Electronic Structures of the Atoms

Main Groups →	I	II											III	IV	V	VI	VII	VIII
All Groups →	**1**	**2**	**3**	**4**	**5**	**6**	**7**	**8**	**9**	**10**	**11**	**12**	**13**	**14**	**15**	**16**	**17**	**18**
Period 1	1 **H** $1s^1$																	2 **He** $1s^2$
Period 2 He core	3 **Li** $2s^1$	4 **Be** $2s^2$											5 **B** $2s^2 2p^1$	6 **C** $2s^2 2p^2$	7 **N** $2s^2 2p^3$	8 **O** $2s^2 2p^4$	9 **F** $2s^2 2p^5$	10 **Ne** $2s^2 2p^6$
Period 3 Ne core	11 **Na** $3s^1$	12 **Mg** $3s^2$	*d* Series of Transition Metals (*d* orbitals fill)										13 **Al** $3s^2 3p^1$	14 **Si** $3s^2 3p^2$	15 **P** $3s^2 3p^3$	16 **S** $3s^2 3p^4$	17 **Cl** $3s^2 3p^5$	18 **Ar** $3s^2 3p^6$
Period 4 Ne core; also 3*d* orbitals in Ga–Kr	19 **K** $4s^1$	20 **Ca** $4s^2$	21 **Sc** $3d^1 4s^2$	22 **Ti** $3d^2 4s^2$	23 **V** $3d^3 4s^2$	24 **Cr** $3d^5 4s^1$	25 **Mn** $3d^5 4s^2$	26 **Fe** $3d^6 4s^2$	27 **Co** $3d^7 4s^2$	28 **Ni** $3d^8 4s^2$	29 **Cu** $3d^{10} 4s^1$	30 **Zn** $3d^{10} 4s^2$	31 **Ga** $4s^2 4p^1$	32 **Ge** $4s^2 4p^2$	33 **As** $4s^2 4p^3$	34 **Se** $4s^2 4p^4$	35 **Br** $4s^2 4p^5$	36 **Kr** $4s^2 4p^6$
Period 5 Kr core; also 4*d* orbitals in In–Xe	37 **Rb** $5s^1$	38 **Sr** $5s^2$	39 **Y** $4d^1 5s^2$	40 **Zr** $4d^2 5s^2$	41 **Nb** $4d^4 5s^1$	42 **Mo** $4d^5 5s^1$	43 **Tc** $4d^5 5s^2$	44 **Ru** $4d^7 5s^1$	45 **Rh** $4d^8 5s^1$	46 **Pd** $4d^{10}$	47 **Ag** $4d^{10} 5s^1$	48 **Cd** $4d^{10} 5s^2$	49 **In** $5s^2 5p^1$	50 **Sn** $5s^2 5p^2$	51 **Sb** $5s^2 5p^3$	52 **Te** $5s^2 5p^4$	53 **I** $5s^2 5p^5$	54 **Xe** $5s^2 5p^6$
Period 6 Xe core; also 4*f* orbitals in Rf–Rn and full 5*d* orbitals in Tl–Rn	55 **Cs** $6s^1$	56 **Ba** $6s^2$	57 **La** $5d^1 6s^2$	72 **Hf** $5d^2 6s^2$	73 **Ta** $5d^3 6s^2$	74 **W** $5d^4 6s^2$	75 **Re** $5d^5 6s^2$	76 **Os** $5d^6 6s^2$	77 **Ir** $5d^7 6s^2$	78 **Pt** $5d^9 6s^1$	79 **Au** $5d^{10} 6s^1$	80 **Hg** $5d^{10} 6s^2$	81 **Tl** $6s^2 6p^1$	82 **Pb** $6s^2 6p^2$	83 **Bi** $6s^2 6p^3$	84 **Po** $6s^2 6p^4$	85 **At** $6s^2 6p^5$	86 **Rn** $6s^2 6p^6$
Period 7 Rn core; also 5*f* orbitals in Unq–Une	87 **Fr** $7s^1$	88 **Ra** $7s^2$	89 **Ac** $6d^1 7s^2$	104 **Unq** $6d^2 7s^2$	105 **Unp** $6d^3 7s^2$	106 **Unh** $6d^4 7s^2$	107 **Uns** $6d^5 7s^2$	108 **Uno** $6d^6 7s^2$	109 **Une** $6d^7 7s^2$									

Inner Transition Metals (*f* orbitals fill)

Lanthanides Xe core →	58 **Ce** $4f^1 5d^1 6s^2$	59 **Pr** $4f^3 5d^0 6s^2$	60 **Nd** $4f^4 5d^0 6s^2$	61 **Pm** $4f^5 5d^0 6s^2$	62 **Sm** $4f^6 5d^0 6s^2$	63 **Eu** $4f^7 5d^0 6s^2$	64 **Gd** $4f^7 5d^0 6s^2$	65 **Tb** $4f^9 5d^0 6s^2$	66 **Dy** $4f^{10} 5d^0 6s^2$	67 **Ho** $4f^{11} 5d^0 6s^2$	68 **Er** $4f^{12} 5d^0 6s^2$	69 **Tm** $4f^{13} 5d^0 6s^2$	70 **Yb** $4f^{14} 5d^0 6s^2$	71 **Lu** $4f^{14} 5d^1 6s^2$
Actinides Rn core →	90 **Th** $5f^0 6d^2 7s^2$	91 **Pa** $5f^2 6d^1 7s^2$	92 **U** $5f^3 6d^1 7s^2$	93 **Np** $5f^4 6d^1 7s^2$	94 **Pu** $5f^6 6d^0 7s^2$	95 **Am** $5f^7 6d^0 7s^2$	96 **Cm** $5f^7 6d^1 7s^2$	97 **Bk** $5f^9 6d^0 7s^2$	98 **Cf** $5f^{10} 6d^0 7s^2$	99 **Es** $5f^{11} 6d^0 7s^2$	100 **Fm** $5f^{12} 6d^0 7s^2$	101 **Md** $5f^{13} 6d^0 7s^2$	102 **No** $5f^{14} 6d^0 7s^2$	103 **Lr** $5f^{14} 6d^1 7s^2$

GENERAL CHEMISTRY

GENERAL CHEMISTRY

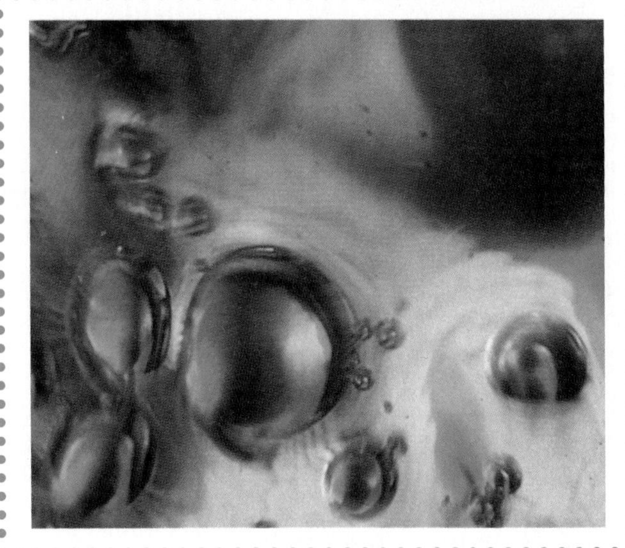

Uno Kask
Towson State University

J. David Rawn
Towson State University

with contributions from

Ronald A. DeLorenzo
Middle Georgia College

Wm. C. Brown Publishers
Dubuque, Iowa • Melbourne, Australia • Oxford, England

Book Team

Editor *Craig S. Marty*
Developmental Editor *Elizabeth M. Sievers*
Publishing Services Coordinator—Production Specialist *Julie Avery Kennedy*
Publishing Services Coordinator—Design Specialist *Barbara J. Hodgson*

Wm. C. Brown Publishers
A Division of Wm. C. Brown Communications, Inc.

Vice President and General Manager *Beverly Kolz*
National Sales Manager *Vincent R. Di Blasi*
Director of Marketing *John W. Calhoun*
Marketing Manager *Christopher T. Johnson*
Advertising Manager *Amy Schmitz*
Director of Production *Colleen A. Yonda*
Manager of Visuals and Design *Faye M. Schilling*
Design Manager *Jac Tilton*
Art Manager *Janice Roerig*
Publishing Services Manager *Karen J. Slaght*
Permissions/Records Manager *Connie Allendorf*

Wm. C. Brown Communications, Inc.

President and Chief Executive Officer *G. Franklin Lewis*
Corporate Vice President, President of WCB Manufacturing *Roger Meyer*
Vice President and Chief Financial Officer *Robert Chesterman*

Cover photo © C: Raymond/Photo Researchers, Inc.
Production by Lachina Publishing Services
Interior/cover design by Marsha Cohen/Parallelogram Graphic Communications
Illustrations by Diphrent Strokes, Inc.

The credits section for this book begins on page 975 and is considered an extension of the copyright page.

A Times Mirror Company

Library of Congress Catalog Card Number: 92–72533

ISBN 0–697–12144–5

Printed in the United States of America by Wm. C. Brown Communications, Inc., 2460 Kerper Boulevard, Dubuque, IA 52001

10 9 8 7 6 5 4 3 2 1

Contents, An Overview

PART 4 Thermodynamics and Kinetics 747

PART 5 Some Special Branches of Chemistry 809

Expanded Contents

PART 1

Elements, Compounds, Reactions, and Energy Changes / 1

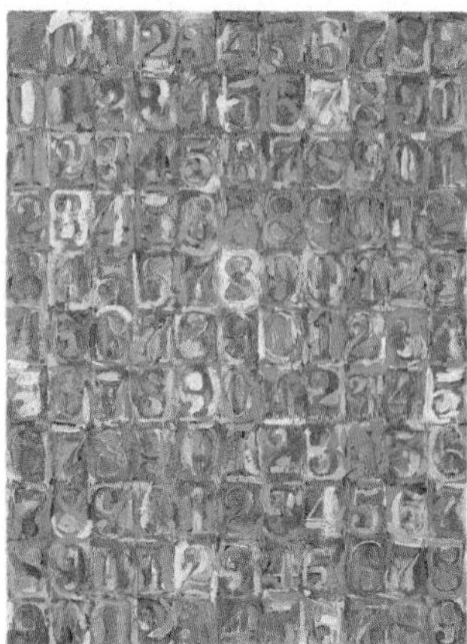

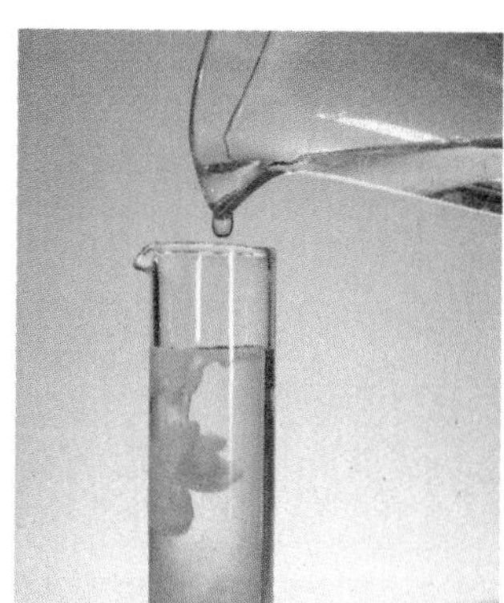

PART 2

Structure and Properties / 201

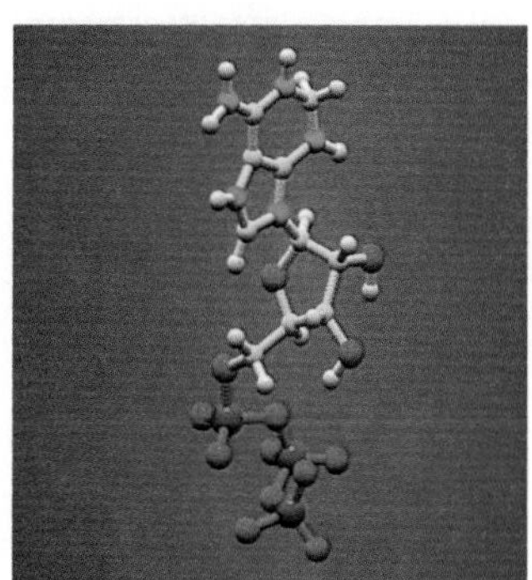

CHAPTER 9 Chemistry of Nonmetals 324

CHAPTER 10 The Gaseous State of Matter 368

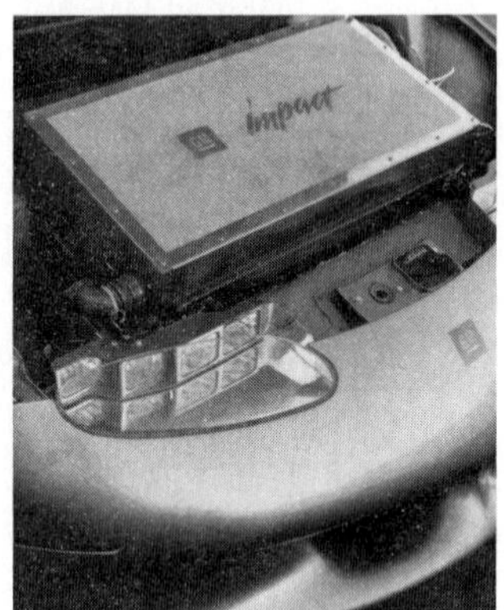
impact

PART 4

Thermodynamics and Kinetics / 747

PART 5

Some Special Branches of Chemistry / 809

Preface

To the Instructor

The Level of This Text Is Suitable for Most General Chemistry Courses

This text is intended for a two-semester general chemistry course for chemistry and other science majors. The text has been written on a ''level'' that will allow it to be used in most general chemistry courses. Textbooks in general chemistry have grown to gargantuan proportions, perhaps hoping to provide topics to satisfy the interests of every conceivable instructor. Thus, general chemistry books contain far more material than can be taught in a two-semester course. This text represents a move in the other direction, toward a more concise treatment. However, we too have included more material than can be covered in a typical course, so that instructors will have some flexibility in designing their courses (the Instructor's Manual lists various orders of topics that can be used). The book assumes no previous knowledge of chemistry, although in keeping with nearly universal standards, a background of high school or college algebra is assumed.

This Book Thoroughly Integrates Chemical Principles with Descriptive Chemistry

The major goal in writing this text has been to thoroughly integrate descriptive chemistry with chemical principles so that students will be able to learn abstract concepts within a firm empirical context. This emphasis on descriptive chemistry also conforms to the recent recommendation by the Committee of Professional Training of the American Chemical Society. We define the term ''descriptive chemistry'' as the study of the properties of substances, their physical and chemical changes, and the practical applications of these properties and changes in other sciences as well as in everyday life.

Most general chemistry textbooks either concentrate on principles only or place descriptive chemistry at the end of the book, long after the principles and theories that underline chemical reactions have been discussed. Therefore, instructors often teach *only* principles either (a) because their text contains virtually no descriptive chemistry, or (b) because there is no time left at the end of the second semester to cover descriptive chemistry. The latter difficulty is nearly universal given the organization of most general chemistry books. The old joke that ''silver chloride is a pale green gas'' (D.A. Davenport, *Journal of Chemical Education*) is perilously close to an accurate depiction of many students' knowledge of chemical facts. We have integrated an extensive amount of descriptive chemistry in many of the chapters of this book in a unique way; these chapters are indicated by a triangle in the table of contents.

▶ Descriptive Chemistry Is Introduced "Early" and Is Incorporated into the Treatment of General Principles Throughout the Book

Chemical elements are the simplest and most fundamental substances. They are as fundamental to chemistry as an alphabet is to a language. We therefore introduce the properties of common elements early in the study of chemistry. We provide a survey of the elements, their classification, and their periodic properties in Chapter 2. Some common reactions of elements are also introduced in Chapter 2. We use these reactions to introduce the writing of chemical equations, a topic students often find difficult. Chemical reactions in aqueous solution—acid–base reactions, precipitation reactions, and oxidation–reduction reactions, as well as net ionic equations for these reactions—are introduced in Chapter 4. Thus, Part 1 of the book is entitled "Elements, Compounds, Reactions, and Energy Changes." Chapter 1 deals with the usual introductory material for all chemistry courses, Chapter 3 covers stoichiometry, and Chapter 5 thermochemistry.

After this five-chapter introduction to descriptive chemistry and chemical principles, Part 2 of the book (Chapters 6 through 12) deals with structure and properties. Theoretical principles—atomic and molecular structure, the theoretical foundation for the periodic table, and so forth—begin to enter the picture. These topics are, of course, covered in all general chemistry texts. Our treatment differs somewhat since we describe theoretical principles within an empirical foundation. Although we have emphasized what we believe to be unique in our approach, we certainly provide a thorough coverage of chemical principles.

We take up more descriptive aspects of chemistry in Chapters 8 and 9. In Chapter 8 we discuss the covalent bonding and the compounds and reactions of nonmetallic elements. The descriptive chemistry of nonmetallic elements is discussed more extensively in Chapter 9 from the viewpoint of covalent bonding and structure. The chemistry of metals is treated in Chapter 18, primarily in the context of oxidation–reduction reactions and the energetics of these reactions.

Various aspects of descriptive chemistry are further interspersed with principles in later chapters. Thus, Chapter 14, in Part 3, "Chemical Equilibria and Their Applications," discusses the periodic trends in the properties of oxides and their corresponding acids and hydroxides. These are important topics, and the students who do not learn them will be hampered in their later study of organic chemistry and biochemistry. The coverage of descriptive chemistry culminates in Chapter 19, a short treatment of aqueous qualitative analysis of selected cations, anions, and salts. In this chapter, almost every major type of chemical reaction is encountered. Students therefore learn to write net ionic equations for these reactions, and they also learn how to use these reactions to identify unknown substances. When coupled with concomitant laboratory work, this topic is challenging and interesting to students. We have taught this material in this way for many years, and it "works."

Besides the descriptive chemistry discussed in dedicated chapters, many descriptive topics relating chemistry to other subjects are blended into every chapter. For example, in Chapter 5 chemical reactions are considered as sources of energy, including the foods consumed by living organisms. In Chapter 10 we describe some aspects of the chemistry of the atmosphere and air pollution, and carbon and nitrogen cycles in the biosphere. In Chapter 11 the relations between properties of various substances and their phase diagrams are described, as well as the relations between intramolecular and intermolecular bonding and physical states of substances. In Chapter 12 water pollution and methods of water purification are discussed. In Chapters 17 and 18 we consider the chemical reactions that cause corrosion and the chemical reactions used in commercial batteries. In Chapter 24 we consider the biological effects of ionizing radiation. As we noted earlier, many students in general chemistry also study biological sciences, and biological applications of chemistry are introduced in many chapters, not as "special topics" (which are seldom covered in class), but as an integral part of "general chemistry." We hope this descriptive part of the book will appeal to students and instructors alike.

Example Problems and Practice Problems Are Found Throughout Each Chapter and Extensive Review Material at the End of Each Chapter

Every chapter includes many example problems and questions that are worked out. Most of these are followed by practice problems of similar type so that students can test their mastery of the subject. Key terms and concepts in each chapter are **boldfaced** for emphasis, and definitions are *italicized.*

At the end of each chapter a list of key terms with reference to the section numbers where the terms first appear is provided. The page numbers where these key terms are first defined are indicated in the index. We have also included many study problems at the end of each chapter. These exercises are preceded by a section called "General Review," in which the goals or objectives of the chapter are formulated as a series of questions that urges the students to recall and actively think through the concepts introduced in the chapter. The answers to the odd-numbered practice problems in each chapter and to the odd-numbered questions and problems at the end of the chapters are found in Appendix A. Complete, worked-out solutions to all odd numbered problems and exercises in the book are given in the Student's Solutions Manual. The complete, worked-out solutions to all even numbered problems and exercises are available in an Instructor's Solutions Manual.

To the Student

Here are some suggestions that help you use this book most effectively in your studies. First, take good notes of the professor's lectures and study them as soon as possible after the lecture, and read the chapter being discussed. Pay particular attention to the **boldfaced** key terms introduced in the chapter, and learn the definitions of these terms, which are *italicized* for emphasis. Carefully study all the example problems in the chapter and work the corresponding practice problems to test your mastery of the topics.

At the conclusion of the chapter, read the summary and the key terms at the end of the chapter, and try to recall the meaning of each term. Then try to answer the questions in the General Review part of the chapter-end exercises which test your ability to recall the important topics covered in the chapter. Finally, do all the exercises assigned to you by your professor at the end of the chapter and check the correctness of your answers given in Appendix A or in the Solutions Manual.

If you cannot do some of the exercises assigned to you, go back and review the material in the chapter dealing with the relevant topics. Try to anticipate various questions and problems you might be asked in a test. For each topic, think about possible questions and problems you would ask if you were an instructor. You may be surprised to see how many similar questions might actually be asked in your examinations. Also, you should realize that there are many ways questions and problems can be worded, and you should learn to recognize or anticipate various alternatives.

Acknowledgments

A project of this magnitude cannot be completed without the help of many people. During the early stages of writing the manuscript for this text, the advice and comments of Professors Nordulf Debye, Alan Pribula, Joseph Topping, and Alan Wingrove of Towson State University made a valuable contribution to the project. The chapters were then chain-delivered back and forth many times by Jane Kask and Don Hooper to a small house near the Pennsylvania border where they were expertly typed by Debby Rouzer using an Eagle computer with Spellbinder software. These files were subsequently converted to WordPerfect by Jerry Bunker, who has an ingenious touch with computers. The garbled conversion product was ungarbled and beautifully typed by Teresa (Grzymala) Hornbach, a wizard at constructing equations and tables with columns and typing subscripts and superscripts one above the other as well as adding subscripts to subscripts.

We are grateful to the late Ed Jaffe of Wm. C. Brown Company for negotiating the project. His skillful negotiating ability, his human touch, and his special chair at a table with coffee and bran muffins will be long remembered. The manuscript was developed to its final form by Liz Sievers, who is responsible for many fine improvements and additions to the text. Julie Kennedy, Barbara Hodgson, and Craig Marty have been in charge of the production, which has been expertly carried out by Lachina Publishing Services, Inc., Cleveland, Ohio. Their efforts have produced a beautiful product.

The following reviewers have made more than the usual contributions to the project and therefore deserve special mention. Our colleague, Professor Alan Pribula of Towson State University, has read the entire manuscript at least three times in its various stages of development. He has given valuable advice and comments both as a reviewer and in consultation whenever the need arose. Professor David Goldberg of the City University of New York, Brooklyn College, must be termed a "super reviewer" because of his unusually thorough and speedy reviews. His attention to the details of both subject matter and language have considerably reduced the errors in the book. Professor Delwin Johnson of St. Louis Community College at Forest Park has been an outstanding advocate of the students' point of view, and has thus helped to produce a book that is as easy for students to understand as possible. The following reviewers have also contributed: Edwin H. Abbott, Montana State University; Thomas J. Greenbowe, Iowa State University of Science and Technology; Larry G. Hargis, University of New Orleans; David C. Hilderbrand, South Dakota State University; R. C. Legendre, University of South Alabama; Joe M. Ross, Central State University; Vernon J. Thielmann, Southwest Missouri State University.

Finally, but most importantly, we want to thank our wives, Jane and Margie, for their support, help, and patience during the many years of writing this book. Our thanks to all who helped us directly or indirectly in this project.

Uno Kask
David Rawn

PART 1

Elements, Compounds, Reactions, and Energy Changes

CHAPTER 1

Philip de Loutherbourg, *Coalbrookdale by Night* (1801). A steel manufacturing company. The application of chemistry present here is the process by which naturally occurring matter that contains iron is converted to steel.

Chemistry, Matter, and Measurement

1.1 Chemistry and the Scientific Method

A Bird's-Eye View of Chemistry

In this section we take a brief look at some of the many aspects of chemistry. Some interesting questions that chemists are trying to answer probe the fundamental nature of things: What are the simplest components of wood, rocks, and living organisms? Can the components in these systems be isolated from one another? Can we explain the changes that occur in such processes as the rusting of iron, the burning of natural gas, or the action of a drug?

Chemists study the nature of material things to improve the quality of life. Such improvements include the synthesis of pharmaceuticals to cure illnesses, the creation of wrinkle-free clothing for easy maintenance, the production of durable paints to protect buildings, and the manufacture of materials suitable for building cars, airplanes, and spacecraft.

Chemists also try to alleviate environmental problems such as the pollution caused by the gaseous exhaust products of internal combustion engines and industrial smokestacks. These gaseous exhaust products are largely responsible for atmospheric pollution—including ''photochemical smog'' and acid rain. Many other aspects of the industrial economies of developed countries involve the application of chemistry for better life.

In a broad sense, **chemistry** is a *science that examines the properties, composition, structure, and changes of matter. Anything that occupies space and has mass* is called **matter**. Matter that has a particular set of characteristics different from the characteristics of another kind of matter is referred to as a *substance*. Thus, oxygen, water, and table salt are examples of different substances.

A given sample of matter has characteristics such as color, texture, relative hardness, odor, and taste. These characteristics of matter are called its *properties*. A sample of matter can consist of two or more components. For example, water consists of hydrogen and oxygen, and table salt consists of sodium and chlorine. Determining the identities of the components and their quantities present in a substance establishes the *composition* of the substance.

Figure 1.1

A cut diamond: an example of a crystalline substance.

The *structure* of matter refers to the shapes of the constituent smallest particles of matter and how these particles are arranged. Later we will see how the properties of matter are related to its structure. Some substances consist of particles arranged in a repeating pattern. This pattern is called a **lattice**. *Substances that consist of particles that have regular geometric shapes* are called **crystalline** substances; a cut diamond is a beautiful example (Figure 1.1). The structures of some substances have no regularly repeating pattern. Such *shapeless substances* are **amorphous**. Chocolate, glass, and soot are examples of amorphous substances (Figure 1.2).

Figure 1.2

Soot: an example of an amorphous substance.

Matter can undergo physical and chemical *changes*. In a **physical change**, the basic nature or the composition of the substance is not altered. Examples of physical change include water turning from a liquid to a solid, from a liquid to a gas, or from a solid to a gas without losing its identity. A change in which the composition of a substance is altered is called a **chemical change**. For example, when natural gas (methane) burns in air, it changes to carbon dioxide and water, which are different from methane.

The study of chemistry is commonly divided into five major areas:

1. The branch of chemistry that deals primarily with substances containing carbon is called **organic chemistry**.
2. Non-carbon-containing substances are the major concern of **inorganic chemistry**.
3. The branch of chemistry concerned with the methods and instruments for determining the composition of matter is called **analytical chemistry**.
4. The laws of physics are applied to the investigation of the structure and changes of matter in the branch of chemistry called **physical chemistry**.
5. The physical and chemical changes that occur in living organisms are the province of **biochemistry**.

These various branches of chemistry are by no means unrelated to one another. For example, a biochemist applies organic, inorganic, physical, and analytical chemistry to the study of the components of living organisms.

The Scientific Method

If this be madness, there's some method in't.

Hamlet

Chemistry, as all other sciences, starts with observations of phenomena, continues with an organization of ideas, and ends with explanations of observed phenomena and predictions of phenomena not yet observed. The *process of experimentation and reasoning that leads from observations to the construction of unifying principles for explanation and prediction* is called the **scientific method**.

The scientific method consists essentially of two processes: (1) A process of reasoning that leads from specific observations to general statement(s) and abstract ideas is called **inductive reasoning**. (2) The reasoning process in which general ideas or theories lead to specific conclusions is called **deductive reasoning**. These reasoning processes are summarized in Figure 1.3.

Inductive reasoning in science is based on experiments in which events are observed and described. Experimental observations may lead to the formation of a *tentative assumption*, or **hypothesis**, that explains the observations.

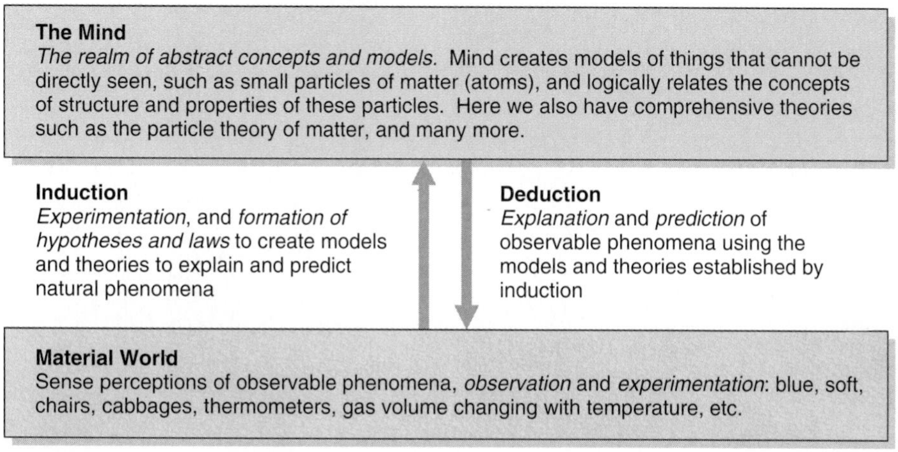

The Mind
The realm of abstract concepts and models. Mind creates models of things that cannot be directly seen, such as small particles of matter (atoms), and logically relates the concepts of structure and properties of these particles. Here we also have comprehensive theories such as the particle theory of matter, and many more.

Induction
Experimentation, and *formation of hypotheses and laws* to create models and theories to explain and predict natural phenomena

Deduction
Explanation and *prediction* of observable phenomena using the models and theories established by induction

Material World
Sense perceptions of observable phenomena, *observation* and *experimentation*: blue, soft, chairs, cabbages, thermometers, gas volume changing with temperature, etc.

Figure 1.3

A diagrammatic outline of induction and deduction in the scientific method of inquiry. Induction involves reasoning from particular observations to form general (abstract) theories and models. The process of induction involves the formation of hypotheses and laws in science. Deduction is the opposite of induction; it consists of reasoning from general theories and mental models to particular observable events that can be predicted from theory.

Sometimes the results of experiments can be expressed in a concise verbal or mathematical statement that expresses the relationship between two or more variable quantities. For example, when the volume of a container with a sample of a gas in it is decreased without changing the temperature, the pressure of the gas in the container increases. If the volume of the container is increased, the pressure of the gas decreases. These observations can be summarized by the following statement: The volume and pressure of a gas vary inversely with each other—that is, when one increases, the other decreases—provided that the temperature remains constant. A *verbal or mathematical statement of a relationship that is always true under certain circumstances* is known as a **scientific law**.

Often hypotheses of the existence of components of matter that cannot be observed directly may be necessary to explain natural phenomena. Such *hypothetical entities* are called **models**. A *network of interrelated hypotheses and models consistent with experimentally established laws* in science is known as a **comprehensive theory**. A comprehensive theory is a unifying idea that can be used to *explain* observed phenomena and *predict* new phenomena.

For example, to *explain* the phenomenon of ice melting, it is theorized that ice consists of small particles (see the models in Section 1.5) that adhere to each other strongly in solid ice. When the solid is heated, the particles are driven apart, and they are assumed to be in constant motion in all directions throughout the liquid water. This motion explains the fluid character of liquid water. The fluid character of all liquids can be explained by assuming that liquids consist of a large number of tiny particles in constant motion in all directions.

The unifying theory that liquids consist of particles in constant motion can be used to *predict* new phenomena. (Recall that both explanation and prediction in science are deductive processes—going from unifying theories to particular observations, as we can see from Figure 1.3.) We can, for example, predict that when a drop of blue ink is placed in water, the ink will disperse in the water until the entire solution is blue. If the particles of the ink and the water are in constant motion in all directions—eventually they can form a uniform mixture. This phenomenon is, indeed, observed (Figure 1.4).

Figure 1.4

Dispersion of ink in water: (a) a drop of ink before making contact with water; (b) the ink-drop starting to disperse; (c) the dispersion (solution) of ink in water.

1.2 Measurement and Conversion of Units

SI Units

Observations and experiments usually involve measurements. For example, we frequently measure mass, volume, length, and temperature. We next discuss the units of these measurements and how the results of these measurements should be recorded to reflect the quality of each measurement.

In 1960, the General Conference of Weights and Measures, an international body of 40 countries (including the United States), adopted a new system of units, the International System of Units. In the French language, this is Le Système International d'Unités; hence they are abbreviated **SI units**.

There are *seven basic SI units*. These units can be converted to other units by either multiplying or dividing a basic unit by the factor of 10^x (where x is an integer), preferably by 10^3 (1000). Each unit that differs from the basic unit by a multiple of 10^x is given a special prefix, for example, kilo- for 10^3 and mega- for 10^6.

Table 1.1 lists the seven basic SI units. Table 1.2 gives the prefixes and symbols for units, as well as the factors with which the basic units must be multiplied to obtain other units. For example, the basic SI unit for length is the meter, which has the symbol m (Table 1.1). Since the symbol for kilo is k, the symbol for kilometer is km (Table 1.2). One meter, the basic unit of length, must be multiplied by the factor of 10^3 to obtain 1 kilometer, or by the factor of 10^6 to obtain 1 megameter (Mm). Thus, 1 km = 10^3 m, and 1 Mm = 10^6 m. It is important to distinguish between upper- and lowercase symbols. Thus, 1 Mm (megameter) is not equal to 1 mm (millimeter).

Table 1.1 Basic SI Units

Quantity	Name of SI Unit	Symbol
Length	meter	m
Mass	kilogram[a]	kg
Time	second	s
Electric current	ampere	A
Temperature	kelvin	K
Luminosity	candela	cd
Quantity of substance	mole	mol

[a] Although the kilogram is the SI unit of mass, the unit used for conversion is the gram. Thus, 1 kilo*gram* = 1×10^3 *grams* = 1×10^6 milli*grams*.

Table 1.2

Prefixes Used in the SI System, Their Symbols, and the Factors with Which the Basic Units Must Be Multiplied to Obtain the Given Units[a]

Prefix	Symbol	Factor
exa	E	10^{18} or 1,000,000,000,000,000,000
peta	P	10^{15} or 1,000,000,000,000,000
tera	T	10^{12} or 1,000,000,000,000
giga	[illegible]	10^{9} or 1,000,000,000
mega	**M**	**10^{6} or 1,000,000**
kilo	**k**	**10^{3} or 1000**
hecto	h	10^{2} or 100
deka	da	10
basic unit	—	**1**
deci	**d**	**10^{-1} or 0.1**
centi	**c**	**10^{-2} or 0.01**
milli	**m**	**10^{-3} or 0.001**
micro	**μ**	**10^{-6} or 0.000001**
nano	**n**	**10^{-9} or 0.000000001**
pico	**p**	**10^{-12} or 0.000000000001**
femto	f	10^{-15} or 0.000000000000001
atto	a	10^{-18} or 0.000000000000000001

[a] The most commonly encountered prefixes and factors are shown in **boldface**.

Conversion of Units

A measurement with a given set of units can be converted to another set of units by using an appropriate conversion factor from the data in Table 1.2. A **conversion factor** is *a ratio that can be used to convert one quantity and its units to another equivalent quantity and its units*. For example, we know that 1 kilometer equals exactly 1000 meters; therefore, the ratio of these two quantities is 1:

$$\frac{1\text{ km}}{1000\text{ m}} = 1$$

This ratio is the conversion factor that converts meters to kilometers. The reciprocal of this ratio must also be 1:

$$\frac{1000\text{ m}}{1\text{ km}} = 1$$

This ratio is the conversion factor that converts kilometers to meters.

A conversion factor can be used to convert meters to kilometers, or kilometers to meters, without changing the value of the measurement. We usually use many units of different sizes simply for convenience. We measure large quantities in large units and small quantities in small units. Such a conversion can conveniently be performed by using the following equation:

(given quantity in its units) × (conversion factor)
= same quantity in sought units

The use of this equation is illustrated in Examples 1.1 and 1.2.

Example 1.1

Using a Conversion Factor in Unit Conversion

The athletes in a marathon race run a distance of 42.186 km. How many meters do they run?

SOLUTION: The given quantity and unit is 42.186 km. The conversion factor is

$$\frac{1000 \text{ m}}{1 \text{ km}}$$

Multiplying the given quantity by this conversion factor gives the desired result:

$$42.186 \cancel{\text{km}} \left(\frac{1000 \text{ m}}{1 \cancel{\text{km}}}\right) = 42{,}186 \text{ m}$$

Note that the unit, km, cancels, and the desired unit, m, remains in the answer. Recall that the value of a conversion factor expressed as a ratio is 1; therefore, multiplication with such a factor leaves the value of a measurement unchanged.

In all simple, one-step unit conversions, *the conversion factor must be chosen so that the numerator of the factor has the same unit as the unit in the answer that is sought.* When this is done, the unit in the denominator of the conversion factor cancels with the identical unit of the quantity being converted. We can see in Example 1.1 that the unit km in the given quantity cancels the km in the denominator of the conversion factor, leaving the unit m in the numerator of the conversion factor as the unit desired for the answer. In many conversions, more than one conversion factor must be used as shown in Example 1.2.

Example 1.2

Using Several Conversion Factors in a Multistep Operation

Your doctor has prescribed a 375-mg capsule of a medicine for you to take once a day. What is the mass of this capsule in pounds? Use the necessary data from Tables 1.2 and 1.3.

SOLUTION: From Table 1.2 we find the relationships between milligrams and grams, and between grams and kilograms. Table 1.3 gives the number of pounds in 1 kilogram. We can perform the required conversion in three steps or in one multistep operation. If we choose to perform this conversion in three different steps, these can be set up as follows:

Step 1. $375 \cancel{\text{mg}} \left(\frac{1 \text{ g}}{1000 \cancel{\text{mg}}}\right) = 0.375 \text{ g}$

Step 2. $0.375 \cancel{\text{g}} \left(\frac{1 \text{ kg}}{1000 \cancel{\text{g}}}\right)$ 0.000375 kg or 3.75×10^{-4} kg

Step 3. $3.75 \times 10^{-4}\ \cancel{\text{kg}}\left(\frac{2.20\ \text{lb}}{1\ \cancel{\text{kg}}}\right) = 8.25 \times 10^{-4}\ \text{lb}$

This conversion is more easily carried out in a single multistep operation using the three conversion factors in a single expression:

$$375\ \cancel{\text{mg}}\left(\frac{1\ \cancel{\text{g}}}{1000\ \cancel{\text{mg}}}\right)\left(\frac{1\ \cancel{\text{kg}}}{1000\ \cancel{\text{g}}}\right)\left(\frac{2.20\ \text{lb}}{1\ \cancel{\text{kg}}}\right) = 8.25 \times 10^{-4}\ \text{lb}$$

Let us next consider some important units that are used to measure *length*, *mass*, *volume*, *energy*, and *temperature*: the properties frequently dealt with in chemistry and other sciences. Additional illustrations of unit conversion will then be considered. Table 1.3 gives the relations between some commonly used units of length, mass, and volume in the SI, metric, and English systems.

Table 1.3

Equivalents of Some SI, Metric, and English Units

Length	Mass	Volume
1 m = 39.37 in.	1 kg = 2.2046 lb	1 m^3 = 10^3 L
1 in. = 2.54 cm (exactly)	1 lb = 453.6 g	1 L = 1.057 qt
1 mile = 1.609 km	1 lb = 16 oz	1 qt = 32 oz
1 mile = 5280 ft	1 oz = 28.35 g	1 oz = 29.57 mL

Length

The basic SI unit of length, the *meter* (m), was originally defined as exactly one ten-millionth of the distance from the equator to the north pole. Later, a reference metal bar 1 meter long was kept in the International Bureau of Weights and Measures at Sèvres, near Paris, France. In 1960 the International Conference of Weights and Measures redefined the meter as being equal to 1650763.73 times the wavelength of the red-orange radiation emitted by the isotope krypton-86. (See Section 1.6 for isotopes and Chapter 7 for the meanings of wavelength and radiation.) This krypton standard was adopted because that particular wavelength of radiation is very accurately reproducible.

The meter is the basic unit of length in both the SI and metric systems: 1 meter equals 100 centimeters, and 1 inch equals 2.54 centimeters. Figure 1.5 compares some frequently used units of length.

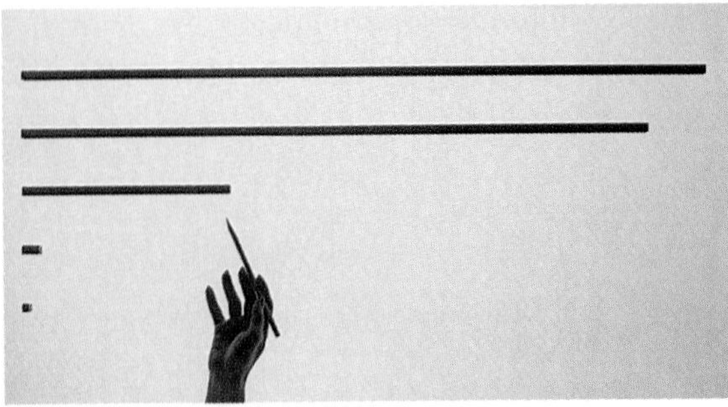

Figure 1.5

A meter, a yard, a foot, an inch, and a centimeter.

Example 1.3

Comparing Lengths Given in Different Units

If you are 5 ft $10\frac{1}{4}$ in. tall, would another person who measures 1800 millimeters be taller or shorter than you are?

SOLUTION: In this problem we compare two heights that are not given in the same units. To solve the problem, we can convert your height to millimeters, or 1800 m to feet and inches. If we choose to convert your height to millimeters, we can first convert 5 ft $10\frac{1}{4}$ in. to inches, followed by the conversion of inches to centimeters, and centimeters to millimeters:

$$\text{your height in inches} = 5\,\cancel{\text{ft}}\left(\frac{12\text{ in.}}{1\,\cancel{\text{ft}}}\right) + 10.25\text{ in.} = 70.25\text{ in.}$$

This number of inches can now be converted to millimeters by a one-step operation using two conversion factors:

$$70.25\,\cancel{\text{in.}}\left(\frac{2.54\,\cancel{\text{cm}}}{1\,\cancel{\text{in.}}}\right)\left(\frac{10\text{ mm}}{1\,\cancel{\text{cm}}}\right) = 1784\text{ mm}$$

Thus, a person whose height is 1800 mm is slightly taller than you are.

There are several other ways or routes to perform such conversions. For example, 5 ft $10\frac{1}{4}$ in. can be converted to feet, the feet to yards, and the yards to meters, meters to centimeters, and centimeters to millimeters.

Example 1.4

Converting Small and Large Length Units

The radius of an average atom is 100 pm. What is the equivalent in meters, centimeters, and nanometers?

SOLUTION: From Table 1.2 we find that 1 pm equals 10^{-12} m. Thus, 1 m is equal to 10^{12} pm. Based on this and other relevant information from Table 1.2, the required conversions can be performed as follows:

$$100\,\cancel{\text{pm}}\left(\frac{1\text{ m}}{10^{12}\,\cancel{\text{pm}}}\right) = 1 \times 10^{-10}\text{ m}$$

$$1 \times 10^{-10}\,\cancel{\text{m}}\left(\frac{100\text{ cm}}{1\,\cancel{\text{m}}}\right) = 1 \times 10^{-8}\text{ cm}$$

$$1 \times 10^{-10}\,\cancel{\text{m}}\left(\frac{10^{9}\text{ nm}}{1\,\cancel{\text{m}}}\right) = 0.1\text{ nm}$$

Practice Problem 1.1: The radius of an aluminum atom is 0.122 nm. What is the equivalent in meters, centimeters, millimeters, and picometers?

Figure 1.6

Different masses of sugar—one kilogram, one pound, one ounce, and one gram.

Mass

The standard mass is the *kilogram* (kg), which equals 1000 g or 2.20 lb. The primary reference kilogram is a platinum–iridium cylinder kept in a vault at the International Bureau of Weights and Measures at Sèvres, France. Replicas of this cylinder are stored for reference by other nations of the world. These replicas are called secondary standards.

In practice, the mass of an object is usually determined by weighing. (Different masses of sugar are shown in Figure 1.6.) The result is, therefore, frequently called the weight of the object. Actually, weight is the gravitational attraction between the object and the earth. Gravitational attraction varies slightly over the earth's surface, but for most practical purposes the change is not important.

Objects are weighed on a device called a *balance*. There are two essentially different types of balances, the beam balance and the spring balance (Figure 1.7). In a beam balance, the mass of an object on one side of a "beam" is "balanced" by a known mass on the other side.

In weighing an object on a beam balance (Figure 1.7a) the mass of the object, m_1, is compared with that of a standard mass, m_2. The weight of an object is its mass, m, times the acceleration of gravity, g. The weight of the object equals the weight of the standard mass when the balance beam is in a horizontal position, the pointer is at the center of the scale, and the balance arms are equal. At that point the masses of m_1 and m_2 are also equal:

$$m_1 g = m_2 g \quad \text{and} \quad m_1 = m_2$$

Most balances used in today's laboratories are much more sophisticated than the one portrayed in Figure 1.7a, but they operate on the same principle of comparing masses.

A spring balance (Figure 1.7b) compares the force of a spring with the

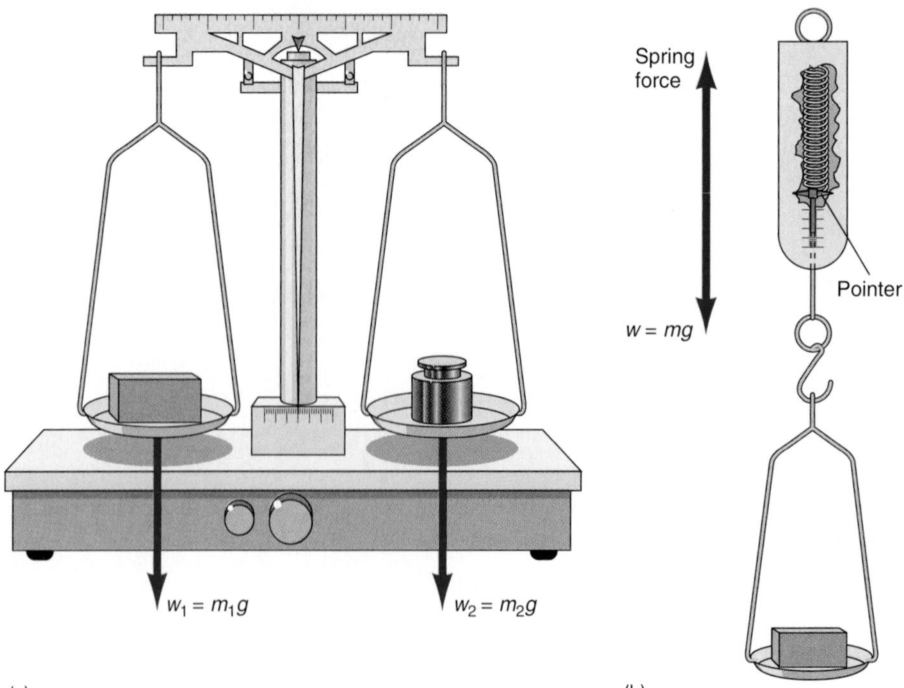

Figure 1.7

Two different types of balances. (a) On a beam balance, the mass of an object is compared with a reference mass. (b) On a spring balance, the weight of an object is dependent on the gravitational attraction by the Earth.

gravitational force acting on an object being weighed. The weight of an object on a spring balance varies with location because the gravitational force varies. The mass of the object, however, is the same at any location.

Example 1.5

Converting Large to Small Mass Units

What is the mass in grams of a person who weighs 158 lb?

SOLUTION

$$158 \text{ lb} \left(\frac{1 \text{ kg}}{2.20 \text{ lb}}\right)\left(\frac{1000 \text{ g}}{1 \text{ kg}}\right) = 7.18 \times 10^4 \text{ g}$$

Volume

The volume of a sample of matter can be expressed in length units cubed (Figure 1.8). Thus, the basic SI unit of volume is the cubic meter (m^3). Smaller, more practical units of volume are the metric unit cubic decimeter (dm^3), which is equal to a *liter* (L), and the unit cubic centimeter (cm^3 or cc), which is equal to a *milliliter* (mL). A liter is slightly larger than a quart (1 L = 1.057 qt) (Figure 1.9).

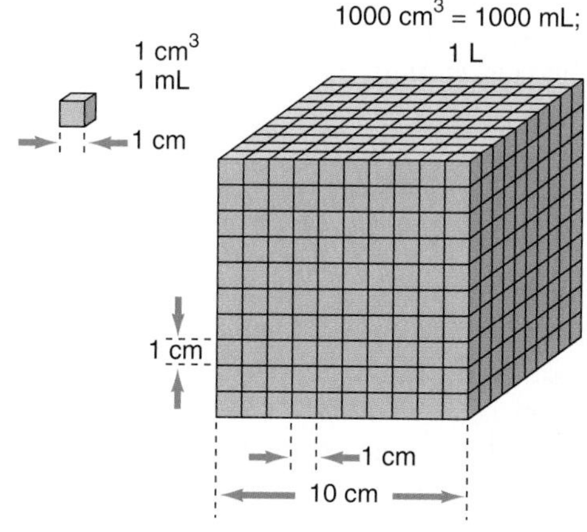

Figure 1.8

The relationship of two volumes to each other: 1 cm^3 = 1 mL; 1000 cm^3 = 1 L (1000 mL).

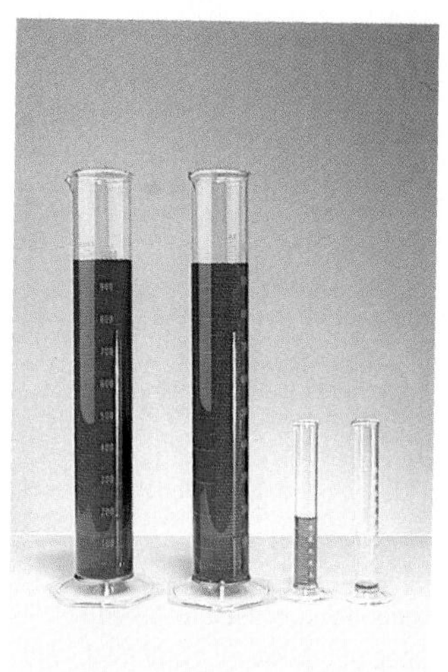

Figure 1.9

Different volumes of fruit juice—one liter, one quart, one fluid ounce, and one milliliter.

Density

The *mass per unit volume of an object* is its **density**. If the volumes of two objects are equal, the one with the greater mass has the higher density (an iron ball has a higher mass than that of a softball of equal size, so the iron ball has a higher density than the softball). The density, d, of an object can be determined by dividing its mass, m, by its volume, V:

$$\text{density} = \frac{\text{mass}}{\text{volume}} \quad \text{or} \quad d = \frac{m}{V} \tag{1.1}$$

The densities of some common materials are listed in Table 1.4.

Table 1.4

Densities of Some Common Materials at 20°C and 1 Atmosphere Pressure

Solids	Density ($g\ cm^{-3}$)	Liquids	Density ($g\ cm^{-3}$)	Gases	Density ($g\ L^{-1}$)
Aluminum	2.699	Ethyl alcohol	0.791	Carbon dioxide	1.977
Copper	8.94	Gasoline	0.67	Chlorine	3.214
Cork	0.24	Glycerine	1.27	Hydrogen	0.089
Gold	19.32	Mercury	13.546	Nitrogen	1.25
Iron	7.874	Water	0.998203	Oxygen	1.429
Platinum	21.37				
Silver	10.4				

Example 1.6

Determining Density to Identify an Element

A street peddler sells you a ring that he claims is made of solid platinum. To check whether the ring really is made of platinum, you go home and determine its density. You first weigh the ring, then place it in water in a graduated cylinder and observe the increase in volume in the cylinder (this is the volume of the ring). The results obtained: the mass of the ring, 9.5 g; the volume of the ring, 1.2 cm^3. From these values, calculate the density of the ring. Using data in Table 1.4, determine whether the ring is made of platinum or of some other metal.

SOLUTION

$$\text{density} = \frac{\text{mass}}{\text{volume}} = \frac{9.5\ \text{g}}{1.2\ \text{cm}^3} = 7.9\ \text{g/cm}^3 \quad \text{or} \quad 7.9\ \text{g cm}^{-3}$$

The units for the result obtained, grams per cubic centimeter, can be written as g/cm^3 or as $g\ cm^{-3}$, because in general $1/x = x^{-1}$, $1/x^2 = x^{-2}$, $1/x^3 = x^{-3}$, and so on (in this case x = cm). Table 1.4 reveals that the density of platinum is 21.37 $g\ cm^{-3}$; therefore, the ring could not be made of platinum. The experimental density compares rather closely with the density of iron, which could, therefore, be the substance of which the ring is made. However, it could also be made of a mixture of two or more metals. Thus, although you may not know what the ring is made of, you can be certain that it is not pure platinum.

Some devices used to measure volumes of liquids—a graduated cylinder, a pipet, a buret, a pycnometer, and a volumetric flask.

If any two of the variables in Equation 1.1 are known, the third can be calculated. For example, if the density and the mass of an object are known, its volume can be calculated. Similarly, if an object's density and volume are known, its mass can be calculated. Examples 1.7 and 1.8 illustrate such calculations.

Example 1.7

Calculating Volume from Given Mass and Density

What is the volume of a metal coin with a density of 5.8 g cm^{-3}, and a mass of 17.4 g?

SOLUTION

$$\text{density} = \frac{\text{mass}}{\text{volume}}, \quad \text{from which}$$

$$\text{volume} = \frac{\text{mass}}{\text{density}} = \frac{17.4 \text{ g}}{5.8 \text{ g cm}^{-3}} = 3.0 \text{ cm}^3$$

This problem can also be solved by using the conversion factor method. In this case the given quantity is 17.4 g and the conversion factor is 1 cm^3/5.8 g:

$$17.4 \cancel{\text{g}} \left(\frac{1 \text{ cm}^3}{5.8 \cancel{\text{g}}}\right) = 3.0 \text{ cm}^3$$

Example 1.8

Calculating Mass from Given Density and Volume

The density of a 38.0 percent (by mass) solution of sulfuric acid (H_2SO_4) is 1.30 g mL^{-1}. How much sulfuric acid (in grams) is in 300 mL of this solution?

SOLUTION

$$d = \frac{m}{V}; \quad \text{from this,} \quad m = Vd = 300 \text{ mL} \left(\frac{1.30 \text{ g}}{1 \text{ mL}}\right) = 390 \text{ g}$$

This is the mass of the entire solution. The mass of sulfuric acid, H_2SO_4, in the solution is 38.0 percent of 390 g or 38 one-hundreths (38/100) of 390 g:

$$390 \cancel{\text{ g solution}} \left(\frac{38.0 \text{ g } H_2SO_4}{100 \cancel{\text{ g solution}}}\right) = 148 \text{ g } H_2SO_4$$

Practice Problem 1.2: A rectangular plate of gold that measures 1.24 cm by 2.06 cm by 5.02 mm has a mass of 0.0247 kg. What is the density of gold in g cm^{-3}? What is the mass of a circular gold coin 2.40 mm thick with a radius of 2.15 cm?

Heat and Temperature

The *ability of matter to do work* is defined as **energy**. **Heat** is a form of energy; other forms of energy include electrical energy, mechanical energy, and the energy of light.

Temperature is the *property of an object that determines whether heat flows to it or away from it spontaneously*. Heat flows spontaneously from a body of higher temperature to a body of lower temperature.

The SI unit of heat and other forms of energy is the joule (J). One **joule** is the *energy of motion possessed by a 2-kilogram mass moving with a velocity of 1 meter per second*. Another unit of energy sometimes used is the calorie (cal). A **calorie** is *the amount of energy required to raise the temperature of 1 gram of water by 1 degree Celsius*. One calorie equals exactly 4.184 J.

The energy content of foods is conventionally given in kilocalories (1 kcal = 1000 cal). These are sometimes called large calories or nutritional calories but are usually referred to as "calories" (Cal) in books on nutrition.

Temperature is measured with a thermometer, which is usually a narrow glass tube connected to a lower reservoir that is filled with mercury or some other liquid. The German physicist Gabriel Fahrenheit (1686–1736) introduced the **Fahrenheit** temperature scale. He designated zero degrees (0°F) as the lowest temperature obtainable using a mixture of ice and salt, 32°F as the freezing temperature of pure water, and 212°F as the temperature at which water boils.

The Swedish astronomer Anders Celsius (1701–1744) originated the **Celsius** scale. On the Celsius scale, the temperature at which water freezes is designated as zero (0°C), and the temperature at which it boils at normal pressure is 100°C. The Celsius scale is often called the **centigrade** scale because the interval between the freezing and boiling points of water is divided into 100 units on the Celsius scale (Figure 1.10).

The international scale of temperature is the **Kelvin** (K) scale. The Fahrenheit, Celsius, and Kelvin thermometer scales are illustrated in Figure 1.10. Note that 1.8 or $\frac{9}{5}$ Fahrenheit units equal 1 Celsius unit, which is also equal to 1 Kelvin unit. As you can see from Figure 1.10, 0°C = 32°F = 273.15 K. In recording Kelvin temperatures, the degree sign is omitted. The relationship

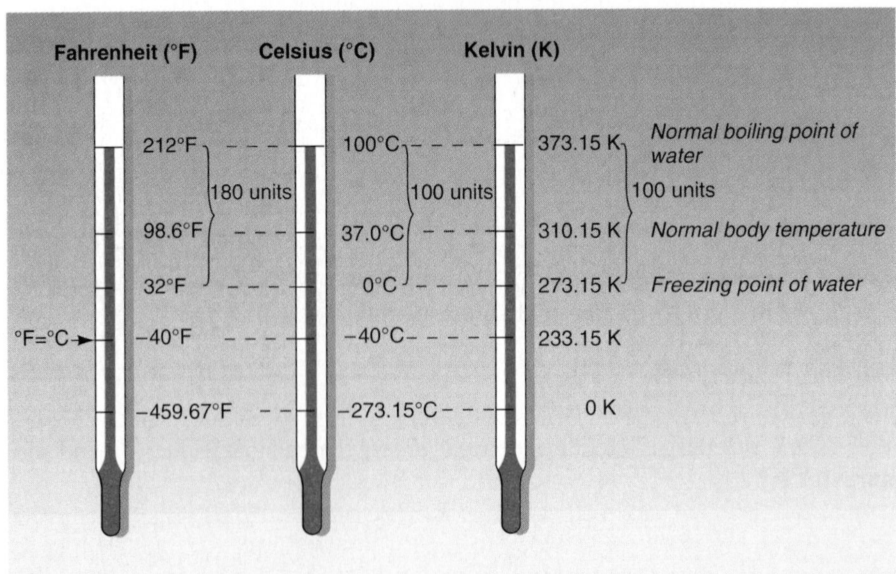

Figure 1.10

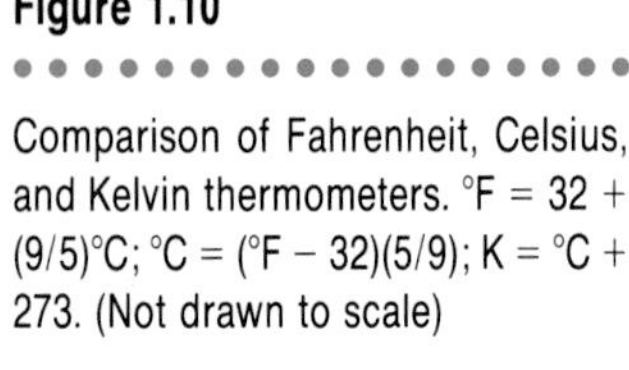

Comparison of Fahrenheit, Celsius, and Kelvin thermometers. °F = 32 + (9/5)°C; °C = (°F − 32)(5/9); K = °C + 273. (Not drawn to scale)

between a reading on the Fahrenheit scale (°F) and a reading on the Celsius scale (°C) is as follows:

$$°F = 32 + 1.8°C \quad \text{or} \quad °F = 32 + \tfrac{9}{5}(°C) \tag{1.2}$$

From this relationship, the Celsius reading corresponding to any given Fahrenheit reading is

$$°C = \frac{°F - 32}{1.8} \quad \text{or} \quad °C = (°F - 32)(\tfrac{5}{9}) \tag{1.3}$$

The relationship between the Kelvin and Celsius scales is

$$K = °C + 273.15 \tag{1.4}$$

However, for most practical purposes,

$$K = °C + 273$$

Example 1.9

Converting Celsius to Fahrenheit and Kelvin

Gold melts when heated to 1065°C. What does this reading correspond to on the Fahrenheit and Kelvin scales?

SOLUTION: One unit on the Celsius scale equals 1.8 or $\tfrac{9}{5}$ units on the Fahrenheit scale, and 1065 Celsius units are equal to $(1065)(\tfrac{9}{5}) = 1917$ Fahrenheit units. The units on the Celsius scale are counted from the 0 mark on that scale, but the Fahrenheit units are counted from the corresponding 32 mark. Thus, the Fahrenheit reading corresponding to the Celsius reading of 1065 is 1917 + 32 = 1949. The same result can be obtained by substituting 1065 for °C in Equation 1.2 and solving for °F:

$$°F = 32 + (\tfrac{9}{5})(1065) = 1949°F$$

The Kelvin reading corresponding to 1065°C is

$$K = 1065 + 273 = 1338 \text{ K}$$

Example 1.10

Converting Fahrenheit to Celsius

If a home thermostat is set at 70°F, what is the equivalent reading on the Celsius scale?

SOLUTION: The Fahrenheit reading of 70 is 70 − 32 = 38 units above the 32 mark. The number of Celsius units corresponding to 38 Fahrenheit units is $(38)/(\tfrac{9}{5}) = (38)(\tfrac{5}{9}) = 21$. This is also the Celsius reading, because the Fahrenheit reading of 32 corresponds to 0 on the Celsius scale. The same result can be obtained by substituting the Fahrenheit reading of 70 into Equation 1.3 and solving for °C:

$$°C = (70 - 32)(\tfrac{5}{9}) = 21$$

Practice Problem 1.3: Lead metal melts at 622°F. What is the equivalent temperature on the Celsius and Kelvin scales?

APPLICATIONS OF CHEMISTRY 1.1
Temperature, Heat Capacity, and Fire Walking

Fire walking refers to an age-old practice in which people walk across red-hot coals whose temperature exceeds 1000°F apparently without pain or physical injury (see photo on right). The mystery of fire walking exists primarily because many people confuse heat with temperature. But they are not the same.

You have probably done something very similar to fire walking without even realizing it. In a hot oven, the oven air, a pan in the oven, and the oven rack are all equally hot because they are all at the same temperature. When you place your hands into the oven briefly, the experience is painless. However, if you were to pick up the pan or touch the oven rack, it would burn you. Although the air, the pan, and the rack are all at the same temperature, the air in the oven contains less heat (it has a lower heat capacity) than does the pan or the rack. Your hands, which are mostly water, have a very high heat capacity (see Chapter 5 for a discussion of heat capacity) compared with air. When heat flows from the hot oven air over your hands, the hot oven air cools quickly. Since your hands have a higher heat capacity, they warm only slightly.

Fire walkers use the same concept. The red-hot embers and pumice rocks that they walk across have much lower heat capacities than those of the firewalker's feet. As heat energy leaves the embers or rock, they cool down quickly, and the fire walker's feet warm up only slightly. As a result, the experience is painless. In fact, one firewalker said that the coals feel like hot peanut shells. You should realize that fire walkers could not walk across a

1000°F hot steel plate for the same reason that you could not pick up a hot pan from the oven. The steel plate has a higher heat capacity than that of embers or pumice rocks.

There are several other reasons why firewalking can be accomplished. One of these is the length of the fire bed. Typical fire beds are only about 10 ft long. Fire walkers cross them in about two or three quick steps. A second factor is that the embers are insulators, not conductors. Once cooled, heat does not quickly flow into the cooled ember. Similarly, hot oven air is an insulator while the oven rack is a conductor.

Warning: Without professional instruction, fire walking can be dangerous. You should not try this on your own.

Source: Ronald DeLorenzo, *Journal of Chemical Education*, November 1986, pages 976–977.

1.3 Recording Measurements

Significant Figures

Measurements of all quantities, regardless of how well the measurements are performed, involve some uncertainties. These uncertainties and the way they are reflected in the recording of the results of measurements are discussed next.

When you repeat a measurement several times with the same measuring instrument, the results you obtain for each measurement will probably be slightly different regardless of how carefully you measure. You can usually see the difference between the measurements in the last digit you record for each measurement. Thus each measurement involves an *uncertainty* that is reflected in the last digit you record for the measurement.

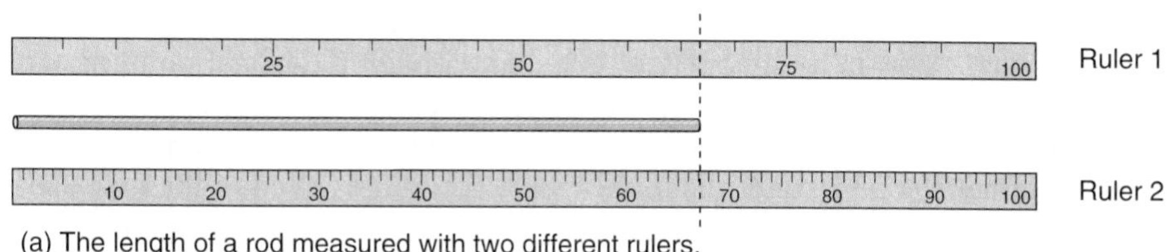

(a) The length of a rod measured with two different rulers.

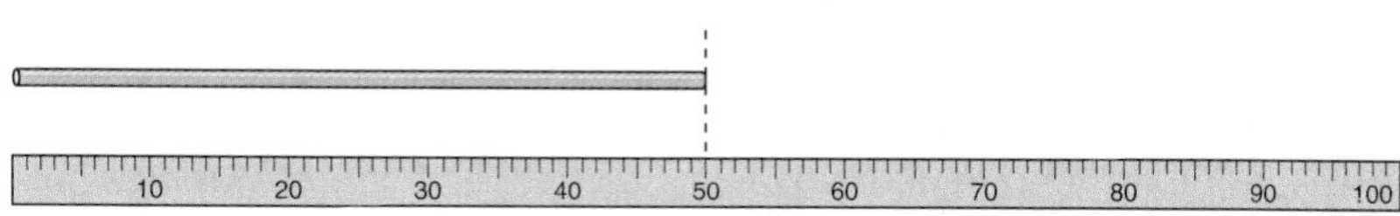

(b) The length of a rod of 50.0 length units measured with ruler 2.

Figure 1.11

Measurement of length with different rulers. (a) Measurements with ruler 1 can be recorded with two digits, but measurements with ruler 2 can be recorded with three digits. (b) If a rod measured with ruler 2 falls exactly at 50 units, the measurement should be expressed in three digits, 50.0.

The *agreement between two or more results* of a measurement is called the **precision** of the measurement. If the generally accepted value of the dimension measured is known, the *agreement between a result of a measurement and the accepted value* is known as the **accuracy** in the measurement. The precision of a measurement depends on the skill of the person conducting the measurement, whereas accuracy depends largely on the quality of the measuring instrument. For example, if the length of a rod is measured with two different rulers, one more finely marked than the other, as shown in Figure 1.11a, two different results should be obtained.

Ruler 1 (Figure 1.11a) indicates that the length of the rod is between 65 and 70 units, possibly 67 units. This measurement involves two digits. Both of these digits are significant in expressing the result of the measurement with that particular ruler. The last digit indicates the *estimated* position of the end of the rod between 65 and 70 unit marks.

According to ruler 2, the length of the rod lies between 67 and 68 units. This measurement involves three digits, the last of which is estimated and involves some uncertainty.

The digits in a recorded measurement that are known with certainty, plus the estimated final digit, are called **significant figures**. If the length of another rod is measured with ruler 2 (Figure 1.11b) and found to be "exactly" 50 units, the result should be recorded as 50.0 units. That number contains three significant figures and correctly expresses the quality of a measurement with that ruler.

In a recorded measurement that has the value of less than one, *all zeros preceding the first nonzero digit* are *not significant* in expressing the quality of a measurement; they only locate the position of the decimal point. For example, 50.0 cm is equal to 0.500 m or 0.000500 km. In each of these values, there are three significant figures: the integer five, the zero that immediately follows it, and the final zero, which involves some uncertainty but is estimated to be zero. Thus, the *final zeros in a measurement that contains a decimal point are always significant*. For example, the number 5.0 contains two significant figures, 0.200 contains three, and 0.008000 contains four.

In a recorded measurement that contains a decimal point, the *nonzero digits*, the *zeros between the nonzero digits*, and the *final zeros* are all significant. Thus, 5.780 contains four significant figures, 50.080 contains five, and

0.050080 also contains five. It is always assumed that only the last digit in a recorded measurement that involves a decimal point is somewhat uncertain.

The final zeros in a number that does not contain a decimal point may or may not be significant, depending on the precision of the measurement. For example, if a distance from one city to another is recorded as 500 km, we do not know whether the uncertainty is in the third, second, or even in the first digit. The number can therefore contain one, two, or three significant figures.

To avoid ambiguity, a measurement like 500 km should be recorded in **exponential notation**. An exponential number consists of two components: a coefficient and a power of 10, 10^x. The number preceding the 10^x component in such a notation is written with only one digit to the left of the decimal point. Thus, 500 km can be written in exponential notation as 5.00×10^2 km, 5.0×10^2 km, or 5×10^2 km, depending on how well the distance is measured. 5.00×10^2 km has three significant figures, 5.0×10^2 km has two, and 5×10^2 km has only one significant figure. Note that the 10^x component of an exponential notation expresses only the magnitude of a measurement and does not indicate the number of significant figures in the measurement.

Practice Problem 1.4: How many significant figures does each of the following measurements contain? **(a)** 10.2 m; **(b)** 0.010 kg; **(c)** 6.02×10^3 g; **(d)** 0.00030 kg; **(e)** 0.03000 kg; **(f)** 1.0003 kg.

Some numbers in conversion factors, and other numbers resulting from such operations as counting objects, are *exact* numbers; they contain no uncertainty and therefore have an infinite number of significant figures. For example, each number in the following statement is an exact number (a number with no uncertainty) that has an infinite number of significant figures: By definition, there are exactly 1000 grams in 1 kilogram, and 1 foot is equal to exactly 12 inches.

Relative Errors

We recall that the last digit of a recorded measurement always involves an uncertainty. Let us consider again a measurement of the length of a rod using the ruler 1 as shown in Figure 1.11a. An uncertainty of ±1 unit is quite likely in this measurement. This error may be expressed as a relative error of 1 part per 67, which is 1.5 parts per 100, or 1.5 percent:

$$\left(\frac{\text{1 unit of uncertainty}}{\text{67 units of measurement}}\right) 100 = 1.5 \text{ percent}$$

In a measurement with ruler 2, as shown in Figure 1.11, the probable uncertainty is ±0.1 unit, and the relative uncertainty is 0.1 part per 67.1, 1 part per 671, or 0.15 percent. All measurements are assumed to have an uncertainty of at least 1 unit in the last digit of the measured quantity.

Repeated measurements of the length of the rod with ruler 2 (Figure 1.11a) produce results that are closer to each other than the results that can be obtained by ruler 1. This means that the measurements of the length of the rod made by ruler 2 are more precise than the measurements made by ruler 1. Thus, you can see that the greater the precision of a measurement, the smaller is the relative error, and the greater is the number of significant figures in the recorded result of the measurement.

Precise measurements are usually accurate; that is, they agree with the generally accepted value. However, sometimes very precise measurements are quite inaccurate. This happens when the same error is incurred in each measurement. Such errors are called "systematic" errors; they can be caused by a faulty balance or by an incorrectly marked ruler. For example, a balance of high precision may have a "built-in" error that causes it to weigh 5 mg too high every time the same object is weighed.

Calculations Involving Significant Figures

A calculation involving measured quantities is recorded to express the same precision as the least precise measurement. This is usually accomplished by applying the following rules:

1. The sum or the difference of two or more numbers should contain as many decimal places as the measurement with the smallest number of decimal places. If an integer is involved in addition or subtraction, the sum or the difference should contain no decimal places unless the integer is an exact number.
2. The product or the quotient of two or more numbers should usually be recorded with as many significant figures as the measurement with the fewest number of significant figures (the least precise measurement). Examples 1.11 and 1.12 illustrate these rules.

Example 1.11

Significant Figures in Adding and Subtracting

(a) Add 5.473 m and 0.0013 m; **(b)** subtract 21.1 g from 76.23 g.

SOLUTION

(a)

$$\begin{array}{ll} 5.473\text{ m} & \leftarrow 3\text{ decimal places} \\ \underline{0.0013\text{ m}} & \underline{\leftarrow 4\text{ decimal places}} \\ 5.474\text{ m} & \leftarrow 3\text{ decimal places} \end{array}$$

(b)

$$\begin{array}{ll} 76.23\text{ g} & \leftarrow 2\text{ decimal places} \\ \underline{-21.1\text{ g}} & \underline{\leftarrow 1\text{ decimal place}} \\ 55.1\text{ g} & \leftarrow 1\text{ decimal place} \end{array}$$

Both the sum and the difference are expressed with the same number of decimal places as in the measurement with the smaller number of decimal places.

Example 1.12

Significant Figures in Multiplying and Dividing

(a) Multiply 0.017 by 35.89; **(b)** divide 0.02586 by 0.250.

SOLUTION

(a) $0.017 \times 35.89 = 0.61$

Of the two numbers multiplied, 0.017 has two significant figures and 35.89 has four. The answer is therefore recorded to two significant figures.

(b) $\frac{0.02586}{0.0250} = 1.03$

Of the two numbers given in the ratio, 0.02586 has four significant figures and 0.0250 has three. Therefore the answer is recorded to three significant figures.

When a product or a quotient obtained on a calculator contains more digits than the necessary number of significant figures, the result has to be shortened or rounded off to the correct number of significant figures. To determine the number of significant figures in a product or a quotient, find the measurement used in the calculation that has the least number of significant figures. This is also the number of digits that should be recorded in the result. A sum or a difference should be recorded with the same number of decimal places as in the measurement that contains the smallest number of decimal places.

Rounding Off Numbers

After the number of significant figures in a calculated result has been determined, the result is rounded off to contain the correct number of significant figures according to the following guidelines:

1. If the first digit to be dropped is less than 5, the preceding digit is not changed. Thus, the number 7.236 rounded off to two significant figures becomes 7.2.
2. If the first digit to be dropped is more than 5, the preceding digit is increased by 1; 5.647 rounded to three significant figures becomes 5.65.
3. If the first digit to be dropped is a 5 that is followed by one or more nonzero digits, the last digit to be retained is increased by 1. For example, both numbers 0.0056 and 0.005001 when rounded to two places to the right of the decimal point become 0.01.
4. If the first digit to be dropped is a 5 that is not followed by any other digits or only by zeros, the last digit to be retained is raised by 1 if odd, and retained unchanged if even. Thus, 6.75 rounds to 6.8, while 5.650 rounds to 5.6.

If measurements are to be added or subtracted, they are sometimes rounded off before the final result is obtained. However, errors can accumulate in rounding several measurements before they are used in a calculation. Therefore, if in doubt, it is best to round only the final answer. The same guideline applies to multiplying and dividing.

Examples 1.13 and 1.14 illustrate the determination of the number of significant figures and rounding of numbers in adding and dividing measured quantities (remember that the rules for subtraction are the same as for addition, and the rules for multiplication are the same as for division).

Example 1.13

Significant Figures and Rounding of Numbers

Three students collected and combined samples of water from a lake for analysis. One student weighed the sample on a precise analytical balance and reported the mass of the sample as 15.0732 g. Another student, using a less precise balance, weighed the sample to the nearest milligram and recorded the mass of the sample as 20.016 g. The third student recorded the mass of the sample as 6.2 g. What is the total mass of the three samples?

SOLUTION: If the addition is carried out with all the numbers as recorded, the result is

$$15.0732 \text{ g} + 20.016 \text{ g} + 6.2 \text{ g} = 41.2892 \text{ g}$$

which rounds to 41.3 g—a result that has only one decimal place, as in 6.2 g.

Practice Problem 1.5: **(a)** What is the total mass of water you would have if you added 5.8 mL of water to 0.5 L of water? Assume the density of the water to be 0.99872 g cm^{-3}. Record your answer to the correct number of significant figures. **(b)** A sample of sugar weighs 25.0 g. How many grams remain after 5 mg is removed? Record your answer to the correct number of significant figures.

Example 1.14

Significant Figures and Rounding of Numbers in a Density Determination

In a density determination, a student weighs an irregularly shaped piece of a metal on a balance and records the mass as 7.0342 g. The volume of the metal is determined by placing the metal into water in a graduated cylinder and observing the increase in the water level (the displacement of water by the metal) as 2.6 mL. Calculate the density of the metal, and record it to the correct number of significant figures.

SOLUTION

$$d = \frac{m}{V} = \frac{7.0342 \text{ g}}{2.6 \text{ mL}}$$

$$= 2.7054615 \text{ g mL}^{-1} \quad \text{(from a calculator)}$$

When the answer is rounded to two significant figures, the result becomes 2.7 g mL^{-1}. This reflects approximately the same relative error as the less precisely known measurement, the volume in this case.

To obtain a density with a greater number of significant figures, a better method of volume determination is needed.

Practice Problem 1.6: In an agricultural experiment, the moisture content of a 15.25-g sample of soil was determined by heating the sample to drive out all water. The mass loss of the sample after heating was 4.28 g. What is the percent moisture in the sample, expressed to the correct number of significant figures?

Example 1.14 shows that a calculator does not determine the number of significant figures. The calculator is merely a computational aid—it reveals nothing about the reliability of calculated results obtained from measurements.

1.4 Matter: Its Properties, Changes, and Classification

The *quantity of matter that an object possesses* is defined as its **mass**. We can also define the mass of an object as a measure of its *resistance to a change in velocity*. For example, the force required to stop an iron ball in its motion is much greater than the force required to stop an equal-sized baseball moving with the same velocity, because the mass of the iron ball is greater than the mass of the baseball.

Properties of Matter

Properties of matter can be physical or chemical. Many **physical properties** of matter *can be observed directly by our senses*; these properties include color, odor, taste, and hardness. Others can be established by certain laboratory operations; these properties include mass, volume, density, melting point, and boiling point. Two different substances may look alike but seldom will have exactly the same physical properties. Thus, different substances can often be identified or distinguished from one another by the physical properties that are characteristic of each.

Properties that determine how a substance changes into one or more other substances as a result of changing conditions, or as a result of bringing a substance into contact with other substances, are **chemical properties**. The *changes of matter in which one or more substances convert into different substances* are called **chemical changes** or **chemical reactions**. For example, a chemical property of oxygen gas is that it converts to water when mixed with hydrogen gas at high temperatures. A chemical property of water is that it decomposes into hydrogen gas and oxygen gas as a result of an applied electric current, a process called electrolysis.

The States of Matter

At normal room temperature and pressure, matter can exist in three different forms or **states of matter**: **solid**, **liquid**, and **gas**. A solid is rigid, and its volume is nearly independent of changes in temperature and pressure. A liquid has a tendency to flow, and it assumes the shape of its container. The volume of a liquid changes only slightly with changes in temperature and pressure. A gas spreads throughout the entire volume of its container, is very compressible, and is capable of indefinite expansion.

To explain the properties of the three states of matter, we assume that all matter consists of small particles in a constant random motion. In the solid state, the particles are relatively tightly packed. They have little or no freedom

APPLICATIONS OF CHEMISTRY 1.2
How Many Days Are There in One Century?

To illustrate the importance of rounding off numbers, let us consider the number of days in a century. There are several ways that we can calculate the number of days in a century, and each of these approaches gives us a slightly different answer. First, let's assume that 1 week contains 7 days, that a month contains 4 weeks, and that a year contains 12 months. Then, using the conversion factor technique that we developed in Section 1.2, we calculate that one century contains 33,600 days.

$$\frac{7 \text{ days}}{\text{week}} \times \frac{4 \text{ weeks}}{\text{month}} \times \frac{12 \text{ months}}{\text{year}} \times \frac{100 \text{ years}}{\text{century}} = \frac{33{,}600 \text{ days}}{\text{century}}$$

In our second approach, let's again use the definition that one week contains seven days. However, this time we will also use the definition that one year contains 52 weeks. Using the conversion factors shown below, we calculate that one century contains 36,400 days.

$$\frac{7 \text{ days}}{\text{week}} \times \frac{52 \text{ weeks}}{\text{year}} \times \frac{100 \text{ years}}{\text{century}} = \frac{36{,}400 \text{ days}}{\text{century}}$$

Why is there a difference in our two answers? In the first calculation, all the ratios are exact conversion factors except the conversion factor 4 weeks/month. One month does not contain exactly 4 weeks. Actually, there are about 4.3 weeks in a 30-day month. Also, some months have more than 30 days, and February alone has only 28 days. The other conversion factors are exact. There are exactly 7 days in 1 week (by definition), there are exactly 12 months per year, and there are exactly 100 years per century.

We can think of the 4 weeks/month as a measurement and not as an exact conversion factor. It is the only nonexact conversion factor in the calculation. Since this nonexact conversion factor contains only one significant digit, we must round our answer off to 3×10^4 days/century.

In our second approach, 52 weeks/year is an inexact conversion factor. This conversion factor contains two significant digits. Therefore, we need to round off our answer to 3.6×10^4 days per century.

Now consider a third approach.

$$\frac{365 \text{ days}}{\text{year}} \times \frac{100 \text{ years}}{\text{century}} = \frac{36{,}500 \text{ days}}{\text{century}}$$

In this approach, 365 days/year is an inexact conversion factor that contains three significant digits. Therefore, we need to round off our answer to 3.65×10^4 days per century.

A more accurate measurement for the number of days in a year is 365.25. (That's why we add one day in leap years.) Using this measurement, we get the following:

$$\frac{365.25 \text{ days}}{\text{year}} \times \frac{100 \text{ years}}{\text{century}} = \frac{36{,}525 \text{ days}}{\text{century}}$$

In this approach, 365.25 days/year is an inexact conversion factor that contains five significant digits. Therefore, our answer should be reported as 36,525 days/century with five significant digits. By the way, a more accurate (but still inexact) value for the number of days/year is 365.242500.

Source: George Lisensky, *Journal of Chemical Education*, July 1990, page 562.

of movement, but they vibrate in their set positions. The particles in the liquid state are free to move throughout the entire sample of the liquid. In the gaseous state, the particles are much farther apart than they are in the liquid state. The expansion of a gas in space can be explained by assuming that the gaseous particles in their random motion spread out as far as possible from one another.

Homogeneous and Heterogeneous Matter

Material substances can be divided into two classes (Figure 1.12). Substances such as water and gasoline that *appear uniform throughout* are called **homogeneous** substances. Homogeneous matter is further divided into pure substances and solutions. **Pure substances** can be either elements or compounds (Figure 1.12). A **solution** is a *homogeneous mixture* such as sugar dissolved in coffee, or iodine dissolved in alcohol (tincture of iodine). A **mixture** can contain two or more different components in varying proportions. A *mixture that consists of two or more visibly different homogeneous components*, such as oil floating on water, is **heterogeneous** (Figure 1.13).

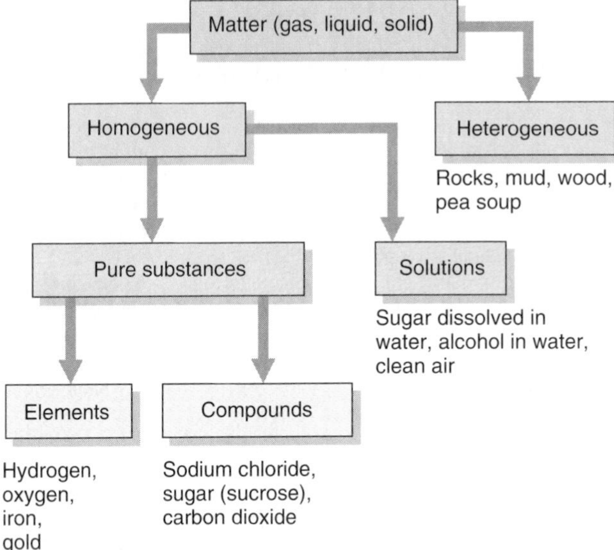

Figure 1.12

Classification of matter.

Figure 1.13

Homogeneous matter: (a) gold (pure), (b) sodium chloride. Heterogeneous matter: (c) rocks showing different types of elements in rock; (d) oil floating in water.

(a)

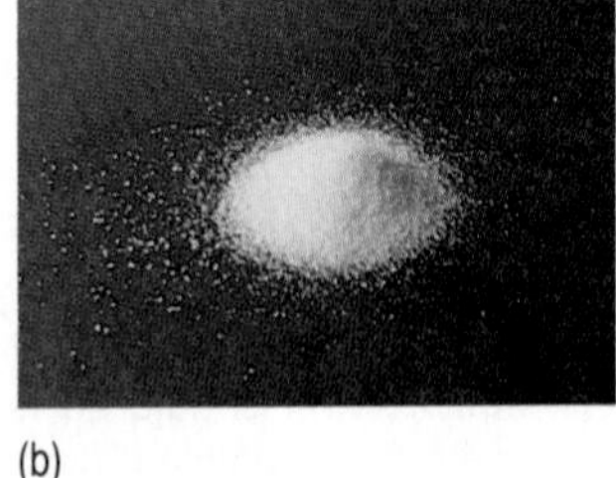

(b)

(c)

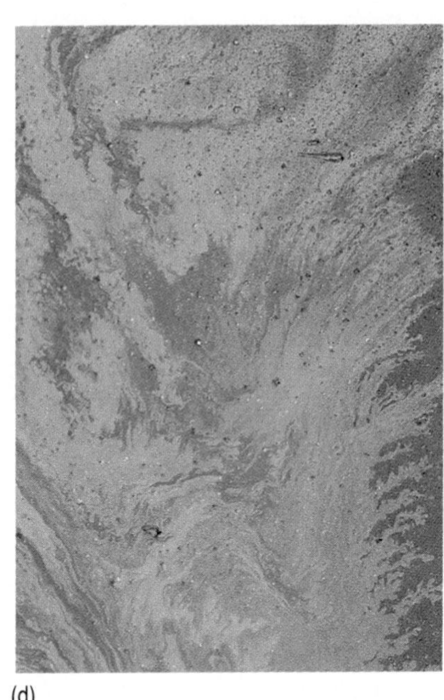

(d)

The different homogeneous portions in a heterogeneous mixture are sometimes called **phases**. For example, a mixture of mercury, water, and oil consists of three different liquid phases (Figure 1.14). A rusty iron nail consists of two different phases of the solid state—the unrusted iron and the rust (an iron oxide).

Elements and Compounds

From ancient times through the nineteenth century, philosophers and scientists have referred to an **element** as a *substance that cannot be converted to other simpler substances*. This definition is still sometimes used today, but see Section 1.6 for a modern definition. Examples of elements include oxygen, nitrogen, mercury, gold, and iodine.

Each element is represented by a symbol containing one, two, or three letters. The *letters and letter combinations that are used to represent elements* are called **symbols**. A symbol can be either one uppercase letter or one uppercase letter followed by one or two lowercase letters. A one-letter symbol is usually the first letter of the name of the element. For example, H is the symbol for hydrogen, O for oxygen, and C for carbon.

The names of some elements, such as sulfur, silicon, strontium, and sodium, start with the same letter. For that reason, one of these elements is usually symbolized by the first letter of its name; others are symbolized by their first letter followed by the second, third, or fourth letters; and most other symbols are derived from their Latin names. Thus, the symbol for sulfur is S, for silicon Si, and for strontium Sr. The symbol for sodium is Na, which is derived from its Latin name, *natrium*. Table 1.5 lists the names, symbols, and physical states of the 15 most abundant elements in the earth's crust, the oceans, and the atmosphere.

Two or more elements can combine chemically; that is, they can undergo a chemical reaction with each other, to form a different substance called a compound. A **compound** is a *pure substance that consists of two or more elements held together by natural forces called chemical bonds*. The properties of a compound are different from the properties of the elements from which the compound is formed. Examples of some common compounds are water, sucrose (table sugar), and sodium chloride (salt).

Figure 1.14

Different phases in a liquid state. When mercury, oil, and water are mixed, mercury falls to the bottom of the container, water forms the middle layer, and oil the top layer. All of these components are in the liquid state, but they are visibly different homogeneous portions, or phases, of this heterogeneous mixture.

Table 1.5

Names, Symbols, and Physical States of the 15 Most Abundant Elements in the Earth's Crust, the Oceans, and the Atmosphere

Name	Symbol	Physical State (Under Ordinary Conditions)	Relative Abundance[a] (percent by mass)
Oxygen	O	Gas	49.4
Silicon	Si	Solid	25.7
Aluminum	Al	Solid	7.5
Iron	Fe	Solid	4.7
Calcium	Ca	Solid	3.4
Sodium	Na	Solid	2.6
Potassium	K	Solid	2.4
Magnesium	Mg	Solid	1.9
Hydrogen	H	Gas	0.87
Titanium	Ti	Solid	0.58
Chlorine	Cl	Gas	0.19
Phosphorus	P	Solid	0.12
Manganese	Mn	Solid	0.09
Carbon	C	Solid	0.08
Sulfur	S	Solid	0.06

[a] In both pure state and in compounds.

(a)

(b)

(c)

Figure 1.15

(a) Elemental chlorine and sodium; (b) reaction of sodium and chlorine; (c) sodium chloride product.

Table 1.6

Sets of Properties for the Reactants and for the Product in the Reaction of Sodium with Chlorine

Reactants				Product
Sodium	+	**Chlorine**	→	**Sodium Chloride**
Silvery solid		Pale green gas		White solid
m.p. 97.81°C		m.p. −100.98°C		m.p. 801°C
b.p. 882.9°C		b.p. −34.6°C		b.p. 1413°C
Reacts with water		Slightly soluble[a] in water		Soluble in water

[a] When a substance is soluble in another substance, the two substances are miscible with one another to form a homogeneous mixture—a solution.

Let us consider the formation of sodium chloride from its elements. Sodium is a solid, and chlorine is a pale green gas. Each of these elements by itself is a dangerous substance. Sodium reacts violently with water, and chlorine is a poisonous gas that killed many soldiers when it was used as a chemical weapon in World War I. Yet when these two dangerous substances are mixed, they combine to form sodium chloride (Figure 1.15), which is common table salt and an ingredient in many foods. Table 1.6 lists some properties for the elements sodium and chlorine and for the compound sodium chloride.

Just as elements are represented by symbols, *compounds are represented by formulas*. A **formula** for a compound is a *combination of symbols of the elements that make up the compound*. Many formulas have subscripts for symbols to describe the exact composition of a compound. (Formulas are discussed in Chapter 2.) The compound sodium chloride has the formula NaCl because this compound consists of the elements sodium (Na) and chlorine (Cl). *Compounds that consist of only two elements* are called **binary compounds**. The names of binary compounds nearly always end with *-ide*. For example, in naming the compound NaCl, the name of the element chlorine is changed to "chlor*ide*."

A chemical reaction is usually described by an equation. A **chemical equation** is a *shorthand method of describing a chemical reaction by using symbols and formulas to designate the substances that change and the new substances that form as a result of the reaction.* The substances that come into contact with each other for a reaction are called **reactants**, and the new substances that form in the reaction are called **products**. In a chemical equation, the reactants are written first, followed by an arrow pointing toward the products. For example, an equation for the reaction that occurs when magnesium is heated with sulfur to form the compound magnesium sulfide is written as follows:

$$\underset{\text{Reactants}}{Mg + S} \longrightarrow \underset{\text{Product}}{MgS}$$

1.5 Development of Atomic Theory of Matter

Ancient Theories About the Nature of Matter

The nature of matter has always held mystery and fascination for human beings. Some ancient philosophers thought that the universe was a continuous whole that could be compressed, expanded, or twisted into different shapes to explain the properties of different material things. This continuous "stuff" was variously named the "boundless," air, fire, and the "one."

In the fifth century B.C., some Greek philosophers suggested that all matter consisted of four "roots": earth, air, fire, and water. Later this idea was largely abandoned for a concept of matter consisting of infinitely divisible, qualitatively different "seeds." Various hard and soft objects of different shapes were explained as mixtures of different seeds packed with varying amounts of empty space between them. Soon thereafter, Leucippus and Democritus (Greek philosophers of the fifth century B.C.) argued that ultimately, the "seeds," as the fundamental "building blocks" of matter, were indivisible. This idea was the first concept of the *fundamental, indivisible material particles* called **atoms**. According to Democritus, all things were composed of atoms having various shapes and sizes. Democritus and his contemporaries thus invented the first **atomic theory** of matter.

The atomic theory of Democritus was scarcely altered until the eighteenth century A.D. With the emerging experimental science in the eighteenth century, a new atomic theory, which to a great extent resembles our modern theory, was developed. This new atomic theory, and the experimentally established laws on which the theory was based, are discussed next.

Figure 1.16

Portrait of Antoine Lavoisier (1743–1794). Lavoisier was executed during the Terror in the French Revolution. Robespierre cryptically remarked that "the revolution has no need for chemists."

Fundamental Laws in Chemistry

One important invention for the development of the new atomic theory in the eighteenth century is credited to the French chemist Antoine Lavoisier (1743–1794) (Figure 1.16). He invented accurate balances to measure the masses of reactants and products in various chemical reactions. He studied the process of burning, or combustion, of substances and concluded that the element oxygen was necessary both for combustion and for processes that supported life. The burning of a candle was one of the reactions studied by Lavoisier. He found that when a candle burns, it combines with oxygen to form water vapor and a substance he called "fixed air," now known as carbon dioxide:

$$\text{candle} + \text{oxygen} \longrightarrow \text{water vapor} + \text{"fixed air"}$$

Careful measurements showed that the total mass of the candle and the oxygen consumed by combustion equals the total mass of water vapor and the "fixed air" produced. The results of this and other similar experiments led to the establishment of the **law of conservation of matter**, also known as the **law of conservation of mass**, which states that *matter can be neither created nor destroyed*. The mass of the reactants in a chemical reaction is equal to the mass of the products formed. The matter originally present in the reactants is still present in the products, but the properties of the products are different from the properties of the reactants.

In 1799, another French chemist, Joseph Proust (1754–1826), analyzed many different compounds and found that *the elements in a compound are present in definite proportions by mass*. This is a statement of the **law of definite proportions**, also known as the **law of constant composition**. For example, sodium chloride always contains 39.32 percent by mass of sodium and 60.68 percent chlorine. Thus, the masses of the elements chlorine and sodium in sodium chloride are always in definite proportions of 60.68 to 39.32 or 1.543 to 1.000. Figure 1.17 illustrates both the law of constant composition and the law of conservation of matter (mass).

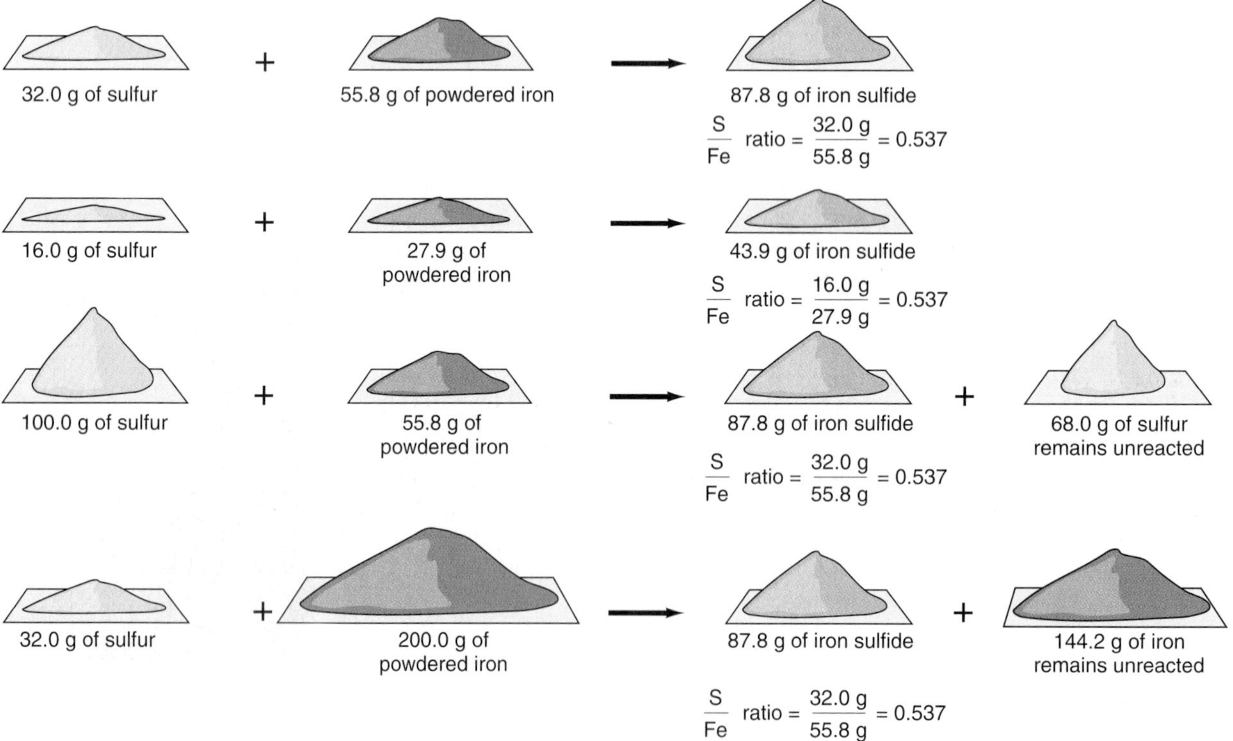

Figure 1.17

Illustrations of the law of conservation of mass and the law of constant composition. The total mass of the reactants equals the mass of the product(s) formed in the reaction (the law of conservation of mass). The ratio of the masses of sulfur and iron that *react* to form iron sulfide is always 0.537 regardless of the amounts of these elements mixed (the law of constant composition).

Example 1.15

The Law of Constant Composition

The compound carbon monoxide, CO, contains 42.9 percent carbon. In accordance with the law of constant composition, how much oxygen would be in combination with 10.0 g of carbon in carbon monoxide?

SOLUTION: The information given above reveals that a 100-g sample of the compound contains 42.9 g of carbon, and 100 g minus 42.9 g, or 57.1 g, of oxygen. The amount of oxygen in combination with 10.0 g of carbon is

$$10.0\,\cancel{\text{g C}}\left(\frac{57.1\text{ g O}}{42.9\,\cancel{\text{g C}}}\right) = 13.3\text{ g O}$$

Practice Problem 1.7: A 3.00-g sample of water, H_2O, contains 2.67 g of oxygen. What is the percent by mass of oxygen and hydrogen in water?

Some elements combine to form two or more different compounds under different experimental conditions. The British scientist John Dalton (1766–1844) found that *in different binary compounds of two elements A and B, the masses of the element B in combination with a fixed mass of A in each of these compounds can be expressed in the ratios of small whole numbers*. This statement is known as the **law of multiple proportions.**

For example, the elements carbon and oxygen form two different compounds, called oxides, under different conditions. Experiments show that if a fixed mass of carbon, for example, 2.4 g, is used to prepare each of these compounds, the mass of oxygen needed to prepare one of the oxides is 3.2 g, and the mass of oxygen needed to prepare the other oxide is 6.4 g:

$$2.4\text{ g C} + 3.2\text{ g O} \longrightarrow 5.6\text{ g oxide 1}$$
$$2.4\text{ g C} + 6.4\text{ g O} \longrightarrow 8.8\text{ g oxide 2}$$

The masses of oxygen in combination with 2.4 g of carbon in these two compounds are in the ratio

$$\frac{6.4\text{ g}}{3.2\text{ g}} = 2.0$$

a small whole number. One oxide is carbon monoxide, with the formula CO, and the other is carbon dioxide, CO_2.

The formula for a compound specifies the relative number of atoms of an element in the compound by a subscript at the right of the symbol for the element (the subscript 1 is omitted). Thus, carbon monoxide, CO, contains one atom of oxygen for each one carbon atom, and carbon dioxide, CO_2, contains two oxygen atoms for each carbon atom.

Figure 1.18

John Dalton.

Dalton's Atomic Theory

In 1805, John Dalton (Figure 1.18) explained the experimentally determined laws of chemical combination in terms of an atomic theory of matter. The main ideas of Dalton's theory are given below.

1. Each element is composed of very small, indivisible particles called atoms that are all alike for a given element but differ from the atoms of other elements.
2. Atoms of two or more elements can combine to form a compound in which the atoms are not changed to different types of atoms or fractions of atoms. Atoms are neither created nor destroyed in a chemical reaction.
3. The relative number of atoms of each element in a given compound is always the same, and is a characteristic of the compound.

Dalton's theory was the first major atomic theory based on experimental evidence. The establishment of this theory was an important cornerstone in the development of chemistry. Modern atomic theory is essentially a refinement of Dalton's original theory. However, the first postulate of Dalton's theory was modified when it was found that atoms of the same element can have slightly different masses.

Atoms of an element that have different masses are known as **isotopes**. It is also now known that atoms are not indivisible but consist of 30 or so smaller particles called subatomic particles (see Section 1.6). However, Dalton was correct in believing that in ordinary chemical reactions, atoms are not divided into fractions of atoms.

Dalton's second postulate is in agreement with the law of conservation of matter. Since the identity or mass of an atom does not change during an ordinary chemical reaction, and since atoms cannot be created or destroyed, the total mass of the elements that combine must equal the mass of the compound formed. Thus, the atoms are "conserved" and the corresponding mass is also "conserved."

The third postulate of Dalton's theory explains both the law of constant composition and the law of multiple proportions. If the relative numbers of atoms of the elements in a compound are in a fixed ratio, the relative masses of the corresponding elements must be in the same ratio.

Molecules

A *neutral group of two or more atoms of the same or different elements held together by natural forces called chemical bonds* is called a **molecule**. Compounds that exist as gases or liquids at room temperature, and many solids as well, consist of molecules. Other solid compounds at room temperature consist of electrically charged particles.

Let us consider the formulas of a few common molecules. A water molecule contains three atoms: two hydrogen atoms and one oxygen atom. Its formula is therefore H_2O. This formula represents the compound water and also a single molecule of water. A molecule of table sugar, sucrose, which has the formula $C_{12}H_{22}O_{11}$, contains 12 carbon atoms, 22 hydrogen atoms, and 11 oxygen atoms, a total of 45 atoms.

The simplest molecules consist of two identical atoms. Such molecules are called *homonuclear diatomic molecules*. Elements that consist of diatomic molecules at room temperature are hydrogen, H_2; nitrogen, N_2; oxygen, O_2; fluorine, F_2; chlorine, Cl_2; bromine, Br_2; and iodine, I_2.

A molecule that consists of three atoms of the same element is called a *homonuclear triatomic molecule*. Three oxygen atoms can bond together to form a molecule of ozone, O_3. Ozone gas is present in the upper atmosphere of

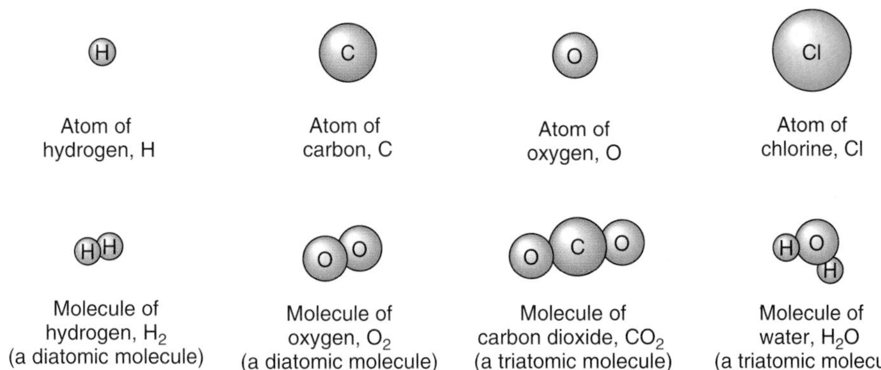

Figure 1.19

Some models of atoms and molecules. (The relative sizes of atoms shown are approximate.) Many types of models are used by chemists to help visualize atoms and the arrangements of atoms in molecules.

the earth, where it helps to block the potentially harmful ultraviolet radiation from the sun. Figures 1.19 and 1.20 illustrate some models of atoms and molecules.

Figure 1.20

Models of molecules (H_2O, CH_4, and CO_2).

1.6 Composition of Atoms

By the early part of the nineteenth century, Dalton's atomic theory became generally accepted, but another century elapsed before it was discovered that atoms have a complex inner structure. We discuss the details of the modern atomic theory in Chapter 6. In the present section we summarize some of the most important features of modern atomic theory.

The Major Components of Atoms

Nearly all of the mass of an atom is concentrated in its center, called the **nucleus**. The major, most important, components of a nucleus are **protons** and **neutrons**. The nucleus of an isotope of hydrogen contains only a single proton. Protons and neutrons are collectively called **nucleons**. A proton has a mass of 1.67265×10^{-24} g and has an electric charge of +1. A neutron has a mass of 1.67495×10^{-24} g and is electrically uncharged, or neutral.

The nucleus is surrounded by negatively charged particles called **electrons**. The electric charge on an electron is −1. The charge on an electron is thus equal and opposite to the charge on the proton. The mass of an electron is 9.10953×10^{-28} g.

The number of protons in a neutral atom equals the number of electrons. Therefore, the total positive charge on the protons and the total negative charge on the electrons cancel.

Atomic Mass Units

Atoms of different elements have different masses. The masses of atoms and their component particles are commonly expressed in atomic mass units (amu). An **atomic mass unit** is *exactly one-twelfth of the mass of a carbon atom that has six protons and six neutrons in its nucleus*. One atomic mass unit equals 1.66056×10^{-24} g.

A carbon atom that has 12 nucleons is called *carbon-12* (symbolized as ^{12}C, or $^{12}_{6}C$, or C-12). The superscript to the left of a symbol of an atom or its component is the number of nucleons in the particle; the subscript is the nu-

Table 1.7

Major Components of Atoms

Name of Particle	Symbol	Symbol with Mass and Charge[a]	Mass (amu) Approx.	Mass (amu) Precise	Mass (g)	Charge[a]
Proton	p	${}^{1}_{1}p$	1	1.00728	1.67265×10^{-24}	+1
Neutron	n	${}^{1}_{0}n$	1	1.00866	1.67495×10^{-24}	0
Electron	e	${}^{0}_{-1}e$	0	0.000548580	9.10953×10^{-28}	−1

[a] Relative to the charge on a proton.

clear charge, which is the number of protons in the nucleus. The masses and charges of the major components of atoms—protons, neutrons, and electrons—are listed in Table 1.7. The charge of an electron is indicated by the subscript −1.

Atomic Number

The *number of protons in the nucleus of an atom* is known as its **atomic number**. All the atoms of a given element have the same number of protons and therefore the same atomic number. Some atoms of an element, however, have different numbers of neutrons and therefore different masses. The *total number of protons and neutrons* (*the number of nucleons*) *in an atom* is the **mass number** of the atom. The number of neutrons in an atom is therefore the difference between the mass number and the atomic number (the number of protons):

$$\text{number of neutrons} = \text{mass number} - \text{atomic number}$$

To indicate both the mass number of an atom and its nuclear charge, the mass number is written as a superscript and the nuclear charge as a subscript to the left of the symbol for the atom. Thus, the symbol ${}^{4}_{2}He$ specifies that the mass number of a helium atom is 4 and its nuclear charge is +2. The nuclear charge of He is due to two protons in the nucleus, and the mass number 4 is the sum of two protons and two neutrons.

Example 1.16

Determining the Number of Protons, Neutrons, and Electrons from Atomic Number and Mass Number

The atomic number of an element is 15. An isotope of this element has a mass number of 31. How many protons and how many neutrons are in the nucleus of an atom of this isotope? How many electrons does an atom of the element have?

SOLUTION: The number of protons is equal to the atomic number, 15. The number of neutrons is the difference between the mass number (which is the total number of nucleons) and the atomic number (the number of protons): 31 − 15 = 16. The number of electrons in a neutral atom is equal to the number of protons in the nucleus of the atom, which is 15. This atom of the element phosphorus can be symbolized as ${}^{31}_{15}P$. This symbol specifies the mass number as a superscript and the nuclear charge as a subscript. The nuclear charge is also the number of protons in the nucleus of the atom: that is, the atomic number for the element to which the atom belongs.

Example 1.17

Determining the Numbers of Nucleons and Neutrons from the Atomic Number and the Mass Number

How many neutrons and how many nucleons does each of the following species contain? **(a)** ${}^{2}_{1}H$; **(b)** ${}^{9}_{4}Be$.

SOLUTION

(a) The number 1 at the lower left of the symbol H is the atomic number of hydrogen, which is also the number of protons in the nucleus of an H atom. The mass number (2) is the number of nucleons—the total number of protons and neutrons. There is one proton, so the number of neutrons is $2 - 1 = 1$.

(b) Using similar reasoning, the number of neutrons in ${}^{9}_{4}Be$ is $9 - 4 = 5$, and the number of nucleons is 9.

Practice Problem 1.8: How many nucleons, how many protons, how many neutrons, and how many electrons are there in each of the following atoms? **(a)** ${}^{11}_{5}B$; **(b)** ${}^{52}_{24}Cr$; **(c)** ${}^{197}_{79}Au$.

Most elements have two or more isotopes. For example, hydrogen has three isotopes, each having a special name. Special names are not normally given to isotopes; the case of hydrogen is an exception. Of the three hydrogen isotopes, the most abundant, ${}^{1}_{1}H$, is simply called hydrogen, sometimes *protium*. A protium atom has a single proton in its nucleus. The elemental hydrogen found in nature is almost entirely protium.

Another isotope of hydrogen is *deuterium*, symbolized ${}^{2}_{1}H$ or D. A deuterium atom contains one proton and one neutron in its nucleus. Only 0.015 percent of elemental hydrogen consists of deuterium. Deuterium, often called *heavy hydrogen*, is a constituent of *heavy water*, D_2O, which is used as a cooling agent in some nuclear reactors.

The third isotope of hydrogen is *tritium*, symbolized ${}^{3}_{1}H$ or T. The nucleus of tritium contains one proton and two neutrons. Tritium is present in naturally occurring hydrogen only in trace amounts. It is unstable and converts slowly to an isotope of helium, ${}^{3}_{2}He$.

Since most elements are mixtures of two or more isotopes, the eighteenth-century definition of an element as a substance that cannot be made simpler (Section 1.4) is not strictly correct. An **element** is now defined as a *substance whose atoms all have the same atomic number*.

Atomic Weight

The *weighted average mass of all the isotopes of an element* is known as the atomic mass, usually referred to as **atomic weight** of the element. The mass of each isotope is weighted according to the natural abundance of the isotope in nature. Atomic weights are conveniently expressed in atomic mass units. We recall that an atomic mass unit is the mass of exactly one-twelfth of the mass of a carbon-12 (${}^{12}C$) atom. Thus, the atomic weights of all the elements are relative to the mass of a carbon-12 atom.

To determine atomic weights, the natural abundance of each isotope of an element must be considered. For example, hydrogen has two naturally occurring isotopes: ${}^{1}H$, which has a mass of 1.0078 amu and constitutes 99.985

percent of all the naturally occurring hydrogen, and ^{2}H, with a mass of 2.0140 amu and with an abundance of 0.015 percent. The fraction of the average mass of the hydrogen atoms due to ^{1}H is therefore

$$(0.99985)(1.0078 \text{ amu}) = 1.0076 \text{ amu}$$

and the fraction due to ^{2}H is

$$(0.00015)(2.0140 \text{ amu}) = 0.00030 \text{ amu}$$

The sum of these two fractions is 1.0079 amu, which is the weighted average mass of the hydrogen atoms, or the atomic weight of hydrogen.

Practice Problem 1.9: Carbon has two naturally occurring isotopes: ^{12}C, which has a mass of 12.0000 amu and an abundance of 98.89 percent, and ^{13}C, which has a mass of 13.00335 amu and an abundance of 1.11 percent. What is the atomic weight of carbon?

Summary

Chemistry and the Scientific Method

Chemistry is a *science dealing with properties, structure, and changes of matter*. The **scientific method** of problem solving involves the processes of *induction* and *deduction*. Induction starts with particular observations that lead to generalizations. Observations are carefully described, and the objects involved are classified. These processes lead to the formation of **hypotheses**, **laws**, **models**, and simple or comprehensive networks of ideas called **theories**. Deduction involves explanation of observable phenomena in terms of models and theories, and predictions of as yet unknown phenomena. In the processes of observation and description, various types of measurements are performed.

Units of Measurement

Measurements involve various units. The **SI** and **metric units** of measurement are used almost exclusively in chemistry today. The abbreviation "SI" represents the *International System of Units*. There are *seven basic SI units*, from which all the other units can be derived. The SI and metric units frequently used in chemistry are *meter*, *centimeter*, *millimeter*, and *nanometer* for length; *kilogram* and *gram* for mass; *liter* and *milliliter* (cubic centimeter) for volume; *Kelvin* and *Celsius* degrees for temperature; and *joule* and *calorie* for energy.

Significant Figures and Recording of Measurements

In a recorded measurement, the *digits that are known with certainty, plus the last digit, which is somewhat uncertain*, are called **significant figures**. A property that is calculated from measured quantities can be known only as well as the least precise measurement. Such a *quantity derived by multiplying or dividing the results of measurements is therefore usually recorded with as many significant figures as in the least precise measurement. A quantity derived by adding or subtracting the results of measurements should be recorded with as many digits to the right of the decimal point as in the measurement with the fewest number of digits to the right of the decimal point.*

Matter: Its Properties, Changes, and Classification

Matter is classified by its appearance and properties as **homogeneous** or **heterogeneous**, and as existing in **solid**, **liquid**, or **gaseous** states. Matter can be described by its **physical** or **chemical properties**. Matter can undergo **physical changes** involving changes only in its form or state, or **chemical changes** that involve the formation of new substances. Chemical changes are also called **chemical reactions**. Matter can consist of **elements**, **compounds** (both referred to as **pure substances**), and **mixtures**. Elements consist of **atoms**. *A neutral unit of two or more atoms bonded together* is called a **molecule**. A **compound** *is a pure substance that consists of two or more elements in definite proportions by mass*. A **molecule of a compound** *contains two or more atoms of different elements*. Mixtures consist of two or more components (elements, compounds, or both) in varying proportions. *Elements are represented by* **symbols**, and *compounds are represented by* **formulas**. A *chemical reaction is described by* a **chemical equation**.

Atomic Theory

The first atomic theory of matter based on experimental evidence was developed by John Dalton. Dalton's theory was based on the observation that *matter can neither be created nor destroyed*—the **law of conservation of matter**—and on the **law of constant composition** (*the composition of a compound is always the same*). According to Dalton's theory, all matter consists of atoms, and the atoms of an element are all alike. In chemical reactions, atoms act as whole entities and not as fractions of atoms. This theory offers a simple explanation for the **law of multiple proportions**: If an atom of an element A combines with one atom of element B in one compound, and with two atoms of element B in another compound, the *masses of the element B per given mass of A in these different compounds are in a ratio of small whole numbers*, 1:2 in this case.

Composition of Atoms

An atom consists of a **nucleus** and one or more **electrons**. The major components of the nucleus of an atom are **protons** and **neutrons**, collectively called **nucleons**. The *number of protons in an atom* is known as the **atomic number**, and the *total number of protons and neutrons in an atom* is the **mass number** for the atom. *Atoms of the same element can have different numbers of neutrons; such atoms have different masses and are called* **isotopes**. The **atomic weight** of an element is *the weighted average mass of all the naturally occurring isotopes* of the element. Atomic weights can be expressed in atomic mass units. An **atomic mass unit** is *exactly one-twelfth of the mass of a carbon-12 atom* (a carbon atom that contains six protons and six neutrons).

Important Relationships

Some important relationships introduced in this chapter are:

1. $d = m/V$; the **density** (d) of an object is the mass (m) of the object divided by its volume (V). From this relationship, any one of these quantities can be calculated from the known values of the other two.
2. $°C = \frac{5}{9}(°F - 32)$; **Celsius temperature** is $\frac{5}{9}$ times the difference between Fahrenheit temperature and 32.
3. $°F = \frac{9}{5}°C + 32$; **Fahrenheit temperature** is Celsius temperature times $\frac{9}{5}$ plus 32.
4. $K = °C + 273$; **Kelvin** temperature is Celsius temperature plus 273.

New Terms

The number in parentheses following each term denotes the section in which the term is introduced and explained.

Accuracy (1.3)
Amorphous (1.1)
Analytical chemistry (1.1)
Atom (1.5)
Atomic mass unit (1.6)
Atomic number (1.6)
Atomic theory (1.5)
Atomic weight (1.6)
Binary compound (1.4)
Biological chemistry (biochemistry) (1.1)
Calorie (1.2)
Celsius temperature (1.2)
Centigrade temperature (1.2)
Chemical change (1.1 and 1.4)
Chemical equation (1.4)
Chemical property (1.1 and 1.4)
Chemical reaction (1.4)
Chemistry (1.1)
Compound (1.4)
Comprehensive theory (1.1)
Conversion factor (1.2)
Crystalline substance (1.1)
Deduction (1.1)
Density (1.2)
Electron (1.6)
Element (1.4 and 1.6)
Energy (1.2)
Exponential notation (1.3)
Fahrenheit temperature (1.2)

Formula (1.4)
Gas (1.4)
Heat (1.2)
Heterogeneous matter (1.4)
Homogeneous matter (1.4)
Hypothesis (1.1)
Induction (1.1)
Inorganic chemistry (1.1)
Isotopes (1.5 and 1.6)
Joule (1.2)
Kelvin temperature (1.2)
Lattice (1.1)
Law of conservation of matter (mass) (1.5)
Law of constant composition (1.5)
Law of definite proportions (1.5)
Law of multiple proportions (1.5)
Liquid (1.4)
Mass (1.2 and 1.4)
Mass number (1.6)
Matter (1.1)
Mixture (1.4)
Model (1.1)
Molecule (1.5)
Neutron (1.6)
Nucleon (1.6)
Nucleus (1.6)
Organic chemistry (1.1)
Phase (1.4)
Physical change (1.1)
Physical chemistry (1.1)
Physical property (1.4)
Precision (1.3)
Product (1.4)
Proton (1.6)
Pure substance (1.4)
Reactant (1.4)
Relative error (1.3)
Scientific law (1.1)
Scientific method (1.1)
Significant figures (1.3)
SI units (1.2)
Solid (1.4)
Solution (1.4)
States of matter (1.4)
Symbol (1.4)
Temperature (1.2)

Exercises

General Review

1. Describe the nature of science and the scientific method.
2. What is chemistry? What are the major branches of chemistry?
3. What is matter?
4. Explain and illustrate each of the following: **(a)** hypothesis; **(b)** law in science; **(c)** theory; **(d)** mental model in science.
5. List the SI units of measurement and their abbreviated symbols for length, mass, time, temperature, and energy.
6. Convert 50.6 m to kilometers, centimeters, millimeters, inches, feet, and miles (1 inch = 2.54 centimeters; 1 mile = 1.6 kilometers).
7. Convert 25.8 kg to grams, milligrams, and pounds (1 kilogram = 2.20 pounds).
8. Convert 12.4 L to milliliters, quarts, and gallons (1 liter = 1.06 quarts). How many liters are equal to 1.75 qt?
9. Convert 80° Fahrenheit to degrees Celsius and to Kelvin.
10. What is the Fahrenheit temperature equal to 30°C? to 300 K?
11. What Celsius temperature corresponds to 77°F?
12. What is the density of a 12.0-g block of a metal that occupies the volume of 1.85 mL?
13. A 15.2-g sample of a liquid has a density of 0.962 g/mL. What is the volume of the sample?
14. Explain in terms of the particle nature of matter why solids, liquids, and gases have different densities.
15. How many significant figures are there in each of the following measurements? **(a)** 15.7 cm; **(b)** 0.056 g; **(c)** 25.00 mL; **(d)** 0.002 m; **(e)** 0.0020 kg; **(f)** 2.0608 g; **(g)** 0.08050 m.
16. Round each of the following measurements to three significant figures: **(a)** 0.8604 m; **(b)** 25.06 mL; **(c)** 0.050752 g; **(d)** 5.285 L.
17. Perform each of the following operations and express the answers to the correct number of significant figures and with the correct units: **(a)** multiply 15.2 m by 0.026 m; **(b)** divide 6.030 g by 2.10 mL; **(c)** add 0.115 g to 25.1 g; **(d)** subtract 0.0028 cm from 32.85 cm.
18. State the laws of conservation of matter, constant composition (definite proportions), and multiple proportions.
19. State the basic ideas of Dalton's atomic theory. How does that theory differ from the modern theory?
20. What is an atomic mass unit?
21. An atom has a mass number of 19 and atomic number of 9. How many protons, neutrons, and electrons does this atom contain?
22. What is the approximate mass (in amu) and the charge of: **(a)** a proton; **(b)** a neutron; **(c)** an electron?
23. What is the difference between the atomic weight of an element and the mass of an atom of the element?

Exponential Numbers and Conversion of Units

24. Express the following numbers in scientific (exponential) notation so that only one nonzero digit precedes the decimal point: **(a)** 1000000; **(b)** 25070; **(c)** 0.000523; **(d)** 0.02073.
25. Express the following numbers in ordinary (nonex-

ponential) decimal form: **(a)** 7.5×10^2; **(b)** 7.51×10^{-2}; **(c)** 1.8×10^{-5}.

26. Carry out the following mathematical operations and express your answers in scientific (exponential) notation: **(a)** $(2.8 \times 10^3)(1.8 \times 10^{-5})$; **(b)** $4.821 \times 10^3 + 6.83 \times 10^2$; **(c)** $[(5.8 \times 10^4)(6.63 \times 10^{-27})]/6.02 \times 10^{23}$; **(d)** $(7.8 \times 10^{-7})^{1/2}$; **(e)** $(7.8 \times 10^{-7})^{1/3}$.

27. Without the help of a conversion table, convert 2.5 m to: **(a)** cm; **(b)** mm; **(c)** km; **(d)** in.; **(e)** ft.

28. Without the help of a conversion table, convert 0.050 kg to: **(a)** g; **(b)** mg; **(c)** lb.

29. Without the help of a conversion table, convert 50 mL to: **(a)** L; **(b)** qt; **(c)** cm^3; **(d)** m^3.

30. Consulting Tables 1.2 and 1.3, convert 9.06 mm to: **(a)** Mm; **(b)** Tm; **(c)** nm; **(d)** am; **(e)** yd.

31. The summer temperature in the United States sometimes reaches 104°F. What are the corresponding readings on the Celsius and Kelvin scales?

32. "Room temperature" is 25°C. What is the corresponding temperature on the Kelvin and Fahrenheit scales?

33. The volume of a fishpond is 15 ft^3. How many gallons of water is required to fill the pond? Consult Table 1.3.

Significant Figures, Percent Error, and Density

34. How many significant figures are there in each of the following quantities? **(a)** 150.0 m; **(b)** 1.5×10^3 m; **(c)** 0.0015 km; **(d)** 0.001500 km; **(e)** the product $(6.02 \times 10^{23})(5785.2)$; **(f)** the quotient 5000.2/2.0; **(g)** the difference $47.552 - 0.22$; **(h)** the result of (4.844)(2.00)/0.020.

35. A measurement of a distance of 1500 m involves an absolute uncertainty of ±2 m. Calculate the percent relative error and the relative error in parts per thousand in this measurement.

36. Perform the following mathematical operations and express the answers to the correct number of significant figures: **(a)** $(47.00)(2.1 \times 10^{20})$; **(b)** $5.2 - 1.8 \times 10^{-5}$; **(c)** $[(2.50)(9.17 \times 10^{-17})]/0.005$; **(d)** $[4.4 + 50]/[3.0 - 0.02]$

37. An object has a mass of four grams and the absolute uncertainty in the weighing is ±0.0001 g. Record the mass of the object to the correct number of significant figures.

38. A piece of metal with the mass of 8.426 g is submerged in water in a graduated cylinder. The water level rises from 15.6 mL to 16.5 mL. Calculate the density of the metal, and record it to the correct number of significant figures.

39. A flask has a mass of 62.520 g when empty and 96.472 g when filled with an unknown liquid. When filled with water, the flask has a mass of 101.728 g. The density of water in the flask is 0.99998 g cm^{-3}. Calculate: **(a)** the volume of the flask; **(b)** the density of the unknown liquid.

40. The density of a metal is 8.94 g cm^{-3}. A sample of this metal has a volume of 50.3 mL. Calculate the mass of the sample.

41. The density of a metal is 11.3 g cm^{-3}. Which of the following quantities contains the greatest mass of that metal: 1 lb, 0.5 kg, or 0.05 L?

42. A solution of sulfuric acid has a density of 1.30 g cm^{-3} and is 38.0 percent sulfuric acid by mass. What is the mass of 30.0 mL of this solution? How many grams of sulfuric acid (H_2SO_4), and how many grams of water, are present in 30.0 mL of the solution? How many milliliters of water is present in the sample? (Assume that the density of water is 0.9987 g mL^{-1}.)

Scientific Method, Chemistry, and Matter

43. Distinguish among hypothesis, law, and theory in science.

44. What is the difference between scientific proof and logical proof?

45. What are the major branches of chemistry? What is the primary concern of each branch?

46. Explain as well as you can how the phenomenon of diffusion (such as the uniform and spontaneous mixing of ink with water) supports the particle theory of matter.

47. Classify each of the following as an element, compound, or mixture: **(a)** water; **(b)** mercury; **(c)** sodium chloride; **(d)** sugar (sucrose); **(e)** toothpaste; **(f)** bread; **(g)** iron, **(h)** cloth; **(i)** ice; **(j)** steam; **(k)** iodine; **(l)** wood; **(m)** pea soup.

48. How many different phases of matter are present in each of the following? **(a)** A cocktail with ice cubes; **(b)** a cocktail in which all the ice has melted; **(c)** a sample of clean air; **(d)** the contents of a cup of coffee with cream and sugar; **(e)** milk with butter fat on the surface; **(f)** a rusty nail; **(g)** salt and pepper.

49. Which of the following changes of matter involve the formation of a new element or compound? **(a)** Chopping wood; **(b)** fermenting sugar; **(c)** evaporating water; **(d)** adding ice to water; **(e)** sharpening a knife; **(f)** adding sodium to water to produce hydrogen gas; **(g)** a reaction of sodium with chlorine?

Symbols, Formulas, and Equations

50. Decide whether each of the following items is a chemical symbol, a chemical formula, a chemical equation, or none of these: **(a)** NaCl; **(b)** water; **(c)** H; **(d)** CaO;

(e) Ca + S → CaS; **(f)** $d = m/V$; **(g)** S_8; **(h)** C; **(i)** carbon; **(j)** CO_2; **(k)** °C = $\frac{5}{9}$(°F − 32).

51. What is the total number of atoms in a molecule of each of the following compounds? **(a)** CO_2; **(b)** H_2SO_4; **(c)** C_3H_8; **(d)** $C_6H_{12}O_6$.

The Laws of Definite and Multiple Proportions

52. An 18.0-g sample of water contains 16.0 g of oxygen and 2.02 g of hydrogen. According to the law of definite proportions, what is the mass in grams of oxygen and of hydrogen in 50.0 g of water?

53. **(a)** In a sample of a gold bromide, 197 g of gold is combined with 80.0 g of bromine. A 218.5-g sample of another gold bromide contains 98.5 g of gold. How many grams of bromine is in combination with 98.5 g of gold in the second sample? **(b)** Show how these data illustrate the law of multiple proportions.

54. Under certain conditions (1), 35.45 g of chlorine combines with 18.62 g of iron, and under different conditions (2), 17.72 g of chlorine combines with 13.96 g of iron. **(a)** What is the mass of chlorine that would combine with 10.00 g of iron under each of these conditions? What laws are illustrated by these data? **(b)** If one atom of iron unites with three atoms of chlorine under the conditions described by case (1) above, how many atoms of chlorine unite with one atom of iron in the second case?

55. Tin and oxygen can combine to form two different oxides. One of these oxides contains 78.77 percent tin, and the other contains 88.12 percent tin. **(a)** How many grams of oxygen would be required to react with 5.00 g of tin to prepare each of these oxides? **(b)** What laws are illustrated by these data?

Composition of Atoms

56. What percent of the mass of an ^{1}H atom is due to the mass of its nucleus? Use the necessary data from Table 1.7.

57. Assume that an atom is spherical in shape and its diameter is 1.0×10^{-10} m. Assume also that the nucleus of the atom is spherical with a diameter of 1.0×10^{-15} m. What percent of the volume of the atom is occupied by its nucleus?

58. What is: **(a)** atomic number; **(b)** mass number; **(c)** atomic weight?

59. What is the mass in amu and the charge on each of the following particles? **(a)** a proton; **(b)** a neutron; **(c)** an electron.

60. The mass number of an atom is 19 and the atomic number of the atom is 9. The atom contains how many: **(a)** nucleons; **(b)** protons; **(c)** electrons; **(d)** neutrons?

61. How many protons, electrons, and neutrons are there in each of the following atoms? **(a)** ^{12}C; **(b)** ^{4}He; **(c)** ^{35}Cl; **(d)** ^{23}Na.

62. What are the sign and the magnitude of the charge on the nucleus of a chromium atom? What are the sign and the magnitude of the charge of all the electrons present in a chromium atom?

63. Using the data in Table 1.7, calculate the mass in grams of: **(a)** a deuterium atom, ^{2}H; **(b)** an oxygen atom, ^{16}O; **(c)** a molecule of heavy water, D_2O.

64. The element nitrogen has two naturally occurring isotopes. One of these has a mass of 14.00308 amu and an abundance of 99.635 percent; the other isotope has a mass of 15.00011 amu and an abundance of 0.365 percent. Calculate the atomic weight of nitrogen.

65. Silver has two naturally occurring isotopes, one with a mass of 106.90509 amu, the other with a mass of 108.9047 amu. The atomic weight of silver is 107.868. Calculate the percent abundance of each of these isotopes of silver.

Miscellaneous Exercises

66. Add the following masses and express the answer in grams to the correct number of significant figures: 5.0 g + 20 mg + 4.000008 kg.

67. The gasoline mileage of an automobile is 17.0 miles/gallon. What is this mileage in kilometers/liter?

68. A graduated cylinder contains 50.0 mL of water. Uniform stones, each weighing 5.000 g and having a density of 2.50 g cm^{-3}, are placed into the graduated cylinder until the water level rises to 130.0 mL. How many stones are in the cylinder?

69. A perfect cube of gold with an edge 2.000 cm long is beaten into a uniform rectangular sheet 160.0 cm long and 5.000 cm wide. Calculate the thickness of the sheet in millimeters, assuming that the density of the gold is unchanged in the process.

70. A pycnometer is a carefully constructed flask of a specific volume with a ground-glass stopper. Pycnometers are used to determine densities of liquids more accurately than is possible with graduated cylinders. In an experiment a pycnometer with a volume of 10.00 mL was filled with water with a density of 1.000 g cm^{-3}. The total mass of the pycnometer and water was 37.245 g. The same pycnometer, containing a pile of small metal pellets and filled with water, weighed 57.835 g. The dry pellets alone weighed 22.697 g.

What is the density of the pellets? Record your answer with the correct units and to the correct number of significant figures.

71. Calculate the density of a solution formed by adding 25.00 mL of carbon tetrachloride (density = 1.6 g cm^{-3}) to 75.00 mL of carbon tetrabromide (density = 3.40 g cm^{-3}). (Assume that total V is 100.00 mL.)

72. A kilogram of meat in a German supermarket costs 12.03 German marks. How many dollars does a pound of this meat cost if 1.00 dollar = 1.82 marks?

73. The density of gold is 19.3 g cm^{-3}. What is the mass in pounds of 1 cubic foot of gold?

74. Convert 3.754 in. to feet and express the answer to the correct number of significant figures.

CHAPTER 2

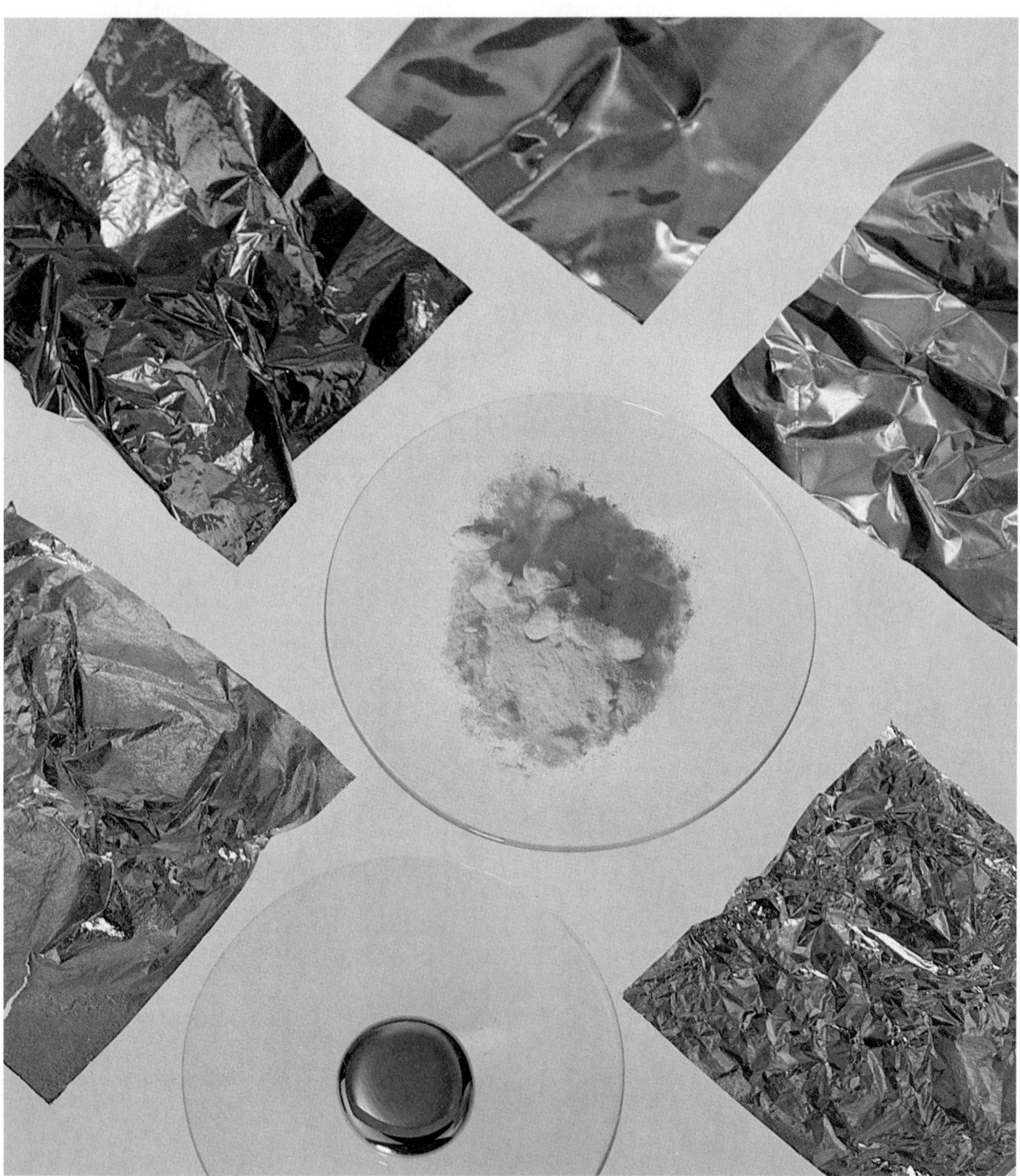

A group of elements. Center: sulfur; clockwise from top: tin, aluminum, gold, mercury, silver, and copper.

Elements, Compounds, and Chemical Reactions

Figure 2.1

Dmitri Mendeleev (1834–1907).

In this chapter we describe some properties of elements, how elements are classified, and how they react with each other to form compounds. We also discuss how chemical reactions can be described by chemical equations. One simple way of classifying chemical reactions is also discussed in this chapter.

2.1 Classification of Elements and Their Periodicity

Mendeleev's Periodic Table of Elements

During the nineteenth century, several attempts were made to classify the 65 elements that had been discovered. The most significant of these efforts were carried out independently by the Russian chemist Dmitri Mendeleev (1834–1907) (Figure 2.1), and the German chemist Lothar Meyer (1830–1895). Mendeleev and Meyer arranged all the known elements into a table. The elements in this table were arranged in the order of their increasing atomic weights. The table has columns designated by Roman numerals; these columns are called **groups**.

Mendeleev received most of the credit for this early classification of elements because he published his table first and because he used his table to make predictions. Mendeleev's table (Figure 2.2) was published in English in 1871. This table is very similar to some forms of the periodic tables used at the present time.

Mendeleev discovered that the properties of the elements vary in a regular and predictable way. He found that the same properties recurred periodically when the elements were arranged in the order of their increasing atomic weights. This generalization was called the **periodic law.**

For example, elements in the same group form similar compounds when they react with oxygen or with chlorine. Table 2.1 shows that the formulas of the oxygen-containing compounds of lithium and sodium in the first group are Li_2O and Na_2O, respectively; and the formulas for the chlorine-containing compounds are LiCl and NaCl. In the second group, beryllium forms BeO and $BeCl_2$, and magnesium forms MgO and $MgCl_2$. Mendeleev designated the elements in each group by the letter R (Table 2.1) and indicated the general formulas of oxygen and chlorine compounds of each element at the head of each group of elements. Mendeleev's classification of the elements was based on his experimental observations of the properties of elements, and it preceded any theory that might have suggested such an arrangement.

When Mendeleev made his original periodic table, it contained some empty spaces. The elements now known to occupy these spaces had not yet been discovered. Mendeleev predicted the existence of these missing elements from their anticipated properties, which he expected would lie between the properties of the element above and below and to either side of the missing element. The atomic weights of these elements were predicted to be 44, 68, and 72. All of these elements were discovered in his lifetime.

Because of its predictive power, Mendeleev's periodic table was an important cornerstone for chemistry and for other sciences as well. However, the unifying theory that explains the periodic law and the corresponding arrange-

Row	Group I R_2O	Group II RO	Group III R_2O_3	Group IV RH_4 RO_2	Group V RH_3 R_2O_5	Group VI RH_2 RO_3	Group VII RH R_2O_7	Group VIII RO_4
1	H = 1							
2	Li = 7	Be = 9.4	B = 11	C = 12	N = 14	O = 16	F = 19	
3	Na = 23	Mg = 24	Al = 27.3	Si = 28	P = 31	S = 32	Cl = 35.5	
4	K = 39	Ca = 40	— = 44	Ti = 48	V = 51	Cr = 52	Mn = 55	Fe = 56, Co = 59, Ni = 59, Cu = 63
5	(Cu = 63)	Zn = 65	— = 68	— = 72	As = 75	Se = 78	Br = 80	
6	Rb = 85	Sr = 87	?Yt = 88	Zr = 90	Nb = 94	Mo = 96	— = 100	Ru = 104, Rh = 104, Pd = 106, Ag = 108
7	(Ag = 108)	Cd = 112	In = 113	Sn = 118	Sb = 122	Te = 125	I = 27	
8	Cs = 133	Ba = 137	?Di = 138	?Ce = 140				
9								
10			?Er = 178	?La = 180	Ta = 182	W = 184		Os = 195, Ir = 197, Pt = 198, Au = 199
11	(Au = 199)	Hg = 200	Tl = 204	Pb = 207	Bi = 208			
12				Th = 231		U = 240		

Figure 2.2

Mendeleev's periodic table, published in English in 1871. The elements are arranged according to their increasing atomic weights relative to the weight of 1 assigned to the hydrogen atom. The letter R in the column headings is the general symbol for an element in the column. The formulas under the group numbers are the formulas for the compounds the element R in each group forms with oxygen and with hydrogen. Elements in parentheses indicate the continuation of the preceding period. Spaces are left beneath B, Al, and Si for elements unknown at that time.

ment of elements in a periodic table was not developed for nearly a half a century, when a new theory of physics, called quantum mechanics, was applied to chemistry.

The Modern Periodic Table and Classification of Elements

In the modern periodic table, the *elements are arranged in the order of their increasing atomic numbers* rather than by atomic weights. This new arrangement of elements gives a better correlation between their properties and their positions in the periodic table.

The modern periodic table also contains more elements than did Mendeleev's version. For example, an entirely new group of elements, called the **noble gases**, was discovered in 1894. These are helium (He), neon (Ne), argon (Ar), krypton (Kr), xenon (Xe), and radon (Rn). Several other elements were discovered in the twentieth century.

Some sections of Mendeleev's table (Table 2.1) correspond to the modern periodic table (Table 2.2). In the modern periodic table, the atomic number of an element is usually indicated above the symbol of the element and the atomic weight usually below. The *rows* in the periodic table are called **periods**, and the *columns* are called **groups** or **families**. The modern periodic table is divided by a heavy zigzag line. To the left of this line are the metals and to the right are the nonmetals. Hydrogen is an exception. Although hydrogen is a nonmetal, it is the first member of group I in most periodic tables because it forms compounds that have formulas similar to the corresponding compounds of other group I elements—HCl, NaCl, H_2O, Li_2O. In some periodic tables, hydrogen is also the first member of group VII because it sometimes forms compounds that have formulas similar to the corresponding compounds of other group VII elements—NaCl, NaH, MgF_2, MgH_2. In other periodic tables, hydrogen is placed in both groups I and VII.

Table 2.1

Properties of Some Elements Listed in Mendeleev's Periodic Table[a]

Group I R_2O, RCl	Group II RO, RCl_2	Group III R_2O_3, RCl_3	Group IV RO_2, RCl_4	Group V R_2O_5, RCl_3	Group VI Na_2R, BeR	Group VII NaR, BeR_2
Li, lithium Soft metal, low density, very reactive, forms Li_2O and LiCl	*Be, beryllium* Much harder than Li, low density, less reactive than Li, forms BeO and $BeCl_2$	*B, boron* Very hard nonmetal, not very reactive, forms B_2O_3 BCl_3	*C, carbon* Brittle nonmetal, unreactive at room temperature, forms CO_2 and CCl_4	*N, nitrogen* Gas, not very reactive, forms N_2O_5 and NCl_3	*O, oxygen* Gas, quite reactive, forms Na_2O and BeO	*F, fluorine* Gas, very reactive, poisonous, forms NaF and BeF_2
Na, sodium Soft metal, low density, very reactive, forms Na_2O and NaCl	*Mg, magnesium* Much harder than Na, low density, less reactive than Na, forms MgO and $MgCl_2$	*Al, aluminum* As hard as Mg, somewhat less reactive than Mg, forms Al_2O_3 and $AlCl_3$	*Si, silicon* Brittle nonmetal, unreactive, forms SiO_2 and $SiCl_4$	*P, phosphorus* Solid, low melting point, reactive, forms P_2O_5 and PCl_3	*S, sulfur* Solid, low melting point, reacts with most elements, forms Na_2S and BeS	*Cl, chlorine* Gas, very reactive, poisonous, forms NaCl and $BeCl_2$
K, potassium Soft metal, low density, very reactive, forms K_2O and KCl	*Ca, calcium* Much harder than K, low density, less reactive than K, forms CaO and $CaCl_2$					

[a] Mendeleev designated an element in each group by the letter R. The formulas of the compounds that the element R in each group forms with oxygen and with chlorine are listed on top of each group I through V. In groups VI and VII, the formulas for sodium and beryllium compounds of R are listed.

Elements in groups I through VIII (Table 2.2) are known as **main group elements**; these are also called **representative elements** because their properties reveal the periodic trends in the most "representative" or predictable way. Elements in columns not numbered by Roman numerals in Table 2.2 are called the **transition elements** or **transition metals**, and **inner transition metals**. The latter include **lanthanides** and **actinides** (also called **lanthanoids** and **actinoids**). The lanthanides are so named because they are preceded by the element lanthanum (La) in the periodic table, and the actinides are preceded by actinium (Ac).

Several elements along the diagonal zigzag line are called **metalloids**. Some properties of metalloids—particularly their electrical conductivity—resemble the properties of metals; other properties resemble nonmetals. The metalloids include boron (B), silicon (Si), germanium (Ge), arsenic (As), antimony (Sb), tellurium (Te), polonium (Po), and astatine (At). Some metalloids, particularly silicon and germanium, are used in virtually every electronic device.

As we move from left to right across a period (except the first one, whose only members are hydrogen and helium), there is a transition from metals to

Table 2.2

Modern Periodic Table of the Elements[a]

Key: Period number; 1 — Atomic Number; Hydrogen — Name; H — Symbol; 1.0079 — Atomic Weight; 1s — Valence electrons

Period		Representative Elements (*s* Series) I 1	II 2	Transition Metals (*d* Series of Transition Elements) 3	4	5	6	7	8	9	10	11	12	Representative Elements (*p* Series) III 13	IV 14	V 15	VI 16	VII 17	VIII 18
1	1s	1 Hydrogen **H** 1.0079																	2 Helium **He** 4.0026
2	2s2p	3 Lithium **Li** 6.941	4 Beryllium **Be** 9.0122											5 Boron **B** 10.811	6 Carbon **C** 12.0112	7 Nitrogen **N** 14.0067	8 Oxygen **O** 15.9994	9 Fluorine **F** 18.9984	10 Neon **Ne** 20.179
3	3s3p	11 Sodium **Na** 22.989	12 Magnesium **Mg** 24.305											13 Aluminum **Al** 26.9815	14 Silicon **Si** 28.086	15 Phosphorous **P** 30.9738	16 Sulfur **S** 32.064	17 Chlorine **Cl** 35.453	18 Argon **Ar** 39.948
4	4s3d4p	19 Potassium **K** 39.098	20 Calcium **Ca** 40.08	21 Scandium **Sc** 44.956	22 Titanium **Ti** 47.90	23 Vanadium **V** 50.942	24 Chromium **Cr** 51.996	25 Manganese **Mn** 54.938	26 Iron **Fe** 55.847	27 Cobalt **Co** 58.933	28 Nickel **Ni** 58.71	29 Copper **Cu** 63.546	30 Zinc **Zn** 65.38	31 Gallium **Ga** 69.723	32 Germanium **Ge** 72.59	33 Arsenic **As** 74.922	34 Selenium **Se** 78.96	35 Bromine **Br** 79.904	36 Krypton **Kr** 83.80
5	5s4d5p	37 Rubidium **Rb** 85.468	38 Strontium **Sr** 87.62	39 Yttrium **Y** 88.905	40 Zirconium **Zr** 91.22	41 Niobium **Nb** 92.906	42 Molybdenum **Mo** 95.94	43 Technetium **Tc** (99)	44 Ruthenium **Ru** 101.07	45 Rhodium **Rh** 102.905	46 Palladium **Pd** 106.4	47 Silver **Ag** 107.868	48 Cadmium **Cd** 112.40	49 Indium **In** 114.82	50 Tin **Sn** 118.69	51 Antimony **Sb** 121.75	52 Tellurium **Te** 127.60	53 Iodine **I** 126.904	54 Xenon **Xe** 131.30
6	6s(4f) 5d6p	55 Cesium **Cs** 132.905	56 Barium **Ba** 137.34	*57 Lanthanum **La** 138.91	72 Hafnium **Hf** 178.49	73 Tantalum **Ta** 180.948	74 Tungsten **W** 183.85	75 Rhenium **Re** 186.2	76 Osmium **Os** 190.2	77 Iridium **Ir** 192.2	78 Platinum **Pt** 195.09	79 Gold **Au** 196.967	80 Mercury **Hg** 200.59	81 Thalium **Tl** 204.37	82 Lead **Pb** 207.19	83 Bismuth **Bi** 208.980	84 Polonium **Po** (209)	85 Astatine **At** (210)	86 Radon **Rn** (222)
7	7s(5f) 6d	87 Francium **Fr** (223)	88 Radium **Ra** (226)	**89 Actinium **Ac** (227)	104 Unnilquadium **Unq** (261)	105 Unnilpertium **Unp** (262)	106 Unnilhexium **Unh** (263)	107 Unnilseptium **Uns**	108 Unniloctium **Uno**	109 Unnilennium **Une**									

Inner Transition Elements (*f* Series)

Lanthanides	4f	58 Cerium **Ce** 140.12	59 Praseodymium **Pr** 140.907	60 Neodymium **Nd** 144.24	61 Promethium **Pm** 144.913	62 Samarium **Sm** 150.35	63 Europium **Eu** 151.96	64 Gadolinium **Gd** 157.25	65 Terbium **Tb** 158.925	66 Dysprasium **Dy** 162.50	67 Holmium **Ho** 164.930	68 Erbium **Er** 167.26	69 Thulium **Tm** 168.934	70 Ytterbium **Yb** 173.04	71 Lutetium **Lu** 174.97
Actinides	5f	90 Thorium **Th** 232.038	91 Protactinium **Pa** (231)	92 Uranium **U** 238.03	93 Neptunium **Np** (237)	94 Plutonium **Pu** 244.064	95 Americium **Am** (243)	96 Curium **Cm** (247)	97 Berkelium **Bk** (247)	98 Californium **Cf** 242.058	99 Einsteinium **Es** (254)	100 Fermium **Fm** 257.095	101 Mendelevium **Md** 258.10	102 Nobelium **No** 259.101	103 Lawrencium **Lr** 260.105

[a] A number in parentheses is the mass number of the isotope of longest known half-life. Main groups are numbered with Roman numerals. All the groups are numbered with Arabic numerals.

Numbers 1–18 are group numbers recommended by the International Union of Pure and Applied Chemistry (IUPAC).

nonmetals. For example, the first two elements of the second period, lithium and beryllium, are metals; the third element, boron, is a metalloid; and the remainder of the elements are nonmetals. Each period ends with a noble gas.

Some groups of elements in the periodic table have been given group names. For example, the metallic elements in group I are called **alkali metals**, a term that originated in the Arabic language to describe the ash (Arabic *alqili*) produced when certain plants are burned. Oxides and hydroxides of these metals are caustic or "alkaline."

The group II metals are called **alkaline earth metals**. The elements in group VII are the **halogens**, a name derived from the Greek language meaning "salt former." The group VIII elements are called **noble gases** because they do not readily undergo chemical reactions. The term "noble gas" also has a historic antecedent since the nobility did not readily "mix" with the lower classes.

2.2 Metals and Nonmetals

In this section we briefly survey the properties of metals and nonmetals. In later sections of this chapter we describe a few reactions of metals with nonmetals. In later chapters, particularly in Chapters 9 and 18, the chemistry of nonmetals and metals, respectively, is considered in greater detail.

Properties of Metals

Metals have several characteristic physical properties that are familiar to all of us from our daily experience. These properties include their colors, electrical and thermal conductivity, hardness, density, and melting point. Most pure metallic elements have a *silvery luster*, except copper and gold, whose colors have entered the language as the names of the metals themselves (Figure 2.3). Metals are generally *good conductors of electricity and heat*. Some metals are soft; others are very hard. For example, the alkali metals (group I) are relatively soft and can be cut with a knife (Figure 2.4). Others, like chromium, are

Figure 2.3

A gold ring, a silver necklace, and a copper bracelet.

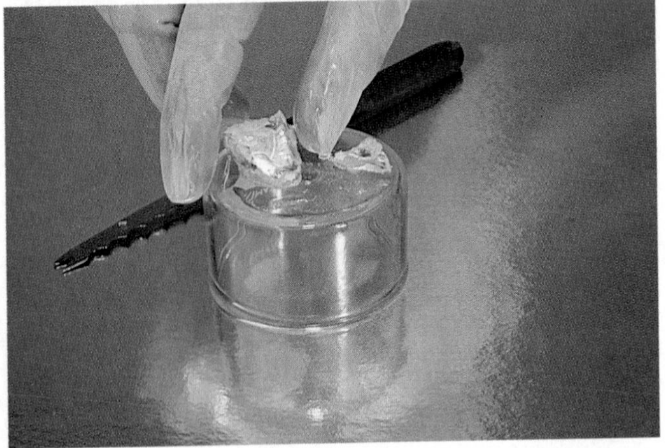

Figure 2.4

A piece of sodium that was cut with a knife.

very hard. Some metals, such as lead and iron, can be hammered to different shapes and are thus said to be **malleable**. Many metals can also be drawn into a wire and are said to be **ductile**.

Metals have densities that vary from just above 0.5 g cm^{-3} for lithium to slightly more than 22 g cm^{-3} for certain transition metals, such as platinum. The densities of most metals, excepting lithium, sodium, and potassium, are considerably higher than that of water (1 g cm^{-3}).

One of the metals, mercury, is a liquid at room temperature; all the others are solids. A metal in its solid state consists of an ordered structure of tightly packed atoms. This structure is disrupted when the metal melts. The melting points of metals range from about room temperature to thousands of degrees Celsius.

Among the main group metals with low melting points are cesium and gallium, which melt at about 29°C and 30°C, respectively, slightly above room temperature. At the other extreme are germanium and beryllium, which melt at 937°C and 1278°C, respectively. The melting points of the transition metals (with the exception of the zinc family) are generally much higher than the melting points of most of the representative metals, ranging from 920°C for lanthanum to 3410°C for tungsten, the metal with the highest melting point.

Properties of Nonmetals

Like metals, the nonmetals have characteristic physical properties. All metals but mercury are solids, in contrast to nonmetals, whose physical states at room temperature (25°C) range from gas to solid. The gaseous nonmetals are hydrogen, nitrogen, oxygen, fluorine, chlorine, and the noble gases. The rest of the nonmetals are solids, except bromine, which is the only liquid at 25°C.

Nonmetals typically lack the luster of metals, and they generally have lower densities than metals. In contrast to metals, solid nonmetals are often brittle and are generally *poor conductors of electricity and heat*. Some nonmetals consist of arrays of molecules containing two or more atoms; others exist as large networks of atoms having definite rigid structures (see Section 2.8).

The common gaseous nonmetals—hydrogen, nitrogen, oxygen, fluorine, and chlorine—as well as bromine (a liquid) and iodine (a solid) exist at room temperature as diatomic molecules (molecules composed of two atoms). These molecules are illustrated in Figure 2.5 in the order in which the corresponding elements appear in the periodic table. The noble gases exist as single atoms.

Binary Metal–Nonmetal Compounds

Most metals react with most nonmetals to form compounds. A reaction of a metal with a nonmetal produces a compound that contains only two elements. A *compound that consists of only two elements* is called a **binary compound**. A binary compound of a metal and a gaseous nonmetal can be prepared in a laboratory using the type of apparatus shown in Figure 2.6. A photograph of magnesium burning in oxygen is shown in Figure 2.7.

Most binary metal–nonmetal compounds are solids at room temperature and normal atmospheric pressure. Most are relatively high melting, and brittle. Some, such as sodium chloride, are soluble in water; others, such as silver chloride, are insoluble. Many binary metal–nonmetal compounds are electrical

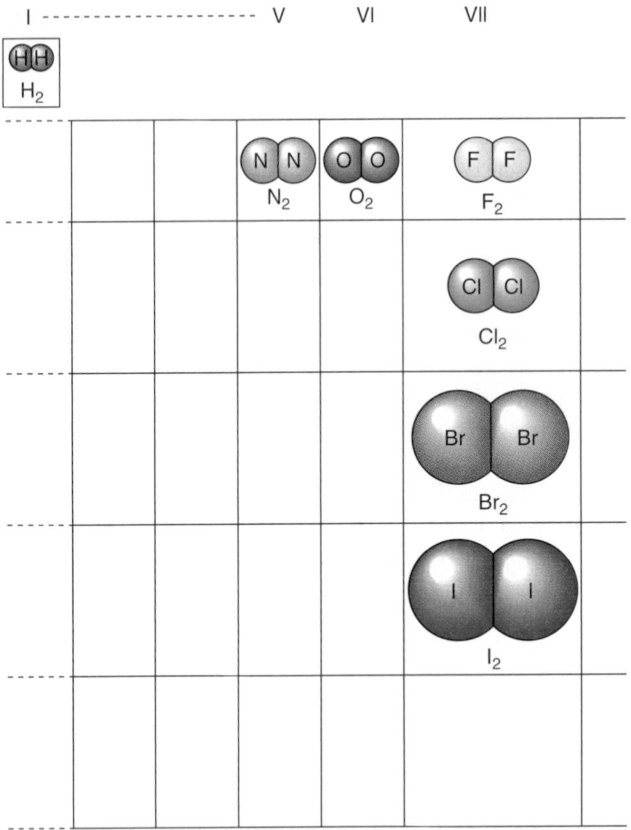

Figure 2.5

Nonmetals that exist as diatomic molecules (as elements) under normal conditions. The elements are arranged according to their positions in the periodic table. The approximate relative sizes of the molecules are also shown.

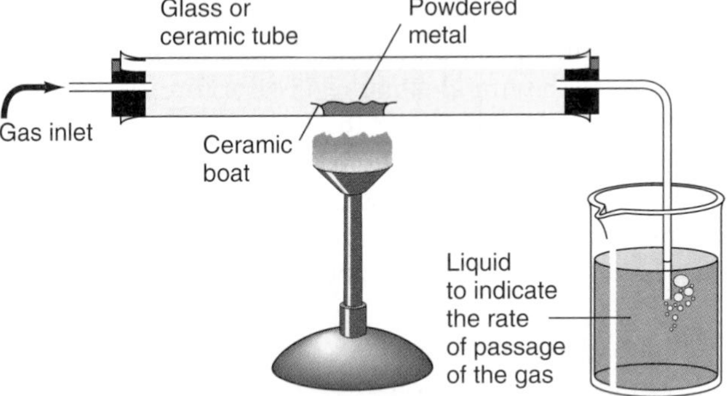

Figure 2.6

A simple method of preparing a binary compound of a metal and a gaseous nonmetal. The gaseous nonmetal replaces the air in the reaction tube and reacts with the metal at a suitable temperature.

conductors in their molten states or in water solution. The electric conductivity of these compounds suggests that they contain electrically charged particles.

An *electrically charged particle formed from an atom or a group of atoms* is called an **ion** (Greek *ion*, to go). *Substances that consist of ions* are called **ionic compounds**. Most ionic compounds are called **salts**. Perhaps the most common salt is "table salt," sodium chloride (NaCl).

Monatomic Ions of Common Metals

In the reaction of a metal such as sodium or magnesium with a nonmetal such as oxygen or chlorine, the metal atom loses electrons and forms a positively charged ion, as shown schematically in Figure 2.8 for sodium and magnesium.

Figure 2.7

Magnesium burning (reacting with oxygen), and the magnesium oxide (white product) that is a product of the reaction.

The electrons lost by the metal atoms are gained by the nonmetal atoms to form negative ions.

An atom is electrically uncharged, or neutral, because the number of (negatively charged) electrons in the atom equals the number of (positively charged) protons in the nucleus of the atom. When an electrically neutral atom loses one or more electrons, the ion that forms has more protons than electrons, and therefore has a net positive charge. The magnitude of the positive charge on the ion equals the number of electrons lost by the neutral atom, or the difference between the number of protons and electrons in the ion.

For example, Figure 2.8 shows that a sodium atom has 11 protons and 11 electrons. When this atom loses an electron to form a sodium ion, the ion has 11 protons and 10 electrons. The charge on the ion is therefore 1+, which is

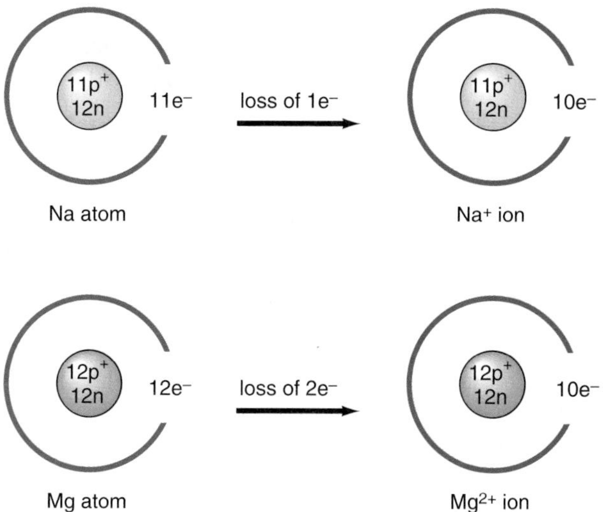

Figure 2.8

Formation of positive ions from neutral atoms. When (negative) electrons are lost by a neutral atom, the ion that forms has a positive charge because it has more protons than electrons.

indicated as a superscript at the right of the symbol for sodium, Na^+. A magnesium atom has 12 protons and 12 electrons. In a chemical reaction with a nonmetal, this atom loses 2 electrons to form an ion that has 12 protons and 10 electrons; the ion has a charge of 2+ and is symbolized by Mg^{2+}.

An atom of most main group elements forms an ion that contains the same number of electrons as that of the noble gas atom whose atomic number is closest to that of the atom from which the ion is derived. We recall that the noble gas atoms are called "noble" because they resist chemical reaction. An ion that has the same number of electrons as its nearest noble gas atom is generally less reactive than its corresponding atom.

For example, the noble gas nearest sodium (atomic number 11) is neon (atomic number 10). Thus, a sodium atom loses one electron to form an ion that has the same number of electrons as a neon atom (10). *Atoms and ions that have the same number and arrangement of electrons* are said to be **isoelectronic**.

The noble gas nearest calcium is argon, whose atomic number is 18. A calcium atom (20 electrons) loses 2 electrons to form a Ca^{2+} ion (18 electrons), which is isoelectronic with an argon atom. Similarly, an aluminum atom (13 electrons) loses 3 electrons to form an Al^{3+} ion, which is isoelectronic with a neon atom (10 electrons). We can see that an atom of a group I metal loses 1 electron to form a 1+ ion, an atom of a group II metal loses 2 electrons to form a 2+ ion, and an atom of a group III metal can lose 3 electrons to form a 3+ ion.

Example 2.1

Writing Symbols and Charges for Metal Ions

Write the symbols and charges for ions of the following metals: **(a)** potassium; **(b)** barium. How many electrons does each of these ions contain? In each case, which noble gas atom is isoelectronic with the ion?

SOLUTION

(a) A potassium atom has 19 electrons, and an atom of its nearest noble gas, argon, has 18 electrons. A potassium atom must lose 1 electron to form a K^+ ion with 18 electrons, which is isoelectronic with an Ar atom.

(b) The noble gas atom nearest to barium is xenon (atomic number 54). A barium atom must lose 2 of its 56 electrons to form a Ba^{2+} ion with 54 electrons, which is isoelectronic with an Xe atom.

Practice Problem 2.1: Write the symbol and charge for an ion of each of the following metals: **(a)** lithium; **(b)** beryllium. The ions should be isoelectronic with a helium atom.

Up to this point we have discussed main group metals that form ions whose charge equals the group number of the metal. A few main group metals and many transition metals form ions with more than one charge.

The first metal in group III, aluminum, forms only 3+ ions, while the heavier members of the group form both 1+ and 3+ ions. The 3+ charge is more common for the lighter members, aluminum and gallium, of the group, and the 1+ charge is more common for the heavier members, indium and thallium.

Table 2.3

Common Monatomic Ions of Metals and Nonmetals. The species in parentheses are less common than others.

I	II											III	IV	V	VI	VII
H^- H^+																
Li^+	Be^{2+}													N^{3-}	O^{2-}	F^-
Na^+	Mg^{2+}											Al^{3+}		P^{3-}	S^{2-}	Cl^-
K^+	Ca^{2+}	Sc^{3+}	Ti^{2+} Ti^{3}	V^{2+} V^{3+}	(Cr^{2+}) Cr^{3+}	Mn^{2+}	Fe^{2+} Fe^{3+}	Co^{2+}	Ni^{2+}	(Cu^+) Cu^{2+}	Zn^{2+}	Ga^{3+}		As^{3-}	Se^{2-}	Br^-
Rb^+	Sr^{2+}	Y^{3+}								Ag^+	Cd^{2+}	In^{3+} In^+	Sn^{2+}		Te^{2-}	I^-
Cs^+	Ba^{2+}	La^{3+}								(Au^+)	Hg_2^{2+}* Hg^{2+}	(Tl^{3+}) Tl^+	Pb^{2+}	Bi^{3+}		

* The commonly encountered diatomic ion for mercury (I)

The metals of group IV form 2+ ions. In group V the only predominantly metallic element is bismuth (Bi), which usually forms 3+ ions. Table 2.3 lists the symbols and charges of ions for common metals.

The members of the first group of transition metals, scandium, yttrium, and lanthanum, form only ions of 3+ charge. This charge can be predicted by assuming that the number of electrons of the ion equals the number of electrons of the preceding noble gas atom. However, the charges of most other transition metal ions cannot be predicted by the same line of reasoning. All the fourth-period transition metals, except scandium, form 2+ ions (Table 2.3). Metal ions, their compounds, and their reactions are discussed in more detail in Chapters 18, 19, and 22.

Monatomic Ions of Nonmetals

We recall that nonmetals form negative ions when they react with metals. A nonmetal atom tends to gain one or more electrons to form an ion that is isoelectronic with the nearest noble gas atom. For example, a fluorine atom (9 electrons) gains one electron to form a 1− ion (10 electrons), which is isoelectronic with a neon atom. An oxygen atom (8 electrons) gains two electrons to form a 2− ion, and a nitrogen atom (7 electrons) gains three electrons to form a 3− ion. Both O^{2-} and N^{3-} ions are also isoelectronic with a neon atom. Three of the nonmetals, boron, carbon, and silicon, have almost no tendency to form ions, and the ions of phosphorus and arsenic are not very common.

A hydrogen atom can lose an electron to form a positive hydrogen ion, H^+, which, however, is very reactive and therefore so short-lived that it does not exist independently. A hydrogen atom can also gain an electron to form a negative ion, H^-, which is isoelectronic with a helium atom.

Naming Ions

Many *positively charged monatomic ions are named as the atoms* from which the ions are formed. For example, Na is sodium atom, and Na^+ is sodium ion. Similarly, Mg is magnesium atom, and Mg^{2+} is magnesium ion.

The *names of negative monatomic ions always end with -ide*. Thus, an F^- ion is called fluoride; an O^{2-} ion, oxide; an H^- ion, hydride; and an N^{3-} ion, nitride. Table 2.3 lists the symbols and charges of the common monatomic ions of nonmetals. The *ions of the halogens*, F^-, Cl^-, Br^-, and I^-, are referred to collectively as *halide ions*.

Example 2.2

Identifying Isoelectronic Species

Which of the following species have the same number of electrons: Ar, Al^{3+}, Sc^{3+}, P^{3-}, Ne?

SOLUTION: An atom of Ar, as well as the Sc^{3+} and P^{3-} ions, have 18 electrons each and are therefore isoelectronic with each other. An atom of Ne and an Al^{3+} ion, with 10 electrons each, constitute an isoelectronic pair.

Practice Problem 2.2: Write the symbol and charge for each of the following nonmetal ions that are isoelectronic with an argon atom: **(a)** phosphorus; **(b)** sulfur; **(c)** chlorine.

2.3 Formulas and Names of Binary Metal–Nonmetal Compounds

Compounds of Metals That Form Ions of Only One Charge

We recall that the elements in group I form only 1+ ions, that those in group II form only 2+ ions, and that aluminum (in group III) forms only 3+ ions. The charges of these ions are identical to the group numbers of the elements. The formulas and names of the compounds of these elements are therefore easiest to learn and so are considered first.

A binary ionic compound has a formula in which the *symbol of the positive ion is written first, followed by the symbol of the negative ion*. No charges are written for the ions in the formula because a *compound* is electrically uncharged, or neutral. The total charge of the positive ions in the compound is balanced by the total charge of the negative ions. The charges of these oppositely charged ions thus sum to zero. The name of a binary ionic compound consists of two parts: *first the name of the positive ion, then the name of the negative ion*. The latter, we recall, *ends with -ide*.

Binary ionic compounds of alkali metals and halogens are collectively called *alkali halides* because they consist of alkali metal ions and halide ions. For example, the compound formed from lithium metal and bromine has the formula LiBr and the name lithium bromide. Since lithium bromide contains equal numbers of Li^+ and Br^- ions, it is electrically neutral.

When an alkali metal reacts with an element of group VI that forms ions of 2− charge, the resulting neutral compound contains two alkali metal ions, with a charge of 1+ for each doubly negative ion of the group VI element. Thus, the formula for lithium oxide is Li_2O, and the formula for potassium sulfide is K_2S. An atom of group V nonmetal forms a 3− ion. Three 1+ ions are needed to balance a 3− ion. Thus, the formula for sodium phosphide is Na_3P.

An alkaline earth metal ion has a 2+ charge. A binary compound of an alkaline earth metal and a halogen has two halide ions of 1− charge for each alkaline earth metal ion of 2+ charge. Thus, the formula for magnesium chloride is $MgCl_2$, and for barium fluoride, BaF_2.

A binary compound of a group II metal and a group VI nonmetal consists of an equal number of positive and negative ions because the 2+ charge on the alkaline earth metal ion equals the 2− charge on the group VI nonmetal ion. For example, the formulas for calcium sulfide and beryllium oxide are CaS and BeO, respectively.

Other metals react with most nonmetals to form binary ionic compounds. The formulas of some of these compounds are a little more complicated, as shown by Example 2.3, but in writing formulas we must remember that *every compound is electrically neutral.* Thus, the total charge of the positive ions equals the total charge of the negative ions. Therefore, *the subscripts for the positive and negative ions in the formula for a binary ionic compound must be chosen to indicate the minimum relative numbers of positive and negative ions required for electrical neutrality.*

Example 2.3

Writing Formulas of Binary Compounds

Write the formulas for magnesium nitride and for aluminum oxide.

SOLUTION: A magnesium ion has a 2+ charge and a nitride ion has a 3− charge. To determine the minimum relative numbers of Mg^{2+} and N^{3-} ions that must be present for the compound to be neutral, find the lowest common multiple of 2 and 3, which is 6. This means that the compound must have three Mg^{2+} ions (a total of 6+ charge) for every two N^{3-} ions (a total of 6− charge). The formula for magnesium nitride is therefore Mg_3N_2.

Similarly, aluminum oxide contains two Al^{3+} ions (a total of 6+ charge) for three O^{2-} ions (a total of 6− charge). Therefore, the formula for aluminum oxide is Al_2O_3.

Practice Problem 2.3: What is the formula for each of the following compounds? **(a)** aluminum fluoride; **(b)** lithium phosphide; **(c)** calcium bromide.

Compounds of Metals That Form Ions of More Than One Charge

We recall that the representative metals in group III (except aluminum), and most transition metals, can form ions of more than one charge (Table 2.3). A metal that forms ions of two different charges can form different compounds with the same nonmetal under different conditions.

Group IV metals form ionic compounds such as $SnCl_2$ and $PbCl_2$ which contain 2+ metal ions. Tin and lead also form $SnCl_4$ and $PbCl_4$, respectively. These are **molecular compounds** (compounds that consist of molecules), although they *appear* to contain 4+ metal ions. The two "tin chlorides" and the two "lead chlorides" clearly need different names.

To name different binary compounds of a metal with the same nonmetal, a method developed by the German chemist Alfred Stock (1876–1946) is used. According to the "Stock method," the metallic element of the compound is named first, followed by its real or apparent charge in Roman numerals enclosed in parentheses. There is no space between the name of the metal and the parentheses. Next, after a space following the parentheses, the nonmetal is named, ending with "-ide." Thus, $PbCl_2$ is lead(II) chloride, $PbCl_4$ is lead(IV) chloride, $SnCl_2$ is tin(II) chloride, and $SnCl_4$ is tin(IV) chloride. Similarly, PbO is lead(II) oxide, and SnO_2 is tin(IV) oxide.

Next, let us name some compounds formed from commonly encountered transition metals. Chromium and iron both form ions of 2+ and 3+ charge (Table 2.3), and copper forms ions of 1+ and 2+ charge. Mercury also forms ions of 2+ charge, and a diatomic ion, Hg_2^{2+}, called mercury(I) ion. In a mercury(I) ion, two Hg^+ ions are bonded together. Below are some examples of naming compounds by the Stock method:

Two oxides of iron (FeO and Fe_2O_3).

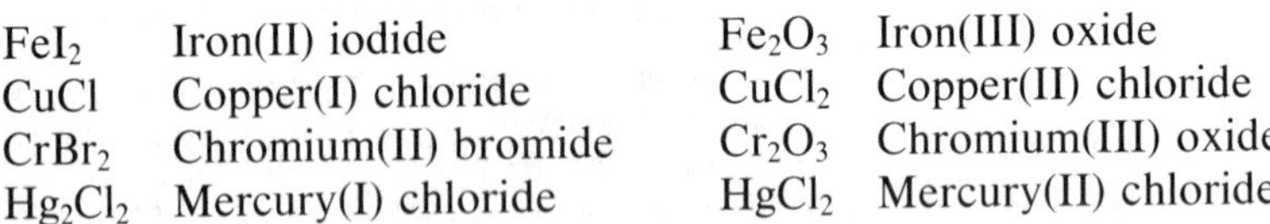

FeI_2	Iron(II) iodide	Fe_2O_3	Iron(III) oxide
CuCl	Copper(I) chloride	$CuCl_2$	Copper(II) chloride
$CrBr_2$	Chromium(II) bromide	Cr_2O_3	Chromium(III) oxide
Hg_2Cl_2	Mercury(I) chloride	$HgCl_2$	Mercury(II) chloride

Although the Stock method of naming metal ions of different charge is now used by most chemists, an older nomenclature system also exists. It is still often used but is no longer the preferred method. In this older method of naming metal ions of different charges, the name of the *ion of lower charge* ends with *-ous*, and the name of the *ion of higher charge* ends with *-ic*. For example, Cr^{2+} ion is named chromous ion, and Cr^{3+} ion is called chromic ion.

Many metals have Latin names. The ions of these metals are named by adding *-ous* or *-ic* to their Latin stems. For example, the symbol Cu for copper is derived from its Latin name cuprum, whose stem is *cupr-*. Thus, Cu^+ is called cuprous ion, and Cu^{2+} is cupric ion. There is an exception to the "Latin stem rule." The Latin name for mercury, Hg, is hydrargyrum, but the Latin stem is not used in naming mercury compounds. Below are some examples of naming compounds by this older method.

SnO	Stannous oxide (from Latin *stannum* for tin)	Cu_2O	Cuprous oxide (from Latin *cuprum* for copper)
SnO_2	Stannic oxide	CuO	Cupric oxide
FeS	Ferrous sulfide (from Latin *ferrum* for iron)	Hg_2Cl_2	Mercurous chloride
$FeCl_3$	Ferric chloride	$HgCl_2$	Mercuric chloride

Example 2.4

Naming Binary Compounds of Transition Metals

Name each of the following compounds using the Stock system of nomenclature: **(a)** CuF_2; **(b)** Cu_2S; **(c)** CrO; **(d)** Cr_2O_3.

SOLUTION

(a) In CuF_2 there are two fluoride ions for each copper ion. The charge on a copper ion must therefore be 2+. Thus, the name of the compound is copper(II) fluoride.

(b) A sulfide ion has a 2− charge, and each of the two copper ions in a Cu_2S unit must therefore have a 1+ charge for the compound to be electrically neutral. The name of the compound is copper(I) sulfide.

(c) From the 2− charge on an oxide ion we conclude that the charge on the chromium ion in CrO is 2+, and the name of the compound is chromium(II) oxide.

(d) In a Cr_2O_3 unit there are three oxide ions of 2− charge amounting to a total of 6− charge in the unit. Each of the two chromium ions in the unit must therefore have a charge of 3+, and the name of the compound is chromium(III) oxide.

Practice Problem 2.4: An oxide of iron is a black powder used in manufacture of green, heat-absorbing glass. The formula for this oxide is FeO. Name this oxide by the Stock method and by the traditional older nomenclature.

Another oxide of iron is called iron(III) oxide. This oxide is the principal ingredient of rust. It is used as a paint pigment. What is the formula for this oxide?

2.4 Ionic and Covalent Bonding

Ionic Bonding

Thus far we have considered ionic compounds of metals and nonmetals. An ionic compound contains positive and negative ions that are held together by an *attractive force* known as **ionic bonding**, also called **electrovalent bonding**. The force of attraction, f, between two oppositely charged ions is directly proportional to the charges on the ions and inversely proportional to the square of the distance, d, between the ions:

$$f = \frac{q_1 q_2}{d^2}$$

where q_1 and q_2 represent the magnitudes of the positive and negative charges on the ions. This equation is a mathematical statement of Coulomb's law, named after the French physicist Charles Augustine de Coulomb (1736–1806).

The ions in a solid ionic compound are arranged or ''packed'' in a characteristic *geometric pattern* called a *lattice*, as shown in Figure 2.9 for sodium chloride, NaCl, as one example.

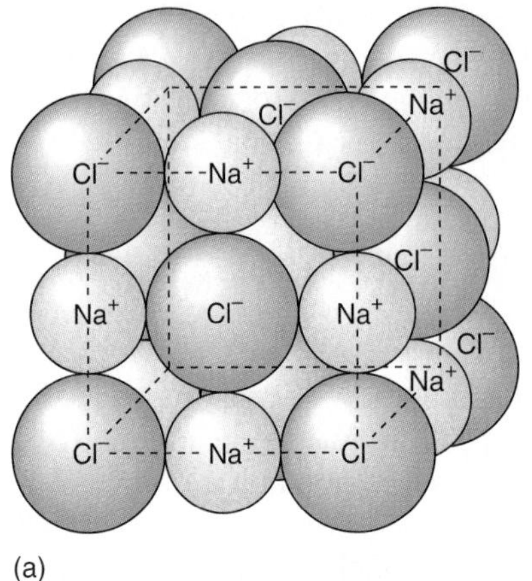

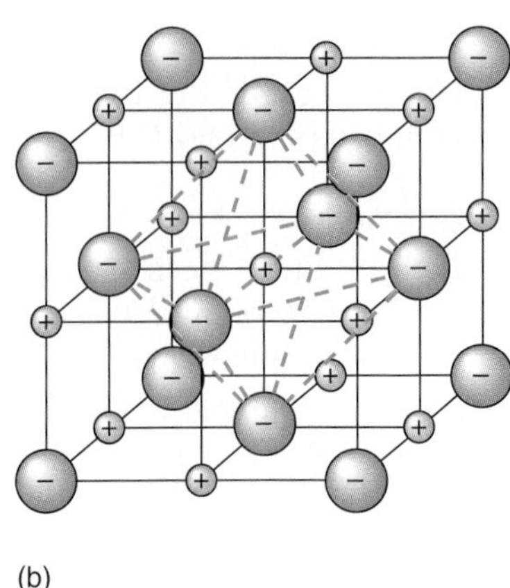

Figure 2.9

Sodium chloride lattice: (a) arrangement of Na^+ and Cl^- ions in NaCl; (b) in this structure the distances between ions are exaggerated to show that each Na^+ ion is surrounded by six Cl^- ions. By extension of this structure it can be shown that each Cl^- ion is surrounded by six Na^+ ions as well.

Covalent Bonding

We recall that nonmetallic elements can form compounds that contain molecules as their basic structural units. We also recall that compounds that consist of molecules are called molecular compounds. Such compounds include liquids such as water and ethyl alcohol, and gases such as ammonia, carbon dioxide, and methane. Molecular models of these compounds are illustrated in Figure 2.10. Many solid compounds also consist of molecules. These compounds include oxides of phosphorus such as tetraphosphorus decoxide and a simple sugar glucose (also called blood sugar) (Figure 2.11), and many organic compounds, such as aspirin and naphthalene (mothballs).

Figure 2.10

Molecular models of liquid and gas molecules (water, ammonia, carbon dioxide, methane, and ethyl alcohol).

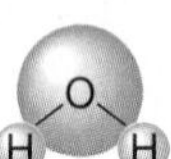

Water, H_2O

A triatomic molecule (a molecule that contains three atoms—two hydrogen atoms and one oxygen atom)

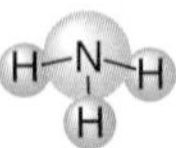

Ammonia, NH_3

A tetraatomic molecule (a molecule that contains four atoms—three hydrogen atoms and one nitrogen atom)

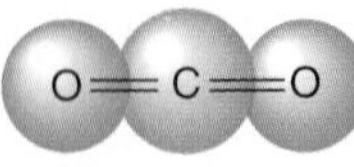

Carbon dioxide, CO_2

A triatomic molecule (this molecule contains one carbon atom and two oxygen atoms)

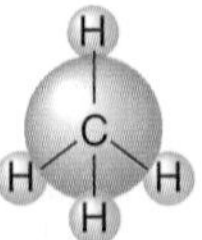

Methane, CH_4

A molecule that contains five atoms—one carbon atom and four hydrogen atoms

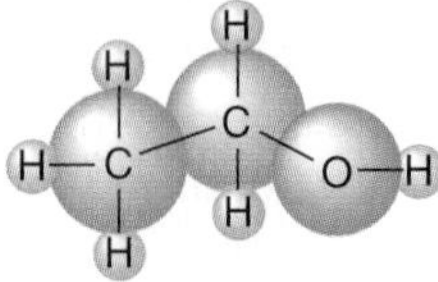

Ethyl alcohol, C_2H_5OH

A molecule containing nine atoms—two carbon atoms, one oxygen atom, and six hydrogen atoms

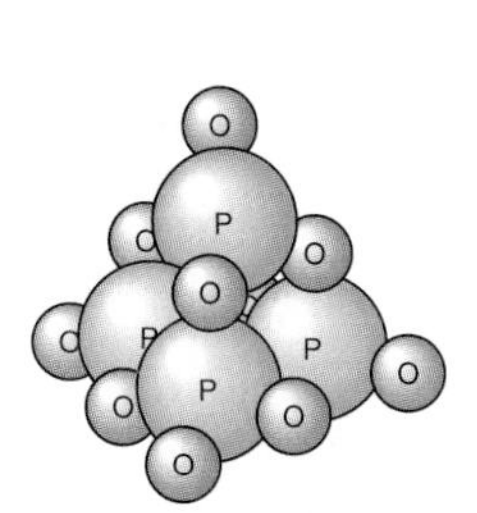

An oxide of phosphorus (tetraphosphorus decoxide)

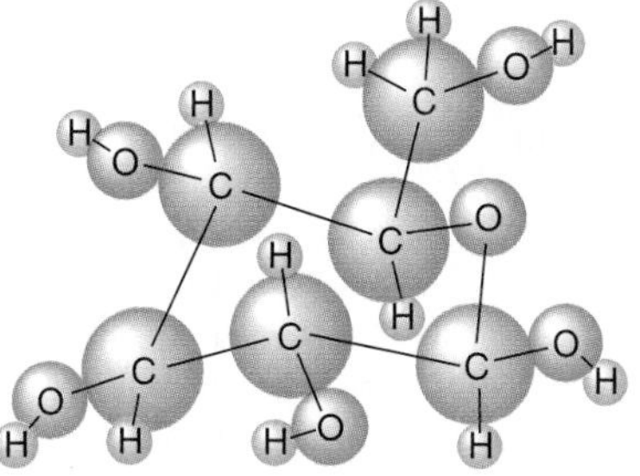

A simple sugar, glucose, $C_6H_{12}O_6$ (Distance between atoms is exaggerated for clarity)

Figure 2.11

Molecular models of tetraphosphorus decoxide and glucose, solid compounds at room temperature.

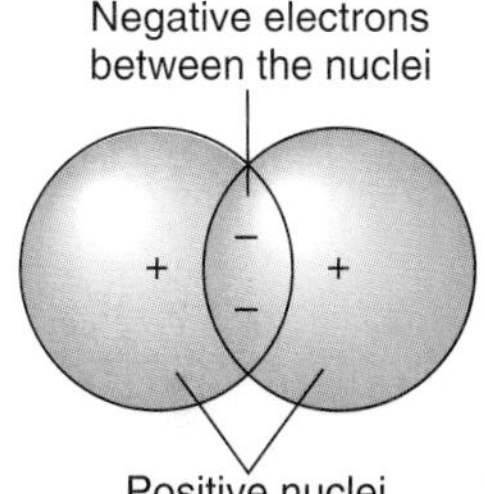

Figure 2.12

A covalent bond between two atoms in a molecule. Positive nuclei are held together by negative electrons that are equally shared by the two atoms.

Let us consider the bonding between nonmetal atoms in a molecule. One of the simplest molecules is H_2. In the formation of an H_2 molecule from two H atoms, there is no transfer of electrons from one atom to another to form a pair of ions. Instead, a pair of electrons is *shared* between the nuclei of the atoms. Each hydrogen atom has one proton in its nucleus. These positively charged nuclei in an H_2 molecule are attracted to the negatively charged electrons between the nuclei. A *bond in which electrons are shared between the bonded atoms* is known as a **covalent bond** (Figure 2.12).

The shared electron pair as a covalent bond between two atoms can be represented by two dots, such as H:H, or by a dash, H—H. Thus, H:H and H—H both represent H_2.

A triatomic molecule with the general formula AB_2, in which two atoms of B are bonded to one atom of A, can be written as B:A:B or B—A—B. Beryllium fluoride, BeF_2, in the gas phase consists of triatomic F—Be—F molecules.

Names and Formulas for Binary Nonmetal–Nonmetal Compounds

A molecule such as H_2 or Cl_2, in which the two bonded atoms are identical, is called a *homonuclear* diatomic molecule. In such a molecule the bonding electrons are shared equally by the atoms.

A diatomic molecule in which the two atoms are not identical is called a *heteronuclear* diatomic molecule. In such a molecule, one of the atoms has a greater share of the bonding electron pair than the other atom. The atom that has the greater share is said to be more *electronegative* than the other atom, which is the more *electropositive* atom. As a general rule, electronegativity increases from left to right across a given period in the periodic table, and decreases from top to bottom within a group. Thus, oxygen is more electronegative than nitrogen, and fluorine is more electronegative than oxygen or chlorine.

In the formulas for binary nonmetal–nonmetal compounds, the symbol of the more electropositive element is written first, followed by the symbol of the more electronegative element. In naming such compounds, the electropositive element is named first, followed by the name of the electronegative element. The latter has the ending *-ide*, as if it were a negative ion. For example, HCl is called hydrogen chloride, and BN is called boron nitride.

Many molecules of binary compounds consist of more than two atoms. We recall that the number of atoms of an element in a molecule of a compound is indicated by a subscript. The numbers shown by the subscripts are named by the following Greek prefixes: *mono-* for 1, *di-* for 2, *tri-* for 3, *tetra-* for 4, *penta-* for 5, *hexa-* for 6, *hepta-* for 7, *octa-* for 8, *nona-* for 9, and *deca-* for 10.

Table 2.4

Examples of Naming Binary Nonmetal–Nonmetal Compounds

Compound	Systematic Name	Common Name
N_2O	Dinitrogen monoxide	Nitrous oxide
NO	Nitrogen monoxide	Nitric oxide
PCl_3	Phosphorus trichloride	
PCl_5	Phosphorus pentachloride	
SF_6	Sulfur hexafluoride	
P_4O_6	Tetraphosphorus hexoxide	Phosphorous oxide
P_4O_{10}	Tetraphosphorus decoxide	Phosphoric oxide

The prefix *mono-* is not used to name the first element in a formula. Thus, CO is named carbon monoxide, and CO_2 is carbon dioxide. Note that for smooth pronunciation, the letters "o" and "a" are often dropped when the second element in a formula starts with a vowel. Therefore, CO is called carbon monoxide instead of carbon monooxide, and N_2O_5 is called dinitrogen pentoxide instead of dinitrogen pentaoxide. Some additional examples of naming binary covalent compounds are listed in Table 2.4.

Although some compounds may be named by both their systematic and common names, water, H_2O, and ammonia, NH_3, are always named by their common names.

Practice Problem 2.5: Name each of the following compounds: **(a)** SO_3; **(b)** SF_2; **(c)** S_2Cl_2; **(d)** IF_7; **(e)** N_2O_4.

Practice Problem 2.6: Write a formula for each of the following compounds: **(a)** carbon tetrachloride; **(b)** diiodine pentoxide; **(c)** tetraphosphorus decasulfide.

2.5 Polyatomic Ions and Their Compounds

Polyatomic Ions

Until now, we have considered only ions containing one electrically charged atom, such as Na^+ and Cl^-. These ions are called **monatomic ions**. We also mentioned one diatomic ion, the mercury(I) ion, Hg_2^{2+}. We will next consider more ions in which two or more atoms are linked by covalent bonds.

When two or more atoms are covalently bonded in a cluster in which the total number of protons does not equal the total number of electrons, the assembly has an electric charge. An *electrically charged group of atoms* is called a **polyatomic ion**. A polyatomic ion has a negative charge if the number of electrons in the group is greater than the number of protons. Conversely, if there are more protons than electrons, the ion has a positive charge. Generally, *positively charged monatomic and polyatomic ions* are called **cations** (pronounced *cat-ions*) and *negatively charged ions* are called **anions** (pronounced *an-ions*).

Formulas and Names of Polyatomic Ions

Most common polyatomic ions are anions containing two or more nonmetal atoms. A few polyatomic anions also contain metal and nonmetal atoms. There are only a few polyatomic cations. One of the most common polyatomic cations consists of one nitrogen atom covalently bonded to four hydrogen atoms. This ion has a net 1+ charge and has the formula NH_4^+; it is called the *ammonium ion*.

Some common polyatomic anions are listed in Table 2.5. The covalent bonds between atoms in these anions are represented by dashes. The *exact* arrangement of atoms in space is not inferred.

Only a few common polyatomic anions are named with an *-ide* ending. Two of these anions, listed in Table 2.5, are cyanide (CN^-) and hydroxide (OH^-) ions. Most of the names of oxygen-containing polyatomic anions end with *-ate* or *-ite*. *Polyatomic anions that contain oxygen* are called **oxoanions** or **oxyanions**.

An oxoanion that consists of three or more atoms usually has a central atom other than oxygen bonded to two or more oxygen atoms (Table 2.5). Some elements form two different oxoanions, one containing more oxygen than the other. The name of the oxoanion that contains *more oxygen* ends with *-ate*, and the name of the one with *less oxygen* ends with *-ite*, as shown below for the oxoanions of nitrogen and sulfur:

Nitrate ion	NO_3^-	Sulfate ion	SO_4^{2-}
Nitrite ion	NO_2^-	Sulfite ion	SO_3^{2-}

These examples show that the endings *-ate* and *-ite* do not specify the *number* of oxygen atoms in an oxoanion. For example, a nitrate ion contains three oxygen atoms, but a sulfate ion contains four.

There are four oxoanions of chlorine. Of these, two are named by the use of the endings *-ate* (greater) and *-ite* (lesser). The anion that contains the least oxygen is named by using the prefix *hypo-* (meaning the "least") with the ending *-ite*. The anion with the most oxygen is named by using the prefix *per-* (meaning the "most") with the ending *-ate*. The formulas and names of the oxoanions of chlorine are as follows:

ClO_2^-	Chlorite ion	ClO^-	Hypochlorite ion
ClO_3^-	Chlorate ion	ClO_4^-	Perchlorate ion

Note that the atom other than oxygen is written first in the formula of an oxoanion. Compounds of these different anions have very different chemical properties. Salts containing the chlorate ions can be safely heated (see Chapter 3). But heating *per*chlorate salts is very dangerous because they are likely to explode.

Sulfate, sulfite, and carbonate ions can combine with a hydrogen ion, H^+, to form a hydrogen sulfate ion, HSO_4^-, a hydrogen sulfite ion, HSO_3^-, and a hydrogen carbonate ion, HCO_3^-, respectively. The common names for these ions are bisulfate, bisulfite, and bicarbonate.

The phosphate ion can combine with one or two hydrogen ions to form

HPO_4^{2-}	Monohydrogen phosphate ion
$H_2PO_4^-$	Dihydrogen phosphate ion

Table 2.5

Some Common Polyatomic Anions

Anions with 1− Charge

Name	Formula	Structure[a]	Name	Formula	Structure[a]
Acetate	$C_2H_3O_2^-$	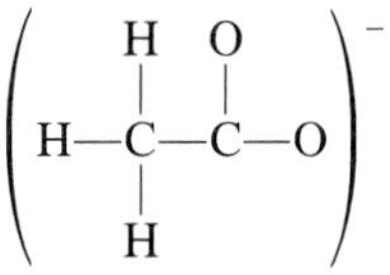 H O H—C—C—O H ( )$^-$	Hypochlorite	ClO^-	(Cl—O)$^-$
			Iodate	IO_3^-	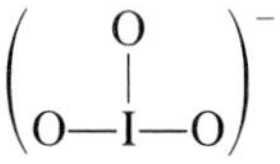 O O—I—O ( )$^-$
Chlorate	ClO_3^-	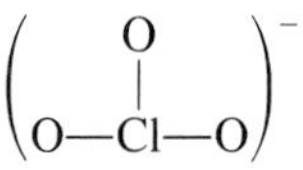 O O—Cl—O ( )$^-$	Nitrate	NO_3^-	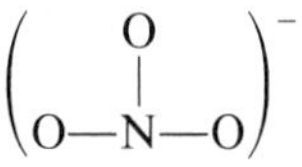 O O—N—O ( )$^-$
Chlorite	ClO_2^-	(O—Cl—O)$^-$	Nitrite	NO_2^-	(O—N—O)$^-$
Cyanide	CN^-	(C—N)$^-$	Perchlorate	ClO_4^-	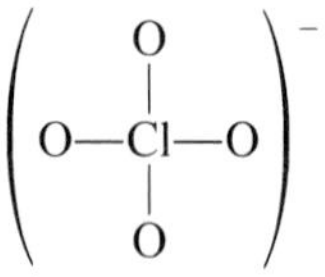 O O—Cl—O O ( )$^-$
Hydrogen carbonate (bicarbonate)[b]	HCO_3^-	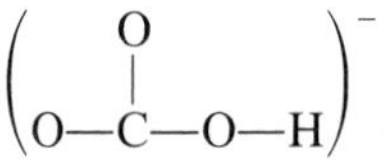 O O—C—O—H ( )$^-$			
Hydrogen sulfate (bisulfate)[b]	HSO_4^-	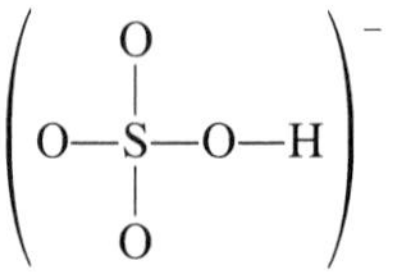 O O—S—O—H O ( )$^-$	Permanganate	MnO_4^-	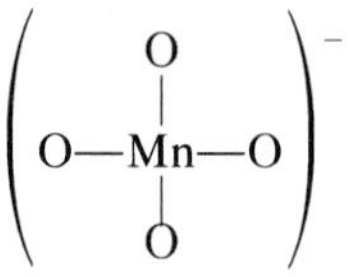 O O—Mn—O O ( )$^-$
Hydroxide	OH^-	(O—H)$^-$			

Anions with 2− Charge

Name	Formula	Structure[a]	Name	Formula	Structure[a]
Carbonate	CO_3^{2-}	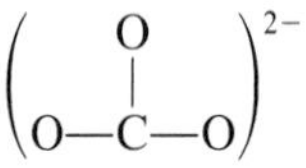 O O—C—O ( )$^{2-}$	Oxalate	$C_2O_4^{2-}$	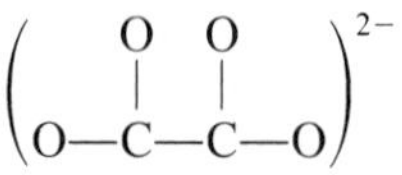 O O O—C—C—O ( )$^{2-}$
Chromate	CrO_4^{2-}	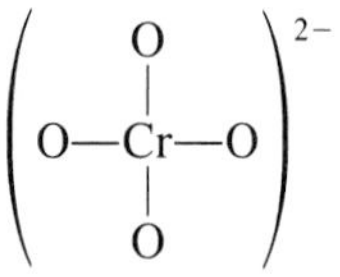 O O—Cr—O O ( )$^{2-}$	Sulfate	SO_4^{2-}	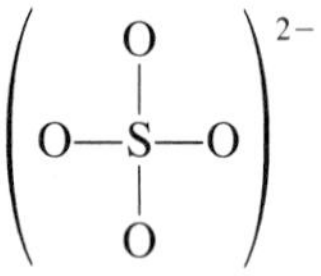 O O—S—O O ( )$^{2-}$
Dichromate	$Cr_2O_7^{2-}$	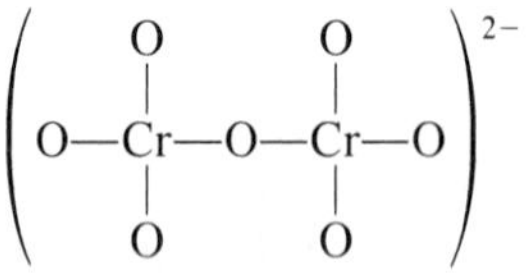 O O O—Cr—O—Cr—O O O ( )$^{2-}$	Sulfite	SO_3^{2-}	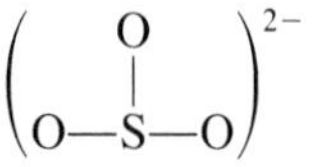 O O—S—O ( )$^{2-}$

Anions with 3− Charge

Name	Formula	Structure[a]	Name	Formula	Structure[a]
Arsenate	AsO_4^{3-}	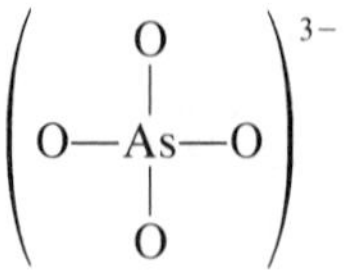 O O—As—O O ( )$^{3-}$	Phosphate	PO_4^{3-}	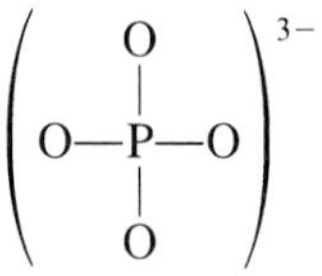 O O—P—O O ( )$^{3-}$
Arsenite	AsO_3^{3-}	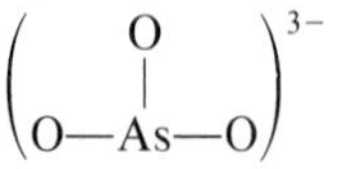 O O—As—O ( )$^{3-}$			

[a] Dashes show the sequence of atom linkages.

[b] Old trivial name still in common use.

The sodium salt of monohydrogen phosphate ion, Na_2HPO_4, is named sodium monohydrogen phosphate. In a similar way, specifying the name of the anion, the sodium salt of dihydrogen phosphate ion, NaH_2PO_4, is sodium dihydrogen phosphate.

Compounds Containing Polyatomic Ions

Polyatomic ions such as carbonates, sulfates, phosphates, and others are found in the earth's crust as salts of many different metal ions. The formulas for salts involving polyatomic ions are written like the formulas for the compounds involving monatomic cations and anions. That is, the formula of any ionic compound must signify the minimum relative numbers of cations and anions for the compound to be electrically neutral. For example, the salt sodium nitrate has the formula $NaNO_3$, which signifies that the Na^+ and NO_3^- ions are present in a 1:1 ratio. Similarly, the salt ammonium chlorate has the formula NH_4ClO_3, in which the NH_4^+ ions and ClO_3^- ions are also present in a 1:1 ratio.

If more than one polyatomic ion is present in the formula for a compound, the formula for the polyatomic ion is enclosed in parentheses and a subscript is added to indicate the number of polyatomic ions in the formula. For example, the Ca^{2+} ion has a charge of 2+ and the NO_3^- ion has a charge of 1−. For the salt to be electrically neutral, the Ca^{2+} and NO_3^- ions must be present in a 1:2 ratio. This ratio is reflected by the formula, which is $Ca(NO_3)_2$. If the formula of a salt contains only one polyatomic cation or anion, the formula for the polyatomic ion is not enclosed in parentheses and no subscript is needed. For example, the formula for ammonium nitrate is NH_4NO_3.

Example 2.5

Writing Formulas for Compounds Involving Polyatomic Ions

Write formulas for the following compounds: **(a)** sodium sulfate; **(b)** aluminum nitrate; **(c)** calcium phosphate; **(d)** ammonium monohydrogen phosphate.

SOLUTION

(a) We recall that a sodium ion has a charge of 1+ and a sulfate ion 2−. Thus, for the compound to be neutral, the Na^+ and the SO_4^{2-} ions must be present in a 2:1 ratio, and the formula is Na_2SO_4.

(b) An aluminum ion has a charge of 3+ and a nitrate ion 1−. The aluminum and nitrate ions are therefore present in a 1:3 ratio, and the formula for aluminum nitrate is $Al(NO_3)_3$. The parentheses indicate that nitrate ions are counted as units, and each unit consists of one nitrogen and three oxygen atoms.

(c) The charge on a calcium ion is 2+, and on a phosphate ion 3−. The lowest common multiple of 2 and 3 is 6. Therefore, three calcium ions are needed for a total of 6+ charge and two phosphate ions for a total of 6− charge. Thus, the formula for calcium phosphate is $Ca_3(PO_4)_2$.

(d) Ammonium ions, NH_4^+, and monohydrogen phosphate ions, HPO_4^{2-}, must be present in a 2:1 ratio in a neutral compound, and the formula for ammonium monohydrogen phosphate is $(NH_4)_2HPO_4$.

Practice Problem 2.7: Name each of the following compounds: **(a)** K_2CO_3; **(b)** $Ba(NO_2)_2$; **(c)** $Al_2(SO_3)_3$; **(d)** Li_3PO_4; **(e)** $Sr(HSO_4)_2$; **(f)** $Mg(ClO_3)_2$; **(g)** $(NH_4)_3AsO_4$; **(h)** $Fe(NO_3)_3$.

Acids

Some *substances produce H^+ ions when dissolved in water*. Such substances are called **acids**. If an acid is placed in water, it produces both H^+ ions and anions characteristic of the acid. The H^+ ions bond to water molecules to form hydronium ions, H_3O^+. An acid whose anion does not contain oxygen is named with a prefix *hydro-* and suffix *-ic*, and the acid is called a **hydroacid**. For example, when gaseous hydrogen chloride, HCl, is dissolved in water, it forms *hydro*chlor*ic* acid. Similarly, gaseous hydrogen bromide, HBr, in water forms hydrobromic acid, and hydrogen sulfide gas, H_2S, forms hydrosulfuric acid. In each of these cases, the gaseous compound and its corresponding acid in water are represented by the same formula.

An acid whose anion contains oxygen is called an **oxoacid** and is named as follows. If the name of the anion ends with *-ate*, the name of the corresponding acid ends with *-ic*; if the name of the anion ends with *-ite*, the name of the acid ends with *-ous*. Table 2.6 lists the formulas and names of some common hydroacids and oxoacids. Gases such as hydrogen chloride, HCl, are obtained from chemical supply houses in metal cylinders. Aqueous acids, such as hydrochloric acid and acetic acid, are stored in glass bottles (Figure 2.13).

Table 2.6

Some Common Hydroacids and Oxoacids

Hydroacids		Oxoacids	
HF	Hydrofluoric acid	HNO_3	Nitric acid
HCl	Hydrochloric acid	HNO_2	Nitrous acid
HBr	Hydrobromic acid	H_2SO_4	Sulfuric acid
HI	Hydroiodic acid	H_2SO_3	Sulfurous acid
H_2S	Hydrosulfuric acid	H_3PO_4	Phosphoric acid
HCN	Hydrocyanic acid	$HC_2H_3O_2$	Acetic acid

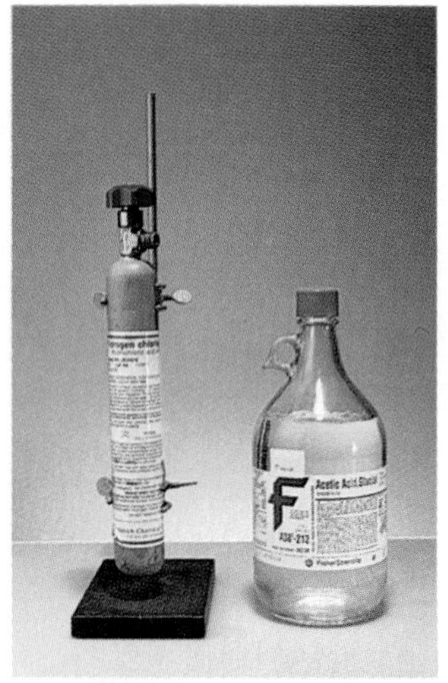

Figure 2.13

Storage of gases and aqueous acids. Left: An HCl gas cylinder. Right: A bottle of aqueous acetic acid. HCl gas is highly toxic, and contact with acetic acid can cause burns.

2.6 Chemical Equations

We recall from Chapter 1 that a chemical reaction can be described by a chemical equation. In an equation the symbols or formulas of the substances mixed for a reaction are written first. These substances are called the **reactants**. After the formulas for the reactants have been written, an arrow is drawn that points to the substances produced in the reaction. These are called the **products**. The general form of a chemical equation is thus

$$\text{reactants} \longrightarrow \text{products}$$

Let us consider the equation for the reaction of sodium with chlorine to form sodium chloride as a specific example. We start by writing the symbols for the reactants, sodium and chlorine (we recall that chlorine consists of diatomic

APPLICATIONS OF CHEMISTRY 2.1
Chemical Nomenclature in Daily Life

You may have encountered many simple, yet interesting chemical compounds for years without being aware of them. For example, did you know that barns are red today because historically they were once preserved with rust (Fe_2O_3)? (See photo on right.) In the mid-nineteenth century, American farmers made an inexpensive, yet long-lasting barn preservative by mixing rust-iron(III) oxide, skim milk, linseed oil, and lime (CaO, calcium oxide). By the late nineteenth century, red had become the traditional color for barns. Barns are no longer preserved with rust and the red color is the result of red paint pigments. But historically, they are red because of rust.

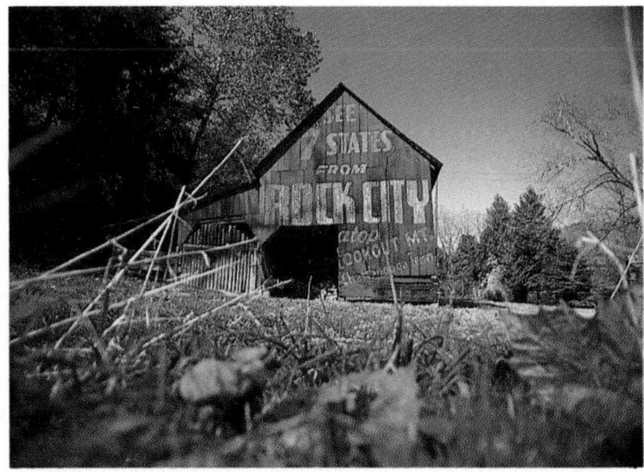

Another use of chemicals in everyday life can be found in the supermarket. What keeps meat at the supermarket looking fresh? The color, flavor, and texture of meats depend on additives such as sodium nitrite ($NaNO_2$) and sodium nitrate ($NaNO_3$). Sodium nitrate also acts as a preservative, and its use has virtually wiped out botulism food poisoning. Despite this, the Food and Drug Administration has been trying to limit the concentration of these two chemicals because excessive amounts may be harmful. Food processors add iron(II) sulfate ($FeSO_4$) and iron(III) phosphate ($FePO_4$) to many breads and cereals as a dietary source of iron.

You probably think that the only chemical in table salt is sodium chloride. But there can be any of 20 other chemicals in table salt. Sodium chloride is the primary ingredient, but others include potassium iodide (KI), which is a nutrient. Sodium hydrogen carbonate ($NaHCO_3$), sodium carbonate (Na_2CO_3), calcium hydroxide [$Ca(OH)_2$], and sodium monohydrogen phosphate (Na_2HPO_4) are also present to help stabilize the potassium iodide. Magnesium carbonate ($MgCO_3$), calcium hydrogen phosphate ($CaHPO_4$), calcium phosphate [$Ca_3(PO_4)_2$], and calcium carbonate ($CaCO_3$) are present as desiccants (drying agents) to help keep the salt crystals from fusing.

Some other chemicals that you may find around your own home include SiO_2 (silicon dioxide, sand), $NaOH$ (sodium hydroxide, lye, used in drain cleaners), and H_2SO_4 (sulfuric acid, battery acid). Also common are $CuSO_4$ [copper(II) sulfate], an algicide, HCl (hydrochloric acid, called muriatic acid, used to clean masonry and bricks), and $NaClO$ (sodium hypochlorite, used in liquid laundry bleach).

molecules, Cl_2). Then we draw an arrow pointing to the formula of the product; and we write the formula for the product, sodium chloride:

$$Na + Cl_2 \longrightarrow NaCl \quad \text{(unbalanced)}$$

The equation above suggests that two atoms of Cl in a molecule of Cl_2 produce only one atom of chlorine in NaCl. But this violates the law of conservation of mass. In other words, the equation is not *balanced.* In a balanced equation, the number of atoms of each element present in the reactants must be equal to the number of atoms of the corresponding elements in the products because atoms are neither created nor destroyed in a chemical reaction.

An equation is balanced by writing appropriate numbers or multipliers, called *coefficients*, in front of the formulas for the reactants and the products. A coefficient preceding a formula multiplies all the atoms in the formula by that number. The coefficients in a chemical equation are chosen to provide the same number of atoms of each element on the left and right of the arrow. The equation for the reaction of sodium with chlorine can be balanced by writing the coefficient 2 preceding the symbol Na and the formula NaCl. Thus, the correctly balanced equation is

$$2Na + Cl_2 \longrightarrow 2NaCl$$

The equation is now balanced, but we can describe the reaction further by indicating the physical states of the reactants and the products. The solid state is indicated by the letter *s* in parentheses following the symbol or formula for the substance, a liquid is shown by the letter *l*, and a gas by a *g*. A substance in water solution is usually indicated by *aq* for the word "aqueous." A complete and balanced equation for the reaction of sodium with chlorine can now be written as follows:

$$2Na(s) + Cl_2(g) \longrightarrow 2NaCl(s)$$

A pictorial representation of this reaction, assuming that only two atoms of sodium combine with a diatomic chlorine molecule, is shown in Figure 2.14.

The *steps involved in writing a chemical equation* may be summarized as follows.

1. Write the symbols or formulas for the reactants, and indicate the physical state of each reactant by using the abbreviations *s*, *l*, *g*, or *aq*.
2. Draw an arrow pointing from the reactants to the products, predict the products, and indicate the physical states of the products. In Section 2.3 we discussed how to predict the products of metal–nonmetal reactions. The prediction of products of other reactions is discussed in Section 2.7.
3. Balance the equation by counting the atoms of each element on the left and right of the arrow. To add atoms where needed, write coefficients *preceding* the symbols or formulas involved in the equation. The coefficients should be the *smallest* set of positive integers.

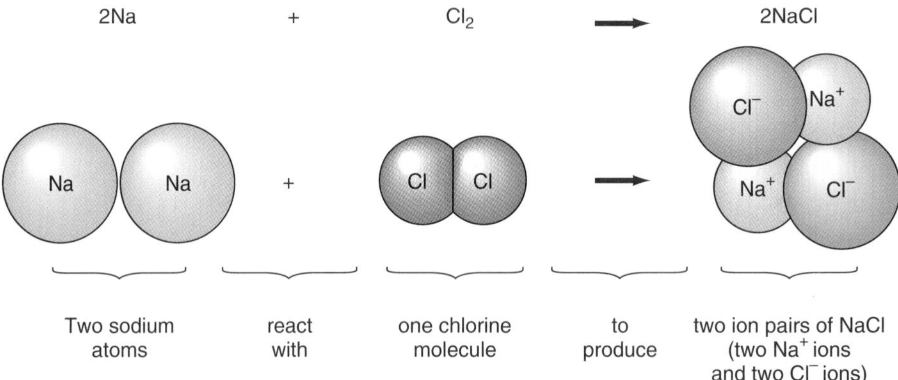

Figure 2.14

A reaction of two atoms of sodium with a diatomic molecule of chlorine. Electrons are transferred from the Na atoms to the two Cl atoms in a Cl_2 molecule. As a result, two positively charged Na^+ ions and two negatively charged Cl^- ions are formed. These ions attract each other to form a lattice of sodium chloride, NaCl, that is electrically uncharged because the total positive charge due to the Na^+ ions equals the total negative charge due to the Cl^- ions. In reality, when observable quantities of sodium and chlorine react, the number of sodium atoms and chlorine molecules involved is enormous. As a result of a reaction, the oppositely charged ions form a three-dimensional array of alternating plus (+) and minus (−) ions as depicted in Figure 2.9.

Example 2.6

Writing an Equation for a Reaction of a Metal with a Nonmetal

Reactions of metals with atmospheric oxygen in moist air, a process called corrosion, are very common. Among these processes are the rusting of iron to form iron(III) oxide (rust) and the corrosion of aluminum by the formation of aluminum oxide, commonly seen as the accumulation of a gray deposit on aluminum kitchen utensils and on aluminum storm doors and window frames. Write a balanced equation for the reaction of aluminum with oxygen.

SOLUTION

Step 1. We write the symbol for aluminum and the formula for diatomic oxygen as the reactants and indicate their respective physical states in parentheses:

$$Al(s) + O_2(g)$$

Step 2. We draw an arrow pointing to the product and write the formula for the product, aluminum oxide, using the principles learned in Section 2.3. Next we indicate the physical state for aluminum oxide. We know that aluminum oxide is a solid because it is an ionic compound and because we have seen it as a solid gray deposit on the surface of objects made of aluminum:

$$Al(s) + O_2(g) \longrightarrow Al_2O_3(s)$$

Step 3. We balance the equation. It is usually easiest to start with the compound containing the largest number of atoms (Al_2O_3 in this case), then go to the elements consisting of polyatomic molecules such as O_2, leaving until last monatomic species such as Al. To balance the preceding equation, count the atoms on both sides of the arrow. There are three O atoms on the right, but only two on the left. The lowest common multiple of two and three is six. Thus, the assignment of the coefficients three for O_2 on the left of the arrow and two for Al_2O_3 to the right provides six O atoms on both sides of the arrow:

$$Al(s) + 3O_2(g) \longrightarrow 2Al_2O_3(s)$$

The coefficient four for Al on the left balances the equation:

$$4Al(s) + 3O_2(g) \longrightarrow 2Al_2O_3(s)$$

Practice Problem 2.8: Write a balanced equation for the reaction that occurs when each of the following pairs of elements combine: **(a)** calcium and fluorine; **(b)** potassium and sulfur, S_8 (the element sulfur exists as S_8 molecules); **(c)** magnesium and nitrogen.

Practice Problem 2.9: Write balanced equations for the formation of the following compounds from their elements: **(a)** lithium oxide; **(b)** magnesium phosphide (the element phosphorus exists as P_4 molecules); **(c)** aluminum sulfide.

2.7 Types of Reactions

In Section 2.6 we considered equations for some reactions in which elements combine to form binary compounds. Many other types of reactions occur in gas, liquid, and solid phases and in solutions. These reactions can be grouped into a few general types. These include (1) **combination reactions**, (2) **decomposition reactions**, and (3) **displacement reactions**; the displacement reactions include single displacements and double displacements.

Combination Reactions

A combination reaction produces one complex substance from two or more simpler substances. Most combination reactions can be divided into the following subcategories: (1) *combination of an element with another element to form a binary compound*, (2) *combination of an element with a compound to form another compound*, or (3) *combination of two compounds to form a new compound.*

The simplest combination reactions are those in which *two elements combine to form a binary compound.* We have already considered a few reactions of metals with nonmetals to form ionic compounds (see Figures 2.6 and 2.7). We also recall that a nonmetal can react with another nonmetal in a combination reaction to form a binary molecular compound. For example, hydrogen can combine with oxygen to form water:

$$2H_2(g) + O_2(g) \longrightarrow 2H_2O(l)$$

or sulfur can combine with oxygen to form sulfur dioxide:

$$S_8(s) + 8O_2(g) \longrightarrow 8SO_2(g)$$

The combination of a metal with another metal to form a compound does not usually occur, although two or more metals can be melted together to form a *solution of metals* called an **alloy**. For example, the alloy consisting of iron and at least 11.5 percent of chromium and other minor components is stainless steel.

The combination of an *element with a compound to form a new compound* can be illustrated by the reaction of oxygen with carbon monoxide to form carbon dioxide:

$$2CO(g) + O_2(g) \longrightarrow 2CO_2(g)$$

Some metal oxides combine with oxygen to form another oxide in which the metal ions have a higher charge. For example, iron(II) oxide reacts with oxygen to form iron(III) oxide, the principal component of rust:

$$4FeO(s) + O_2(g) \longrightarrow 2Fe_2O_3(s)$$

A very common example of the combination of an element with a compound occurs in the burning of hydrocarbons. **Hydrocarbons** are *binary compounds of hydrogen and carbon.* Natural gas is a mixture of hydrocarbons. When a hydrocarbon burns, it reacts with oxygen to form carbon dioxide and water as shown below for the reaction of methane, CH_4, the major component of natural gas:

$$CH_4(g) + 2O_2(g) \longrightarrow CO_2(g) + 2H_2O(l)$$

Practice Problem 2.10: Write a balanced equation for each of the following combination reactions: **(a)** sulfur dioxide reacts with oxygen to form sulfur trioxide; **(b)** iron(II) chloride reacts with chlorine to form iron(III) chloride; **(c)** a component of natural gas, ethane, C_2H_6, burns in the atmosphere.

The *reaction of a metal oxide with a nonmetal oxide* is an example of the *combination of two compounds to form a new compound.* Most metal oxides are ionic compounds, and nonmetal oxides are molecular compounds. When a metal oxide reacts with a nonmetal oxide, the product is an ionic compound that consists of the cations of the metal and polyatomic oxoanions of the nonmetal. The equation for the reaction of calcium oxide (lime) with carbon dioxide to form calcium carbonate (limestone) is

$$CaO(s) + CO_2(g) \longrightarrow CaCO_3(s)$$

Example 2.7

Combination Reaction of a Metal Oxide with a Nonmetal Oxide

Write an equation for the reaction of sodium oxide with sulfur dioxide.

SOLUTION: Sodium oxide is a metal oxide, and sulfur dioxide is a nonmetal oxide. These two oxides combine to form the salt sodium sulfite (this reaction is similar to the reaction of calcium oxide with carbon dioxide shown above):

$$Na_2O(s) + SO_2(g) \longrightarrow Na_2SO_3(s)$$

Most metal oxides combine with water to yield metal hydroxides. Calcium oxide reacts with water to produce calcium hydroxide:

$$CaO(s) + H_2O(l) \longrightarrow Ca(OH)_2(s)$$

Calcium hydroxide is called slaked lime (used in mortar, plaster, and cement).

Metal hydroxides belong to a class of substances called **bases**. When a base dissolves in water it produces hydroxide ions. Oxides that dissolve in water to produce hydroxides are called **basic oxides**.

Most nonmetal oxides combine with water to form acids. Sulfur dioxide forms sulfurous acid:

$$SO_2(g) + H_2O(l) \longrightarrow H_2SO_3(aq)$$

and sulfur trioxide forms sulfuric acid:

$$SO_3(g) + H_2O(l) \longrightarrow H_2SO_4(aq)$$

Oxides that dissolve in water to produce acids are called **acidic oxides**.

Practice Problem 2.11: **(a)** Write an equation for the reaction of sodium oxide with water to produce sodium hydroxide. **(b)** Write an equation for the reaction of dinitrogen pentoxide with water to produce nitric acid.

Decomposition Reactions

A process in which a substance breaks down—usually upon heating or electrolysis (a decomposition as a result of an applied electric current)—into two or more simpler products is called a decomposition reaction. The *products of a decomposition reaction* can be *elements*, *compounds*, or *elements and compounds*.

Oxides of relatively nonreactive metals such as platinum, gold, silver, and mercury *decompose upon heating to their component elements*. For example, heating mercury(II) oxide produces mercury and oxygen:

$$2HgO(s) \longrightarrow 2Hg(l) + O_2(g)$$

Figure 2.15

Left: A vial containing HgO. Right: A vial containing Hg.

Figure 2.15 illustrates a vial that contains HgO before heating and a vial after heating, which now contains only liquid Hg.

The decomposition of hydrogen peroxide, H_2O_2, a disinfectant and bleaching agent, is an example of a *decomposition reaction that produces an element and a compound*. Hydrogen peroxide decomposes to water and oxygen:

$$2H_2O_2(aq) \longrightarrow 2H_2O(l) + O_2(g)$$

Heat and light accelerate this reaction; therefore, a bottle of hydrogen peroxide should be kept in a cool, dark environment. A photo of a watch glass that contains H_2O_2 with a drop of blood in it showing the evolution (foaming) of oxygen is shown in Figure 2.16.

Figure 2.16

Watch glass of H_2O_2 with a drop of blood showing the production of oxygen (foaming).

An example of a *decomposition reaction that produces two different compounds as products* is the decomposition of calcium carbonate, the principal ingredient of *limestone*. When calcium carbonate is heated to a high temperature, it decomposes into calcium oxide (*lime*) and carbon dioxide:

$$CaCO_3(s) \longrightarrow CaO(s) + CO_2(g)$$

The industrial decomposition of limestone is carried out in large "lime kilns" to make lime for the construction industry. Figure 2.17 shows a sample of limestone and lime produced from the limestone by heating.

Figure 2.17

Left: Limestone (calcium carbonate). Right: Lime (calcium oxide) as a byproduct of heating limestone.

Displacement Reactions

Many chemical reactions involve the displacement of a component of a reactant by another reactant. Displacement reactions can be divided into **single displacement** and **double displacement** reactions. The latter are also called **metathesis** reactions.

In a single displacement reaction, one element replaces another element in a compound. The metals of groups I and II, as well as aluminum and zinc, are much more reactive than transition metals such as copper, silver, gold, mercury, and platinum. The more reactive metals are often referred to as **active metals**. The less reactive metals are sometimes called **noble metals**. *Relatively nonreactive metals can be displaced from their compounds by more active metals*. For example, when copper(II) oxide is heated in the presence of magnesium, copper is displaced from its oxide and magnesium oxide is produced:

$$Mg(s) + CuO(s) \longrightarrow MgO(s) + Cu(s)$$

In this single displacement reaction, magnesium replaces copper from its oxide.

Single displacement reactions can also be carried out in water solution. In such a reaction, the ions of a less reactive metal in solution are displaced by a more reactive metal. For example, when magnesium metal is placed in copper(II) sulfate solution, the magnesium atoms convert to magnesium ions that displace the copper(II) ions in the solution. The copper(II) ions convert to copper atoms that adhere to the unreacted magnesium metal as a copper coating. The reaction occurs according to the following equation:

$$Mg(s) + CuSO_4(aq) \longrightarrow MgSO_4(aq) + Cu(s)$$

Figure 2.18 shows Mg foil and blue $CuSO_4$ solution separately, and the Mg foil dipped into the $CuSO_4$ solution and being coated with elemental Cu.

Example 2.8

Predicting the Products and Writing an Equation for a Single Displacement Reaction

What would you expect to observe if you placed an aluminum nail into a solution of silver nitrate, $AgNO_3$? Write an equation for the reaction.

SOLUTION: Aluminum is an active metal that can replace silver ions from the solution of silver nitrate. Therefore, a deposit of elemental silver can be seen on the nail. The balanced equation for the reaction is

$$Al(s) + 3AgNO_3(aq) \longrightarrow Al(NO_3)_3(aq) + 3Ag(s)$$

Practice Problem 2.12: **(a)** Propose a chemical reaction that would enable you to recover silver from its oxide, Ag_2O. Write an equation for the reaction you would carry out in this recovery. (*Note*: Ag_2O is not soluble in water.) **(b)** How would you recover copper metal from a solution of copper(II) sulfate? Write an equation for the reaction you would carry out in this recovery process.

Another important type of single displacement reaction involves the *displacement of hydrogen from an acid by an active metal.* The reaction of an active metal with an acid can be used to prepare small quantities of hydrogen in a laboratory. Some examples of such reactions are:

$$Mg(s) + 2HCl(aq) \longrightarrow H_2(g) + MgCl_2(aq)$$
$$2Al(s) + 3H_2SO_4(aq) \longrightarrow 3H_2(g) + Al_2(SO_4)_3(aq)$$

Less active metals such as copper, silver, mercury, and gold do not react with hydrochloric acid. Gold does not react with any acid except with a mixture of concentrated hydrochloric acid and nitric acid.

Practice Problem 2.13: Write an equation for the "dissolution" reaction of aluminum in hydrochloric acid.

When two reactants, AB and CD, react to form products, AC and BD, by an exchange of partners, the reaction is called a double displacement or metathesis reaction. Many metathesis reactions occur in aqueous solution in which one of the products is insoluble and settles to the bottom of the solution in which it forms. The insoluble solid is called a **precipitate**. The solution that remains above the precipitate is called the **supernatant solution**, or simply the

A typical example of a double displacement reaction is a reaction that occurs when solutions of silver nitrate and sodium chloride are mixed. Upon mixing, the cations and the anions in the original solutions exchange partners to form a solution of sodium nitrate and a white precipitate of silver chloride,

(a)

(b)

Figure 2.18

(a) Magnesium foil and blue copper(II) sulfate solution in beaker; (b) magnesium foil dipped into the $CuSO_4$ solution.

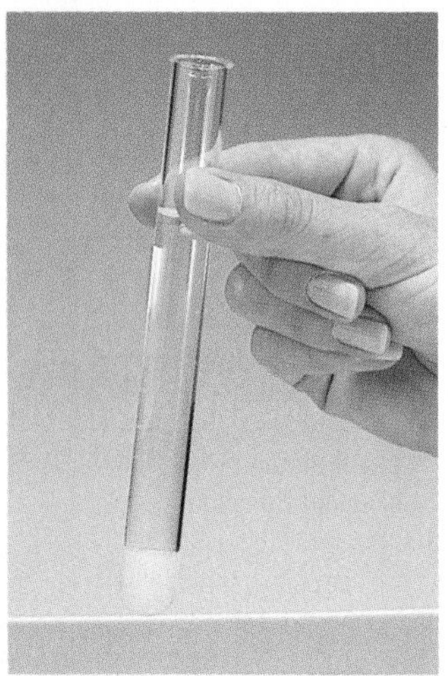

Figure 2.19

A precipitate of silver chloride.

which settles to the bottom of the container. Silver chloride precipitates because it is insoluble in water. The equation for this reaction is

$$NaCl(aq) + AgNO_3(aq) \longrightarrow NaNO_3(aq) + AgCl(s)$$

In this reaction, silver ions replace sodium ions to form the precipitate of silver chloride (Figure 2.19), and nitrate ions remain in the solution with sodium ions.

Practice Problem 2.14: When solutions of barium chloride and sodium sulfate are mixed, a precipitate of barium sulfate forms. Write an equation for this double displacement reaction.

Double displacement reactions can be used to synthesize compounds such as silver chloride and barium sulfate that are known to be insoluble in water. We examine many more double displacement reactions in Chapter 4.

Summary of General Reaction Types

Here are some general equations that provide introductory guidelines for predicting products of reactions and for writing equations for these reactions.

1. Metal + nonmetal → binary compound of metal and nonmetal
2. Metal oxide + nonmetal oxide → salt that consists of the metal ions and oxoanions of the nonmetal
3. Most metal oxides + water → metal hydroxide
4. Most nonmetal oxides + water → oxoacid
5. Hydrocarbon + oxygen → carbon dioxide + water
6. Most metal carbonates + heat → metal oxide + carbon dioxide
7. Active metal + acid → salt + hydrogen gas
8. Active metal + salt of a less active metal → salt of the active metal + less active metal
9. $Salt_1$ solution + $salt_2$ solution → $salt_3$ solution + precipitate of $salt_4$ (if $salt_4$ is insoluble in water)

2.8 A Brief Survey of Elements

In Section 2.2 we described some properties common to most metals and those common to most nonmetals. In this section we consider some important properties characteristic of selected metals and nonmetals. Various uses of the elements are also listed in this section.

The Representative Metals

Most metals exist in nature only as compounds such as oxides, chlorides, sulfides, sulfates, and phosphates. About the only metals commonly found in elemental form are gold and platinum. We recall from Section 2.2 that most pure metals are shiny substances with a silvery color of different shades ranging from white to gray. The exceptions are the "yellow" metals, copper and gold. The actual surface color of a metal may lack the luster of the pure substance if it has reacted with atmospheric oxygen or with another nonmetal.

APPLICATIONS OF CHEMISTRY 2.2
Some Interesting Consequences of Acid Rain

For centuries, acid rain has caused substantial environmental damage. Acid rain comes from many sources, both human-made and natural. We hear primarily about problems in North America, but the problems are worldwide.

One primary human-made source of acid rain is the burning of fuel that contains sulfur impurities. Along with the fuel, the sulfur impurities burn to form sulfur dioxide. Sulfur dioxide then reacts with oxygen in the air to form sulfur trioxide, and the sulfur trioxide reacts with rainwater to form sulfuric acid. All three of these reactions are combination reactions.

$S_8(s) + 8O_2(g) \longrightarrow 8SO_2(g)$ combination
$2SO_2(g) + O_2(g) \longrightarrow 2SO_3(g)$ combination
$H_2O(l) + SO_3(g) \longrightarrow H_2SO_4(aq)$ combination

Other nonmetal oxides, such as CO_2 and NO_2, can react with rainwater to form nitric acid (HNO_3) and carbonic acid (H_2CO_3). The decomposition of dead plants is one major natural source of carbon dioxide. Volcanoes are another source of nonmetal oxides. (See Applications of Chemistry 3.1 for a very unusual source of acid rain.)

For centuries buildings and statues have been constructed of marble and limestone. Both of these building materials are composed primarily of calcium carbonate. Acids react with calcium carbonate in a double displacement reaction to form carbon dioxide, thereby breaking down the very structures of buildings and statues (see photo on right). In the United States, acid rain decays exterior building stonework at an estimated cost of $2 billion. The reaction of nitric acid with calcium carbonate is

$$2HNO_3(aq) + CaCO_3(s) \longrightarrow Ca(NO_3)_2(aq) + CO_2(g) + H_2O(l)$$
double displacement

Acids also react with iron in a single displacement reaction. A specific example of this is the reaction of nitric acid in acid rain with iron in outdoor bells in Holland.

$$Fe(s) + 2HNO_3(aq) \longrightarrow Fe(NO_3)_2(aq) + H_2(g)$$
single displacement

The acid (HNO_3) in rain causes the thinning of the bells' walls which changes their natural frequency and makes the bells go out of tune. Smaller bells go out of tune faster than larger bells. By filing down the larger bell's interior, their pitch parallels that of the smaller bells.

A more serious problem occurs when acid rain affects our drinking water and makes lakes so acidic that they become lifeless. Acid rain dissolves toxic metals in the soil causing them to be leached into drinking water. Normally soil ties up many toxic metals as insoluble carbonates. Acid rain unties these toxic metals by reacting with carbonates in a double displacement reaction. The reaction of nitric acid with lead(II) carbonate is

$$2HNO_3(aq) + PbCO_3(s) \longrightarrow Pb(NO_3)_2(aq) + CO_2(g) + H_2O(l)$$
double displacement

The resulting lead nitrate is soluble in water, and it enters our drinking water supplies.

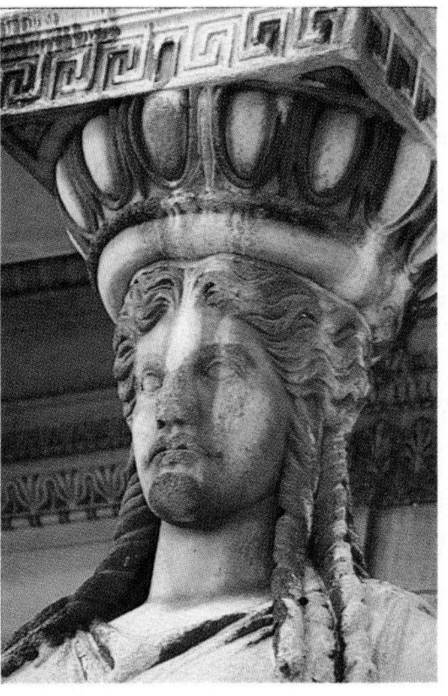

Sources: *Chemical and Engineering News*, October 12, 1981, page 7; *Science and Children*, October 1985, page 4.

In Figure 2.20, the nonradioactive representative metals are highlighted in the periodic table and some common ones are pictured below the table.

Let us begin with group I, whose members are called *alkali metals*. We often think of metals as hard substances, but the alkali metals are so soft that most can be cut with a knife (see Figure 2.4). The densities of the alkali metals are among the lowest of all metals. Three of them, lithium, sodium, and potas-

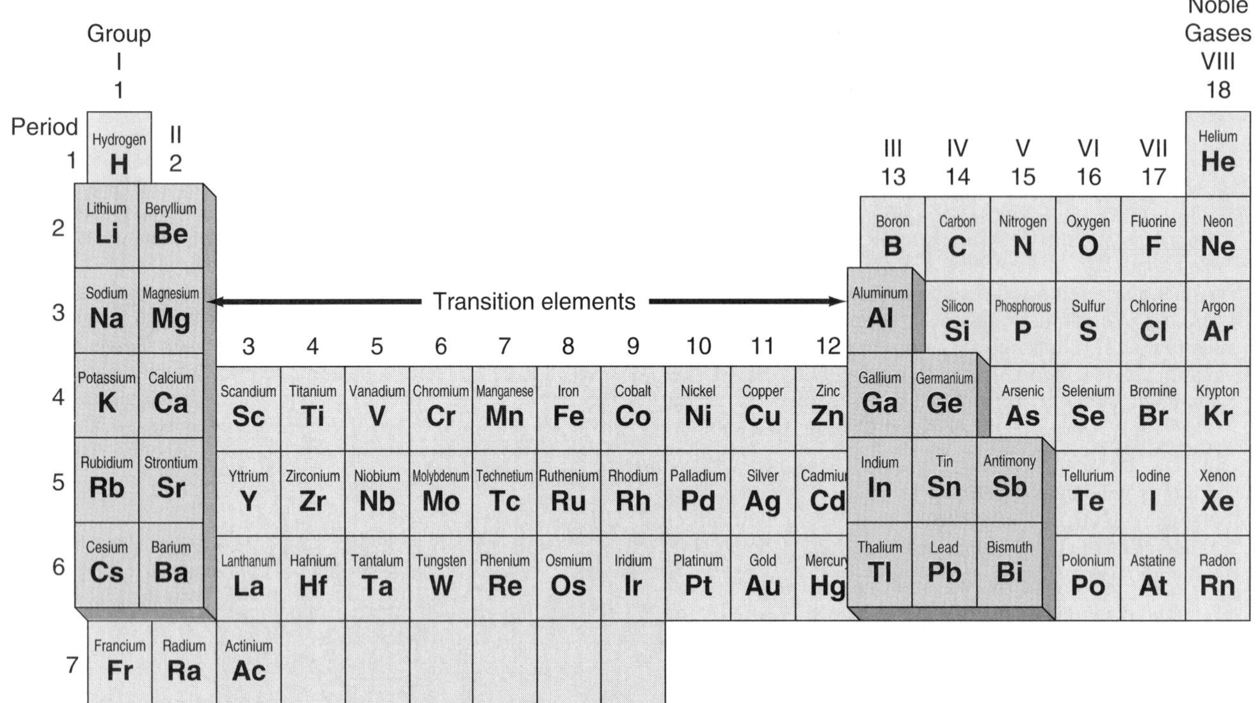

Sodium kept in oil to prevent combustion

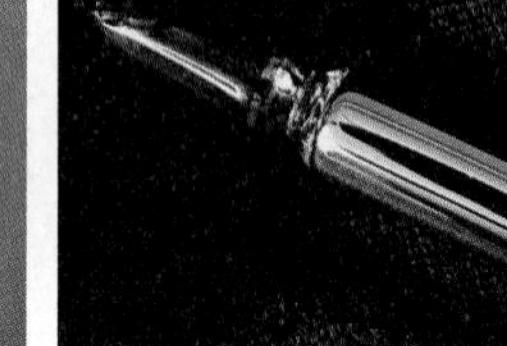
Potassium

Aluminum

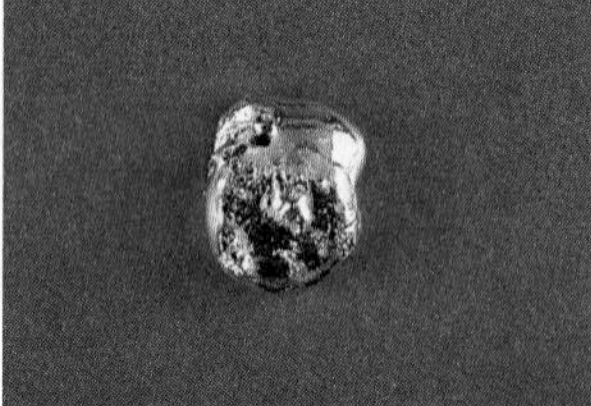
Gallium

Figure 2.20

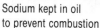

The nonradioactive representative metals (main group metals) in the periodic table and pictures of some common ones.

sium, have densities lower than 1 g cm^{-3}, which means that these metals would float in water if they did not react with it.

Alkali metals react vigorously with nonmetals, with acids, and with water. In the latter reaction, an alkali metal displaces hydrogen in water and produces the metal hydroxide as another product. The reaction of sodium with water is

$$2Na(s) + 2H_2O(l) \longrightarrow 2NaOH(aq) + H_2(g) + \text{heat}$$

The vigorous reaction of an alkali metal with water releases a large amount of heat. As a result, the temperature can rise to a point at which the hydrogen gas that is formed in the reaction reacts with the oxygen gas from the surrounding atmosphere to form water vapor:

$$2H_2(g) + O_2(g) \longrightarrow 2H_2O(g)$$

This reaction often occurs violently and releases so much heat that surrounding material may catch on fire. The exception to this generalization is lithium, which reacts only sluggishly with water.

Some uses of alkali metals in their elemental forms are listed in Table 2.7. The use of rubidium or cesium in photoelectric cells is illustrated in Figure 2.21.

Table 2.7

Some Uses of the Alkali Metals in Their Elemental Form

Element	Uses
Lithium	In the manufacture of alloys. Lithium added to aluminum increases the corrosion resistance of aluminum. A lithium–lead alloy is used for flexible cable sheaths.
Sodium	Sodium mixed with potassium forms a liquid alloy that is used as a coolant in some nuclear reactors. In the vapor state the metal is used in sodium lamps, primarily in outdoor areas and roadways, where they are relatively economical to operate.
Potassium	Alloyed with sodium as a medium of heat transfer and as a coolant.
Rubidium	In photoelectric cells (see Figure 2.21).
Cesium	In photoelectric cells and in the atomic clock. Cesium as a time-measuring element is based on the precise movement of one of its electrons around its axis. The accuracy of this movement is 5 seconds in 300 years. The radioactive isotope ^{137}Cs is used in medicine.

Sodium ions and potassium ions are found in all living cells, and the relative amounts of these ions stay fairly constant in healthy organisms.

A reaction of sodium with water, causing fire.

Practice Problem 2.15: Write an equation for each of the following reactions: **(a)** lithium with oxygen; **(b)** rubidium with water.

The group II metals—*alkaline earth metals*—are somewhat harder and most are less reactive than the alkali metals, but like alkali metals, they react readily with most nonmetals, acids, and water.

Practice Problem 2.16: Write an equation for each of the following reactions: **(a)** beryllium with bromine; **(b)** calcium with hydrochloric acid; **(c)** barium with water; **(d)** magnesium with nitrogen.

The commonly encountered alkaline earth metals are *magnesium* and *calcium*. *Magnesium* has a relatively low density, and it is used as a component in lightweight alloys with other metals. Alloys containing magnesium are found in the frames of automobiles and aircraft. Magnesium ions are present in all living cells, where they are essential for many chemical reactions.

Calcium salts are abundant in nature. The salt of calcium called hydroxyapatite is a major component of bones and teeth. Calcium ions are also found in all living cells, and processes such as muscle contractions are triggered by transient changes in the concentration of calcium ions. Some uses of alkaline earth metals are listed in Table 2.8.

In groups III, IV, and V, the commonly encountered metals are *aluminum* in group III and *tin* and *lead* in group IV. Aluminum has a relatively low density; tin and lead have much higher densities but are much softer than aluminum. Like the alkali metals, they can be cut with a knife.

Aluminum is the third most abundant element in the earth's crust. Aluminum can be drawn to a wire, rolled into a sheet as a wrapping material, and pressed into any shape. The low density of aluminum and its resistance to corrosion make it extremely useful in the construction industry.

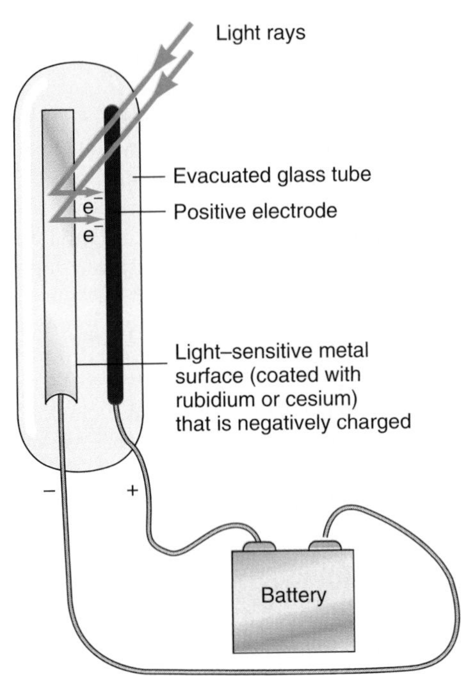

Figure 2.21

A photoelectric cell. When light shines on the negative electrode coated with rubidium or cesium, electrons are ejected from the surface of this light- sensitive electrode and are attracted to the positive electrode. As long as light shines to the cell, an electric current flows through the circuit. When the light beam is interrupted, the current flow is stopped; this may trigger a device that opens or closes a door or a parking gate.

Table 2.8

Some Uses of the Alkaline Earth Metals

Element	Uses
Beryllium	Neutron moderator in nuclear reactors and as an ingredient in beryllium–copper alloy that is as hard as steel.
Magnesium	As a constituent of lightweight alloys; in flashbulbs, flares, and in intense signal lights (magnesium burns with a brilliant flame).
Calcium	Primarily as an ingredient in an alloy with silicon, used in high-temperature furnaces for steel-making; also, in an alloy with cerium, to make flints for cigarette and gas lighters.
Strontium	In fireworks and in red signal flares (strontium and its compounds produce a brilliant red color when heated to high temperatures). The artificial isotope of strontium, ^{90}Sr, liberates heat as it disintegrates to one-half of its original amount in 28 years; this isotope is therefore considered as a possible source of electric power.
Barium	The alloys of barium with aluminum and magnesium are used to make spark plugs and electronic vacuum tubes.
Radium	Radium is radioactive and is used in making x-ray photographs of metals.

Aluminum is chemically much more reactive than are tin and lead. It reacts readily with most nonmetals, acids, and even to a small extent with hot water:

$$2Al(s) + 6H_2O(l) \longrightarrow 2Al(OH)_3(s) + 3H_2(g)$$

The surfaces of aluminum cooking utensils and siding on houses initially react with water and oxygen to form a coating of aluminum hydroxide and aluminum oxide. This coating adheres very strongly to the surface of the metal and prevents further reaction.

Table 2.9

Some Uses of the Metals in Groups III, IV, and V

Element	Uses
Aluminum	As an ingredient in lightweight alloys; in aluminum siding and cooking utensils. Aluminum is used extensively in the building industry, aircraft, and automobile construction, shipbuilding, and in thin sheets as a wrapping "material" in the food industry.
Gallium, indium and thallium	No large-scale industrial uses.
Tin	Mainly for tin plating of cans (tin cans) and other containers and equipment. *Bronze* is an alloy containing tin, lead, copper, zinc, and phosphorus in varying proportions. *Pewter* is about 85 percent tin and the remainder lead, or a mixture of copper, zinc, and antimony. Tin has also been put to use in organ pipes. Pipes of tin are noted for their wonderful tone and for their ability to be precisely tuned.
Lead	Tank linings and other equipment handling corrosive chemicals, in protective shields for x-ray and other radiation, and in plates for automobile batteries.
Antimony	In alloys for bullets and bearing metal, and in fireworks.
Bismuth	For making drugs and various alloys with other metals.

Tin and *lead* are malleable metals with relatively low melting points (232°C and 328°C, respectively). Both metals are relatively unreactive to air, acids, and water. Tin is used as a plating material for cans. Lead is used in protective shields for x-ray and other radiation. Lead compounds are poisonous. Prolonged exposure to lead compounds can lead to severe mental retardation. Thus, lead compounds are no longer used in paints, gasoline, pipes, or solder. Table 2.9 lists some uses of the metals in groups III, IV, and V.

Practice Problem 2.17: Write a balanced equation for each of the following reactions: **(a)** aluminum metal reacts with oxygen; **(b)** tin metal reacts with sulfur to form tin(II) sulfide; **(c)** lead(II) chloride reacts with chlorine to form lead(IV) chloride.

Common Transition Metals

In the first row of transition metals, *titanium*, *chromium*, *iron*, *nickel*, *copper*, and *zinc* are particularly important. *Titanium* is stronger and has a lower density than steel. It is very corrosion-resistant, even in the presence of acids and seawater. Titanium is therefore used in the construction of jet engines and chemical plants, where corrosion resistance is important. Figure 2.22 highlights nonradioactive transition metals in the periodic table and pictures some common ones.

Another metal that resists corrosion is *chromium*. This hard, strong metal is used as a protective coating on some automobile bumpers and other steel products. *Iron* is a relatively soft, malleable, and ductile metal. It is the principal component in *steel*, an alloy of iron that contains from 0.1 to 1.3 percent carbon and varying amounts of chromium, vanadium, nickel, manganese, tung-

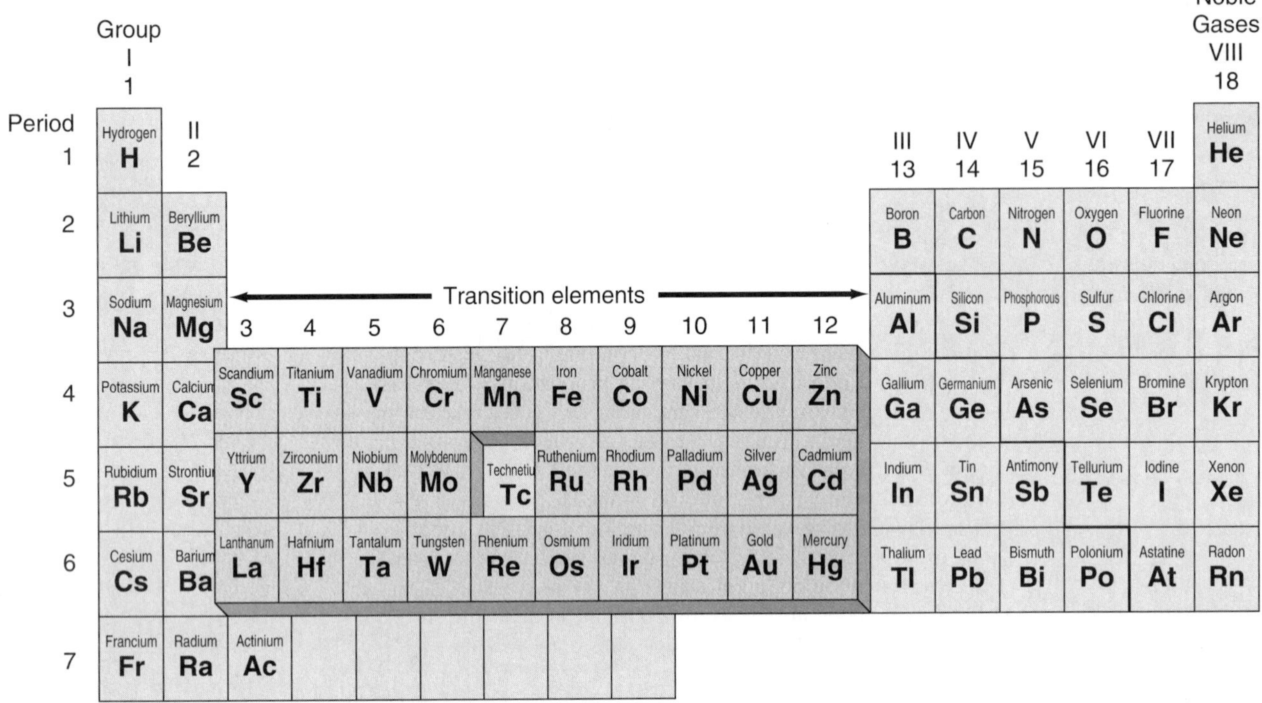

Cobalt

Copper

Tungsten

Platinum

Figure 2.22

The nonradioactive transition metals in the periodic table with photos of some common ones.

sten, and other metals. An alloy of iron, chromium, and tungsten is used as a steel for high-speed tools that stay hard and sharp even when red hot.

Nickel is used primarily to make coins, which contain varying proportions of nickel and copper as their principal ingredients. *Copper* is relatively unreactive and a good conductor of electricity. It is therefore used to make tubing for pipes and wires for electrical conductors. *Zinc* is a soft, malleable, corrosion-resistant metal with a relatively low melting point (420°C). It is used as a coating for iron to make *galvanized iron*, which is used in applications where resistance to corrosion is important.

Some of the commonly encountered transition metals of the second and third rows include *silver*, *gold*, *platinum*, and *mercury*. Silver and gold are used extensively in jewelry and coinage. A silver alloy that contains at least 92.5 percent silver is called *sterling silver*. Pure gold is too soft to be used alone for most purposes and is therefore alloyed with other metals, such as nickel and zinc. The purity of gold alloys is expressed in *carats*. A 24-carat gold consists of pure gold, whereas a 14-carat gold contains 14/24 parts (58 percent) of gold by mass.

Platinum is used as an agent that increases the rates of many chemical reactions. Such substances are called **catalysts**. Platinum catalysts are found in

APPLICATIONS OF CHEMISTRY 2.3
Buckminsterfullerene: A New Form of Carbon

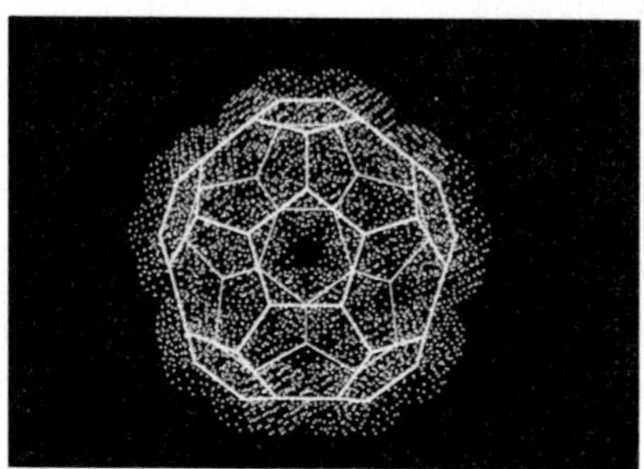

In this chapter we have described two structural forms of carbon: diamond and graphite. A diamond is a form of carbon in which each carbon atom is bonded to four others in a regular tetrahedron; graphite consists of stacked sheets of carbon in hexagonal arrays (see Figure 2.24). Diamond and graphite often contain some atoms other than carbon and are therefore not "pure" forms of carbon. Recently, a new, pure form of carbon has been discovered that consists of discrete molecules, the most stable of which is C_{60}. The name given to C_{60} is *buckminsterfullerene,* after the American inventor Buckminster Fuller, who is most famous for inventing the architectural form known as a geodesic dome (see photo, top right).

Buckminsterfullerene has the structure shown in the lower right photograph. It has an icosahedral structure. That is, it is a 60-sided molecule made of 12 pentagons and 20 hexagons. We can think of a C_{60} molecule as a hollow, molecular soccer ball. Perhaps for this reason, buckminsterfullerene is known colloquially as a "buckyball."

Buckyballs can easily be made in the laboratory simply by burning a piece of paraffin in the flame of a bunsen burner. This process produces a sooty material which turns out to be composed almost entirely of buckyballs and smaller amounts of other molecules having similar structures. Thus soot, which conjures images of diesel exhaust and the black residue that lines the insides of fireplaces, consists mostly of buckyballs.

Buckyballs have extraordinary properties. For example, since buckyballs are almost completely "round," they are likely to be exceptionally good lubricants and should provide a virtually friction-free contact between surfaces. Buckyballs have been chemically converted to compounds that contain metal atoms. One example is K_3C_{60}. This material conducts electricity without electrical resistance at a temperature of 18 K; that is, it is a *superconductor.* (See Applications of Chemistry 6.2 for a description of a "superfluid.") Buckyballs are not merely laboratory curiosities, they have a tremendous range of potential commercial applications.

automobile catalytic converters. Platinum is also used as a chemically inert material for laboratory ware and in jewelry.

Mercury is the only metal that is a liquid at room temperature; it is used in barometers and thermometers. Mercury dissolves many metals (with the exception of iron) to form mercury alloys called *amalgams*. One such amalgam, an alloy of mercury, silver, and tin, is used by dentists for filling cavities in teeth.

The *inner transition metals* are less often encountered than those we described above. Two of the inner transition metals, *uranium* and *plutonium*, are used as fuels for nuclear energy production. The elements starting with the element of atomic number 93 (neptunium) do not exist in nature. They can be produced only by nuclear reactions in nuclear reactors.

Metallic elements with an atomic number of 84 (polonium) or higher have unstable nuclei that disintegrate spontaneously with concomitant emission of energy and subatomic particles. These elements are said to be *radioactive*.

Practice Problem 2.18: Consulting Table 2.3 if necessary, write formulas and names for each of the following compounds: **(a)** a binary compound of silver and oxygen; **(b)** two sulfates of iron; **(c)** two binary compounds of copper and chlorine; **(d)** two binary compounds of mercury and chlorine; **(e)** a binary compound of zinc and phosphorus.

Nonmetals

Nonmetals exist at room temperature as solids or gases, except bromine, which is liquid. The gaseous nonmetals are hydrogen, nitrogen, oxygen, fluorine, chlorine, and all of the noble gases. We recall from Section 2.2 that the noble gases consist of single atoms. In contrast, hydrogen, nitrogen, oxygen, and the halogens consist of diatomic molecules. Figure 2.23 highlights the nonradioactive nonmetals in the periodic table and displays photos of some common ones.

Figure 2.23

The nonradioactive nonmetals in the periodic table and photos of some common ones.

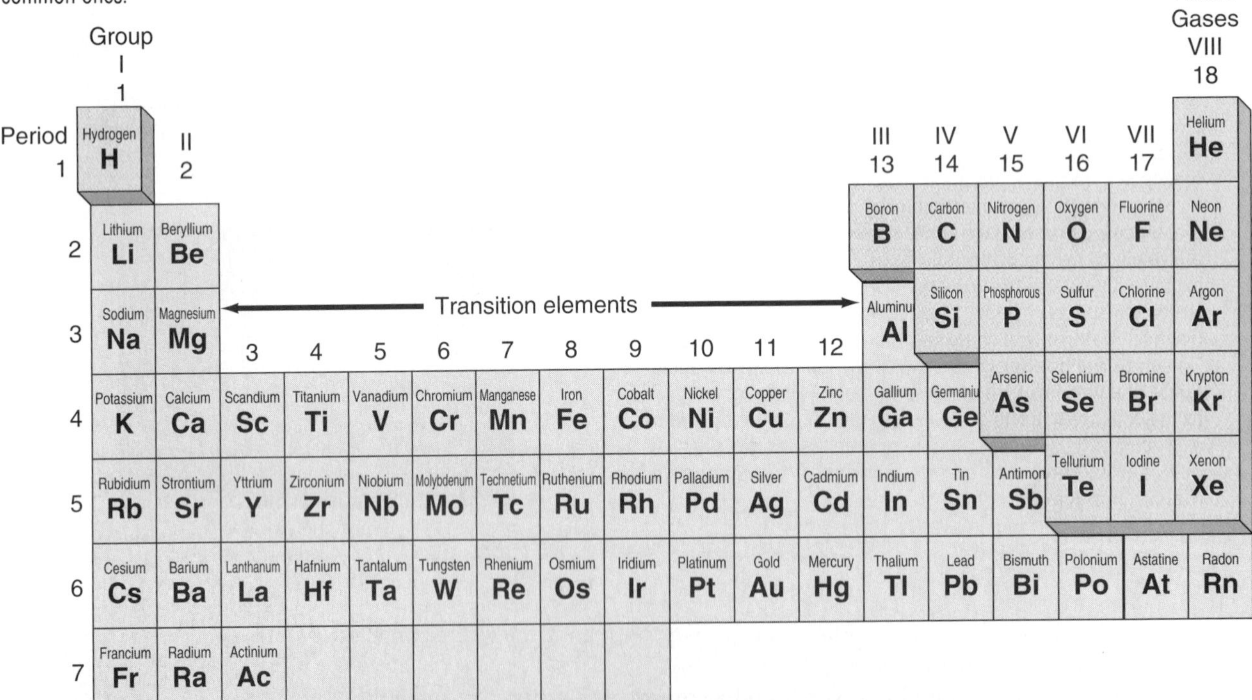

Boron

Silicon

Sulfur

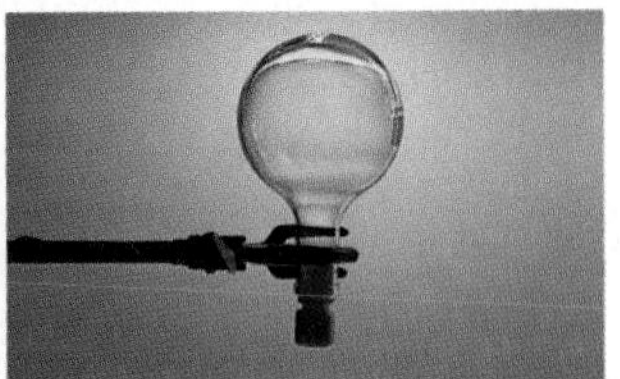

Xenon gas

The common nonmetals that exist as solids at room temperature consist of polyatomic molecules, or arrays of covalently bonded atoms. The only common solid nonmetal that consists of diatomic molecules at room temperature and pressure is iodine.

Some nonmetals can exist in two or more different arrangements of atoms. The *different forms of an element in the same physical state and phase* are called **allotropic forms**, or **allotropes**. Some common elements that are found in two or more allotropic forms include carbon, oxygen, phosphorus, and sulfur.

Amorphous form of carbon (soot) and two crystalline forms of carbon (diamond and graphite).

Carbon exists in two common crystalline allotropic forms, *diamond* and *graphite* (Figure 2.24), and in an amorphous form called soot. The carbon atoms in a diamond are bonded to each other in a large network that resists any pressure exerted on it. Thus, diamond is very hard. The carbon atoms in graphite are arranged in layers of hexagonal networks that can easily slide past one another. This arrangement makes graphite slippery. The slippery character of graphite makes it an excellent lubricant.

Another common nonmetal, *phosphorus*, consists of tetratomic molecules, P_4 (Figure 2.25), at 25°C. Phosphorus is found in two common allotropic forms, known as *white* phosphorus and *red* phosphorus. In white phosphorus, the P_4 molecules are not bonded to each other, whereas in the red allotrope they are covalently bonded to one another in long chains (Figure 2.26). White phosphorus is very poisonous, ignites spontaneously at 35 to 40°C, causes painful

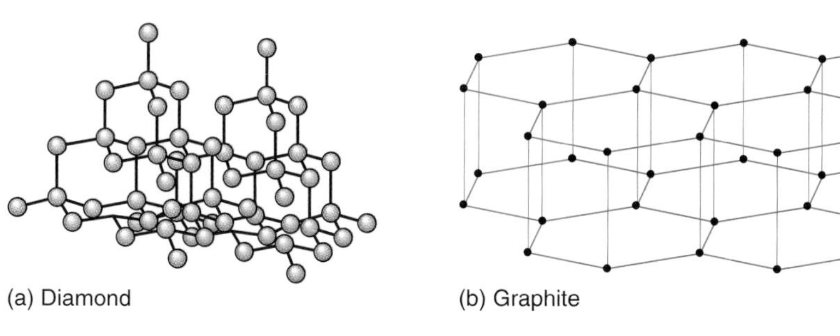

Figure 2.24

Two crystalline allotropic forms of carbon. (a) *Diamond*: each carbon atom in diamond is connected to four other carbon atoms in a rigid, giant network of atoms. Distances between atoms are exaggerated for clarity. (b) *Graphite*: layers of carbon atoms can slide on top of one another making graphite slippery to the fingers.

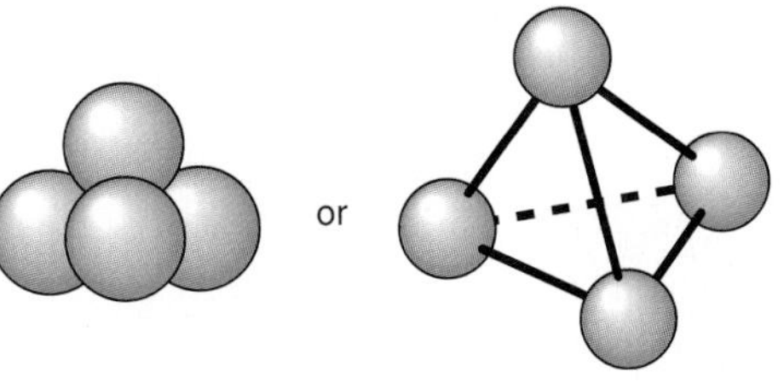

Figure 2.25

The structure of a P_4 molecule. At the left is a model of a P_4 molecule in which the atoms are touching one another. At the right, the distances between atoms are exaggerated for clarity, and lines between atoms are drawn to show that the atoms occupy the corners of a tetrahedron (a polyhedron with four equilateral triangular sides).

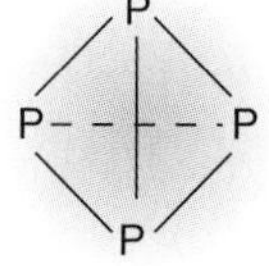

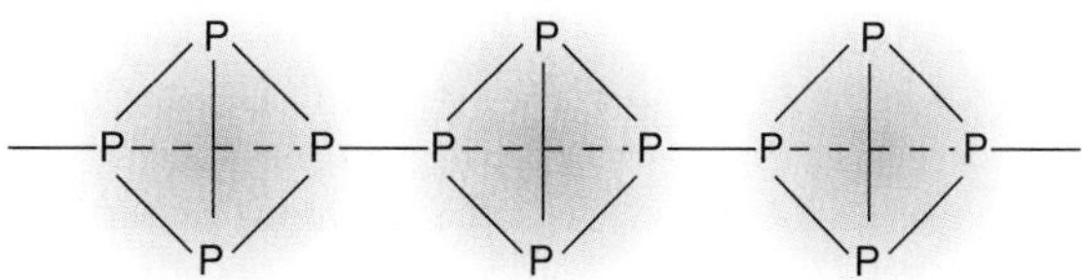

Figure 2.26

Two allotropic forms of phosphorus: (a) white phosphorus and (b) red phosphorus.

Left: Crystalline sulfur. Right: Amorphous sulfur.

burns in contact with skin, and reacts with most metals and nonmetals. The red variety is much less reactive and safer to handle; it is used in making matches.

Elemental *sulfur*, at 25°C, consists of molecules that contain eight atoms, octatomic molecules, S_8 (Figure 2.27). Sulfur exists in several crystalline allotropic forms plus an amorphous modification. Large deposits of sulfur are found in Louisiana and Texas. Sulfur is mined by a method called the Frasch process (Figure 2.28).

Oxygen is the only gaseous element that exists in allotropic forms. The allotrope consisting of diatomic oxygen, O_2, usually called "oxygen," is by far the more abundant. It is a colorless and odorless gas. Oxygen is a very reactive element that combines with many elements and compounds. This general combination reaction is called **combustion**, or burning. Combustion reactions emit heat and sometimes light. Oxygen is required by humans and animals for respiration. It is produced by plants as a by-product of photosynthesis.

The other allotrope of oxygen is *ozone*, O_3. Ozone is a pale blue gas with a characteristic odor that one associates with operating electrical equipment. Ozone is found in the upper atmosphere, where it is formed from diatomic oxygen, O_2, when it absorbs the sun's ultraviolet rays. Ozone has an important biological function. It prevents most ultraviolet solar radiation from reaching the earth's surface. Ultraviolet radiation causes changes in the structure of the

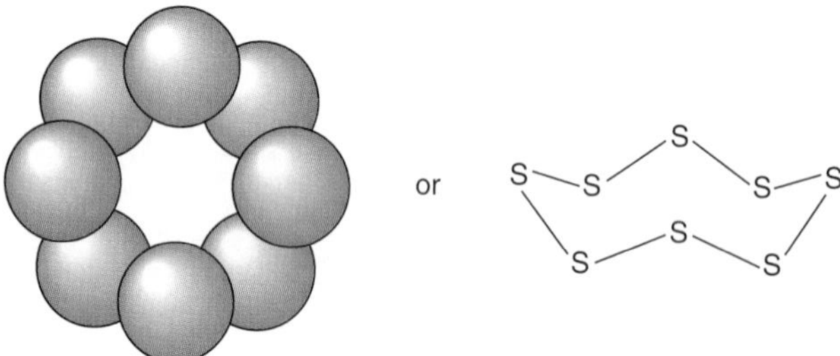

Figure 2.27

The structure of an octatomic sulfur molecule. Note that the sulfur atoms in an S_8 molecule do not lie in one plane but form a puckered ring in which four atoms lie in a plane, and four other atoms lie in a parallel plane.

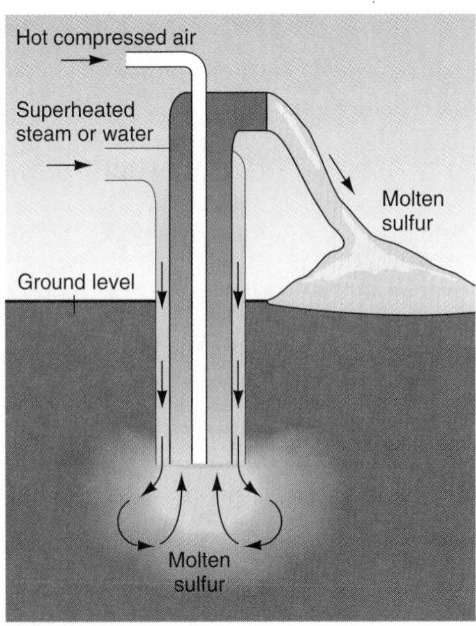

Figure 2.28

The Frasch process for sulfur extraction. Superheated steam is compressed down through the outer pipe to melt the sulfur in the ground. Hot compressed air is pumped down the inner pipe to form a foamy mixture of sulfur, water, and air. This mixture rises to the surface through the middle pipe. After removal of water, the sulfur is stored in large bins.

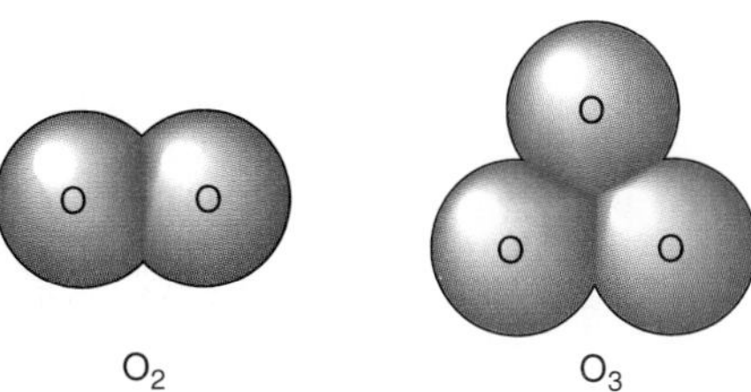

Figure 2.29

Models of O_2 and O_3 molecules. The oxygen atoms in an ozone molecule do not lie along a straight line but are located at the corners of a triangle.

genetic material, DNA, and long exposure to this radiation can cause cancer. Various atmospheric pollutants are presently suspected of decreasing the ozone layer, especially above the south and north poles. Models of O_2 and O_3 molecules are shown in Figure 2.29.

Like carbon, *boron* and *silicon* also exist in both crystalline and amorphous forms. Highly purified silicon is used in the construction of electronic devices such as microcomputers and solar cells. In nature, silicon is found mainly in the form of silicon dioxide, SiO_2, an inert solid that is the principal ingredient of sand. The usual crystalline form of silicon dioxide is *quartz*, a hard, brittle, clear, and colorless solid. Most of the rocks in the earth's crust consist of minerals that contain silicon; these are called silicates (see Chapter 9).

Arsenic has a metallic appearance and exists in crystalline forms similar to some forms of phosphorus. Elemental arsenic and its compounds are poisonous. The "arsenic" favored as a poison in detective stories is diarsenic trioxide, As_2O_3. (The terrible ravages of arsenic poisoning are described with morbid brilliance near the conclusion of Gustave Flaubert's novel *Madame Bovary*.)

To summarize the discussion of the nonmetals, let us consider some trends in the periodic table. The nonmetals in groups III and IV are all solids at room temperature. In group V, the first element, *nitrogen*, is a colorless and odorless gas, and the other elements are solids. In group VI, as in group V, the first element, *oxygen*, is a colorless and odorless gas, while the other elements are solids. The first element in group VII, *fluorine*, is a poisonous, pale yellow gas with a choking odor. *Chlorine* is a poisonous, greenish-yellow gas with a pungent odor. *Bromine* is also very poisonous; it is a dark red liquid that vaporizes easily at room temperature. *Iodine* is a black-to-dark violet solid that easily converts to a violet vapor upon warming. The halogens are very reactive and poisonous substances. When working with bromine or iodine, be careful to avoid skin contact.

The nonmetals located at the *upper right* of the periodic table (except the noble gases) are chemically *most reactive*. They include fluorine, oxygen, and chlorine, in that order. These elements react readily with most metals and nonmetals. Noble gases, as we noted earlier, are extremely unreactive.

Hydrogen is a colorless and odorless gas that reacts with alkali metals and with alkaline earth metals to form binary compounds called hydrides, such as sodium hydride, NaH, and barium hydride, BaH_2. In these compounds, hydrogen is present in the form of hydride ions, H^-.

Boron, carbon, nitrogen, and silicon are the *least reactive* of the nonmetals. Some uses of the nonmetallic elements are listed in Tables 2.10 and 2.11.

Table 2.10

Some Uses of the Nonmetallic Elements of Groups III, IV, and V

Element	Uses
Boron	As neutron absorber in nuclear reactors, and as an ingredient in metal alloys to harden metals.
Carbon	Diamond is used in jewelry, and because of its hardness, in polishing, grinding, cutting glass, and in making bearings for delicate instruments. Graphite is used in "lead" pencils, as a lubricant, and as a moderator in nuclear reactors. Amorphous carbon is a good absorbent of gases and is widely used as a filtering element. Because of its color, amorphous carbon is also used in making ink and in painting.
Silicon	In the manufacture of transistors (electronic devices that amplify electric current), radios, televisions, computers, and calculators.
Nitrogen	For manufacture of ammonia. As an inert atmosphere for preservation of materials that would spoil in a normal atmosphere containing oxygen and moisture. Liquid nitrogen is used as a coolant and as a medium that provides low temperature for laboratory experiments (liquid nitrogen boils at about −196°C and freezes to a solid at about −210°C under normal pressure).
Phosphorus	In the manufacture of fertilizers, matches, smoke bombs, and fireworks.
Arsenic	As an additive to lead and copper to increase the hardness of these metals. Also in the manufacture of certain types of glass.

Table 2.11

Some Uses of the Nonmetals in Groups VI, VII, and VIII

Element	Uses
Oxygen	In oxyhydrogen and oxyacetylene flame for welding metals, as a propellant for rockets, and in medicine.
Sulfur	Mainly for synthesis of compounds such as sulfuric acid and sulfur-containing drugs ("sulfa" drugs), in the vulcanization (treatment of material to give it useful properties) of rubber, and in the manufacture of gunpowder, matches, and fungicidal fruit and flower spray.
Selenium	In the xerographic process used in copying machines and as a pigment to make colored glass.
Tellurium	As a coloring agent of glass and porcelain.
Fluorine	Used mainly to prepare fluorine compounds called fluorocarbons, which are useful as plastics (Teflon), refrigerants (Freon), and insecticides (agents that kill insects).
Chlorine	For water purification, as a disinfectant in swimming pools, and for the manufacture of chlorinated lime for bleaching fabrics.
Bromine	For synthesis of the antiknock compound, ethylene bromide, for gasoline; for bleaching fibers and silk; and for making bromine-containing drugs.
Iodine	For manufacture of iodine compounds and as an antiseptic agent in an alcohol solution (tincture of iodine).
Helium	In gas balloons, in inert-atmosphere boxes for research laboratories, and in a mixture with oxygen for use by divers to permit easier breathing when ascending to the surface.
Neon	In illuminating signs ("neon signs").
Argon, krypton and xenon	In gas-filled incandescent lamps.

Summary

Classification of Elements

The 109 elements known today are arranged in the periodic table according to their increasing atomic numbers. In this table the *rows* are called **periods**, and the *columns* are **groups**. The physical and chemical properties of the elements recur periodically from one period to another so that the *elements in each group have roughly similar properties*.

The elements are classified into **metals** and **nonmetals**. Some elements that have properties on the borderline between metals and nonmetals are called **metalloids**. The elements in groups I through VIII of the periodic table, as shown in Table 2.2, are called **main group elements** or **representative elements**. The remaining elements are **transition metals** and **inner transition metals**. The metallic elements in group I are called **alkali metals**, and those in group II are called **alkaline earth metals**. The nonmetals in group VII are called **halogens**, and those in group VIII are called **noble gases**.

Ions of Metals and Nonmetals; Ionic and Molecular Compounds

In a reaction of most metals with nonmetals, the metal atoms lose electrons to form positive ions, and the atoms of the nonmetal gain electrons to form negative ions. The ions of most main group elements have the same number of electrons as those of their nearest noble gas atoms. Some exceptions to this rule exist for the lower members of groups III and IV.

Compounds that consist of ions are called **ionic compounds**, and those that consist of molecules are **molecular compounds**. The ions in an ionic compound are held together by **ionic bonding**, which is the electrostatic attraction between positively charged ions (**cations**) and negatively charged ions (**anions**). The atoms in a molecule of a molecular compound are held together by **covalent bonding**, in which the electrons are shared by the bonded atoms.

Formulas and Names of Binary Metal–Nonmetal Compounds

The subscripts in the formula of most ionic compounds specify the lowest relative numbers of cations and anions in the compound. The total positive charge of the cations in a compound must be equal to the total negative charge of the anions because the compound is electrically neutral.

The *names of binary compounds* generally *end with -ide*. In naming compounds of metals that form ions of more than one charge, the charge is indicated by a Roman numeral in parentheses following the name of the metal. The Roman numeral is not used in the names of the compounds of metals that form ions of only one charge.

Naming Oxoanions

In naming two **oxoanions** of the same element, the ending *-ate* is used for the anion that contains *more oxygen*, and the ending *-ite* for the anion with *less oxygen*.

Naming Binary Nonmetal–Nonmetal Compounds

In naming binary compounds consisting of two nonmetals, the Greek prefixes *mono-*, *di-*, *tri-*, and so on, are used to designate the number of atoms of an element in a molecule of the compound.

Naming Acids

Substances that produce hydrogen ions in water are called **acids**. The acids that do not contain oxygen are called **hydroacids**. They are named with the prefix *hydro-* and suffix *-ic*. The acids that contain oxygen are **oxoacids**. The names of oxoacids are related to the names of their oxoanions. When the name of the oxoanion of an acid ends with *-ate*, the name of the acid ends with *-ic*; when the name of the anion ends with *-ite*, the name of the acid ends with *-ous*.

Chemical Equations and Types of Reactions

A chemical equation is a shorthand description of a chemical reaction. A chemical equation is **balanced** to satisfy the law of conservation of matter. Chemical reactions can generally be classified into three types: **combination**, **decomposition**, and **displacement** reactions.

Combination reactions can involve the *combination of two elements* to form a binary compound, the *combination of an element with a compound* to form a new compound, or the *combination of two compounds* to form a new compound. Examples of the latter type of combination reactions are the reactions of **acidic oxides** with **basic oxides** to form **salts**.

Common decomposition reactions include decomposition of some metal oxides and metal carbonates. *Oxides of relatively unreactive metals* such as gold, silver, and mercury *decompose into their elements* upon heating. Metal *carbonates decompose to form two different compounds, the metal oxide and carbon dioxide. Displacement reactions* can be divided into *single and double displacement.* In a single displacement reaction, *a more active metal displaces a less active metal from its compound*, or *an active metal can displace hydrogen from an acid.* A *double displacement reaction involves an exchange of components of the two reactants.* Such a reaction usually occurs in water solution to form a **precipitate**.

A Brief Survey of Elements

Metals are solids at room temperature (except mercury, which is a liquid), range from soft to very hard, have a characteristic metallic luster, are generally good conductors of electricity and heat, and many are malleable and ductile. Some nonmetals under normal conditions are gases, some are solids, and one (bromine) is a liquid.

The metals and the noble gases consist of atoms as their basic structural units. Hydrogen, nitrogen, oxygen, and the halogens exist as diatomic molecules. The molecules of white phosphorus are tetratomic, and the molecules of sulfur are octatomic. Several nonmetals exhibit **allotropic forms** such as diamond and graphite for carbon and the white and red varieties of phosphorus.

New Terms

The number in parentheses following each term denotes the section in which the term is introduced.

Acid (2.5)
Acidic oxide (2.7)
Actinide (2.1)
Active metal (2.7)
Alkali metal (2.1)
Alkaline earth metal (2.1)
Allotropic forms or allotropes (2.8)
Amalgam (2.8)
Anion (2.5)
Base (2.7)
Basic oxide (2.7)
Binary compound (2.2)
Carat (2.8)
Catalyst (2.8)
Cation (2.5)
Combination reaction (2.7)
Combustion (2.8)
Covalent bond (2.4)
Decomposition reaction (2.7)
Displacement reaction (2.7)
Ductile (2.2)
Group in a periodic table (2.1)
Halide ion (2.2)
Halogen (2.1)
Heteronuclear diatomic molecule (2.4)
Homonuclear diatomic molecule (2.4)
Hydroacid (2.5)
Hydrocarbon (2.7)
Ion (2.2)
Ionic bond (2.4)
Ionic compound (2.2)
Isoelectronic (2.2)
Lanthanide (2.1)
Lattice (2.4)
Malleable (2.2)
Metal (2.1 and 2.2)
Metalloid (2.1)
Molecular compound (2.3)
Monatomic ion (2.5)
Noble gas (2.1)
Noble metal (2.7)
Nonmetal (2.1 and 2.2)
Oxoacid (2.5)
Oxoanion (2.5)
Period in a periodic table (2.1)
Periodic law (2.1)
Polyatomic ion (2.5)
Precipitate (2.7)
Product of a reaction (2.6)
Reactant of a reaction (2.6)
Representative element (2.1)
Salt (2.2)
Stainless steel (2.7)
Supernatant solution (2.7)
Transition metal (2.1)

Exercises

General Review

1. Describe the classification scheme of elements by Mendeleev.
2. In what order did Mendeleev arrange the elements in his periodic table?
3. What is the meaning of the word "periodic" in periodic table?
4. What is the order in which the elements are arranged in the modern periodic table?
5. Illustrate the meaning of the periodicity of properties of elements in terms of the periodic table.
6. What are periods, groups, and families in a periodic table?

7. What are the group names of the elements in groups I, II, VII, and VIII of the periodic table?
8. Point out the location of each of the following types of elements in a periodic table: **(a)** main group elements; **(b)** transition elements; **(c)** inner transition elements; **(d)** noble gases; **(e)** representative elements; **(f)** lanthanides; **(g)** actinides; **(h)** metals; **(i)** nonmetals; **(j)** metalloids.
9. How many main group elements are in each of the first six periods in a periodic table?
10. How many transition elements (not including the inner transition elements) are in each of the first six periods in a periodic table?
11. How many inner transition elements are in each of the lanthanide and actinide series?
12. Describe some distinguishing properties of the elements in each group of the representative elements, and show how these properties differ from the properties of the elements in other groups.
13. What are some general characteristics of metals and of nonmetals?
14. In a reaction of a typical metal with a nonmetal, do **(a)** the atoms of the metal and **(b)** the atoms of the nonmetal generally lose or gain electrons? **(c)** What is the sign of the charge on metal ions? on nonmetal ions?
15. How can you generally arrive at the magnitude of the charge on an ion of: **(a)** a representative metal; **(b)** a nonmetal?
16. Which common representative metals form more than one compound with the same nonmetal? Give examples by writing formulas and names for such compounds.
17. What is the ending of the name of a monatomic negative ion?
18. What is the general name of all of the common ions of the group VII elements?
19. What is a "binary compound"?
20. In writing the formula for a binary compound of aluminum and oxygen, how can you decide what the subscripts for Al and for O should be?
21. What is: **(a)** ionic bonding; **(b)** covalent bonding?
22. Write formulas and names for several polyatomic ions.
23. Write the names and formulas for three common acids.
24. Write a balanced equation as an example of each of the following reactions: **(a)** combination reaction; **(b)** decomposition reaction; **(c)** single displacement reaction; **(d)** double displacement reaction.
25. List all the elements that exist as gases and those that exist as liquids at room temperature.
26. Which elements exist as diatomic molecules, as tetratomic molecules, and as octatomic molecules?
27. Discuss and give examples of allotropy of elements.
28. Write the formulas and names for three molecular compounds and for three ionic compounds.
29. Explain how you decide what the formula of the binary compound of magnesium and phosphorus should be. What is the name of this compound?
30. Write a formula for each of the compounds that would result from the combination of phosphate ions with: **(a)** sodium ions; **(b)** magnesium ions; **(c)** aluminum ions; **(d)** ammonium ions.
31. What is the meaning of the name: **(a)** "noble" gas; **(b)** "noble" metal?

Monatomic Ions of Elements

32. Write the symbol, charge, and the name for the ion that each of the following elements is likely to form: **(a)** Rb, an alkali metal; **(b)** Sr, an alkaline earth metal; **(c)** phosphorus; **(d)** selenium; **(e)** iodine.
33. **(a)** If an atom has an atomic number of 42 and the mass number of 96, how many neutrons does the nucleus of this atom contain? **(b)** How many electrons does the atom contain? **(c)** What is the number of electrons in a 2+ ion of this atom? **(d)** What is the number of protons in a 2+ ion of this atom?
34. An atom with a mass number of 31 contains 16 neutrons in its nucleus. **(a)** How many protons does a 3− ion of this atom contain? **(b)** How many electrons?
35. In which of the following categories are all of the species isoelectronic with each other? **(a)** Li^+, Na^+, K^+; **(b)** O^{2-}, F^-, Ne; **(c)** Ar, K^+, Ca^{2+}; **(d)** P^{3-}, Ar, K^+; **(e)** As^{3-}, Kr, Sr^{2+}.
36. How many protons and how many electrons does each of the following species contain? **(a)** An atom of bromine; **(b)** a bromide ion; **(c)** an aluminum atom; **(d)** an aluminum ion.
37. In some periodic tables, hydrogen appears in both groups I and VII. Explain.
38. Write the formulas for all of the ions that each of the following elements is likely to form: **(a)** tin; **(b)** lead; **(c)** iron; **(d)** copper; **(e)** chromium; **(f)** silver; **(g)** mercury.
39. Two representative metals A and B belong to groups II and V, respectively. What is the most likely charge of the ion derived from an atom of A, and of B?

Names and Formulas of Acids and Polyatomic Ions

40. Name each of the following ions: **(a)** NO_3^-; **(b)** NO_2^-; **(c)** SO_4^{2-}; **(d)** SO_3^{2-}; **(e)** $H_2PO_4^-$; **(f)** CN^-; **(g)** CO_3^{2-}; **(h)** HCO_3^-.
41. Write the formula for each of the following ions: **(a)** ammonium; **(b)** arsenate; **(c)** chlorate; **(d)** perchlorate; **(e)** permanganate; **(f)** dichromate.
42. Write the names and formulas for the acids that correspond to the following anions: **(a)** NO_3^-; **(b)** NO_2^-; **(c)** SO_4^{2-}; **(d)** SO_3^{2-}; **(e)** CN^-; **(f)** F^-; **(g)** chlorate; **(h)** perchlorate; **(i)** acetate; **(j)** chlorite; **(k)** hypochlorite; **(l)** sulfide; **(m)** phosphate.

Formulas and Names of Compounds

43. Write a formula for each of the following compounds: **(a)** an alkali halide; **(b)** calcium bromide; **(c)** lithium nitride; **(d)** magnesium phosphide; **(e)** aluminum oxide; **(f)** hydrogen bromide; **(g)** potassium permanganate; **(h)** magnesium phosphate; **(i)** ammonium sulfate; **(j)** cesium dihydrogen phosphate; **(k)** tin(II) fluoride; **(l)** tin(IV) selenide; **(m)** calcium hydroxide; **(n)** chromium(III) sulfide; **(o)** chromium(VI) oxide; **(p)** manganese(II) nitrate; **(q)** iron(III) chromate; **(r)** scandium iodide; **(s)** zinc arsenide; **(t)** titanium(IV) chloride; **(u)** carbon tetrachloride; **(v)** diarsenic trisulfide; **(w)** tetraphosphorus hexoxide.

44. Name each of the following compounds using the Stock system: **(a)** $Na_2Cr_2O_7$; **(b)** $NH_4H_2AsO_4$; **(c)** $Mg_3(AsO_4)_2$; **(d)** $Al(NO_2)_3$; **(e)** PbO; **(f)** PbO_2; **(g)** $FeCl_2$; **(h)** $Fe(NO_3)_3$; **(i)** $Ba(ClO)_2$; **(j)** $Sr(HSO_4)_2$; **(k)** Cs_2Se; **(l)** $Ca(C_2H_3O_2)_2$; **(m)** Rb_2CrO_4; **(n)** Ca_3P_2; **(o)** K_3As; **(p)** K_3AsO_3; **(q)** K_3AsO_4; **(r)** $Mg(ClO_4)_2$; **(s)** $HgSO_4$; **(t)** Hg_2Br_2; **(u)** Cu_2O; **(v)** $Cu(NO_3)_2$; **(w)** PF_3; **(x)** N_2O_4; **(y)** As_2O_5.

Bonding

45. Differentiate between covalent and ionic bonding.

46. What type of bonding holds the atoms together in a molecule?

47. The formula for sodium chloride is NaCl. Does this mean that sodium chloride consists of NaCl molecules, of Na^+ and Cl^- ion pairs as distinct units, or other units? Explain.

Equation Balancing

48. Balance each of the following equations: **(a)** $Fe(s) + Cl_2(g) \rightarrow FeCl_3(s)$; **(b)** $Mg(s) + HCl(aq) \rightarrow MgCl_2(aq) + H_2(g)$; **(c)** $NaOH(aq) + H_2SO_4(aq) \rightarrow Na_2SO_4(aq) + H_2O(l)$; **(d)** $KOH(aq) + H_3PO_4(aq) \rightarrow K_3PO_4(aq) + H_2O(l)$; **(e)** $Ca(OH)_2(aq) + H_3PO_4(aq) \rightarrow Ca_3(PO_4)_2(s) + H_2O(l)$; **(f)** $(NH_4)_2Cr_2O_7(s) \rightarrow Cr_2O_3(s) + N_2(g) + H_2O(g)$; **(g)** $BiCl_3(s) + NH_3(g) + H_2O(l) \rightarrow Bi(OH)_3(s) + NH_4Cl(aq)$; **(h)** $C_4H_{10}(g) + O_2(g) \rightarrow CO_2(g) + H_2O(g)$.

Equation Writing and Balancing

49. Complete and balance each of the following equations for the reactions of metals with nonmetals: **(a)** $K(s) + F_2(g) \rightarrow$; **(b)** $Ba(s) + I_2(s) \rightarrow$; **(c)** $Ca(s) + P_4(s) \rightarrow$; **(d)** $Al(s) + S_8(s) \rightarrow$; **(e)** $Li(s) + N_2(g) \rightarrow$; **(f)** $Al(s) + Br_2(l) \rightarrow$; **(g)** $Mg(s) + H_2(g) \rightarrow$.

50. Write a balanced equation for each of the following reactions: **(a)** magnesium is heated in nitrogen gas; **(b)** a piece of lithium reacts with water; **(c)** potassium is heated with sulfur; **(d)** magnesium powder is heated with phosphorus; **(e)** oxygen gas is passed through a glass tube containing hot powdered aluminum; **(f)** a piece of magnesium ribbon is burned in oxygen. **(g)** barium metal is "dissolved" in water; **(h)** hydrogen reacts with nitrogen to form ammonia; **(i)** powdered calcium metal is heated with arsenic; **(j)** powdered iron is heated with oxygen gas to produce iron(III) oxide.

Types of Reactions

51. Identify each of the following reactions as a combination, decomposition, single displacement, or double displacement reaction: **(a)** $2NH_3(g) \rightarrow N_2(g) + 3H_2(g)$; **(b)** $H_2(g) + Cl_2(g) \rightarrow 2HCl(g)$; **(c)** $4FeO(s) + O_2(g) \rightarrow 2Fe_2O_3(s)$; **(d)** $CaO(s) + H_2O(l) \rightarrow Ca(OH)_2(s)$; **(e)** $3Ba(s) + Cr_2O_3(s) \rightarrow 3BaO(s) + 2Cr(s)$; **(f)** $MgO(s) + CO_2(g) \rightarrow MgCO_3(s)$; **(g)** $Hg_2(NO_3)_2(aq) + 2KCl(aq) \rightarrow Hg_2Cl_2(s) + 2KNO_3(aq)$; **(h)** $Zn(s) + NiSO_4(aq) \rightarrow ZnSO_4(aq) + Ni(s)$; **(i)** $2KClO_3(s) + \text{heat} \rightarrow 2KCl(s) + 3O_2(g)$.

52. Complete and balance each of the following reactions and indicate what type of a reaction is involved in each case: **(a)** $H_2O_2(aq) \rightarrow O_2(g) +$ _____; **(b)** $CO(g) + O_2(g) \rightarrow$ _____; **(c)** $BaCO_3(s) \rightarrow CO_2 +$ _____; **(d)** $Ag_2O(s) \rightarrow$ _____ + _____; **(e)** $Mg(s) + Fe(NO_3)_3(aq) \rightarrow$ _____ + _____; **(f)** $Pb(NO_3)_2(aq) + KCl(aq) \rightarrow PbCl_2(s) +$ _____; **(g)** $Sr(s) + CuSO_4(aq) \rightarrow Cu(s) +$ _____; **(h)** _____ $+ H_2O(l) \rightarrow Mg(OH)_2(s)$; **(i)** $CrO(s) +$ _____ $\rightarrow Cr_2O_3(s)$; **(j)** $SO_3(g) +$ _____ $\rightarrow H_2SO_4(aq)$; **(k)** $ZnO(s) + H_2O(l) \rightarrow$ _____; **(l)** $SO_2(g) + H_2O(l) \rightarrow$ _____; **(m)** $Al(s) + H_2SO_4(aq) \rightarrow$ _____ + _____.

53. Write a balanced equation for each of the following reactions, and indicate what type of a reaction is involved in each case: **(a)** magnesium metal is heated strongly with solid copper(II) oxide; **(b)** solutions of barium bromide and potassium sulfate are mixed to produce a precipitate of barium sulfate and another product in solution; **(c)** copper(I) oxide is treated with oxygen to produce copper(II) oxide; **(d)** lithium oxide is treated with carbon dioxide; **(e)** platinum(IV) oxide is heated strongly; **(f)** a solution of calcium nitrate is mixed with a solution of sodium phosphate to form a precipitate of calcium phosphate as one of the two products; **(g)** a piece of zinc metal is placed in hydrochloric acid; **(h)** an aluminum nail is placed into a beaker of sulfuric acid.

Survey of Elements

54. Name each of the following elements: **(a)** F; **(b)** Mg; **(c)** Mn; **(d)** Ti; **(e)** Fe; **(f)** Ag; **(g)** Hg; **(h)** K; **(i)** P; **(j)** Sb.

55. How does the hardness of the alkali metals as a group differ from the hardness of the alkaline earth metals?

56. Which element in group III of the representative elements is the most widely used in commerce and industry? What are some of the properties and uses of this element?

57. In which one group in the periodic table can you find uncombined elements of which two exist as solids, one as a liquid, and two as gases under normal conditions? What is the group name of these elements?

58. Discuss some physical and chemical properties of carbon, oxygen, chlorine, bromine, and iodine.

59. Give an example of a metal that reacts vigorously with water, and write an equation for the reaction. Give an example of another metal that does not react with water.

60. Give an example of a metal that reacts with hydrochloric acid, and write an equation for the reaction. Give an example of another metal that does not react with hydrochloric acid.

61. Which metal and which nonmetal are the basic components of steel?

62. Identify some important transition metals and discuss their uses.

63. List the major allotropes of carbon, phosphorus, and oxygen, and describe their properties.

64. What is the percent by mass of gold in a 14-carat gold alloy?

65. What is an amalgam?

66. Discuss some uses of two of the allotropes of carbon.

67. For each of the metals in the third period of the periodic table, write a formula for its bromide and of its sulfide.

68. Write the formula for an oxide of: **(a)** silicon; **(b)** iron; **(c)** sulfur; **(d)** copper; **(e)** aluminum.

69. Write formulas for the two common chlorides of each of the following elements: **(a)** iron; **(b)** copper; **(c)** mercury; **(d)** tin; **(e)** lead; **(f)** chromium.

CHAPTER 3

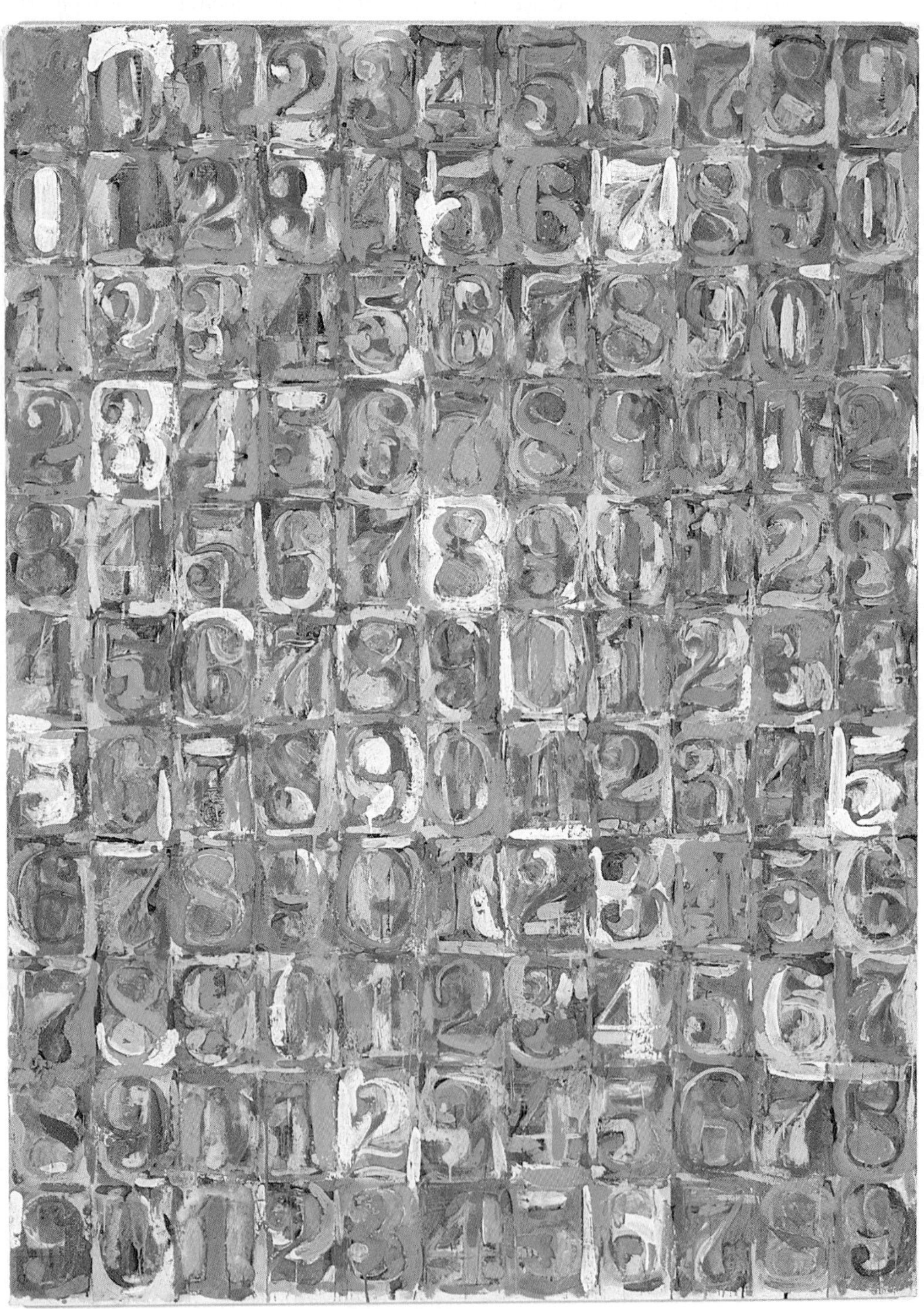

Jasper Johns, *Numbers in Color* (1958–59). This chapter discusses a variety of numbers: numbers of grams, moles, molecules, and atoms.

Stoichiometry

In Chapter 2 we considered chemical formulas and equations, which together make up the "vocabulary" and "language" of chemistry. In this chapter we discuss how the formulas of compounds are related to their quantitative compositions. We also discuss how a balanced chemical equation for a reaction can be used to determine the relative quantities of the substances involved in the reaction. These quantitative aspects of chemistry are brought together under the heading ***stoichiometry*** *(from the Greek* stoicheion, *element, and* metron, *measure).*

3.1 Quantity of Matter

The quantity of a given sample of a substance can be specified in terms of its *mass*, its *volume*, or the *number of atoms, ions, or molecules* that are present. The most commonly used unit of the quantity of matter in chemistry is the mass of a definite number of atoms, ions, or molecules.

The Mole

The *amount of matter that contains as many atoms, ions, or molecules as the number of atoms in exactly 12 grams of carbon-12 isotope* is the *SI unit of quantity of matter*, called the **mole** (from Latin *moles*, meaning a mass or a pile of a substance). A mole is a quantity of matter that can be weighed conveniently in a laboratory. Single atoms or molecules cannot be weighed on a laboratory balance because of their extremely small size, but the masses of atoms and molecules can be determined by an instrument called a **mass spectrometer** (Figures 3.1 and 3.2).

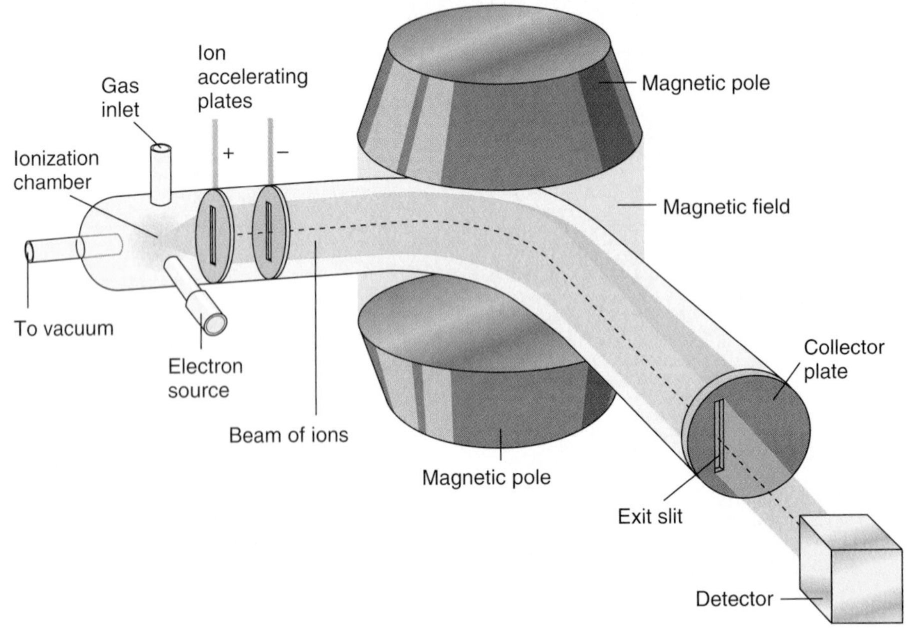

Figure 3.1

Schematic diagram of one type of mass spectrometer. A gaseous substance is injected into an evacuated chamber of a mass spectrometer and the atoms or molecules of the gas are ionized (converted to positive ions). The ions are accelerated through an electric field and pass through two slits to produce a narrow beam. The beam is then deflected in the magnetic field toward a collector plate. The extent to which the ions are deflected depends on the masses and charges of the ions. The greater the positive charge and the smaller the mass of an ion, the more it is deflected. The magnetic field refocuses the ion beams so that all the ions of the same charge-to-mass ratio strike the same spot on a photographic plate. The magnitude of the field allows the calculation of the masses of the ions. Modern mass spectrometers have electronic detectors and recorders that automatically record the intensity of the signal that corresponds to the mass number of the atoms involved. The intensity of the signal is proportional to the relative number of atoms of a particular ion. The resulting graph is called a mass spectrum. An example of a mass spectrum for neon is illustrated in Figure 3.2.

Note that the definition of a mole is based on a 12-g mass of ^{12}C. This isotope of carbon is also the standard for the atomic weights of all the elements. Various modern experimental methods have shown that the number of atoms in 12 g of ^{12}C is 6.0221×10^{23}. This large number is known as **Avogadro's number**, in honor of the Italian scientist Amadeo Avogadro (1776–1856).

Amadeo Avogadro (1776–1856).

Just as the mass of a mole of atoms is based on the mass of a mole of ^{12}C atoms (12 g), the mass of a single atom of an element in atomic mass units is based on the mass of a single ^{12}C atom (12 amu). A mole of naturally occurring carbon contains several isotopes, and the mass of 1 mole (mol) of naturally occurring carbon is 12.011 g. A mole of naturally occurring hydrogen atoms contains 6.0221×10^{23} H atoms and has a mass of 1.00794 g, and a mole of oxygen atoms contains 6.0221×10^{23} O atoms with a mass of 15.9994 g. In every case, a mole of atoms contains Avogadro's number of atoms.

Avogadro's number is so large that it virtually defies comprehension in the context of familiar objects such as pennies or dollars. A mole of pennies is 6.02×10^{23} pennies, which equals 6.02×10^{21} dollars, or 6.02×10^{12} billion dollars. This number is still too large to comprehend. If a person could count at a steady rate of 2 pennies a second for an entire year, the number of pennies the person could count is

$$\left(\frac{2 \text{ pennies}}{1 \text{ s}}\right)\left(\frac{60 \text{ s}}{1 \text{ min}}\right)\left(\frac{60 \text{ min}}{1 \text{ h}}\right)\left(\frac{24 \text{ h}}{1 \text{ day}}\right)\left(\frac{365 \text{ days}}{1 \text{ yr}}\right) = \frac{6 \times 10^7 \text{ pennies}}{1 \text{ yr}}$$

Assuming the human population of the world is about 5 billion (5×10^9) people, the number of pennies all these people could count in one year at the steady rate of 6×10^7 pennies per year is

$$\left(\frac{6 \times 10^7 \text{ pennies}}{1 \text{ yr}}\right)(5 \times 10^9) = \frac{3 \times 10^{17} \text{ pennies}}{1 \text{ yr}}$$

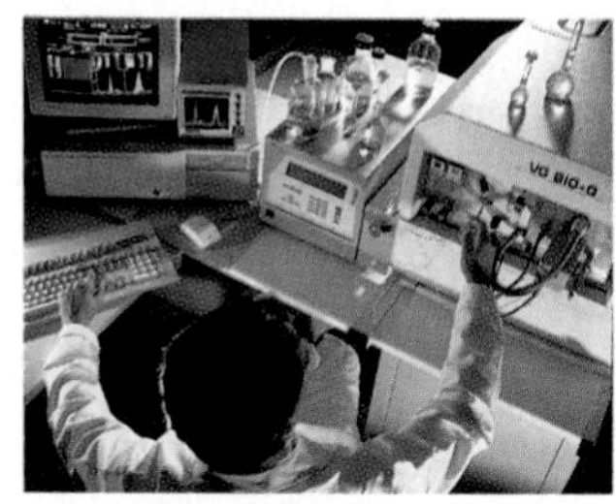

Mass spectrometer.

The number of years needed for the entire human population to count a mole of pennies (assuming a mole to be approximately 6×10^{23}) is

$$(6 \times 10^{23} \text{ pennies})\left(\frac{1 \text{ yr}}{3 \times 10^{17} \text{ pennies}}\right) = 2 \times 10^6 \text{ (or 2 million) yr!}$$

The enormous value of the number of atoms in a mole reflects the very small size of an atom. For example, 1 mol of iron has a mass of 55.8 g, and 1 mol of aluminum, 27.0 g. These quantities represent relatively small samples of these metals, as you can see from Figure 3.3.

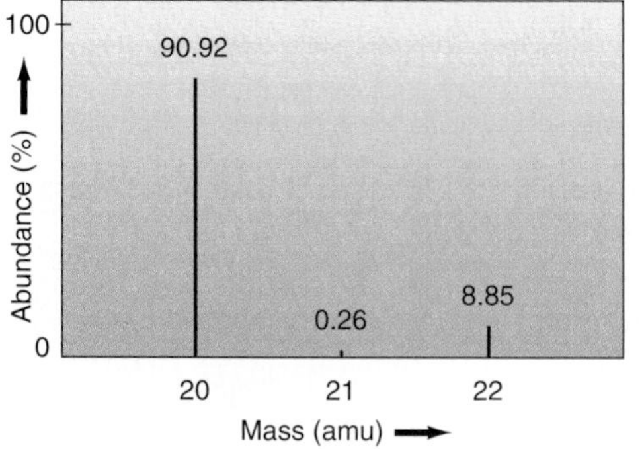

Figure 3.2

Mass spectrum of neon. The electronic detector of a mass spectrometer automatically draws a vertical line on the graph paper to indicate the masses of the ions striking the collector plate at various points. The length of each vertical line is a measure of the percent abundance of each type of particle. The three lines in the mass spectrum of neon indicate that there are three naturally occurring isotopes of neon, with mass numbers of 20, 21, and 22. The exact masses of these isotopes are 19.99 amu, 20.99 amu, and 21.99 amu, respectively. The relative abundances of the isotopes are indicated on the vertical axis. The percent of abundance is shown on top of the line for each isotope. From the mass spectrum data, the atomic weight of neon can be calculated as follows: Atomic weight = (20.00 amu)(0.9092) + (21.00 amu)(0.0026) + (22.00 amu)(0.0882) = 20.18 amu.

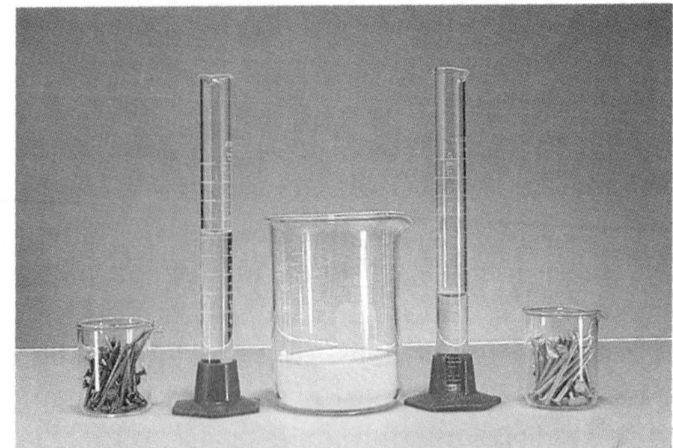

Figure 3.3

One-mole samples of iron, water, table sugar, carbon tetrachloride, and aluminum.

We can use the mole concept to calculate (1) the number of moles in any given mass of an element, (2) the number of atoms in any given mass of an element, (3) the mass of a single atom, and (4) the mass in grams of one atomic mass unit. Such calculations are illustrated by the following examples.

Example 3.1

Calculating the Number of Moles of Atoms of an Element and the Number of Atoms of the Element from Given Mass of the Element

Calculate the number of moles of carbon atoms, and the number of individual carbon atoms in 20.0 g of carbon.

SOLUTION: One mole of carbon atoms has a mass of 12.0 g, and the number of moles of atoms in 20.0 g of carbon is

$$20.0 \text{ g C} \left(\frac{1 \text{ mol C}}{12.0 \text{ g C}}\right) = 1.67 \text{ mol C}$$

One mole of atoms consists of 6.02×10^{23} atoms, and the number of atoms in 1.67 mol is

$$1.67 \text{ mol} \left(\frac{6.02 \times 10^{23} \text{ atoms}}{1 \text{ mol}}\right) = 1.01 \times 10^{24} \text{ atoms}$$

Example 3.2

Calculating the Mass of an Atom in Grams and in Atomic Mass Units

What is the mass in grams of one ^{12}C atom? of one atomic mass unit?

SOLUTION: Exactly 12 g of ^{12}C contains 6.02×10^{23} ^{12}C atoms. The mass of one atom is

$$\frac{12.0 \text{ g}}{6.02 \times 10^{23} \text{ atoms}} = 1.99 \times 10^{-23} \text{ g/atom}$$

Since one ^{12}C atom has a mass of exactly 12 amu, 1 amu has a mass of one-twelfth of the mass of one ^{12}C atom:

$$\left(\frac{1.99 \times 10^{-23} \text{ g}}{1 \text{ atom}}\right)\left(\frac{1 \text{ atom}}{12.0 \text{ amu}}\right) = 1.66 \times 10^{-24} \text{ g/amu}$$

Practice Problem 3.1: What is the mass in grams of an average hydrogen atom (the lightest of all atoms)?

Example 3.3

Calculating the Number of Atoms of an Element from Its Mass and Atomic Weight

Look up the atomic weight of aluminum and calculate the number of aluminum atoms in 10.0 g of aluminum.

SOLUTION: The number of moles of aluminum atoms in 10.0 g is

$$10.0 \text{ g Al} \left(\frac{1 \text{ mol Al}}{27.0 \text{ g Al}}\right) = 0.370 \text{ mol Al}$$

The number of atoms is

$$0.370 \text{ mol Al} \left(\frac{6.02 \times 10^{23} \text{ atoms Al}}{1 \text{ mol Al}}\right) = 2.23 \times 10^{23} \text{ atoms Al}$$

This problem can be more conveniently solved by a two-step operation as follows:

$$10.0 \text{ g Al} \left(\frac{1 \text{ mol Al}}{27.0 \text{ g Al}}\right)\left(\frac{6.02 \times 10^{23} \text{ atoms Al}}{1 \text{ mol Al}}\right) = 2.23 \times 10^{23} \text{ atoms Al}$$

Practice Problem 3.2: If you had 4.52×10^{18} silver atoms, how many moles of silver and how many grams of silver would you have? What would be the cost of that much silver if silver were valued at \$6.00 an ounce?

Molecular Weight and Molar Mass

The mass of a molecule equals the sum of the masses of the atoms in the molecule. Thus, the *sum of the atomic weights of the atoms in a molecule of a compound* is called its **molecular weight**. For example, the molecular weight of water, H_2O, equals twice the atomic weight of hydrogen plus the atomic weight of oxygen:

$$(2)(1.01 \text{ amu}) + 16.0 \text{ amu} = 18.0 \text{ amu}$$

A mole of water consists of 2 mol of hydrogen and 1 mol of oxygen. Thus, the mass of 1 mol of water equals the mass of 2 mol of hydrogen atoms (2.02 g) plus the mass of 1 mol of oxygen atoms (16.0 g), a total of 18.0 g. A mole of water also contains 6.02×10^{23} molecules and has a volume of 18.0 mL at 3.98°C (the temperature at which water has its highest density of exactly 1 g mL^{-1}). Just as a mole of atoms of an element equals the atomic weight of the element expressed in grams, a mole of molecules of a compound equals the molecular weight of the compound expressed in grams.

For ionic compounds, the sum of the atomic weights of the atoms in a formula unit is called the **formula weight**, a term that is sometimes used for molecular compounds as well. Thus, the formula weight of NaCl equals the sum of the atomic weights of Na and Cl: 23.0 amu + 35.5 amu = 58.5 amu. A mole of an ionic compound is the formula weight of the compound expressed in grams.

The *mass of 1 mol of any substance* is called **molar mass** or **molar weight** of the substance. Molar mass is a convenient term because it refers to the mass of

1 mol of elements as well as molecular or ionic compounds. Thus, the molar mass of hydrogen atoms is 1.01 g/mol H, of hydrogen molecules 2.02 g/mol H_2, of water molecules 18.0 g/mol H_2O, and of sodium chloride 58.5 g/mol NaCl. One-mole samples of different substances are illustrated in Figure 3.3.

Example 3.4

Calculating the Mass in Grams of a Given Number of Moles of a Substance

What is the mass in grams of each of the following quantities? **(a)** 0.328 mol of silver atoms; **(b)** 1.72 mol of oxygen molecules (O_2); **(c)** 0.250 mol of magnesium perchlorate (a good drying agent).

SOLUTION: From the periodic table we find the atomic weights to three significant figures for all the elements involved in this problem: Ag = 108 amu, O = 16.0 amu, Mg = 24.3 amu, and Cl = 35.5 amu.

(a) One mole of Ag atoms is 108 g, and

$$0.328 \text{ mol Ag} \left(\frac{108 \text{ g Ag}}{1 \text{ mol Ag}}\right) = 35.4 \text{ g Ag}$$

(b) One mole of O_2 molecules has a mass of (2)(16.0 g) = 32.0 g; therefore,

$$1.72 \text{ mol } O_2 \left(\frac{32.0 \text{ g } O_2}{1 \text{ mol } O_2}\right) = 55.0 \text{ g } O_2$$

(c) The formula for magnesium perchlorate is $Mg(ClO_4)_2$ (Chapter 2). One mole of $Mg(ClO_4)_2$ consists of 1 mol of Mg atoms (24.3 g), 2 mol of Cl atoms (2 × 35.5 g = 71.0 g), and 8 mol of O atoms (8 × 16.0 g = 128 g). The molar mass for $Mg(ClO_4)_2$ is 24.3 g + 71.0 g + 128 g = 223 g, and

$$0.250 \text{ mol } Mg(ClO_4)_2 \left(\frac{223 \text{ g } Mg(ClO_4)_2}{1 \text{ mol } Mg(ClO_4)_2}\right) = 55.8 \text{ g } Mg(ClO_4)_2$$

Practice Problem 3.3: If the price of gold were \$500.00 a troy ounce (31.10 g), how many moles of gold atoms could you buy for \$50.00? how many gold atoms?

Example 3.5

Calculating the Mass of a Molecule from Molar Mass

One mole of water molecules has a mass of 18.0 g. **(a)** What is the mass of one water molecule in grams; **(b)** what is the mass of hydrogen in 18.0 g of water?

SOLUTION

(a)
$$\left(\frac{18.0 \text{ g } H_2O}{1 \text{ mol } H_2O}\right)\left(\frac{1 \text{ mol } H_2O}{6.02 \times 10^{23} \text{ molecules } H_2O}\right)$$
$$= 2.99 \times 10^{-23} \text{ g/molecule } H_2O$$

(b) One mole of H_2O contains 2 mol of H atoms. One mole of H atoms has a mass of 1.01 g, and 2 mol of H atoms has a mass of 2.02 g.

Practice Problem 3.4: If you drink a quart (0.946 L) of water, how many moles of water would you drink? How many molecules? Assume the density of water to be 1.00 g mL^{-1}.

Example 3.6

Calculating the Molecular Weight, the Mass of a Molecule, and the Number of Moles as Well as the Number of Molecules in a Given Mass of a Compound

Dinitrogen pentoxide (one of the major oxides of nitrogen) has the formula N_2O_5. Look up the necessary atomic weights and calculate: **(a)** the molecular weight of dinitrogen pentoxide; **(b)** the mass in grams of one N_2O_5 molecule; **(c)** the number of moles of N_2O_5 in 2.00 g of N_2O_5; **(d)** the number of molecules in 2.00 g of N_2O_5.

SOLUTION

(a) The atomic weights of nitrogen and oxygen are 14.0 amu and 16.0 amu, respectively. The molecular weight of N_2O_5 is

$$(2)(14.0 \text{ amu}) + (5)(16.0 \text{ amu}) = 108.0 \text{ amu}$$

(b) Since one molecule of N_2O_5 has a mass of 108.0 amu, 1 mol of N_2O_5 has a mass of 108 g. One mole of N_2O_5 contains 6.02×10^{23} molecules of N_2O_5. Therefore, the mass of one molecule of N_2O_5 is

$$\left(\frac{108.0 \text{ g}}{1 \text{ mol } N_2O_5}\right)\left(\frac{1 \text{ mol } N_2O_5}{6.02 \times 10^{23} \text{ molecules } N_2O_5}\right)$$

$$= 1.79 \times 10^{-22} \text{ g/molecule } N_2O_5$$

(c) $2.00 \text{ g } N_2O_5 \left(\frac{1 \text{ mol } N_2O_5}{108 \text{ g } N_2O_5}\right) = 0.0185 \text{ mol } N_2O_5$

(d) The number of molecules of N_2O_5 is

$$0.0185 \text{ mol } N_2O_5 \left(\frac{6.02 \times 10^{23} \text{ molecules } N_2O_5}{1 \text{ mol } N_2O_5}\right)$$

$$= 1.11 \times 10^{22} \text{ molecules } N_2O_5$$

3.2 Percent Composition of Compounds

We recall that a compound has a definite characteristic composition (Chapter 1). Below we consider how the formula of a compound is used to calculate the percent composition of the compound.

The formula of a molecular compound specifies the number of atoms of each element in a single molecule; it also specifies the number of *moles of atoms* in a *mole of the compound*. From the known molar masses of the atoms in a compound and the molar mass of the compound, the mass percent of each element in a compound can be calculated. For example, 1 mol of water (18.0 g) consists of 2 mol of hydrogen atoms (2.02 g) and 1 mol of oxygen atoms (16.0 g). From these known masses, the percent by mass composition of water is

$$\text{percent H} = \frac{2.02\ \text{g}}{18.0\ \text{g}} \times 100 = 11.2\ \text{percent}$$

$$\text{percent O} = \frac{16.0\ \text{g}}{18.0\ \text{g}} \times 100 = 88.9\ \text{percent}$$

The percentages of H and O thus obtained do not add up to exactly 100 because of the rounding off of the atomic weights to three significant figures.

Example 3.7

Calculating the Percent Composition of a Compound

What is the percent composition of aspirin (acetyl salicylic acid), $C_9H_8O_4$?

SOLUTION: The atomic weights involved are C = 12.0, H = 1.01, and O = 16.0. The molar mass of the compound is

$$(9)(12.0\ \text{g/mol}) + (8)(1.01\ \text{g/mol}) + (4)(16.0\ \text{g/mol}) = 180.1\ \text{g/mol}$$

The percent composition of aspirin is

$$\text{percent C} = \frac{(9\ \text{mol C})(12.0\ \text{g/mol C})}{180.1\ \text{g}} \times 100 = 60.0\ \text{percent}$$

$$\text{percent H} = \frac{(8\ \text{mol H})(1.01\ \text{g/mol H})}{180.1\ \text{g}} \times 100 = 4.49\ \text{percent}$$

$$\text{percent O} = \frac{(4\ \text{mol O})(16.0\ \text{g/mol O})}{180.1\ \text{g}} \times 100 = 35.5\ \text{percent}$$

$$\text{Total} = 100.0\ \text{percent}$$

Practice Problem 3.5: What is the percent by mass of each element in nitroglycerin (used in explosives and in medicine), $C_3H_5(NO_3)_3$?

The formula of an ionic compound specifies the number of moles of *ions* present in a mole of the compound. The number of moles of polyatomic ions in a mole of a compound is indicated by a subscript outside of the parentheses that enclose the ion. For example, 1 mol of aluminum nitrate, $Al(NO_3)_3$ (213 g), contains 1 mol of Al^{3+} ions (27.0 g) and 3 mol of NO_3^- ions (3 × 62.0 g = 186 g). Since there are 3 mol of oxygen atoms in 1 mol of nitrate ions, and since 1 mol of aluminum nitrate contains 3 mol of nitrate ions, a total of 9 mol of oxygen atoms are in 1 mol of aluminum nitrate; this amounts to 144 g of oxygen in 1 mol (213 g) of aluminum nitrate.

Example 3.8

Determining the Mass Percent of Elements and Ions in an Ionic Compound

Aluminum sulfate, $Al_2(SO_4)_3$, is used for fireproofing and waterproofing cloth, and as an antiperspirant. What is the percent of each element in $Al_2(SO_4)_3$? What is the percent by mass of sulfate ions in $Al_2(SO_4)_3$? Atomic weights: Al = 27.0 amu, S = 32.1 amu, O = 16.0 amu.

SOLUTION: One mole of $Al_2(SO_4)_3$ contains 2 mol of Al, 3 mol of S, and 12 mol of O. The corresponding masses are

$$2 \text{ mol Al} \left(\frac{27.0 \text{ g Al}}{1 \text{ mol Al}}\right) = 54.0 \text{ g Al}$$

$$3 \text{ mol S} \left(\frac{32.1 \text{ g S}}{1 \text{ mol S}}\right) = 96.3 \text{ g S}$$

$$12 \text{ mol O} \left(\frac{16.0 \text{ g O}}{1 \text{ mol O}}\right) = 192.0 \text{ g O}$$

The mass of 1 mol of $Al_2(SO_4)_3$ = 54.0 g + 96.3 g + 192.0 g = 342.3 g:

$$\text{percent Al} = \frac{54.0 \text{ g}}{342.3 \text{ g}} \times 100 = 15.8 \text{ percent}$$

$$\text{percent S} = \frac{96.3 \text{ g}}{342.3 \text{ g}} \times 100 = 28.1 \text{ percent}$$

$$\text{percent O} = \frac{192 \text{ g}}{342.3 \text{ g}} \times 100 = 56.1 \text{ percent}$$

One mole of sulfate ions contains 1 mol of S atoms (32.1 g) and 4 mol of O atoms (4 × 16.0 g = 64.0 g). The molar mass of sulfate ions is therefore 32.1 g + 64.0 g = 96.1 g. Since there are 3 mol of sulfate ions in 1 mol of aluminum sulfate, the percent of sulfate ions in aluminum sulfate is

$$\text{percent } SO_4^{2-} = \frac{(3)(96.1 \text{ g})}{342.3 \text{ g}} \times 100 = 84.2 \text{ percent}$$

The percent by mass of SO_4^{2-} ions in this compound is also equal to the sum of the percent S and the percent O in the compound (28.1 percent + 56.1 percent = 84.2 percent).

Practice Problem 3.6: Nitrogen, phosphorus, and potassium are important plant nutrients. The percentages of nitrogen, phosphorus, and potassium compounds present in commercially available garden fertilizers are listed on fertilizer bags as three numbers, such as 10–10–10, respectively. Ammonium phosphate is a compound that supplies both nitrogen and phosphorus. What is the percent nitrogen and the percent phosphorus in ammonium phosphate?

3.3 Derivation of Formulas

In Section 3.2 we learned how the percent composition of a compound can be calculated from its formula. In this section we consider the derivation of the formula of a compound from its percent composition.

The formula of a compound can be derived in the following way. First, the constituent elements in the compound are identified by chemical tests. Then the mass percent of each element in the compound is determined. *From the mass percent data, the number of moles of atoms of each element in a given sample of the compound can be calculated. The simplest formula of the compound can then be obtained from the mole ratios of the elements, which are the same as their atomic ratios.*

The formula that specifies the *smallest* whole numbers of atoms of the constituent elements in a compound is known as the **simplest formula** or **empirical formula** for the compound. In most cases, the empirical formula accurately describes the formula of an ionic compound. However, the empirical formula may not correctly represent the actual number of atoms in a molecule of a molecular compound. *The formula that specifies both the relative and the actual numbers of atoms in a molecule* is called the **molecular formula**. For example, the gaseous organic compound ethylene has the empirical formula CH_2, but its molecular formula is C_2H_4. The empirical formula for another organic compound, propylene, is also CH_2, but its molecular formula is C_3H_6, which is three times the empirical formula.

To obtain the molecular formula for a compound, its molecular weight is compared with the formula weight of the empirical formula. If the two agree, the empirical formula is the molecular formula. If the molecular weight is a multiple of the formula weight of the empirical formula, the molecular formula is that same multiple of the simplest formula.

Example 3.9 illustrates how the principles outlined above can be applied to derive an empirical and a molecular formula of a compound from its percent composition. A method of determining the carbon and hydrogen content in an organic compound is depicted in Figure 3.4.

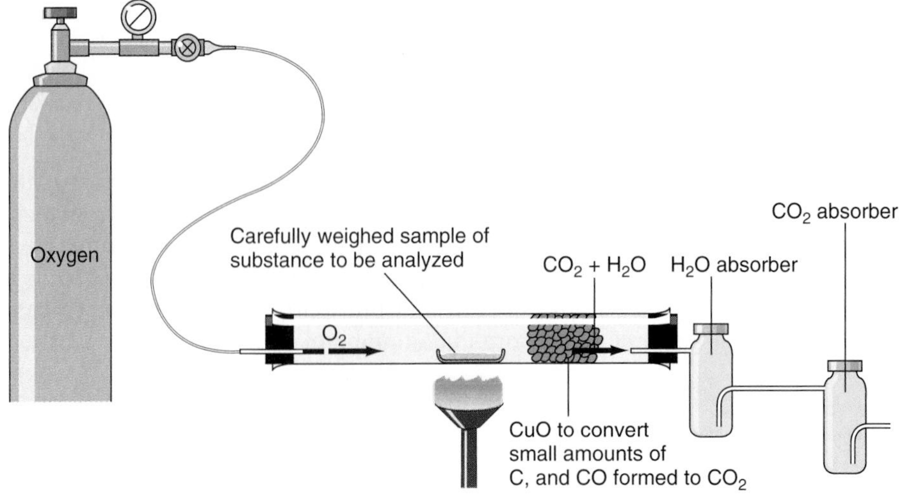

Figure 3.4

Determination of percent carbon and hydrogen in a compound. The incoming oxygen reacts with the compound at the temperature of the flame and converts carbon to carbon dioxide, and hydrogen to water vapor. The increase in mass of each of the absorbers gives the mass of water and carbon dioxide. The mass percent of elemental hydrogen and carbon can then be calculated.

Example 3.9

Deriving the Simplest Formula and Molecular Formula from Mass Percent Data and Molecular Weight

An organic compound contains 40.0 percent by mass of carbon, 6.73 percent hydrogen, and 53.3 percent oxygen. The molecular weight of the compound is 181 amu/molecule. Derive the simplest formula and the molecular formula for the compound.

SOLUTION: The first step is to convert the percent composition to the number of moles of atoms of each element. The mass percent of an element in a compound is the mass in grams of the element in 100 g of the compound. Thus, we calculate the number of moles of atoms of each element in 100 g of the compound:

$$\text{moles C} = 40.0 \text{ g C} \left(\frac{1 \text{ mol C}}{12.0 \text{ g C}}\right) = 3.33 \text{ mol C}$$

$$\text{moles H} = 6.73 \text{ g H} \left(\frac{1 \text{ mol H}}{1.01 \text{ g H}}\right) = 6.66 \text{ mol H}$$

$$\text{moles O} = 53.3 \text{ g O} \left(\frac{1 \text{ mol O}}{16.0 \text{ g O}}\right) = 3.33 \text{ mol O}$$

Next we determine the C:H:O mole ratio. The lowest number of moles found is 3.33 for both C and O. The H:C mole ratio is 6.66/3.33 = 2, and the C:O mole ratio is 3.33/3.33 = 1. Thus, the C:H:O mole ratio is 1:2:1, and the empirical formula for the compound is CH_2O.

Many substances have the empirical formula CH_2O. One of these is formaldehyde, a compound with many uses: it is used to make some kinds of plastics; its water solution is called formalin, which is used as a disinfectant and in embalming fluids. Another compound with the formula CH_2O is the sugar glucose, sometimes called blood sugar.

To determine the molecular formula for the compound, the experimental molecular weight must be compared with the formula weight of the empirical formula, which for CH_2O is 30.0 amu:

$$\frac{181 \text{ amu}}{30.0 \text{ amu}} \approx 6 \qquad (\text{“}\approx\text{” means approximately equal})$$

The molecular formula is therefore six times the empirical formula, $C_6H_{12}O_6$. This is the molecular formula for glucose (as well as the formula for several other sugars).

Practice Problem 3.7: A compound in a household fuel contains 85.70 percent by mass of carbon and 14.30 percent hydrogen. The molecular weight of the compound is 56.1 amu. What is the simplest formula and the molecular formula for this compound?

For derivation of the formula of a compound, its composition can be given in either *mass percent*, or in terms of the *mass* of each element *in a given mass of the compound*. If the masses of the constituent elements in a sample of a compound are given, the number of moles of each element in the given mass of the sample can be calculated (see Example 3.10).

Example 3.10

Deriving a Formula from Given Masses of the Elements Present in a Sample of a Compound

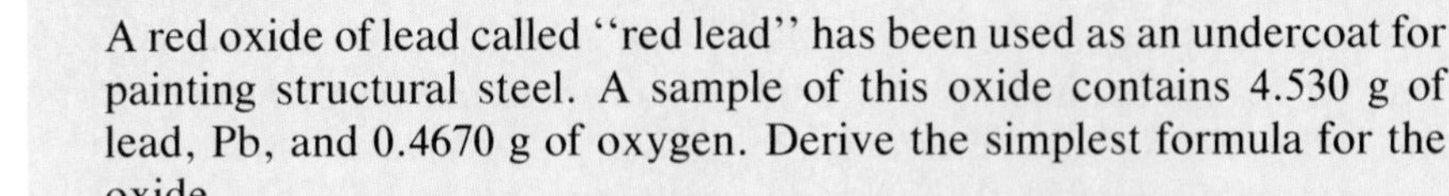

A red oxide of lead called "red lead" has been used as an undercoat for painting structural steel. A sample of this oxide contains 4.530 g of lead, Pb, and 0.4670 g of oxygen. Derive the simplest formula for the oxide.

SOLUTION: The atomic weights of lead and oxygen (to four significant figures) are 207.2 and 16.00, respectively. The number of moles of Pb and O atoms in the sample of red lead are

$$\text{moles Pb} = 4.530 \text{ g Pb} \left(\frac{1 \text{ mol Pb}}{207.2 \text{ g Pb}}\right) = 0.02186 \text{ mol Pb}$$

$$\text{moles O} = 0.4670 \text{ g O} \left(\frac{1 \text{ mol O}}{16.00 \text{ g O}}\right) = 0.02919 \text{ mol O}$$

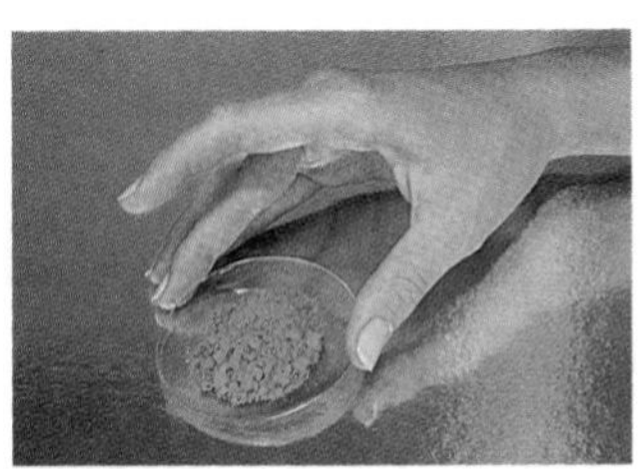

Pb_3O_4 ("red lead").

According to these data, the number of moles of O atoms per mole of Pb atoms in the sample of the compound is

$$\frac{0.02919 \text{ mol O}}{0.02186 \text{ mol Pb}} = \frac{1.335 \text{ mol O}}{1.000 \text{ mol Pb}}$$

We know that the formula of a compound is expressed with whole-number subscripts. If we multiply the numerator and the denominator of the O:Pb mole ratio by 3, we obtain 4.005 mol O:3.000 mol Pb. Thus, within a small experimental error, the formula for red lead is Pb_3O_4.

Example 3.10 shows that it is important to express calculated results to the correct number of significant figures to reflect the precision of the measurements used in the calculation. For example, if the number of moles of O and Pb atoms in Example 3.10 had been expressed to only one significant figure, an incorrect formula would be obtained:

$$\frac{0.03 \text{ mol O}}{0.02 \text{ mol Pb}} \quad \text{or} \quad \frac{3 \text{ mol O}}{2 \text{ mol Pb}}$$

which corresponds to the incorrect formula Pb_2O_3.

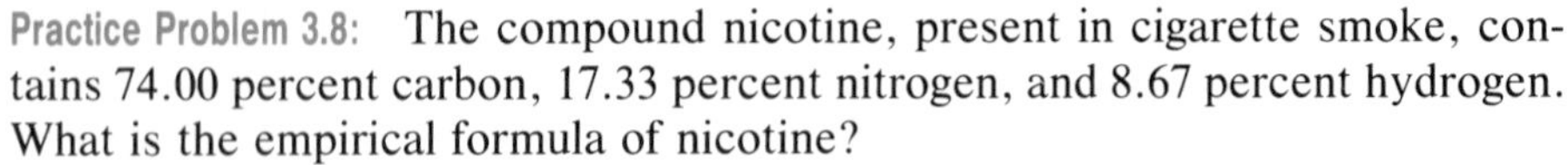

Practice Problem 3.8: The compound nicotine, present in cigarette smoke, contains 74.00 percent carbon, 17.33 percent nitrogen, and 8.67 percent hydrogen. What is the empirical formula of nicotine?

Figure 3.5

Left: Hydrated copper(II) sulfate, $CuSO_4(H_2O)_5$. Right: Anhydrous $CuSO_4$ residue after heating.

We can use the same method to determine formulas of salts that are chemically bound to water molecules. These salts are called **hydrates**. The absorption of water by a salt to form a hydrate is called **hydration**. Some hydrates include $MgSO_4(H_2O)_7$ (Epsom salt) and $Na_2CO_3(H_2O)_{10}$ (washing soda).

The water contained in a hydrate—called the **water of hydration**—can often be driven out by heating. The residue that remains is the **anhydrous compound** (Figure 3.5). The mass of the water and the mass of the anhydrous residue can be converted to moles, from which the formula for a hydrate can be derived as shown in Example 3.11.

Example 3.11

Deriving the Formula of a Hydrate

Heating a 2.576-g sample of a blue hydrated copper(II) sulfate produces 1.646 g of white anhydrous $CuSO_4$. Derive the formula for the hydrated salt.

SOLUTION: The mass of water driven out from the hydrate is

$$2.576 \text{ g} - 1.646 \text{ g} = 0.930 \text{ g}$$

$$\text{moles } CuSO_4 \text{ in sample} = 1.646 \text{ g } CuSO_4 \left(\frac{1 \text{ mol } CuSO_4}{159.6 \text{ g } CuSO_4}\right)$$

$$= 0.01031 \text{ mol } CuSO_4$$

$$\text{moles } H_2O \text{ in sample} = 0.930 \text{ g } H_2O \left(\frac{1 \text{ mol } H_2O}{18.0 \text{ g } H_2O}\right)$$

$$= 0.0517 \text{ mol } H_2O$$

The mole ratio of water to anhydrous copper(II) sulfate is

$$\frac{0.0517 \text{ mol } H_2O}{0.01031 \text{ mol } CuSO_4} = \frac{5.01 \text{ mol } H_2O}{1.00 \text{ mol } CuSO_4}$$

and the formula for the hydrate is $CuSO_4(H_2O)_5$.

3.4 Calculations Based on Chemical Equations

Quantitative Meaning of a Chemical Equation

In Chapter 2 we learned how to write balanced equations. A chemical equation provides much quantitative information. Table 3.1 lists some important data that can be obtained from a chemical equation.

We can see from Table 3.1 that in a balanced chemical equation, the total masses of the reactants and products are equal, as required by the law of conservation of mass. However, the total number of moles of the reactants may or may not equal the total number of moles of the products because the molar masses of different substances are usually not equal.

Similarly, on the microscale of atoms and molecules, the total number of atoms of the reactants equals the total number of atoms of the products in a reaction. But the total number of molecules of the reactants and products may not be equal because molecules can have different numbers of atoms. For example, two molecules of H_2 react with one molecule of O_2 to produce two molecules of H_2O:

$$2H_2(g) + O_2(g) \longrightarrow 2H_2O(l)$$

In this reaction the number of atoms of hydrogen and oxygen in the reactants equals the number of atoms of hydrogen and oxygen in the products, but the number of molecules of the reactants is not equal to the number of molecules of the products.

Table 3.1

Information That Can Be Obtained from a Balanced Chemical Equation[a]

Macroscale information:

$$4NH_3(g) + 5O_2(g) \longrightarrow 4NO(g) + 6H_2O(g)$$

$$4 \text{ mol } NH_3 + 5 \text{ mol } O_2 \longrightarrow 4 \text{ mol NO} + 6 \text{ mol } H_2O$$

$$\underset{NH_3}{(4)(17.0 \text{ g})} + \underset{O_2}{(5)(32.0 \text{ g})} \longrightarrow \underset{NO}{(4)(30.0 \text{ g})} + \underset{H_2O}{(6)(18.0 \text{ g})}$$

228 g of reactants ⟶ 228 g of products

Microscale information:

$$\underset{NH_3}{4 \text{ molecules}} + \underset{O_2}{5 \text{ molecules}} \longrightarrow \underset{NO}{4 \text{ molecules}} + \underset{H_2O}{6 \text{ molecules}}$$

$$4\text{N atoms} + 12\text{H atoms} + 10 \text{ O atoms} \longrightarrow 4\text{N atoms} + 4 \text{ O atoms} + 12\text{H atoms} + 6 \text{ O atoms}$$

26 atoms of reactants ⟶ 26 atoms of products

$$\underset{NH_3}{(4)(17.0 \text{ amu})} + \underset{O_2}{(5)(32.0 \text{ amu})} \longrightarrow \underset{NO}{(4)(30.0 \text{ amu})} + \underset{H_2O}{(6)(18.0 \text{ amu})}$$

228 amu of reactants ⟶ 228 amu of products

[a] According to the law of conservation of mass, the total mass of the reactants equals the total mass of the products. The total number of atoms of the reactants must also be equal to the total number of atoms of the products, because atoms are neither created nor destroyed in a chemical reaction.

A chemical equation also gives the mole ratios of the reactants and products of a reaction. For example, in the reaction of ammonia with oxygen (Table 3.1), the mole ratio of O_2 to NH_3 is 5:4. This ratio specifies that 5 mol of O_2 can react with 4 mol of NH_3, or that $\frac{5}{4}$ mol of O_2 reacts with 1 mol of NH_3. In the same reaction, the mole ratio of H_2O to NH_3 is 6:4, or $1\frac{1}{2}$. This means that $1\frac{1}{2}$ mol of H_2O can be obtained from 1 mol of NH_3.

Practice Problem 3.9: How many moles of O_2 are needed to completely react with 2.50 mol of NH_3 according to the equation for the reaction described in Table 3.1?

Calculation of the Amounts of Reactants and Products

The mole ratios provided by a balanced chemical equation are very important in many stoichiometric calculations. For example, a common type of a stoichiometric problem requires calculating the mass of a product that can be obtained from a given mass of a reactant. To solve this type of problem, we first *convert the given mass of the reactant to moles*. Then we *convert the moles of reactant to moles of product using the mole ratio from the balanced equation*. Next we *convert moles of product to mass of the product*. This sequence of steps can be written as

mass reactant ⟶ moles reactant ⟶ moles product ⟶ mass product

An application of this sequence is illustrated in Example 3.12.

Example 3.12

Calculating the Mass of a Reaction Product from a Given Mass of a Reactant

Consider the reaction of ammonia with oxygen to produce nitrogen monoxide and water according to the equation

$$4NH_3(g) + 5O_2(g) \longrightarrow 4NO(g) + 6H_2O(g)$$

If 2.85 g of NH_3 reacts completely with excess oxygen, what mass of water can be obtained?

SOLUTION: A sequence of steps to solve this problem is

$$\text{mass } NH_3 \xrightarrow{1} \text{moles } NH_3 \xrightarrow{2} \text{moles } H_2O \xrightarrow{3} \text{mass } H_2O$$

These steps can be carried out as follows:

Step 1.

$$2.85 \text{ g } NH_3 \left(\frac{1 \text{ mol } NH_3}{17.0 \text{ g } NH_3}\right) = 0.168 \text{ mol } NH_3$$

Step 2. According to the balanced equation, 6 mol of H_2O can be obtained from 4 mol of NH_3; therefore,

$$0.168 \text{ mol } NH_3 \left(\frac{6 \text{ mol } H_2O}{4 \text{ mol } NH_3}\right) = 0.252 \text{ mol } H_2O$$

Step 3.

$$0.252 \text{ mol } H_2O \left(\frac{18.0 \text{ g } H_2O}{1 \text{ mol } H_2O}\right) = 4.54 \text{ g } H_2O$$

These three steps can be combined for a single multistep operation on a calculator that stores the answers to the first two steps, so you do not need to write them down:

$$2.85 \text{ g } NH_3 \left(\frac{1 \text{ mol } NH_3}{17.0 \text{ g } NH_3}\right)\left(\frac{6 \text{ mol } H_2O}{4 \text{ mol } NH_3}\right)\left(\frac{18.0 \text{ g } H_2O}{1 \text{ mol } H_2O}\right) = 4.53 \text{ g } H_2O$$

The small difference between this answer (4.53 g) and the one obtained from step 3 (4.54 g) results from rounding off the answers in steps 1 and 2. When a calculator is used in one multistep operation, all the intermediate answers are stored in the calculator to the maximum number of digits characteristic of the calculator.

The reaction of NH_3 with O_2 using a Pt catalyst. The NH_3 gas that escapes from its concentrated solution reacts with O_2 gas in the atmosphere. The reaction releases heat that causes the platinum wire to glow.

Practice Problem 3.10: If 3.00 g of hydrogen were consumed in a reaction to produce water, how many moles of water could be obtained? how many molecules of water? how many grams of water?

If the total mass of either the reactants or the products is known, the mass of one reactant, or product, can be calculated from the known mass of the other using the law of conservation of mass (Example 3.13).

Example 3.13

Using the information calculated in Example 3.12, determine the mass of nitrogen monoxide that can be prepared from 2.85 g of ammonia that consumes 6.71 g of O_2 in this reaction.

SOLUTION: According to the law of conservation of mass, the total mass of the reactants in a chemical reaction equals the total mass of the products. The total mass of the reactants in this reaction is

$$2.85 \text{ g} + 6.71 \text{ g} = 9.56 \text{ g}$$

Since this is also the total mass of the products, and since the mass of one of the products, H_2O, is 4.54 g, the mass of the other product, NO, is

$$9.56 \text{ g} - 4.54 \text{ g} = 5.02 \text{ g}$$

To check this answer, we obtain

$$2.85 \text{ g NH}_3 \left(\frac{1 \text{ mol NH}_3}{17.0 \text{ g NH}_3}\right)\left(\frac{4 \text{ mol NO}}{4 \text{ mol NH}_3}\right)\left(\frac{30.0 \text{ g NO}}{1 \text{ mol NO}}\right) = 5.03 \text{ g NO}$$

A chemist in a laboratory often needs to know the mass of one reactant that should be mixed with a given mass of another reactant for a reaction that will yield a desired quantity of product. This type of problem is illustrated in Example 3.14.

Example 3.14

Calculating the Mass of a Reactant Consumed by a Given Mass of Another Reactant

Aluminum reacts with oxygen according to the equation

$$4\text{Al}(s) + 3\text{O}_2(g) \longrightarrow 2\text{Al}_2\text{O}_3(s)$$

How much oxygen must be present for a complete reaction of 2.85 g of aluminum?

SOLUTION: An outline of the steps that can be used to solve this problem is

$$\text{mass Al} \longrightarrow \text{moles Al} \longrightarrow \text{moles O}_2 \longrightarrow \text{mass O}_2$$

Carrying out these steps in one multistep operation yields

$$2.85 \text{ g Al} \left(\frac{1 \text{ mol Al}}{27.0 \text{ g Al}}\right)\left(\frac{3 \text{ mol O}_2}{4 \text{ mol Al}}\right)\left(\frac{32.0 \text{ g O}_2}{1 \text{ mol O}_2}\right) = 2.53 \text{ g O}_2$$

Many industrial preparations are carried out in several steps. Calculation of the amount of a product obtained in such a multistep operation is illustrated in Example 3.15.

Example 3.15

Calculating the Mass of a Product Obtained in a Process Involving a Series of Steps

In an industrial operation, 2.6 metric tons of calcium carbonate (limestone) is decomposed by heating. The solid decomposition product (lime, or unslaked lime) is then treated with water to produce calcium hydroxide (slaked lime) to be used in making mortar for the building industry. Write a balanced equation for the decomposition of limestone and another for the production of calcium hydroxide from calcium oxide and water. How much calcium hydroxide can be obtained in this process assuming ideal conditions, that is, no loss of material during the reactions?

SOLUTION: Decomposition of calcium carbonate by heating is a typical example of a decomposition of a metal carbonate (Chapter 2) to produce metal oxide and carbon dioxide:

$$CaCO_3(s) \longrightarrow CaO(s) + CO_2(g)$$

Most metal oxides react with water to produce hydroxides. Such reactions are typical combination reactions that we discussed in Chapter 2. Calcium oxide reacts with water according to the following equation:

$$CaO(s) + H_2O(l) \longrightarrow Ca(OH)_2(s)$$

According to the equations above, the mole ratios of $CaCO_3$ to CaO to $Ca(OH)_2$ are 1:1:1, or

$$\text{mol } CaCO_3 = \text{mol } CaO = \text{mol } Ca(OH)_2$$

Based on this information, a sequence of steps leading to the solution of this problem is

$$\text{metric tons } CaCO_3 \longrightarrow \text{g } CaCO_3 \longrightarrow \text{mol } CaCO_3 \longrightarrow$$
$$\text{mol } CaO \longrightarrow \text{mol } Ca(OH)_2 \longrightarrow \text{g } Ca(OH)_2 \longrightarrow \text{metric tons } Ca(OH)_2$$

Carrying out these steps yields

$$(2.6 \text{ tons } CaCO_3)\left(\frac{10^6 \text{ g } CaCO_3}{1 \text{ ton } CaCO_3}\right)\left(\frac{1 \text{ mol } CaCO_3}{100 \text{ g } CaCO_3}\right)\left(\frac{1 \text{ mol } CaO}{1 \text{ mol } CaCO_3}\right)$$
$$\times \left(\frac{1 \text{ mol } Ca(OH)_2}{1 \text{ mol } CaO}\right)\left(\frac{74 \text{ g } Ca(OH)_2}{1 \text{ mol } Ca(OH)_2}\right)\left(\frac{1 \text{ ton } Ca(OH)_2}{10^6 \text{ g } Ca(OH)_2}\right)$$
$$= 1.9 \text{ tons } Ca(OH)_2$$

Practice Problem 3.11: A 5.258-g sample of a mixture of calcium carbonate and calcium oxide is heated to decompose the calcium carbonate. Upon heating, the mass loss is 1.892 g. What is the percent of calcium carbonate in the mixture?

Practice Problem 3.12: Nitric acid is prepared commercially by the Ostwald process. This process involves three steps. In the first step, ammonia gas reacts with oxygen to produce nitrogen monoxide and water (the reaction featured in Table 3.1):

$$4NH_3(g) + 5O_2(g) \longrightarrow 4NO(g) + 6H_2O(g)$$

APPLICATIONS OF CHEMISTRY 3.1
Environmental Pollution from Outer Space

On June 30, 1908, people in China and Russia saw a fireball crossing the sky, now known as the Tunguska meteorite. The fireball landed and exploded in the Tunguska river valley in central Siberia. The explosive force, which was equivalent to that of a 10-megaton nuclear detonation, completely leveled 1300 square miles of forest (by comparison, Maryland is 10,000 square miles).

The popular press became interested in the Tunguska meteorite when investigating teams found no crater. Rumors and popular press stories attributed the explosion to an alien spacecraft. A more recent and plausible theory suggests that a comet or rocky meteorite exploded as it passed through the earth's atmosphere. It broke up into pieces that were too small to produce a crater, but its effects were still environmentally devastating.

Scientists have determined that the Tunguska meteorite generated enough heat to cause oxygen and nitrogen present in the atmosphere to react. As shown in Equation 1, the reaction formed nitrogen monoxide (NO). According to their calculations, the meteorite produced 30 million metric tons of nitrogen monoxide (NO).

$$N_2(g) + O_2(g) \longrightarrow 2NO(g) \qquad (1)$$

This nitrogen monoxide reacted with more oxygen and water to form nitric acid, as shown by the combination reaction

$$4NO(g) + 3O_2(g) + 2H_2O(l) \longrightarrow 4HNO_3(aq) \qquad (2)$$

Let's calculate the maximum number of tons of nitric acid that can form from the 30 million metric tons of nitrogen monoxide produced by the Tunguska meteorite. (Nitric acid is a primary contributor to acid rain. See Applications of Chemistry 2.2 for more details.) According to Equation 2, the mole ratio of NO to O_2 to H_2O to HNO_3 is 4:3:2:4. Based on this information, a sequence of steps leading to the solution of this problem is

$$\text{metric tons NO} \longrightarrow \text{g NO} \longrightarrow \text{mol NO} \longrightarrow \text{mol HNO}_3 \longrightarrow \text{g HNO}_3 \longrightarrow \text{metric tons HNO}_3$$

Carrying out these steps yields

$$(3 \times 10^7 \text{ tons NO})\left(\frac{10^6 \text{ g NO}}{1 \text{ ton NO}}\right)\left(\frac{1 \text{ mol NO}}{30 \text{ g NO}}\right) \times \left(\frac{4 \text{ mol HNO}_3}{4 \text{ mol NO}}\right)\left(\frac{63 \text{ g HNO}_3}{1 \text{ mol HNO}_3}\right)\left(\frac{1 \text{ ton HNO}_3}{10^6 \text{ g HNO}_3}\right) = 6.3 \times 10^7 \text{ tons HNO}_3$$

This calculation shows that from the 30 million metric tons of NO produced by the Tunguska meteorite, 63 million metric tons of nitric acid formed in the earth's atmosphere. In comparison, the United States produces about 25 million metric tons of nitrogen monoxide waste per year from sources such as automobile exhaust.

Source: *Sky and Telescope*, December 1986, pages 577–578.

In the second step, the nitrogen monoxide is converted to nitrogen dioxide:

$$2NO(g) + O_2(g) \longrightarrow 2NO_2(g)$$

and in the third step, the nitrogen dioxide is absorbed by water to form nitric acid solution and more nitrogen monoxide:

$$3NO_2(g) + H_2O(l) \longrightarrow 2HNO_3(aq) + NO(g)$$

The NO produced in this step is recycled. If 2.50 kg of ammonia is used for a reaction with excess oxygen, what mass of nitric acid can be obtained?

Limiting Reactant

When two substances are mixed for a reaction, usually only one of them reacts completely. Part of the second reagent does not react completely because it is present in excess. The amount of the *reactant that is totally consumed* in the reaction limits the amounts of the products formed; this reactant is therefore called the **limiting reactant**.

To illustrate the limiting reactant concept, let us consider the reaction of hydrogen with oxygen to form water:

$$2H_2(g) + O_2(g) \longrightarrow 2H_2O(l)$$

In this reaction, each mole of O_2 requires 2 mol of H_2 for both reactants to be completely consumed.

If 1 mol of O_2 (32 g) and 1 mol of H_2 (2.0 g) are mixed for a reaction, all of the oxygen cannot react because 2 mol of hydrogen is not available. However, 1 mol of hydrogen reacts completely because it needs only $\frac{1}{2}$ mol of oxygen. Hydrogen is therefore the limiting reactant, and $\frac{1}{2}$ mol of oxygen remains unreacted.

The limiting reactant determines the quantity of a product that can form. Therefore, we can identify the limiting reactant by determining which of the given quantities of the reactants produces the smaller quantity of a product. Let us consider 1 mol of H_2 and 1 mol of O_2 mixed for a reaction to produce H_2O. To determine which of the two reactants is the limiting reactant, we first calculate the number of moles of water that can be obtained from a complete reaction with 1 mol of hydrogen. Then we calculate the amount of water that can be produced from a complete reaction of 1 mol of oxygen. In each case we assume that an excess of the other reactant is present:

$$1 \text{ mol } H_2 \left(\frac{2 \text{ mol } H_2O}{2 \text{ mol } H_2}\right) = 1 \text{ mol } H_2O$$

$$1 \text{ mol } O_2 \left(\frac{2 \text{ mol } H_2O}{1 \text{ mol } O_2}\right) = 2 \text{ mol } H_2O$$

The calculations show that 1 mol of hydrogen produces a smaller quantity of water than 1 mol of oxygen does. Hydrogen is therefore the limiting reactant.

Example 3.16

Determining the Limiting Reactant

Most metals are found in natural deposits in which the metals are in combination with various nonmetals. Metal sulfides are common in these deposits, and they are also easy to prepare in a laboratory. In an experiment, 10.0 g of aluminum is heated with 50.0 g of sulfur to form aluminum sulfide. Which of the two reactants is the limiting reactant? How many grams of aluminum sulfide is formed? How much of one of the reactants remains unreacted?

SOLUTION: The balanced equation for the reaction is:

$$2Al(s) + 3S(s) \longrightarrow Al_2S_3(s)$$

(For simplicity, sulfur is often written as S instead of S_8.) To determine whether aluminum or sulfur is the limiting reactant, we calculate which one gives the smaller quantity of the product, assuming in each case that the other reactant is present in excess:

$$10.0 \text{ g Al} \left(\frac{1 \text{ mol Al}}{27.0 \text{ g Al}}\right)\left(\frac{1 \text{ mol Al}_2\text{S}_3}{2 \text{ mol Al}}\right) = 0.185 \text{ mol Al}_2\text{S}_3$$

$$50.0 \text{ g S} \left(\frac{1 \text{ mol S}}{32.1 \text{ g S}}\right)\left(\frac{1 \text{ mol Al}_2\text{S}_3}{3 \text{ mol S}}\right) = 0.519 \text{ mol Al}_2\text{S}_3$$

Since 10.0 g of aluminum gives a smaller quantity of the product than 50.0 g of sulfur, aluminum is the limiting reactant.

The amount of aluminum sulfide formed in the reaction is determined by the amount of aluminum (the limiting reactant) consumed. We have already calculated the number of moles of Al_2S_3 formed from 10.0 g of aluminum. The mass in grams of Al_2S_3 formed is

$$0.185 \text{ mol Al}_2\text{S}_3 \left(\frac{150 \text{ g Al}_2\text{S}_3}{1 \text{ mol Al}_2\text{S}_3}\right) = 27.8 \text{ g Al}_2\text{S}_3$$

The mass of sulfur remaining unreacted is the mass present minus the mass that reacts. The mass of sulfur that reacts is

$$10.0 \text{ g Al} \left(\frac{1 \text{ mol Al}}{27.0 \text{ g Al}}\right)\left(\frac{3 \text{ mol S}}{2 \text{ mol Al}}\right)\left(\frac{32.1 \text{ g S}}{1 \text{ mol S}}\right) = 17.8 \text{ g S}$$

The mass of sulfur originally present is 50.0 g. The mass that remains unreacted is the mass present minus the mass that reacts:

$$50.0 \text{ g} - 17.8 \text{ g} = 32.2 \text{ g}$$

Practice Problem 3.13: Gold does not react with most nonmetals, but it does react with chlorine gas at 200°C to form gold(III) chloride, $AuCl_3$. If 0.625 g of powdered gold and 0.872 g of chlorine are mixed for a reaction, which of the two reactants is the limiting reactant? How much of the other reactant present in excess remains unreacted? How many grams of gold(III) chloride can be obtained?

Percent Yield

In the stoichiometric calculations we have considered until now, we have assumed that the reactions proceeded to completion. The *amount of a product calculated with this assumption* is called the **theoretical yield**.

The *experimentally measured yield of the product of a reaction*, the **actual yield**, is often less than the theoretical yield. Various factors can decrease the theoretical yield. For example, the limiting reactant in a reaction may not completely react, or some of the product may be lost in the separation and purification procedures. Also, other reactions—called side reactions—may occur and consume some of the limiting reactant or the product.

The **percent yield** is defined as

$$\text{percent yield} = \left(\frac{\text{actual yield}}{\text{theoretical yield}}\right) 100$$

Example 3.17 illustrates how the percent yield of a reaction can be calculated.

APPLICATIONS OF CHEMISTRY 3.2
Life and Death Struggle: A Real-World Limiting Reactant

In the past, human beings used soap to launder their clothes. Today, most people use laundry detergents. Detergents contain phosphates, which soften the wash water by removing calcium and magnesium ions. Without phosphates or phosphate substitutes, calcium and magnesium salts in the wash water would come out of solution and deposit on clothes as a grimy residue. Although phosphate substitutes exist, none are quite as effective as phosphates. However, one-third of U.S. communities ban phosphate use in laundry detergents because of hazards to the environment.

Recall that a limiting reactant determines the amount of product that forms. This is true both in simple chemical reactions as well as in living organisms. Most plants need an adequate supply of several nutrients (called essential elements), such as phosphorus, nitrogen, potassium, hydrogen, sulfur, calcium, and magnesium, to sustain life. Not only do plants need an ample supply of these essential elements, but they also need to have these elements available at all times.

Consider your own bodily needs as an analogy. Among other nutrients in your diet, you need protein and carbohydrates. Without protein, you would have trouble remaining alive even if you had the entire world's supply of carbohydrates available to you. Similarly, if plants had large supplies of all essential elements except phosphorus available to them, they would die.

Most plant nutrients except phosphorus are abundant. Air contains ample supplies of oxygen and nitrogen. Carbon dioxide and sulfur dioxide, also present in the air, provide plentiful sources of carbon and sulfur. Lake waters provide large reserves of hydrogen. Only phosphorus is scarce in nature. Phosphorus is recycled into the environment by the death and decay of plants and animals. Therefore, phosphorus is the limiting reactant in maintaining plant life. It keeps plants such as algae from growing without bound. Therefore, the limiting reactant phosphorus actually controls the amount of plant life (the final product) that can be sustained in a habitat.

When plants begin to grow without limit because of the availability of phosphates from detergents, eventually they cover the water's surface, keeping sunlight from reaching plants at the bottom of the lake (see photo above). When plants at the bottom of the lake die, they decompose, which consumes oxygen that is needed by fish and other organisms in the lake. Eventually, the lake can no longer support life.

Example 3.17

Determining the Percent Yield

Cisplatin, $Pt(NH_3)_2Cl_2$, is used to treat certain types of cancer. It is prepared by the following reaction:

$$K_2PtCl_4(aq) + 2NH_3(aq) \longrightarrow 2KCl(aq) + Pt(NH_3)_2Cl_2(aq)$$

If you start with 8.50 g of K_2PtCl_4 and isolate 5.72 g of $Pt(NH_3)_2Cl_2$, what are the theoretical yield and the percent yield?

SOLUTION: The molar masses are: For K_2PtCl_4, 415 g/mol; and for $Pt(NH_3)_2Cl_2$, 300 g/mol. The theoretical yield is

$$8.50 \text{ g } K_2PtCl_4 \left(\frac{1 \text{ mol } K_2PtCl_4}{415 \text{ g } K_2PtCl_4}\right)\left(\frac{1 \text{ mol } Pt(NH_3)_2Cl_2}{1 \text{ mol } K_2PtCl_4}\right)$$

$$\times \left(\frac{300 \text{ g } Pt(NH_3)_2Cl_2}{1 \text{ mol } Pt(NH_3)_2Cl_2}\right) = 6.14 \text{ g } Pt(NH_3)_2Cl_2$$

The percent yield is

$$\text{percent yield} = \left(\frac{\text{actual yield}}{\text{theoretical yield}}\right) 100 = \left(\frac{5.72 \text{ g}}{6.14 \text{ g}}\right) 100$$

$$= 93.2 \text{ percent}$$

Practice Problem 3.14: A 0.582-g sample of mercury(II) oxide is decomposed by heating to elemental mercury and oxygen. The mass of oxygen recovered is 0.0341 g. What are the theoretical yield and the percent yield in this process?

3.5 Solution Stoichiometry

We recall that a solution is a homogeneous mixture of two or more components (Chapter 1). The solutions prepared in laboratories are often two-component mixtures. In a two-component solution, the component present in larger quantity is usually designated as the **solvent**, the other component as the **solute**.

If water is one of the components of a solution, it is usually designated as the solvent in which another substance, the solute, is dissolved. A solution in which water is the solvent is said to be **aqueous** (Latin *aqua*, water). Aqueous solutions are among the most common type of solutions, and they will be considered in this section.

Concentration of Solutions

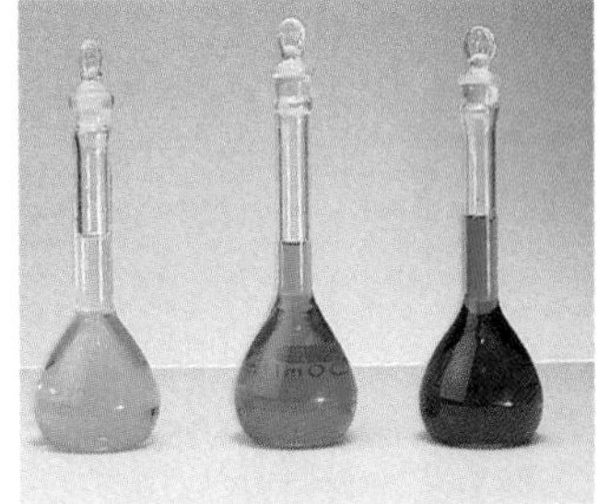

Figure 3.6

Colored solutions. Left: Dilute solution. Middle: Concentrated solution. Right: Saturated solution.

The *amount of solute dissolved in a given amount of solvent* determines the **concentration** of a solution. If a "small" amount of a solute is dissolved in a relatively "large" amount of solvent, the solution is said to be **dilute**. If a solution contains a relatively "large" amount of a solute in a given quantity of solvent, the solution is said to be **concentrated**. The borderline between a dilute and a concentrated solution is arbitrary. If the amount of a solute in a solvent is increased until no more solute dissolves at a given temperature, the solution is **saturated** (Figure 3.6).

The concentration of a solution can be expressed in several different quantitative ways. Concentration can be expressed as **percent by mass**, which indicates the *mass of solute per 100 mass units of solution*. Concentration can also be given in terms of the *volume of solute per 100 volume units of solution*, that is, as **percent by volume**. The concentration of alcohol in alcohol-containing beverages is expressed as "proof," a unit of concentration that equals twice the percent by volume. A 100 proof whisky therefore contains 50% alcohol by volume.

In chemistry, one very common way of specifying the concentration of a solution is in terms of its **molarity** (M), which equals the *number of moles of solute per liter of solution*:

$$\text{molarity } (M) = \frac{\text{moles of solute}}{\text{liters of solution}} \qquad (3.1)$$

The following examples illustrate the concept of molarity and its application.

Example 3.18

Calculating Molarity from a Given Mass of Solute and Volume of Solution in Liters

A 15.7-g sample of NaCl is dissolved in water to make 2.00 L of solution. Calculate the molarity of the solution.

SOLUTION: First we determine the number of moles in 15.7 g of NaCl. The molar mass of NaCl = 23.0 g mol^{-1} Na + 35.5 g mol^{-1} Cl = 58.5 g mol^{-1} NaCl:

$$\text{moles NaCl} = 15.7 \text{ g NaCl} \left(\frac{1 \text{ mol NaCl}}{58.5 \text{ g NaCl}}\right) = 0.268 \text{ mol NaCl}$$

The number of moles of solute in 1 L of solution is found by applying Equation 3.1:

$$\frac{0.268 \text{ mol}}{2.00 \text{ L}} = 0.134 \text{ mol L}^{-1} \quad \text{or} \quad 0.134\ M$$

Example 3.19

Calculating Molarity from a Given Mass of Solute and Volume of Solution in Milliliters

What is the molarity of a sodium chloride solution that contains 6.82 g of NaCl in 450 mL of solution?

SOLUTION: The number of moles of NaCl in 6.82 g is

$$6.82 \text{ g NaCl} \left(\frac{1 \text{ mol NaCl}}{58.5 \text{ g NaCl}}\right) = 0.117 \text{ mol NaCl}$$

The number of liters is 450 mL is

$$450 \text{ mL} \left(\frac{1 \text{ L}}{1000 \text{ mL}}\right) = 0.450 \text{ L}$$

Since a liter equals 1000 mL, we need only divide the given number of milliliters by 1000 to convert to liters. This operation is done by moving the decimal point three places to the left. If there is no decimal point, simply place one preceding the third digit from the end of the number.

Applying Equation 3.1 we obtain

$$\text{molarity} = \frac{0.117 \text{ mol}}{0.450 \text{ L}} = 0.260 \text{ mol L}^{-1} \quad \text{or} \quad 0.260\ M$$

Practice Problem 3.15: What is the molarity of a glucose, $C_6H_{12}O_6$, solution that contains: **(a)** 25.6 g of glucose in 3.75 L of solution; **(b)** 0.108 mol of glucose in 750 mL of solution?

The number of moles of solute in a solution of known volume and molarity can be calculated by solving Equation 3.1 for moles of solute:

$$\text{moles of solute} = (\text{liters of solution})(M) \tag{3.2}$$

Since molarity is the number of moles of solute in 1 L of solution, the number of liters times the molarity must be the *total* number of moles of the solute in the solution.

Example 3.20

Calculating the Number of Moles of Solute from a Given Volume and Molarity of Solution

What is the number of moles of ammonia in 500 mL of 2.00 *M* solution of household ammonia?

SOLUTION: Converting the given volume from milliliters to liters and using Equation 3.2, we obtain

$$0.500\ \text{L}\left(\frac{2.00\ \text{mol}}{1\ \text{L}}\right) = 1.00\ \text{mol}$$

Practice Problem 3.16: How many liters of a 2.00 *M* solution of alcohol would contain 3.20 mol of alcohol?

In preparing solutions, a measured amount of the solute is dissolved in a small amount of the solvent, and the resulting solution is transferred to a volumetric flask (Chapter 1) of appropriate size. Solvent is then added to dissolve the solute and water is added until the total volume reaches the etched mark on the neck of the flask. By this method, the total volume of the solution can be determined without having to worry about any change in the volume of the components when they are mixed (Figure 3.7).

(a)

(b)

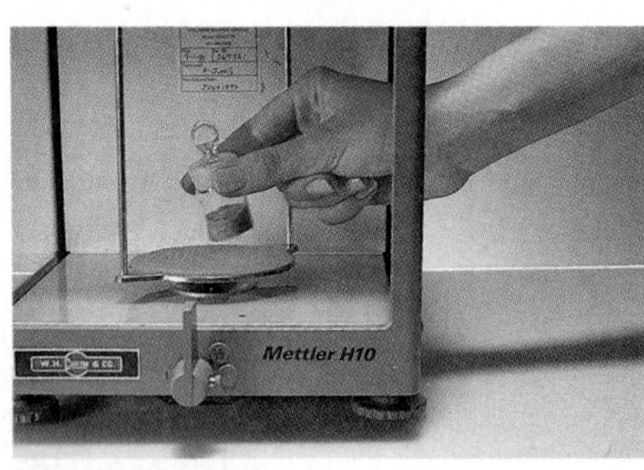

(c)

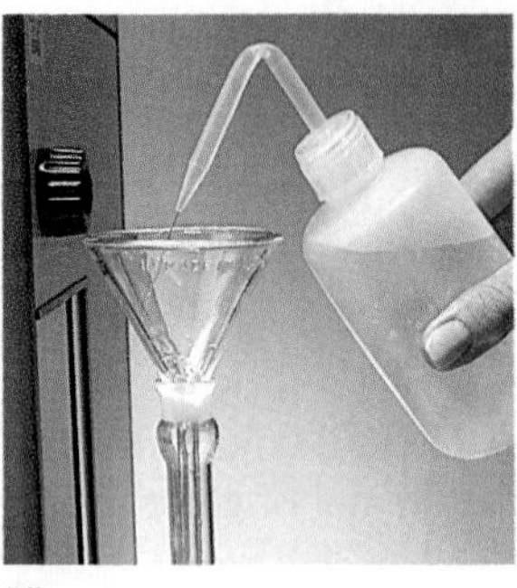

(d)

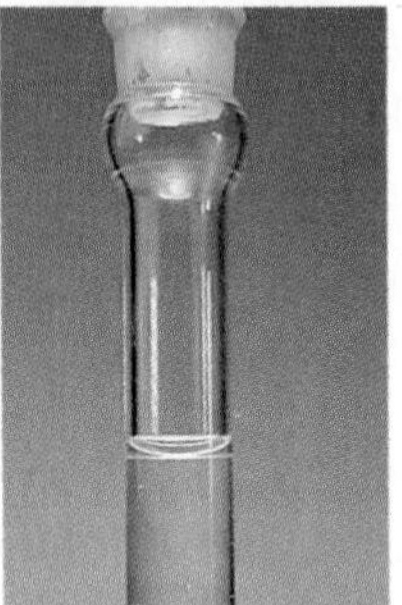

(e)

Figure 3.7

Preparing a solution of a given molarity: (a) weighing a weighing bottle containing a solid, and a volumetric flask with water nearby; (b) adding the solute from the weighing bottle into the volumetric flask; (c) weighing the weighing bottle with remaining solute; (d) rinsing the funnel with a wash bottle and adding water from wash bottle into the volumetric flask; (e) the solution in the volumetric flask brought to the mark on the stem.

Example 3.21

Calculating the Mass of Solute to Make a Solution of a Given Volume and Molarity

How many grams of sucrose, $C_{12}H_{22}O_{11}$ (table sugar), is required to make 2.00 L of a 0.300 *M* solution? Describe the steps you would take to make such a solution.

SOLUTION: The number of moles of sucrose needed can be calculated by applying Equation 3.2:

$$\text{moles sucrose} = 2.00 \text{ L soln} \left(\frac{0.300 \text{ mol sucrose}}{1 \text{ L soln}}\right)$$
$$= 0.600 \text{ mol sucrose}$$

The conversion of moles to grams yields

$$0.600 \text{ mol sucrose} \left(\frac{342 \text{ g sucrose}}{1 \text{ mol sucrose}}\right) = 205 \text{ g sucrose}$$

The mass of sucrose needed can be calculated more effectively by a single multistep operation by using the following sequence of conversions:

$$\text{liters solution} \longrightarrow \text{moles sucrose} \longrightarrow \text{mass sucrose}$$

$$2.00 \text{ L soln} \left(\frac{0.300 \text{ mol sucrose}}{1 \text{ L soln}}\right)\left(\frac{342 \text{ g sucrose}}{1 \text{ mol sucrose}}\right) = 205 \text{ g sucrose}$$

To prepare this solution, 205 g of sucrose should be weighed and dissolved in less than 2.00 L of water (the minimum volume needed to easily dissolve the sugar). Water should then be added to bring the total volume of the solution to the 2.00-L mark in a volumetric flask. The solution in the flask is then thoroughly mixed (see Figure 3.7).

Practice Problem 3.17: If you had 13.2 g of solid ammonium sulfate (used in fireproofing of cloth and paper), how many milliliters of 0.250 *M* solution could you make?

Dilution of Solutions

Many solutions are purchased from chemical supply houses. These solutions—called stock solutions—are often very concentrated. The stock solution can be diluted by adding solvent to achieve any desired lower concentration. Adding water to a solution does not change the total number of moles of the solute in the solution, but it does change the number of moles of the solute per liter of the solution. Thus, adding water decreases the molarity of the solution.

The number of moles of a solute in a solution is given by Equation 3.2 as the volume in liters (L) times the molarity (M): moles solute = (L)(M). If we let V_c and M_c be the volume and the molarity of the concentrated stock solution, the number of moles of the solute in the solution is given by

$$\text{moles solute} = V_c M_c$$

If V_d and M_d are the volume and molarity of the dilute solution, the number of moles of solute in the dilute solution is

$$\text{moles solute} = V_d M_d$$

Since the number of moles of solute does not change by dilution,

$$V_c M_c = V_d M_d \tag{3.3}$$

The dilution formula given by Equation 3.3 can be applied to molarity or any other unit of concentration. For most practical purposes, the volume of water to be added to dilute an aqueous solution is the difference between the volume of the diluted solution (V_d) and the volume of the original concentrated solution (V_c):

$$\text{volume of } H_2O \text{ to add} \approx V_d - V_c \tag{3.4}$$

When a solution is mixed with water or with another solution, the total volume obtained may not exactly equal the sum of the two volumes mixed. Sometimes the volume decreases upon mixing; other times it increases, depending on the interaction between the solute and solvent molecules. However, for most practical purposes we can *assume*, as in Equation 3.4, that the volume of a diluted solution equals the sum of the volumes of the concentrated solution and the solvent added.

Example 3.22

Calculating the Volume of Water to Add to Dilute a Solution

Hydrochloric acid is one of the most commonly used reagents in chemistry laboratories. A stock solution of concentrated hydrochloric acid is 12 *M*. To what volume must 0.20 L of this solution be diluted to prepare 0.50 *M* solution? How much water must be added to accomplish this, assuming no increase or decrease in volume upon mixing?

SOLUTION: The number of moles of HCl in the stock solution = moles HCl in the diluted solution. Solving Equation 3.3 for V_d (the total volume of the diluted solution), we obtain

$$V_d = \frac{V_c M_c}{M_d} = \frac{(0.20 \text{ L})(12 \, M)}{0.50 \, M} = 4.8 \text{ L}$$

According to Equation 3.4,

$$\text{volume } H_2O \text{ to add} = V_d - V_c = 4.8 \text{ L} - 0.20 \text{ L} = 4.6 \text{ L}$$

Practice Problem 3.18: What volumes of 6.0 *M* sodium hydroxide solution and water must be mixed to make 400 mL of 0.20 *M* solution, assuming no expansion or contraction of volume upon mixing?

Practice Problem 3.19: A 25.0-mL sample of a 3.00 *M* stock solution of nitric acid is diluted with water to a total volume of 100.0 mL. What is the molarity of the resulting solution?

Concentrations of Ions in Solution

Some substances dissolve in water to form solutions that conduct an electric current. Such substances are called **electrolytes**. A solution that conducts an electric current contains mobile positively and negatively charged ions. Substances whose solutions do not conduct an electric current are **nonelectrolytes**.

The electrical conductance of a solution can be tested by an ordinary light bulb as shown in Figure 3.8. The bulb lights up when the electrodes are placed

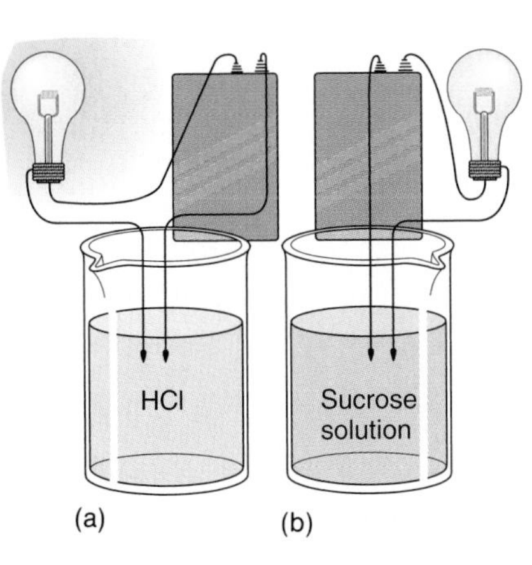

(c)

(d)

(e)

Figure 3.8

Electrical conductance as shown by a light bulb. (a) HCl (strong electrolyte) solution; the bulb lights up brightly; (b) sucrose (nonelectrolyte) solution; the bulb doesn't light up; (c) water (nonelectrolyte); the bulb doesn't light up; (d) acetic acid (a weak electrolyte) solution; the bulb lights up with a dim glow; (e) solution of sodium chloride (strong electrolyte); the bulb lights up brightly.

into a solution of an electrolyte such as hydrochloric acid or sodium chloride, but it does not light up in pure water or in a solution of a nonelectrolyte such as sucrose. Thus, the solutions of hydrochloric acid and sodium chloride contain ions, whereas pure water and an aqueous sucrose solution consist mainly of uncharged molecules.

The light bulb glows brightly in some electrolyte solutions, but only a dim glow can be seen in other electrolyte solutions of equal concentration (Figure 3.8). We conclude that the bulb glows brightest in the solution that contains the highest concentration of ions.

A solution in which the bulb glows brightly contains substances that are extensively dissociated into ions in solution. *Substances that are nearly completely dissociated into ions* are called **strong electrolytes**. In contrast, *substances that dissociate into ions only to a small degree* are called **weak electrolytes**.

Salts, acids, and bases are electrolytes. Of these, sodium chloride, hydrochloric acid, and sodium hydroxide are examples of strong electrolytes. Acetic acid (the major nonaqueous ingredient of vinegar) is a weak electrolyte. A 0.1 *M* solution of acetic acid is only about 1 percent dissociated in aqueous solution. Thus, 0.1 *M* aqueous solution of acetic acid consists of approximately 99 percent acetic acid molecules and only 1 percent of ions derived from the molecules. Below we consider how the various ion concentrations in solutions of strong electrolytes can be calculated from appropriate data.

When a mole of an ionic solid such as sodium chloride dissolves in water, a mole of sodium ions and a mole of chloride ions are produced:

$$\underset{1\text{ mol}}{\mathrm{NaCl}(s)} \xrightarrow{\mathrm{H_2O}} \underset{1\text{ mol}}{\mathrm{Na^+}(aq)} + \underset{1\text{ mol}}{\mathrm{Cl^-}(aq)}$$

In the equation above, the formula H_2O on the arrow indicates that solid sodium chloride is dissolved in water. The notation *aq* following the symbols for ions signifies that the ions are in water solution where each ion is surrounded by water molecules that are weakly bonded to the ion. An ion that is bonded to water molecules is called a **hydrated ion**. (Recall from Section 3.3 that some salts are also hydrated in their solid state.)

Figure 3.9

Dissociation and hydration of ions in dissolving of solid sodium chloride in water.

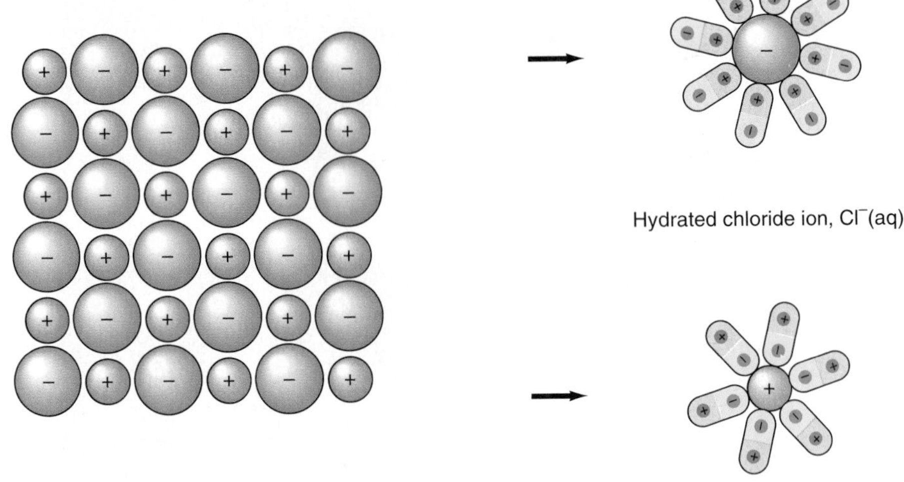

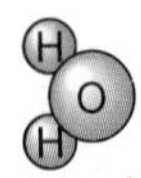
A water molecule

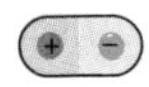
A simple representation of a water molecule in which the electron-rich oxygen side is marked with "–" and the electron-deficient hydrogen side "+".

Figure 3.9 illustrates the dissolution process of an ionic solid, sodium chloride, and the hydration of its ions. The electrons in a water molecule are concentrated around the oxygen atom, leaving the hydrogen atoms more electropositive. Thus, a positive sodium ion attracts the oxygen atom of a water molecule, and a negative chloride ion attracts the hydrogen atoms (Figure 3.9).

Let us consider a salt in which the cations and anions are not in a 1:1 ratio. An example of such a salt is calcium nitrate. A mole of calcium nitrate, $Ca(NO_3)_2$, consists of 1 mol of Ca^{2+} ions and 2 mol of NO_3^- ions. Thus, 1 mol of calcium nitrate produces a total of 3 mol of ions in solution, assuming complete dissociation of the salt. When the salt dissolves, the ions separate from the solid lattice and become bound to water molecules as hydrated ions:

$$\underset{1\text{ mol}}{Ca(NO_3)_2(s)} \xrightarrow{H_2O} \underset{1\text{ mol}}{Ca^{2+}(aq)} + \underset{2\text{ mol}}{2NO_3^-(aq)}$$

Example 3.23

Calculating the Concentration of Ions in Solution from a Given Molarity of a Salt

Barium chloride solution is used as a heart stimulant. What is the molar concentration of Ba^{2+} and of Cl^- ions in a 0.300 *M* barium chloride solution? How many moles of barium ions and how many moles of chloride ions does 50.0 mL of 0.300 *M* barium chloride solution contain?

SOLUTION: When barium chloride dissolves in water, it converts to aqueous barium ions and chloride ions:

$$BaCl_2(s) \xrightarrow{H_2O} Ba^{2+}(aq) + 2Cl^-(aq)$$

Each mole of barium chloride that dissolves produces 1 mol of barium ions and 2 mol of chloride ions. Thus, a 0.300 *M* solution of $BaCl_2$ is 0.300 *M* in Ba^{2+} ions and 0.600 *M* in Cl^- ions. This means that 1 L of a 0.300 *M* solution of $BaCl_2$ contains 0.300 mol of Ba^{2+} ions and 0.600 mol of Cl^- ions. The number of moles of Ba^{2+} ions in 50.0 mL (0.0500 L) of 0.300 *M* $BaCl_2$ solution is

$$0.0500\text{ L}\left(\frac{0.300\text{ mol Ba}^{2+}}{1\text{ L}}\right) = 0.0150\text{ mol Ba}^{2+}$$

and the number of moles of Cl^- ions is

$$0.0500\text{ L}\left(\frac{0.600\text{ mol Cl}^{-}}{1\text{ L}}\right) = 0.0300\text{ mol Cl}^{-}$$

Example 3.24

Calculating Individual Ion Concentrations and the Total Ion Concentration in a Solution

A 12.5-g sample of aluminum sulfate is dissolved in sufficient water to make 1.80 L of solution. What is the molar concentration of aluminum ions and of sulfate ions in this solution, assuming complete dissociation? What is the total ion concentration?

SOLUTION: Each mole of aluminum sulfate dissolves in water to produce 2 mol of aluminum ions and 3 mol of sulfate ions:

$$Al_2(SO_4)_3(s) \longrightarrow 2Al^{3+}(aq) + 3SO_4^{2-}(aq)$$

The number of moles of aluminum sulfate dissolved is

$$12.5\text{ g Al}_2(\text{SO}_4)_3\left(\frac{1\text{ mol Al}_2(\text{SO}_4)_3}{342\text{ g Al}_2(\text{SO}_4)_3}\right) = 0.0365\text{ mol Al}_2(\text{SO}_4)_3$$

The number of moles of aluminum ions formed in the solution is

$$0.0365\text{ mol Al}_2(\text{SO}_4)_3\left(\frac{2\text{ mol Al}^{3+}}{1\text{ mol Al}_2(\text{SO}_4)_3}\right) = 0.0730\text{ mol Al}^{3+}$$

and the molarity of aluminum ions is

$$M\text{ of Al}^{3+} = \frac{0.0730\text{ mol}}{1.80\text{ L}} = 0.0406\text{ mol L}^{-1}$$

Similarly for sulfate ions:

$$0.0365\text{ mol Al}_2(\text{SO}_4)_3\left(\frac{3\text{ mol SO}_4^{2-}}{1\text{ mol Al}_2(\text{SO}_4)_3}\right) = 0.110\text{ mol SO}_4^{2-}$$

$$M\text{ of SO}_4^{2-} = \frac{0.110\text{ mol}}{1.80\text{ L}} = 0.0611\text{ mol L}^{-1}$$

The total number of moles of ions in the solution is

$$0.0730\text{ mol Al}^{3+} + 0.110\text{ mol SO}_4^{2-} = 0.183\text{ mol ions}$$

The total molar concentration of ions is

$$\frac{0.183\text{ mol}}{1.80\text{ L}} = 0.102\text{ mol L}^{-1}$$

Practice Problem 3.20: What volume of 0.45 *M* Na_3PO_4 contains 2.5 mol of sodium ions? Assume complete dissociation of Na_3PO_4.

Summary

Stoichiometry is the *branch of chemistry that deals with the calculation of relative quantities of substances as components of compounds or as reactants and products in chemical reactions.*

Quantity of Matter

The *SI unit for quantity of matter* is the **mole**. A mole is defined as that *quantity of a substance which contains as many characteristic units of the substance as the number of atoms in exactly 12 g of carbon-12.* This number, called **Avogadro's number**, is **6.022 × 10^{23}**. The masses of molecules are conveniently expressed in terms of **molecular weight**, which is the *sum of the atomic weights of all the atoms in a molecule*. The *mass of a mole of atoms or molecules* in grams is called **molar mass**. The molar mass of atoms of an element *equals the atomic weight of the element in grams*. The molar mass of a compound *is the formula weight of the compound expressed in grams.* The **formula weight** for a molecular or an ionic compound *is the sum of all the atomic weights of the atoms specified by the formula for the compound.*

Percent Composition of Compounds

The formula for a compound specifies the relative numbers of atoms in a molecule or formula unit of the compound, as well as the relative numbers of *moles of atoms in a mole of the compound*. The *mass percent* of an element in a compound can be calculated by *converting the number of moles* of the element in a mole of the compound *to mass*. Then the ratio of this mass to the molar mass of the compound, times 100, is the mass percent of the element in the compound.

Derivation of Formulas

If the mass percent of each element in a compound is known, the number of moles of the elements in a given quantity of the compound can be calculated. A *comparison of the number of moles of each element in a sample of the compound* to the element with the lowest number of moles present leads to the **simplest** or **empirical formula** of the compound. The **molecular formula** of a molecular compound *can be obtained by comparing the experimentally determined molecular weight with the formula weight corresponding to the simplest formula.*

Calculations Based on Chemical Equations

From a given mass of a reactant in a chemical reaction, the mass of another reactant needed for a complete reaction, or the mass of a product formed, can be calculated. This can be done by the following sequence of steps in which the reactant is represented by A, and the other reactant, or the product, by B:

$$\text{grams A} \longrightarrow \text{moles A} \longrightarrow \text{moles B} \longrightarrow \text{grams B}$$

The number of moles B that can be obtained from moles A is determined by the B/A mole ratio in the balanced equation for the reaction.

In a chemical reaction, the *reactant that reacts completely* is called the **limiting reactant**. From the masses of two reactants mixed for a reaction, the limiting reactant can be identified in the following way: (1) Convert the masses of the reactants to moles. (2) Calculate the number of moles of a product formed from each of the reactants assuming that it reacts completely with an excess of the other reactant. (3) The reactant that forms a lower number of moles of the product is the limiting reactant.

The *measured yield of a product* of a reaction is called the **actual yield**. The *calculated yield of the product* is the **theoretical yield**. The actual yield is often less than the theoretical yield because of a possible incomplete reaction of the limiting reactant, loss of some of the product during its isolation, and so on. The **percent yield** is

$$\text{percent yield} = \frac{\text{actual yield}}{\text{theoretical yield}} \times 100$$

Solution Stoichiometry

The concentration of a solution can be expressed in different ways, including **percent by mass**, **percent by volume**, and **molarity**. *Molarity* (M) specifies the *number of moles of solute per liter of solution*. To calculate the molarity, M, of a solution from a given mass of the solute and the volume of the solution, convert the mass to moles and divide the number of moles by the volume, V, in liters:

$$M = \frac{\text{moles solute}}{\text{liters solution}}$$

To calculate the volume of a solvent needed to dilute a given volume of a concentrated stock solution we use the equation

$$V_c M_c = V_d M_d$$

where V_c is the volume of the concentrated solution, M_c the molarity of the concentrated solution, V_d the volume of the diluted solution, and M_d the molarity of the diluted solution. *The volume of solvent to add for the dilution* can, for most practical purposes, be determined by use of the equation

$$V_{\text{solvent}} = V_d - V_c$$

Substances that conduct electricity in water solution are called **electrolytes**. *Substances that dissociate almost completely into ions* in solution are called **strong electrolytes**, *those which dissociate only to a small extent* are **weak electrolytes**. *Substances that do not dissociate into ions in solution* are **nonelectrolytes**. The molarity of the ions in the solution of a strong electrolyte equals the molarity of the electrolyte times the number of moles of the ion present in 1 mol of the electrolyte.

New Terms

Actual yield (3.4)
Anhydrous compound (3.3)
Aqueous (3.5)
Avogadro's number (3.1)
Concentrated solution (3.5)
Concentration (3.5)
Dilute solution (3.5)
Electrolyte (3.5)
Empirical formula (3.3)
Formula weight (3.2)
Hydrate (3.3)
Hydrated ion (3.5)
Hydration (3.3)
Limiting reactant (3.4)
Mass spectrometer (3.1)
Mass spectrum (3.1)
Molarity (3.5)
Molar mass (3.2)
Molar weight (3.2)
Mole (3.1)
Molecular formula (3.3)
Molecular weight (3.2)
Nonelectrolyte (3.5)
Percent by mass (3.5)
Percent by volume (3.5)
Percent yield (3.4)
Saturated solution (3.5)
Simplest formula (3.3)
Solute (3.5)
Solvent (3.5)
Stoichiometry (introduction)
Strong electrolyte (3.5)
Theoretical yield (3.4)
Water of hydration (3.3)
Weak electrolyte (3.5)

Exercises

General Review

1. What is the atomic weight of an element? How does the atomic weight of an element differ from the mass number of an isotope?

2. Which isotope is used today as the standard for the determination of atomic weights?

3. How is the isotope in question 2 related to the atomic mass unit? Define atomic mass unit.

4. How are atomic weights determined?

5. How is the atomic weight of an element related to the masses of the isotopes of the element and the abundance of these isotopes?

6. What is the SI unit for quantity of matter?

7. **(a)** How many atoms does a mole of atoms contain? **(b)** How many molecules are in a mole of molecules?

8. What is Avogadro's number?

9. What advantage is there in using a mole as a unit for quantity of matter rather than a kilogram or a pound?

10. What is: **(a)** molecular weight; **(b)** formula weight; **(c)** molar mass?

11. **(a)** How many H atoms does a molecule of acetic acid, $HC_2H_3O_2$, contain? **(b)** How many moles of H atoms does a mole of acetic acid contain?

12. **(a)** What is the molecular weight of acetic acid? **(b)** What is the mass in grams of a mole of acetic acid?

13. How many grams of hydrogen does a mole of acetic acid contain?

14. What is the percent of hydrogen in acetic acid?

15. How would you calculate the mass in grams of an average atom of an element from the atomic weight of the element?

16. How can the mass in grams of 1.00 amu be calculated from the mass of a mole of ^{12}C atoms?

17. How many grams is: **(a)** a mole of H atoms; **(b)** a mole of H_2O; **(c)** a mole of $(NH_4)_2SO_4$; **(d)** a mole of O_3; **(e)** a mole of $Cr_2O_7^{2-}$ ions?

18. What is the percent by mass nitrogen in ammonium nitrate?

19. How would you proceed to calculate the simplest formula for a compound from the given percent composition of the compound? Outline a sequence of steps you would follow in this procedure.

20. What additional information, other than the percent composition, would you need to determine the molecular formula for a compound?
21. What is a hydrate?
22. List all the quantitative information that you can get from the equation $N_2(g) + 3H_2(g) \rightarrow 2NH_3(g)$.
23. What is a limiting reactant in a reaction? Why is it called the limiting reactant?
24. What is: **(a)** the actual yield of a reaction; **(b)** the theoretical yield; **(c)** the percent yield?
25. What is: **(a)** a dilute solution; **(b)** a concentrated solution?
26. List some quantitative ways to express the concentration of a solution.
27. List the steps you would take to prepare 500 mL of 1.80 *M* sodium chloride solution. Include the mass of sodium chloride you need for this purpose.
28. If you had 25.0 mL of 5.0 *M* sodium hydroxide solution, how many milliliters of water must be added to this solution to dilute it to 2.0 *M*?
29. What is: **(a)** an electrolyte; **(b)** a strong electrolyte; **(c)** a weak electrolyte; **(d)** a nonelectrolyte? Give an example of each.
30. What is the chloride ion concentration of a 0.060 *M* calcium chloride solution, assuming complete dissociation of the solute?

Moles, Molecules, and Atoms

31. What is the mass of 1 mol of each of the following substances? **(a)** Helium; **(b)** H_2; **(c)** silver; **(d)** silver chloride, AgCl; **(e)** aluminum sulfide; **(f)** sodium carbonate; **(g)** $Mg_3(PO_4)_2$.
32. What is the mass in grams equal to 2.70 mol of each of the following substances? **(a)** Sodium chloride; **(b)** O_2; **(c)** aluminum sulfate; **(d)** S_8; **(e)** dinitrogen pentoxide.
33. How many moles of the following substances is present in a 10.0-g sample of each? **(a)** Neon; **(b)** calcium nitrate; **(c)** water; **(d)** sucrose, $C_{12}H_{22}O_{11}$; **(e)** ammonium chloride.
34. Calculate the number of carbon atoms in 0.0120 g of carbon.
35. How many moles of O_2 is represented by 2.0×10^{12} molecules of O_2? How many moles of O atoms can be obtained from 2.0×10^{12} molecules of O_2?
36. Calculate the mass in grams of an average boron atom.
37. How many molecules are present in: **(a)** 20.0 g of carbon dioxide; **(b)** 20.0 g of glucose, $C_6H_{12}O_6$?
38. How many oxygen atoms and how many hydrogen atoms are needed to form 10.0 g of water?
39. How many moles of carbon atoms and how many moles of oxygen atoms are required to form 3 mol of carbon dioxide?

Percent Composition

40. Calculate the percent composition of the elements in each of the following compounds: **(a)** sulfuric acid; **(b)** calcium phosphate; **(c)** ammonium dichromate; **(d)** $Ca_5(PO_4)_3OH$; **(e)** $CuSO_4(H_2O)_5$; **(f)** aspirin, $C_9H_8O_4$.
41. Nitrogen is an important plant nutrient. Calculate the mass of nitrogen that can be obtained: **(a)** from 1.00 kg of ammonium nitrate; **(b)** from 1.00 kg of ammonium sulfate.
42. Calcium phosphate is a plant nutrient that supplies phosphorus. **(a)** What is the percent by mass phosphorus in calcium phosphate? **(b)** What is the mass percent phosphate ions in calcium phosphate? **(c)** What is the mass percent calcium ions?

Derivation of Formulas

43. Elemental analysis of a compound gives the following percent by mass composition: K = 26.57 percent, Cr = 35.36 percent, and O = 38.07 percent. Derive the simplest (empirical) formula for the compound.
44. A gaseous compound contains 20.11 percent hydrogen and 79.89 percent carbon. The molecular weight of the compound is 30.07 g mol^{-1}. What is the molecular formula for the compound?
45. A compound consists of 19.30 percent sodium, 26.91 percent sulfur, and 53.80 percent oxygen. The molar mass of the compound is 237.9 g mol^{-1}. Derive the molecular formula for the compound.
46. A 1.600-g sample of an oxide of chromium contains 1.095 g of chromium. What is the simplest formula for the oxide?
47. A 1.000-g sample of a compound loses 0.450 g of oxygen when it decomposes upon heating. The solid residue that remains is sodium chloride. Derive the simplest formula for the compound.
48. A 0.501-g sample of a compound that is known to contain calcium, carbon, and oxygen decomposes upon heating to 0.280 g of calcium oxide and 0.221 g of carbon dioxide. Derive the simplest formula for the compound.
49. A 3.136-g sample of a hydrated sodium carbonate (washing soda) is heated to drive out its water of hydration. The mass loss of the sample upon heating is 1.970 g. Derive the formula for the hydrate.
50. A compound consists of 10.15 percent carbon and 89.85 percent chlorine. The molecular weight of the compound is 236.7. What is the molecular formula for this compound?
51. When a 1.214-g sample of a compound consisting only of carbon and hydrogen is burned, 4.059 g of carbon dioxide and 0.9494 g of water are produced. Calculate the percent of carbon and the percent of hydrogen in the compound, and derive the empirical formula for the compound.

52. When fish decay on beaches, many chemical compounds are formed. One of these is cadaverine, which has a strong odor of decaying fish. Cadaverine consists of carbon, hydrogen, and nitrogen. When a sample of 0.03560 g of cadaverine is completely burned in excess oxygen, it produces 0.07665 g of CO_2 and 0.04392 g of H_2O. What is the empirical formula for cadaverine?

53. Ammonium nitrate is used in fireworks and as a fertilizer. When ammonium nitrate is gently heated (*Caution!* Explosion may occur upon strong heating), water vapor is formed along with another gaseous product that consists of 63.60 percent nitrogen and the remainder oxygen. The molecular weight of this nitrogen-containing gaseous product is 44.0. Derive the molecular formula for this product, and write a balanced equation that occurs when ammonium nitrate is heated.

Calculations Based on Equations

54. Hydrogen gas reacts with oxygen gas to form water vapor according to the equation

$$2H_2(g) + O_2(g) \longrightarrow 2H_2O(g)$$

Supply the missing information in the following table, assuming complete reaction:

	H_2	O_2	H_2O
(a)	5.60 g	__ g	__ g
(b)	2.80 mol	__ mol	__ mol
(c)	1.69×10^{24} molecules	__ molecules	__ molecules

55. Based on the equation

$$4NH_3(g) + 5O_2(g) \longrightarrow 4NO(g) + 6H_2O(g)$$

complete the following table assuming complete reaction:

	NH_3	O_2	NO	H_2O
(a)	2.02 g	__ mol	__ g	__ mol
(b)	0.0364 mol	__ molecules	__ mol	__ g
(c)	7.82×10^{23} molecules	__ g	__ molecules	__ mol

56. What mass of iron(III) oxide (the principal ingredient of rust) is produced by complete reaction of a 100-g bar of iron according to the equation

$$4Fe(s) + 3O_2(g) \longrightarrow 2Fe_2O_3(s)?$$

57. Potassium chlorate (used in making matches and dyes) decomposes upon heating to produce potassium chloride and oxygen. Write a balanced equation for this reaction. If 12.26 g of potassium chlorate is completely decomposed, how many grams of oxygen is produced? How many moles of oxygen? How many molecules of oxygen?

58. Butane, C_4H_{10}, is used as a fuel in portable cigarette lighters. Butane can also be used as a rocket fuel. The energy for rocket launching comes from the reaction of butane with liquid oxygen; the products of this reaction are carbon dioxide and water vapor. Write a balanced equation for this reaction and calculate the mass of liquid oxygen needed for combustion of 1.00 kg of butane.

59. When ethane, C_2H_6, burns in air, it reacts with oxygen from the atmosphere to produce carbon dioxide and water vapor. Write a balanced equation for this reaction and calculate: **(a)** the number of moles of CO_2 produced from 3.00 mol of C_2H_6; **(b)** the number of moles of H_2O produced from 3.00 mol of C_2H_6; **(c)** the mass of CO_2 formed from 0.450 mol of C_2H_6; **(d)** the number of moles of CO_2 formed from 15.0 g of C_2H_6.

60. When cane sugar, $C_{12}H_{22}O_{11}$, is burned in air, it reacts with oxygen to produce carbon dioxide and water. What is the mass of water and of carbon dioxide that can be obtained by burning 0.342 g of cane sugar?

61. Phosphorus can be converted to phosphoric acid, H_3PO_4, by treatment with nitric acid:

$$3P_4(s) + 20HNO_3(aq) + 8H_2O(l) \longrightarrow 12H_3PO_4(aq) + 20NO(g)$$

What mass of 30 percent nitric acid solution must be used for a complete conversion of 1.0 kg of phosphorus to phosphoric acid if 50 percent excess of nitric acid is actually required for this process over the amount theoretically calculated? What mass of phosphoric acid can be obtained in the process?

62. For each of the following reactions, write a balanced equation (review the relevant sections of Chapter 2 if necessary), and perform the specified calculation in each case. **(a)** Calculate the mass of the product formed when 0.825 g of aluminum completely reacts with oxygen. **(b)** What mass of sodium oxide must completely react with excess carbon dioxide to produce 5.28 g of sodium carbonate? **(c)** When a 5.00-g sample of calcium oxide completely reacts with water, how much calcium hydroxide can be obtained? **(d)** What mass of potassium nitrate must be decomposed to produce 2.48 g of potassium nitrite? **(e)** How many grams of barium metal must react with excess

iron(III) sulfate to make 5.00 g of barium sulfate? **(f)** A solution containing 1.50 g of barium chloride reacts completely with a solution of sodium sulfate. What is the maximum number of grams of solid barium sulfate that can be obtained in this process? **(g)** How many grams of bismuth(III) oxide must react with excess zinc metal to produce 3.52 g of zinc oxide?

63. A thirsty chemist wanders in the Sahara desert and comes across an abandoned cache of gasoline apparently left over from the World War II Africa campaign. The chemist cleverly burns some of the gasoline to produce water vapor, trapping the liquid by condensation in a long piece of tubing. Assume that the gasoline is C_8H_{18}, the combustion reaction is

$$2C_8H_{18}(l) + 25O_2(g) \longrightarrow 16CO_2(g) + 18H_2O(g)$$

Assuming also that the chemist and the process are 100 percent efficient and that 180 mL of liquid H_2O with a density of 1.00 g/mL is collected, calculate: **(a)** the number of moles of H_2O collected; **(b)** the number of moles of C_8H_{18} burned; **(c)** the number of grams of C_8H_{18} burned; **(d)** the number of milliliters of C_8H_{18} burned (the density of C_8H_{18} is 0.700 g/mL).

64. Sulfuric acid can be prepared from the mineral pyrite (FeS_2) by the "lead chamber process," in which the following reactions occur:

$$2FeS_2(s) + 5O_2(g) \longrightarrow 2FeO(s) + 4SO_2(g)$$
$$SO_2(g) + NO_2(g) \longrightarrow SO_3(g) + NO(g)$$
$$2NO(g) + O_2(g) \longrightarrow NO_2(g)$$
$$SO_3(g) + H_2O(l) \longrightarrow H_2SO_4(aq)$$

What mass of 60 percent by mass H_2SO_4 solution can be produced from 7.5 metric tons of pyrite that is 91.3 percent pure FeS_2, assuming that 3.8 percent of SO_2 is lost during the process? The other 8.7 percent of the pyrite is inert material.

Limiting Reactant

65. Powdered aluminum reacts with solid iodine to produce aluminum iodide (used in organic syntheses) according to the equation

$$2Al(s) + 3I_2(s) \longrightarrow 2AlI_3(s)$$

If 12.7 g of iodine is heated with 12.7 g of aluminum to produce aluminum iodide, which reactant is the limiting reactant? How much of the other reactant remains unreacted? How many moles of aluminum iodide is produced? How many grams of aluminum iodide is produced?

66. If 10.0 g of calcium is heated with 10.0 g of phosphorus, how much calcium phosphide can theoretically be obtained? Which is the limiting reactant here, and how much of the other reactant remains unreacted?

67. The refrigerant Freon-12, CCl_2F_2, is prepared by the reaction of carbon tetrachloride with antimony trifluoride:

$$3CCl_4(l) + 2SbF_3(s) \longrightarrow 3CCl_2F_2(g) + 2SbCl_3(s)$$

If 0.400 kg of CCl_4 is mixed with 0.350 kg of SbF_3 for a reaction, how many grams of CCl_2F_2 can be formed?

68. Acetylene gas, C_2H_2 (used as a fuel in portable carbide lanterns, in welding, and for the production of some plastics), can be prepared by the following sequence of reactions:

(1) $CaCO_3(s) \longrightarrow CaO(s) + CO_2(g)$

(2) $CaO(s) + C(s) \longrightarrow CaC_2(s) + CO(g)$
(CaC_2 is calcium carbide)

(3) $CaC_2(s) + H_2O(l) \longrightarrow Ca(OH)_2(s) + C_2H_2(g)$

(a) Balance the equations above if not already balanced. **(b)** In a preparation of acetylene, 15.0 kg of calcium carbonate and 5.00 kg of carbon are used with excess water. Determine the limiting reactant and calculate the mass of acetylene that can be produced in this process.

Percent Yield

69. Boron trifluoride, BF_3, used as a catalyst in organic synthesis, can be prepared by the reaction of diboron trioxide, B_2O_3 (used in eyewash), with hydrofluoric acid according to the equation

$$B_2O_3(s) + 6HF(aq) \longrightarrow 2BF_3(g) + 3H_2O(l)$$

In a reaction of 10.0 g of B_2O_3 with 18.5 g of HF, 10.2 g of BF_3 is recovered. Calculate the theoretical yield and the percent yield of BF_3 in this reaction.

70. Copper metal reacts in concentrated sulfuric acid to form hydrated copper sulfate, $CuSO_4(H_2O)_5$, according to the equation

$$Cu(s) + 2H_2SO_4(aq) + 3H_2O(l) \longrightarrow CuSO_4(H_2O)_5(s) + SO_2(g)$$

How many grams of hydrated copper sulfate can be obtained in this reaction when 20.0 g of 98.0 percent sulfuric acid reacts with excess copper, assuming that the yield is 85.0 percent?

Derivation of Formula, Equation Writing, and Percent Yield

71. In an experiment, 1.90 g of solid chromium(VI) oxide is heated with excess sulfur at 250°C. The resulting reaction produces 1.70 g of a black solid and sulfur dioxide gas. The black solid contains 54.9 percent chromium and the remainder oxygen. **(a)** Derive the simplest formula for the black solid. **(b)** Write a balanced equation for the reaction of chromium(VI) oxide with sulfur at 250°C. **(c)** Calculate the percent yield of the solid product.

Solution Stoichiometry

72. A 0.476-g sample of magnesium chloride is dissolved in water and the total volume of the solution is brought to 1.60 L. What is the molarity of the solution? What is the molarity of chloride ions in this solution? What is the molarity of magnesium ions?
73. How many grams of potassium hydrogen phthalate, $KHC_8H_4O_4$, must be used to make 250.0 mL of 0.400 *M* solution?
74. How many moles of barium hydroxide is present in 300.0 mL of 0.00501 *M* solution of barium hydroxide? How many moles of hydroxide ions is present in this solution? How many moles of barium ions? Assume complete dissociation of the solute.
75. A 2.86 *M* solution of sucrose contains 1.24 mol of sucrose. What is the volume of the solution?
76. What volume of 6.00 *M* sodium hydroxide solution is needed to prepare 400.0 mL of 0.100 *M* solution?
77. What volume of water should be added to 12.5 mL of 8.00 *M* hydrochloric acid to dilute it to 1.20 *M*?
78. What is the molarity of a 40.0-mL sample of sodium chloride solution that yields a 0.450-g residue after evaporation of the water?
79. A 25.2-mL sample of a 5.24 *M* solution of nitric acid is diluted to 40.5 mL. What is the molarity of the dilute solution?
80. A 1.285-g sample of magnesium nitrate is dissolved in water and the total volume of the solution is brought to 250.0 mL. How many moles of magnesium nitrate does a 10.00-mL portion of this solution contain? How many moles of nitrate ions does the 10.00-mL portion contain?
81. How many milliliters of 0.600 *M* H_3PO_4 is needed to react with 37.0 g of KOH according to the equation

$$H_3PO_4(aq) + 3KOH(s) \longrightarrow K_3PO_4(aq) + 3H_2O(l)$$

82. Aqueous lead(II) nitrate reacts with aqueous potassium chloride according to the equation

$$Pb(NO_3)_2(aq) + 2KCl(aq) \longrightarrow PbCl_2(s) + 2KNO_3(aq)$$

How many grams of $PbCl_2$ is produced by the reaction of 40.0 mL of 0.100 *M* KCl with excess $Pb(NO_3)_2$?

CHAPTER 4

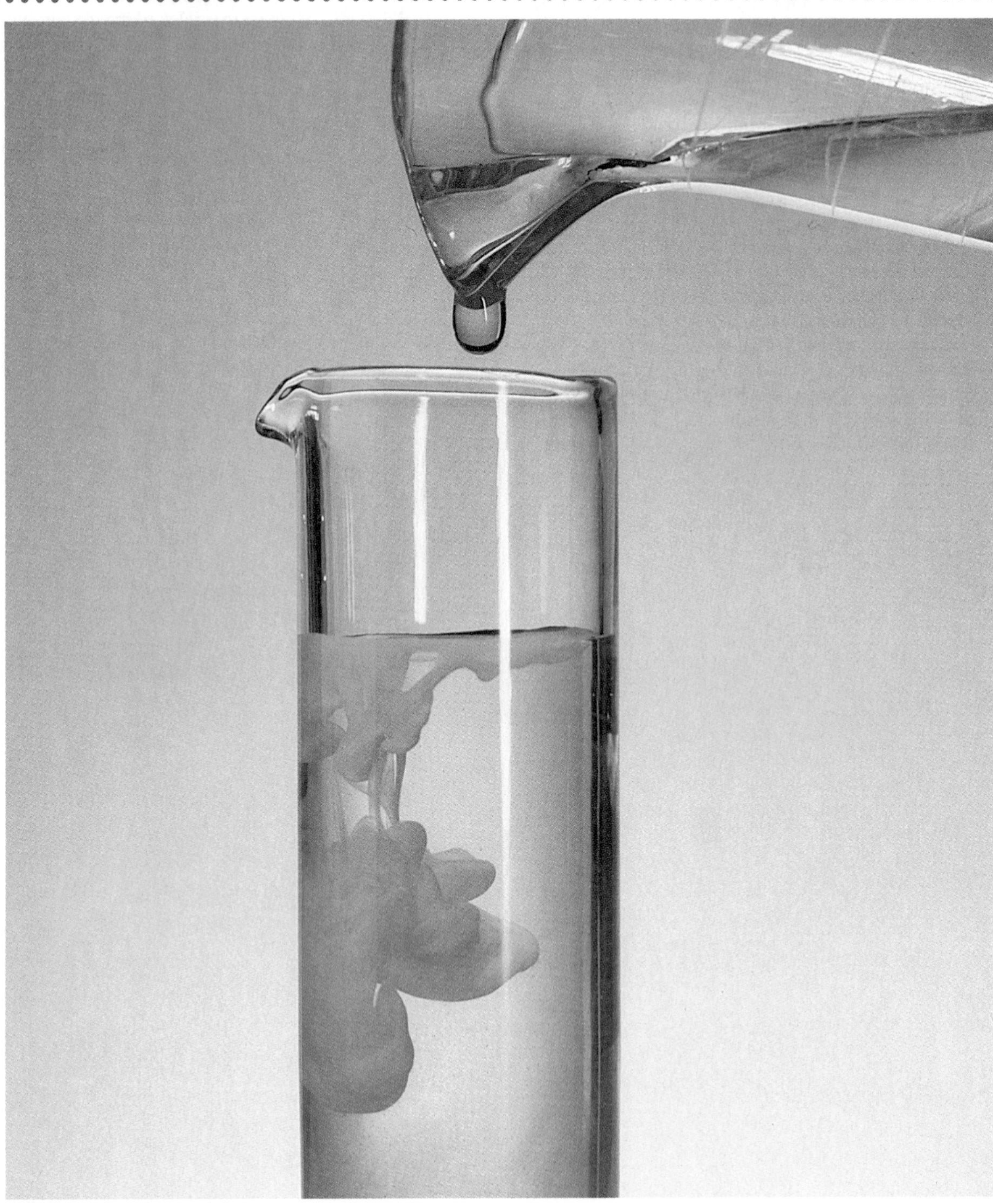

A reaction in a solution: mixing the solutions of lead(II) nitrate and sodium iodide produces yellow solid lead(II) iodide.

Reactions in Solution

In the preceding chapters we described some properties of solutions. In this chapter we consider some of the major types of reactions that occur in solution. These reactions span almost the entire range of chemistry, from the chemical reactions that occur in the tiny volume of a living cell to reactions that occur in the vast oceans. The study of chemical reactions in solutions is thus part of the study of such diverse fields as ecology, physiology, and air pollution.

4.1 Acids and Bases

Many substances can be described as "acids" or "bases." When these substances are dissolved in water, the resulting solution is either "acidic" or "basic." Some common examples of acidic solutions are vinegar and grapefruit juice; a commonly encountered basic substance is "household ammonia."

We can identify some solutions as acidic or basic by our senses of taste and touch. Thus, acidic solutions have a *sour taste*. In contrast, *basic solutions* have a *bitter taste* and they feel *slippery*. Acidic and basic solutions also undergo characteristic chemical reactions. For example, acidic solutions react with active metals to produce hydrogen gas. They also react with a dye called *litmus*, causing its color to change from blue to red. Basic solutions also react with litmus, changing its color from red to blue. Table 4.1 lists some common acidic and basic substances, and Figure 4.1 illustrates some properties of acids and bases.

Table 4.1

Common Acidic and Basic Substances

Acidic Substances	Basic Substances
Apples and citrus fruits	Washing soda
Aspirin	Detergents
Carbonated beverages	Drano
Vinegar	Household ammonia
Vitamin C	Soap
	Milk of magnesia

(a)

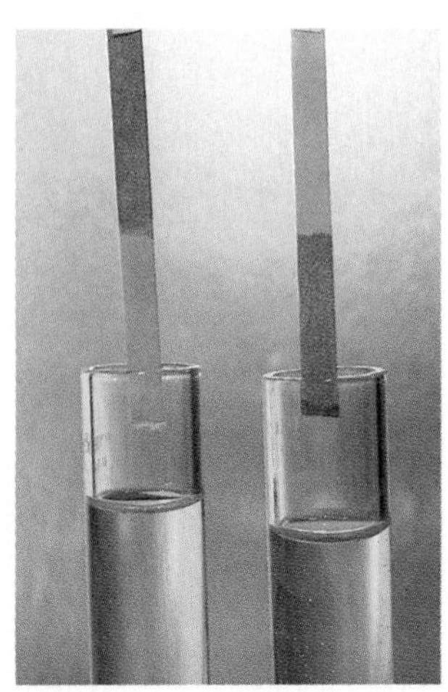

(b)

Figure 4.1

Some properties of acids and bases: (a) an active metal reacting with an acid forming bubbles of H_2 gas and (b) an acid and a base on litmus paper.

Arrhenius Acids and Bases

Next let us consider more precise definitions of the terms **acid** and **base**. Definitions for acids and bases were proposed by the Swedish chemist Svante Arrhenius in the 1880s. An **Arrhenius acid** is a substance that produces *hydrogen* (H^+) *ions in water solution* plus negative ions characteristic of the acid.

An H^+ ion in water is bonded to a water molecule to form a *hydronium ion*, H_3O^+. Thus, the hydronium ion is a hydrated H^+ ion. For simplicity, this ion is often represented by $H^+(aq)$. Hydrogen ions are not the only ions that are bound to water molecules in solution; this property is typical of all ions.

An **Arrhenius base** produces *hydroxide* (OH^-) *ions in water solution* plus positive ions characteristic of the base. For example, solid sodium hydroxide consists of sodium ions and hydroxide ions that separate from the solid lattice when the solid dissolves:

$$NaOH(s) \longrightarrow Na^+(aq) + OH^-(aq)$$

Brønsted–Lowry Acids and Bases

According to the Arrhenius definition, substances are acids or bases only in water solution. Broader definitions for an acid and for a base were given independently by the Danish chemist Johannes Brønsted (1879–1947) and the English chemist Thomas Lowry (1874–1936). According to the **Brønsted–Lowry definition**, an **acid** *donates protons* (H^+ *ions*) *in a chemical reaction*, and a **base** *accepts protons in a chemical reaction*. We will call the Brønsted–Lowry definition of acid and base the Brønsted definition for brevity.

Brønsted acids and bases can be solids, liquids, or gases, or they can be dissolved in water or other solvents. Even water itself can act as a Brønsted acid or base. For example, hydrogen chloride gas reacts with water by donating protons to water molecules to produce hydronium ions and chloride ions:

$$\underset{\text{Acid}}{HCl(g)} + \underset{\text{Base}}{H_2O(l)} \longrightarrow H_3O^+(aq) + Cl^-(aq)$$

In this reaction, hydrogen chloride gas is an acid and water is a base.

Ammonia reacts with water to produce ammonium ions and hydroxide ions. In this reaction, ammonia accepts protons from water and is therefore a Brønsted base; water acts as an acid because it donates protons to ammonia:

$$\underset{\text{Acid}}{H_2O(l)} + \underset{\text{Base}}{NH_3(g)} \longrightarrow NH_4^+(aq) + OH^-(aq)$$

Common Acids

The process of ion formation in solution is variously called dissociation and ionization. **Dissociation** is a process in which *particles break down into smaller particles*. The *dissociation of uncharged particles into ions as a result of a chemical reaction* is called **ionization**. Thus, hydrogen chloride gas dissociates or ionizes when it reacts with water to produce hydronium ions and chloride ions:

$$HCl(g) + H_2O(l) \longrightarrow H_3O^+(aq) + Cl^-(aq)$$

This dissociation of HCl into ions in water solution is virtually 100 percent complete. An acid that is *virtually 100 percent dissociated into ions* in aqueous solution is called a **strong acid**. An acid that *partly dissociates* is a **weak acid**. Table 4.2 lists six common strong acids and the anions derived from these acids, as well as the molarities of the commercially available concentrated solutions of the acids.

Table 4.2

Common Strong Acids and Their Anions

Acid Name	Formula	Molarity of Concentrated Solution	Anion	
Hydrochloric acid	HCl	12	Cl^-	Chloride ion
Hydrobromic acid	HBr	7	Br^-	Bromide ion
Hydroiodic acid	HI	8	I^-	Iodide ion
Nitric acid	HNO_3	15	NO_3^-	Nitrate ion
Sulfuric acid	H_2SO_4	18	HSO_4^-	Hydrogen sulfate ion
			SO_4^{2-}	Sulfate ion
Perchloric acid	$HClO_4$	12	ClO_4^-	Perchlorate ion

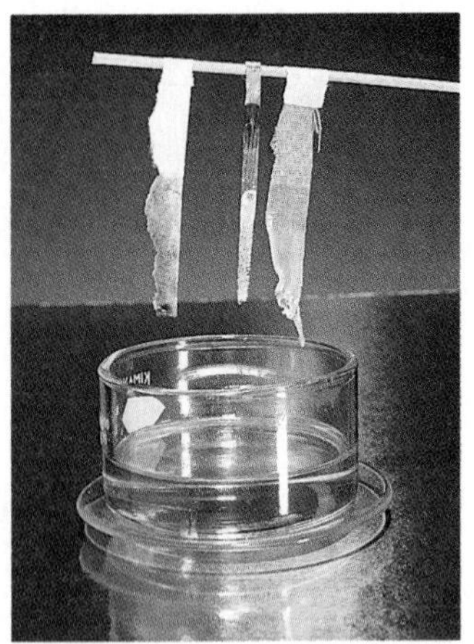

Figure 4.2

Action of sulfuric acid on paper, cloth and metals.

Concentrated solutions of all the strong acids listed in Table 4.2 cause traumatic damage to human skin and they can destroy clothing in a few seconds. Wood, metals, and even concrete are affected by contact with concentrated solutions of these acids (Figure 4.2). Concentrated perchloric acid reacts explosively when heated in the presence of organic matter.

Acids sold by commercial chemical supply houses are concentrated solutions whose concentrations are usually given in terms of percent by mass and density rather than molarity. The calculation of the molarity of an acid from its percent composition and density is illustrated in Example 4.1.

Example 4.1

Calculating the Molarity of an Acid from a Given Percent by Mass and Density

A bottle of hydrochloric acid contains 37.0 percent of HCl by mass and has a density of 1.19 g cm^{-3}. Calculate the molarity of the acid.

SOLUTION: This problem can be solved by the following sequence of conversions. First, we convert the density in grams per milliliter to grams per liter. Since the solution is 37.0 percent HCl by mass, we multiply the mass of 1 L by 37/100 to get the mass of HCl in 1 L of the solution. Finally, we convert the mass of HCl per liter of solution to moles per liter, which is the molarity. Carrying out these steps yields

$$\left(\frac{1.19\text{ g soln}}{1\text{ cm}^3\text{ soln}}\right)\left(\frac{1000\text{ cm}^3\text{ soln}}{1\text{ L soln}}\right)\left(\frac{37.0\text{ g HCl}}{100\text{ g soln}}\right)\left(\frac{1\text{ mol HCl}}{36.5\text{ g HCl}}\right)$$

$$= 12.1\text{ mol HCl/L soln} \quad \text{or} \quad 12.1\ M$$

Practice Problem 4.1: A solution of concentrated sulfuric acid is 98.0 percent by mass H_2SO_4 and has a density of 1.84 g cm^{-3}. **(a)** What is the molarity of the solution? **(b)** What volume of the solution contains 50.0 g of H_2SO_4?

Table 4.2 shows that some acids, such as HCl, contain one H atom per formula unit. These acids ionize to *produce one mole of protons per mole of acid*. They are known as **monoprotic acids**. Acids such as H_2SO_4 can *produce 2 mol of protons per mole of acid*. They are **diprotic acids**. Similarly, an acid that can furnish 3 mol of protons per mole of acid is a **triprotic acid**. An important example is phosphoric acid, H_3PO_4. Acids that can produce 2 mol or more of protons per mole of acid are called **polyprotic acids**.

The dissociation of a diprotic acid such as sulfuric acid in water occurs in two steps, as shown in the following equations:

Step 1. $H_2SO_4(l) + H_2O(l) \longrightarrow H_3O^+(aq) + HSO_4^-(aq)$

Step 2. $HSO_4^-(aq) + H_2O(l) \longrightarrow H_3O^+(aq) + SO_4^{2-}(aq)$

For H_2SO_4 the first step goes essentially to completion in a dilute solution. However, only about 10 percent of the HSO_4^- ions formed in the first step react with water. Thus, a dilute solution of sulfuric acid contains water molecules, hydronium ions, hydrogen sulfate ions, and sulfate ions. The approximate concentrations of ions present in a 0.10 *M* solution of H_2SO_4 are as follows:

$$\begin{array}{ccccc} H_3O^+ & > & HSO_4^- & > & SO_4^{2-} \\ 0.11\ M & & 0.09\ M & & 0.01\ M \end{array}$$

Phosphoric acid is a moderately weak triprotic acid that dissociates in three steps:

Step 1. $H_3PO_4(l) + H_2O(l) \longrightarrow H_3O^+(aq) + H_2PO_4^-(aq)$

Step 2. $H_2PO_4^-(aq) + H_2O(l) \longrightarrow H_3O^+(aq) + HPO_4^{2-}(aq)$

Step 3. $HPO_4^{2-}(aq) + H_2O(l) \longrightarrow H_3O^+(aq) + PO_4^{3-}(aq)$

In the first step, about 27 percent of the H_3PO_4 molecules in a 0.10 *M* solution react with water. The other two steps proceed only to a very small extent. The second step proceeds to about 2.6×10^{-4} percent, and the third step to only 1.5×10^{-9} percent. Thus, a 0.10 *M* solution of phosphoric acid contains H_3PO_4 molecules, H_3O^+ ions, $H_2PO_4^-$ ions, HPO_4^{2-} ions, and PO_4^{3-} ions. The approximate concentrations of the phosphoric acid molecules and its ions in a 0.10 *M* solution are as follows:

$$\begin{array}{ccccccccc} H_3PO_4 & > & H_3O^+ & \approx & H_2PO_4^- & \gg & HPO_4^{2-} & \gg & PO_4^{3-} \\ 0.073\ M & & 0.027\ M & & 0.027\ M & & 7.0 \times 10^{-8}\ M & & 1.0 \times 10^{-18}\ M \end{array}$$

Table 4.3 lists some common weak acids and acids of intermediate strength. Perhaps the most common of the weak acids is acetic acid, $HC_2H_3O_2$, which is the major acid in *vinegar*. A typical vinegar solution contains about 5 percent by mass of acetic acid. In a 0.10 *M* aqueous solution of acetic acid, the acid is only about 1.3 percent dissociated into H_3O^+ and $C_2H_3O_2^-$ (acetate) ions:

$$HC_2H_3O_2(aq) + H_2O(l) \longrightarrow H_3O^+(aq) + C_2H_3O_2^-(aq)$$

Thus, a 0.10 *M* solution of acetic acid contains 0.0013 mol/L (1.3 percent of 0.10) of H_3O^+ ions and 0.0013 mol/L of $C_2H_3O_2^-$ ions. Since acetic acid dissociates only to a small extent, it is a weak electrolyte.

Practice Problem 4.2: What is the molar concentration of acetic acid molecules in a 0.10 *M* solution of acetic acid?

We recall that an acid which contains oxygen is called an oxoacid. Usually, the more oxygen atoms bonded to the central atom of an acid the stronger the acid. That is, an oxoacid whose name ends with *-ic* is usually stronger than the corresponding acid whose name ends with *-ous* (Tables 4.2 and 4.3). Thus,

Table 4.3

Common Weak and Intermediate-Strength Acids

Acid Name	Formula	Percent Dissociated in 0.1 *M* Solution	Anions	
Acetic acid	$HC_2H_3O_2$	1.3	$C_2H_3O_2^-$	Acetate ion
Formic acid	$HCHO_2$	4.2	CHO_2^-	Formate ion
Hydrocyanic acid	HCN	7×10^{-3}	CN^-	Cyanide ion
Hydrofluoric acid	HF	8.2	F^-	Fluoride ion
Hydrosulfuric acid	H_2S	0.10	HS^-	Hydrogen sulfide ion
			S^{2-}	Sulfide ion
Nitrous acid	HNO_2	6.7	NO_2^-	Nitrite ion
Oxalic acid	$H_2C_2O_4$	77	$HC_2O_4^-$	Hydrogen oxalate ion
			$C_2O_4^{2-}$	Oxalate ion
Phosphoric acid	H_3PO_4	27	$H_2PO_4^-$	Dihydrogen phosphate ion
			HPO_4^{2-}	Monohydrogen phosphate ion
			PO_4^{3-}	Phosphate ion
Phosphorous acid	H_2PHO_3	32	$HPHO_3^-$	Dihydrogen phosphite ion
			PHO_3^{2-}	Monohydrogen phosphite ion
Sulfurous acid	H_2SO_3	41	HSO_3^-	Hydrogen sulfite ion
			SO_3^{2-}	Sulfite ion

nitric acid (HNO_3) is a strong acid that is completely dissociated into ions in a 0.10 *M* solution, whereas nitrous acid (HNO_2) is a weak acid that is only 6.7 percent dissociated in a 0.10 *M* solution. Similarly, sulfuric acid is a very strong acid, but sulfurous acid (H_2SO_3) is an acid of intermediate strength.

Common Bases

Among the most common **strong bases** are the hydroxides of alkali metals and alkaline earth metals. These hydroxides are ionic solids that dissociate in water to metal ions and hydroxide ions in solution. Examples are shown below for sodium hydroxide and for barium hydroxide:

$$NaOH(s) \longrightarrow Na^+(aq) + OH^-(aq)$$
$$Ba(OH)_2(s) \longrightarrow Ba^{2+}(aq) + 2OH^-(aq)$$

Hydroxides of the alkali metals are all very soluble in water. Hydroxides of the alkaline earth metals are only sparingly soluble; their solubility increases from $Mg(OH)_2$ to $Ba(OH)_2$.

Practice Problem 4.3: What is the molar concentration of hydroxide ions in: **(a)** 0.020 *M* potassium hydroxide; **(b)** 0.0030 *M* barium hydroxide?

Figure 4.3

Glass corroded by NaOH.

Concentrated solutions of strong bases are corrosive. A solution of a strong base such as sodium hydroxide reacts slowly with glass (Figure 4.3) and should therefore be stored in a plastic bottle. Solutions of strong bases also cause severe damage to skin and clothing. Protective eye covering should be worn by anyone who works with strong acids and bases.

Perhaps the most commonly encountered **weak base** is ammonia, NH_3, an ingredient in a cleaning solution called household ammonia. Ammonia reacts slightly with water to produce ammonium ions and hydroxide ions:

$$NH_3(g) + H_2O(l) \longrightarrow NH_4^+(aq) + OH^-(aq)$$

This reaction goes only about 1.3 percent to completion in a 0.10 *M* solution. Thus, a 0.10 *M* aqueous solution of ammonia contains mostly ammonia molecules and a small concentration of ammonium ions and hydroxide ions. The presence of ammonia in a solution can easily be detected by the pungent odor of the gas above the solution. Since ammonia is only 1.3% ionized in a 0.10 *M* solution, it is a weak electrolyte.

4.2 Aqueous Acid–Base Reactions

Chemical reactions between acids and bases are collectively called acid–base reactions. When a solution of a strong acid reacts with a solution of a strong base so that the number of hydrogen ions produced by the acid equals the number of hydroxide ions produced by the base, the resulting solution is neither acidic nor basic, it is *neutral*. Therefore, acid–base reactions are also called **neutralization reactions**.

A solution of a strong acid contains hydronium ions and anions characteristic of the acid. A solution of a strong base contains hydroxide ions and cations characteristic of the base. When these two solutions are mixed, the hydronium ions react with the hydroxide ions to form water:

$$H_3O^+(aq) + OH^-(aq) \longrightarrow 2H_2O(l)$$

For simplicity, this reaction is often written as

$$H^+(aq) + OH^-(aq) \longrightarrow H_2O(l)$$

The cations derived from the base and the anions derived from the acid most often remain in the solution that results from an acid–base reaction. These cations and anions can be recovered as a salt if the water is allowed to evaporate from the solution. Sometimes the cation and the anion react to form an insoluble salt that can be separated from the solution by filtration.

Equations for Acid–Base Reactions: Total Equations, Ionic Equations, and Net Ionic Equations

Aqueous acid–base reactions can be described by three different kinds of chemical equations. Let us first consider an equation for the reaction of a *strong acid with a strong base*. We write the formulas of the acid and the base that react, and the formulas for the salt and water produced. For example, the equation for the reaction of hydrochloric acid with sodium hydroxide in solution is

$$HCl(aq) + NaOH(aq) \longrightarrow NaCl(aq) + H_2O(l)$$

This equation does not explicitly specify that there are ions in solution, although the notation *aq* implies it.

An *equation in which the formulas of all the reactants and products are written, regardless of whether they are present as ions or molecules in solution*, is called a **total equation**. Total equations identify the reagents used to make the solutions to be mixed for a reaction.

A total equation also specifies the mole ratio in which an acid and base react. For example, sulfuric acid reacts with sodium hydroxide in 1:2 mole ratio:

$$H_2SO_4(aq) + 2NaOH(aq) \longrightarrow Na_2SO_4(aq) + 2H_2O(l)$$

The mole ratios of the reactants in solution are important in stoichiometric calculations that we will examine later in this chapter.

An *equation that specifies the major ionic species as reactants or products in an aqueous reaction* is called an **ionic equation**. We recall that a hydrochloric acid solution contains $H^+(aq)$ and $Cl^-(aq)$ ions, sodium hydroxide solution contains $Na^+(aq)$ and $OH^-(aq)$ ions, and sodium chloride solution contains $Na^+(aq)$ and $Cl^-(aq)$ ions. Thus, an ionic equation for the reaction of an HCl solution with an NaOH solution is

$$H^+(aq) + Cl^-(aq) + Na^+(aq) + OH^-(aq) \longrightarrow H_2O(l) + Na^+(aq) + Cl^-(aq)$$

Aqueous reactants and products that do not contribute ions in solution to any significant extent *are not written in ionic form in an ionic equation*. These include *insoluble solids, pure liquids, gases, nonelectrolytes, and weak electrolytes*. Thus, water consists predominantly of molecules and is therefore written in molecular form. Soluble or insoluble solids that are added to a solution for a reaction are also not written in ionic form, although they may be ionic solids (see Example 4.5).

When we consider the ionic equation for the aqueous reaction of HCl with NaOH, we see that sodium ions and chloride ions are present in the solution before and after the reaction. These ions are not changed during the reaction. They are called **spectator ions**. The spectator ions in an ionic equation are present in equal numbers on the left and right of the arrow.

The only chemical reaction that occurs when HCl and NaOH solutions are mixed is the reaction of aqueous hydrogen ions with hydroxide ions to form water. We can write an equation for a neutralization reaction in solution by omitting the spectator ions. An *ionic equation in which the spectator ions are omitted* is called a **net ionic equation**. The net ionic equation for the reaction of aqueous hydrochloric acid with aqueous sodium hydroxide is

$$H^+(aq) + OH^-(aq) \longrightarrow H_2O(l)$$

This is the net ionic equation for neutralization of *any* strong acid by *any* strong base.

Example 4.2

Writing the Total Equation, Ionic Equation, and Net Ionic Equation for a Reaction of a Strong Acid with a Strong Base

Write a total equation, ionic equation, and net ionic equation for the reaction that occurs when equal volumes of equimolar solutions (solutions of equal molarities) of nitric acid and potassium hydroxide are mixed. (Potassium ions do not react with nitrate ions.)

SOLUTION: First, we write the formulas for nitric acid and potassium hydroxide as the reactants, each followed by the notation *aq* in parentheses to indicate that it is in solution. To predict the products, we pair the cation of the base with the anion of the acid and combine hydrogen ion with hydroxide ion, thus obtaining the formulas for potassium nitrate and water. For potassium nitrate we add the notation *aq* to indicate that it is in solution, and for water we add the notation *l* for liquid state. Thus, the total equation is

$$HNO_3(aq) + KOH(aq) \longrightarrow KNO_3(aq) + H_2O(l)$$

In the ionic equation we write the formulas for all ions present before and after the reaction. Aqueous nitric acid consists of hydrogen ions and nitrate ions, potassium hydroxide solution contains potassium ions and hydroxide ions, potassium nitrate solution contains potassium ions and nitrate ions, and water consists of molecules. Thus, the ionic equation is

$$H^+(aq) + NO_3^-(aq) + K^+(aq) + OH^-(aq) \longrightarrow K^+(aq) + NO_3^-(aq) + H_2O(l)$$

Nitrate ions and potassium ions appear in equal numbers on both sides of the arrow. They are therefore the spectator ions. Omitting the spectator ions, the net ionic equation is

$$H^+(aq) + OH^-(aq) \longrightarrow H_2O(l)$$

We thus find that the net ionic equation for this reaction is the same as that for the reaction of hydrochloric acid with sodium hydroxide in solution.

Practice Problem 4.4: Write a total equation and a net ionic equation for the reaction that occurs when a solution of hydrobromic acid is mixed with a solution of cesium hydroxide. (Cesium ions and bromide ions do not react with each other.)

Next, let us consider the reaction of a diprotic acid, such as sulfuric acid, H_2SO_4, with a base. A mole of hydrogen ions neutralizes a mole of hydroxide ions to produce a mole of water. A diprotic acid such as sulfuric acid can furnish 2 mol of hydrogen ions per mole of the acid. Therefore, 2 mol of sodium hydroxide completely neutralize 1 mol of sulfuric acid. The total equation for the reaction is

$$H_2SO_4(aq) + 2NaOH(aq) \longrightarrow Na_2SO_4(aq) + 2H_2O(l)$$

Sulfuric acid is a strong acid that dissociates completely in water into hydrogen ions and hydrogen sulfate ions. The number of sulfate ions in a sulfuric acid solution is relatively small (Section 4.1). Therefore, the ionic equation for the reaction of sulfuric acid solution with sodium hydroxide solution can be written to include only the *major* ionic species—the $H^+(aq)$ and $HSO_4^-(aq)$—as the reactants in the sulfuric acid solution:

$$H^+(aq) + HSO_4^-(aq) + 2Na^+(aq) + 2OH^-(aq) \longrightarrow 2H_2O(l) + 2Na^+(aq) + SO_4^{2-}(aq)$$

This equation shows that both hydrogen ions and hydrogen sulfate ions from sulfuric acid react with the hydroxide ions from sodium hydroxide. The hydrogen sulfate ion reacts by a transfer of a proton to a hydroxide ion to form a water molecule and a sulfate ion. The spectator ions in this reaction are the sodium ions. Thus, the net ionic equation for the reaction is

$$H^+(aq) + HSO_4^-(aq) + 2OH^-(aq) \longrightarrow 2H_2O(l) + SO_4^{2-}(aq)$$

Practice Problem 4.5: Write a total equation, an ionic equation, and a net ionic equation for the reaction between potassium hydroxide solution and aqueous sulfuric acid.

Neutralization reactions of weak acids and weak bases follow the same general principles as the reactions between strong acids and bases described above. There is, however, one significant difference. The *major species in solutions of weak acids or weak bases are undissociated molecules. Thus, only the molecular forms of these substances are written in ionic equations* for their neutralization reactions. Examples 4.3 and 4.4 illustrate how to write equations for these reactions.

Example 4.3

Writing the Equations for the Reactions of a Weak Acid with a Strong Base, and a Strong Acid with a Weak Base

Write a total equation, an ionic equation, and a net ionic equation for the reaction of: **(a)** aqueous acetic acid with aqueous sodium hydroxide; **(b)** aqueous nitric acid with aqueous ammonia.

SOLUTION

(a) The total equation for the reaction is

$$HC_2H_3O_2(aq) + NaOH(aq) \longrightarrow NaC_2H_3O_2(aq) + H_2O(l)$$

In the ionic equation, we write the molecular formula for the weak acid, acetic acid. Aqueous sodium hydroxide contains sodium ions and hydroxide ions. Sodium acetate is a soluble salt that exists in solution as sodium ions and acetate ions. Thus, the ionic equation is

$$HC_2H_3O_2(aq) + Na^+(aq) + OH^-(aq) \longrightarrow Na^+(aq) + C_2H_3O_2^-(aq) + H_2O(l)$$

Sodium ions are spectator ions in this reaction, so the net ionic equation is

$$HC_2H_3O_2(aq) + OH^-(aq) \longrightarrow C_2H_3O_2^-(aq) + H_2O(l)$$

(b) For the reaction of nitric acid with ammonia, the total equation is

$$HNO_3(aq) + NH_3(aq) \longrightarrow NH_4NO_3(aq)$$

In an ionic equation, ammonia, a weak base, is written in molecular form. Nitric acid and ammonium nitrate are strong electrolytes and are therefore written in ionic form:

$$H^+(aq) + NO_3^-(aq) + NH_3(aq) \longrightarrow NH_4^+(aq) + NO_3^-(aq)$$

The only chemical change is the reaction of ammonia with hydrogen ions to form ammonium ions. Nitrate ions are spectator ions, so the net ionic equation is

$$H^+(aq) + NH_3(aq) \longrightarrow NH_4^+(aq)$$

Example 4.4

Writing the Equations for a Reaction of a Weak Acid with a Weak Base

Write the total equation, the ionic equation, and the net ionic equation for the reaction of acetic acid solution with aqueous ammonia.

SOLUTION: In this reaction, the acetic acid molecules donate protons to ammonia molecules to form aqueous ammonium acetate. Thus, the total equation for the reaction is

$$HC_2H_3O_2(aq) + NH_3(aq) \longrightarrow NH_4C_2H_3O_2(aq)$$

In the ionic equation, both acetic acid and ammonia are written in molecular form because they are weak electrolytes. Ammonium acetate is a soluble salt that consists of ions in solution and is therefore written in ionic form:

$$HC_2H_3O_2(aq) + NH_3(aq) \longrightarrow NH_4^+(aq) + C_2H_3O_2^-(aq)$$

There are no spectator ions in this reaction, and the ionic and net ionic equations are identical (and only slightly different from the total equation).

Example 4.5

Writing the Equations for Reactions of Strong Electrolytes, Weak Electrolytes, Solids, and Gases

Write total equations and net ionic equations for the reactions of: **(a)** aqueous nitric acid with aqueous barium hydroxide; **(b)** aqueous nitrous acid with aqueous ammonia; **(c)** solid calcium hydroxide with a solution of sulfuric acid (this is a violent reaction!); **(d)** hydrogen chloride gas with potassium hydroxide solution. In all parts of this problem, there is no reaction between the cations of the base and the anions of the acid.

SOLUTION: We need to know how to write formulas of common strong and weak acids and the hydroxides of representative metals. These substances are discussed in Sections 2.5 and 4.1.

(a) When we pair the cations of the base with the anions of the acid, we get barium nitrate as one of the products. Thus, the total equation is

$$2HNO_3(aq) + Ba(OH)_2(aq) \longrightarrow Ba(NO_3)_2(aq) + 2H_2O(l)$$

Nitric acid and barium hydroxide are strong electrolytes that are completely dissociated in solution. Nitrate ions are present in nitric acid solution before the reaction and in barium nitrate solution after the reaction. Similarly, barium ions are present in barium hydroxide solution before the reaction and in barium nitrate solution after the reaction. Thus, nitrate ions and barium ions are the spectator ions. Omitting the spectator ions, the net ionic equation is

$$2H^+(aq) + 2OH^-(aq) \longrightarrow 2H_2O(l)$$

We divide both sides of this equation by 2 to arrive at the net ionic equation in its final form:

$$H^+(aq) + OH^-(aq) \longrightarrow H_2O(l)$$

(b) In this reaction, protons are transferred from nitrous acid (a weak acid) to ammonia to produce aqueous ammonium nitrite as the product. The total equation is

$$HNO_2(aq) + NH_3(aq) \longrightarrow NH_4NO_2(aq)$$

In the ionic equation, both nitrous acid and ammonia are written in molecular form because they are weak electrolytes that exist essentially as molecules in water solution. However, aqueous ammonium nitrite is a strong electrolyte and exists in the form of ammonium ions and nitrite ions. Therefore, the ionic equation is

$$HNO_2(aq) + NH_3(aq) \longrightarrow NH_4^+(aq) + NO_2^-(aq)$$

There are no spectator ions in this reaction, and the ionic and net ionic equations are identical.

(c) The total equation for the reaction of sulfuric acid solution and solid calcium hydroxide is

$$H_2SO_4(aq) + Ca(OH)_2(s) \longrightarrow CaSO_4(aq) + 2H_2O(l)$$

To write an ionic equation, we recall that sulfuric acid solution consists mostly of hydrogen ions and hydrogen sulfate ions. We also recall that solid reactants that are added to a solution for a reaction are not written in ionic form. Following these guidelines, the ionic equation is

$$H^+(aq) + HSO_4^-(aq) + Ca(OH)_2(s) \longrightarrow Ca^{2+}(aq) + SO_4^{2-}(aq) + 2H_2O(l)$$

Since there are no spectator ions, the ionic and net ionic equations are the same.

(d) The total equation for the reaction of hydrogen chloride gas with potassium hydroxide solution is

$$HCl(g) + KOH(aq) \longrightarrow KCl(aq) + H_2O(l)$$

In the ionic equation, hydrogen chloride gas is written in molecular form because gases consist of molecules:

$$HCl(g) + K^+(aq) + OH^-(aq) \longrightarrow K^+(aq) + Cl^-(aq) + H_2O(l)$$

When we omit potassium ions (they are spectator ions) the net ionic equation is

$$HCl(g) + OH^-(aq) \longrightarrow H_2O(l) + Cl^-(aq)$$

Practice Problem 4.6: Write a total equation and a net ionic equation for the reaction that occurs when: **(a)** a solution of hydrofluoric acid (a weak acid) is mixed with a solution of barium hydroxide; **(b)** a solution of hydrocyanic acid (a weak acid) is added to aqueous ammonia; **(c)** solid aluminum hydroxide is added to sulfuric acid solution.

Since an Arrhenius acid–base reaction produces a salt and water, the salt can be recovered from its solution by evaporating the water. Thus, acid–base reactions can be used to synthesize salts. For example, sodium nitrate can be prepared by mixing sodium hydroxide solution with aqueous nitric acid:

$$NaOH(aq) + HNO_3(aq) \longrightarrow NaNO_3(aq) + H_2O(l)$$

Some salts are insoluble in water and precipitate as they form (see Section 4.4). These salts can be recovered by pouring the liquid from a precipitate through a semipermeable paper, called filter paper, that retains the precipitate. This process is called **filtration**. The following practice problems illustrate additional syntheses of salts.

Practice Problem 4.7: Potassium iodide is a water-soluble electrolyte that is used as an ingredient in table salt to provide the iodine needed for the human body. Magnesium nitrate is also a water-soluble salt that is used in fireworks. Write the formula and the name of the acid and the base whose solutions you would mix for a reaction to produce: **(a)** potassium iodide; **(b)** magnesium nitrate. Write the total equation and the net ionic equation for each of these reactions.

Practice Problem 4.8: What salt can be recovered when water is evaporated from the solution that results from mixing of equimolar quantities of aqueous nitrous acid and aqueous potassium hydroxide? Write the net ionic equation for this reaction.

Practice Problem 4.9: Ammonium sulfate is used in flameproofing fabrics and paper. What acid and what base solutions would you mix to prepare ammonium sulfate? Write the total and the net ionic equations for the reaction.

4.3 Acid–Base Titration

Acid–base reactions, or neutralization reactions, are commonly used to determine the concentrations of acids (or bases) in solutions. If the concentration and volume of one of the reactants in a neutralization reaction is known, the concentration of the second solution can be determined if its volume is known. This procedure requires the measurement of volumes, and is therefore called **volumetric analysis**.

Let us see how volumetric analysis is used to determine the concentration of an acid solution using a base of known concentration. The base solution is added slowly from a buret to a measured volume of the acid solution until all of the acid is neutralized (Figure 4.4). The point at which neutralization occurs is usually detected by the change in color of an organic dye such as litmus or phenolphthalein, which has one color in acidic solution and a different color in basic solution. The organic dye used for this purpose is called an **indicator**. This type of volumetric analysis is known as **titration**. Figure 4.4 illustrates a setup for an acid–base titration.

The procedure we just described, in which the concentration of one solution is determined by titration with another solution of known concentration, is called **standardization**. In a standardization procedure, the solution whose concentration is accurately known is called a **standard solution**.

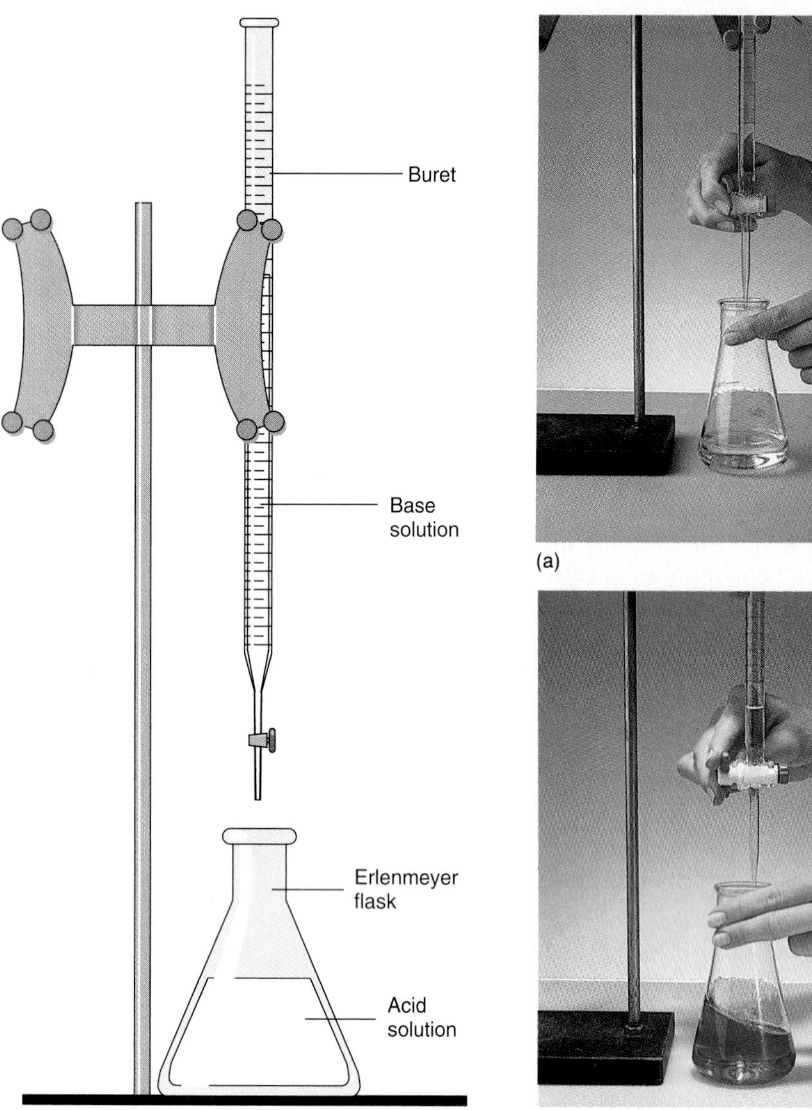

Figure 4.4

A setup for an acid-base titration.

Figure 4.5

Titration of HCl with NaOH using phenolphthalein as indicator: (a) before reaching the endpoint and (b) after reaching the endpoint.

Let us consider the standardization of a sodium hydroxide solution with a standard hydrochloric acid solution. An accurately measured volume of the standard hydrochloric acid solution is delivered with a pipet into a conical flask called an Erlenmeyer flask (Figure 4.4). A few drops of an indicator solution are also added to the flask. The sodium hydroxide solution is then slowly added from a buret into the acid solution, with swirling, until just enough of the base is added to neutralize the acid. This point is called the **neutralization point** or the **equivalence point**. The indicator changes color at or very near the neutralization point and signals the **endpoint** for the titration (Figure 4.5).

In the titration just described, the volume of the NaOH solution used in the titration is measured as the difference between the final and initial buret readings of the liquid level in the buret. From the volume and the molarity of the standard HCl solution, we can calculate the number of moles of HCl present by use of the equation

$$(\text{volume of HCl in liters})(M\ \text{HCl}) = \text{number of moles HCl}$$

The number of moles of HCl is also the number of moles of NaOH used in the titration because HCl and NaOH react in a 1:1 mole ratio. The molarity of NaOH is the number of moles of NaOH divided by the volume of NaOH solution in liters:

$$M \text{ NaOH} = \frac{\text{number of moles NaOH}}{\text{volume NaOH solution in liters}}$$

This type of calculation is illustrated in Example 4.6.

Example 4.6

Calculating the Molarity of NaOH from the Volume and Molarity of a Standard HCl Solution

A 25.00-mL sample of 0.1208 *M* standard HCl solution is titrated with 28.52 mL of NaOH solution to the endpoint. What is the molarity of the NaOH solution?

SOLUTION: We convert the volume of HCl in milliliters to liters by moving the decimal point three places to the left to obtain 0.02500 L. The number of moles of HCl is the volume of the solution (soln) in liters times the molarity:

$$0.02500 \text{ L soln} \left(\frac{0.1208 \text{ mol HCl}}{1 \text{ L soln}}\right) = 3.020 \times 10^{-3} \text{ mol HCl}$$

At the equivalence point of the titration, the number of moles of HCl equals the number of moles of NaOH because HCl reacts with NaOH in 1:1 mole ratio, as we see from the balanced equation for the titration reaction:

$$\text{HCl}(aq) + \text{NaOH}(aq) \longrightarrow \text{NaCl}(aq) + \text{H}_2\text{O}(l)$$

Therefore, the number of moles HCl = the number of moles NaOH = 3.020×10^{-3} mol. The volume of NaOH used in the titration is 28.52 mL or 0.02852 L. The molarity of NaOH is the number of moles of NaOH divided by its volume in liters:

$$M \text{ soln} = \frac{3.020 \times 10^{-3} \text{ mol NaOH}}{0.02852 \text{ L NaOH soln}}$$

$$= 0.1059 \text{ mol L}^{-1} \quad \text{or} \quad 0.1059\ M$$

The steps outlined above can be carried out on a calculator in a *single multistep operation*:

$$\frac{0.02500 \text{ L soln} \left(\frac{0.1208 \text{ mol HCl}}{1 \text{ L soln}}\right)\left(\frac{1 \text{ mol NaOH}}{1 \text{ mol HCl}}\right)}{0.02852 \text{ L soln}}$$

$$= 0.1059 \text{ mol L}^{-1} \quad \text{or} \quad 0.1059\ M$$

Practice Problem 4.10: In a titration, 50.00 mL of 0.1204 *M* hydrochloric acid requires 48.54 mL of sodium hydroxide solution for neutralization. What is the molarity of the sodium hydroxide solution?

In Example 4.6, the acid and base react in a 1:1 mole ratio. But a complete neutralization of 1 mol of a polyprotic acid requires more than 1 mol of sodium hydroxide. For example, the complete neutralization of 1 mol of sulfuric acid, a diprotic acid, requires 2 mol of sodium hydroxide:

$$H_2SO_4(aq) + 2NaOH(aq) \longrightarrow Na_2SO_4(aq) + 2H_2O(l)$$

Similarly, the complete neutralization of 1 mol of phosphoric acid, a triprotic acid, requires 3 mol of sodium hydroxide:

$$H_3PO_4(aq) + 3NaOH(aq) \longrightarrow Na_3PO_4(aq) + 3H_2O(l)$$

When an acid–base titration is carried out using a standard solution of NaOH to neutralize a polyprotic acid, the calculations resemble those outlined in Example 4.6. The only difference is in the conversion of the number of moles of the base to the number of moles of the acid. The conversion factor is the acid/base mole ratio given by the balanced equation for the reaction, as shown by Example 4.7.

Example 4.7

Calculating the Molarity of a Diprotic Acid from the Volume and Molarity of a Base

Sulfuric acid is sold by chemical supply houses in an approximately 18 *M* solution. In an experiment, the 18 *M* stock solution is diluted to approximately 0.3 *M*. A 25.00-mL aliquot (or portion) of this diluted solution of sulfuric acid requires 32.58 mL of 0.5000 *M* sodium hydroxide for a complete neutralization. What is the molarity of the sulfuric acid solution?

SOLUTION: We follow the general method outlined in Example 4.6. The only difference is in the conversion of the number of moles of base to the number of moles of acid. From the balanced equation for the reaction we see that 1 mol of the acid reacts with 2 mol of the base:

$$H_2SO_4(aq) + 2NaOH(aq) \rightarrow Na_2SO_4(aq) + 2H_2O(l)$$

Thus, the molarity of H_2SO_4 solution can be obtained by the following multistep operation:

$$\frac{0.03258 \text{ L soln} \left(\frac{0.5000 \text{ mol NaOH}}{1 \text{ L soln}}\right)\left(\frac{1 \text{ mol } H_2SO_4}{2 \text{ mol NaOH}}\right)}{0.02500 \text{ L soln}}$$

$$= 0.3258 \text{ mol } H_2SO_4/\text{L soln} \quad \text{or} \quad 0.3258\ M$$

Practice Problem 4.11: **(a)** How many milliliters of 0.2056 *M* sodium hydroxide is required to completely neutralize 25.00 mL of 0.1000 *M* phosphoric acid solution? **(b)** How many milliliters of 0.600 *M* H_3PO_4 is completely neutralized by 37.0 g of solid KOH?

Practice Problem 4.12: Hydrochloric acid is an ingredient of gastric juice. A 12.58-mL sample of gastric juice is neutralized with 15.72 mL of 0.04209 *M* barium hydroxide solution. What is the molarity of hydrochloric acid in gastric juice? How many milliliters of 0.06800 *M* barium hydroxide solution would be required to neutralize all the hydrochloric acid in the same sample of gastric juice?

As we have seen, a solution of a base can be standardized with a solution of a standard acid. But where do we get the standard acid? Several acids are available in a pure state. One that is frequently used to standardize base solutions is potassium hydrogen phthalate, $KHC_8H_4O_4$, abbreviated KHPh. Pure KHPh is a salt that dissolves in water to form a solution of potassium ions and hydrogen phthalate ions, $HC_8H_4O_4^-$:

$$KHC_8H_4O_4(s) \longrightarrow K^+(aq) + HC_8H_4O_4^-(aq)$$

Hydrogen phthalate ion is a weak acid that reacts with a hydroxide ion in a titration to produce a phthalate ion, $C_8H_4O_4^{2-}$, and water:

$$HC_8H_4O_4^-(aq) + OH^-(aq) \longrightarrow C_8H_4O_4^{2-}(aq) + H_2O(l)$$

We will use the abbreviations HPh^- and Ph^{2-} to represent hydrogen phthalate ions and phthalate ions, respectively. The overall ionic equation for the reaction of aqueous potassium hydrogen phthalate with aqueous sodium hydroxide is

$$K^+(aq) + HPh^-(aq) + Na^+(aq) + OH^-(aq) \longrightarrow K^+(aq) + Na^+(aq) + Ph^{2-}(aq) + H_2O(l)$$

and the net ionic equation is

$$HPh^-(aq) + OH^-(aq) \longrightarrow Ph^{2-}(aq) + H_2O(l)$$

The equations above show that KHPh(aq) and NaOH(aq) react in 1:1 mole ratio. Thus, the number of moles of a base used in a titration of a KHPh solution equals the number of moles of the KHPh neutralized in the titration.

To standardize an NaOH solution with KHPh, a known mass of KHPh is dissolved in water, and an indicator is added to the solution. This solution is then titrated with the NaOH solution to the endpoint signaled by the color change of the indicator. The molarity of the NaOH solution can be calculated using the following sequence of conversions:

$$\text{mass KHPh} \longrightarrow \text{moles KHPh} \longrightarrow \text{moles NaOH} \longrightarrow M\ \text{NaOH}$$

Experimental procedures in this process are illustrated in Figure 4.6, and calculation of the molarity of NaOH from experimental data is shown in Example 4.8.

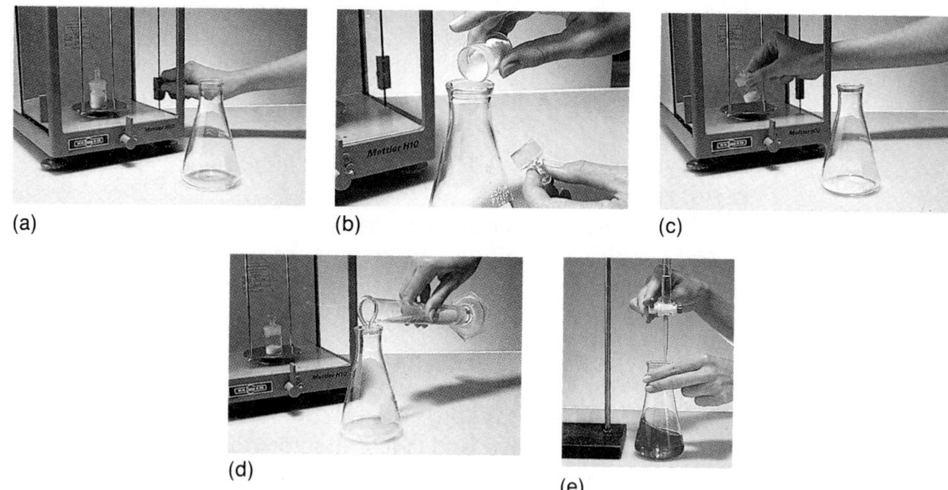

(a) (b) (c) (d) (e)

Figure 4.6

Sequence of steps involved in a standardization of a base solution using KHPh as a standard acid: (a) weighing a weighing bottle with KHPh; (b) tapping some KHPh from weighing bottle into an Erlenmeyer flask; (c) weighing the weighing bottle with the remaining KHPh; (d) the solution of KHPh with added phenolphthalein indicator; (e) the solution of KHPh titrated to the endpoint.

Example 4.8

Calculating the Molarity of NaOH from the Mass and Molar Mass of Potassium Hydrogen Phthalate

In a standardization of an NaOH solution, 0.7284 g of KHPh is dissolved in water. This KHPh solution requires 38.58 mL of the NaOH solution for neutralization. Calculate the molarity of the sodium hydroxide solution.

SOLUTION: The molar mass of KHPh to four significant figures is 204.2 g mol^{-1}. Using this molar mass, we convert the given mass of the sample of KHPh to the number of moles of KHPh. We know from the balanced equation for the titration reaction that KHPh reacts with NaOH in 1:1 mole ratio. Therefore, the number of moles KHPh = the number of moles NaOH. The molarity of NaOH is the number of moles of NaOH divided by the volume in liters of its solution used in the titration. This sequence of steps can be summarized as follows:

$$\text{mass KHPh} \longrightarrow \text{moles KHPh} \longrightarrow \text{moles NaOH} \longrightarrow M\ \text{NaOH}$$

Carrying out these steps yields

$$\frac{0.7284\ \text{g KHPh}\left(\dfrac{1\ \text{mol KHPh}}{204.2\ \text{g KHPh}}\right)\left(\dfrac{1\ \text{mol NaOH}}{1\ \text{mol KHPh}}\right)}{0.03858\ \text{L}}$$

$$= 0.09246\ \text{mol L}^{-1} \quad \text{or} \quad 0.09246\ M$$

Practice Problem 4.13: In a titration, 36.58 mL of 0.1208 *M* sodium hydroxide is required to neutralize a solution of potassium hydrogen phthalate. What mass of potassium hydrogen phthalate was in the solution?

4.4 Precipitation Reactions

A *reaction in which an insoluble solid forms* when two aqueous solutions are mixed is called a **precipitation reaction**. In a precipitation reaction, the ions of the two reactants exchange partners and form a combination of ions that precipitates as an insoluble salt. Thus, precipitation reactions fall into the category of double displacement reactions we discussed in Chapter 2. A precipitated salt is "insoluble" in water. But no salt is absolutely insoluble. The term "insoluble" generally refers to a "sparingly soluble" solid that dissolves to an extent less than about 0.1 g per 100 mL of water.

Precipitation reactions can be used to prepare insoluble salts. They can also be used to precipitate certain ions from a solution by other ions that form an insoluble salt with the ions in the solution. For example, barium ions can be precipitated from a solution by sulfate ions as barium sulfate:

$$Ba^{2+}(aq) + SO_4^{2-}(aq) \longrightarrow BaSO_4(s)$$

The ions precipitated from a solution can be quantitatively determined by weighing the precipitate and calculating the ion content in the precipitate. This procedure is a branch of quantitative analysis known as **gravimetric analysis**, so called because a precipitate is weighed after it has been isolated from the solution.

We can also use precipitation reactions to identify ions in solution, a technique called **qualitative analysis**, which is a branch of analytical chemistry dealing with the identification of the components present in a compound or a mixture. We consider some specific applications of precipitation reactions later in this section.

Equations for Precipitation Reactions

When we mix solutions of sodium nitrate and calcium chloride, no precipitate forms. However, when we mix a solution of silver nitrate with a solution of calcium chloride, a white precipitate appears (Figure 4.7).

To explain these observations, let us consider the composition of the solutions being mixed. Sodium nitrate solution contains $Na^+(aq)$ and $NO_3^-(aq)$ ions, and calcium chloride solution contains $Ca^{2+}(aq)$ and $Cl^-(aq)$ ions. No precipitate forms when these two solutions are mixed because no combination of the cations and the anions in this solution can produce a salt that is insoluble in water.

Silver nitrate solution contains $Ag^+(aq)$ ions and $NO_3^-\ (aq)$ ions. The combination of $Ag^+(aq)$ ions from $AgNO_3$ and $Cl^-(aq)$ ions from $CaCl_2$ gives a white precipitate of $AgCl(s)$, which we know is insoluble in water. The total equation for the reaction that occurs when we mix solutions of silver nitrate and calcium chloride is

$$2AgNO_3(aq) + CaCl_2(aq) \longrightarrow 2AgCl(s) + Ca(NO_3)_2(aq)$$

The ionic equation for the same reaction is

$$2Ag^+(aq) + 2NO_3^-(aq) + Ca^{2+}(aq) + 2Cl^-(aq) \longrightarrow 2AgCl(s) + Ca^{2+}(aq) + 2NO_3^-(aq)$$

The net ionic equation is

$$Ag^+(aq) + Cl^-(aq) \longrightarrow AgCl(s)$$

The calcium ions and nitrate ions remain in solution as spectator ions. This net ionic equation reveals that when we mix any soluble silver salt solution with a solution containing chloride ions, silver chloride will precipitate.

Dozens of other reactions result in the formation of precipitates. For example, when we mix solutions of potassium sulfate and barium chloride, a white precipitate forms. This precipitate is barium sulfate (Figure 4.8). This precipitation reaction has the following total equation:

$$BaCl_2(aq) + K_2SO_4(aq) \longrightarrow BaSO_4(s) + 2KCl(aq)$$

and the net ionic equation is

$$Ba^{2+}(aq) + SO_4^{2-}(aq) \longrightarrow BaSO_4(s)$$

(a)

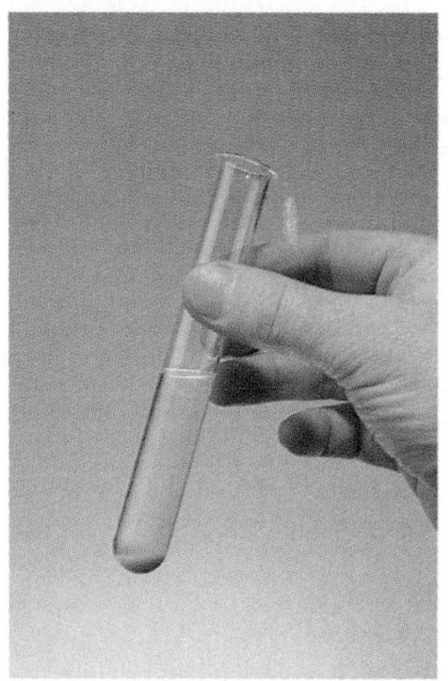

(b)

Figure 4.7

Precipitation of silver chloride: (a) adding a solution of chloride ions to a solution of silver ions and (b) a precipitate of silver chloride.

Practice Problem 4.14: Lead(II) salts in aqueous solution contain Pb^{2+} ions. When a solution of hydrochloric acid is added to a solution of lead(II) nitrate, a white precipitate of lead(II) chloride forms. Write a total equation and a net ionic equation for this reaction.

(a)

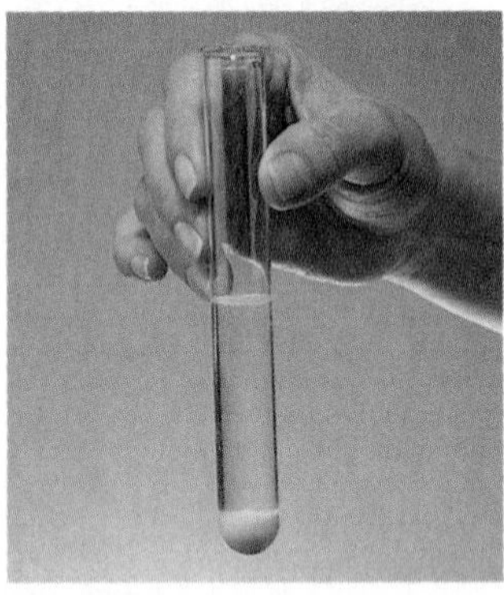

(b)

Figure 4.8

Precipitation of barium sulfate: (a) adding potassium sulfate solution to a solution of barium chloride and (b) a precipitate of barium sulfate.

Solubility Rules

It is difficult for a beginning student to predict which salts are soluble or insoluble in water. But it is important in qualitative and quantitative analysis to know the solubilities of some common groups of compounds. General guidelines about the solubilities of some common compounds are listed below:

Soluble Compounds

1. Most salts and bases of the *alkali metals* and *ammonium* ions, as for example, KCl, $(NH_4)_2SO_4$, and NaOH.
2. Most metal *nitrates*, *acetates*, and *chlorates*.
3. Most metal *chlorides*, *bromides*, and *iodides*, *except* those of Ag^+, Hg_2^{2+}, and Pb^{2+} ($HgCl_2$, $HgBr_2$, and HgI_2 are slightly soluble). Thus, $Pb(NO_3)_2$ is soluble, but $PbCl_2$ is not.
4. Most metal *sulfates*, *except* those of Ba^{2+}, Sr^{2+}, and Pb^{2+} ($CaSO_4$ and Ag_2SO_4 are slightly soluble).

Insoluble Compounds

1. Most metal *carbonates*, *phosphates*, and *sulfides*, *except* those of the alkali metals, and ammonium ion.
2. Most metal *hydroxides*, *except* those of the alkali metals [$Ca(OH)_2$, $Sr(OH)_2$, and $Ba(OH)_2$ are slightly soluble].

These rules are only general guidelines. Many exceptions and borderline cases exist. *Some* borderline cases such as $HgCl_2$ and Ag_2SO_4 are listed above in parentheses.

The solubilities of many compounds in water and in other solvents have been experimentally determined. Table 4.4 gives the solubility data for some commonly known compounds. The solubility rules can be used to predict the products of many precipitation reactions as shown in Example 4.9.

Table 4.4

Solubilities of Some Common Ionic Compounds in Water at 25°C[a]

	NO_3^-, $C_2H_3O_2^-$, ClO_3^-	Cl^-, Br^-, I^-	SO_4^{2-}	CO_3^{2-}, $C_2O_4^{2-}$	OH^-	S^{2-}
Na^+, K^+, Rb^+, Cs^+, NH_4^+	s	s	s	s	s	s
Mg^{2+}	s	s	s	i	i	d
Ca^{2+}	s	s	sl s	i	i	d
Sr^{2+}	s	s	i	i	i	i
Ba^{2+}	s	s	i	i	sl s	d
Al^{3+}	s	s	s	i[b]	i	d
Pb^{2+}	s	i	i	i	i	i
Ag^+	s	i	sl s	i	i	i
Hg_2^{2+}	s	i	sl s	i	—	—
Hg^{2+}	s	sl s	sl s	i	i	i

[a] s, soluble; sl s, slightly soluble; i, insoluble; d, decomposes in or reacts with water; —, compound does not exist.

[b] $Al_2(CO_3)_3$ does not exist.

Example 4.9

Predicting Products and Writing Net Ionic Equations for Precipitation Reactions

Predict the compound, if any, that precipitates when each of the following pairs of aqueous solutions are mixed and write the net ionic equation for the reaction: **(a)** sodium phosphate and calcium nitrate; **(b)** magnesium chloride and lead(II) chlorate; **(c)** barium nitrate and sodium carbonate; **(d)** potassium acetate and strontium nitrate.

SOLUTION

(a) A solution of sodium phosphate contains sodium ions and phosphate ions, and a solution of calcium nitrate contains calcium ions and nitrate ions. The two possible combinations of cations and anions that can produce a precipitate upon mixing of these solutions are sodium ions with nitrate ions and calcium ions with phosphate ions. According to the solubility rules, alkali metal nitrates are soluble, but calcium phosphate is insoluble and therefore precipitates. The net ionic equation for the reaction is

$$3Ca^{2+}(aq) + 2PO_4^{3-}(aq) \longrightarrow Ca_3(PO_4)_2(s)$$

(b) The two possible combinations of ions that might produce a precipitate are magnesium ions with chlorate ions and lead ions with chloride ions. According to the solubility rules, most chlorates are soluble and most chlorides are soluble, except lead chloride, silver chloride, and mercury(I) chloride. Thus, lead chloride precipitates, and the net ionic equation for the reaction is

$$Pb^{2+}(aq) + 2Cl^-(aq) \longrightarrow PbCl_2(s)$$

(c) Following similar reasoning, we find that sodium ions do not combine with nitrate ions because sodium nitrate is soluble, but barium ions combine with carbonate ions to form insoluble barium carbonate:

$$Ba^{2+}(aq) + CO_3^{2-}(aq) \longrightarrow BaCO_3(s)$$

(d) In this case, we see that neither of the possible combinations of ions—potassium ions with nitrate ions and strontium ions with acetate ions—can occur because the products are soluble. Therefore, no reaction occurs.

Practice Problem 4.15: Write a net ionic equation for the reaction that occurs when each of the following pairs of solutions are mixed: **(a)** hydrochloric acid and mercury(I) nitrate; **(b)** magnesium chloride and sodium hydroxide.

APPLICATIONS OF CHEMISTRY 4.1
Adolf Hitler's Diaries: A Modern Con Game

In the early 1980s, publishers of *Stern,* a West German weekly magazine, paid about 4 million dollars for what they believed were Adolf Hitler's diaries. *Stern* purchased the 64 volumes of diaries after handwriting experts had analyzed the documents for possible forgery. The forged diaries appeared so genuine, however, that they fooled those experts. Chemical analysis is another method that can be used to test for authenticity, and chemical tests showed that the diaries were fraudulent. The chemical analysis of the diaries relied on simple chemical tests. As you learned in this section, most chlorides are soluble except lead chloride, mercury(I) chloride, and silver chloride. This simple solubility rule was used in one of the tests performed on the Hitler diaries.

The ink used in the Hitler diaries contained chloride ions. When you apply such an ink to paper, the chloride ions migrate through the paper away from the original ink lines for about two years. The maximum migration distance is only about 3 mm from the original point of ink application.

The pages of the Hitler diaries were placed into a silver nitrate solution to precipitate the chloride ions. Since silver chloride is insoluble, it does not wash off the paper when the paper is placed in the silver nitrate solution. When freshly made, silver chloride is white. When exposed to light, the white silver chloride turns black as it decomposes into silver atoms and chlorine molecules. (All silver halides—chloride, bromide, and iodide—do this. We take advantage of this when we use silver halides in photographic film.)

Using a microscope, chemists carefully examined the lines of ink in the diaries that were treated with silver nitrate. They saw small black specks of silver surrounding the original ink lines. All black specks of silver were less than 3 mm away from the original ink lines. Since it takes two years for the chloride ions to migrate 3 mm away from the original ink lines, the diaries that the chemists examined had to be less than two years old. Hitler died in 1945; therefore, the diaries should have been several decades old and the chloride ions in the ink would have migrated the 3-mm distance from the ink lines.

Think what might have happened without those chemical tests. Historians would have accepted the diaries as historical documents, and researchers would have wasted countless hours studying them. Your history courses would contain incorrect information and interpretations.

Source: Elizabeth and Henry Urrows, *Chem Matters*, October 1989, pages 13–15.

Precipitation Reactions in Qualitative Analysis and Syntheses of Compounds

Many laboratory experiments require the identification of the ions present in a solution. We can use our knowledge of precipitation reactions and solubility rules to identify the ions in a given solution, as shown in Example 4.10.

Example 4.10

Identifying Ions in Solution

A student prepared sodium chloride, sodium carbonate, and sodium nitrate solutions in three different bottles. The student went to lunch and forgot to label the bottles. All three of the solutions are colorless. How can these solutions be identified using only silver nitrate and barium chloride solutions as test reagents?

SOLUTION: We add a small quantity of each test reagent to a sample of each of the unknown solutions. Assume that the first randomly chosen bottle is the sodium chloride solution. Since silver chloride is insoluble, adding a few drops of silver nitrate solution produces a white precipitate of silver chloride:

$$Ag^+(aq) + Cl^-(aq) \longrightarrow AgCl(s)$$

Adding barium chloride solution to the sodium chloride solution does not produce a precipitate. That is, the ions mix without a reaction. Thus, the solution is sodium chloride.

If the solution tested is sodium carbonate, adding silver nitrate solution precipitates silver carbonate:

$$2Ag^+(aq) + CO_3^{2-}(aq) \longrightarrow Ag_2CO_3(s)$$

Adding barium chloride solution precipitates barium carbonate:

$$Ba^{2+}(aq) + CO_3^{2-}(aq) \longrightarrow BaCO_3(s)$$

Thus, if a precipitate is obtained with both of the test reagents, the solution is sodium carbonate.

Sodium nitrate solution gives no precipitate with either silver nitrate or barium chloride because no combination of ions yields an insoluble salt. Thus, if no precipitate is observed, the solution is sodium nitrate.

Practice Problem 4.16: A colorless solution is either potassium chloride or lead(II) nitrate. What reagent would you use to identify the solution? Write the net ionic equation for the reaction that occurs when you add the reagent, and explain how the reaction confirms your conclusion.

Precipitation reactions can also be used to separate ions from a mixture of solutions, as illustrated in Example 4.11.

Example 4.11

Separating Ions from an Aqueous Mixture

A solution is an aqueous mixture of magnesium nitrate and lead(II) nitrate. How can the magnesium ions be separated from the lead(II) ions in this solution using only hydrochloric acid and sodium hydroxide solutions? Consult Table 4.4 if necessary. Write a net ionic equation for each reaction.

SOLUTION: According to Table 4.4, magnesium chloride is soluble and magnesium hydroxide is insoluble. Lead(II) chloride and lead(II) hydroxide are both insoluble. Adding hydrochloric acid to the mixture of magnesium nitrate and lead(II) nitrate precipitates lead(II) chloride and thus removes the lead(II) ions from the solution:

$$Pb^{2+}(aq) + 2Cl^-(aq) \longrightarrow PbCl_2(s)$$

The lead(II) chloride precipitate can be removed from the solution by filtration. Sodium hydroxide solution is then added to the remaining solution. The hydroxide ions in the sodium hydroxide first neutralize any excess acid added in the first step:

$$H^+(aq) + OH^-(aq) \longrightarrow H_2O(l)$$

and more OH^- ions precipitate the magnesium ions in the solution as magnesium hydroxide:

$$Mg^{2+}(aq) + 2OH^-(aq) \longrightarrow Mg(OH)_2(s)$$

Precipitation reactions can be used to prepare inorganic compounds that are insoluble in water or in other solvents. For example, silver chloride can be prepared by mixing a solution containing silver ions with a solution containing chloride ions. The silver chloride precipitate is collected by filtration, dried, and stored in a suitable container. Similarly, barium sulfate can be prepared by mixing solutions containing barium ions and sulfate ions.

Practice Problem 4.17: Write the names and formulas for two compounds you would dissolve in water to make solutions which, upon mixing, would produce each of the following compounds as a precipitate: **(a)** lead(II) sulfate; **(b)** calcium phosphate; **(c)** silver sulfide. Write a net ionic equation for each of these precipitation reactions.

When we synthesize a compound by a precipitation reaction, we can determine the mass of the precipitate if we know the number of moles of each of the reactants in solution. The limiting reactant is determined by using the balanced equation for the reaction. Once the limiting reactant and its quantity are known, the mass of the product can be calculated as described in Chapter 3 and illustrated in Example 4.12.

Example 4.12

Calculating the Mass of a Precipitate

Lead(II) chloride is a white solid that was used in paint pigments; it is also a component in some types of solder. What mass of lead(II) chloride precipitates when 500 mL of a 2.00 *M* solution of lead(II) nitrate is mixed with 450 mL of a 2.10 *M* solution of sodium chloride? Assume that the precipitate is 100 percent insoluble.

SOLUTION: The total equation for the reaction is

$$Pb(NO_3)_2(aq) + 2NaCl(aq) \longrightarrow PbCl_2(s) + 2NaNO_3(aq)$$

First we determine which of the two reactants is the limiting reactant. We do this by calculating which reactant would produce the lower number of moles of the product, assuming that an excess of the other reactant is present:

$$0.500 \text{ L soln} \left(\frac{2.00 \text{ mol Pb(NO}_3)_2}{1 \text{ L soln}}\right)\left(\frac{1 \text{ mol PbCl}_2}{1 \text{ mol Pb(NO}_3)_2}\right) = 1.00 \text{ mol PbCl}_2$$

$$0.450 \text{ L soln} \left(\frac{2.10 \text{ mol NaCl}}{1 \text{ L soln}}\right)\left(\frac{1 \text{ mol PbCl}_2}{2 \text{ mol NaCl}}\right) = 0.472 \text{ mol PbCl}_2$$

Since NaCl gives the lower number of moles of $PbCl_2$, NaCl is the limiting reactant. The mass of $PbCl_2$ that can be obtained is

$$0.472 \text{ mol PbCl}_2 \left(\frac{278 \text{ g PbCl}_2}{1 \text{ mol PbCl}_2}\right) = 131 \text{ g PbCl}_2$$

Practice Problem 4.18: **(a)** How many grams of $PbCl_2$ is produced by the reaction of 40.0 mL of 0.100 *M* KCl with excess $Pb(NO_3)_2$ solution, assuming that $PbCl_2$ is 100 percent insoluble? **(b)** Barium sulfate is used in x-ray diagnosis. What volume of 0.500 *M* barium chloride solution is needed to precipitate 2.84 g of barium sulfate from a solution containing excess sulfate ions?

Gravimetric Analysis

In gravimetric analysis, a precipitate is filtered, dried, and weighed. From the mass of the precipitate obtained, the mass of a component can be calculated. Example 4.13 illustrates how the silver content in a coin can be determined by this method.

Example 4.13

An Application of a Precipitation Reaction in Gravimetric Analysis

A 0.528-g sample of a silver-containing coin is dissolved in an acid. The solution is treated with excess sodium chloride solution to precipitate the silver ions as silver chloride. After filtering and drying, the precipitate weighs 0.288 g. What is the percent silver in the coin?

SOLUTION: The net ionic equation for the precipitation reaction is

$$Ag^+(aq) + Cl^-(aq) \longrightarrow AgCl(s)$$

This equation shows that 1 mol of silver ions is removed from the solution per mol of silver chloride formed. Based on the 1:1 mole ratio of AgCl to Ag, we can calculate the mass of silver removed from the solution. To do this, we convert the mass of silver chloride obtained to moles of silver chloride, the moles of silver chloride to moles of silver, and the moles of silver to mass of silver:

$$0.288 \text{ g AgCl} \left(\frac{1 \text{ mol AgCl}}{144 \text{ g AgCl}}\right)\left(\frac{1 \text{ mol Ag}}{1 \text{ mol AgCl}}\right)\left(\frac{108 \text{ g Ag}}{1 \text{ mol Ag}}\right) = 0.216 \text{ g Ag}$$

Since the mass of an electron is negligibly small, the mass of a silver ion equals the mass of a silver atom. The percent of silver in the coin is

$$\text{percent Ag} = \frac{0.216 \text{ g Ag}}{0.528 \text{ g sample}} \times 100 = 40.9\%$$

Practice Problem 4.19: Barium carbonate is an ingredient in rat poison. A 2.582-g sample of a rat poison is dissolved in an acid. This solution is treated with a solution of excess sodium sulfate to precipitate the barium ions as barium sulfate. After filtering and drying, the precipitate has a mass of 1.018 g. What is the percent of barium carbonate in the rat poison? How many milliliters of 0.1052 *M* sodium sulfate solution is needed to precipitate the barium ions from the solution of the rat poison?

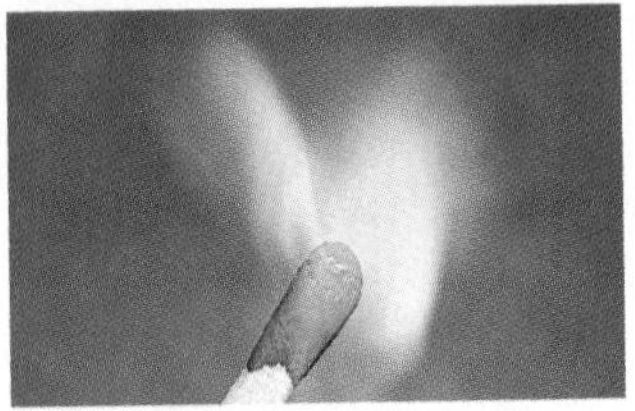
Burning of a match is an example of a redox reaction.

Physiological processes of respiration and metabolism are accompanied by a series of redox reactions.

4.5 Oxidation–Reduction Reactions

Place a shiny new iron nail outside a windowsill; after a few days, the nail will have a brownish color. Strike a match; within a few seconds the stem is charred and smoke drifts to the ceiling. Take a deep breath, oxygen enters your lungs; breathe out, carbon dioxide leaves. These different processes belong to a class of reactions known as **oxidation–reduction**, or **redox** reactions. The first is an example of corrosion, the second is combustion, and the third is related to human respiration. Some of the reactions we discussed in Chapter 2 also fall under the heading of redox reactions. These include the combination reactions of metals with nonmetals, many decomposition reactions, and single displacement reactions as well.

Let us first approach redox reactions by comparing them with acid–base reactions. We recall that Brønsted acid–base reactions occur by *transfer of protons* from the acid to the base. Most oxidation–reduction reactions occur by *transfer of electrons from one reactant to another*. For example, in a combination reaction of calcium with fluorine,

$$Ca(s) + F_2(g) \longrightarrow CaF_2(s)$$

electrons are transferred from calcium to fluorine.

An example of a redox reaction that is also a single displacement reaction is the displacement of hydrogen in an acid by an active metal. For example, when magnesium reacts with aqueous hydrochloric acid, hydrogen gas is evolved, and the magnesium metal is converted to Mg^{2+} ions:

$$Mg(s) + 2HCl(aq) \longrightarrow MgCl_2(aq) + H_2(g)$$

In this reaction, electrons are transferred from Mg atoms to $H^+(aq)$ ions in solution. The Mg atoms convert to Mg^{2+} ions in solution, and the $H^+(aq)$ ions convert to H_2 gas. The net ionic equation for this reaction is

$$Mg(s) + 2H^+(aq) \longrightarrow Mg^{2+}(aq) + H_2(g)$$

Oxidation States

The examples of redox reactions we have considered thus far have involved ions or ionic compounds as reactants or products. But there are redox reactions in which neither the reactants nor products are ions or ionic compounds. An example of such a redox reaction is the combination of two different nonmetals to form a binary compound that does not contain ions because the atoms are covalently bonded. However, since the component elements in the product are not identical, one has a greater share of the bonding electrons than the other. Such an element can be said to have an *apparent* negative charge, and the other element is said to have an apparent positive charge. *The real or apparent charge on an atom, assigned according to a set of rules*, is called the **oxidation state**, or the **oxidation number**, of the element.

Oxidation states are assigned to atoms according to the following set of rules:

1. The oxidation state of an atom *in its elemental form is 0*. Thus, atoms of monatomic helium, He, the atoms in H_2, and P_4 molecules all have oxidation states of 0.

2. The *oxidation state of a monatomic ion equals the charge on the ion.* For example, the oxidation state of sodium in the form of an Na^+ ion is +1. Similarly, in the binary ionic compound Na_2S, the oxidation states of Na and S are +1 and −2, respectively.
3. *Oxygen* has an oxidation state of *−2 in most of its compounds, except in peroxides, in which it is* −1. Thus, the oxidation state of oxygen is −2 in H_2O and in H_2SO_4 but −1 in H_2O_2 and in Na_2O_2.
4. *Hydrogen* has an oxidation state of *+1 in most of its compounds, except in binary compounds with metals, in which it is −1.* Thus, in H_2O, HNO_3, and NaOH, the oxidation state of hydrogen is +1, but in NaH and AlH_3, it is −1.
5. The *sum of the oxidation states of all the atoms in a neutral compound equals 0. In a polyatomic ion, the sum of oxidation states equals the charge on the ion.*
6. In a *binary compound of two nonmetals* (excluding the noble gases), the *element that lies above or to the right* of the other element in the periodic table *has the negative oxidation state.* The magnitude of this negative oxidation state equals the charge this element would have as an ion in a binary ionic compound. For example, the oxidation state of Cl in PCl_3 is −1. The oxidation state of P must therefore be +3 for the compound to be neutral. In IF_5, fluorine has an oxidation state of −1, and iodine +5.

If the oxidation states for all but one of the atoms in a compound or a polyatomic ion are known, the unknown oxidation state can be determined, as shown in Examples 4.14 to 4.16.

Example 4.14

Determining the Oxidation State of an Atom in a Neutral Compound

What is the oxidation state of sulfur in sulfuric acid?

SOLUTION: We know from the rules above that the oxidation states of H and O in H_2SO_4 are +1 and −2, respectively. To determine the oxidation state of S, we assign it the value of x. Since H_2SO_4 is a neutral compound, the sum of all the oxidation states is zero, and the value of x can be calculated from the equation: $x + 2(+1) + 4(-2) = 0$. Solving this equation, we find that $x = +6$, which is the oxidation state for S in H_2SO_4.

Example 4.15

Determining the Oxidation State of an Atom in a Polyatomic Ion

What is the oxidation state of phosphorus in a pyrophosphate, $P_2O_7^{4-}$, ion?

SOLUTION: The oxidation state of oxygen is −2. The sum of the oxidation states of all the atoms in the ion must equal the charge on the ion, that is, −4. Letting the oxidation state of phosphorus be x, we can write $2x + 7(-2) = -4$, from which $x = +5$. Thus, the oxidation state of P in $P_2O_7^{4-}$ is +5.

Example 4.16

Determining the Oxidation States of Atoms in an Ionic Compound Consisting of Polyatomic Ions

What are the oxidation states of nitrogen and of sulfur in ammonium sulfate, $(NH_4)_2SO_4$?

SOLUTION: This compound consists of NH_4^+ and SO_4^{2-} ions. The oxidation states of N and S in these ions are the same as in the neutral compound made of these ions. In an NH_4^+ ion, the oxidation state of H is +1 (Rule 4). Letting x be the oxidation state of N, we can write $x + 4(+1) = +1$, from which $x = -3$.

In an SO_4^{2-} ion, the oxidation state of O is −2. If y is the oxidation state of S, $y + 4(-2) = -2$, from which $y = +6$.

Practice Problem 4.20: What is the oxidation state of: **(a)** S in Na_2SO_3; **(b)** P in PF_3; **(c)** Cr in CrO_3; **(d)** Cr in $Cr_2O_7^{2-}$; **(e)** P in $(NH_4)_3PO_4$?

Oxidation and Reduction

When a reactant loses electrons in a redox reaction, its oxidation state increases and it is **oxidized**. *When a reactant gains electrons, its oxidation state decreases* and it is **reduced**. The processes of oxidation and reduction occur simultaneously. For example, magnesium reacts with chlorine to give magnesium chloride:

$$Mg(s) + Cl_2(g) \longrightarrow MgCl_2(s)$$

In this reaction, magnesium loses electrons, and its oxidation state increases from 0 in the elemental state to +2 in the product, $MgCl_2$. Magnesium is therefore *oxidized*. Chlorine simultaneously gains electrons and is *reduced* from an oxidation state of 0 in the element to −1 state in $MgCl_2$. In all redox reactions, the total number of electrons lost by the species that is oxidized equals the total number gained by the species that is reduced.

A *substance that causes the oxidation of another reactant* is called an **oxidizing agent**. A *reactant that causes the reduction of another substance* is a **reducing agent**. Therefore, in a redox reaction, the *oxidizing agent is reduced*, and the *reducing agent is oxidized*. In the reaction of magnesium with chlorine, chlorine is the oxidizing agent because it oxidizes magnesium (it is itself reduced in the process). Magnesium is the reducing agent because it reduces chlorine (it is itself oxidized in the process).

Balancing Redox Equations by the Oxidation State Method

Oxidation states can be used to balance equations of redox reactions. The "oxidation state method" or the "oxidation number method" of balancing equations for redox reactions is based on the principle that the gain in the oxidation number in one reactant equals the loss in the oxidation number of the other reactant. In other words, the number of electrons lost by one reactant equals the number of electrons gained by another reactant. This method of balancing equations is outlined below.

Step 1. Determine the increase in the oxidation number by the oxidized species and the decrease in the oxidation number by the reduced species.

Step 2. Balance the gain and loss in the oxidation numbers by assigning coefficients to the reactant side of the equation.

Step 3. Assign coefficients to the products to balance the equation.

Example 4.17 illustrates an application of these steps.

Example 4.17

Balancing a Redox Equation by the Oxidation State Method and Determining the Oxidizing and Reducing Agents

Balance the equation

$$NH_3 + O_2 \longrightarrow N_2 + H_2O$$

by the oxidation state method, and indicate which reactant is the oxidizing agent and which is the reducing agent.

SOLUTION: To balance this equation, we follow the steps outlined above.

Step 1. The oxidation number of nitrogen increases by 3, from −3 in NH_3 to 0 in N_2. Nitrogen is therefore oxidized. The oxidation number of oxygen decreases by 2 from 0 in O_2 to −2 in H_2O. Oxygen is thus reduced as shown below.

Increase in the oxidation number of an N atom by 3

$$\overset{-3}{N}H_3 + \overset{0}{O}_2 \longrightarrow \overset{0}{N}_2 + H_2\overset{-2}{O}$$

Decrease in the oxidation number of an O atom by 2 and of two O atoms in an O_2 molecule by 4

(The oxidation state of hydrogen does not change; it is +1 in NH_3 and +1 in H_2O.)

Step 2. To balance the gain of three units and the loss of four units in the oxidation numbers, we assign coefficients of 4 to NH_3 and 3 to O_2 to obtain

$$4NH_3 + 3O_2 \longrightarrow N_2 + H_2O$$

There is a total oxidation number loss of 12 by 4 N atoms, and a total oxidation number gain of 12 by six O atoms in three O_2 molecules.

Step 3. Having balanced the loss and gain in the oxidation numbers in the reactants, we next assign coefficients to the products to balance the equation. We count four nitrogen atoms on the left of the arrow, but only two on the right. A coefficient 2 for N_2 on the right balances the nitrogen atoms. There are six oxygen atoms on the left, but only one on the right. A coefficient 6 for H_2O on the right balances the oxygen atoms. Now there are 12 hydrogen atoms on both sides of the arrow, and the equation is balanced:

$$4NH_3 + 3O_2 \longrightarrow 2N_2 + 6H_2O$$

Practice Problem 4.21: Using the oxidation state method, balance the equation

$$HNO_3 + H_2S \longrightarrow NO + S + H_2O$$

Balancing Redox Equations by the Half-Reaction Method

Many redox reactions occur in an aqueous solution that contains ions. Equations for these reactions can best be balanced by the *half-reaction method* in which we separate an unbalanced equation into oxidation and reduction *half-reactions*. These half-reactions are balanced separately and then combined to obtain the overall equation. The following steps can be used to balance a redox equation by the half-reaction method.

Step 1. Write the redox equation in net ionic form and separate it into oxidation and reduction half-reactions. Each half-reaction must include the oxidized and reduced species of the same element. For example, SO_3^{2-} and SO_4^{2-} are reduced and oxidized species of sulfur.

Step 2. Balance all the atoms, except oxygen and hydrogen, in each half-reaction.

Step 3. Balance the oxygen atoms by adding water molecules to the side deficient in oxygen.

Step 4. Balance the hydrogen atoms as follows:

(a) If the reaction occurs in acid solution, add H^+ ions to the side deficient in hydrogen.

(b) If the reaction occurs in basic solution, add a water molecule for each H atom to the side deficient in hydrogen, then add an equal number of hydroxide ions to the opposite side.

Step 5. Balance the charge by adding electrons to the side that has the higher total positive charge or the lower total negative charge.

Step 6. Multiply the half-reactions, if necessary, with the lowest possible coefficients to provide an equal number of electrons in each half-reaction. Then combine the half-reactions and cancel any identical species on the left and right of the arrow.

The overall balanced equation must not contain electrons because they are neither reactants nor products, they are only transferred from one reactant to another in the reaction. Balancing a redox reaction in acid solution by the half-reaction method is illustrated in Example 4.18.

Example 4.18

Using the Half-Reaction Method for Balancing an Equation for a Redox Reaction in Acid Solution

Complete and balance the following redox equation in acid solution using the half-reaction method:

$$Cr_2O_7^{2-} + HSO_3^- \longrightarrow Cr^{3+} + SO_4^{2-}$$

SOLUTION

Step 1. We separate the unbalanced equation into two skeletal half-reactions, each containing the same element on both sides of the arrow. One half-reaction involves chromium-containing species and the other half-reaction includes sulfur-containing species:

$$Cr_2O_7^{2-} \longrightarrow Cr^{3+}$$
$$HSO_3^- \longrightarrow SO_4^{2-}$$

We follow the steps above to complete and balance the first half-reaction, and then we balance the other one in a similar way.

Step 2. The half-reaction involving chromium has two Cr atoms on the left of the arrow and one on the right. We assign the coefficient 2 for Cr^{3+} to balance the Cr atoms:

$$Cr_2O_7^{2-} \longrightarrow 2Cr^{3+}$$

Step 3. We now have 7 O atoms on the left, but none on the right. To balance the O atoms, we add 7 H_2O to the right:

$$Cr_2O_7^{2-} \longrightarrow 2Cr^{3+} + 7H_2O$$

Step 4. The equation in step 3 has 14 H atoms on the right and none on the left. To balance the H atoms, we add 14 H^+ ions to the left:

$$Cr_2O_7^{2-} + 14H^+ \longrightarrow 2Cr^{3+} + 7H_2O$$

All the atoms in this equation are balanced.

Step 5. In the equation obtained in step 4, the total charge on the left of the arrow is $-2 + 14 = +12$. The total charge on the right of the arrow is $(2)(+3) = +6$. We balance the charge by adding 6 electrons to the left, which yields the net charge of $+6$ on both sides of the arrow. The equation for this half-reaction is now balanced:

$$Cr_2O_7^{2-} + 14H^+ + 6e^- \longrightarrow 2Cr^{3+} + 7H_2O \qquad \text{(reduction)}$$

This half-reaction represents reduction because electrons are gained by the $Cr_2O_7^{2-}$ ions to reduce the oxidation state of chromium from $+6$ to $+3$.

We complete and balance the second half-reaction in a similar way. We add 1 H_2O to the left to balance the O atoms, and we add 3 H^+ ions to the right to balance the H atoms. The charge is then balanced by adding two electrons to the right:

$$HSO_3^- + H_2O \longrightarrow SO_4^{2-} + 3H^+ + 2e^- \qquad \text{(oxidation)}$$

This half-reaction represents oxidation because electrons are lost by the HSO_3^- ions to increase the oxidation state of sulfur from $+4$ to $+6$.

Step 6. We multiply the second half-reaction by 3 to provide an equal number of electrons for both half-reactions. Adding the half-reactions yields

$$Cr_2O_7^{2-} + 14H^+ + 6e^- \longrightarrow 2Cr^{3+} + 7H_2O$$
$$3HSO_3^- + 3H_2O \longrightarrow 3SO_4^{2-} + 9H^+ + 6e^-$$
$$Cr_2O_7^{2-} + 14H^+ + 6e^- + 3HSO_3^- + 3H_2O \longrightarrow 2Cr^{3+} + 7H_2O + 3SO_4^{2-} + 9H^+ + 6e^-$$

Canceling $9H^+$, $6e^-$, and $3H_2O$ from both sides of the arrow yields the balanced equation in its final form, which does not contain electrons:

$$Cr_2O_7^{2-} + 5H^+ + 3HSO_3^- \longrightarrow 2Cr^{3+} + 4H_2O + 3SO_4^{2-}$$

In the equation above, both the atoms and the charge are balanced. Each side of the arrow has 2 Cr, 3 S, 16 O, and 8 H atoms, as well as the net charge of zero. The charge happens to be zero in this case; it could be any number as long as it is the same on both sides of the arrow.

The only difference in balancing equations for redox reactions in acidic and basic solutions is in step 4, where the H atoms are balanced. Example 4.19 illustrates this difference.

Example 4.19

Using the Half-Reaction Method for Balancing an Equation for a Redox Reaction in Basic Solution

Write a balanced equation for the reaction of zinc metal with chromate ions, CrO_4^{2-}, to produce $Zn(OH)_4^{2-}$ ions and $Cr(OH)_4^-$ ions in basic solution.

SOLUTION: The unbalanced, incomplete equation is

$$Zn + CrO_4^{2-} \longrightarrow Zn(OH)_4^{2-} + Cr(OH)_4^-$$

The skeletal half-reactions involve zinc- and chromium-containing species:

$$Zn \longrightarrow Zn(OH)_4^{2-}$$
$$CrO_4^{2-} \longrightarrow Cr(OH)_4^-$$

In the zinc-containing half-reaction, we balance the O atoms by adding 4 H_2O to the left to obtain

$$Zn + 4H_2O \longrightarrow Zn(OH)_4^{2-}$$

Since this equation contains 8 H atoms on the left and 4 on the right, there is a deficiency of 4 H atoms on the right. Adding *4 H_2O to the right* and *4 OH^- to the left* balances the H atoms (the balance of the O atoms is undisturbed):

$$Zn + 4H_2O + 4OH^- \longrightarrow Zn(OH)_4^{2-} + 4H_2O$$

Canceling 4 H_2O molecules from both sides yields

$$Zn + 4OH^- \longrightarrow Zn(OH)_4^{2-}$$

The charge is balanced by adding two electrons to the right:

$$Zn + 4OH^- \longrightarrow Zn(OH)_4^{2-} + 2e^- \quad \text{(oxidation)}$$

In the chromium-containing half-reaction, the Cr and O atoms are already balanced:

$$CrO_4^{2-} \longrightarrow Cr(OH)_4^-$$

There is a deficiency of 4 H atoms on the left. We therefore add 4 H_2O to the left and 4 OH^- to the right to balance the H atoms:

$$CrO_4^{2-} + 4H_2O \longrightarrow Cr(OH)_4^- + 4OH^-$$

To balance the charge, we add three electrons to the left to obtain a complete and balanced half-reaction:

$$CrO_4^{2-} + 4H_2O + 3e^- \longrightarrow Cr(OH)_4^- + 4OH^- \quad \text{(reduction)}$$

We now have two balanced half-reactions; but if we add them at this point, we will find that the electrons do not cancel out. A balanced overall equation must not contain electrons because they are neither created nor destroyed in the reaction. Multiplying the oxidation half-reaction by 3 and the reduction half-reaction by 2 gives six electrons in both half-reactions. We know that these electrons will cancel when the half-reactions are added so we need not write the electrons in the overall equation:

$$\begin{array}{r} 3Zn + 12OH^- \longrightarrow 3Zn(OH)_4^{2-} + 6e^- \\ 2CrO_4^{2-} + 8H_2O + 6e^- \longrightarrow 2Cr(OH)_4^- + 8OH^- \\ \hline 3Zn + 12OH^- + 2CrO_4^{2-} + 8H_2O \longrightarrow \qquad\qquad \\ 3Zn(OH)_4^{2-} + 2Cr(OH)_4^- + 8OH^- \end{array}$$

We now have 12 OH^- on the left of the equation and 8 OH^- on the right. We can therefore cancel 8 OH^- from both sides to obtain the final form of the equation:

$$3Zn + 2CrO_4^{2-} + 4OH^- + 8H_2O \longrightarrow 3Zn(OH)_4^{2-} + 2Cr(OH)_4^-$$

Practice Problem 4.22: Complete and balance each of the following redox equations by the half-reaction method:

(a) $MnO_4^- + Cl^- \longrightarrow Mn^{2+} + Cl_2$ (in acid solution)
(b) $Bi_2O_3 + OCl^- \longrightarrow BiO_3^- + Cl^-$ (in basic solution)

Redox Titration

Redox reactions are applied in many branches of chemical research and industry. For example, the percent of iron in an iron ore can be determined by a titration that involves a redox reaction.

Iron ore contains either Fe_2O_3, Fe_3O_4, or a mixture of the two, depending on the source. Fe_3O_4 is a combination of FeO and Fe_2O_3. A sample of iron ore is weighed and dissolved in an acid. The solution usually contains both Fe^{2+} and Fe^{3+} ions. The Fe^{3+} ions are reduced to Fe^{2+} ions, and the solution is titrated

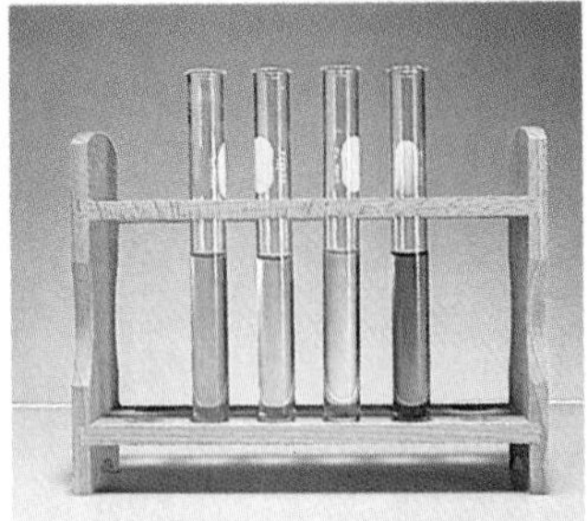

Figure 4.9

The colors of equimolar aqueous solutions of Fe^{2+}, Fe^{3+}, Mn^{2+}, and MnO_4^-.

with a standard solution of a strong oxidizing agent such as $KMnO_4$ or $K_2Cr_2O_7$. The net ionic equation for the reaction of Fe^{2+} ions in acid solution with MnO_4^- ions is

$$5Fe^{2+}(aq) + MnO_4^-(aq) + 8H^+(aq) \longrightarrow Mn^{2+}(aq) + 4H_2O(l) + 5Fe^{3+}(aq)$$

A solution of aqueous permanganate ions is purple and a solution of Fe^{2+} ions is slightly green. When the titration is started, the purple MnO_4^- solution is added from a buret to the slightly greenish Fe^{2+} solution that is swirled. The purple color of the MnO_4^- ions disappears instantly as the redox reaction occurs. When all of the Fe^{2+} ions in the sample have been converted to Fe^{3+} ions and a slight excess of the permanganate solution is added, the color of the solution turns permanently purple, signaling the endpoint of the titration. The reaction of permanganate produces very pale pink Mn^{2+} ions, as well as pale yellow Fe^{3+} ions. However, these pale colors are completely masked by the intense purple color of the slight excess of MnO_4^- ions (Figures 4.9 and 4.10).

The percent iron in the ore sample is calculated by converting the molarity and the volume of the permanganate solution used in the titration to the number of moles of the permanganate ions. The number of moles of the permanganate ions is converted to the number of moles of the Fe^{2+} ions in the ore sample by using the Fe^{2+}/MnO_4^- mole ratio from the balanced equation for the reaction. The number of moles of Fe^{2+} ions is converted to mass and mass percent of iron in the ore. Example 4.20 illustrates the details in these calculations.

Example 4.20

Calculating the Percent of Iron in Iron Ore

To determine the percent of iron in an ore, a 2.032-g sample of the ore is dissolved in an acid, and all of the iron present is converted to Fe^{2+} ions. The sample is titrated with 32.40 mL of 0.1000 *M* $KMnO_4$ solution to the endpoint. Calculate the percent of iron in the ore sample.

SOLUTION: Carrying out the steps outlined above yields

$$0.03240 \text{ L soln} \left(\frac{0.1000 \text{ mol } MnO_4^-}{1 \text{ L soln}}\right)\left(\frac{5 \text{ mol } Fe^{2+}}{1 \text{ mol } MnO_4^-}\right)\left(\frac{55.85 \text{ g } Fe^{2+}}{1 \text{ mol } Fe^{2+}}\right) = 0.9048 \text{ g } Fe^{2+}$$

$$\text{percent Fe} = \left(\frac{0.9048 \text{ g}}{2.032 \text{ g}}\right) 100 = 44.53 \text{ percent}$$

Practice Problem 4.23: Sodium thiosulfate, $Na_2S_2O_3$, is a reducing agent that reduces halogens to their respective halide ions as shown below for the reduction of iodine to iodide ions by thiosulfate ions in solution:

$$2S_2O_3^{2-}(aq) + I_2(aq) \longrightarrow S_4O_6^{2-}(aq) + 2I^-(aq)$$

A 100.0-mL sample of water contaminated by iodine is titrated with 20.2 mL of 0.0100 *M* $Na_2S_2O_3$ solution to convert all the iodine to iodide ions. Calculate the number of milligrams of iodine present in each milliliter of this sample of water.

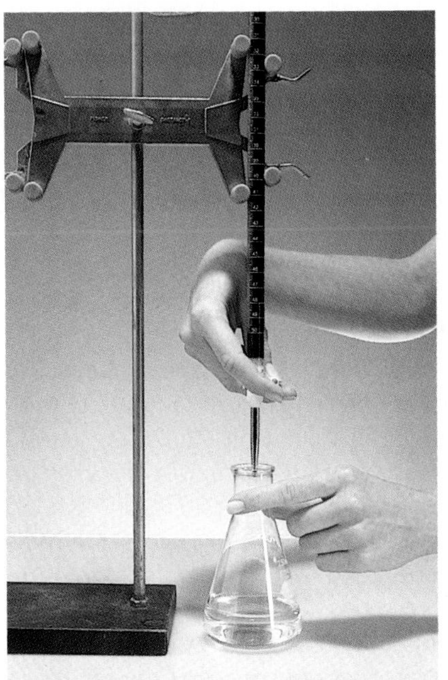

(a)

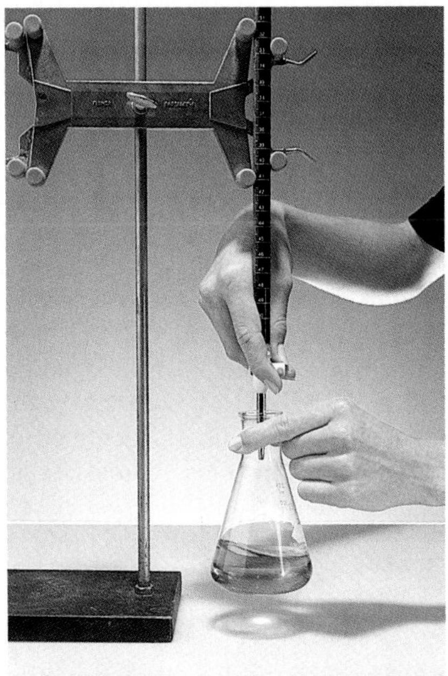

(b)

Figure 4.10

Titration of $Fe^{2+}(aq)$ with $MnO_4^-(aq)$: (a) before the endpoint is reached; (b) after the endpoint is reached.

Summary

In this chapter we discuss three important types of reactions in solution. These are *acid–base reactions*, *precipitation reactions*, and *oxidation–reduction reactions*.

Acids and Bases

There are several definitions for an acid and a base. An *Arrhenius acid is a substance that produces H^+ ions in water solution*, and an *Arrhenius base is a substance that produces OH^- ions in water solution*. A *Brønsted acid is a proton donor*, and a *Brønsted base is a proton acceptor*. A Brønsted acid–base reaction is a proton transfer reaction.

A strong acid dissociates in water solution almost completely into $H^+(aq)$ ions and anions characteristic of the acid. A strong base is completely dissociated into $OH^-(aq)$ ions and cations characteristic of the base. Common strong acids include hydrochloric acid, HCl; nitric acid, HNO_3; and sulfuric acid, H_2SO_4. The hydroxides of the alkali metals and of the alkaline earth metals are strong bases.

Weak acids and bases ionize only to a small extent in water solution. Common weak acids include acetic acid, $HC_2H_3O_2$, and nitrous acid, HNO_2. Phosphoric acid, H_3PO_4, is a moderately weak acid. The most common weak base found in most laboratories is aqueous ammonia, $NH_3(aq)$.

Equations for Acid–Base Reactions

Reactions in solution can be described by three types of equations. In one type, the formulas of all the reactants and products are written without regard to whether they are present as molecules or ions in solution. This type of equation is called a **total equation**. An equation that specifies the major ionic species present in a reaction in solution is called an **ionic equation**. Ions in solution that do not react are called **spectator ions**. An ionic equation in which the spectator ions are omitted is a **net ionic equation**. Substances that are not written in ionic form in an ionic equation are *undissolved solids*, *pure liquids*, *gases*, *nonelectrolytes*, and *weak electrolytes*.

Acid–Base Titration

Acid–base reactions can be used to determine the concentration of an acid or a base in solution. This determination belongs to a branch of quantitative analysis called **volumetric analysis**. In volumetric analysis, the volume of a reactant solution is measured by a procedure known as **titration**. In calculating the results of a titration, the known volume and molarity of a reactant are converted to moles of the reactant. The number of moles of the other reactant is determined by the mole ratio in the balanced equation for the titration reaction. The molarity of this reactant is obtained by dividing the number of moles by the volume in liters of the reactant used in the titration. These steps in the calculation can be outlined as follows:

$$V \text{ and } M \text{ reactant 1} \longrightarrow \text{mol reactant 1} \longrightarrow \text{mol reactant 2} \longrightarrow M \text{ reactant 2}$$

Acid–base reactions can also be used to synthesize compounds that contain the cation of the base and the anion of the acid.

Precipitation Reactions

A *reaction in which a water-insoluble solid*, a **precipitate**, *forms* is called a **precipitation reaction**. The products of a precipitation reaction can be predicted by solubility rules. Precipitation reactions can be used in qualitative and quantitative analysis, and to synthesize insoluble salts.

Oxidation–Reduction Reactions

Most **oxidation–reduction**, or **redox**, **reactions** are *electron transfer reactions*. These reactions can be identified by changes in oxidation states. An **oxidation state** is *either a real or an apparent charge on an atom assigned according to a set of rules*. Redox reactions involve the simultaneous processes of oxidation and reduction. In **oxidation**, the *oxidation state of an element increases*. In **reduction**, the *oxidation state of an element decreases*. The *reactant that oxidizes another reactant* is called the **oxidizing agent**, and the *reactant that reduces the other reactant* is the **reducing agent**.

New Terms

Acid (4.1)
Acid–base reaction (4.2)
Arrhenius acid (4.1)
Arrhenius base (4.1)
Base (4.1)
Brønsted–Lowry acid (4.1)
Brønsted–Lowry base (4.1)
Diprotic acid (4.1)
Dissociation (4.1)
Endpoint (4.3)
Equivalence point (4.3)
Filtration (4.2)
Gravimetric analysis (4.4)
Indicator (4.3)
Ionic equation (4.2)
Ionization (4.1)
Monoprotic acid (4.1)
Net ionic equation (4.2)
Neutralization point (4.3)
Neutralization reaction (4.2)
Oxidation (4.5)
Oxidation number (4.5)
Oxidation–reduction (4.5)
Oxidation–reduction reaction (4.5)
Oxidation state (4.5)
Oxidizing agent (4.5)
Polyprotic acid (4.1)
Precipitation reaction (4.4)
Qualitative analysis (4.4)
Redox reaction (4.5)
Reducing agent (4.5)
Reduction (4.5)
Spectator ion (4.2)
Standardization (4.3)
Standard solution (4.3)
Strong acid (4.1)
Strong base (4.1)
Titration (4.3)
Total equation (4.2)
Triprotic acid (4.1)
Volumetric analysis (4.3)
Weak acid (4.1)
Weak base (4.1)

Exercises

General Review

1. What are the Arrhenius definitions of an acid and a base?
2. What are some common properties of acids and bases?
3. What are the Brønsted–Lowry definitions of an acid and a base?
4. What is a strong acid and what is a strong base?
5. What is a weak acid and what is a weak base?
6. Write the names and formulas of three common strong acids and of three common strong bases.
7. Write the names and formulas of two common weak acids and of one common weak base.
8. Write a total equation for the reaction of: **(a)** a strong acid with a strong base; **(b)** a strong acid with a weak base; **(c)** a weak acid with a strong base; **(d)** a weak acid with a weak base.
9. Write an ionic equation for each of the reactions listed in question 8.
10. Write a net ionic equation for each of the reactions listed in question 8.
11. Explain how you would determine the concentration of an acid solution using a standard solution of a base. What data do you need to obtain in this procedure?
12. Explain how you would determine the concentration of a base solution using a pure solid, water-soluble acid of known composition. List the data you need to obtain in this procedure.
13. Write equations for the different stages of neutralization of a polyprotic acid with a base.
14. Write a total equation for each of the following reactions: **(a)** hydrochloric acid is added to a solution of lead(II) nitrate to precipitate lead(II) chloride; **(b)** potassium sulfate solution is added to barium nitrate solution to precipitate barium sulfate.
15. Write an ionic equation for each of the reactions listed in question 14.
16. Write a net ionic equation for each of the reactions listed in question 14.
17. How can you distinguish an oxidation–reduction reaction from other types of reactions?
18. Define oxidation state?
19. What is: **(a)** oxidation; **(b)** reduction?
20. What is: **(a)** an oxidizing agent; **(b)** a reducing agent?
21. Write an equation for a simple redox reaction, and indicate the changes in the oxidation states involved. Show which one of these changes occurs as a result of a loss of electrons, and which one involves a gain of electrons.
22. Indicate which reactant in question 21 is the oxidizing agent and which is the reducing agent.

Acids and Bases

23. Write the formula and the name for: **(a)** a monoprotic acid; **(b)** a diprotic acid; **(c)** a triprotic acid.
24. Write equations for the two steps of dissociation of sulfuric acid in water. Name the products of each step.
25. Write equations for the three steps of dissociation of phosphoric acid in water. Name the products of each step.
26. In each of the following reactions, indicate which of the reactants is Brønsted acid and which is Brønsted base: **(a)** $NH_4^+(aq) + OH^-(aq) \rightarrow NH_3(aq) + H_2O(l)$; **(b)** $H_2O(l) + HCl(g) \rightarrow H_3O^+(aq) + Cl-(aq)$
27. Which is the stronger acid? **(a)** $HC_2H_3O_2$ or HNO_3; **(b)** HNO_2 or HCl?
28. Which one of the following acids is a weak acid? HF(*aq*); HCl(*aq*); HBr(*aq*); HI(*aq*).

Equations for Acid–Base Reactions

29. Write a balanced total equation for each of the following reactions: **(a)** hydrochloric acid is added to a solution of sodium hydroxide; **(b)** hydrochloric acid is added to a solution of barium hydroxide; **(c)** solid magnesium hydroxide is added to hydrochloric acid; **(d)** a solution of potassium hydroxide is neutralized with a solution of sulfuric acid; **(e)** solid aluminum hydroxide is neutralized with a solution of sulfuric acid; **(f)** a solution of barium hydroxide is neutralized with a solution of acetic acid; **(g)** solid calcium hydroxide is neutralized with a solution of acetic acid; **(h)** aqueous ammonia is neutralized with a solution of nitrous acid; **(i)** aqueous ammonia is neutralized with a solution of sulfuric acid.
30. Write a net ionic equation for each of the reactions in question 29.

Acid–Base Titration

31. In an acid–base titration, what is the meaning of each of the following words? "Neutralization," "equivalence point," and "endpoint."
32. How many milliliters of 0.200 *M* sodium hydroxide is required to neutralize 40.0 mL of 0.0500 *M* HCl solution?
33. How many milliliters of 0.200 *M* sodium hydroxide is required to completely neutralize: **(a)** 40.0 mL of 0.0500 *M* H_2SO_4; **(b)** 40.0 mL of 0.0500 *M* H_3PO_4?
34. How many grams of potassium hydrogen phthalate, $KHC_8H_4O_4$ (molar mass 204.2), is required to make 250.0 mL of 0.4000 *M* solution?

35. A 0.1205-g sample of potassium hydrogen phthalate, $KHC_8H_4O_4$, requires 42.00 mL of a sodium hydroxide solution for neutralization. What is the molarity of the sodium hydroxide?

36. How many milliliters of 0.00100 M barium hydroxide solution is required to completely neutralize 40.0 mL of 0.0300 M phosphoric acid?

37. How many grams of solid barium hydroxide is required to make enough 0.00100 M solution that would completely neutralize 40.0 mL of 0.0300 M phosphoric acid?

38. A 0.0100-g sample of potassium hydrogen phthalate requires 50.0 mL of barium hydroxide solution for neutralization. What is the molarity of the barium hydroxide solution?

39. A 42.60-mL sample of 0.08500 M hydrochloric acid requires 37.94 mL of a sodium hydroxide solution for neutralization. Calculate the molarity of the sodium hydroxide solution.

40. A 35.78-mL volume of nitric acid solution is neutralized by 40.54 mL of 2.583×10^{-3} M barium hydroxide solution. What is the molarity of the nitric acid solution?

41. 42.80 mL of phosphoric acid solution is neutralized by 0.662 g of solid calcium hydroxide. What is the molarity of the phosphoric acid solution?

42. A 25.00-mL sample of 0.1208 M hydrochloric acid is neutralized by 28.26 mL of a sodium hydroxide solution. 24.65 mL of this sodium hydroxide solution neutralizes a 1.025-g sample of a solid monoprotic acid. Calculate the molar mass of the acid.

43. In a standardization of a sodium hydroxide solution, 42.58 mL of the solution is added to a solution containing 1.208 g of potassium hydrogen phthalate (molar mass 204.2). The volume of sodium hydroxide added is in excess to that needed to neutralize all the acid. The excess sodium hydroxide is neutralized by 10.40 mL of 0.01126 M hydrochloric acid. Calculate the molarity of the sodium hydroxide solution.

44. A solution containing 1.15 g of a mixture of NaOH and KOH requires 12.1 mL of 1.00 M H_2SO_4 for neutralization. Calculate the mass of NaOH and of KOH in the solution. (*Hint*: Use two equations with two unknowns.)

45. The nitrogen content in a fertilizer (in the form of NH_3 or its salts) is determined as follows. A 0.196-g sample of the fertilizer is treated with sulfuric acid to convert all of the nitrogen to ammonium sulfate. The latter is converted to ammonia gas by treatment with sodium hydroxide. The ammonia produced is absorbed and neutralized by 25.0 mL of 0.100 M HCl solution, of which a part remains unreacted. The unreacted HCl in the solution is neutralized by 12.3 mL of 0.0500 M $Ba(OH)_2$ solution. Calculate the percent of nitrogen in the fertilizer.

Precipitation Reactions

46. Explain how silver ions can be separated from barium ions in an aqueous solution of silver nitrate and barium nitrate. Write a net ionic equation for each reaction that occurs in the separation scheme you propose.

47. What substance, if any, precipitates when each of the following pairs of 1 M solutions are mixed? **(a)** $AgNO_3$ and RbCl; **(b)** $Pb(NO_3)_2$ and KCl; **(c)** $Hg_2(NO_3)_2$ and HCl; **(d)** $CaCl_2$ and Na_2CO_3; **(e)** $Mg(NO_3)_2$ and $CaCl_2$; **(f)** K_2SO_4 and $BaCl_2$.

48. Write a balanced total equation and a net ionic equation for each of the reactions that occurs when the solutions are mixed as described in question 47.

49. What two substances in aqueous solution would you mix to prepare each of the following compounds by precipitation from the solution? **(a)** barium sulfate; **(b)** silver chloride; **(c)** calcium carbonate; **(d)** barium phosphate.

50. Write total equations and net ionic equations for the reactions that would occur between the substances you chose in each part of question 49.

51. If you were given three different solid substances, barium chloride, sodium nitrate, and silver chloride, how would you tell them apart? Include equation(s) for the reaction(s) that you would carry out, if any, in each case.

52. If you had two unlabeled solutions and you knew that one of them is a solution of lead nitrate and the other a solution of potassium sulfate, how would you determine which is which? Write a net ionic equation for each reaction that you would carry out in your testing.

53. A 0.5000-g sample of a soluble metal sulfate gives a precipitate of barium sulfate that weighs 0.4620 g. Calculate the percent of sulfur in the sample.

54. A 1.5280-g sample of a coin that contains silver is dissolved in an acid and the silver ions are precipitated from the solution as silver chloride. After filtering and drying, the precipitate has a mass of 1.7367 g. Calculate the percent of silver in the coin.

55. To determine the amount of phosphorus in a rock, a 2.48-g sample of the rock is crushed and dissolved in an acid. All of the phosphorus is precipitated from this solution as magnesium ammonium phosphate, $MgNH_4PO_4$. The latter is ignited to magnesium pyrophosphate, $Mg_2P_2O_7$, which has a mass of 0.928 g. Calculate the percent of phosphorus in the rock.

56. A 0.500-L sample of well water that contains dissolved calcium sulfate is treated with sodium carbonate solution to precipitate the calcium ions present. The precipitate is dissolved in 30.0 mL of 0.100 M HCl, which converts the precipitate to aqueous calcium chloride, water, and carbon dioxide. The CO_2 is

boiled off. A part of the HCl remains unreacted and is neutralized by 9.80 mL of 0.100 *M* sodium hydroxide. What is the mass of calcium sulfate present in a liter of the well water?

57. How many grams of silver is present in 30.0 mL of 0.100 *M* silver nitrate?

58. A 40.5-mL portion of a sodium carbonate (washing soda) solution is treated with an excess of calcium chloride solution. The precipitate formed is strongly ignited, leaving 0.235 g of calcium oxide as a residue after cooling. Calculate the molarity of the sodium carbonate solution.

Redox Reactions

59. Write a balanced equation for each of the following redox reactions: **(a)** calcium metal reacts with hydrochloric acid by displacing hydrogen in the acid (review single displacement reactions in Chapter 2 if necessary); **(b)** potassium metal reacts with sulfuric acid (*Caution*: This reaction is violent!); **(c)** zinc metal reduces iron(III) chloride in solution to elemental iron while the zinc metal is oxidized to $ZnCl_2$ in the solution; **(d)** magnesium metal reacts with a solution of silver nitrate producing silver metal and a solution of magnesium nitrate.

60. For each reaction specified in question 59, identify the substance oxidized, the substance reduced, and the oxidizing and reducing agents.

61. Write a balanced net ionic equation for each of the reactions in question 59.

62. What is the oxidation state of the boldfaced element in each of the following compounds? **(a)** $K\mathbf{N}O_3$; **(b)** $\mathbf{Cr}_2O_3$; **(c)** $K_2\mathbf{Cr}_2O_7$; **(d)** $\mathbf{N}_2O$; **(e)** $(NH_4)_3\mathbf{P}O_4$; **(f)** $K_4\mathbf{Fe}(CN)_6$; **(g)** $\mathbf{P}_4O_6$; **(h)** $Ca_2\mathbf{P}_2O_7$; (i) S_2O_3; **(j)** $K_3\mathbf{Fe}(CN)_6$; **(k)** $\mathbf{Cr}_3O_8$; **(l)** $\mathbf{I}F_7$.

63. What is the oxidation state of the boldfaced element in each of the following ions? **(a)** $\mathbf{Cr}_2O_7^{2-}$; **(b)** $\mathbf{P}_2O_7^{4-}$; **(c)** $\mathbf{Fe}(CN)_6^{3-}$; **(d)** $\mathbf{N}O_3^-$; **(e)** $\mathbf{N}H_4^+$; **(f)** $\mathbf{S}_2O_3^{2-}$.

64. Balance the following redox equations by the oxidation state method. In each case indicate the oxidation state of the element oxidized and reduced before and after the reaction. Also, in each case, identify the oxidizing agent and the reducing agent: **(a)** $V_2O_3(s) + H_2(g) \rightarrow VO(s) + H_2O(g)$; **(b)** $PH_3(g) + N_2O(g) \rightarrow H_3PO_4(l) + N_2(g)$; **(c)** $FeS(s) + O_2(g) \rightarrow Fe_2O_3(s) + SO_2(g)$; **(d)** $Ca_3(PO_4)_2(s) + SiO_2(s) + C(s) \rightarrow CaSiO_3(s) + CO(g) + P_4(s)$; **(e)** $K_2Cr_2O_7(aq) + S_8(s) + H_2O(l) \rightarrow SO_2(g) + KCr(OH)_4(aq)$; **(f)** $Cr_2O_7^{2-}(aq) + I^-(aq) + H^+(aq) \rightarrow Cr^{3+}(aq) + I_2(s) + H_2O(l)$; **(g)** $Cr(OH)_4^-(aq) + H_2O_2(aq) + OH^-(aq) \rightarrow CrO_4^{2-}(aq) + H_2O(l)$.

65. Using the half-reaction method, balance the following equations for redox reactions in acid solution: **(a)** $Cl^-(aq) + MnO_4^-(aq) \rightarrow Mn^{2+}(aq) + Cl_2(g)$; **(b)** $Cr_2O_7^{2-}(aq) + HNO_2(aq) \rightarrow Cr^{3+}(aq) + NO_3^-(aq)$; **(c)** $Fe^{2+}(aq) + H_2O_2(aq) \rightarrow Fe^{3+}(aq) + H_2O(l)$; **(d)** $MnO_4^-(aq) + I^-(aq) \rightarrow Mn^{2+}(aq) + IO_3^-(aq)$; **(e)** $CdS(s) + NO_3^-(aq) \rightarrow Cd^{2+}(aq) + S_8(s) + NO_2(g)$; **(f)** $H_4IO_6^-(aq) + Cr^{3+}(aq) \rightarrow IO_3^-(aq) + Cr_2O_7^{2-}(aq)$; **(g)** $HgCl_2(aq) + Sn^{2+}(aq) + Cl^-(aq) \rightarrow Hg_2Cl_2(s) + SnCl_6^{2-}(aq)$; **(h)** $I_2(s) + S_2O_3^{2-}(aq) \rightarrow S_4O_6^{2-}(aq) + I^-(aq)$.

66. Using the half-reaction method, complete and balance the following equations for redox reactions in basic solution: **(a)** $Bi(OH)_3(s) + SnO_2^{2-}(aq) \rightarrow Bi(s) + SnO_3^{2-}(aq)$; **(b)** $N_2O_4(g) + Br^-(aq) \rightarrow NO_2^-(aq) + BrO_3^-(aq)$; **(c)** $Al(s) + HPO_3^{2-}(aq) \rightarrow Al(OH)_4^-(aq) + HPHO_2^-(aq)$; **(d)** $Fe(CN)_6^{4-}(aq) + CrO_4^{2-}(aq) \rightarrow Fe(CN)_6^{3-}(aq) + Cr_2O_3(s)$; **(e)** $V(s) \rightarrow HV_{10}O_{28}^{5-}(aq) + H_2(g)$.

Redox Titration

67. Potassium permanganate solution is commonly used as a good oxidizing agent. Potassium permanganate solution can be standardized by titrating with a solution of oxalic acid, $H_2C_2O_4$. The unbalanced equation for the titration reaction is

$$MnO_4^-(aq) + H_2C_2O_4(aq) \longrightarrow Mn^{2+}(aq) + CO_2(g)$$

If 24.78 mL of a potassium permanganate solution is required to completely react with 0.1062 g of oxalic acid, what is the molarity of the permanganate solution?

68. In a determination of iron in an iron ore, a 0.5000-g sample of iron ore is dissolved in a dilute sulfuric acid. All the iron present is reduced to Fe^{2+} ions, and the resulting solution is titrated with 42.00 mL of 0.02000 *M* potassium dichromate solution. The unbalanced net ionic equation for the reaction in acid solution is

$$Cr_2O_7^{2-}(aq) + Fe^{2+}(aq) \longrightarrow Cr^{3+}(aq) + Fe^{3+}(aq)$$

Calculate the percent of iron in the ore.

69. Thiosulfate ions reduce iodine to iodide ions according to the following equation:

$$S_2O_3^{2-}(aq) + I_2(aq) \longrightarrow S_4O_6^{2-}(aq) + I^-(aq) \quad \text{(unbalanced)}$$

The iodine, a contaminant in a sample of water, is titrated with 40.2 mL of 0.104 *M* sodium thiosulfate solution. How many grams of iodine is present in the sample?

CHAPTER 5

Old Faithful geyser, Yellowstone National Park, Wyoming.

Thermochemistry

In the preceding chapters we have focused our attention on the properties of elements, compounds, and their reactions. In this chapter we consider the role of energy *in chemical reactions. Every process in the physical world—whether it occurs in stars, in geysers, in living cells, or in test tubes—is accompanied by an energy change. Some processes release energy, others require energy. But in all cases, the total amount of energy in the universe remains constant.*

We restrict the discussion in this chapter to the changes of heat energy that accompany chemical reactions, a topic called ***thermochemistry****. The study of all types of energy changes associated with chemical and physical changes is known as* ***thermodynamics****.*

5.1 Energy and Work

In Chapter 1 we defined energy as the ability to do work. In this section we consider some major types of energy and then examine the relation between work and energy.

Types of Energy

The energy of an object can be classified as either kinetic energy or potential energy. **Kinetic energy** is possessed by bodies in motion. The kinetic energy of an object equals one half of the product of its mass (m) and the square of its velocity (v):

$$\text{kinetic energy} = \tfrac{1}{2}mv^2$$

Potential energy is the energy that an object possesses by virtue of its position or composition. The potential energy of an object can be determined by its mass and position relative to some reference point. For example, the gravitational potential energy of an object is determined by its mass and its distance from the center of the earth: potential energy = mgh, where m is the mass of the object, g the gravitational acceleration, and h the height or position of the object relative to a reference point.

Potential energy is also called "stored energy." The amount of potential energy stored in an object depends on its composition, the bonding of its atoms in molecules, the attractive forces between molecules, and the attractive forces between ions in an ionic lattice.

The SI unit of energy is the joule (J). One calorie (cal) equals 4.184 J, which is also the heat required to raise the temperature of 1 g of water by 1°C. We can think of a joule in "everyday" terms as approximately the energy expended by a 1-lb rabbit jumping upward 3 ft (Figure 5.1).

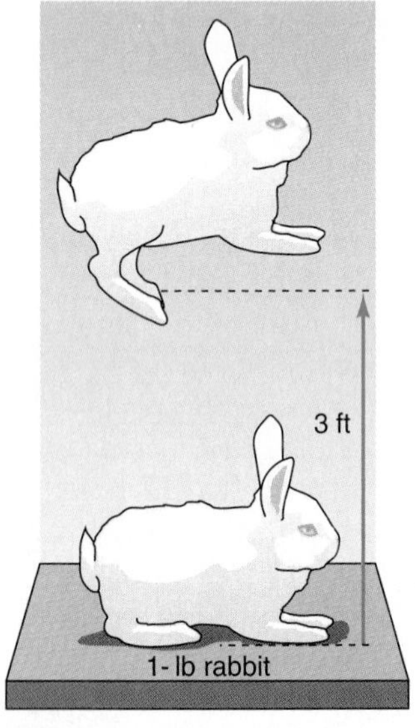

Figure 5.1

The SI unit of energy is the joule (J). A one pound rabbit that jumps 3 feet upward expends approximately 1 J of energy.

During a chemical reaction, the atoms of the reactants separate and recombine to form the products. If the products have lower energy than the reactants, energy is released by the reaction. Chemical *reactions that release energy* are called **exothermic reactions** (Greek, *exo*, out, + *therm-*, heat). *Reactions that absorb energy* are **endothermic** (Greek, *endo-*, in). We discuss endothermic and exothermic reactions in more detail in Section 5.3.

Other kinds of energy include *heat*; energy associated with mechanical work, or *mechanical energy*; *electrical energy*; *nuclear energy*; and *radiant energy*. Radiant energy itself can be subdivided into many categories (see Chapter 6).

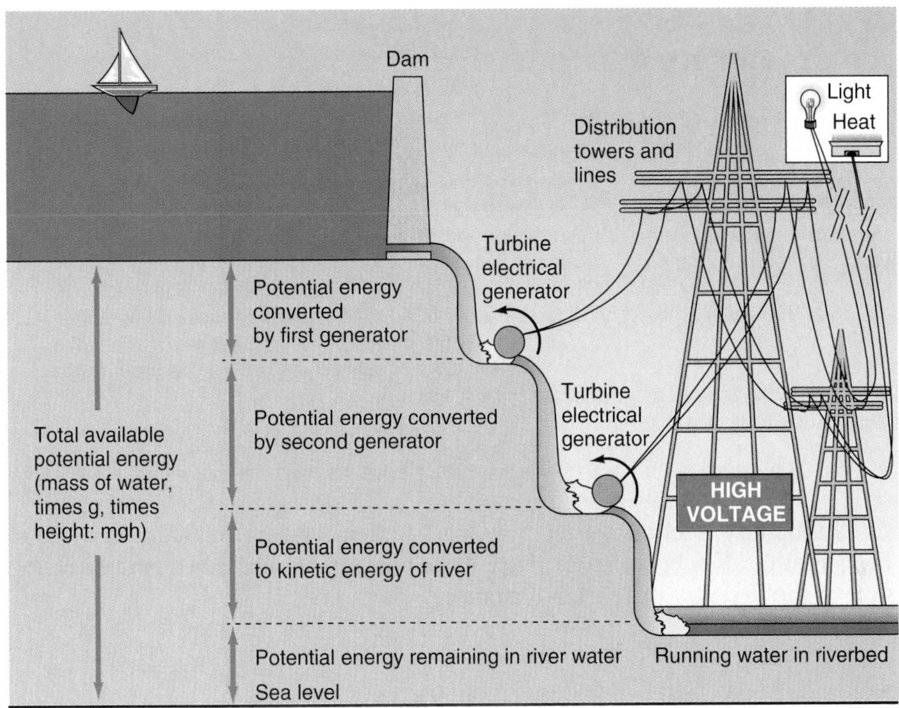

Figure 5.2

Conversion of energies: When the gates of the dam are opened and the water is allowed to fall, the potential energy of the water behind the dam is converted to kinetic energy of running water and rotating turbines. The kinetic energy is then converted to electrical energy, then light energy and heat energy.

Energy can be converted from one form to another. For example, the potential energy of water harnessed by a dam can be converted to kinetic energy by allowing the water to fall to a lower level. This kinetic energy can in turn be used to do electrical, mechanical, or other kinds of work (Figure 5.2).

Work

Work is done by an object when it moves against a force. **Work** is defined as the force times the distance moved:

$$w = f \times d$$

where w is work, f is force, and d is distance.

When an object does work by moving against a force, energy is consumed. Thus, work is measured in units of energy. For example, energy is spent for lifting an object against the force of gravity, or for removing a positively charged particle from the vicinity of a negatively charged particle.

The English scientist Benjamin Thompson noticed in 1798 that the temperature of the barrel of a cannon increases as it is being bored. This temperature increase is caused by friction between the drill bit and the metal. The drill bit is doing work as it bores through the metal. As the work proceeds, heat is released. When the cannon and the drilling apparatus were immersed in water, the water started to boil. This demonstrated that the energy expended to do work can be converted to heat—a form of energy.

After Thompson's observations, James **Joule** carried out many experiments in which he carefully measured the heat released when work of various kinds was done. The experiments conducted by Thompson and Joule established the *law of conservation of energy*, which states that *the energy content of the universe is constant*. That is, *energy can be neither created nor destroyed* (Chapter 1). The law of conservation of energy is also known as the **first law of thermodynamics**.

APPLICATIONS OF CHEMISTRY 5.1
James Joule: A Biographical Sketch

James Joule was born on Christmas Eve, 1818, into a northern English family that owned a prosperous brewery. He died in 1889. Early illness left him with a weakened spine, and Joule was a hunchback. His deformity made him extremely shy, and he shunned public appearances.

Joule worked for part of his youth in the family brewery, but eventually he established his own laboratory, and began a life-long effort to accurately measure the relationship of heat to work. His research showed that work could be completely converted into heat, thus demonstrating the law of conservation of energy. Among his many experiments, Joule showed that if the work done in raising 1 pound of water through 772 ft is converted to heat, the temperature of the water will increase by 1°F. (These units are, of course, not metric.) He also found that the temperature of a liquid could be increased by agitating it. This simple phenomenon explains why the water that falls 49 m over Niagara Falls is 0.11°C warmer at the bottom than at the top.

In the next section we see how work and energy are related to the law of conservation of energy. We will use these concepts to provide a precise definition of the first law of thermodynamics.

5.2 The First Law of Thermodynamics

The grand statement of the law of conservation of energy must be formulated in a way that will allow us to measure the energy changes that accompany chemical reactions. For that purpose we will divide the world into two parts. One part we call the thermodynamic **system**; everything else constitutes the **surroundings**.

We can define our thermodynamic system in any way that is convenient for our purposes. A thermodynamic system for an astronomer might be the solar system, a galaxy, or even the whole observable universe. A thermodynamic system for a chemist might be a test tube and its contents; for a biologist, a cell.

A system can be *open*, *closed*, or *isolated*. A system is *open* if matter and energy can be exchanged between the system and its surroundings. An example of an open system is a living organism. A system is *closed* if matter cannot cross its boundaries, but energy can. An example of a closed system is a sealed flask. If neither matter nor energy can cross the boundary between the system and its surroundings, the system is *isolated*. It is almost impossible to construct a truly isolated system. We can keep matter in, but despite all precautions, energy inevitably leaks across the boundary between the system and the surroundings in one way or another.

A system is characterized in terms of a set of variables called **state functions**, which define the **state** of a system. Examples of state functions of a closed system are the temperature, pressure, the number of moles of material in the system, and the energy of the system. When one or more of the state functions of a system change from their initial values to new values, the system is said to change from state 1, the **initial state**, to state 2, the **final state**. *The magnitude of change of a state function depends only on the difference between the initial and final states of a system, not on the way in which the system changes*. For example, the temperature difference between two states of a system does not depend on whether the temperature change is brought about by heating the system or by agitating it.

According to the first law of thermodynamics, the total energy of the universe is constant. Therefore, any change of energy in a thermodynamic system is exactly balanced by an opposite change in the surroundings. Thus, if the energy of a system decreases, the energy of the surroundings increases by the same amount. We can write this statement as a simple equation

$$-\Delta E_{\text{system}} = +\Delta E_{\text{surroundings}}$$

where the Greek capital letter delta, Δ, means "difference." Similarly, if the energy of a system increases, the energy of the surroundings decreases by the same amount:

$$+\Delta E_{\text{system}} = -\Delta E_{\text{surroundings}}$$

Internal Energy Change

The **internal energy** of a system is the total of all possible types of energy present in the system. When the energy of a system changes from some initial energy, E_{initial}, to final energy, E_{final}, the internal energy change, ΔE, is

$$\Delta E = E_{\text{final}} - E_{\text{initial}}$$

When E_{final} is larger than E_{initial} ($E_{\text{final}} > E_{\text{initial}}$), ΔE has a positive sign, which indicates that the energy of the system has increased during the process. Thus, energy has flowed from the surroundings *into* the system. In contrast, if the final energy of the system is lower than the initial energy, the system has lost energy—energy has flowed from the system to the surroundings.

We recall that Thompson's "cannon drilling experiment" showed that heat and work are different aspects of energy. Thus, when a thermodynamic system changes from one state to another, heat may flow into or out of the system, and work may be done on the system or by the system. Therefore, the **first law of thermodynamics** is formulated in terms of internal energy, heat, and work: *The change of the internal energy, ΔE, of a system is the sum of the heat, q, and work, w, added to the system*:

$$\Delta E = q + w \tag{5.1}$$

If the value of q in Equation 5.1 is *positive*, *heat flows into the system* from the surroundings. If the value of w is positive, *work is done on the system* by the surroundings. For example, the surroundings can do work on the system by forcing the volume of the system to decrease. Thus, the internal energy change of a system that gains 50 J of heat and has 5 J of work done on it by the surroundings is: 50 J + 5 J = 55 J.

If the value of q is *negative*, the *system loses heat* to the surroundings. If the value of w is negative, the *system does work on the surroundings*. For example, a system can do work by expanding against the external pressure. When the volume of the system increases, the system does work on the surroundings.

Figure 5.3

PV work done by a system at constant pressure. PV work can be done by a chemical reaction whose volume increases during the reaction (the total volume of the products is larger than the total volume of the reactants).

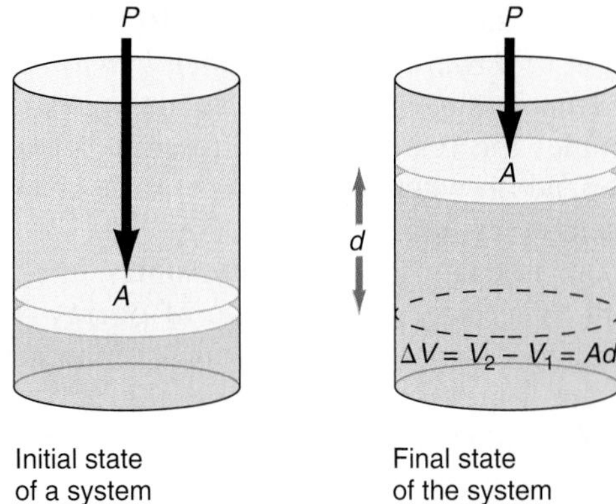

Work can be done in many ways. If the work that causes a volume change is carried out at constant pressure, the work is called *pressure–volume*, or *PV*, *work* (Figure 5.3). Pressure, P, is defined as a force, f, per unit area, A. Thus,

$$P = \frac{f}{A}$$

Let us consider a gas expanding in a cylinder against a frictionless piston with a cross-sectional area, A. As the gas expands, it pushes the piston against a constant atmospheric pressure, P, through a distance, d. The increase in the volume, ΔV, of the gas is therefore Ad (Figure 5.3). Since work is defined as a force acting through a distance, the product of the pressure and the change in volume is work:

$$P\,\Delta V = \frac{f}{A}(Ad) = fd = -w$$

Here w is negative because the system does work on the surroundings. When a system does work on the surroundings, $P\,\Delta V$ (and hence w) is negative. When PV work is done by the surroundings on the system, $P\,\Delta V$ (and hence w) is positive.

Example 5.1

Determining the Internal Energy Change of a System and Its Surroundings

A system absorbs 1000 J of heat *from* the surroundings and does 2 J of work *on* the surroundings. What is ΔE for the system and for the surroundings?

SOLUTION: The system absorbs heat, and q is positive; w is negative because the system loses energy due to the work it does on the surroundings:

$$\Delta E_{\text{system}} = q + w = 1000\text{ J} + (-2\text{ J}) = 998\text{ J}$$
$$\Delta E_{\text{surroundings}} = -\Delta E_{\text{system}} = -1000\text{ J} + 2\text{ J} = -998\text{ J}$$

Enthalpy Change

Many chemical reactions are carried out in open containers at a constant atmospheric pressure. Under these conditions, the volume of a reaction mixture can change. Therefore, PV work can be done and heat can flow to or from the system.

The symbol for the heat transferred at constant pressure is q_p. Therefore, for reactions carried out at constant pressure, which do only PV work,

$$\Delta E = q_p + P\Delta V \tag{5.2}$$

where q_p is the heat transferred to the system from the surroundings, and PV work, $+P\,\Delta V$, is done by the surroundings on the system at constant pressure.

If a system at constant pressure does work on the surroundings, w has a negative sign. If the system does only PV work at constant P,

$$\Delta E = q_p + w = q_p + (-P\Delta V)$$

from which

$$q_p = \Delta E + P\Delta V \tag{5.3}$$

If a chemical reaction involves only PV work, the *heat absorbed or evolved by the reaction at constant pressure*, q_p, is an important quantity known as the **enthalpy change**, ΔH, for the reaction (Greek *enthalpein*, to heat). Thus, the *enthalpy change for a reaction, or any system, equals the internal energy change of the system plus any PV work done by the system at constant pressure*,

$$\Delta H = q_p = \Delta E + P\Delta V \tag{5.4}$$

Since the enthalpy change, ΔH, for a reaction is measured as the heat, q_p, it is also often called the **heat of reaction**, or the **enthalpy of reaction**.

The amount of work done by a chemical reaction at constant pressure depends on the volume change that accompanies the reaction. There is no appreciable volume change in reactions in which the reactants and products are solids or liquids. For such reactions, no PV work is done because $\Delta V = 0$. Therefore, $P\Delta V = 0$, and $\Delta H = \Delta E$.

A reaction in which gases are involved can have different volumes for reactants and products. Such a reaction may be accompanied by PV work. However, the $P\Delta V$ term is usually small compared to ΔE even for gaseous reactions.

5.3 Enthalpy Changes of Chemical Reactions

We have seen that the change in internal energy of a system is the difference between the final and initial energy states of the system. Similarly, the enthalpy change for a system equals the difference between the enthalpies of the final and initial states of the system:

$$\Delta H = H_{\text{final}} - H_{\text{initial}}$$

If we consider a chemical reaction as a system that changes from an initial state consisting of the reactants to a final state consisting of the products, the enthalpy change for the reaction, ΔH_{rxn}, is

$$\Delta H_{rxn} = H_{\text{products}} - H_{\text{reactants}} \tag{5.5}$$

If the total enthalpy of the products of a reaction is greater than the total enthalpy of the reactants, ΔH is positive because $H_{\text{final}} - H_{\text{initial}} > 0$. A positive ΔH indicates that the heat flows from the surroundings to the system during the reaction. A reaction that absorbs heat from the surroundings is an *endothermic*

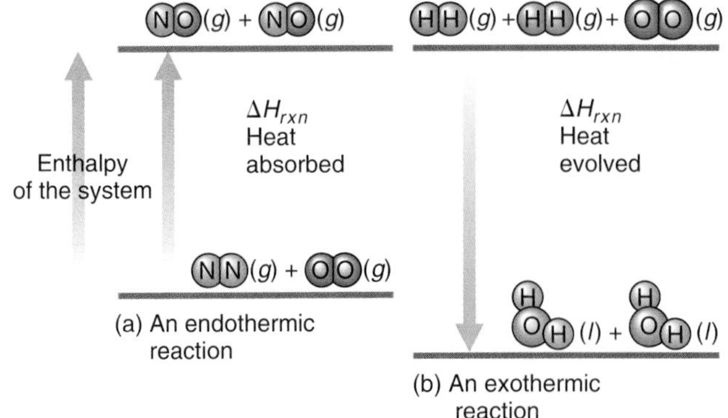

Figure 5.4

Enthalpy changes for endothermic and exothermic reactions.

reaction (Figure 5.4). The reaction of nitrogen with oxygen to form nitrogen monoxide is an example of an endothermic reaction:

$$N_2(g) + O_2(g) + \text{heat} \longrightarrow 2NO(g)$$

This reaction occurs at high temperature in internal combustion engines and contributes to the formation of the air pollutants known collectively as "smog."

If the total enthalpy of the products is less than the total enthalpy of the reactants, that is, if $H_{\text{final}} - H_{\text{initial}} < 0$, heat flows from the system to the surroundings. A reaction that releases heat to the surroundings is an *exothermic* reaction. An example of an exothermic reaction is the reaction of hydrogen with oxygen to form water:

$$2H_2(g) + O_2(g) \longrightarrow 2H_2O(g) + \text{heat}$$

Combination reactions of metals with nonmetals to form binary ionic compounds are also exothermic. The enthalpy changes for endothermic and exothermic reactions are illustrated in Figure 5.4.

Factors That Affect Enthalpy Change

The enthalpy change for a reaction depends on the conditions under which the reaction occurs. The effects of various conditions on enthalpy changes are discussed below.

1. The enthalpy change of a reaction is an *extensive property* of the system, that is, it *depends on the quantities of reactants and products.* Consider the enthalpy change for the reaction of hydrogen with oxygen at one atmosphere pressure and 298 K:

$$2H_2(g) + O_2(g) \longrightarrow 2H_2O(g) \qquad \Delta H = -484 \text{ kJ}$$

According to this equation, the enthalpy change for the reaction of 2 mol of hydrogen gas with 1 mol of oxygen gas to form 2 mol of water vapor is −484 kJ. The enthalpy change for the reaction of 1 mol of hydrogen with $\frac{1}{2}$ mol of oxygen to form 1 mol of water vapor under the same conditions is one-half of the enthalpy of formation of 2 mol of water:

$$H_2(g) + \tfrac{1}{2}O_2(g) \longrightarrow H_2O(g) \qquad \Delta H = -242 \text{ kJ}$$

This equation contains a fractional coefficient to emphasize the formation of 1 mol of water.

2. *The enthalpy change for a reaction written from "left to right" is equal in magnitude but opposite in sign to the enthalpy change for the reverse reaction.*

In the preceding paragraph we saw that the formation of 1 mol of water vapor from 1 mol of hydrogen and $\frac{1}{2}$ mol of oxygen is an exothermic reaction having an enthalpy change of -242 kJ mol^{-1}. The reverse reaction, that is, the decomposition of 1 mol of water vapor to 1 mol of H_2 gas and $\frac{1}{2}$ mol of O_2 gas is an *endothermic reaction* whose enthalpy change is +242 kJ:

$$H_2O(g) \longrightarrow H_2(g) + \tfrac{1}{2}O_2(g) \qquad \Delta H = +242 \text{ kJ}$$

3. *The enthalpy change for a reaction depends on the physical states of the reactants and products.* For example, the formation of 1 mol of gaseous hydrogen iodide from *gaseous iodine* and hydrogen gas is an exothermic process:

$$\tfrac{1}{2}I_2(g) + \tfrac{1}{2}H_2(g) \longrightarrow HI(g) \qquad \Delta H = -36.5 \text{ kJ}$$

In contrast, the formation of gaseous hydrogen iodide from *solid iodine* and hydrogen gas is an endothermic process:

$$\tfrac{1}{2}I_2(s) + \tfrac{1}{2}H_2(g) \longrightarrow HI(g) \qquad \Delta H = +25.9 \text{ kJ}$$

The reason for this difference is that the reaction of solid iodine with hydrogen involves a preliminary *endothermic* step in which the solid *sublimes* to the gaseous state:

$$\tfrac{1}{2}I_2(s) \longrightarrow \tfrac{1}{2}I_2(g) \qquad \Delta H = 62.4 \text{ kJ}$$

Sublimation is an endothermic process because energy must flow from the surroundings to the system to disrupt the lattice forces of the solid.

Similarly, the enthalpy change for the formation of 1 mol of *liquid* water from its gaseous elements is -286 kJ. The enthalpy change for the formation of 1 mol of *gaseous* water from its gaseous elements is -242 kJ. The ΔH for the formation of liquid water is slightly more negative than that for gaseous water because heat is evolved when water vapor condenses to a liquid:

$$H_2O(g) \longrightarrow H_2O(l) \qquad \Delta H = -44 \text{ kJ mol}^{-1}$$

4. *The enthalpy change for a reaction depends on temperature and pressure.* Most enthalpy values are tabulated for chemical reactions that occur at 1 atmosphere (atm) pressure and 298 K. The conditions of *1 atm and 298 K* are therefore referred to as **standard conditions**. The *physical state of a substance in which it is most stable under standard conditions* is called its **standard state**. For example, the standard state for sodium chloride is the solid state, and for diatomic oxygen the gaseous state.

The *enthalpy change of a reaction that occurs under standard conditions and involves only substances in their standard states* is called the **standard enthalpy change of reaction**, or simply the **standard enthalpy of reaction**. The standard enthalpy of reaction is represented by $\Delta H°$.

Hess's Law of Heat Summation

In the preceding section we saw that the reaction of solid iodine with gaseous hydrogen consists of two related processes. First, solid iodine sublimes; second, it reacts with gaseous hydrogen. The enthalpy change for the *net reaction* is the *sum* of the enthalpy changes for the individual steps:

Step 1:	$\tfrac{1}{2}I_2(s) \longrightarrow \tfrac{1}{2}I_2(g)$	$\Delta H = +62.4$ kJ
Step 2:	$\tfrac{1}{2}I_2(g) + \tfrac{1}{2}H_2(g) \longrightarrow HI(g)$	$\Delta H = -36.5$ kJ
Sum:	$\tfrac{1}{2}I_2(s) + \tfrac{1}{2}H_2(g) \longrightarrow HI(g)$	$\Delta H = +25.9$ kJ

This example can be expanded to a general principle: *If a reaction occurs in two or more steps, the enthalpy for the reaction is the sum of the enthalpies of the individual steps*. This statement is known as **Hess's law of heat summation**, formulated by Germain Henri Hess (a Swiss-born Russian chemist, 1802–1850).

Let us consider another example of Hess's law. The reaction of solid carbon with gaseous oxygen to form carbon dioxide can be carried out in two steps. In the first step, carbon reacts with oxygen to form carbon monoxide. In the second step, carbon monoxide reacts with more oxygen to form carbon dioxide. These two steps and their corresponding enthalpies can be added to obtain the enthalpy for the net reaction:

$$\begin{array}{ll} C(s) + \frac{1}{2}O_2(g) \longrightarrow CO(g) & \Delta H = -111 \text{ kJ} \\ CO(g) + \frac{1}{2}O_2(g) \longrightarrow CO_2(g) & \Delta H = -283 \text{ kJ} \\ \hline CO(s) + O_2(g) \longrightarrow CO_2(g) & \Delta H = -394 \text{ kJ} \end{array}$$

We can use Hess's law to determine the enthalpy for a reaction that is difficult or impossible to carry out in a laboratory. This can be done if some related reactions can be performed whose sum equals the reaction of interest. These related reactions can be multiplied by certain factors, or written in reverse to obtain the net reaction whose enthalpy we want to determine.

When we *multiply an equation* for a reaction by a factor, we must also remember to *multiply the* ΔH of the reaction by the same factor. When we *reverse a reaction*, we must remember to *change the sign* of the ΔH of the reaction. If a reaction is exothermic when written from left to right, it is endothermic when written from right to left. Example 5.2 illustrates an application of Hess's law.

Example 5.2

Using Hess's Law to Determine the Enthalpy of a Reaction from Given Enthalpies of Several Related Reactions

Under ordinary conditions, carbon does not react with hydrogen to any appreciable extent. However, the enthalpy change for the reaction

$$C(s) + 2H_2(g) \longrightarrow CH_4(g)$$

can be determined by carrying out the following series of related reactions and measuring their enthalpy changes:

$$\begin{array}{lll} \textbf{(1)} & C(s) + O_2(g) \longrightarrow CO_2(g) & \Delta H = -393.51 \text{ kJ} \\ \textbf{(2)} & H_2(g) + \frac{1}{2}O_2(g) \longrightarrow H_2O(l) & \Delta H = -285.83 \text{ kJ} \\ \textbf{(3)} & CH_4(g) + O_2(g) \longrightarrow CO_2(g) + 2H_2O(l) & \Delta H = -890.37 \text{ kJ} \end{array}$$

From these data, calculate the enthalpy for the reaction of solid carbon with hydrogen gas to form methane gas, CH_4, at 25°C.

SOLUTION: Reaction (1) has the $C(s)$ that we need as a reactant, so this reaction can be used in its present form in the summation process. Reaction (2) has $H_2(g)$ as a reactant, but we need $2H_2(g)$, so we multiply reaction (2) by 2. Reaction (3) has CH_4 as a reactant, so this reaction must be reversed to give $CH_4(g)$ as a product. Adding these revised reactions and their enthalpies yields:

$$\begin{array}{ll} C(s) + O_2(g) \longrightarrow CO_2(g) & \Delta H = -393.51 \text{ kJ} \\ 2H_2(g) + O_2(g) \longrightarrow 2H_2O(l) & \Delta H = -571.66 \text{ kJ} \\ \underline{CO_2(g) + 2H_2O(l) \longrightarrow CH_4(g) + 2O_2(g)} & \underline{\Delta H = +890.37 \text{ kJ}} \\ \text{Sum: } C(s) + 2H_2(g) \longrightarrow CH_4(g) & \Delta H = -74.80 \text{ kJ} \end{array}$$

Practice Problem 5.1: The decomposition of water vapor to the atoms of its elements, hydrogen and oxygen,

$$H_2O(g) \longrightarrow 2H(g) + O(g)$$

is difficult to carry out because the atoms released by the decomposition of water immediately combine to form diatomic molecules. Calculate the enthalpy change for the foregoing reaction from the following data:

$$\begin{array}{ll} H_2(g) + \frac{1}{2}O_2(g) \longrightarrow H_2O(g) & \Delta H = -242 \text{ kJ} \\ H_2(g) \longrightarrow 2H(g) & \Delta H = +436 \text{ kJ} \\ O(g) \longrightarrow \frac{1}{2}O_2(g) & \Delta H = -249 \text{ kJ} \end{array}$$

Enthalpy of Formation

The *enthalpy change observed when one mole of a compound in its standard state forms from its elements in their standard states at 1 atm and 298 K* is called the **standard molar enthalpy of formation** or, unless otherwise specified, the **standard enthalpy of formation**, denoted by ΔH_f°. Table 5.1 lists the standard enthalpies of formation for some common compounds. We discussed the standard physical states of the elements in Section 2.8.

Example 5.3

Writing an Equation to Represent a Standard Molar Enthalpy of Formation

Which of the following equations represents the correct conditions for determining the standard molar enthalpy of formation, ΔH_f°, of gaseous hydrogen iodide? (1) $\frac{1}{2}H_2(g) + \frac{1}{2}I_2(g) \rightarrow HI(g)$; (2) $\frac{1}{2}H_2(g) + \frac{1}{2}I_2(s) \rightarrow HI(g)$; (3) $\frac{1}{2}H_2(g) + \frac{1}{2}I_2(s) \rightarrow HI(s)$; (4) $H(g) + I(g) \rightarrow HI(g)$; (5) $H_2(g) + I_2(s) \rightarrow 2HI(g)$.

SOLUTION: Equation (2) is the correct one. In equation (1), I_2 is written as a gas, which is not the standard state of iodine. In equation (3), HI is written as a solid, but the standard state is a gas. In equation (4), gaseous H and I are written in atomic forms, which are not the standard states for these elements. Equation (5) refers to the formation of 2 mol of HI, which is not consistent with the definition of ΔH_f°.

Table 5.1

Standard Enthalpies of Formation (kJ mol^{-1}) of Some Common Compounds by Periodic Groups

Compound	ΔH_f°
Group I	
$LiF(s)$	−612.1
$LiCl(s)$	−408.8
$LiI(s)$	−271.0
$NaF(s)$	−569.0
$NaCl(s)$	−411.0
$NaI(s)$	−288.0
$KCl(s)$	−435.89
$CsCl(s)$	−433.0
$Na_2O(s)$	−416
$Cs_2O(s)$	−318
$LiH(s)$	−90.4
$Na_2CO_3(s)$	−1131
$KClO_3(s)$	−391.2
Group II	
$MgF_2(s)$	−1102
$MgCl_2(s)$	−641.8
$MgO(s)$	−601.83
$CaF_2(s)$	−1215
$CaH_2(s)$	−189
Group III	
$BF_3(g)$	−1135.95
$B_2O_3(s)$	−1273.5
$Al_2O_3(s)$	−1669.8
$Al_2S_3(s)$	−508.8
Group IV	
$CH_4(g)$ methane	−74.848
$C_2H_6(g)$ ethane	−84.667
$C_3H_8(g)$ propane	−103.847
$C_8H_{18}(l)$ octane	−250.0
$C_2H_4(g)$ ethene	52.283
$C_2H_2(g)$ ethyne	226.75
$C_6H_6(l)$ benzene	49.028
$CH_3OH(l)$ methanol	−238.64
$C_2H_5OH(l)$ ethanol	−277.63
$C_6H_{12}O_6(s)$ glucose	−1260
$C_{12}H_{22}O_{11}(s)$ sucrose	−2221
$CCl_4(l)$	−139.3
$CO(g)$	−110.54
$CO_2(g)$	−393.51
$SiO_2(s)$	−859.4

Compound	ΔH_f°
Group V	
$NH_3(g)$ ammonia	−45.94
$NO(g)$	90.4
$NO_2(g)$	33.8
Group VI	
$H_2O(g)$	−241.814
$H_2O(l)$	−285.830
$SO_2(g)$	−296.81
$SO_3(g)$	−395.2
$H_2S(g)$	−20.15
Group VII	
$HF(g)$	−269
$HCl(g)$	−92.30
$HBr(g)$	−36.2
$HI(g)$	25.9
Transition metals	
$Fe_2O_3(s)$	−822.2
$AgCl(s)$	−127.0
$TiO_2(s)$	−912.1

We have defined the standard enthalpy of formation of chemical compounds. But what is the standard enthalpy of formation of an element? Let us consider the formation of a mole of O_2 gas from the same element, $O_2(g)$:

$$O_2(g) \longrightarrow O_2(g)$$

The initial and final states are identical, their enthalpies are equal, and according to Equation 5.5, ΔH_{rxn} is zero:

$$\Delta H_{rxn} = H[O_2(g)] - H[O_2(g)] = 0$$

Therefore, the *standard enthalpy of formation of an element in its standard state is zero.*

Enthalpies of formation provide us with information about the stabilities of compounds relative to their elements. The word "stability" here means "thermodynamic stability." If the standard enthalpy of formation of a compound is highly exothermic, the reverse reaction, thermal decomposition of the compound, is highly endothermic. Such a compound must absorb a considerable amount of heat before it decomposes, and is therefore described as thermodynamically stable.

Enthalpies of formation of compounds also enable us to calculate the enthalpies of reactions involving these compounds. We discuss this type of calculation below.

Enthalpies of Reaction from Enthalpies of Formation

Once we have determined the enthalpies of formation of many compounds, we can use these data to calculate the enthalpy changes for various reactions. Suppose, for example, that we wish to calculate the amount of heat required to convert a mole of liquid water to its gaseous state, a quantity called the *heat of vaporization*. The process can be written as

$$H_2O(l) \longrightarrow H_2O(g) \qquad \Delta H_{\text{vaporization}} = ?$$

To solve this problem we use the standard enthalpies of formation of liquid and gaseous H_2O, and we can apply Hess's law:

$$\begin{aligned} H_2(g) + \tfrac{1}{2}O_2(g) &\longrightarrow H_2O(g) \qquad \Delta H_f^\circ = -242 \text{ kJ mol}^{-1} \\ H_2(g) + \tfrac{1}{2}O_2(g) &\longrightarrow H_2O(l) \qquad \Delta H_f^\circ = -286 \text{ kJ mol}^{-1} \end{aligned}$$

If we reverse the second reaction and then add the reactions and their corresponding ΔH values, we obtain the equation for the vaporization and the value for $\Delta H_{\text{vaporization}}$:

$$\begin{aligned} H_2(g) + \tfrac{1}{2}O_2(g) \longrightarrow H_2O(g) &\qquad \Delta H^\circ = -242 \text{ kJ} \\ \underline{H_2O(l) \longrightarrow H_2(g) + \tfrac{1}{2}O_2(g)} &\qquad \underline{\Delta H^\circ = +286 \text{ kJ}} \\ H_2O(l) \longrightarrow H_2O(g) &\qquad \Delta H_{\text{vaporization}} = +44 \text{ kJ} \end{aligned}$$

This result is identical to the value obtained by subtracting the standard enthalpy of formation of liquid water (the reactant in the original reaction) from that of gaseous water (the product):

$$-242 \text{ kJ} - (-286 \text{ kJ}) = 44 \text{ kJ}$$

Similarly, the *enthalpy change for a chemical reaction* equals *the sum of the enthalpies of formation of all the products minus the sum of the enthalpies of formation of all the reactants*:

$$\Delta H^\circ_{rxn} = \Sigma\, \Delta H_f^\circ(\text{products}) - \Sigma\, \Delta H_f^\circ(\text{reactants}) \tag{5.6}$$

(The Greek uppercase sigma, Σ, is used to symbolize summation.) When we sum the enthalpies of formation, *the enthalpy of formation of each substance must be multiplied by its coefficient in the balanced equation* for the reaction. Example 5.4 illustrates this type of calculation.

Example 5.4

Calculating ΔH°_{rxn} from Known ΔH°_f Values

Use the data from Table 5.1 to calculate the standard enthalpy change for the photosynthesis reaction in which plants convert carbon dioxide and water to the sugar glucose, $C_6H_{12}O_6(s)$, and oxygen gas. Is this reaction exothermic or endothermic?

SOLUTION: The balanced equation for the reaction is

$$6CO_2(g) + 6H_2O(l) \longrightarrow C_6H_{12}O_6(s) + 6O_2(g)$$

Applying Equation 5.6, we obtain

$$\begin{aligned}\Delta H^\circ_{rxn} &= [\Delta H^\circ_f \text{ of } C_6H_{12}O_6(s) + 6\Delta H^\circ_f \text{ of } O_2(g)] \\ &\quad - [6\Delta H^\circ_f \text{ of } CO_2(g) + 6\Delta H^\circ_f \text{ of } H_2O(l)] \\ &= [(1 \text{ mol})(-1260 \text{ kJ mol}^{-1}) + 0] \\ &\quad - [(6 \text{ mol})(-394 \text{ kJ mol}^{-1}) + (6 \text{ mol})(-286 \text{ kJ mol}^{-1})] \\ &= +2820 \text{ kJ}\end{aligned}$$

The positive sign for ΔH°_{rxn} indicates that the reaction is endothermic. The energy required to drive the synthesis of glucose from carbon dioxide and water is provided by light.

Practice Problem 5.2: Propane gas, C_3H_8, is used as a fuel in rural areas. When propane is ignited, the combustion of propane occurs. We recall from Chapter 2 that combustion is a reaction in which a substance combines with oxygen and that the complete combustion of hydrocarbons produces carbon dioxide and water. Assume that the water forms as a liquid, and use the necessary data from Table 5.1 to calculate the standard enthalpy of combustion of propane.

Practice Problem 5.3: Octane is a component of gasoline. The standard enthalpy of combustion of octane, $C_8H_{18}(l)$, to produce CO_2 gas and liquid water, is -5470 kJ mol^{-1}. Use data from Table 5.1 to calculate the standard enthalpy of formation of octane.

5.4 Enthalpy Changes and Bond Energies

In the preceding section we saw that when a chemical reaction is exothermic, the products are more stable than the reactants. During the course of a chemical reaction of molecular compounds, the bonds between the atoms of the reactant molecules break, and the atoms rearrange to form new bonds in the product molecules. Since we know that the products of an exothermic reaction are more stable than the reactants, it follows that the products have *stronger bonds* or *more bonds* than the reactants.

We can use the enthalpy change of a chemical reaction to determine bond strengths, or **bond energies**. The bond energy is defined as the *energy required to break a bond* (Figure 5.5). Since energy is required to break a bond, energy is

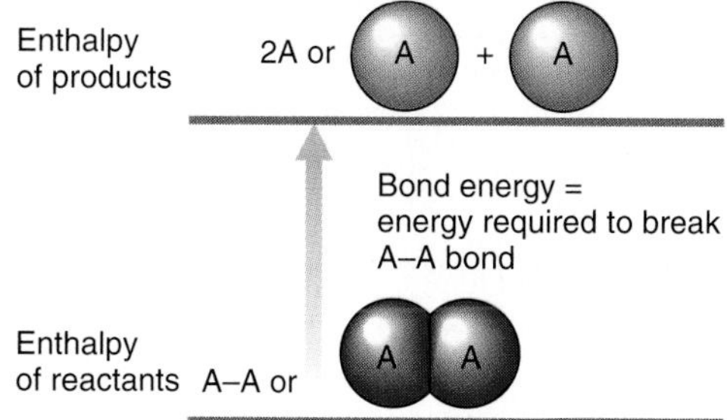

Figure 5.5

Bond energy. The energy required to break the bond in the molecule A—A is called the bond energy. This process is endothermic.

Table 5.2

Single Bond Energies (kJ mol^{-1}) in Diatomic and Polyatomic Molecules

H	C	N	O	F	Si	P	S	Cl	Br	I	
436	413	391	459	569	318	322	363	432	366	299	**H**
	346	305	358	485	318	264	272	327	285	213	**C**
		167	201	283				313			**N**
			142	190	452	335	532	218	201	201	**O**
				157	565	490	284	250	234	280	**F**
					222		293	381	310	234	**Si**
						201		326	264	184	**P**
							226	255	217		**S**
								243	219	211	**Cl**
									194	179	**Br**
										152	**I**

released when a bond forms. Furthermore, since heat is a form of energy, bond energies can be measured in terms of enthalpy changes. For example, the energy of a mole of H—H bonds equals the enthalpy change for the dissociation of 1 mol of H_2 gas into 2 mol of H atoms:

$$H_2(g) \longrightarrow H(g) + H(g) \qquad \Delta H^\circ = +436 \text{ kJ}$$

In contrast, the formation of 1 mol of H_2 molecules from H atoms is an exothermic process:

$$2H(g) \longrightarrow H_2(g) \qquad \Delta H^\circ = -436 \text{ kJ}$$

Thus, the energy required to break a mole of H—H bonds is 436 kJ. In other words, the H—H bond energy is 436 kJ/mol. Some common bond energies are listed in Table 5.2.

Let us consider the reaction of $\frac{1}{2}$ mol of H_2 with $\frac{1}{2}$ mol of Cl_2 to form 1 mol of HCl:

$$\tfrac{1}{2}H_2(g) + \tfrac{1}{2}Cl_2(g) \longrightarrow HCl(g)$$

Mentally, we can divide this reaction into two processes, considering first the bonds that are broken, and then the bonds that are formed. Finally, we can add the enthalpy changes for these processes to obtain the enthalpy change for the net reaction. These steps are diagrammed in Figure 5.6 and outlined below.

In this reaction, let us first consider the endothermic processes of breaking $\frac{1}{2}$ mol of H—H bonds and $\frac{1}{2}$ mol of Cl—Cl bonds:

$$\tfrac{1}{2}(\text{H—H})(g) \longrightarrow H(g) \qquad \Delta H^\circ = +218 \text{ kJ}$$
$$\tfrac{1}{2}(\text{Cl—Cl})(g) \longrightarrow Cl(g) \qquad \Delta H^\circ = +122 \text{ kJ}$$

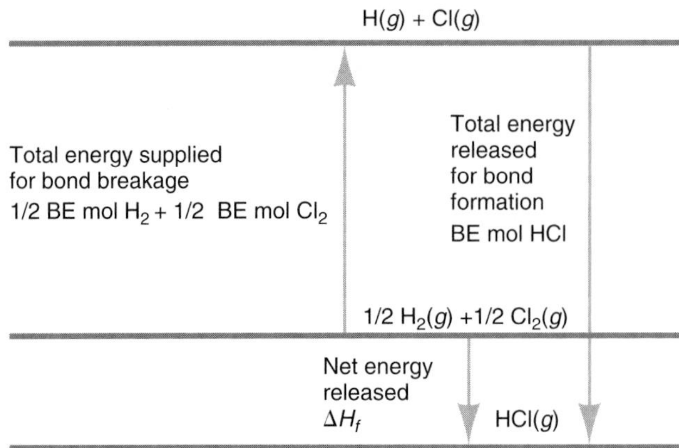

Figure 5.6

Energetics of the reaction $\frac{1}{2}H_2(g) + \frac{1}{2}Cl_2(g) \rightarrow HCl(g)$.

The separated H and Cl atoms then recombine to form H—Cl molecules. The formation of the H—Cl bonds is an exothermic process. The heat released by the formation of a mole of H—Cl bonds is −432 kJ:

$$H(g) + Cl(g) \longrightarrow HCl(g) \qquad \Delta H^\circ = -432 \text{ kJ}$$

The sum of the enthalpy changes of the bond-breaking and bond-forming processes equals the enthalpy change for the net reaction:

$$\Delta H^\circ_{rxn} = 218 \text{ kJ} + 122 \text{ kJ} - 432 \text{ kJ} = -92 \text{ kJ}$$

This example reveals a general relationship between ΔH of a reaction and the bond energies of its molecules. *For any chemical reaction, the enthalpy change is the sum of all the bond energies of the reactants minus the sum of all the bond energies of the products*:

$$\Delta H^\circ_{rxn} = \Sigma(\text{BE of reactants}) - \Sigma(\text{BE of products}) \qquad (5.7)$$

where BE means bond energy. Example 5.5 illustrates an application of Equation 5.7.

Example 5.5

Using Bond Energies to Calculate the Enthalpy of a Reaction

Methane, CH_4, gas is the major component of natural gas. When methane reacts with excess chlorine, one of the products of the reaction is carbon tetrachloride, CCl_4. Using the data from Table 5.2, calculate ΔH° for the reaction

$$CH_4(g) + 4Cl_2(g) \longrightarrow CCl_4(g) + 4HCl(g)$$

Hint: In a methane molecule, the C atom is bonded to four H atoms, and in a carbon tetrachloride molecule, the C atom is bonded to four Cl atoms:

```
      H                  Cl
      |                  |
   H—C—H    and    Cl—C—Cl
      |                  |
      H                  Cl
    (CH4)              (CCl4)
```

SOLUTION: The bond energies of the reactant molecules are as follows:

For 4 mol C—H bonds: (4 mol)(413 kJ mol^{-1}) = 1652 kJ
For 4 mol Cl—Cl bonds: (4 mol)(243 kJ mol^{-1}) = 972 kJ
Total = 2624 kJ

The bond energies of the product molecules are:

For 4 mol C—Cl bonds: (4 mol)(327 kJ mol^{-1}) = 1308 kJ
For 4 mol H—Cl bonds: (4 mol)(432 kJ mol^{-1}) = 1728 kJ
Total = 3036 kJ

Application of Equation 5.7 yields

$$\Delta H^\circ_{rxn} = 2624 \text{ kJ} - 3036 \text{ kJ} = -412 \text{ kJ}$$

This is an exothermic reaction because ΔH_{rxn} is negative.

Practice Problem 5.4: The standard enthalpy change for the reaction

$$CH_4(g) + 4F_2(g) \longrightarrow CF_4(g) + 4HF(g)$$

is −1918 kJ. Use this value and the data from Table 5.2 to calculate the C—F bond energy.

In the present chapter we emphasize the enthalpies of reactions of molecular compounds in which the atoms are covalently bonded. Ionic compounds consist of ions held together by electrostatic attraction in characteristic geometric structures or lattices. These compounds contain no molecules and thus no covalent bonds. The equivalent of bond energy in an ionic compound is the *energy required to separate the ions from a solid lattice*. This energy is called the **lattice energy**. The enthalpy of formation of an ionic solid therefore equals the enthalpy change for forming a mole of a solid lattice from its elements. For example, when 1 mol of sodium chloride forms from 1 mol of sodium and $\frac{1}{2}$ mol of chlorine, the enthalpy change is

$$Na(s) + \tfrac{1}{2}Cl_2(g) \longrightarrow NaCl(s) \qquad \Delta H^\circ_{lattice} = -410 \text{ kJ mol}^{-1}$$

The enthalpy of a reaction is also the energy that can be obtained from a fuel in its combustion, or the energy (calories) that can be derived from a food in the process of metabolism. We discuss fuels and foods in Section 5.6. In Section 5.5 we consider some methods used to measure enthalpy changes.

5.5 Measurement of Enthalpy Changes: Calorimetry

The heat transferred by a chemical or physical change can be measured in an insulated container called a **calorimeter** (Figure 5.7). The measurement of heat transfer in a calorimeter is called **calorimetry**. The first calorimeter was built in the 1780s by Antoine Lauren Lavoisier—who, we recall, also proposed the law of conservation of mass (Chapter 1)—and Pierre Simon de Laplace. These workers studied the energy changes in a guinea pig's respiration, a process they described as combustion.

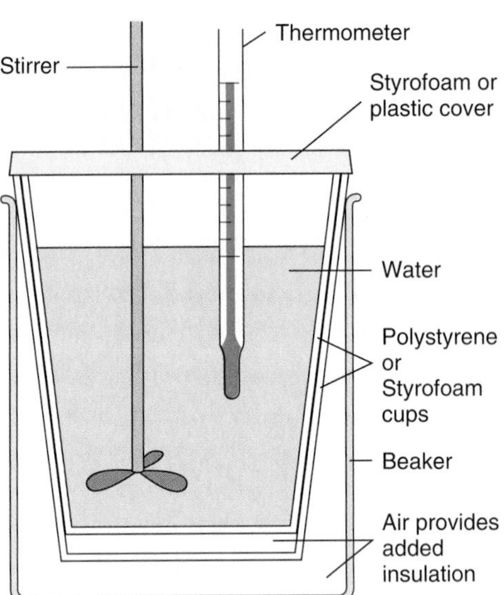

Figure 5.7

A simple constant pressure calorimeter. A chemical reaction or other process carried out in this type of a calorimeter occurs at the constant pressure of the surrounding atmosphere. The amount of heat, q_p, absorbed or released by the process is calculated as the product of the observed temperature change and the known heat capacity of the water and the calorimeter.

Figure 5.7 illustrates an inexpensive calorimeter. It is an insulated container filled with water and fitted with a thermometer and a stirrer. When a chemical reaction is carried out in this calorimeter, the temperature of the water rises if heat is released by the reaction. The temperature drops if the reaction absorbs heat from the water. The amount of heat absorbed or released by a reaction in a calorimeter can be calculated from the known quantity of the water in the calorimeter and its temperature change during the reaction. We describe how this is done in the next section.

The simple calorimeter illustrated in Figure 5.7 is not tightly sealed, and reactions carried out in it occur at constant external atmospheric pressure. The device is therefore called a **constant pressure calorimeter**. A constant pressure calorimeter is generally used to measure the enthalpy changes of reactions that occur in aqueous solution or those that involve solid or liquid reactants and products. A constant pressure calorimeter is not usually used for reactions involving gaseous reactants or products.

Heat Capacity

The *amount of heat required to raise the temperature of an object by 1 degree Celsius (or 1 K)* is its **heat capacity**. The *heat capacity of 1 mole* of a substance is the **molar heat capacity**. The *heat capacity of 1 gram* of a substance is its **specific heat**. Table 5.3 lists the specific heats of some common substances.

The specific heat of water at 15°C is 1.000 cal g^{-1} $°C^{-1}$, or 4.184 J g^{-1} $°C^{-1}$. Thus, the definition of the specific heat of water is also a definition of the calorie (see Chapter 1). The molar heat capacity of water is

$$(1.000 \text{ cal g}^{-1}\ °\text{C}^{-1})(18.02 \text{ g mol}^{-1}) = 18.02 \text{ cal mol}^{-1}\ °\text{C}^{-1}$$
$$(18.02 \text{ cal mol}^{-1}\ °\text{C}^{-1})(4.184 \text{ J cal}^{-1}) = 75.40 \text{ J mol}^{-1}\ °\text{C}^{-1}$$

If the heat capacity of an object is known, the amount of heat it gains when heated equals the product of its heat capacity and its temperature change (Δt)

Table 5.3

Specific Heats of Some Common Substances at 25°C[a]

Substance	Specific Heat ($J\ g^{-1}\ °C^{-1}$)
Aluminum	0.900
Carbon tetrachloride	0.8624
Ethyl alcohol	2.4558
Ice	2.06[b]
Iron	0.444
Iron oxide (Fe_2O_3)	0.653
Mercury	0.1389
Sand	0.787
Sulfur	0.732
Water	4.1796

[a] Specific heats vary with temperature. For example, the specific heat of water is 4.2177 at 0°C, 4.1840 at 17°C, and 4.2160 at 100°C.

[b] At −5°C.

during the process. Similarly, if an object is cooled, the heat lost is also the product of its heat capacity and the temperature change. These statements can be summarized by the equation

$$\text{heat gained or lost} = (\text{heat capacity})(\Delta t) \qquad (5.8)$$

Example 5.6

Using the Heat Capacity of an Object to Calculate the Amount of Heat Transferred

The heat capacity of a coffee mug is 112 J °C^{-1}. When the mug is filled with hot coffee, the temperature of the mug increases from 24.7°C to 56.8°C. How much heat is gained by the mug?

SOLUTION: The temperature change of the mug is

$$\Delta t = t_{\text{final}} - t_{\text{initial}} = 56.8°C - 24.7°C = 32.1°C$$

Applying Equation 5.8, we have

$$(112\ J\ °C^{-1})(32.1°C) = 3.60 \times 10^3\ J$$

Practice Problem 5.5: The heat capacity of water in a cup is 418 J °C^{-1}. How much heat is lost by the water when it cools from 30.0°C to 24.8°C?

The heat capacity of an object is its mass in grams times its specific heat:

$$\text{heat capacity} = (\text{mass in grams})(\text{specific heat}) \qquad (5.9)$$

Thus, if the mass and the specific heat of a substance are known, the amount of heat it gains or loses can be calculated from its temperature change, Δt, during the process:

$$\text{heat gained or lost} = (\text{grams of substance})(\text{specific heat})(\Delta t) \qquad (5.10)$$

Example 5.7

Using the Specific Heat of a Substance to Calculate the Amount of Heat Transferred

The specific heat of iron is 0.444 J g^{-1} $°C^{-1}$. If a 25.0-g iron bolt is cooled from 26.58°C to 20.45°C, how much heat is lost by the bolt?

SOLUTION: The temperature change upon cooling is

$$\Delta t = t_{\text{final}} - t_{\text{initial}} = 20.45°\text{C} - 26.58°\text{C} = -6.13°\text{C}$$

Applying Equation 5.10, we have

$$\text{heat lost} = (25.0\text{ g})(0.444\text{ J g}^{-1}\,°\text{C}^{-1})(-6.13°\text{C}) = -68.0\text{ J}$$

Practice Problem 5.6: A 42.5-g sample of an alloy (a mixture of metals) is heated to 94.20°C and then placed into water, where it cools to 26.54°C. The amount of heat lost by the alloy sample upon cooling is 870 J. What is the specific heat of the alloy?

We can see from Table 5.3 that water has the highest specific heat of all the substances listed. Thus, if equal masses of water and, say, aluminum, absorb the same quantity of heat, the temperature of aluminum increases more than the temperature of water. The high heat capacity of water has important biological consequences. Since the human body is about 60 percent of water by mass, the temperature of the human body does not change significantly when it absorbs or loses a moderate quantity of heat. Of course, various other mechanisms in the human body also help to maintain a constant temperature.

Heat Capacity and Calorimetry

Having defined heat capacity, we can now use this concept to determine heats of reactions in a calorimeter. As a preliminary step, the mass and temperature of the calorimeter's water bath are measured. When a reaction occurs in the calorimeter, the temperature change of the water in the calorimeter is Δt. The mass of the water in the bath, of course, remains constant.

One calorie (4.184 J) is absorbed by 1 gram of water if its temperature rises by 1 degree Celsius. Similarly, 1 calorie of heat is lost by 1 gram of water for a 1-degree decrease in the Celsius temperature. Thus, according to Equation 5.10, the heat of a reaction can be calculated by multiplying the mass in grams of the water in a calorimeter by the specific heat of water and by the temperature change, Δt, of water:

$$\text{heat lost or gained by H}_2\text{O} = (\text{grams H}_2\text{O})(\text{specific heat H}_2\text{O})(\Delta t)$$

Example 5.8 illustrates an application of this equation.

APPLICATIONS OF CHEMISTRY 5.2
The Amazing Water Diet

According to medical specialists, the average American daily diet is about 2500 Calories per day. (Nutritionists use a capital C to spell the word "calorie." Later in this box we discuss the difference between a calorie and a Calorie.) This means that the food we consume oxidizes in our bodies to produce on average 2500 Cal. Our bodies use these 2500 Cal to power the various biochemical reactions that keep us alive. These reactions include breaking down nutrients to produce cellular energy, conversion of nutrients into compounds that our bodies need, and maintaining a constant body temperature, to name a few.

Medical authorities recommend that the average American drink six glasses of water each day for overall good health. Six glasses of water equals 1.5 qt or about 1.5 L. The heat from our bodies warms the water we drink until the water reaches body temperature. How much heat would our bodies have to provide to warm 1.5 L of ice-cold (0°C) water to body temperature (37°C)?

The temperature change upon warming is 37°C. Applying Equation 5.10, we have the following:

$$\begin{aligned}\text{heat lost} &= (1500\text{ g})(1.000\text{ cal g}^{-1}\,{}^\circ\text{C}^{-1})(37^\circ\text{C})\\ &= 5.5\times 10^{4}\text{ cal}\end{aligned}$$

Does this answer make sense? Does it make sense that each day you consume enough food to produce 2500 Cal and can drink six glasses of water that consumes 55,000 cal? What an amazingly easy way to lose weight. Unfortunately, it does not work.

In the area of nutrition, a Calorie (with a capital C) is what other scientists refer to as a kilocalorie (that is, 1000 calories). The average American diet is actually 2,500,000 calories per day (2500 kcal/day). This nutritional jargon of referring to a kilocalorie as a Calorie is very confusing and can lead to the type of misunderstanding that we have just encountered.

One and a half liters of ice-cold water does absorb 55,000 calories from our bodies, but that's only 55 kcal of the 2500 kcal supplied by our daily diets. However, drinking cold beverages can be a worthwhile activity. Fifty-five kcal is significant when we realize that we have to walk more than 2 miles to burn 55 kcal of food or body fat.

Example 5.8

Determining the Heat of Reaction Assuming No Gain or Loss of Heat by the Calorimeter Walls

A chemical reaction is carried out in 0.500 kg of water in a calorimeter. As a result of the reaction, the temperature of the water rises from 15.5°C to 17.7°C. Calculate the amount of heat released by the reaction, assuming that all the heat is absorbed by the water and only a negligible amount is absorbed by the calorimeter walls. Is this reaction endothermic or exothermic?

SOLUTION: The temperature change (Δt) of the water is

$$\Delta t = t_{\text{final}} - t_{\text{initial}} = 17.7^\circ\text{C} - 15.5^\circ\text{C} = 2.2^\circ\text{C}$$

Each gram of water absorbs 4.184 J per 1°C temperature increase. The mass of 0.500 kg (5.00×10^2 g) of water absorbs $(5.00 \times 10^2)(4.184)$ joules of heat for each 1°C temperature increase. The total amount of heat absorbed by the water is therefore the product of the mass of water, its specific heat, and the temperature change (Equation 5.10):

$$\begin{aligned}\text{heat absorbed by water} &= -\text{heat released by the reaction}\\ &= (\text{mass of water})(\text{specific heat of water})(\Delta t)\\ &= (5.00\times 10^{2}\text{ g})(4.184\text{ J g}^{-1}\,{}^\circ\text{C}^{-1})(2.2^\circ\text{C})\\ &= 4.6\times 10^{3}\text{ J} = 4.6\text{ kJ}\end{aligned}$$

The temperature increase of water is caused by the heat released by the reaction. The reaction is therefore exothermic, and the heat of the reaction = $\Delta H_{rxn} = -4.6$ kJ.

In our preceding calculation we neglected any heat gained or lost by the walls of the calorimeter itself. The amount of heat needed to raise or lower the temperature of the calorimeter by 1°C is the heat capacity of the calorimeter. This heat capacity is either determined by the manufacturer or by the user of the calorimeter.

Example 5.9

Determining the Heat of a Reaction by Considering the Heat Absorbed by Both the Water and the Calorimeter

A 50.0-mL sample of a 0.500 *M* HCl solution is mixed with 50.0 mL of a 0.500 *M* NaOH solution in a calorimeter. The heat capacity of the calorimeter is 335 J °C^{-1}. The temperature of the calorimeter and each of the solutions before mixing is 25.20°C. The temperature of the calorimeter and the solution after mixing is 26.40°C. Assuming that the density and the specific heat of the mixture are 0.9970 g mL^{-1} and 4.184 J g^{-1} °C^{-1}, respectively, calculate the heat of the reaction. Is this reaction endothermic or exothermic?

SOLUTION: The reaction in the calorimeter is

$$HCl(aq) + NaOH(aq) \longrightarrow NaCl(aq) + H_2O(l)$$

Because equal numbers of moles of the acid and base are mixed, the product is a solution of sodium chloride having a total volume of 100 mL. The mass of the solution is

$$100 \text{ mL} \left(\frac{0.9970 \text{ g}}{1 \text{ mL}}\right) = 99.70 \text{ g}$$

The temperature change is the same for both the solution and the calorimeter walls:

$$\Delta t = 26.40°\text{C} - 25.20°\text{C} = 1.20°\text{C}$$

The positive sign for Δt indicates that the reaction is exothermic. The heat released by the reaction increases the temperature of the solution and the calorimeter. We apply Equation 5.10 to calculate the amount of heat, q_p(soln), released by the reaction that increases the temperature of the solution:

$$\begin{aligned} q_p(\text{soln}) &= -(99.70 \text{ g})(4.184 \text{ J g}^{-1} \text{ °C}^{-1})(1.20°\text{C}) \\ &= -5.01 \times 10^2 \text{ J} = -0.501 \text{ kJ} \end{aligned}$$

The amount of heat released by the reaction that increases the temperature of the calorimeter, q_p(cal), is the heat capacity of the calorimeter times the temperature change:

$$q_p(\text{cal}) = -(335 \text{ J °C}^{-1})(1.20°\text{C}) = -402 \text{ J} = -0.402 \text{ kJ}$$

The total amount of heat, $q_p(rxn)$, released by the reaction is the sum q_p(soln) and q_p(cal):

$$q_p(rxn) = q_p(\text{soln}) + q_p(\text{cal}) = -(0.501 \text{ kJ} + 0.402 \text{ kJ}) = -0.903 \text{ kJ}$$

This is also the ΔH for the reaction since the reaction occurs at constant pressure.

Practice Problem 5.7: In an experiment, 5.00 g of an unknown salt is dissolved in 50.0 mL of water both at 25.00°C. As a result of the solution process, the temperature decreases to 23.80°C. The density of water is 0.997 g mL^{-1}, the specific heat of the solution is 4.05 J g^{-1} °C^{-1}, and the heat capacity of the calorimeter is 32.3 J °C^{-1}. Calculate the heat of solution per gram of salt (the heat absorbed or released for each gram of the salt dissolved).

A constant pressure calorimeter can also be used to determine specific heats of solids, as shown in Example 5.10.

Example 5.10

Determining the Specific Heat of a Solid

In a determination of the specific heat of lead, 75.5 g of lead at 99.20°C is added to 76.0 g of water at 25.20°C in a constant pressure calorimeter. The final temperature of the system is 27.20°C. The heat capacity of the calorimeter is 29.1 J °C^{-1}. Calculate the specific heat of lead.

SOLUTION: Heat flows from an object of higher temperature (lead) to an object of lower temperature (water and the calorimeter). According to the first law of thermodynamics, the heat gained by water and the calorimeter equals the heat lost by lead:

$$q_{\text{water}} + q_{\text{calorimeter}} = -q_{\text{lead}}$$

The temperature change of water is the same as the temperature change of the calorimeter:

$$\Delta t_{\text{water}} = \Delta t_{\text{calorimeter}} = 27.20°\text{C} - 25.20°\text{C} = 2.00°\text{C}$$

The temperature change for lead is

$$\Delta t_{\text{lead}} = 27.20°\text{C} - 99.20°\text{C} = -72.00°\text{C}$$

The heat absorbed by water is

$$q_{\text{water}} = (76.0\text{ g})(4.184\text{ J g}^{-1}\text{ °C}^{-1})(2.00°\text{C}) = 636\text{ J}$$

and the heat absorbed by the calorimeter is

$$q_{\text{calorimeter}} = (29.1\text{ J °C}^{-1})(2.00°\text{C}) = 58.2\text{ J}$$

The heat lost by lead is

$$q_{\text{lead}} = -(75.5\text{ g})(x)(-72.00°\text{C})$$

where x represents the specific heat of lead. When we equate the heat gained by water and the calorimeter with the heat lost by lead, we have

$$636\text{ J} + 58.2\text{ J} = -(75.5\text{ g})(x)(-72.00°\text{C})$$

from which

$$x = \frac{636\text{ J} + 58.2\text{ J}}{(75.5\text{ g})(72.00°\text{C})} = 0.128\text{ J g}^{-1}\text{ °C}^{-1}$$

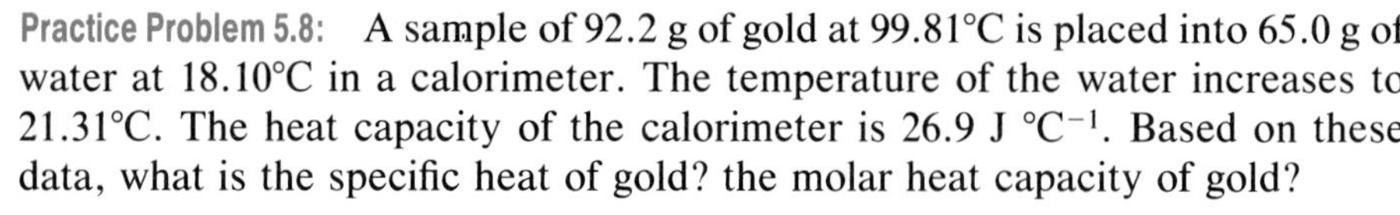

Practice Problem 5.8: A sample of 92.2 g of gold at 99.81°C is placed into 65.0 g of water at 18.10°C in a calorimeter. The temperature of the water increases to 21.31°C. The heat capacity of the calorimeter is 26.9 J °C^{-1}. Based on these data, what is the specific heat of gold? the molar heat capacity of gold?

Bomb Calorimetry

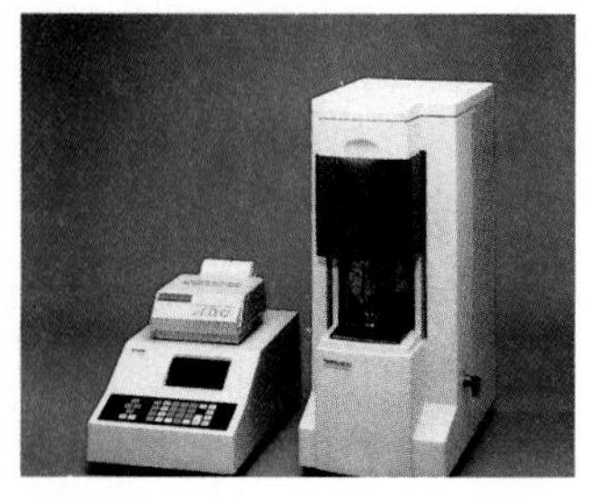

A commercial bomb calorimeter.

Enthalpy changes of reactions in which the reactants, products, or both are gases cannot be studied in a constant pressure calorimeter since it is open to the atmosphere. Enthalpy changes for gaseous reactions are measured in a **bomb calorimeter** (Figure 5.8).

Reactions in a bomb calorimeter occur at *constant volume*. A bomb calorimeter is therefore also called a **constant volume calorimeter**. In bomb calorimetry, the reactants are placed in a steel cylinder, the "bomb," that is sealed and lowered into a measured quantity of water in a calorimeter. The reaction is usually started by an electric current, and the temperature change of the water is recorded.

We recall that the change in internal energy, ΔE, of a system is the sum of heat, q, and work, w: $\Delta E = q + w$. If w represents only PV work, $\Delta E = q + P\Delta V$. Since the volume of a bomb calorimeter does not change, no PV work is done. Therefore, the heat, q_v, measured in a bomb calorimeter equals the internal energy change of the system,

$$q_v = \Delta E$$

But as we discussed in Section 5.2, the difference between ΔH and ΔE for even gaseous reactions is often quite small.

The heat absorbed or evolved in a bomb calorimeter is calculated as in constant pressure calorimetry, as shown in the following example.

Example 5.11

Determining the Heat of Combustion in a Bomb Calorimeter

A 1.000-g sample of hydrazine, N_2H_4, a rocket fuel, is burned with excess oxygen in a bomb calorimeter. The temperature in the calorimeter rises from 25.01°C to 28.56°C. The total heat capacity of the calorimeter and its contents is 5859 J °C^{-1}. What is the heat of combustion of hydrazine in joules per gram and in joules per mole of hydrazine?

SOLUTION: The temperature rise in the calorimeter indicates that the reaction evolves heat. The temperature change is

$$\Delta t = 28.56°\text{C} - 25.01°\text{C} = 3.55°\text{C}$$

$$\begin{aligned} -\text{heat released by the reaction} &= \text{heat gained by the calorimeter} \\ &= (5859\ \text{J °C}^{-1})(3.55°\text{C}) \\ &= 2.08 \times 10^4\ \text{J for } 1.00\ \text{g } N_2H_4 \end{aligned}$$

$$\left(\frac{2.08 \times 10^4\ \text{J}}{1.00\ \text{g } N_2H_4}\right)\left(\frac{32.0\ \text{g } N_2H_4}{1\ \text{mol } N_2H_4}\right) = 6.66 \times 10^5\ \text{J mol}^{-1}$$

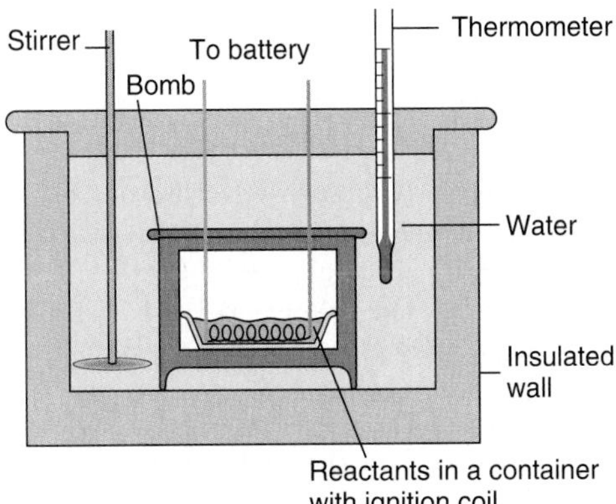

Figure 5.8

A sketch of a bomb calorimeter. A reaction in this type of a calorimeter occurs at constant volume (the volume of the bomb does not change). The amount of heat, q_V, absorbed or released by the reaction in a constant volume calorimeter is measured in the same way as in a constant pressure calorimeter—as the product of the observed temperature change and the known heat capacity of the calorimeter and its contents.

Practice Problem 5.9: A 1.60-g sample of magnesium is ignited electrically in a bomb calorimeter containing excess oxygen. The magnesium burns completely to form magnesium oxide. The temperature in the calorimeter rises from 25.8°C to 36.4°C. The heat capacity of the calorimeter is 1213 J °C^{-1}. What is the heat of combustion of magnesium in joules per gram and in joules per mole of magnesium?

5.6 Fuels and Foods

Combustion reactions of fuels are usually carried out in bomb calorimeters because such reactions have gaseous reactants and products. The amount of heat evolved by a given quantity, such as a gram, of a fuel is called its *fuel value*. In this section we discuss some data obtained by calorimetric measurements of fuel values, as well as the ''calories'' we gain from various foods.

Fuels

In this section we consider the heat released from the combustion of substances used to produce heat and other forms of energy for human consumption. Such substances are called **fuels**. The enthalpy of a combustion reaction is called the **heat of combustion**. In our search for fuels as energy sources, we may ask whether all exothermic chemical reactions are useful. The answer is ''no.'' We can use only combustible substances that are readily and economically available. Equally important, the combustion products of these substances must not be environmental pollutants. For example, the combustion of sulfur is an exothermic reaction. Furthermore, sulfur is relatively cheap and readily available. But when we burn sulfur, we obtain oxides of sulfur, which are air pollutants that contribute to acid rain. In fact, sulfur is an undesirable contaminant in oil and coal. Table 5.4 lists the amounts of energy that can be derived from the reaction of 1 g of various elements.

Table 5.4

Heats Released by 1.00 g of Various Elements in Their Reaction as Indicated (25°C and 1 atm)

Element	Reaction	Energy (kJ)
Li	$Li(s) + \frac{1}{2}F_2(g) \longrightarrow LiF(s)$	88.2
Na	$Na(s) + \frac{1}{2}Cl_2(g) \longrightarrow NaCl(s)$	17.9
Mg	$Mg(s) + F_2(g) \longrightarrow MgF_2(s)$	45.3
Al	$2Al(s) + 1\frac{1}{2}O_2(g) \longrightarrow Al_2O_3(s)$	30.9
B	$2B(s) + 1\frac{1}{2}O_2(g) \longrightarrow B_2O_3(s)$	58.9
C	$C(s) + O_2(g) \longrightarrow CO_2(g)$	32.8
Si	$Si(s) + O_2(g) \longrightarrow SiO_2(s)$	30.6
S_8	$\frac{1}{8}S_8(s) + O_2(g) \longrightarrow SO_2(g)$	9.26
H_2	$H_2(g) + \frac{1}{2}O_2(g) \longrightarrow H_2O(g)$	120
H_2	$H_2(g) + \frac{1}{2}O_2(g) \longrightarrow H_2O(l)$	142
H_2	$\frac{1}{2}H_2(g) + \frac{1}{2}F_2(g) \longrightarrow HF(g)$	266
H_2	$\frac{1}{2}H_2(g) + \frac{1}{2}Cl_2(g) \longrightarrow HCl(g)$	91.4

Of the elements listed in Table 5.4, only carbon and hydrogen are suitable as fuels. Carbon is the major component of coal. The reaction of hydrogen with oxygen is explosive but can be controlled. Since the combustion of hydrogen produces only water vapor as a by-product, hydrogen is environmentally safe and may find increasing use as a fuel in the near future.

Among the most common fuels are binary compounds of carbon and hydrogen—the hydrocarbons. The simplest of these is methane, CH_4, which is a major ingredient in natural gas. When methane (or any hydrocarbon) reacts completely with oxygen, the products are carbon dioxide and water (Chapter 2), as shown for the reaction of methane.

$$CH_4(g) + 2O_2(g) \longrightarrow CO_2(g) + 2H_2O(g) \qquad \Delta H^\circ = -891 \text{ kJ}$$

Methane is a gas at room temperature and 1 atm pressure, but many hydrocarbons are liquid at room temperature and pressure. These liquid fuels are mixtures of higher-molecular-weight hydrocarbons than those in natural gas.

Combustion reactions of hydrocarbons occur with breakage of C—H bonds in the hydrocarbon and O—O bonds in oxygen, and with the formation of C—O bonds in carbon dioxide and O—H bonds in water. Since these combustion reactions are exothermic, as we saw above for methane, we know that the products are more stable than the reactants.

The enthalpy change of a reaction corresponds to the difference in bond energies of the reactants and products. Therefore, the C—O bonds and the H—O bonds in the combustion products are collectively stronger than the O—O and C—H bonds in the reactants. Table 5.5 lists the fuel values of some common fuels.

Table 5.5

Approximate Fuel Values of Some Common Fuels

Fuel	Fuel Value (kJ g^{-1})
Coal (bituminous)	32
Gasoline	48
Natural gas	49
Oil (crude, from Texas)	45
Wood (pine)	18

Present Energy Sources

Figure 5.9

An oil refinery. (Photograph courtesy of Standard Oil Company)

Modern industrial societies consume vast amounts of energy. Our major energy sources today are *coal*, *oil* (petroleum), *natural gas*, and *nuclear energy*. Coal, oil, and natural gas are known as *fossil fuels* since they were formed from the decomposition of plants and microorganisms over a long period of time (10 to 300 million years). Once fossil fuels have been burned, they cannot easily be replaced. Thus, fossil fuels are sometimes called nonrenewable resources.

Coal is a combustible solid that contains more than 50 percent carbon by mass, and over 70 percent carbon-containing material by volume. Coal can be subdivided into several types or grades based on the carbon content of the sample. The highest-grade coal is *anthracite*, or hard coal, which contains 92 to 98 percent carbon. Lower-grade coals include semianthracite coal, semibituminous coal, and *bituminous* coal. Bituminous coal is also called soft coal because of its high content of soft, volatile matter. It contains 69 to 78 percent carbon. The lowest-grade coals are *lignite* and *peat*. The latter is a deposit of semicarbonized plant matter typically found in swamps and bogs.

Another fossil fuel is petroleum, or crude oil. Crude oil is a mixture of liquid hydrocarbons that are produced in the absence of air under conditions of high temperature and pressure deep beneath the earth's surface. Many petroleum deposits were formed from immense masses of microorganisms. The study of the hydrocarbons of these crude oil deposits reveals information about the chemical reactions that occurred in this decomposing matter in the remote past. Crude oil undergoes an immense metamorphosis in oil refineries, which produce everything from gasoline to heavy industrial heating oils and various high-molecular-weight lubricants, colloquially known as "greases." Varying amounts of gasoline are produced by oil refineries (Figure 5.9), ranging from 6 to 35 percent by volume of the crude oil, depending on the source of the oil.

Natural gas consists of low-molecular-weight gaseous hydrocarbons, mainly methane, CH_4 (70 to 90 percent by volume), and smaller percentages of ethane, C_2H_6, propane, C_3H_8, and butane, C_4H_{10}, in that order of decreasing percentage. Natural gas is found underground at various depths, as well as in solution with crude oil deposits. Methane is produced by decaying plant matter in areas of stagnant water, where the gas is known as *marsh gas*.

Fossil fuels are enormously important in industrial economies. Countries that control significant deposits of coal and oil possess both economic and political power. The United States consumes large amounts of fossil fuels and depends on foreign countries for a significant fraction of its oil.

It is inevitable that alternative sources of energy will be found to replace fossil fuels as these become increasingly scarce and expensive and as the environmental damage of burning fossil fuels becomes more and more intolerable. The alternative energy sources available today include solar, nuclear, and geothermal energies.

Foods

Human beings consume three classes of foods: *proteins*, *carbohydrates*, and *fats*. Of these three, proteins play a major role in building muscles, cartilage, hair, and skin. Most human energy is derived from carbohydrates and fats. Carbohydrates are the source of the quickest energy. This is why some athletes consume large amounts of carbohydrates in the form of honey or other sugar-containing foods shortly before competition. Table sugar, sucrose, $C_{12}H_{22}O_{11}$, is broken down in the body to simpler sugars, glucose and fructose, which both have the formula $C_6H_{12}O_6$. Glucose is known as blood sugar; it is soluble in blood and is transported by the blood to various tissues, where it is oxidized to form carbon dioxide, water, and energy. The net reaction for the complete oxidation of glucose is

$$C_6H_{12}O_6(s) + 6O_2(g) \longrightarrow 6CO_2(g) + 6H_2O(l) \qquad \Delta H^\circ_{rxn} = -2815 \text{ kJ}$$

This equation indicates that 1 mol of glucose produces 2815 kJ of energy in its reaction with oxygen. The molar mass of glucose is 180 g mol^{-1}; thus, the energy produced by 1 g of glucose is

$$\left(\frac{2816 \text{ kJ}}{1 \text{ mol}}\right)\left(\frac{1 \text{ mol}}{180 \text{ g}}\right) = \frac{15.6 \text{ kJ}}{1.00 \text{ g}}$$

This equals 3.73 kcal per gram of glucose.

Nutrition books commonly list the energies derived from foods in kilocalories, which are called "large calories" (Cal). Most foods are mixtures of proteins, fats, carbohydrates, water, and insoluble fiber. Table 5.6 lists the compositions of some common foods and the energies that they can provide. As a rule of thumb, the heats of combustion of carbohydrates, proteins, and fats are 4, 4, and 9 Cal g^{-1}, respectively.

Energy requirements of human bodies vary considerably with activity, body weight, and age. For example, from a small physically inactive adult to a large active athlete, the daily energy requirement may vary from 6000 kJ (1400 kcal) to 15,000 kJ (3600 kcal).

It has been estimated that an average 25-year-old male weighing 154 lb (70 kg) needs 2259 kJ (540 kcal) of energy for basic life activities (essentially, the

Table 5.6

Approximate Composition and Energy Equivalents of Some Common Foods

	Approximate Percent			Approximate Energy Value	
Food	**Protein**	**Fat**	**Carbohydrate**	**kJ g^{-1}**	**kcal g^{-1}**
Apples (raw)	0.2	0.6	14	2.3	0.56
Bacon (fried)	30	52	3.2	26	6.1
Beef	13–30	15–41	0	8.4–29	2–7
Beer	0.3	0	3.8	1.8	0.42
Bread (white, enriched)	8.7	3.2	50.4	11	2.7
Butter	0.6	81	0.4	30	7.2
Candy (chocolate)	4.4	35	58	22	5.3
(hard)	0	1.1	97	16	3.9
Cheese	8–36	4–38	1.6–8.2	4–18	1–4
Chicken (broiled)	24	3.8	0	5.9	1.4
Eggs (hard-cooked)	13	12	0.9	6.7	1.6
Fish	18	2–3	2	4–8	1–2
Ice cream	10	2.4	78	16	3.8
Milk	3.5	3.5	4.9	2.7	0.65
Peanuts (raw with skins)	26	48	19	24	5.6
Popcorn (plain)	12.7	5.0	77	16	3.9
Potatoes (baked)	2.6	0.1	21	3.9	0.93
(French fried)	4.3	13	36	11	2.7

Table 5.7

Energy Requirements by an Average Man (154 lb or 70 kg) and Woman (128 lb or 58 kg) of Age 25 for Various Activities

Type of Activity	Time (h)	Energy [kJ (kcal)]	
		Man	Woman
Sleeping	8	2259 (540)	1841 (440)
Sitting (with activities such as reading, writing, driving, etc.)	1	452 (108)	314 (75.0)
Standing (with some occasional walking)	1	628 (150)	377 (90.0)
Walking 2.5–3.0 mph, shopping, golf, sailing	1	637 (200)	628 (150)
Walking 3.5–4.0 mph, weeding and hoeing, scrubbing floors, moderate cycling	1	1255 (300)	1004 (240)
Walking with load uphill, work with pick and shovel, swimming, climbing, competitive tennis, skiing	1	2510 (600)	2008 (480)

basal metabolism) during 8 hours of sleeping. An average 25-year-old woman weighing 128 lb (58 kg) needs 1841 kJ (440 kcal) during her sleep. Energy requirements for various activities are listed in Table 5.7.

Summary

Types of Energy, Work, and Thermochemistry

All chemical and physical changes of matter are associated with energy changes. Energy can be classified as **kinetic energy**, the energy of motion, and **potential energy**, the energy of position or composition. **Work** is a movement of an object against some force and is defined as the *force times the distance moved.*

Investigation of *heat changes associated with chemical reactions* is called **thermochemistry**. Study of *all energy changes associated with chemical or physical changes* is called **thermodynamics**. In thermodynamics, a *sample of matter, or a chemical reaction under investigation*, is called a **system**, and *everything else* is referred to as the **surroundings**. The **state** of a system is described by a *set of properties* of the system such as its energy, temperature, and volume.

The First Law of Thermodynamics

The **first law of thermodynamics** states that *energy is neither created nor destroyed.* The energy lost by a system is gained by the surroundings, and vice versa:

$$+\Delta E_{\text{system}} = -\Delta E_{\text{surroundings}}$$

The first law of thermodynamics can be stated in terms of the energy change of a system, ΔE, which equals the sum of the heat, q, absorbed by the system, and the work, w, that is done on the system by the surroundings:

$$\Delta E = q + w$$

The work associated with a chemical reaction occurring at constant pressure is called pressure–volume or PV work. PV work is defined as the product of the constant pressure, P, and the change in volume, ΔV:

$$\text{pressure–volume work} = P\Delta V$$

Enthalpy Change

The *sum of the internal energy change of a system and its pressure–volume work* is called the **enthalpy change**, ΔH, for the system:

$$\Delta H = \Delta E + P\Delta V \quad \text{(at constant pressure)}$$

The enthalpy change for a system is also the heat, q_p, gained or lost by the system at constant pressure:

$$\Delta H = q_p$$

For most chemical reactions, particularly for those involving only solid and liquid reactants and products, the $P\,\Delta V$ term is relatively small compared with ΔE, and $\Delta H \approx \Delta E$.

The enthalpy change for a chemical reaction is also known as the **enthalpy of reaction** or **heat of reaction**. A *reaction that absorbs heat* is called an **endothermic reaction**. The ΔH for an endothermic reaction has a positive sign. A *reaction that releases heat* is **exothermic**. An exothermic reaction has a negative ΔH.

Hess's Law

Hess's law states that the *enthalpy changes of different steps of a reaction can be added to obtain the enthalpy change for the reaction.*

Enthalpy of Formation

A special case of the enthalpy of reaction is the **standard enthalpy of formation**, or **standard heat of formation**, ΔH_f°. Standard enthalpy of formation is the enthalpy of a reaction in which *1 mol of a compound in its standard state forms from its elements in their standard states at 1 atm pressure and 298 K.*

Calculation of the Enthalpy of a Reaction from Enthalpies of Formation

The enthalpy change of a reaction is related to the enthalpies of formation of the reactants and the products by the equation

$$\Delta H_{rxn} = \Sigma\ \Delta H_f(\text{products}) - \Sigma\ \Delta H_f(\text{reactants})$$

The enthalpy of a reaction is a measure of the relative bond strengths in reactant and product molecules.

Calorimetry

Enthalpy changes in chemical reactions can be measured in a device called a **calorimeter**. An important concept used in such measurements is heat capacity. The **heat capacity** of an object is the *amount of heat required to increase its temperature by 1 degree Celsius*. A special case of heat capacity is **specific heat,** which is the *amount of heat required to increase the temperature of 1 gram of a substance by 1 degree Celsius*. The amount of heat absorbed or released by a reaction in a calorimeter equals the total heat capacity of the calorimeter and its contents times the temperature change. The heat, q_p, of a reaction measured at constant pressure equals the enthalpy change, ΔH, of the reaction. The heat, q_v, of a reaction measured at constant volume equals the internal energy change, ΔE, of the reaction. Thus,

$$q_p = \Delta H \quad \text{and} \quad q_v = \Delta E$$

The difference between ΔH and ΔE is relatively small even for gaseous reactions. The energy values of fuels and foods are determined by calorimetric measurements.

Important Relationships

Heat gained by a body
= (heat capacity of the body)(Δt)
= (mass)(specific heat)(Δt)

The first law of thermodynamics: $\Delta E = q + w$

Enthalpy change, $\Delta H = \Delta E + P\,\Delta V = q_p$

Hess's law: $\Delta H_{rxn} = \Delta H_{\text{step 1}} + \Delta H_{\text{step 2}} + \Delta H_{\text{step 3}} + \cdots$

$\Delta H^\circ_{rxn} = \Sigma\ \Delta H_f^\circ(\text{products}) - \Sigma\ \Delta H_f^\circ(\text{reactants})$

$\Delta H^\circ_{rxn} = \Sigma(\text{bond energies of reactants}) - \Sigma(\text{bond energies of products})$

New Terms

Bomb calorimeter (5.5)
Bond energy (5.4)
Calorimeter (5.5)
Calorimetry (5.5)
Constant pressure calorimeter (5.5)
Constant volume calorimeter (5.5)
Endothermic reaction (5.1)
Enthalpy (5.2)
Enthalpy change (5.2)
Enthalpy of formation (5.3)
Enthalpy of reaction (5.2)
Exothermic reaction (5.1)
Final state (5.2)
First law of thermodynamics (5.1)
Fuel (5.6)
Heat capacity (5.5)
Heat of combustion (5.6)
Heat of formation (*see* Enthalpy of formation)
Heat of reaction (*see* Enthalpy of reaction)
Hess's law of heat summation (5.3)
Initial state (5.2)
Internal energy (5.2)
Joule (5.1)
Kinetic energy (5.1)
Lattice energy (5.4)
Molar heat capacity (5.5)
Potential energy (5.1)
Specific heat (5.5)
Standard conditions (5.3)
Standard enthalpy of formation (5.3)
Standard molar enthalpy of formation (5.3)
Standard state (5.3)
State (5.2)
State function (5.2)
State of a system (5.2)
Surroundings (5.2)
System (5.2)
Thermochemistry (introduction)
Thermodynamics (introduction)
Work (5.1)

Exercises

General Review

1. What is: **(a)** an exothermic reaction; **(b)** an endothermic reaction? Give two examples of each.
2. Discuss some different forms of energy and give examples of their interconversions.
3. **(a)** How many joules equal 1 calorie? **(b)** How many kilojoules equal 1 kilocalorie?
4. Write an expression for a joule in terms of kilograms, meters, and seconds. Give a verbal explanation of this expression.
5. Define a calorie in terms of a quantity of water and its temperature change.
6. Define "thermochemistry" and "thermodynamics."
7. Give a verbal explanation of the equation $\Delta E = q + w$. Be sure to clearly state the sign conventions for q and w values.
8. What is meant by a "system" in thermodynamics?
9. What is meant by the "state" of a system?
10. Explain how the sum of the internal energy change of a system and that of the surroundings can be equal to zero.
11. For what type of reactions is pressure–volume work negligibly small?
12. Define "enthalpy change for a system."
13. Under what conditions does the measured heat of a reaction equal the enthalpy change for the reaction?
14. **(a)** What is the meaning of a positive sign for the enthalpy of a reaction? **(b)** What is the meaning of a negative sign?
15. What is the difference, if any, between "enthalpy of formation" and "enthalpy of reaction?"
16. Explain how the heat of formation can be considered as a special case of heat of reaction.
17. What is the difference, if any, between ΔH_f and ΔH_f°?
18. Write an equation relating the heat of a reaction and the heats of formation of the reactants and the products of the reaction.
19. What is the relation between bond strengths between the atoms in a molecule of a compound and the thermodynamic stability of the compound?
20. Write an equation relating the enthalpy of a reaction with the bond strengths in the reactant and product molecules.
21. Why is it not appropriate to use the concept "bond strength" or "bond energy" for ionic compounds?
22. Define "lattice energy."
23. Define: **(a)** "heat capacity"; **(b)** "specific heat."
24. Describe how the heat capacity of a metal at constant pressure can be measured in a calorimeter. List the measurements that should be performed, and show how these measurements can be used to calculate the heat capacity.
25. Is the heat of a reaction that is measured in a constant pressure calorimeter the ΔE or the ΔH of the reaction?
26. Write an equation for a reaction for which you expect q_p and q_v to be equal. Write an equation for a reaction for which you think q_p and q_v may not be equal. Do you expect a large difference?
27. How is a bomb calorimeter different from a constant pressure calorimeter?

28. What are the three basic types of foods? Which one of these provides the quickest energy to a human body?

29. List some major fuels available today.

30. Discuss some of our present energy problems and suggest what could be done to improve the situation.

Energy Conversions

31. A reaction absorbs 15.0 J of heat. How many calories is this? How many kilocalories?

32. How many joules of heat is required to warm 30 g of water from 20°C to 50°C?

33. How many joules of heat is released when 2.0 L of water is cooled from 40°C to 15°C?

The First Law of Thermodynamics

34. A reaction absorbs 500 J of heat from the surroundings, and 5 J of work is done by the surroundings on the reaction as a system. Calculate ΔE for the reaction and for the surroundings.

35. The internal energy change for a system changing from state 1 to state 2 is −150 J. During this change, the system does 6 J of work on its surroundings. Does the system absorb heat from the surroundings or does it release heat to the surroundings? How many joules? What is ΔE for the surroundings? Show how your answer illustrates the first law of thermodynamics.

Enthalpy Changes

36. The temperature of 100 g of water is increased from 20.0°C to 30.0°C at constant pressure. Calculate ΔH for this change.

37. If the heat of a reaction is listed as −20 J, is this reaction exothermic or endothermic? Are the products thermodynamically less or more stable than the reactants?

38. For the reaction $H_2(g) + \frac{1}{2}O_2(g) \rightarrow H_2O(l)$, $\Delta H = -286\ \text{kJ mol}^{-1}$ of H_2O. Calculate: **(a)** the amount of heat liberated for each gram of water formed; **(b)** the amount of heat liberated for each gram of hydrogen consumed.

39. For the metabolism of sugar, $C_{12}H_{22}O_{11}$, in the body, $\Delta H = -5645\ \text{kJ mol}^{-1}$. How many kilocalories (Cal) of energy would your body gain from the sugar in a candy bar that has the mass of 150.0 g and contains 60.0 percent sugar and no other nutrients?

40. When 1.00 g of oxygen difluoride reacts with water vapor according to the equation

$$OF_2(g) + H_2O(g) \longrightarrow O_2(g) + 2HF(g)$$

at 1.00 atm and 25.0°C, 5.98 kJ of heat is liberated. Calculate ΔH for this reaction per mole of OF_2.

Hess's Law of Heat Summation

41. Given the following thermochemical equations:

$$C(s) + O_2(g) \longrightarrow CO_2(g) \qquad \Delta H = -394\ \text{kJ}$$
$$H_2(g) + \tfrac{1}{2}O_2(g) \longrightarrow H_2O(l) \qquad \Delta H = -286\ \text{kJ}$$
$$HCOOH(l) + \tfrac{1}{2}O_2(g) \longrightarrow CO_2(g) + H_2O(l) \qquad \Delta H = -275\ \text{kJ}$$

Calculate the enthalpy of formation of formic acid, HCOOH, from its elements:

$$H_2(g) + O_2(g) + C(s) \longrightarrow HCOOH(l)$$

42. Burning 12.0 g of amorphous carbon (soot) in oxygen liberates 408 kJ of heat. When the same amount of diamond is burned, 394 kJ of heat is evolved. In both cases the combustion product is CO_2 gas. Calculate the enthalpy of conversion of 12.0 g of amorphous carbon to diamond. What does your answer indicate about the relative thermal stabilities of amorphous carbon and diamond?

43. Given the following reactions and their enthalpy changes:

$$N_2(g) + 2O_2(g) \longrightarrow 2NO_2(g) \qquad \Delta H = 67.8\ \text{kJ}$$
$$N_2(g) + 2O_2(g) \longrightarrow N_2O_4(g) \qquad \Delta H = 9.67\ \text{kJ}$$

Calculate the enthalpy change for the conversion of $NO_2(g)$ to 1 mol of $N_2O_4(g)$.

44. Calculate ΔH for the reaction

$$Ca(s) + C(s) + 1\tfrac{1}{2}O_2(g) \longrightarrow CaCO_3(s)$$

from the following data:

(1) $Ca(s) + 2C(s) \longrightarrow CaC_2(s) \qquad \Delta H = -62.8\ \text{kJ}$

(2) $CO_2(g) \longrightarrow C(s) + O_2(g) \qquad \Delta H = 394\ \text{kJ}$

(3) $CaCO_3(s) + CO_2(g) \longrightarrow CaC_2(s) + \tfrac{5}{2}O_2(g) \qquad \Delta H = 1538\ \text{kJ}$

45. Calculate ΔH for the reaction

$$2N_2O_3(g) \longrightarrow 2N_2(g) + 3O_2(g)$$

from the following data:

(1) $N_2O_3(g) \longrightarrow NO(g) + NO_2(g) \qquad \Delta H = 39.7\ \text{kJ}$

(2) $\tfrac{1}{2}N_2(g) + \tfrac{1}{2}O_2(g) \longrightarrow NO(g) \qquad \Delta H = 90.4\ \text{kJ}$

(3) $\tfrac{1}{2}N_2(g) + O_2(g) \longrightarrow NO_2(g) \qquad \Delta H = 33.8\ \text{kJ}$

Enthalpy of Formation and Enthalpy of Reaction

46. For which of the following reaction(s) does the ΔH of the reaction represent the enthalpy of formation, and for which reaction(s) does it represent the standard enthalpy of formation? (The standard states of elements are discussed in Chapter 2. The standard state of a binary compound of a metal and a nonmetal is a solid.) **(a)** $K(s) + \frac{1}{2}Br_2(g) \rightarrow KBr(s)$; **(b)** $K(s) + \frac{1}{2}Br_2(l) \rightarrow KBr(s)$; **(c)** $Rb(s) + \frac{1}{2}I_2(g) \rightarrow RbI(s)$; **(d)** $CaO(s) + CO_2(g) \rightarrow CaCO_3(s)$; **(e)** $Ca(s) + \frac{1}{2}O_2(g) \rightarrow CaO(s)$; **(f)** $H_2O(l) + SO_2(g) \rightarrow H_2SO_3(aq)$

47. The combustion of benzene is represented by the equation

$$C_6H_6(l) + 7\tfrac{1}{2}O_2(g) \longrightarrow 6CO_2(g) + 3H_2O(l)$$

Using the data from Table 5.1 to four significant figures, calculate the standard enthalpy of combustion of benzene.

48. The reaction of aluminum with iron(III) oxide, Fe_2O_3, is called the "thermite reaction." It occurs according to the equation

$$2Al(s) + Fe_2O_3(s) \longrightarrow Al_2O_3(s) + 2Fe(l)$$

This reaction is so exothermic that the elemental iron produced in the reaction is a liquid and can be used in welding iron rails. Using the data from Table 5.1, calculate the heat released by this reaction per mole of iron formed, and per kilogram of iron formed, under standard conditions.

49. The $\Delta H°$ for the reaction

$$PbS(s) + H_2(g) \longrightarrow Pb(s) + H_2S(g)$$

is 78.9 kJ. The standard enthalpy of formation of $H_2S(g)$ is -20.2 kJ mol^{-1}. Calculate ΔH_f° for $PbS(s)$.

50. The combustion of octane occurs according to the equation

$$2C_8H_{18}(l) + 25O_2(g) \longrightarrow 16CO_2(g) + 18H_2O(l)$$

The standard enthalpy for this reaction as written is -1.094×10^4 kJ. Using the data from Table 5.1, calculate the standard enthalpy of formation of octane.

51. Use the data from Table 5.1 to three significant figures to calculate $\Delta H°$ for the reaction

$$C_2H_4(g) + H_2(g) \longrightarrow C_2H_6(g)$$

52. For the reaction $SiH_4(g) + 4HCl(g) \rightarrow SiCl_4(g) + 2H_2(g)$, ΔH is -275 kJ. **(a)** How many grams of SiH_4 must react in this reaction with excess HCl to release 900 kJ of heat? **(b)** How many joules of heat is evolved if 35.0 g of HCl reacts with excess SiH_4?

53. For the reaction, $NH_3(g) + HCl(g) \rightarrow NH_4Cl(s)$, $\Delta H°$ is -177 kJ mol^{-1}. Using the data in Table 5.1, calculate ΔH_f° for $NH_4Cl(s)$.

54. **(a)** Calculate the standard enthalpy change for each of the following reactions:

$$SiO_2(s) + 4HF(g) \longrightarrow SiF_4(g) + 2H_2O(g)$$
$$SiO_2(s) + 4HCl(g) \longrightarrow SiCl_4(g) + 2H_2O(g)$$

ΔH_f° for $SiF_4(g)$ and for $SiCl_4(g)$ are -1548 kJ mol^{-1} and -609.6 kJ mol^{-1}, respectively. Additional data can be found in Table 5.1. **(b)** Silicon dioxide, SiO_2, is one of the major ingredients of glass. Suggest an explanation for the fact that hydrofluoric acid, $HF(aq)$, attacks glass, but hydrochloric acid, $HCl(aq)$, does not.

Bond Energies

55. Using the data from Table 5.2, calculate $\Delta H°$ for the reaction $H_2(g) + Cl_2(g) \longrightarrow 2HCl(g)$.

56. For the reaction

$$2Cl_2(g) + 2H_2O(g) \longrightarrow 4HCl(g) + O_2(g)$$

$\Delta H°$ is 96.0 kJ. Using the necessary bond energy values from Table 5.2, calculate the oxygen–oxygen bond energy in an O_2 molecule.

57. The nitrogen–nitrogen bond energy in an N_2 molecule is 945 kJ mol^{-1}. Using additional data from Table 5.2, calculate $\Delta H°$ of the reaction

$$N_2(g) + 3H_2(g) \longrightarrow 2NH_3(g)$$

58. From the bond energy data in Table 5.2, calculate $\Delta H°$ for the reaction

```
    H                              H
    |                              |
H—N—H(g) + Cl—Cl(g) ⟶          H—N—Cl(g) + H—Cl(g)
```

59. Calculate the N—H bond energy in NH_3 from the following information: the ΔH_f° of 1 mol of $NH_3(g)$ is -45.9 kJ mol^{-1}; the ΔH_f° of 1 mol of $N(g)$ from $\frac{1}{2}$ mol of $N_2(g)$ is 474 kJ; and the ΔH_f° of 1 mol of $H(g)$ from $\frac{1}{2}$ mol of $H_2(g)$ is 218 kJ.

Calorimetry

60. The specific heat of aluminum is 0.878 J g^{-1} °C^{-1}. Calculate the heat capacity of 15.0 g of aluminum.

61. The specific heat capacity of copper is 0.38 J g^{-1} °C^{-1}. What is the temperature increase of 2.0 kg of copper after absorbing 8.4 kJ of heat?

62. Dissolving 4.058 g of a certain salt in 50.00 g of water at 25.72°C decreases the temperature to 23.58°C in a calorimeter. The specific heat of the solution of the salt in the calorimeter is 33.910 J g^{-1} $°C^{-1}$. Is the process exothermic or endothermic? Assuming that no heat is gained or lost by the calorimeter walls, calculate the heat of solution of the salt in joules per gram of the salt.

63. A 10.0-kg sample of an unknown solid at 65.0°C is placed in 1.00 kg of water at 25.0°C in a calorimeter. The temperature of the water reaches 46.4°C. The heat capacity of the calorimeter is 11.2 J $°C^{-1}$. Calculate the specific heat of the unknown solid.

64. A chemical reaction is carried out in 500 g of water in a constant pressure calorimeter. As a result, the temperature of the water drops from 24.68°C to 20.12°C. The specific heat of the solution of the reaction products in water is 4.03 J g^{-1} $°C^{-1}$, and the heat capacity of the calorimeter is 1.21×10^3 J $°C^{-1}$. Calculate ΔH for this reaction. Is this reaction endothermic or exothermic?

65. A 2.000-g sample of benzoic acid, $C_7H_6O_2$, is burned in a bomb calorimeter containing 1000 g of water. As a result, the temperature of the water increases by 5.26°C. The heat of combustion of benzoic acid is 2.6×10^4 J g^{-1}. What is the heat capacity of the calorimeter?

66. **(a)** Burning a 1.500-g sample of methane, CH_4, in a bomb calorimeter increases the temperature in the calorimeter by 2.190°C. The heat of combustion of methane is 890.3 kJ mol^{-1}. Calculate the heat capacity of the calorimeter in kJ $°C^{-1}$. **(b)** When 2.266 g of an unknown compound is burned in the same calorimeter, the temperature of the calorimeter increases by 1.059°C. Calculate the heat of combustion of the unknown compound in kJ g^{-1}.

67. The combustion of a 5.000-g sample of boron carbide, B_4C, with excess oxygen in a bomb calorimeter increases the temperature in the calorimeter by 1.685°C. The heat capacity of the calorimeter and its contents is 153.1 kJ $°C^{-1}$. Calculate the molar heat of combustion of B_4C.

68. A 2.27-g sample of the explosive, nitroglycerin, $C_3H_5(NO_3)_3$, is allowed to explode in a bomb calorimeter at standard conditions. The amount of heat evolved is 15.1 kJ. The explosion of nitroglycerin occurs according to the equation

$$2C_3H_5(NO_3)_3(l) \longrightarrow 6CO_2(g) + 5H_2O(l) + 3N_2(g) + \tfrac{1}{2}O_2(g)$$

Using the necessary data from Table 5.1, calculate ΔH_f° for nitroglycerin.

69. In a calorimetric determination of the heat of solution of hydrogen chloride gas in water, a sample of the gas is introduced into water in a calorimeter. The temperature of the water increases by 0.951°C. The total heat capacity of the calorimeter and the water in the calorimeter is 4.05×10^3 J $°C^{-1}$. The HCl solution in the calorimeter is titrated to the endpoint by 54.0 mL of a 1.00 *M* solution of NaOH. Calculate the molar heat of solution of HCl.

Foods and Fuels

70. A typical fat, glyceryl trioleate, is completely oxidized by the body according to the following equation:

$$C_{57}H_{104}O_6(s) + 80O_2(g) \longrightarrow 57CO_2(g) + 52H_2O(l)$$

For this reaction, $\Delta H = -3.35 \times 10^4$ kJ. **(a)** How much heat is evolved when 1.00 g of this fat is metabolized? **(b)** What is the calorie equivalent of work you would have to do to reduce 1 lb of this fat from your body if you are an average woman weighing 128 lb? (See Table 5.7.) **(c)** Approximately how many hours would you have to work with pick and shovel, or play competitive tennis, to get rid of 1 lb of this fat, assuming that you are an average man weighing 154 lb? (See Table 5.7.) **(d)** Approximately how many miles would you have to walk at the rate of 4.0 miles an hour to get rid of 1 lb of this fat, assuming that you are an average woman weighing 128 lb? (See Table 5.7.)

71. **(a)** Propane gas is used as a fuel for household heating. Using the data from Table 5.1 to three significant figures, calculate the standard heat of combustion of propane in kJ mol^{-1}. Assume that gaseous water is formed as a combustion product. **(b)** Calculate the number of cubic feet of propane (density 1.80 g L^{-1}) that would be required to heat 150 gallons of water from 60.0°F to 190°F. Use the following data: 1 ft^3 = 28.3 L; 1 gallon = 3.78 L; the density of water = 0.998 g cm^{-3}, and the specific heat of water = 4.18 J g^{-1}. **(c)** What would be the answer to part **(b)** if the gas were methane (density 0.654 g L^{-1}) instead of propane?

PART 2

Structure and Properties

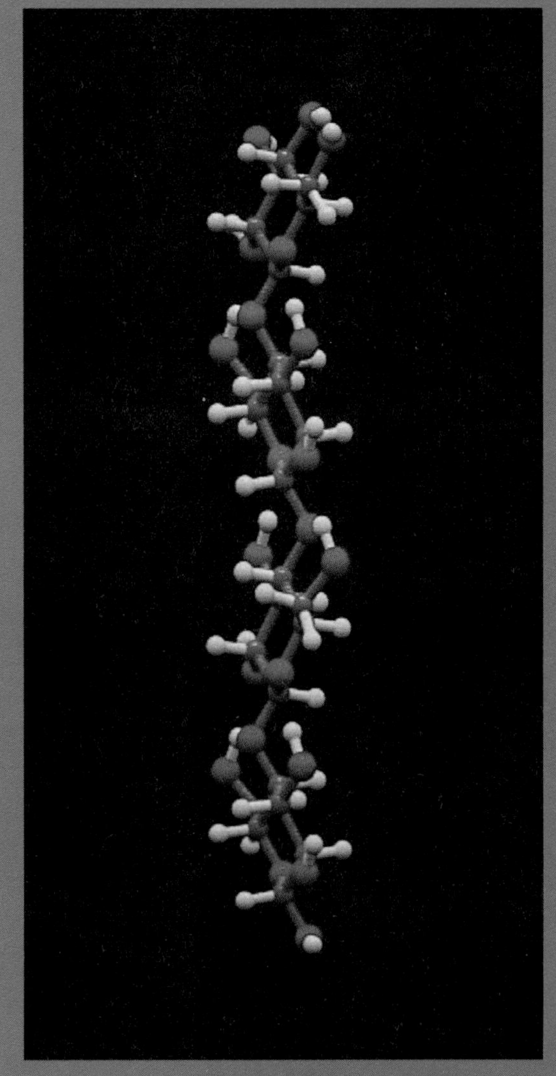

CHAPTER 6

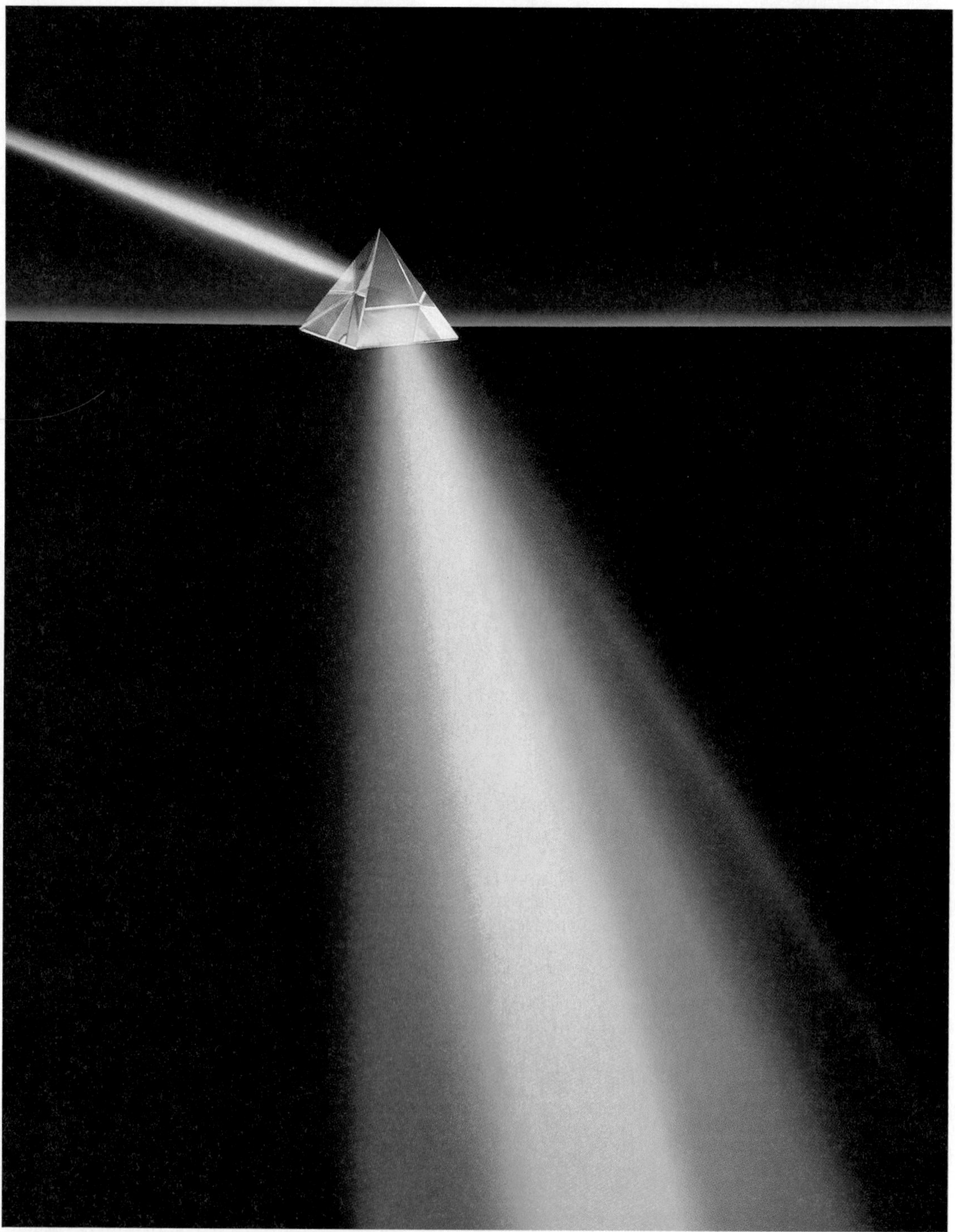

A continuous spectrum of white light.

Atomic Structure and the Periodic Table

Up to this point we have discussed elements, compounds, chemical reactions, and the energy changes associated with these reactions. We turn next to the theories of atomic structure, which enable us to explain many of the phenomena we described in previous chapters.

Dalton's atomic theory became widely accepted by scientists in the nineteenth century. His theory explained the laws of conservation of matter, constant composition, and multiple proportions. However, it did not explain chemical bonding. For example, Dalton's theory could not explain why an oxygen atom combines with two hydrogen atoms and a nitrogen atom combines with three hydrogen atoms. The only way to account for such facts was to postulate that atoms possess an inner structure.

The details of atomic structure were formulated from a series of experiments during the latter part of the nineteenth century and the beginning of the twentieth century. This was an exciting period of discovery, important experiments followed one another in rapid succession. The modern theory of atomic structure is based on the results of these experiments. A brief overview of these experiments is presented below.

6.1 Composition of Atoms

Electrical Nature of Matter

People have always been fascinated by electricity. Much of our knowledge about the composition and structure of atoms was gained by experiments in which high-voltage electricity was passed through different gases which had been introduced at low pressures into sealed glass tubes (Figure 6.1).

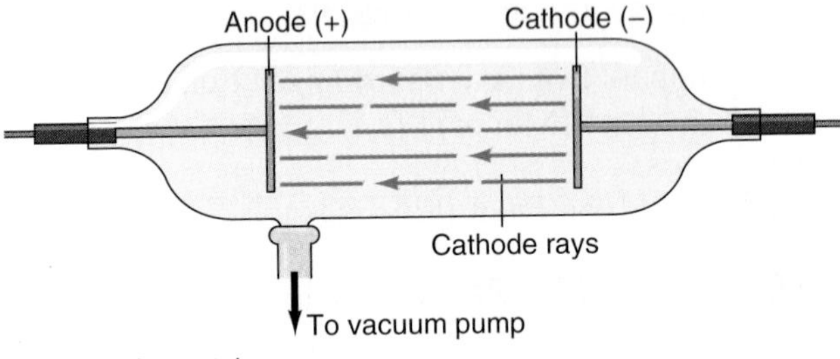

(a) Cathode ray tube

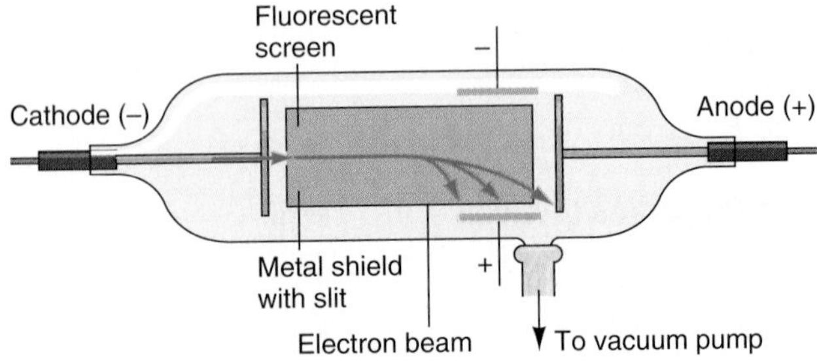

(b) Deflection of cathode rays (electrons) in an electric field

Figure 6.1

Cathode ray tubes. The rays can be made visible through their impact on a fluorescent material.

In these experiments, a glass tube containing a gas was sealed with electrodes inserted in each end. One electrode was negatively charged; it was called the **cathode**. The other electrode was positively charged; it was called the **anode**. An electric current entered the tube through the cathode and emerged at the anode. The electricity emanating from the cathode was seen as a glowing *beam* or *ray*. The entire apparatus was called a **cathode ray tube**.

Some cathode ray tubes were constructed with two additional parallel plates as additional electrodes. When electric current was applied to the tube, the rays moved from the negative plate and moved toward the additional positive plate (Figure 6.1b). This observation indicates that the beam is negatively charged (Figure 6.1b). The English scientist Sir William Crookes suggested in 1879 that the beam consisted of a stream of negative particles which he called **cathode rays**.

Crooke's hypothesis was supported by the experiments of another English scientist, Joseph John Thomson. In 1897, Thomson was able to determine the charge-to-mass ratio of a cathode ray particle. He found that this ratio was the same for all the particles no matter what type of metal was used for the electrodes or what type of gas was used in the cathode ray tube. The cathode ray particles were later termed **electrons**. Since identical electrons are produced regardless of the nature of the materials used in a cathode ray tube, Thomson concluded that electrons must be present in *all* matter.

In 1909, the U.S. physicist Robert A. Millikan (at the University of Chicago) determined the charge on an electron. Once Millikan had determined the charge on an electron, he used Thomson's charge-to-mass ratio to calculate its mass. Millikan found that the mass of an electron is 0.000548 amu. This is about one eighteen hundredth of the mass of a hydrogen atom. The electron is much lighter than an atom. It is a **subatomic particle**.

Protons

Electrons are not the only charged subatomic particles. In 1886 the German physicist Eugen Goldstein designed a cathode ray tube whose cathode had holes in it. Goldstein noticed that not only were cathode rays emitted from the cathode, but there were other rays which originated from the anode. These rays moved toward the cathode and passed through the holes in the cathode. These rays were termed **canal rays** because they passed through "canals"—the openings in the cathode (Figure 6.2). Because canal rays are attracted to the negative electrode, Goldstein concluded that they consist of positively charged particles.

Canal rays are produced when cathode rays knock some electrons off neutral gaseous atoms or molecules in the tube. The charge of a canal ray particle was found to be equal to that of the electron or to some whole-number

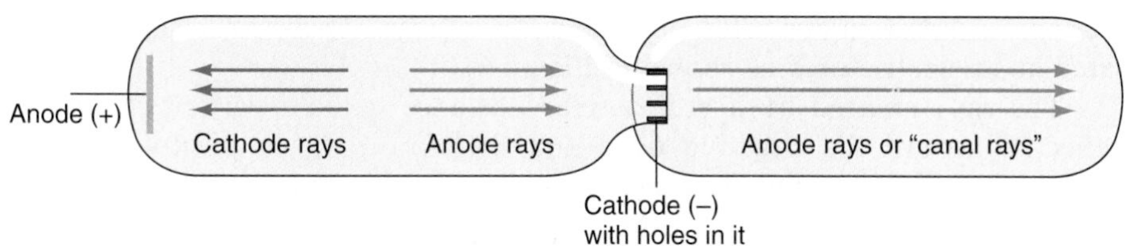

Figure 6.2

A schematic diagram of a canal ray tube.

multiple of that charge, but opposite in sign. The charge-to-mass ratio of these particles varied with the gas used in the tube. The masses of the particles were found to be the lowest when hydrogen was used in the tube. The positive ions produced from hydrogen atoms were called **protons**. The mass of a proton is 1.0073 amu. Protons are now known to be fundamental components in the nuclei of all atoms.

Neutrons

In 1932 the English physicist James Chadwick discovered a subatomic particle when he bombarded beryllium and other elements with nuclei of helium atoms called **alpha (α) particles**. When beryllium is bombarded with alpha particles, electrically neutral particles are emitted. Chadwick named the neutral particles **neutrons**. A neutron was found to have a mass of 1.0087 amu, which is only slightly higher than the mass of a proton. Neutrons are stable species within atoms, but outside an atom a neutron is unstable and disintegrates into a proton and an electron.

Radioactivity

In 1896 the French physicist Henri Becquerel discovered an entirely new facet of atomic structure. Late that winter Becquerel was studying the emission of light by potassium uranyl sulfate [$K_2UO_2(SO_4)_2(H_2O)_2$], an example of a phenomenon called phosphorescence. He wrapped a glass photographic plate with two pieces of black paper. On top of this package he placed another glass plate that was coated with a thin film of the uranium salt. He placed a metal object between the glass plates and exposed the "sandwich" to sunlight. After a day in the sun, the photographic plate was developed and the image of the object could be seen on the resulting photograph. On February 26 he set up another such experiment, but the weather in Paris was cloudy and he placed the experiment in the cupboard. On Sunday, March 1, he removed the plate from the cupboard and for some reason decided to develop it. Amazingly, the photograph showed a picture of the object upon a white background. The radiation responsible for the phenomenon was released from the uranyl salt in the absence of light. Becquerel had discovered **radioactivity**.

The rays emitted by the uranium salt—originally called uranic rays or Becquerel rays—were studied extensively by Marie Sklodowska Curie (1867–1934), born in Poland, and her husband Professor Pierre Curie of France, who demonstrated that radioactivity is a property of certain atoms. Today we know that radioactivity results from the disintegration of atoms. Energy is released in this process. The emission or transmission of energy in the form of waves, particles, or both, is called **radiation**. The type of energy that is transmitted as waves is known as **radiant energy**.

Uranic rays were also studied by the British physicist Ernest Rutherford (1871–1937). Rutherford found that the radiation emitted from such elements as uranium and radium consisted of three different rays. These rays can be separated in an electric field as shown in Figure 6.3.

The rays emitted by a radioactive substance contain particles that are deflected toward the negative plate and therefore carry a positive charge (Figure 6.3). These positively charged particles have a charge and mass equal to those of helium ions, He^{2+}, called alpha (α) particles. Rays that consist of

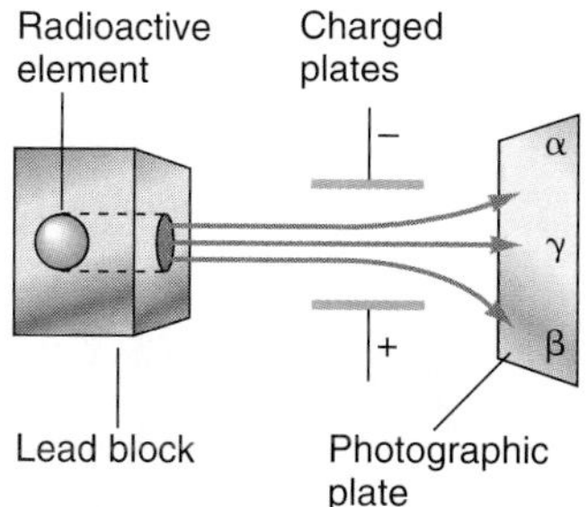

Figure 6.3

Rays from a radioactive element in an electric field. The alpha rays are positively charged and are, therefore, attracted toward the negative plate. The beta rays are negatively charged and are attracted toward the positive plate. The gamma rays are uncharged and are not affected by the electric field.

alpha particles are called **alpha rays**. Alpha particles are easily absorbed by matter and can be stopped by a few centimeters of air.

The rays from a radioactive substance also contain a component consisting of highly energetic particles that are deflected toward the positive plate. These negatively charged particles are called **beta** (β) **particles**. Beta particles were found to be high-velocity electrons. High-energy beta particles can penetrate a sheet of aluminum metal several millimeters thick.

A third component of the rays studied by Rutherford was not affected by the electric field. These neutral rays were termed **gamma** (γ) **rays**. Gamma radiation was the most penetrating radiation of the three. It was found to be even more energetic than x-rays, which had recently been discovered by the German physicist Wilhelm Conrad Röntgen.

The composition and properties of the radiations emitted by radioactive substances are listed in Table 6.1. The masses and charges of common subatomic particles are given in Table 6.2.

Table 6.1

Natural Radioactivity

Name	Symbol	Mass (amu)	Charge (Relative to Electron)
Alpha particle	α	4	+2
Beta particle	β	1/1837	−1
Gamma ray	γ	0	0

Table 6.2

Masses and Charges of Common Subatomic Particles

Name of Particle	Symbol	Mass (amu) Approx.	Mass (amu) Exact	Electrical Charge (Based on the Charge of an Electron)	Symbol with Mass Number and Charge
Proton	p	1	1.007276470	+1	${}^{1}_{1}\mathrm{p}$
Neutron	n	1	1.008665012	0	${}^{1}_{0}\mathrm{n}$
Alpha particle	α	4	4.0015061	+2	${}^{4}_{2}\alpha$
Electron or beta particle	e or β	1/1800	0.00054858026	−1	${}^{0}_{-1}\mathrm{e}$ or ${}^{0}_{-1}\beta$

APPLICATIONS OF CHEMISTRY 6.1
Preservation Through Radiation

One-third of all food grown in the world today rots before it reaches the table. As a result of this and other factors, within the next hour about 1100 people (mostly children) will starve to death, and another 50,000 undernourished people will suffer permanent mental or physical injuries.

The primary ways that we preserve food in the United States, for example, refrigeration and canning, are too expensive for citizens of third-world nations. Even if refrigerators were available to these people, their governments cannot afford to build and maintain the electric generating plants needed to operate the refrigerators. Consider the following: Californians need two full-time (24 hours/day, 7 days/week) power plants to supply the electricity needed to operate just their refrigerators. Therefore, an alternative method of preserving food is needed for the conditions found in many third-world countries.

To develop a new means of preserving food we need to understand how or why food rots. Food rots because microorganisms invade the food and release waste products. Heating food kills these microorganisms, and refrigerating food slows their growth. Radiation also kills these microorganisms, and radiation is being used more and more to preserve food (see photo).

Gamma radiation is a primary source of radiation in food preservation. Gamma radiation is so energetic that it removes electrons from the atoms that make up the microorganisms. These damaged microorganisms then die, and packaging protects the irradiated food from further contamination.

Preserving food with radiation is about 80 percent cheaper than freezing or canning. Today, about 50 countries treat their food with radiation. They use radiation to preserve meats, fruits, and vegetables, and once they package the food, it stays fresh for over a year. Our astronauts on space missions eat some food preserved this way.

A similar radiation process also helps to preserve pieces of fine art. Works of art decay for the same reasons that food rots. Microorganisms and insects such as termites damage canvases, tapestries, wood frames, and wood sculptures. Gamma radiation effectively kills the microorganisms and insects in artwork without harming it. The aging process for paintings and tapestries stops after radiation treatment has destroyed the harmful microorganisms.

Source: Ronald DeLorenzo, *Journal of Chemical Education*, August 1983, page 671.

Atomic Nuclei

Although the experiments just described demonstrated the existence of various subatomic particles, they did not describe how these particles might be arranged in an atom. An important insight into the architecture of atoms was gained from an experiment of Ernest Rutherford. In this experiment, alpha particles (obtained from the disintegration of polonium) were passed through a thin metal foil (Figure 6.4). Most of these particles went straight through the foil and a few were only slightly deflected. But about one in 1 million were greatly deflected, some more than 90°, from their straight paths.

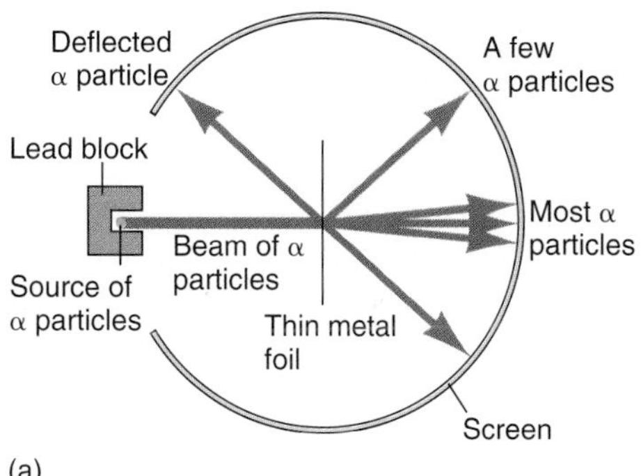

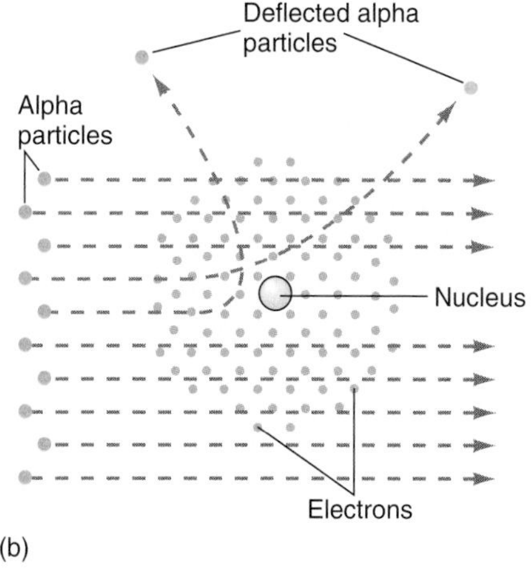

Figure 6.4

Rutherford's experiment: (a) a general view of the experiment and (b) a model of a metal atom being bombarded by alpha particles.

Rutherford (Figure 6.5) was utterly amazed when he learned that some alpha particles were greatly deflected from their straight paths. He wrote about his surprise: "It was quite the most incredible event that has ever happened to me in my life. It was almost as incredible as if you fired a 15-inch shell at a piece of tissue paper and it came back and hit you."

Rutherford's experiments led to a new model of the atom. Since most alpha particles went straight through the foil, Rutherford proposed that the mass of an atom is concentrated in a very small, dense, positively charged **nucleus**. Alpha particles that move straight toward a nucleus, or come very close to a nucleus, are deflected by electrostatic repulsion between the positively charged alpha particle and the positively charged nucleus. The much lighter electrons around the nucleus of an atom are relatively far from each other and from the nucleus. Massive alpha particles are not deflected by electrons.

It is now known that the diameter of the nucleus of an average atom is about 10^{-12} cm. This is one ten-thousandth of the diameter of an atom, which is about 10^{-8} cm, or 0.1 nm.

Figure 6.5

Ernest Rutherford.

6.2 Light and Optical Spectra

Our knowledge of the structure of atoms and molecules is intimately connected with the interactions of matter with electromagnetic radiation, or "light." Thus, as a prelude to our discussion of atomic structure, we describe some of the fundamental properties of light. After we have considered the nature of light, we discuss the interaction of light with atoms.

Radiant Energy

There are many forms of radiant energy, also called *electromagnetic radiation*. Visible light, x-rays, and radio waves are examples of radiant energy. Radiant energy can be described in terms of waves. A wave is characterized by its length, amplitude, and frequency. The length of a wave, or **wavelength**, is the *distance from a point in one wave to an equivalent point in the next wave* (Figure 6.6). Wavelength is symbolized by the Greek lowercase letter lambda, λ. The **amplitude** of a wave is *one-half of the vertical distance between a crest and a trough*, as shown in Figure 6.6. The **frequency**, ν (Greek lowercase letter nu), of electromagnetic radiation equals the *number of waves that pass a given reference point in a unit time*, such as a second. The SI unit for frequency is the **hertz**, Hz, which equals one wave per second. Thus, 1 Hz = 1 s^{-1}.

The rate at which a light wave travels through space, that is, the time it takes for a given point in a wave pattern to travel a specified distance, is the **velocity**, c, of the wave. The speed of all forms of light in a vacuum is 2.998×10^8 meters per second.

The wavelength and frequency of radiant energy are inversely proportional to each other. Thus, a radiation with short wavelength has high frequency. The wavelength, λ, frequency, ν, and speed, c, of light and other forms of radiant energy are related as follows:

$$\lambda\nu = c \tag{6.1}$$

Figure 6.7 compares the wavelengths and frequencies of several radiant energies.

Example 6.1

Calculating the Wavelength from a Given Frequency

What is the wavelength in nanometers of an x-ray radiation that has a frequency of 9.4×10^{17} Hz?

SOLUTION: The frequency of 9.4×10^{17} Hz = 9.4×10^{17} s^{-1}. Solving Equation 6.1 for λ, and substituting the known values for ν and c yields

$$\lambda = \frac{c}{\nu} = \frac{3.0 \times 10^8 \text{ m s}^{-1}}{9.4 \times 10^{17} \text{ s}^{-1}} = 3.2 \times 10^{-10} \text{ m}$$

This result in nanometers is

$$3.2 \times 10^{-10} \text{ m} \left(\frac{10^9 \text{ nm}}{1 \text{ m}}\right) = 0.32 \text{ nm}$$

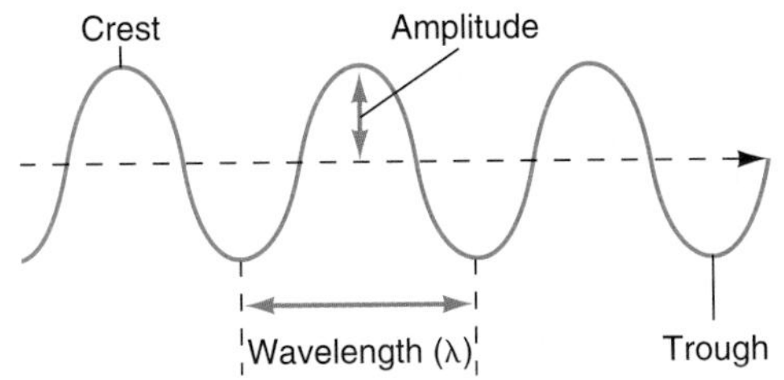

Figure 6.6

Wavelength and amplitude of a wave.

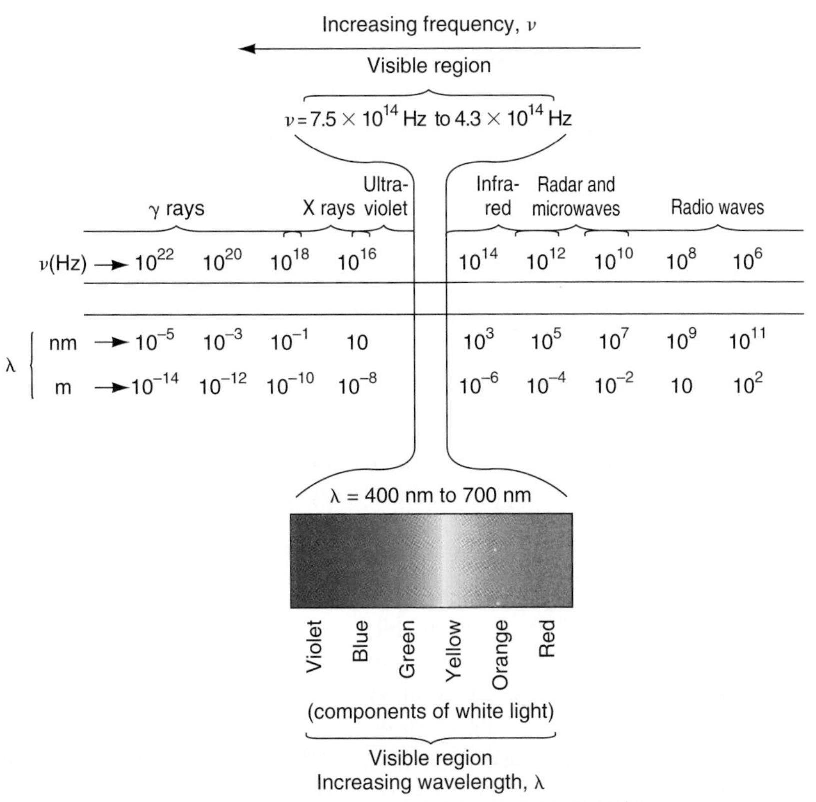

Figure 6.7

Wavelengths and frequencies of spectral regions.

Practice Problem 6.1: The wavelength at which a radio station transmits its programs is 3.11 m. What is the frequency of this transmission in megahertz (MHz)?

Albert Einstein, a German-born physicist (1879–1955), proposed that electromagnetic radiation can be viewed as a "stream" of particles called **photons**. The energy, E, of a photon of any form of a radiation is proportional to its frequency, and is given by the relationship

$$E = h\nu \tag{6.2}$$

where h is an experimentally measured physical constant, known as Planck's constant, which has the value of 6.626×10^{-34} joule second (J s). Equation 6.2 is known as Planck's equation, or Planck's law, named after the German physicist Max Planck (1858–1947), who announced it in 1900.

Example 6.2

Calculating the Wavelength of a Photon from Its Energy and Determining Its Spectral Region

What is the wavelength in nanometers of a photon whose energy, E, is 3.82×10^{-19} J? In which spectral region shown in Figure 6.7 can this radiation be detected?

SOLUTION: We first solve Equation 6.2 for ν, then we solve Equation 6.1 for λ, in each case substituting the given values. From Equation 6.2 we have

$$\nu = \frac{E}{h} = \frac{3.82 \times 10^{-19}\ \text{J}}{6.626 \times 10^{-34}\ \text{Js}} = 5.77 \times 10^{-14}\ \text{s}^{-1}$$

Equation 6.1 yields

$$\lambda = \frac{c}{\nu} = \frac{3.00 \times 10^{8}\ \text{m s}^{-1}}{5.77 \times 10^{-14}\ \text{s}^{-1}} = 5.20 \times 10^{-7}\ \text{m}$$

$$5.20 \times 10^{-7}\ \text{m} \left(\frac{10^{9}\ \text{nm}}{1\ \text{m}}\right) = 520\ \text{nm}$$

The radiation with the wavelength of 520 nm falls in the visible range, which is between 400 and 700 nm (Figure 6.7).

Practice Problem 6.2: The wavelength of a certain microwave radiation is 1.25×10^{5} nm. What are the frequency and the energy of this radiation?

Optical Spectra

Most of us are familiar with a rainbow. The continuously changing series of colors from red to yellow to green to blue to violet in a rainbow is called a **continuous spectrum** of colors. Such a spectrum can also be obtained when white light passes through a glass prism (Figure 6.8).

The prism disperses the component wavelengths of white light as shown in Figure 6.8. The wavelength corresponding to the red color is the longest and is bent least by the prism. Next in order of decreasing wavelength are orange, yellow, green, blue, and violet.

The human eye perceives "visible" light, which has wavelengths in the approximate range 400 to 700 nm. The radiation responsible for red color has a

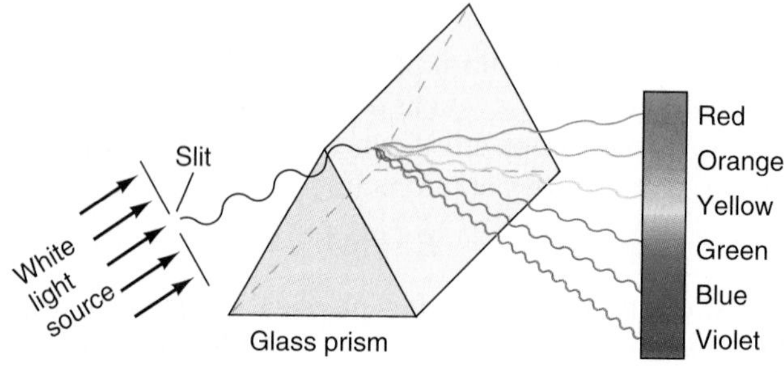

Figure 6.8

Spectrum of white light.

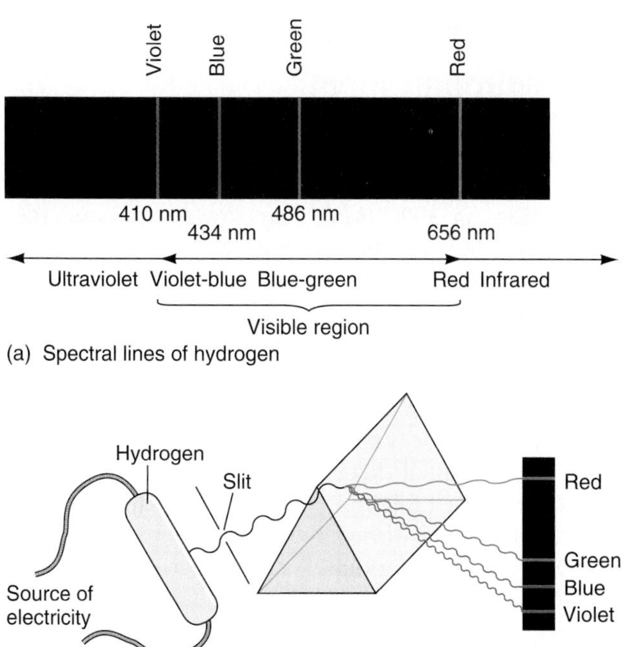

Figure 6.9

Visible line spectrum of hydrogen. Note that the wavelength of the radiation causing red color is longer than the wavelength of the radiation causing violet color.

wavelength of slightly less than 700 nm, blue corresponds to a wavelength of slightly above 400 nm, and violet corresponds to a wavelength of approximately 400 nm. Radiations having wavelengths less than 400 nm or greater than 700 nm cannot be detected by the human eye. The spectral region from approximately 200 to 400 nm is called the **ultraviolet** (UV) region, and the region from approximately 700 to 1500 nm is the **infrared** (IR) region.

The interaction of atoms and molecules with light is a very powerful tool for analyzing atomic and molecular structure. The first experiments to reveal the interaction of light with atoms were made over a century ago. In 1853, the Swedish physicist Anders Jonas Ångström placed hydrogen in a sealed tube with electrodes at each end. When he passed an electric current through the tube and viewed the tube through a prism, he observed three thin, bright lines: a red one, a green one, and a violet one; a second violet line was observed shortly thereafter.

The pattern of colored lines observed by Ångström is an example of a **line spectrum**. Each element in the gas phase has its own characteristic line spectrum. This spectrum can be used as an ''atomic fingerprint'' to identify the element. A line spectrum of hydrogen is shown in Figure 6.9.

The line spectrum of hydrogen contained one of the first clues that atoms possess an internal structure. However, line spectra were not understood until the beginning of the twentieth century, when a new theory of physics, called the quantum theory, was developed. We give a brief account of the quantum theory in the next section.

6.3 Quantum Theory and the Bohr Model of the Hydrogen Atom

Quantum Theory

Almost a century ago, in 1900, Max Planck was studying the emission of radiation through a tiny hole of a heated, hollow, spherical solid. Such a heated sphere emits radiation from its dark interior. This phenomenon is called blackbody radiation. Planck found the distribution of colors of the spectrum produced by the hot solid to be puzzling, because when a solid is heated, it turns red, and as the temperature increases, it turns yellow and, finally, white. The radiation that Planck observed did not have enough blue and violet. This absence of blue and violet color (known as the "ultraviolet catastrophe") could not be explained by traditional theory.

To explain blackbody radiation, Planck made the radical proposal that radiant energy is emitted or absorbed by the solid in multiples of *small, discrete units*. Each unit was called a **quantum**. As noted earlier, Einstein had proposed that light and other radiant energies consist of streams of photons. Each photon represents a quantum of energy. The magnitude of a quantum is proportional to the frequency of light. As given by Equation 6.2 (Planck's equation), the energy, E_{quantum}, of a photon equals the frequency, ν, times Planck's constant, h (6.626×10^{-34} J s):

$$E_{\text{quantum}} = h\nu$$

Photons of higher-frequency radiation have higher energies than photons of lower-frequency radiation.

Thus, according to the quantum theory of Planck, transfer of radiant energy, ΔE, can occur only in integral multiples of $h\nu$: ΔE can equal $h\nu$, $2h\nu$, or $3h\nu$ . . . , never a fractional multiple of $h\nu$. Since radiant energy appears in small "packages" called quanta, it is said to be **quantized**.

The Bohr Model of the Hydrogen Atom

In 1913 the Danish physicist Niels Henrik David Bohr (1885–1962) proposed that the energy states of hydrogen atoms are quantized. In so doing he ushered in the modern theory of atomic structure. Bohr's theory provided a link between atomic spectra and Planck's constant. Rutherford's experiments strongly suggested that a hydrogen atom consists of a proton bound to one electron. In 1913 this assumption had not been proven, but it provided Bohr with a starting point for his theory.

The laws of physics before quantum theory suggested that when a negatively charged electron moves around a positively charged nucleus, it should gradually "spiral" into the nucleus. If this spiraling occurred, the electron would lose energy to the surroundings. However, atoms do not normally emit energy.

Bohr leaped right over this difficulty by proposing that the single electron in a hydrogen atom moves around the nucleus in *well-defined circular paths* called **orbits**. He also defined his model to conform to the experimental observation that *an atom in its lowest energy state, called the ground state, does not emit energy.*

APPLICATIONS OF CHEMISTRY 6.2
Helium Turns Funny

After reading the section on quantum theory you may think that you can observe quantum states only at the atomic level. Although this is usually the case, there is an interesting exception: liquid helium at 2 K.

When cooled to about 4.2 K, helium turns into a liquid. If liquid helium is cooled even further, to about 2.2 K, it begins to flow as if it had no apparent friction. Any liquid that behaves in this way is called a **superfluid**. To visualize the properties of a superfluid, imagine filling a bowl with helium at 2.2 K and attempting to stir it. You will find that stirring the helium is almost like trying to stir an empty beaker. Although there is some friction between your stirring rod and the superfluid helium, you would probably not be able to detect it.

To emphasize the drastic change that takes place in helium as it cools from 4 K to 2.2 K, superfluid helium is given its own name: helium II. Helium above 2.2 K is referred to as helium I.

If helium II exhibited no friction, you would not be able to move the liquid inside a container, but because helium II does exhibit a very slight friction, it is possible to place superfluid helium into a doughnut-shaped container and start it rotating. If you wanted the helium to rotate faster or slower, you would find that you could rotate helium II only at certain speeds. You would have to put in or take away just the right amount of energy to change the rotation speed. Because helium II rotates only at certain speeds, the rotation energies are said to be quantized, just like the energies of an electron in a hydrogen atom.

Helium II has another interesting property: It climbs the walls of its container. If you were to partially fill an empty glass with helium II, you would see that the helium climbs the inside walls of the glass and pours down the outside of the glass. Helium II does this because of the van der Waals attraction between the helium atoms and the atoms of the glass. (See Chapter 11 for more information on van der Waals attractions.)

Source: William Steele, *Science World*, April 1, 1983, pages 6–7.

According to Bohr, an electron in a hydrogen atom ordinarily resides in the orbit closest to the nucleus. This orbit is called the **ground state** orbit. If precisely the right amount of energy is supplied, the electron in the ground state can move to a larger, **excited state** orbit, farther from the nucleus.

We recall from Chapter 5 that the gravitational potential energy of a body increases with the distance between the body and a reference point. Similarly, the potential energy of an electron relative to the nucleus increases with the distance between the electron and the nucleus. Therefore, the potential energy of an electron is higher in the larger, excited state orbit than in the smaller, ground state orbit. The electron can occupy only certain orbits and is never found between these orbits. This is a condition described by saying that the energy states of an electron in a hydrogen atom are *quantized*. The orbits of an electron correspond to its energy levels or **quantum levels**.

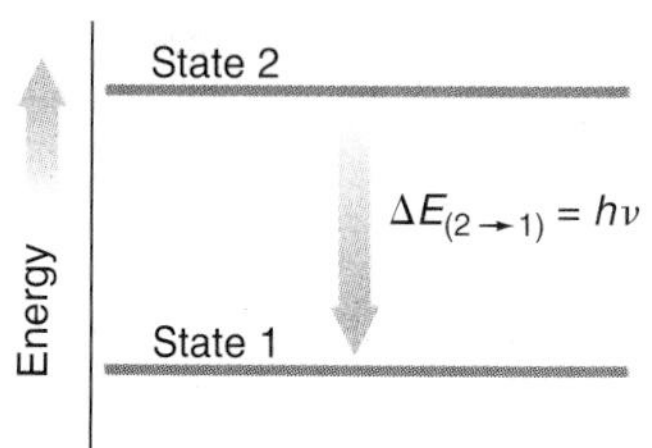

Figure 6.10

A simple energy level diagram.

The Line Spectrum of Hydrogen and the Bohr Model

We saw earlier that a line spectrum is produced when an electric current passes though a tube containing hydrogen gas. The electric current "excites" the hydrogen atoms, driving them to higher-energy states in which their electrons are in higher orbits. When the energy of a hydrogen atom changes from a high-energy state—the excited state—to a lower energy state, the energy difference between the two levels is emitted as light. The energy change that accompanies a transition from an excited state to the ground state is illustrated by the energy level diagram shown in Figure 6.10. The energy of the emitted light equals the difference in energies, ΔE, between the higher energy, E_h, of an electron in the atom and the lower energy, E_l, of the electron. This energy difference equals $h\nu$ (Equation 6.2):

$$\Delta E = E_h - E_l = h\nu$$

The packet of energy, $h\nu$, released by the transition of one electron from an excited state to a lower energy state is a photon. The transition of electrons in a sufficiently large number of atoms from a given excited state directly to the ground state is detected as a discrete line in the spectrum of hydrogen.

The line spectrum of hydrogen can thus be interpreted in terms of the internal structure of the atom. A hydrogen atom consists of a single proton and an electron. The different energy states of the hydrogen atom correspond to different energy levels or orbits of the electron. In the absence of an outside influence, nearly all hydrogen atoms are in their ground states. When the hydrogen atoms are excited in some way, many of them enter excited states. When the electrons of many hydrogen atoms "fall" from various excited states to orbits closer to the nucleus, light is emitted. This light produces a line spectrum which we discuss in more detail below.

When an electron in a hydrogen atom moves from a higher-energy orbit to a lower energy orbit, it emits a quantum of radiant energy of a specific wavelength and frequency corresponding to a spectral line. We know that this transition does not occur gradually by a spiraling process inward toward the nucleus because such a process would emit energy of gradually changing wavelength and would result in a continuous spectrum. The electron "leaps" virtually instantaneously from one orbit to another (Figure 6.11).

In Bohr's model, each orbit in a hydrogen atom is assigned an integer, n, known as the **principal quantum number**, which can have values of 1, 2, 3, and

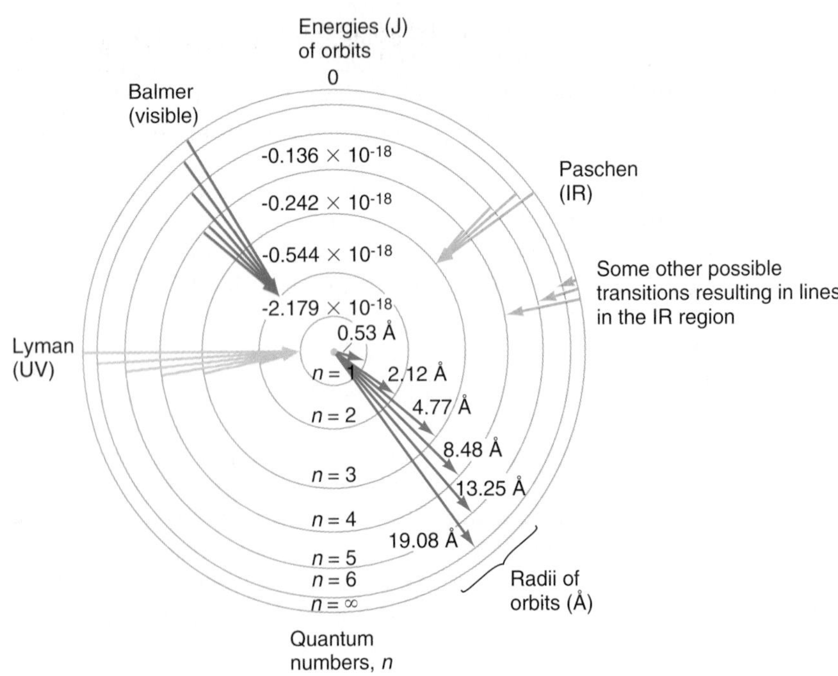

Figure 6.11

A summary of the electron transitions and spectra associated with Bohr's theory of the hydrogen atom. The electron transitions to the first orbit or quantum level ($n = 1$) produce lines in the ultraviolet region of the spectrum; these lines are called the Lyman series of lines. The transitions from higher orbits to $n = 2$ level (the second orbit or energy level) are the only transitions that produce lines in the visible region of the spectrum; these lines are called the Balmer series of lines. All the other transitions down to $n = 3$ (the Paschen series) and to higher levels give rise to lines in the infrared region of the spectrum.

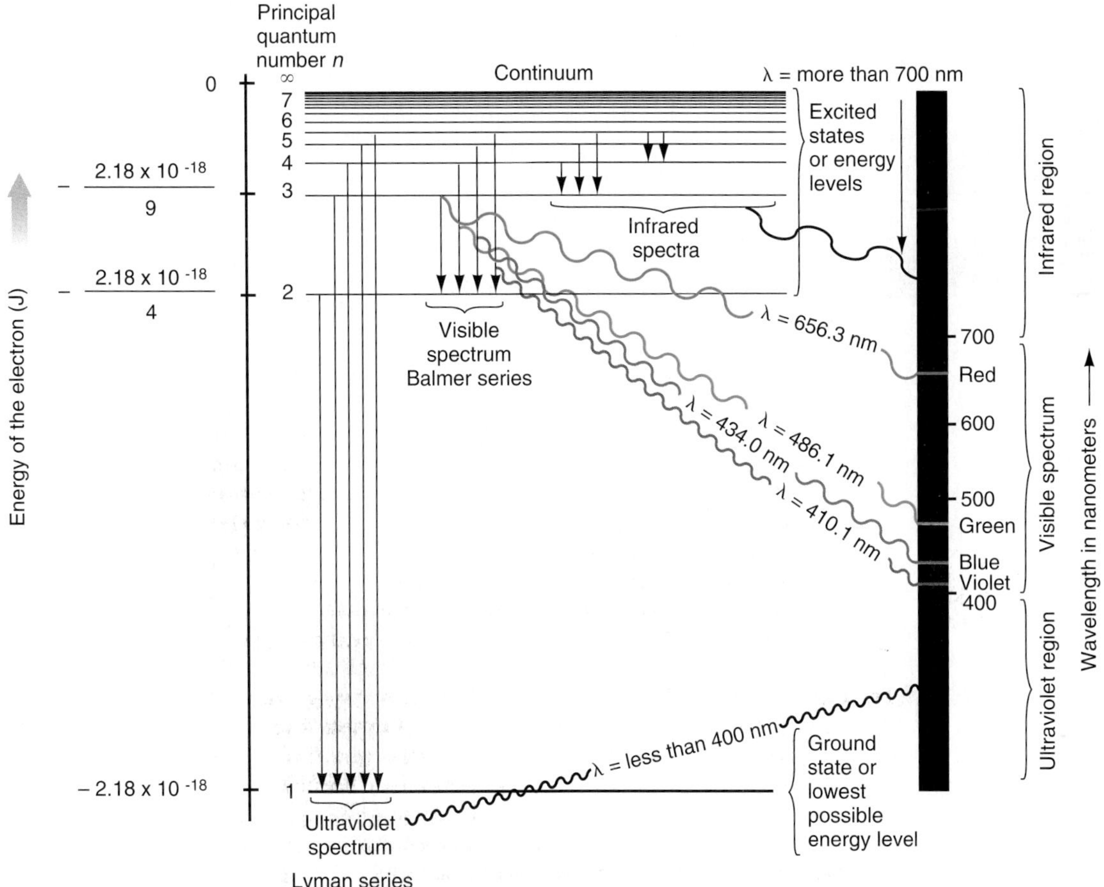

Figure 6.12

Electron transitions and the corresponding spectral lines of hydrogen. Note how the energies of the higher orbits (energy levels) are closer together at the higher values of n. As a result, the energy differences between two adjacent energy levels also decrease toward the higher values of n. As the energy differences between adjacent energy levels become smaller, the corresponding radiations emitted have successively longer wavelengths, and the spectral lines shift from the ultraviolet toward the visible and infrared regions. As n approaches infinity (∞), the energy levels form a continuum (a series of energies with indistinguishably small differences).

so on, up to infinity, as shown in Figures 6.11 and 6.12. The ground state orbit, which has the lowest energy and the smallest radius, has a principal quantum number of $n = 1$. Higher-energy orbits are assigned n values of 2, 3, 4, and so on, in the order of increasing energy and radius of the orbits.

Considerable work went into the derivation of Bohr's mathematical model of the hydrogen atom. Without going into the details of this derivation, we present the results of this work. Bohr showed that the quantized energies of the electron are given by the relationship

$$E_n = -R_H\left(\frac{1}{n^2}\right) \tag{6.3}$$

where E_n is the energy of the electron in any given orbit or quantum level n, and R_H is a constant, called the Rydberg constant, after the Swedish physicist Johannes Rydberg (1854–1919). The numerical value for this constant is 2.179×10^{-18} J.

The negative sign in Equation 6.3 indicates that the energy of an electron bound to the nucleus of an atom is lower than the energy of a "free" electron that is not bound to the nucleus. By convention, the energy of a free electron at rest is given the value of zero.

According to Equation 6.3, the energy of the electron in the first orbit is $-R_H$, the energy in the second orbit is $-\frac{1}{4}R_H$, in the third orbit $-\frac{1}{9}R_H$, and so on. As the electron moves from any given orbit to higher orbits, its energy

Niels Bohr.

increases (becomes less negative) until it reaches the value of zero when the electron becomes completely separated from the nucleus. At that point, $n = \infty$, and according to Equation 6.3,

$$E_\infty = -R_H\left(\frac{1}{\infty^2}\right) = 0$$

The energy emitted by the transition of an electron from a higher orbit or quantum level, n_h, to a lower level, n_l, is the difference between the energy of the electron in the higher quantum level, E_h, and the energy at the lower level, E_l:

$$\Delta E = E_h - E_l = -R_H\left(\frac{1}{n_h^2}\right) - \left[-R_H\left(\frac{1}{n_l^2}\right)\right]$$

This equation simplifies to

$$\Delta E = R_H\left(\frac{1}{n_l^2} - \frac{1}{n_h^2}\right) \qquad (6.4)$$

Example 6.3 illustrates an application of Equation 6.4.

Example 6.3

Calculating the Energy, Frequency, and Wavelength of the Radiation Emitted by an Excited Hydrogen Atom

Calculate the energy, the frequency, and the wavelength of the radiation emitted by the electron transition from the fifth to the second quantum level in a hydrogen atom. In which spectral region can a line corresponding to this transition be detected? What is the color of the line? (See Figure 6.12.)

SOLUTION: According to Equation 6.4, the energy of the radiation is

$$\Delta E_{5\to 2} = \mathrm{R_H}\left(\frac{1}{2^2} - \frac{1}{5^2}\right)$$

$$= 2.179 \times 10^{-18}\ \mathrm{J}\left(\tfrac{1}{4} - \tfrac{1}{25}\right)$$

$$= 4.58 \times 10^{-19}\ \mathrm{J}$$

Solving Equation 6.2 for the frequency, we obtain

$$\nu = \frac{\Delta E}{h} = \frac{4.58 \times 10^{-19}\ \mathrm{J}}{6.63 \times 10^{-34}\ \mathrm{J\ s}} = 6.91 \times 10^{14}\ \mathrm{s^{-1}}$$

The wavelength of the radiation can be obtained from Equation 6.1:

$$\lambda = \frac{c}{\nu} = \frac{3.00 \times 10^8\ \mathrm{m\ s^{-1}}}{6.91 \times 10^{14}\ s^{-1}} = 4.34 \times 10^{-7}\ \mathrm{m}$$

In nanometers this wavelength is 434 nm. The spectral line corresponding to this transition lies in the visible region of the spectrum, which extends from approximately 400 to 700 nm. Figure 6.12 indicates that the electron transition from $n = 5$ to $n = 2$ produces a blue line.

Practice Problem 6.3: Calculate the energy, frequency, and wavelength of the radiation emitted by the following electron transitions in hydrogen atoms: **(a)** from $n = 2$ to $n = 1$; **(b)** from $n = 3$ to $n = 2$; **(c)** from $n = 4$ to $n = 3$; **(d)** from $n = 5$ to $n = 3$. Compare your calculated energies and state whether they are increasing or decreasing. In which spectral region would each of these transitions be detected?

Practice Problem 6.4: In the visible series of lines of the hydrogen spectrum, a line is observed at a wavelength of 486.1 nm. Calculate the upper quantum level of the electron in the hydrogen atom whose transition to the lower level ($n = 2$) is responsible for this line. What is the color of this line?

The Bohr model of the hydrogen atom allows us to calculate the energies required to excite the electron from lower to higher quantum levels in the atom. Thus, we can calculate the energy required to remove the electron from the atom by taking the electron from its ground state of $n = 1$ to infinity (∞). This energy is known as the **ionization energy** because removing an electron from a hydrogen atom produces a hydrogen ion, H^+.

Example 6.4

Calculating the Ionization Energy of Hydrogen

Calculate the ionization energy of hydrogen in its ground state in joules per atom and in joules per mole of atoms.

SOLUTION: When a hydrogen atom is in its ground state, its electron is in the first orbit, for which $n = 1$. The ionization energy for an atom is the energy required to take the electron out of the atom (from $n = 1$ to $n = \infty$):

$$\text{ionization energy per atom} = R_H\left(\frac{1}{1^2} - \frac{1}{\infty^2}\right) = 2.179 \times 10^{-18}\ \text{J}$$

The ionization energy per mole of atoms is

$$\left(\frac{2.179 \times 10^{-18}\ \text{J}}{1\ \text{atom}}\right)\left(\frac{6.023 \times 10^{23}\ \text{atoms}}{1\ \text{mol}}\right) = 1.312 \times 10^{6}\ \text{J mol}^{-1}$$

The Bohr theory explains some very important properties of hydrogen atoms. However, attempts to apply the Bohr theory to atoms with more than one electron did not explain their spectra. A more sophisticated theory was required. The Bohr theory, however, provided the basis for applying quantum theory to atomic structure. The principal quantum number, n, of Bohr's theory remains as a part of the modern atomic theory, although the notion of an electron moving around the nucleus along a clearly defined orbit has been abandoned.

6.4 Wave Theory

De Broglie's Hypothesis and the Heisenberg Uncertainty Principle

Prior to 1900 it was believed that light was exclusively wavelike in character. The work of Planck and Einstein, however, suggested that in some circumstances light behaves as if it consists of particles, called photons. In 1923 the French physicist Louis de Broglie made a surprising suggestion that if radiation has particlelike properties, then particles in motion should have wavelike properties. He showed that the wavelength, λ, of a moving particle equals the Planck's constant, h, divided by its momentum (the product of the mass, m, of the particle and its velocity, v): $\lambda = h/mv$.

We can see from this equation that when the mass of a particle is large, its wavelength is small. Thus, heavy objects have immeasurably small wavelengths, but subatomic particles such as electrons have wavelengths that can be determined experimentally.

In the world of familiar macroscopic objects, it is easy to determine the position and momentum of a moving object such as a tennis ball or a billiard ball. Furthermore, if we know the position and velocity of a billiard ball moving across a table at a given instant, we can predict the position of the ball at any future time if it is not acted on by an external force.

In 1928 the German physicist Werner Heisenberg (1901–1976) showed that we cannot accurately measure both the position and momentum of a subatomic particle such as an electron. That is, if we manage to determine the position of an electron precisely, we are not able *simultaneously* to determine its precise momentum. This principle is known as the **Heisenberg uncertainty principle**.

We can never invent a measuring device that will evade the dictates of the uncertainty principle. We can therefore speak only in terms of a given probability of finding an electron at a particular location in an atom. Since Bohr's theory placed an electron along a circular orbit, it was evident that Bohr's model of an atom had to be modified.

The Schrödinger Wave Equation

The ideas of Planck, Bohr, de Broglie, Heisenberg, and others were carried further by the Austrian physicist Erwin Schrödinger (1887–1961). In 1926 he developed an equation to describe the waves associated with electron motions. This equation is now known as the Schrödinger equation. The Schrödinger equation treats electrons in atoms as standing waves having certain allowed, or quantized energies.

When the Schrödinger equation is solved mathematically, it yields a set of equations called *wave functions*. Each of these is designated by the Greek lowercase letter psi, ψ. These wave functions are mathematical equations relating the amplitude, or the intensity of an electron wave, to its energy. The energies of the electron waves are related to a set of three quantum numbers, including a quantum number, n, which also emerges from the Bohr model of the hydrogen atom.

The solutions to the Schrödinger equation give us the energies of electrons in atoms, but these solutions do not provide us directly with any information about the location of these electrons in space. In 1926, Max Born, a German physicist, showed that the *square of the wave function, ψ^2, is proportional to the probability of finding the electron at a given point in an atom.* This probability is often referred to as the **electron density**. In the next section we consider the shapes of these electron density regions in space, which are usually called "orbitals."

6.5 Quantum Numbers and Atomic Orbitals

Quantum Numbers from the Schrödinger Equation

The solutions to the Schrödinger equation consist of a set of wave equations where electron energies are defined in terms of a set of three quantum numbers designated n, l, and m_l. Each electron in an atom is described by a set of all three of these quantum numbers. The principal quantum number, n, has the same meaning in both the Schrödinger equation and the Bohr model of the atom. The quantum numbers l and m_l have no counterpart in the Bohr model of the atom. We will consider n, l, and m_l in greater detail in the following paragraphs.

The numerical value of n is related to the energy of an electron and its most probable distance from the nucleus. According to both the Bohr theory and the wave theory, n can have only integral values of 1, 2, 3, and so on. These values correspond to **principal energy levels** or **quantum levels** for an electron in an atom. Principal quantum levels are also called **shells** around the nucleus of an atom. The shells are labeled as *K*, *L*, *M*, *N*, and so on, corresponding to the following n values:

Value of n	1	2	3	4	. . .
Designation of shell	*K*	*L*	*M*	*N*	. . .

Relation of Quantum Numbers to Orbitals

The electrons in a given shell can occupy different **subshells** in atoms with two or more electrons. These subshells have slightly different energies. An electron in a given subshell occupies a *spatial region* called an **orbital** *where the probability of finding an electron is greatest.* The quantum number l is called the **azimuthal** or **secondary quantum number**, but for our purposes it is more informative to call it the **subshell quantum number**.

The numerical value of the quantum number l specifies the subshell in any given shell. The possible values of l depend on the value of the principal quantum number, n. For a given value of n, l can have integral values from 0 to $(n - 1)$. For example, if $n = 1$, l can be only 0; if $n = 2$, l can have values of 0 *or* 1; and if $n = 3$, l can be 0, 1, or 2.

The l values are also designated by letters:

Value of l	0	1	2	3
Letter designation	*s*	*p*	*d*	*f*

Figure 6.13 illustrates the shapes of *s* and *p* orbitals. We can see from Figure 6.13 that the *s* orbital has a spherical shape, and each *p* orbital has a dumbbell shape with two lobes on opposite sides of the nucleus.

The probability of finding an electron increases as the distance from the nucleus increases, comes to a maximum, then decreases and approaches zero farther from the nucleus. The variation of electron density with the distance from the nucleus is illustrated in Figure 6.14 for the *s* orbital electrons in the first, second, and third quantum levels (the 1*s*, 2*s*, and 3*s* electrons). The surfaces where the electron density is zero are called *nodes*.

Figures 6.13 and 6.14 suggest that an *s* orbital does not have a definite boundary. The orbital models shown represent about ninety percent of the electron density. The greatest electron densities for 1*s*, 2*s*, and 3*s* electrons lie successively farther from the nucleus. However, both 2*s* and 3*s* electrons have small electron density regions rather close to the nucleus.

The quantum number m_l is called the **magnetic quantum number**. The values of m_l determine the spatial orientations of orbitals relative to the *x*, *y*, and *z* coordinates in the space around the nucleus. Just as the value of l depends on the value of n, the possible values of m_l depend on the value of l.

For a given l, m_l can have integral values from $-l$ to $+l$, including 0. For $l = 0$, $m_l = 0$; for $l = 1$, m_l has three possible values, -1, 0, and $+1$; and for $l = 2$, m_l has five possible values, -2, -1, 0, $+1$, and $+2$. There are $(2l + 1)$ values of m_l for any given value of l.

The total number of possible m_l values for a given value of l specifies the number of orbitals with that l value in a given principal quantum level (shell). For example, for the first principal quantum level, for which $n = 1$, l can be only 0. The l value of 0 represents an *s* orbital. For $l = 0$, m_l can be only 0. Thus there is only one value of m_l if $l = 0$. This means that the first principal quantum level contains only one orbital, the 1*s* orbital. In the orbital designation "1*s*" the number preceding the letter specifies the principal quantum level for the orbital.

In the second principal quantum level ($n = 2$), l can be 0 or 1. These l values correspond to *s* and *p* orbitals, respectively. Just as the first principal quantum level, the second level can only have one *s* orbital. The *s* orbital in the second principal quantum level is called the 2*s* orbital.

Figure 6.13

The shapes of *s* and *p* orbitals.

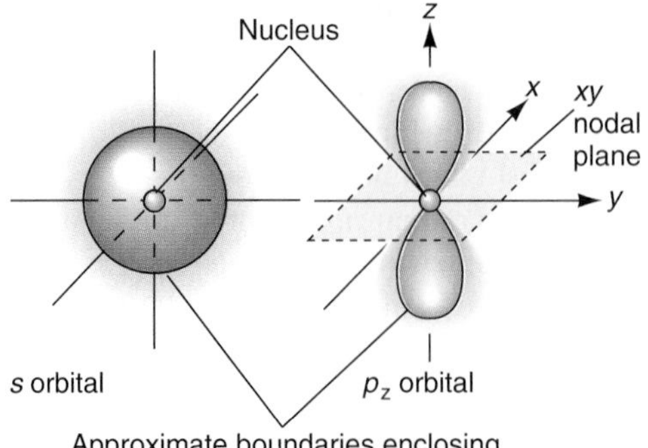

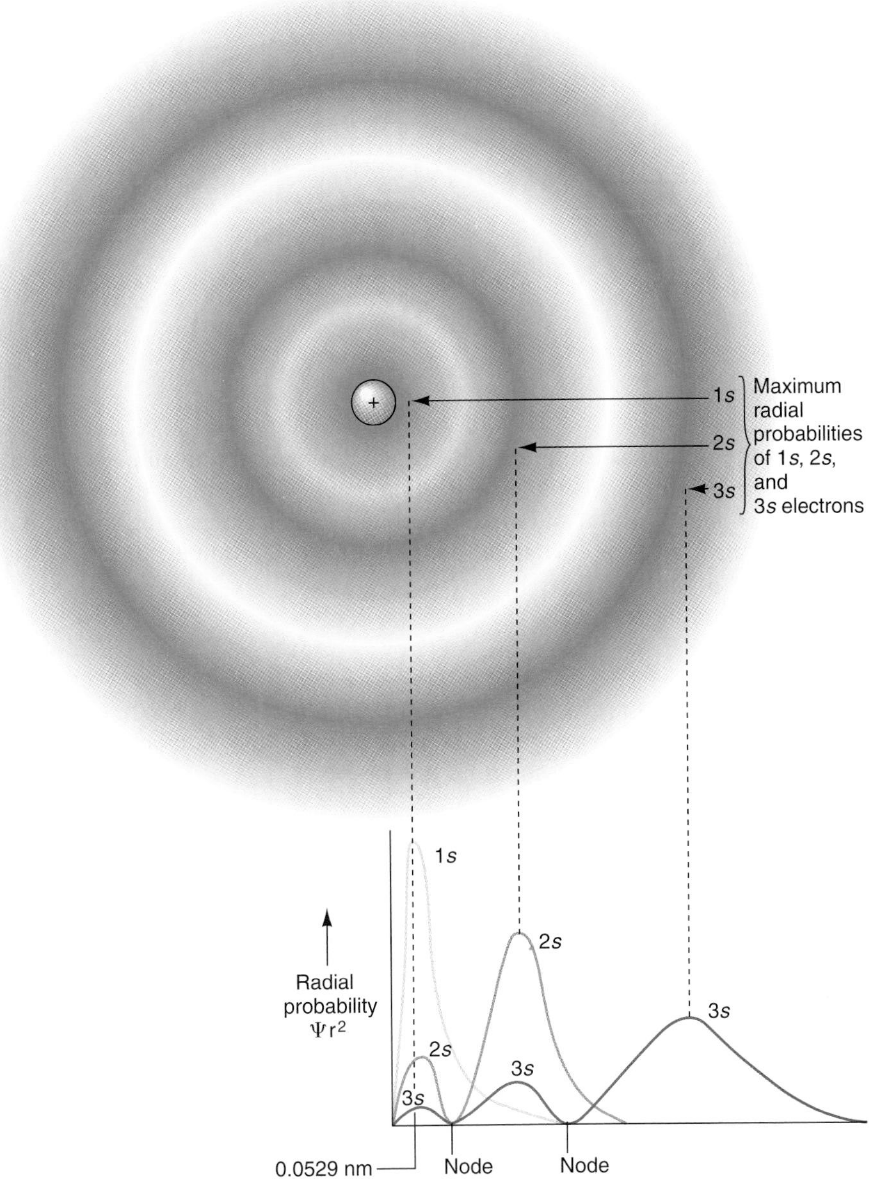

Figure 6.14

Radial probabilities of *s* orbitals. The maximum probability of finding the electron in the 1*s* orbital of a hydrogen atom is at the distance r = 0.0529 nm, which is also the radius of Bohr's first orbit.

The l value of 1 corresponds to p orbitals. When $l = 1$, m_l can have three different values. This means that there are three p orbitals in a principal quantum level. This is true for all principal quantum levels except the $n = 1$ level, for which $l = 0$. Each of the three p orbitals has a different m_l value and lies along a different coordinate axis, x, y, and z. The p orbital that lies along the x axis is the p_x orbital, the orbital along the y axis is the p_y orbital, and the one along the

Figure 6.15

Spatial orientations of *s*, *p*, and *d* orbitals.

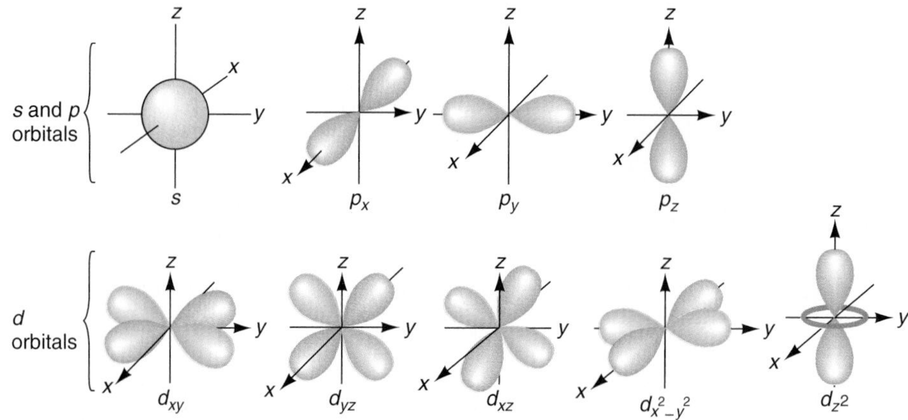

z axis is the p_z orbital (Figure 6.15). Thus, all the p orbitals in a shell lie at 90° angles to each other. We can now see that the $n = 2$ quantum level (the L shell) consists of one $2s$ orbital and three $2p$ orbitals. Similar reasoning shows that the third quantum level (the M shell) consists of one $3s$ orbital, three $3p$ orbitals, and five $3d$ orbitals.

Example 6.5

Determining the Relations Between Quantum Numbers and Orbitals

(a) What are the possible values of l for $n = 4$? **(b)** What are the possible values of m_l for $l = 3$? **(c)** What is the total number of orbitals in the fourth principal quantum level?

SOLUTION

(a) For $n = 4$, l can have values from 0 to 3: 0, 1, 2, 3.

(b) For $l = 3$, there are $(2)(3) + 1 = 7$ values of m_l: $-3, -2, -1, 0, +1, +2, +3$. These numbers refer to 7 different f orbitals.

(c) The fourth principal quantum level consists of one $4s$ orbital, three $4p$ orbitals, five $4d$ orbitals, and seven $4f$ orbitals, for a total of 16 orbitals.

The maximum number of electrons in a quantum level with principal quantum number n is $2n^2$. Thus, the $n = 1$ level contains a maximum of $2(1^2) = 2$ electrons, and the $n = 2$ level contains a maximum of $2(2^2) = 8$ electrons. Thus, the maximum number of electrons in quantum levels 1 through 4 are 2, 8, 18, and 32, respectively.

Example 6.6

Determining Quantum Numbers and Orbital Designations

What is the designation of the orbital defined by each of the following sets of values for n and l? **(a)** $n = 1, l = 0$; **(b)** $n = 2, l = 1$; **(c)** $n = 3, l = 0$; **(d)** $n = 3, l = 2$.

SOLUTION

(a) The values $n = 1$ and $l = 0$ refer to an *s* orbital in the first principal quantum level, the 1*s* orbital. Note that the number 1 preceding the letter designation of the orbital is the *n* value for the principal quantum level to which the orbital belongs.

(b) The values of $n = 2$ and $l = 1$ represent a *p* orbital in the second principal quantum level, a 2*p* orbital. Using similar reasoning, the remaining answers are: **(c)** 3*s* and **(d)** 3*d*.

Practice Problem 6.5: What are the values for the quantum numbers *n* and *l* for a 4*d* orbital?

The shapes of *s*, *p*, and *d* orbitals, and their spatial orientations are illustrated in Figure 6.15. Four of the five *d* orbitals have a double dumbbell shape, and one has the shape resembling a *p* orbital with a doughnut-shaped region of electron density around its center (Figure 6.15).

The *d* orbital that lies with its lobes along the *x* and *y* axes is labeled the $d_{x^2-y^2}$ (read *d x* squared minus *y* squared) orbital. The lobes of the remaining three *d* orbitals of double dumbbell shape lie between the coordinate axes. The one that lies in the *x*–*y* plane, with its lobes between the *x* and *y* axes, is the d_{xy} orbital. The orbital that lies in the *x*–*z* plane with its lobes between the *x* and *z* axes is the d_{xz} orbital, and the one in the *y*–*z* plane with the lobes between *y* and *z* axes is the d_{yz} orbital. The single dumbbell-shaped *d* orbital is customarily designated to lie along the *z* axis and is labeled the d_{z^2} orbital.

Table 6.3 summarizes the relations between quantum numbers and orbitals for the first four principal quantum levels. We can see that the fourth level (the *N* shell) contains one *s* orbital, three *p* orbitals, five *d* orbitals, and seven *f* orbitals. The shapes of the *f* orbitals are complex, and they are not discussed here.

Table 6.3

Quantum Numbers and Orbitals

Value of *n*	Shell	Possible Values of *l*	Possible Values of m_l	Number and Type of Orbitals in Shell
1	*K*	0	0	One 1*s* orbital
2	*L*	0	0	One 2*s* orbital
		1	−1, 0, 1	Three 2*p* orbitals
3	*M*	0	0	One 3*s* orbital
		1	−1, 0, 1	Three 3*p* orbitals
		2	−2, −1, 0, 1, 2	Five 3*d* orbitals
4	*N*	0	0	One 4*s* orbital
		1	−1, 0, 1	Three 4*p* orbitals
		2	−2, −1, 0, 1, 2	Five 4*d* orbitals
		3	−3, −2, −1, 0, 1, 2, 3	Seven 4*f* orbitals

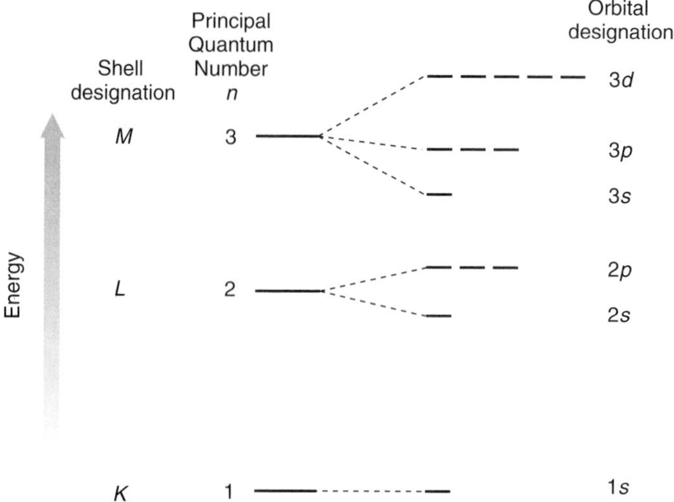

Figure 6.16

Relative orbital energies of the first three principal quantum levels in an atom of many electrons.

Orbital Energies

We have seen that each electron in an atom occupies an orbital that is characterized by a set of quantum numbers, and that the energy of an electron is related to the principal quantum number n. The energy of an electron increases (becomes less negative) with increasing n. Furthermore, in each shell with $n >$ 1, various subshells have slightly different energies. Within a given shell, the orbitals with higher l value have higher energy than the orbitals with lower l value. Thus, the p orbitals ($l = 1$) in a shell have higher energy than the s orbital for which l is 0. The order of orbital energies in a given principal quantum level, n, is as follows:

$$E_{ns} < E_{np} < E_{nd} < E_{nf}$$

In any given shell, orbitals with the same l value have equal energy. *Orbitals with the same energy* are said to be **degenerate**. Thus, all the p orbitals within a given shell are degenerate, all the d orbitals in that shell are degenerate, and all the f orbitals are also degenerate. The energy of the p orbitals in a shell is lower than the energy of the d orbitals, and the energy of the d orbitals is lower than the energy of the f orbitals. Figure 6.16 illustrates the approximate relative orbital energies of the first three principal quantum levels.

Electron Spin and the Pauli Principle

Knowledge of atomic structure gained from the Schrödinger equation was developed further by Dutch graduate students Samuel Goudsmit and George Uhlenbeck at the University of Leyden, Holland. They suggested that an electron spins around its axis like the earth spins around its axis.

The spin of an electron can be described by a quantum number that is called the **spin quantum number**, m_s. The spin quantum number can have one of only two possible values, $+\frac{1}{2}$ or $-\frac{1}{2}$. One of these values designates direction of the spin, say clockwise. The other value designates the opposite spin—counterclockwise (Figure 6.17). If two electrons in an atom spin in the same direction, they both have the same m_s value, either $+\frac{1}{2}$ or $-\frac{1}{2}$, and they are said to be *parallel*. In a given subshell, two electrons with parallel spin occupy different orbitals. If two electrons spin in opposite directions, one of them has the m_s value of $+\frac{1}{2}$ and the other $-\frac{1}{2}$. When two electrons are in the same orbital, they *must* have opposite spins; they are said to be *paired*.

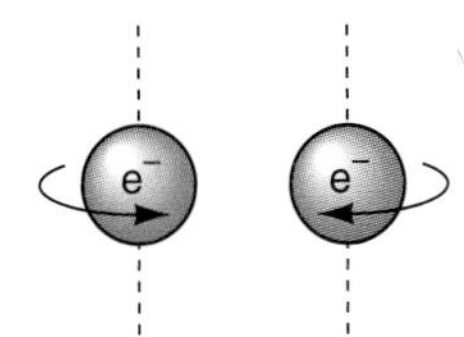

Figure 6.17

The opposite directions of the spin of an electron about its axis, clockwise and counterclockwise. The m_s value for one of the electrons is $+\frac{1}{2}$ and for the other $-\frac{1}{2}$.

The spin quantum number and a principle developed by the Austrian physicist Wolfgang Pauli (1900–1958) are important in developing our description of atomic structures. According to Pauli, *no two electrons in an atom can have the same set of four quantum numbers*. This statement is known as the **Pauli exclusion principle**.

Electrons in the same orbital of an atom have the same values of n, l, and m_l. According to the Pauli principle, they must have different values of m_s. Since there are only two values of m_s, an *orbital can hold a maximum of only two electrons. Moreover, two electrons in the same orbital must spin in opposite directions* (*that is, they must be paired*).

6.6 Atomic Structure and the Periodic Table

The Aufbau Process

In this section we describe the structures of isolated atoms in their ground energy states. We use a hypothetical atom building process known by its German name as the **aufbau** ("buildup") **process**. In this process we start with the smallest atom, the hydrogen atom, and imagine that we add protons, neutrons, and electrons to this atom to build larger atoms. Since the chemical properties of atoms mostly depend on electrons, we restrict our discussion here to adding only electrons. Although protons and neutrons must also be added, we do not mention them explicitly.

Building Hydrogen and Helium Atoms

The electron in a hydrogen atom in its ground state resides in the first principal quantum level, which contains only one occupied orbital, the 1*s* orbital. In "building" a helium atom in its ground state from a hydrogen atom in its ground state, another electron is placed outside the nucleus.

The electron that builds a helium atom out of a hydrogen atom enters the lowest possible energy level, which is still the 1*s* orbital. According to the Pauli principle, two electrons in the same orbital must spin in opposite directions; that is, their spins are *paired*. Thus, one electron in a helium atom spins in one direction and the other spins in the opposite direction. The following orbital diagram shows the electronic configurations of hydrogen and helium atoms (the orbitals are represented by dashes and electrons by arrows, with the arrowheads indicating the relative directions of spin):

	H		He
$n = 1$, 1*s* orbital . . .	↑	. . . .	↑ ↓

A shorthand method may also be used to describe electronic structures. In this method, atomic orbitals are represented by their corresponding letters *s*, *p*, *d*, and *f*. Each letter designation is preceded by a number that specifies the principal quantum level of the orbital. The number of electrons in each orbital is indicated by a superscript on the letter representing the orbital. Thus, the electronic structure of a hydrogen atom is $1s^1$, and the electronic structure of a helium atom is $1s^2$. Hydrogen and helium constitute the entire first period in the periodic table.

In succeeding sections we generate the periodic table by ''building'' the electronic configurations of the atoms of each period. In this process we use the noble gas electronic configuration of one period as the inner core upon which the electronic configurations of the atoms in the next period are assembled.

The Atoms of the Second Period

The elements of the second period are Li, Be, B, C, N, O, F, and Ne. We will build the atoms of these elements on the ''foundation'' of a helium atom. Figure 6.18 gives the electronic structures of all the second period atoms.

The first element in the second period is the alkali metal lithium. A lithium atom has three protons in its nucleus and three electrons. The two lowest energy electrons have quantum numbers identical to those of the helium ''core.'' No more electrons can be added to the *K* shell without violating the Pauli principle. Therefore, the third electron that builds the lithium atom must enter the second shell, for which $n = 2$. The lowest-energy orbital (sublevel) available for this electron is the 2*s* orbital. Thus, the third electron that builds the lithium atom enters the 2*s* orbital.

The next element in the second period is beryllium. A beryllium atom has four electrons. The fourth electron that builds a beryllium atom from a lithium atom enters the 2*s* orbital that already contains an electron. According to the Pauli principle, two electrons in the same orbital spin in opposite directions—

Figure 6.18

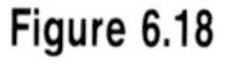

Electronic configurations of the second period atoms in their ground states. The orbitals are represented by dashes and the electrons by arrows.

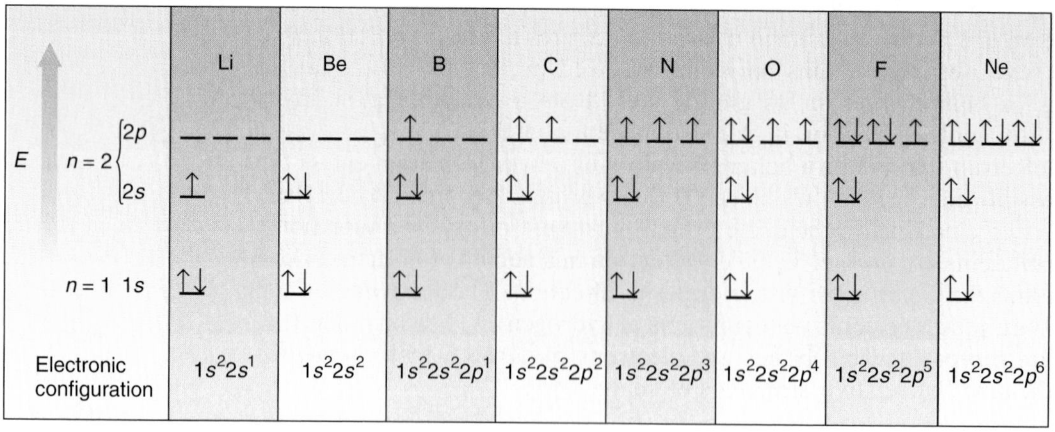

E		Li	Be	B	C	N	O	F	Ne
$n = 2$	2*p*	— — —	— — —	↑ — —	↑ ↑ —	↑ ↑ ↑	↑↓ ↑ ↑	↑↓ ↑↓ ↑	↑↓ ↑↓ ↑↓
	2*s*	↑	↑↓	↑↓	↑↓	↑↓	↑↓	↑↓	↑↓
$n = 1$	1*s*	↑↓	↑↓	↑↓	↑↓	↑↓	↑↓	↑↓	↑↓
Electronic configuration		$1s^22s^1$	$1s^22s^2$	$1s^22s^22p^1$	$1s^22s^22p^2$	$1s^22s^22p^3$	$1s^22s^22p^4$	$1s^22s^22p^5$	$1s^22s^22p^6$

their spins are paired. The complete electron configurations of lithium and beryllium atoms in their ground states are given below by both the shorthand symbolism and by the use of orbital diagrams:

	Li: $1s^2 2s^1$	Be: $1s^2 2s^2$
$n = 2$, $2s$ orbital . . .	↑	↑↓
$n = 1$, $1s$ orbital . . .	↑↓	↑↓

Note that once we have made beryllium we have filled the $2s$ subshell.

The fifth electron that builds a boron atom enters a p orbital of the second shell (the second principal quantum level). As we move from boron to carbon, we add one more electron. But does this electron enter a $2p$ orbital that already has an electron, or does it enter an empty $2p$ orbital? This question was answered by the German physicist Frederick Hund (1896–). According to Hund, *electrons occupy orbitals of the same subshell singly, with their spins parallel, before any pairing occurs*. This rule is known as **Hund's rule**.

Hund's rule has a simple explanation. We know that electrons repel one another. If two electrons are in the same $2p$ orbital (say, $2p_x$), the energy of the system is higher than the energy of the system in which one electron is in a $2p_x$ orbital and the other in a $2p_y$ or $2p_z$ orbital because two electrons in the same orbital repel each other more than two electrons in separate orbitals.

The spins of electrons in two degenerate p orbitals are parallel rather than paired because energy is required to pair electrons. Therefore, the $2p$ electrons in carbon and nitrogen atoms occupy different $2p$ orbitals and have parallel spins. The $2p$ orbital configurations of boron, carbon, and nitrogen atoms are shown below:

	B			C			N		
$2p$ orbitals . . .	↑			↑	↑		↑	↑	↑

We see that a boron atom has one unpaired electron, a carbon atom two, and a nitrogen atom three. The complete electronic configurations of boron, carbon, and nitrogen atoms are:

	B: $1s^2 2s^2 2p^1$			C: $1s^2 2s^2 2p^2$			N: $1s^2 2s^2 2p^3$		
$2p$ orbitals . . .	↑			↑	↑		↑	↑	↑
$2s$ orbitals . . .	↑↓			↑↓			↑↓		
$1s$ orbitals . . .	↑↓			↑↓			↑↓		

The electronic configurations of oxygen, fluorine, and neon atoms are obtained by adding one electron at a time. In this aufbau process, each added electron enters a half-filled p orbital. We can see from the orbital diagrams of the first and second period atoms that the spins of two electrons in any given orbital are paired.

Filling the $2s$ and $2p$ orbitals with electrons completes the second principal quantum level—the L shell. The L shell holds a maximum of eight electrons:

Table 6.4

Electronic Structure of the Atoms[a]

Main Groups →	I	II											III	IV	V	VI	VII	VIII
All Groups →	1	2	3	4	5	6	7	8	9	10	11	12	13	14	15	16	17	18
Period 1	1 **H** $1s^1$																	2 **He** $1s^2$
Period 2 He core	3 **Li** $2s^1$	4 **Be** $2s^2$											5 **B** $2s^2 2p^1$	6 **C** $2s^2 2p^2$	7 **N** $2s^2 2p^3$	8 **O** $2s^2 2p^4$	9 **F** $2s^2 2p^5$	10 **Ne** $2s^2 2p^6$
Period 3 Ne core	11 **Na** $3s^1$	12 **Mg** $3s^2$	*d* Series of Transition Metals (*d* orbitals fill)										13 **Al** $3s^2 3p^1$	14 **Si** $3s^2 3p^2$	15 **P** $3s^2 3p^3$	16 **S** $3s^2 3p^4$	17 **Cl** $3s^2 3p^5$	18 **Ar** $3s^2 3p^6$
Period 4 Ne core; also 3*d* orbitals in Ga–Kr	19 **K** $4s^1$	20 **Ca** $4s^2$	21 **Sc** $3d^1 4s^2$	22 **Ti** $3d^2 4s^2$	23 **V** $3d^3 4s^2$	24 **Cr** $3d^5 4s^1$	25 **Mn** $3d^5 4s^2$	26 **Fe** $3d^6 4s^2$	27 **Co** $3d^7 4s^2$	28 **Ni** $3d^8 4s^2$	29 **Cu** $3d^{10} 4s^1$	30 **Zn** $3d^{10} 4s^2$	31 **Ga** $4s^2 4p^1$	32 **Ge** $4s^2 4p^2$	33 **As** $4s^2 4p^3$	34 **Se** $4s^2 4p^4$	35 **Br** $4s^2 4p^5$	36 **Kr** $4s^2 4p^6$
Period 5 Kr core; also 4*d* orbitals in In–Xe	37 **Rb** $5s^1$	38 **Sr** $5s^2$	39 **Y** $4d^1 5s^2$	40 **Zr** $4d^2 5s^2$	41 **Nb** $4d^4 5s^1$	42 **Mo** $4d^5 5s^1$	43 **Tc** $4d^5 5s^2$	44 **Ru** $4d^7 5s^1$	45 **Rh** $4d^8 5s^1$	46 **Pd** $4d^{10}$	47 **Ag** $4d^{10} 5s^1$	48 **Cd** $4d^{10} 5s^2$	49 **In** $5s^2 5p^1$	50 **Sn** $5s^2 5p^2$	51 **Sb** $5s^2 5p^3$	52 **Te** $5s^2 5p^4$	53 **I** $5s^2 5p^5$	54 **Xe** $5s^2 5p^6$
Period 6 Xe core; also 4*f* orbitals in Rf–Rn and full 5*d* orbitals in Tl–Rn	55 **Cs** $6s^1$	56 **Ba** $6s^2$	57 **La** $5d^1 6s^2$	72 **Hf** $5d^2 6s^2$	73 **Ta** $5d^3 6s^2$	74 **W** $5d^4 6s^2$	75 **Re** $5d^5 6s^2$	76 **Os** $5d^6 6s^2$	77 **Ir** $5d^7 6s^2$	78 **Pt** $5d^9 6s^1$	79 **Au** $5d^{10} 6s^1$	80 **Hg** $5d^{10} 6s^2$	81 **Tl** $6s^2 6p^1$	82 **Pb** $6s^2 6p^2$	83 **Bi** $6s^2 6p^3$	84 **Po** $6s^2 6p^4$	85 **At** $6s^2 6p^5$	86 **Rn** $6s^2 6p^6$
Period 7 Rn core; also 5*f* orbitals in Unq–Une	87 **Fr** $7s^1$	88 **Ra** $7s^2$	89 **Ac** $6d^1 7s^2$	104 **Unq** $6d^2 7s^2$	105 **Unp** $6d^3 7s^2$	106 **Unh** $6d^4 7s^2$	107 **Uns** $6d^5 7s^2$	108 **Uno** $6d^6 7s^2$	109 **Une** $6d^7 7s^2$									

Inner Transition Metals (*f* orbitals fill)

Lanthanides Xe core →	58 **Ce** $4f^1 5d^1 6s^2$	59 **Pr** $4f^3 5d^0 6s^2$	60 **Nd** $4f^4 5d^0 6s^2$	61 **Pm** $4f^5 5d^0 6s^2$	62 **Sm** $4f^6 5d^0 6s^2$	63 **Eu** $4f^7 5d^0 6s^2$	64 **Gd** $4f^7 5d^0 6s^2$	65 **Tb** $4f^9 5d^0 6s^2$	66 **Dy** $4f^{10} 5d^0 6s^2$	67 **Ho** $4f^{11} 5d^0 6s^2$	68 **Er** $4f^{12} 5d^0 6s^2$	69 **Tm** $4f^{13} 5d^0 6s^2$	70 **Yb** $4f^{14} 5d^0 6s^2$	71 **Lu** $4f^{14} 5d^1 6s^2$
Actinides Rn core →	90 **Th** $5f^0 6d^2 7s^2$	91 **Pa** $5f^2 6d^1 7s^2$	92 **U** $5f^3 6d^1 7s^2$	93 **Np** $5f^4 6d^1 7s^2$	94 **Pu** $5f^6 6d^0 7s^2$	95 **Am** $5f^7 6d^0 7s^2$	96 **Cm** $5f^7 6d^1 7s^2$	97 **Bk** $5f^9 6d^0 7s^2$	98 **Cf** $5f^{10} 6d^0 7s^2$	99 **Es** $5f^{11} 6d^0 7s^2$	100 **Fm** $5f^{12} 6d^0 7s^2$	101 **Md** $5f^{13} 6d^0 7s^2$	102 **No** $5f^{14} 6d^0 7s^2$	103 **Lr** $5f^{14} 6d^1 7s^2$

[a] This table lists the symbols, atomic numbers, and the outermost electron structures of the atoms of the elements.

two are in the 2*s* orbital and six occupy the three 2*p* orbitals. Thus, a neon atom has a completely filled *L* shell and neon is stable and chemically unreactive. It has virtually no tendency to gain, lose, or share electrons.

The Third Period

The third period elements include sodium through argon. The electronic configuration of neon provides the core electronic structure for third period atoms. As we proceed across the third period, we fill the 3*s* and 3*p* orbitals. The sodium atom has one 3*s* electron and the magnesium atom has two 3*s* electrons.

The 3*p* orbitals in the third period are filled from aluminum through argon in the same way as the 2*p* orbitals are filled from boron to neon. Thus, like a neon atom, an argon atom has a total of eight electrons in its outer shell.

Each period in the periodic table (Table 6.4) ends with a noble gas. The electronic configuration of any atom in the second and later periods can be built by adding electrons to the noble gas atom of the preceding period. The electronic structures of atoms can therefore be conveniently described by listing the symbol of the noble gas from the preceding period, followed by the shorthand symbolism we have already described. The noble gas symbol is usually enclosed in square brackets, and the symbol with its brackets is called the **noble gas core**. For example, the electronic configuration of a beryllium atom is [He]$2s^2$, and the electronic configuration of a chlorine atom is [Ne]$3s^2 3p^5$. Thus, a beryllium atom consists of a helium core plus two $2s$ electrons, and a chlorine atom consists of a neon core plus two $3s$ electrons and five $3p$ electrons.

The electronic structures of the elements in the first three periods are shown in Figure 6.19. The approximate relative orbital energies in each quantum level are also indicated in Figure 6.19. The differences between the s and p orbital energies in the same principal quantum level are relatively small compared with the energy differences between principal quantum levels.

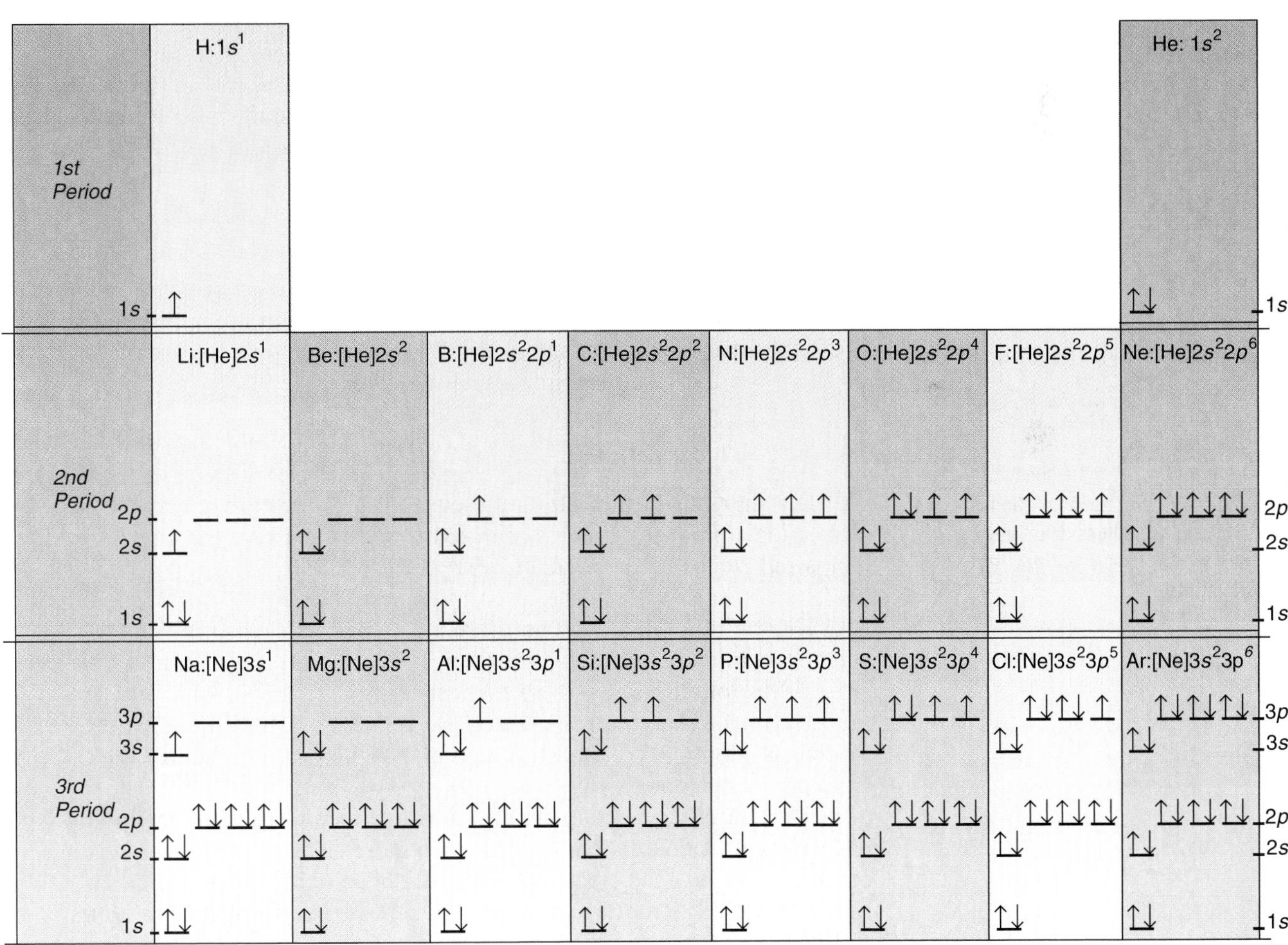

Figure 6.19

Electronic configurations of the elements in the first three periods.

Example 6.7

Periods and Quantum Levels

How many elements are in each of the first three periods of the periodic table? How is the number of elements in each of these periods related to the maximum number of electrons in the highest principal quantum level in the period?

SOLUTION: The first period consists of two elements. Only two electrons are required to fill the entire first principal quantum level, or *K* shell. Thus, the number of elements in the first period equals the number of electrons that completely fill the first principal quantum level.

The second period consists of eight elements. The atoms of these elements are built from the helium atom by successively adding two 2*s* electrons and six 2*p* electrons to complete the second principal quantum level, the *L* shell. The number of elements in the second period equals the maximum number of electrons in the second principal quantum level.

The third period also consists of eight elements. The number of elements in the third period equals the number of electrons which completely fill the 3*s* and 3*p* orbitals. These electrons make up only a part of the third principal quantum level which also contains five as yet unoccupied 3*d* orbitals.

Practice Problem 6.6: How many *s* orbitals does any given principal quantum level contain? How many *p* orbitals? How many electrons are required to fill all the *s* orbitals in a principal quantum level? How many electrons are required to fill all the *p* orbitals in a principal quantum level?

Example 6.8

Shorthand Description of Atomic and Ionic Structures, and Determining the Number of Unpaired Electrons

Write a shorthand description of the electronic structure of each of the following atoms and ions in their ground states and give the number of unpaired electrons on each: Be, Be^{2+}, O, O^{2-}, Na, P.

SOLUTION: Be: $[He]2s^2$. There are no unpaired electrons. The two electrons in each of the 1*s* and 2*s* orbitals are paired. To form a Be^{2+} ion, a Be atom loses two of its highest-energy electrons, which are the 2*s* electrons. The electronic structure of a Be^{2+} ion is therefore $1s^2$, which is identical to the structure of a helium atom. There are no unpaired electrons.

O: $[He]2s^22p^4$. There are two unpaired electrons, each occupying a 2*p* orbital. An oxide ion, O^{2-}, has two more electrons than an oxygen atom. Therefore, the electronic structure of an oxide ion is $[He]2s^22p^6$. This electronic structure is identical to the structure of a neon atom. All the electrons are paired.

Na: $[Ne]3s^1$. One unpaired electron in the 3*s* orbital.

P: $[Ne]3s^23p^3$. Three unpaired electrons, each occupying a separate 3*p* orbital.

Practice Problem 6.7: Write a shorthand description of the electronic structure of the following atoms and ions in their ground states: magnesium atom, magnesium ion, nitrogen atom, nitride ion, chlorine atom. How many unpaired electrons does each of these species contain?

The Long Periods

In atoms beyond the first three periods, the energies of the outer main quantum levels get closer to one another, the subshells within each shell become more diverse, and some overlapping occurs as shown in Figure 6.20. In the beginning of the fourth period, the energy of the 4*s* orbital is below the energy of the 3*d* orbitals at potassium (atomic number 19) and at calcium (atomic number 20). At scandium (atomic number 21), the 4*s* and 3*d* orbital energies are almost equal. As the atomic number increases beyond scandium, the energy of the 3*d* orbitals becomes progressively lower than that of the 4*s* orbitals (Figure 6.20).

In accord with the orbital energy variations discussed above, the fourth period begins with the element potassium, followed by calcium. In these atoms, the 4*s* orbital is filled because its energy is lower than the 3*d* orbital energy at atomic numbers 19 and 20 (Figure 6.20). Then the five 3*d* orbitals are filled in atoms of scandium through zinc. These 10 elements are called the **3*d* series of transition elements**, also called the first transition series. The fourth period is then completed by filling the three 4*p* orbitals in the six atoms from gallium through krypton. The fourth period thus consists of a total of 18 elements.

Just as the *p* orbitals in any given shell are degenerate, so are the *d* orbitals. Therefore, the *d* orbitals in a shell fill with electrons according to Hund's rule. Each of the five *d* orbitals will be occupied with one electron before any pairing occurs. For example, a titanium atom has two unpaired *d* electrons, a vanadium atom three, and a manganese atom five.

The electronic configurations of transition elements can be described by either of two methods. One of these methods follows the aufbau order of orbital filling, the other specifies the actual orbital energies so that the orbitals of highest energy are written last. For example, according to the aufbau order of orbital filling, the structure of the zinc atom is $[Ar]4s^2 3d^{10}$. However, since the 4*s* orbital energy of a zinc atom is higher than the 3*d* orbital energy, the structure of the zinc atom that expresses this order of energies is $[Ar]3d^{10}4s^2$. In this book we follow the convention that expresses the actual order of orbital energies.

The actual order of orbital energies is consistent with the fact that all the transition metals in the 3*d* series except scandium form 2+ ions. When an ion forms from an atom, the highest energy electrons are lost first. When an atom in the 3*d* series of transition metals loses two 4*s* electrons, a 2+ ion forms. In scandium, the energies of the 4*s* and the 3*d* electrons are about the same, so the scandium atom loses all three of these electrons to form a 3+ ion (see Table 2.3).

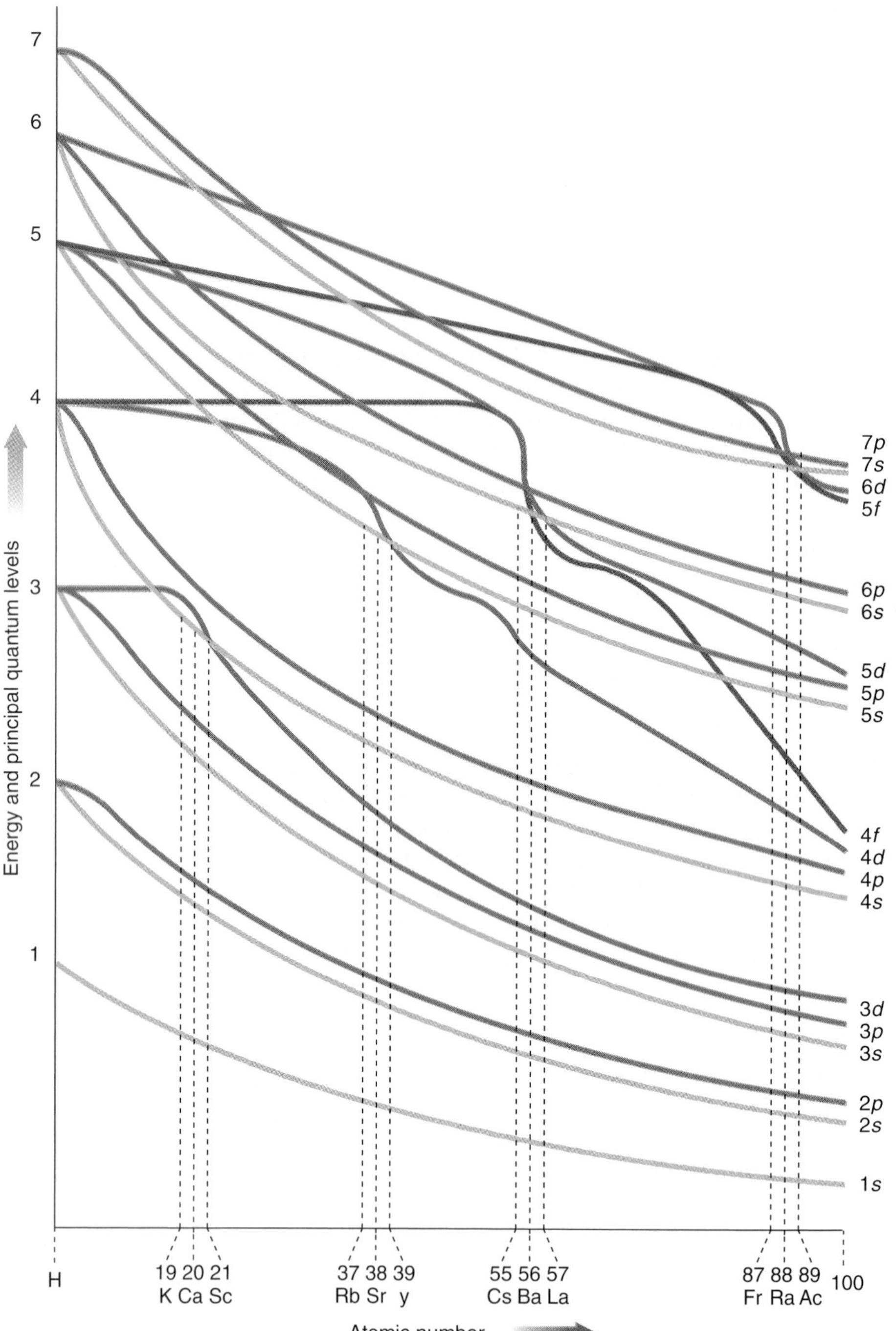

Figure 6.20

Variation of orbital energies with atomic number in isolated atoms. The *d* orbital energy falls below the *s* orbital energy at the beginning of each of the *d* series of transition metals.

Example 6.9

Electronic Structures of the 3*d* Series of Transition Metals

Write a shorthand description of the structure of a vanadium atom, an iron atom, and an iron(III) ion in their ground states. How many unpaired electrons do a vanadium atom and an iron atom contain?

SOLUTION: The vanadium atom is the third atom in the 3*d* series and has three 3*d* electrons. The structure of the vanadium atom is therefore $[Ar]3d^3 4s^2$. The three 3*d* electrons are unpaired. The iron atom is the sixth atom in the 3*d* series and thus contains six 3*d* electrons. The structure of the iron atom is $[Ar]3d^6 4s^2$. Of the six 3*d* electrons, four are unpaired, and two occupy a *d* orbital as a pair. When an iron(III) ion forms from an atom, the atom loses its highest energy electrons—the two 4*s* electrons—plus one 3*d* electron. The structure of the Fe^{3+} ion is $[Ar]3d^5$. All five of the 3*d* electrons are unpaired.

Practice Problem 6.8: Write the shorthand descriptions of the ground state electron configurations of a titanium atom, a nickel atom, and a nickel(II) ion. How many unpaired electrons does each of these species contain?

Two of the 3*d* series of transition metals have electronic configurations that cannot be predicted from the regular trend in the aufbau process. These metals are chromium and copper. For chromium, one would expect the structure $[Ar]3d^4 4s^2$, and for copper $[Ar]3d^9 4s^2$. However, the actual structures are $[Ar]3d^5 4s^1$ for chromium and $[Ar]3s^{10} 4s^1$ for copper. It has been found that structures that have half-full and full *d* orbital levels, that is, d^5 and d^{10} electronic configurations, are more stable than d^4 and d^9 electronic configurations. Chromium and copper atoms thus have the d^5 and d^{10} structures, respectively, at the expense of the 4*s* level.

The fifth period begins with the elements rubidium and strontium. In these atoms the most energetic electrons occupy 5*s* orbitals. Just as the 4*s* orbital energy is below 3*d* at potassium and calcium in the fourth period, the energy of the 5*s* orbital is lower than 4*d* at rubidium and strontium. After the 5*s* orbital is full and the 4*d* electrons start to enter, the energy of the 4*d* orbitals falls below 5*s* (Figure 6.20). The 4*d* orbitals are then filled with 10 electrons in the ***4d* series of transition elements** from yttrium through cadmium (a total of 10 elements). Like chromium and copper in the 3*d* series, the structures of the molybdenum and silver atoms in the 4*d* series are $[Kr]4d^5 5s^1$ and $[Kr]4d^{10} 5s^1$, respectively. The fifth period is completed by filling the 5*p* orbitals from indium through xenon. As in the fourth period, there are a total of 18 elements in the fifth period.

At the beginning of the sixth period, the 6*s* orbital is filled at cesium and barium (atomic numbers 55 and 56, respectively). At atomic number 57 (lanthanum) the energies of the 6*s*, 5*d*, and 4*f* orbitals are very close together as shown in Figure 6.20. With increasing atomic number, the 5*d* and 4*f* orbital energies become progressively lower than the 6*s* orbital energy but stay close to one another until about atomic number 70 when the energy of the 4*f* orbitals sinks abruptly below the energy of the 5*d* orbitals. These energy variations determine the order of orbital filling after barium in the sixth period.

The element after barium in the sixth period is lanthanum. A lanthanum atom has 57 electrons. The fifty-seventh electron that builds a lanthanum atom is the first 5*d* electron. Following this, 4*f* orbital filling occurs from cerium through lutetium (atomic numbers 58 through 71, respectively). The 14 elements from cerium through lutetium are known as the **lanthanides** or **lanthanoids** because they follow the element lanthanum. The lanthanides are also called the **4*f* series of inner transition metals**. After the 4*f* orbitals are filled, the 5*d* orbitals are then completed from hafnium through mercury. This is followed by the filling of the 6*p* orbitals from thallium through radon to complete the sixth period. Thus, the sixth period contains a total of 32 elements.

The seventh and last period begins with the filling of the 7*s* orbitals in francium and radium. Like the order of orbital filling in the sixth period, the first 6*d* orbital is filled at actinium (atomic number 89) in the seventh period. After actinium, the seven 5*f* orbitals are filled from thorium through lawrencium (atomic numbers 90 through 103, respectively). The elements from thorium through lawrencium are called the **actinides** or **actinoids** because they follow actinium; they are also called the **5*f* series of inner transition metals.** After the actinides, the filling of the 6*d* orbitals continues from element unnilquadium (atomic number 104) to unnilennium (atomic number 109), the most recently discovered elements. The highest-energy electron in an atom of the element to be discovered next will most likely be a 6*d* electron.

The highest-energy orbitals in the atoms of each period of the periodic table, as well as the order of orbital filling in the aufbau process, are shown in Figure 6.21. There are several irregularities in the general trend of orbital filling (for example, chromium and copper). The number of irregularities increases toward the larger atoms, as shown in Table 6.4. The reasons for some of these irregularities are presently not well understood, and some details of the electronic configuration of heavy atoms remain uncertain.

Example 6.10

Electronic Structures of Metal Atoms and Numbers of Unpaired Electrons

Write the shorthand description of the electronic structure for each of the following atoms and indicate the number of unpaired electrons in each: cobalt, chromium, silver, mercury, and lead.

SOLUTION: Co: $[Ar]3d^7 4s^2$. Three unpaired electrons in the 3*d* orbitals. Cr: $[Ar]3d^5 4s^1$. Six unpaired electrons, five in the 3*d* orbitals and one in the 4*s* orbital. Ag: $[Kr]4d^{10} 5s^1$. One unpaired electron in the 5*s* orbital. Hg: $[Xe]4f^{14} 5d^{10} 6s^2$. No unpaired electrons. Pb: $[Xe]4f^{14} 5d^{10} 6s^2 6p^2$. Two unpaired electrons in the 6*p* orbitals.

Practice Problem 6.9: Write the shorthand descriptions of the ground state atoms of each of the following elements, and indicate the number of unpaired electrons in each: **(a)** the element of the second group and fifth period; **(b)** the noble

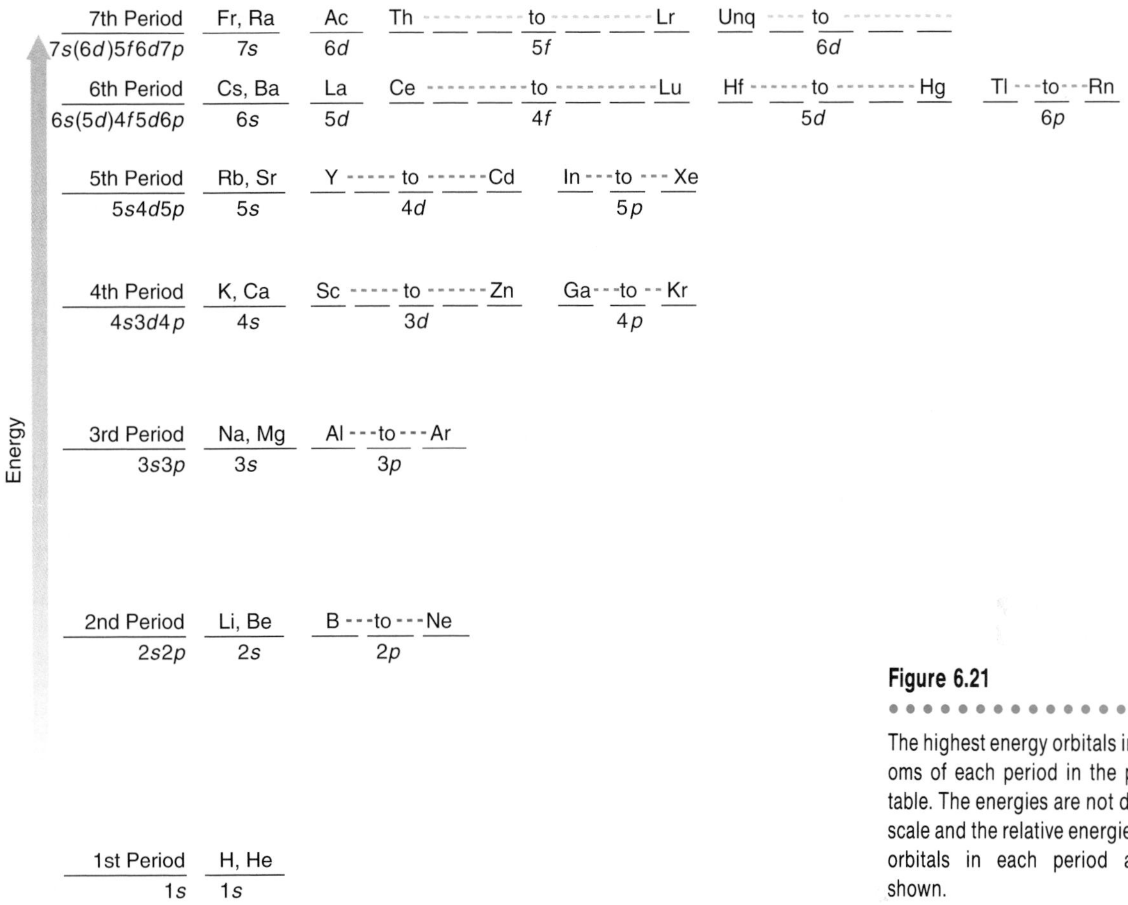

Energy ↑

Period					
7th Period 7s(6d)5f6d7p	Fr, Ra 7s	Ac 6d	Th - - - to - - - Lr 5f	Unq - - - to - - - 6d	
6th Period 6s(5d)4f5d6p	Cs, Ba 6s	La 5d	Ce - - - to - - - Lu 4f	Hf - - - to - - - Hg 5d	Tl - - - to - - - Rn 6p
5th Period 5s4d5p	Rb, Sr 5s	Y - - - to - - - Cd 4d	In - - - to - - - Xe 5p		
4th Period 4s3d4p	K, Ca 4s	Sc - - - to - - - Zn 3d	Ga - - - to - - - Kr 4p		
3rd Period 3s3p	Na, Mg 3s	Al - - - to - - - Ar 3p			
2nd Period 2s2p	Li, Be 2s	B - - - to - - - Ne 2p			
1st Period 1s	H, He 1s				

Figure 6.21

The highest energy orbitals in the atoms of each period in the periodic table. The energies are not drawn to scale and the relative energies of the orbitals in each period are not shown.

gas of the fourth period; **(c)** the halogen of the fourth period; **(d)** the alkali metal of the fifth period; **(e)** manganese; **(f)** copper; **(g)** zinc; **(h)** the alkaline earth metal of the sixth period.

When we consider the complete periodic table (Table 6.4), we see that in the building of the atoms of groups I and II in the aufbau process, the *s* orbitals are being filled. The *elements in groups I and II* are therefore called the **s-block elements**. The filling of *p* orbitals occurs in groups III through VIII. The *elements of groups III through VIII* are therefore called the ***p*-block elements**. Similarly, the *d series of transition elements* are the ***d*-block elements**, and the *f-series of inner transition elements* are the ***f*-block elements**.

Valence Electrons

For the main group (the *s*- and *p*-block) elements, the electrons in the outermost principal quantum level of an atom are called **valence electrons**. The valence electrons of an atom participate in the formation of bonds with other atoms. In the shorthand description of electronic configuration, valence electrons are written beyond the noble gas core. For example, the electron configuration of magnesium is [Ne]$3s^2$; the valence electrons are the 3*s* electrons. For phosphorus, [Ne]$3s^23p^3$, the 3*s* and 3*p* electrons are the valence electrons. Similarities among the valence shell configurations of atoms account for similarities in their chemical properties.

Table 6.4 shows that *the number of valence electrons of a main group element is given by its group number* (*the Roman numeral in Table 6.4*). Thus, magnesium is in group II and its atom has two valence electrons—the 3*s* electrons. Phosphorus is in group V, and its atom has five valence electrons—two 3*s* electrons and three 3*p* electrons. Thus, the periodic table is an arrangement of atoms that places elements with the same number of valence electrons in the same group. Valence electrons will be important when we discuss chemical bonding in Chapters 7 and 8.

Example 6.11

Valence Electrons

What is the number of valence electrons in a carbon atom and in a bromine atom? How many valence electrons are unpaired in each case?

SOLUTION: Carbon is in group IV and a carbon atom has four valence electrons, two 2*s* electrons and two 2*p* electrons. The two 2*p* electrons are unpaired. Bromine is in group VII, and a bromine atom has seven valence electrons, two 4*s* electrons and five 4*p* electrons. Only one of the 4*p* electrons is unpaired.

With the knowledge of atomic structure we have now acquired, we can turn our attention to some periodic properties of atoms that depend on their electronic configurations. As we recall, periodic properties vary in a similar way in each period of the periodic table.

6.7 Periodic Properties of Atoms

Atomic and Ionic Radii

Although we cannot be certain about the shapes of atoms, we will assume that atoms are spherical. The covalent radius of an atom is one-half of the distance between the nuclei of two identical atoms bonded by a single covalent bond. The radii of metal atoms—the metallic radii—are obtained from internuclear distances in metals. Ionic radii can be calculated from internuclear distances and lattice structures of ionic solids. The atomic and ionic radii of the main group elements are illustrated in Figure 6.22.

We can see from Figure 6.22 that the radii of atoms generally decrease from left to right in a period and increase as we move down a group. To explain

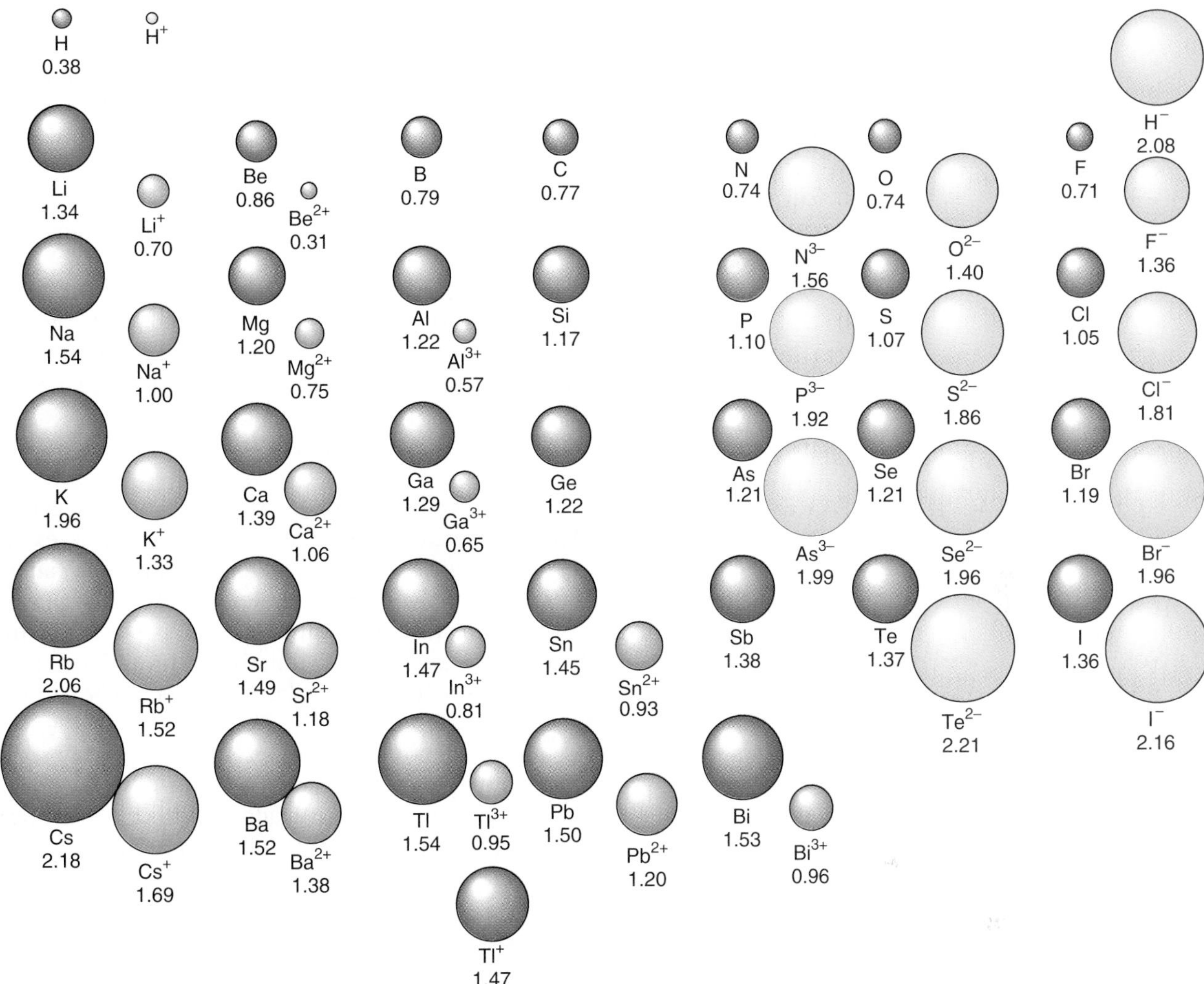

Figure 6.22

Atomic and ionic radii (in Å) of the main group elements. 1 Å = 0.1 nm = 10^{-10} m.

the trend of decreasing atomic radii within a period, we recall that the outermost electrons of the main group atoms in the same period have the same value of the principal quantum number, n, and belong to the same shell. We also recall that the value of n designates the most probable distance of an electron from the nucleus of an atom. As we proceed from the first atom at the left of a period toward the right of the period, the nuclear charge of the atoms increases. The increasing nuclear charge attracts the valence electrons closer to the nucleus, decreasing the atomic radii.

As we move down a group, the valence electrons are in successively higher principal quantum levels, which represent larger shells. Also, the valence electrons become more *shielded* from the nucleus by the increasing number of electron shells between the nucleus and the valence shell. As a result, the valence electrons do not ''feel'' the full nuclear charge. Thus, as we move down a group, the successively increasing value of n for larger shells and the shielding factor act in concert to cause an increase in atomic radii despite the increasing nuclear charge.

The atomic and ionic radii of the *d* series of transition metals are shown in Figure 6.23. We can see that the trends in the variation of the atomic radii of the transition metals within periods are irregular. In the three periods shown, there is a small decrease in the radii to the middle of the period, followed by a small irregular variation through the latter part of the period. The reason for this variation is that the outer *s* electrons are effectively shielded from the increased nuclear charge by the electrons in the inner *d* subshell. There is only a small increase of the radii toward the heavier elements in each group of transition metals.

When a neutral atom loses an electron, it becomes a positive ion. The positive ions of main group and transition metals are smaller than the atoms from which the ions are formed (Figures 6.22 and 6.23). When a positive ion forms from an atom, the electrons lost are those of highest energy and greatest distance from the nucleus. The electrons remaining in the ion occupy lower energy levels closer to the nucleus. Also, interelectronic repulsion is decreased.

Removing more electrons from a positive ion decreases the total negative charge of the electrons, while the nuclear charge remains constant. As a result, the remaining electrons are pulled closer to the nucleus and the ionic radius decreases even further. Thus, for any given element, a cation of higher charge is smaller than the cation of lower charge. For example, we can see in Figure 6.22 that a Tl^{3+} ion is smaller than a Tl^{+} ion. Similarly, in the 3*d* series of transition elements, an Fe^{3+} ion is smaller than an Fe^{2+} ion, and a Cu^{2+} ion is smaller than a Cu^{+} ion (Figure 6.23).

When a neutral atom gains an electron, it becomes a negative ion. Negative ions are larger than their corresponding atoms (Figure 6.22). An anion is larger than its corresponding atom because electrons added to the atom to form the anion increase the mutual repulsion between the electrons. The electrons "spread out" into a larger volume to minimize the repulsion.

Figure 6.23

Atomic and ionic radii (in Å) of transition metals.

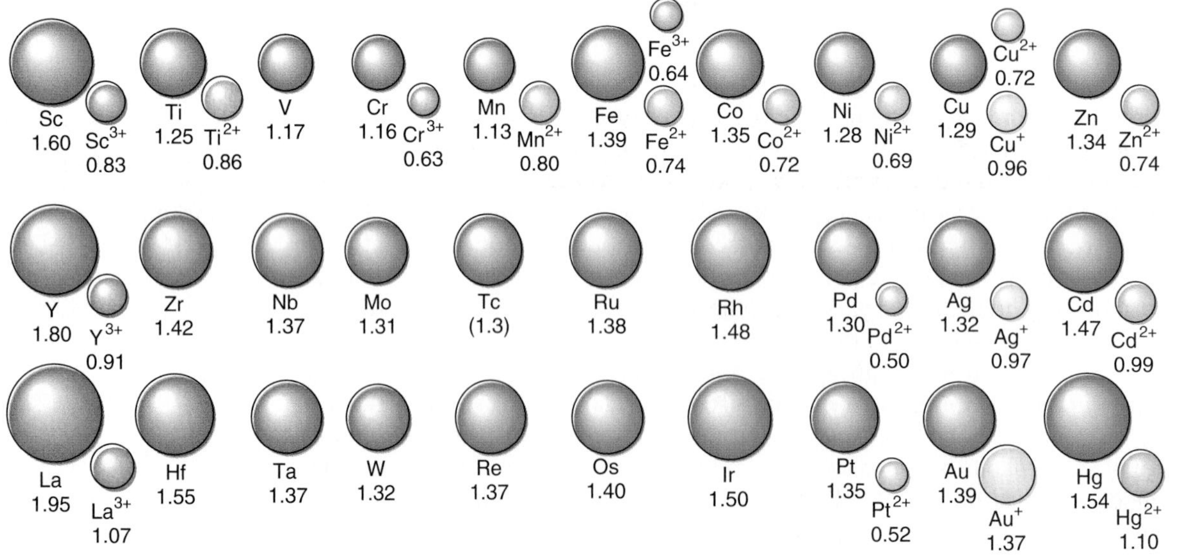

Figure 6.22 reveals that the radii of the negative ions from N^{3-} to F^- decrease, as do the radii of the corresponding atoms from left to right within a period. Nitride, N^{3-}, oxide, O^{2-}, and fluoride, F^-, ions have identical electronic structures. Thus, they are said to be *isoelectronic*. The valence shell configurations of nitride, oxide, and fluoride ions are all $2s^22p^6$. The decrease in the radii from N^{3-} to F^- again reflects the increasing nuclear charge from N^{3-} to F^-. A neon atom is also isoelectronic with N^{3-}, O^{2-}, and F^- ions, as well as with Na^+, Mg^{2+}, and Al^{3+} ions. The radii from N^{3-} through Al^{3+} decrease in that order because of the increasing nuclear charge from N^{3-} to Al^{3+}.

Within a group of elements in the periodic table, the radii of both the cations and anions increase as we move toward the heavier elements, paralleling the trend in atomic radii. In heavier atoms, the outer electrons belong to increasingly larger shells that are shielded from the nuclear charge by the electrons in the inner shells. The increasing distance of outer electrons in a larger shell, and the shielding factor, outweigh the effect of the increasing nuclear charge in determining both the atomic and ionic radii in a group of elements.

Example 6.12

Relative Sizes of Isoelectronic Species

Which of the following atoms and ions has the largest radius, and which has the smallest radius: K^+, Ca^{2+}, Sc^{3+}, S^{2-}, Cl^-, Ar?

SOLUTION: All the species given are isoelectronic—they all have the Ar structure. Because the nuclear charge in an S^{2-} ion is the lowest in this group, the S^{2-} ion has the largest radius. The nuclear charge in the Sc^{3+} ion is the highest in this group, and the Sc^{3+} ion has the smallest radius.

Ionization Energy

Earlier in this chapter we defined the ionization energy of hydrogen as the energy required to take the electron from its ground state ($n = 1$) completely out of the atom (to $n = \infty$). In an atom with more than one electron, the electrons have different energies. These atoms can have more than one ionization energy. The *energy required to remove the highest energy electron from an isolated atom* is its **first ionization energy**. After one electron is removed from an atom, the *energy needed to remove the highest energy electron from the 1+ ion* is the **second ionization energy**, and so on. We can illustrate the first and second ionization energies for any isolated (gaseous) atom X as follows:

$$X(g) + \text{first ionization energy} \longrightarrow X^+(g) + e^-$$
$$X^+(g) + \text{second ionization energy} \longrightarrow X^{2+}(g) + e^-$$

The second ionization energy of an atom is higher than the first (Table 6.5) because it is more difficult to remove a negatively charged electron from a positively charged ion than from a neutral atom. Each successive ionization energy is considerably higher than the preceding one because of the increasing positive charge on the ion.

Table 6.5

First and Second Ionization Energies of Elements (kJ mol^{-1})

Key: Li — 520 ← First ionization energy; 7297 ← Second ionization energy

H 1311																	**He** 2372 5250
Li 520 7297	**Be** 899 1757											**B** 801 2427	**C** 1087 2353	**N** 1402 2855	**O** 1313 3388	**F** 1681 3376	**Ne** 2081 3963
Na 495 4563	**Mg** 737 1450											**Al** 577 1816	**Si** 786 1577	**P** 1011 1903	**S** 1000 2258	**Cl** 1255 2297	**Ar** 1521 2665
K 419 3070	**Ca** 590 1145	**Sc** 631 1235	**Ti** 658 1310	**V** 650 1414	**Cr** 652 1591	**Mn** 717 1509	**Fe** 762 1561	**Co** 758 1645	**Ni** 736 1751	**Cu** 745 1958	**Zn** 906 1733	**Ga** 579 1979	**Ge** 760 1537	**As** 947 1798	**Se** 941 2075	**Br** 1143 2084	**Kr** 1351 2370
Rb 403 2653	**Sr** 549 1064	**Y** 616 1180	**Zr** 660 1267	**Nb** 664 1382	**Mo** 685 1558	**Tc** 703 1473	**Ru** 710 1617	**Rh** 720 1744	**Pd** 804 1874	**Ag** 731 2073	**Cd** 868 1631	**In** 588 1820	**Sn** 708 1412	**Sb** 834 1592	**Te** 869 1795	**I** 1008 1842	**Xe** 1171 2046
Cs 375 2422	**Ba** 503 965	**La** 541 1103	**Hf** 676 1438	**Ta** 760 1563	**W** 770 1708	**Re** 759 1602	**Os** 840 1640	**Ir** 868	**Pt** 868 1795	**Au** 888 1978	**Hg** 1006 1809	**Tl** 590 1971	**Pb** 716 1450	**Bi** 703 1610	**Po** 813 1834	**At** 917 1940	**Rn** 1037
Fr 370 2123	**Ra** 510 976	**Ac** 666 1168															

Ce 540 1187	**Pr** 529	**Nd** 531	**Pm**	**Sm** 540	**Eu** 547	**Gd** 594	**Tb** 577	**Dy** 656	**Ho**	**Er** 587	**Tm** 561	**Yb** 600	**Lu** 593
Th 671	**Pa**	**U** 587	**Np**	**Pu** 560	**Am** 579	**Cm**	**Bk**	**Cf**	**Es**	**Fm**	**Md**	**No**	**Lr**

Ionization energy generally increases moving from left to right across a period, as shown in Figure 6.24. This increase is due to the increasing nuclear charge and the consequent decrease in atomic radii. It is more difficult to pull an electron out of a small atom with a high nuclear charge than from a larger atom of lower nuclear charge.

In each of the second, third, and fourth periods of the representative elements, there are two exceptions to the generally increasing trend of ionization energy. In each of these periods, the exceptions occur in Groups III and VI. In Group III, there is a small decrease from the generally increasing trend at boron, aluminum, and gallium atoms (Figure 6.24). In Group VI, this decrease occurs at oxygen, sulfur, and selenium atoms.

The valence shell electron structures of B, Al, and Ga atoms are $2s^22p^1$, $3s^23p^1$, and $4s^24p^1$, respectively. Thus, in each of these atoms, the highest-energy electron is the single *p* electron in the second, third, and fourth principal quantum levels, respectively. The lone *p* electron is easier to remove from an atom than the lower-energy *s* electron of the same principal quantum level. The valence shell structures of the O, S, and Se atoms are $2s^22p^4$, $3s^23p^4$, and $4s^24p^4$, respectively. The highest-energy electron in each of these atoms is the fourth *p* electron in the outer shell. This electron is somewhat higher in energy and therefore easier to remove from an atom than the other *p* electrons because the fourth electron has to share an orbital with another electron, and two electrons in the same orbital repel each other.

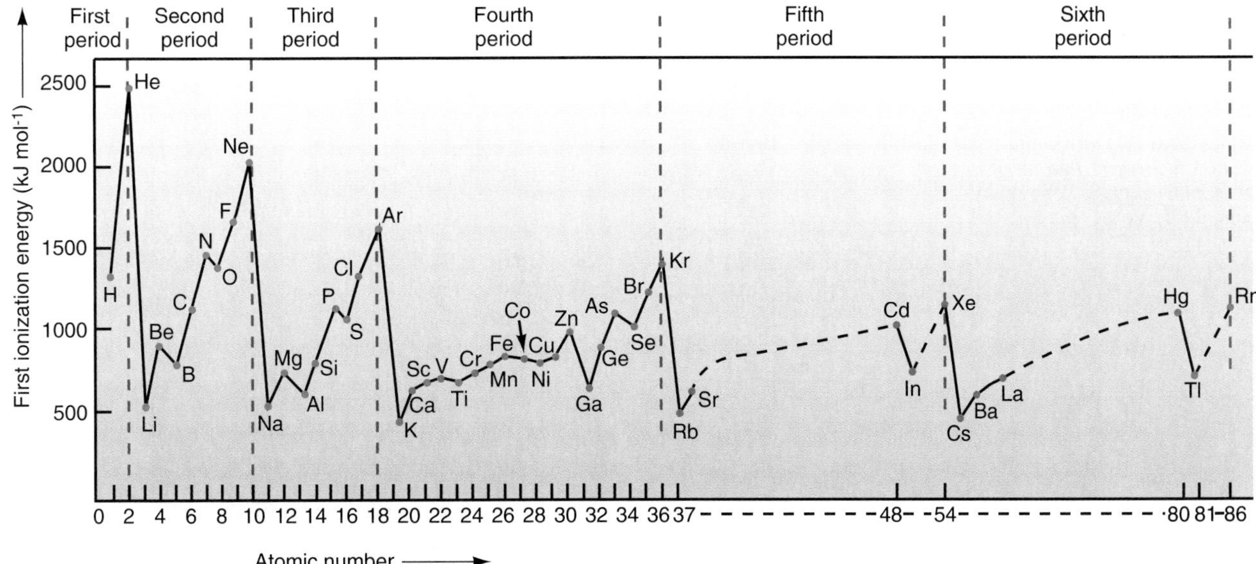

Figure 6.24

Periodic variation of first ionization energies with atomic number. Note that in each period the ionization energies of the alkali metals are the lowest, and those of the noble gases are the highest.

The ionization energy of the main group elements decreases toward the heavier atoms within a group because of the increasing atomic radii and the shielding effect that we described earlier. The valence electrons in the larger atoms of elements near the bottom of a group are farther from the nucleus than the valence electrons of the smaller atoms in the top of the group. Also, the larger atoms have more core electrons and these shield the outer electrons from the nucleus. Table 6.5 lists the first and second ionization energies of most of the elements.

Figure 6.24 and Table 6.5 show that ionization energies are the lowest for the metallic elements at the lower left of the periodic table and highest for the nonmetals at the upper right of the periodic table. Therefore, atoms of the metallic elements commonly form positive ions because the energy needed to remove the valence electrons from a metal atom is relatively low. The nonmetals do not form positive ions in ordinary chemical reactions because the ionization energy of a nonmetal atom is relatively high.

Electron Affinity

Another important periodic property of atoms is the relative ability of a gaseous atom to acquire an electron to form a negative ion. The *energy change that occurs when an electron adds to an isolated atom to form a negative ion* is called **electron affinity**. If the energy is measured as heat energy at constant pressure, electron affinity is defined as the *enthalpy change for adding an electron to an isolated atom*. Electron affinity values are usually given per mole of atoms. For example, for chlorine

$$\mathrm{Cl}(g) + \mathrm{e}^- \longrightarrow \mathrm{Cl}^-(g) \qquad \Delta H^\circ = -349 \text{ kJ mol}^{-1}$$

In most cases, the addition of an electron to an isolated neutral atom is an exothermic process (Table 6.6). Adding an electron to a negative ion is always an endothermic process because an electron and a negative ion repel each other.

Table 6.6

Electron Affinities of Main Group Elements ($kJ\ mol^{-1}$)

Values from H. Hotop and W. C. Lineberger, *Journal of Physical Chemistry Reference Data*, 1975, Vol. 4, p. 539.

H −73							He 0
Li −60	Be 0	B −27	C −122	N +7	O −141	F −328	Ne 0
Na −53	Mg 0	Al −44	Si −134	P −71.7	S −200	Cl −349	Ar 0
K −48	Ca 0	Ga −29	Ge −120	As −77	Se −195	Br −325	Kr 0
Rb −47	Sr 0	In −29	Sn −121	Sb −101	Te −190	I −295	Xe 0
Cs −45	Ba 0	Tl −30	Pb −110	Bi −110	Po −180	At −270	Rn 0

Table 6.6 reveals that nonmetals, except the noble gases, have more negative electron affinities than metals. Like ionization energy, electron affinity generally increases (becomes more negative) from left to right across a period and decreases (becomes less negative or more positive) from the top to the bottom of a group. However, as Table 6.6 reveals, there are several exceptions to this rule (see Example 6.13).

Example 6.13

Dependence of Electron Affinity on Valence Shell Structure

The electron affinity of silicon is $-134\ kJ\ mol^{-1}$. Phosphorus lies to the right of silicon in the periodic table, yet the electron affinity of phosphorus is $-72\ kJ\ mol^{-1}$, which is lower than that of silicon. Explain this anomaly in terms of the valence shell electron configurations of silicon and phosphorus atoms.

SOLUTION: The valence shell electron configuration of silicon is $3s^23p^2$, and that of phosphorus is $3s^23p^3$. An electron added to a silicon atom enters an empty p orbital. An electron added to a phosphorus atom enters a half-filled p orbital. There is repulsion between the electron already in the orbital and the added electron. The energy needed to overcome this repulsive force decreases the tendency of a phosphorus atom to acquire an extra electron. Thus, the electron affinity of a silicon atom is greater than that of a phosphorus atom.

Practice Problem 6.10: Arrange the atoms Mg, Ca, P, and N in the order of: **(a)** increasing atomic radii; **(b)** increasing ionization energy; **(c)** increasing electron affinity.

The structures and periodic properties of atoms discussed in this chapter will help us understand the theories of chemical bonding that we discuss in the following chapters.

Summary

Composition of Atoms

In the latter part of the nineteenth century and in the beginning of the twentieth century, experiments conducted by Thomson, Millikan, Rutherford, and others established the gross structures of atoms. An atom consists of a dense nucleus surrounded by much lighter electrons spread out in a relatively large volume of space. The nucleus consists of positively charged protons and electrically uncharged neutrons.

Bohr's Model of the Hydrogen Atom

According to quantum mechanics, the energies of electrons in atoms are quantized; that is, they can occupy only certain discrete energy levels. According to **Bohr's model of the hydrogen atom**, the electron in a hydrogen atom can exist only in discrete **orbits**. The transition of electrons from higher to lower orbits releases radiant energy, giving rise to *line spectra*.

Wave Theory, Quantum Numbers, and Orbitals

The **wave theory** asserts that electrons have properties of both waves and particles. According to **Heisenberg's uncertainty principle**, we cannot simultaneously determine both the position and momentum of an electron. We can only determine the probability that it will occupy a given volume of space when it has a certain quantized energy. The solution to **Schrödinger's wave equation** yields three quantum numbers that specify the energies of electrons in atoms.

The **principal quantum number**, n, designates the *principal energy level* or **shell** of an electron, and the *distance of the maximum probability region of the electron from the nucleus*. The azimuthal or **subshell quantum number**, l, *specifies the shape and the energy* of an **orbital** within a given shell of an atom. The value of l can equal any integer from 0 up to $(n - 1)$. The **magnetic quantum number**, m_l, depends on the value of l. For any given value of l, there are always $(2l + 1)$ values of m_l. This number equals the number of the orbitals in a given subshell of an atom. Individual values of m_l relate to the spatial orientations of the orbitals relative to the x, y, and z coordinates about the nucleus of the atom.

The fourth quantum number, the **spin quantum number**, m_s, was developed later to specify the *relative direction of the spin of an electron*. According to the **Pauli principle**, *no two electrons in an atom can have identical sets of four quantum numbers*.

Aufbau Process and Periodic Table

The quantum numbers and structures of atoms are related to the periodic table by a *hypothetical atom-building process* called the **aufbau process**. In this process, atoms are theoretically "built" by placing electrons in successively higher energy orbitals. The occupancy of **degenerate (equal energy) orbitals** occurs according to **Hund's rule**: *Degenerate orbitals are occupied first by a single electron in each orbital, with spins parallel, before any electron pairing occurs.*

The *first period* of the periodic table contains *two elements*, hydrogen and helium. The electronic structure of the helium atom is "built" by adding one electron to a hydrogen atom. Two electrons fill the first principal quantum level.

The *second period* has *eight elements*. The atoms of these elements are built by filling the 2*s* and 2*p* orbitals in the second shell, the *L* shell (the second principal quantum level). The *third period* also contains *eight elements*. In this period, the 3*s* and 3*p* orbitals in the third shell are being filled.

The *fourth* and *fifth periods* contain *eighteen elements* each. In the fourth period, the 4*s*, 3*d*, and 4*p* orbitals are being filled. In the fifth period, the filling of the 5*s*, 4*d*, and 5*p* orbitals is completed.

The *sixth period* has *32 elements*, corresponding to the occupancy of 6*s*, 4*f*, 5*d*, and 6*p* orbitals. The *seventh period* is presently incomplete. For the known elements in this period, the highest-energy electrons are in the 7*s*, 5*f*, and 6*d* orbitals.

The *electrons of an atom that can be involved in chemical bonding* are called **valence electrons**. The number of valence electrons of a main group atom equals its group number (a Roman numeral in this book) in the periodic table.

Periodic Properties of Atoms

Important periodic properties of atoms that are related to their electronic structures include the **atomic and ionic radii**, **ionization energy**, and **electron affinity**. Ionization energy is the *energy needed to remove a highest-energy electron from an isolated atom*. Electron affinity is the *energy change that occurs when an electron adds to an isolated atom*.

The *atomic radii* generally *decrease from left to right within a period* of elements in the periodic table and *increase down a group* of elements. The *ionization energy* generally *increases toward the right of each period* and *decreases down a group*. The *electron affinity* generally *increases from left to right in a period* and *decreases down a group*.

Important Relationships

1. The product of the frequency, ν, of a radiation and its wavelength, λ, equals the velocity of light, c:

$$\nu\lambda = c$$

2. Energy, E, of a photon equals Planck's constant, h, times the frequency, ν:

$$E = h\nu$$

3. The energy, E_n, of an electron in the nth orbital of a hydrogen atom is given as the negative of the Rydberg constant, R_H, in joules, divided by the square of the quantum number, n:

$$E_n = -\frac{R_H}{n^2}$$

4. The energy emitted, ΔE, by an electron transition from a higher quantum level, n_h, to a lower level, n_l, is given by the equation

$$\Delta E_{nh \rightarrow nl} = R_H\left(\frac{1}{n_l^2} - \frac{1}{n_h^2}\right)$$

5. The correlations between the principal quantum number n and shell designations are

n	1, 2, 3, 4, 5 . . .
Shell	K, L, M, N, O . . .

6. The correlations between the values of the quantum number l and orbital designations are

l	0, 1, 2, 3
Orbital	s, p, d, f

New Terms

Actinides (6.6)
Actinoids (6.6)
Alpha particle (6.1)
Alpha rays (6.1)
Amplitude (of a wave) (6.2)
Anode (6.1)
Aufbau process (6.6)
Azimuthal quantum number (6.5)
Beta particles (6.1)
Canal rays (6.1)
Cathode (6.1)
Cathode rays (6.1)
Cathode ray tube (6.1)
Continuous spectrum (6.2)
d-Block elements (6.6)
Degenerate orbitals (6.5)
Electron (6.1)
Electron affinity (6.7)
Electron density (6.4)
Excited state (6.3)
f-Block elements (6.6)
Frequency (of a wave) (6.2)
Gamma rays (6.1)
Ground state (6.3)
Heisenberg uncertainty principle (6.4)
Hertz (6.2)
Hund's rule (6.6)
Infrared spectrum (6.2)
Inner transition metals (6.6)
Ionization energy (6.3)
Ionization energy (first, second) (6.7)
Lanthanides (6.6)
Lanthanoids (6.6)
Line spectrum (6.2)
Magnetic quantum number (6.5)
Neutron (6.1)
Noble gas core (6.6)
Nucleus (6.1)
Orbit (6.3)
Orbital (6.5)
p-Block elements (6.6)
Paired spin (6.6)
Parallel spin (6.6)
Pauli exclusion principle (6.5)
Photon (6.2)
Principal energy levels (6.5)
Principal quantum number (6.3)
Proton (6.1)
Quantized (6.3)
Quantum (6.3)
Quantum levels (6.3)
Quantum number (6.5)
Radiant energy (6.1)
Radiation (6.1)
Radioactivity (6.1)
s-Block elements (6.6)
Schrödinger's wave equation (6.4)
Shell (6.5)
Spectrum (6.2)
Spin quantum number (6.5)
Subatomic particle (6.1)
Subshell (6.5)
Subshell quantum number (6.5)
Superfluid (6.3)
Ultraviolet spectrum (6.2)
Unpaired electrons (6.6)
Valence electrons (6.6)
Velocity (6.2)
Wave function (6.4)
Wavelength (6.2)
Wave theory (6.4)

Exercises

General Review

1. Cite some evidence for the presence of electrons in atoms.
2. What evidence exists to support the theory that atoms contain nuclei that are very small and dense compared with an atom as a whole?
3. What are the relative masses and charges of a proton, a neutron, an electron, and an alpha particle?
4. How were protons and neutrons discovered?
5. Cite an important contribution to science by each of the following scientists: Curie, Rutherford, Bohr, de Broglie, Heisenberg, Schrödinger, Pauli.
6. Define each of the following terms: cathode rays, alpha rays, beta rays, gamma rays, radioactive substance.
7. Describe the natures of alpha, beta, and gamma rays.
8. Write an equation showing how the frequency, ν, of radiant energy is related to its wavelength, λ, and the velocity of light, c. Include the appropriate units for each.
9. Write an equation to show how the energy of an electromagnetic radiation is related to its frequency and Planck's constant, h. Include the appropriate units for each.
10. Write an equation to show the relationship between the energy of an electromagnetic radiation and its wavelength.
11. What is "quantum theory"?
12. What is the difference between a continuous spectrum and a line spectrum?
13. Explain in terms of the quantum theory how line spectra of elements arise.
14. What are the two spectral regions immediately adjacent to the visible region? What are the approximate limits of the wavelengths of radiation of the visible spectrum?
15. Convert any wavelength of radiant energy, such as 500 nm, to frequency and to energy associated with the radiation. Show how any one of these three variables can be calculated from the known value of either of the other two.
16. In terms of Bohr's theory, explain why hydrogen produces a line spectrum.
17. According to Bohr's theory, which is the lowest possible orbit of a hydrogen atom that is involved in producing: **(a)** the visible spectrum; **(b)** the ultraviolet spectrum?
18. Write an equation for the energy of an electron in any given Bohr orbit (quantum level) in a hydrogen atom.
19. Write an equation for the energy emitted as a result of the transition of an electron from a higher to a lower orbit in a hydrogen atom.
20. How does the wavelength of a radiation producing a red line in a visible spectrum compare with: **(a)** the wavelength of a radiation producing a yellow line; **(b)** a radiation in the infrared region of the spectrum; **(c)** a radiation in the ultraviolet region of the spectrum?
21. What are the possible numerical values for the quantum numbers n, l, m_l, and m_s for each of the valence electrons of a boron atom in its ground energy state?
22. Explain and illustrate how the numerical values 0, 1, and 2 of the quantum number, l, are related to the letter designations and the shapes of the corresponding orbitals.
23. How many of each of the s, p, d, and f orbitals can be found in each of the principal quantum levels from $n = 1$ to $n = 4$?
24. What is the maximum number of electrons that can occupy an orbital?
25. What is the maximum number of electrons that can occupy each of the following orbital energy levels (subshells) within a given principal quantum level (shell)? s, p, d, f.
26. By drawing sketches, show how an s orbital, three p orbitals, and five d orbitals can be oriented in space relative to the x, y, and z axes. Label each orbital.
27. Arrange the s, p, d, and f orbitals in order of increasing energy within a given principal quantum level.
28. Write a shorthand description ($1s^2$, $2s^2$, etc.) of the ground state structures of the first, second, and fourth period atoms.
29. What are the highest energy (outermost) electrons ($1s$, $2p$, etc.) in the ground state atoms of: **(a)** B to Ne; **(b)** Na to Mg; **(c)** K to Ca; **(d)** Ti to Zn; **(e)** Ga to Kr?
30. What are valence electrons?
31. Which are the highest-energy electrons in the atoms of the $4d$ series of transition metals?
32. What is the relationship between the number of valence electrons in an atom of a representative element and the group number of the element?
33. How many unpaired electrons are there in each of the ground state atoms of the second period? of the fourth period?
34. How do the atomic radii generally vary: **(a)** within a group; **(b)** within a period?
35. How does the ionization energy generally vary: **(a)** within a group; **(b)** within a period?
36. How does the electron affinity generally vary: **(a)** within a group; **(b)** within a period?

Meaning of Terms

37. Briefly define or explain the meaning of each of the following terms and illustrate by drawing or by giving examples of each: **(a)** orbital; **(b)** s orbital; **(c)** p_y or-

bital; **(d)** d_{xy} orbital; **(e)** d_{z^2} orbital; **(f)** $d_{x^2-y^2}$ orbital; **(g)** principal or main energy level in an atom; **(h)** visible, ultraviolet, and infrared spectrum; **(i)** Pauli exclusion principle; **(j)** Hund's rule; **(k)** atomic shell; **(l)** atomic subshell; **(m)** atomic radius; **(n)** unpaired electron; **(o)** ionization energy; **(p)** electron affinity.

Radiant Energy, Spectra, and Bohr Theory

38. Answer the following questions with respect to the radiant energies responsible for blue and red colors. **(a)** Which has a longer wavelength? **(b)** Which is more energetic? **(c)** Which is deflected more by a glass prism? **(d)** Which forms a line toward the higher-energy region in the spectrum? **(e)** Which is caused by a higher-energy transition of electrons?

39. How does the wavelength of x-ray radiation compare with that responsible for blue color?

40. Which is most energetic and penetrating, x-ray radiation, radiation responsible for blue color, radiation responsible for red color, or infrared radiation?

41. Which one of the radiations in question 40 has the longest wavelength and is in the lower-energy side of the spectrum?

42. Explain how a gaseous element, when subjected to electric discharge, produces a spectrum of lines instead of a continuous spectrum (like a rainbow).

43. Discuss a possible use of line spectra in the identification of elements.

44. What is the meaning of "quantum"?

45. What evidence is there for the argument that electrons in atoms occupy discrete energy levels?

46. What evidence is there to support the view that an electron in an atom can move from one energy level to another only by a complete "jump" and not by a gradual spiraling process?

47. What is the nature of light?

48. What is the nature of the rainbow?

49. When an electron falls from the second to the first quantum level in a hydrogen atom, much more energetic radiation is produced than that emitted when the electron falls from the third to the second, or from the fourth to the third quantum level. Explain.

50. Calculate the wavelength and energy of radiation with a frequency of 1.14×10^3 kHz. The velocity of light is 3.00×10^8 m s^{-1}, and Planck's constant, h, is 6.63×10^{-34} J s.

51. Calculate the energy of an electron in the second principal quantum level in a hydrogen atom. The Rydberg constant, R_H, is 2.179×10^{-18} J.

52. Calculate the wavelength and the energy of the radiation emitted by the transition of an electron from the fifth to the second quantum level in a hydrogen atom. R_H is 2.179×10^{-18} J. What color is the spectral line produced by this transition? Find it in Figure 6.12.

53. Calculate the frequency and the energy of the radiation that produces a violet spectral line at 410 nm in the hydrogen spectrum.

Quantum Numbers, Orbitals, and the Periodic Table

54. Give the name, symbol, and meaning of each of the four quantum numbers used to designate electrons in atoms.

55. Give one possible value for each of the four quantum numbers (n, l, m_l, and m_s) for one of the highest-energy electrons of each of the following atoms: Ca, Zn, Ga. Which subshell (s, p, d, or f) does the electron occupy in each case?

56. Give the numerical values for the quantum numbers n and l for the two highest-energy electrons of a carbon atom. Which type of orbital do these electrons occupy?

57. Explain the fact that there are no d orbitals in the second main energy level.

58. Explain the fact that there are no f orbitals in the third main energy level.

59. Draw pictures of the comparative sizes of $2s$ and $3s$ orbitals; $2p$ and $3p$ orbitals.

60. What is the relationship between the quantum number n and periods in the periodic table, and between the quantum number l and the various blocks of elements in the periodic table?

61. What quantum number(s), excluding the spin quantum number, is (are) the same for the last or highest-energy electron(s) of each of the following atoms? **(a)** Alkali metals; **(b)** alkaline earth metals; **(c)** halogens.

62. Why does the first period have two elements?

63. Why does the second period have eight elements?

64. The third main energy level in an atom can be filled with 18 electrons. Why does the third period have only eight elements?

Electronic Structures of Atoms

65. Describe the electronic structure of each of the atoms in the first three periods by drawing dashes for the orbitals in their relative order of energies, and arrows for electrons indicating the direction of spins of the electrons.

66. How many unpaired (spins parallel) electrons are there: **(a)** in a nitrogen atom in its ground energy state; **(b)** in an oxygen atom; **(c)** in a chromium atom; **(d)** in a zinc atom?

67. How many valence electrons does each of the following atoms have: nitrogen, magnesium, sulfur, scandium, iodine? In each case, specify the number of *s* electrons, *p* electrons, *d* electrons, and *f* electrons, if any, in the outer shell.

68. The electronic structures of chromium, molybdenum, copper, and silver differ from those predicted by the aufbau process. Explain. Write the valence shell structures for the atoms of each of these elements.

69. How many *s* electrons, *p* electrons, *d* electrons, and *f* electrons are there in the outer shell of: **(a)** a bromine atom; **(b)** an aluminum atom; **(c)** a silver atom?

70. How many of each of the types of electrons specified in question 69 are there in the outer shell of: **(a)** a bromide ion; **(b)** an aluminum ion; **(c)** a sulfide ion; **(d)** a scandium ion, Sc^{3+}; **(e)** an iron(II) ion?

71. Which two ions in question 70 are isoelectronic with each other? What is their electronic configuration?

72. Write a shorthand description ($1s^22s^2$ or $[He]2s^2$, etc.) of the ground state electron population of each of the following species: **(a)** aluminum atom; **(b)** aluminum ion; **(c)** sodium atom; **(d)** sodium ion; **(e)** oxygen atom; **(f)** oxide ion; **(g)** fluoride ion; **(h)** scandium atom; **(i)** copper atom; **(j)** manganese atom; **(k)** arsenic atom; **(l)** tin atom; **(m)** zinc ion. Which four of the above species are isoelectronic with one another?

73. How many unpaired electrons are there in each of the species in question 72?

Periodic Properties of Atoms

74. Account for the general increasing trend of the atomic radii from top to bottom within a group of elements in the periodic table.

75. The ionization energies of boron and oxygen deviate from the regular increasing trend in the second period. This deviation also occurs at aluminum and sulfur in the third period, and for gallium and selenium in the fourth period. Explain.

76. Given the following atoms: Na, Li, S, Se, Cl. Arrange these atoms in the order of: **(a)** increasing atomic radius; **(b)** increasing ionization energy; **(c)** increasing electron affinity.

77. Arrange the following ions in the order of decreasing radius: calcium ion, scandium ion, chloride ion, phosphide ion.

78. Arrange calcium, strontium, arsenic, bromine, and chlorine atoms in the order of increasing first ionization energies.

79. Which one of the following species has the largest radius and which has the smallest radius? Fe, Fe^{2+}, Fe^{3+}. Explain.

80. Which one of the following species has the largest and which has the smallest radius? N^{3-}, N, F^-, F, Na^+, Na, Mg^{2+}, Mg. Explain.

81. Arrange the following atoms in the order of increasing electron affinity: Li, K, F, I.

CHAPTER 7

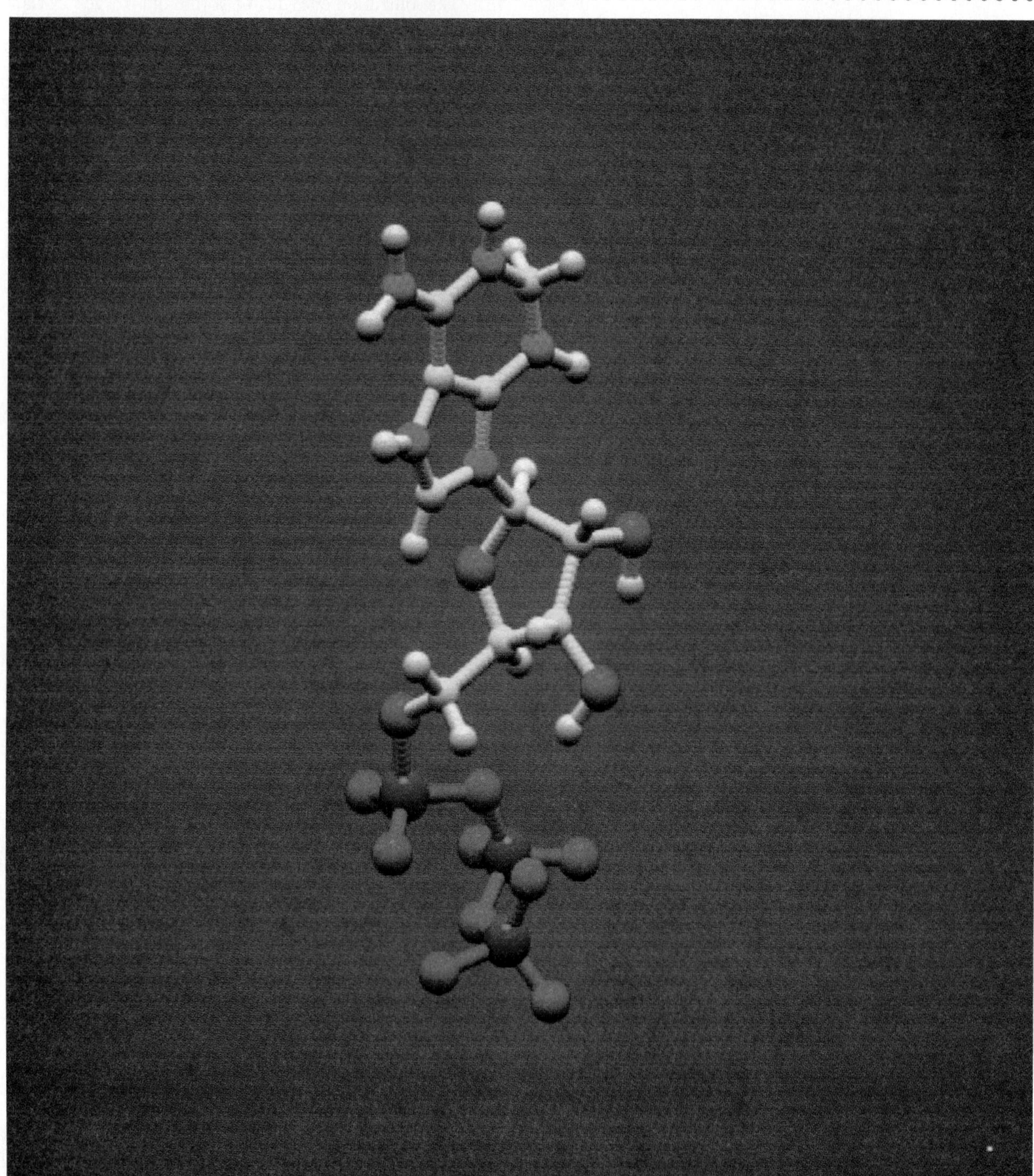

The molecular structure of ATP.

Chemical Bonding and Structure

Figure 7.1

Democritus (470 – 380 B.C.).

As we recall from Chapter 2, the Greek philosophers Leucippus and Democritus (Figure 7.1) of the fifth century B.C. *believed that the universe was made of tiny particles called atoms. They could not explain how atoms are held together in different substances such as water and sand. They attempted to explain the bulk properties of matter by assuming that different states of matter were made of different kinds of atoms. Thus, they imagined that atoms of water are smooth and round, and are therefore able to roll over one another without sticking, whereas the atoms of iron are rough, jagged, and uneven, clinging together in a solid body.*

More than 2000 years elapsed before the speculative views of the ancient Greeks began to be replaced by the observations of empirical science. We recall that John Dalton formulated the first atomic theory that was based on experimentally established laws. Like the ancient theory, Dalton's theory had to account for the existence of many different kinds of substances with very different properties made from relatively few kinds of atoms. To explain this immense diversity, Dalton invented the concept of atomic clusters or "compound atoms" around 1800.

In a similar vein, Stanislao Cannizzaro of Italy suggested in 1858 that gases consist of tiny aggregates of atoms called molecules, and he insisted on the distinction between molecular weights and atomic weights. The originators of atomic theory, however, could not explain how atoms are bonded to one another in chemical compounds. Theories of chemical bonding were developed much later.

In this and the following chapters we discuss the modern theories that describe how atoms are linked to one another in chemical compounds. The subject of this discussion is called ***chemical bonding****. We also discuss the geometric arrangement of different atoms in various molecules. This topic is called* ***molecular structure****.*

7.1 Ionic and Covalent Bonding

In Chapter 2 we briefly discussed ionic and covalent bonding. We recall that ionic bonding results from the electrostatic attraction between oppositely charged ions. Ionic bonding is found in *solids* such as the binary metal–nonmetal compounds, and also in metal nitrates, sulfates, and carbonates. We also remember that in a covalent bond, electrons are "shared" by the bonded atoms. The atoms in molecules are covalently bonded. Molecular compounds include all compounds that are gaseous or liquid at room temperature, and many solid compounds as well.

Ionic compounds typically have much higher melting points than molecular compounds. Table 7.1 lists the melting points of various compounds. This table reveals that the melting points of most binary metal–nonmetal compounds and metal sulfates are relatively high. Most of these substances are ionic solids. A few solids that we might imagine to be ionic—including beryllium hydride, beryllium chloride, and aluminum chloride—have low melting points. In fact, the bonding in these compounds is essentially covalent.

Table 7.1

Melting Points of Compounds (°C). The notations *g*, *l*, and *s* refer to the Physical States of the Compounds at 25°C

Compounds of the second-period elements	$LiH(s)$	680	$BeH_2(s)$	125[a]	$B_2H_6(g)$	−166	$CH_4(g)$	−182	$NH_3(g)$	−77.7	$H_2O(l)$	0	$HF(g)$	−83.1
	$LiCl(s)$	605	$BeCl_2(s)$	405	$BCl_3(l)$	−107	$CCl_4(l)$	−23.0	$NCl_3(l)$	−40	—		—	
	$Li_2O(s)$	1700	$BeO(s)$	2530	$B_2O_3(s)$	450	$CO_2(g)$	−56.6[b]	$NO_2(g)$	−11.2	—		—	
	$Li_2SO_4(s)$	843	$BeSO_4(s)$	550[a]	—		—		—		—		—	
Compounds of the third-period elements	$NaH(s)$	800[a]	$MgH_2(s)$	280[a]	$AlH_3(s)$	150–200[a]	$SiH_4(g)$	−185	$PH_3(g)$	−133	$H_2S(g)$	−85.5	$HCl(g)$	−115
	$NaCl(s)$	801	$MgCl_2(s)$	714	$AlCl_3(s)$	190[c]	$SiCl_4(l)$	−70	$PCl_3(l)$	−112	$SCl_2(l)$	−78	—	
	$Na_2O(s)$	1275[d]	$MgO(s)$	2852	$Al_2O_3(s)$	2072	$SiO_2(s)$	1703	$P_4O_{10}(s)$	300[d]	$SO_2(g)$	−72.7	$Cl_2O(g)$	−20
	$Na_2SO_4(s)$	884	$MgSO_4(s)$	1124[a]	$Al_2(SO_4)_3(s)$	770[a]	—		—		—		—	
Organic compounds	Acetic acid(*l*)			16.6	Ethanol(*l*)			−117						
	Benzoic acid(*s*)			122	Naphthalene(*s*)			357						
	Camphor(*s*)			179	Sucrose(*s*)			185						

[a] Decomposes. [b] At 5.2 atm. [c] At 2.5 atm. [d] Sublimes.

Electronic Structures of Ions in Ionic Compounds

The ions of most main group elements are isoelectronic with their nearest noble gas atoms. Noble gas atoms are stable and so are ions that are isoelectronic with noble gas atoms.

We recall that metals in general have low ionization energies (they lose electrons readily), and nonmetals have high electron affinities (they gain electrons readily). Therefore, when a metal reacts with a nonmetal, metal atoms lose electrons to form positive ions, and nonmetal atoms gain electrons to form negative ions. The electrons lost by a metal atom lie outside its noble gas core. These are the valence electrons. Nonmetal atoms gain electrons to acquire their nearest noble gas electronic configuration.

A few ions of main group metals do not have noble gas electronic configurations. These ions are derived from main group metals whose atoms have either filled *d* subshells or filled *d* and *f* subshells. For example, lead forms Pb^{2+} ions. To explain the charge on this ion, let us consider the electronic structure of a lead atom, which is $[Xe]4f^{14}5d^{10}6s^2 6p^2$. When a neutral lead atom loses its two 6*p* electrons, a lead(II) ion forms. Its electronic configuration is $[Xe]4f^{14}5d^{10}6s^2$. This is not a noble gas configuration. A very large amount of energy would be needed to remove 26 more electrons from a Pb^{2+} ion to produce the core electronic configuration of the noble gas xenon. In an ordinary chemical reaction, only enough energy is available to remove the highest-energy electrons.

If a Pb^{2+} ion lost two 6*s* electrons, a Pb^{4+} ion would form. But a Pb^{4+} ion is not stable. Although both lead(II) and lead(IV) compounds are known, most lead(IV) compounds are molecular compounds, which can easily be converted to lead(II) compounds. For example, both $PbCl_2$ and $PbCl_4$ exist. Of these, $PbCl_2$ is a stable ionic solid at room temperature and pressure, with a melting point of 501°C, while $PbCl_4$ is a liquid that freezes at −15°C. Similarly, tin(II) chloride is a solid with a melting point of 246°C, but tin(IV) chloride is a liquid with a freezing point of −33°C.

Practice Problem 7.1: In terms of the electronic configuration of a thallium atom, explain how it can form both Tl^{+} and Tl^{3+} ions.

Atoms of transition metals contain *d* electrons, or both *d* and *f* electrons. Scandium, yttrium, and lanthanum atoms have two *s* electrons and one *d* electron as their highest-energy electrons. In these atoms the outer *s* and *d* electrons are of approximately equal energy. When an atom of scandium, yttrium, or lanthanum forms an ion, it loses all three of these highest-energy electrons to form a 3+ ion.

As we proceed to the right in a series of transition elements, it becomes increasingly more difficult to remove more *d* electrons. After scandium, the remaining elements in the 3*d* series of transition metals form 2+ ions. Most of these ions are formed by losing two 4*s* electrons, which are the highest-energy electrons in these atoms. The atoms of two elements in the 3*d* series of transition metals, chromium and copper, have only one 4*s* electron. To form a 2+ ion, each of these atoms loses its 4*s* electron and one of its 3*d* electrons.

Most transition metals form ions of more than one charge. For example, chromium forms Cr^{2+} and Cr^{3+} ions, iron forms Fe^{2+} and Fe^{3+} ions, and copper forms Cu^{+} and Cu^{2+} ions. Zinc is the last element in the 3*d* series of transition metals, it forms only 2+ ions (see Table 2.3 for a more complete list of the ions of the main group elements and the transition metals). We discuss the chemistry of transition metals further in Chapters 18 and 22.

The Covalent Bond

A covalent bond consists of the simultaneous attraction of two nuclei for the same pair of electrons. We will illustrate this fundamental principle by considering the formation of a covalent bond between two hydrogen atoms to give a hydrogen molecule, H_2.

The formation of an H—H covalent bond is depicted in Figure 7.2. The two nuclei, called H_A and H_B, are initially associated with electrons called e_1 and e_2. As the isolated hydrogen atoms approach each other, two sets of repulsions arise: the two hydrogen nuclei repel each other and the two electrons repel each other. *At the same time*, however, *four* sets of attractions emerge: H_A is attracted to *both* e_1 and e_2, and H_B is attracted to *both* e_1 and e_2. Thus there are twice as many electrostatic attractions as electrostatic repulsions, and the result is a net bonding interaction.

Let us now look at the formation of the covalent bond in H_2 in terms of the interactions of the atomic orbitals of the two hydrogen atoms (Figure 7.3). The same attractions and repulsions exist that we described in our model in Figure 7.2, but now we will consider the interactions of atomic orbitals. *A covalent bond exists when electron "clouds" of two atomic orbitals overlap*. This overlap occurs when two nuclei approach each other so that the electrostatic attractions are more powerful than the electrostatic repulsions. The nuclei of both hydrogen atoms are simultaneously attracted to both electrons in the overlapping electron "cloud" that lies between them.

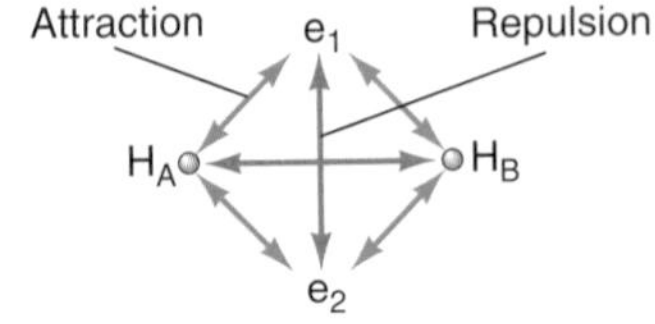

Figure 7.2

Attractive and repulsive forces in a hydrogen atom.

The nuclei of the two H atoms in an H_2 molecule move toward and away from each other, as if they were two balls at the end of a spring. At a given temperature, the two nuclei are separated by an average or equilibrium distance

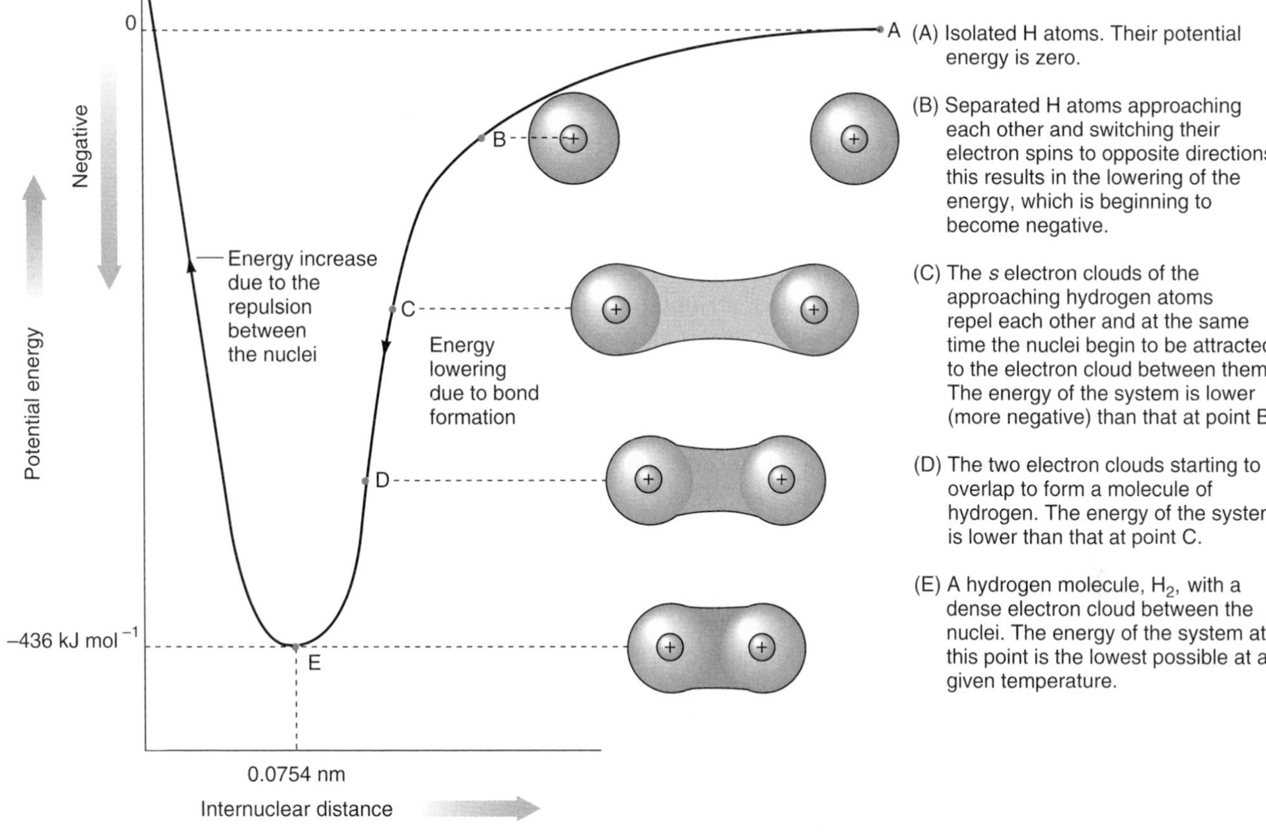

0
-436 kJ mol^{-1}
Potential energy
Negative
Energy increase due to the repulsion between the nuclei
Energy lowering due to bond formation
A
B
C
D
E
0.0754 nm
Internuclear distance
(A) Isolated H atoms. Their potential energy is zero.
(B) Separated H atoms approaching each other and switching their electron spins to opposite directions; this results in the lowering of the energy, which is beginning to become negative.
(C) The s electron clouds of the approaching hydrogen atoms repel each other and at the same time the nuclei begin to be attracted to the electron cloud between them. The energy of the system is lower (more negative) than that at point B.
(D) The two electron clouds starting to overlap to form a molecule of hydrogen. The energy of the system is lower than that at point C.
(E) A hydrogen molecule, H_2, with a dense electron cloud between the nuclei. The energy of the system at this point is the lowest possible at a given temperature.

Figure 7.3

Model of H—H bond formation.

from each other, so that the entire system of the two nuclei and the two electrons is at the lowest possible energy. The distance between H_A and H_B is the H—H bond length (r_{AB}). The two atoms are linked by a *covalent bond*. The formation of an H—H bond occurs with the energy changes shown in Figure 7.3.

Figure 7.3 shows that the energy of the molecule is lower than that of the isolated atoms. Figure 7.3 also shows that when two identical atoms unite to form a molecule, the most probable location of the two bonding electrons is between the nuclei. They are equally shared by the nuclei.

Since a single covalent bond contains two electrons, it is often called an *electron pair bond*. The bonded nuclei are both attracted to the bonding electron pair. The two electrons in a bonding pair have opposite spins, just as they do when occupying the same atomic orbital.

The internuclear distance in an H_2 molecule is 0.0754 nm. We mentioned in Chapter 6 that the covalent radius of an atom is one-half of the internuclear distance between two identical covalently bonded atoms. The experimentally measured potential energy of an H—H bond is -436 kJ mol^{-1} (-104 kcal mol^{-1}). This is the *bond energy*, or the amount of energy released when 1 mol of H—H bonds forms. An equal amount of energy must be supplied for the dissociation of 1 mol of H_2 molecules into 2 mol of H atoms.

7.2 Polar and Nonpolar Diatomic Molecules

Polar Molecules and Polar Bonds

Homonuclear diatomic molecules, such as H_2, and heteronuclear diatomic molecules, such as HCl, behave differently when they are placed between positively and negatively charged metal plates or poles. Homonuclear diatomic molecules are not much affected by an electric field. Such molecules are **nonpolar**. The electron density of the bonding pair in a nonpolar diatomic molecule is symmetrically distributed, equidistant between the nuclei.

However, heteronuclear diatomic molecules tend to line up in an electric field so that one side of a molecule is attracted to the positive pole and the other side to the negative pole (Figure 7.4). A diatomic molecule that aligns in an electric field has its bonding electron pair closer to one of the atoms than to the other. As a result, one end of the molecule has a slight negative charge relative to the other end. The other end of the molecule, having lost some electron density, has a slight positive charge. *Diatomic molecules in which the electron density is more concentrated on one side of a molecule than on the other side* are called **polar diatomic molecules**. A *bond in which the bonding electron pair is shifted toward one atom* is called a **polar bond**.

Polar molecules act as miniature *dipoles*. The more electron-rich, or negative side of a polar diatomic molecule, is attracted to the positive pole in an electric field, and the less electron-rich atom, or positive side of the molecule, is attracted to the negative pole (Figure 7.4).

Dipole Moment

In a polar molecule such as HF, the center of the partial positive charge on the H atom is separated by a certain distance from the center of the partial negative

Figure 7.4

Representation of polarity of molecules.

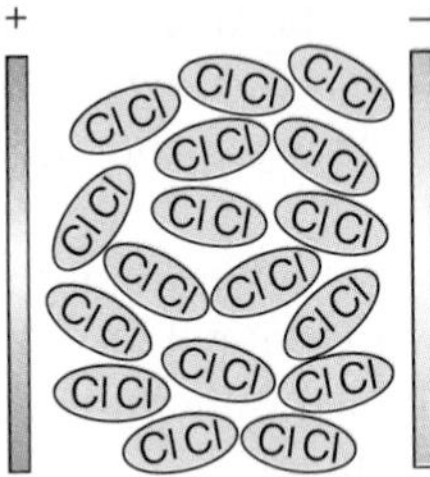

Nonpolar Cl_2 molecules in a random arrangement in an electric field

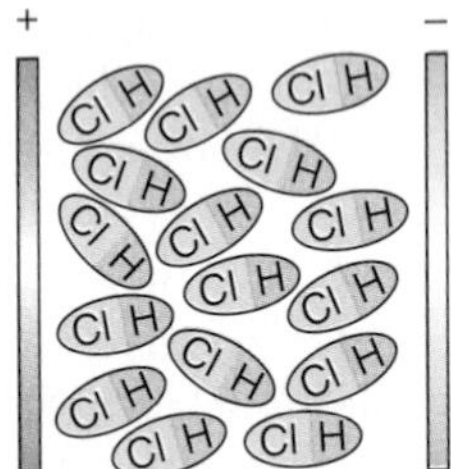

Polar HCl molecules oriented in an electric field so that the more electron-rich chlorine side of a molecule is pointed to the positive pole and the relatively electron-deficient hydrogen side is pointed to the negative pole

APPLICATIONS OF CHEMISTRY 7.1
"Like Dissolves Like"

Polar solvents such as water dissolve ionic solutes such as sodium chloride. Similarly, nonpolar solvents such as oil dissolve nonpolar solutes such as iodine. Chemists summarize this by saying that "like dissolves like." On the other hand, polar solvents such as water do not dissolve nonpolar solutes such as oil.

The premise that like dissolves like is used extensively by chemists. Carbon dioxide has some unique properties that are utilized by food and oil industries. Carbon dioxide molecules are nonpolar. Although carbon dioxide is a gas, it begins to act more and more like a liquid when it is subjected to high pressure. Under pressures of about 30,000 kPa (see Chapter 10 for units of pressure), it turns into a supercritical fluid (SCF). Under high-pressure conditions, a supercritical fluid is a gas that acts as if it were a liquid.

The food industry uses SCF carbon dioxide to remove oils from food, thereby reducing fat content and calories. For example, potato chips are usually around 50 percent oil by mass. Since like dissolves like, when SCF carbon dioxide meets an oily potato chip, the nonpolar carbon dioxide dissolves the nonpolar oil in the potato chips. This removes some of the oil from the potato chips. Cholesterol is also nonpolar. The food-processing industry removes cholesterol from food using SCF carbon dioxide.

In many instances, when oil companies abandon an oil well, they leave much oil behind. Oil consists of many substances, some more viscous than others, but most are nonpolar. The oil extracted from oil wells is the least viscous portion, which is easiest to remove. The remaining viscous oil is too difficult and expensive to recover. One way to recover the remaining viscous oil is to make it less viscous by pumping carbon dioxide into the well. Since like dissolves like, carbon dioxide injected into abandoned wells readily dissolves in the viscous oil. Because the resulting solution of oil and carbon dioxide is less viscous, it is easier to remove. There are more than 300 billion barrels of viscous oil left in exhausted U.S. wells. This translates into trillions of dollars for those who recover this oil.

Sources: Chemical & Engineering News, December 22, 1986, page 44: *Science*, August 26, 1983, page 815.

charge on the F atom. The product of the distance, d, and the charge, q, is called the **dipole moment**, μ:

$$\mu = dq$$

The magnitudes of the partial positive charge on the H atom and the partial negative charge on the F atom are equal. In calculating the dipole moment, the positive value of the charge is used to obtain a positive value for the dipole moment. Figure 7.5 illustrates the meaning of dipole moment.

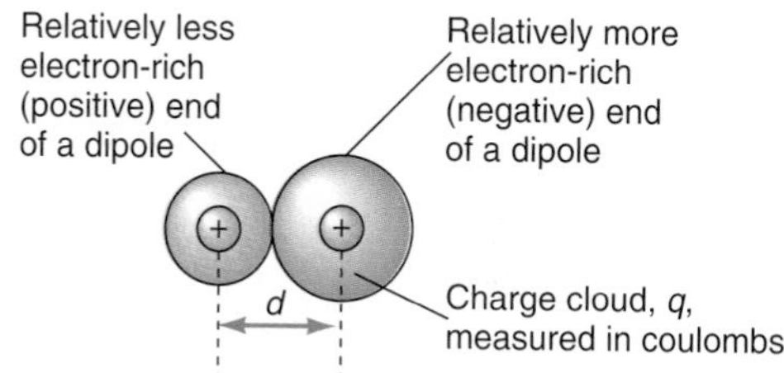

Figure 7.5

Dipole moment of a polar diatomic molecule. The dipole moment, μ, is the product of the distance, d, between the two nuclei, and the partial charge, q, on a charge center: $\mu = dq$.

Figure 7.6

Linus Pauling (1901–) has won two Nobel Prizes. He was awarded the Nobel Prize in chemistry in 1954. He won the Nobel Peace Prize in 1962 for his contribution to the adoption of the nuclear test ban treaty.

7.3 Electronegativity and Ionic Character of Bonding

Electronegativity

Atoms with different sizes and nuclear charges have different abilities to attract bonding electron pairs. The *relative ability of an atom in a molecule to attract bonding electron pairs* is called its **electronegativity**. Atoms having high ionization energies and high electron affinities usually also have high electronegativities. Let us consider the electronegativities of the elements in the second period of the periodic table from lithium to neon. As we move from left to right, the ionization energies and electron affinities increase. Electronegativity also increases from left to right. Thus, in the second period, the order of electronegativities is

$$\text{Li} < \text{Be} < \text{B} < \text{C} < \text{N} < \text{O} < \text{F}$$

As we move down a group in the periodic table, we recall that the ionization energy and electron affinity decrease. Thus, electronegativity also decreases toward the heavier atoms in a group. In summary, atoms toward the lower left in the periodic table have low electronegativities, and those toward the upper right (omitting the noble gases) have high electronegativities.

Electronegativities have been calculated in several different ways. One of the most commonly used methods was developed by the U.S. chemist Linus Pauling (Figure 7.6). The values of the electronegativities on the Pauling scale are adjusted to give fluorine the highest value of 4.0 (Figure 7.7). A more complete list of Pauling's electronegativities for the representative elements, and for the *d* series of transition metals, is given in Table 7.2.

Figure 7.7

Electronegativities of the representative elements. The elements of the same group are connected with broken lines.

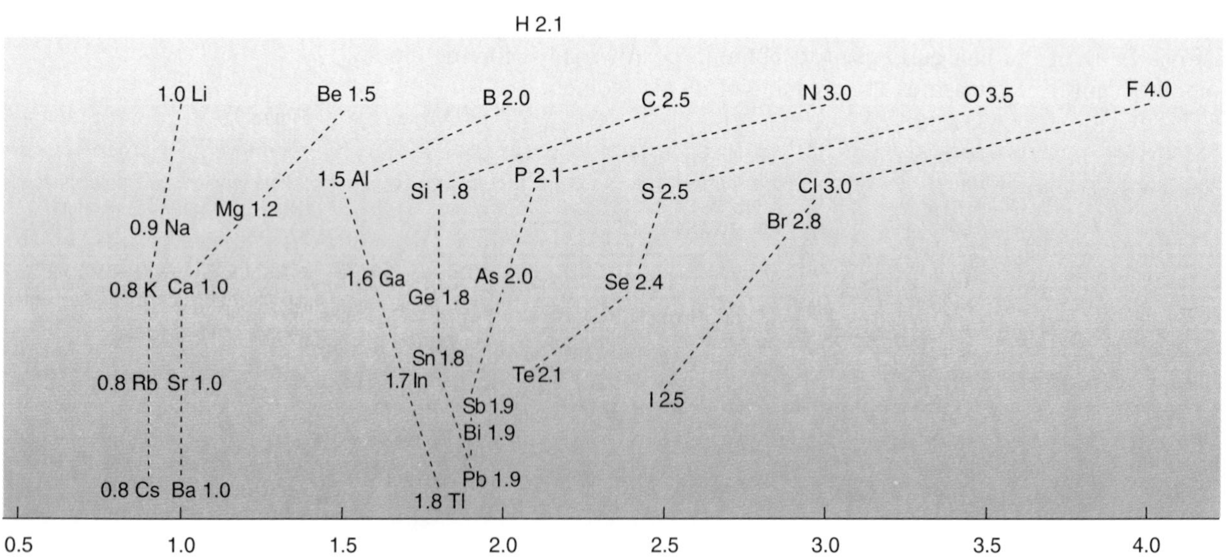

Table 7.2

Pauling's Electronegativities of Elements

H 2.1																
Li 1.0	Be 1.5											B 2.0	C 2.5	N 3.0	O 3.5	F 4.0
Na 0.9	Mg 1.2											Al 1.5	Si 1.8	P 2.1	S 2.5	Cl 3.0
K 0.8	Ca 1.0	Sc 1.3	Ti 1.5	V 1.6	Cr 1.6	Mn 1.5	Fe 1.8	Co 1.9	Ni 1.9	Cu 1.9	Zn 1.6	Ga 1.6	Ge 1.8	As 2.0	Se 2.4	Br 2.8
Rb 0.8	Sr 1.0	Y 1.2	Zr 1.3	Nb 1.6	Mo 1.8	Tc 1.9	Ru 2.2	Rh 2.2	Pd 2.2	Ag 1.9	Cd 1.7	In 1.7	Sn 1.8	Sb 1.9	Te 2.1	I 2.5
Cs 0.8	Ba 1.0	La 1.1	Hf 1.3	Ta 1.5	W 1.7	Re 1.9	Os 2.2	Ir 2.2	Pt 2.2	Au 2.4	Hg 1.9	Tl 1.8	Pb 1.9	Bi 1.9	Po 2.0	At 2.2
Fr 0.8	Ra 1.0	Ac 1.1	Th 1.3	Pa 1.4												

Bond Polarity and Ionic Character

Electronegativity can be used to explain the polarity of a covalent bond. In a homonuclear diatomic molecule, the bonded atoms are identical, their electronegativities are identical, and the bonding pair is shared equally by the nuclei. Such a molecule is nonpolar.

In a heteronuclear diatomic molecule, one atom has a higher electronegativity than the other. The electron density of the bonding pair is concentrated somewhat toward the nucleus of the more electronegative atom. A difference in the electronegativities of a pair of bonded atoms results in a polar bond. Thus, the polarity of a covalent bond depends on the electronegativity difference, Δ(EN), of the two bonded atoms.

As the bond in a heteronuclear diatomic molecule becomes more polar, the molecule begins to resemble an ion pair. A *polar covalent bond* therefore has **partial ionic character**. As the electronegativity difference between two bonded atoms increases, the partial ionic character of the bond increases and the bonded atoms become more like ions.

Figure 7.8 shows the effect of the electronegativity difference of two bonded atoms on bond polarity, and thus on the partial ionic character, of the bond. First, consider a homonuclear, diatomic molecule such as Cl_2. The Cl—Cl bond in Cl_2 has a Δ(EN) value of 0 and is therefore a nonpolar covalent bond. Next, consider the heteronuclear, diatomic molecule aluminum phosphide, AlP. The Al—P bond in aluminum phosphide has a Δ(EN) of 0.6 and has a small degree of ionic character. As the Δ(EN) increases, a bond becomes

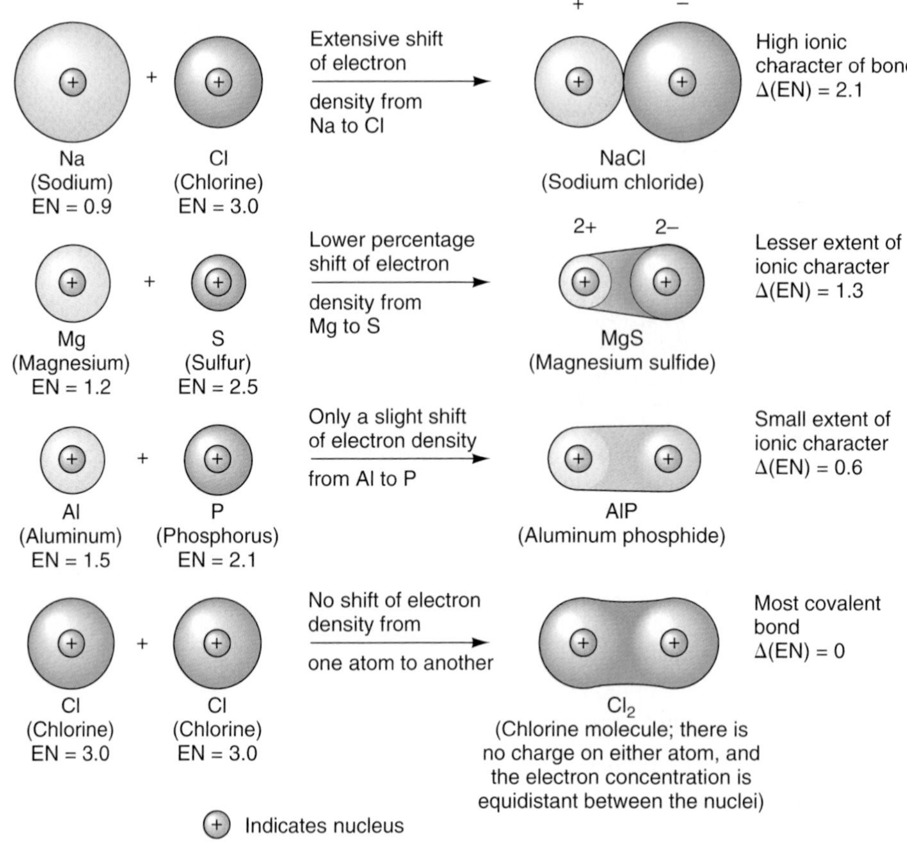

Figure 7.8

Electronegativity difference, Δ(EN), and ionic character of bonds. As Δ(EN) between two atoms in the covalent bond increases, the ionic character of the bond increases, and bonding electrons are pulled closer to the more electronegative atom.

more polar. Thus, in magnesium sulfide, MgS, the Mg—S bond has a Δ(EN) value of 1.3 and has therefore a greater degree of ionic character than the Al—P bond. The bond between sodium and chlorine is an ionic bond, with a Δ(EN) value of 2.1.

Figure 7.8 shows that there is no sharp borderline between ionic and covalent bonding. There is some covalent character in ionic bonds, and most covalent bonds have some ionic character. The ionic character of a covalent bond increases with increasing Δ(EN) of the bonded atoms.

Pauling has suggested the following *approximate* guidelines to determine the percent of ionic character of a bond:

Electronegativity Difference	Percent Ionic Character
1.0	20
1.5	40
2.0	60
2.5	80

As a rule of thumb, bonds that have 0 to 40 percent ionic character are called covalent, and bonds that have above 60 percent ionic character are called ionic. Bonds that have 40 to 60 percent ionic character cannot be classified simply as "covalent" or "ionic."

Example 7.1

Using Electronegativity Difference to Predict the Extent of Ionic Character of a Bond

Arrange the following bonds in order of increasing ionic character: K—Br, Cl—F, H—O.

SOLUTION: Of the three bonds given, the K—Br bond has the highest degree of ionic character because it is a bond between a metal and a nonmetal. The electronegativity of potassium is 0.8 and that of bromine is 2.8. The Δ(EN) in KBr is 2.0.

The Cl—F and H—O bonds link nonmetals and these bonds have little ionic character. The Δ(EN) for Cl—F bond is 1.0 and for H—O bond 1.4. Therefore, the H—O bond has more ionic character than the Cl—F bond. Thus, the order of increasing ionic character is:

$$\text{Cl—F} < \text{H—O} < \text{K—Br}$$

Practice Problem 7.2: Arrange the following bonds in the order of increasing polarity: H—S, S—Cl, Br—Cl, F—F.

Ionic Character and Bond Strength

Pauling showed that the bond energy of a diatomic molecule is related to the ionic character of the bond. Table 7.3 shows the relationship between the electronegativity difference, the percent ionic character, and the bond energy of hydrogen halides. The bond energies of hydrogen halides increase with increasing ionic character. This is true for many other diatomic molecules.

If the bond in HF were 100 percent covalent, the H—F bond energy would equal the sum of one-half of the H—H bond energy and one-half of the F—F bond energy. The H—H bond energy is 436 kJ mol^{-1} and the F—F bond energy is 157 kJ mol^{-1}. The sum of 218 kJ mol^{-1} and 78 kJ mol^{-1} is 296 kJ mol^{-1}, considerably less than the actually observed bond energy of 569 kJ mol^{-1} for HF. According to Pauling, the difference between the observed and calculated values is due to the ionic character of the bond.

Table 7.3

Electronegativity Difference, Δ(EN); Percent Ionic Character; and Bond Energy of Hydrogen Halides

Compound	Δ(EN)	Percent Ionic Character	Bond Energy (kJ mol^{-1})
HF	1.9	45	569
HCl	0.9	17	432
HBr	0.7	12	366
HI	0.4	5	299

Example 7.2

Comparing Bond Energies Based on Electronegativity Differences

Based only on electronegativity differences, which of the following diatomic molecules is expected to have the strongest bond? BrCl, HCl, or HBr.

SOLUTION: The electronegativity difference is greatest in HCl. Therefore, the H—Cl bond has the greatest ionic character and is expected to be the strongest of those given.

Practice Problem 7.3: Based on the electronegativity differences, which one of the bonds in Example 7.2 is the weakest?

7.4 Lewis Structures

Writing Lewis Structures

Thus far we have considered only single covalent bonds in diatomic molecules. Polyatomic molecules contain two or more covalent bonds. The U.S. chemist Gilbert Lewis (1875–1946), of the University of California at Berkeley, devised a method to describe the role of valence electrons in atoms and molecules.

Lewis described the valence shell structures of atoms by writing the symbols for the atoms surrounded by dots to represent their valence electrons. The *symbol of an atom surrounded by dots which represent its valence electrons* is called the **Lewis symbol** or the **Lewis dot symbol**. Table 7.4 lists the Lewis symbols for the representative elements. A *combination of Lewis symbols that represents a molecule* is called **Lewis dot formula** or **Lewis structure**.

Table 7.4

Lewis Symbols and Valence Shell Electron Configurations for the Representative Elements

H· $1s^1$								He: $1s^2$
Li· $2s^1$	Be: $2s^2$		B· $2s^22p^1$	C· $2s^22p^2$	·N· $2s^22p^3$	:O· $2s^22p^4$	:F· $2s^22p^5$	:Ne: $2s^22p^6$
Na· $3s^1$	Mg: $3s^2$		Al· $3s^23p^1$	Si· $3s^23p^2$	·P· $3s^23p^3$	:S· $3s^23p^4$	:Cl· $3s^23p^5$	:Ar: $3s^23p^6$
K· $4s^1$	Ca: $4s^2$		Ga· $4s^24p^1$	Ge· $4s^24p^2$	·As· $4s^24p^3$	:Se· $4s^24p^4$	:Br· $4s^24p^5$	:Kr: $4s^24p^6$
Rb· $5s^1$	Sr: $5s^2$		In· $5s^25p^1$	Sn· $5s^25p^2$	·Sb· $5s^25p^3$	:Te· $5s^25p^4$	:I· $5s^25p^5$	:Xe: $5s^25p^6$
Cs· $6s^1$	Ba: $6s^2$		Tl· $6s^26p^1$	Pb· $6s^26p^2$	·Bi· $6s^26p^3$	:Po· $6s^26p^4$	:At· $6s^26p^5$	:Rn: $6s^26p^6$
Fr· $7s^1$	Ra: $7s^2$							

Example 7.3

Writing Lewis Symbols

Without looking at Table 7.4, write Lewis symbols for each of the following atoms. Write the paired valence electrons as dot pairs and the unpaired valence electrons as single dots. **(a)** As; **(b)** Si; **(c)** Ca.

SOLUTION

(a) Arsenic is in group V in the periodic table, and an As atom has five valence electrons, with an electronic structure of $4s^24p^3$. The $4s$ electrons are paired and the $4p$ electrons are unpaired. Thus, the Lewis symbol for an As atom is

$$:\dot{\underset{\cdot}{\mathrm{As}}}\cdot$$

(b) Silicon is in the fourth group, and an atom of silicon has four valence electrons with an electronic structure of $3s^23p^2$. The two $3s$ electrons are present as a pair, and the two $3p$ electrons are unpaired:

$$:\underset{\cdot}{\mathrm{Si}}\cdot$$

(c) Calcium is in the second group. A calcium atom has two paired $4s$ electrons as its valence electrons:

$$\mathrm{Ca}:$$

Lewis structures for molecules and polyatomic ions are based on the premise that in chemical reactions *nonmetal atoms, except hydrogen and helium, tend to acquire a full octet of electrons in their valence shells*. This generalization is known as Lewis's **octet rule**.

We recall that noble gas atoms and most monatomic ions of main group elements have eight electrons in their valence shells. Similarly, nonmetal atoms that combine to form covalent bonds tend to acquire eight electrons in their valence shells by sharing electrons. Exceptions to the octet rule are hydrogen and helium, which can never have more than two electrons.

A Lewis structure for a molecule is constructed from the Lewis symbols for the atoms of which the molecule is composed. For example, a molecule of hydrogen chloride can be represented by combining the Lewis symbols for hydrogen and chlorine atoms:

$$\mathrm{H}\cdot + \cdot\ddot{\underset{\cdot\cdot}{\mathrm{Cl}}}: \longrightarrow \mathrm{H}:\ddot{\underset{\cdot\cdot}{\mathrm{Cl}}}:$$

A pair of dots between two atoms in a molecule is usually represented by a dash:

$$\mathrm{H}—\ddot{\underset{\cdot\cdot}{\mathrm{Cl}}}:$$

Two dots or a dash between two atoms in a Lewis structure represents a covalent bond. The H—Cl bond is formed by sharing the $1s$ electron of the H atom with the unpaired $3p$ electron of the Cl atom. An *electron pair in a covalent bond* is called the **bonding pair**. *Electron pairs in a molecule that do not participate in bonding* are variously called **lone pairs**, **unshared pairs**, or **nonbonding pairs**. The Lewis structure for HCl indicates that a molecule of hydrogen chloride has a covalent bond between the hydrogen and chlorine atoms and also has three lone pairs on the chlorine atom.

The bonding electrons in a covalent bond are contributed equally or nearly equally by the two bonded atoms. Bonding electrons therefore belong to the valence shells of *both* of the bonded atoms. Thus, the chlorine atom in an HCl molecule has a full octet of electrons, and the hydrogen atom has two electrons.

Let us consider the Lewis structure for a water molecule. The oxygen atom has two unpaired electrons that can be used to form the two O—H bonds in an H_2O molecule:

$$:\ddot{\underset{\cdot}{O}}\cdot + 2H\cdot \longrightarrow :\ddot{\underset{\underset{\displaystyle H}{|}}{O}}—H$$

The resulting structure for H_2O shows that a water molecule contains two H—O bonds and two lone pairs on the oxygen atom. The oxygen atom has a full octet of electrons, and each of the two hydrogen atoms has two electrons. The following rules will help us to write Lewis structures.

1. *Draw a skeletal structure for the molecule or ion, and connect the atoms with dashes to represent bonding electron pairs.* For a binary compound whose molecule consists of one atom of an element and two or more atoms of another element, the single atom is *usually* in the center of the molecule, with the other atoms attached to it. An example of such a molecule is methane, CH_4. In a CH_4 molecule, the single carbon atom is in the center of the molecule, and the four hydrogen atoms are connected to the carbon atom:

```
    H
    |
H—C—H
    |
    H
```

When a molecule contains atoms of more than two elements, the atom with the lowest electronegativity is *usually* the central atom of the molecule. For example, in a $POCl_3$ molecule, the phosphorus atom is the central atom because its electronegativity is lower than that of chlorine or oxygen:

```
      O
      |
Cl—P—Cl
      |
      Cl
```

The following guidelines are useful when writing Lewis structures:

(a) Hydrogen and fluorine atoms are never the central atoms in molecules.
(b) A carbon atom is usually linked to other atoms by four bonding pairs, nitrogen *usually* has three bonding pairs, and oxygen two. There are a few exceptions to these rules. For example, in some molecules and polyatomic ions, carbon has three bonding pairs, as in CO, nitrogen has four bonding pairs in NH_4^+ ion, and oxygen has three bonding pairs in H_3O^+ ion.

2. *Count the total number of valence electrons of all the atoms in a molecule.* For a polyatomic anion, add as many electrons as needed to equal the charge on the anion. For a polyatomic cation, subtract the number of electrons equal to the positive charge on the ion.

3. *Subtract two electrons for each single bond in the skeletal structure drawn in step 1 from the total number of valence electrons counted.*
4. *Distribute the remaining electrons as lone pairs so that each atom has eight electrons, if possible.* If not enough electrons are available as lone pairs to satisfy the octet rule for all the atoms, convert some lone pairs to bonding pairs by creating double or triple bonds where necessary. *Two dashes* represent two electron pairs or a **double bond**, and *three dashes* represent three electron pairs or a **triple bond**. Double and triple bonds are called *multiple bonds*. The *number of bonding electron pairs* between any two atoms is known as the **bond order** for these atoms.

The guidelines for writing Lewis structures are illustrated in Example 7.4.

Example 7.4

Writing Lewis Structures for Molecules and Polyatomic Ions

Write a Lewis structure for each of the following species: **(a)** CO_2; **(b)** CH_2O; **(c)** ClO_3^-; **(d)** NO^+.

SOLUTION.

(a) A plausible skeletal structure for CO_2 is

O—C—O

The carbon atom is in the center because it is the only atom of its kind and because carbon is less electronegative than oxygen. The number of valence electrons in the molecule is:

Contributed by the C atom (group IV):	4
Contributed by the two O atoms (group VI):	12
	16

We subtract four electrons for the two single C—O bonds leaving 12 electrons to be distributed as unshared pairs, or for multiple bonding if necessary.

An attempt to distribute the 12 electrons as unshared pairs can give several structures such as the following:

(1) :Ö—C̤̈—Ö (2) :Ö̤—C—Ö̤: (3) :Ö—C̤̈—Ö:

In structure (1) one of the O atoms lacks a full octet; in structure (2) the C atom has only four electrons; and in structure (3) each of the two oxygen atoms has only six electrons. However, when two unshared pairs in structure (2) or (3) are converted to bonding pairs to create two carbon–oxygen double bonds, an acceptable Lewis structure for CO_2 is obtained:

:Ö=C=Ö:

(b) Skeletal structures for CH_2O include the following possibilities:

(1) H—C—O—H (2) C—O(—H)—H (3) C—H—O—H (4) H—C(—H)—O

Structure (1) is not acceptable because the carbon atom has only two bonds. Multiple bonding is not possible because that would give more than one bond to hydrogen or more than two bonds to oxygen. Structure (2) is not acceptable because oxygen has three bonds. Structure (3) is not acceptable because hydrogen has two bonds.

Structure (4) is best because the central carbon atom is less electronegative than the oxygen atom. Although hydrogen is less electronegative than carbon, the hydrogen atom can never be the central atom because it can form only one bond. Structure (4) would be acceptable with a double bond between the carbon and oxygen atoms, a possibility we test below.

The number of valence electrons contributed by two H atoms is 2, by one C atom, 4, and by one O atom, 6, a total of 12. We subtract six electrons for the three bonds in structure (4), which leaves six electrons to be distributed as lone pairs, or for multiple bonding. If all the six electrons were placed as lone pairs, two possible structures would be obtained:

$$\begin{array}{c}\mathrm{H}\\ |\\ \mathrm{H{-}\underset{\cdot\cdot}{C}{-}\ddot{O}:}\end{array} \quad \text{or} \quad \begin{array}{c}\mathrm{H}\\ |\\ \mathrm{H{-}C{-}\underset{\cdot\cdot}{\ddot{O}}:}\end{array}$$

Neither of these structures is acceptable because a full octet for both carbon and oxygen is not obtained. A double bond between carbon and oxygen atoms gives the correct structure:

$$\begin{array}{c}\mathrm{H}\\ |\\ \mathrm{H{-}C{=}\ddot{O}:}\end{array}$$

This is the structure for formaldehyde.

(c) The skeletal structure for ClO_3^- ion is

$$\left[\begin{array}{c}\mathrm{O}\\ |\\ \mathrm{O{-}Cl{-}O}\end{array}\right]^-$$

The chlorine atom is in the center because it is the only atom of its kind and because it is less electronegative than oxygen. The number of valence electrons contributed by one Cl atom (group VII) is 7, by three O atoms (group VI) 18, plus one more for the 1− charge on the ion, a total of 26 electrons. After subtracting 6 electrons for the three Cl—O bonds, there are 20 electrons available for lone pairs. The resulting structure is

$$\left[\begin{array}{c}\mathrm{:\ddot{O}:}\\ |\\ \mathrm{:\underset{\cdot\cdot}{\ddot{O}}{-}\underset{\cdot\cdot}{Cl}{-}\underset{\cdot\cdot}{\ddot{O}}:}\end{array}\right]^-$$

(d) The starting skeletal structure for the NO^+ ion is

$$[\mathrm{N{-}O}]^+$$

The nitrogen atom has five valence electrons and the oxygen atom has six, minus one for the 1+ charge, leaving a total of 10 electrons on the NO^+ ion. Subtracting two for the N—O bond leaves eight electrons to be distributed as lone pairs. Some possible structures are

(1) $[:\ddot{N}—\ddot{O}:]^+$ (2) $[:\ddot{\underset{..}{N}}—O:]^+$ (3) $[:N—\ddot{\underset{..}{O}}:]^+$

Structure (1) is unacceptable because both the N and O atoms lack a full octet. Structure (2) is not acceptable because the O atom has only four electrons. Similarly, structure (3) is not acceptable because the N atom has only four electrons. However, when we convert an electron pair from each of the N and O atoms of structure (1) to multiple bonding, an acceptable Lewis structure is obtained:

$$[:N\equiv O:]^+$$

Practice Problem 7.4: Write Lewis structures for $AsCl_3$ and SO_3^{2-}.

Example 7.4 shows that the Lewis structures for CO_2 and CH_2O are written with a double bond for each C—O linkage. In a double bond, two electron pairs are shared by the bonded nuclei. We recall that the number of bonding electron pairs between any two atoms is called the bond order for these atoms. Thus, the C—O bond order in both CO_2 and in CH_2O is two, but the C—H bond order in CH_2O is one.

Example 7.5

Lewis Structures and Bond Order

Draw a Lewis structure for **(a)** disulfur dichloride, S_2Cl_2 (which has a sulfur-to-sulfur bond) and for **(b)** carbon monoxide, CO. What is the S—S bond order in S_2Cl_2 and the C—O bond order in CO?

SOLUTION

(a) Since S is less electronegative than Cl, the skeletal structure for S_2Cl_2 involving sulfur-to-sulfur bond can be written as

$$Cl—S—S—Cl$$

The number of valence electrons contributed by two Cl atoms is 14, and the number contributed by two S atoms is 12, for a total of 26. After subtracting 6 for the three bonds drawn, we have 20 electrons to distribute as unshared pairs, or for multiple bonding. The distribution of these electrons as unshared pairs gives an acceptable structure

$$:\ddot{\underset{..}{Cl}}—\ddot{\underset{..}{S}}—\ddot{\underset{..}{S}}—\ddot{\underset{..}{Cl}}:$$

Note that if we use any of the unshared pairs to make a double bond, one of the atoms will have 10 electrons. There is a single dash between the sulfur atoms, which means that the S—S bond order is one.

(b) The only skeletal structure that can be written for CO is

$$\mathrm{C{-}O}$$

The total number of valence electrons is 4 + 6 = 10. Subtracting two for the C—O bond leaves eight electrons to be distributed as lone pairs or for multiple bonding. Distributing these as lone pairs creates Lewis structures such as

$$:\ddot{\mathrm{C}}{-}\ddot{\mathrm{O}}: \quad \text{or} \quad :\mathrm{C}{-}\ddot{\underset{\cdot\cdot}{\mathrm{O}}}: \quad \text{or} \quad :\ddot{\underset{\cdot\cdot}{\mathrm{C}}}{-}\mathrm{O}:$$

None of these structures satisfy the octet rule for both of the atoms, but if we convert two lone pairs to multiple bonding and leave one lone pair for each of the carbon and oxygen atoms, an acceptable Lewis structure that satisfies the octet rule for both carbon and oxygen atoms is obtained:

$$:\mathrm{C{\equiv}O}:$$

The three bonding pairs between carbon and oxygen atoms show that the C—O bond order is three.

Practice Problem 7.5: Write the Lewis structure for each of the following: nitrogen molecule, N_2; ethene, C_2H_4; and ethyne, C_2H_2. Predict the N—N bond order in N_2, the C—C bond order in C_2H_4, and the C—C bond order in C_2H_2.

Certain atoms can combine in two or more different ways to form different stable compounds. *Compounds having the same molecular formula but different structural formulas* are called **isomers** (Example 7.6).

Example 7.6

Lewis Structures for Isomers

Draw Lewis structures for the isomers corresponding to the molecular formula C_2H_6O.

SOLUTION: For C_2H_6O, two reasonable Lewis structures can be drawn:

```
     H  H                    H      H
     |  |         ..         |  ..  |
  H—C—C—O—H   and   H—C—O—C—H
     |  |         ..         |  ..  |
     H  H                    H      H
```

These structures represent two isomers with the formula C_2H_6O. The first of these structures, simplified as CH_3CH_2OH, is the structure of ethyl alcohol (ethanol), and the other, CH_3OCH_3, represents dimethyl ether.

Practice Problem 7.6: Write Lewis structures for three isomers of C_3H_8O.

Resonance Structures

Sometimes two or more equivalent Lewis structures can be written for a molecule or ion. For example, two different Lewis structures can be written for sulfur dioxide, SO_2. In each of these structures, a different oxygen atom is double-bonded to the sulfur atom:

$$:\ddot{\underset{..}{O}}-\ddot{S}=\ddot{O}: \longleftrightarrow :\ddot{O}=\ddot{S}-\ddot{\underset{..}{O}}:$$

Lewis structures that differ only by the number of bonding electron pairs between a given pair of atoms are called **resonance structures**.

The resonance structures for a molecule or a polyatomic ion are usually separated by double-headed arrows. Although the structures for an SO_2 molecule we have written are acceptable Lewis structures, neither of them represents the known structure of an SO_2 molecule. Experiments show that the S—O bond strengths in an SO_2 molecule are equal. Therefore, one of the two S—O bonds cannot be a single bond, while the other is a double bond. The S—O bond strength in an SO_2 molecule is greater than that expected for a single bond but weaker than the S=O double bond.

Since an SO_2 molecule cannot be adequately represented by either of the resonance structures, the real structure for the molecule is a *hybrid* of the two resonance structures. This *hybrid* is called a **resonance hybrid**.

A resonance structure for SO_2 shows that there are a total of three bonding pairs per two S—O linkages. Thus, there are $\frac{3}{2}$ or $1\frac{1}{2}$ bonding electron pairs for each S—O linkage. Therefore, the S—O bond order in SO_2 is $1\frac{1}{2}$. *Partial or fractional bonding* between two atoms is sometimes represented by a broken or dashed line above or below a solid line for a single bond. Using this method, the resonance hybrid for SO_2 can be drawn as

$$O\overset{---}{—}S\overset{---}{—}O$$

The dashed lines in this structure show that the additional partial bonding is equally distributed between the two S—O bonds, which are equivalent in all respects. Theories that explain partial bonding are discussed in Chapters 8 and 9.

Example 7.7

Drawing Lewis Resonance Structures and Predicting Bond Order

Draw all the resonance structures for a carbonate ion, CO_3^{2-}, and determine the C—O bond order in a CO_3^{2-} ion.

SOLUTION: Using the guidelines previously outlined for drawing Lewis structures, three resonance structures for a CO_3^{2-} ion can be obtained:

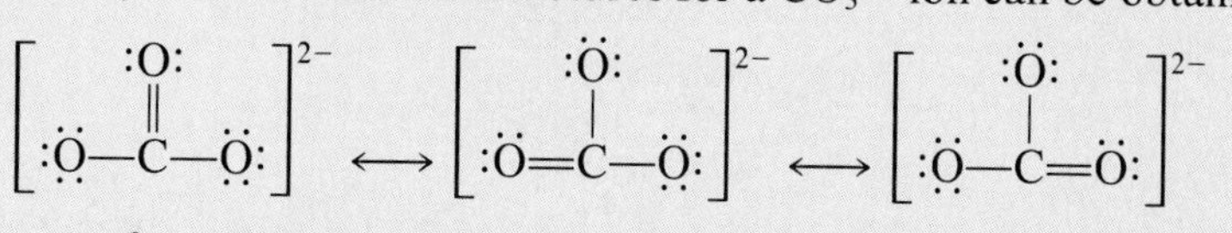

$$\left[\begin{array}{c}:O:\\ \| \\ :\ddot{\underset{..}{O}}-C-\ddot{\underset{..}{O}}:\end{array}\right]^{2-} \longleftrightarrow \left[\begin{array}{c}:\ddot{O}:\\ | \\ :\ddot{O}=C-\ddot{\underset{..}{O}}:\end{array}\right]^{2-} \longleftrightarrow \left[\begin{array}{c}:\ddot{O}:\\ | \\ :\ddot{\underset{..}{O}}-C=\ddot{O}:\end{array}\right]^{2-}$$

There are four electron pairs and three C—O linkages in a CO_3^{2-} ion; therefore, the C—O bond order is $\frac{4}{3}$ or $1\frac{1}{3}$.

Practice Problem 7.7: Write all the Lewis resonance structures for a sulfur trioxide molecule, for a nitrate ion, and for an ozone molecule, O_3. Predict the S—O bond order in sulfur trioxide, the N—O bond order in a nitrate ion, and the O—O bond order in ozone.

Exceptions to the Octet Rule

Exceptions to the octet rule are observed in three situations: (1) the central atom in a molecule has fewer than eight electrons, (2) the central atom has more than eight electrons, (3) the total number of electrons in a molecule is an odd number.

Examples of molecules in which the central atom has fewer than an octet of electrons are provided by certain gaseous or liquid compounds of beryllium (in group II) and of boron and aluminum (in group III). Such molecules are said to be *electron deficient*. First, we consider beryllium chloride, which can be made by heating beryllium oxide and carbon tetrachloride to very high temperatures.

$$2BeO(s) + CCl_4(g) \xrightarrow{800°C} 2BeCl_2(g) + CO_2(g)$$

The Lewis structure of beryllium chloride is shown below. In this structure, the beryllium atom has only four electrons:

$$:\ddot{\underset{\cdot\cdot}{Cl}}—Be—\ddot{\underset{\cdot\cdot}{Cl}}:$$

We might try moving a pair of electrons from each of the two chlorine atoms around beryllium in this molecule to create double bonds. However, the electronegativity of beryllium is low and double bonds do not form.

Another exception to the octet rule is boron trichloride, BCl_3. The Lewis structure of boron trichloride is

$$\begin{array}{c} :\ddot{Cl}: \\ | \\ :\ddot{\underset{\cdot\cdot}{Cl}}—B—\ddot{\underset{\cdot\cdot}{Cl}}: \end{array}$$

In this structure, the boron atom has only six electrons. We can provide a full octet for the boron atom by converting an electron pair of one of the chlorine atoms to a bonding pair. This way we could write three resonance structures for the BCl_3 molecule. But because of the low electronegativity of boron, the electron pairs remain with the chlorine atoms and do not participate in a double bond.

When the central atom of a molecule is in the third period or beyond, it can be surrounded by more than eight electrons. This is possible because the atoms of the third and later periods have unfilled *d* orbitals that can accommodate bonding electron pairs. An example of a molecule in which the central atom has more than eight electrons is phosphorus pentachloride, PCl_5. In a PCl_5 molecule, there are five P—Cl bonds, and the central phosphorus atom has 10 electrons. The Lewis structure of PCl_5 is

$$\begin{array}{c} :\ddot{Cl}: \quad \ddot{Cl}: \\ | \diagup \\ :\ddot{\underset{\cdot\cdot}{Cl}}—P \\ | \diagdown \\ :\underset{\cdot\cdot}{Cl}: \quad \ddot{Cl}: \end{array}$$

Another exception to the octet rule is found in molecules having an odd number of electrons. An example of a molecule with an odd number of electrons is nitrogen dioxide, NO_2. This molecule has 17 valence electrons; therefore, one electron is unpaired. The Lewis structure for NO_2 can be written with the odd electron on either the oxygen or the nitrogen atom:

$$:\underset{\cdot\cdot}{\ddot{O}}-\dot{N}=\ddot{O}\cdot \quad \text{or} \quad :\underset{\cdot\cdot}{\ddot{O}}-\dot{N}=\ddot{O}:$$

The most probable position of the unpaired electron can be deduced by considering the chemical reactions of NO_2. Two NO_2 molecules react to form dinitrogen tetroxide (N_2O_4), which has an N—N covalent bond. We conclude that the odd electron in NO_2 is concentrated mostly on the nitrogen atom so that it is available to form the N—N covalent bond in N_2O_4:

$$O_2N\cdot + \cdot NO_2 \longrightarrow O_2N{-}NO_2 \quad \text{or} \quad N_2O_4 \quad \text{(dinitrogen tetroxide)}$$

Practice Problem 7.8: Write Lewis structures for: **(a)** beryllium hydride; **(b)** boron trifluoride; **(c)** sulfur hexafluoride; **(d)** nitrogen monoxide, NO. A sulfur atom in its ground state has only two unpaired electrons. Account for the formation of six S—F bonds in a molecule of sulfur hexafluoride.

The Coordinate Covalent Bond

A molecule in which the central atom is electron deficient, such as BCl_3, reacts with species that can donate a pair of electrons to its central atom to complete its octet. For example, a chloride ion, $[:\underset{\cdot\cdot}{\ddot{Cl}}:]^-$, has four electron pairs in its valence shell. It can donate one electron pair to complete the octet of an electron-deficient boron atom of BCl_3. A molecule of BCl_3 reacts with a chloride ion to form the polyatomic anion BCl_4^-:

$$\begin{array}{c} :\ddot{Cl}: \\ | \\ :\underset{\cdot\cdot}{\ddot{Cl}}-B \\ | \\ :\underset{\cdot\cdot}{Cl}: \end{array} + [:\underset{\cdot\cdot}{\ddot{Cl}}:]^- \longrightarrow \left[\begin{array}{c} :\ddot{Cl}: \\ | \\ :\underset{\cdot\cdot}{\ddot{Cl}}-B-\underset{\cdot\cdot}{\ddot{Cl}}: \\ | \\ :\underset{\cdot\cdot}{Cl}: \end{array}\right]^-$$

A covalent bond in which one of the bonding atoms (or ions) contributes both bonding electrons is called a **coordinate covalent bond**. Of course, all electrons are identical, and once a coordinate covalent bond has formed, it becomes equivalent to any other covalent bond. Thus, all the B—Cl bonds in a BCl_4^- ion are equivalent.

Two commonly encountered polyatomic cations have coordinate covalent bonds. These are ammonium ion, NH_4^+, and hydronium ion, H_3O^+. Ammonium ions are formed when ammonia gas is passed through a solution that contains hydrogen ions:

$$NH_3(g) + H^+(aq) \longrightarrow NH_4^+(aq)$$

In this reaction the nitrogen atom of the ammonia molecule furnishes a pair of electrons to form a coordinate covalent bond with the hydrogen ion:

$$\underset{\text{Ammonia molecule}}{\begin{array}{c} H \\ | \\ H-N: \\ | \\ H \end{array}} + H^+ \longrightarrow \underset{\text{Ammonium ion}}{\left[\begin{array}{c} H \\ | \\ H-N-H \\ | \\ H \end{array}\right]^+}$$

A hydronium ion forms when a nonbonding electron pair of the oxygen atom of a water molecule donates a pair of electrons to a hydrogen ion:

$$\underset{\text{Water molecule}}{\mathrm{H{-}\overset{\overset{\displaystyle H}{|}}{\underset{\cdot\cdot}{O}}{:}}} + H^+ \longrightarrow \underset{\text{Hydronium ion}}{\left[\mathrm{H{-}\overset{\overset{\displaystyle H}{|}}{\underset{\cdot\cdot}{O}}{-}H}\right]^+}$$

7.5 Structures of Molecules with No Lone Electron Pairs on the Central Atom

Lewis structures do not represent the three-dimensional structures of molecules. However, Lewis structures can help us to predict three-dimensional structures. In this section we learn how to predict three-dimensional structures from Lewis structures. Three-dimensional structures are important because they determine many physical and chemical properties of compounds.

The arrangement of atomic nuclei in a molecule can be determined experimentally. When we know how the atomic nuclei are arranged, we also know the positions of the bonding electron pairs, because the bonding electron pairs are concentrated between the nuclei.

The *positions of the nuclei in a molecule* determine its **molecular structure**. The *arrangement and angles between all valence shell electron pairs around a central atom in a molecule* is called the **electron pair structure**. If the central atom has no unshared electrons, the electron pair structure of the molecule and the molecular structure are identical. The locations of any lone electron pairs cannot be experimentally determined but can be inferred from the molecular structure.

Valence Shell Electron Pair Repulsion Theory

Molecular structure is conveniently described in terms of **bond angles**. A bond angle is the *angle between two lines drawn from the central atom of a molecule to two adjacent atoms bonded to the central atom*. N. V. Sidgwick and R. J. Nyholm in England, and R. J. Gillespie of Canada have developed a theory that makes it possible for us to predict the molecular and electron pair structures of molecules and polyatomic ions from their formulas. Their theory is called the **valence shell electron pair repulsion** (VSEPR) **theory**.

The central hypothesis of VSEPR theory is that *valence shell electron pairs of the central atom in a molecule adopt a spatial arrangement in which their mutual repulsion is minimized*. This repulsion is minimal when all electron groups (both bonding and nonbonding) are as far as possible from one another in the space around the central atom.

We first consider the structures of molecules that have no lone electron pairs on the central atom and in which all the bonds are single bonds. Figure 7.9 illustrates the geometrical arrangements of different numbers of bonding electron pairs represented by lines drawn from the central atom. The central atom is represented by a dot in the center of a sphere.

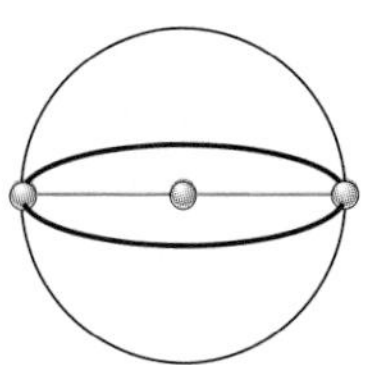

(a) Linear arrangement (2 electron pairs)

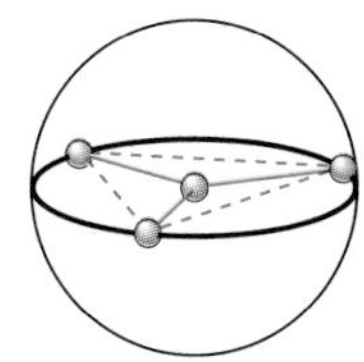

(b) Trigonal planar arrangement (3 electron pairs)

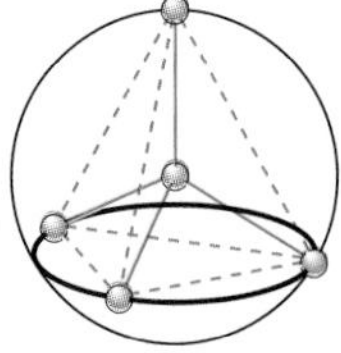

(c) Tetrahedral arrangement (4 electron pairs)

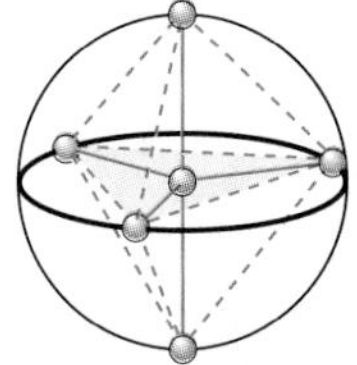

(d) Trigonal bipyramidal arrangement (5 electron pairs)

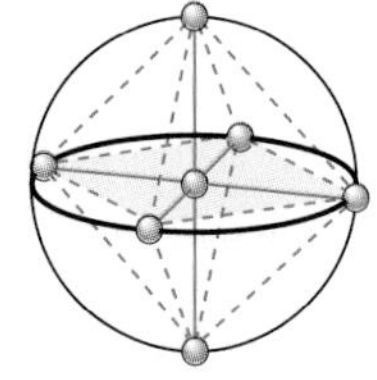

(e) Octahedral arrangement (6 electron pairs)

Figure 7.9

Various structural models of molecules. The dot in the center of each sphere represents the central atom in the molecule. The lines drawn from the central atom to the surface of the sphere represent the electron pairs on the central atom. The dashed lines describe the imaginary sides of the structures with corners toward which the electron pairs are pointed.

If the central atom, A, of a molecule is bonded to two other atoms, each represented as B, in a molecule AB_2, there are *two* bonding electron pairs around the central atom. Electron pair repulsion is minimized when the bonding electrons lie on a straight line on opposite sides of the central atom. The BAB bond angle is 180° and the molecule has a **linear structure**:

180°
B—A—B

If a molecule has *three* bonding pairs around the central atom (an AB_3 molecule), electron pair repulsion is minimized when the bonding electrons lie in a plane and each BAB bond angle is 120°. Thus, the bonded atoms (the three B's) lie at the corners of an equilateral triangle. Such a molecule has a **trigonal planar structure** or **triangular planar structure**:

120° B
B—A
B

If a molecule has *four* bonding pairs around the central atom (AB_4), electron pair repulsion is minimized when the bonding electrons point at the corners of a *tetrahedron*. (A tetrahedron is a polyhedron whose sides are four equilateral triangles.) Such a molecule has a **tetrahedral structure**. Each BAB bond angle in a tetrahedral structure is 109.5°:

B
109.5° A B
B
B

If a molecule has *five* bonding pairs around the central atom (AB_5), electron pair repulsion is minimized when three bonding pairs lie in a trigonal planar structure and the remaining two lie on opposite sides of the central atom along a straight line perpendicular to the trigonal plane. Since the trigonal plane is a common base for two imaginary pyramids on opposite sides of the plane, the structure is called **trigonal bipyramidal**:

If a molecule has *six* bonding pairs around the central atom (AB_6), electron pair repulsion is minimized when the bonding electrons point at the corners of an *octahedron*. (An octahedron is a polyhedron whose sides are eight equilateral triangles.) A molecule with such geometry has an **octahedral structure**:

Some molecules that represent each of these structures are discussed below.

Linear Molecules

We shall now consider the structures of some polyatomic molecules from the viewpoint of the VSEPR theory. We will begin with linear molecules. Linear molecules whose central atoms have only two covalent bonding pairs, and no lone pairs, are formed from group II atoms which have two valence electrons. Group II elements are metals, which generally form ionic bonds with most nonmetals. However, the first element in group II, beryllium, has the highest ionization energy and also the highest electronegativity in its group. Beryllium hydride, BeH_2, and beryllium chloride, $BeCl_2$, exist in the vapor state as molecules whose bonds have considerable covalent character.

Beryllium hydride, BeH_2, has two bonding pairs on the central beryllium atom. VSEPR theory predicts that the two bonding pairs must be as far apart as possible to minimize their mutual repulsion. This condition is satisfied in a linear structure having a bond angle of 180°, as shown in Figure 7.10. The

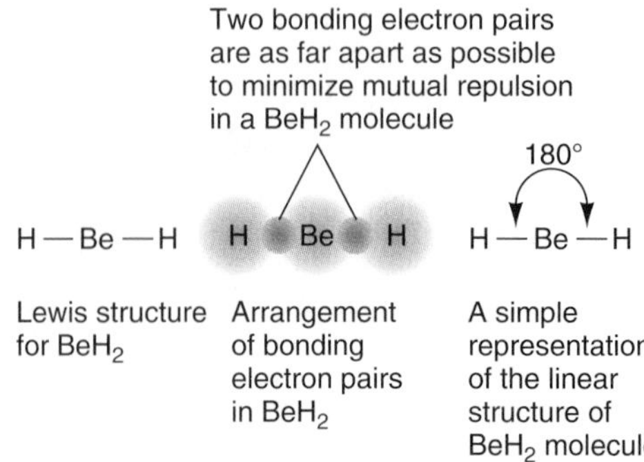

Figure 7.10

A diagrammatic representation of the positions of electron pairs in a BeH_2 molecule according to the VSEPR theory.

structure of beryllium hydride has been determined experimentally. In accord with the prediction of VSEPR theory, it is a linear molecule.

A beryllium hydride molecule has two polar Be—H bonds, each having a dipole moment. Since hydrogen is more electronegative than beryllium, the negative end of the dipole points toward the hydrogen (Figure 7.11). Because the dipole moments of the two Be—H bonds are equal in magnitude and point in opposite directions, BeH_2 is a nonpolar molecule.

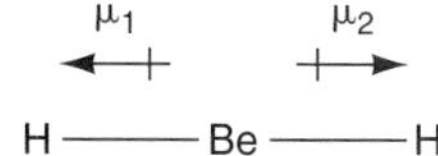

Figure 7.11

Bond dipole moments in a BeH_2 molecule. Since the magnitude of μ_1 equals the magnitude of μ_2, the molecule is nonpolar.

Now, let us consider a linear molecule in which the central atom has two double bonds. When we use VSEPR theory to predict molecular structure, double and triple bonds are considered in the same way as single bonds since they are groups of electrons, or regions of electron density, between atomic nuclei. For example, carbon dioxide contains two carbon–oxygen double bonds and no lone electrons on the central carbon atom. It is linear. Its Lewis structure is shown in Example 7.4. A carbon dioxide molecule has two polar C—O bonds. As in BeH_2, the dipole moments of the individual bonds cancel, and the molecule is nonpolar.

Trigonal Planar Molecules

Next we consider some examples of molecules whose central atom is connected to three atoms of another element. To form three covalent bonds, an atom needs three unpaired valence electrons. To find atoms with three valence electrons that can form three bonds, we look in group III of the periodic table. An example of a molecule whose central atom has three bonding pairs and trigonal planar geometry is boron trichloride, BCl_3. In a trigonal planar molecule, all the atoms lie in the same plane, and all the bond angles are 120° (Figure 7.12). In this geometry, all the bonding pairs are as far as possible from one another to minimize their mutual repulsion.

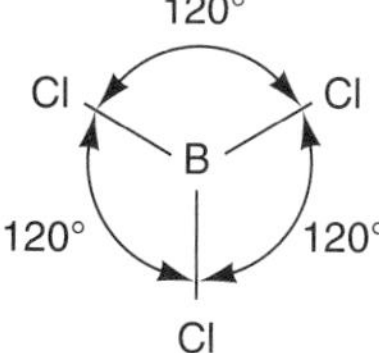

Figure 7.12

The structure of the trigonal planar BCl_3 molecule.

A BCl_3 molecule has three polar B—Cl bonds with dipole moments of equal magnitude. Each of these dipole moments is directed to an electronegative chlorine atom. Since all the ClBCl bond angles are equal, and all the atoms lie in a plane, the bond dipole moments cancel. Therefore, the BCl_3 molecule has no net dipole moment and is nonpolar.

Sulfur trioxide, SO_3, is also a trigonal planar molecule. It can be represented by three Lewis resonance structures (Practice Problem 7.7). There are no lone electrons on the central atom, and the S—O bond order is $1\frac{1}{3}$. Since there are no lone pairs on the sulfur atom, VSEPR theory predicts a trigonal planar structure for SO_3. The three polar S—O bonds are symmetrically distributed around the central atom. Therefore, as in BCl_3, these bond dipole moments cancel, and the molecule is nonpolar.

Example 7.8

Predicting the Polarity of a Trigonal Planar Molecule with No Lone Electrons on the Central Atom

Is a BCl_2F molecule polar or nonpolar? Explain.

SOLUTION: BCl_2F is a polar molecule because of an unsymmetric distribution of electron density. Fluorine is more electronegative than chlorine. Therefore, the dipole moment of a B—F bond is greater in magnitude than that of a B—Cl bond. Consequently, a molecule of BCl_2F has a small net dipole moment directed toward the fluorine atom.

Tetrahedral Molecules

In a tetrahedral molecule, the central atom has four bonds. To form four covalent bonds, an atom needs four unpaired valence electrons. To find atoms with four valence electrons, we look in group IV of the periodic table. An example of a molecule whose central atom has four bonding pairs and tetrahedral geometry is methane, CH_4.

For a group IV atom such as carbon (or silicon) to form four covalent bonds, it must rearrange its electrons by unpairing them. We consider this process in Chapter 8. For now, we need only apply the rule that minimal electron repulsion is achieved when the electrons are farthest apart. This means that the four covalent bonds will point at the corners of a tetrahedron, as shown for methane in Figure 7.13. The HCH bond angle in methane is 109.5°. In this arrangement, repulsion between bonding pairs is minimized. Thus, the VSEPR model correctly predicts the geometry of methane.

The electronegativity difference between carbon and hydrogen is so small (0.4) that C—H bonds are almost nonpolar. Because of a symmetrical distribution of the bonding pairs about the central carbon atom, the small C—H bond dipole moments cancel, and a molecule of CH_4 is nonpolar.

A molecule of carbon tetrachloride, CCl_4, is also tetrahedral and nonpolar. Although there are four polar C—Cl bonds in a CCl_4 molecule, the molecule is nonpolar because the dipole moments of the four bonds are distributed symmetrically around the central atom and therefore cancel one another (Figure 7.14).

Figure 7.13

Structure of methane, a tetrahedral molecule. All the HCH bond angles are 109.5°.

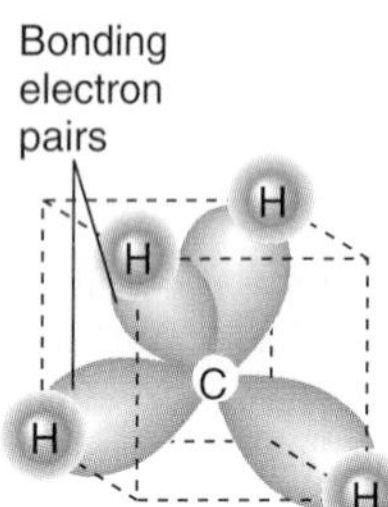

(a) Tetrahedral structure: Each C—H bond is directed to each of the four opposite corners of an imaginary cube

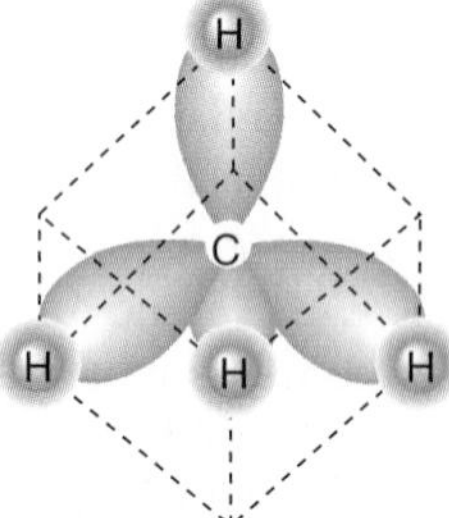

(b) Identical to structure (a) resulting from the rotation of the cube to one of its corners

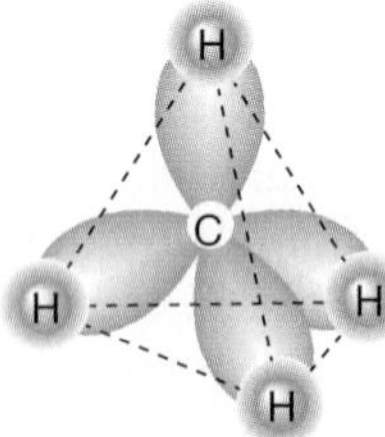

(c) Identical to structures (a) and (b) but inscribed in an imaginary tetrahedron

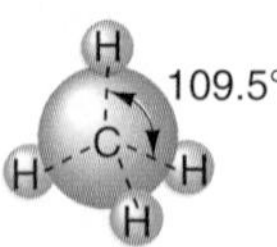

(d) Tetrahedral molecule of methane

H
H—C—H
H

(e) A simple way of showing the structure of a tetrahedral molecule such as methane

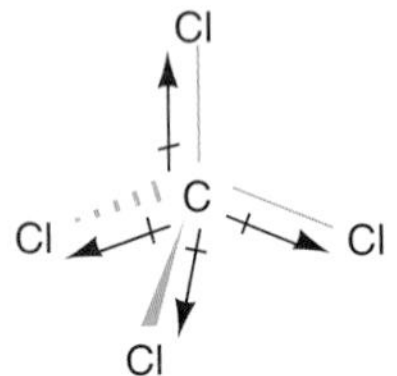

Figure 7.14

CCl_4 has four polar C—Cl bonds. However, the dipole moments of the molecule is zero because the dipole moments of the individual bonds cancel.

Example 7.9

Predicting the Structure and Polarity of a Tetrahedral Molecule with No Lone Electrons on the Central Atom

The formula for chloroform (a good solvent for fats, oils, and rubber, formerly used as an anesthetic) is $CHCl_3$. What is the structure of chloroform? Is a molecule of chloroform polar or nonpolar?

SOLUTION: The Lewis structure for $CHCl_3$ is

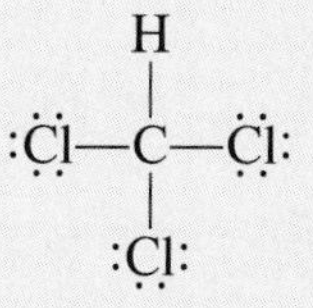

```
        H
        |
  :Cl—C—Cl:
        |
      :Cl:
```

From the Lewis structure we can see that there are four bonding pairs on the central atom. Although the molecule can be predicted to be tetrahedral, we might expect a slight distortion from that structure because all of the bonding pairs are not equivalent.

Due to the electronegativity difference of C and Cl atoms, there are three bond dipole moments directed from the central C atom toward the Cl atoms situated at the corners of a tetrahedron. There is also a small H—C dipole moment directed toward the central C atom. As a result, the molecule possesses a net dipole moment and is polar (Figure 7.15).

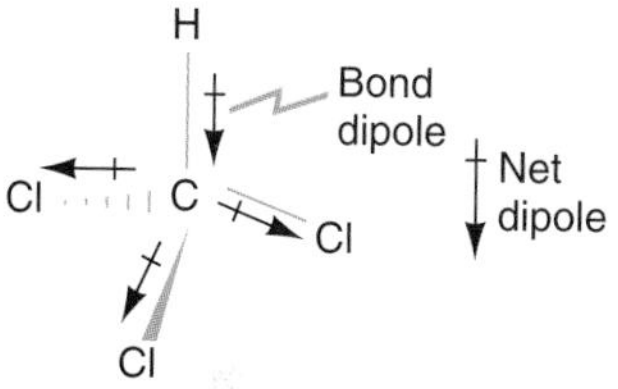

Figure 7.15

Chloroform $CHCl_3$ is a polar molecule.

When we proceed to the right from carbon in the second period, we find that nitrogen in neutral molecules *usually* forms three bonds because a nitrogen atom has three unpaired valence electrons. Oxygen in neutral molecules forms two bonds because an oxygen atom has two unpaired valence electrons. A fluorine atom has only one unpaired valence electron, and it can therefore form only one bond.

Trigonal Bipyramidal and Octahedral Molecules

Trigonal bipyramidal molecules have five bonds to the central atom. To form five covalent bonds, an atom needs five valence electrons. Atoms with five valence electrons are in group V of the periodic table. First, let us consider the group V element phosphorus. We recall from Section 7.4 that phosphorus can form a pentachloride, which in its vapor phase consists of PCl_5 molecules. In a PCl_5 molecule, all five of the electron pairs in the valence shell of the phosphorus atom are bonding pairs. These five bonding pairs are arranged around the central phosphorus atom in a trigonal bipyramidal structure (Figure 7.16). The PCl_5 molecule is nonpolar because the electron pairs are distributed symmetrically around the central atom.

The Cl atoms in a PCl_5 molecule are situated at the corners of the trigonal bipyramid (Figure 7.16). The P atom lies in the center of an equilateral triangle. Each Cl—P—Cl bond angle in the central trigonal plane is 120°. The three Cl atoms at the corners of the central trigonal plane are called *equatorial* Cl atoms. Two of the Cl atoms lie above and below the central plane at 90° angles with that plane. These atoms occupy *axial* positions. To remember the meaning of these terms, think of the equatorial atoms as pointing toward the equator, and the axial atoms as pointing along the north–south axis of the earth.

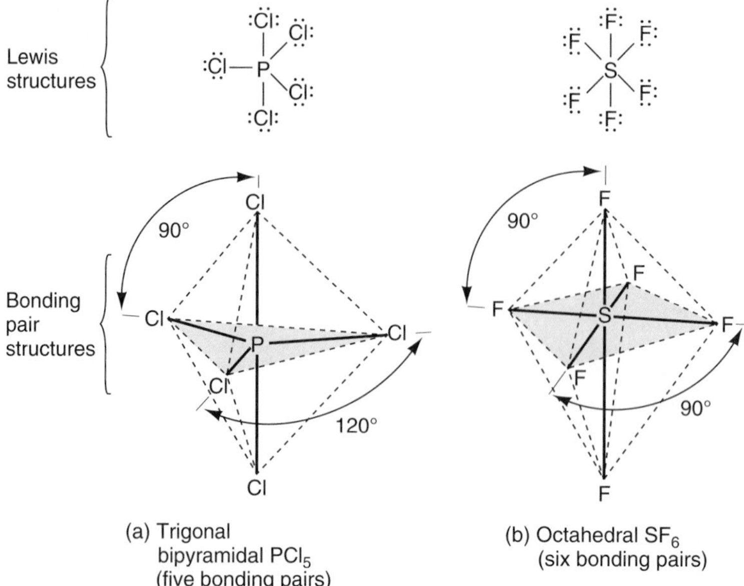

Figure 7.16

Trigonal bipyramidal and octahedral molecules.

Other pentahalides of the group V elements can also have the same geometry. Thus, PF_5 and AsF_5 are nonpolar, trigonal bipyramidal molecules.

Next, let us examine a group VI atom, sulfur. A sulfur atom has six valence electrons and can therefore form six bonds in molecules such as SF_6. In an SF_6 molecule, the central atom has no lone electron pairs. The six bonding pairs form an octahedral structure (Figure 7.16). Because of the symmetric arrangement of the electron pairs, SF_6 is nonpolar.

The central atom in an SF_6 molecule lies in the center of an imaginary square, with four other atoms at the corners of the square, and two more atoms on opposite sides of the square at 90° angles with the square. In this configuration, all of the bond angles are 90°, as shown in Figure 7.16 for SF_6. In an octahedron, all of the faces, edges, corners, and angles are equivalent. Table 7.5 summarizes the structural types of molecules in which the central atom is surrounded by bonding pairs only.

To use the VSEPR theory to predict the structure of a polyatomic molecule or ion, first write the Lewis structure for the species. If the Lewis structure has no unshared electron pairs on the central atom, use the information in Table 7.5 to predict the geometry of the molecule. Multiple bonds are considered the same as single bonds in determining geometry (see Example 7.10c).

Table 7.5

Structures of Molecules with Only Bonding Electron Pairs on the Central Atom

Type of Molecule[a]	Example	Structure of Molecule	Bond Angles at the Central Atom
AB_2	BeH_2	Linear	180°
AB_3	BCl_3	Trigonal planar	120°
AB_4	CH_4	Tetrahedral	109.5°
AB_5	PCl_5	Trigonal bipyramidal	120° and 90°
AB_6	SF_6	Octahedral	90°

[a] A is the central atom and B an atom attached to it.

Example 7.10

Predicting the Structures of Polyatomic Molecules and Ions

Predict the structure of each of the following polyatomic molecules and ions: **(a)** silicon tetrachloride; **(b)** ammonium ion; **(c)** nitrate ion; **(d)** PF_6^- ion.

SOLUTION

(a) The Lewis structure for $SiCl_4$ is

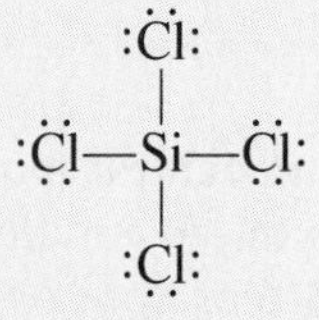

```
         ..
        :Cl:
         |
   ..    |    ..
  :Cl—Si—Cl:
   ..    |    ..
         |
        :Cl:
         ..
```

The Lewis structure shows that the central atom has no unshared electron pairs. There are four bonding pairs and four chlorine atoms connected to the central silicon atom. The four equivalent electron pairs on the central atom are arranged in a tetrahedral structure.

(b) The Lewis structure for an NH_4^+ ion is

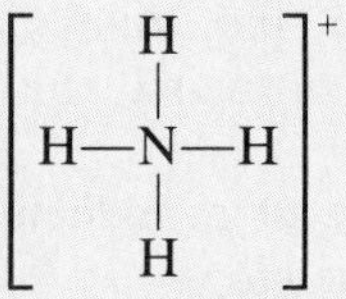

```
[      H     ]+
       |
   H—N—H
       |
[      H     ]
```

The central atom has four bonding pairs and no lone pairs. The structure is therefore tetrahedral.

(c) A Lewis resonance structure for nitrate ion is

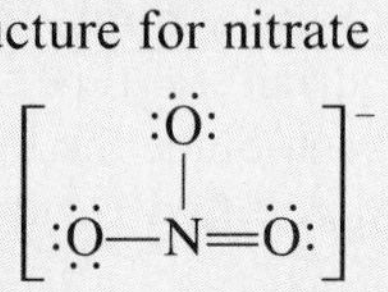

```
         ..
[       :O:       ]-
         |
   ..         ..
[ :O—N=O: ]
   ..
```

Remember that one resonance structure does not represent the real structure of a molecule. The real structure is a hybrid of all the resonance structures. The resonance hybrid of NO_3^- contains three equivalent sets of electrons for the three N—O bonds. From these three regions of bonding electron density around the central nitrogen atom, we predict a trigonal planar structure for an NO_3^- ion.

(d) For the PF_6^- ion, the Lewis structure is

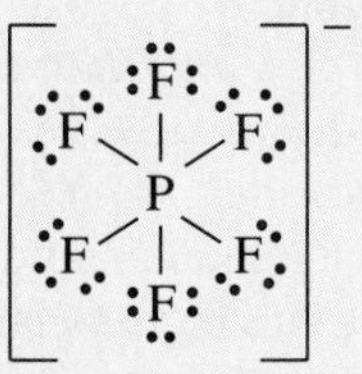

```
[            :F:          ]-
      :F\    |    /F:
               P
      :F/    |    \F:
[            :F:          ]
```

For each of the six bonding pairs to have maximum space between them, the structure is octahedral.

Practice Problem 7.9: Predict the structure and polarity of each of the following molecules: **(a)** beryllium chloride; **(b)** Cl_3PO; **(c)** arsenic pentafluoride; **(d)** ethane, C_2H_6 (predict the structure at each of the C atoms).

7.6 Molecules with Unshared Electron Pairs on the Central Atom

Unshared electron pairs on the central atom of a molecule play an important part in determining its structure. A bonding electron pair is concentrated between two bonded nuclei and therefore occupies a somewhat smaller volume than a lone pair, which is associated with only one nucleus. The repulsion between two lone pairs (lp) is therefore greater than the repulsion between a lone pair and a bonding pair (bp). The order of repulsions between electron pairs is

$$\text{lp–lp} > \text{lp–bp} > \text{bp–bp}$$

Although *exact* bond angles cannot be predicted when lone pairs are present on the central atom, molecules can readily be assigned to various structural types on the basis of their Lewis structures and the VSEPR theory, as we discuss below.

To predict the structure of a molecule with lone electron pairs on the central atom, we follow these steps:

1. Write the Lewis structure of the molecule to see how many bonding and lone pairs are attached to the central atom.
2. From Table 7.5 find the structural type that corresponds to the *total* number of bonding electron sets. Since the lone pairs and bonding pairs are not equivalent, the structure you find is not *exactly* as predicted from Table 7.5.
3. From the central atom, draw both the bonding and lone pairs as dashes pointing approximately toward the corners of the predicted structure.

Following these rules leads us to the *electron pair geometry*, which describes the approximate arrangement of both the bonding electron pairs and lone pairs around the central atom according to the VSEPR model. Remember that experimentally only the positions of the nuclei in a molecule can be determined, not the positions of lone electron pairs. The positions of lone pairs can be inferred from the VSEPR model.

The *arrangement of the atoms and therefore the bonding pairs* around the central atom in a molecule is the *molecular structure* that describes the positions of the atoms in the molecule. In the following discussion we use the VSEPR model to predict the molecular and the electron pair structures of molecules and polyatomic ions having unshared electron pairs on the central atom.

Triatomic Molecules with One Lone Pair on the Central Atom

First, we consider molecules of the type ($:AB_2$). These are triatomic molecules having one lone pair (shown as a dot pair) on the central atom A and two other atoms B attached to A. An example of such a molecule is sulfur dioxide, SO_2. It has two Lewis resonance structures:

$$:\ddot{O}—\ddot{S}=\ddot{O}: \longleftrightarrow :\ddot{O}=\ddot{S}—\ddot{O}:$$

These structures show that the central sulfur atom in an SO_2 molecule is surrounded by two sets of bonding electrons and one lone pair. If all three sets of electrons on the central sulfur atom in an SO_2 molecule were bonding pairs, they would be distributed in a trigonal planar geometry, at 120° angles (Table 7.5). We expect the O—S—O bond angle to be somewhat less than 120° because lone pair-bonding pair repulsion is stronger than bonding pair-bonding pair repulsion. The O—S—O bond angle in SO_2 is actually 119°. Thus, the *molecular geometry* is **bent** or **angular** (Figure 7.17).

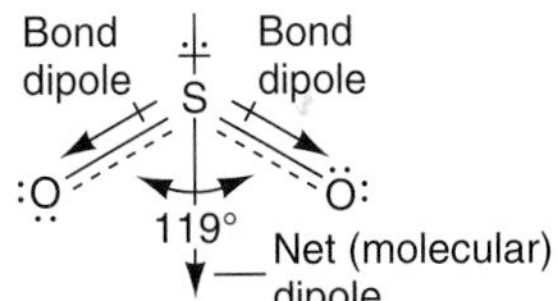

Figure 7.17

SO_2 has an angular geometry. The OSO bond angle is 119°. The molecule has a net dipole moment, and is polar.

An SO_2 molecule has two S—O bond dipole moments directed toward the more electronegative O atoms. Because the molecule is bent, these dipole moments do not cancel, and the molecule has a net dipole moment and is polar.

Practice Problem 7.10: An SO_2 molecule is bent but CO_2 and BeH_2 are linear. Explain why.

Tetratomic Molecules with One Lone Pair on the Central Atom

An example of a tetratomic molecule (a molecule with four atoms) with one lone pair and three bonding pairs on the central atom is ammonia ($:NH_3$). The four electron pairs around the central N atom in an NH_3 molecule have nearly tetrahedral geometry. The HNH bond angle in ammonia is about 107°. This is slightly *less* than the tetrahedral bond angle of 109.5° in a tetrahedral AB_4 structure (Figure 7.13). The bond angle in ammonia is slightly less than the bond angle in a perfectly tetrahedral molecule because of lone pair-bonding pair repulsion, which is stronger than the bonding pair-bonding pair repulsion.

The nearly tetrahedral *electron pair* distribution around the N atom in NH_3 (Figure 7.18) must not be confused with the arrangement of atoms in space in the *molecular geometry* of the molecule. Ammonia has **trigonal pyramidal** molecular geometry. This structure is shown in Figure 7.18c where the N atom and its lone electron pair are at the apex of a trigonal pyramid. The three bonding pairs and the H atoms form the three "legs" of the pyramid.

An ammonia molecule has polar N—H bonds. Each has a dipole moment whose negative end is on the nitrogen atom. When the dipole moments are added, a net dipole results, and ammonia is therefore a polar molecule.

Note the difference between tetrahedral and trigonal pyramidal structures. A tetrahedral structure has *four* equivalent faces and an atom in the center,

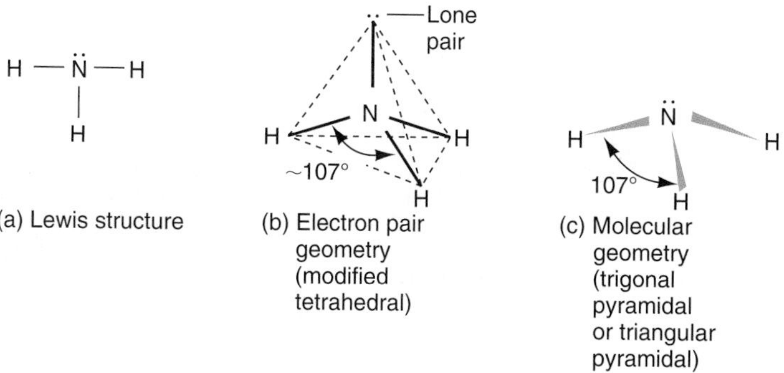

Figure 7.18

The structure of an NH_3 molecule.

whereas a trigonal pyramid has *three* equivalent faces with a base that can be different from the faces. In a trigonal pyramidal structure, the central atom is at the apex, and there is an atom at each of the corners of the base.

Other tetratomic molecules with one lone pair on the central atom include NCl_3, PCl_3, and PH_3. The structures of these molecules are similar to the structure of NH_3.

Triatomic Molecules with Two Lone Pairs on the Central Atom

An example of a molecule with two lone pairs and two bonding pairs on the central atom is water. Again we predict that the four electron pairs will point approximately to the corners of a tetrahedron. As in ammonia, we expect the HOH bond angle to be less than 109.5° because lone pair-bonding pair repulsion is stronger than bonding pair-bonding pair repulsion. The experimental results confirm this prediction. The HOH bond angle in water is about 105°, and the molecular structure for water is bent, or angular, as illustrated in Figure 7.19. The bond angle in H_2O is slightly less than that in NH_3 because an H_2O molecule has two lone pairs, whereas an NH_3 molecule has only one. Two lone pairs repel the bonding pairs more strongly than one lone pair, forcing the bonding pairs closer together, and decreasing the bond angle.

Water is a polar molecule. The two H—O dipole moments in a water molecule are both directed to the highly electronegative oxygen atom. These dipole moments do not cancel because the molecule is bent.

Other triatomic molecules with two lone pairs on the central atom include H_2S, SCl_2, and Cl_2O. The structures of these molecules are similar to the structure of H_2O.

Molecules with Five Electron Pairs on the Central Atom

Molecules with five electron pairs on the central atom can have *zero*, *one*, *two*, or *three lone pairs*. In Section 7.5 we saw that molecules having the general formula AB_5 have trigonal bipyramidal *molecular geometry*. Thus, in molecules having five electron pairs distributed around the central atom, we expect the *electron pair geometry* to be approximately trigonal bipyramidal (Figure 7.20).

An example of a molecule with five electron pairs around the central atom, including one lone pair, is sulfur tetrafluoride, SF_4. In an SF_4 molecule, the four bonding pairs are somewhat displaced from the corners of the bipyramid because of the lp–bp repulsion. The molecular geometry of SF_4 resembles a "seesaw" (Figure 7.20).

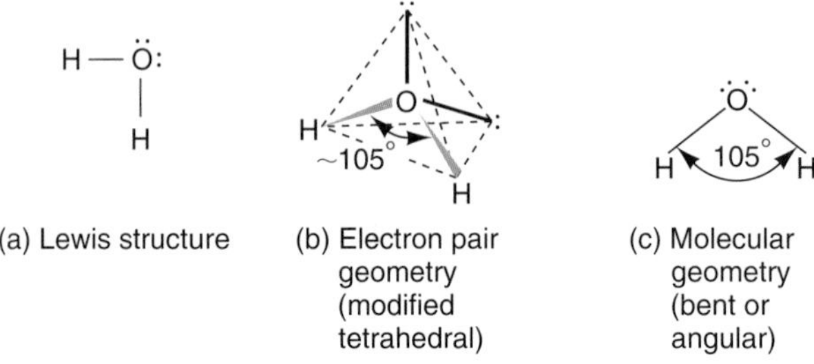

Figure 7.19

The structure of an H_2O molecule.

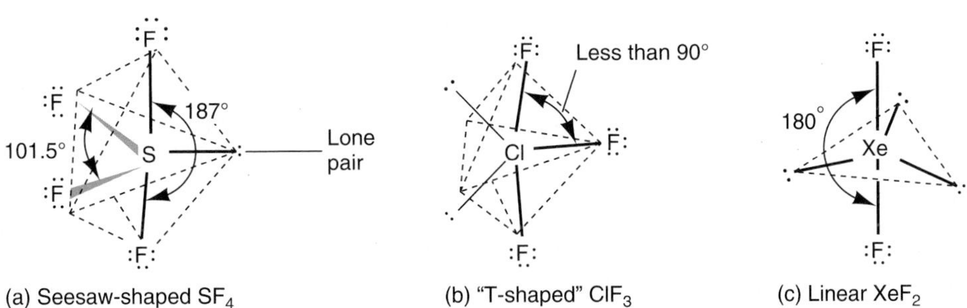

Figure 7.20

Structures of molecules with five electron pairs on the central atom.

An example of a molecule with five electron pairs around the central atom, including two lone pairs, is ClF_3 (Figure 7.20). The electron pair structure of the ClF_3 is a modified trigonal bipyramid. The *molecular structure* of ClF_3 is "T-shaped."

An example of a molecule with five electron pairs around the central atom, including three lone pairs, is xenon difluoride, XeF_2. In the XeF_2 molecule the three lone pairs occupy equatorial positions, and the two bonding pairs are in axial positions. A xenon difluoride molecule is therefore linear (Figure 7.20).

We can see in Figure 7.20 that the lone electron pairs occupy the equatorial rather than the axial positions in all of the molecules shown. Experiments show that SF_4 is indeed seesaw-shaped, ClF_3 is "T-shaped," and XeF_2 is linear. Geometrical analysis of these structures shows that the repulsive forces between the lone pairs in equatorial positions are minimal compared with other arrangements.

Molecules with Six Electron Pairs on the Central Atom

Molecules in which the central atom has six electron pairs fall into three commonly encountered types. One type has six bonding pairs and *no lone pair* on the central atom; a second type has five bonding pairs and *one lone pair*; and a third type has four bonding pairs and *two lone pairs*. In Section 7.5 we discussed sulfur hexafluoride, SF_6, as an example of a molecule with six bonding pairs and no lone pair on the central atom.

An example of a molecule with one lone pair and five bonding pairs is iodine pentafluoride, IF_5. A molecule with two lone pairs and four bonding pairs is represented by xenon tetrafluoride, XeF_4. The structures of IF_5 and XeF_4 are illustrated in Figure 7.21.

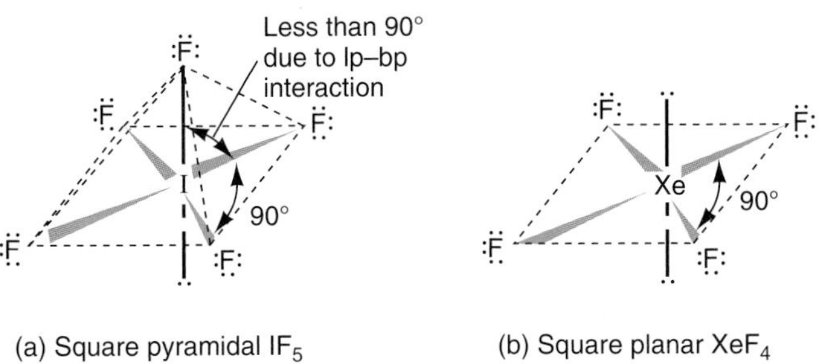

Figure 7.21

Structures of square pyramidal and square planar molecules.

We saw in Section 7.5 that six electron pairs around a central atom are positioned in an octahedral electron pair structure. Thus, the electron pair structures of both IF_5 and XeF_4 are approximately octahedral. The electron pair structures deviate from pure octahedral because all the electron pairs are not equivalent. In an IF_5 molecule the iodine atom is at the center of the base of an imaginary pyramid. One of the fluorine atoms is at the apex of the pyramid and the other four fluorine atoms occupy the corners of the square base of the pyramid. The molecular geometry of IF_5 is therefore **square pyramidal.**

In an XeF_4 molecule, the xenon atom is in the center of a square, and the fluorine atoms occupy the corners of the square. Since all of the atoms lie in a plane, the molecular structure of xenon tetrafluoride is called **square planar**.

The structures of molecules with lone electron pairs on the central atom that have been discussed in this section are summarized in Table 7.6.

In summary, to predict the structure of a molecule that has lone electron pairs on the central atom, first draw the Lewis structure for the molecule. This will show the number of bonding pairs and lone pairs on the central atom. Remember to count multiple bonds as single groups of electrons. Then, derive the most likely *electron pair geometry* of the molecule from Table 7.5. The *molecular geometry* is obtained by considering the arrangement of the bonding electrons and the repulsion between the bonding electrons and the lone pairs.

Table 7.6

Structures of Molecules with Lone Electron Pairs on the Central Atom

Total Number of Electron Groups on the Central Atom	Type of Molecule[a]	Example	Approximate Electron Pair Structure of Molecule	Molecular Structure
3	AB_2E	SO_2	Trigonal planar	Bent
4	AB_3E	NH_3	Tetrahedral	Trigonal pyramidal
	AB_2E_2	H_2O	Tetrahedral	Bent
5	AB_4E	SF_4	Trigonal bipyramidal	Irregular "seesaw"
	AB_3E_2	ClF_3	Trigonal bipyramidal	Irregular "T-shape"
	AB_2E_3	XeF_2	Trigonal bipyramidal	Linear
6	AB_5E	IF_5	Octahedral	Square pyramidal
	AB_4E_2	XeF_4	Octahedral	Square planar

[a] A is the central atom, B is an atom attached to the central atom, and E a lone pair on the central atom.

Example 7.11

Predicting the Structures of Molecules and Polyatomic Ions

Using the VSEPR model, predict the geometry of each of the following species: **(a)** hydrogen sulfide; **(b)** sulfite ion; **(c)** nitrite ion; **(d)** ethene (ethylene), C_2H_4; **(e)** bromine trifluoride. Predict the polarity of each molecule.

SOLUTION

(a) The Lewis structure for hydrogen sulfide is

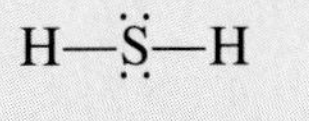

H—S̤̈—H

From the Lewis structure we see that the central sulfur atom has four electron pairs. The electron pair structure of an H_2S molecule is therefore modified tetrahedral. The molecular structure is bent because of lone pair-bonding pair repulsion. Because of the bent structure, the two H—S dipole moments do not cancel, and the molecule is polar.

(b) The Lewis structure for SO_3^{2-} ion is

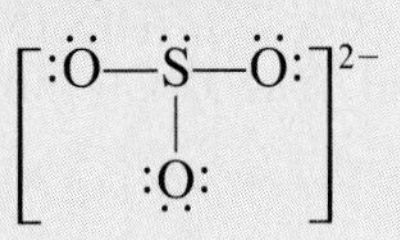

[:Ö—S̈—Ö:
    |
   :O̤:]$^{2-}$

There are four electron pairs at the sulfur atom. This suggests a modified tetrahedral electron pair distribution. A lone pair at the sulfur atom at one of the corners of the imaginary tetrahedron causes the molecular structure to be trigonal pyramidal (Figure 7.22). This structure is analogous to the structure of ammonia (see Figure 7.18).

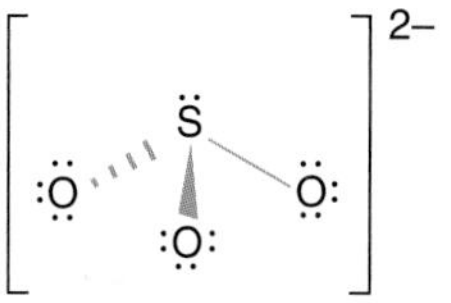

Figure 7.22

Sulfite ion, SO_3^{2-}, has a pyramidal structure analogous to that of ammonia.

(c) For an NO_2^- ion, one Lewis resonance structure is

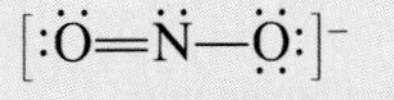

[:Ö=N̈—Ö̤:]$^-$

This is one of the two possible resonance structures. The other structure reverses the single and double bonds. The resonance hybrid has two equal electron density regions that form the two N—O bonds. There is a lone pair at the central atom, which causes the molecular structure to be bent. The electron pair structure is modified trigonal planar.

(d) The Lewis structure for ethene can be written as

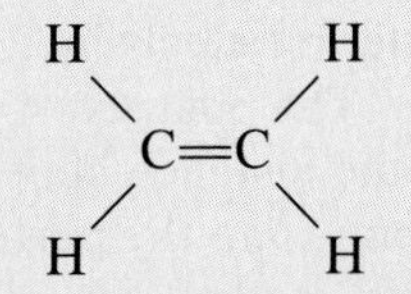

```
H       H
 \     /
  C=C
 /     \
H       H
```

Counting the double bond as a single region of electron density, each carbon atom has three groups of bonding electrons and no lone pairs. Therefore, the structure at each of the C atoms is trigonal planar.

According to the trigonal planar geometry at each carbon atom, the HCH bond angle in C_2H_4 is expected to be about 120°.

The experimentally determined HCH bond angle is 117°. The volume occupied by the double bond forces the C—H bonding pairs closer together, reducing the bond angle from the expected 120° for a perfect trigonal planar structure. Because of the symmetrical electron distribution in the molecule, an ethene molecule is nonpolar.

(e) The Lewis structure for $BrCl_3$ is

```
:C̤̈l—B̈r:
    /   \
 :C̤l:    :C̤l:
```

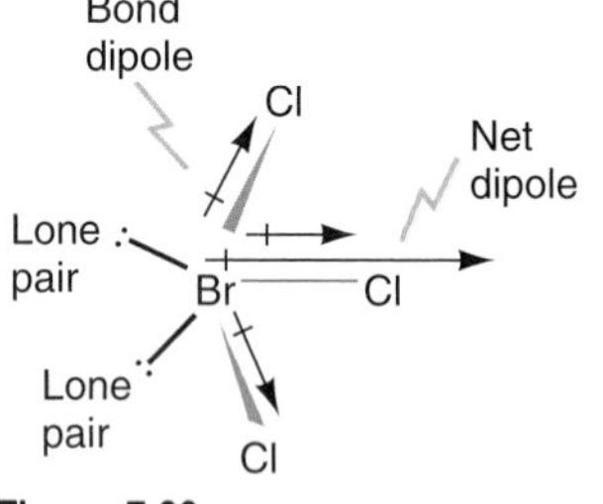

Figure 7.23

$BrCl_3$ has an irregular, "T-shape." It is polar.

From the total of five nonequivalent electron pairs on the Br atom, the electron pair geometry of the molecule can be predicted to be modified trigonal bipyramidal. The two lone pairs are situated in equatorial positions. One of the bonding pairs is in the third equatorial position, and the two remaining bonding pairs are in axial positions. The molecular structure is therefore irregular or "T-shaped," (Figure 7.23). The three Br—Cl bond dipole moments do not cancel. The $BrCl_3$ molecule is therefore polar.

Practice Problem 7.11: Predict the electron pair structure, the molecular structure, and the polarity of each of the following molecules: **(a)** arsenic trifluoride; **(b)** ethyne (acetylene), C_2H_2; **(c)** selenium hexafluoride.

Summary

Ionic and Covalent Bonding

Ionic solids consist of oppositely charged ions that are held together by an electrostatic attraction called **ionic bonding**. Other compounds consist of molecules in which atoms are covalently bonded. In a **covalent bond** between two atoms, each atom contributes an electron to the bonding electron pair, which is shared between the atoms. The nuclei of both atoms are simultaneously attracted to the bonding electron pair. When one of the two bonded atoms contributes both electrons to the bond, the bond is called a **coordinate covalent bond**.

Ionic compounds are relatively high melting solids. All gaseous and liquid compounds at room temperature and pressure, and some low-melting solid compounds, are molecular compounds.

In the formation of both ionic and covalent bonds, atoms tend to acquire a full octet of electrons in their valence shells. That is, they tend to become isoelectronic with their nearest noble gas atoms.

Polar Molecules

Some molecules are aligned between positively and negatively charged electrodes in an electric field. Such molecules are called **polar molecules**. Molecules that are unaffected by an electric field are **nonpolar**. In a polar diatomic molecule, the bonding electron pair is shifted toward the nucleus of the more **electronegative** atom. A bond between two atoms of different electronegativities is called a **polar bond**.

The polarity of a bond is expressed as a product of the distance between the nuclei of the two bonded atoms, and the partial charge on either of the atoms. This product is called **dipole moment**. A diatomic molecule with a polar bond is a polar molecule. In a *homonuclear* diatomic molecule, the bonding electron pair is equidistant between the nuclei, and the bond, like the molecule, is nonpolar. Increasing polarity of a bond increases its **ionic character** and the bond strength.

Lewis Structures

Atoms and monatomic ions can be represented by **Lewis symbols**; molecules and polyatomic ions can be represented by **Lewis structures** or **Lewis dot formulas**. A Lewis symbol for an atom represents its valence electrons as dots. The Lewis structure for a molecule is obtained by combining the Lewis symbols of its atoms. A Lewis structure describes how the atoms in a molecule are linked. According to the **octet rule**, nonmetal atoms other than hydrogen and helium tend to have eight electrons in their valence shells. The octet rule for all the atoms in the molecule is obeyed, if possible.

When writing a Lewis structure of a polyatomic species, the peripheral atoms are positioned symmetrically around the central atom. Lines are then drawn between the atoms to represent covalent bonds. From the total number of valence electrons of the atoms in the species, two are subtracted for each covalent bond drawn. The remaining electrons are then distributed as **lone pairs**. If necessary, some lone pairs are converted to multiple bonds to give each atom a full octet.

For some molecules and polyatomic ions, several acceptable Lewis structures, called **resonance structures**, can be drawn. The real structure for such a species is some combination or hybrid of resonance structures that is called the **resonance hybrid**.

Prediction of Three-Dimensional Molecular Structures

Valence shell electron pair repulsion (VSEPR) **theory** enables us to predict the three-dimensional structures of molecules and polyatomic ions. According to VSEPR theory, electron pairs in the valence shell of the central atom of a molecule are arranged as far as possible from one another to minimize repulsion. If a molecule has only bonding electron groups, the number of these groups and their distributions around the central atom are as follows:

Number of Bonding Groups	Distribution
Two	Linear
Three	Trigonal planar
Four	Tetrahedral
Five	Trigonal bipyramidal
Six	Octahedral

In molecules whose central atom has lone electron pairs, we distinguish between the **electron pair structure** and the **molecular structure**. The electron pair structure is determined by the total number of both bonding and lone electron sets. Molecular structure is the arrangement of atoms in a molecule; it excludes any lone pairs.

A polyatomic molecule is polar if its bond dipole moments do not cancel. In this case the bond dipole moments combine to give a net molecular dipole moment. A polar molecule has an unsymmetric distribution of electron pairs around the central atom.

New Terms

Angular molecule (7.6)
Axial atom (7.5)
Bent molecule (7.6)
Bond angle (7.5)
Bond order (7.4)
Bonding pair (7.4)
Chemical bonding (introduction)
Coordinate covalent bond (7.4)
Covalent bond (7.1)
Dipole (7.2)
Dipole moment (7.2)
Double bond (7.4)
Electron pair structure (7.5)
Electronegativity (7.3)
Equatorial atom (7.5)
Ionic character of bond (7.3)
Isomers (7.4)
Lewis dot symbol (7.4)
Lewis dot formula (7.4)
Lewis structure (7.4)
Lewis symbol (7.4)
Linear structure (7.5)
Lone electron pair (7.4)
Molecular structure (7.5)
Nonbonding electron pair (7.4)
Nonpolar (7.2)
Octahedral structure (7.5)
Octet rule (7.4)
Partial ionic character (7.3)
Polar bond (7.2)
Polar diatomic molecule (7.2)
Resonance hybrid (7.4)
Resonance structure (7.4)
Shared electron pair (7.4)
Square planar structure (7.6)
Square pyramidal structure (7.6)
Tetrahedral structure (7.5)
Trigonal bipyramidal structure (7.5)
Trigonal planar structure (7.5)
Trigonal pyramidal structure (7.6)
Triple bond (7.4)
Unshared electron pair (*see* Lone electron pair)
Valence shell electron pair repulsion (VSEPR) theory (7.5)

Exercises

General Review

1. What physical properties are common to ionic compounds? How do these properties differ from those generally characteristic of molecular compounds?
2. What are the smallest structural units in ionic compounds? What are the smallest structural units in compounds that exist as liquids and gases at room temperature and pressure?
3. What is meant by the term "completion of an octet"?
4. When a metal reacts with a nonmetal to form an ionic compound, how do the atoms of the elements tend to obtain an octet of outer electrons?
5. When two elements react to form covalent bonds, how do the atoms of the elements tend to complete their octets?
6. Which main group atoms can never gain a full octet?
7. What is a polar molecule? Give examples of some polar diatomic molecules.
8. Explain how a polar bond in a diatomic molecule differs from a nonpolar bond in terms of the location of the bonding electron pair relative to the nuclei of the bonded atoms.
9. What is dipole moment?
10. What element is most electronegative? Next-to-the most electronegative? Where in the periodic table are the least electronegative elements located?
11. What are the approximate Pauling electronegativity values for: **(a)** fluorine, **(b)** oxygen; **(c)** the alkali metals; **(d)** hydrogen?
12. Define electronegativity and electron affinity and explain the difference.
13. Discuss the relationship between electronegativity, ionic character of a bond, and bond energy.
14. What is a Lewis symbol? Give examples of some Lewis symbols.
15. What is a Lewis dot formula or Lewis structure? Give examples.
16. What is meant by "resonance structures" and "resonance hybrid"?
17. What is VSEPR theory?
18. How can VSEPR theory be used to determine the structures of molecules with no lone electrons on the central atom?
19. When the central atom of a molecule has no lone electrons, what is the structure of the molecule that corresponds to each of the following numbers of bonding electron pairs on the central atom? **(a)** Two; **(b)** three; **(c)** four; **(d)** five; **(e)** six.
20. What is the difference, if any, between "electron pair structure" and "molecular structure"?
21. Which one of the structure types in question 20 can be determined experimentally?
22. For what type of molecules is the electron pair structure identical to the molecular structure?
23. How can the VSEPR theory be used to predict the *electron pair structure* of a molecule that has one or more lone pairs on the central atom?
24. How can the VSEPR theory be used to predict the *molecular structure* of a molecule that has one or more lone pairs on the central atom?
25. Give an example of a polar and a nonpolar triatomic molecule.
26. Give an example of a polar and a nonpolar tetratomic molecule.
27. Give three examples of nonpolar molecules with polar bonds.

Ionic and Covalent Bonding

28. Which of the following species are isoelectronic with a sulfide ion: Na, Ar, K^+, F^-, Ca^{2+}?
29. Given the following atoms: Na, O, N, B, He. Which two of these atoms would react with each other to form a bond with greatest ionic character? Would you classify this bond as ionic or covalent? Why?
30. Explain the difference between a covalent bond and an ionic bond in terms of the distribution of electron density in the bond.
31. Given the following atoms: Na, Al, O, S, Te, Rb, Zn. Which two react with each other to form a bond with the greatest ionic character? Would you classify this bond as ionic or covalent? Why?
32. Which of each of the following pairs of compounds would you expect to have the higher melting point? **(a)** $CaCl_2$ or SCl_2; **(b)** $FeCl_3$ or NCl_3; **(c)** AlN or AlF_3.

Bond Polarity, Electronegativity, and Ionic Character

33. Which of the following molecules would you expect to be polar? **(a)** H_2; **(b)** N_2; **(c)** NO; **(d)** BrCl; **(e)** I_2; **(f)** HI.
34. Arrange the following elements in the order of increasing electronegativity: Cl, Br, I, P, Al, Na.
35. Which one of the following molecules would you expect to be most polar? HCl, HF, F_2, H_2, BrCl.
36. Which is more polar, a bond that has 20 percent ionic character or another bond that has 50 percent ionic character (assume the internuclear distances to be equal)?

37. The F—F and Br—Br bond energies are 157 and 194 kJ mol^{-1}, respectively. Calculate the Br—F bond energy assuming no ionic character. The experimentally determined Br—F bond energy is 234 kJ mol^{-1}. How much "extra" energy does the Br—F bond possess due to its ionic character?

38. Which one of the following molecules would you expect to have the highest percent of ionic character: HCl, HI, I_2, O_2, IBr?

39. Which two elements are likely to form the most ionic bond in their reaction with each other? Explain.

40. Give examples of atoms that are likely to form the most covalent bond when they combine with each other.

Lewis Structures

41. Write a Lewis structure for each of the following molecules or ions: **(a)** CCl_4; **(b)** NCl_3; **(c)** BrCl; **(d)** SCl_2; **(e)** H_2NNH_2; **(f)** BCl_4^-; **(g)** HCO_2^-; **(h)** H_2CO; **(i)** CH_3OH; **(j)** CH_3NH_2; **(k)** NO_2; **(l)** NO_2^-; **(m)** H_2O_2; **(n)** SF_6; **(o)** Cl_2SO; **(p)** ONONO (N_2O_3); **(q)** PCl_4^+; **(r)** PCl_6^-; **(s)** NO_2^+; **(t)** H_2NOH; **(u)** CH_3CO_2H; **(v)** IF_3; **(w)** I_3^-.

42. Write all the resonance structures for each of the following molecules or ions: **(a)** SO_3; **(b)** SO_3^{2-}; **(c)** NNO (N_2O); **(d)** $HCOO^-$.

43. What is the S—O bond order in SO_2 and the C—O bond order in HCO_2^-?

44. Write Lewis structures for all the isomers of each of the following molecules: **(a)** C_2H_6O; **(b)** C_4H_8; **(c)** C_4H_6.

Structures of Molecules

45. Write Lewis structures for each of the following molecules and ions, and predict the electron pair structure and the molecular structure of each: **(a)** boron trifluoride; **(b)** silicon tetrachloride; **(c)** ammonium ion; **(d)** phosphonium ion, PH_4^+; **(e)** propane, C_3H_8 (predict the structure at each of the carbon atoms); **(f)** acetylene, C_2H_2; **(g)** carbon dioxide; **(h)** formaldehyde, H_2CO; **(i)** PF_6^- ion; **(j)** sulfur hexafluoride; **(k)** iodine pentafluoride; **(l)** carbonate ion; **(m)** nitrate ion; **(n)** nitrite ion; **(o)** nitrogen trichloride; **(p)** oxygen difluoride.

46. What are the approximate bond angles at the central atom of each of the species in question 45?

47. Write Lewis structures for each of the following molecules and ions, and predict both the electron pair structure and the molecular structure for each: **(a)** sulfur dioxide; **(b)** dichlorine monoxide; **(c)** hydrogen sulfide; **(d)** phosphine, PH_3; **(e)** hydronium ion, H_3O^+; **(f)** diethyl ether, $H_3CCH_2OCH_2CH_3$ (predict the geometry at each of the carbon atoms and at the oxygen atom); **(g)** methyl alcohol, CH_3OH (geometries at both the C and O atoms); **(h)** methyl amine, CH_3NH_2 (geometries at both the C and N atoms); **(i)** sulfite ion; **(j)** sulfur tetrafluoride; **(k)** sulfur dichloride; **(l)** triiodide ion, I_3^-; **(m)** iodine trifluoride; **(n)** xenon difluoride; **(o)** xenon tetrafluoride.

Polarity of Molecules

48. Which of the molecular species listed in questions 45 and 47 are polar and which are nonpolar?

CHAPTER 8

A spider web is a natural wonder of bonding.

Theories of Covalent Bonding

In Chapter 7 we discussed the general nature of covalent bonds and used Lewis structures to predict molecular geometries. In this chapter we focus on two theories of covalent bonding: valence bond theory and molecular orbital theory.

In our discussions of covalent bonding in Chapter 7 we introduced the rudiments of **valence bond** *(VB)* **theory**. *We recall that a covalent bond is formed when two atoms are attracted simultaneously to the same pair of electrons, and that this simultaneous attraction corresponds to the overlap of two atomic orbitals. These overlapping orbitals share the same region in space and provide a high electron density between the nuclei of the bonded atoms. We represent the shared electron pair between two bonding atoms as a "dash" when we write Lewis structures for molecules.*

In valence bond theory the "dash" by which we represent a bonding electron pair in a Lewis structure indicates that the electrons in the bond are localized *between the two bonded atoms. In contrast, the second major theory of covalent bonding—***molecular orbital** *(MO)* **theory***—considers a molecule as a set of nuclei in which electrons belong to* molecular orbitals.

8.1 Valence Bond Theory

We will begin by using valence bond theory to describe the bonding of second-period elements in their binary compounds with hydrogen and halogens. The elements lithium and beryllium have only 2*s* electrons in their valence shells. The elements boron through neon have 2*s* and 2*p* electrons. Atoms of *p*-block elements of the third and later periods have *s* and *p* electrons, as well as empty *d* orbitals available for bonding. Elements of later periods can therefore form some types of compounds not found with the second-period elements.

Below we consider binary hydrogen compounds of second-period elements and binary halogen compounds of second-period elements. We recall that in a covalent bond the difference between the electronegativities, Δ(EN), of the bonded atoms is small. The electronegativity of hydrogen is 2.1, which is in the intermediate range of electronegativities (0.8 to 4.0). The Δ(EN) between hydrogen and most other elements is therefore small, and the atoms in most binary hydrogen compounds are covalently bonded. The exceptions to this generalization are alkali metal hydrides such as NaH, and hydrides of alkaline earth metals other than beryllium and magnesium.

Binary halogen compounds of second-period elements will be included in our discussion because a halogen atom, like a hydrogen atom, has only one unpaired electron available for bonding. Most alkali and alkaline earth metal halides are ionic compounds. Beryllium is the only metal in the second period that forms predominantly covalent bonds with hydrogen and halogens.

Orbital Overlap Theory

Let us review the bonding in a hydrogen molecule. We recall from Section 7.1 that an H_2 molecule forms when two hydrogen atoms approach each other until their 1*s* atomic orbitals overlap. In this configuration the nuclei of the two H atoms are simultaneously attracted to the electron pair, which is concentrated

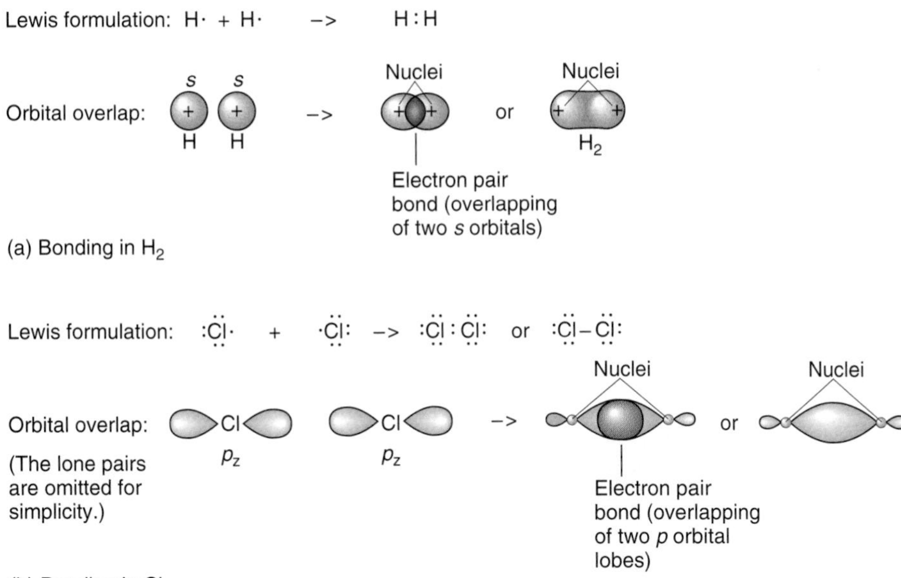

Figure 8.1

Bonding in (a) H_2 and (b) Cl_2.

between the nuclei (Figure 8.1a). Note that the electron density extends somewhat beyond the nuclei. After the bond has formed, the spherical electron clouds of the original 1*s* orbitals have given way to an oval-shaped cloud.

A covalent bond between two chlorine atoms forms when the singly occupied 2*p* orbitals of the two chlorine atoms overlap (Figure 8.1b). The overlapping *p*-orbital lobes form an oval-shaped electron cloud between the nuclei, and the nonoverlapping *p*-orbital lobes on the opposite sides of the internuclear axis decrease in size, indicating regions in which the probability of finding an electron is relatively low.

Hybrid Orbitals; Covalent Bonding in Beryllium Compounds

We now extend the foregoing discussion to molecules of the second-period elements, starting with beryllium. The electronegativity of beryllium is 1.5. Thus, beryllium forms covalent bonds with hydrogen (EN 2.1) and with chlorine (EN 3.0). For example, the Δ(EN) value between Be and H is $2.1 - 1.5 = 0.6$, whereas Δ(EN) between Ba (barium) and H is $2.1 - 1.0 = 1.1$. Thus, the Be—H bond is far more covalent than the Ba—H bond.

Beryllium hydride, $BeH_2(s)$, can be prepared by treating beryllium chloride with lithium hydride:

$$BeCl_2(s) + 2LiH(s) \longrightarrow BeH_2(s) + 2LiCl(s)$$

Beryllium reacts with chlorine at high temperature to form beryllium chloride:

$$Be(s) + Cl_2(g) \longrightarrow BeCl_2(s)$$

Both BeH_2 and $BeCl_2$ in their solid states exist as continuous networks of covalently bonded Be and H, or Be and Cl atoms, respectively. When either BeH_2 or $BeCl_2$ is heated in vacuum, it vaporizes. In the vapor phase, both BeH_2 and $BeCl_2$ exist as linear, triatomic molecules.

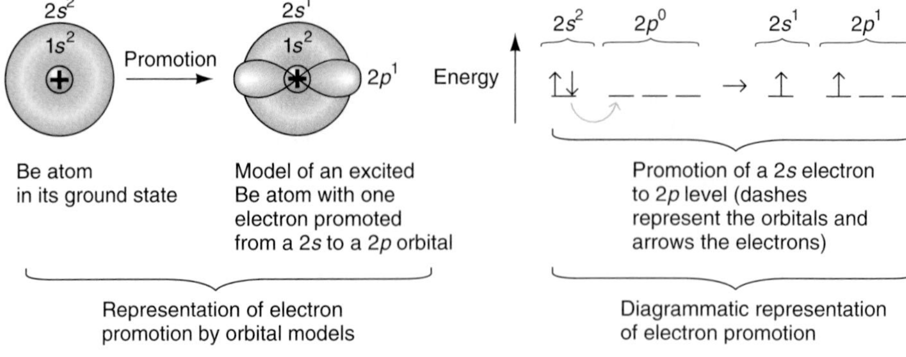

Figure 8.2

Promotion of a 2*s* electron to the 2*p* level in a Be atom.

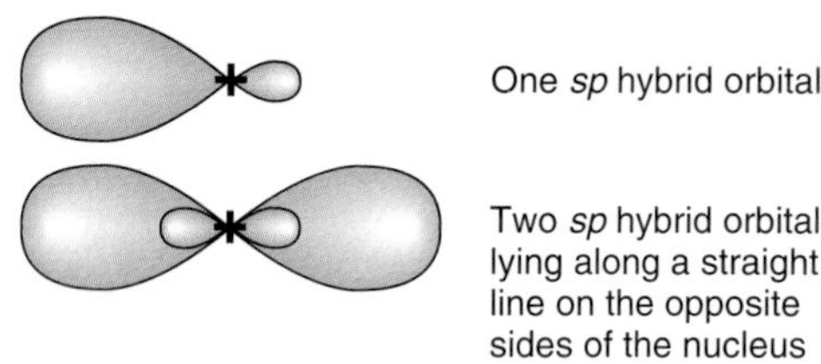

Figure 8.3

Models of *sp* hybrid orbitals.

Let us consider the bonding in gaseous BeH_2 molecules. A beryllium atom has two valence electrons. The electronic structure of the valence shell of the beryllium atom is $2s^2$. When an atom forms two covalent bonds, it must provide one unpaired electron for each. But the two electrons of a beryllium atom are paired. To form two bonds, they must be unpaired. To do this, energy must be supplied to "promote" one electron from the 2*s* orbital to a 2*p* orbital (Figure 8.2). This "promotion" results in two unpaired valence electrons.

At this point we might be tempted to make one Be—H bond using the half-occupied 2*s* orbital and the other bond using the half-occupied 2*p* orbital. That would create two Be—H bonds having different bond energies and bond lengths. But the Be—H bonds are known to be identical. Therefore, beryllium must have two electrons in identical orbitals. According to valence bond theory, these identical orbitals arise from mixing or **hybridizing** the 2*s* and 2*p* orbitals. *Mixing an s and a p orbital* generates two identical orbitals called ***sp* hybrid orbitals**. Each contains one unpaired electron. The *sp* hybrid orbitals of a beryllium atom have energies midway between the energies of the 2*s* and 2*p* atomic orbitals.

Models of *sp* hybrid orbitals are illustrated in Figure 8.3. Note that the two hybrid *sp* orbitals and the Be nucleus lie on a straight line. Thus, triatomic molecules made from *sp* hybrid orbitals are linear. The H—Be—H bond angle is therefore 180°.

The overlap of the *s* orbital of a hydrogen atom with an *sp* hybrid orbital of the beryllium atom constitutes a Be—H bond (Figure 8.4). A *covalent bond in which orbitals overlap along the axis between the bonded nuclei* is called a **sigma (σ) bond**. The two Be—H σ bonds in a BeH_2 molecule correspond to the Lewis electron pair bond representation of H:Be:H, or H—Be—H.

According to valence bond theory, bonding in $BeCl_2$ is similar to bonding in BeH_2. Each of the *sp* hybrid orbitals of the beryllium atom overlaps with a 3*p* orbital of a chlorine atom to form two Be—Cl bonds (Figure 8.5).

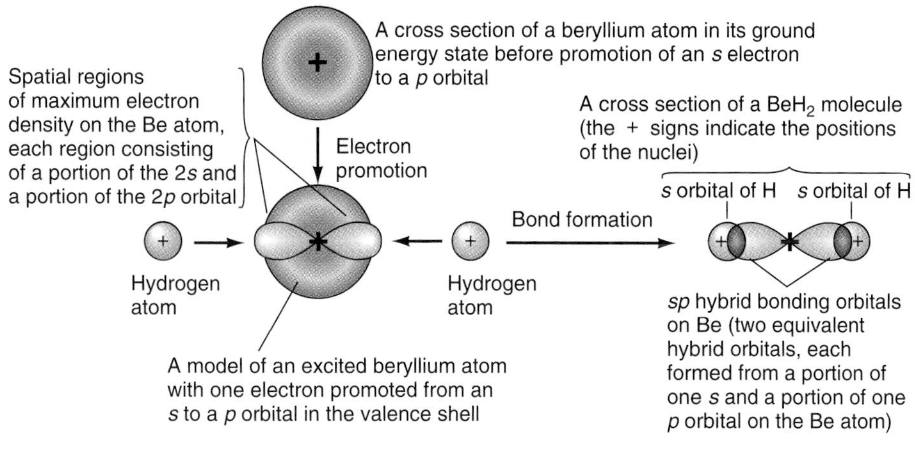

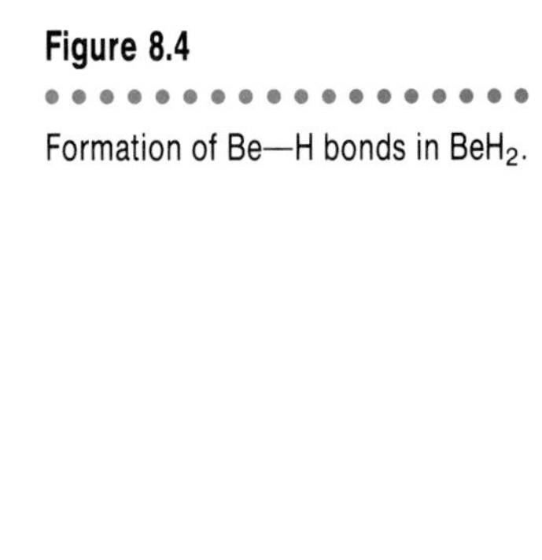

Figure 8.4

Formation of Be—H bonds in BeH_2.

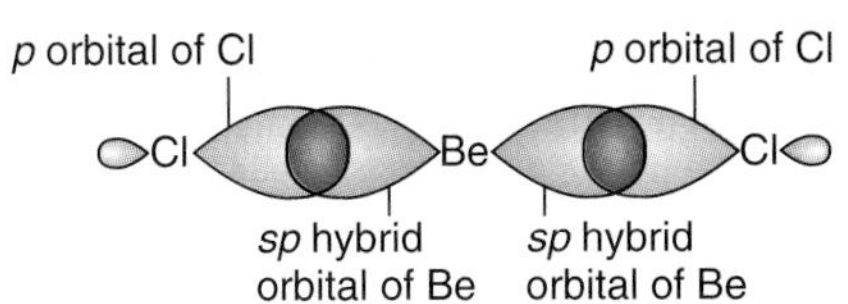

Figure 8.5

Bonding in a molecule of $BeCl_2$.

Bonding in Boron Trichloride

The binary compound of boron and hydrogen with the formula "BH_3" does not exist. However, boron halides exist, and we will consider boron trichloride, BCl_3 (a colorless liquid that boils at 12.5°C). Boron reacts with chlorine at an elevated temperature to yield boron trichloride:

$$2B(s) + 3Cl_2(g) \longrightarrow 2BCl_3(l)$$

A boron atom in its ground state has the following electronic configuration:

$$\underset{2s^2}{\underline{\uparrow\downarrow}} \quad \underbrace{\underline{\uparrow} \quad \underline{\quad} \quad \underline{\quad}}_{2p^1}$$

Thus, the ground state boron atom has one unpaired electron, but three are needed for the three B—Cl bonds in a BCl_3 molecule. To obtain the three unpaired electrons, a 2*s* electron is promoted to the 2*p* level, as shown below.

$$\underset{2s^2}{\underline{\uparrow\downarrow}} \quad \underbrace{\underline{\uparrow} \quad \underline{\quad} \quad \underline{\quad}}_{2p^1} \xrightarrow{\text{"Promotion"}} \underset{2s^1}{\underline{\uparrow}} \quad \underbrace{\underline{\uparrow} \quad \underline{\uparrow} \quad \underline{\quad}}_{2p^2}$$

Ground state B atom — Excited B atom

BCl_3 is a trigonal planar molecule with three identical B—Cl bonds. To generate this geometry, the half-occupied 2*s* orbital mixes with the two half-occupied 2*p* orbitals to form three identical ***sp*² hybrid orbitals** (Figure 8.6). We read sp^2 as "*s*-*p*-two," <u>*not*</u> "*s*-*p*-squared." The superscript 2 means that two *p* orbitals are mixed with one *s* orbital (the superscript 1 is omitted). The sp^2 hybrid orbitals are coplanar and are separated by 120°, as is characteristic of a trigonal planar molecule. Each sp^2 hybrid orbital overlaps with a *p* orbital of a chlorine atom to form a B—Cl bond (Figure 8.6).

Figure 8.6

A diagram of the formation of B—Cl bonds in a BCl_3 molecule.

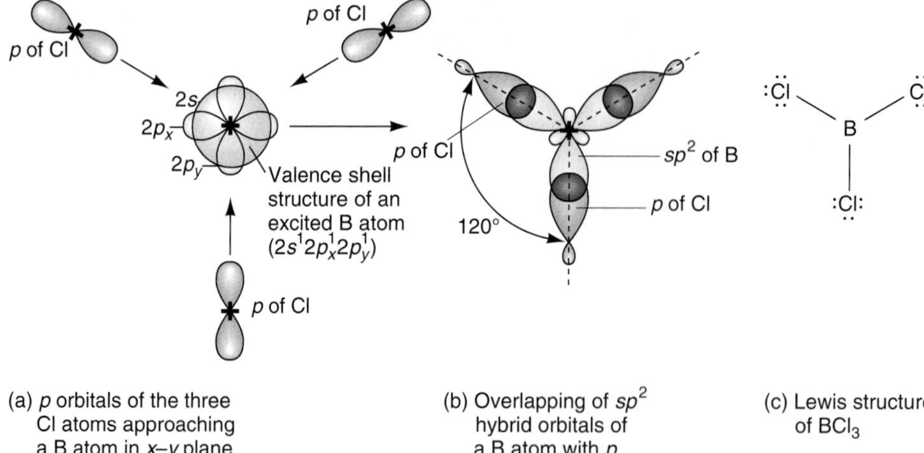

(a) p orbitals of the three Cl atoms approaching a B atom in x–y plane

(b) Overlapping of sp^2 hybrid orbitals of a B atom with p orbitals of Cl atoms in a BCl_3 molecule

(c) Lewis structure of BCl_3

Example 8.1

Predicting the Shape of a Molecule and the Hybridization of Its Central Atom

Aluminum chloride is a solid at room temperature, but when heated to temperatures above 800°C, it converts to a gas consisting of $AlCl_3$ molecules. **(a)** What is the structure of an $AlCl_3$ molecule? **(b)** What is the hybridization of the aluminum atom in this molecule?

SOLUTION: Aluminum is in the same group as boron.

(a) Since a BCl_3 molecule is trigonal planar, we also expect $AlCl_3$ to be trigonal planar.

(b) Just as the boron atom in BCl_3 is sp^2 hybridized, the aluminum atom in $AlCl_3$ is sp^2 hybridized.

Practice Problem 8.1: What orbital of an aluminum atom and what orbital of a bromine atom do you expect to overlap to form an Al—Br bond in $AlBr_3$?

From the preceding discussion we can see that valence bond theory is actually an extension of Lewis theory. Valence bond theory explains covalent bonds in terms of overlapping atomic orbitals.

8.2 Bonding in Simple Carbon Compounds

In this section we consider bonding in some simple binary compounds of carbon with hydrogen and with chlorine.

Methane and Carbon Tetrachloride

Methane, CH_4, is the major constituent (up to 97 percent) of **natural gas**. It is also produced in marshy areas by decaying organic material, hence its common name, **marsh gas**. Let us consider the bonding in a methane molecule. The valence shell electron configuration of a carbon atom in its ground state is $2s^22p^2$:

$$\underset{2s^2}{\underline{\uparrow\downarrow}} \quad \underbrace{\underline{\uparrow} \quad \underline{\uparrow} \quad \underline{}}_{2p^2}$$

A carbon atom has two unpaired electrons, but four unpaired electrons are required to make the four C—H bonds in methane. To obtain four unpaired electrons, the electrons in the 2*s* orbital of the ground state carbon atom must be unpaired. This can be accomplished by "promoting" one of the 2*s* electrons to a 2*p* orbital (Figure 8.7).

It is known experimentally that a molecule of methane is tetrahedral, and that its four C—H bonds are identical. Using one 2*s* and three 2*p* electrons for bonding by the carbon atom would not produce a tetrahedral molecule because a 2*s* orbital is spherical and the three 2*p* orbitals of an atom lie at right angles to one another. To generate the required tetrahedral geometry, the 2*s* orbital mixes with the three 2*p* orbitals to produce four identical ***sp*³ hybrid orbitals**. Each half-filled sp^3 hybrid orbital of the carbon atom overlaps with the half-filled 1*s* orbital of a hydrogen atom to form a C—H σ bond. A diagram of the promotion and hybridization process, and the formation of a CH_4 molecule, is presented in Figure 8.7.

Next, let us consider the bonding in carbon tetrachloride, CCl_4. Like methane, we expect carbon tetrachloride to be tetrahedral. As in methane, the carbon atom in a CCl_4 molecule is sp^3 hybridized. A C—Cl bond is formed by the overlap of an sp^3 hybrid orbital of the carbon atom with the *p* orbital which contains the unpaired electron of a chlorine atom.

Figure 8.7

Schematic diagram of a reaction of carbon with hydrogen to form CH_4.

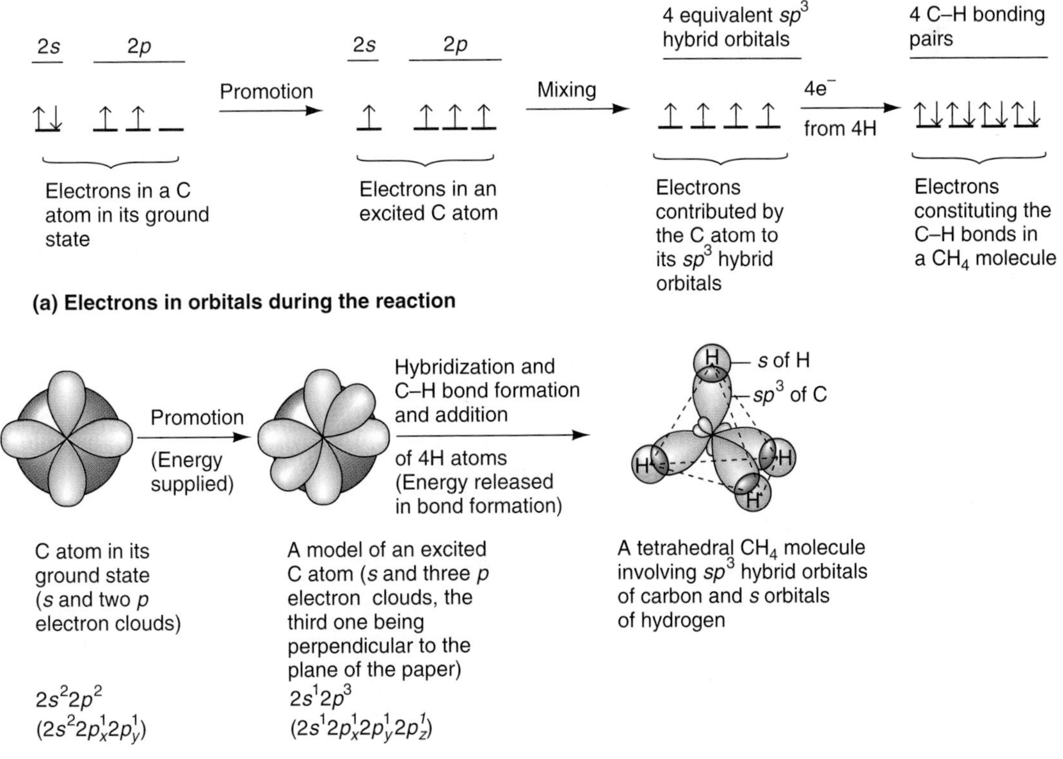

Practice Problem 8.2: Silicon tetrachloride is a liquid that fumes in moist air, producing a dense smoke. What is the structure of a silicon tetrachloride molecule? What is the hybridization of silicon in silicon tetrachloride? What orbitals of silicon and chlorine overlap to form a silicon–chlorine σ bond in silicon tetrachloride?

Alkanes

Carbon atoms are bonded to one another and to hydrogen atoms in binary compounds called **hydrocarbons**. If all the carbon atoms in a hydrocarbon are linked by single bonds, the molecule is called an **alkane**. Alkanes contain the maximum number of hydrogen atoms for a given number of carbon atoms and are therefore known as **saturated hydrocarbons**. Alkanes have the general formula C_nH_{2n+2}, where n is an integer. Among the alkanes are fuels such as methane, CH_4, ethane, C_2H_6, propane, C_3H_8, butane C_4H_{10}, and octane, C_8H_{18}.

Let us consider the bonding in ethane, C_2H_6. We know that a hydrogen atom can form only one bond. Therefore, the carbon atoms of ethane *must* be linked to each other. Each carbon atom has four valence electrons with which to form one C—C bond and three C—H bonds. The molecular formula of ethane is shown below.

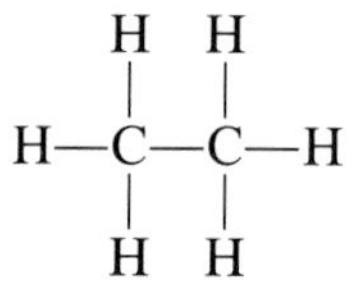

```
    H   H
    |   |
H — C — C — H
    |   |
    H   H
```

Ethane

When we examine the molecular formula, we see that each carbon atom is bonded to four other atoms. From our previous example of bonding in methane, we expect each carbon atom to be tetrahedral and sp^3 hybridized. The structure of an ethane molecule thus corresponds to two tetrahedra that share a corner. The two carbon atoms of ethane are bonded by overlapping sp^3 hybrid orbitals. The remaining three sp^3 hybrid orbitals of each carbon atom overlap with the s orbitals of hydrogen atoms to form C—H bonds, as shown in Figure 8.8.

Bonding in propane, C_3H_8, is just like bonding in ethane: each carbon atom is tetrahedral and sp^3 hybridized (Figure 8.8). As in ethane, each of the carbon atoms in a propane molecule has four bonding pairs arranged in a tetrahedral structure.

Practice Problem 8.3: Silicon forms binary silicon and hydrogen compounds called silanes with the general formula Si_nH_{2n+2}. The largest known silane molecule is hexasilane, Si_6H_{14}. What is the most probable bonding orbital structure and hybridization on each of the silicon atoms in a molecule of Si_2H_6, called disilane? What orbitals of the silicon and the hydrogen atoms overlap to form each of the Si—H bonds in this molecule?

Alkenes: Sigma and Pi Bonds

When an alkane is heated at high temperature in the absence of air and in the presence of a finely divided metal such as palladium (a process known as

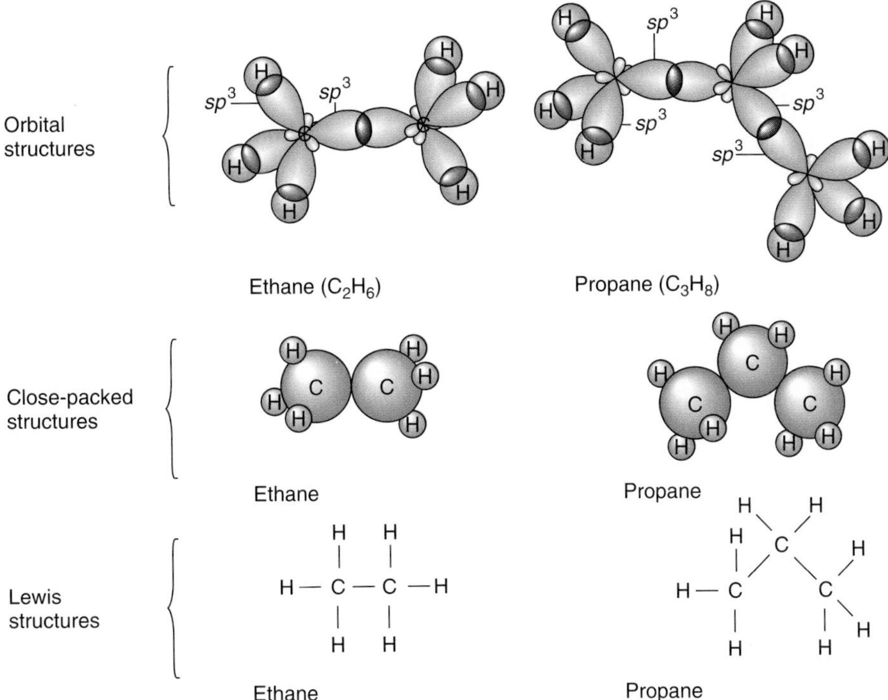

Figure 8.8

Structures of ethane and propane.

pyrolysis or cracking), it loses hydrogen. For example, heating ethane with a palladium catalyst in the absence of air produces **ethene** and hydrogen:

$$\underset{\text{Ethane}}{C_2H_6} + \text{heat} \xrightarrow{\text{Pd}} \underset{\text{Ethene}}{C_2H_4} + H_2$$

Each carbon atom in an ethene molecule is surrounded by three bonding electron groups and therefore has a trigonal planar structure:

```
H         H
 \       /
  C ═ C          sometimes written     CH2═CH2
 /       \
H         H
  Ethene
```

Ethene is the simplest member of a series of hydrocarbons, called **alkenes**. All alkenes contain double-bonded carbon atoms as in $CH_3CH{=}CH_2$ and $CH_3CH{=}CHCH_3$. Although an ethene molecule has two carbon atoms, it has fewer hydrogen atoms than ethane, and is an example of an **unsaturated hydrocarbon**. Alkenes have the general formula C_nH_{2n}.

Let us consider the bonding in ethene. Each carbon atom in an ethene molecule is attached to three other atoms. By analogy with BCl_3, we expect each carbon atom in C_2H_4 to have a trigonal planar structure and to be sp^2 hybridized. To form three σ bonds, we need three unpaired electrons. To obtain the three unpaired electrons, a 2*s* electron is promoted to the 2*p* level as shown in the diagram below.

Ground state C atom					Excited C atom			
$\uparrow\downarrow$	$\uparrow$	$\uparrow$		$\longrightarrow$	$\uparrow$	$\uparrow$	$\uparrow$	$\uparrow$
$2s$	$2p_x$	$2p_y$	$2p_z$		$2s$	$2p_x$	$2p_y$	$2p_z$

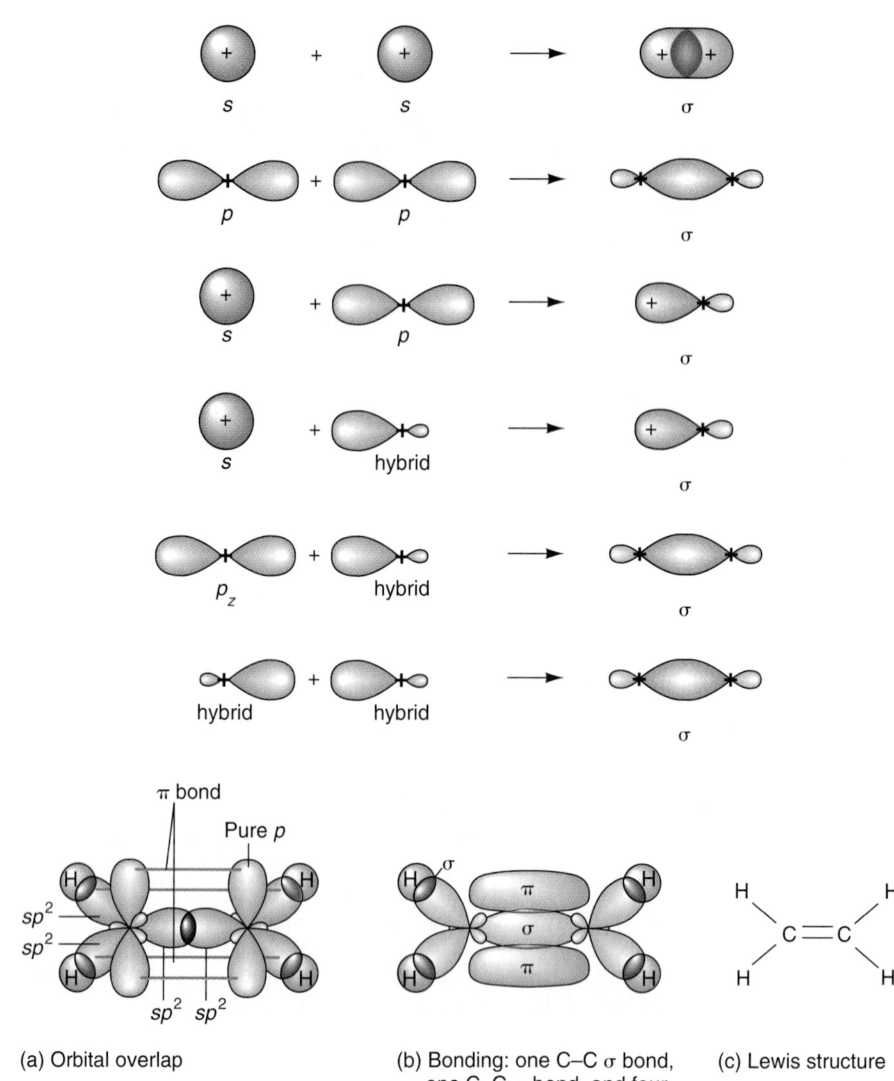

Figure 8.9

Orbital overlaps to form sigma bonds.

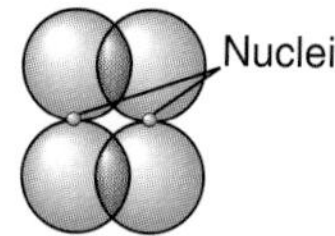

Figure 8.10

Sideways overlapping of *p* orbitals in a π bond.

Figure 8.11

Structure and bonding in ethene.

Mixing the 2*s* orbital with *two* 2*p* orbitals of the carbon atom in its excited state gives three sp^2 hybrid orbitals. Each of the two sp^2 hybrid orbitals overlaps with a 1*s* orbital of a hydrogen atom to form a carbon–hydrogen σ bond. The third sp^2 orbital overlaps with an sp^2 orbital of the other carbon atom to form a carbon–carbon σ bond. Each of these overlapping orbitals of a carbon atom contains one electron for the bond. The various orbital overlaps to form σ bonds in ethene and other molecules are illustrated in Figure 8.9.

If the two *p* orbitals used by a carbon atom for the σ bonds in an ethene molecule are the $2p_x$ and $2p_z$ orbitals, the $2p_y$ orbital remains unhybridized as a *pure p* orbital. This $2p_y$ orbital contains one unpaired electron. It overlaps side by side with a $2p_y$ orbital of the other carbon atom, which also contains one unpaired electron. The *side-by-side overlap of p orbitals* of adjacent atoms forms a covalent bond called a **pi (π) bond** (Figure 8.10). Orbital overlap in a π bond occurs on opposite sides of the internuclear axis, not directly between the nuclei (Figure 8.10).

The bonding and structure of ethene are summarized in Figures 8.11 and 8.12.

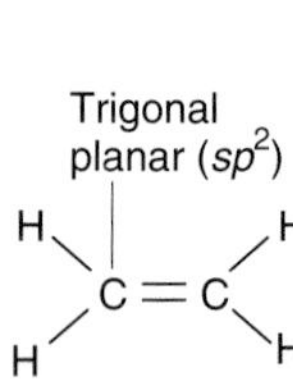

Figure 8.12

Trigonal planar structure of each of the two carbon atoms in an ethene molecule.

Example 8.2

Determining the Structure and Hybridization of Carbon Atoms in an Alkene Molecule

The formula for 1-butene, an alkene, is $CH_3CH_2CH{=}CH_2$. Determine the structure and hybridization of each of the four carbon atoms in 1-butene.

SOLUTION: The Lewis structure for 1-butene is

```
                H           H
                 \ 2     1 /
         H   H    C=====C
         |   |  /          \
   H —4C—3C                H
         |   |
         H   H
```

In this structure we see that C-4 and C-3 are each surrounded by four σ bonding pairs. C-4 and C-3 are therefore tetrahedral, and sp^3 hybridized.

C-2 and C-1 are linked by a double bond that consists of one σ and one π bond. One p orbital of each of these two carbon atoms is used to form the π bond. Therefore, C-2 and C-1 each have one s electron and two p electrons with which to form three sp^2 hybrid orbitals for three σ bonds. The three σ bonding pairs are arranged in a trigonal planar structure. C-2 uses two of its sp^2 hybrid orbitals for the two C—C σ bonds and one for the C—H σ bond. C-1 uses one of its sp^2 hybrid orbitals for the C—C σ bond and the two remaining sp^2 hybrid orbitals for C—H σ bonds.

Practice Problem 8.4: Predict the structure and hybridization of each of the carbon atoms in $CH_3CH{=}CHCH_3$ (2-butene).

Alkynes

Removing two hydrogen atoms from a molecule of ethene results in a molecule of **ethyne**, also called **acetylene**, C_2H_2. Ethyne is the first member of a class of unsaturated hydrocarbons called **alkynes**. Other alkynes include $CH_3C{\equiv}CH$ (propyne), $CH_3CH_2C{\equiv}CH$ (1-butyne), and many other compounds that contain a triple bond. Alkynes have the general formula C_nH_{2n-2}.

The Lewis structure for ethyne has a triple bond between the carbon atoms: H—C≡C—H. The valence bond model of ethyne is similar to that of beryllium hydride. We recall from Section 8.1 that BeH_2 is a linear molecule in which the beryllium atom is sp hybridized. Each carbon atom in C_2H_2 is also sp hybridized.

To obtain an sp-hybridized carbon atom, one of the $2s$ electrons must be promoted to the $2p$ level. Then the $2s$ orbital that contains one unpaired elec-

APPLICATIONS OF CHEMISTRY 8.1
Alkenes Help Plants Outfox Animals and Fruit Growers Outfox Nature

Imagine an isolated strawberry patch. In this patch, each strawberry plant ripens randomly, independently of the others and in random order. Think of the results. It would be possible for a single animal to eat all the fruit that ripens on the first day. Then that animal could come back on the next day to eat the fruit that ripened on that day. Continuing in this way, a single animal could eat all the berries. No ripened strawberries would remain to sow seeds for the next generation of strawberry plants.

In real life, this does not happen. Nature has built in a protection system that makes all the fruit ripen at the same time. When fruit on the first strawberry plant ripens, it releases ethene into the air, which causes nearby fruit to ripen. Thus, the first plant that ripens unleashes a chain reaction that causes all the fruit to ripen at about the same time. Hence, it is more difficult for a single animal to wipe out the entire crop of strawberries. An animal can eat only so many strawberries before walking away full. So some of the strawberries will reach the ground to begin seeding the next generation.

Fruit growers exploit this natural process to assure the customer of fresh produce. Fruit growers pick their produce unripened because ripened fruit does not ship well. Unripened fruit is firmer than ripened fruit, and better survives the physical hardships of shipping. Once the unripened fruit reaches grocery stores, grocery clerks spray ethene on the shipment to ripen it.

Fruit growers discovered the secret of ripening fruit quite by accident. In the 1920s, fruit growers tried to protect their crops from frost by placing lighted smudge pots in their fields to generate heat. Smudge pots (see photo) use kerosene, which is primarily a mixture of alkanes. Alkanes undergo a reaction at high temperatures and in the absence of a good supply of oxygen in which they lose hydrogen and produce alkenes. This incomplete combustion of kerosene in the smudge pots produced ethene and other alkenes. Whenever smudge pots were placed in the fields, the growers noticed that all the plants flowered at about the same time, which made it easier to time the harvesting of the fruit, which would later ripen simultaneously. This led to the discovery that ethene stimulates the ripening of fruit.

Source: *Chem Matters*, April 1989, pages 11–13.

tron mixes with a $2p$ orbital that contains another unpaired electron to give two sp hybrid orbitals, as shown by the following scheme:

↑	↑	↑	↑		↑	↑	↑	↑
$2s$	$2p_z$	$2p_x$	$2p_y$	$\xrightarrow{\text{Mixing one } s \text{ orbital with one } p \text{ orbital yields two } sp \text{ hybrid orbitals}}$			$2p_x$	$2p_y$
					Two sp_z orbitals for the σ bonds		Two p orbitals for the π bonds	
Electrons in the valence shell of an excited C atom					Electrons in the orbitals of an sp-hybridized C ready for bonding			

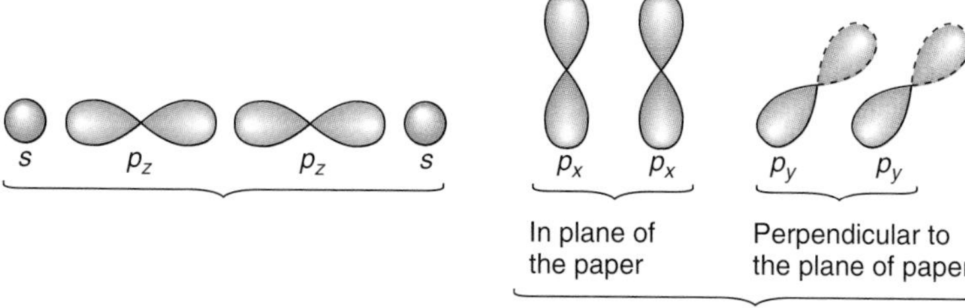

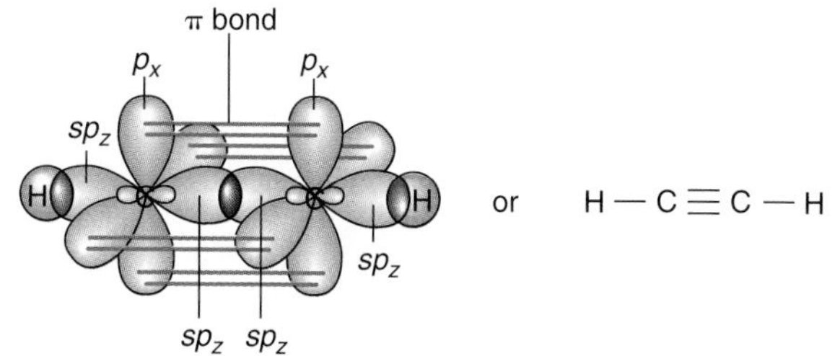

Figure 8.13

Bonding in ethyne.

One of the *sp* hybrids on a carbon atom is used to make a σ bond to the other carbon atom, the second *sp* hybrid is used to make a σ bond to a hydrogen atom (Figure 8.13). One of the π bonds in an ethyne molecule arises from a side-by-side overlap of the $2p_x$ orbitals of the two carbon atoms. The other π bond results from a similar side-by-side overlap of the $2p_y$ orbitals. The two π bonds, like the *p* orbitals from which they originate, are at right angles to each other, as shown in Figure 8.13. Thus, an ethyne molecule has a total of three σ bonds and two π bonds.

We have seen that the electron pair geometry and the hybridization of an atom in a molecule depend on its σ bonds. We recall that a double bond consists of one σ and one π bond, and that a triple bond consists of one σ and two π bonds.

To determine the structure and hybridization of a carbon atom in a hydrocarbon molecule, count the total number of electron groups attached to the atom (each single or a multiple bond counts as one electron group). This number of electron groups equals the number of σ bonds on the atom. Each single bond is a σ bond, and a multiple bond contains one σ bond. The number of σ bonds equals the number of *s* and *p* orbitals that are mixed to form hybrid orbitals.

For example, a carbon atom in an H—C≡C—H molecule is surrounded by two electron groups. Two electron groups include two σ bonds. Two σ bonds lie on opposite sides of the carbon atom along a straight line. The two *sp* hybrid orbitals for the σ bonds form from mixing two atomic orbitals—the *s* and *p* orbitals.

Example 8.3

Determining the Structure and Hybridization of the Carbon Atoms in an Alkyne Molecule

What is the structure and hybridization of each of the carbon atoms in a propyne molecule, $CH_3C{\equiv}CH$?

SOLUTION: The Lewis structure for propyne is

```
        H
       3|   2   1
     H—C—C≡C—H
        |
        H
```

The C-3 atom in the Lewis structure is surrounded by four σ bonding pairs. Four σ bonds are distributed in tetrahedral geometry. To form four σ bonds, four sp^3 hybrid orbitals are used by the C-3 atom.

C-2 and C-1 are each surrounded by two groups of electrons, or two σ bonds that lie on a straight line. The two σ bonds are formed from sp hybrid orbitals on C-2 and C-1.

Practice Problem 8.5: Predict the structure and hybridization of each of the carbon atoms in a $CH_3C{\equiv}CCH_3$ (2-butyne) molecule.

8.3 Bonding in Other Molecules of Second- and Third-Period Elements

In this section we consider molecules in which nitrogen, oxygen, phosphorus, or a halogen is the central atom.

The Nitrogen Molecule, Ammonia, and the Ammonium Ion

The Lewis structure for N_2 is

$$:N{\equiv}N:$$

The nitrogen atoms in an N_2 molecule are linked by one σ bond and two π bonds. The valence shell electronic structure of a ground state nitrogen atom is $2s^22p^3$, which can be represented by the following orbital diagram:

↑↓	↑	↑	↑
$2s^2$	$2p^3$		

The s orbital and one of the half-full p orbitals hybridize to produce two sp hybrid orbitals. An sp hybrid orbital of one nitrogen atom overlaps with an sp hybrid orbital of the second nitrogen atom to form the N—N σ bond. The other

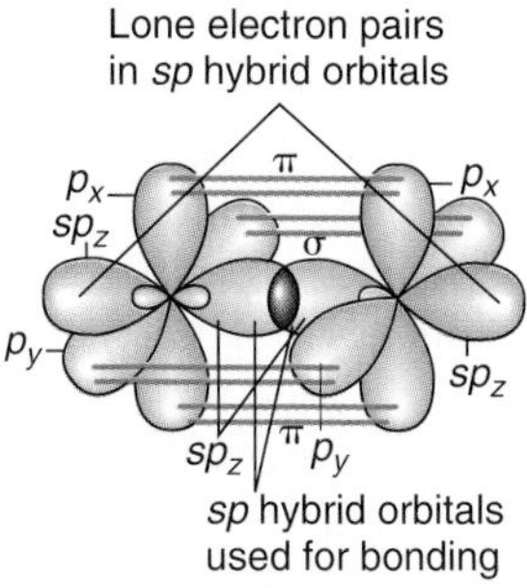

(a) Orbital overlap for σ and π bonds

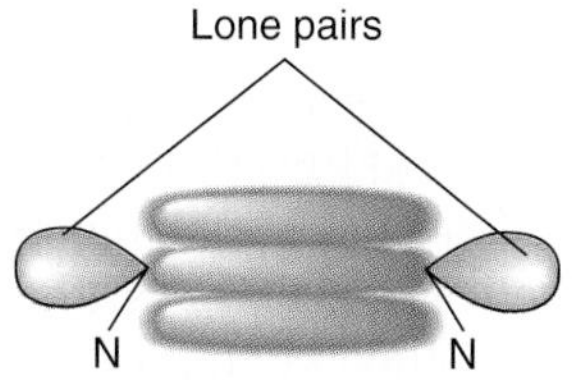

(b) π bonding electron cloud in the molecule, and the two lone pairs

:N⋮⋮N: or :N≡N:

(c) Lewis structures

Figure 8.14

Structure of N_2 molecule.

sp hybrid orbital on each N atom contains a lone electron pair as shown in Figure 8.14. Two remaining *p* orbitals of each N atom in an N_2 molecule are used for the two π bonds in the triply bonded molecule.

Ammonia, NH_3, is a colorless gas with a suffocating odor. It is an important industrial product that is made by the direct combination of nitrogen and hydrogen at high temperature and pressure in the *Haber process*:

$$N_2(g) + 3H_2(g) \longrightarrow 2NH_3(g)$$

Ammonia and ammonium salts are used as fertilizers. Aqueous ammonia is a household cleaning liquid. Ammonia is also used in the production of nitric acid.

We recall that the Lewis structure for ammonia is

$$\begin{array}{c} \text{H—}\ddot{\text{N}}\text{—H} \\ | \\ \text{H} \end{array}$$

We also recall that ammonia is a pyramidal molecule whose H—N—H bond angle is approximately 107°, close to the tetrahedral bond angle of 109.5°. The approximately tetrahedral H—N—H bond angle and the fact that the central atom is surrounded by four electron pairs suggest that the nitrogen is sp^3 hybridized. Of the four sp^3 hybrid orbitals, three are used for N—H bonds, and one contains the lone pair (Figure 8.15).

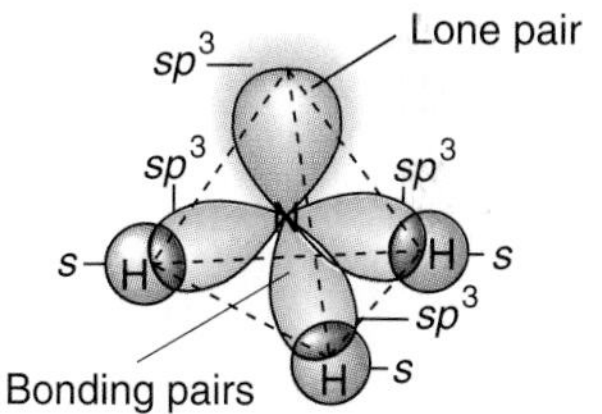

Figure 8.15

Orbital structure of NH_3.

The nitrogen atom of ammonia, with its lone pair of electrons, is the center of electron density in the molecule. The unshared electron pair of ammonia can form a coordinate covalent bond with a proton (H^+), generating an ammonium ion, NH_4^+. Thus, when ammonia is passed through an aqueous solution that contains H^+ ions, the NH_3 molecules convert to ammonium ions, NH_4^+:

$$NH_3(g) + H^+(aq) \longrightarrow NH_4^+(aq)$$

When a proton attaches to the unshared electron pair of the ammonia molecule, the resulting positively charged ammonium ion has a tetrahedral structure. The NH_4^+ ion has the same geometry as a methane molecule. Like the C atom in CH_4, the N atom in NH_4^+ is sp^3 hybridized.

Example 8.4

Predicting the Structure, Polarity, and Hybridization of Nitrogen in a Trihalide of Nitrogen

At room temperature, nitrogen trichloride is a liquid that explodes at 95°C. It was prepared in 1811 by P. L. Dulong (who lost three fingers and an eye in an accident while studying its properties). Nitrogen tribromide explodes even at −100°C. Nitrogen triiodide has not been prepared as a pure compound, but it is believed to exist as a solid that is bonded to ammonia; this solid explodes when touched even with a feather. Nitrogen trifluoride, NF_3 (a colorless gas at room temperature), is a stable trihalide of nitrogen. **(a)** Write a Lewis structure for NF_3. **(b)** Predict the shape of an NF_3 molecule. Is the molecule polar or nonpolar? **(c)** What is the hybridization of nitrogen in NF_3?

SOLUTION

(a) The Lewis structure for NF_3 is similar to that of ammonia:

:F̤—N̈—F̤:
|
:F̤:

(b) The NF_3 molecule is trigonal pyramidal because the central atom has an unshared electron pair. Since the electronegativity of fluorine is higher than that of nitrogen, the N—F bond dipole moments are directed toward the F atoms. The NF_3 molecule is polar because it possesses a net dipole moment.

(c) The nitrogen atom in an NF_3 molecule is surrounded by four electron pairs in four hybrid orbitals that are obtained by mixing one *s* orbital and three *p* orbitals. Thus, the hybridization of nitrogen in NF_3 is sp^3. Of the four sp^3 hybrid orbitals, three are bonding and one contains the lone pair.

Water and the Hydronium Ion

Proceeding along the second-period elements, we next come to oxygen. An oxygen atom forms two covalent bonds with hydrogen or with halogen atoms in its binary compounds. One of these compounds, water, is essential for life on earth. The properties of water depend on the structure of water molecules.

We recall that the Lewis structure for water is

H—Ö̤—H

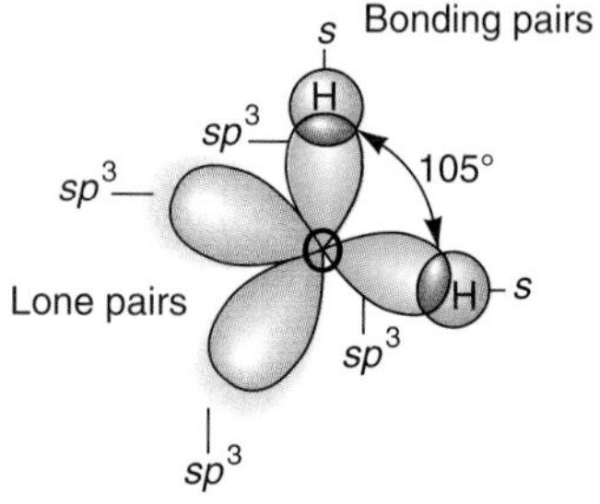

Figure 8.16

Orbital structure of H_2O.

The H—O—H bond angle is 105°, only slightly less than the H—N—H bond angle in NH_3 (107°), and still close to the tetrahedral bond angle of 109.5° of methane. As in methane and ammonia, there are four electron pairs around the central atom. Since the bond angle in water is close to the tetrahedral bond angle, we assume that the four electron pairs on the oxygen atom are in sp^3 hybrid orbitals. Of the four sp^3 hybrid orbitals in a water molecule, two are bonding and two contain the lone pairs (Figure 8.16).

The lone pairs on the oxygen atom of a water molecule attract protons. A water molecule can add one proton to one of its lone pairs. Attaching of a proton to a water molecule forms a coordinate covalent O—H bond, and converts the bent water molecule to a trigonal pyramidal hydronium ion, H_3O^+:

$$H_2O(l) + H^+(aq) \longrightarrow H_3O^+(aq)$$

The geometry of an H_3O^+ ion is similar to that of an ammonia molecule.

Example 8.5

Predicting the Structure and Polarity of the Molecule of a Binary Compound of Oxygen and Chlorine, and the Hybridization on the Central Atom

An oxide of chlorine has the formula Cl_2O. **(a)** What is the structure of this molecule? **(b)** Is it polar or nonpolar? **(c)** What is the hybridization of oxygen in Cl_2O?

SOLUTION

(a) The Lewis structure for Cl_2O is similar to that of H_2O:

$$:\ddot{\underset{..}{Cl}}—\ddot{\underset{..}{O}}—\ddot{\underset{..}{Cl}}:$$

The Cl_2O molecule is bent because the lone pairs on the central oxygen atom repel each other and the bonding pairs.

(b) The Cl—O bond is polar. Since oxygen is more electronegative than chlorine, the negative end of the bond dipole is directed to the O atom. The Cl_2O molecule is polar because it is bent.

(c) As in water, the hybridization of oxygen in Cl_2O is sp^3.

Practice Problem 8.6: Predict the structure and polarity of the SF_2 molecule. What is the hybridization on the sulfur atom in SF_2?

The bonding in an oxygen molecule, O_2, cannot be accurately represented by a Lewis structure. Although valence bond theory correctly accounts for the double bond in O_2, it fails to predict the behavior of O_2 in an applied magnetic field. Molecules can either be attracted or repelled by an applied magnetic field. If they are attracted to the applied field they are said to be **paramagnetic**; if repelled, they are called **diamagnetic**. Molecules with unpaired electrons are paramagnetic, whereas those without unpaired electrons are diamagnetic.

Valence bond theory predicts an electronic structure for O_2 in which there are no unpaired electrons. Therefore, O_2 should be diamagnetic. But it is paramagnetic. Thus, valence bond theory is unable to explain the paramagnetism of oxygen. Since valence bond theory cannot explain the paramagnetism of O_2, it also cannot explain the bonding in O_2. In Section 8.4 we account for the paramagnetism and bonding of O_2 by using molecular orbital theory.

Hydrogen Halides, Halogens, and Interhalogens

Hydrogen gas reacts directly with halogens, X_2, to produce hydrogen halides, HX, as shown by the equation for the reaction of hydrogen with fluorine:

$$H_2(g) + F_2(g) \longrightarrow 2HF(g)$$

The hydrogen halides are noxious gases with pungent odors.

The H—F bond in hydrogen fluoride and the H—X bond in other hydrogen halides form when the half-filled $1s$ orbital of a hydrogen atom overlaps with a half-filled p orbital of a halogen atom.

The bond between two halogen atoms in an X_2 molecule is formed by the overlap of a half-filled p orbital of one halogen atom with the half-filled p orbital of the other. The formation of a chlorine molecule is shown below by Lewis symbols:

$$:\ddot{\underset{..}{Cl}}\cdot + \cdot\ddot{\underset{..}{Cl}}: \longrightarrow :\ddot{\underset{..}{Cl}}:\ddot{\underset{..}{Cl}}:$$

Atoms of two different halogens can also react with each other to form binary compounds. A *binary compound consisting of two different halogens* is called an **interhalogen** compound, such as ClF and ICl. The bonding in diatomic interhalogen molecules resembles that in Cl_2.

Summary of Bonding Involving *s* and *p* Orbitals

Table 8.1 summarizes our discussion of molecular structure, hybridization, and bonding to this point. The structures in Table 8.1 include those atoms that use only *s* and *p* orbitals for bonding. Atoms having only *s* and *p* orbitals in their valence shells are, of course, second-period atoms.

The following rules enable us to deduce quickly the correct electron pair geometry and hybridization of an atom that uses only *s* and *p* electrons for bonding. Both the electron pair geometry and hybridization are determined by the total number of electron groups, which can be single bonds, multiple bonds, or lone pairs. Molecular structure is determined by the positions of atomic nuclei only.

1. If the central atom is surrounded by two electron groups, the central atom has a linear electron pair geometry and is *sp* hybridized. Examples of such atoms include Be in H—Be—H, C in H—C≡C—H, and N in :N≡N:.
2. If the central atom is surrounded by three electron groups, it has a trigonal planar or nearly trigonal planar electron pair geometry and is sp^2 hybridized. Examples of such atoms include B in BCl_3, S in SO_2, and C in $CH_2{=}CH_2$.
3. If the central atom is surrounded by four electron groups, it has tetrahedral or modified tetrahedral electron pair geometry and is sp^3 hybridized. Examples of such atoms include C in CH_4, N in NH_3, and O in H_2O.

If the central atom in a molecule has lone pairs, the molecular geometry can be deduced by considering the bonding electron groups and the lone pair-bonding pair repulsion. Thus, the S atom in an SO_2 molecule has two bonding electron groups and one lone pair. The SO_2 molecule is therefore bent. An O atom in an H_2O molecule has two bonding electron groups and two lone pairs. The H_2O molecule is also bent. The N atom in an NH_3 molecule has three bonding electron groups and one lone pair; it is trigonal pyramidal.

Table 8.1

Structures of Molecules and Hybrid Orbitals on the Central Atom Having Only *s* and *p* Orbitals in the Valence Shell

Type of Molecule[a]	Examples	Electron Pair Structure	Molecular Structure	Bond Angles	Number of Electron Pairs on Central Atom (Lone and Bonding)	Atomic Orbitals Mixed to Form Hybrid Orbitals	Designation of Hybrid Orbitals
AB_2	BeH_2, $BeCl_2$	Linear	Linear	180°	2	s, p_z	sp
AB_3	BCl_3, BF_3	Trigonal planar	Trigonal planar	120°	3	s, p_x, p_y	sp^2
AB_4	CH_4, CCl_4	Tetrahedral	Tetrahedral	109.5°	4	s, p_x, p_y, p_z	sp^3
AB_3E	NH_3, NCl_3	Modified tetrahedral	Trigonal pyramidal	Less than 109.5°	4	s, p_x, p_y, p_z	sp^3
AB_2E_2	H_2O, Cl_2O	Modified tetrahedral	Bent or angular	Less than 109.5°	4	s, p_x, p_y, p_z	sp^3

[a] A is the central atom, B a peripheral atom, and E a lone electron pair.

The Role of *d* Orbitals in Bonding

We consider next the bonding in some molecules in which phosphorus and sulfur are the central atoms. These atoms can use not only *s* and *p* orbitals for bonding, they can also employ *d* orbitals, giving rise to geometries not observed for second-period elements. For example, nitrogen forms nitrogen trichloride, NCl_3, but phosphorus forms both phosphorus trichloride, PCl_3, and phosphorus pentachloride, PCl_5.

The central atoms of NCl_3 and PCl_3 are surrounded by four electron pairs. The central atoms in these molecules are therefore sp^3 hybridized, and their molecular geometries are trigonal pyramidal. In PCl_5, however, the phosphorus atom is surrounded by five electron pairs. We recall from Chapter 7 that the geometry of phosphorus pentachloride is trigonal bipyramidal. The phosphorus atom must have five unpaired valence electrons to form five P—Cl bonds. In its ground state, a phosphorus atom has only three unpaired valence electrons:

$$\begin{array}{c|ccc} \uparrow\downarrow & \uparrow & \uparrow & \uparrow \\ 3s^2 & & 3p^3 & \end{array}$$

To get five unpaired electrons, one of the 3*s* electrons is promoted to an empty 3*d* orbital:

$$\begin{array}{c|ccc|ccccc} \uparrow\downarrow & \uparrow & \uparrow & \uparrow & \_ & \_ & \_ & \_ & \_ \\ 3s^2 & & 3p^3 & & & & 3d^0 & & \end{array} \longrightarrow$$

$$\begin{array}{c|ccc|ccccc} \uparrow & \uparrow & \uparrow & \uparrow & \uparrow & \_ & \_ & \_ & \_ \\ 3s^1 & & 3p^3 & & & & 3d^1 & & \end{array}$$

As a result of this "promotion," five unpaired electrons are in five different orbitals. These orbitals mix to form five identical ***sp*³*d*** (read *s*-*p*-three-*d*) **hybrid orbitals**. These are arranged in a trigonal bipyramidal geometry around the phosphorus atom (Figure 8.17).

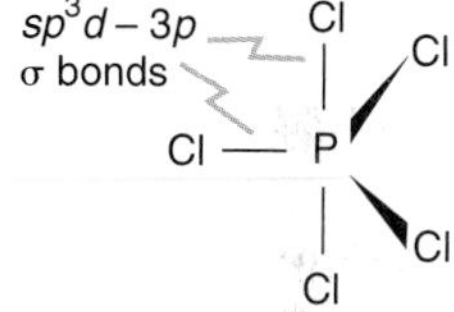

Figure 8.17

Structure and bonding in PCl_5.

Now let us turn to compounds of sulfur. A sulfur atom in its ground state has six valence electrons whose configuration is $3s^23p^4$:

$$\begin{array}{c|ccc} \uparrow\downarrow & \uparrow\downarrow & \uparrow & \uparrow \\ 3s^2 & & 3p^4 & \end{array}$$

In a molecule such as SF_2, the sulfur atom is surrounded by two σ bonding pairs and two lone pairs like the oxygen atom in H_2O. The sulfur atom in SF_2 is sp^3 hybridized.

Sulfur also forms a tetrafluoride, SF_4. In this compound the sulfur atom is surrounded by five electron pairs; four of these are σ bonding pairs and one is a lone pair. To obtain four unpaired electrons for the four S—F bonds, the 3*p* electron pair must be unpaired and one electron promoted to an empty *d* orbital as shown below.

$$\begin{array}{c|ccc|ccccc} \uparrow\downarrow & \uparrow\downarrow & \uparrow & \uparrow & \_ & \_ & \_ & \_ & \_ \\ 3s^2 & & 3p^4 & & & & 3d^0 & & \end{array} \longrightarrow$$

$$\begin{array}{c|ccc|ccccc} \uparrow\downarrow & \uparrow & \uparrow & \uparrow & \uparrow & \_ & \_ & \_ & \_ \\ 3s^2 & & 3p^3 & & & & 3d^1 & & \end{array}$$

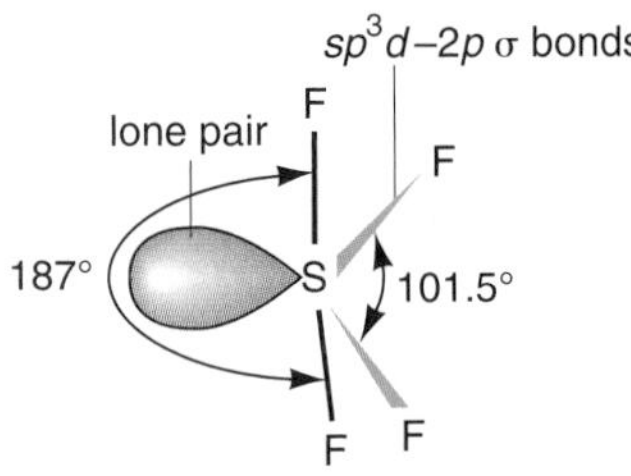

Figure 8.18

Structure and bonding in SF_4.

Hybridization of one *s* orbital, three *p* orbitals, and one *d* orbital yields five sp^3d hybrid orbitals. Of these, four are bonding orbitals and one houses the lone electron pair. As we recall from Chapter 7, the structure of sulfur tetrafluoride is "seesaw-shaped" (Figure 8.18).

Sulfur reacts readily with excess fluorine to give sulfur hexafluoride, SF_6:

$$S_8(s) + 24F_2(g) \longrightarrow 8SF_6(g)$$

To form six S—F bonds, the sulfur atom must have six unpaired electrons. These can be provided by unpairing both the 3*s* and the 3*p* electron pairs on the sulfur atom and promoting two electrons to two vacant 3*d* orbitals:

$$\underset{3s^2}{\underline{\uparrow\downarrow}} \quad \underbrace{\underline{\uparrow\downarrow}\ \underline{\uparrow}\ \underline{\uparrow}}_{3p^4} \quad \underbrace{\underline{\ \ }\ \underline{\ \ }\ \underline{\ \ }\ \underline{\ \ }\ \underline{\ \ }}_{3d^0} \longrightarrow$$

$$\underset{3s^1}{\underline{\uparrow}} \quad \underbrace{\underline{\uparrow}\ \underline{\uparrow}\ \underline{\uparrow}}_{3p^3} \quad \underbrace{\underline{\uparrow}\ \underline{\uparrow}\ \underline{\ \ }\ \underline{\ \ }\ \underline{\ \ }}_{3d^2}$$

Mixing one *s*, three *p*, and two *d* orbitals results in six $\boldsymbol{sp^3d^2}$ hybrid orbitals. These orbitals are all bonding orbitals. We recall from Chapter 7 that six bonding orbitals are distributed around the central atom in octahedral geometry (Figure 8.19).

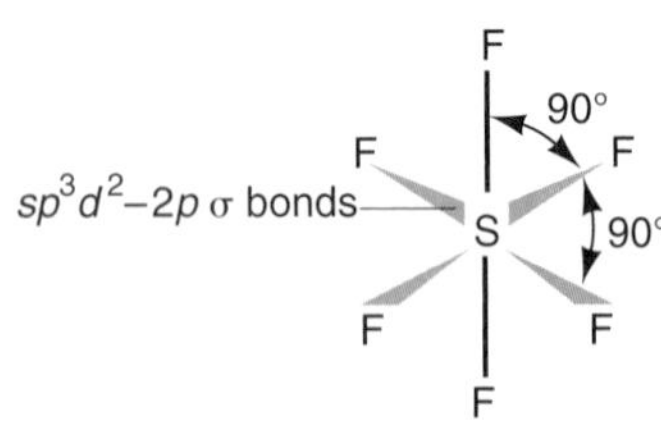

Figure 8.19

Structure and bonding in SF_6.

Practice Problem 8.7: What are the electron pair structure, the molecular structure, and the hybridization on the central atom of each of the following molecules? **(a)** BrF_3; **(b)** IF_5; **(c)** XeF_2; **(d)** XeF_4.

8.4 Molecular Orbital Theory

We now turn our attention to a theory of bonding called **molecular orbital** (MO) **theory**. MO theory explains many phenomena which cannot be accounted for by valence bond theory. For example, molecular orbital theory predicts the existence of diatomic molecules such as Li_2 and C_2, which exist in the gas phase. It explains why the noble gases do not form diatomic molecules, and it also explains the magnetic properties and the bond orders in O_2 and other molecules. Molecular orbital theory also provides much more quantitative information than valence bond theory.

According to valence bond theory, a covalent bond exists when the atomic orbitals of two valence electrons overlap. These bonding electrons are *localized* in the region of space between the two bonded nuclei. In contrast, molecular orbital (MO) theory treats a molecule as a "multinuclear atom" in which all the electrons of the bonded atoms belong to **molecular orbitals** that extend over the entire molecule.

Molecular orbitals are combinations of atomic orbitals. A given number of atomic orbitals combine to produce an equal number of molecular orbitals. We recall that atomic orbitals can be arranged in the order of their increasing energies. These orbitals are occupied by electrons one at a time starting with the lowest-energy orbital. Molecular orbitals are filled with electrons in a similar way, according to Hund's rule. Let us see how these ideas are applied to diatomic molecules.

Hydrogen Through Beryllium

Consider the bonding in H_2. According to molecular orbital theory, when two H atoms combine to form an H_2 molecule, the 1*s* atomic orbitals (AOs) of the two H atoms combine to form two molecular orbitals (MOs). We recall that electrons can be regarded as waves. If we add two waves, they can combine by either constructive or destructive interference. Adding atomic orbitals by constructive interference, which is the same as adding the two waves *in phase*, gives a molecular orbital in which electron density is concentrated between the nuclei. This MO is called a **bonding molecular orbital**.

Adding two electron waves by destructive interference—which is the same as adding the two waves *out of phase*—gives a higher-energy orbital called an **antibonding molecular orbital**. Electrons in an antibonding MO cancel the bonding effect of an equal number of electrons in the corresponding bonding MO.

Electrons that reside in an antibonding MO are not located between the nuclei. Thus an antibonding MO has higher energy than its corresponding bonding MO. Figure 8.20 shows the shapes and the relative energies of the molecular orbitals in an H_2 molecule. Figure 8.20 shows that a bonding MO is lower in energy than its corresponding antibonding MO, which consists of two spatial regions on opposite sides of the nuclei. As in atomic orbitals, the maximum occupancy of a molecular orbital is two electrons. When two electrons occupy a given MO, they have paired, or opposite spins.

The two electrons of the H_2 molecule occupy the lower-energy (bonding) MO. This bonding MO is labeled σ_{1s} because it results from adding the two 1*s* orbitals in phase so that the greatest electron density lies along the axis between the bonded nuclei. Adding two 1*s* orbitals out of phase gives the antibonding MO, which is designated σ^*_{1s} (read "sigma-star"). An antibonding MO is indicated by a star at the upper right of its symbol.

In MO theory, the **bond order** is one-half of the difference between the number of electrons in bonding molecular orbitals (B) and the number of electrons in antibonding molecular orbitals (A). The bond order is

$$\text{bond order} = \frac{\text{B} - \text{A}}{2}$$

For example, in an H_2 molecule, two electrons are in the bonding σ_{1s} MO and no electrons in the antibonding σ^*_{1s} MO. Thus, the bond order in an H_2 molecule is 1.

The electrons in the bonding MO in an H_2 molecule are paired with their spins opposed. Hydrogen gas is therefore diamagnetic and is slightly repelled by an applied magnetic field.

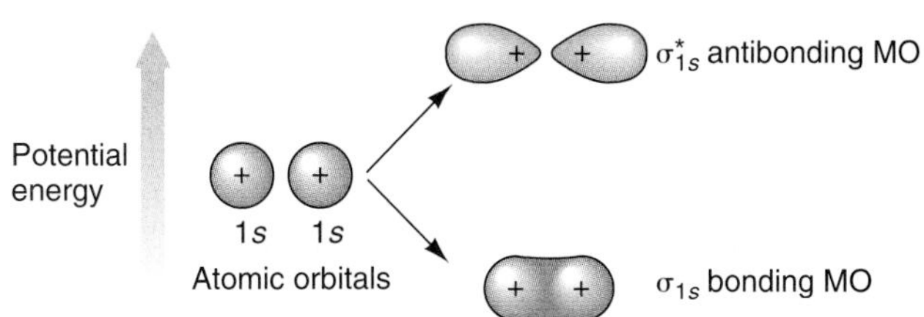

Figure 8.20

The molecular orbitals resulting from a combination of two 1*s* atomic orbitals.

Example 8.6

Predicting the Bond Order and Magnetism of a Diatomic Ion

(a) What is the bond order in the hydrogen molecule ion, H_2^+? **(b)** Is the species diamagnetic or paramagnetic?

SOLUTION

(a) There is only one electron in H_2^+. This electron is in the bonding MO of H_2^+. The bond order is 0.5.

(b) The species is paramagnetic since it has an unpaired electron.

MO theory explains why helium consists of atoms rather than diatomic molecules. If a diatomic helium molecule existed, it would contain four electrons. Two of them would be in the σ_{1s} MO and two in the σ^*_{1s} MO. The bonding contributed by the electrons in the σ_{1s} MO would be completely canceled by the two electrons in the σ^*_{1s} MO. Consequently, He_2 molecules are not stable. Figure 8.21 illustrates the relative energies and the electron populations of the MOs in H_2 and the hypothetical He_2 molecule.

Next we consider the diatomic molecule Li_2. Each Li atom has the electronic configuration $1s^22s^1$. The four electrons of the two $1s$ orbitals occupy σ_{1s} and σ^*_{1s} MOs. As in He_2, these MOs make no net contribution to bonding. The $2s$ orbitals of the two lithium atoms combine to form a σ_{2s} bonding MO and a σ^*_{2s} antibonding MO. Since the $2s$ AOs are higher in energy than the $1s$ AOs, the σ_{2s} and the σ^*_{2s} MOs are higher in energy than the σ_{1s} and σ^*_{1s} MOs, respectively.

The two $2s$ electrons of the lithium atoms occupy the σ_{2s} MO as a pair (Figure 8.22). There are no electrons in the σ^*_{2s} MO. The total number of electrons in the bonding MOs is 4, and in antibonding MOs 2. The bond order (BO) of an Li_2 molecule is therefore 1:

$$BO = \frac{4-2}{2} = 1$$

MO theory thus predicts that Li_2 is a stable molecule. It exists in the gas phase, but condenses to larger aggregates in the solid phase. Each Li_2 molecule contains six electrons. Since all the electrons in the bonding and antibonding MOs are paired, Li_2 is diamagnetic.

The bond order of a homonuclear diatomic molecule can be determined by considering only the *valence* electrons of the atoms in their corresponding MOs because electrons in the lower complete shells of atoms occupy bonding and antibonding MOs in equal numbers so there is no net bonding effect. Thus, the bond order of a diatomic molecule equals one half of the difference between the

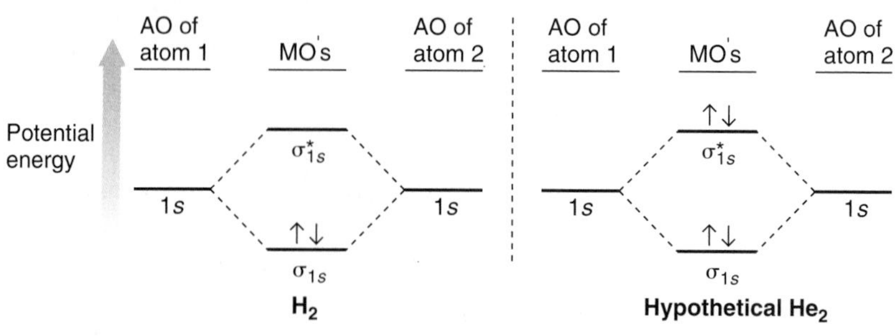

Figure 8.21

Molecular orbital energies and electron populations for H_2 and hypothetical He_2 molecules.

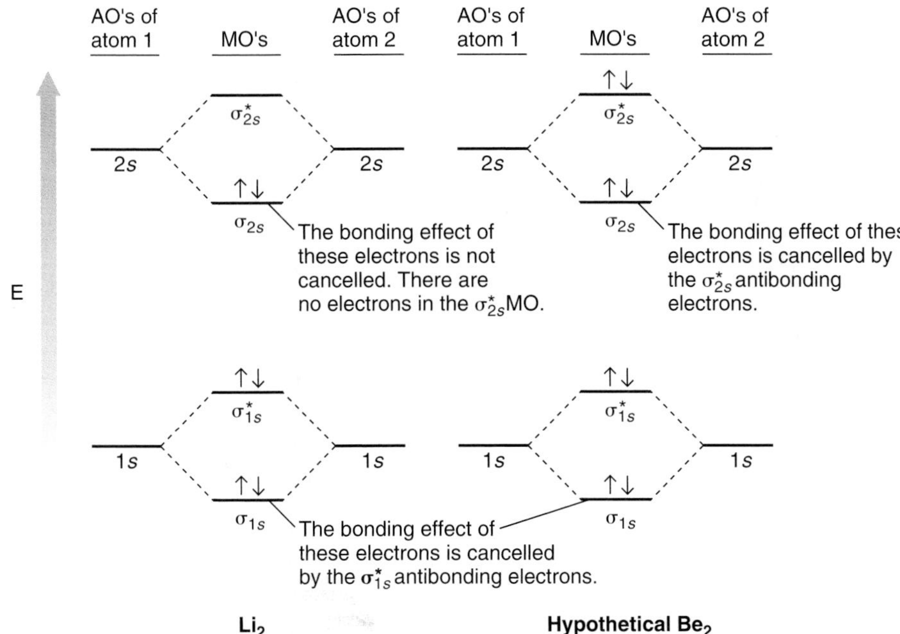

Figure 8.22

Relative MO energies and electron populations for Li_2 and for hypothetical Be_2.

number of valence electrons occupying bonding MOs and the number of valence electrons occupying antibonding MOs. If we designate the number of valence electrons in bonding MOs by VB and the number of valence electrons in antibonding MOs by VA, the bond order, BO, is

$$\text{BO} = \frac{\text{VB} - \text{VA}}{2}$$

We have seen that Li_2 is predicted to be stable and that it actually exists in the gas phase at high temperatures. But does a Be_2 molecule exist? MO theory predicts that Be_2 is not stable since the four valence electrons of the two Be atoms would occupy σ_{2s} and σ^*_{2s} MOs (Figure 8.22). Thus, the contribution to bonding of two electrons in the σ_{2s} bonding MO would be canceled by two electrons in the σ^*_{2s} antibonding MO. Indeed, no Be_2 molecules have been found.

The population of molecular orbitals of a molecule can be described by a shorthand symbolism similar to that used to describe atomic structures. MO designations are written in parentheses, and the number of electrons in the MOs are indicated as superscripts to the right of the parentheses. The MOs are listed in the order of their increasing energies. For example, the MO population of an Li_2 molecule is $(\sigma_{1s})^2(\sigma^*_{1s})^2(\sigma_{2s})^2$. This description means that in an Li_2 molecule there are two electrons in the σ_{1s} MO, two in the σ^*_{1s} MO, and two in the σ_{2s} MO. The MO populations, bond orders (BOs), and magnetic properties of the homonuclear diatomic molecules from H_2 to hypothetical Be_2 may be summarized:

H_2: $(\sigma_{1s})^2$; BO = 1; the molecule is diamagnetic.
He_2: $(\sigma_{1s})^2\ (\sigma^*_{1s})^2$; BO = 0; the molecule does not exist.
Li_2: $(\sigma_{1s})^2\ (\sigma^*_{1s})^2\ (\sigma_{2s})^2$; BO = 1; the molecule is diamagnetic.
Be_2: $(\sigma_{1s})^2\ (\sigma^*_{1s})^2\ (\sigma_{2s})^2\ (\sigma^*_{2s})^2$; BO = 0; the molecule does not exist.

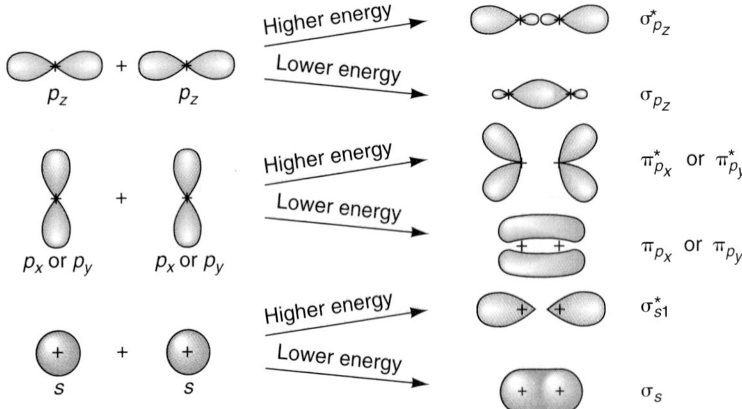

Figure 8.23

Molecular orbitals resulting from the combination of *s* and *p* atomic orbitals.

Boron Through Neon

The atoms of the *p*-block elements of the second period—boron through neon—have both *s* and *p* orbitals in their valence shells. Figure 8.23 illustrates the various ways that *s* and *p* orbitals can combine to produce molecular orbitals. The end-to-end combination of p_z orbitals produces a σ_p bonding MO and a σ_p^* antibonding MO. (The *z* axis is conventionally designated as the internuclear axis.) The side-by-side combination of two 2*p* orbitals results in a π-bonding MO and a π^*-antibonding MO. The bonding and antibonding π MOs resulting from the p_x orbitals are perpendicular to the MOs resulting from the p_y orbitals (if one MO is in the plane of this page, the other is above and below the plane of this page).

A π-bonding MO consists of two spatial regions on opposite sides of the internuclear axis. These *two* spatial regions together constitute *one* molecular orbital. The corresponding antibonding, π^* MO also consists of two spatial regions of two lobes each, for a total of four regions of space.

The relative MO energies for the homonuclear diatomic molecules from H_2 to N_2 are as follows:

$$\sigma_{1s} < \sigma^*_{1s} < \sigma_{2s} < \sigma^*_{2s} < \pi_{2p_x} = \pi_{2p_y} < \sigma_{2p_z} < \pi^*_{2p_x} = \pi^*_{2p_y} < \sigma^*_{2p_z}$$

Figure 8.24 illustrates the shapes and relative energies of the MOs for the second-period homonuclear diatomic molecules from Li_2 to N_2.

In a shorthand description of the MO population of a molecule consisting of atoms of the second or later periods, the 1*s* electrons are omitted because they do not affect the bond order. For example, the MO population of an Li_2 molecule may be described as $(\sigma_{2s})^2$.

As in the aufbau process (Chapter 6), molecular orbitals are filled with electrons entering the most stable, lowest-energy MO first. Then, successively higher energy MOs are occupied. Each MO can contain a maximum of two electrons. MOs of equal energy are occupied according to Hund's rule; one electron enters each of the orbitals (with spins parallel) before any pairing occurs.

At room temperature, elemental boron and carbon exist in the solid state as various arrays of atoms (Chapter 2). However, in the vapor state at high temperature, diatomic molecules B_2 and C_2, respectively, have been found. MO theory predicts the existence of these diatomic species.

When two B atoms combine to form a B_2 molecule, the six valence electrons of the two B atoms fill the σ_{2s} and σ^*_{2s} MOs in pairs. The next energy level

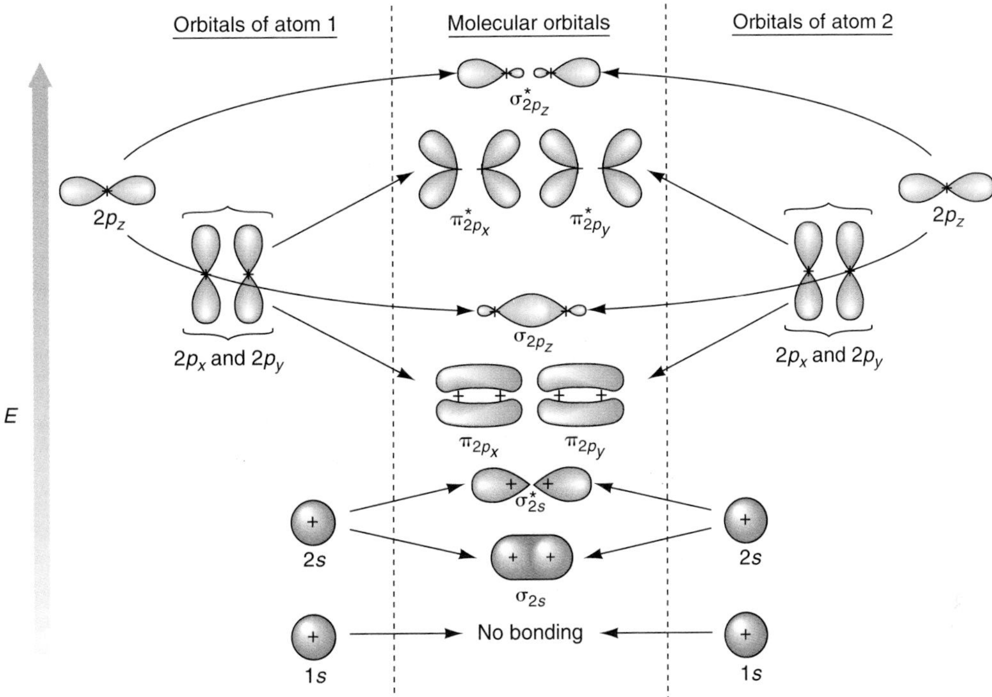

Figure 8.24

The shapes and relative energies of the MOs from Li_2 to N_2.

is π_{2p}. One of the remaining electrons occupies one of the π_{2p} MOs, and the other electron occupies the other π_{2p} MO according to Hund's rule. The bonding effect of the two valence electrons in the σ_{2s} MOs is canceled by two electrons in the σ^*_{2s}-antibonding MO. The bond order of B_2 is therefore

$$BO = \frac{4 - 2}{2} = 1$$

The bond order of a B_2 molecule equals 1 because of the two electrons in the π_{2p} MOs. There are no electrons in the corresponding π^*_{2p}-antibonding MOs. Since there are two unpaired electrons in the π_{2p} MOs, the B_2 molecule is paramagnetic.

The molecular orbital populations and their relative energies for B_2, C_2, and N_2 are described in Figure 8.25. The bond orders, bond energies, bond lengths, and magnetic properties for these molecules are also listed. The letters "KK" preceding each shorthand description of the MO population refer to the filled σ_{1s} and σ^*_{1s} MOs that result from the combination of atomic 1*s* orbitals in the *K* shells of the atoms.

The orbital energies decrease from B_2 to N_2 because the nuclear charge increases from boron to nitrogen. The increased nuclear charge pulls the electrons closer to the nuclei, thereby decreasing the energy of the electrons and making the orbitals more stable.

One result of the increasing nuclear charge and the resulting decrease of MO energies as we move from left to right across the second period is that the energy of the σ_{2p} MO moves successively closer to the π_{2p} level from B_2 through N_2, and drops below the π_{2p} level in O_2 and F_2 molecules. This switching of the MO energies occurs because σ_{2p}-bonding MOs are concentrated directly between the nuclei and are therefore attracted by an increasing nuclear charge more than the π_{2p} MOs.

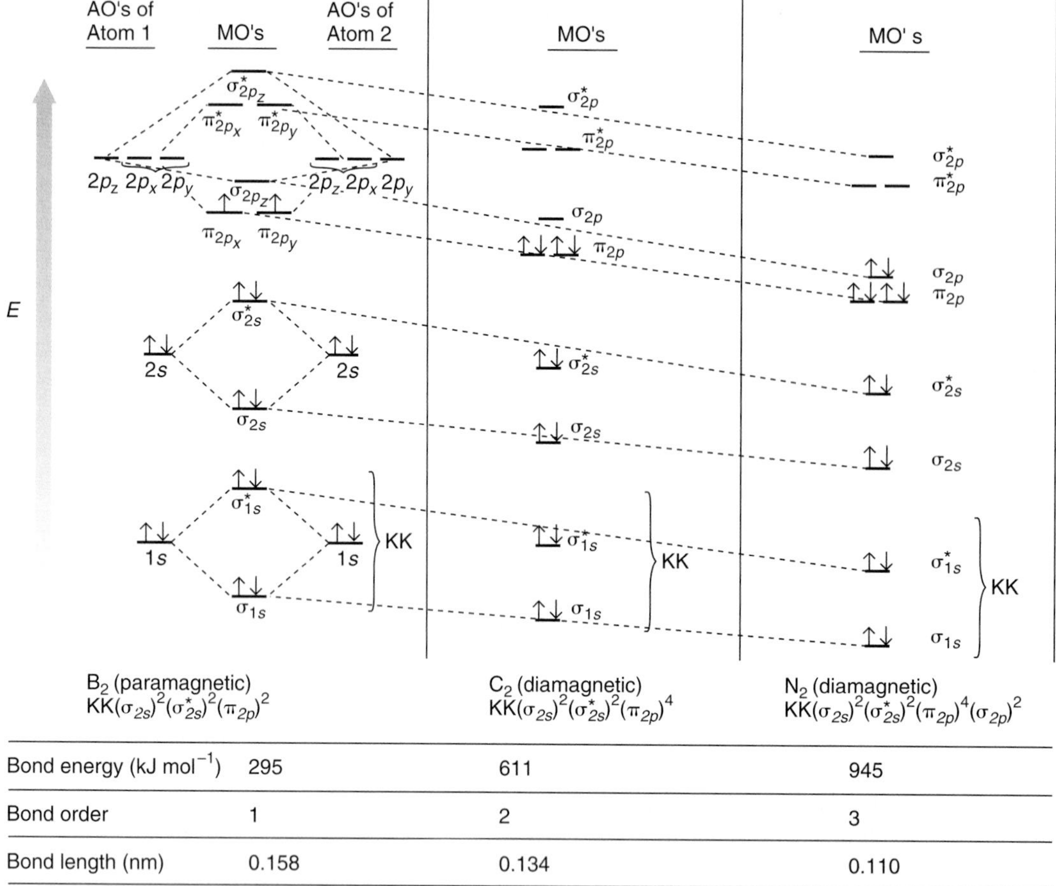

	B_2	C_2	N_2
Bond energy (kJ mol^{-1})	295	611	945
Bond order	1	2	3
Bond length (nm)	0.158	0.134	0.110

Figure 8.25

Molecular orbital populations and relative energies for B_2, C_2, and N_2.

Figure 8.26 describes the MO populations and their relative energies for O_2 and F_2 molecules, as well as for the hypothetical Ne_2 molecule. The bond orders, bond lengths, and magnetic properties for these molecules are also listed.

Figures 8.25 and 8.26 show that the bond energy increases with increasing bond order, and bond length decreases with increasing bond order. Bond lengths and bond energies for these molecules are summarized in Table 8.2.

From the MO description of a diatomic molecule or ion, the bond order and magnetic property can be predicted. Of the molecules shown in Figures 8.25 and 8.26, only B_2 and O_2 have unpaired electrons. These molecules are therefore paramagnetic; the others are diamagnetic.

From the foregoing discussion we can see that MO theory corresponds to VB theory in predicting the bond orders for simple homonuclear diatomic molecules like N_2, O_2, and F_2. In VB theory, the bond orders in these molecules are 3, 2, and 1, respectively, because the atoms in N_2 are triple-bonded, in O_2 double-bonded, and in F_2 single-bonded. These bond orders are the same as predicted by MO theory.

However, according to VB theory, N_2, O_2, and F_2 should all be diamagnetic. Experimentally, this is true for N_2 and F_2 but not for O_2. Unlike VB theory, MO theory can also explain the paramagnetism of O_2. MO theory also explains why helium, neon, and other noble gases are monatomic. Namely, the bond order for a diatomic noble gas molecule is zero. Therefore, diatomic noble gas molecules do not exist.

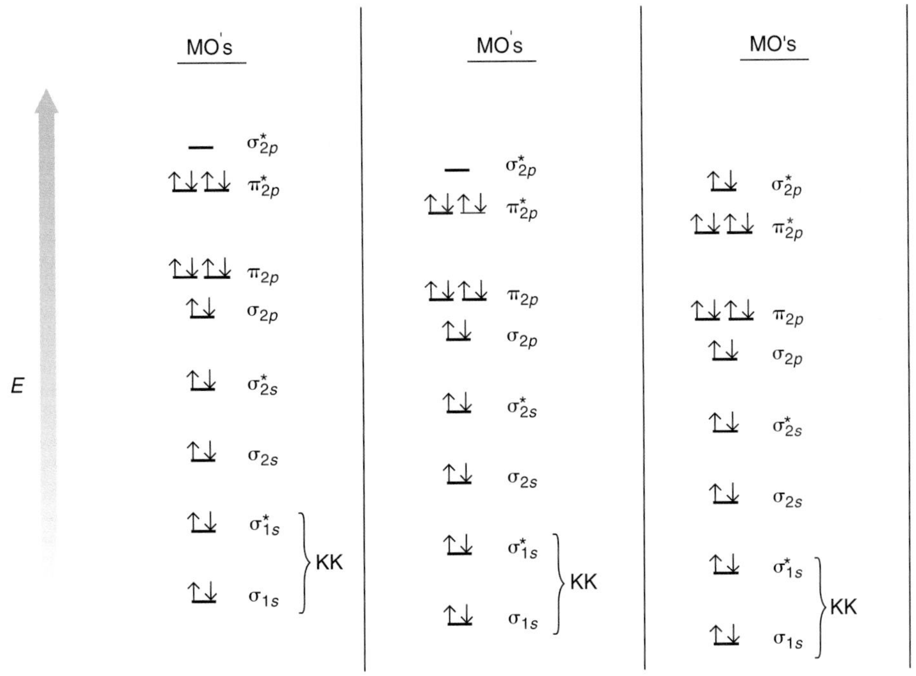

MO's		MO's		MO's	
—	σ^*_{2p}	—	σ^*_{2p}	↑↓	σ^*_{2p}
↑↓↑↓	π^*_{2p}	↑↓↑↓	π^*_{2p}	↑↓↑↓	π^*_{2p}
↑↓↑↓	π_{2p}	↑↓↑↓	π_{2p}	↑↓↑↓	π_{2p}
↑↓	σ_{2p}	↑↓	σ_{2p}	↑↓	σ_{2p}
↑↓	σ^*_{2s}	↑↓	σ^*_{2s}	↑↓	σ^*_{2s}
↑↓	σ_{2s}	↑↓	σ_{2s}	↑↓	σ_{2s}
↑↓	σ^*_{1s} KK	↑↓	σ^*_{1s} KK	↑↓	σ^*_{1s} KK
↑↓	σ_{1s}	↑↓	σ_{1s}	↑↓	σ_{1s}

E

O_2 (paramagnetic)
$KK(\sigma_{2s})^2(\sigma^*_{2s})^2(\sigma_{2p})^2(\pi_{2p})^4(\pi^*_{2p})^2$

F_2 (diamagnetic)
$KK(\sigma_{2s})^2(\sigma^*_{2s})^2(\pi_{2p})^4(\pi^*_{2p})^4$

Hypothetical Ne_2
$KK(\sigma_{2s})^2(\sigma^*_{2s})^2(\pi_{2p})^4(\pi^*_{2p})^4(\sigma^*_{2p})^2$

	O_2	F_2	Ne_2
Bond energy (kJ mol^{-1})	498	159	——
Bond order	2	1	0
Bond length (nm)	0.116	0.142	——

Figure 8.26

Molecular orbital populations and relative energies for O_2, F_2, and hypothetical Ne_2.

Example 8.7

Describing the MO Population of a Diatomic Ion and Predicting Its Bond Order and Magnetism

(a) Describe the MO population of the peroxide ion, O_2^{2-}, and predict its bond order. **(b)** Is a peroxide ion diamagnetic or paramagnetic?

SOLUTION

(a) A peroxide ion has two more electrons than an O_2 molecule (Figure 8.26). Each oxygen atom has six valence electrons. Adding two more electrons to give the 2− charge on the peroxide ion yields a total of 14 valence electrons. These 14 electrons completely fill the σ_{2s}, σ^*_{2s}, σ_{2p}, π_{2p}, and π^*_{2p} MOs. The total number of electrons in the bonding MOs that result from the valence electrons of the O atoms is 8. The total number of antibonding electrons is 6. Therefore, the bond order in an O_2^{2-} ion is

$$BO = \frac{8 - 6}{2} = 1$$

(b) Since it has no unpaired electrons, peroxide ion is diamagnetic.

Table 8.2

Single and Multiple Bond Energies and Bond Lengths

(a) Average Bond Energies and Bond Lengths in Various Molecules

Bond	Bond Energy ($kJ\ mol^{-1}$)	Bond Length (nm)
C—C	346	0.154
C=C	602	0.134
C≡C	835	0.120
C—H	413	0.109
C—Cl	327	0.177
C—O	358	0.143
C=O	799	0.120
C≡O	1072	0.113
C—N	305	0.147
C=N	615	0.129
C≡N	887	0.116
N—N	167	0.148
N=N	418	0.125
N≡N	942	0.110
N—O	201	0.140
N=O	607	0.121
O—O	142	0.148

(b) Bond Energies and Bond Lengths in Some Specific Molecules

Bond	Molecule	Bond Energy ($kJ\ mol^{-1}$)	Bond Length (nm)
H—H	H_2	436	0.0746
H—O	H_2O	498	0.0958
H—C	CH_4	435	0.1091
C—C	C_2H_6	368	0.1536
C=C	C_2H_4	720	0.134
C≡C	C_2H_2	962	0.120
N≡N	N_2	945	0.10975
O=O	O_2	498	0.1208
F—F	F_2	157	0.1417

Practice Problem 8.8: **(a)** Describe the MO population and predict the bond order of a superoxide ion, O_2^-. **(b)** Is an O_2^- ion diamagnetic or paramagnetic?

In the third and later periods, the *s* and *p* atomic orbitals can combine to form molecular orbitals in a manner similar to that in the second period. The bond orders and magnetic properties of diatomic molecules and ions consisting of atoms of any given period can thus be determined by the MO theory as shown by Example 8.8.

Example 8.8

Predicting the Bond Order and Magnetic Properties of the Third-Period Homonuclear Diatomic Molecules and Ions

Predict the bond order and magnetic properties of: **(a)** an Al_2 molecule; **(b)** an S_2^+ ion.

SOLUTION

(a) Only the electrons in the valence shells of the Al atoms need to be considered. The number of valence electrons in an Al atom is 3 ($3s^23p^1$), and an Al_2 molecule has a total of six valence electrons. The MO population by these electrons is $(\sigma_{3s})^2(\sigma_{3s}^*)^2(\pi_{3p})^2$. From this description, the bond order is 1:

$$\text{BO} = \frac{4 - 2}{2} = 1$$

The two π_{3p} MOs are occupied by one electron each, with parallel spins. Consequently, the molecule is paramagnetic. Thus, we see that an Al_2 molecule is similar to a B_2 molecule with respect to its bond order and paramagnetism.

(b) A sulfur atom has six valence electrons, so an S_2 molecule has a total of 12 valence electrons. Subtracting one electron to give the 1+ charge on the S_2^+ ion leaves 11 electrons to be distributed in the MOs. The MO population of these 11 electrons of an S_2^+ ion is $(\sigma_{3s})^2(\sigma_{3s}^*)^2(\sigma_{3p})^2(\pi_{3p})^4(\pi_{3p}^*)^1$. From this, the bond order is $2\frac{1}{2}$:

$$\text{BO} = \frac{8 - 3}{2} = 2\tfrac{1}{2}$$

The ion is paramagnetic because it has one unpaired electron in the π_{3p}^* orbital.

Practice Problem 8.9: Describe the MO population and predict the bond order of the N_2^+ ion. Is this ion diamagnetic or paramagnetic?

Heteronuclear Diatomic Molecules

The molecular orbital treatment of heteronuclear diatomic molecules is similar to that of homonuclear diatomic molecules. An important difference is that the equivalent atomic orbitals of each contributing atom do not have the same energies. The energies of two combining atomic orbitals should be as close as possible to give a maximum overlapping for a strong bond. If the energy difference is too large, the orbitals cannot combine effectively and remain nonbonding orbitals, or the molecule cannot form at all.

To illustrate the molecular orbital treatment of a heteronuclear molecule, we use nitrogen monoxide, NO. The energies of N and O atomic orbitals are not the same because O is more electronegative than N. The energies of the

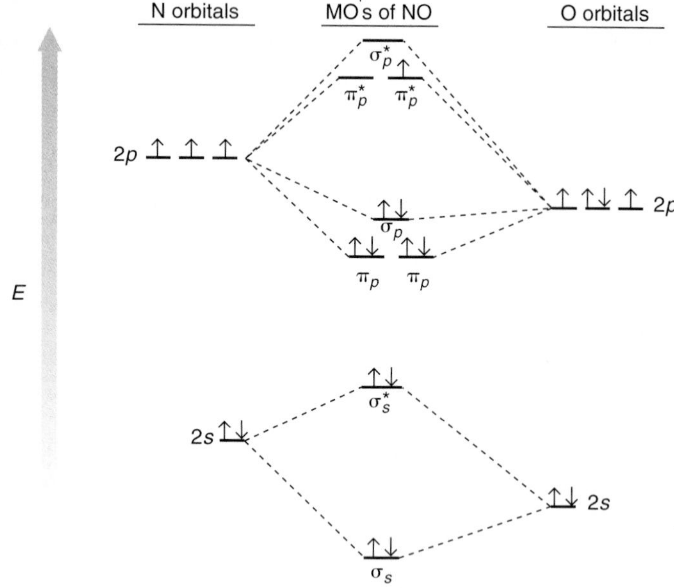

Figure 8.27

Relative atomic and molecular orbital energies of an NO molecule.

atomic orbitals of atoms N and O relative to the corresponding MO energies of an NO molecule are shown in Figure 8.27.

The bond orders and magnetic properties of heteronuclear diatomic molecules and ions can also be predicted by using MO diagrams, or shorthand descriptions, of the electron populations of these molecules. For example, nitrogen monoxide, NO, has an electron structure in which the octet on one of the atoms cannot be satisfied in the Lewis resonance structures:

$$:\dot{N}=\ddot{O}: \longleftrightarrow :\ddot{N}=\dot{O}:$$

Since one atom in either of these structures has an unpaired electron, an NO molecule is paramagnetic. According to VB theory, NO has a bond order of 2. MO theory agrees with VB theory that the NO molecule is paramagnetic, but MO theory predicts a bond order of $2\frac{1}{2}$ for NO as shown in Figure 8.27. Eight of the total of eleven valence electrons of the N and O atoms in an NO molecule are in the bonding MOs, and three are in the antibonding MOs (Figure 8.27). The bond order is therefore $2\frac{1}{2}$:

$$\text{BO} = \frac{8 - 3}{2} = 2\tfrac{1}{2}$$

The molecule is paramagnetic since it has one unpaired electron in a π_p^* antibonding MO. The bond order of $2\frac{1}{2}$ for NO is reasonable because the N$\overset{\cdots}{=}$O bond energy of 607 kJ mol^{-1} (Table 8.2) in NO is intermediate between the N≡N bond energy of 945 kJ mol^{-1} in N_2 and the O=O bond energy 498 kJ mol^{-1} in O_2.

Example 8.9

Predicting the Relative Bond Lengths in Heteronuclear Diatomic Species

Which of the following species has the shortest bond and which has the longest bond: **(a)** NO; **(b)** NO^-; **(c)** NO^+?

SOLUTION: The NO^- ion contains one more electron than the NO molecule, which has one more electron than the NO^+ ion. The valence shell MO population of an NO^- ion is $(\sigma_s)^2(\sigma_s^*)^2(\pi_p)^4(\sigma_p)^2(\pi_p^*)^2$, which makes it isoelectronic with an O_2 molecule. In both the NO^- ion and the O_2 molecule, the bond order is 2:

$$BO = \frac{8 - 4}{2} = 2$$

The NO molecule has one fewer electrons in the π_p^* MO than does the NO^- ion. Therefore, the bond order in an NO molecule is $2\frac{1}{2}$. The NO^+ ion has no electrons in the π_p^* MO and therefore has a bond order of 3. The greater the bond order, the stronger and shorter the bond. Thus, the bond in NO^+ is the shortest and the bond in NO^- the longest.

We can now summarize some advantages and disadvantages of the MO and VB theories. One advantage of MO theory is its ability to predict the bond orders and magnetic properties of diatomic molecules and ions. MO theory can also be extended to polyatomic molecules.

VB theory is simpler than MO theory. VB theory can be used in conjunction with the VSEPR theory to predict molecular structures. However, VB theory cannot predict the existence of diatomic molecules such as Li_2, B_2, and C_2, nor can it predict the paramagnetism of O_2. Furthermore, VB theory cannot predict the bond orders in molecules that contain an odd number of electrons.

Summary

Covalent Bonding Theories

Two major theories describe covalent bonding: **valence bond** (VB) **theory** and **molecular orbital** (MO) **theory**. Valence bond theory describes a covalent bond between two atoms as an attraction of two nuclei for the same pair of electrons, which are localized between the nuclei. Electrons are "shared" by atoms when the valence shell orbitals of the atoms overlap. The bonding electrons in a molecule belong to their parent atoms, although they are shared equally or nearly equally between the bonded atoms. In contrast, molecular orbital theory treats a molecule as a "multinuclear atom" in which all electrons of the bonded atoms belong to molecular orbitals that extend over the entire molecule.

Valence Bond Theory

When two atoms combine to form a diatomic molecule, the electron clouds of atomic orbitals or hybrid orbitals overlap to form the bond. Orbitals that result from mixing two or more atomic orbitals are called **hybrid orbitals**. In polyatomic molecules, the bonding orbitals of the central atom are hybrids of two or more atomic orbitals. Mixing one s and one p orbital results in two ***sp*** hybrid orbitals. Mixing one s with two p orbitals results in three ***sp*²** hybrid orbitals, and so on, as shown below:

One s orbital + one p orbital $\longrightarrow$ two ***sp*** hybrid orbitals
One s orbital + two p orbitals $\longrightarrow$ three ***sp*²** hybrid orbitals
One s orbital + three p orbitals $\longrightarrow$ four ***sp*³** hybrid orbitals
One s orbital + three p orbitals + one d orbital $\longrightarrow$ five ***sp*³*d*** hybrid orbitals
One s orbital + three p orbitals + two d orbitals $\longrightarrow$ six ***sp*³*d*²** hybrid orbitals

If the central atom in a molecule has no lone electron pairs, all of its hybrid orbitals are bonding orbitals. If the central atom has any lone pairs, these occupy nonbonding hybrid orbitals. The hybrid orbitals are distributed as far as possible from each other in space, in accordance with VSEPR theory. The number of electron groups (single bonds, multiple bonds, or lone pairs) on the central atom, the electron pair geometry of the central atom, and the hybridization on the central atom are related as follows:

Two groups	Linear	sp
Three groups	Trigonal planar	sp^2
Four groups	Tetrahedral	sp^3
Five groups	Trigonal bipyramidal	sp^3d
Six groups	Octahedral	sp^3d^2

According to valence bond theory, a double bond consists of one **sigma** (σ) **bond** and one **pi** (π) **bond**; a triple bond consists of one σ bond and two π bonds. The electron density in a σ bond is concentrated along the internuclear axis between two bonded nuclei. If the z axis is the internuclear axis, a σ bond can be formed by any of the following types of s and p orbital overlaps: s–s, s–p_z, p_z–p_z, s–hybrid, p_z–hybrid, or hybrid–hybrid *along* the internuclear axis. A π bond is formed by a side-by-side overlap of two p_x or two p_y orbitals. The electron density in a π bond is concentrated on opposite sides of the internuclear axis.

Molecular Orbital Theory

According to molecular orbital theory, when atoms unite to form a molecule, the electrons occupy **molecular orbitals** (MOs). Two atomic orbitals combine to form a **bonding** and an **antibonding** MO. Electrons in an antibonding MO cancel the bonding effect of an equal number of electrons in the corresponding bonding MO. The **bond order** in a diatomic molecule equals one half of the difference between the number of valence electrons occupying bonding MOs and the number of electrons occupying antibonding MOs. If all of the electrons in molecular orbitals are paired, the molecule is **diamagnetic**; if any electrons are unpaired, the molecule is **paramagnetic**.

New Terms

Acetylene (8.2)
Alkane (8.2)
Alkene (8.2)
Alkyne (8.2)
Antibonding molecular orbital (8.4)
Bond order (8.4)
Bonding molecular orbital (8.4)
Diamagnetic substance (8.3)
Ethene (8.2)
Ethyne (8.2)
Hybrid orbital (8.1)
Hybridizing (8.1)
Hydrocarbon (8.2)
Interhalogen compound (8.3)
Marsh gas (8.2)
Molecular orbital (8.1 and 8.4)
Molecular orbital (MO) theory (8.1 and 8.4)
Natural gas (8.2)
Paramagnetic substance (8.3)
Pi (π) bond (8.2)
Saturated hydrocarbon (8.2)
Sigma (σ) bond (8.1)
sp hybrid orbital (8.1)
sp^2 hybrid orbital (8.1)
sp^3 hybrid orbital (8.2)
sp^3d hybrid orbital (8.3)
sp^3d^2 hybrid orbital (8.3)
Unsaturated hydrocarbon (8.2)
Valence bond (VB) theory (8.1 to 8.3)

Exercises

General Review

1. What is the major difference between valence bond theory and molecular orbital theory?
2. What is a hybrid orbital?
3. What is the relationship between the electron pair geometry of a molecule and the hybridization on the central atom? Give examples of this relationship.
4. From a given Lewis structure of a molecule, how would you predict the geometry of the molecule and the hybridization on the central atom?
5. What is a saturated hydrocarbon?
6. What is an unsaturated hydrocarbon?
7. What is a σ bond?
8. What is a π bond?
9. How many σ bonds and how many π bonds does a double bond contain? A triple bond?
10. What are isomers? Give examples of isomers.
11. Explain why nitrogen forms only a trichloride, whereas phosphorus can form both a trichloride and a pentachloride.
12. What is a reasonable explanation for the nonexistence of phosphorus tetrachloride?
13. Sulfur forms a difluoride, a tetrafluoride, and a hexafluoride, but no trifluoride or pentafluoride. Explain.
14. Sketch the cross sections of the shapes of each of the following molecular orbitals: **(a)** σ_{1s}; **(b)** σ^*_{1s}; **(c)** σ_{2p}; **(d)** σ^*_{2p}; **(e)** π_{2p}; **(f)** π^*_{2p}.
15. How is the bond order of a diatomic molecule predicted by molecular orbital theory?
16. How does molecular orbital theory explain the magnetic properties of a diatomic molecule?

Hybrid Orbitals

17. How many atomic orbitals contribute to a set of three sp^2 hybrid orbitals? Which orbitals are they?
18. What is the geometry of the distribution of: **(a)** sp hybrid orbitals of an atom; **(b)** sp^2 hybrid orbitals; **(c)** sp^3 hybrid orbitals; **(d)** sp^3d hybrid orbitals; **(e)** sp^3d^2 hybrid orbitals?
19. What is the structure and hybridization on each of the carbon atoms in the molecules of: **(a)** methane; **(b)** ethane; **(c)** ethene (ethylene); **(d)** ethyne (acetylene)?
20. What is the electron pair structure, molecular structure, and hybridization on the boldfaced atom in each of the following molecules or ions? **(a)** $\mathbf{B}F_3$; **(b)** $\mathbf{Si}F_4$; **(c)** $\mathbf{N}H_4^+$; **(d)** $\mathbf{C}O_2$; **(e)** $\mathbf{N}O_3^-$; **(f)** $\mathbf{C}O_3^{2-}$; **(g)** $\mathbf{N}O_2^+$; **(h)** $H_2\mathbf{C}O$; **(i)** $\mathbf{P}F_5$; **(j)** $\mathbf{I}F_5$; **(k)** $\mathbf{S}F_4$; **(l)** $\mathbf{S}F_6$.
21. What is the hybridization on the boldfaced atom or atoms in each of the following molecules or ions? **(a)** $\mathbf{N}H_3$; **(b)** $H_3\mathbf{O}^+$; **(c)** $\mathbf{O}F_2$; **(d)** $H_3\mathbf{CO}H$; **(e)** $H_3\mathbf{CN}H_2$; **(f)** $\mathbf{C}_3H_8$; **(g)** $\mathbf{C}_2Cl_2H_2$; **(h)** $CH_3\mathbf{C}H{=}\mathbf{C}H_2$; **(i)** $CH_3\mathbf{C}{\equiv}CH$; **(j)** $H_2\mathbf{N}NH_2$; **(k)** $\mathbf{N}O_2^-$; **(l)** $\mathbf{S}O_3^{2-}$.

Sigma and Pi Bonds

22. According to valence bond theory, how many σ bonds and how many π bonds (if any) does each of the following molecules contain? **(a)** Fluorine; **(b)** nitrogen; **(c)** ethene; **(d)** ethyne; **(e)** propane; **(f)** germanium tetrachloride; **(g)** phosphine, PH_3; **(h)** hydrogen selenide; **(i)** HCO_2H.
23. Would you predict a σ bond to be stronger or weaker than a π bond on the basis of the location of the electron density of the bond relative to the internuclear axis?

Equation Writing

24. Write a balanced equation for each of the following reactions: **(a)** the formation of beryllium chloride from its elements at high temperature; **(b)** the formation of boron trichloride from its elements at high temperature; **(c)** carbon reacts with chlorine at high temperature; **(d)** the formation of ammonia from nitrogen and hydrogen; **(e)** preparation of hydrogen fluoride from hydrogen and fluorine.

Molecular Orbital Theory

25. Draw a relative energy-level diagram for the molecular orbitals of an N_2 molecule. Label each molecular orbital. How is this order of molecular orbital energies different from those for O_2 and F_2?
26. By drawing relative energy levels for molecular orbitals as dashes and writing arrows as electrons to indicate their relative direction of spin, describe the molecular orbital population of the following diatomic molecules and ions: **(a)** C_2; **(b)** O_2; **(c)** O_2^-; **(d)** O_2^{2-}; **(e)** O_2^+; **(f)** CO; **(g)** CO^+; **(h)** CO^-; **(i)** CN^-; **(j)** F_2^+.
27. Write an abbreviated description of the molecular orbital population of each of the species in question 26 [for example, Li_2: $(\sigma_{2s})^2$].
28. What is the bond order in each species in question 26?
29. For each species in question 26, state whether it is diamagnetic or paramagnetic.
30. When an electron is removed from an N_2 molecule, the bond energy decreases and the internuclear distance increases. The removal of an electron from an O_2 molecule causes the opposite effects. Explain.

CHAPTER 9

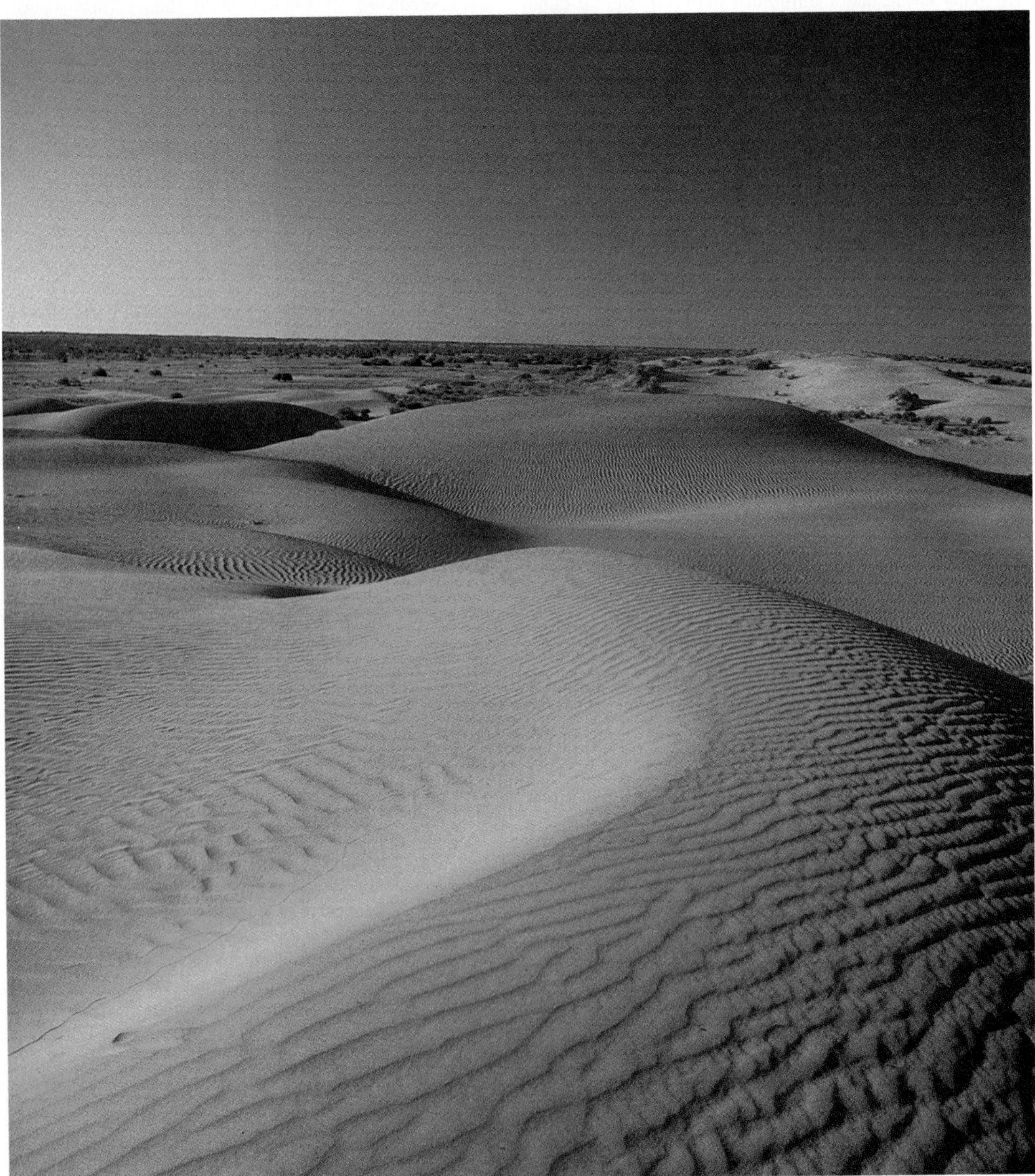

Sand dune of the Australian outback. The vast expanse of land mass and its surrounding atmosphere contain most of the nonmetals and their compounds. Sand is particularly rich in silicon dioxide.

Chemistry of Nonmetals

In this chapter we discuss the chemistry of nonmetals. In compounds, nonmetals form predominantly covalent bonds with one another. Thus, this chapter is an extension of the covalent structures and bonding theories we have just completed in Chapters 7 and 8. In Chapter 8 we discussed the valence bond and molecular orbital theories of covalent bonding, taking most of the examples from compounds whose central atom was a second-period element. We recall that atoms of period 3 and beyond have d *orbitals available for bonding. Therefore, even though nonmetallic elements in a given group form similar compounds, the heavier elements in a group form some compounds that have no counterparts in the first member of the group. In this chapter we examine many molecules where* d *orbitals are required for bonding.*

9.1 Binary Compounds of Hydrogen

We begin with binary compounds of hydrogen. Hydrogen has an electronegativity of 2.1 on the Pauling scale. This value is in the middle range of electronegativity values of the elements (0.8 for cesium to 4.0 for fluorine). There is a small electronegativity difference between hydrogen and most elements, and hydrogen forms predominantly covalent bonds except in compounds of alkali metals and the alkaline earth metals (calcium, strontium, and barium).

Hydrides of Group I, II, and III Elements

Hydrogen typically forms similar hydrides with all the elements in a group, with a few exceptions. Group I and group II metals react with hydrogen gas at elevated temperatures to form metal hydrides. For example, the reaction of sodium with hydrogen gives sodium hydride:

$$2\text{Na}(s) + \text{H}_2(g) \longrightarrow 2\text{NaH}(s)$$

Like most alkali metal hydrides, sodium hydride is a reactive ionic solid that consists of sodium ions, Na^+, and hydride ions, H^-.

Hydrides of beryllium and magnesium, BeH_2 and MgH_2, respectively, are solids at room temperature that consist of covalently bonded metal and hydrogen atoms (Figure 9.1). Hydrides of the heavier elements of group II are ionic solids.

We recall that the group III element boron does not form a stable compound with the formula BH_3, although we might expect it to do so since a boron atom has three valence electrons. Instead, boron forms a family of binary hydrogen compounds called **boranes**. The simplest of these is *diborane*, B_2H_6, which is a gas at room temperature. Other boranes include tetraborane, B_4H_{10}, which is a liquid below 18°C; pentaborane(9), B_5H_9, a liquid that boils at 48°C; pentaborane(11), B_5H_{11}, a liquid boiling at 63°C; and decaborane(14), $B_{10}H_{14}$, a solid melting at 99.7°C. The structures of B_5H_9 and B_6H_{10} are shown in Figure 9.2.

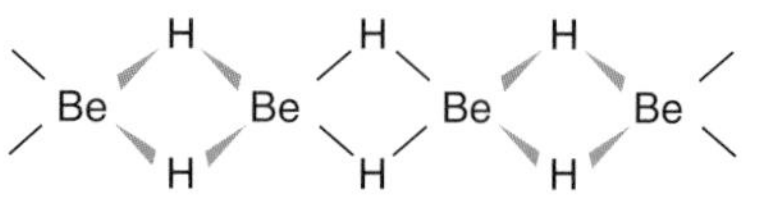

Be H Be H Be H Be (bridging H atoms above and below)

Figure 9.1

Structure of beryllium hydride.

The boranes have no large-scale uses, but their unique structures and bonding have been of great interest to research chemists. For example, borane molecules do not contain enough valence electrons to explain the bonding in terms of conventional valence bond theory.

Of the group III metallic elements, only aluminum forms a hydride, AlH_3. At 25°C this compound is a white solid. In the vapor state, AlH_3 consists of trigonal planar molecules in which the aluminum atoms are sp^2 hybridized. Thus, the structure of this molecule and the hybridization on the central atom are similar to those of BCl_3 discussed in Chapter 8.

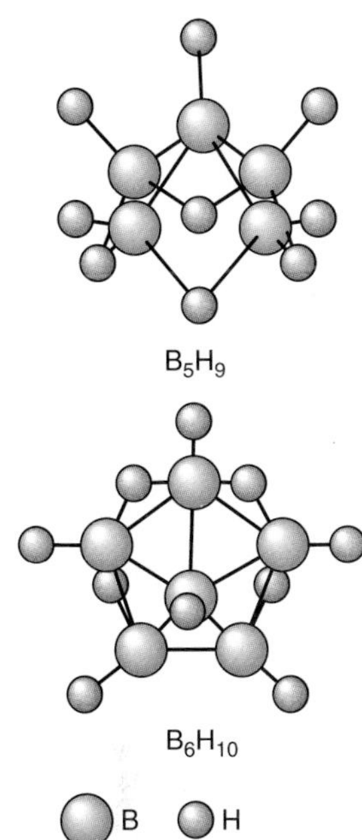

Figure 9.2

Structures of B_5H_9 and B_6H_{10}.

Hydrogen Compounds of Group IV Elements

Group IV elements form binary hydrogen compounds with the general formula MH_4, where M is a group IV atom. Carbon forms methane, CH_4, and silicon forms monosilane, SiH_4. Hydrides of germanium, tin, and lead also exist.

Like the alkane series (Chapter 8) of hydrocarbons, silicon forms molecules called **silanes**. Figure 9.3 illustrates the structural similarities of alkanes and silanes. In the alkane series there is no known limit of the number of carbon atoms per molecule. The largest silane molecule is hexasilane, Si_6H_{14}. Silanes are used in the manufacture of many commercial products including silicone rubber. The silanes are much more reactive than the alkanes.

From the discussion above, we can see that the formulas for the binary hydrogen compounds of the elements in groups I through IV (except boron) can be predicted from the number of valence electrons on the central atom in each of these groups. For the elements of groups V, VI, and VII, the formulas of binary hydrogen compounds can be predicted from the number of *unpaired* valence electrons.

Hydrogen Compounds of Group V Elements

The atoms of group V elements have three unpaired valence electrons. Thus, nitrogen forms NH_3, phosphorus forms PH_3, and arsenic forms AsH_3. PH_3 is called phosphine, and AsH_3 is arsine. Phosphine and arsine are both extremely poisonous gases. Each molecule of NH_3, PH_3, and AsH_3 has three bonding

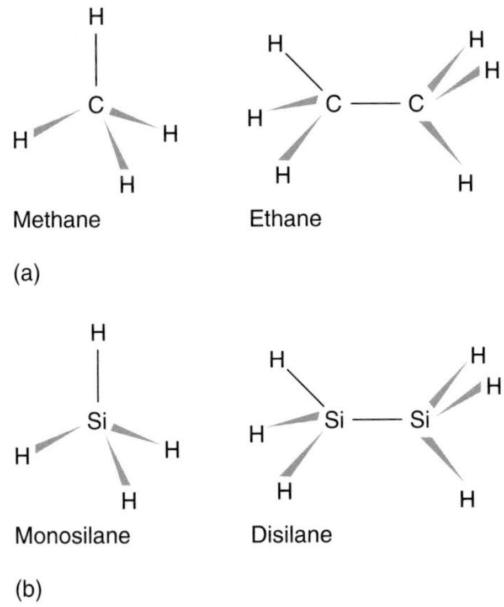

Methane   Ethane

(a)

Monosilane   Disilane

(b)

Figure 9.3

Structures of the first two members of (a) alkanes, (b) silanes.

electron pairs and one lone pair. Thus, each of these molecules has a trigonal pyramidal structure. The hydrides of antimony (SbH_3) and bismuth (BiH_3) are unstable and have been obtained only in trace quantities.

Arsine can easily be decomposed to its elements upon heating. This property of arsine is the basis for a qualitative test for arsenic called the *Marsh test*, which is used in criminal proceedings when arsenic poisoning is suspected. Arsenic compounds in the human body are converted to arsenic acid, H_3AsO_4. In a Marsh test, a sample of stomach contents is treated with hydrogen gas, which reacts with any arsenic acid in the sample to form arsine:

$$4H_2(g) + H_3AsO_4(aq) \longrightarrow AsH_3(g) + 4H_2O(l)$$

Gaseous arsine is passed through a heated glass tube, where it decomposes and forms a metallic arsenic mirror:

$$2AsH_3(g) \longrightarrow 2As(s) + 3H_2(g)$$

Arsenic poisons often contain arsenate, AsO_4^{3-}, and arsenite, AsO_3^{3-} ions. Arsenite ions react with sulfhydryl, $-SH$, groups in biological catalysts called enzymes. One arsenite ion reacts with two sulfhydryl groups that are connected to carbon atoms in enzymes (Figure 9.4). The result of this reaction can be fatal.

Hydrogen Compounds of Group VI Elements

Oxygen (in group VI) forms two important binary compounds with hydrogen. One is water, the other is hydrogen peroxide. Hydrogen peroxide in 3 percent aqueous solution is used as a germicide and bleach. Pure hydrogen peroxide is a pale blue liquid that freezes at $-0.89°C$ and boils at $152°C$. Hydrogen peroxide decomposes exothermally to water and oxygen:

$$2H_2O_2(l) \longrightarrow 2H_2O(l) + O_2(g) \qquad \Delta H° = -5.7 \text{ kJ}$$

The heavier elements in the oxygen family—sulfur, selenium, tellurium, and (radioactive) polonium—all form binary hydrogen compounds consisting of bent molecules with structures similar to that of water. In contrast with water, which is a liquid at room temperature, hydrogen sulfide (H_2S), hydrogen selenide (H_2Se), and hydrogen telluride (H_2Te) are all toxic gases that have unpleasant odors. Hydrogen sulfide, the most common of these compounds, has the odor of rotten eggs.

Hydrogen Halides

The binary hydrogen compounds of the halogens are all gases at room temperature. They can be prepared by direct reaction of hydrogen and the halogen, X_2:

$$H_2(g) + X_2(g) \longrightarrow 2HX(g)$$

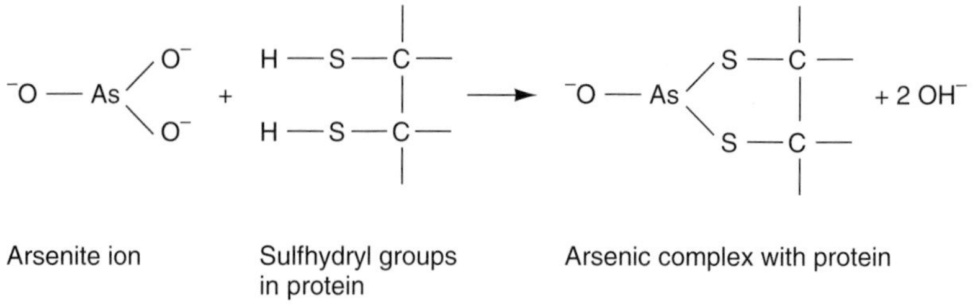

Figure 9.4

Reaction of arsenite ion with sulfhydryl groups.

Arsenite ion + Sulfhydryl groups in protein → Arsenic complex with protein + 2 OH^-

Hydrogen halides have pungent irritating odors. They damage the respiratory tract, and their aqueous solutions are acidic. The aqueous solution of hydrogen chloride is hydrochloric acid. Hydrochloric acid is sold in hardware stores under the name muriatic acid (Latin *murus*, wall). It is commonly used to clean cement walls, patios, driveways, and bricks.

Hydrogen fluoride and hydrogen chloride can be prepared by the action of concentrated sulfuric acid on calcium fluoride and sodium chloride, respectively:

$$CaF_2(s) + H_2SO_4(l) \longrightarrow 2HF(g) + CaSO_4(s)$$

$$NaCl(s) + H_2SO_4(l) \longrightarrow HCl(g) + NaHSO_4(s)$$

Hydrogen bromide and hydrogen iodide cannot be prepared this way because sulfuric acid oxidizes bromide and iodide ions to their respective elements, bromine and iodine, as shown below for the reactions of sodium bromide and sodium iodide with sulfuric acid:

$$2NaBr(s) + 2H_2SO_4(l) \longrightarrow Br_2(l) + SO_2(g) + Na_2SO_4(s) + 2H_2O(l)$$

$$8NaI(s) + 9H_2SO_4(l) \longrightarrow 4I_2(s) + H_2S(g) + 8NaHSO_4(s) + 4H_2O(l)$$

Note that NaBr reduces H_2SO_4 to SO_2, but NaI reduces it to H_2S.

HBr and HI can be prepared from their sodium salts by reactions with a nonoxidizing acid such as concentrated phosphoric acid, H_3PO_4:

$$NaBr(s) + H_3PO_4(l) \longrightarrow HBr(g) + NaH_2PO_4(s)$$

$$NaI(s) + H_3PO_4(l) \longrightarrow HI(g) + NaH_2PO_4(s)$$

Hydrogen halides (HX) can also be prepared by the reaction of phosphorus trihalides (PX_3) with water:

$$PX_3(g, l, s) + 3H_2O(l) \longrightarrow 3HX(g) + H_3PO_3(aq)$$

The binary hydrogen compounds of the second- and third-period elements are listed in Figure 9.5. The hydrides of the first- and second-group metals, as well as aluminum hydride, are solids at room temperature. The remaining compounds listed in Figure 9.5 are molecular compounds. With the exception of water, they are gases under ordinary conditions.

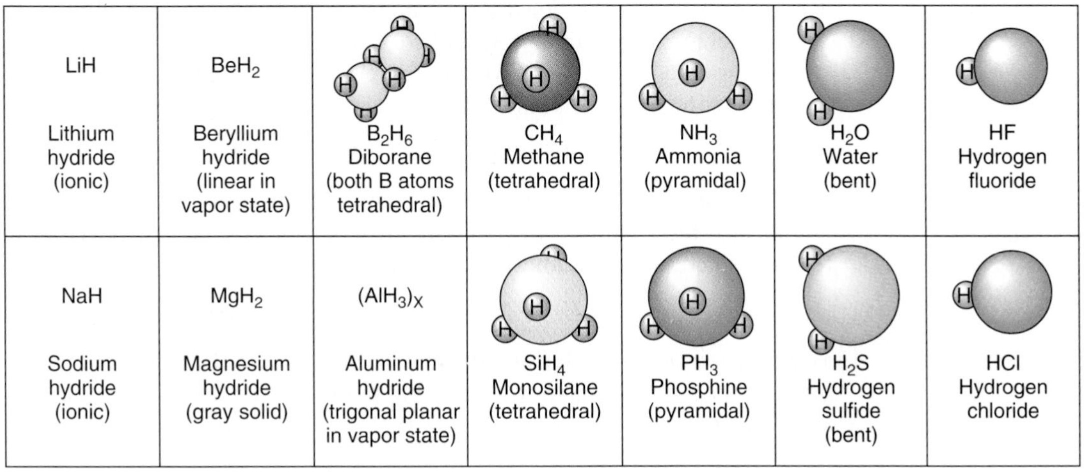

Figure 9.5

Binary hydrogen compounds of the second and third period elements. All the hydrides of the first and second group metals, as well as aluminum, are solids at room temperature, water is a liquid, and the rest of the compounds shown are gases.

LiH Lithium hydride (ionic)	BeH_2 Beryllium hydride (linear in vapor state)	B_2H_6 Diborane (both B atoms tetrahedral)	CH_4 Methane (tetrahedral)	NH_3 Ammonia (pyramidal)	H_2O Water (bent)	HF Hydrogen fluoride
NaH Sodium hydride (ionic)	MgH_2 Magnesium hydride (gray solid)	$(AlH_3)_X$ Aluminum hydride (trigonal planar in vapor state)	SiH_4 Monosilane (tetrahedral)	PH_3 Phosphine (pyramidal)	H_2S Hydrogen sulfide (bent)	HCl Hydrogen chloride

9.2 Binary Halogen Compounds

Halides of Group I, II, and III Elements

The halides of group I and II metals are stable ionic solids with high melting points. Sodium chloride is familiar to all of us as table salt. Sodium bromide is used as a sedative, and calcium chloride as a drying agent, to mention just a few of the common uses of metal halides. Of the trihalides of boron, BF_3 is a gas at room temperature; BCl_3 is a liquid below 12.5°C; BBr_3 is a liquid that boils at about 91°C, and BI_3 is a solid with a melting point of 49.9°C.

Aluminum fluoride is a solid that melts at 1290°C, suggesting a high degree of ionic character in bonding. Aluminum chloride is a solid that melts at 192.4°C. This relatively low melting point indicates that $AlCl_3$ is not an ionic solid. Aluminum chloride can be prepared by a reaction of aluminum with chlorine:

$$2Al(s) + 3Cl_2(g) \longrightarrow 2AlCl_3(s)$$

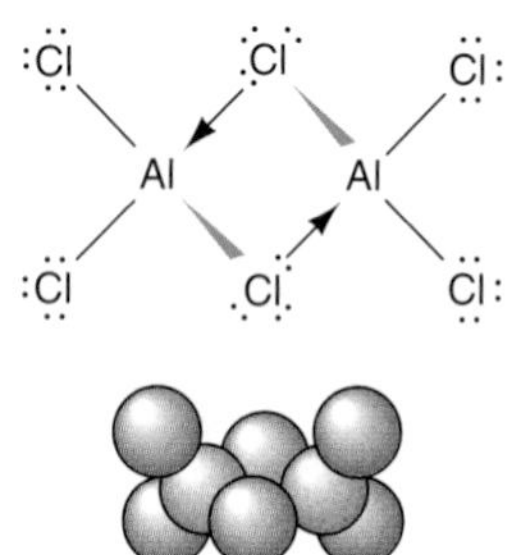

Figure 9.6

The structure of Al_2Cl_6 in its liquid and vapor state. The arrows indicate coordinate covalent bonds between the Al and Cl atoms.

Aluminum chloride in its solid state consists of layers of covalently bonded aluminum and chlorine atoms. When $AlCl_3$ dissolves in water, the solution contains Al^{3+} and Cl^- ions.

In its liquid and gaseous states, aluminum chloride consists of Al_2Cl_6 molecules (Figure 9.6). At high temperature, the vapor of aluminum chloride contains trigonal planar $AlCl_3$ molecules similar to BCl_3. Aluminum also forms Al_2Br_6 and Al_2I_6, which are molecular compounds.

Two $AlCl_3$ units combine to give a molecule of Al_2Cl_6. A *molecule that consists of several identical characteristic units* is called a **polymer**. The *smallest repeating structural unit in a polymer* is called a **monomer**. A polymer that consists of *two monomers* is a **dimer**, *three monomers* make a **trimer**, and *four monomers* comprise a **tetramer**. For example, Al_2Cl_6 is a dimer consisting of $AlCl_3$ monomers.

In the molecule of Al_2Cl_6 shown in Figure 9.6, a lone pair from a Cl atom in each $AlCl_3$ monomer is donated to the Al atom of the other monomer to form two coordinate covalent bonds, shown by arrows in Figure 9.6.

Halides of the Carbon Family

The group IV elements form tetrahalides that consist of tetrahedral molecules. The tetrachloride of each group IV element can be made by heating the element with chlorine gas. Carbon tetrachloride can also be prepared by the reaction of chlorine with carbon disulfide, CS_2:

$$CS_2(l) + 3Cl_2(g) \longrightarrow CCl_4(l) + S_2Cl_2(l)$$

Carbon tetrachloride and the tetrachlorides of other group IV elements are liquids at room temperature. Silicon tetrachloride reacts vigorously with water to form silicon dioxide and hydrochloric acid:

$$SiCl_4(l) + 2H_2O(l) \longrightarrow SiO_2(s) + 4HCl(aq)$$

The reaction of silicon tetrachloride with moisture in the atmosphere is used to produce smoke screens in warfare. When dispersed in air, silicon dioxide forms a dense smoke.

Halides of Group V Elements

Group V elements form halides whose molecules contain three or five halogen atoms. When elemental phosphorus is treated with chlorine gas, phosphorus trichloride, a low-boiling liquid, is obtained:

$$P_4(s) + 6Cl_2(g) \longrightarrow 4PCl_3(l)$$

This product is the phosphorus analog of NCl_3. In contrast to NCl_3, which explodes easily, PCl_3 is stable. Like nitrogen, phosphorus and the other elements of group V form trifluorides and trichlorides. But unlike nitrogen, the other elements in this group also form tribromides and triiodides (see Table 9.1).

Nitrogen is the smallest group V atom. A nitrogen atom can be surrounded by three fluorine or chlorine atoms in NF_3 or NCl_3, but it cannot combine with three bromine or iodine atoms because they are larger than chlorine or fluorine atoms. The *spatial limitation that makes it difficult, or impossible, to fit large atoms or groups of atoms into a given space* is called **steric hindrance**.

When PCl_3 is treated with *excess* chlorine, phosphorus pentachloride, a white crystalline solid, is formed:

$$PCl_3(l) + Cl_2(g) \longrightarrow PCl_5(s)$$

PCl_5 can also be prepared by the reaction of excess chlorine with elemental phosphorus:

$$P_4(s) + 10Cl_2(g) \longrightarrow 4PCl_5(s)$$

Solid pentachloride of phosphorus consists of PCl_4^+ and PCl_6^- ions. In the vapor state, however, PCl_5 consists of trigonal bipyramidal molecules in which the phosphorus atom is sp^3d hybridized. There is no nitrogen analog for PCl_5 because a nitrogen atom, which does not have d orbitals for bonding, cannot form five bonds (recall Section 8.3).

The pentafluorides of phosphorus, arsenic, antimony, and bismuth, PF_5, AsF_5, SbF_5, and BiF_5, have all been prepared. We might also expect pentachlorides of these elements to exist based on the similarities of the valence shell structures of the group V atoms. Yet only PCl_5 and $SbCl_5$ are known. Only one pentabromide, PBr_5, exists. No pentaiodide exists for any group V element, probably because of steric hindrance. Iodine atoms are larger than chlorine or fluorine atoms, so that five iodine atoms cannot fit around a phosphorus atom. Table 9.1 lists the major halides of the group V elements, as well as their melting and boiling points.

Example 9.1

Silicon Halides

Although a silicon atom has d orbitals available for bonding, only silicon tetrahalides exist, but not pentahalides analogous to PCl_5 and PF_5. Explain.

SOLUTION: A silicon atom has four valence electrons with the configuration $3s^23p^2$. The promotion of an s electron to a p level produces a total of four unpaired electrons capable of forming four bonds with halogen atoms. A phosphorus atom has five valence electrons with the configuration $3s^23p^3$. The promotion of a $3s$ electron to a $3d$ level creates five unpaired electrons for five P—Cl or P—F bonds. Although a silicon atom has d orbitals, it does not have sufficient electrons to promote them to d orbital.

Table 9.1

Halides of Group V Elements

		Fluorides			Chlorides		
Element	Oxidation State	Compound	m.p. (°C)	Normal b.p. (°C)	Compound	m.p. (°C)	Normal b.p. (°C)
N	+3	NF_3	−207	−129	NCl_3	<−40	<71
P	+3	PF_3	−151.5	−101.5	PCl_3	−112	75.5[a]
	+5	PF_5	−83	−75	PCl_5	162 Sublimes at 166.8 Decomposes at 167	
As	+3 +5	AsF_3 AsF_5	−8.5 −80	−63[b] −53	$AsCl_3$	−83	63[b]
Sb	+3 +5	SbF_3 SbF_5	292 7	319 Sublimes at 150	$SbCl_3$ $SbCl_5$	73.4 2.8	283 79[c]
Bi	+3 +5	BiF_3 BiF_5	727 <160	 230	$BiCl_3$	230	447

		Bromides			Iodides		
Element	Oxidation State	Compound	m.p. (°C)	Normal b.p. (°C)	Compound	m.p. (°C)	Normal b.p. (°C)
N	+3						
P	+3	PBr_3	−40	170.9	PI_3	61 Decomposes at >200	
+5	PF_5	PBr_5	Decomposes at <100				
As	+3 +5	$AsBr_3$	32.8	221	AsI_3	146	403
Sb	+3 +5	$SbBr_3$	96.6	280	SbI_3	170	401
Bi	+3 +5	$BiBr_3$	218	453	BiI_3	408	~500

[a] At 749 mm Hg pressure.

[b] At 752 mm Hg pressure.

[c] At 22 mm Hg pressure.

Halides of the Sulfur Family

When elemental sulfur is heated with chlorine, disulfur dichloride forms:

$$S_8(s) + 4Cl_2(g) \longrightarrow 4S_2Cl_2(l)$$

Disulfur dichloride, a yellow liquid with a revolting odor, is used to vulcanize rubber. Vulcanizing is a chemical treatment of rubber that increases its strength, stability, and elasticity.

Further chlorination of S_2Cl_2 at room temperature yields sulfur dichloride, a red liquid with a pungent odor:

$$S_2Cl_2(l) + Cl_2(g) \longrightarrow 2SCl_2(l)$$

The Lewis structures of S_2Cl_2 and SCl_2, respectively, are

$$:\ddot{\underset{..}{Cl}}-\ddot{\underset{..}{S}}-\ddot{\underset{..}{S}}-\ddot{\underset{..}{Cl}}: \quad \text{and} \quad :\ddot{\underset{..}{Cl}}-\ddot{\underset{..}{S}}-\ddot{\underset{..}{Cl}}:$$

Treatment of SCl_2 with chlorine at $-78°C$ produces sulfur tetrachloride, a yellow solid (stable only below $-30°C$):

$$SCl_2(l) + Cl_2(g) \longrightarrow SCl_4(s)$$

SCl_6 is not known, but sulfur hexafluoride, SF_6, can be prepared by heating elemental sulfur with excess fluorine:

$$S_8(s) + 24F_2(g) \longrightarrow 8SF_6(g)$$

By different methods, the lower fluorides, SF_4, SF_2, and S_2F_2 can also be made. The fluorides of sulfur are gases at room temperature.

To form an SF_4 (or SCl_4) molecule, a sulfur atom needs four unpaired electrons. These can be obtained by promoting one of the paired $3p$ electrons to a $3d$ level, as discussed in Chapter 8:

$$3s^2 3p^4 \xrightarrow{\text{promotion}} 3s^2 3p^3 3d^1$$

One $3s$, three $3p$, and one $3d$ orbitals mix to give five sp^3d hybrid orbitals. Four of these are bonding, and one contains the unshared electron pair.

To form an SF_6 molecule, six unpaired electrons are needed on the S atom. These can be obtained by promoting both a $3s$ and a $3p$ electron to the $3d$ level to obtain six unpaired electrons for the six sp^3d^2 hybrid orbitals. These orbitals are arranged in an octahedral structure around the central sulfur atom.

Sulfur hexafluoride is colorless, odorless, tasteless, and nontoxic. Therefore, SF_6 can be inhaled without ill effect, provided that sufficient oxygen is also present. Thus, it can be administered during an x-ray examination of the lungs to produce a more detailed picture. SF_6 is also used as an insulator in high-voltage generators and other electrical equipment.

Tetrafluorides and hexafluorides of selenium and tellurium have been prepared. However, no hexachloride, hexabromide, or hexaiodide exists for any group VI element. There is simply not enough space around the central atom to accommodate six large halogen atoms. Tellurium, the largest stable group VI atom, is the only one that forms a tetraiodide, and only selenium and tellurium form tetrabromides (Table 9.2). Thus, steric hindrance is an important factor in the stability of group VI halides.

Table 9.2

Halides of Sulfur, Selenium, and Tellurium[a]

Element	Oxidation State	Fluorides: Compound	Fluorides: m.p. (°C)	Fluorides: Normal b.p. (°C)	Chlorides: Compound	Chlorides: m.p. (°C)	Chlorides: Normal b.p. (°C)
S	+1	S_2F_2	−120.5	38.5	S_2Cl_2	−80	135.6
	+2	SF_2			SCl_2	−78	59
	+4	SF_4	−124	−40	SCl_4	−30	−15
	+6	SF_6	−50.5	Sublimes at −63.8			
Se	+1				Se_2Cl_2		
	+2				$SeCl_2$	Vapor decomposes at 305	
	+4	SeF_4	−13.8	106	$SeCl_4$	Sublimes	191
	+6	SeF_6	−39	−34.5			
Te	+2				$TeCl_2$	209	327
	+4	TeF_4	Sublimes	>97	$TeCl_4$	225	380
	+6	TeF_6	−36	35.5			

Element	Oxidation State	Bromides: Compound	Bromides: m.p. (°C)	Bromides: Normal b.p. (°C)	Iodides: Compound	Iodides: m.p. (°C)	Iodides: Normal b.p. (°C)
S	+1	S_2Br_2	−40				
	+2						
	+4						
	+6						
Se	+1	Se_2Br_2	Vapor decomposes at 227				
	+2	$SeBr_2$	Vapor decomposes				
	+4	$SeBr_4$	Exists only as a solid which decomposes at 75				
	+6						
Te	+2	$TeBr_2$	210	339			
	+4	$TeBr_4$	380 Decomposes at 421		TeI_4	280 Decomposes at >300	
	+6						

[a] In some cases the melting and boiling points have not been established.

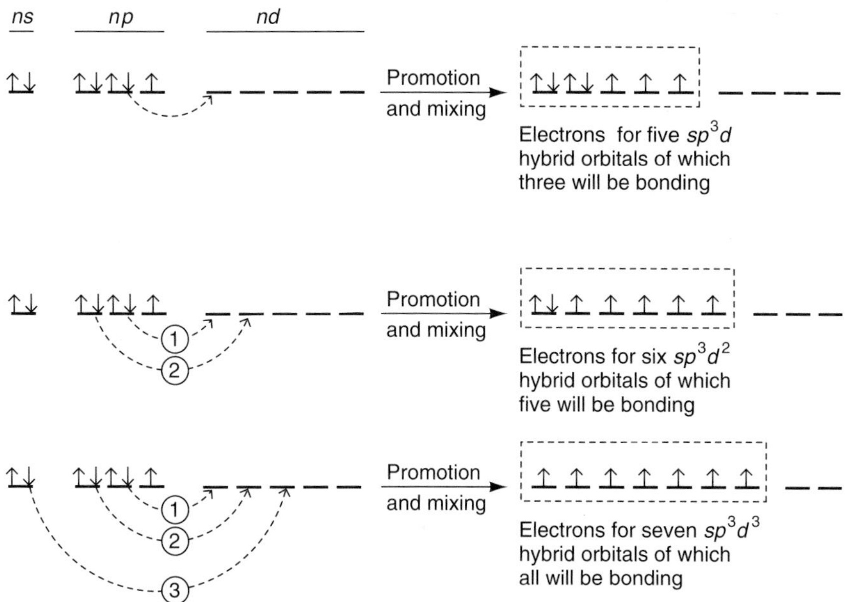

Figure 9.7

Promotion of *ns* and *np* electrons of a halogen atom to produce three, five, and seven unpaired electrons for bonding.

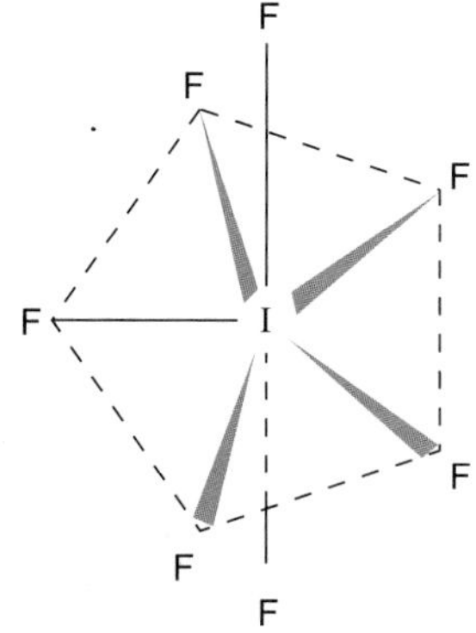

Figure 9.8

Structure of iodine heptafluoride.

Practice Problem 9.1: Explain why sulfur forms SF_2, SF_4, and SF_6, but phosphorus forms PF_3 and PF_5.

Interhalogen Compounds

Interhalogen compounds are binary compounds consisting of two different halogens, such as ICl and ClF_3. Interhalogens have properties intermediate between the properties of their constituent elemental halogens. A halogen atom, X, has seven valence electrons, of which only one is unpaired. To form more than one bond, the valence shell electrons of the halogen atom must be unpaired and promoted to *d* orbitals. Promoting *s* and *p* valence shell electrons to the *d* orbitals in the same shell can produce three, five, or seven unpaired electrons, as shown in Figure 9.7. Fluorine is an exception since, like the other elements in the second period, it does not have *d* orbitals in its valence shell.

Seven unpaired electrons on a halogen atom are in sp^3d^3 hybrid orbitals that are arranged in a pentagonal bipyramidal geometry. A pentagonal bipyramid is a polyhedron in which two pyramids, each having five triangular sides, are joined together by a common pentagonal base. Iodine heptafluoride, IF_7, consists of pentagonal bipyramidal molecules (Figure 9.8).

A halogen atom can have oxidation states of +1, +3, +5, or +7 when it reacts with another more electronegative halogen or oxygen. In an interhalogen compound, the oxidation state of the less electronegative halogen atom equals the number of more electronegative halogen atoms bonded to it. For example, the binary interhalogens of iodine and fluorine have formulas IF, IF_3, IF_5, and IF_7. In these compounds, the iodine atom has oxidation numbers of +1, +3, +5, and +7, respectively.

Although each halogen can form interhalogen compounds, the number of atoms that can be packed around the central atom is at least in part determined by steric hindrance. For example, we have seen that an iodine atom can be surrounded by as many as seven fluorine atoms. But bromine and chlorine atoms are smaller than iodine. Therefore, no heptafluorides for bromine or chlorine exist. The structures of BrF_5 and ClF_3 are shown in Figure 9.9.

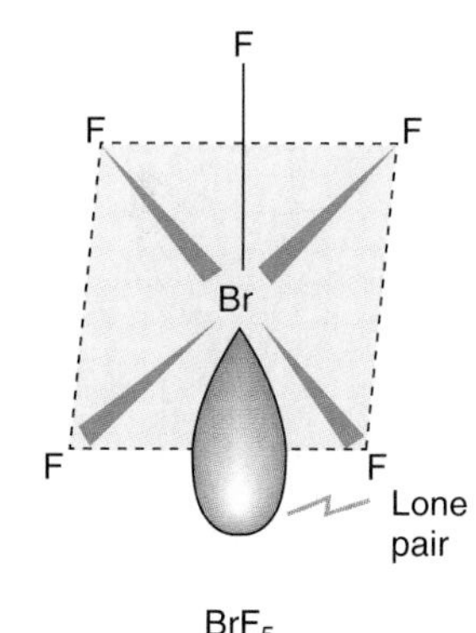

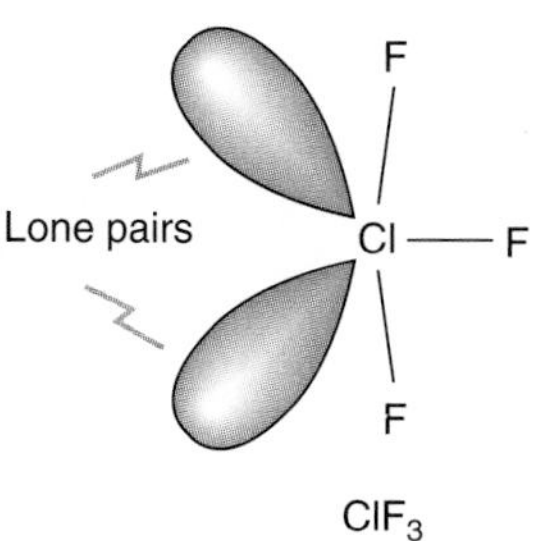

Figure 9.9

Structures of BrF_5 and ClF_3.

Table 9.3

Interhalogen Compounds

		Fluorides			Chlorides			Bromides		
Element	Oxidation State	Compound	m.p. (°C)	Normal b.p. (°C)	Compound	m.p. (°C)	Normal b.p. (°C)	Compound	m.p. (°C)	Normal b.p. (°C)
Cl	+1	ClF	−157	−100						
	+3	ClF_3	−83	11.3						
Br	+1	BrF	−33	−20	BrCl	Unstable; pure compound not prepared				
	+3	BrF_3	8.8	135						
	+5	BrF_5	−61.3	40.5						
I	+1	IF			ICl	27.2	97.4	IBr	42	Decomposes at 116
	+3	IF_3	Decomposes above −28		ICl_3	101[a]	Sublimes at 64			
	+5	IF_5	9.6	98						
	+7	IF_7	5.5	Sublimes at 4.5						

[a] At 16 atm pressure.

Even the large iodine atom cannot be surrounded by more than three chlorine atoms. Thus, ICl and ICl_3, exist, but ICl_5 and ICl_7 do not. Furthermore, iodine can be bonded to only one bromine atom (in IBr), and bromine can be bonded to only one chlorine (in BrCl). The known interhalogen compounds are listed in Table 9.3.

Interhalogen compounds are usually made simply by mixing the elements in a nickel tube, which acts as a catalyst, using appropriate stoichiometric quantities of the reactants. Most interhalogens are quite reactive, and they are often used as halogenating agents (compounds supplying halogen atoms) in the synthesis of other compounds. For example, ClF_3 or BrF_3 can be used to prepare metal fluorides from various metals or metal oxides. The trifluorides are liquids (BrF_3 at room temperature and ClF_3 below 12°C) and are therefore more convenient to store and handle than fluorine gas.

9.3 Noble Gas Compounds

Before 1962, compounds of noble gases were unknown, and these gases were often called "inert" gases because it was thought that they could not combine chemically with other elements. Since 1962, however, various noble gas compounds have been prepared. These are mainly compounds of xenon with fluorine and oxygen, but a few compounds of krypton and radon have also been made.

Let us consider the structures and bonding in some compounds of xenon, beginning with xenon difluoride, XeF_2. The valence shell electron configuration of a xenon atom is $5s^25p^6$. All these electrons are paired. To form an XeF_2

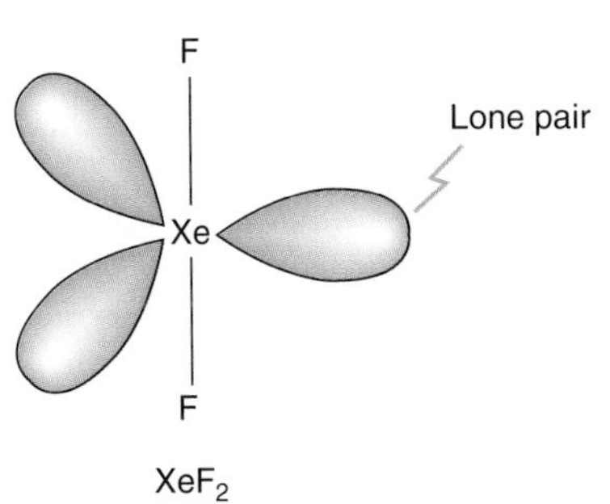

Figure 9.10

Structure of XeF_2.

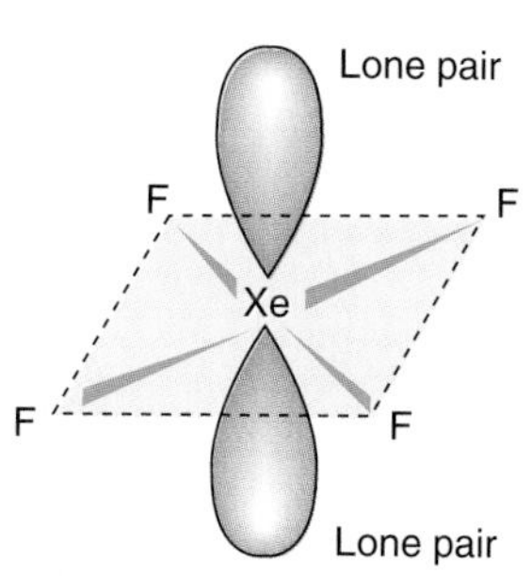

Figure 9.11

Structure of XeF_4.

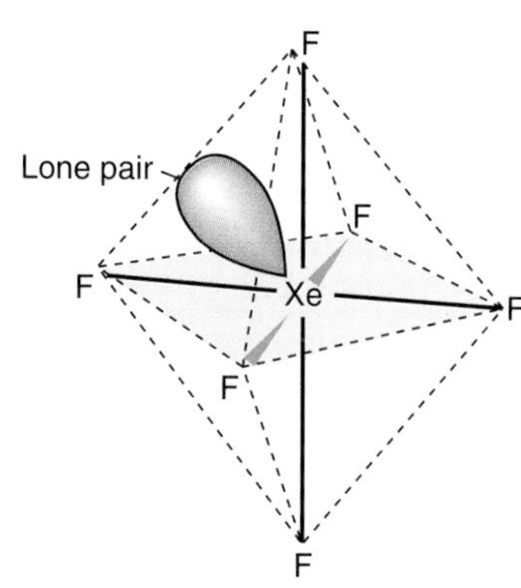

Figure 9.12

Structure of XeF_6.

molecule, a pair of electrons on a xenon atom must be unpaired to get two unpaired electrons for the two Xe—F bonds:

$$\underbrace{\uparrow\downarrow}_{5s^2}\quad \underbrace{\uparrow\downarrow\ \ \uparrow\downarrow\ \ \uparrow\downarrow}_{5p^6}\quad \underbrace{\_\ \ \_\ \ \_\ \ \_\ \ \_}_{5d^0} \longrightarrow$$

$$\underbrace{\uparrow\downarrow}_{5s^2}\quad \underbrace{\uparrow\downarrow\ \ \uparrow\downarrow\ \ \uparrow}_{5p^5}\quad \underbrace{\uparrow\ \ \_\ \ \_\ \ \_\ \ \_}_{5d^1}$$

The Xe—F bonds in an XeF_2 molecule are identical; therefore, the likely bonding orbitals on xenon are sp^3d, which arise from mixing one $5s$, three $5p$, and one $5d$ orbitals. Of these five hybrid orbitals, two are bonding and three contain the lone pairs. The lone pairs occupy the equatorial positions at the corners of a triangular plane bisecting the central xenon atom. The bonding electron pairs occupy axial positions perpendicular to the triangular plane. Thus, as we recall from Chapter 7, an XeF_2 molecule has a linear structure (Figure 9.10).

To form XeF_4 and XeF_6 molecules, additional electron pairs on the xenon atom must be unpaired. To form four Xe—F bonds in XeF_4, two $5p$ electrons on the xenon atom must be promoted to two empty $5d$ orbitals. The valence shell orbitals then hybridize to produce six sp^3d^2 hybrid orbitals. Of these, the four that contain unpaired electrons become bonding orbitals; the two that remain are nonbonding because they each have a pair of electrons. The lone pairs occupy axial positions on opposite sides of the central xenon atom. The bonding pairs are in equatorial positions pointing toward the corners of a square. As we recall from Chapter 7, xenon tetrafluoride has a square planar structure (Figure 9.11). Xenon hexafluoride is believed to have a distorted octahedral structure, with one lone pair on the central atom (Figure 9.12).

The fluorides of xenon—XeF_2, XeF_4, and XeF_6—are among the most stable noble gas compounds. All have negative enthalpies of formation and are thermodynamically stable. In contrast to the xenon fluorides, the oxides of xenon—xenon(VI) oxide, XeO_3, and xenon(VIII) oxide, XeO_4—have positive enthalpies of formation and are thermodynamically unstable. For example, XeO_3 decomposes violently on the slightest provocation. Table 9.4 lists the physical states and the enthalpies of formation of some noble gas compounds.

Table 9.4

Physical States and Enthalpies of Formation of Some Well-Known Noble Gas Compounds

Compound	Physical State at 25°C	ΔH_f° (kJ mol^{-1})
XeF_2	Colorless solid	−109
XeF_4	Colorless solid	−218
XeF_6	Colorless solid	−293
XeO_3	Colorless to white solid	+418
KrF_2	White solid	+59

Example 9.2

Structures of Isoelectronic Species

The valence electron configuration of a triiodide ion, I_3^-, is isoelectronic with that of an XeF_2 molecule. Write the Lewis dot formula for the I_3^- ion and predict the most likely geometry for the ion.

SOLUTION: Both the XeF_2 molecule and the I_3^- ion have 22 valence electrons. The Lewis structure for I_3^- is

$$[:\ddot{\underset{..}{I}}-\ddot{\underset{..}{I}}-\ddot{\underset{..}{I}}:]^-$$

The I_3^- ion and the XeF_2 molecule each has a trigonal bipyramidal electron pair geometry and linear molecular geometry. The three lone pairs are directed toward the corners of an equilateral triangle. The two bonding pairs are arranged axially, perpendicular to the plane of the triangle (see Figure 9.10).

9.4 Oxygen Compounds of Boron, Carbon, and Silicon

Oxygen-containing compounds of nonmetals are abundant in nature. Most *nonmetal oxides react with water to form acid solutions*. Such oxides are called **acidic oxides** or **acidic anhydrides**. For example, sulfur dioxide reacts with water to form sulfurous acid:

$$SO_2(g) + H_2O(l) \longrightarrow H_2SO_3(aq)$$

In contrast, most *metal oxides react with water to produce basic* (*alkaline*) *solutions*. Such oxides are known as **basic oxides** or **basic anhydrides**. For example, barium oxide produces barium hydroxide:

$$BaO(s) + H_2O(l) \longrightarrow Ba(OH)_2(aq)$$

In this section we discuss nonmetal oxides, their corresponding acids, and their salts.

Boron

☐ Boron occurs in nature as a water-soluble mineral, sodium *tetraborate* or *borax*, $Na_2B_4O_7(H_2O)_{10}$, which is a hydrated salt of tetraboric acid, $H_2B_4O_7$.

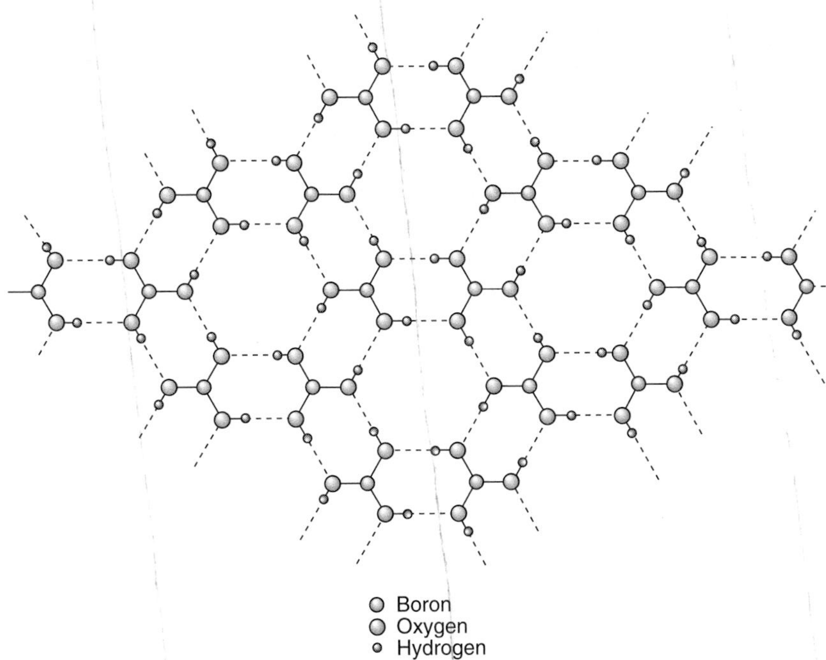

Figure 9.13

Structure of boric acid H_3BO_3.

Borax is sometimes used as a cleaning agent in laundry detergent mixtures. It is also used in welding because of its ability, in the molten state, to dissolve metal oxides, leaving a clean surface for welding or soldering.

When borax is treated with an aqueous solution of sulfuric acid, slightly soluble, weakly acidic *boric acid*, H_3BO_3, is produced:

$$Na_2B_4O_7(H_2O)_{10}(s) + H_2SO_4(aq) \longrightarrow Na_2SO_4(aq) + 5H_2O(l) + 4H_3BO_3(s)$$

Aqueous boric acid is a mild antiseptic that is often used as an eyewash. Boric acid is also used in ointments to soothe mild skin irritations.

Boric acid consists of trigonal planar H_3BO_3 units that are polymerized in sheetlike structures (Figure 9.13). The anhydride of boric acid is boron(III) oxide or boric oxide, B_2O_3. This oxide can be obtained from boric acid by strong heating:

$$2H_3BO_3(s) \longrightarrow B_2O_3(s) + 3H_2O(g)$$

Quick cooling of molten boric oxide yields a solid with properties of glass. ◘ The glass in windowpanes and bottles is made by fusing the oxides of silicon, sodium, and calcium in an approximate mole ratio of 7:1:1. The fusion product is a mixture of sodium silicate and calcium silicate. This mixture, called *soda-lime glass*, cracks rather easily upon sudden temperature changes or when subjected to mechanical shock. When glass is made with lesser amounts of sodium and calcium oxides, but with added B_2O_3, the resulting *borosilicate glass* resists thermal and mechanical shock much better than does soda-lime glass. Borosilicate glasses such as Pyrex and Kimax are used as ovenware and laboratory glassware.

Boron has a +3 oxidation state in virtually all of its common compounds. This is also true of aluminum. However, the heavier elements in group III can also have an oxidation state of +1. For example, when Tl_2O_3 is heated to about 100°C, it loses oxygen and forms Tl_2O, in which the element thallium is in a +1 oxidation state.

Carbonates and Hydrogen Carbonates

Carbon forms two common oxoanions; one is carbonate ion and the other is hydrogen carbonate ion, commonly named bicarbonate ion. Examples of naturally occurring metal carbonates include $CaCO_3$, Na_2CO_3, $MgCO_3$ (magnesite), and $FeCO_3$ (siderite).

Sodium carbonate, or *soda ash*, is a constituent of some soaps and soap powders. Sodium carbonate decahydrate, $Na_2CO_3(H_2O)_{10}$, is known as *washing soda*. Like soda ash, it is used in laundering. Sodium hydrogen carbonate, $NaHCO_3$, is *baking soda*, a principal component of baking powders. Magnesium carbonate is used in toothpaste, in talcum powder, and for glass manufacture.

Calcium carbonate is found principally as *limestone*, but it is also a component in other minerals, including *marble*, *calcite*, *pearls*, *coral*, and *chalk*. Limestone is not soluble in water, but if it is treated with water and carbon dioxide, it slowly converts to a solution of calcium hydrogen carbonate:

$$CaCO_3(s) + H_2O(l) + CO_2(g) \longrightarrow Ca^{2+}(aq) + 2HCO_3^{-}(aq)$$

This reaction is mainly responsible for the formation of limestone caves. When water percolates through rock containing calcium carbonate in the ground above a cavern, some of the salt dissolves, and the solution of calcium hydrogen carbonate may start dripping from the ceiling of the cavern. As water evaporates from this solution in an open cavern, the solution of calcium hydrogen carbonate converts back to solid calcium carbonate:

$$Ca^{2+}(aq) + 2HCO_3^{-}(aq) \longrightarrow CaCO_3(s) + H_2O(g) + CO_2(g)$$

The solid $CaCO_3$ grows down from the ceiling to form *stalactites* and up from the floor to make *stalagmites*. The formation of stalactites and stalagmites is shown schematically in Figure 9.14. A photograph of a cavern with stalagmites and stalactites is shown in Figure 9.15.

A carbonate ion is trigonal planar. It can be represented by the following three Lewis resonance structures:

$$\left[\begin{matrix} & :\ddot{O} & \\ & \| & \\ :\ddot{\underset{..}{O}}- & C & -\ddot{\underset{..}{O}}: \end{matrix}\right]^{2-} \longleftrightarrow \left[\begin{matrix} & :\ddot{O}: & \\ & | & \\ :\ddot{\underset{..}{O}}- & C & =\underset{..}{O}: \end{matrix}\right]^{2-} \longleftrightarrow \left[\begin{matrix} & :\ddot{O}: & \\ & | & \\ :\underset{..}{O}= & C & -\ddot{\underset{..}{O}}: \end{matrix}\right]^{2-}$$

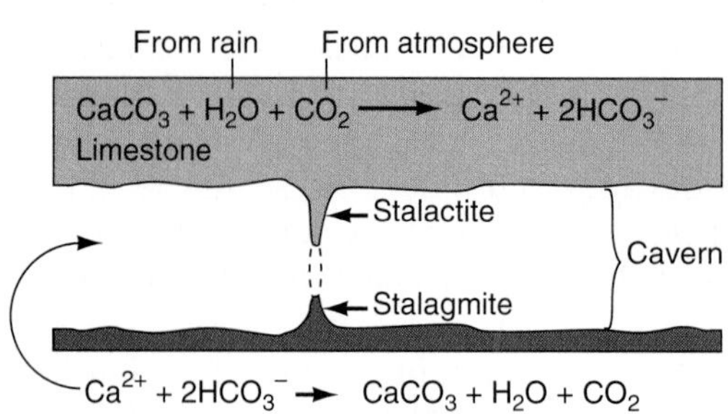

Figure 9.14

Formation of stalactites and stalagmites.

Figure 9.15

A cavern with stalagmites and stalactites.

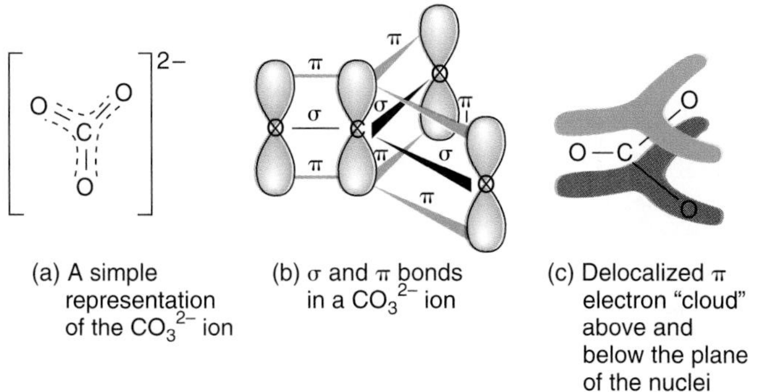

(a) A simple representation of the CO_3^{2-} ion

(b) σ and π bonds in a CO_3^{2-} ion

(c) Delocalized π electron "cloud" above and below the plane of the nuclei

Figure 9.16

Models of σ bonding and delocalizing π bonding in a CO_3^{2-} ion.

The actual electronic structure of the carbonate ion is a resonance hybrid to which each of the resonance forms shown above makes an equal contribution (Figure 9.16). Since each resonance structure makes an equal contribution to the electronic structure of a carbonate ion, the three C—O bonds are identical. Each is somewhat stronger than a C—O single bond but not as strong as a double bond.

Figure 9.16 illustrates various models of bonding in a carbonate ion. Figure 9.16a is a simple representation of the bonding in a carbonate ion showing the σ bonds by solid lines and delocalized π bonding by dashed lines on opposite sides of the plane bisecting the nuclei. Figure 9.16b emphasizes the role of p orbitals in π bonding. Figure 9.16c shows that the π-bonding electrons are *delocalized* throughout the entire ion rather than localized between each carbon and oxygen atom.

The delocalized π electrons in a carbonate ion provide partial π bonding equal to one-third of a π bond for each C—O linkage in a CO_3^{2-} ion. This can be seen from any single resonance structure since each has four bonding electron pairs divided among three C—O linkages. The C—O bond order therefore equals the number of bonding electron pairs divided by the number of C—O linkages per ion:

$$\text{C—O bond order} = \tfrac{4}{3} = 1\tfrac{1}{3}$$

The bond order of $1\frac{1}{3}$ means that each C—O bond in a CO_3^{2-} ion consists of one σ bond and partial bonding equivalent to one-third of a π bond. The structure of CO_3^{2-} is trigonal planar.

Carbon Oxides and Carbonic Acid

Carbon forms two common oxides, carbon monoxide, CO, and carbon dioxide, CO_2. They are colorless, odorless gases. Carbon monoxide is a poisonous gas produced by burning carbon or carbon-containing substances such as wood, coal, or oil in a limited supply of air or oxygen:

$$2C(s) + O_2(g) \longrightarrow 2CO(g)$$

In excess air or oxygen, carbon reacts to form carbon dioxide, CO_2, either directly,

$$C(s) + O_2(g) \longrightarrow CO_2(g)$$

or by the conversion of the initially produced carbon monoxide to dioxide:

$$2CO(g) + O_2(g) \longrightarrow 2CO_2(g)$$

The bonding in CO and CO_2 is illustrated in Figure 9.17.

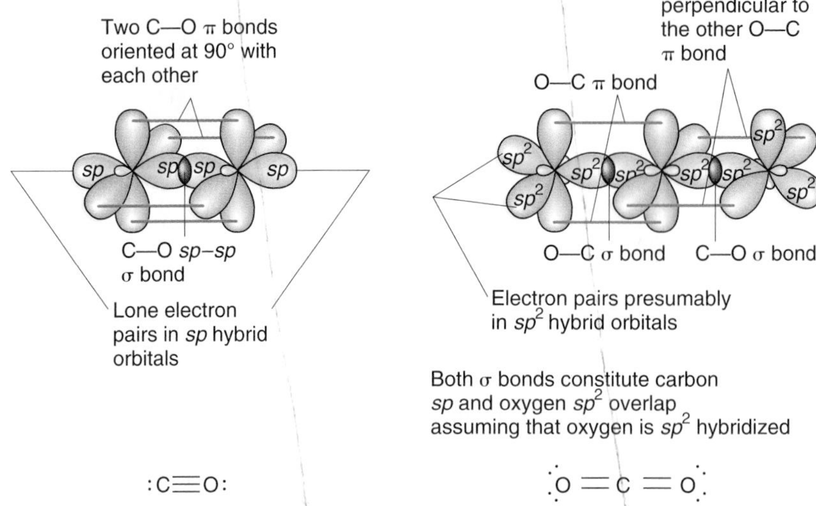

Figure 9.17

Orbital structures and bonding in CO and CO_2.

Carbon monoxide is toxic because a CO molecule forms a coordinate covalent bond by donating a lone electron pair to an Fe^{2+} ion in a protein called hemoglobin. Hemoglobin is found in red blood cells. It transports O_2 to bodily tissues. Carbon monoxide bonds to Fe^{2+} in hemoglobin more strongly than does oxygen. As a result, hemoglobin cannot transport oxygen from the lungs to bodily tissues. The result can be suffocation and death. If carbon monoxide poisoning is detected in time, its effects can be reversed by administering oxygen.

Carbon dioxide dissolves in water to form an acidic solution that contains mostly CO_2 and H_2O molecules plus a small amount of *carbonic acid*, H_2CO_3, a relatively weak acid. A carbonic acid solution contains hydronium ions, H_3O^+, hydrogen carbonate ions, HCO_3^-, and carbonate ions, CO_3^{2-}. Carbonated beverages contain carbon dioxide, carbonic acid, and the ions listed above. Since carbonic acid is a weak acid, the number of ions in its solution is much less than the number of CO_2 molecules.

Silicon Dioxide

In contrast to carbon, which has two common gaseous oxides, the only stable oxide of silicon is silicon dioxide, commonly called *silica*. Silicon dioxide is an extremely stable solid. Silicon dioxide consists of a tetrahedral network of covalently bonded silicon and oxygen atoms in which no individual SiO_2 units or molecules can be distinguished (Figure 9.18). This structure is analogous to that of diamond.

Silicon dioxide is present in *quartz, sand, sandstone, flint, jasper, amethyst* (Figure 9.19), and *agate*. Silicon dioxide dissolves in water very slowly, and only to a slight extent to form orthosilicic acid, H_4SiO_4, or metasilicic acid, H_2SiO_3. Orthosilicic acid is hydrated metasilicic acid, $H_2SiO_3(H_2O)$. When either of these acids is completely dehydrated, silica forms. This dehydration reaction is the source of the silica in petrified wood and in the skeletons of small one-celled animals called diatoms (Figure 9.20).

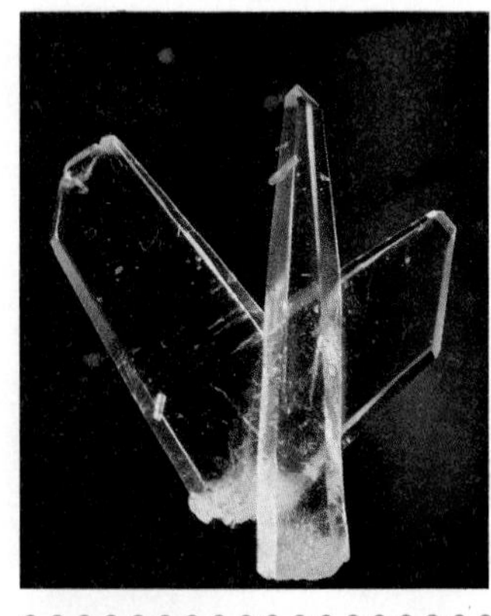

Quartz crystal.

Pure quartz crystals and some other crystalline solids generate an electric potential if they are mechanically deformed in a certain direction. Substances having this property are **piezoelectric** (Greek, *piezin*, to press). Piezoelectric crystals also contract if an electric potential is applied to them. Thus, crystalline quartz can transform mechanical motion into an electric potential differ-

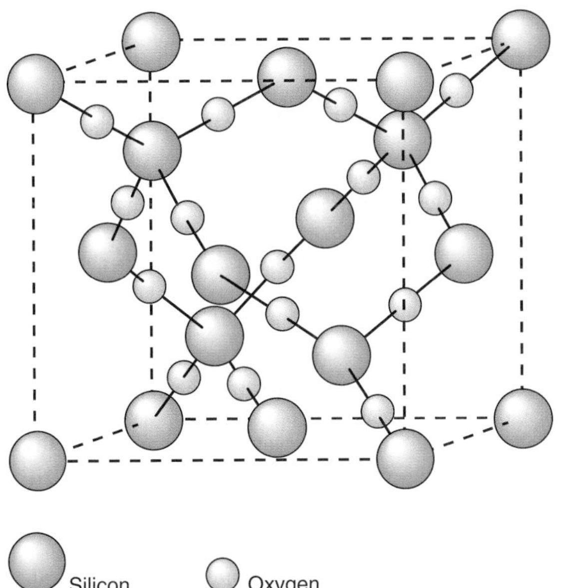

Figure 9.18

Tetrahedral network of silicon and oxygen atoms in SiO_2.

Figure 9.19

An amethyst.

Figure 9.20

Diatoms.

Figure 9.21

Etched glass.

ence, and vice versa. Piezoelectricity has many applications. For example, a quartz plate of appropriate thickness will vibrate in exact resonance with an applied alternating current of a particular frequency. This property is the basis for the operation of quartz watches, where the crystal vibration of a precise frequency is used as the standard of time. Piezoelectric devices are also used in police and citizens' band radio transmitters.

Silicon dioxide reacts with only a few substances, including hydrofluoric acid, fluorine, and very strong hot alkalies such as a concentrated solution of sodium hydroxide. The reaction of silicon dioxide with hydrofluoric acid is

$$SiO_2(s) + 4HF(aq) \longrightarrow SiF_4(g) + 2H_2O(l)$$

Since silicon dioxide is an ingredient of glass, hydrofluoric acid also reacts with glass. This reaction is used to etch glass. The surface of a piece of glass is covered with a layer of wax because wax does not react with hydrofluoric acid. The wax is then removed from certain areas. These exposed areas are treated with hydrofluoric acid, which etches the glass as shown in Figure 9.21.

APPLICATIONS OF CHEMISTRY 9.1
Piezoelectric Eyes, Ears, and Legs for the Handicapped

As you just learned, quartz and certain other crystals exhibit the piezoelectric effect. Such crystals vibrate when in contact with an electric current and they generate electric potentials when they are mechanically deformed. This property has profound applications in medicine and physical therapy.

Piezoelectric material can enable blind persons to "see." Special eyeglasses have been made that convert light waves into electric signals. Those signals are sent to piezoelectric material contained in a large patch measuring about $1\frac{1}{2}$ ft by $1\frac{1}{2}$ ft. The patch is worn by a blind person on his or her back. The piezoelectric material in the patch vibrates when it receives electric signals from the glasses. When the patch receives the electric signals and vibrates, the vibrations form a tactile impulse on the wearer's back. The first time that someone wore the glasses and the piezoelectric patch, something unexpected happened: The wearer thought that he was seeing images in his head. Somehow, the brain converted the tactile vibrations from the patch into mental images.

When wearing this device, a blind person has enough visual capabilities to find articles in a room and read meters; he or she can correctly identify wave patterns displayed on oscilloscopes, and can assemble objects as small as microcircuits.

A similar device exists for the deaf. Like the special glasses for the blind, this device connects a hearing aid to a piezoelectric back patch. The hearing aid converts sound waves into electric signals and sends the signals to the piezoelectric patch. The piezoelectric patch converts the electrical signal into vibrations. Wearers are able to recognize such sounds as ringing bells and barking dogs as well as individually spoken words and short sentences. One 3-year-old wearer can recognize her own name, respond to questions, and has a hearing vocabulary of over 100 words. Surgery is not required to install, this tactile hearing aid.

Piezoelectric materials also help people with artificial limbs to function more normally. When piezoelectric material is placed on the feet of artificial legs, the wearer senses pressure on the feet. While walking, the piezoelectric material on the bottom of the feet converts tactile pressure into electrical signals. The electric signals travel up wires in the artificial leg and connect to the wearer's thighs. The wearer feels these electric signals as tingling sensations, and the brain acts on these sensations as if a real foot had sent them. The brain then directs the movement of the rest of the body so that running and walking are more natural.

Sources: *Science World*, April 5, 1991, page 3, and November 2, 1984, pages 13–15; *Science 81*, March 1981, pages 38–43.

Silicates and Silicones; Polymeric Network Structures

Silica (SiO_2) reacts with sodium carbonate at about 1300°C to form sodium silicate, Na_2SiO_3:

$$Na_2CO_3(s) + SiO_2(s) \longrightarrow Na_2SiO_3(s) + CO_2(g)$$

Many metal *silicates* exist in nature. Silicate ions have polymeric network structures in which each silicon atom is surrounded by four oxygen atoms in a tetrahedral arrangement as shown in Figure 9.22. The SiO_4 tetrahedra share

Figure 9.22

Structure of a silicate chain.

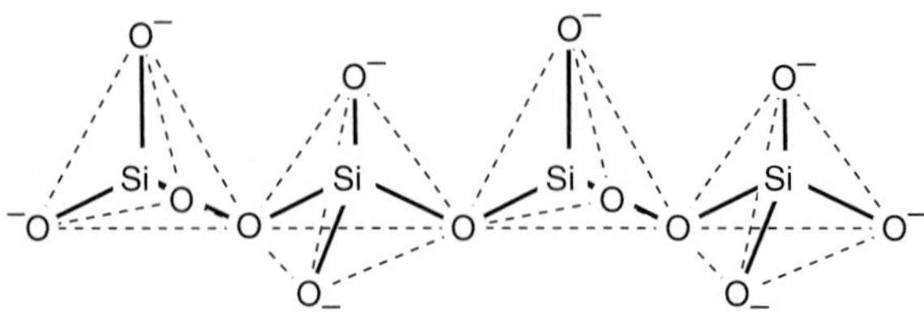

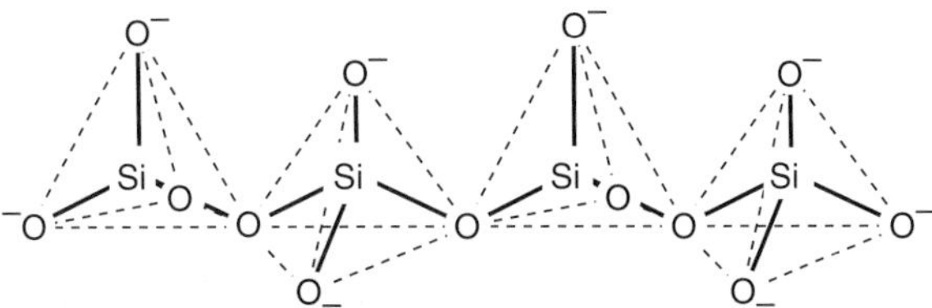

Figure 9.22

Structure of a silicate chain.

$CaSiO_3$. When molten glass is poured on a smooth layer of pure molten tin, it hardens to *plate glass*. The surface of the molten tin is perfectly smooth and so is the glass floating on it. Plate glass produced in this way does not need to be polished after hardening.

When silica is dissolved in an aqueous solution of NaOH, a mixture called *water glass* is obtained:

$$SiO_2(s) + 2NaOH(aq) \longrightarrow \text{``}Na_2SiO_3(aq)\text{''} + H_2O(l)$$

The quotation marks indicate that water glass is not a pure compound but a mixture in which the sodium and silicon content are variable. Water glass is used as a fire retardant on fabrics, as an adhesive in the manufacture of cardboard cartons, and as an egg preservative that acts by sealing the pores in egg shells.

When an aqueous solution of a metal silicate is treated with strong aqueous acid, a semisolid substance called *silica gel* is formed. When silica gel is dried, a porous solid is obtained that is widely used as a drying agent and as an absorbent for certain gases.

Hundreds of silicate minerals also contain aluminum. These minerals are called aluminosilicates. They include such minerals as *feldspars* and *clays*. There are dozens of other silicate minerals of general interest, such as mica and various gems, including *zircon*, *topaz*, *emerald*, *aquamarine*, and *tourmaline* (Figure 9.23). *Micas* are silicates in which the molecules form layers of SiO_4 tetrahedra, so mica has a layer structure.

Clay contains a considerable amount of water. When the water is driven off by strong heating, a glasslike material called a *ceramic* is produced. Ceramics have many applications, from fine porcelain vases and china to the tiles that cover space shuttles to protect them from the fierce heat they generate when reentering the earth's atmosphere.

Another important class of oxygen-containing compounds of silicon are the *silicones*. These are synthetic polymeric substances containing alternating silicon and oxygen atoms in —O—Si—O— chains. Hydrocarbon groups, such as methyl groups ($—CH_3$), are attached to the two remaining bonding positions of each silicon atom. Silicones can be represented by the formula

$$—O—\underset{\displaystyle R}{\overset{\displaystyle R}{\overset{|}{\underset{|}{Si}}}}—\left[O—\underset{\displaystyle R}{\overset{\displaystyle R}{\overset{|}{\underset{|}{Si}}}}\right]_n—O—\underset{\displaystyle R}{\overset{\displaystyle R}{\overset{|}{\underset{|}{Si}}}}—O—$$

where R is a hydrocarbon group and n signifies an indefinite number of the units shown in the brackets. Silicones are chemically inert, water repellant, heat resistant, and good electric insulators. These properties make them useful as protective coatings, lubricants, insulators, sealants, and adhesives. Silicones are biologically inert. They are therefore used cosmetically for certain surgical implants.

(a)

(b)

(c)

Figure 9.23

(a) Aquamarine, (b) topaz, and (c) beryl-emerald.

APPLICATIONS OF CHEMISTRY 9.2
Ceramics and Supermetals

Iraq's invasion of Kuwait in 1990 and the ensuing Desert Storm crisis caused renewed concern by Americans about the accessibility of oil that is controlled by potentially hostile states. A large percentage of our oil is imported and many oil-exporting countries are unfriendly or unstable. There is an equally important crisis of which few Americans are aware: the supermetal crisis. Supermetals are metals that are essential for the operation of an industrial economy.

There are about 35 supermetals. The United States imports 90 to 100 percent of many of them, including manganese (Mn), cobalt (Co), tantalum (Ta), niobium (Nb), aluminum (Al), chromium (Cr), platinum (Pt), rhodium (Rh), and titanium (Ti), and we import them from countries that are also unfriendly or unstable. The U.S. economy and U.S. industry rely heavily on the availability of supermetals. For example, every commercial jet engine contains about 8000 lb of supermetals. One supermetal alone, titanium, accounts for 5000 lb of a typical commercial jet engine's weight. We also use supermetals to make items such as nuclear power plants, submarine hulls, stainless steel, missiles, and electronic devices.

Supermetals are valuable investments. Over one eight-year period, tantalum increased in value by 1480%. At this rate, one dollar's worth of tantalum purchased 20 years ago would be worth about $1,250,000,000 today. Not bad for investors!

To ease our dependence on supermetals, some scientists are trying to create ceramics that will replace the metals. Ceramics consist primarily of forms of aluminum and silicon. The human race has been turning clay and sand (silicates) into ceramic objects for more than 10,000 years. Unlike supermetals, ceramic materials come from dirt, rocks (except limestone and dolomite), and many mineral compounds, most of which are inexpensive and abundant in nature.

Ceramic objects can be easily shaped and, once fired, they can withstand high temperatures without corroding. The tiles on space shuttles are made from ceramics because they are good insulators and are very lightweight. The tiles protect space shuttles from excessive heat during reentry into the earth's atmosphere (see photo at right).

Conventional metal automobile engines need lubrication and coolants to prevent excessive wear and tear. Because of the high heat resistances of ceramics, scientists have been able to make ceramic automobile engines that run without lubrication or coolant. These engines have performed reliably in tests even at temperatures as high as 2200°F. More research is under way into the practical uses of ceramics. In time, the United States may no longer be dependent on foreign supermetal suppliers.

9.5 Oxygen Compounds of Nitrogen

Oxides of Nitrogen

Nitrogen forms several oxides. Their formulas, names, and physical properties are listed in Table 9.5. Nitrogen is in its +1 oxidation state in *dinitrogen monoxide*, also called *nitrous oxide*, N_2O. This oxide can be prepared by carefully heating ammonium nitrate:

$$NH_4NO_3(s) \longrightarrow N_2O(g) + 2H_2O(g)$$

This is a dangerous reaction because ammonium nitrate decomposes explosively if heated above 300°C in the presence of oxidizable substances.

Dinitrogen monoxide is a colorless gas that is sometimes called *laughing gas* because of its exhilarating effect when inhaled. It was one of the first anesthetics. When used as an anesthetic, N_2O must be administered with oxygen; otherwise, suffocation will result. Dinitrogen monoxide was discovered by Joseph Priestly in 1771. He found that a candle burned in an N_2O atmosphere, which suggests that N_2O somehow supplies oxygen. But Priestly found that a mouse could not live in an atmosphere consisting entirely of N_2O.

If N_2O is a source of oxygen for combustion, why doesn't it support living processes, which also require oxygen? The answer to this question is rather subtle. The formation of N_2O from its elements is an endothermic process. The decomposition of N_2O to its elements is therefore exothermic. At the temperature of the flame of a candle, N_2O decomposes exothermally into its elements:

$$2N_2O(g) \longrightarrow 2N_2(g) + O_2(g) \qquad \Delta H° = -164.1 \text{ kJ}$$

Table 9.5

Oxides of Nitrogen

Formula	Name	Physical State at 25°C and 1 atm
N_2O	Nitrogen(I) oxide, dinitrogen monoxide, or nitrous oxide (laughing gas)	Colorless gas
NO	Nitrogen(II) oxide, nitrogen monoxide, or nitric oxide	Colorless gas
N_2O_3	Nitrogen(III) oxide or dinitrogen trioxide (nitrogen trioxide)	Exists in pure form only as a blue solid below −102°C, as a liquid between −102 and 3.5°C; decomposes above 3.5°C to NO and NO_2
NO_2	Nitrogen(IV) oxide or nitrogen dioxide	Brown gas
N_2O_4	Dinitrogen tetroxide	Exists together with NO_2
N_2O_5	Nitrogen(V) oxide or dinitrogen pentoxide (nitrogen pentoxide)	Colorless solid, melts at 30°C and decomposes at 47°C

The liberated oxygen supports combustion, and a candle burns even more brightly in N_2O than in air because a unit volume of N_2O can supply more oxygen than an equal volume of air, which contains only one-fifth oxygen. However, the decomposition of N_2O does not occur at the physiological temperature of a human or a mouse.

Dinitrogen monoxide has a linear structure. It can be considered as a resonance hybrid of three structures:

$$:\dot{\text{N}}=\text{N}=\ddot{\text{O}}: \longleftrightarrow :\text{N}\equiv\text{N}-\ddot{\underset{..}{\text{O}}}: \longleftrightarrow :\dot{\underset{..}{\text{N}}}-\text{N}\equiv\text{O}:$$

The central nitrogen atom in N_2O is *sp* hybridized.

The oxide of nitrogen in which nitrogen is in its +2 oxidation state is called *nitrogen(II) oxide* or *nitric oxide*, NO. Nitrogen(II) oxide is formed by the reaction of atmospheric nitrogen and oxygen at high temperatures:

$$N_2(g) + O_2(g) \longrightarrow 2NO(g) \qquad \Delta H° = +180.8 \text{ kJ}$$

Since the formation of nitrogen(II) oxide is endothermic, it is favored by high temperatures. Nitrogen(II) oxide is a major air pollutant that is produced in every automobile combustion chamber, where air (a mixture of N_2 and O_2) is heated at a high temperature.

The Lewis resonance structures for NO are

$$:\dot{\text{N}}=\ddot{\text{O}}: \longleftrightarrow :\ddot{\text{N}}=\dot{\text{O}}:$$

Nitrogen(II) oxide has an unpaired electron and is therefore paramagnetic.

An oxide of nitrogen in which nitrogen has a +4 oxidation state is *nitrogen dioxide*, NO_2, a brown gas at room temperature. Nitrogen dioxide is formed by the reaction of nitrogen(II) oxide with oxygen:

$$2NO(g) + O_2(g) \longrightarrow 2NO_2(g)$$

A molecule of NO_2, like that of NO, has an unpaired electron which is associated primarily with the nitrogen atom, as shown in the following resonance structures:

$$:\ddot{\underset{..}{\text{O}}}-\dot{\text{N}}=\ddot{\text{O}}: \longleftrightarrow :\ddot{\text{O}}=\dot{\text{N}}-\ddot{\underset{..}{\text{O}}}:$$

Two nitrogen dioxide molecules can react to form *dinitrogen tetroxide*, N_2O_4. Each nitrogen atom of NO_2 contributes an unpaired electron to the N—N covalent bond in an N_2O_4 molecule (Figure 9.24), and the oxidation state of nitrogen in the conversion of nitrogen dioxide to dinitrogen tetroxide remains unchanged. The formation of N_2O_4 is readily reversible, and NO_2 and N_2O_4 are usually found associated with each other.

Figure 9.24

Formation of N_2O_4 from NO_2.

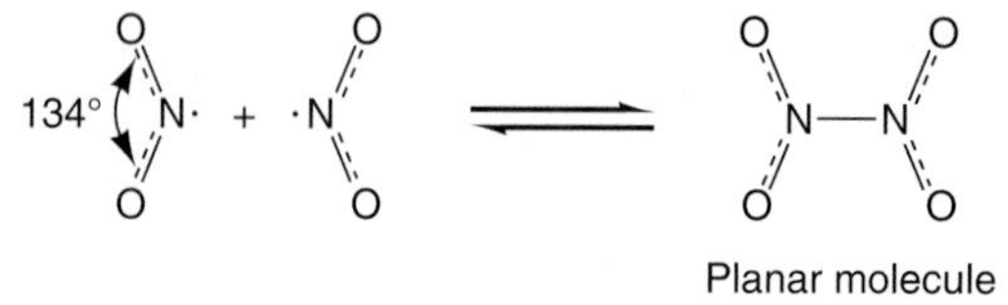

134° · N· + ·N · ⇌ N—N

Planar molecule

Example 9.3

Effect of Unshared Electrons on the Bond Angle of NO_2

In an NO_2 molecule, there is a single unshared electron on the N atom. Using VSEPR theory, explain why NO_2 is bent and not linear. Is the ONO bond angle in NO_2 less or more than what it would be if the N atom had two unshared electrons?

SOLUTION: VSEPR theory predicts that NO_2 is bent because there is an unshared electron on the central nitrogen atom. We expect the ONO bond angle to be 120° or more. The single unshared electron on the nitrogen atom allows the two bonding pairs to be farther apart than they would be if there were a lone *pair* of electrons on the nitrogen atom. Indeed, the experimental bond angle for NO_2 is 134°, as shown in Figure 9.24.

Practice Problem 9.2: Why is the CO_2 molecule linear rather than bent like NO_2 and SO_2?

A nitrite ion, NO_2^-, contains one more electron than an NO_2 molecule and is structurally similar to an NO_2 molecule. The central nitrogen atom in an NO_2^- ion has an unshared electron pair, but there is only one lone electron in an NO_2 molecule. The extra electron on the central nitrogen atom reduces the ONO bond angle to 115° in a nitrite ion compared with 134° in the NO_2 molecule. In both the NO_2 molecule and the NO_2^- ion, the nitrogen atom is surrounded by three sp^2 hybrid orbitals. Two of these are bonding. The third sp^2 hybrid orbital contains one electron in an NO_2 molecule and two electrons in an NO_2^- ion.

Example 9.4

Predicting the Structure of NO_2^+

What is the geometry of an NO_2^+ ion?

SOLUTION: The Lewis structure for an NO_2^+ ion is

$$[:\ddot{O}{=}N{=}\ddot{O}:]^+$$

An NO_2^+ ion is linear because the central nitrogen atom has no unshared electrons. The Lewis structure for NO_2^+ is similar to that for CO_2, which is also a linear species.

The oxide of nitrogen in which the nitrogen is in its +3 oxidation state is *dinitrogen trioxide*, N_2O_3. This oxide is the anhydride of *nitrous acid*, HNO_2. Nitrous acid is formed by the reaction of N_2O_3 with water:

$$N_2O_3(l) + H_2O(l) \longrightarrow 2HNO_2(aq)$$

Nitrous acid cannot be isolated from solution as a pure substance. When a solution of nitrous acid is left standing or is warmed, it decomposes to nitric acid, nitrogen monoxide, and water:

$$3HNO_2(aq) \longrightarrow HNO_3(aq) + 2NO(g) + H_2O(l)$$

In this reaction, nitrogen is both oxidized and reduced. A *reaction in which the same element is both oxidized and reduced* is called a **disproportionation reaction**.

Example 9.5

Structure and Hybridization on N and O Atoms in N_2O_3

What is the structure and hybridization on each of the nitrogen atoms and at the central oxygen atom of a molecule of N_2O_3 (ONONO)?

SOLUTION: An acceptable Lewis structure for an N_2O_3 molecule is

:Ö=N̈—Ö—N̈=Ö:

This structure shows that each of the nitrogen atoms has an unshared electron pair, two σ bonds and a π bond. The lone electron pair and the two σ-bonding electron pairs occupy sp^2 hybrid orbitals on the nitrogen atom. (Recall that the orbitals holding electrons in a π bond are unhybridized.) Thus, the electron pair geometry at each of the nitrogen atoms is trigonal planar, but the molecular geometry is bent.

The two bonding pairs and two lone pairs on the central oxygen atom suggest a modified tetrahedral electron pair geometry, sp^3 hybridization, and a bent molecular structure at that atom (Figure 9.25).

Practice Problem 9.3: Write an acceptable Lewis structure for N_2O_4 and predict the structure and hybridization on each of the nitrogen atoms in the molecule.

In dinitrogen pentoxide, N_2O_5, nitrogen is in its highest, +5, oxidation state. Dinitrogen pentoxide is a solid that melts at 30°C. It reacts with water to produce *nitric acid*:

$$N_2O_5(s) + H_2O(l) \longrightarrow 2H^+(aq) + 2NO_3^-(aq)$$

Dinitrogen pentoxide is therefore the anhydride of nitric acid.

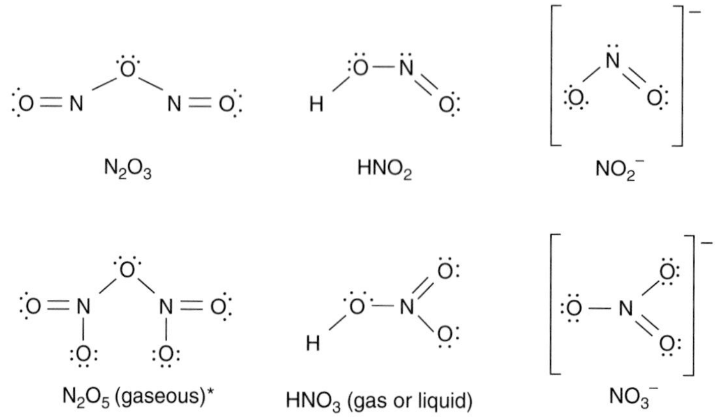

Figure 9.25

Structures of N_2O_3, N_2O_5, their corresponding acids, and the oxoanions derived from the acids.

N_2O_3 HNO_2 NO_2^-

N_2O_5 (gaseous)* HNO_3 (gas or liquid) NO_3^-

Nitric Acid, Nitrates, and Nitrites

Nitric acid, a colorless liquid (m.p. −42°C, b.p. 83°C), can be prepared by treating KNO_3 with 100% H_2SO_4 at 0°C:

$$KNO_3(s) + H_2SO_4(l) \longrightarrow KHSO_4(s) + HNO_3(l)$$

Nitric acid in aqueous solution is a strong acid and it is also a relatively good oxidizing agent. It reacts with most metals and with many of their compounds (see Chapters 17 to 19).

Nitric acid and several nitrates are of great importance in industry. Major industrial uses of nitric acid include the synthesis of ammonium nitrate for use as a fertilizer and in explosives. Ammonium nitrate is potentially dangerous because it can decompose explosively in the presence of oxidizable material or at high temperatures:

$$2NH_4NO_3(s) \xrightarrow{>300°C} 2N_2(g) + 4H_2O(g) + O_2(g)$$

At lower temperatures, ammonium nitrate decomposes smoothly into dinitrogen oxide and water as discussed in the preceding section.

Large quantities of nitric acid are prepared industrially by the *Ostwald process*. First ammonia is converted to nitric oxide by reaction with oxygen in the presence of platinum metal, which acts as a catalyst to increase the rate of the reaction:

$$4NH_3(g) + 5O_2(g) \xrightarrow{Pt} 4NO(g) + 6H_2O(g)$$

The NO produced is then treated with O_2 to form NO_2, which is absorbed by water to produce nitric acid:

$$2NO(g) + O_2(g) \longrightarrow 2NO_2(g)$$

$$3NO_2(g) + H_2O(l) \longrightarrow 2HNO_3(aq) + NO(g)$$

The NO gas produced in this reaction is used as a reactant in the prior reaction to form more NO_2.

Nitric acid can be converted into salts called nitr*ates*; nitrous acid can be converted into salts called nitr*ites*. The structures of N_2O_3 and N_2O_5, their corresponding acids, HNO_2 and HNO_3, and of the nitrite and nitrate ions are shown in Figure 9.25.

Common nitrates of industrial importance include sodium nitrate, $NaNO_3$, used in preserving meats; potassium nitrate, KNO_3, used in fertilizers and for making gunpowder; silver nitrate, $AgNO_3$, used to prepare photographic film and paper; and ammonium nitrate, NH_4NO_3, used in fertilizers and explosives.

The most common *nitrites* are those of the alkali and alkaline earth metals. One of these compounds, sodium nitrite, $NaNO_2$, is used as a food additive. It prevents the growth of *Clostridium botulinum*, a bacterium that causes the food poisoning known as botulism. Sodium nitrite is also added to meat to preserve the red color associated with fresh meat.

However, nitrites are potentially dangerous food additives. They can be converted in the stomach to organic derivatives called nitrosamines (R_2NNO, where R is an organic group). Nitrosamines are carcinogens; that is, they cause cancer. They are also mutagens. That is, they cause changes in the genetic material, DNA. These changes can adversely affect cellular processes.

9.6 Oxygen Compounds of Phosphorus and Its Family

Oxides of Phosphorus

We have seen that most of the oxides of nitrogen are gases under ordinary conditions. In contrast, the oxides of phosphorus, arsenic, antimony, and bismuth are solids. Phosphorus forms two major oxides, *tetraphosphorus hexoxide*, P_4O_6, and *tetraphosphorus decoxide*, P_4O_{10}. The former is usually called *phosphorus trioxide* and the latter *phosphorus pentoxide*, reflecting the empirical formulas of these oxides, P_2O_3 and P_2O_5. The oxides of phosphorus can also be named to reflect the oxidation state of phosphorus. Thus, P_4O_6 is *phosphorus(III) oxide*, and P_4O_{10} is *phosphorus(V) oxide*. The structures of these oxides, and of the P_4 molecule for comparison, are shown in Figure 9.26.

Phosphorus(III) oxide forms when elemental phosphorus burns in a limited supply of oxygen:

$$P_4(s) + 3O_2(g) \longrightarrow P_4O_6(s)$$

Burning phosphorus in excess oxygen produces phosphorus(V) oxide:

$$P_4(s) + 5O_2(g) \longrightarrow P_4O_{10}(s)$$

Phosphorus(V) oxide has a great affinity for water and is among the most powerful dehydrating (drying) agents known.

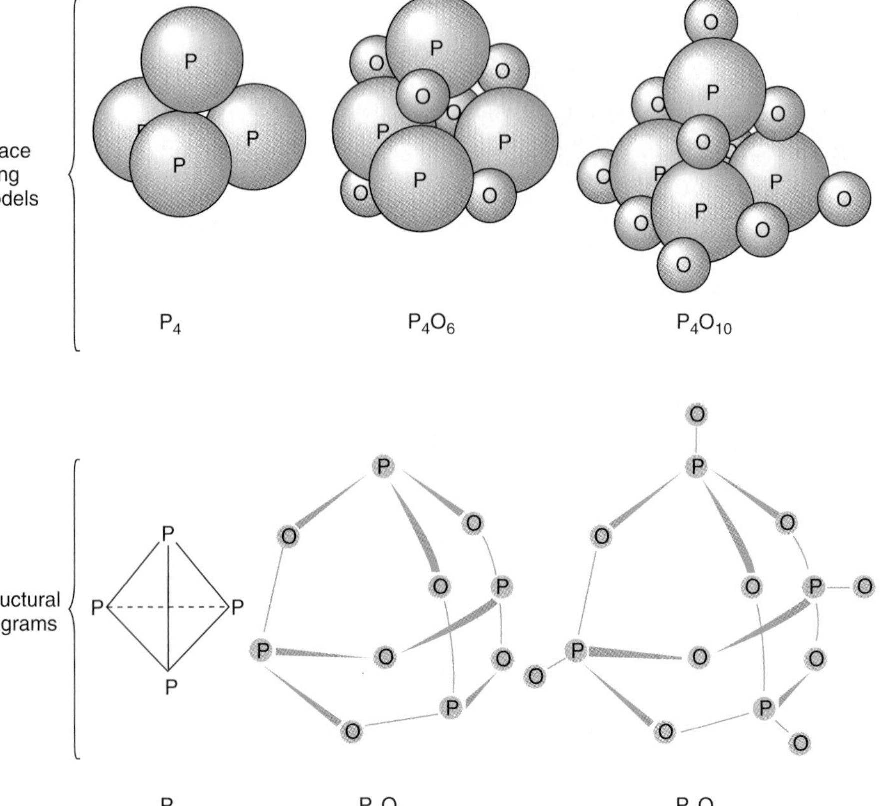

Figure 9.26

Structures of P_4, P_4O_6 and P_4O_{10} molecules.

Common Acids and Salts of Phosphorus

Phosphorus(III) oxide is the anhydride of *phosphorous acid*, H_2PHO_3, and phosphorus(V) oxide is the anhydride of *orthophosphoric acid*, commonly called *phosphoric acid*, H_3PO_4. (Note the difference in the spelling of the element phosphor*us* and the acid, phosphor*ous* acid.) The reactions of these anhydrides with water to produce the corresponding acids are as follows:

$$P_4O_6(s) + 6H_2O(l) \longrightarrow 4H_2PHO_3(aq)$$

$$P_4O_{10}(s) + 6H_2O(l) \longrightarrow 4H_3PO_4(aq)$$

Phosphorous acid can also be prepared by the reaction of phosphorus trichloride with water:

$$PCl_3(l) + 3H_2O(l) \longrightarrow H_2PHO_3(aq) + 3HCl(g)$$

The resulting solution is heated to expel the hydrogen chloride and to evaporate excess water.

Phosphorous acid is a colorless solid that melts at 70.1°C. It is a diprotic acid, and its formula can best be written as H_2PHO_3 to emphasize that fact. In a molecule of phosphorous acid, a hydrogen atom is attached to each of the two oxygen atoms, and the third hydrogen atom is attached to the phosphorus atom (Figure 9.27). The hydrogen atoms attached to the oxygen atoms are available as protons in a reaction with a base:

$$H_2PHO_3(aq) + OH^-(aq) \longrightarrow HPHO_3^-(aq) + H_2O(l)$$

$$HPHO_3^-(aq) + OH^-(aq) \longrightarrow PHO_3^{2-}(aq) + H_2O(l)$$

Thus, monohydrogen phosphite ion has the formula PHO_3^{2-}, and dihydrogen phosphite ion has the formula $HPHO_3^-$ (Figure 9.27).

When phosphorous acid reacts with a base such as NaOH, two kinds of salts can be obtained. When equimolar quantities of H_2PHO_3 and NaOH are mixed in aqueous solution and the water is allowed to evaporate, *sodium dihydrogen phosphite* is obtained:

$$H_2PHO_3(aq) + NaOH(aq) \longrightarrow NaHPHO_3(aq) + H_2O(l)$$

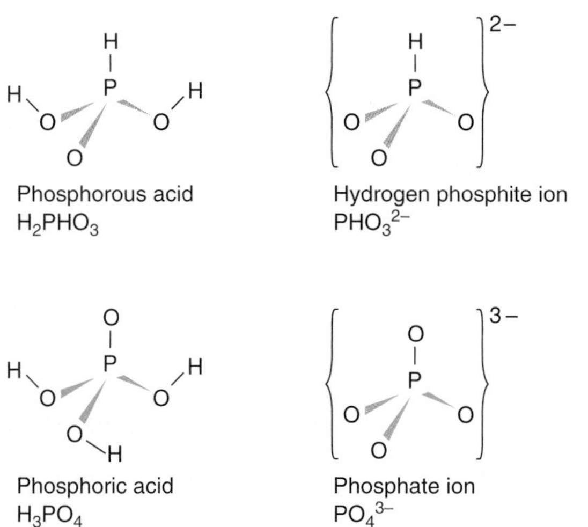

Figure 9.27
Structures of phosphorous acid, phosphoric acid, and their oxoanions.

The reaction of 1 mol of H_2PHO_3 with 2 mol of NaOH yields sodium monophosphite:

$$H_2PHO_3(aq) + 2NaOH(aq) \longrightarrow Na_2PHO_3(aq) + 2H_2O(l)$$

However, when 1 mol of H_2PHO_3 is mixed with 3 mol of NaOH, the salt that forms is Na_2PHO_3 (not Na_3PO_3) because the hydrogen atom in PHO_3^{2-} is tightly held to the phosphorus atom and the excess base does not react.

Phosphoric acid is a colorless, crystalline solid that melts at 42.35°C. The acid is made industrially in large quantities as 85% "syrupy" phosphoric acid by the reaction of calcium phosphate, present in *phosphate rock*, with sulfuric acid:

$$Ca_3(PO_4)_2(s) + 3H_2SO_4(aq) \longrightarrow 2H_3PO_4(aq) + 3CaSO_4(s)$$

All three hydrogen atoms in a molecule of phosphoric acid are linked to the oxygen atoms and are available for reaction with base. Since a molecule of phosphoric acid contains three acidic hydrogens, it is called a *triprotic acid.* Three sodium salts can be obtained from the reaction of H_3PO_4 with NaOH:

NaH_2PO_4	Sodium dihydrogen phosphate	Na_3PO_4	Sodium orthophosphate or sodium phosphate
Na_2HPO_4	Sodium monohydrogen phosphate		

The structures of phosphoric acid and orthophosphate ion are shown in Figure 9.27.

Example 9.6

Writing Equations for Preparations of Different Salts of a Polyprotic Acid

Write equations for reactions of phosphoric acid with sodium hydroxide solution showing how the sodium hydrogen phosphates and sodium orthophosphate can be prepared.

SOLUTION: To prepare NaH_2PO_4 from H_3PO_4, a given quantity of H_3PO_4 is treated with an equal number of moles of NaOH:

$$H_3PO_4(aq) + NaOH(aq) \longrightarrow NaH_2PO_4(aq) + H_2O(l)$$

When water is evaporated from the solution, solid NaH_2PO_4 is obtained.

To prepare Na_2HPO_4, the solution of a given number of moles of H_3PO_4 can be treated with twice as many moles of NaOH:

$$H_3PO_4(aq) + 2NaOH(aq) \longrightarrow Na_2HPO_4(aq) + 2H_2O(l)$$

Evaporation of water from the Na_2HPO_4 solution gives solid Na_2HPO_4.

To obtain sodium orthophosphate, Na_3PO_4, a solution of a given number of moles of H_3PO_4 can be treated with three times as many moles of NaOH:

$$H_3PO_4(aq) + 3NaOH(aq) \longrightarrow Na_3PO_4(aq) + 3H_2O(l)$$

Evaporation of water from the Na_3PO_4 solution yields solid Na_3PO_4.

Practice Problem 9.4 Write equations for the reaction of phosphorous acid with sodium hydroxide to prepare: **(a)** sodium hydrogen phosphite; **(b)** sodium phosphite.

Condensed Acids of Phosphorus and Condensed Phosphates

When orthophosphoric acid is heated to 250°C, it loses water and condenses to *pyrophosphoric acid*, also called *diphosphoric acid*, $H_4P_2O_7$, a white crystalline solid:

$$2H_3PO_4(s) \xrightarrow{250^\circ C} H_4P_2O_7(s) + H_2O(g)$$

This reaction involves the elimination of water from two molecules of H_3PO_4, as shown schematically by the following equation:

```
        H                     H                         H        H
        |                     |                         |        |
        O                     O                         O        O
        |                     |                         |        |
H—O—P—O—H + H—O—P—O—H → H—O—P—O—P—O—H + O—H
        |                     |                         |        |            |
        O                     O                         O        O            H
```

Similarly, three molecules of H_3PO_4 can condense to form *triphosphoric acid*, $H_5P_3O_{10}$. A whole series of such *polyphosphoric acids* with the general formula $H_{n+2}P_nO_{3n+1}$ can be made.

Pyrophosphoric acid (diphosphoric acid) can be converted to salts called *pyrophosphates*. An example is magnesium pyrophosphate, $Mg_2P_2O_7$, which is insoluble in water. The insolubility of magnesium pyrophosphate can be exploited in quantitative determination of Mg^{2+} ions in a solution. First, soluble ammonium pyrophosphate is added to the solution to form insoluble magnesium ammonium phosphate, $MgNH_4PO_4(H_2O)_6$. Ammonium ions are removed from this precipitate by heating:

$$2MgNH_4PO_4(H_2O)_6(s) \xrightarrow{\text{heat}} Mg_2P_2O_7(s) + 2NH_3(g) + 13H_2O(g)$$

The solid $Mg_2P_2O_7$ is then weighed and the amount of magnesium in the solid is calculated (see questions 31 and 32 in the exercise section).

Another important class of condensed phosphoric acids, collectively known as *metaphosphoric acids*, can be prepared by heating orthophosphoric acid at 400°C. One mole of water is lost per mole of H_3PO_4, as shown by the equation

$$nH_3PO_4(s) \longrightarrow (HPO_3)_n(s) + nH_2O(g)$$

where n equals the number of moles of water lost. The product, $(HPO_3)_n$, is a viscous, sticky mass that solidifies to a glasslike substance called *glacial phosphoric acid*.

In all phosphoric acids the phosphorus atom is surrounded by four oxygen atoms. Each PO_4 unit therefore has a tetrahedral structure, as predicted by the VSEPR theory. These tetrahedra are connected by oxygen atoms at corners, as shown in Figure 9.28 for pyrophosphoric acid (diphosphoric acid), triphosphoric acid, and metaphosphoric acid.

Figure 9.28

Structures of pyrophosphoric (diphosphoric), triphosphoric, and metaphosphoric acids. The PO_4 tetrahedra (shown by dotted lines) are connected by oxygen atoms at the corners of the tetrahedra.

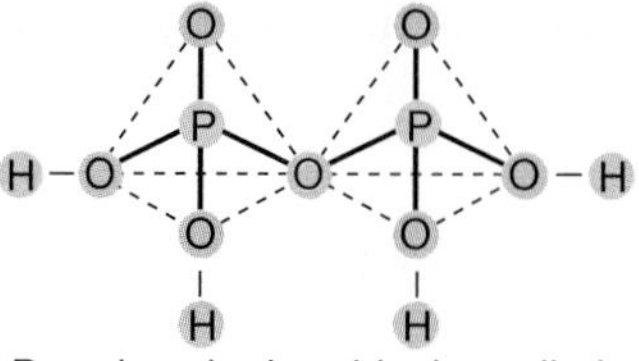

Pyrophosphoric acid, also called diphosphoric acid, $H_4P_2O_7$

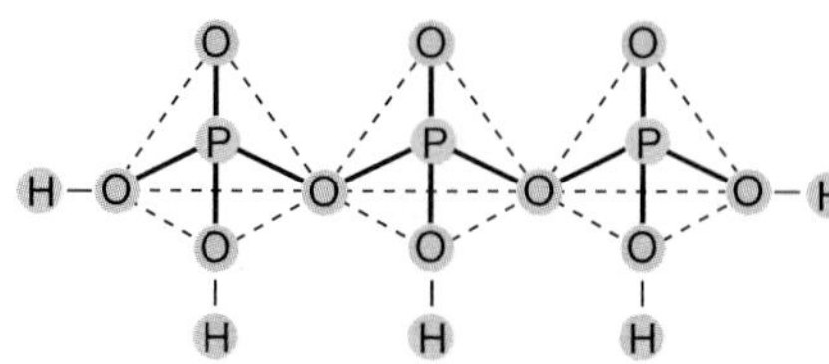

Triphosphoric acid, $H_5P_3O_{10}$

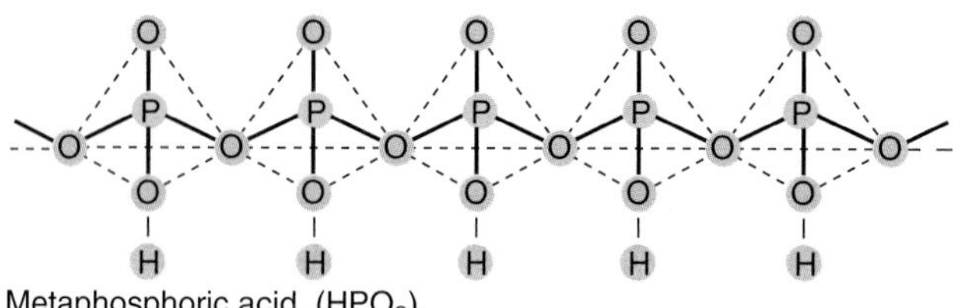

Metaphosphoric acid, $(HPO_3)_n$

Phosphorus is a required nutrient in all living organisms. An important phosphorus-containing fertilizer for plants is *superphosphate*, which is a mixture of calcium dihydrogen phosphate, $Ca(H_2PO_4)_2$, and hydrated calcium sulfate (gypsum), $CaSO_4(H_2O)_2$.

Oxygen Compounds of Arsenic, Antimony, and Bismuth

Arsenic and antimony form oxides similar to those of phosphorus. In these oxides, arsenic and antimony exhibit oxidation states of +3 or +5. In the +3 oxidation state, arsenic and antimony form As_4O_6 and Sb_4O_6, respectively. These oxides have structures similar to that of P_4O_6. In the +5 oxidation states arsenic and antimony form the oxides As_2O_5, and Sb_2O_5. The most common oxide of bismuth is Bi_2O_3.

When As_4O_6 is dissolved in water, the aqueous solution obtained is called *arsenous acid* and is given the formula H_3AsO_3. Arsenous acid is a triprotic acid that can form three types of anions: monohydrogen *arsenite* ion, dihydrogen arsenite ion, and arsenite ion.

Arsenic acid, H_3AsO_4, is a white solid with a structure similar to that of H_3PO_4. Like phosphoric acid, arsenic acid is also a triprotic acid.

Arsenic compounds are poisonous. Even skin contact with these compounds should be avoided. A dose of 0.1 g of As_4O_6 (the "arsenic" referred to in detective stories) is lethal for an average human being. Arsenic compounds cause degeneration of the lining of the digestive tract and the tissues of other internal organs.

Arsenic compounds are present in herbicides and pesticides. Some of these compounds include calcium arsenate, used to kill boll weevils and crabgrass; lead arsenate, which destroys fruit pests; sodium arsenate, a weed killer, and a pesticide used to protect sheep from disease carriers such as ticks. These substances must be used carefully to prevent toxic effects to people and animals.

9.7 Oxygen Compounds of the Sulfur Family

The Oxides of Sulfur

Sulfur forms two major oxides, sulfur dioxide, SO_2, and sulfur trioxide, SO_3, in which sulfur is in +4 and +6 oxidation states, respectively. Sulfur dioxide is a colorless gas with a pungent odor. It is produced by burning sulfur or sulfur-containing substances in air:

$$S_8(s) + 8O_2(g) \longrightarrow 8SO_2(g)$$

Sulfur dioxide is converted to sulfur trioxide when heated with oxygen in the presence of a catalyst such as vanadium pentoxide, V_2O_5, or finely divided Pt:

$$2SO_2(g) + O_2(g) \xrightarrow{V_2O_5} 2SO_3(g)$$

The product, SO_3, first appears as a dense white smoke with an extremely choking odor, which then condenses on the reaction vessel as the temperature decreases to room temperature.

The structure of SO_2 is similar to that of ozone, O_3. Oxygen and sulfur are in the same group of the periodic table. In comparing ozone and SO_2 we see that each of the molecules can be represented by two resonance structures (Figure 9.29). In an SO_2 molecule the central sulfur atom is sp^2 hybridized and the molecule has a delocalized π electron cloud, as shown in Figure 9.29. Note that the structures of the ozone and sulfur dioxide molecules are similar to that of nitrite ion, NO_2^-. Each of these species contains 18 valence shell electrons. Since they are isoelectronic, we expect them to have similar structures. The bond angles in O_3, SO_2, and NO_2^- are 117°, 119°, and 115°, respectively.

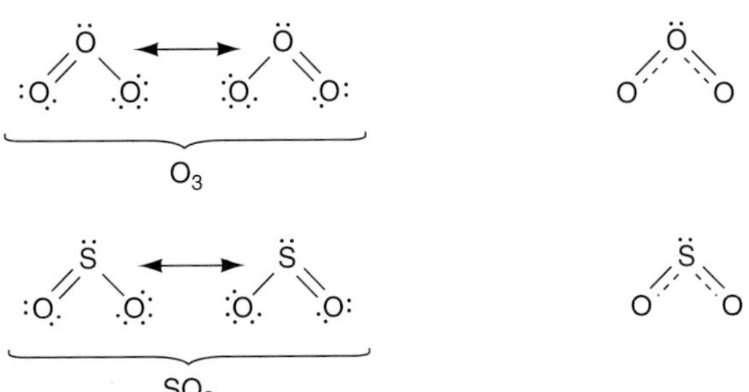

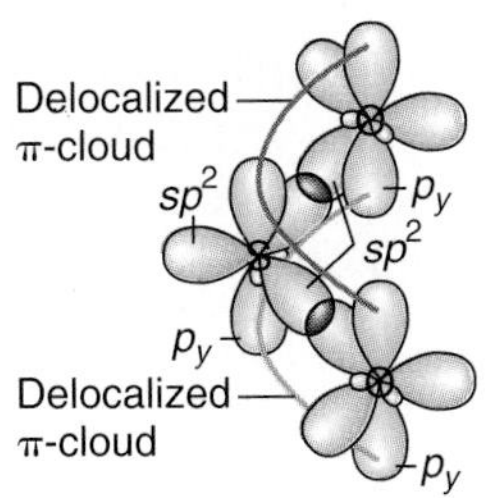

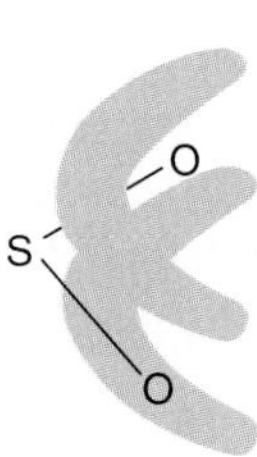

Figure 9.29

Structures of SO_2 and O_3.

Table 9.6

Selected Bond Orders and Bond Energies of Single Bonds, Double Bonds, Triple Bonds, and Nonintegral Bonds

Bond	Molecule or Ion	Bond Order	Bond Energy (kJ mol^{-1})
C—O	H_3COCH_3	1	310
$\mathrm{C}\overset{\cdots}{-}\mathrm{O}$	CO_3^{2-}	$1\frac{1}{3}$	735
C=O	CO_2	2	803
C≡O	CO	3	1076
N—O	H_2NOH	1	163
$\mathrm{N}\overset{\cdots}{-}\mathrm{O}$	NO_2	$1\frac{1}{2}$	305
N=O	NOF	2	594
$\mathrm{N}\underset{\cdots}{=}\mathrm{O}$	NO	$2\frac{1}{2}$	632
N≡O	NO^+	3	1054
S—O	SO_4^{2-}	1	469
$\mathrm{S}\underset{\cdots}{-}\mathrm{O}$	SO_3	$1\frac{1}{3}$	532
$\mathrm{S}\underset{\cdots}{-}\mathrm{O}$	SO_2	$1\frac{1}{2}$	552

The structure of SO_3 can be represented by three resonance structures:

$$:\ddot{\mathrm{O}}{-}\underset{}{\overset{\overset{:\ddot{\mathrm{O}}}{\|}}{\mathrm{S}}}{-}\ddot{\mathrm{O}}: \longleftrightarrow :\ddot{\mathrm{O}}{-}\overset{\overset{:\ddot{\mathrm{O}}:}{|}}{\mathrm{S}}{=}\ddot{\mathrm{O}}: \longleftrightarrow :\ddot{\mathrm{O}}{=}\overset{\overset{:\ddot{\mathrm{O}}:}{|}}{\mathrm{S}}{-}\ddot{\mathrm{O}}:$$

These resonance structures indicate that the central atom has no unshared electron pairs. Thus, the structure of SO_3 is trigonal planar, and the sulfur atom is sp^2 hybridized. There are three S—O σ bonds and a delocalized π-bonding network. SO_3 molecule is isoelectronic and isostructural with NO_3^- and CO_3^{2-} ions. Each has 24 valence electrons and each has a trigonal planar structure (see the structure for CO_3^{2-} in Figure 9.16).

When the structure of a molecule or a polyatomic ion is a resonance hybrid, the species has σ bonds and partial π bonds. Table 9.6 lists the bond orders and bond energies in some molecules. The data show that increasing the bond order increases the bond strength.

Example 9.7

Determining the Bond Orders in Resonance Hybrids

What is the S—O bond order in an SO_2 molecule and in an SO_3 molecule?

SOLUTION: The bond order between two atoms in a molecule equals the number of shared electron pairs per bond. In an SO_2 molecule, there are three bonding electron pairs and two S—O bonds (see the resonance structures for SO_2 in Figure 9.29). For each S—O linkage, the number of electron pair bonds is $\frac{3}{2}$ or $1\frac{1}{2}$. This is the S—O bond order. In an SO_3 molecule, there are four bonding electron pairs and three S—O linkages. Therefore, the S—O bond order is $1\frac{1}{3}$.

Practice Problem 9.5: What is the O—O bond order in ozone?

Example 9.8

Comparing Bond Angles

Is the OSO bond angle larger in SO_2 or in SO_3? Explain your answer.

SOLUTION: In an SO_3 molecule, the three sets of bonding electrons (as described by the Lewis resonance structures) are distributed in a trigonal planar structure in which each bond angle is 120°. The lone pair-bonding pair repulsion in an SO_2 molecule is responsible for the slightly lower 119° bond angle than the expected 120° bond angle.

Practice Problem 9.6: Which is larger, the O—S—O bond angle in an SO_3 molecule or in an SO_4^{2-} ion? Explain.

Sulfur Acids and Their Salts

Sulfur dioxide is the anhydride of *sulfurous acid*, H_2SO_3, and sulfur trioxide is the anhydride of *sulfuric acid*, H_2SO_4. The reactions of these oxides with water are

$$SO_2(g) + H_2O(l) \longrightarrow H_2SO_3(aq)$$

$$SO_3(s) + H_2O(l) \longrightarrow H_2SO_4(aq)$$

Sulfurous acid is a relatively weak acid that exists only in aqueous solution. It cannot be isolated as a stable compound. In contrast, sulfuric acid is a colorless liquid that freezes at 10.4°C. Both sulfurous and sulfuric acids are diprotic acids, and two series of salts can be derived from each. The salts of sulfurous acid are *hydrogen sulfites* and *sulfites*. The salts of sulfuric acid are *hydrogen sulfates* and *sulfates*. Hydrogen sulfite is also called *bisulfite*, or acid sulfite, and hydrogen sulfate is called *bisulfate*, or acid sulfate.

The reaction of 1 mol of sulfurous acid with 1 mol of sodium hydroxide produces sodium hydrogen sulfite, $NaHSO_3$, and water:

$$H_2SO_3(aq) + NaOH(aq) \longrightarrow NaHSO_3(aq) + H_2O(l)$$

One mole of sulfurous acid reacts with 2 mol of sodium hydroxide to yield sodium sulfite:

$$H_2SO_3(aq) + 2NaOH(aq) \longrightarrow Na_2SO_3(aq) + 2H_2O(l)$$

Similar reactions of sulfuric acid and sodium hydroxide produce sodium hydrogen sulfate and sodium sulfate:

$$H_2SO_4(aq) + NaOH(aq) \longrightarrow NaHSO_4(aq) + H_2O(l)$$

$$H_2SO_4(aq) + 2NaOH(aq) \longrightarrow Na_2SO_4(aq) + 2H_2O(l)$$

Example 9.9

Writing the Equations for a Multistep Synthesis of a Compound

Write equations for the reactions you would carry out to synthesize sodium sulfate from elemental sulfur as a starting material.

SOLUTION: There are several ways to effect this type of synthesis. Using reactions we have discussed in this chapter, we can carry out the following steps. Sulfur can be burned to produce sulfur dioxide. Sulfur dioxide can be converted to sulfur trioxide in the presence of a catalyst such as V_2O_5 or platinum. Sulfur trioxide can then be treated with water to produce sulfuric acid, which can be treated with sodium hydroxide to form an aqueous solution of sodium sulfate. This salt can be recovered in its solid state by evaporation of water. Equations for the foregoing reactions are

$$S_8(s) + 8O_2(g) \longrightarrow 8SO_2(g)$$

$$2SO_2(g) + O_2(g) \xrightarrow{V_2O_5} 2SO_3(g)$$

$$SO_3(g) + H_2O(l) \longrightarrow H_2SO_4(aq)$$

$$H_2SO_4(aq) + 2NaOH(aq) \longrightarrow Na_2SO_4(aq) + 2H_2O(l)$$

The reaction of SO_3 with water is vigorous and difficult to control. Therefore, in industry sulfuric acid is prepared in a slightly different way, which we discuss below.

Practice Problem 9.7: Write equations for the reactions you would carry out to synthesize a solution of sulfurous acid from elemental sulfur.

Sulfuric acid and metal and ammonium sulfates are important industrial chemicals. Sulfuric acid is produced in larger quantity than any other synthetic chemical. Of the millions of tons of H_2SO_4 produced in the United States and other countries each year, the largest share, nearly a third, goes to the manufacture of phosphate fertilizers. (See Section 9.6 for the conversion of phosphate rock to phosphoric acid by use of sulfuric acid.) Smaller amounts are used in the manufacture of paint pigments, ammonium sulfate fertilizers, and other chemicals.

Most of the sulfuric acid in the United States is prepared by a method called the *contact process*. In this process, gaseous SO_2 and O_2 are brought into contact with solid V_2O_5, which catalyzes the conversion of SO_2 to SO_3. The SO_3 is then passed into 98 percent H_2SO_4 to form pyrosulfuric (disulfuric) acid, $H_2S_2O_7$:

$$SO_3(g) + H_2SO_4(l) \longrightarrow H_2S_2O_7(l)$$

Although SO_3 reacts with water to give sulfuric acid, its reaction with sulfuric acid is less exothermic and is easier to control. Pyrosulfuric acid formed in the reaction of SO_3 with H_2SO_4 is added to water to produce sulfuric acid:

$$H_2S_2O_7(l) + H_2O(l) \longrightarrow 2H_2SO_4(aq)$$

Concentrated sulfuric acid has a great affinity for water, and mixing sulfuric acid and water is a highly exothermic process. When a concentrated solution of sulfuric acid is diluted, so much heat is generated that a container made of soda-lime glass can be shattered. Therefore, when diluting a concentrated solution of sulfuric acid, the concentrated solution should be slowly added to water in a Pyrex container, with continuous stirring to avoid rapid localized heating and spattering.

Concentrated sulfuric acid is a powerful dehydrating agent. The affinity of the concentrated acid for water is so great that it can remove water from many substances. For example, cane sugar ($C_{12}H_{22}O_{11}$), and cellulose $(C_6H_{10}O_5)_n$ in paper, which contain hydrogen and oxygen in 2:1 mole proportions, are both converted to carbon and water (are *charred*) by concentrated sulfuric acid (Figure 9.30). The equation for the dehydration of sucrose is shown below:

$$C_{12}H_{22}O_{11}(s) \xrightarrow{H_2SO_4} 12C(s) + 11H_2O(g)$$

Sulfuric acid dehydrates human skin and flesh. If you ever spill concentrated sulfuric acid on yourself, immediately flood the exposed surfaces with a large volume of cold water. *Always* wear safety glasses and protective clothing when working with sulfuric acid!

Metal sulfates have several commercial uses. As we recall, most metal sulfates are water soluble, but some are insoluble. One of the important insoluble sulfates is barium sulfate, $BaSO_4$, a white solid. A suspension of $BaSO_4$ is administered to persons undergoing x-ray examination of the intestinal tract. As the suspension of the barium salt passes through the intestine, it can be traced by x-rays to reveal the location of any lesions.

Another important, sparingly soluble sulfate is calcium sulfate dihydrate or *gypsum*, $CaSO_4(H_2O)_2$, which occurs in large natural deposits. Gypsum is an important material in the building industry. When gypsum is heated to drive off a portion of its water of hydration, a hemihydrate called *plaster of paris*, $(CaSO_4)_2H_2O$, is formed:

$$2CaSO_4(H_2O)_2(s) \longrightarrow (CaSO_4)_2H_2O(s) + 3H_2O(g)$$

Plaster of paris is used to make plaster casts. In this process, water is added to plaster of paris, causing a slow reformation of gypsum. As water is absorbed, the volume of the material increases, forming a hard, white mass. Plaster of paris is also an important constituent of cement.

The commercially important water soluble sulfates include sodium sulfate decahydrate, sometimes called *Glauber's salt*, $Na_2SO_4(H_2O)_{10}$, used in fireproofing fabrics, as a cathartic, and in textile dyeing. Another common soluble sulfate is magnesium sulfate. Magnesium sulfate heptahydrate, $MgSO_4(H_2O)_7$, called Epsom salt, is used as a cathartic and in dyeing, ceramics, and cosmetic lotions. It is also used in motion pictures to simulate snow. Epsom salt produces blizzards in Hollywood (Figure 9.31).

Figure 9.30

Top: A piece of paper being dipped into sulfuric acid. Bottom: The same paper charred by the sulfuric acid.

Figure 9.31

Use of Epsom salt to make "snow."

Figure 9.32

Left: Hydrated copper(II) sulfate. Right: Anhydrous copper(II) sulfate.

Anhydrous copper(II) sulfate is a white powder that efficiently absorbs moisture from air and becomes a blue pentahydrate (Figure 9.32). It is therefore used to maintain a dry atmosphere for moisture-sensitive substances.

Thiosulfate

The sulfur atom of the sulfate ion is surrounded by four oxygen atoms. If one of these oxygen atoms is replaced by a second sulfur atom, the resulting ion is called *thiosulfate*. The prefix "thio-" in "thiosulfate" indicates that one of the oxygen atoms in a sulfate ion is replaced by a sulfur atom. The formula of thiosulfate ion is $S_2O_3^{2-}$. Thiosulfate ions can be prepared by boiling a solution of sulfite ions with elemental sulfur:

$$S_8(s) + 8SO_3^{2-}(aq) \longrightarrow 8S_2O_3^{2-}(aq)$$

A sulfite ion has a trigonal pyramidal structure with the sulfur atom at the apex and the three oxygen atoms forming the "legs" of the pyramid. When a sulfur atom becomes attached to a sulfite ion to form thiosulfate, the pyramidal sulfite ion converts to a modified tetrahedral thiosulfate ion. This reaction is shown below by use of Lewis structures:

$$\left[\begin{array}{c} :\ddot{O}-\ddot{S}-\ddot{O}: \\ | \\ :\underset{..}{O}: \end{array}\right]^{2-} + :\dot{\ddot{S}}\cdot \longrightarrow \left[\begin{array}{c} :\ddot{S}: \\ | \\ :\ddot{O}-S-\ddot{O}: \\ | \\ :\underset{..}{O}: \end{array}\right]^{2-}$$

Pyramidal   Tetrahedral

Sodium thiosulfate, known to photographers as "hypo," is used in the photographic fixing process. Black-and-white photographic film is usually coated with a mixture of silver halides, AgX (where X is any halide ion). Sodium thiosulfate solution dissolves any unreacted silver halides that remain on a photographic film after exposure. A thiosulfate solution dissolves silver halides according to the equation

$$AgX(s) + 2S_2O_3^{2-}(aq) \longrightarrow Ag(S_2O_3)_2^{3-}(aq) + X^-(aq)$$

In the $Ag(S_2O_3)_2^{3-}$ ion each $S_2O_3^{2-}$ ion is linked to the Ag^+ ion through a sulfur atom that provides a pair of electrons for a coordinate covalent bond between the silver ion and sulfur atoms:

$$\left[\begin{array}{c} O \\ | \\ O-S-S \\ | \\ O \end{array}\right]^{2-} + Ag^+ + \left[\begin{array}{c} O \\ | \\ S-S-O \\ | \\ O \end{array}\right]^{2-} \longrightarrow \left[\begin{array}{c} O \qquad\qquad\qquad\qquad O \\ | \qquad\qquad\qquad\qquad | \\ O-S-S-Ag-S-S-O \\ | \qquad\qquad\qquad\qquad | \\ O \qquad\qquad\qquad\qquad O \end{array}\right]^{3-}$$

9.8 Halogen Oxides, Oxoacids, and Their Salts

Oxides

The halogens form several oxides and oxoacids. The oxidation states of halogens in these compounds range from -1 for fluorine in OF_2 to $+7$ for chlorine in Cl_2O_7. The formulas and physical properties of binary halogen–oxygen compounds are listed in Table 9.7.

Table 9.7

Binary Halogen–Oxygen Compounds

Compound	Melting Point (°C)	Normal Boiling Point (°C)	Physical Properties
OF_2	−223.8	−144.8	Colorless gas, decomposes at 250°C
O_2F_2	−163.5	−57.0 (estimated) Decomposes above −100	Brown gas
Cl_2O	−20	3.8	Brown gas, red-brown liquid; explodes on heating or shock
Cl_2O_3	Decomposes at −45		Dark-brown solid
ClO_2	−59.0	11.0	Orange-red gas, red liquid, orange-red solid; explodes when disturbed
Cl_2O_6	3.5	~203.0	Orange solid, red liquid; explodes in contact with organic matter
Cl_2O_7	−91.5	82	Colorless liquid; can explode on shock or heating
Br_2O	−17.5	Decomposes at −17.5	Brown liquid and solid
BrO_2	Decomposes at 0		Yellow solid
Br_2O_6	Decomposes at −80		White solid
I_2O_4	Decomposes at 130		Yellow crystals
I_4O_9	Decomposes at 75		Pale yellow solid
I_2O_5	Decomposes at 275		White solid

Hypohalous Acids and Hypohalites

The halogens (X_2) are slightly soluble in water. Chlorine, bromine, and iodine also react with water to form a *hydrohalic acid* and a *hypohalous acid*. For example, chlorine reacts with water to a slight extent to form hydrochloric acid and *hypochlorous acid*:

$$Cl_2(g) + H_2O(l) \longrightarrow \underbrace{H^+(aq) + Cl^-(aq)}_{\text{Hydrochloric acid}} + \underbrace{HOCl(aq)}_{\text{Hypochlorous acid}}$$

Hydrochloric acid is a strong acid that is almost entirely dissociated to $H^+(aq)$ and $Cl^-(aq)$ ions in water solution. In contrast, an aqueous solution of hypochlorous acid consists mostly of undissociated HOCl molecules. Bromine and iodine react with water in a similar way except that they produce mostly halic acid and relatively less hypohalous acid. The hypohalous acids of chlorine, bromine, or iodine exist only in aqueous solution.

Salts of hypochlorous acid, the *hypochlorites*, are good bleaching agents. Sodium hypochlorite can be prepared as a pentahydrate by the reaction of chlorine with an ice-cold solution of sodium hydroxide:

$$2NaOH(aq) + Cl_2(g) + 4H_2O(l) \longrightarrow NaOCl(H_2O)_5(s) + NaCl(s)$$

The bleaching agent called Clorox is a 5.25 percent solution of NaOCl. Bleaches that contain NaOCl must never be mixed with household ammonia. NaOCl reacts with excess ammonia to produce toxic hydrazine, N_2H_4:

$$2NH_3(aq) + NaOCl(aq) \longrightarrow N_2H_4(aq) + NaCl(aq) + H_2O(l)$$

Hydrazine is explosive in oxygen.

The only known oxoacid of fluorine is *hypofluorous acid*, HOF. This acid was not prepared until 1971. HOF is a white solid below −117°C. It melts to a pale yellow liquid at −117°C.

Halous Acids and Halites

The only *halous acid* definitely known to exist is *chlorous acid*, HOClO or $HClO_2$. Like the hypohalous acids, chlorous acid is a weak acid. It cannot be isolated as a pure substance, and it exists only in aqueous solution. The salts of chlorous acid are known as *chlorites*.

Halic Acids and Halates

Of the halic acids, only *iodic acid*, HIO_3, has been isolated as a pure compound. It is a colorless solid that melts with decomposition at 110°C. Chloric acid, $HClO_3$, and bromic acid, $HBrO_3$, exist only in aqueous solution. Halic acids are moderately strong acids. Their solutions can be prepared by the reaction of sulfuric acid with a solution of barium halate:

$$Ba^{2+}(aq) + 2XO_3^-(aq) + H^+(aq) + HSO_4^-(aq) \longrightarrow BaSO_4(s) + \underbrace{2H^+(aq) + 2XO_3^-(aq)}_{\text{Halic acid}}$$

The insoluble $BaSO_4$ can be separated from the halic acid solution by filtration. Halic acids can be converted to salts called *halates*. We recall from Chapter 3 that potassium chlorate, $KClO_3$, decomposes upon heating (with MnO_2 as a catalyst) to produce potassium chloride and oxygen:

$$2KClO_3(s) \xrightarrow[\text{heat}]{MnO_2} 2KCl(s) + 3O_2(g)$$

Perhalic Acids and Perhalates

Perhalic acids are compounds with the general formula HXO_4. Perhalic acids are very strong acids that react readily, often explosively, with many substances. *Perchloric acid*, $HClO_4$, is the most common perhalic acid. It can be prepared by the reaction of a metal perchlorate, such as $KClO_4$, with H_2SO_4:

$$KClO_4(s) + H_2SO_4(aq) \longrightarrow KHSO_4(aq) + HClO_4(aq)$$

Perchloric acid can be isolated from the reaction mixture by distillation at low pressure and low temperature. Anhydrous $HClO_4$ is a colorless liquid that freezes at −112°C and boils at 19°C under 111 mm Hg pressure. The pure acid is an extremely dangerous substance that decomposes explosively upon heating (above 92°C), and it reacts with explosive violence with many substances, particularly organic compounds.

Aqueous solutions of perchloric acid at concentrations up to 70 percent by mass are much safer than the pure acid but must still be handled with extreme caution. A 70% aqueous solution of perchloric acid can easily explode if it is heated with organic compounds. Perchloric acid can be dehydrated with a strong dehydrating agent such as phosphorus(V) oxide, P_4O_{10}, to form the anhydride, Cl_2O_7. Perchloric anhydride is even more unstable and dangerous than perchloric acid. In an aqueous solution, perchloric acid and its salts, *perchlorates*, are used as reagents in analytical chemistry. Perchlorates are explosive and must be handled very gingerly. The names and formulas of halogen oxoacids and their anions are listed in Table 9.8. The structures of the anions of these acids are illustrated in Figure 9.33.

Table 9.8

Oxoacids and Oxoanions of the Halogens

Oxidation State of Halogen	Formula of Acid F	Cl	Br	I	Name of Acid	Name and General Formula of the Anion (X = halogen)
1	HOF	HOCl	HOBr	HOI	Hypohalous	Hypohalite, XO^-
3		$HClO_2$			Halous[a]	Halite, XO_2^-
5		$HClO_3$	$HBrO_3$	HIO_3	Halic	Halate, XO_3^-
7		$HClO_4$	$HBrO_4$	HIO_4	Perhalic[b]	Perhalate,[c] XO_4^-
				H_5IO_6	Paraperhalic[d]	Paraperhalate, XO_6^{5-}

[a] The only halous acid known to exist is chlorous acid, $HClO_2$.

[b] Metaperiodic.

[c] Metaperiodate.

[d] The only paraperhalic acid known to exist is paraperiodic acid, H_5IO_6.

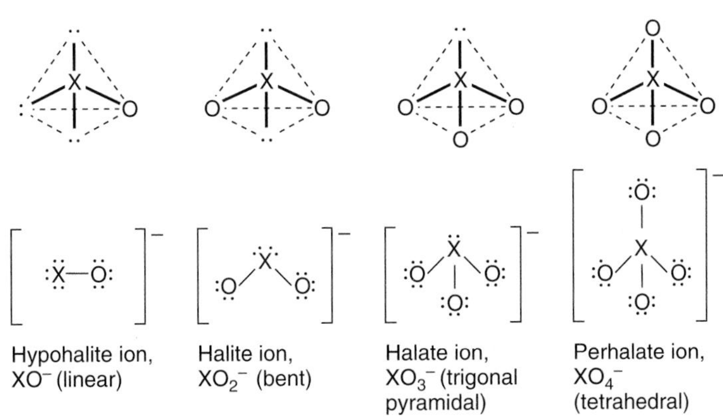

Figure 9.33

Structures of the anions of oxoacids of halogens.

Summary

Elements within a group of the periodic table form similar compounds whose formulas, structures, and bonding can be predicted from the valence shell electronic structure of the atoms in the group. However, an element in the third or a later period can form some compounds for which the second-period element in the same group has no analogs. Elements in the third or later periods have *d* orbitals available for bonding. Elements in groups VI and VII can have higher oxidation states than those of the second-period elements.

Binary Hydrogen Compounds

The formulas of *binary hydrogen compounds* of the representative elements in groups I through IV can generally be predicted from the *number of valence electrons* on the atoms in each group. An exception to this rule is boron, which does not form a stable BH_3 molecule, as might be expected from the number of valence electrons on the B atom. The binary hydrogen compounds of the elements in groups V, VI, and VII can be predicted from the number of *unpaired* valence electrons on the atoms in these groups.

Binary Halogen Compounds

The formulas of many *binary halogen compounds* of the representative elements can generally be predicted by the rules discussed above for hydrogen compounds. Thus, the formulas of most of the halides of the elements in groups I through IV can be predicted from the number of valence electrons on the atoms in these groups. The heavier metallic elements in groups III and IV form additional halides that cannot be predicted from this rule.

The formulas of some of the halides of nitrogen and oxygen can be predicted from the number of unpaired valence electrons. For the third- and later-period elements of these groups, additional halides exist. Their formulas can be predicted by unpairing their valence shell electrons.

Thus, boron forms BCl_3, carbon forms CCl_4, nitrogen forms NCl_3, and oxygen forms Cl_2O. However, phosphorus forms both a *trichloride* (PCl_3) and a *pentachloride* (PCl_5). The formula for the trichloride can be predicted from the number of unpaired valence electrons on a phosphorus atom, and the pentachloride from the *total number of valence electrons*. Similarly, sulfur forms SF_2 (there are two unpaired electrons on an S atom), and also SF_4 and SF_6. The formulas for SF_4 and SF_6 can be predicted by successive unpairing of valence shell electrons of the S atom.

Oxygen Compounds of Nonmetals

Most oxides of nonmetals react with water to give acidic solutions. These oxides are *acidic oxides* or *acidic anhydrides*. Most nonmetals form two or more oxides. One of these usually corresponds to the maximum oxidation state of the nonmetal, and one other oxide generally corresponds to the oxidation state that is equal to the number of unpaired electrons on the nonmetal atom.

Common oxoacids of nonmetals include H_3BO_3, HNO_3, HNO_2, H_3PO_4, H_3PO_3, H_3AsO_4, H_3AsO_3, H_2SO_4, H_2SO_3, $HClO_4$, $HClO_3$, $HClO_2$, and $HClO$.

New Terms

Acidic anhydride (9.4)
Acidic oxide (9.4)
Basic anhydride (9.4)
Basic oxide (9.4)
Borane (9.1)
Dimer (9.2)
Disproportionation reaction (9.5)
Monomer (9.2)
Piezoelectric (9.4)
Polymer (9.2)
Silane (9.1)
Silicate (9.4)
Silicone (9.4)
Steric hindrance (9.2)
Tetramer (9.2)
Trimer (9.2)

Exercises

General Review

1. Write the formulas and names of binary compounds of a representative metal of group I with a nonmetal from each of groups VII, VI, and V. Repeat this for a representative metal from each of groups II, III, IV, and V.
2. What general guideline can be used to predict the formula for a binary hydrogen compound of: **(a)** a group I, II, III, or IV element; **(b)** a group V, VI, or VII element?
3. Can the same general rules stated in your answer to question 2 be used to predict the formulas of binary halogen compounds of the elements in: **(a)** group I, II, III, or IV; **(b)** group V, VI, or VII? Explain.
4. Nitrogen forms NF_3 and oxygen forms OF_2. Write the formulas for the fluorides of phosphorus and of sulfur reviewed in this chapter.
5. What is the electron pair geometry and the hybridization on the central atom in each of the fluorides of phosphorus? Of the fluorides of sulfur? What is the molecular structure of each of these molecules?
6. An atom of xenon has a full octet of electrons in its valence shell. Explain how a xenon atom can react with fluorine atoms.
7. Write the formulas and names for all the fluorides of xenon.
8. What is the structure and the hybridization on the central atom of each of the fluorides in question 7?
9. What is a polymer?
10. Aluminum chloride in its liquid state consists of dimeric Al_2Cl_6 molecules. Sketch the structure of an Al_2Cl_6 molecule and explain the bonding involved.
11. Explain why sulfur forms a hexafluoride but no hexaiodide.
12. Write the formulas and names for different fluorides of iodine. Sketch the molecular structures of each of the fluorides, and indicate the hybridization on iodine in each of these molecules.
13. Write the formulas and names for all the well-known oxides of boron, carbon, silicon, nitrogen, phosphorus, and sulfur.
14. Write the formulas and names for all the common acids of boron, carbon, silicon, nitrogen, phosphorus, sulfur, and chlorine.

15. Write the formulas and names for all the possible sodium salts that can be obtained from each of the acids in question 14.

Bond Angles

16. Which of the following species do you predict to have the largest bond angle: NO_2, NO_2^-, or NO_2^+? Explain.
17. Which do you predict to be the larger, the FSF bond angle in SF_6 or the FXeF bond angle in XeF_6? Explain.
18. Which can be predicted to be smaller, the FIF bond angle in IF_3, or the FPF bond angle in PF_3? Explain by drawing sketches.

Steric Hindrance

19. Explain why IF_7 exists but BrF_7 does not exist. Would you expect chlorine to form ClF_7? Explain.
20. Selenic and telluric acids have the formulas H_2SeO_4 and $Te(OH)_6$, respectively. What is the oxidation state of the central element in each? Sketch the structures of these molecules. Account for the fact that the selenium atom in H_2SeO_4 is surrounded by four oxygen atoms, but the tellurium atom in $Te(OH)_6$ is surrounded by six oxygen atoms.
21. Arsenate ion has the formula AsO_4^{3-}, and antimonate ion $Sb(OH)_6^-$. What is the oxidation state of the central element in each? Sketch the structures of these anions and give a reason for the difference.
22. Pentafluorides of phosphorus, arsenic, and antimony exist, but no pentaiodides have been prepared for these elements. Suggest an explanation.

Equation Writing

23. Write a balanced equation for each of the following reactions: **(a)** phosphorus(III) chloride is heated with chlorine gas; **(b)** elemental sulfur is heated with chlorine gas to form disulfur dichloride; **(c)** elemental sulfur is heated with oxygen gas without a catalyst; **(d)** 1 mol of bromine is heated with 3 mol of fluorine; **(e)** sulfur dioxide is bubbled through water; **(f)** calcium oxide is treated with water; **(g)** dehydration of boric acid; **(h)** limestone (calcium carbonate) is treated with carbon dioxide and water; **(i)** aqueous calcium hydrogen carbonate is warmed in a cavern to form stalagmites and stalactites; **(j)** a quartz crystal is "dissolved" in hydrofluoric acid; **(k)** silicon dioxide is treated with aqueous sodium hydroxide; **(l)** ammonium nitrate is decomposed by heating to form "laughing gas" and water; **(m)** colorless nitrogen monoxide is heated in the open atmosphere to form a brown gas in which nitrogen is in the +4 oxidation state; **(n)** dinitrogen trioxide is treated with water; **(o)** the aqueous product from part **(n)** is treated with lime water (calcium hydroxide); **(p)** phosphorus(III) oxide is treated with water; **(q)** aqueous phosphorous acid is treated with excess sodium hydroxide solution; **(r)** phosphoric acid is treated with excess sodium hydroxide solution; **(s)** sulfuric acid is heated to produce a dense white smoke with an extremely choking odor; **(t)** plaster of paris, $(CaSO_4)_2H_2O$, is mixed with water to form gypsum (calcium sulfate dihydrate); **(u)** phosphorus trichloride is treated with water.
24. Write equations to show how each of the following compounds can be prepared: **(a)** hydrogen chloride from its elements; **(b)** hydrogen chloride by the reaction of calcium chloride with sulfuric acid; **(c)** hydrogen iodide by the reaction of sodium iodide with phosphoric acid (Why can't sulfuric acid be used?); **(d)** hydrogen iodide by the reaction of phosphorus triiodide with water; **(e)** phosphorus trifluoride from its elements; **(f)** tetraphosphorus decoxide from its elements; **(g)** nitric acid by the Ostwald process, which starts with the oxidation of ammonia to NO, followed by the oxidation of NO to NO_2, which is then treated with water; **(h)** sulfurous acid starting with sulfur and oxygen; **(i)** ammonia from its elements; **(j)** ammonium chloride starting with nitrogen, hydrogen, and other chemicals.

General Problems

25. Write a formula for a chlorine compound in which the element exhibits each of the following oxidation states: −1, +1, +3, +4, +5, +6, and +7.
26. Write a formula for a nitrogen compound in which the element exhibits each of the following oxidation states: −3, +1, +2, +3, +4, and +5.
27. Sulfur forms SF_2, SF_4, and SF_6, but no SF, SF_3, or SF_5. Explain.
28. Bromine forms BrF, BrF_3, and BrF_5, but no BrF_2, BrF_4, or BrF_6. Explain.
29. Nitrogen and phosphorus are both in group V, which suggests that the maximum oxidation state of each of the elements should be +5. Indeed, both elements exhibit this oxidation state in the oxides, N_2O_5, and P_4O_{10}. However, phosphorus forms PF_5, but nitrogen forms only NF_3, and no NF_5. Explain why.
30. What is a possible hybridization on nitrogen in N_2O_5? (Refer to Figure 9.25.)
31. A salt of an organic acid was found to contain sodium, carbon, phosphorus, and oxygen. An analysis of the salt produced the following data: A 0.283-g sample gave 0.102 g of Na. Another 0.499-g sample gave 0.114 g of CO_2 on burning. A third 0.440-g sample gave 0.255 g of $Mg_2P_2O_7$. Derive the empirical formula of the salt.
32. A 0.5255-g sample of impure Epsom salt, $MgSO_4(H_2O)_7$, is dissolved in water and the magnesium is precipitated from the solution as $Mg_2P_2O_7$, which has a mass of 0.2337 g after drying. Calculate: **(a)** the percent of magnesium in the sample; **(b)** the percent purity of the Epsom salt.

CHAPTER 10

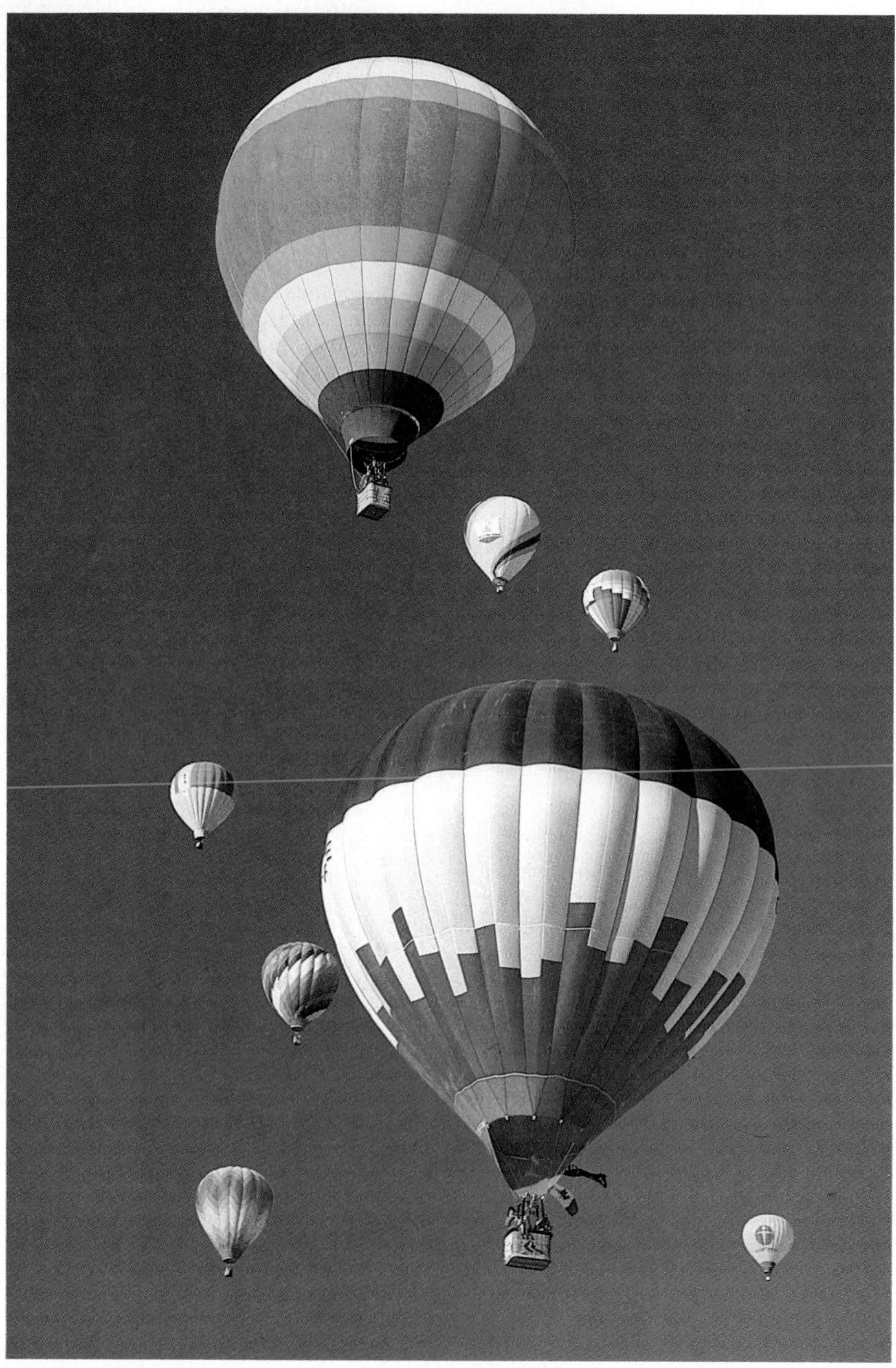

When the air enclosed in a hot air balloon is heated, it expands and becomes less dense than the cooler surrounding atmosphere. The balloon rises until its density equals that of the surrounding atmosphere.

The Gaseous State of Matter

In this chapter and the next we consider some physical properties of gases, liquids, and solids. The atoms or molecules in the solid, liquid, and gaseous states have different degrees of order and mobility. Solids and liquids are condensed states *of matter. In a crystalline solid, the individual atoms, molecules, or ions are closely packed, and they occupy fixed positions in a highly ordered structure. In a liquid, the molecules are still closely packed, but they can move easily in all directions. As a result, a liquid flows and adapts to the shape of its container. In the gaseous state, the molecules are much farther apart, and they are free to travel in any direction. As a result, a gas expands indefinitely in space. Both liquids and gases tend to flow or conform to the shape of their container; they are called* fluids. *Figure 10.1 illustrates the nature of the solid, liquid, and gaseous states.*

In the present chapter we concentrate on the properties of gases. In Chapter 11 we consider liquids, solids, and phase changes.

10.1 Gas Pressure

One of the most fundamental and easily measured properties of a gas is its pressure. Pressure, P, is defined as force, F, divided by area, A. Thus, $P = F/A$. The SI unit of force is the newton (N),* and the SI unit of area is the square meter (m^2). Thus, the SI unit of pressure is a newton per square meter (N/m^2). This unit is also called the **pascal** (Pa), after the French mathematician, scientist, and philosopher Blaise Pascal (1623–1662):

$$1 \text{ Pa} = \frac{1 \text{ N}}{1 \text{ m}^2}$$

Pressures in automobile tires and gas cylinders are commonly measured in pounds per square inch (psi). One psi equals 6893 Pa.

* 1 newton is the force that can accelerate a 1 kg mass 1 m per second each second: $1 \text{ N} = 1 \text{ kg m s}^{-2}$.

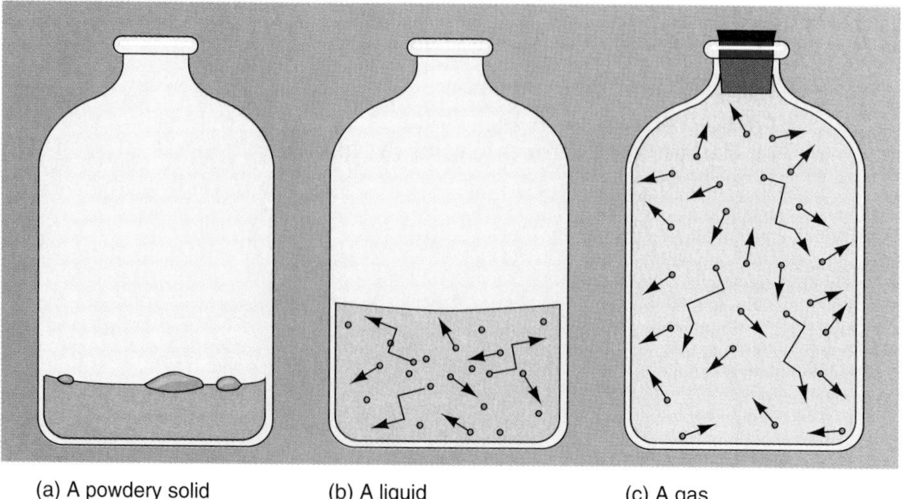

Figure 10.1

Solid, liquid, and gaseous states. Dots represent the particles (molecules, atoms, or ions) in each state. In a solid, the particles are close together and unable to move from one place to another. In most liquids, the particles move around randomly as shown by the arrows in part (b). In a gas, the average distance between the particles is great. The particles of a gas move around in all directions (as shown by arrows in part c) and exert a uniform pressure on the walls of the container.

Gas pressure was first measured in 1643 by the Italian scientist Evangelista Torricelli (1608–1647), a student of the famous Italian scientist and mathematician Galileo Galilei (1564–1642). Torricelli showed that the atmosphere exerts a pressure. The method he used to measure this pressure is still used today. Torricelli filled a long glass tube, sealed at one end, with mercury and inverted the tube in an open vessel of mercury so that the open end of the tube was beneath the surface of the mercury (Figure 10.2).

When the tube is inverted, some of the mercury flows out of the tube into the open vessel, and the level of mercury in the closed end of the tube drops. Since all of the mercury does not flow out of the tube, there must be a force acting on the surface of the mercury in the open vessel. This force is transmitted in all directions throughout the fluid mercury, and it supports the weight of the column of mercury remaining in the tube. The force acting on a unit area of the surface of mercury in the open vessel equals the pressure of the atmosphere in contact with the surface. The height of the mercury column in the vertical tube indicates the magnitude of the pressure.

The simple arrangement shown in Figure 10.2 is a rudimentary form of an instrument that measures atmospheric pressure called a **barometer**. The magnitude of the pressure exerted by the atmosphere is measured by the difference in height, h, between the mercury level in the tube and the level in the open vessel.

At sea level, the average atmospheric pressure supports a column of mercury 760 mm (29.9 inches) high. A *barometric pressure* equivalent to 760 mm (76.0 cm) of mercury is called a pressure of 1 *atmosphere* (atm). A gas pressure that supports 1 millimeter of mercury is called 1 *torr* (after Torricelli).

The atmosphere and the torr are not SI units, but they are commonly used by chemists in the United States. A pressure of 1 atm equals 101,325 pascals (101.325 kilopascals), or 14.7 pounds per square inch.

If we place the barometer shown in Figure 10.2 in a closed container and remove the air from the container, the mercury column will drop until the level of the mercury inside the tube is the same as the level in the dish. Under these conditions, the pressure in the container corresponds to zero millimeters of mercury ($h = 0$). If gas is again admitted into the container, mercury rises in the

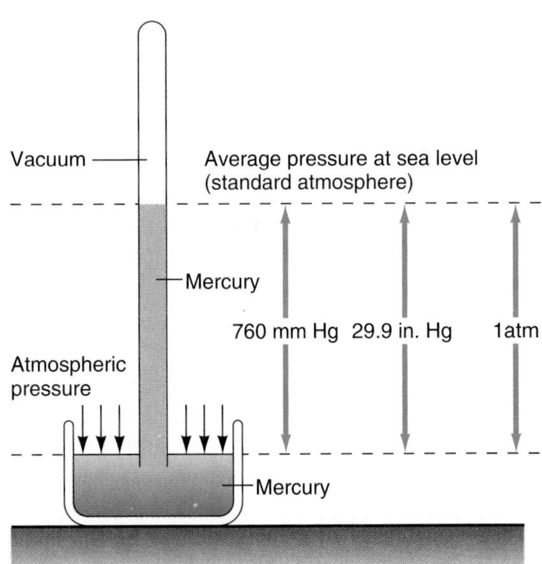

Figure 10.2

Mercury barometer. The arrows on the surface of mercury represent gas molecules bombarding the surface of mercury and exerting the pressure that sustains the mercury column in the tube. 760 mm Hg = 1.00 atmosphere = 101325 pascals = 29.9 in. Hg.

tube, indicating that the pressure in the container increases as more gas is introduced. The column of mercury in the barometer tube is supported by the pressure of the gas on the surface of the mercury and not pulled up by the near-vacuum that exists above the mercury column.

Example 10.1

Comparing the Heights of Mercury and Water Columns Supported by Atmospheric Pressure

If the barometer shown in Figure 10.2 contained water instead of mercury, what height of the water column would correspond to 76.0 cmHg? The density of mercury is 13.6 g cm^{-3} and that of water is 1.00 g cm^{-3}.

SOLUTION: Since the density of water is less than that of mercury, the atmosphere supports a taller column of water than mercury. From the densities of mercury and water, we can see that any given volume of mercury is 13.6 times as heavy as the same volume of water. A column of water that exerts the same pressure at its base as 76.0 cm of mercury must be 13.6 times as high as the mercury column:

$$h_{\text{water}} = h_{\text{Hg}} \times \frac{d_{\text{water}}}{d_{\text{Hg}}} = 76.0 \text{ cm} \times \frac{13.6 \text{ g cm}^{-3}}{1.00 \text{ g cm}^{-3}} = 1.03 \times 10^3 \text{ cm}$$

which is 33.8 ft.

Figure 10.3

Drinking a soda with a straw.

The pressure exerted by the atmosphere is not obvious to us because the inward pressure of the atmosphere upon one's body is counterbalanced by an equal outward pressure. But the presence of atmospheric pressure is demonstrated by our ability to drink soda through a straw or to pump water out of a well. Sucking air out of a straw creates a vacuum, which is filled as the soda is pushed up the straw by pressure of the atmosphere (Figure 10.3). A similar principle is involved when water is pumped out of a well. However, the air pressure of 1 atm cannot raise water higher than 33.8 ft, as shown by Example 10.1. Deeper wells require mechanical pumps.

The average atmospheric pressure at sea level and at 0°C is 1 atm, which equals 760 mmHg or 760 torr. These conditions of temperature and pressure, *0°C and 1 atm*, are defined as the **standard conditions of temperature and pressure** (STP).

The earth's atmosphere is concentrated near the surface both as a result of gravitational attraction and because gases are compressible. The variation of the atmospheric pressure with altitude (Figure 10.4) has many practical consequences. For example, mountain climbers often find it difficult to breathe because of a lack of air, and the cabins of high-flying jet planes are pressurized to provide a cabin pressure similar to that of the internal pressure of human bodies to prevent blood vessels and cells from rupturing.

The atmospheric pressure also changes with weather conditions. As the humidity in the atmosphere increases, more water is present in the vapor state. Water molecules (molar mass 18) are lighter than the nitrogen molecules (molar mass 28) and oxygen molecules (molar mass 32). At the same total pressure, a given volume of moist air is therefore lighter than an equal volume of dry air because moist air contains more water molecules than an equal volume of dry air. Atmospheric pressure therefore decreases with increasing humidity.

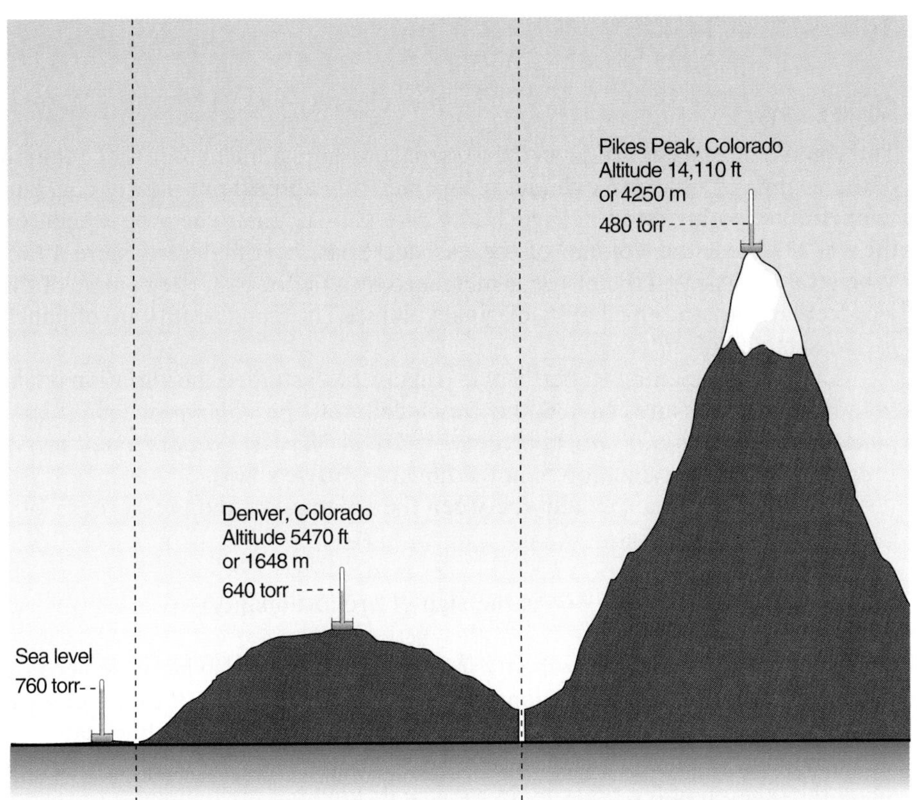

Figure 10.4

Change of atmospheric pressure with altitude. The atmospheric pressure decreases with increasing altitude, because air is mostly concentrated on the surface of the Earth and becomes less dense farther from the surface.

Changes in barometric pressure are thus related to changing weather conditions and are used in weather forecasting.

The pressure of a sample of a gas in a container in a laboratory is usually measured by a device called a **manometer** (Figure 10.5). In an open-end manometer (Figure 10.5a), the height of the mercury column, h_1, represents the difference between the gas pressure, P_1, and the atmospheric pressure, P_{atm}.

In a closed manometer (Figure 10.5b), the height of the mercury column, h_2, is a direct measure of the gas pressure, P_2. The pressure exerted by the mercury vapor in the closed end of the tube at room temperature (25°C) is negligible for most purposes. It is only about 0.002 mm. A closed-end manometer is used to measure low pressures.

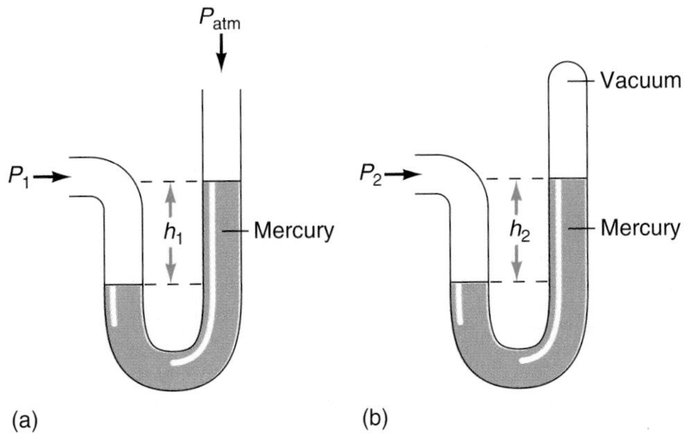

Figure 10.5

Manometers: (a) Open-end; (b) Closed-end. In the open-end manometer, the pressure of a gas is measured as the difference between the gas pressure, P_1, and the atmospheric pressure, P_{atm}. In the case shown, $P_1 = P_{atm} + h_1$. In the closed-end manometer, the height of the mercury column, h_2, is a direct measure of the gas pressure, P_2.

10.2 Elementary Gas Laws

Boyle's Law

The volume of a gas depends on the external pressure applied to it. The relation of the volume of a gas in a cylinder to an externally applied pressure at constant temperature is illustrated in Figure 10.6. We can see that when the weight on the gas doubles, the volume of the gas decreases by one-half (Figure 10.6). When the pressure on the piston is increased by a factor of 3, the volume of the gas decreases to one-third of its original volume. The same results are obtained for *all* samples of *all* gases.

The British scientist Robert Boyle (Figure 10.7) studied the variation of gas volumes with pressure. In 1660 he concluded that *the volume occupied by a given mass of a gas at constant temperature is inversely proportional to the pressure*. This generalization is now known as **Boyle's law**.

The inverse proportionality between the volume, V, and pressure, P, of a gas can be expressed as

$$V \propto \frac{1}{P} \qquad (\propto \text{ is the sign of proportionality})$$

This relationship means that when V increases, $1/P$ must also increase because V and $1/P$ are directly proportional to each other. But for $1/P$ to increase, P must decrease since $1/P$ and P are inversely proportional to each other.

The volume of a gas also varies with temperature. Therefore, when we study the relation of the pressure of a gas to its volume, the temperature is kept constant. For a given number of moles of a gas at constant temperature, Boyle's law can be formulated mathematically as

$$PV = k \quad \text{(at constant temperature)} \tag{10.1}$$

where k is a proportionality constant. The constant k is the pressure–volume product, a number that does not change with changes in P and V, as shown in Table 10.1. The pressure–volume variation of a gas at constant temperature is shown graphically in Figure 10.8.

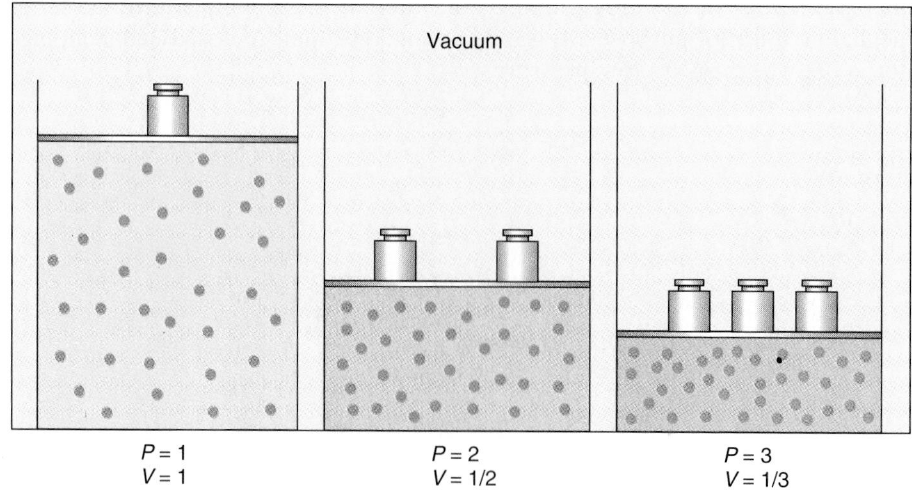

Figure 10.6

Variation of the volume of a gas with pressure at constant temperature. The dots represent gas molecules. When the pressure of the gas is doubled, the volume decreases to one half of its original volume, and when the pressure is tripled, the volume decreases to one third of its original volume.

If we apply increasingly higher pressures on the sample of the gas, as shown in Figure 10.6, the gas eventually condenses to a liquid, and even to a solid, if the temperature is low enough. Boyle's law does not apply to liquids or solids. As the pressure on the gas increases, the molecules become more crowded in the smaller volume and start repelling one another. As the pressure increases close to the point where the gas turns to a liquid, the volume of the gas does not decrease as much as would be predicted by Boyle's law. *Boyle's law is an approximation that applies only at moderate or low pressures and moderate or high temperatures.* At very high pressure or very low temperature, the behavior of most gases deviates significantly from the ideal behavior described by Boyle's law. Boyle's law is applicable for most purposes when we need answers only to three significant figures.

Figure 10.7

Robert Boyle.

Charles' Law

When gases are heated at constant pressure, they expand; when they are cooled, they contract. The French scientist Jacques Charles (1746–1823) found that *the volume, V, of a given mass of a gas at constant pressure is directly proportional to the Kelvin temperature.* This statement is known as **Charles' law**. It can be expressed algebraically as

$$V \propto T$$

This statement of proportionality can be expressed by a mathematical equation for Charles' law as follows:

$$\frac{V}{T} = k' \text{ at constant pressure} \qquad (10.2)$$

Table 10.1

Pressure–Volume Variation of a Sample of a Gas at Constant Temperature

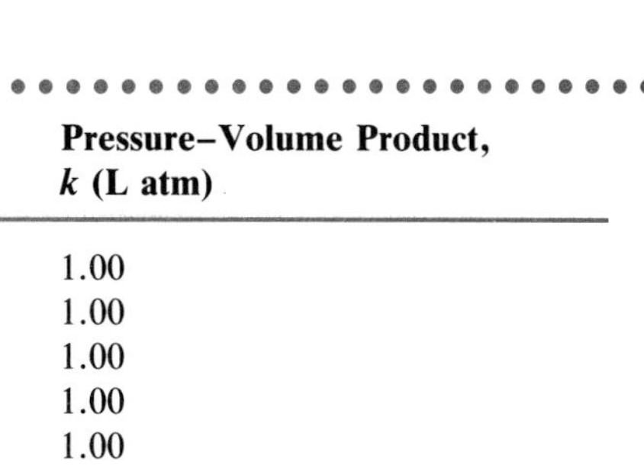

Volume, V (L)	Pressure, P (atm)	Pressure–Volume Product, k (L atm)
10.00	0.100	1.00
3.00	0.333	1.00
2.00	0.500	1.00
1.00	1.00	1.00
0.500	2.00	1.00
0.333	3.00	1.00

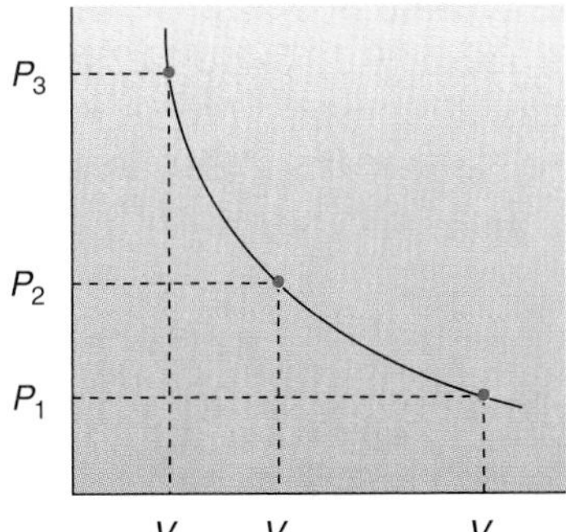

Figure 10.8

Graphical illustration of the variation of gas volume with pressure at constant temperature. If we start with a sample of a gas with volume V_1 and pressure P_1, and if we increase the pressure to P_2 and then to P_3, the volume will decrease to V_2 and V_3, respectively.

APPLICATIONS OF CHEMISTRY 10.1
The Loch Ness Monster and Boyle's Law

The Loch Ness Monster has been an object of curiosity since A.D. 565 (see photo at right). In that year St. Columba, while on a mission to convert Scotland to Christianity, first reported seeing the Loch Ness Monster. Thousands of other people have also reported seeing the Loch Ness Monster.

The scientific community began to take these sightings somewhat seriously as the reported sightings continued. In 1972, a team led by MIT physics graduate Robert Rines searched for the monster. A year later, Japanese scientists funded a \$500,000 Loch Ness Monster expedition, and the *New York Times* sponsored another search in 1976. Scientists conducted a \$1.6 million investigation in 1987.

Monsters have been seen in three of the 500 freshwater lochs (lakes) in Scotland. These three lakes are very deep, and each is surrounded by Scotch pines, which are not present around the other lakes. These facts suggest the following theory.

Scotch pines have much more resin than other pines. If you have ever had a Scotch pine Christmas tree, you probably noticed that the presents under the tree become covered with sap. When a Scotch pine dies and falls into a lake, it sinks to the bottom and the wood releases its resin. The resin, clinging to the dead tree, traps decomposition gases from the decaying wood. Trapped by the resin, the gases form blisters on the tree. As the blisters increase in size, they become large enough to buoy the tree to the surface. Since water pressure is greater on the bottom of the loch than at the top, the pressure on the blisters decreases continuously as the tree rises to the top. Boyle's law tells us that as gas pressure decreases, its volume increases. So as the tree rises, the blisters continue to inflate. The more the blisters inflate, the greater the buoyant force becomes, and as the buoyancy increases, the tree rises faster and faster. Eventually, the blisters burst and the tree sinks quickly back to the bottom of the loch. Every once in a while, a tree reaches the surface, where it raises its "monster-like" head out of the water before quickly diving again out of sight.

Scientists have shown that about 80 percent of the monster sightings occurred during surface temperature inversions. A surface temperature inversion exists when the temperature of a layer of air increases with altitude. Temperature inversions produce atmospheric image distortions, and these distortions can make protruding trees appear to move, grow, shrink, curve back, and expand vertically.

Source: Ronald De Lorenzo, *Journal of Chemical Education*, July 1989, page 570.

The relation between the volume of a gas and its Kelvin temperature at constant pressure is given in Table 10.2. Table 10.2 indicates that as the temperature of a sample of a gas at constant pressure increases, the volume also increases proportionately, so that the V/T ratio stays constant. Like Boyle's law, Charles' law applies only at moderate or low pressures and moderate or high temperatures.

Table 10.2

Variation of the Volume of a Sample of a Gas with Temperature at Constant Pressure

Volume (L)	Temperature (K)	V/T, k (L/K)
1.00	100	1.00×10^{-2}
2.00	200	1.00×10^{-2}
3.00	300	1.00×10^{-2}
4.00	400	1.00×10^{-2}
5.00	500	1.00×10^{-2}

If a gas is trapped in a container whose volume cannot change, an increase in temperature increases the pressure of the gas, and a decrease in temperature decreases the pressure. Just as the volume of a gas at constant pressure is directly proportional to Kelvin temperature, the pressure of a gas at constant volume is directly proportional to Kelvin temperature. This pressure–temperature variation of a gas is shown in Equation 10.3:

$$\frac{P}{T} = k'' \text{ at constant volume} \qquad (10.3)$$

where P is the pressure of a gas and T the Kelvin temperature.

Charles' law is illustrated by experiments in which the temperature of a gas is gradually decreased. If we begin with a 273-mL sample of a gas at 273 K and 1 atm pressure, and decrease the temperature to 272 K, the volume decreases to 272 mL. If the temperature is further decreased to 271 K, the volume of the sample decreases to 271 mL. In other words, for each 1-K decline in Kelvin temperature, the volume of a gas decreases by 1 mL from its original volume of 273 mL at constant pressure. Lowering the temperature by 1 K on any volume of a gas at constant pressure, decreases the volume by 1/273 of its original volume at 273 K. A graphic representation of these results is illustrated in Figure 10.9.

It appears from Figure 10.9 that the gas should occupy zero volume if the temperature could be lowered to 0 K. This does not and cannot happen, because the disappearance of matter would violate the law of conservation of mass. Before a temperature of 0 K is reached, the gas condenses to a liquid. Gas laws do not apply to liquid and solid states.

The temperature of 0 K (−273.15°C to five significant figures) is of fundamental significance, even though that temperature can never be achieved. This temperature is defined as **absolute zero** on the *absolute* or *Kelvin temperature scale* (Chapter 1). This scale is named for the British physicist William Thomson, known as Lord Kelvin (1824–1907).

Scientists using very ingenious experiments have produced temperatures as low as a fraction of a degree from absolute zero. At temperatures near absolute zero, matter exhibits some very unusual properties. Gases such as nitrogen and oxygen are solids at that temperature and 1 atm pressure, while helium is a liquid.

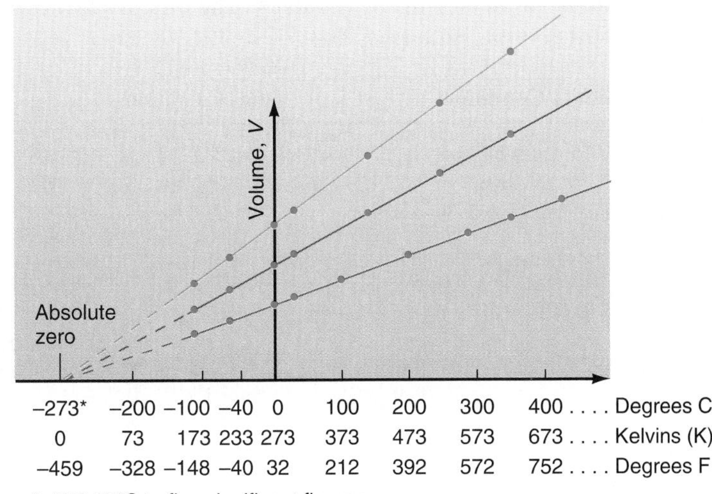

Figure 10.9

Variation of gas volume with temperature at constant pressure. At a constant pressure, a given volume of any gas at a given temperature decreases with decreasing temperature until the gas turns to a liquid. An extrapolation of this decreasing trend to zero volume (shown by dotted lines) ends up at absolute zero temperature (zero K) at zero volume.

The Combined Gas Law

We have seen that the volume, pressure, and temperature of a sample of a gas are interrelated quantities. If we change any one of these quantities, one or both of the others will also change. The change in one of these variables (P, V, or T) can be calculated from the known changes of the other two from an equation that results from the combination of Equations 10.1 ($PV = k$), 10.2 ($V/T = k'$), and 10.3 ($P/T = k''$). If V/T is constant and P/T is constant, the ratio PV/T must also be a constant:

$$\frac{PV}{T} = k'''$$

If a sample of a gas has a pressure, P_1, volume, V_1, and temperature, T_1, we can write

$$\frac{P_1V_1}{T_1} = k'''$$

For a given sample of a gas, this ratio, PV/T, is a constant even if one or more of the variables is changed. Thus, if the new pressure, volume, and temperature are P_2, V_2, and T_2, respectively, we can write

$$\frac{P_2V_2}{T_2} = k'''$$

and therefore

$$\frac{P_1V_1}{T_1} = \frac{P_2V_2}{T_2} \qquad (10.4)$$

Equation 10.4 is often called the **combined gas law equation**. Example 10.2 illustrates an application of this equation.

Example 10.2

An Application of the Combined Gas Law Equation

In a laboratory, 2.50 L of nitrogen gas is prepared at 30.0°C and 750 torr. Calculate the volume of this sample of nitrogen at standard conditions of temperature and pressure (STP), 0.00°C and 760 torr.

SOLUTION: A combined gas law problem is conveniently solved by first tabulating the values for all the variables, and by identifying them as the initial and final values for these variables. The unknown variable is then obtained by solving Equation 10.4.

Initial Conditions	**Final Conditions**
$V_1 = 2.50$ L	$V_2 = ?$
$P_1 = 750$ torr	$P_2 = 760$ torr
$T_1 = 30.0 + 273 = 303$ K	$T_2 = 273$ K

Solving Equation 10.4 for V_2, and substituting the given values into the equation, we obtain

$$V_2 = \frac{V_1P_1T_2}{P_2T_1} = \frac{(2.50\text{ L})(750\text{ torr})(273\text{ K})}{(760\text{ torr})(303\text{ K})} = 2.22\text{ L}$$

Practice Problem 10.1: A student prepared 1.26 L of oxygen gas at 77.0°F and 1.00 atm. Soon afterward the weather turned cold and rainy, the temperature in the unheated lab dropped to 45.0°F, and the barometric pressure dropped to 746 torr. What is the new volume of oxygen under these conditions?

Reactions of Gases; Gay-Lussac's Law of Combining Volumes

The French chemist Joseph Gay-Lussac (1778–1850) studied reactions of gases. In 1809 he made the surprising observation that *the volumes of gaseous reactants and products at the same temperature and pressure can be expressed in whole-number ratios*. This law is known as **Gay-Lussac's law of combining volumes**. For example, if a 57.0-mL sample of hydrogen gas were chosen for a reaction with chlorine gas at the same temperature and pressure, 57.0 mL of chlorine would be required to react with all the hydrogen, and 114 mL of hydrogen chloride (measured at the same temperature and pressure) would be produced:

$$\begin{array}{ccccc} H_2(g) & + & Cl_2(g) & \longrightarrow & 2HCl(g) \\ 57.0\text{ mL} & & 57.0\text{ mL} & & 114\text{ mL} \end{array}$$

The volumes of H_2, Cl_2, and HCl in this reaction are in the ratio 1:1:2, respectively.

Similarly, if 1.00 L of hydrogen reacts with 0.500 L of oxygen at the same temperature and pressure, 1.00 L of water vapor is produced:

$$\begin{array}{ccccc} 2H_2(g) & + & O_2(g) & \longrightarrow & 2H_2O(g) \\ 1.00\text{ mL} & & 0.500\text{ L} & & 1.00\text{ L} \end{array}$$

In this reaction, the volumes of H_2, O_2, and H_2O are in the ratio 2:1:2. The implications of Gay-Lussac's law will become clear in the next subsection.

Avogadro's Law

The Italian physicist Amedeo Avogadro (for whom Avogadro's number is named) became interested in the findings of Gay-Lussac. In 1811, Avogadro proposed that *equal volumes of all gases at the same temperature and pressure contain equal numbers of molecules*. This hypothesis became known as **Avogadro's law**.

Avogadro's law implies that 1 liter of hydrogen gas (H_2) contains as many molecules as 1 liter of oxygen gas (O_2) at the same temperature and pressure. Doubling the number of molecules (or moles) of either hydrogen or oxygen also doubles the volume of the gas if the temperature and pressure remain constant. Thus, the volume of a gas and the number of moles of the gas are directly proportional to each other, and Avogadro's law can be expressed mathematically as

$$V = kn \qquad (10.5)$$

where V is the volume of a gas, k the proportionality constant, and n the number of moles of the gas.

It follows from Avogadro's law that since hydrogen and chlorine react in 1:1 mole ratio at constant temperature and pressure, they must also react in 1:1 volume ratio because equal numbers of moles of two gases occupy the same volumes under the same conditions of temperature and pressure. Similarly, hydrogen reacts with oxygen in 2:1 mole ratio, and therefore also in 2:1 volume ratio at any given temperature and pressure.

Table 10.3

Molar Volumes of Some Common Gases at 0°C and 1 atm

Gas	Molar Volume (L)
Hydrogen (H_2)	22.433
Oxygen (O_2)	22.397
Nitrogen (N_2)	22.402
Carbon dioxide (CO_2)	22.260
Ammonia (NH_3)	22.079
Helium (He)	22.434

One mole of any gas at STP occupies a volume of approximately 22.4 L. This volume is known as the **molar volume**. Gases actually have slightly different molar volumes (Table 10.3) because molecules have different sizes and different intermolecular attractions. However, the average distance between gas molecules is so great compared with the sizes of the molecules that molecular size and the intermolecular interactions play a relatively small role in determining the volume of a gas under standard conditions.

Gases that have the smallest molecules, such as hydrogen and helium, obey the gas laws better than other gases do over a wide range of temperatures and pressures. Figure 10.10 illustrates the relation between the molar volumes and the numbers of moles, molecules, and masses of different gases.

Example 10.3

Stoichiometry Involving Gas Volumes at STP

Ammonia gas can be prepared by the reaction of hydrogen with nitrogen. Write a balanced equation for this reaction and calculate the volume of nitrogen, measured at STP, that would be needed for a complete reaction of 2.56 mol of hydrogen.

SOLUTION: The balanced equation for the reaction is

$$3H_2(g) + N_2(g) \longrightarrow 2NH_3(g)$$

Using the coefficients in the balanced equation, the number of moles of nitrogen needed for the reaction of 2.56 mol of hydrogen is

$$2.56 \text{ mol } H_2 \left(\frac{1 \text{ mol } N_2}{3 \text{ mol } H_2}\right) = 0.853 \text{ mol } N_2$$

Since 1 mol of a gas at STP has a volume of 22.4 L, the volume of nitrogen needed is

$$0.853 \text{ mol } N_2 \left(\frac{22.4 \text{ L } N_2}{1 \text{ mol } N_2}\right) = 19.1 \text{ L at STP}$$

Practice Problem 10.2: High-voltage electric discharge can convert oxygen, O_2, to ozone, O_3. How many liters of ozone, measured at STP, can be obtained: **(a)** from 1.28 mol of oxygen; **(b)** from 1.28 L of oxygen?

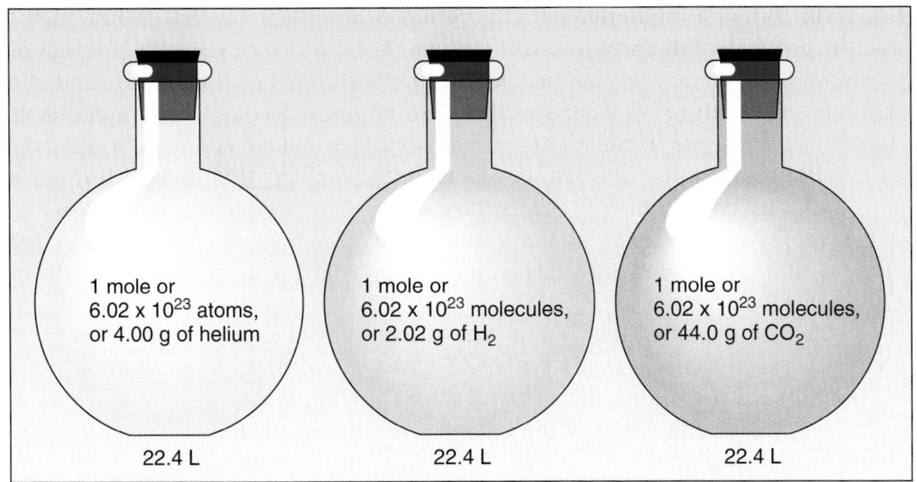

Figure 10.10

Molar volumes of gases. Approximately 22.4 L of any gas at 0°C (273 K) and 1 atm contains 1 mole of the gas. The masses of a mole of different gases differ because of the different masses of molecules, but the average intermolecular distances are the same at the same temperature and pressure. A change of temperature or pressure changes the intermolecular distances and thus the volumes, but not the number of molecules or the mass of matter in a container.

10.3 The Ideal Gas Equation and Its Applications

The Ideal Gas Equation

Up to this point we have discussed several experimentally established gas laws. According to Boyle's law, the volume of a gas is inversely proportional to pressure at constant temperature:

$$V \propto \frac{1}{P} \qquad \text{(at constant } T\text{)}$$

Charles's law states that the volume of a gas is directly proportional to the absolute temperature, at constant pressure:

$$V \propto T \qquad \text{(at constant } P\text{)}$$

Avogadro's law states that the volume of a gas is directly proportional to the number of moles, n, of the gas at constant temperature and pressure:

$$V \propto n \qquad \text{(at constant } T \text{ and } P\text{)}$$

Since the volume is proportional to $1/P$, T, and n, it is also proportional to the product of these variables:

$$V \propto \left(\frac{1}{P}\right)(T)(n)$$

This relation between the volume of a gas, the reciprocal of its pressure, its temperature, and the number of moles, can be made into an equality by introducing a proportionality constant, R:

$$V = R\left(\frac{nT}{P}\right)$$

This equation is usually written as

$$PV = nRT \qquad (10.6)$$

Equation 10.6 is a mathematical expression of the **ideal gas law**, which states that the product of the pressure and volume (PV) of a gas equals the product of the number of moles of the gas, n, the proportionality constant R, and the absolute temperature, T. Equation 10.6 is sometimes also called the **equation of state** for an **ideal gas** because the variables P, V, n, and T completely describe its condition, or state. An ideal gas is an imaginary gas that obeys all the gas laws under all conditions of temperature and pressure (see also Section 10.5). If the values of any three of the variables in the ideal gas law equation are known, the fourth can only have one value, which can be calculated from Equation 10.6.

Although the ideal gas law is rigorously correct only for ideal gases, it gives reasonably good results with real gases at or near standard conditions (273 K and 1 atm). Using the value of 22.4 L as the molar volume of a gas at 1 atm and 273 K, the value of the constant R can be calculated from Equation 10.6:

$$R = \frac{PV}{nT} = \frac{(1.00 \text{ atm})(22.4 \text{ L})}{(1.00 \text{ mol})(273 \text{ K})} = 0.0821 \text{ L atm mol}^{-1} \text{ K}^{-1}$$

If the pressure is expressed in torr and the volume in milliliters, the value of R is

$$R = \frac{(760 \text{ torr})(2.24 \times 10^4 \text{ mL})}{(1.00 \text{ mol})(273 \text{ K})} = 6.24 \times 10^4 \text{ mL torr mol}^{-1} \text{ K}^{-1}$$

In most applications of the ideal gas equation, the pressure of the gas is expressed in atmospheres, volume in liters, and temperature in kelvin. The value of R is then 0.0821 L atm mol^{-1} K^{-1}.

Applications of the Ideal Gas Equation

The ideal gas law has several practical applications, as shown by the following examples.

Example 10.4

Calculating the Pressure of a Gas from a Given Number of Moles, Volume, and Temperature

A 4.00-L container is filled with 0.200 mol of helium at 107°C. What is the pressure of the gas?

SOLUTION: We apply Equation 10.6, the ideal gas equation. It is important to express the given variables in units that correspond to the value and the units of the constant R we choose. We choose R as 0.0821 L atm mol^{-1} K^{-1}. The volume must therefore be expressed in liters, the temperature in kelvin, and pressure in atmospheres. The volume given in this problem is in liters, but the temperature is given in Celsius degrees. 107°C converted to kelvin is

$$T = 107 + 273 = 380 \text{ K}$$

Solving the equation $PV = nRT$ for P and substituting the foregoing data into the equation yields

$$P = \frac{nRT}{V} = \frac{(0.200 \text{ mol})(0.0821 \text{ L atm mol}^{-1} \text{ K}^{-1})(380 \text{ K})}{4.00 \text{ L}} = 1.56 \text{ atm}$$

Practice Problem 10.3: A sample of 1.48 mol of a gas has a pressure of 2.58 atm at 105°C. What is the volume of the gas?

Example 10.5

Using the Ideal Gas Equation to Calculate the Number of Moles and the Mass of a Sample of a Gas

In a chemical reaction, 50.0 mL of hydrogen gas is prepared at 750 torr and 297 K. How many moles and how many grams of hydrogen are obtained?

SOLUTION: We first list the variables and make sure that their units are compatible with the units of R. We will continue using R as 0.0821 L atm mol^{-1} K^{-1}. We see that the pressure is given in torr and volume in milliliters. So we convert the pressure from torr to atmospheres, and the volume from milliliters to liters:

$$P = (750 \text{ torr})\left(\frac{1 \text{ atm}}{760 \text{ torr}}\right) = 0.987 \text{ atm} \qquad V = 0.0500 \text{ L} \qquad n = ?$$

We next solve the equation $PV = nRT$ for the number of moles, n, and then convert the number of moles to mass in grams:

$$n = \frac{PV}{RT} = \frac{(0.987 \text{ atm})(0.0500 \text{ L})}{(0.0821 \text{ L atm mol}^{-1} \text{ K}^{-1})(297 \text{ K})} = 0.00202 \text{ mol}$$

$$= 2.02 \times 10^{-3} \text{ mol}$$

$$\text{mass of H}_2 = 2.02 \times 10^{-3} \text{ mol} \left(\frac{2.02 \text{ g}}{1 \text{ mol}}\right) = 4.08 \times 10^{-3} \text{ g}$$

Practice Problem 10.4: A balloon filled with 6.68 g of helium is allowed to ascend to a certain height where it has the volume of 52.5 L at a pressure of 553 torr. Assume that the material of the balloon has a negligible resistance to the expansion of the gas. What is the Celsius temperature of the helium in the balloon?

The ideal gas equation is frequently applied in stoichiometric calculations in reactions with gaseous reactants or products, as illustrated by Example 10.6.

Example 10.6

Using the Ideal Gas Equation in Stoichiometry

Potassium chlorate ($KClO_3$) is a convenient source of oxygen gas in a laboratory. A sample of potassium chlorate decomposes upon heating to produce 120 mL of oxygen gas at 303 K and 800 torr according to the equation

$$2KClO_3(s) \longrightarrow 2KCl(s) + 3O_2(g)$$

What mass of $KClO_3$ decomposes?

SOLUTION: In this problem we work backward from the volume of the gaseous product to the mass of the solid reactant. We use the ideal gas law to calculate the number of moles of O_2 produced. From the number of moles of oxygen, we calculate the number of moles and the mass of potassium chlorate using the methods outlined in Chapter 3. These steps are:

$$(P, V, T) \text{ of } O_2 \longrightarrow n \text{ of } O_2 \longrightarrow n \text{ of } KClO_3 \longrightarrow \text{mass of } KClO_3$$

Before solving for the number of moles of O_2, we convert pressure from torr to atmospheres and volume from milliliters to liters:

$$P = 800 \text{ torr} \left(\frac{1 \text{ atm}}{760 \text{ torr}}\right) = 1.05 \text{ atm} \qquad V = 0.120 \text{ L}$$

Next we determine the number of moles of O_2 produced:

$$\text{moles } O_2 \text{ produced} = n = \frac{PV}{RT} = \frac{(1.05 \text{ atm})(0.120 \text{ L})}{(0.0821 \text{ L atm mol}^{-1} \text{ K}^{-1})(303 \text{ K})}$$

$$= 5.07 \times 10^{-3} \text{ mol}$$

Finally, we use the stoichiometry of the reaction to convert moles of O_2 produced to grams of $KClO_3$ in the original sample.

$$5.07 \times 10^{-3} \text{ mol } O_2 \left(\frac{2 \text{ mol } KClO_3}{3 \text{ mol } O_2}\right)\left(\frac{122.5 \text{ g } KClO_3}{1 \text{ mol } KClO_3}\right) = 0.414 \text{ g } KClO_3$$

Practice Problem 10.5: A 0.196-g sample of impure limestone (calcium carbonate, $CaCO_3$) is treated with hydrochloric acid, which reacts with the calcium carbonate in the sample to produce carbon dioxide according to the following equation:

$$CaCO_3(s) + 2HCl(aq) \longrightarrow CaCl_2(aq) + CO_2(g) + H_2O(l)$$

Assume that the impurities do not react with HCl. When all the calcium carbonate is consumed by this reaction, 32.5 mL of carbon dioxide is collected at 23.6°C and 745 torr. Calculate the percent calcium carbonate in the limestone sample.

Practice Problem 10.6: When silver nitrate is heated, it decomposes to solid silver oxide, nitrogen dioxide gas, and oxygen gas. Write a balanced equation for this reaction, and calculate the volume in liters of nitrogen dioxide and of oxygen produced by the decomposition of 2.82 g of silver nitrate at 450°C and 772 torr.

The ideal gas equation can also be used to determine the density of a gas under any conditions of temperature and pressure. For that purpose we rearrange the ideal gas equation slightly. We know that density equals the ratio of mass to volume. The volume of a gas appears explicitly as V in the equation $PV = nRT$. The number of moles, n, of a substance can be calculated by dividing the mass, m, in grams of the substance by its molar mass, MM:

$$n = \frac{m}{\text{MM}}$$

Substituting m/MM for n in $PV = nRT$, we obtain

$$PV = \frac{mRT}{\text{MM}} \tag{10.7}$$

Since density, d, is mass divided by volume, Equation 10.7 can be rearranged to express the density of a gas by first dividing both sides of the equation by V:

$$P = \left(\frac{m}{V}\right)\left(\frac{RT}{\text{MM}}\right)$$

Since $m/V = d$, we can write

$$P = (d)\left(\frac{RT}{\text{MM}}\right)$$

from which

$$d = \frac{P(\text{MM})}{RT} \qquad (10.8)$$

Example 10.7

Using the Ideal Gas Equation to Determine the Density of a Gas

What is the density of nitrogen gas at 1.00 atm and 346 K?

SOLUTION: The molar mass (MM) of N_2 = 28.0 g mol^{-1}. Substituting the above data into Equation 10.8, we obtain:

$$d = \frac{(1.00 \text{ atm})(28.0 \text{ g mol}^{-1})}{(0.0821 \text{ L atm mol}^{-1}\text{ K}^{-1})(346 \text{ K})} = 0.986 \text{ g L}^{-1}$$

Practice Problem 10.7: What is the density of methane, CH_4, the principal ingredient of natural gas, at: **(a)** STP; **(b)** 815 torr and 325 K?

The ideal gas law can also be used to determine the molar mass of a gas or a volatile liquid. For that purpose, the ideal gas law can best be expressed in the form of Equation 10.7. The molar mass can be calculated from the known mass of a gaseous sample, its volume, pressure, and temperature, using Equation 10.7. Such a calculation is illustrated in Example 10.8.

Example 10.8

Determining the Molar Mass of a Gas

A sample of a gaseous compound has a mass of 0.316 g. At 100°C, the sample occupies 125 mL at 755 torr. Calculate the molar mass of the gas.

SOLUTION: We first convert temperature from Celsius to kelvin, volume from milliliters to liters, and pressure from torr to atmospheres:

$$T = 100 + 273 = 373 \text{ K} \qquad P = 755 \text{ torr}\left(\frac{1 \text{ atm}}{760 \text{ torr}}\right) = 0.993 \text{ atm}$$

$$V = 0.125 \text{ L}$$

We next solve Equation 10.7 for the molar mass, MM:

$$\text{MM} = \frac{mRT}{PV} = \frac{(0.316 \text{ g})(0.0821 \text{ L atm mol}^{-1}\text{ K}^{-1})(373 \text{ K})}{(0.993 \text{ atm})(0.125 \text{ L})}$$

$$= 78.0 \text{ g mol}^{-1}$$

Practice Problem 10.8: A 0.172-g sample of an unknown gaseous compound occupies 261 mL at 100°C and 752 torr. What is the molar mass of the gas?

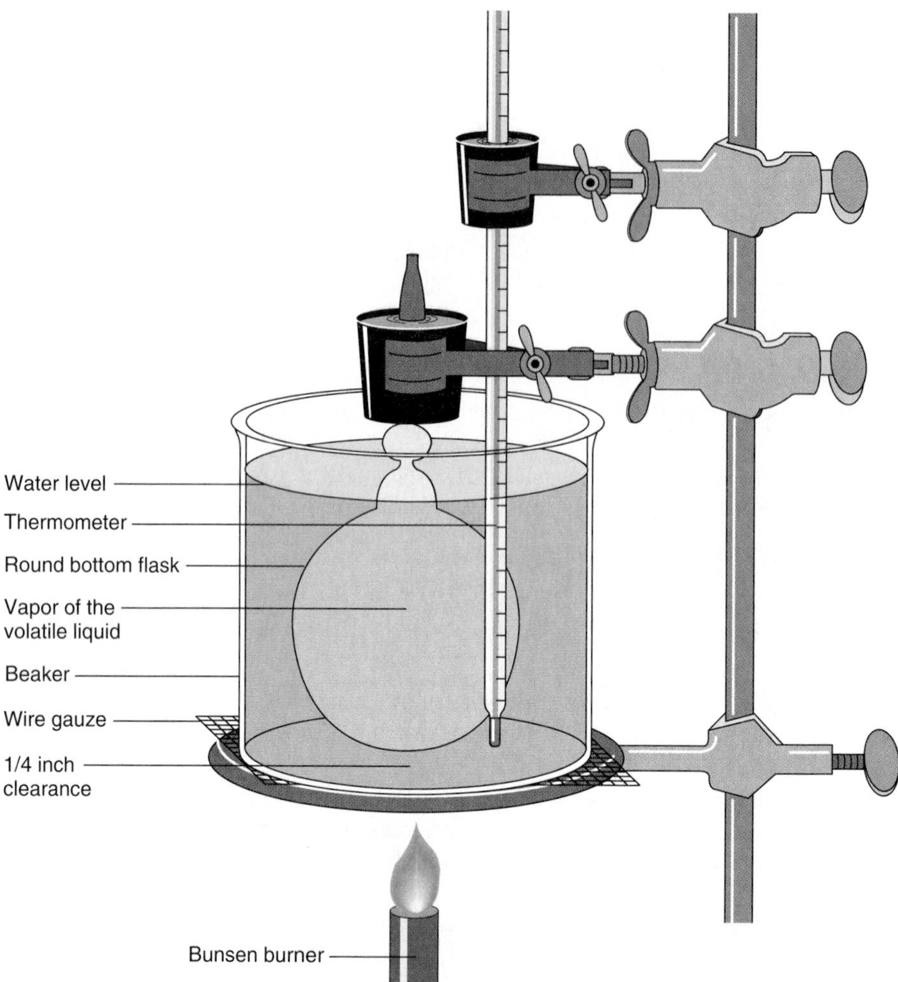

Figure 10.11

Determination of the molar mass of a volatile liquid. A small sample of the liquid is vaporized in a boiling water bath. Some vapor escapes through a small pinhole. The vapor remaining in the flask is weighed with the flask. The volume of the flask is the volume of the gas, the temperature of the boiling water is the temperature of the gas, and the barometric pressure is the pressure of the gas. From these data, the molar mass of the gas (the original volatile liquid) is calculated from the equation MM = gRT/PV, where MM is the molar mass and g the mass of the gas in grams.

The molar mass of a liquid that can easily be converted to a gas can be determined by heating the liquid in a water or oil bath (Figure 10.11) to convert the liquid completely to a gas. Heating drives some of the vapor out of the flask. The vapor that remains occupies the volume of the flask and is at the temperature of the bath. The vapor is allowed to cool to room temperature. When it cools, it condenses to a liquid again, and is then weighed with the flask. The mass of the liquid is obtained by subtracting the mass of the empty flask from that of the flask and the liquid.

Practice Problem 10.9: To determine the molar mass of a volatile liquid, excess liquid is placed into a flask fitted with a stopper containing a pinhole. The flask is then heated in a water bath to convert the liquid to vapor and to drive the excess vapor out through the pinhole. The flask is then cooled to room temperature. During the cooling process, the vapor remaining in the flask condenses to a liquid. From the following data in this experiment, calculate the molar mass of the liquid:

Mass of empty flask: 125.256 g
Mass of flask and condensed liquid after cooling: 125.568 g
Volume of the flask: 101 mL
Temperature of the water bath: 100°C
Barometric pressure: 758 torr

APPLICATIONS OF CHEMISTRY 10.2
Nuclear Power Plant Accidents and the Gas Laws

There have been two major accidents at nuclear power plants. One occurred at Three Mile Island, near Harrisburg, Pennsylvania in 1979. A second, catastrophic accident occurred at Chernobyl in Ukraine, near Kiev, in 1986 (see photo). The Three Mile Island and Chernobyl accidents had much in common. Each began when cooling pumps in the reactor failed and temperatures rose to at least 337°C. At these high temperatures, the zirconium used in reactor rod casings reacts with water. (Zirconium is used for making corrosion-resistant steel and as a hardening agent in steel alloys. Zirconium is also used in the tubes for cladding uranium oxide fuel because it is corrosion-resistant and does not absorb thermal neutrons.) The displacement reaction that occurred when the temperature in the reactor increased was

$$Zr(s) + 2H_2O(g) \longrightarrow ZrO_2(s) + 2H_2(g) \qquad (1)$$

The Three Mile Island accident released hydrogen gas (a product of the zirconium displacement reaction), which collected above the radioactive core as a large 1000-ft^3 bubble. The presence of the hydrogen bubble prevented cooling water from entering the core. Three problems were associated with the hydrogen bubble: (1) the potential of a hydrogen explosion and (2) the large pressure (70 atm) inside the core. When combined with oxygen, hydrogen explodes in the presence of a spark. [The space shuttle *Challenger* (1986) and the *Hindenburg* explosion (1937) are tragic demonstrations of the explosive force of hydrogen.] (3) The hydrogen gas was radioactive.

Knowing Boyle's law, the scientists did not attempt to reduce the pressure inside the core. Had they reduced the pressure, say from 70 atm to 35 atm, the hydrogen bubble would have doubled in size to 2000 ft^3. The large bubble would have further prevented cooling water to reach the core, causing the core to overheat even more. To compound that problem, according to Charles' law, an increase in core temperature would further increase the size of the gas bubble. To circumvent the first two problems, radioactive hydrogen gas was released slowly into the atmosphere.

We can use this information to calculate the number of moles of hydrogen that formed during the Three Mile Island accident. Use dimensional analysis to convert 1000 ft^3 to 28,000 L. Then convert 337°C to 500 K. Finally, use the ideal gas equation to solve for the number of moles of hydrogen.

$$n = \frac{(70 \text{ atm})(28{,}000 \text{ L})}{(500 \text{ K})(0.0821 \text{ L atm/mol K})}$$

$$= 50{,}000 \text{ mol of hydrogen}$$

As a final step, we can get a feeling for the amount of structural damage that took place at Three Mile Island by calculating how much zirconium reacted. From Equation 1 we can calculate the mass of zirconium that reacted to produce 50,000 mol of hydrogen. (The correct answer is 2 tons of zirconium.) The percent zirconium in the rod casings and the structural steel used in nuclear reactors is classified information. However, if we estimate it to be 10 percent, 20 tons of reactor rod casings and structural steel had to have been destroyed if 2 tons of zirconium was consumed by the displacement reaction. This gives us a vivid feeling for the amount of structural damage that took place at Three Mile Island.

Source: Earl F. Pearson, Curtis C. Wilkins, and Norman W. Hunter, *Journal of Chemical Education*, August 1988, page 718.

10.4 Gas Mixtures

Dalton's Law of Partial Pressures

Boyle's and Charles' laws apply to any gas or mixture of gases such as air. Gases that do not react with one another mix in all proportions, giving homogeneous mixtures that obey Boyle's and Charles' laws. Each gas in a mixture behaves independently of the others. Each gas occupies the entire volume of the container and exerts a pressure that is independent of the pressures exerted by the other gases in the mixture.

John Dalton observed that the total pressure of a mixture of gases equals the sum of the pressures of the components in the mixture. The contribution each gas makes to the total pressure is known as its **partial pressure**. In 1801 Dalton found that *the total pressure of a mixture of gases equals the sum of the partial pressures exerted by the individual gases in the mixture*. This generalization is known as **Dalton's law of partial pressures**. Dalton's law can be written algebraically as:

$$P_t = P_1 + P_2 + P_3 + \cdot \cdot \cdot + P_n \qquad (10.9)$$

where P_t is the total pressure, and P_1, P_2, P_3, and P_n are the partial pressures of the first, second, third, and the last components, respectively, of a gaseous mixture. The partial pressure of the component, P_1, can be obtained from the equation $P_1V = n_1RT$, where n_1 is the number of moles of component 1:

$$P_1 = \frac{n_1RT}{V}$$

Dalton's law for a mixture of three gases is expressed as follows:

$$P_t = \frac{n_1RT}{V} + \frac{n_2RT}{V} + \frac{n_3RT}{V}$$

from which

$$P_t = (n_1 + n_2 + n_3)\left(\frac{RT}{V}\right) \qquad (10.10)$$

From this equation we can calculate the total pressure of any number of components.

Example 10.9

Calculating the Total Pressure of a Mixture of Gases from Their Masses

What is the total pressure of a gaseous mixture containing 1.00 g of He, 2.00 g of H_2, and 3.00 g N_2 in a 10.0-L container at 300 K?

SOLUTION: First, we determine the number of moles of each component, then we calculate the total pressure by using Equation 10.10.

$$n_{He} = 1.00 \text{ g He} \left(\frac{1 \text{ mol He}}{4.00 \text{ g He}}\right) = 0.250 \text{ mol He}$$

$$n_{H_2} = 2.00 \text{ g } H_2 \left(\frac{1 \text{ mol } H_2}{2.02 \text{ g } H_2}\right) = 0.990 \text{ mol } H_2$$

$$n_{N_2} = 3.00 \text{ g } N_2 \left(\frac{1 \text{ mol } N_2}{28.0 \text{ g } N_2}\right) = 0.107 \text{ mol } N_2$$

The total number of moles of the components, n_t, is

$$n_t = 0.250 \text{ mol} + 0.990 \text{ mol} + 0.107 \text{ mol} = 1.347 \text{ mol}$$

Solving Equation 10.10 for P_t, we obtain

$$P_t = \frac{(n_t)RT}{V}$$

$$= \frac{(1.347 \text{ mol})(0.0821 \text{ L atm mol}^{-1} \text{ K}^{-1})(300 \text{ K})}{10.0 \text{ L}}$$

$$= 3.32 \text{ atm}$$

Practice Problem 10.10: Deep-sea divers breathe a mixture of oxygen and helium so they can ascend to the surface quickly and safely. A mixture of 0.0826 mol of oxygen and some helium occupies a volume of 2.52 L at 72.5°F and 768 torr. How many moles of helium does the mixture contain?

Collecting a Gas over Water

When a gas is prepared in a laboratory, it is often collected by displacement of water, as illustrated in Figure 10.12. If the water levels inside and outside of the graduated gas collecting tube shown in Figure 10.12 are made equal by raising or lowering the tube, the gas pressure inside the tube equals the atmospheric pressure outside the tube. The atmospheric pressure can be read on a barometer. The volume of the gas in the tube can be read directly on the tube at the

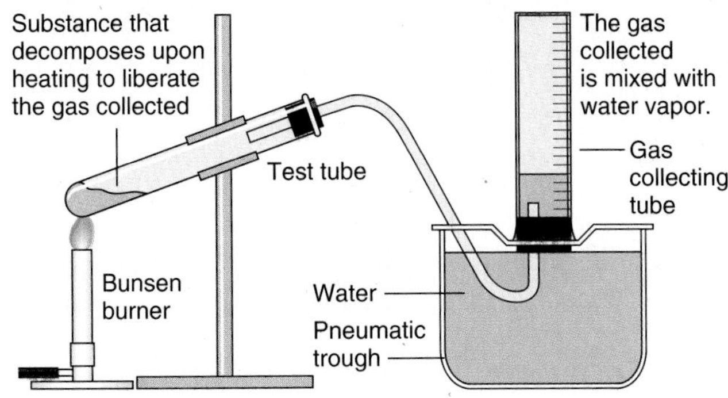

Figure 10.12

Preparation of a gas by displacement of water. The gas is produced by the decomposition of a suitable solid. The pressure of the gas generated forces some water out of the collecting tube. The volume of water displaced by the gas is the volume of the gas collected.

point where the water levels inside and outside the tube are equal. This is the volume of the mixture which consists of the gas being prepared, and of water vapor, which is always present above liquid water. *The vapor pressure of water in air which is saturated with water vapor depends only on temperature.*

Example 10.10

Calculating the Number of Moles of a Gas Collected over Water

Hydrogen gas can be prepared by the reaction of magnesium with hydrochloric acid. In an experiment, 45.0 mL of H_2 gas is collected over water at 25.0°C and 754 torr. The vapor pressure of water (the partial pressure of water vapor in air saturated with water vapor) at 25.0°C is 23.8 torr. How many moles of hydrogen is collected?

SOLUTION: To solve this problem, we first calculate the partial pressure of the H_2 collected. Then we convert pressure to atmospheres, volume to liters, and temperature to kelvin. Finally, we solve the ideal gas equation for the number of moles of H_2 produced.

In this experiment the hydrogen produced by the reaction is mixed with water vapor. The total pressure of the mixture is 754 torr:

$$P_t = 754 \text{ torr} = P_{H_2} + P_{H_2O}$$

$$P_{H_2} = 754 \text{ torr} - 23.8 \text{ torr} = 730 \text{ torr}$$

$$730 \text{ torr} \left(\frac{1 \text{ atm}}{760 \text{ torr}}\right) = 0.961 \text{ atm}$$

The temperature in kelvins is 25.0 + 273.15 = 298.2 K.

The ideal gas equation can now be written in terms of the pressure of hydrogen and the number of moles of hydrogen:

$$(P_{H_2})V = (n_{H_2})RT$$

Solving this equation for the number of moles of hydrogen yields

$$n_{H_2} = \frac{(P_{H_2})V}{RT}$$

$$= \frac{(0.961 \text{ atm})(0.0450 \text{ L})}{(0.0821 \text{ L atm mol}^{-1} \text{ K}^{-1})(298.2 \text{ K})}$$

$$= 1.77 \times 10^{-3} \text{ mol } H_2$$

10.5 Kinetic Molecular Theory, Effusion and Diffusion

The Kinetic Molecular Theory

Thus far we have described the behavior of gases in terms of experimentally established laws. Let us now consider how these laws and the properties of gases, such as their ability to expand in space and be compressed under pressure, can be explained in terms of their molecular motion.

Ideas about molecular motion were developed into the **kinetic molecular theory** by three physicists, Rudolf Clausius (1822–1888) of Germany, James Maxwell (1831–1879) of England, and Ludwig Boltzmann (1844–1906) of Austria. The kinetic molecular theory was published in its complete form by Clausius in 1857 and may be summarized as follows.

1. *All gases consist of molecules that are in random motion with varying speeds.* The molecules collide with one another and with the walls of the container. Collisions with the walls exert a force that constitutes pressure. *The molecules have negligible volume compared to the volume of the container.*
2. *Collisions between molecules, and between the molecules and the walls of the container, are perfectly elastic.* This means that the molecules collectively experience no *net* loss or gain of kinetic energy when they collide with one another or with the walls of the container, although there may be an *exchange* of energy between colliding molecules. The average kinetic energy of the molecules does not change as long as the temperature of the gas stays constant. The difference between elastic and inelastic collisions can be understood by comparing collisions of billiard balls and collisions of putty balls. Collisions between billiard balls are nearly elastic—the balls bounce off each other and continue moving. A collision between two putty balls is inelastic; as the balls collide, they stop moving.
3. *The average kinetic energy of gas molecules is proportional to the absolute temperature.* The kinetic energy of a gas molecule changes when it collides with other molecules. This means that the concept of a single, unvarying, kinetic energy for a gas molecule makes little sense. It is more useful to discuss the average kinetic energy, $(KE)_{av}$, of a collection of molecules. The average kinetic energy of gas molecules, $(KE)_{av}$, is directly proportional to the absolute temperature, T:

$$(KE)_{av} \propto T$$

Although the molecules of a gas at a given temperature have varying speeds, the distribution of these speeds at any given temperature is constant (Figure 10.13). A few molecules have very low speeds and a few have extremely high speeds. The largest fraction of the molecules has the *most probable* speed at that given temperature (T_2 in Figure 10.13). A somewhat smaller fraction has the speed that is the *average* for all the gas molecules at that temperature. The distribution of the speeds of molecules is shown in Figure 10.13 for a gas at three different temperatures.

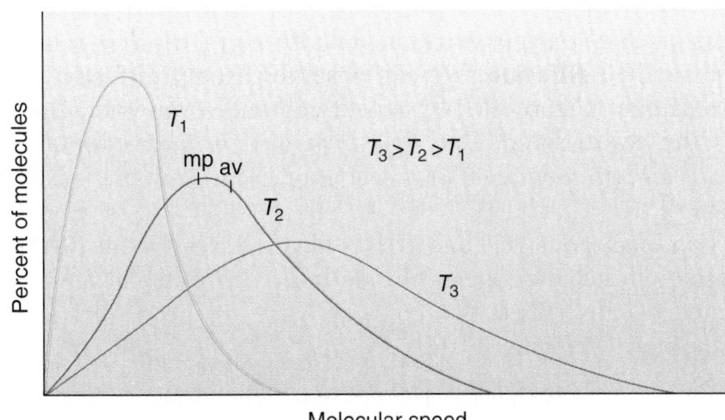

Figure 10.13

Distribution of molecular speeds of a gas at different temperatures. Point mp represents the most probable speed at temperature T_2, and point av represents the average speed at that temperature. We can see from this figure that an increase in temperature increases the average speed of gas molecules. An increase in temperature also "widens" the distribution of molecular speeds. That is, at a higher temperature, there are more molecules having much lower than the average speed and also more molecules having much higher than the average speed compared with the corresponding numbers at a lower temperature.

We see that at temperature T_2 in Figure 10.13 the average speed (marked by the point "av" on the curve) of the gas molecules is greater than the average speed at a lower temperature, T_1. At higher temperatures, there is a wider distribution of speeds among the molecules than at lower temperatures. This finding is in accord with the kinetic molecular theory. As the temperature increases, the average kinetic energy of gas molecules increases, and the molecular speed increases.

Increasing the average speed of gas molecules increases both the frequency and vigor with which the molecules strike a unit area of the container. This means that an increase in temperature increases gas volume at constant pressure, or it increases the pressure at constant volume. Thus, the kinetic molecular theory provides a simple explanation of Charles' law.

The kinetic molecular theory also explains Boyle's law. Consider a sample of a gas whose pressure is increased at constant temperature. When the pressure increases, the molecules of the gas are forced closer together, and the volume of the gas decreases. The frequency of gas molecules striking a unit area of the container walls in a smaller volume increases, and pressure increases. Thus, decreasing the volume increases the pressure.

If molecules are sufficiently close to each other, they experience mutual attraction. (We describe this phenomenon in Chapter 11.) As a result of this intermolecular attraction at close range, a smaller number of molecules strike the container walls, and the pressure and volume variation of the gas deviates from that predicted from Boyle's law. At high pressure, the molecules of a gas are relatively close together, and any further increase in pressure decreases the volume of the gas more than predicted by Boyle's law because as they are forced closer to one another, the molecules attract one another more strongly. A further increase in pressure at a sufficiently low temperature forces the molecules so close together that the gas condenses to a liquid.

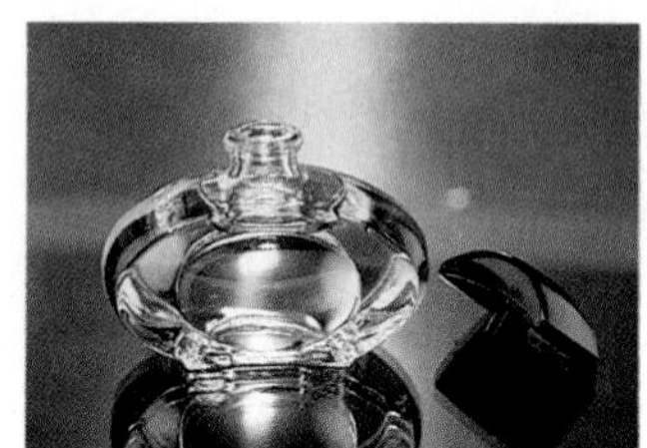

Figure 10.14

The vapor of perfume diffuses into the surrounding air.

At low temperatures, the average speed of gas molecules is low. When molecules move past each other slowly, they are more likely to be attracted to one another than at higher temperatures and higher speeds. Thus, at a low temperature, the attraction between gas molecules is more significant than at a higher temperature. This intermolecular attraction at low temperatures decreases the volume of a gas by an amount greater than that predicted by Charles' law.

Now we can better understand why the equation $PV = nRT$ is called the "ideal gas equation." Only an **ideal gas** behaves exactly as described by the gas laws discussed in this chapter. The molecules of an ideal gas occupy no volume and do not interact with one another.

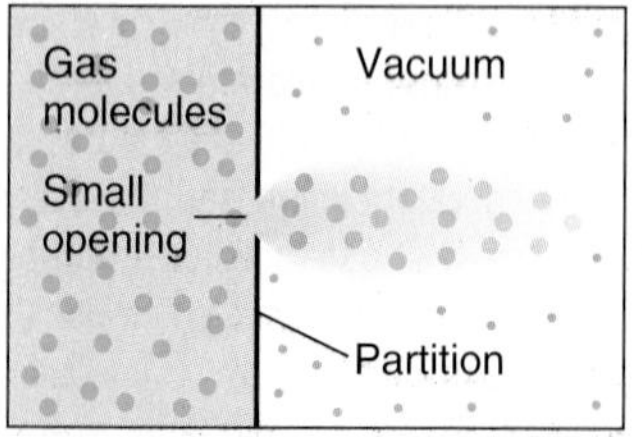

Figure 10.15

A gas flow through a small opening from a region of high pressure to a vacuum, or a region, of lower pressure.

Diffusion and Effusion of Gases

The *process in which one gas mixes with another as a result of random molecular motion* is called **diffusion**. We can observe gaseous diffusion by opening a bottle of perfume (Figure 10.14). As its contents evaporate, the odor soon permeates the entire room. The *flow of a gas through a small opening or porous material into a vacuum or a region of lower pressure* is called **effusion** (Figure 10.15).

Diffusion of gaseous HCl and NH_3 is illustrated in Figure 10.16. The diffusion of ammonia and hydrogen chloride into each other in air can easily be demonstrated because these two colorless gases react to form a white solid, ammonium chloride (Figure 10.16):

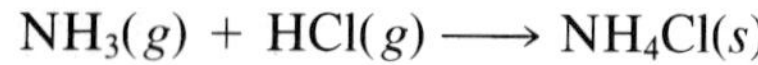

$$NH_3(g) + HCl(g) \longrightarrow NH_4Cl(s)$$

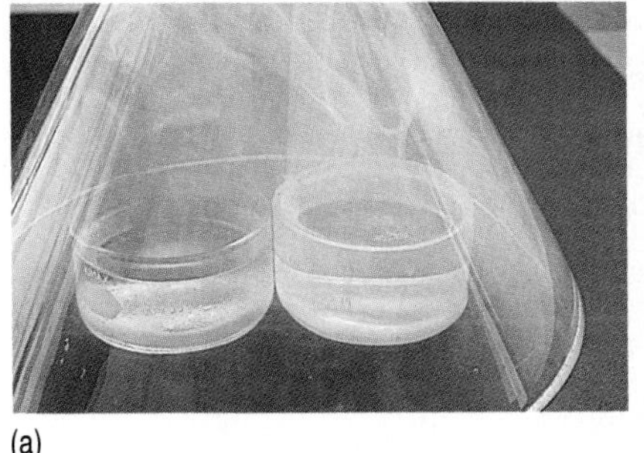
(a)

(b)

Figure 10.16

Diffusion of NH_3 and HCl to produce white solid NH_4Cl. (a) Gaseous diffusion; and (b) white solid NH_4Cl has settled on the surrounding desktop.

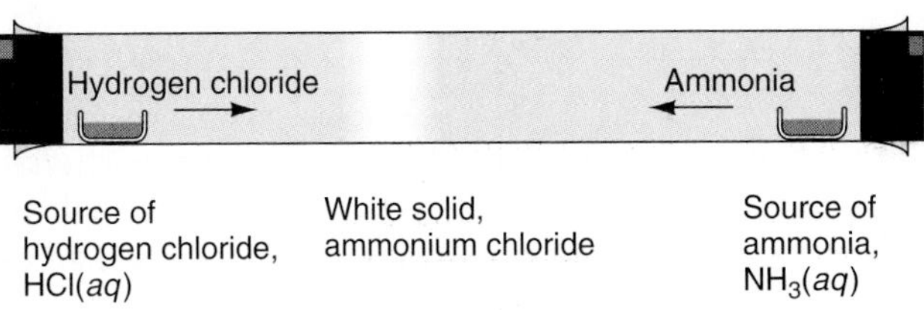

Figure 10.17

An experiment showing the relative rates of diffusion of hydrogen chloride and ammonia. The two gases meet to form a white solid, ammonium chloride. This solid forms at a point closer to the source of hydrogen chloride than the source of ammonia, showing that the rate of diffusion of ammonia is greater than the rate of diffusion of hydrogen chloride.

When a suitable source of ammonia is placed at one end of a long horizontal glass tube and simultaneously, a source of hydrogen chloride is placed at the other end (Figure 10.17), each gas diffuses toward the center of the tube. The point at which the two gases meet can be observed easily as the white solid ammonium chloride forms. Interestingly enough, the white fog of ammonium chloride does not appear at the midpoint of the tube, but somewhat nearer the end containing the source of hydrogen chloride. Since both gases begin diffusing down the tube at the same time, ammonia diffuses more rapidly than hydrogen chloride. In the next section we see why this is so.

Graham's Law of Effusion

The speed at which a gas effuses or diffuses depends on its molar mass and therefore on its density. Lower-density gases effuse faster than higher-density gases. Effusion was extensively studied by the Scottish-born British chemist Thomas Graham (1805–1869), who found that *the rate of effusion or diffusion of a gas is inversely proportional to the square root of its molar mass or density*. This statement is known as **Graham's law of effusion** or **Graham's law of diffusion**. It can be expressed algebraically as

$$v \propto \frac{1}{\sqrt{\text{MM}}}$$

where v is the effusion speed and MM is the molar mass of the gas.

Graham's law is consistent with the kinetic molecular theory. The average kinetic energy of gas molecules depends only on the absolute temperature. This means that the average kinetic energy of gas 1 (KE_1) at a given temperature equals the average kinetic energy of gas 2 (KE_2) at the same temperature, or

$$KE_1 = KE_2 \text{ at any given temperature}$$

If the mass and average speed of a molecule of gas 1 are m_1 and v_1, and for gas 2, m_2 and v_2, respectively, we can write

$$KE_1 = \tfrac{1}{2}m_1v_1^2 \quad \text{and} \quad KE_2 = \tfrac{1}{2}m_2v_2^2$$

If the gases are at the same temperature, their kinetic energies are equal:

$$\tfrac{1}{2}m_1v_1^2 = \tfrac{1}{2}m_2v_2^2$$

Multiplying both sides of this equation by 2 gives

$$m_1v_1^2 = m_2v_2^2$$

Rearranging this equation yields

$$\frac{v_1^2}{v_2^2} = \frac{m_2}{m_1}$$

The mass of the average molecule of a compound is proportional to its molar mass. Thus, replacing the masses, m, of gas molecules by their molar masses, MM, and taking the square root of both sides of the equation gives one form of the Graham's law equation:

$$\frac{v_1}{v_2} = \sqrt{\frac{MM_2}{MM_1}} \tag{10.11}$$

Since gas densities are proportional to their molecular weights or molar masses, MM, Graham's law can also be written

$$\frac{v_1}{v_2} = \sqrt{\frac{d_2}{d_1}} \tag{10.12}$$

Example 10.11

Determining the Relative Rates of Effusion of Gases

Calculate the relative rates of effusion of H_2 and O_2.

SOLUTION: The molar masses of oxygen and hydrogen are 32.0 g mol^{-1} and 2.0 g mol^{-1}, respectively. Thus, according to Graham's law of effusion, the ratio of the relative average rates of effusion of oxygen, v_{O_2}, to hydrogen, v_{H_2}, molecules may be expressed by the equation

$$\frac{v_{H_2}}{v_{O_2}} = \sqrt{\frac{32.0}{2.0}} = \sqrt{\frac{16.0}{1.0}} = \frac{4.0}{1.0}$$

This result means that hydrogen molecules effuse (or diffuse) on average four times faster than oxygen molecules at any given temperature.

Practice Problem 10.11: At a certain temperature, the rate of effusion of helium is twice the rate of effusion of a gaseous compound X. The compound X is a binary compound of carbon and hydrogen. What is the molar mass of X and the formula of X?

Graham's law and the kinetic molecular theory can be used to explain the diffusion of HCl and NH_3 shown in Figure 10.16. The molar mass of NH_3 is 17.0 g mol^{-1} and that of HCl is 36.5 g mol^{-1}. Thus, the NH_3 molecules travel faster than HCl molecules and the two meet closer to the source of HCl than that of NH_3 (see question 68 in the exercise section).

10.6 Real Gases and Deviations from the Ideal Gas Law

Ideal and Real Gas Behavior

We noted earlier that the ideal gas laws apply to real gases only under moderate or low pressure and moderate or high temperatures. Under these conditions, molecules of a real gas behave almost as noninteracting particles that have negligible volume. We know that the molecules of real gases have discrete dimensions and that they interact with one another because real gases can be condensed to the liquid or solid state under appropriate conditions. When we condense a gas to a liquid or solid, we can see that the condensed state occupies space. Therefore, the molecules must have a volume greater than zero. The behavior of real gases deviates from the behavior of ideal gases mostly at high pressures and low temperatures.

We can decide whether or not a gas is "ideal" by observing how well it obeys the ideal gas law. We know that for an ideal gas, $PV = nRT$. For 1 mol of a gas,

$$n = \frac{PV}{RT} = 1$$

Thus, for 1 mol of an ideal gas the ratio PV/RT should equal 1 at all pressures. Figure 10.18 shows the variation of the PV/RT ratio for different gases at 25°C and at various pressures. We can see from Figure 10.18 that at pressures close to 1 atm, the PV/RT ratio does not deviate appreciably from 1. With increasing pressure, however, the ratio falls below 1 for N_2 and CH_4, then increases, crossing the value of 1, and then increases considerably above 1 at higher pressures.

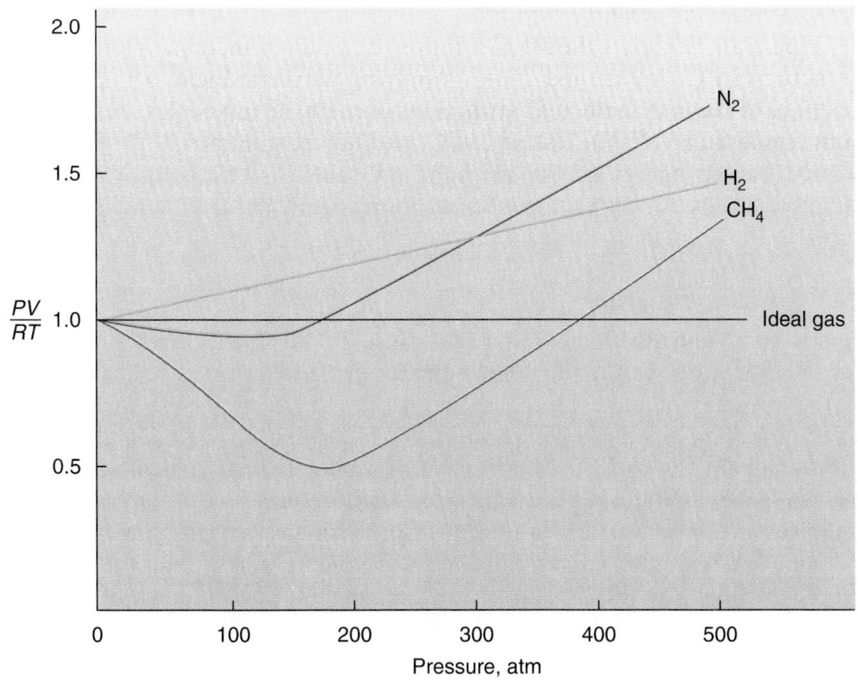

Figure 10.18

Plots of PV/RT for one mole of nitrogen, N_2, one mole of hydrogen, H_2, and one mole of methane, CH_4, as a function of pressure at 25°C. We see that nitrogen behaves pretty much like an ideal gas up to nearly 2 atm pressure. The behavior of hydrogen deviates slightly from the ideal behavior at low pressures; this deviation increases as the pressure increases. The behavior of methane deviates considerably from ideal behavior at all but very low pressures and 4 atmospheres.

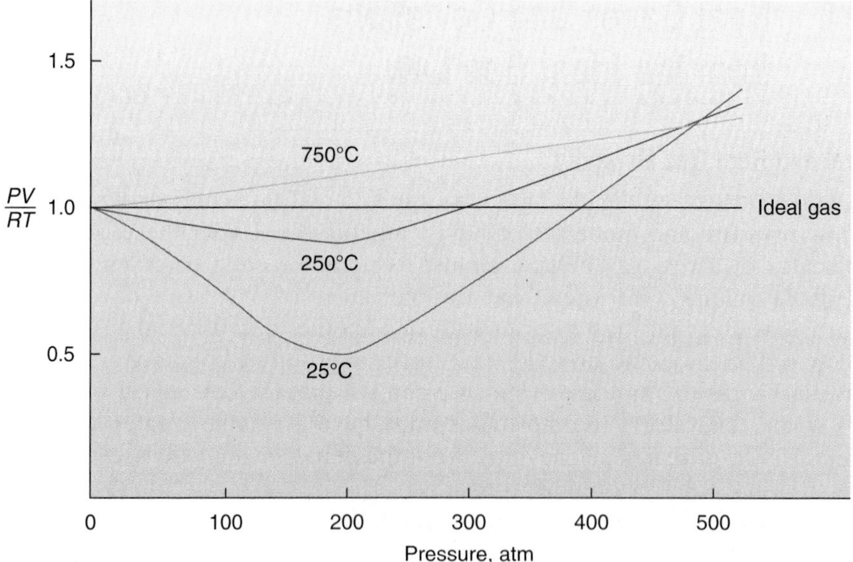

Figure 10.19

Plots of PV/RT for one mole of methane, CH_4, as a function of pressure at three different temperatures. The deviation of the behavior of methane from ideal behavior is less pronounced at higher temperatures than at room temperature.

The curve for H_2 is simplest. For H_2, $PV/RT > 1$ at all pressures, which implies that H_2 molecules repel each other slightly. This repulsion drives the molecules apart, increasing the volume of the gas. The behavior of N_2 and CH_4 is more complex. At pressures less than about 2 atm, $PV/RT < 1$ for N_2 and CH_4, indicating that at low pressures these molecules attract one another. Their intermolecular attractions decrease the volume of the gas. At pressures between 1 and 2 atm, the PV/RT ratio for CH_4 is less than that for N_2, suggesting that the attractive forces among CH_4 molecules are stronger than the attractive forces among N_2 molecules. A further increase in pressure crowds the molecules so close together that they start repelling one another. As a result, the volume decreases with increasing pressure less than expected, and the PV/RT ratio rises steeply.

Figure 10.19 shows the PV/RT ratios for 1 mol of methane plotted against pressure at 25°C, 250°C, and 750°C. Applying the same logic, we can see that the effect of the intermolecular attractions of methane molecules are greatest at room temperature (25°C), less at 250°C, and nonexistent at 750°C. At 25°C the average kinetic energy of the methane molecules is relatively low, and the attractive forces between molecules are more significant than at higher temperatures.

Van der Waals' Equation

Figures 10.18 and 10.19 show that real gases do not strictly obey the ideal gas law. Modifications to the ideal gas equation have been proposed that account for the volumes of molecules and the attractive and repulsive forces among molecules. A modified gas law equation was introduced by the Dutch physicist J. D. van der Waals in 1873. Van der Waals assumed that the ideal volume, V^*, of a real gas was different from the measured volume, V, of the gas, and the ideal pressure, P^*, was also different from the measured pressure, P.

Van der Waals reasoned that the ideal volume is less than the measured volume because the ideal volume does not include the volumes of the molecules. For 1 mol of a gas, the ideal volume is less than the real, or measured, volume by a constant b. This constant is related to the molecular size of the gas and is characteristic of the gas:

$$V^* = V - b$$

The ideal pressure of a gas, P^*, is greater than the measured pressure because the molecules of an ideal gas do not attract each other. In a real gas, the molecules near the wall of a container are attracted by molecules toward the interior of the container. Therefore, they are not able to strike the wall as frequently or as forcefully as they would if they were not attracted by other molecules. This is true for all the molecules of a gas. Consequently, the measured pressure of a real gas is less than the ideal pressure. The extent of intermolecular attraction depends on the nature of the molecules (Chapter 11) and the volume of the gas. If the volume is large, the molecules are relatively far apart, and their attractive forces are not as significant as they would be in a smaller volume. The ideal pressure, P^*, of a gas is given by the expression

$$P^* = P + \frac{a}{V^2}$$

where P is the measured pressure, a is a constant related to the attractive forces between molecules, and V is the volume of the container. This expression indicates that the ideal pressure of a gas is greater than the measured pressure by the quantity a/V^2.

The complete van der Waals' equation for 1 mol of a real gas is

$$\left(P + \frac{a}{V^2}\right)(V - b) = RT \qquad (10.13)$$

For n moles of a gas, the equation is

$$\left(P + \frac{n^2a}{V^2}\right)(V - nb) = nRT \qquad (10.14)$$

Some values of the constants a and b for common gases are listed in Table 10.4. We can see from Table 10.4 that the value of the constant b (which is related to molecular size) generally increases with increasing masses and sizes of the molecules. The value of the constant a is related to intermolecular forces. These forces, as we will learn in Chapter 11, are dependent on both the structure and the size of molecules.

Table 10.4

Van der Waals' Constants for Some Common Gases

Gas	a (L^2 atm mol^{-2})	b (L mol^{-1})
He	0.0341	0.0237
Xe	4.19	0.0510
H_2	0.244	0.0266
O_2	1.36	0.0318
H_2O	5.46	0.030
Cl_2	6.49	0.0562
CO_2	3.59	0.0427
SO_2	6.71	0.056
CH_4	2.25	0.0428

10.7 The Atmosphere and Air Pollution

Composition of the Atmosphere

The earth's atmosphere contains mostly nitrogen (about 78 percent by volume) and oxygen (about 21 percent by volume), as well as other gases in much smaller proportions. Natural processes occurring in the atmosphere, the earth, and the oceans provide a fairly constant composition of the air around us, as shown in Table 10.5.

The composition of dry air does not vary greatly with location on the earth's surface or with altitude. However, air density and pressure vary greatly with altitude. One-half of the total atmosphere lies below 5500 m. The average pressure at sea level and 45° latitude is 760 torr; at 5000 m it is about 400 torr; at 15,000 m about 40 torr; and at 50,000 m only about 0.1 torr. The lower pressure of air at higher elevations on the earth makes it harder to breathe there than at sea level.

Relative Humidity and Dew Point

The amount of water vapor in air varies from as high as 5 percent in the tropics to less than 0.01 percent in the antarctic. The amount of water vapor in air can be expressed as the **relative humidity**, which is defined as the *ratio of the partial pressure of water vapor in air to the vapor pressure of water in saturated air at the same temperature*. This ratio times 100 is the **percent relative humidity**:

$$\text{percent relative humidity} = \left(\frac{P_{H_2O}}{P^\circ_{H_2O}}\right) 100$$

where P_{H_2O} is the partial pressure of water vapor in air at a given temperature, and $P^\circ_{H_2O}$ is the partial pressure of water vapor in saturated air at the same temperature. In saturated air, water evaporates and the vapor condenses to liquid at the same rate. That is, the vapor is in equilibrium with liquid water.

The vapor pressure of water in saturated air depends only on temperature. Vapor pressure increases with increasing temperature. As temperature in-

Table 10.5

Composition of Dry Air at Sea Level

Component[a]	Average Percent by Volume	Average Percent by Mass	Average Mole Percent
Nitrogen, N_2	78.05	75.50	78.08
Oxygen, O_2	21.00	23.21	20.95
Argon, Ar	0.94	1.28	0.93
Carbon dioxide, CO_2	0.03 (variable)	0.04	0.03
Neon, Ne	0.0015	0.0011	0.0018
Helium, He	5×10^{-4}	7×10^{-5}	5×10^{-4}
Krypton, Kr	1×10^{-4}	3×10^{-4}	1×10^{-4}

[a] There are also traces of hydrogen, xenon, oxides of nitrogen and sulfur, ammonia, hydrogen sulfide, and other gases.

creases, the average kinetic energy of the molecules increases, and fewer molecules attract each other strongly enough to be converted to a liquid. Thus, a sample of air at high temperature can hold more water vapor than the same sample of air at a lower temperature. That is why air is less humid in winter than in summer. When an air mass having 100 percent relative humidity at a certain temperature cools, some of the water vapor condenses to a liquid and it rains, leaving the air saturated with water vapor at the lower temperature. However, at the lower temperature, the saturated air contains less total water vapor than it held at the higher temperature.

Example 10.12

Calculating Relative Humidity

One day in New York it was found that the temperature was 20.0°C and the partial pressure of water vapor in the air was 10.3 torr. The vapor pressure of water at 20.0°C is 17.6 torr. What is the relative humidity?

SOLUTION

$$\text{Relative humidity} = \left(\frac{10.3 \text{ torr}}{17.6 \text{ torr}}\right) 100 = 58.5\%$$

Practice Problem 10.12: The vapor pressure of water at 25.0°C is 23.8 torr. What is the partial pressure of water vapor in the air at 25.0°C when the relative humidity is 42.5 percent?

When the air temperature drops, the air may become saturated with water vapor, and any excess moisture condenses. The drops of water that sometimes condense on a glass of cold water are caused by the cooling of air next to the glass. The *temperature at which water just begins to condense out of the air* is called the **dew point**. Air is saturated with water vapor at the dew point temperature.

The Carbon and Nitrogen Cycles in Nature

Two major natural processes account for the rather constant oxygen, nitrogen, and carbon dioxide content in air. One is the *oxygen–carbon dioxide cycle* and the other is the *nitrogen cycle*. The carbon dioxide content in the atmosphere does not vary much despite the constant use and reuse of carbon compounds in nature. Green plants (Figure 10.20) consume carbon dioxide in the process of *photosynthesis* to produce glucose and oxygen:

$$6CO_2(g) + 6H_2O(l) \xrightarrow[\text{photosynthesis}]{\text{sunlight}} C_6H_{12}O_6(s) + 6O_2(g)$$

Figure 10.20

Green plants convert carbon dioxide and water to sucrose and molecular oxygen.

The rate at which carbon dioxide is removed from the air by photosynthesis and by dissolving in natural waters is nearly equal to the rate at which carbon dioxide is produced by plant, animal, and human respiration, fermenta-

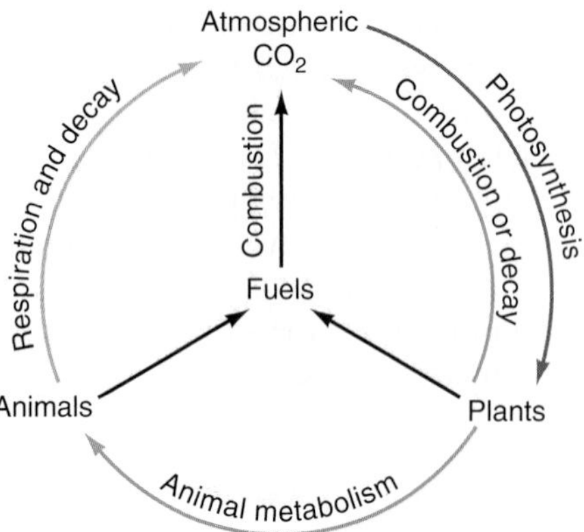

Figure 10.21

The carbon cycle in nature. Carbon dioxide is produced by burning of carbon-containing substances and by respiration and decay. The gas is used up by photosynthesis at approximately the same rate at which it is produced.

tion (conversion of sugar into alcohol and carbon dioxide), and burning of carbon compounds. Figure 10.21 illustrates the carbon cycle in nature.

Recent evidence indicates that the carbon dioxide content in the atmosphere has increased with the use of fossil fuels and the depletion of tropical forests. It is feared that the accumulation of carbon dioxide and other gases, such as methane, might trap too much heat in the earth's atmosphere, a phenomenon called the *greenhouse effect* because it is similar to the effect created by a greenhouse. The increased temperature on the earth that might result from increasing the concentration of atmospheric carbon dioxide could conceivably melt the ice in glaciers and at the north and south poles. If the worst predictions come to pass, most of the coastal cities in the world will be flooded and the earth's climate will change dramatically. The extent of the greenhouse effect is at present uncertain and is a subject of research and debate.

Nitrogen in the atmosphere is "fixed" to oxygen by lightning during thunderstorms to produce oxides of nitrogen. These oxides are washed to the ground by rain and are converted to other simple compounds of nitrogen which are used as nutrients by plants. Atmospheric nitrogen is also "fixed" by bacteria that live on the roots of plants like clover, peas, and beans. Plants convert the simple nitrogen compounds to plant proteins. Animals eat the plants and produce animal proteins. When animals die, their proteins are broken down by denitrifying bacteria to simple nitrogen compounds or to elemental nitrogen to complete the nitrogen cycle (Figure 10.22).

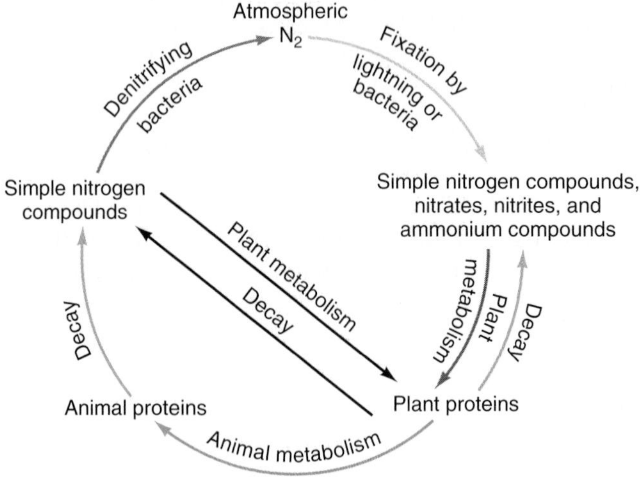

Figure 10.22

The nitrogen cycle in nature. Nitrogen in the atmosphere is produced by the decay of plant and animal proteins. The atmospheric nitrogen gas is converted to simple compounds of nitrogen by lightning and by bacterial action. The simple nitrogen compounds are converted to plant and animal proteins.

Air Pollution

Many components of the atmosphere are produced by human activities. Many of these components are *air pollutants* that adversely affect human health. The major air pollutants are *carbon monoxide*, *oxides of nitrogen*, *oxides of sulfur*, *gaseous hydrocarbons*, and various *small liquid or solid* particles suspended in air. Other pollutants are created by the action of sunlight on hydrocarbons and ozone. These are called *photochemical pollutants*. They include oxygen-containing organic compounds such as aldehydes, ketones, and peroxyacyl nitrates. Photochemical pollutants irritate the eyes, the skin, and the respiratory organs. They constitute a major part of the *smog* that plagues many large cities (Figure 10.23).

Figure 10.23

Smog in Los Angeles.

Carbon monoxide is produced by cars, airplanes, and boats when gasoline is incompletely burned (Figure 10.24). The chemical reaction for complete combustion of isooctane, C_8H_{18} (a typical component of gasoline) is

$$C_8H_{18}(l) + 25O_2(g) \longrightarrow 16CO_2(g) + 18H_2O(g)$$

Figure 10.24

Pollution and smoke produced by automobiles and buses.

If the combustion reaction proceeds to completion, no carbon monoxide is produced. However, some of the fuel is not burned, and most is burned only partially to produce carbon monoxide, CO, instead of carbon dioxide, CO_2.

At the temperature of internal combustion engines, nitrogen and oxygen react to form nitrogen oxides and, indirectly, ozone, all of which are pollutants. The oxides of nitrogen react with water to produce highly corrosive acids such as nitrous acid, HNO_2, and nitric acid, HNO_3. Ozone reacts with hydrocarbons in sunlight to produce photochemical pollutants. Some reactions that produce oxides of nitrogen and nitric acid are:

$$N_2(g) + O_2(g) \longrightarrow 2NO(g)$$

$$2NO(g) + O_2(g) \longrightarrow 2NO_2(g)$$

$$3NO_2(g) + H_2O(l) \longrightarrow 2HNO_3(aq) + NO(g)$$

Nitrogen dioxide decomposes in sunlight to produce oxygen atoms that react with oxygen molecules to give ozone (O_3):

$$NO_2(g) \xrightarrow{\text{sunlight}} NO(g) + O(g)$$

$$O(g) + O_2(g) \longrightarrow O_3(g)$$

Most of the oxides of sulfur (SO_2 and SO_3) are produced by burning sulfur-containing fuels. Coal contains about 0.5 to 3.0 percent sulfur, and its burning accounts for most of the sulfur dioxide (SO_2) in the atmosphere. Sulfur dioxide pollution is also produced by the combustion of oil and gasoline and by various industrial processes such as the roasting of sulfide ores. Sulfur dioxide reacts with water vapor in the atmosphere to give sulfurous acid (H_2SO_3):

$$SO_2(g) + H_2O(g) \longrightarrow H_2SO_3(aq)$$

Sulfur dioxide can also be converted to sulfuric acid (H_2SO_4) by reaction with water vapor and oxygen:

$$2SO_2(g) + 2H_2O(g) + O_2(g) \longrightarrow 2H_2SO_4(aq)$$

Both sulfurous and sulfuric acids in the atmosphere are produced as fine droplets that collectively contribute to smog. The oxides of sulfur and their acids irritate the upper part of the human respiratory tract, corrode metals, and damage textiles, leather, paper, paint, brick, and stone.

Figure 10.25

Cleopatra's Needle, Central Park, NY, before and after corrosion by acid rain.

Table 10.6

Approximate Annual Emission of Pollutants in the United States (Millions of Tons)

Source	Carbon Monoxide	Nitrogen Oxides	Sulfur Oxides	Hydro-carbons	Particulates (Minute Particles)
Transportation	64	8	1.0	17	1
Fuel combustion in industry, home, power generation, etc.	2	10	24	1	9
Industry	10	0.2	7	5	8
Waste disposal	8	1	0.1	2	1
Miscellaneous (agricultural burning, etc.)	17	2	1	9	10

The fine droplets of sulfurous and sulfuric acids produced in the atmosphere are washed to the ground when it rains. This *acid rain* kills fish in freshwater lakes, and it can cause structural damage to vehicles, buildings, and stone statues (Figure 10.25). Acid rain is prevalent in industrial areas where sulfur-containing fuel is used in power plants. Acid rain has killed or severely damaged forests in the northeastern United States, parts of Canada, central Europe, and Scandinavia. Table 10.6 gives a summary of the major air pollutants and their sources in the United States.

Summary

Gas Laws

The relation of the pressure of a gas to its volume at constant temperature is summarized by **Boyle's law**, which states that the volume of a gas is inversely proportional to the pressure at constant temperature:

$$PV = \text{constant} \quad (\text{at constant } T)$$

The relation of the volume of a gas to its temperature at constant pressure is given by **Charles' law**, which states that the volume of a gas is directly proportional to the *absolute* temperature at constant pressure:

$$\frac{V}{T} = \text{constant} \quad (\text{at constant } P)$$

The pressure of a gas also varies directly with absolute temperature at constant volume:

$$\frac{P}{T} = \text{constant} \quad \text{(at constant } V\text{)}$$

A useful equation for calculating the volume, pressure, or temperature of a gas resulting from a change of one or two of these variables on a sample of a gas is the *combined gas law equation*:

$$\frac{P_1V_1}{T_1} = \frac{P_2V_2}{T_2}$$

According to **Gay-Lussac's law of combining volumes**, the volumes of gaseous reactants and products are in whole-number ratios at constant temperature and pressure. This law is explained by **Avogadro's law**, which states that equal volumes of all gases under the same conditions of temperature and pressure contain equal numbers of molecules. This principle also explains why 1 mol of any gas occupies the same volume under the same conditions of temperature and pressure. At 25°C and 1 atm, this volume is 22.4 L and is called the **molar volume**.

An equation that relates the volume, pressure, and temperature of a gas with the number of moles of the gas is the **ideal gas equation**:

$$PV = nRT$$

The ideal gas equation yields reasonably good results for most real gases near **standard temperature and pressure** (STP), 273 K and 1 atm, and at higher temperatures and lower pressures.

The ideal gas equation can be applied to calculate the pressure, volume, temperature, or the number of moles of a gas from the known values of the other three variables. This law can also be used to determine the molar mass and density of a gas, as well as in stoichiometric problems involving gaseous reactants and products. To determine the molar mass, a useful modification of the ideal gas equation is

$$PV = \frac{gRT}{\text{MM}}$$

where g is the mass in grams and MM is the molar mass; thus, $g/\text{MM} = n$. For density determination, the ideal gas equation can be rearranged as

$$d = \frac{P(\text{MM})}{RT}$$

Dalton's law of partial pressures, which applies to gaseous mixtures, states that the total pressure of a mixture of gases equals the sum of the partial pressures of the components:

$$P_t = P_1 + P_2 + P_3 + \cdot\cdot\cdot + P_n$$

A useful combination of the ideal gas equation and Dalton's law is

$$P_tV = (n_1 + n_2 + n_3 + \cdot\cdot\cdot + n_n)RT$$

Kinetic Molecular Theory and Graham's Law of Effusion

The gas laws, gas pressure, and gaseous diffusion can be explained by the **kinetic molecular theory** of gases. According to this theory, all gases consist of particles of negligible volume that are in constant random motion. Collisions among these particles are perfectly elastic, and the average kinetic energy of the particles is proportional to the absolute temperature.

The relative speeds of effusion of gases are given by **Graham's law of effusion**:

$$\frac{v_1}{v_2} = \sqrt{\frac{\text{MM}_2}{\text{MM}_1}} = \sqrt{\frac{d_2}{d_1}}$$

This law agrees with a postulate of the kinetic molecular theory specifying that the average kinetic energy of gas molecules is proportional to the absolute temperature. When applied to the diffusion, this law is known as **Graham's law of diffusion**.

Real Gases

A modification of the ideal gas equation that takes into account the interaction between gas molecules and the volume occupied by the molecules is the *van der Waals equation:*

$$\left(P + \frac{n^2a}{V^2}\right)(V - nb) = nRT$$

The Atmosphere and Air Pollution

The amount of water vapor in the atmosphere can be expressed in terms of relative humidity. The **percent relative humidity** is 100 times the ratio of the partial pressure of water vapor in air, P_{H_2O}, to the vapor pressure of water in saturated air, $P^\circ_{H_2O}$, at the same temperature:

$$\text{percent relative humidity} = \frac{P_{H_2O}}{P^\circ_{H_2O}} \times 100$$

The content of nitrogen, oxygen, and carbon dioxide in the atmosphere is fairly constant due to the *carbon cycle* and the *nitrogen cycle*. In the carbon cycle, carbon dioxide is produced by burning carbon-containing substances and by human and animal respiration. However, carbon dioxide is consumed by photosynthesis at approximately the same rate at which it is produced. Oxygen is also produced by photosynthesis and consumed by human and animal respiration.

In the nitrogen cycle, atmospheric nitrogen is converted to nitrogen compounds by lightning and bacteria. These compounds are plant nutrients and are converted by plants to plant proteins. Animals eat the plants and convert the plant proteins to animal proteins. When animals die, their proteins are broken down to simple nitrogen compounds, and to elemental nitrogen, to complete the cycle.

Many of the variable components of the atmosphere are *pollutants* that are toxic to humans, animals, and plants. These substances include carbon monoxide, the oxides of nitrogen and sulfur, gaseous hydrocarbons, and photochemical pollutants.

New Terms

Absolute zero (10.2)
Avogadro's law (10.2)
Barometer (10.1)
Boyle's law (10.2)
Charles' law (10.2)
Combined gas law equation (10.2)
Dalton's law of partial pressures (10.4)
Dew point (10.7)
Diffusion (10.5)
Effusion (10.5)
Equation of state (10.3)
Gay Lussac's law of combining volumes (10.2)
Graham's law of diffusion (10.5)
Graham's law of effusion (10.5)
Ideal gas (10.3 and 10.5)
Ideal gas equation (10.3)
Kinetic molecular theory (10.5)
Manometer (10.1)
Molar volume (10.2)
Partial pressure (10.4)
Pascal (10.1)
Percent relative humidity (10.7)
Relative humidity (10.7)
Standard conditions of temperature and pressure (10.1)
Van der Waals' equation (10.6)

Exercises

General Review

1. Define each of the following terms: gas pressure; torr; 1 atm pressure; atmospheric pressure; barometer; manometer; Boyle's law; Charles' law; ideal gas; Gay-Lussac's law; Avogadro's law; Dalton's law of partial pressures; Graham's law of effusion; effusion; diffusion; equation of state.
2. Describe some different methods of gas pressure measurement and list the commonly used units of pressure.
3. If water were used instead of mercury in a barometer tube, would the water column be higher or lower than a mercury column at the same atmospheric pressure? How many times higher or lower? The density of mercury is 13.6 g cm^{-3}, and the density of water is 1.00 g cm^{-3}.
4. Write a mathematical expression for each of the following laws: Boyle's law; Charles' law; combined gas law; ideal gas law; Dalton's law of partial pressures.
5. Explain the meaning of absolute temperature.
6. Write the ideal gas equation and show how you would rearrange the equation to solve for each of the following quantities: pressure, volume, temperature, number of moles, mass of the gas, density of the gas, and molar mass of the gas. What are the limitations of this equation?
7. Write an equation for Dalton's law of partial pressures and show how this equation can be used to calculate the partial pressure of a component in a mixture of gases.
8. Explain how Dalton's law of partial pressures can be used to calculate the number of moles of a gas collected over water.
9. Write an equation for a chemical reaction involving only gaseous reactants and products other than the equations used in this chapter. On the basis of a given *mass* of a limiting reactant in this reaction, show how you would calculate the volume of a gaseous product at any given temperature and pressure. Repeat this calculation using a given *volume* of the gaseous limiting reactant.

10. Explain how the ideal gas equation gives results that deviate from those observed for real gases, particularly under extreme conditions of low temperature, high pressure, or both.
11. Explain how van der Waals modified the ideal gas equation to make it more applicable for real gases.
12. Write a mathematical expression for Graham's law of diffusion and show how this expression can be used to determine the relative rates of diffusion of two gases from the molar masses, or densities, of the gases.
13. Describe how Avogadro's principle explains Gay-Lussac's law of combining volumes.
14. List the major tenets of the kinetic molecular theory of gases, and explain the compressibility, expandability, and the pressure of a gas in terms of the kinetic molecular theory.
15. What is the qualitative composition of the atmosphere?
16. What is relative humidity?
17. What is "dew point"?
18. List some major air pollutants and explain how they get into the atmosphere. Suggest some ways to decrease air pollution without affecting industrial output or consumer prices.
19. Suppose that you are given three 1-L containers at 0°C and 1 atm pressure. One contains carbon dioxide, another oxygen, and the third, nitrogen. Compare the contents of these containers with respect to each of the following quantities: **(a)** the number of moles of the gas; **(b)** the mass of the gas; **(c)** the average kinetic energy of the molecules; **(d)** the average speed of the molecules; **(e)** the number of molecules.

Gas Pressure

20. In a mercury barometer, the glass tube has a cross-sectional area of 1.00 cm^2 and the height of the mercury column is 76.0 cm. In another mercury barometer, the cross-sectional area of the tube is 2.00 cm^2. What is the height of the mercury column in this second barometer at the same atmospheric pressure?
21. What is the mass of mercury of each of the mercury columns in the two barometers of problem 20? The density of mercury is 13.6 g cm^{-3}.
22. A student collected a gas over water in the manner illustrated in Figure 10.12. When the reaction was complete, the student found that the water level in the gas collecting tube was 15.4 cm higher than the level in the trough outside the tube. **(a)** Was the pressure of the gas in the tube higher or lower than atmospheric pressure? **(b)** By how many millimeters of Hg? **(c)** By how many atmospheres? The density of mercury is 13.6 g cm^{-3} and that of water is 1.00 g cm^{-3}. Refer to Figure 10.2 for a comparison of pressure units if necessary.

Pressure, Volume, and Temperature Variation

23. A sample of krypton at 273 K and 700 torr occupies 15.0 mL. What is its volume at STP?
24. A 100-mL sample of a gas at 291 K and 760 torr is heated to 303 K. Calculate: **(a)** the new volume of the gas if the pressure is kept constant; **(b)** the new pressure of the gas if the volume is kept constant.
25. A 1.00-L sample of hydrogen gas at STP is subjected to a pressure of 20.0 atm and a temperature of 400°C. Calculate the new volume of the gas.
26. Calculate the temperature to which 200 mL of a gas at 273 K and 750 torr must be heated to increase the volume to 250 mL at 740 torr.
27. A "tennis ball saver" is a cylindrical device that keeps tennis balls under pressure to maintain their bounce longer. When open, this device has a volume of 36.0 $in.^3$. When closed and pressurized, the volume is decreased to 18.2 $in.^3$. Assuming 1.00 atm pressure when the device is open, what is the pressure inside when it is closed? Assume no temperature change during the pressurizing process.
28. A hydrogen-filled balloon has a volume of 3.20 L at 752 torr and 303 K. When allowed to ascend to an altitude where the pressure of the gas is 715 torr, the balloon has the volume of 3.32 L. What is the temperature at that altitude? Assume that the balloon offers negligible resistance to the expansion of the gas.

Pressure, Volume, Moles, Mass, and Temperature

29. Calculate the temperature of 2.00 mol of a gas that occupies 5.00 L at 2.20 atm.
30. If a sample of a gas at STP contains 1.20×10^{23} molecules, how many moles and how many liters are present?
31. How many liters does the sample in question 30 occupy at 50.0°C and 740 torr?
32. A sample of nitrogen gas occupies a volume of 3.50 L at 750 torr and 298 K. How many moles and how many grams of nitrogen does the sample contain?
33. An empty rubber balloon was filled with 20.0 g of argon gas (atomic weight 39.9). The pressure in the balloon was 0.800 atm, and its volume was 20.0 L. Calculate the temperature of the argon gas in the balloon.
34. What volume does 5.04 g of hydrogen occupy at 1.52 atm and 300 K?
35. A sample of oxygen gas at 268 K and 738 torr has a volume of 509 mL. Calculate: **(a)** its volume at STP; **(b)** its mass.
36. What is the mass of: **(a)** 50.0 L of hydrogen at 293 K and 770 torr; **(b)** 50.0 L of helium at 293 K and 770 torr; **(c)** 50.0 L of air at 293 K and 770 torr? The density of air under these conditions is 1.293 g L^{-1}. **(d)** Which gas has a greater lifting power in a balloon, hydrogen or helium? Explain.
37. What volume does 0.771 g of ammonia, NH_3, occupy at 700 torr and 300 K?

Molar Mass and Density

38. Exactly 1 L of a gaseous compound has a mass of 1.429 g at STP. What is the molar mass of the gas? What is the mass of one molecule of the gas in amu?

39. A 125-mL sample of a gaseous compound has a mass of 0.156 g at STP. What is the molar mass of the gas?

40. If the density of a gaseous compound at STP is 1.96 g L^{-1}, what is the molar mass of the gas?

41. The density of a gaseous compound at 740 torr and 300 K is 1.75 g L^{-1}. Calculate the molar mass of the gas.

42. What is the density of NH_3 at STP?

43. What is the density of water vapor at 700 torr and 100°C?

44. If 280.0 mL of a gaseous compound at STP weighs 0.550 g, what is the molar mass of the gas?

45. A 2.00-L sample of a gaseous compound at 1.02 atm and 298 K has a mass of 6.50 g. Calculate the molar mass of the gas.

46. One liter of carbon dioxide at STP has a mass of 1.96 g. At what pressure would 1 L of carbon dioxide have a mass of 1.50 g if the temperature were kept constant? Assume ideal behavior.

47. A gaseous compound is found to contain 80.0 percent carbon and 20.0 percent hydrogen. A 1-L sample of the gas has a mass of 1.339 g at 760 torr and 273 K. Calculate: **(a)** the molar mass of the gas; **(b)** the empirical formula of the gas; **(c)** the molecular formula of the gas.

48. Calculate: **(a)** the density of oxygen at 298 K and 1.00 atm; **(b)** the density of nitrogen at 298 K and 1.00 atm; **(c)** the density of dry air at 298 K and 1.00 atm assuming that the air is composed of 78.0 percent by volume nitrogen and 22.0 percent oxygen.

49. In a determination of the molar mass of a volatile liquid compound by the vapor density method (see Figure 10.11), a sample of the liquid is vaporized in a glass bulb immersed in an oil bath at 100°C. The following data are obtained:

Mass of the glass bulb: 42.418 g
Mass of the bulb filled with water at 20.0°C: 366.800 g
Mass of the bulb + condensed liquid compound: 43.658 g (The liquid is condensed from the vapor in the bulb as the bulb cools from 100°C to 20.0°C.)
Barometric pressure: 786 torr
Density of water at 20.0°C: 0.99820 g mL^{-1}

Calculate the molar mass of the compound from these data.

Gas Mixtures and Dalton's Law of Partial Pressures

50. If 1.00 mol of nitrogen and 4.00 mol of carbon dioxide are mixed in a 10.0-L tank at 300 K, what is the total pressure in the tank (assuming that the tank is completely evacuated before mixing)?

51. A mixture of 5.00 g N_2, 5.00 g H_2, and 10.0 g NH_3 is placed in a 50.0-L container at 298 K. Calculate the total pressure in the container.

52. What volume will 5.00 g of oxygen occupy at 300 K and 1.00 atm when collected over: **(a)** mercury; **(b)** water? The vapor pressure of water at 300 K is 26.7 torr.

53. A 10-L cylinder at 90.0°C contains 2.0 mol of oxygen, 2.0 mol of hydrogen, and 3.0 mol of argon. What is the total pressure in the cylinder?

54. When a spark is passed through the mixture as described in problem 53, causing the oxygen and hydrogen to react, some water vapor is formed and the temperature rises to 550 K. What is the partial pressure of water vapor? What is the new total pressure in the cylinder?

55. In a laboratory preparation of hydrogen, the gas is collected over water. The total volume of hydrogen and water vapor is 1.00 L at 300 K and 727 torr. The vapor pressure of water at 300 K is 26.7 torr. Calculate: **(a)** the partial pressure of hydrogen in the mixture; **(b)** the number of moles of hydrogen collected.

56. A calibrated tube is filled with water and inverted into a beaker of water. Nitrogen gas is collected in the upper end of this tube by replacement of water. The volume of nitrogen in the tube is 18.6 mL at 288 K. The barometric pressure (the air pressure outside the tube) is 765 torr. The water level inside the tube is 12.5 cm higher than the water level in the beaker outside the tube. The vapor pressure of water at 288 K is 12.7 torr, and the density of mercury is 13.6 g cm^{-3}. Calculate the mass of the nitrogen gas collected.

Stoichiometry Involving Gases

57. Oxygen gas can be prepared in the laboratory by heating potassium chlorate and collecting the oxygen over water:

$$2KClO_3(s) \longrightarrow 2KCl(s) + 3O_2$$

How many grams of $KClO_3$ is needed to prepare 82.0 mL of oxygen over water at 295 K and 780 torr? The vapor pressure of water at 295 K is 19.8 torr.

58. Calculate the volume of O_2 required to completely burn 100 cm^3 of each of the following gases (measured at the same temperature and pressure): **(a)** methane, CH_4; **(b)** ethene, C_2H_4; **(c)** propane, C_3H_8. Assume that the final products of combustion are carbon dioxide and water vapor. What is the volume of carbon dioxide produced in each case?

59. Magnesium reacts with hydrochloric acid according to the equation:

$$Mg(s) + 2H^+(aq) \longrightarrow Mg^{2+}(aq) + H_2(g)$$

If 0.0228 g of Mg reacts completely with hydrochloric acid, and if the H_2 is collected over water at 300 K and 757 torr, what is the volume of H_2 collected? The vapor pressure of water at 300 K is 26.7 torr.

60. What is the minimum number of liters of chlorine gas at STP needed to convert 10.0 g of iron(II) chloride to iron(III) chloride? If the chlorine were measured at 283 K and 758 torr, what volume would be needed?

61. Heating copper(II) oxide with hydrogen gas produces copper metal and water vapor. Write a balanced equation for this reaction. In an experiment, 2.50 L of hydrogen at 150°C and 758 torr is passed through a heated tube containing copper(II) oxide at 150°C. What is the decrease in the mass of the tube? Assume that all of the hydrogen reacts.

62. A gaseous sample of an oxide of chlorine with a mass of 0.135 g that occupies 48.4 mL at 289 K and 745 torr is passed into a solution of sulfur dioxide. All the chlorine in the sample is converted to HCl. The solution is then treated with excess silver nitrate solution producing a precipitate of silver chloride which, after drying, weighs 0.2867 g. Determine the molar mass and the formula of the oxide.

Kinetic Molecular Theory; Molecular Speed

63. Arrange the noble gases, He, Ne, Ar, Kr, and Xe, in the order of increasing average speed of atoms at the same temperature.

64. What is the average speed of argon atoms at 25.0°C compared with the average speed of helium atoms at 25.0°C?

65. Oxygen molecules have an average speed of 5.0×10^4 cm s^{-1} at 25°C. What is the average speed of hydrogen molecules at 25°C?

66. Calculate the relative average speeds of chlorine and hydrogen molecules at the same temperature.

67. Argon effuses through a small opening at a rate of 8 mL s^{-1}. How fast will helium effuse through the same opening under the same conditions?

68. Hydrogen chloride gas reacts with ammonia to form white solid ammonium chloride. When these two gases are simultaneously introduced at opposite ends of a glass tube (Figure 10.17) 100 cm long, a white band of solid ammonium chloride forms in the tube where the two gases first meet. How far is this band from the end where HCl is introduced?

CHAPTER 11

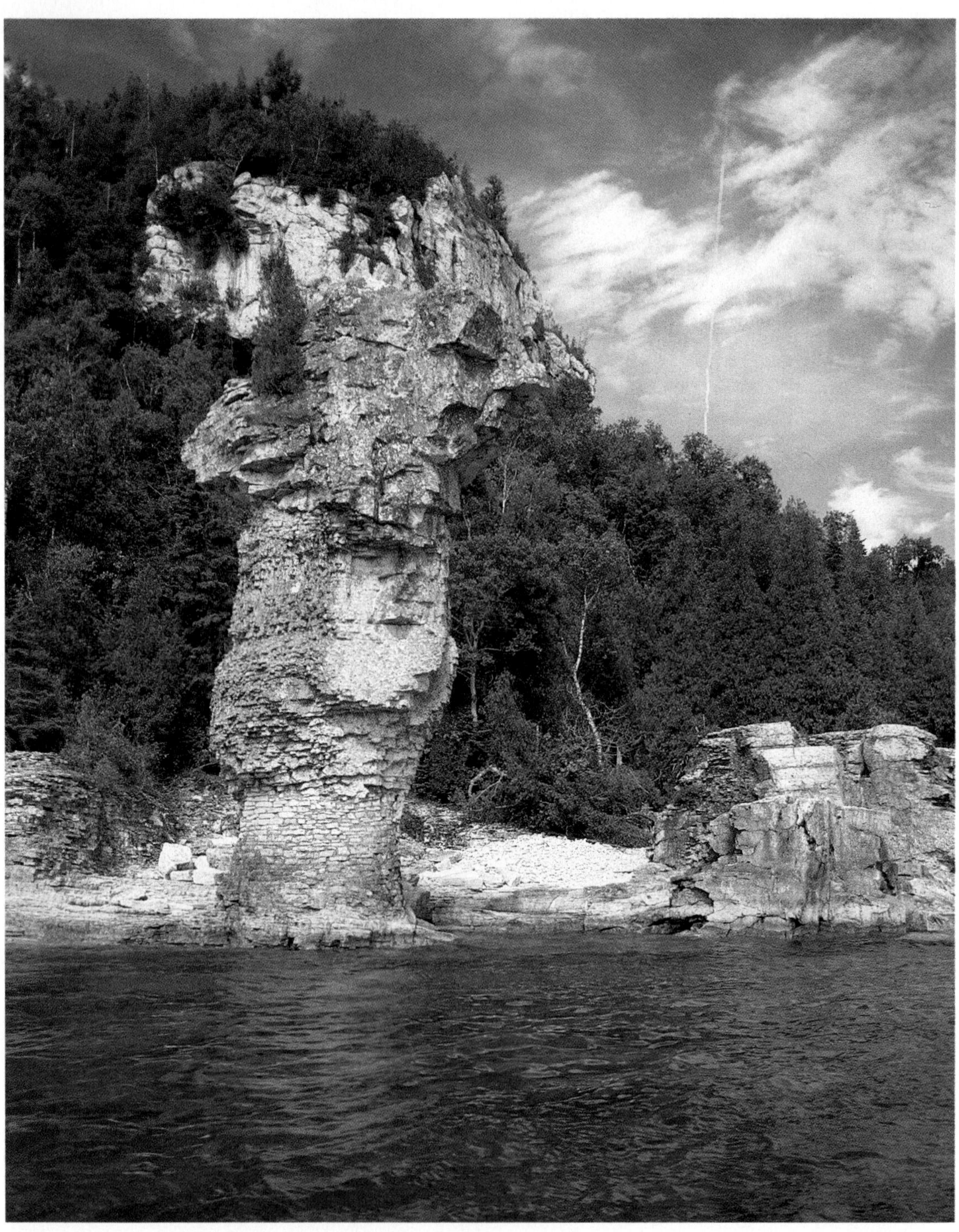

Solid, liquid, and gaseous states of matter in nature. These states coexist and interconvert with changing climate conditions of temperature and pressure.

Liquids, Solids, and Phase Changes

Having discussed the gaseous state in Chapter 10, we now turn to liquid and solid states. We are more familiar with liquids and solids than with gases, which are hard to detect unless they have color, odor, or are in rapid windlike motion. In this chapter we discuss the nature of the liquid and solid states. The ancient Greek atomists struggled to explain how the nature of "atoms" might be responsible for some things being smooth like gases, others wet and flowing like liquids or hard and solid like rocks. In this chapter we describe the modern theories that explain how molecules interact with one another and how this interaction is responsible for gaseous, liquid, or solid states of matter. We also discuss the interconversion of the three states of matter.

11.1 The Nature of Liquids

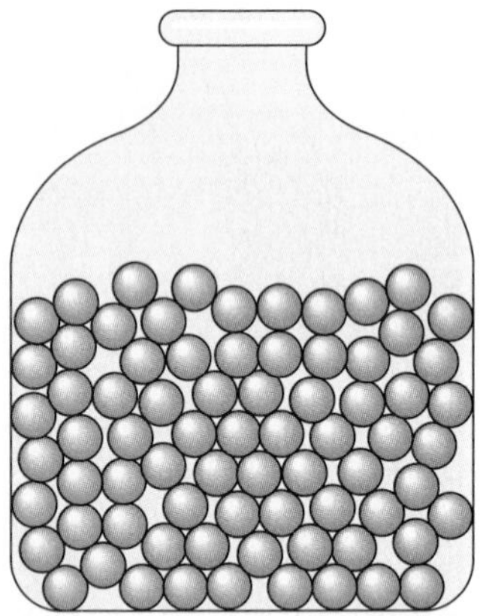

Figure 11.1

A model of a liquid.

In the gas phase, which we considered in Chapter 10, molecules are separated by large distances and intermolecular interactions are relatively weak. In contrast, the liquid state is a *condensed phase* in which the molecules are closely packed. Since the molecules in a liquid are closely packed, intermolecular forces are much stronger in a liquid than in the gas phase. We can make a physical model that explains many properties of liquids. Simply fill an irregularly shaped container with small glass beads (Figure 11.1). The beads slip into every nook and cranny of the container, leaving almost no bead-sized holes. If we shake the container, the beads move from one place to another. In a liquid the molecules are mobile, like the beads. In a solid they are not.

Since the molecules in a liquid are in a condensed phase, it is very difficult to compress a liquid. On the other hand, compressing a gas is relatively easy because it merely decreases the empty space between molecules.

A liquid can flow from one place to another because its molecules are mobile. A liquid's *resistance to flow* is called its **viscosity**. Liquids that flow slowly, such as molasses at room temperature, have high viscosities. Viscosity can be determined by measuring the time it takes for a given amount of a liquid to flow through a small-diameter tube of a specified length under the influence of gravity.

Figure 11.2

Oil can with SAE markings. As the SAE value increases, the viscosity of the oil increases.

Lubricating oils flow more slowly than water and they have higher viscosities than water. The Society of Automotive Engineers (SAE) assigns numbers from 5 to 50 to motor oils to indicate their relative viscosities (Figure 11.2). The higher the number, the greater the viscosity. Highly viscous oils are used in summer; oils of lower viscosity are winter-weight oils.

The spherical shape of a free-falling drop of a liquid shows that the molecules in the drop attract one another (Figure 11.3). A molecule on the surface of a drop is attracted by other molecules on the surface and by molecules in the interior of the drop. These forces tend to pull molecules on the surface toward the interior. In contrast, a molecule in the bulk of the liquid is attracted equally on all sides and experiences no net force in any direction. The inward pull on the molecules on the surface of the drop gives the drop its spherical shape. The energy required to overcome this inward pull, which would *increase the surface area* of the drop, is known as the **surface tension** of the liquid.

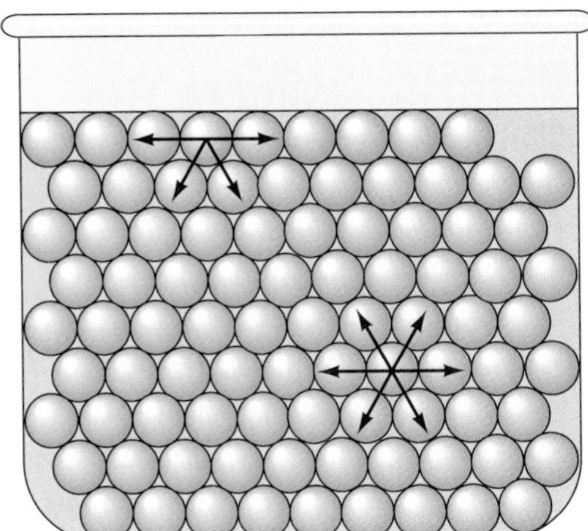

Figure 11.3

Attraction of molecules in the interior and on the surface of a liquid.

Small insects can walk on the surface of the water in a pond as if the water had a plastic cover on it. This apparently plastic or membranelike cover on the water is caused by its surface tension (Figure 11.4).

Figure 11.4

A small bug on the surface of water. An insect can walk on the surface of water because water has a high surface tension.

The surface tension of a liquid is related to its ability to spread out or "wet" a solid surface. Liquids that have low surface tensions readily wet a solid surface. Such liquids include ethyl alcohol, ether, and gasoline. In contrast, water, which has a high surface tension, tends to form droplets on a solid surface rather than to wet it (Figure 11.5). The surface tension of water can be lowered by adding soap or detergent. The resulting solution wets solid surfaces much more readily than does water itself. Table 11.1 gives the surface tensions of some common liquids.

Figure 11.5

Droplets of water on a window form spherical "beads" as the result of the high surface tension of water.

Table 11.1

Surface Tensions of Some Common Liquids at 20°C

Liquid	Formula	Surface Tension ($J\ m^{-2}$)
Benzene	C_6H_6	2.89×10^{-2}
Diethyl ether	$C_2H_5OC_2H_5$	1.70×10^{-2}
Ethyl alcohol	C_2H_5OH	2.23×10^{-2}
Glycerine	$C_3H_5(OH)_3$	6.34×10^{-2}
Mercury	Hg	46.00×10^{-2}
Water	H_2O	7.26×10^{-2}

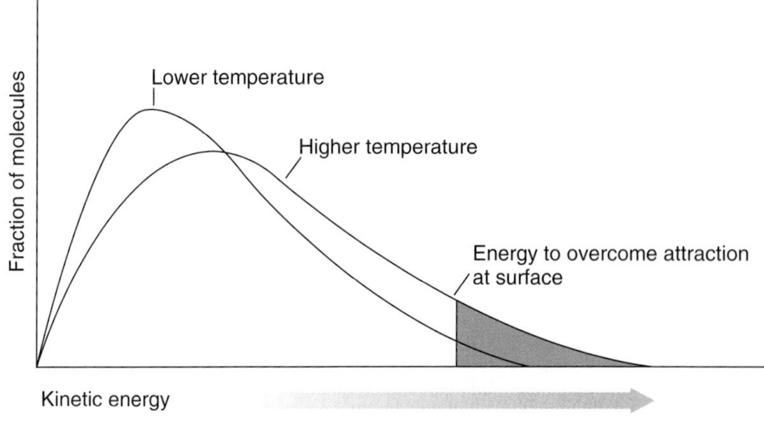

Figure 11.6

Distribution of kinetic energies among molecules of a liquid. The colored areas represent the fractions of the total number of molecules with sufficient energy to escape from the surface of the liquid to the gas phase at two different temperatures.

11.2 Evaporation and Vapor Pressure

Evaporation

The process in which a liquid spontaneously converts to a gas is called *evaporation*. An extension of kinetic molecular theory (Section 10.5) explains evaporation. The molecules in a liquid have a rather wide range of kinetic energies and velocities. These are distributed as shown in Figure 11.6. The curves in Figure 11.6 are quite similar to those shown in Figure 10.13 for molecules in the gaseous state. A molecule in the liquid phase can "escape" to the gas phase if it has sufficient kinetic energy.

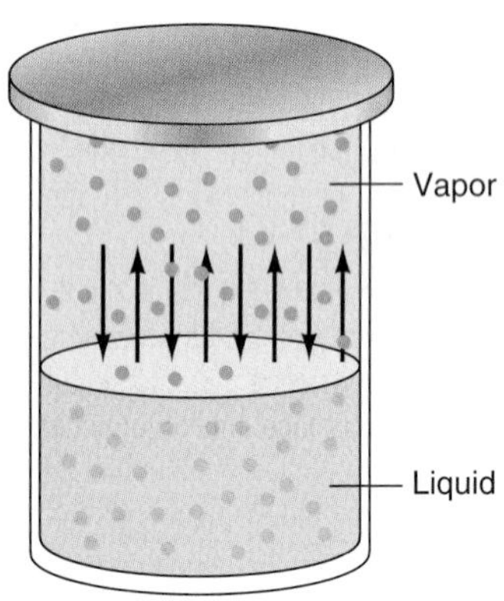

Figure 11.7

Dynamic equilibrium. The number of molecules leaving the liquid state equals the number returning to the liquid in a closed container at a given temperature.

Vapor Pressure

If a liquid is placed in a closed container, some energetic, fast-moving molecules escape from the surface of the liquid into the vapor phase. At the same time, some of the less energetic molecules in the gaseous state return to the liquid by condensation. Evaporation and condensation continue until the number of molecules leaving the liquid per unit time equals the number of molecules returning to the liquid in the same time interval. At this point the amounts of the liquid and the vapor in the container remain constant, and the system is in a state of **equilibrium** (Figure 11.7). An equilibrium in which two opposite processses occur at equal rates is also called a **dynamic equilibrium** to emphasize that the two processes occur continuously. The *pressure exerted by a vapor at equilibrium with its liquid* at a given temperature is the **equilibrium vapor pressure** at that temperature.

The vapor pressure of a liquid depends only on temperature and not on the amount of the liquid as long as *some* liquid is present. When the temperature of a liquid is changed, the average kinetic energy and speed of the molecules change, and the vapor pressure changes. Table 11.2 lists the vapor pressure of water at different temperatures.

Table 11.2

Vapor Pressure of Water at Different Temperatures

Temperature (°C)	Vapor Pressure (mm Hg)	Temperature (°C)	Vapor Pressure (mm Hg)
0	4.58	29	30.04
5	6.54	30	31.82
10	9.21	31	33.70
11	9.84	32	35.66
12	10.52	33	37.73
13	11.23	34	39.90
14	11.99	35	42.18
15	12.79	40	55.32
16	13.63	45	71.88
17	14.53	50	92.51
18	15.48	55	118.04
19	16.48	60	149.38
20	17.54	65	187.54
21	18.65	70	233.7
22	19.83	75	289.1
23	21.07	80	355.1
24	22.38	85	433.6
25	23.76	90	525.8
26	25.21	95	633.9
27	26.74	100	760.0
28	28.35		

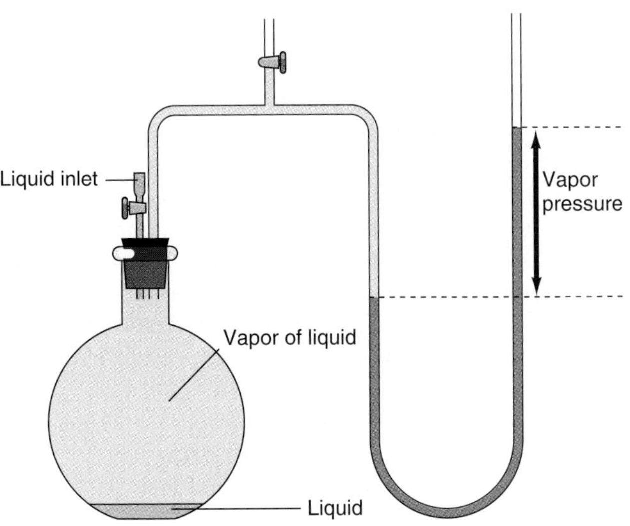

Figure 11.8

A method of measuring the vapor pressure of a liquid.

The vapor pressure of a liquid can be measured in an open-ended manometer (Figure 11.8) which was introduced in Chapter 10. A small amount of a liquid is inserted into the flask shown. The vapor pressure of the liquid is the difference between the mercury levels in the system and in the open arm.

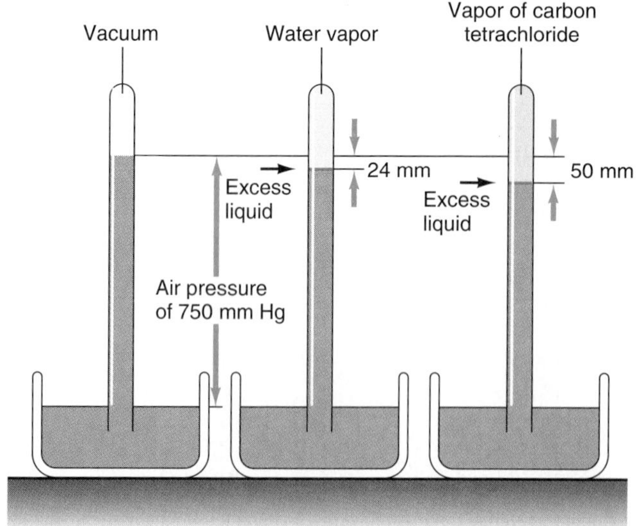

Figure 11.9

Relative vapor pressures of water and carbon tetrachloride at 25°C. A small amount of a liquid is inserted into the vacuum above the mercury column where the liquid vaporizes. Only a small excess of the liquid is used so its mass does not affect the reading.

At a given temperature, each liquid has a characteristic vapor pressure as shown in Figure 11.9. Thus, the vapor pressure of a liquid depends on the *nature of the liquid* as well as on the *temperature*.

Figure 11.9 shows that water has a lower vapor pressure than carbon tetrachloride at 25°C. This indicates that the attraction between water molecules is stronger than the attraction between carbon tetrachloride molecules (see Section 11.4).

Enthalpy of Vaporization

Energy must be supplied to evaporate a liquid. This energy is called the **enthalpy of vaporization**, ΔH_{vap}, when measured at 1 atm pressure. The enthalpy of vaporization of 1 g of a liquid is known as the *specific heat of vaporization*. The specific heat of vaporization for water is 2.26 kJ g^{-1}. The enthalpy change for vaporization of 1 mol of a liquid is the *molar heat of vaporization*. Table 11.3 lists the molar heats of vaporization of some common liquids.

A high value of ΔH_{vap} indicates that the molecules in the liquid are strongly attracted to one another. The heat required to vaporize the liquid overcomes these intermolecular attractions. The **heat of vaporization** of a liquid is thus a measure of the strength of the intermolecular forces in the liquid state.

Table 11.3

Molar Heats of Vaporization and Boiling Points of Some Common Liquids

Liquid	Formula	ΔH_{vap} (kJ mol^{-1})	Boiling Point (°C at 1 atm)
Benzene	C_6H_6	30.8	80.1
Bromine	Br_2	30.0	58.8
Carbon tetrachloride	CCl_4	30.0	76.7
Ethyl alcohol	C_2H_5OH	39.2	78.5
Diethyl ether	$C_2H_5OC_2H_5$	26.0	34.6
Mercury	Hg	59.3	356.9
Water	H_2O	40.7	100.0

Example 11.1

Calculating the Amount of Heat Absorbed from the Surroundings by Evaporation of a Liquid

The molar heat of vaporization of ethyl alcohol is 39.2 kJ mol^{-1}. How much heat would be removed from the surroundings when 1.00 kg of ethyl alcohol, C_2H_5OH, evaporates?

SOLUTION: We first convert the 1.00 kg mass of ethanol to moles. Then we use the molar heat of vaporization to calculate the amount of heat absorbed by the evaporation of the liquid. The molar mass of ethyl alcohol is 46.0 g mol^{-1}. The number of moles in 1.00 kg is:

$$1000 \text{ g} \left(\frac{1 \text{ mol}}{46.0 \text{ g}}\right) = 21.7 \text{ mol}$$

The amount of heat required to vaporize 21.7 mol of ethyl alcohol is

$$21.7 \text{ mol} \left(\frac{39.2 \text{ kJ}}{1 \text{ mol}}\right) = 851 \text{ kJ}$$

Water has a high heat of vaporization. Thus, when a swimmer emerges from water, the evaporation of water from skin causes the skin temperature to drop. Similarly, the evaporation of perspiration from skin removes heat from the body and helps to regulate body temperature on a hot day. On a humid hot day, the rate of evaporation through the skin is slow because of the high concentration of water vapor in the atmosphere. As a result, one feels uncomfortable.

11.3 Boiling and Freezing Points

Boiling Point

When a liquid is heated in an open container, its vapor pressure increases until it becomes equal to the external pressure of the atmosphere. The *temperature at which the vapor pressure of a liquid equals atmospheric pressure* is the **boiling point** of the liquid. *If the external pressure is 1 atm* (760 torr), the boiling temperature is called the **normal boiling point**. Figure 11.10 illustrates the variation of vapor pressure with temperature for some common liquids. All compounds whose normal boiling points are below 25°C are gases above 25°C.

The variation of the boiling point of water with atmospheric pressure has some practical consequences. The atmospheric pressure on a mountain top is lower than at sea level. Therefore, the temperature at which the vapor pressure of water becomes equal to the atmospheric pressure—the boiling point of water—is lower on the mountain top than at sea level. Mountain dwellers find that a "3-minute egg" takes more than 3 minutes to cook; in a cave below sea level it takes less than 3 minutes. The temperature determines the rate of cooking, not the fact that the water is boiling.

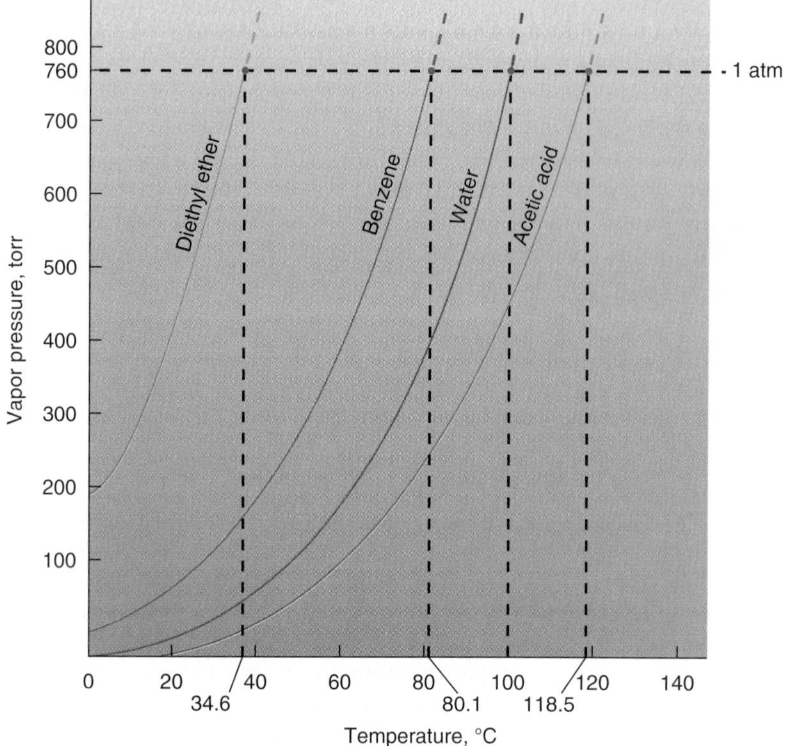

Figure 11.10

Variation of vapor pressures of some common liquids with temperature, and the vapor-pressure dependence of boiling points. The normal boiling points (°C) are the numbers indicated at the crossing points of the vapor pressure curves and the 760 torr line.

Freezing Point

When the temperature of a liquid decreases, the average kinetic energy and velocity of the molecules decrease. Eventually, a temperature is reached at which the average kinetic energy of some of the molecules is so low that the liquid begins to solidify, or freeze. At the freezing temperature of a liquid, the number of molecules entering the solid phase equals the number leaving the solid phase and entering the liquid phase in the same time period. At that temperature the liquid and the solid phases are in dynamic equilibrium.

The *temperature at which the rate of liquid converting to solid equals the rate of solid converting to liquid at 1 atm pressure* is the **normal freezing point** of the liquid. This temperature is also known as the **normal melting point** of the solid. We generally omit the word "normal" when referring to the freezing point of a liquid or melting point of a solid at a pressure of 1 atm. The variation of freezing point with atmospheric pressure is relatively small over pressure ranges ordinarily encountered in laboratories.

When a liquid freezes, heat is released to the surroundings. An equal amount of heat is absorbed from the surroundings to melt the solid. The amount of heat required to melt a specified amount of a solid at its melting point is called **heat of fusion**. The specific heat of fusion of ice at 0°C and 1 atm is 333 J g^{-1}.

11.4 Intermolecular Forces

We noted earlier that the vapor pressure and boiling point of a liquid depend on intermolecular forces. *Inter*molecular forces should be distinguished from *intra*molecular forces, which consist of the covalent bonds between the atoms *within* a molecule. Intermolecular forces are generally much weaker than covalent bonds. The strengths of intermolecular forces range from about 0.1 to 40 kJ mol^{-1}, whereas most of the covalent single bond energies are between 100 and 500 kJ mol^{-1}.

There are two basic types of intermolecular forces: (1) **London dispersion forces** and (2) **dipole–dipole interaction**, of which **hydrogen bonding** is a special case. These are discussed below.

London Dispersion Forces

Let us consider an interaction between nonpolar molecules. As the molecules approach each other, the electrons in one molecule repel the electrons in the other, and vice versa. This mutual distortion or *polarization* of the electron distribution of adjacent molecules induces *temporary polarity* in molecules. The relatively electron-rich side of one temporarily polar molecule is weakly attracted to the relatively electropositive side of another temporarily polar molecule (Figure 11.11). Intermolecular forces due to such *induced dipole–induced dipole interaction* are called **London dispersion forces** or simply **London forces** (after the German physicist Fritz London, 1900–1954). They are also sometimes called **van der Waals forces** (after the Dutch physicist J. D. van der Waals). Strictly speaking, van der Waals forces include *all* types of forces of attraction between molecules.

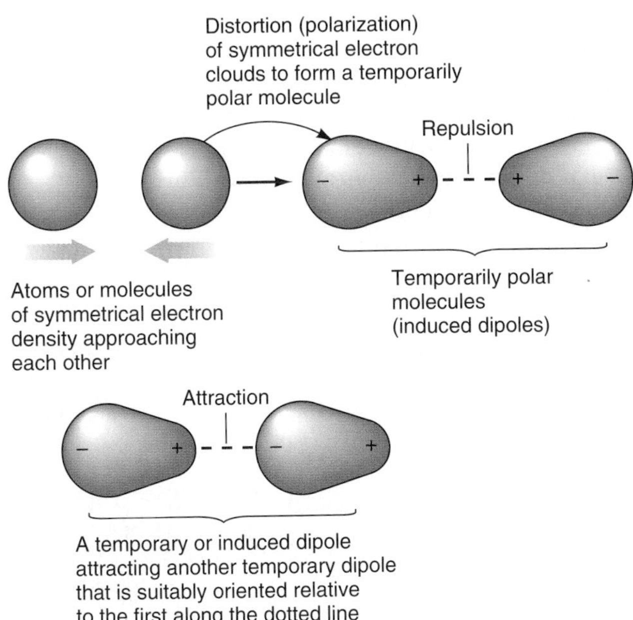

Figure 11.11

Temporary dipoles and London forces. The positive signs indicate centers of positive charge. The negative signs show the relatively high concentrations of electrons, or centers of negative charge.

The strength of London forces depends on the extent to which the electron distribution in the interacting molecules can be distorted or polarized. The polarizability of molecules is related to the number of electrons in a molecule, or its size. For example, I_2 is more polarizable than F_2. In a large molecule, such as I_2, the outer electrons are far from the nucleus. These outer electrons are therefore highly mobile and can easily be distorted by the electric field of an adjacent molecule. In a small molecule, such as F_2, the electrons are strongly attracted to the nucleus and are therefore less easily polarized by the electric field of a nearby molecule. Thus, induced dipole–induced dipole interactions are stronger in I_2 than in F_2.

Figure 11.12

Gaseous Cl_2, liquid Br_2, and solid I_2.

One of the factors that determine the physical state of a substance is molecular size, and we have seen that the strengths of London forces are also related to molecular size. Fluorine, F_2, and chlorine, Cl_2, are gases at 25°C because F_2 and Cl_2 are small molecules that interact by relatively weak London forces. Molecules in the liquid state experience stronger intermolecular attractions than molecules in the gaseous state. Bromine, Br_2, molecules are larger than chlorine molecules and are therefore attracted to one another by stronger London forces than chlorine molecules. Bromine is a liquid at 25°C. Iodine molecules, I_2, are larger than bromine molecules and experience stronger London forces. Iodine is a solid at room temperature (Figure 11.12).

Dipole–Dipole Interaction

In the preceding section we saw that nonpolar molecules attract one another by induced dipole–induced dipole interactions. The *attraction between permanently polar molecules* is a **dipole–dipole interaction**. The partially positively charged end of one polar molecule attracts the partially negatively charged end of another polar molecule. When the dipoles are oriented so that their oppositely charged ends are as close to each other as possible, the maximum interaction occurs. This principle is illustrated below for sulfur dichloride.

Sulfur dichloride, SCl_2, is a liquid at room temperature. Sulfur dichloride is a polar molecule whose structure is similar to that of H_2O. Figure 11.13 illustrates the attraction among SCl_2 molecules. The strongest attraction between polar SCl_2 molecules occurs when the molecules are oriented so that the partially negative end of one molecule is aligned with the partially positive end of another molecule. Sulfur dichloride is a liquid at room temperature and pressure because of strong dipole–dipole interaction.

Hydrogen Bonding

Polar molecules that contain F—H, O—H, or N—H groups attract one another so that the relatively electropositive hydrogen atom of one molecule becomes attracted to an F, O, or N atom of another molecule. This type of

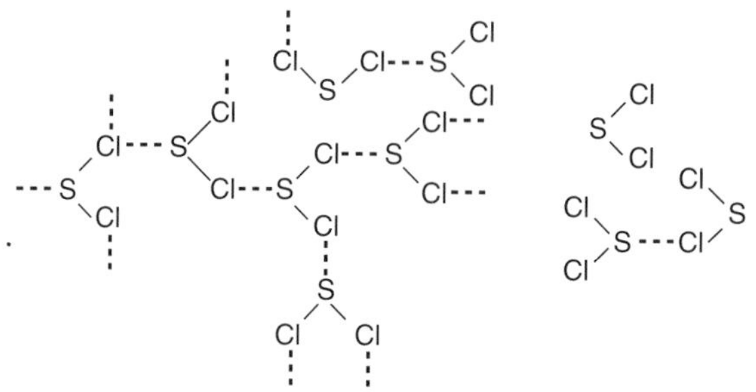

Figure 11.13

Attraction between polar molecules (dipole–dipole interaction). The solid lines represent intramolecular covalent bonds and the dotted lines represent the intermolecular forces between polar molecules.

attraction is a special case of dipole–dipole attraction called a **hydrogen bond** (or **H-bond**), as shown in Figure 11.14 for water molecules.

Hydrogen bonds are generally stronger than other intermolecular forces because the F—H, O—H, and N—H bonds are very polar. The strength of hydrogen bonding accounts for the liquid state of water. Without hydrogen bonding, water would be a gas with a boiling point lower than that of H_2S, which is a gas at temperatures as low as −60.7°C at 1 atm pressure.

The strength of hydrogen bonds decreases in the order H—F···H—F > H—O···H—O > H—N···H—N. The hydrogen bond in HF is the strongest of the three because fluorine is the most electronegative element, and the H—F bond is more polar than H—O or H—N bonds.

Hydrogen fluoride, HF, has a normal boiling point of 19.4°C, higher than the boiling points of other hydrogen halides (see Figure 11.16). This difference is due to strong hydrogen bonding in HF and the lack of hydrogen bonding in other hydrogen halides. The HF molecules attract one another so that the H atom of one of the molecules becomes strongly attracted to the F atom of another molecule (Figure 11.15). The small size of the H atom allows the H and the two F atoms to get relatively close to each other in the network ···H—F···H—F···. Chlorine, bromine, and iodine atoms are larger and much less electronegative than a fluorine atom. Thus, almost no hydrogen bonding exists in HCl, and none in HBr or HI.

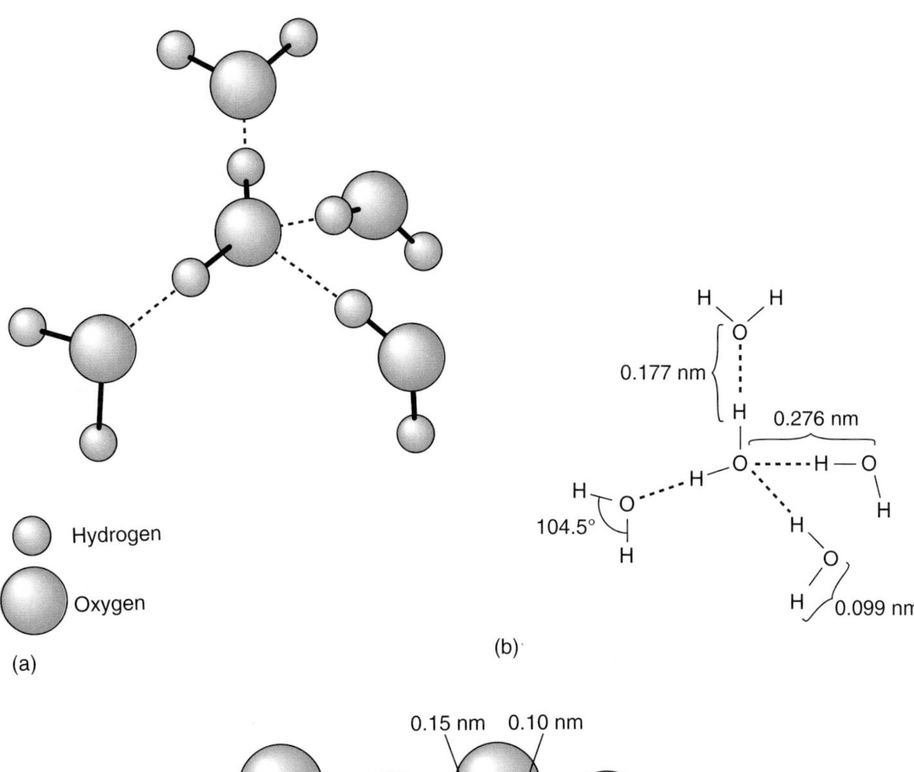

Figure 11.14

Hydrogen bonding in liquid water. (a) Water can form up to four hydrogen bonds. The bonds are extended toward the corners of a tetrahedron. (b) Lengths of hydrogen bonds in a hydrogen-bonded cluster of water molecules.

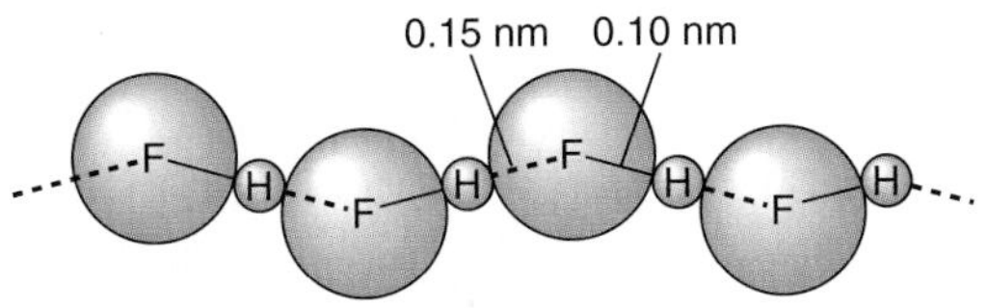

Figure 11.15

Hydrogen bonding in HF. The solid lines represent intramolecular covalent bonds and the dotted lines hydrogen bonds.

Table 11.4

Energies of the Intermolecular Attractive Forces in Some Simple Substances in Their Solid States

Substance	Attractive Energies (kJ mol^{-1})	
	Dipole–Dipole and/or H-Bonding	London Forces
H_2O	36.4	9.0
NH_3	13.3	14.7
HCl	3.31	16.8
HBr	0.69	21.9
HI	0.025	27.9
Ar	0	8.49

We have now considered London forces, dipole–dipole interactions, and hydrogen bonds. Of these, hydrogen bonds are generally the strongest and London forces the weakest. Although London forces are weak among small molecules, their *net* effect among large molecules may be greater than the attractions due to dipole–dipole forces and hydrogen bonds. The energies of some intermolecular forces are listed in Table 11.4.

Example 11.2

Explaining the Liquid State for Water and the Gaseous State for Hydrogen Fluoride at Room Temperature

Explain why water is a liquid at room temperature, whereas HF is a gas, although a hydrogen bond between two HF molecules is stronger than a hydrogen bond between two H_2O molecules.

SOLUTION: A water molecule can form four hydrogen bonds since it has two lone electron pairs and two H atoms (see Figure 11.14). An HF molecule cannot have as many hydrogen bonds as a water molecule.

Intermolecular Forces and Boiling Point

Every liquid has a characteristic boiling point. We recall that the boiling point of a liquid is related to its vapor pressure. Liquids with high vapor pressures have low boiling points. In this section we see how boiling points are related to the strengths of intermolecular forces in a liquid. Figure 11.16 is a plot of normal boiling points of the noble gases and the binary hydrogen compounds of the nonmetals in periods 2 to 5.

The lowest boiling points in Figure 11.16 are those of noble gases. Noble gases interact only by weak, easily disrupted, London forces. The London forces are weakest for (small) helium atoms, but increase in strength with the increasing size of the noble gas atom from helium to xenon, as shown in Figure 11.16. As the intermolecular forces increase, the vapor pressure of a liquid decreases and the boiling point increases.

Figure 11.16 also includes the boiling points of binary hydrogen compounds of the fourth-group elements from CH_4 to SnH_4. These are nonpolar molecules. The only intermolecular forces between nonpolar molecules are London forces. The increasing trend in the boiling points from CH_4 to SnH_4 can therefore be attributed to the increasing strength of London forces with the increasing sizes of the molecules from CH_4 to SnH_4.

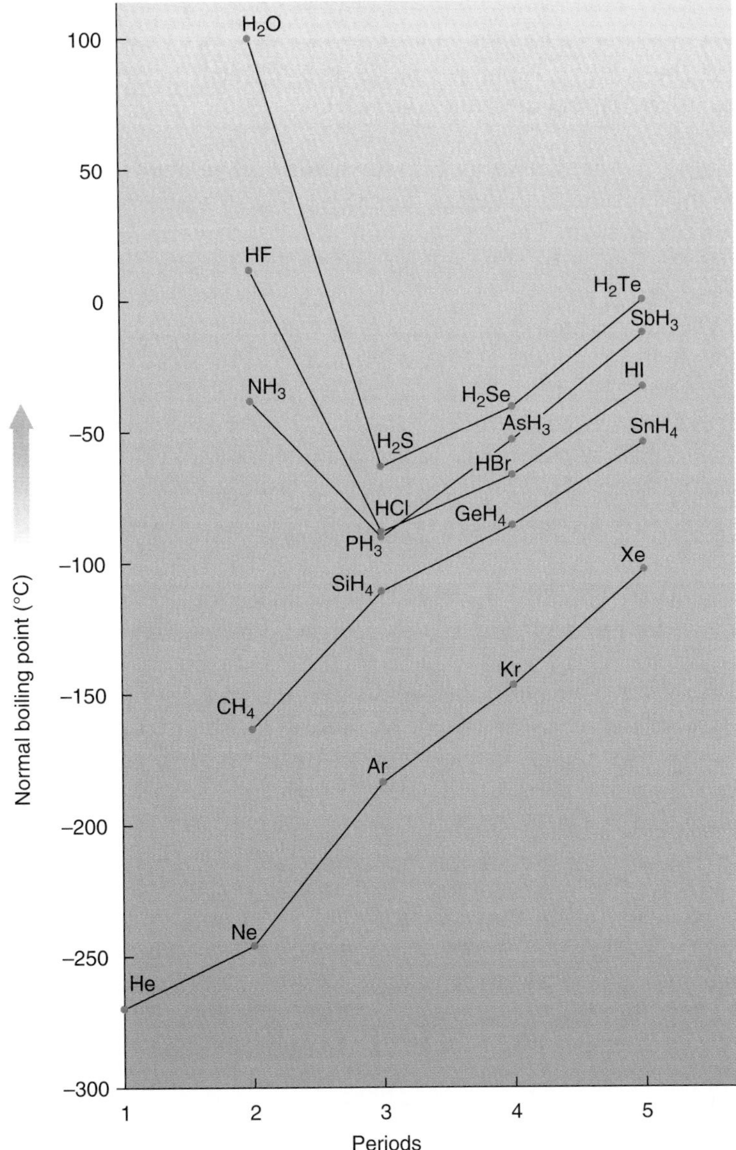

Figure 11.16

Effect of intermolecular forces on the normal boiling points of some substances.

Figure 11.16 also has some anomalous data. The boiling points of H_2O, HF, and NH_3 are abnormally high. These molecules all form intermolecular hydrogen bonds, which must be disrupted before the liquid can boil. We learned in Example 11.2 that a water molecule can form four hydrogen bonds and that these hydrogen bonds stabilize the liquid state. Hydrogen bonding in NH_3 is weaker than in H_2O or in HF because nitrogen is less electronegative than fluorine and oxygen. Thus, the boiling point of NH_3 is lower than the boiling point of H_2O or HF. Water is the only compound shown in Figure 11.16 whose boiling point is above room temperature.

The boiling points of H_2S, HCl, and PH_3, which do not interact by hydrogen bonding, are much lower than the boiling points of H_2O, HF, and NH_3, respectively. The slight increase in the boiling points in each of the series of H_2S to H_2Te, HCl to HI, and PH_3 to SbH_3 results from the increase in the sizes of the molecules and from the consequent increase in London forces.

Example 11.3

Predicting the Relative Vapor Pressures and Boiling Points of Liquids

Predict the relative vapor pressures at 25°C, and the normal boiling points, of water and methanol, CH_3OH.

SOLUTION: To predict vapor pressures and boiling points, we first examine the substances to determine the relative strengths of intermolecular forces in each. The substance in which the intermolecular forces are the weakest must have the highest vapor pressure and the lowest boiling point.

When we examine the formulas of water and methyl alcohol, we see that hydrogen bonding is possible in each. But since a water molecule has two O—H groups and a methyl alcohol molecule only one, water can form more hydrogen bonds and has the lower vapor pressure and higher boiling point. The boiling point of water is 100°C and that of methanol is only 58°C.

Practice Problem 11.1: Predict the relative vapor pressures at 25°C and the normal boiling points for ethane (CH_3CH_3), water, and ethanol, CH_3CH_2OH.

Practice Problem 11.2: Which of the following compounds has the highest vapor pressure, and which the lowest vapor pressure at the same temperature? Which compound has the highest normal boiling point, and which has the lowest normal boiling point? Methanol, CH_3OH; ethanol, CH_3CH_2OH; or fluoroethane, CH_3CH_2F.

The effects of polarity and molecular size on the normal boiling points of some compounds are illustrated in Table 11.5. Substances that have normal boiling points less than 0°C exist as gases above 0°C. In each row of Table 11.5, a polar and a nonpolar substance of approximately equal molar mass are chosen to illustrate the effect of molecular polarity on boiling points. The effect of the size of the molecules may be seen by considering the vertical trends in both polar and nonpolar groups of the substances listed in Table 11.5.

Table 11.5

Normal Boiling Points of Some Polar and Nonpolar Substances

Polar			Nonpolar		
Formula	Molar Mass	Normal b.p. (°C)	Formula	Molar Mass	Normal b.p. (°C)
CO	28	−192	N_2	28	−196
PH_3	34	−85	SiH_4	32	−112
AsH_3	78	−55	GeH_4	77	−90
SCl_2	103	59	BCl_3	117	18
ICl	162	97	Br_2	160	59

11.5 The Solid State

There are four major types of solids: **ionic solids**, **molecular solids**, **covalent or network solids**, and **metallic solids**. We discuss each of these types next.

Ionic Solids

An *ionic solid* consists of positively and negatively charged *ions* that are held together by *electrostatic attraction*. Ionic solids are generally hard and brittle, and they have *high melting points*. They are poor conductors of electricity because the ions are not free to move in the solid state. However, when melted, they become good conductors. Many ionic solids are soluble in water, but not in nonpolar solvents such as carbon tetrachloride.

An ionic solid has a structure in which the ions are closely packed with a minimum of empty space between the oppositely charged ions. Ionic solids have well-defined *crystal lattices*. The geometry of the crystal lattice depends on the relative sizes and numbers of the cations and anions in the compound.

The crystal structure of a solid can be described in terms of its *smallest characteristic structural unit,* called a **unit cell**. Replicating the unit cell in three dimensions generates the crystal lattice of the solid. Figure 11.17 illustrates three different unit cells that have a cubic structure.

Many ionic solids consist of cubic unit cells that have *an ion at each of the corners of the cube and the same type of ion in the center of each face of the cube*. This unit cell is called a **face-centered cubic cell** (Figure 11.17c). The unit cells of most of the alkali halides—except the cesium salts, CsCl, CsBr, and CsI—have a face-centered cubic structure, as shown in Figure 11.18 for sodium chloride.

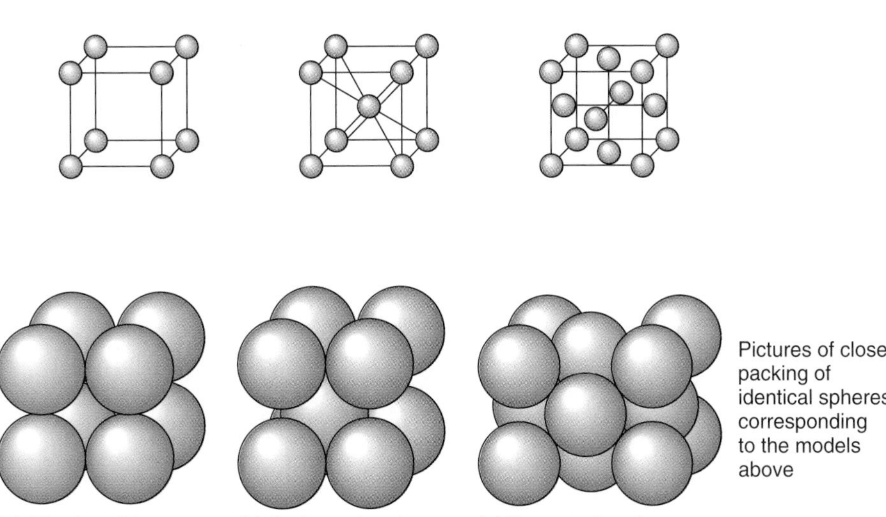

Figure 11.17

Cubic lattices. (a) Simple cubic, (b) body-centered cubic, and (c) face-centered cubic. A simple cubic cell has a lattice point at each corner of the cube. A body-centered cubic cell has a lattice point at the center of the cube and one at each corner of the cube. A face-centered cubic lattice has a lattice point at each corner and one in the center of each face of the cube.

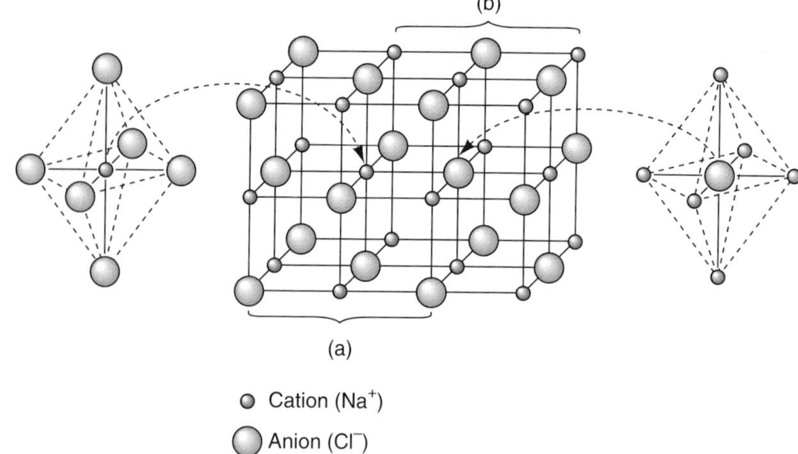

Figure 11.18

Face-centered cubic lattice of NaCl. Two interpenetrating face-centered unit cells are shown: (a) a face-centered cube of anions; (b) a face-centered cube of cations. The structure on the left shows that a cation is surrounded by six anions (is in an octahedral "hole" made by the six anions). The structure on the right shows that an anion is surrounded by six cations.

In a face-centered cubic geometry, each cation has six nearest neighbor anions, and each anion has six nearest cations. For an ion in an ionic solid, the *number of its nearest neighbor oppositely charged ions* is called the **coordination number** for that ion. Thus, in sodium chloride, each sodium and chloride ion has a coordination number of six. Each cation and anion is said to lie in an octahedral "hole" formed by the oppositely charged ions in the lattice. The hole is octahedral because the nearest neighbors of any given ion in such a hole occupy the positions that are equivalent to the corners of an octahedron.

In cesium chloride, simple cubic cells of cations interpenetrate simple cubic cells of anions (Figure 11.19) so that each cation is surrounded by eight anions and each anion is surrounded by eight cations. This type of a structure is called a cesium chloride structure; cesium bromide and cesium iodide also crystallize in this structure.

A cubic cell that has a *lattice point in each corner of the cube and one in the center of the cube* is called a **body-centered cubic cell** (Figure 11.17b). Many metals, such as sodium and iron, have body-centered cubic structures.

Many additional crystal structures exist for various ionic solids, metals, and other crystalline solids as well. The various known crystal lattice types are illustrated in Figure 11.20.

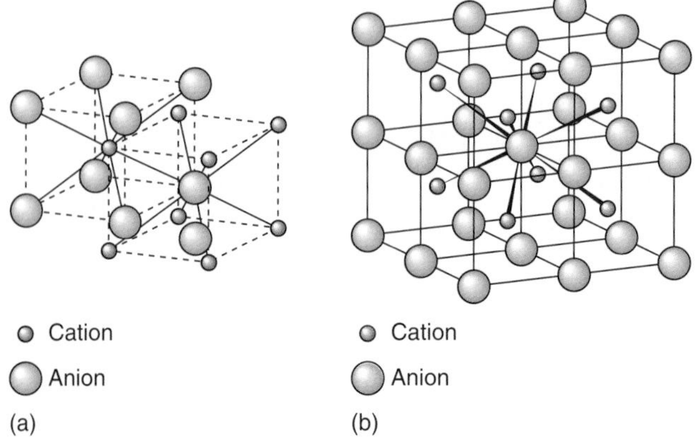

Figure 11.19

Simple cubic CsCl structure. In this structure, simple cubic cells of cations interpenetrate simple cubic cells of anions. (a) A cube of anions has penetrated a cube of cations so that a cation is surrounded by eight anions and an anion is surrounded by eight cations. (b) Eight unit cubes of anions are stacked beside and on top of each other. The eight cations in the centers of these cubes are shown to surround the central anion of the structure.

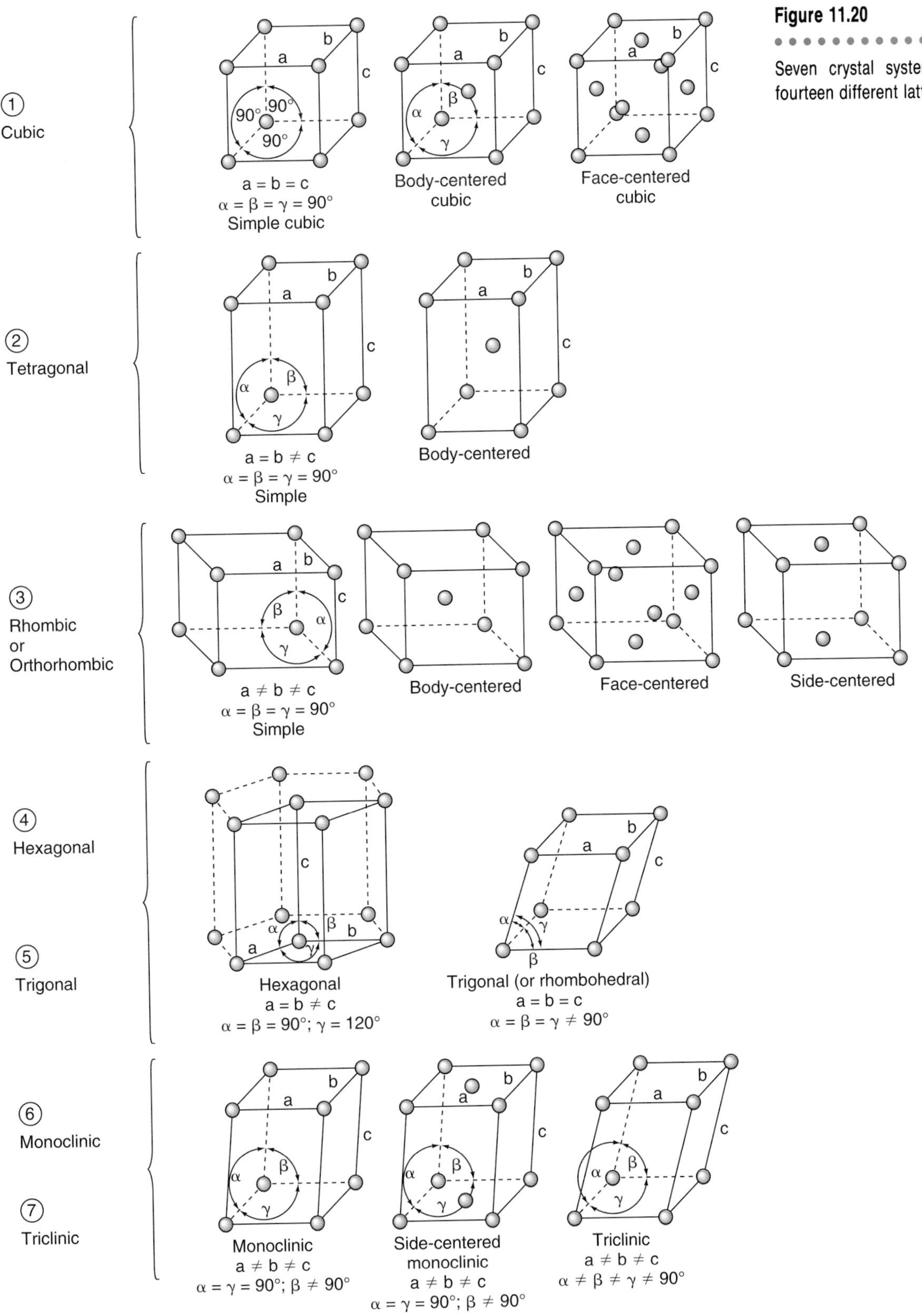

Figure 11.20

Seven crystal systems containing fourteen different lattice types.

Molecular Solids

Solids that consist of molecules are *molecular solids*. Nonpolar molecules are held together by London forces, polar molecules by dipole–dipole attractions and London forces. In special cases, hydrogen bonding is the strongest intermolecular attraction.

There are thousands of molecular solids. Some examples of molecular solids are sucrose, $C_{12}H_{22}O_{11}$; sulfur, S_8; white phosphorus, P_4; iodine, I_2; ice, and solid CO_2 (dry ice).

Molecular solids are generally quite soft, and they have low to moderately high melting points, usually below 300°C, because their intermolecular forces are relatively weak compared with ionic bonding. Molecular solids are poor conductors of electric current.

Covalent or Network Solids

The atoms in a covalent solid are covalently bonded to one another in a continuous network that extends throughout the entire solid. Thus, a covalent solid is also called a *network solid*. Diamond is a typical covalent solid. A diamond may be considered a giant molecule in which many carbon atoms are linked to one another in a continuous network structure.

Each carbon atom in a diamond is covalently bonded to four other carbon atoms in a tetrahedral structure (Figure 11.21). Such a large network of covalently bonded atoms is rigid, making diamond a very hard, though brittle, solid. The whole network of carbon-to-carbon bonds in a diamond resists applied pressure. A strong pressure or a blow on a diamond may break it into many irregularly shaped smaller pieces. However, because of the regular crystalline structure of diamond, a professional can cut a large stone along crystal planes to smaller, regularly shaped pieces.

Figure 11.21

Tetrahedral structure of diamond and hexagonal structure of graphite.

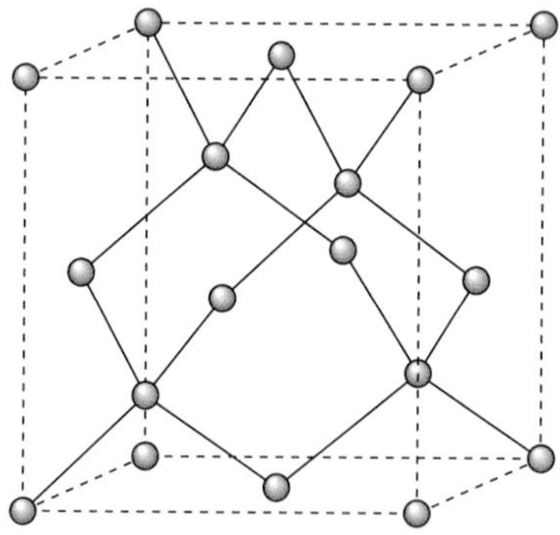

Diamond

Each carbon atom in diamond is connected to four other carbon atoms in a rigid, giant network of atoms. Distances between atoms are exaggerated for clarity.

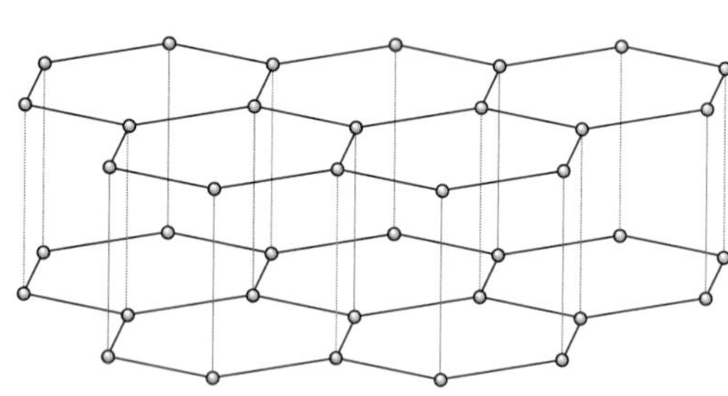

Graphite

Layers of carbon atoms can slide on top of one another, which makes graphite an excellent lubricant.

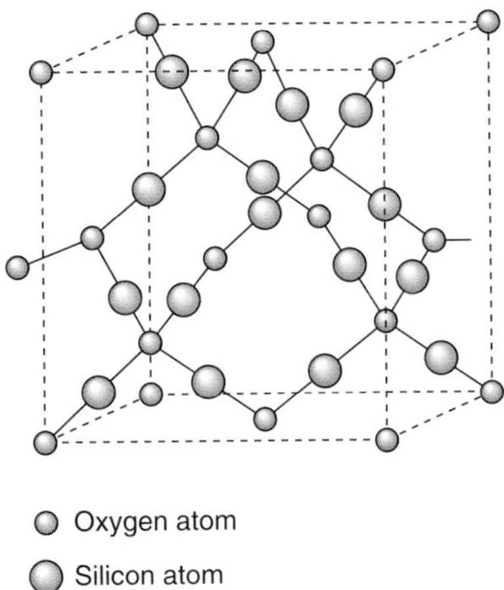

Figure 11.22

The network structure of SiO_2. The SiO_2 structure is similar to the structure of diamond, but it has alternating silicon-oxygen and oxygen-silicon bonds.

Another form of carbon is graphite, which consists of parallel sheets of carbon atoms. Each sheet is a hexagonal array of carbon atoms linked by covalent bonds. Adjacent sheets are held together by relatively weak London forces (Figure 11.21). Graphite is slippery because the adjacent sheets slide easily over each other. This property makes graphite an excellent lubricant.

Other examples of covalent solids are silicon dioxide or quartz, SiO_2, and silicon carbide, or carborundum, SiC. Silicon carbide is a very hard substance whose structure is like that of diamond. Carborundum is an abrasive used to sharpen knives and polish glass. Silicon dioxide is a continuous network of alternating silicon and oxygen atoms in which no individual SiO_2 units or molecules can be distinguished (Figure 11.22).

Covalent solids are very hard (graphite is an exception), they have *very high melting points*, and most are *poor electrical and thermal conductors*. Both the hardness and the high melting points of covalent solids reflect the collective strength of a network of covalent bonds that must be broken to free individual atoms from the crystal lattice of the solid. The melting points of some typical covalent solids are: diamond, >3550°C; silicon dioxide (quartz), 1610°C; and silicon carbide (SiC), 2700°C.

Covalent solids are insoluble in most common solvents because the forces among atoms in the network are much stronger than forces between atoms in the network and the solvent. Covalent solids are also chemically inert.

Metallic Solids

In a *metallic solid* the constituent particles are *metal atoms*. The melting points and hardness of metals are highly variable. Metals are *excellent electrical and thermal conductors*, and many are quite *malleable*, meaning that they can be hammered into sheets, and *ductile*, so that they can be drawn to a wire. Metals are ductile and malleable because the atoms in a metallic lattice can slide with respect to one another.

The bonding of atoms in a metal is called *metallic bonding*. According to a simple theory of metallic bonding, a metal consists of positively charged metal ions or *kernels* embedded in a "sea" of mobile valence electrons. Thus, metallic bonding is unlike either ionic or covalent bonding.

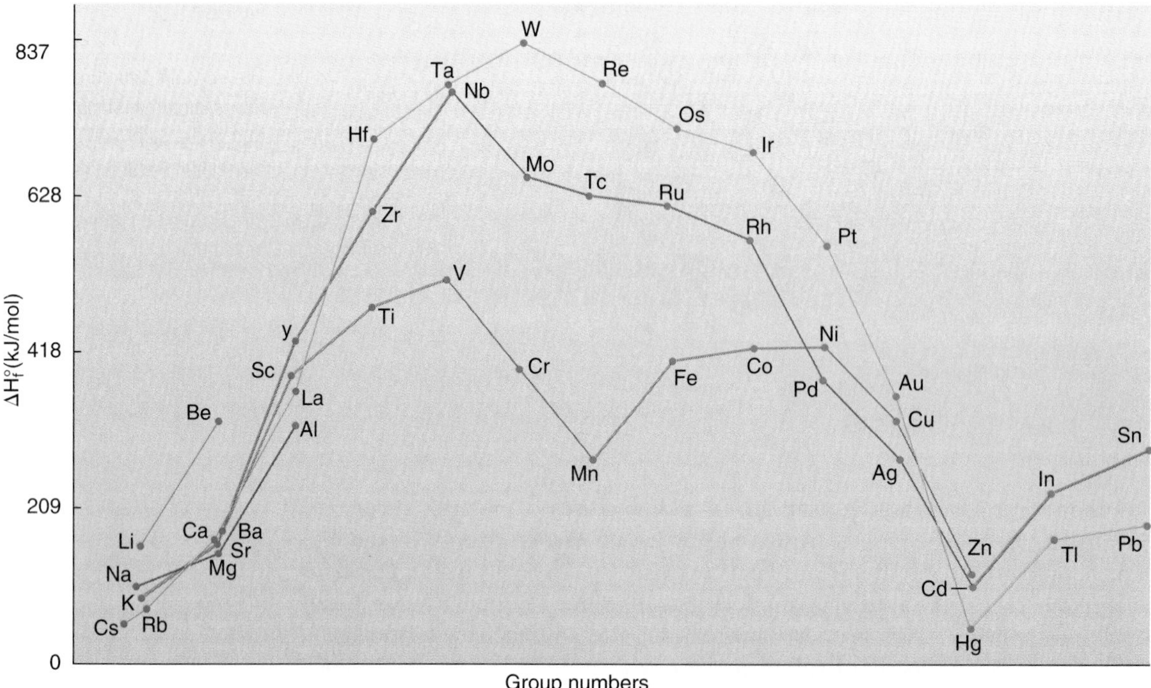

Figure 11.23

Standard enthalpies of formation of gaseous metal atoms from the solid state (enthalpies of atomization).

Some physical and chemical properties of a metal are related to the number of valence electrons of the metal. Figure 11.23 illustrates the periodic variation of the enthalpy change for converting a mole of a solid metal to gas. This property is called the **enthalpy of atomization**. The enthalpy of atomization is related to the strength of metallic bonding. We see that the enthalpy of atomization of metals is small at the beginning of a period, then generally increases to a maximum in the middle of a *d* series of transition metals, falls toward the end of a *d* series, and starts rising again in the *p* series. This trend corresponds to the number of unpaired valence electrons in metal atoms. Since the enthalpy of atomization is a measure of metallic bond strength, the number of unpaired valence electrons is also related to the bond strength. The melting points of metals show similar trends as shown by Figure 11.24.

We noted above that a metal can be regarded as a lattice of positively charged kernels in a "sea" of valence electrons. Metals conduct electricity because the electrons in the sea are mobile. A voltage applied to a metal causes the valence electrons to flow to complete the circuit.

The arrangement or "packing" of atoms in a metal lattice is like an orderly stack of billiard balls. As in the packing of ions in ionic solids, the atoms in many metals are packed to allow a minimum of empty space between them. Figure 11.25 illustrates how metal atoms or any equal-sized spheres can be packed in layers. The layer shown in Figure 11.25a has less empty space between spheres than in the layer shown in Figure 11.25b.

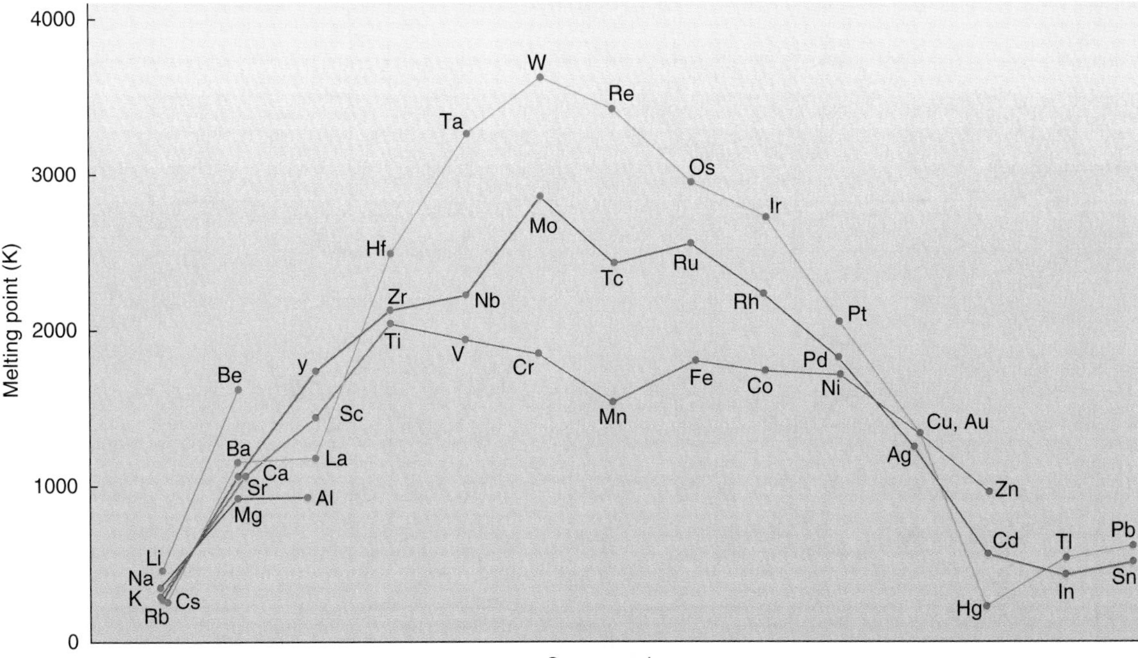

Figure 11.24

Variation of the melting points of metals with atomic number.

The layers of spheres shown in Figure 11.25a can be stacked up in two different ways. These are illustrated in Figure 11.26. In each of these structures, layer B is placed over layer A so that three spheres of layer B touch the same spheres of layer A. The third layer can be placed on top of B in one of two ways. In the first, each atom of the third layer lies directly above an atom in the first layer. This arrangement is repeated in the sequence ABABAB . . . and is called **hexagonal close packing** (Figure 11.26a).

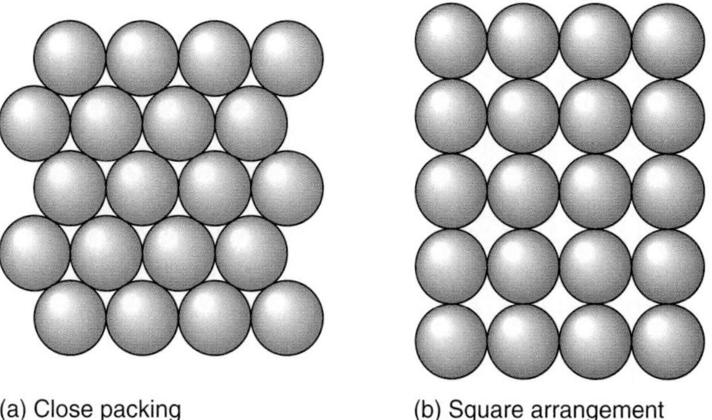

Figure 11.25

Two arrangements of equal-sized spheres. The empty spaces between the spheres are minimal in the close packing arrangement (a).

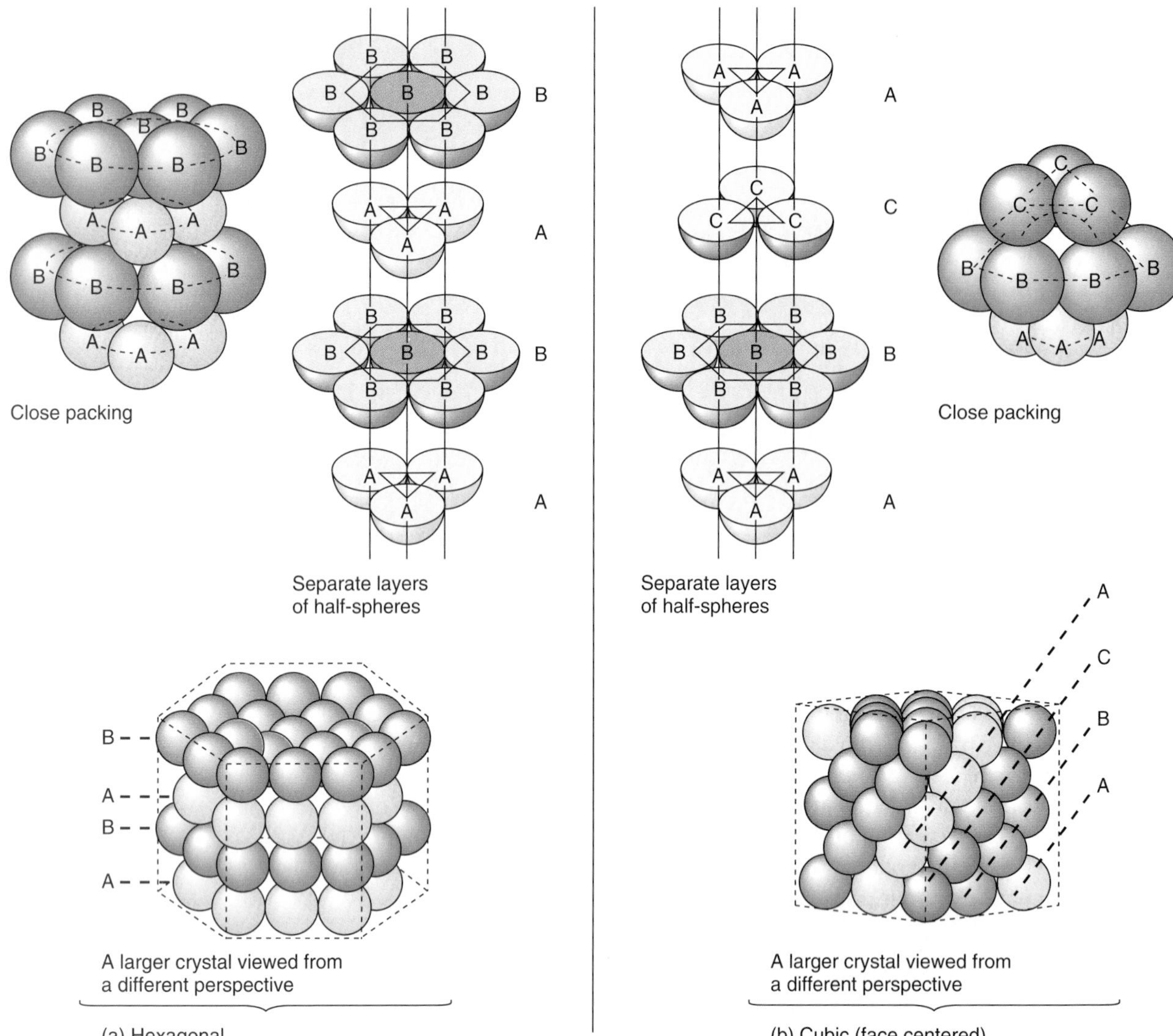

Figure 11.26

Hexagonal (a) and cubic (b) close packing. In each type, an atom has 12 nearest neighbors.

Another way of adding the third layer C on the second layer B is so that the atoms of layer C are not directly above those of either layer A or B. This arrangement is called **cubic close packing**. In this type of packing, the layers are arranged in an ABCABCABC . . . sequence (Figure 11.26b). Extensions of hexagonal and cubic close packings generate larger crystals whose shapes are illustrated in the lower part of Figure 11.26, where they are shown from a different perspective.

In both hexagonal and cubic close packing, each sphere touches six other spheres in a layer, plus three more spheres in the layer above and three more in the layer below. Thus, each sphere has 12 nearest neighbors. So a metal atom in either hexagonal or cubic close packing has a coordination number of twelve.

Most metals crystallize in hexagonal or cubic close packing structures. In these types of packing, approximately 74 percent of the available space is occupied by the atoms, and only 26 percent is empty space between the atoms.

Table 11.6

Types of Solids

	Ionic	Molecular	Covalent	Metallic
Units at lattice sites	Ions	Molecules	Atoms	Atoms
Type of bonding	Electrostatic attraction	London forces, dipole–dipole, hydrogen bonds	Covalent bonds	Metallic bonding
Some properties	High melting, hard and brittle	Low melting, soft	High melting, very hard	Low to high melting, soft to hard, good conductors of electricity
Examples	NaCl, $CaSO_4$, $Mg(NO_3)_2$,	Ice (H_2O), dry ice (CO_2), I_2, P_4, S_8	Diamond (C), silicon dioxide (SiO_2)	Fe, Cu, Na, Ag, Au

About one-third of the metals crystallize in a body-centered cubic geometry in which each atom has a coordination number of eight. In this type of packing, about 68 percent of the space is occupied by the atoms, and 32 percent is empty space. The simple cubic structure (Figure 11.17a) is unstable for uncharged atoms because it contains a relatively large amount of empty space. In this structure, only about 52 percent of the space is occupied by the atoms, and 48 percent is empty space between the atoms. The radioactive metal polonium is known to have the simple cubic geometry.

There is a fifth type of solid, which, however, is rare. At very low temperature and high pressure, a noble gas can be converted to a solid. This type of a solid can be called an **atomic solid** because it consists solely of atoms held together by London forces. As an example, helium exists as a solid at or below −272.2°C and at or above 26 atm! Xenon atoms are larger than helium atoms, so the London forces in xenon are stronger than in helium. Therefore, xenon can exist in a solid state at higher temperature than helium. The freezing point of xenon is −111.9°C. Solid "noble gases" are very soft, and they are poor thermal and electrical conductors. Table 11.6 summarizes the types of solids discussed in this section.

11.6 Phase Changes

Heating and Cooling Curves

The solid, liquid, and gaseous states of matter can be interchanged under different conditions of temperature and pressure. The processes through which one state is changed to another are shown in Figure 11.27.

Let us consider what happens to a solid when it is heated at constant rate and constant 1 atm pressure. As an example, we will start with 1 g of ice at −20°C (Figure 11.28).

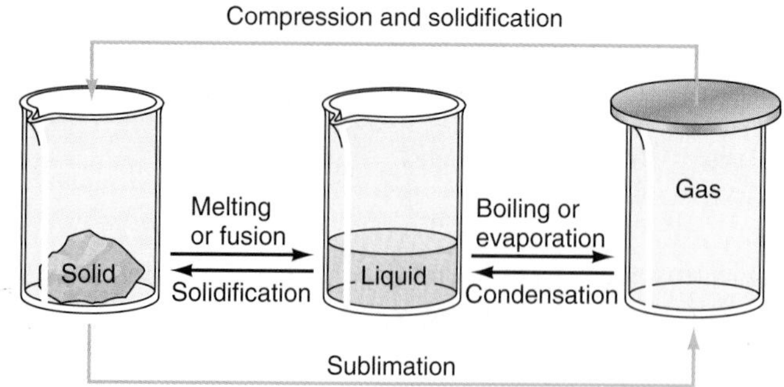

Figure 11.27

Processes by which states of matter can be interconverted.

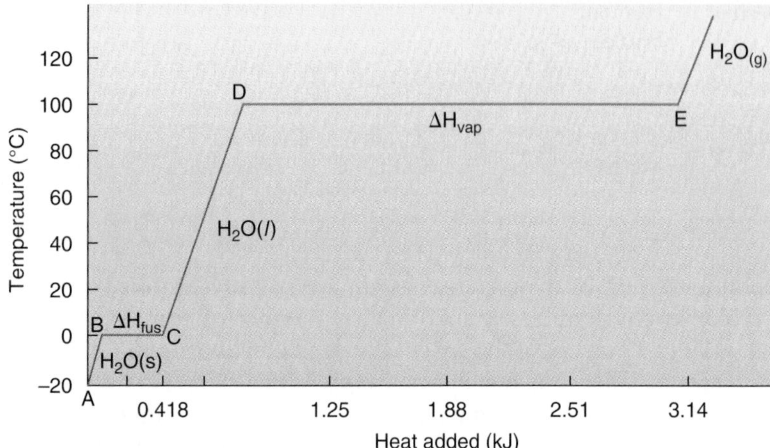

Figure 11.28

Heating curve for 1 gram of water.

The thermal energy supplied to a solid from point A to point B (Figure 11.28) is absorbed by the solid without changing its state. As the temperature of the solid increases, the rate of vibration of molecules increases as the molecules absorb thermal energy.

When the vibrational energy is greater than the forces holding the lattice together, the solid starts to melt. This occurs at point B (Figure 11.28). The temperature does not rise during the melting process from point B to point C because the average kinetic energy of the molecules does not change until all of the solid is changed to a liquid.

After all the solid has changed to a liquid, continued heating increases the kinetic energy of the molecules in the liquid, and the temperature of the liquid increases. As a result of the increasing molecular motion in the liquid state, more molecules escape from the liquid to the surrounding gaseous atmosphere in any given time. Thus, the vapor pressure of the liquid increases with increasing temperature.

When the vapor pressure of the liquid equals the atmospheric pressure, at point D, the boiling point of the liquid is reached. While the liquid is boiling, its temperature does not rise. The added heat is used to overcome the intermolecular forces that maintain the molecules in the liquid state. While the liquid is boiling (point D to point E), the kinetic energy of the molecules does not increase because temperature is proportional to kinetic energy. After all of the liquid has changed to a gas, the temperature of the gas increases again as the kinetic energy of the gas molecules increases.

A cooling curve similar to that shown in Figure 11.28 can be obtained the same way by cooling 1 g of water vapor so that it condenses to a liquid and eventually solidifies.

Example 11.4

Calculating the Amount of Heat Needed to Convert Ice to Water Vapor

How much heat is required to change 10.0 kg of ice at 0.0°C to water vapor at 100°C and 1 atm? The heat of fusion of ice at 0.0°C is 333.5 J g^{-1}, and the heat of vaporization of water is 2257.5 J g^{-1}.

SOLUTION: When we convert a solid at its melting point to a gas at the boiling point of its liquid, we must consider three stages of heating. During the first stage the solid melts. Second, the temperature of the liquid increases from its freezing point to its boiling point. Third, heat must be supplied to convert the liquid at its boiling point to vapor.

The heat needed to melt 1 g of ice at 0.0°C, the heat of fusion, is 333.5 J g^{-1} or 333.5 kJ kg^{-1}. The heat needed to melt 10.0 kg of ice is

$$(10.0 \text{ kg})(333.5 \text{ kJ kg}^{-1}) = 3.34 \times 10^3 \text{ kJ}$$

To increase the temperature of 1 g of water by 1°C, heat equal to the specific heat of water, 4.184 J g^{-1} $°C^{-1}$, is needed. This is equivalent to 4.184 kJ kg^{-1} $°C^{-1}$. The amount of heat needed to increase 10.0 kg of water from 0.0°C to 100°C is

$$(10.0 \text{ kg})(4.184 \text{ kJ kg}^{-1}\ °\text{C}^{-1})(100°\text{C}) = 4.18 \times 10^3 \text{ kJ}$$

The amount of heat required to change 1 g of liquid water to vapor at the boiling point of water, a quantity called the specific heat of vaporization of water, is 2257.5 J g^{-1}. This is equivalent to 2257.5 kJ kg^{-1}. The amount of heat required to convert 10.0 kg of liquid water to vapor at 100°C is

$$(10.0 \text{ kg})(2257.5 \text{ kJ kg}^{-1}) = 2.26 \times 10^4 \text{ kJ}$$

The total heat required for the entire process is

$$3.34 \times 10^3 \text{ kJ} + 4.18 \times 10^3 \text{ kJ} + 2.26 \times 10^4 \text{ kJ} = 3.01 \times 10^4 \text{ kJ}$$

Practice Problem 11.3: Calculate the amount of heat required to convert 1 mol of solid methanol (methyl alcohol) at its melting point (−97.3°C) to vapor at its boiling point (64.96°C). The heat of fusion of solid methanol at its melting point is 3.17 kJ mol^{-1}, the molar heat capacity of methanol is 81.6 J mol^{-1} $°C^{-1}$, and the molar heat of vaporization of methanol at its boiling point is 39.2 kJ mol^{-1}.

Example 11.5

Calculating the Amount of Water That Can Be Cooled a Given Number of Degrees by Melting a Given Quantity of Ice

A 50.0-g ice cube at 0.00°C is placed in water originally at 30.0°C. How many grams of water can be cooled to 0.00°C by the melting of the ice cube? The heat of fusion of ice is 333.5 J g^{-1}.

SOLUTION: This problem deals with a situation in which heat is gained by one substance and lost by another. The heat gained by the ice equals the heat lost by the water, assuming that no heat is transferred elsewhere in the process. We can calculate the heat gained by ice in the melting process since we are given the mass of ice and its heat of fusion. This quantity of heat is also the heat lost by water, which equals the mass of water times its specific heat times the temperature change. From this relationship we can calculate the mass of water that is cooled. Heat gained by the ice cube in melting is

$$(50.0 \text{ g})(333.5 \text{ J g}^{-1}) = 1.67 \times 10^4 \text{ J}$$

Heat lost by water, q_{H_2O} (-1.67×10^4 J), equals the mass of water, m_{H_2O}, times the specific heat of water, C_{H_2O}, times the temperature change of water, Δt_{H_2O}:

$$q_{H_2O} = (m_{H_2O})(C_{H_2O})(\Delta t_{H_2O}) = -1.67 \times 10^4 \text{ J}$$

The temperature change for water is

$$t_{\text{final}} - t_{\text{initial}} = 0.00°\text{C} - 30.0°\text{C} = -30.0°\text{C}$$

Solving this equation for the mass of water gives

$$m_{H_2O} = \frac{q_{H_2O}}{(C_{H_2O})(\Delta t_{H_2O})} = \frac{-1.67 \times 10^4 \text{ J}}{(4.18 \text{ J g}^{-1}\,°\text{C}^{-1})(-30.0°\text{C})} = 133 \text{ g}$$

Practice Problem 11.4: An ice cube at 0.00°C was placed into 70.2 g of water at 24.0°C. The melting of the ice lowered the temperature of water to 0.00°C. How many grams of ice melted?

Phase Diagrams

A diagram of pressure versus temperature that shows the conditions at which a substance exists as a solid, liquid, or gas is called a **phase diagram** (Figure 11.29). A phase diagram typically has a region in which the substance is a gas, one in which it is a liquid, and another in which it is a solid. The line that divides two states of a substance *at equilibrium* is called a *phase boundary*. Thus, there is a liquid–solid phase boundary, a liquid–gas phase boundary, and a solid–gas phase boundary. The last of these may seem surprising, but most solids, like liquids, have detectable vapor pressures. For example, we can smell the vapors emanating from moth balls, and ice can enter the gas phase directly without melting. The process in which a solid converts directly to its vapor is called *sublimation*.

The vapor pressure of a liquid or a solid increases with increasing temperature. The lower portion of the curve in Figure 11.29 represents a plot of the vapor pressure of ice versus temperature. When the temperature reaches 0.01°C, the vapor pressure of ice is 4.58 torr. At this point, all three phases of

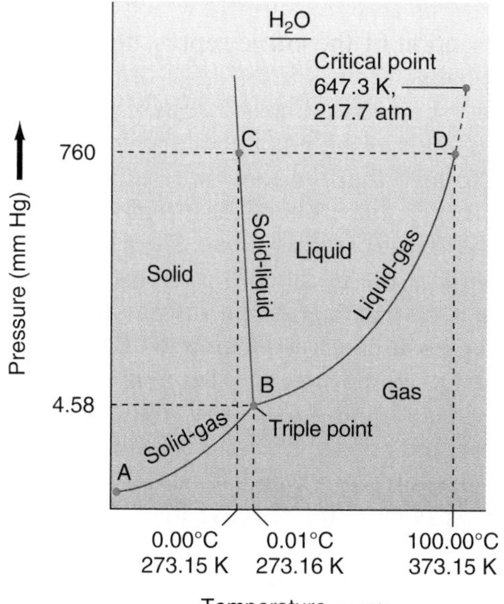

Figure 11.29

Phase diagram for water. The diagram is not drawn to scale. Along any boundary, the two phases indicated exist in equilibrium with each other. At the triple point, three phases may be at equilibrium with one another.

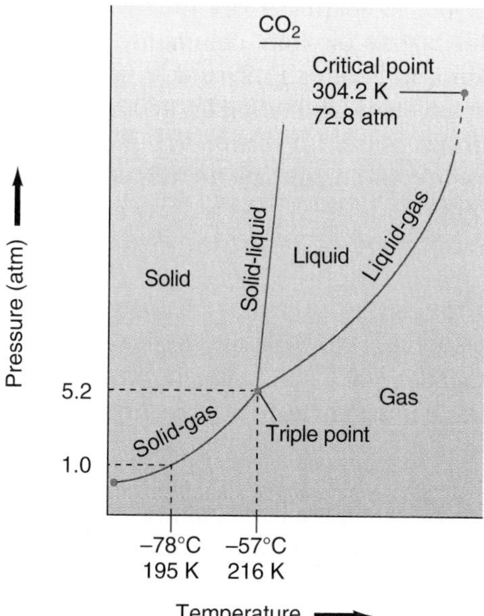

Figure 11.30

Phase diagram for CO_2.

water—ice, liquid water, and water vapor—can exist simultaneously at equilibrium with one another. The temperature and pressure at which all three phases of a substance are at equilibrium is called the **triple point**. The triple point for carbon dioxide is at −57°C and 5.2 atm (Figure 11.30).

When the vapor pressure of liquid water is plotted against temperature, the curve in the upper right of Figure 11.29 is generated. Similarly, plotting the points at which solid ice is at equilibrium with liquid water generates the upper left line of Figure 11.29.

Curve A–B in Figure 11.29 represents the temperatures and pressures at which ice is at equilibrium with water vapor. Curve B–D represents the variation of vapor pressure of liquid water at various temperatures. Curve B–C outlines the temperatures and pressures at which ice exists at equilibrium with liquid water. Each section of the curve represents a phase boundary between the corresponding phases.

The liquid–vapor boundary line in a phase diagram (Figure 11.29) continues upward until it ends at the point above which the average kinetic energy of the molecules is so high that *the substance exists only as a gas that cannot be liquefied under any pressure*. The highest temperature at which a gas can be liquefied is called the **critical temperature**. The pressure required to liquefy a gas at its critical temperature is called the **critical pressure**.

A phase diagram reveals significant information about a substance. For example, below the pressure corresponding to its triple point, a substance can exist only in the solid or gaseous state. The temperature corresponding to the point where the horizontal line drawn from 760 torr pressure crosses the line B–C in Figure 11.29 is the normal freezing point of water. When the horizontal line from 760 torr pressure is extended to line B–D, the temperature that corresponds to the crossing point of these lines is the normal boiling point of water.

Figure 11.29 shows that the solid–liquid boundary line tilts to the left, or has a negative slope. This means that with increasing pressure, the melting point of ice decreases. In carbon dioxide (Figure 11.30) the solid–liquid boundary has a positive slope, indicating that an increase in pressure increases the melting point of solid carbon dioxide, commonly called dry ice.

The phase diagrams in Figures 11.29 and 11.30 also show that ice cannot be changed to liquid water below 4.58 mmHg, and dry ice cannot be changed to liquid below 5.2 atm pressure. At 1 atm, ice melts at 0°C, whereas dry ice sublimes at that pressure and maintains the temperature of −78°C until all the solid is changed to gas.

Example 11.6

Using the Phase Diagram to Predict the Physical State of Carbon Dioxide at a Given Temperature and Pressure

Using Figure 11.30, determine the physical state of a sample of carbon dioxide at 6 atm and −100°C, and explain what happens to the sample when its temperature is increased at constant pressure.

SOLUTION: Figure 11.30 reveals that at 6 atm and −100°C, carbon dioxide is a solid. As we move horizontally to the right from that point parallel to the temperature axis, the sample turns to a liquid as we reach the solid–liquid boundary line at about −57°C. As we continue to the right along that path, we reach the liquid–gas boundary line. At the temperature corresponding to this point, the sample turns to a gas. Any further increase in temperature increases the kinetic energy of the gas molecules. Thus, at 6 atm and 100°C, carbon dioxide is a gas.

Practice Problem 11.5: What is the physical state of water under each of the following conditions? **(a)** −5°C and 0.5 atm; **(b)** 5°C and 4 torr; **(c)** 15°C and 4 torr; **(d)** 15°C and 0.5 atm; **(e)** 100°C and 1.3 atm.

The phase diagram for water shows that the triple point of water is 0.01 degree higher than the normal freezing point of water. The normal freezing point is the temperature corresponding to 1 atm total pressure. However, at the triple point, the substance exists under its own vapor pressure only.

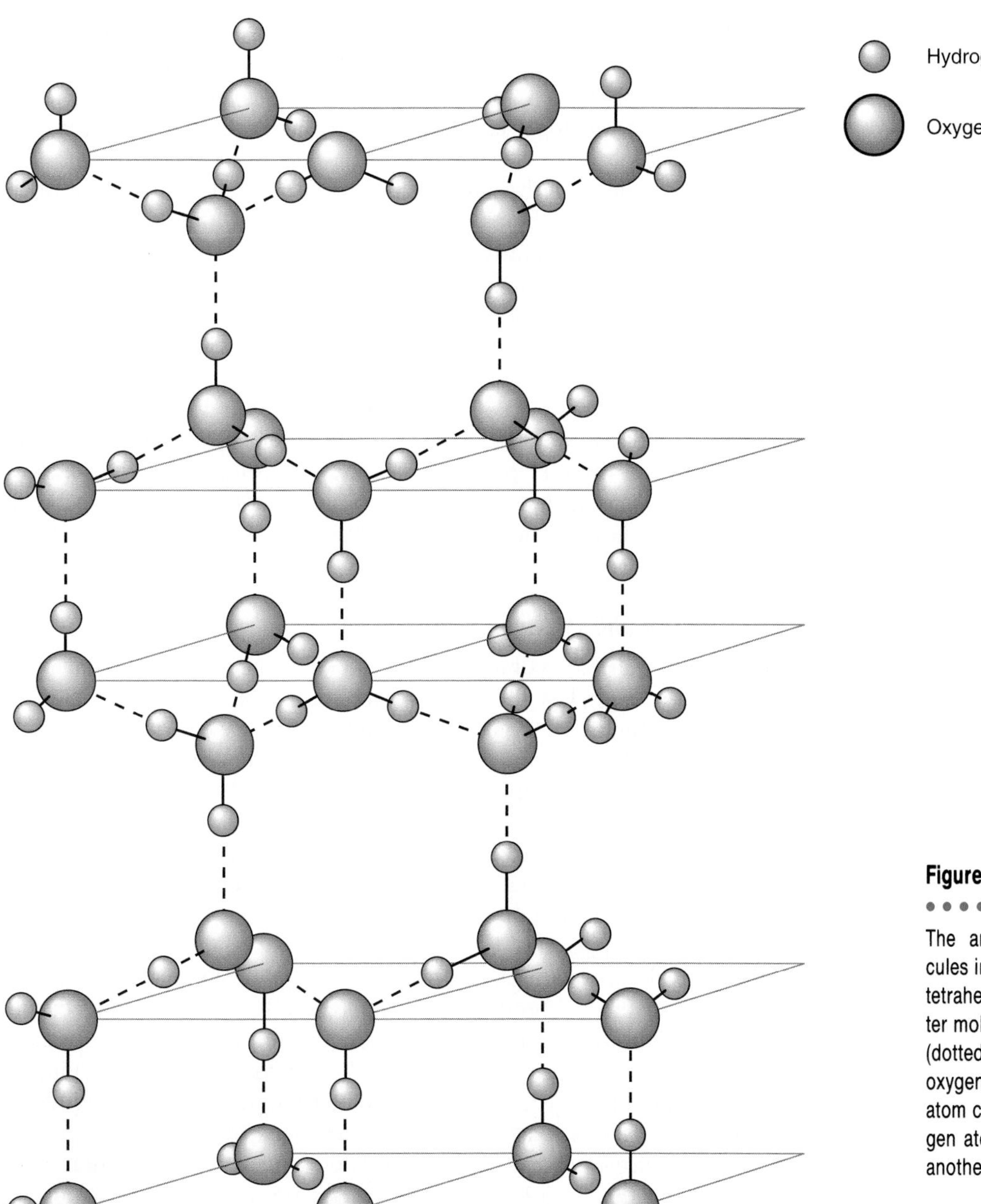

Figure 11.31

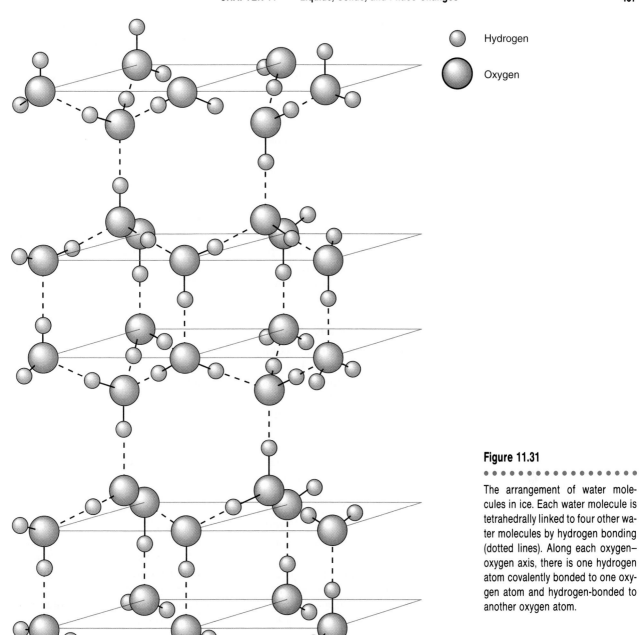

The arrangement of water molecules in ice. Each water molecule is tetrahedrally linked to four other water molecules by hydrogen bonding (dotted lines). Along each oxygen–oxygen axis, there is one hydrogen atom covalently bonded to one oxygen atom and hydrogen-bonded to another oxygen atom.

Phase Diagrams, Structure, and Properties

The phase diagrams for water and carbon dioxide show that increasing the pressure raises the freezing point of carbon dioxide, but lowers it for water (Figures 11.29 and 11.30). This difference can be explained by considering the structures of ice and dry ice. Both are molecular solids. In ice, the principal intermolecular forces are relatively strong hydrogen bonds, whereas in carbon dioxide these forces are relatively weak London forces. The hydrogen bonds in ice hold the molecules in an open crystal lattice (Figure 11.31). In fact, water molecules are farther apart in ice than in liquid water. Thus, ice has a lower density than liquid water. The crystalline ice structure is broken when pressure is exerted on it; as a result, the ice melts.

The molecules of dry ice are nonpolar and are packed with less empty space than in ice. Thus in dry ice, the molecules are closer together than the molecules in liquid carbon dioxide. An increase in pressure on dry ice forces the molecules even closer together and therefore increases the temperature at which the solid melts.

▢ Skating on ice is possible because of the open crystalline structure of the solid. The pressure exerted by the skate blades causes the ice structure to collapse, and a thin film of liquid water forms under the blades, permitting the skates to slide easily over the surface. Skating would be much more difficult on the surface of dry ice, or on another substance displaying a solid–liquid boundary line with a positive slope in the phase diagram. For these substances, an increase in pressure increases the melting point.

We recall that the expansion of the volume of water upon freezing is due to the strength of hydrogen bonding which binds the molecules in fixed positions with cavities between them (Figure 11.31). The tendency of water to expand when it freezes is so great that metal automobile radiators and engine blocks containing water without antifreeze often crack in winter when the water freezes.

Critical Temperature and Intermolecular Forces

We recall that above the critical temperature, a substance remains a gas at any applied pressure. At or below the critical temperature of a gas, intermolecular forces are strong enough so that an increase in the pressure can cause the gas to liquefy. Below the critical temperature, the pressure required to liquefy the gas decreases with decreasing temperature, until the normal boiling point for the substance is reached at 1 atm.

Substances that have relatively strong intermolecular forces, such as water, sulfur dioxide, and ammonia, have high critical temperatures. They tend to exist as liquids at rather high temperatures. Substances with weak intermolecular forces, such as hydrogen and helium, have low critical temperatures. Table 11.7 lists the critical temperatures and pressures for some common substances.

Gases that have critical temperatures above 298 K (25°C), such as carbon dioxide, propane, ammonia, and sulfur dioxide (Table 11.7), are commercially available as liquids under high pressure in metal cylinders. In such a cylinder, the liquid is in equilibrium with its vapor. When the valve of the cylinder is opened, some gas escapes, and more liquid vaporizes to replace the gas that escaped. The pressure of the gas in the cylinder returns to its original value, since the vapor pressure of a liquid depends only on temperature, not on the

Table 11.7

Critical Temperatures and Pressures for Some Common Substances

Substance	Critical Temperature (K)	Critical Pressure (atm)
Hydrogen	33.2	12.8
Helium	5.2	2.3
Nitrogen	126.0	33.5
Oxygen	154.3	49.7
Carbon dioxide	304.2	72.8
Propane	370	41.8
Ammonia	405.5	111.5
Sulfur dioxide	430.3	77.7
Water	647.1	217.7

APPLICATIONS OF CHEMISTRY 11.1
The Dangers of Eating Snow for Emergency Water

If you ever find yourself stranded in a snowstorm, you may be tempted to eat snow as an emergency water source (see photo on right). Although it is reasonable to assume that snow would be an adequate substitute for water, eating snow could burn many valuable calories that you need to keep warm and to survive.

One calorie (or 4.184 J) will raise the temperature of 1 gram of water by 1 degree Celsius. However, 1 cal will not raise the temperature of 1 g of snow (or ice) at all. If you ate 1 g of snow, your body would have to burn 80 cal (335 J) just to melt the snow, assuming it is at 0°C. While your body supplies the 80 cal to melt the snow, the temperature of the melted snow remains unchanged at 0°C. Your body must still provide additional heat to warm the melted snow from 0°C to normal body temperature, which is 37°C.

Let's see how many calories your body would have to burn if you consumed 1000 g of snow or ice (about 1 quart). First, let's calculate how many calories you would need just to melt the snow:

$$(1 \times 10^3 \text{ g snow})(80 \text{ cal absorbed/g snow}) = 8 \times 10^4 \text{ cal}$$

or

$$(1 \times 10^3 \text{ g snow})(335 \text{ J/g snow}) = 3 \times 10^5 \text{ J}$$

In addition to the energy you need to supply to melt the snow, you also need energy to warm the cold water to your body temperature. It takes 3.7×10^4 cal to raise the temperature of 1 L of 0°C water to 37°C. Therefore, your body would have to burn a total of 1.2×10^5 cal to both melt and warm the 1000 g of snow.

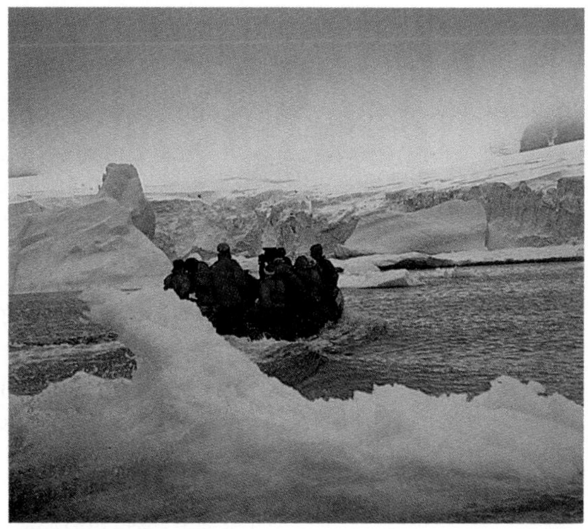

Fortunately, there are several simple ways to get your water from snow and conserve valuable calories so that you do not freeze to death. As part of their car winter emergency kit, some people carry a candle and a metal container such as an empty coffee can in which they can melt and warm the snow. Another approach is to place snow in a container on a dark background and place this in the sun. The heat radiated from the dark background would melt the snow and warm the resulting water.

Source: Ronald DeLorenzo, *Problem Solving in General Chemistry*, 2nd ed., Wm. C. Brown, Publishers, Dubuque, Iowa, 1993.

amount of liquid. Thus, the pressure gauge of the cylinder indicates the same value at a given temperature as long as any liquid is present. After all the liquid has vaporized, the pressure drops, and the gauge on the tank indicates the pressure of the gas that remains in the tank.

Aerosol cans contain a liquid "propellant" such as propane. The propellant has a critical temperature above 298 K. The propellant in the aerosol can is a liquid in equilibrium with its vapor. The can also contains the ingredient to be dispensed, such as paint, hair conditioner, or mosquito repellant. When the valve of the can is pressed open, the propellant gas escapes from the can and delivers a desired quantity of the active ingredient. Some of the residual liquid propellant then vaporizes to replace the gas that escaped, and the pressure in the can returns to its original value. As the liquid phase of the propellant is used up, the pressure of the propellant decreases noticeably, and the can is "empty."

11.7 Physical States of Elements

In this section we give a short survey of the physical states of the elements at 25°C and 1 atm.

General Trends

Most metals have high boiling and melting points. Bonding is strong in metals, and all metals, except mercury, are solids at 25°C and 1 atm. The sublimation energies and melting points of metals generally increase with an increasing number of unpaired valence electrons, as shown in Figures 11.24 and 11.25.

Nonmetals have highly variable melting and boiling points. Many nonmetals consist of small atoms or molecules. They are gases at room temperature and pressure, because the intermolecular forces among small molecules are relatively weak. The nonmetals that exist as gases under normal conditions are located toward the upper right of the periodic table, except hydrogen, which is at the upper left. The gaseous nonmetals at 25°C include hydrogen, nitrogen, oxygen, fluorine, chlorine, and all the group VIII elements (the noble gases).

Larger atoms of nonmetals tend to form polyatomic molecules, such as I_2, P_4, As_4, S_8, and Se_8. These elements exist as solids at 25°C and 1 atm. Table 11.8 lists the standard physical states of the nonmetals. We can see from Table 11.8 that only one of the nonmetals, bromine, is a liquid. The others are either solids or gases.

The elemental halogens exist only in the form of diatomic molecules. The physical states of the halogens therefore depend on London forces, which increase down the group. Thus, the physical states of the halogens range from gaseous fluorine and chlorine, to liquid bromine, to solid iodine.

Table 11.8

Physical States of the Nonmetals at 25°C and 1 atm

Solid Liquid Gas

III	IV	V	VI	VII	VIII
				H_2	He
B	C	N_2	O_2	F_2	Ne
	Si	P_4	S_8	Cl_2	Ar
		As_4	Se_8	Br_2	Kr
			Te	I_2	Xe
				At_2	Rn

Carbon and Nitrogen

Carbon commonly exists as a solid—diamond, graphite, or amorphous soot. But nitrogen, next to carbon in the periodic table, is a gas. This may seem puzzling. Why should such a great difference exist in the physical states of these two elements whose atomic radii are relatively close to each other? A carbon atom has four valence electrons and can link to four other carbon atoms in a continuous tetrahedral network as in diamond.

The solid state and the hardness of diamond are due to the large network of C—C bonds acting more or less as an entire unit against any external stress. Even a single C—C bond is relatively strong (about 346 kJ mol^{-1}). The C—C internuclear distance in diamond is 0.154 nm.

In graphite (see Figure 11.21), the C—C bond length in the planar network of carbon rings is 0.142 nm, but the distance between the layers is 0.335 nm. The macromolecular layers are held together by London forces, which are weak compared with C—C covalent bonding. Hence, graphite is soft, has low density, and is "slippery."

Nitrogen consists of small, triple-bonded N_2 molecules. Because these molecules are small and nonpolar, the London forces between them are weak, and nitrogen is a gas under normal conditions.

Oxygen and Sulfur

Like N_2 molecules, oxygen molecules, O_2, are small and nonpolar. Therefore, oxygen is also a gas at room temperature and pressure.

Sulfur consists of relatively large S_8 molecules. The London forces between such large molecules are sufficiently strong so that sulfur is a solid under normal conditions.

11.8 Physical States of Compounds

We recall that most oxides and halides of metallic elements are ionic solids at room temperature. The oxides and halides of nonmetallic elements range from gases to solids (Tables 11.9 and 11.10).

Oxides of the Boron and Carbon Families

The major oxide of boron, boric oxide, B_2O_3, is a network solid. It contains a continuous single-bonded three-dimensional network of boron–oxygen atoms rather than individual O=B—O—B=O molecules.

Carbon monoxide, CO, and carbon dioxide, CO_2, are colorless and odorless gases. As we discussed in Chapter 9, carbon monoxide consists of triple-bonded C≡O molecules, and carbon dioxide consists of double-bonded O=C=O molecules. These substances are gases at room temperature because the atoms are multiple bonded in small molecules. Attraction between small molecules is weak, unless they can form intermolecular hydrogen bonds.

Although silicon is directly below carbon in the periodic table, silicon dioxide, of which silica and quartz are made, is an inert network solid that melts at about 1700°C. This stability can be attributed to the single-bonded three-dimensional network of silicon–oxygen atoms (see Figure 11.22).

Table 11.9

Physical States of Some Major Oxides of Nonmetals at 25°C and 1 atm

Solid | Liquid | Gas

III	IV	V	VI	VII	VIII
				H_2O	
B_2O_3	CO CO_2	N_2O NO NO_2 N_2O_5		OF_2 O_2F_2	
	SiO_2	P_4O_6 P_4O_{10}	SO_2 SO_3[a]	Cl_2O ClO_2 Cl_2O_7	
		As_4O_6 As_4O_{10}	SeO_2 SeO_3	Br_2O BrO_2	
			TeO_2 TeO_3	IO_2 I_2O_5	XeO_3 XeO_4

[a] The liquid form of SO_3 consists of cyclic trimers (Figure 11.33); the solid form consists of a polymeric network of SO_4 tetrahedra.

Oxides of the Nitrogen and Sulfur Families

Of the common oxides of nitrogen, three are gases at room temperature. They include dinitrogen monoxide, N_2O (called "laughing gas"), nitrogen monoxide, NO, and nitrogen dioxide, NO_2. A gaseous oxide molecule contains multiple bonds, as discussed in Chapter 9. Multiple bonds, as we have seen, favor the formation of small molecules, and hence the gaseous state.

Dinitrogen pentoxide N_2O_5 (generally referred to simply as nitrogen pentoxide) is a colorless ionic solid at 25°C and 1 atm. It is composed of NO_2^+ and NO_3^- ions. The liquid state of N_2O_5 (above 30°C) and its gaseous state both consist of molecules.

The oxides of the heavier nonmetals of the nitrogen family, P_4O_6, P_4O_{10}, As_4O_6, and As_2O_5, are solids. This is expected since these oxides consist of large molecules with relatively strong intermolecular attraction. The oxides of antimony, Sb_4O_6 and Sb_2O_5, are also solids. The only oxide of bismuth is Bi_2O_3. The latter is also a solid, as are all the oxides of metals, since metal–oxygen bonds have considerable ionic character.

Table 11.10

Physical States of Some Major Binary Halogen Compounds of Nonmetals at 25°C and 1 atm

Solid | Liquid | Gas

III	IV	V	VI	VII	VIII
				HF HCl HBr HI	
BF_3 BCl_3 BBr_3 BI_3	CF_4 CCl_4 CBr_4 CI_4	NF_3 NCl_3 NBr_3 NI_3	See group VII of Table 11.9		
	SiF_4 $SiCl_4$ $SiBr_4$ SiI_4	PF_3 PCl_3 PBr_3 PI_3	S_2F_2 SCl_2 S_2Cl_2 S_2Br_2 SF_6	ClF ClF_3	
		AsF_3 $AsCl_3$ $AsBr_3$ AsI_3	Se_2Cl_2 SeF_4 $SeCl_4$ $SeBr_4$ SeF_6	BrF BrF_3 BrF_5	KrF_2
			$TeCl_2$ TeF_4 $TeCl_4$ $TeBr_4$ TeI_4 TeF_6	ICl ICl_3 IF_5 IF_7	XeF_2 XeF_4 XeF_6

Only one of the major oxides of the sulfur group, sulfur dioxide, SO_2, is a gas at room temperature and pressure. Although SO_2 molecules are polar, they are relatively small, so the intermolecular attraction is not strong enough for SO_2 to be a liquid.

The other major oxide of sulfur, sulfur trioxide, SO_3, exists in different forms: one is a white solid that melts at 62.4°C, another is a solid below about 17°C, a liquid between 17 and 45°C, and a gas above 45°C. The solid form of SO_3

Figure 11.32

Solid and liquid sulfur trioxide, SO_3.

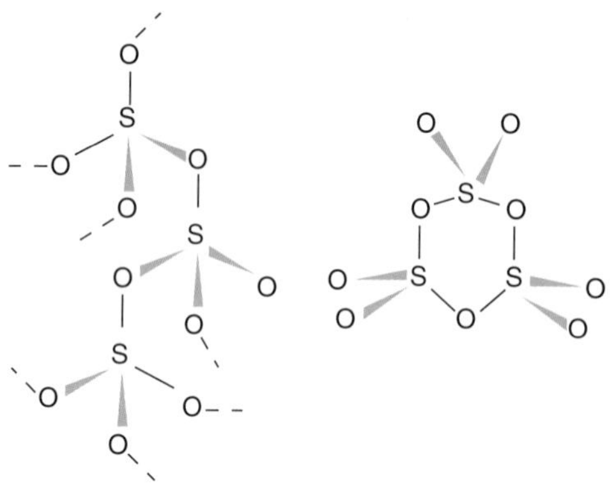

that melts at 62.4°C is a polymeric network of single-bonded SO_4 tetrahedra (Figure 11.32). The solid that melts at 17°C consists of cyclic trimers, as shown in Figure 11.32. The liquid form of SO_3 consists of a mixture of SO_3 molecules and cyclic trimers.

These structures once again illustrate that larger atoms, such as silicon and sulfur, tend to form large molecules or polymeric networks of atoms which exist in solid state at room temperature and pressure. The oxides of the heavier elements of the sulfur group are solids at room temperature and pressure.

General Trends

The physical states of the oxides and halides of the nonmetallic elements at room temperature and pressure (Tables 11.9 and 11.10) reveal some general trends. Substances that consist of small molecules are gases. Compounds of intermediate size molecules are liquids, and those composed of larger molecules, or of polymeric networks of atoms, are solids. There are exceptions, however, depending on the geometry or polarity of the molecules, and other factors determining the strength of intermolecular and intramolecular bonding.

Summary

Liquids

Various properties of liquids such as **viscosity**, **surface tension**, **rate of evaporation**, **vapor pressure**, **boiling point**, and **freezing point** depend on the strengths of **intermolecular forces**. Intermolecular forces depend on the kinds of atoms in a molecule and on its geometry. The strongest intermolecular forces are **hydrogen bonds**, a special type of **dipole–dipole interaction**. The weakest intermolecular forces are **London forces**.

The boiling point of a liquid in an open vessel depends on the nature of the liquid and on the atmospheric pressure. *At the boiling temperature, the vapor pressure of the liquid equals the atmospheric pressure*. The vapor pressure of a liquid depends only on temperature, not on the amount of the liquid. The *freezing point of a liquid* (or the *melting point of the solid*) depends only slightly on pressure. It may increase or decrease with pressure, depending on the nature of the substance.

Solids

Solid substances may be divided into four types in terms of the nature of the particles occupying lattice points and the forces holding the particles together. These four types of solids are **ionic solids**, **molecular solids**, **covalent solids**, and **metallic solids**. The particles of crystalline solids are arranged in definite geometrical patterns, or lattices. Some common lattices are **face-centered cubic** and **body-centered cubic** structures.

Phase Changes

The solid, liquid, and gaseous states of matter can be interconverted by changing the temperature and pressure. A **phase diagram** is a plot of pressure versus temperature that shows the solid–vapor, solid–liquid, and liquid–vapor boundary lines for a substance. Information about the conditions of temperature and pressure under which a substance exists as a gas, liquid, or solid, may be gained from a phase diagram.

The **critical temperature** of a substance is the temperature above which it can exist only as a gas. The liquid–vapor boundary line in a phase diagram ends at the point corresponding to the critical temperature. The pressure needed to liquefy the gas at its critical temperature is the **critical pressure**. In phase changes, the enthalpies of vaporization, fusion, and sublimation are measures of the relative strengths of the intermolecular forces in the substances involved.

Physical States of Substances

Physical states of elements and compounds at 25°C and 1 atm depend on intramolecular bonding and intermolecular forces. Small nonmetallic atoms tend to form diatomic molecules, many of which contain multiple bonds. Elements and compounds that consist of small molecules are usually gases at room temperature and pressure. This is particularly true when the molecules are nonpolar and do not form hydrogen bonds. Substances of intermediate-sized molecules, or polar molecules, tend to exist as liquids. Substances of very large molecules, or substances in which the atoms link together to form ''giant molecules'' such as diamond and silicon dioxide, are solids.

New Terms

Atomic solid (11.5)
Body-centered cubic cell (11.5)
Boiling point (11.3)
Coordination number (11.5)
Covalent solid (11.5)
Critical pressure (11.6)
Critical temperature (11.6)
Cubic close packing (11.5)
Cubic structure (11.5)
Dipole–dipole interaction (11.4)
Dynamic equilibrium (11.2)
Enthalpy of atomization (11.5)
Enthalpy of vaporization (11.2)
Equilibrium (11.2)
Equilibrium vapor pressure (11.2)
Face-centered cubic cell (11.5)
Freezing point (11.3)
Heat of fusion (11.3)
Heat of vaporization (11.2)
Hexagonal close packing (11.5)
Hydrogen (H-) bond (11.4)
Hydrogen bonding (11.4)
Intermolecular forces (11.4)
Ionic solid (11.5)
London dispersion forces (11.4)
London forces (11.4)
Metallic bonding (11.5)
Metallic solid (11.5)
Molecular solid (11.5)
Network solid (11.5)
Normal boiling point (11.3)
Normal freezing point (11.3)
Normal melting point (11.3)
Phase diagram (11.6)
Surface tension (11.1)
Triple point (11.6)
Unit cell (11.5)
Van der Waals forces (11.4)
Viscosity (11.1)

Exercises

General Review

1. Define each of the following terms: viscosity; surface tension; equilibrium vapor pressure; boiling point; freezing point; hydrogen bonding; dipole–dipole interaction; London forces; ionic solid; molecular solid; covalent solid; unit cell; phase diagram; critical temperature; triple point.
2. Discuss the nature of liquids and solids in terms of the relative mobility of particles.
3. Discuss the viscosity and surface tension of a liquid in terms of its intermolecular attractions.
4. How may the viscosity of a liquid be measured?
5. Explain how the vapor pressure of a liquid depends on the intermolecular attraction and temperature.
6. Differentiate between intermolecular and intramolecular forces.
7. Describe each of the following intermolecular forces: dipole–dipole interaction, hydrogen bonding, London forces. Name some liquids in which each of these forces is the strongest intermolecular attraction.
8. In terms of the molecular motion, explain and illustrate the meaning of "equilibrium vapor pressure" of a liquid.
9. What is the relationship between the heat of vaporization and the strength of intermolecular forces of a liquid?
10. Define the boiling point of a liquid in terms of the vapor pressure of the liquid. Define the term "normal boiling point."
11. Discuss the dependence of the boiling point of a liquid on the intermolecular forces of the liquid.
12. Show how the various types of intermolecular forces of a liquid are related to the structure of the molecules of the liquid.
13. Explain how the boiling point of a liquid depends on the atmospheric pressure.
14. Explain how the freezing point of a liquid depends on the external pressure.
15. List the four types of solids discussed in this chapter. Name the characteristic particles of each solid, and the forces holding the particles together. Give some examples of each type of solid.
16. What are some typical physical properties of each type of solids discussed in this chapter?
17. Name some common crystal structures of solids, and describe the unit cells in each.
18. What factors determine the crystal structure of an ionic solid?
19. Describe the difference between simple cubic cells, body-centered cubic cells, and face-centered cubic cells.
20. Give an example of a lattice structure in metallic solids.
21. Explain what happens to the temperature and the kinetic energy of the particles of a solid molecular substance that is heated: **(a)** up to its melting point; **(b)** at its melting point; **(c)** from its melting to its boiling point; **(d)** at its boiling point; **(e)** beyond its boiling point.
22. Draw the phase diagram for water, label the axes and the regions in which water is solid, liquid, and gas. Label the triple point, normal boiling point, and critical point.
23. Explain how the critical temperature of a substance is related to the intermolecular forces of the substance.
24. Which metal is a liquid at 25°C and 1 atm? Which nonmetal is a liquid, and which nonmetals are gases at that temperature and pressure?
25. Make a list of the major (most common) oxides of nonmetals that are gases and those that are liquids at 25°C and 1 atm.
26. Explain and give examples of the possible relationships between the sizes of molecules and the physical states of elements and compounds at 25°C and 1 atm.
27. Explain and give examples of the possible relations between the atomic size and the physical state of an element at 25°C and 1 atm.

Liquids and Intermolecular Forces

28. What forces are responsible for the nearly spherical shape of a drop of water?
29. What are the forces that keep carbon tetrachloride molecules together in a drop of carbon tetrachloride?
30. Which has a higher rate of evaporation at a given temperature, water or carbon tetrachloride? Explain.
31. Which has the higher boiling point at 1 atm pressure, water or carbon tetrachloride? Explain.
32. Why do eggs cook faster below sea level than above sea level?

33. Why do vegetables cook faster in a pressure cooker than in an open pot?

34. Explain why liquid water cannot be heated above 100°C at 1 atm pressure.

35. A carbon disulfide, CS_2, molecule is linear like that of carbon dioxide. At room temperature and 1 atm pressure, carbon disulfide is a liquid, as is water. On the basis of this information, answer the following questions. **(a)** Are carbon disulfide molecules polar or nonpolar? **(b)** What forces keep the molecules of carbon disulfide together in a drop of carbon disulfide? **(c)** Which liquid is likely to have stronger intermolecular forces, water or carbon disulfide? Explain. **(d)** Which substance has stronger intermolecular forces, carbon disulfide or carbon dioxide? Explain. **(e)** Explain which of the two liquids, water or carbon disulfide, is more likely to have the higher vapor pressure at a given temperature. **(f)** Explain which of the two liquids in part **(e)** is more likely to have the higher normal boiling point. **(g)** Do you expect sodium chloride to dissolve in carbon disulfide? Explain. **(h)** Which of the two liquids, water or carbon disulfide, do you expect to have a larger molar heat of vaporization? Explain.

36. Explain how a small needle whose density is higher than that of water can "float" on the surface of water.

37. Do the following properties of liquids increase in magnitude, decrease in magnitude, or remain unaffected with an increase in the strength of intermolecular forces? **(a)** Viscosity; **(b)** surface tension; **(c)** vapor pressure; **(d)** normal boiling point; **(e)** normal melting point; **(f)** heat of vaporization.

38. What is the principal type of intermolecular forces in each of the following substances in their liquid states? **(a)** NH_3; **(b)** PH_3; **(c)** $HC_2H_3O_2$; **(d)** CF_4; **(e)** SO_2; **(f)** CO_2; **(g)** CO; **(h)** H_2.

39. Arrange the following substances in the order of decreasing normal boiling point: **(a)** $HC_2H_3O_2$; **(b)** CO_2; **(c)** SO_2; **(d)** H_2; **(e)** PH_3

Solids

40. Complete the following table by filling in all the blank boxes.

Type of Solid	Example	Units That Occupy Lattice Points	Binding Forces Between Unit Particles	Properties
Metallic				
	Dry ice, sugar, sulfur			
		Atoms		
			Electrostatic attraction	

41. Which of the following solids would you expect to be soluble in water, and which would you expect to be soluble in carbon tetrachloride? **(a)** Sodium sulfate; **(b)** iodine; **(c)** phosphorus; **(d)** potassium nitrate.

42. Does any solid exert a vapor pressure?

Structures of Solids

43. Examine Figures 11.17b and 11.19. If the body-centered cubic cell were reproduced in all directions to build a larger lattice, each corner atom would be shared between eight identical cubes, and the particle in the center of the cube would belong entirely to that cube. How many particles belong to each unit cube in such a lattice?

44. As in question 43, examine Figures 11.17c and 11.18, and determine the number of particles that belong to a face-centered cubic unit cell.

45.** The atoms in potassium metal are packed in a body-centered cubic lattice, and the radius of a potassium atom is 0.231 nm. **(a)** Determine the number of atoms in a unit cell of potassium. **(b)** Calculate the length of an edge of the unit cell, assuming that all atoms are perfect spheres and adjacent atoms are in contact with one another. [*Hint for part* ***(b): Determine the number of atoms, including any fractions, from one corner of a unit cell diagonally to the opposite corner, through the center of the cube.] **(c)** Calculate the volume of the unit cell. **(d)** Calculate the volume of 1 mol (molar volume) of potassium.

***46.** Potassium bromide crystallizes in a face-centered cubic geometry. The edge of a unit cell (Figures 11.17c and 11.18) is 0.658 nm. The density of KBr is 2.75 g cm^{-3}. On the basis of the data above, calculate Avogadro's number.

Phase Changes

47. Draw a phase diagram for water or carbon dioxide, label the axes and all the boundaries, and explain the information that can be obtained from this diagram.

48. Explain whether you expect carbon dioxide, ammonia, or helium to have the highest critical temperature.

49. Explain why ice is less dense than liquid water. Also explain why we can skate on ice but not on glass.

50. Explain in terms of the structure of ice why the solid–liquid boundary line in the phase diagram for water shifts toward lower temperatures at higher pressures. Explain the opposite trend of the solid–liquid boundary line in the phase diagram for carbon dioxide.

51. The normal boiling point of benzene is 353 K, its triple point is 278 K at 21 torr, and its critical point is 562 K at 48 atm. The density of liquid benzene is 0.894 g cm^{-3}, and the density of solid benzene is 1.005 g cm^{-3}. Sketch a phase diagram for benzene in the region of 273 to 573 K.

52. Using Figure 11.30, describe the phase changes that occur, and give the approximate pressures at which these changes occur, when the pressure on carbon dioxide is gradually increased at a constant temperature of: **(a)** −58°C; **(b)** −56°C.

53. **(a)** How much heat is required to melt 1.00 mol of ice at 0°C? **(b)** How much heat is required to change 1 mol of water at 100°C to steam at 100°C? Look up the necessary data.

54. The heat of vaporization of liquid ammonia at its normal boiling point is 23.3 kJ mol^{-1}. **(a)** How much heat would be removed from the surroundings by the vaporization of 100 g of liquid ammonia? **(b)** How much water at 25.0°C would be frozen at 0°C by the cooling action of the liquid ammonia in part **(a)**?

55. What mass of steam at 100°C should be condensed to liquid water at 100°C to supply enough heat to raise the temperature of 500 g of copper from 25.0°C to 30.0°C? The molar heat capacity of copper is 24.7 J mol^{-1} K^{-1}.

Physical States of Elements and Compounds

56. Explain why fluorine and chlorine are gases at 25°C and 1 atm, bromine is a liquid and iodine is a solid.

57. Explain why oxygen is a gas under ordinary conditions, but sulfur is a solid.

58. Explain why nitrogen forms diatomic molecules but phosphorus does not.

59. Explain why OF_2, Cl_2O, and ClO_2 are gases, Br_2 and BrO_2 are liquids, but IO_2 is a solid at room temperature.

60. Explain why water is a liquid at room temperature although its molecules are smaller than the gaseous molecules listed in question 59.

* Denotes problems of greater than average difficulty.

CHAPTER 12

The world's oceans are complex solutions whose most abundant dissolved salt is sodium chloride. About 27 g of NaCl are dissolved in a kilogram of sea water.

Properties of Solutions

We have discussed some general characteristics of solutions in previous chapters. In Chapter 1 we described several types of solutions, and in Chapter 4 we learned about concentrations of solutions. Major types of reactions in solution were also discussed in Chapter 4. In this chapter we consider general dissolution processes and various properties of solutions in more detail.

12.1 Dissolution Processes

Dissolving Solids in Liquids

We recall that a solution is a homogeneous mixture of two or more substances. We begin by considering solutions in which ionic solids are dissolved in water. When an ionic solid dissolves in water, polar water molecules interact with ions on the surface of the solid and pull them out of the crystal lattice. The detached ions then *become surrounded by water molecules*—a process called **hydration**. (See Chapter 3 for a discussion of the hydration of undissolved salts.) Water molecules bound to an ion are not mobile and not part of the bulk solvent. They are often called the *hydration sphere* of the ion.

A water molecule has an electronegative oxygen atom that is attracted to a cation by electrostatic forces. A water molecule is attracted to an anion through its electropositive hydrogen atoms (Figure 12.1). The bonding that keeps an ion and its surrounding water molecules together as a *hydrated ion* is called an *ion–dipole interaction*. The hydrated ions and water molecules form a solution.

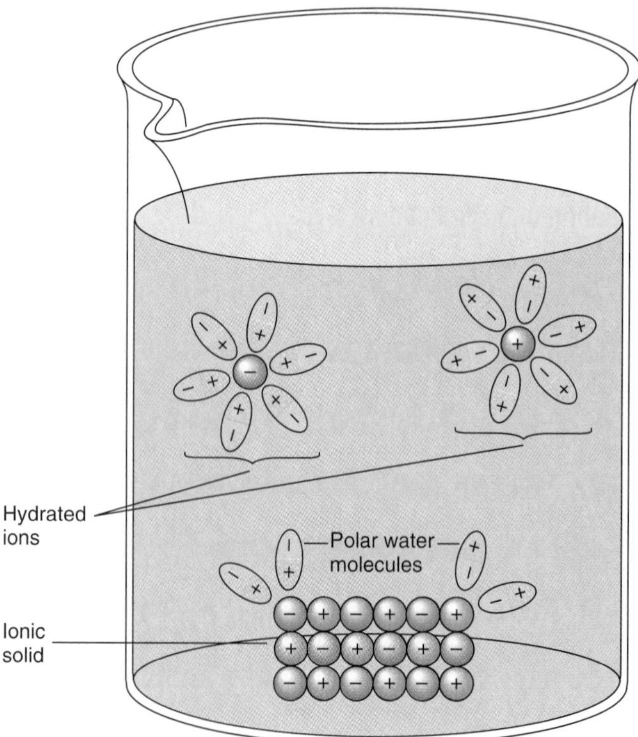

Figure 12.1

Schematic representation of dissolution of an ionic solid in water and hydration of the ions.

When an ionic solid dissolves, the crystal lattice is disrupted. This process requires energy and is therefore *endothermic* (ΔH is positive). On the other hand, energy is released by the hydration of the ions. Therefore, hydration is an *exothermic* process. (This energy, called the *hydration energy*, $\Delta H_{\text{hydration}}$, is negative.) If the absolute value of the hydration energy is greater than that of the lattice energy, heat is released by the system. On the other hand, if the solid dissolves and if the absolute value of the hydration energy is less than that of the lattice energy, heat is absorbed by the system.

Solids consisting of nonpolar molecules are usually not water soluble, particularly when the molecules are large. Naphthalene (mothballs), $C_{10}H_8$, is an example of such a solid. Nonpolar naphthalene molecules are large and therefore easily polarizable. They are attracted to one another by London forces. Water molecules are attracted to one another by hydrogen bonds. There is little solute–solvent interaction, and consequently, little mixing of the solute and the solvent. On the other hand, nonpolar naphthalene dissolves readily in a nonpolar solvent such as the hydrocarbon pentane ($CH_3CH_2CH_2CH_2CH_3$). In this case the attractions between solute (naphthalene) and solvent (pentane) are about the same as the solute–solute and solvent–solvent attractions. A rather reliable rule of thumb is "like dissolves like." That is, polar and ionic substances tend to dissolve in polar solvents, and nonpolar substances tend to dissolve in nonpolar solvents.

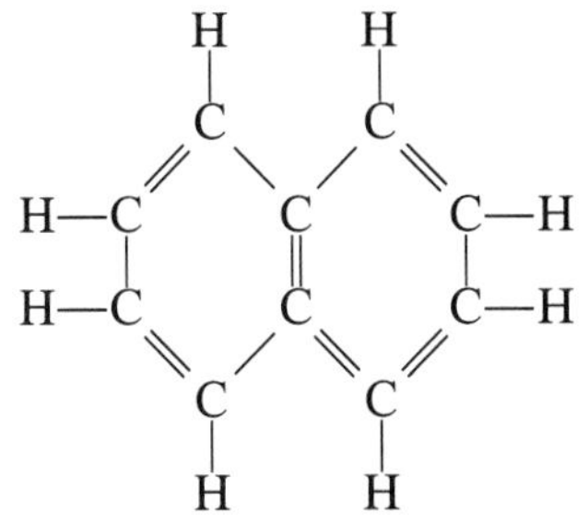

Naphthalene

Covalent solids such as diamond and silicon dioxide do not dissolve in water or in other common solvents. We recall that covalent solids are giant networks of atoms linked by covalent bonds. The interaction between a solvent and a covalent solid is far too weak to disrupt the network of covalent bonds.

Example 12.1

Predicting the Water Solubility of Solids

Predict whether S_8 or $SrCl_2$ will be soluble in water.

SOLUTION: To predict solubility, we consider the nature of the solute and the solvent. If the solid solute is ionic, its solubility in water can be predicted according to the solubility rules discussed in Chapter 4. Solids that consist of large nonpolar molecules are usually insoluble in water.

S_8 molecules are relatively large and nonpolar. Therefore, sulfur does not dissolve in water. However, one of the crystalline forms of sulfur has been found to be soluble in carbon disulfide, CS_2, a nonpolar solvent. (CS_2 is a foul-smelling liquid with a molecular structure similar to that of CO_2.) Strontium chloride is an ionic solid that is soluble in water according to the solubility rules given in Chapter 4.

Practice Problem 12.1: Do you expect barium nitrate and methyl alcohol, CH_3OH, to be water soluble?

Dissolving Liquids in Liquids

The principles that govern the solubility of solids in liquids also apply to the solubility of liquids in other liquids. Water is a good solvent for liquids consisting of small polar molecules that can form hydrogen bonds with water molecules. Some water-soluble liquids include low-molecular-weight alcohols and organic acids. Alcohol molecules contain the polar —O—H group of atoms, and organic acids contain the $-\overset{\overset{O}{\|}}{C}-O-H$ group, which is also polar. For example, methanol, CH_3—**O—H**, and ethanol, CH_3CH_2—**O—H**, are completely miscible with water, as are organic acids such as formic acid, $H-\overset{\overset{\mathbf{O}}{\|}}{\mathbf{C}}-\mathbf{O}-\mathbf{H}$, and acetic acid, $CH_3-\overset{\overset{\mathbf{O}}{\|}}{\mathbf{C}}-\mathbf{O}-\mathbf{H}$. In each of these molecules the polar sites (*boldfaced*) include a hydrogen atom linked to an oxygen atom. Thus, the —O—H groups of methanol, ethanol, formic acid, and acetic acid can form hydrogen bonds to water. Formic acid and acetic acid also contain an oxygen linked to carbon by a double bond. This oxygen has two lone electron pairs that can also form hydrogen bonds with water (Figure 12.2).

Water-soluble alcohols also include propanol, $CH_3CH_2CH_2OH$, and butanol, $CH_3CH_2CH_2CH_2OH$. However, alcohols containing five or more carbon atoms per molecule are insoluble, or only slightly soluble, in water.

Organic acids containing five or fewer carbon atoms per molecule are soluble in water. These include (besides formic and acetic acid) propanoic acid, butanoic acid, and pentanoic acid. Their structures are shown below:

$$CH_3CH_2-\overset{\overset{O}{\|}}{C}-O-H$$

Propanoic acid

$$CH_3CH_2CH_2-\overset{\overset{O}{\|}}{C}-O-H$$

Butanoic acid

$$CH_3CH_2CH_2CH_2-\overset{\overset{O}{\|}}{C}-O-H$$

Pentanoic acid

Organic acids containing six or more carbon atoms per molecule are only slightly soluble in water.

A high-molecular-weight organic acid or alcohol consists of a long nonpolar hydrocarbon "tail" and a small polar "head." The large nonpolar tails of such molecules attract one another by London forces, and the polar heads form

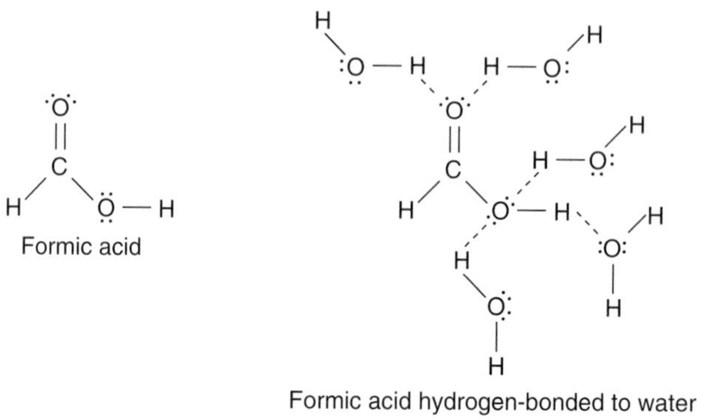

Figure 12.2

Hydrogen bonding between formic acid and water.

Formic acid

Formic acid hydrogen-bonded to water

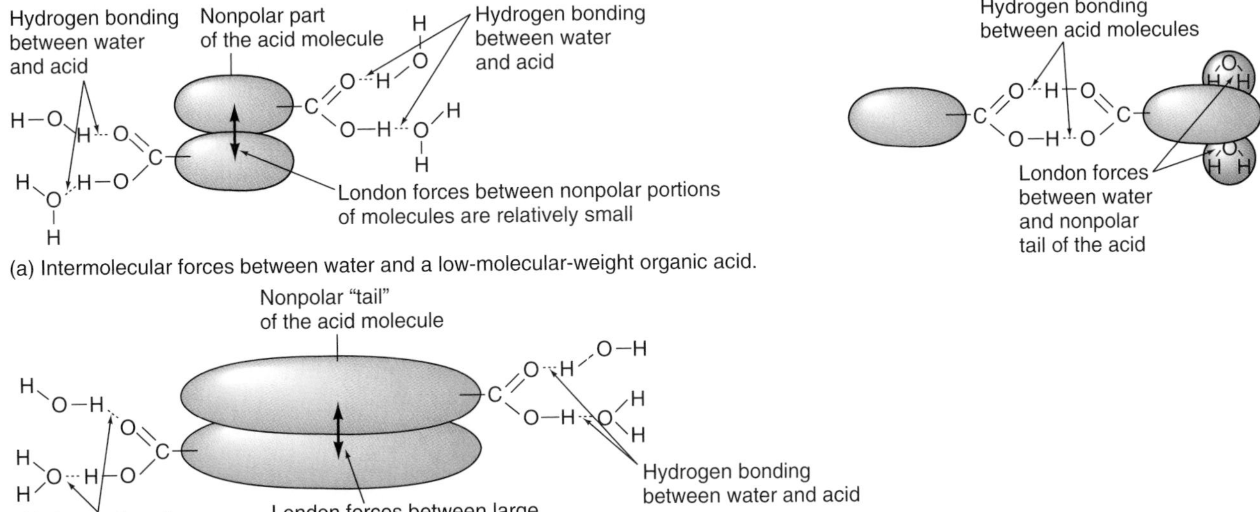

(a) Intermolecular forces between water and a low-molecular-weight organic acid.

(b) Intermolecular forces between water and a high-molecular-weight organic acid. The relatively strong London forces between large nonpolar portions of the acid molecules are responsible for the insolubility of high-molecular-weight acids in water.

Figure 12.3

Intermolecular forces between water and both low and high molecular weight organic acids determine the solubility of the acids in water. The solvent–solvent interaction (hydrogen bonding in water) is not shown. (a) Intermolecular forces between water and a low molecular weight organic acid. (b) Intermolecular forces between water and a high molecular weight organic acid. The relatively strong London forces between large nonpolar portions of the acid molecules is responsible for the insolubility of high molecular weight acids (and alcohols) in water.

hydrogen bonds to water molecules. As the size of the nonpolar tails of an organic acid increases, the London forces between the tails also increase. Thus, even though the $-\overset{\mathbf{O}}{\overset{\|}{\mathbf{C}}}-\mathbf{OH}$ "head" of the acid can form hydrogen bonds to water, this attraction is not strong enough to pull the high-molecular-weight acid into solution (Figure 12.3).

Liquids consisting of nonpolar molecules are generally insoluble in water. Such liquids include carbon tetrachloride (CCl_4), octane (C_8H_{18}), and carbon disulfide. In these cases the interactions of water molecules with one another through hydrogen bonds are much stronger than the interactions of water molecules with the nonpolar solute. However, as we noted above, nonpolar solutes dissolve in nonpolar solvents. Thus, carbon tetrachloride dissolves in octane (like dissolves like).

Solubility of Gases in Water

The principles governing the solubility of solids in liquids and liquids in liquids apply to dissolving gases in water, provided that the gas does not react with water. Thus, polar gases are more soluble in water than nonpolar gases (Table 12.1). The water solubility of the nonpolar molecules tends to increase with increasing molecular weight because the larger gas molecules are polarizable (Table 12.1b and c). Thus, the permanent dipole of water induces a transient dipole in the gas. The resulting dipole-induced dipole interaction enables the gas to dissolve in water.

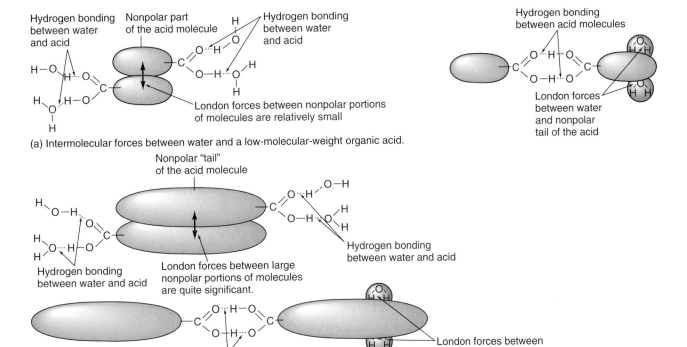

Table 12.1

Solubility of Gases in Water[a]

(a) Gases Consisting of Polar Molecules

Gas	Molar Mass	Solubility[a]
CO	28	3.50
NO	30	7.34
N_2O	44	130
NO_2[b]	46	Soluble
SO_2[b]	48	Soluble

(b) Gases Consisting of Nonpolar Molecules

Gas	Molar Mass	Solubility[a]
H_2	2.0	2.14
N_2	28	2.33
O_2	32	4.89
CO_2[b]	44	171
Cl_2[b]	71	461

(c) Noble Gases

Gas	Atomic Weight	Solubility[a]
He	4.0	0.94
Ne	20	1.47 (at 20°C)
Ar	40	5.6
Kr	84	11.0
Xe	131	24.1

[a] In cm^3 per 100 mL H_2O at 0°C and 1 atm partial pressure of the gas unless otherwise indicated.
[b] Reacts with water.

Four of the gases listed in Table 12.1 react with water when they dissolve. Nitrogen dioxide, sulfur dioxide, carbon dioxide, and chlorine react slightly according to the following equations:

$$3NO_2(g) + H_2O(l) \longrightarrow 2HNO_3(aq) + NO(g)$$

$$SO_2(g) + H_2O(l) \longrightarrow H_2SO_3(aq)$$

$$CO_2(g) + H_2O(l) \longrightarrow H_2CO_3(aq)$$

$$Cl_2(g) + H_2O(l) \longrightarrow HCl(aq) + HOCl(aq)$$

Saturation and Solubility

In most cases, as a solute is added to a given amount of a solvent at a given temperature and with good stirring, a point is eventually reached when no more solute dissolves in the solvent. At that point, the solution is said to be **saturated**. *In a saturated solution, a dynamic equilibrium exists between dissolved and undissolved solute.* At equilibrium, the rate of the solute going into the solution equals the rate of the solute returning from the solution. The condition for a dynamic equilibrium in a saturated solution of a solid in a liquid is illustrated in Figure 12.4.

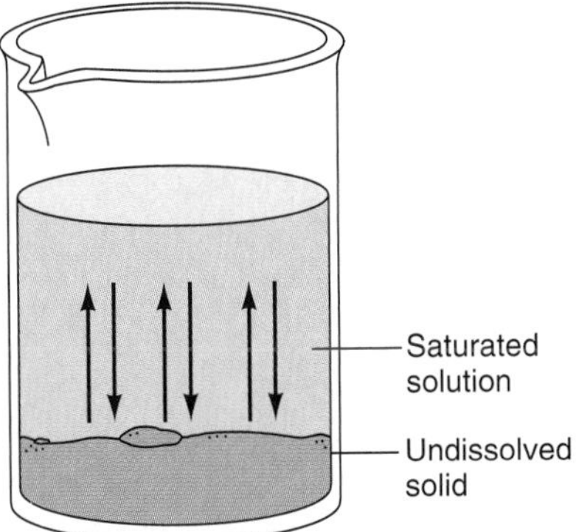

Figure 12.4

Dynamic equilibrium in a saturated solution of a solid in a liquid with excess solid. The rate of the solute going into the solution is equal to the rate of the solute returning from the solution.

The *amount of solute needed to make a saturated solution in a given amount of solvent at a given temperature* is called the **solubility** of the solute under these conditions. Many solid substances are more soluble at higher temperatures than at lower temperatures. Usually, when a saturated solution of such a substance is cooled, excess solid precipitates. However, sometimes a solution contains more solute than is indicated by the equilibrium solubility. Such a solution is called **supersaturated**. A supersaturated solution can often be prepared by slowly cooling a saturated solution. A supersaturated solution is unstable and readily reverts to the more stable saturated system if disturbed by shaking or by adding a seed crystal. The excess solute precipitates in the process.

12.2 Effect of Temperature and Pressure on Solubility

Thus far we have discussed the effect of the nature of solute and solvent on the solubility of a solute in a given solvent. The solubility of a given solute also depends on *temperature* and *pressure.*

Effect of Temperature on Solubility

The solubility of most solids in liquids *increases* with an increase in temperature. The solubility of most liquids in liquids, and gases in liquids, *decreases* with an increase in temperature. These observations can be understood in terms of **Le Châtelier's principle**, which states that *when a stress is applied to a system at equilibrium, the system changes to minimize the effect of the stress.* To understand Le Châtelier's principle, we need to consider how heat energy affects dissolution processes.

An endothermic dissolution process that leads to a saturated solution at equilibrium is described by the following expression:

$$\text{solute} + \text{solvent} + \text{heat} \rightleftharpoons \text{solution}$$

The double arrows represent an equilibrium system in a saturated solution. When the temperature is raised on such a system, the equilibrium is disturbed and the "stress of adding heat" causes more solute to dissolve until a new

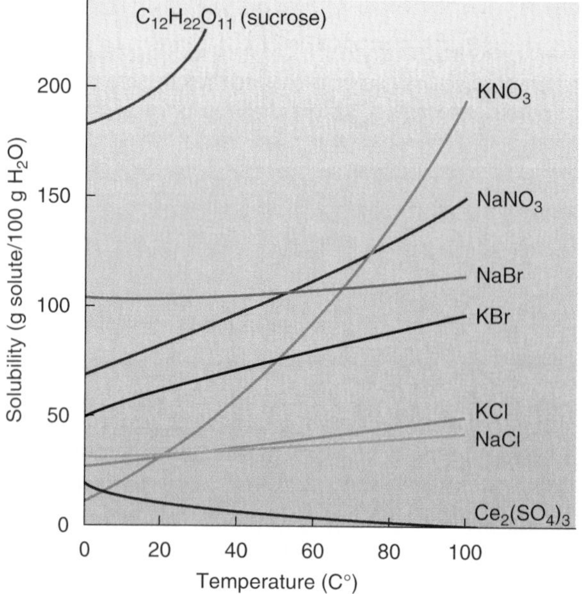

Figure 12.5

Variation of the solubility of common solids in water as a function of temperature.

equilibrium is reached. This new system at equilibrium is a saturated solution that is more concentrated than the original solution. Thus, an increase in temperature increases the solubility of the solute, as shown graphically for some ionic solids and for the molecular solid sucrose in Figure 12.5. A decrease in temperature results in a saturated solution that is more dilute. That is, a decrease in temperature decreases the solubility. The solubility of most ionic solids in water increases as the temperature increases.

An exothermic dissolution process at equilibrium can be described as follows:

$$\text{solute} + \text{solvent} \rightleftharpoons \text{solution} + \text{heat}$$

When the temperature of such a system increases, the solubility of the solute decreases. Conversely, when the temperature decreases, the solubility of the solute increases.

The dissolution of most liquids and gases in water is an exothermic process because solute–solute interactions in many liquids and gases are weaker than solute—solvent interaction. The solubility of liquid and gaseous solutes in water therefore decreases as the temperature increases.

Figure 12.6

On hot summer days fish retreat to cool, deep water where the oxygen concentration is greater than near the surface.

The solubility of oxygen in water decreases from 0.070 g per liter of water at 0°C to 0.030 g per liter at 50°C. In hot summer weather, the temperature of surface layers of lakes and rivers increases, and the water contains relatively small amounts of dissolved oxygen. Dissolved oxygen is required by fish and other marine organisms for respiration. On hot days these organisms retreat to cool deep water where the oxygen concentration is greater (Figure 12.6).

Effect of Pressure on Solubility

The solubility of solids in liquids or liquids in liquids is not much altered by changes in pressure. However, a change in the partial pressure of a gas in equilibrium with its saturated solution markedly changes the solubility of the gas. *The solubility of a gas in a liquid is directly proportional to the partial pressure of the gas above the solution.* This statement is known as **Henry's law**, which can be expressed mathematically as follows:

$$S = kP \tag{12.1}$$

where S is the solubility, k a proportionality constant, and P is the pressure of the gas over the solution. The value of k depends on the gas, the solvent, and the temperature.

If the solubility of a gas at a given pressure is known, the value of the constant, k, can be calculated and used to determine the solubility of the gas at another pressure.

Example 12.2

Calculating the Solubility of Oxygen Using Henry's Law

The solubility of oxygen in water at 20°C and 1.0 atm pressure of oxygen is 1.4×10^{-3} mol L^{-1}. Calculate the solubility of oxygen at 20°C and 0.21 atm (the partial pressure of oxygen in air).

SOLUTION: We calculate the proportionality constant, k, and solve Equation 12.1 for the solubility, S. The constant, k, from Henry's law is

$$k = \frac{S}{P} = \frac{1.4 \times 10^{-3} \text{ mol L}^{-1}}{1.0 \text{ atm}} = 1.4 \times 10^{-3} \text{ mol L}^{-1} \text{ atm}^{-1}$$

The solubility of oxygen at the partial pressure of 0.21 atm is

$$S = kP = (1.4 \times 10^{-3} \text{ mol L}^{-1} \text{ atm}^{-1})(0.21 \text{ atm}) = 2.9 \times 10^{-4} \text{ mol L}^{-1}$$

Carbonated beverages and sparkling wines are bottled under a carbon dioxide pressure of about 4 atm. When a bottle is opened, the beverage becomes ''flat'' because the pressure of the carbon dioxide drops as the gas escapes into the surrounding atmosphere (Figure 12.7).

Figure 12.7

Carbon dioxide escaping from a sparkling beverage.

Deep-sea divers who use compressed air sometimes feel discomfort, and may even suffer from a condition known as ''the bends'' if they ascend too rapidly (Figure 12.8). Oxygen from inhaled air is consumed by respiration and nitrogen dissolves in body fluids. If the diver's ascent is too rapid, the external pressure may drop so fast that the dissolved nitrogen bubbles out of body fluids, blocking blood capillaries, and causing severe pain in muscles and joints, fainting, paralysis, and even death.

Helium is considerably less soluble in body fluids than nitrogen and is therefore often substituted for nitrogen in gas tanks used by divers. The partial pressure of oxygen in the mixture of oxygen and helium must be kept at about 0.2 atm, as it is in air. A higher partial pressure of oxygen results in lower rate of breathing and the consequent accumulation of carbon dioxide in the body, leading to CO_2 poisoning.

Figure 12.8

Scuba diver must ascend slowly to avoid ''the bends.''

12.3 Vapor Pressure and Raoult's Law

In a solution of two liquid components, the vapor pressure of each component is less than the vapor pressure of the component in its pure state at the same temperature. In a solution of a solid solute of negligible vapor pressure, called a nonvolatile solute, the vapor pressure of the solvent is also less than the vapor pressure of the pure solvent at the same temperature.

Figure 12.9

Lowering the vapor pressure of a solvent by adding a nonvolatile solute.

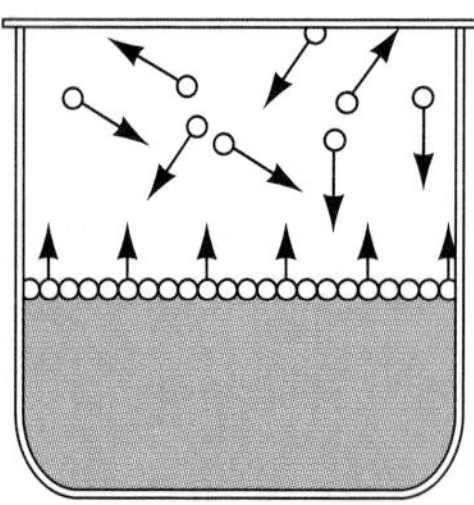

Pure solvent

Circles represent surface molecules. Molecules in the interior of the solvent are omitted for simplicity.

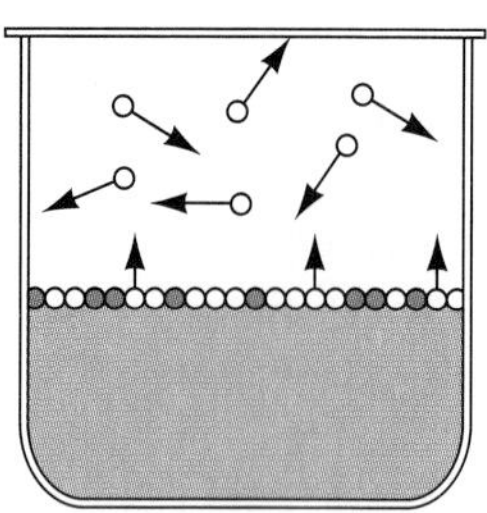

Solution

Shaded circles represent solute molecules. The number of solvent molecules converting to vapor in given time interval is lower than that in the pure solvent.

The lowering of the vapor pressure of a liquid solvent in a solution can be explained in terms of solute–solvent interaction. Solute molecules attract solvent molecules, decreasing the number of solvent molecules that can escape from the liquid to the gaseous state in a given time period (Figure 12.9).

For ideal solutions, the decrease in the vapor pressure of a solvent depends on the number of solute particles in the solution, but not on the nature of the solute. *Properties of dilute solutions that depend only on the number of solute particles* and not on the nature of the solute are known as **colligative properties**.

The *ratio of the number of moles of one component in a solution to the total number of moles of all the components in the solution* is called the **mole fraction** of that component. Thus, the mole fraction of a component i, X_i, can be expressed as

$$X_i = \frac{n_i}{n_1 + n_2 + \cdots + n_i} \tag{12.2}$$

where n_i represents the number of moles of the component i, n_1 the number of moles of component 1, and n_2 the number of moles of component 2.

When the concentration of a solution increases, the mole fraction of solute increases, and the mole fraction of solvent decreases. A solution in which the solute–solvent interaction does not differ from solute–solute or solvent–solvent interaction is called an *ideal solution*. For an ideal solution, *the vapor pressure of a given component in solution is directly proportional to the mole fraction of that component*. This statement is known as **Raoult's law**, after the French chemist François Raoult (1830–1901). This law can be expressed mathematically as follows:

$$P_i = X_i P_i^\circ \tag{12.3}$$

where P_i is the vapor pressure of component i in solution at a given temperature, X_i the mole fraction of component i, and P_i° the vapor pressure of pure component i at the same temperature as the solution.

For a solution of two liquid components A and B, the vapor pressure, P_A, of component A is

$$P_A = X_A P_A^\circ$$

and the vapor pressure, P_B, of component B is

$$P_B = X_B P_B^\circ$$

The vapor pressure of the solution, P_S, is the sum of the vapor pressures of the components:

$$P_S = P_A + P_B = X_A P_A^\circ + X_B P_B^\circ$$

Raoult's law, and other laws dealing with colligative properties of solutions, apply exactly only for ideal solutions. Raoult's law is to solutions as the ideal gas law is to gases.

Example 12.3

Calculating the Vapor Pressure of a Solution of Two Liquid Components

What is the vapor pressure of a solution that contains 3.00 mol of heptane (C_7H_{16}) and 5.00 mol of octane (C_8H_{18}) at 40°C, assuming ideal behavior? The vapor pressure of pure heptane at 40°C is 92 torr, and the vapor pressure of pure octane at 40°C is 31 torr.

SOLUTION: When we calculate the vapor pressure of a solution of two or more liquid components, we first determine the mole fraction of each of the components. Then we calculate the vapor pressure of each of the components from Equation 12.3. The vapor pressure of the solution is the sum of the vapor pressures of all the components.

The mole fraction of heptane in the solution is

$$X_{\text{heptane}} = \frac{3.00 \text{ mol}}{3.00 \text{ mol} + 5.00 \text{ mol}} = 0.375$$

Since the sum of the mole fractions of the components of a solution is 1,

$$X_{\text{octane}} = 1.00 - 0.375 = 0.625$$

From Equation 12.3, the vapor pressure of heptane in the solution is

$$P_{\text{heptane}} = (X_{\text{heptane}})(P^\circ_{\text{heptane}}) = (0.375)(92 \text{ torr}) = 34 \text{ torr}$$

The vapor pressure of octane in the solution is

$$P_{\text{octane}} = (X_{\text{octane}})(P^\circ_{\text{octane}}) = (0.625)(31 \text{ torr}) = 19 \text{ torr}$$

The vapor pressure of the solution is

$$P_{\text{solution}} = 34 \text{ torr} + 19 \text{ torr} = 53 \text{ torr}$$

Practice Problem 12.2: What is the vapor pressure of a solution that consists of 700 g of water and 250 g of ethanol, C_2H_5OH, at 323 K? The vapor pressure of water at that temperature is 0.122 atm, and the vapor pressure of ethanol is 0.292 atm.

If a nonvolatile solute is dissolved in a liquid, the vapor pressure of the solution is due to the vapor pressure of the solvent alone. The vapor pressure of the solvent in such a solution is lower than that of the pure solvent.

Example 12.4

Calculating of the Vapor Pressure of a Solution with a Solid, Nonvolatile Solute

What is the vapor pressure of a solution containing 75.0 g of sucrose, $C_{12}H_{22}O_{11}$, in 500 g of water at 25.0°C? The vapor pressure of water at 25.0°C is 23.8 torr.

SOLUTION: The vapor pressure of a solution that contains a nonvolatile solute is due entirely to the vapor pressure of the solvent. First we calculate the solvent mole fraction. The vapor pressure of the solvent is the mole fraction of the solvent times the vapor pressure of the pure solvent at the given temperature. The molar masses of sucrose and water are 342 and 18.0 g mol^{-1}, respectively. The mole fraction of water is

$$X_{water} = \frac{n_{H_2O}}{n_{C_{12}H_{22}O_{11}} + n_{H_2O}}$$

$$= \frac{500\ g/18.0\ g\ mol^{-1}}{500\ g/18.0\ g\ mol^{-1} + 75.0\ g/342\ g\ mol^{-1}}$$

$$= 0.992$$

The vapor pressure of the solution is

$$P_{solution} = (X_{water})(P^{\circ}_{water}) = (0.992)(23.8\ torr) = 23.6\ torr$$

Practice Problem 12.3: Calculate the vapor pressure of a solution that consists of 205 g of water and 25.0 g of glucose, $C_6H_{12}O_6$, at 70.0°C. The vapor pressure of water at 70.0°C is 0.308 atm.

The difference in the vapor pressures of two liquid components in a solution can be used to separate the components by a process called **distillation** (Figure 12.10). Distillation involves heating a solution to its boiling point, condensing the vapors to liquid, and collecting the condensed liquid, called the *distillate*. The vapor above the solution is richer in the more volatile compo-

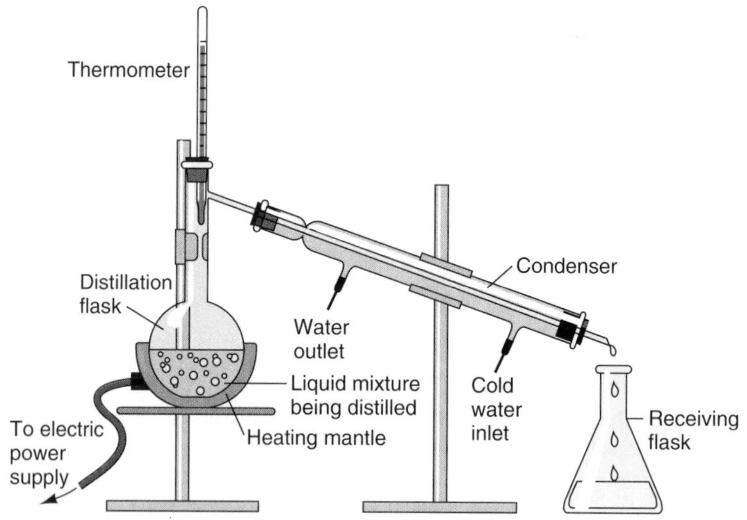

Figure 12.10

Apparatus used for distillation.

nent, and the distillate is also rich in that component. Repeating the distillation several times yields a successively purer component.

Several successive distillations can be accomplished in a one-step operation called **fractional distillation**. In this process, a long glass tube packed with small glass beads or helices is placed between the solution being distilled and the condenser (Figure 12.11). This tube is called a fractionating column. The glass beads in the fractionating column provide a large surface area on which vapors condense. The vapor phase over a solution of two volatile components is richer in the more volatile component than is the liquid solution. When this vapor condenses and reevaporates in the lower section of a fractionating column, the vapor is even richer in the more volatile component. This condensation–evaporation process is repeated many times as the vapors make their way up through the fractionating column. Depending on the length and exact design of the column, a relatively good separation of components can be achieved even if the boiling points of the components are close together.

Heavy water (D_2O or deuterium oxide, ^2H_2O) is used in nuclear power plants as a coolant and a moderator to slow down neutrons in nuclear reactions. Heavy water can be obtained by fractional distillation of regular water, which contains only about 0.015 percent of deuterium, mostly in the form of HDO. The normal boiling point of regular water is 100.00°C and that of heavy water 101.42°C. Heavy water distillation plants have over 300 successive distillation stages and require an input of more than a thousand metric tons of regular water to obtain 1 kg of D_2O. Canada uses heavy water extensively in nuclear power plants and has two D_2O production plants with a combined output capability of about 1600 tons per year.

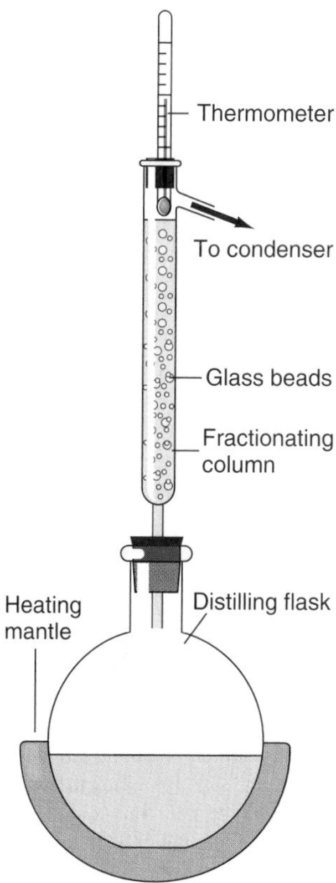

Figure 12.11

Distilling flask with fractionating column.

12.4 Boiling Point Elevation and Freezing Point Depression

We discussed the boiling points and freezing points of liquids in Chapter 11. We recall that the normal boiling point of a liquid is the temperature at which the vapor pressure of the liquid equals 1.00 atm. The freezing point is the temperature at which the solid and liquid phases of a substance are at equilibrium. At the freezing point, the vapor pressure of the solid equals the vapor pressure of the liquid.

Boiling Point Elevation and Vapor Pressure

When a nonvolatile solute dissolves in a liquid, the vapor pressure of the solution is lower than that of the solvent. As a result, the boiling point of the solution is higher than the boiling point of the solvent, and the freezing point of the solution is lower than the freezing point of the solvent. These facts are illustrated in Figure 12.12 by the phase diagrams for pure water and for an aqueous solution.

APPLICATIONS OF CHEMISTRY 12.1
Why Do Joggers Perspire Excessively?

People perspire to maintain constant body temperature. But why do people perspire excessively? (see photo) Water evaporates when the faster-moving liquid water molecules at the surface escape the attractive forces of the other liquid water molecules and become gaseous water molecules. The liquid water molecules left behind have a lower average kinetic energy after the faster-moving molecules escape and are therefore at a lower temperature. Since evaporation is a cooling process, the body perspires to maintain a constant temperature. This mechanism is pushed to the limit under some conditions. For example, as the humidity of the surrounding air increases, the rate of evaporation and therefore the cooling rate decreases. When the humidity is very high, evaporation is severely retarded and the perspiration may be so plentiful that it drips off the body before it has a chance to evaporate and cool the body.

From a physiological standpoint it would be better if our bodies released just enough perspiration to keep the skin moist. Since nature is normally very conservative, we must wonder what advantage exists in producing excessive perspiration.

To answer this question, we need to examine what would happen if our bodies did produce just enough perspiration to keep the skin moist. Perspiration contains solutes such as sodium chloride and potassium chloride. These solutes remain on the skin as the water evaporates. As the body produces more perspiration, the solutes deposited on the skin accumulate.

We just learned that the vapor pressure of a solution is directly proportional to the solute concentration. As the solute concentration increases, the vapor pressure and the evaporation rate of the perspiration decrease. Calculations have shown that increasing the concentration of solute in perspiration can decrease its vapor pressure by 20 percent.

So now we see that there is an advantage to excessive perspiration. As the body perspires excessively, the perspiration rinses off the excess solutes and prevents them from becoming too concentrated. This in turn increases the vapor pressure of the perspiration and cooling is more efficient.

Source: Ronald DeLorenzo, *Journal of Chemical Education*, November 1986, page 977.

The boiling point of water depends on the atmospheric pressure. A pot of boiling water stops boiling when a salt is added to it, indicating that the vapor pressure of the solution is less than the atmospheric pressure. Upon further heating, the temperature rises to the boiling point of the solution and the water boils again. Both the vapor pressure depression and boiling point elevation are colligative properties. That is, in a dilute solution, they are proportional to the concentration of solute particles.

Thus far we have expressed concentration in units of molarity. However, the molarity of a solution depends on temperature since a solution expands when heated and contracts when cooled. A unit of concentration that does not change with temperature is expressed in *moles of solute per kilogram of sol-*

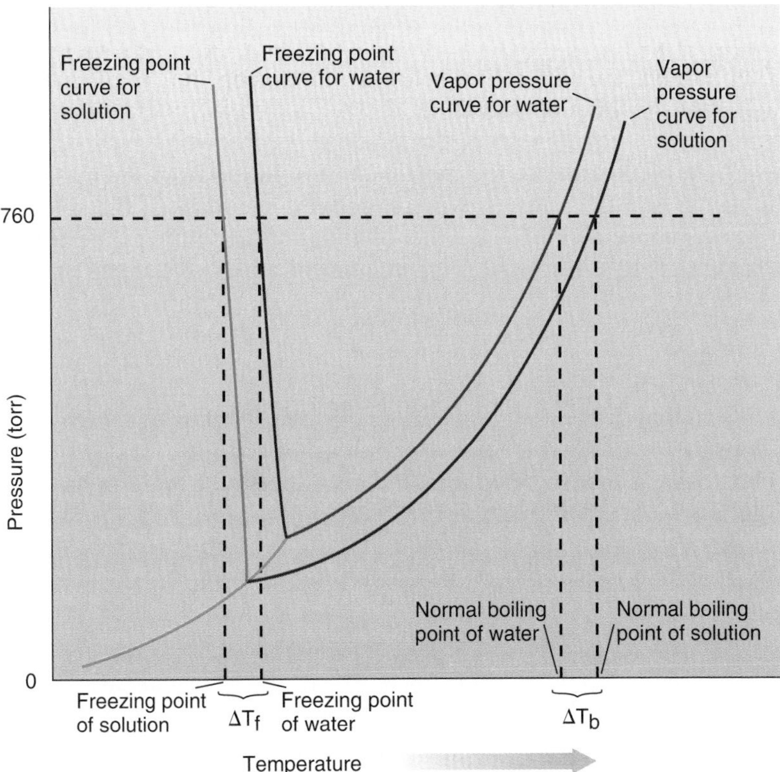

Figure 12.12

Phase diagrams for water and for a water solution (not drawn to scale).

vent. This concentration unit is called **molality** or **molal concentration**. The molality, m, of a solution is given by the equation

$$m = \frac{\text{moles of solute}}{\text{kg of solvent}} \tag{12.4}$$

We recall that the number of moles of a substance is the ratio of the mass of the substance in grams, g, and its molar mass, MM:

$$\text{moles} = \frac{g}{\text{MM}}$$

Substituting this expression for the number of moles in Equation 12.4, the molality of a solution is

$$m = \frac{\text{grams of solute}}{(\text{MM of solute})(\text{kg of solvent})} \tag{12.5}$$

The molality and molarity of dilute aqueous solutions are nearly equal, because the density of a dilute solution nearly equals the density of pure solvent. Thus, 1 L of a dilute aqueous solution has nearly the same mass as 1 L of pure water, that is, 1 kg.

When the concentration of a solution increases, the deviation between molality and molarity increases. When a solute is added to a liter (1 kg) of water, the molality of the solution equals the number of moles of the solute added. However, the molarity does not equal the number of moles of solute added because the volume of the solution changes when the solute is added. The molality of a solution does not change with temperature because molality is expressed in terms of masses of the solute and the solvent, and the masses do not change with temperature.

Example 12.5

Calculating the Molality of a Solution

What is the molality of a solution containing 100 g of sucrose, $C_{12}H_{22}O_{11}$, in 500 g of water?

SOLUTION: To calculate the molality of a solution from the masses of the solute and the solvent, we first convert the mass of the solute to moles. Then we divide the number of moles of the solute by the number of kilograms of the solvent. The number of moles of sucrose is

$$100 \text{ g sucrose} \left(\frac{1 \text{ mol sucrose}}{342 \text{ g sucrose}}\right) = 0.292 \text{ mol sucrose}$$

$$m = \frac{\text{mol solute}}{\text{kg solvent}} = \frac{0.292 \text{ mol}}{0.500 \text{ kg}} = 0.584 \text{ mol kg}^{-1}$$

The molality can also be determined by a one-step operation by use of Equation 12.5 as follows:

$$m = \frac{\text{mass solute in grams}}{(\text{MM solute})(\text{kg solvent})} = \frac{100 \text{ g}}{(342 \text{ g mol}^{-1})(0.500 \text{ kg})}$$

$$= 0.585 \text{ mol kg}^{-1}$$

Practice Problem 12.4: What is the molality of a solution that contains 15.0 g of a solute of molar mass 40.0 g mol^{-1} in 250 g of solvent?

We have seen that the boiling point of a solvent increases when solute is added. The boiling point elevation, ΔT_b, of an ideal solution is proportional to the molal concentration of the solution, and can be expressed by the equation

$$\Delta T_b = K_b m \qquad (12.6)$$

where K_b is the **boiling point constant** and m is the molality of the solution. The boiling point constant equals the *boiling point elevation observed for a one molal solution*. This means that the boiling point constant, K_b, for a 1 molal solution equals ΔT_b. The boiling point constant, K_b, is therefore often referred to as the **molal boiling point constant** or the **molal elevation of the boiling point**. The magnitude of K_b is characteristic of the solvent and does not depend on the nature of the solute. For water, K_b has the value of 0.512°C m^{-1}. For the K_b values of other solvents, see Table 12.2.

Table 12.2

Properties of Some Common Solvents

Solvent	Normal b.p. (°C at 1 atm)	K_b (°C m^{-1})	f.p. (°C)	K_f (°C m^{-1})
Acetic acid	117.9	3.07	16.604	3.90
Benzene	80.1	2.53	5.5	4.90
Carbon tetrachloride	76.54	5.03	−22.99	2.98
Ethanol (ethyl alcohol)	78.5	1.22	−117.3	0.199
Phenol	181.75	3.56	43	7.40
Water	100	0.512	0.000	1.86

Example 12.6

Calculating the Boiling Point of a Solution from Given Masses of the Solute and Solvent

What is the boiling point of the solution containing 75.0 g of sucrose, $C_{12}H_{22}O_{11}$, in 300 g of water at 1.00 atm pressure?

SOLUTION: In problems of this type, it is convenient first to calculate the molality, m, of the solution. The boiling point elevation can then be calculated from Equation 12.6. The boiling point elevation is added to the boiling point of the pure solvent to get the boiling point of the solution.

The molality, m, of the solution is

$$m = \frac{75.0 \text{ g}}{(342 \text{ g mol}^{-1})(0.300 \text{ kg})} = 0.731 \text{ mol kg}^{-1}$$

The boiling point elevation from Equation 12.6 is

$$\Delta T_b = K_b m = (0.512°\text{C } m^{-1})(0.731\ m) = 0.374°\text{C}$$

The boiling point of the solution is the boiling point of the solvent plus the boiling point elevation:

$$100.000°\text{C} + 0.374°\text{C} = 100.374°\text{C}$$

Practice Problem 12.5: A solution is made by dissolving 25.0 g of a solid, molar mass 58.0 g mol^{-1}, in 500 g of water. The boiling point of pure water at the prevailing atmospheric pressure is 99.825°C. Calculate the boiling point of the solution.

Freezing Point Depression

When a nonvolatile solute dissolves in a liquid at its freezing point, the vapor pressure of the liquid at equilibrium with its solid phase decreases. The vapor pressure of the solid phase remains unchanged, and is now higher than the vapor pressure of its liquid phase containing the solid. As a result, the solid melts, and the temperature must decrease for the solution to freeze again. Thus, the freezing point of a solution is lower than the freezing point of the pure solvent regardless of the nature of the solute (Figure 12.12).

Antifreeze added to an automobile radiator lowers the vapor pressure of the solution in the radiator. As a result, the freezing point of the solution is lowered, and the boiling point is increased.

When salt is added to ice on roads, the salt dissolves in a thin layer of water on the surface of the ice. As a result, the equilibrium between ice and its aqueous layer is disturbed because the vapor pressure of the solution in contact with ice is lower than the vapor pressure of water. Some of the ice melts to establish a new equilibrium. This process continues until all of the ice eventually melts or the temperature decreases to the new freezing point.

Freezing point depression is directly proportional to the number of solute particles added to a given mass of the solvent. The *freezing point depression* for a solution can be expressed by Equation 12.7:

$$\Delta T_f = K_f m \tag{12.7}$$

In this equation, ΔT_f is the freezing point depression and K_f the **freezing point constant**, or the **molal freezing point depression**. K_f is the freezing point depression observed when 1 mol of a nonvolatile and nonionizing solute dissolves in 1 kg of solvent. K_f is characteristic of the solvent but does not depend on the nature of solute. The freezing point constant for water is 1.86°C m^{-1}. Values of K_b and K_f for some common solvents are listed in Table 12.2, as well as the boiling and freezing points for these solvents.

Example 12.7

Calculating the Mass of Solute Required to Attain a Given Freezing Point Depression, Assuming Ideal Behavior

What mass of the antifreeze ethylene glycol (HO—CH_2CH_2—OH) must be added to 10.0 kg of water to make a solution that freezes at −4.00°C? Assume ideal conditions.

SOLUTION: First, we determine the molality of the solution needed to lower the freezing point by 4.00°C. From the molality, we determine the mass of the antifreeze needed. The freezing point lowering, ΔT_f, is the freezing point of the solvent minus the freezing point of the solution:

$$\Delta T_f = 0.00 - (-4.00°\text{C}) = 4.00°\text{C}$$

Solving Equation 12.7 for molality, we obtain

$$m = \frac{\Delta T_f}{K_f} = \frac{4.00°\text{C}}{1.86°\text{C kg mol}^{-1}} = 2.15 \text{ mol kg}^{-1}$$

Note that the units for K_f can be obtained by dividing the units for ΔT_f by the units of m:

$$K_f = \frac{\Delta T_f}{m} = \frac{°\text{C}}{\text{mol kg}^{-1}} = °\text{C kg mol}^{-1}$$

From Equation 12.5, the mass of the solute is the product of the molality of the solution, m, the molar mass of the solute, MM, and the mass of the solvent in kilograms:

$$\begin{aligned} \text{g solute} &= (m)(\text{MM solute})(\text{kg solvent}) \\ &= (2.15 \text{ mol kg}^{-1})(62.0 \text{ g mol}^{-1})(10.0 \text{ kg}) \\ &= 1.33 \times 10^3 \text{ g} \quad \text{or} \quad 1.33 \text{ kg} \end{aligned}$$

This answer is approximate only because the properties of a 2.15 m solution deviate somewhat from those calculated from Raoult's law.

Practice Problem 12.6: A 4.50-g sample of a solute with the molar mass of 180 g mol^{-1} is dissolved in 135 g of water. What is the freezing point of the solution?

Ethylene glycol is miscible with water in all proportions. To determine the concentration of an antifreeze solution, we can use a **hydrometer** to determine its density. (A hydrometer consists of a glass tube filled with a heavy metal at one end. The other end of the tube has a smaller diameter and its length is calibrated.) A hydrometer sinks into a liquid in an upright position with its heavy end down until the density of the immersed end equals the density of the liquid. Hydrometers used to test antifreeze solutions are supplied with charts listing the freezing points of an antifreeze solution corresponding to various depths to which the hydrometer sinks. Ethylene glycol (density 1.1 g cm^{-3}) is more dense than water, so increasing the ethylene glycol content of an antifreeze solution *decreases* the depth to which the hydrometer sinks.

"Antifreeze" should be used in radiators even in the summer. The coolant in a radiator that contains antifreeze not only freezes at temperatures lower than 0°C, it also boils at a higher temperature than pure water.

Molar Mass Determination

The principle of freezing point depression or boiling point elevation can be used to determine the molar mass of a solute. Methods relying on freezing point lowering are generally used because they are easier to carry out experimentally. To determine the molar mass of a solute, a weighed sample of the substance is dissolved in a known mass of a solvent, and the freezing point depression is determined (Figure 12.13). The molar mass of the solute can then be calculated from Equation 12.7 as illustrated in Example 12.8.

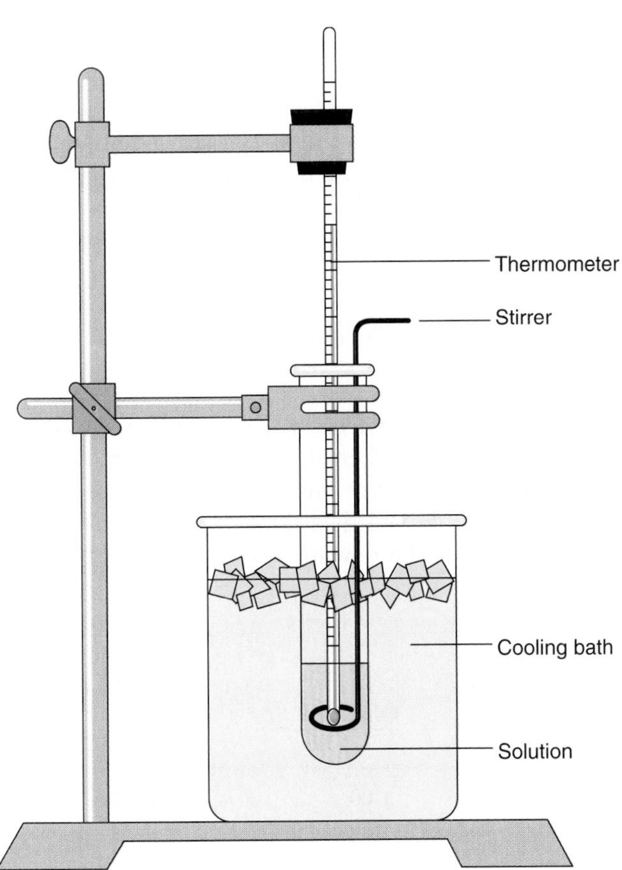

Figure 12.13

A simple apparatus for molar mass determination by freezing point depression.

Example 12.8

Determining the Molar Mass of an Unknown Substance by Freezing Point Depression

A 2.000-g sample of an unknown substance is dissolved in 20.00 g of naphthalene. The freezing point of the solution is 77.52°C. The melting point of pure naphthalene is 80.20°C, and its molal freezing point depression (the freezing point constant) is 6.85°C m^{-1}. Calculate the molar mass of the unknown substance.

SOLUTION: We first calculate ΔT_f and the molality of the solution. From the known molality, the molar mass can be obtained.

$$\Delta T_f = 80.20°\text{C} - 77.52°\text{C} = 2.68°\text{C}$$

The molality of the solution can be determined from the values of ΔT_f and K_f by use of Equation 12.7:

$$m = \frac{\Delta T_f}{K_f} = \frac{2.68°\text{C}}{6.85°\text{C}\ m^{-1}} = 0.391\ m$$

The molar mass, MM, can now be calculated from Equation 12.5,

$$\text{MM} = \frac{\text{g solute}}{(m)(\text{kg solvent})} = \frac{2.000\ \text{g}}{(0.391\ \text{mol kg}^{-1})(0.02000\ \text{kg})}$$

$$= 256\ \text{g mol}^{-1}$$

Practice Problem 12.7: A 2.00-g sample of an unknown compound is dissolved in 10.0 g of water. The normal freezing point of water is depressed to −1.22°C. What is the molar mass of the unknown compound?

12.5 Electrolyte Solutions

So far, our discussion of colligative properties has focused on solutions of nonelectrolytes. Solutions of electrolytes have larger elevations of boiling points and larger depressions of freezing points than do solutions of nonelectrolytes of equal molality. Table 12.3 lists some freezing point depression data for various electrolytes and for the nonelectrolyte sucrose for comparison.

Table 12.3

Freezing Point Depression Data for Sucrose and for Some Common Electrolytes in Water Solution[a]

Substance	Observed ΔT_f (°C) of 1.0 m Solution	Observed ΔT_f / 1.86	Observed ΔT_f (°C) of 0.0010 m Solution	Observed ΔT_f / 0.00186
Sucrose	1.86	1.00	0.00186	1.00
$HC_2H_3O_2$	1.87	1.005	0.00214	1.15
NH_3	1.87	1.005	0.00214	1.15
HCl	(3.4)	(1.8)	0.00368	1.98
NaCl	3.29	1.77	0.00366	1.97
$MgSO_4$	2.02	1.09	0.00339	1.82
K_2SO_4	(4.1)	(2.2)	0.00528	2.84

[a] The data in parentheses are approximate.

The freezing point depression for each of the three strong electrolytes HCl, NaCl, and $MgSO_4$ is considerably greater than that for the weak electrolytes, $HC_2H_3O_2$ and NH_3 (Table 12.3). This is particularly true in dilute solution. The freezing point depression for each of the weak electrolytes listed in Table 12.3 is only slightly more than the freezing point depression for a sucrose solution of equal molality.

We expect a strong electrolyte such as NaCl to dissociate completely into Na^+ and Cl^- ions when it dissolves in water:

$$\mathrm{NaCl}(s) \xrightarrow{\mathrm{H_2O}} \mathrm{Na^+}(aq) + \mathrm{Cl^-}(aq)$$

1 mol added — 1 mol Na^+, 1 mol Cl^- — 2 mol of ions produced

The observed freezing point depression of a 1.0 *m* NaCl solution should therefore be twice that observed for a 1.0 *m* solution of a nonelectrolyte. The data in Table 12.3 indicate that the observed freezing point depression for a 1.0 *m* solution of NaCl (3.29°C) is only 1.77 times the freezing point depression of a 1.0 *m* solution of a nonelectrolyte (1.86°C):

$$\frac{\text{observed } \Delta T_f}{K_f m} = \frac{3.29°\text{C}}{(1.86°\text{C } m^{-1})(1.0\ m)} = 1.77$$

However, for a 0.0010 *m* solution of NaCl, the observed freezing point depression is very close to twice the value for a 0.0010 *m* solution of a nonelectrolyte:

$$\frac{\text{observed } \Delta T_f}{K_f m} = \frac{0.00366°\text{C}}{(1.86°\text{C } m^{-1})(0.0010\ m)} = 1.97$$

These results indicate that in a concentrated solution of an electrolyte, the solute does *not* completely dissociate. The failure of strong electrolytes to dissociate completely at high concentrations can be explained by the Debye–Hückel **interionic attraction theory** (after Peter Debye, a U.S. chemist of Dutch origin, and Erich Hückel, a Swiss chemist, who proposed the theory in 1923).

This theory asserts that as the concentration of ions in solution increases, the ability of the solvent water molecules to screen the ionic charges decreases because the number of the solvent molecules decreases relative to the number of ions present. As a result, some oppositely charged ions are attracted to each other to form ion pairs (Na^+Cl^-) or ion trios ($K^+SO_4^{2-}K^+$), depending on the nature of the solute. The degree of dissociation increases with dilution, as shown by the data in Table 12.3. As a solution becomes more dilute, the average distance between the ions increases, and the opportunity for interionic attraction decreases.

A mole of K_2SO_4 might be expected to dissociate into three moles of ions:

$$\mathrm{K_2SO_4}(s) \xrightarrow{\mathrm{H_2O}} 2\mathrm{K^+}(aq) + \mathrm{SO_4^{2-}}(aq)$$

1 mol — 2 mol K^+, 1 mol SO_4^{2-} — 3 mol of ions

The observed freezing point depression for a 1.0 *m* solution of K_2SO_4 is only 2.2 times $K_f m$, but for a 0.0010 *m* solution it is $2.84K_f m$. With increasing dilution, this value approaches $3K_f m$. If we assume that an electrolyte dissociates completely in a dilute solution, we can calculate its boiling point or freezing point from its molality, as shown by Example 12.9.

Example 12.9

Calculating the Boiling Point of an Electrolyte Solution from a Given Molality

Calculate the normal boiling point of a 0.012 *m* potassium nitrate solution assuming complete dissociation of the electrolyte.

SOLUTION: The molality of an electrolyte solution equals the number of moles of solute per kilogram of solvent. To calculate the freezing point or boiling point of such a solution, we must first determine the total molality of all the molecules and ions of the solute in the solution. From the formula of a strong electrolyte, we determine the number of moles of ions that can be produced from each mole of the electrolyte, assuming complete dissociation in a dilute solution. The total molal concentration of all the ions in solution gives the value of ΔT_b from Equation 12.6. The boiling point of the solution is the sum of the boiling point of the solvent and ΔT_b.

Each mole of potassium nitrate that completely dissociates gives a total of 2 mol of ions:

$$KNO_3(s) \longrightarrow K^+(aq) + NO_3^-(aq)$$

Thus, the molality of the ions in solution is twice the molality of the salt. Application of Equation 12.6 gives

$$\Delta T_b = K_b m = (0.512\ ^\circ\text{C kg mol}^{-1})[(2)(0.012\ \text{mol kg}^{-1})] = 0.012^\circ\text{C}$$

The normal boiling point of the solution is

$$100.000^\circ\text{C} + 0.012^\circ\text{C} = 100.012^\circ\text{C}$$

Practice Problem 12.8: What is the freezing point of a 0.015 *m* magnesium chloride solution assuming complete dissociation of the electrolyte?

The value for the freezing point depression or the boiling point elevation of an electrolyte is a measure of its degree of dissociation in a solution. A solution of a weak acid, HA, consists of undissociated HA molecules in equilibrium with H_3O^+ and A^- ions:

$$HA(l) + H_2O(l) \rightleftharpoons H_3O^+(aq) + A^-(aq)$$

If we let x equal the number of moles of HA that dissociates into ions in a 0.100 *m* solution, the number of moles of HA undissociated is $0.100 - x$. The number of moles of H_3O^+ ions in the solution is x, and the number of moles of A^- ions also equals x. The total number of moles of H_3O^+ ions, A^- ions, and undissociated HA molecules in each kilogram of the solvent is

$$(0.100 - x) + x + x = 0.100 + x$$

The *ratio of the number of moles of an electrolyte that dissociates to ions in solution to the number of moles present before dissociation* is called the **degree of dissociation** of the electrolyte. The *degree of dissociation of an electrolyte times 100* is the **percent dissociation**. Thus, the percent dissociation of a 0.100 *m* solution of HA is:

$$\text{percent dissociation} = \left(\frac{x}{0.100}\right) 100$$

Freezing point depression data can be used to estimate the percent dissociation of a weak electrolyte as shown in Example 12.10.

Example 12.10

Determining the Percent Dissociation of a Weak Electrolyte from Its Freezing Point Depression

The freezing point of a 0.100 *m* solution of acetic acid is −0.188°C. Calculate the percent dissociation of acetic acid in a 0.100 *m* aqueous solution.

SOLUTION: First, we calculate the total molality of all the molecular and ionic species in the solution. The difference between this total molality and the molality of the solution before dissociation equals the number of moles of electrolyte that dissociates per kilogram of solvent. One hundred times the number of moles of electrolyte that dissociates divided by the number of moles before dissociation is the percent dissociation.

The freezing point depression of the solvent is the freezing point of the solvent minus the freezing point of the solution:

$$\Delta T_f = 0.000°\text{C} - (-0.188°\text{C}) = 0.188°\text{C}$$

The total molality of the molecules and ions in acetic acid solution can be calculated from ΔT_f:

$$m = \frac{\Delta T_f}{K_f} = \frac{0.188°\text{C}}{1.86°\text{C}\ m^{-1}} = 0.101\ m$$

An acetic acid solution is an equilibrium mixture of acetic acid molecules, hydronium ions, acetate ions, and water. We let x equal the number of moles of acetic acid that dissociates per kilogram of water. The following equation represents the equilibrium mixture of acetic acid molecules and its ions. Underneath each of the species for acetic acid in this equation, we indicate the number of moles present at equilibrium per kilogram of solvent:

$$\underset{0.100 - x}{HC_2H_3O_2(l)} + H_2O(l) \rightleftharpoons \underset{x}{H_3O^+(aq)} + \underset{x}{C_2H_3O_2^-(aq)}$$

The total number of moles of all the particles of solute present per kilogram of water is

$$(0.100 - x) + x + x = 0.100 + x$$

This sum equals the total molality of the solution:

$$0.100 + x = 0.101$$

From this equality we calculate x, the number of moles of acetic acid dissociated per kilogram of solvent:

$$x = 0.101 - 0.100 = 0.001$$

$$\text{percent dissociation} = \frac{0.001}{0.100} \times 100 = 1 \text{ percent}$$

More precise measurements show that 0.100 *m* acetic acid is 1.3 percent dissociated.

12.6 Osmosis and Osmotic Pressure

Every living cell is surrounded by a thin *membrane* about 6 nm thick which separates the cell from its environment (Figure 12.14). This cell membrane, also called the plasma membrane, permits water to pass through, but blocks the passage of most solutes. Such a membrane is called a **semipermeable membrane**. The diffusion of a solvent through a semipermeable membrane is called **osmosis**. If two solutions are separated by a semipermeable membrane, solvent flows through the membrane into the solution having the higher concentration. Many natural and artificial films serve as semipermeable membranes.

Figure 12.15 illustrates the process of osmosis using a cellophane bag as a semipermeable membrane. A 1 *m* sucrose solution in a glass tube fitted with a cellophane bag is placed into pure water in one of the containers and into sucrose solutions of different concentrations in other containers. The liquid levels inside and outside the tube are initially adjusted to be equal. To discuss the direction of solvent flow across a semipermeable membrane, we use three new terms to describe the relative concentrations of two solutions. If two solutions have the same concentration, they are *isotonic* ("iso-" means the same). If one of the two solutions has a higher concentration than another, the one of higher concentration is called *hypertonic* ("hyper-" means more), and the one having the lesser concentration is *hypotonic* ("hypo-" means less).

Figure 12.15a shows that the level of the solution in the tube has risen and the water level outside the tube is lower than the initial level because some water moved across the semipermeable membrane into the solution in the tube.

When the tube with a 1 *m* sucrose solution is placed in a 0.1 *m* sucrose solution (Figure 12.15b), the liquid level in the tube again rises, but not as much as shown in part (a). When the two solutions are isotonic, there is no change in the liquid levels, as shown in Figure 12.15c.

Figure 12.15d illustrates a model of pure solvent molecules on one side of a semipermeable membrane, and solute and solvent molecules on the other side. In osmosis, solvent flows across the membrane into the more concentrated, hypertonic solution in an effort to equalize the concentrations of the solutions. The greater the difference between these concentrations, the greater is the tendency of the solvent molecules to pass through the membrane. The *minimum pressure required to prevent osmosis* is called **osmotic pressure**.

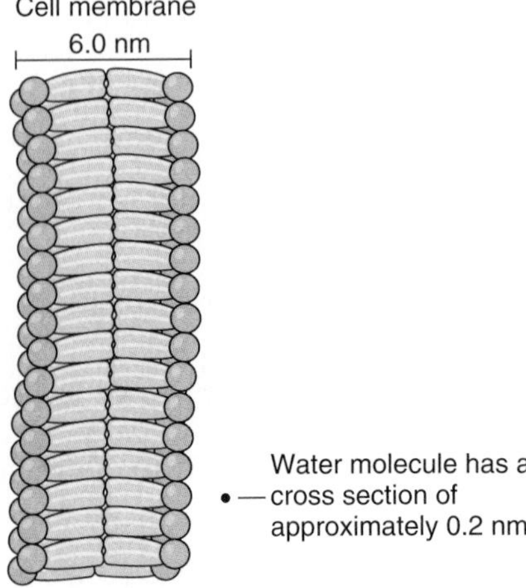

Figure 12.14

The cross-section of the membrane of a living cell.

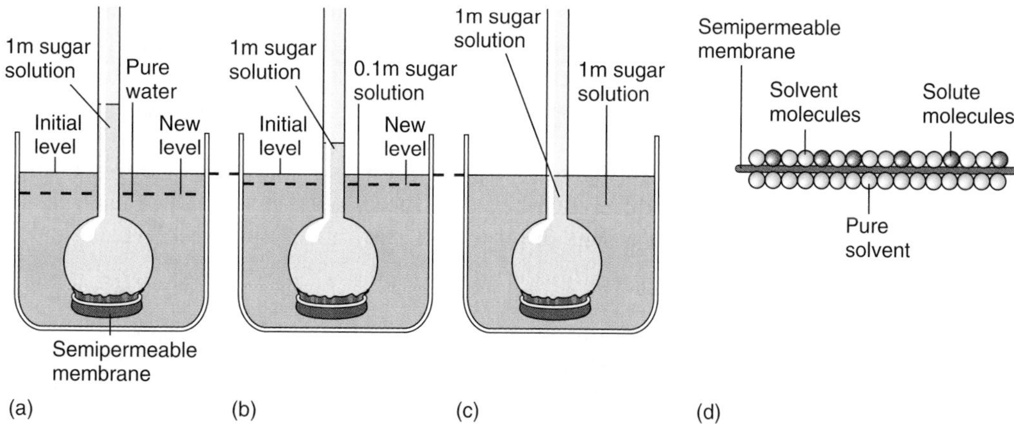

Figure 12.15

Illustration of osmosis. (a) A sugar solution separated from pure solvent (water) by a semipermeable membrane; (b) two sugar solutions of different concentrations separated by a semipermeable membrane; (c) two sugar solutions of equal concentrations separated by a semipermeable membrane; (d) a close-up view of relative numbers of solvent and solute molecules on both sides of the membrane as in part (a).

The osmotic pressure is greater in the case shown in Figure 12.15a than in Figure 12.15b because the difference between 1 m solution and pure solvent is greater than the difference between 1 m and 0.1 m solutions. In the case shown in Figure 12.15c, the solutions inside and outside the tube have equal concentrations, and no osmosis occurs.

Osmotic pressure is proportional to the concentration of solute particles. Osmotic pressure is therefore a colligative property of solutions. The osmotic pressure, π, in a dilute solution is related to the number of moles, n, of solute, the volume, V, of the solution in liters, and the absolute temperature, T, by the following equation:

$$\pi V = nRT \tag{12.8}$$

where R is the ideal gas constant (0.0821 L atm mol^{-1} K^{-1}). The osmotic pressure in atmospheres is obtained by solving Equation 12.8 for π:

$$\pi = \left(\frac{n}{V}\right) RT$$

and since $n/V = M$,

$$\pi = MRT \tag{12.9}$$

where M is the molarity of the solution.

Example 12.11

Calculating the Osmotic Pressure of a Solution of Given Molarity

If red blood cells were placed in pure water, what pressure must be exerted to just prevent water from entering them at normal body temperature of 37°C? Assume that blood cells have a total solute concentration of 0.30 M.

SOLUTION: The osmotic pressure can be calculated directly from Equation 12.9:

$$\pi = MRT = (0.30 \text{ mol L}^{-1})(0.0821 \text{ L atm mol}^{-1} \text{ K}^{-1})(310 \text{ K}) = 7.6 \text{ atm}$$

This result shows that osmotic pressure is amazingly high for a solution as dilute as 0.30 M.

Practice Problem 12.9: Seawater is a solution of many solutes. The average molarity of all seawater is 1.12 *M*. If seawater were separated from pure water by a semipermeable membrane, what pressure must be exerted at 25.0°C to just prevent the pure water from diffusing into the seawater?

The osmotic pressure between two solutions of different concentrations equals the difference between the osmotic pressures of each of the solutions. Calculation of this difference is shown in Example 12.12.

Example 12.12

Calculating the Osmotic Pressure Difference Between Two Solutions of Unequal Concentrations

What pressure must be exerted to just prevent the flow of water through a semipermeable membrane from a 0.10 *M* to a 0.20 *M* solution at 25°C?

SOLUTION: Since osmotic pressure is directly proportional to molar concentration, the difference in osmotic pressures of two solutions is proportional to the difference in the molarities of the solutions. Thus, applying Equation 12.9 yields

$$\begin{aligned}\Delta\pi &= (\Delta M)RT \\ &= (0.20 \text{ mol L}^{-1} - 0.10 \text{ mol L}^{-1})(0.0821 \text{ L atm mol}^{-1}\text{ K}^{-1})(298 \text{ K}) \\ &= 2.4 \text{ atm}\end{aligned}$$

Osmotic pressure can be used to determine the molar masses of polymers and proteins. The molar masses of these substances are so high that the number of moles present in a sample is too small to be detected by freezing point depression or boiling point elevation. Example 12.13 illustrates how osmotic pressure can be used to determine the molar mass of a protein.

Example 12.13

Determining the Molar Mass of a Protein from the Osmotic Pressure of Its Solution

Hemoglobin is the oxygen transport protein of red blood cells. A 35.0-g sample of hemoglobin is dissolved in water and the total volume of the solution is brought to 1.00 L. The osmotic pressure of the solution is 10.0 torr at 25.0°C. Calculate the molar mass of hemoglobin, Hb.

SOLUTION: First, we convert the osmotic pressure from torr to atmospheres. Then we calculate the number of moles of the protein from Equation 12.8. From the number of moles and the mass of protein in the sample we calculate the molar mass.

$$10.0 \text{ torr}\left(\frac{1 \text{ atm}}{760 \text{ torr}}\right) = 0.0132 \text{ atm}$$

Solving Equation 12.8 for the number of moles, *n*, we obtain

$$n = \frac{\pi V}{RT} = \frac{(0.0132 \text{ atm})(1.00 \text{ L})}{(0.0821 \text{ L atm mol}^{-1}\text{ K}^{-1})(298 \text{ K})} = 5.40 \times 10^{-4} \text{ mol}$$

$$\text{molar mass} = \frac{\text{mass of Hb}}{\text{moles of Hb}} = \frac{35.0 \text{ g}}{5.40 \times 10^{-4} \text{ mol}} = 6.48 \times 10^{4} \text{ g mol}^{-1}$$

Practice Problem 12.10: A 10.0-g sample of an unknown protein is dissolved in water at 298 K so that the total volume of the solution is 1.00 L. The osmotic pressure of the solution is 9.25 torr. What is the molar mass of the protein?

If the method of freezing point depression were used to determine the molar mass of hemoglobin, the difference of the freezing points of water and the very dilute solution of hemoglobin would be almost immeasurably small. For example, the freezing point depression for a 5.40×10^{-4} *m* solution is

$$\Delta T_f = K_f m = (1.86°\text{C } m^{-1})(5.40 \times 10^{-4}\ m) = 0.00100°\text{C}$$

Such a small temperature difference is difficult to measure, and a large experimental error would be involved in determining the molar mass of hemoglobin by freezing point depression.

Osmotic pressure has important biological consequences, especially for organisms living in water. Many single-celled organisms live in seawater, which has a concentration of about 1.2 *M*. These organisms are normally isotonic with the solution in which they live, so they bypass the problem of osmosis altogether. We saw in Example 12.11 that red blood cells have a total solute concentration of about 0.30 *M*. Red blood cells are isotonic with the blood plasma in which they are suspended.

Most plant cells are hypertonic relative to their environment, and thus water flows into plant cells by osmosis. As water is absorbed, the cells swell, but the cell walls prevent the cells from bursting. The osmotic pressure is opposed by the mechanical strength of the cell walls. If a plant is placed in a hypertonic medium, say salt water, water flows out of the plant cells into the salt water. As a result, the plant wilts (Figure 12.16).

(a)

(b)

Figure 12.16

Plant in (a) pure water and (b) salt water.

APPLICATIONS OF CHEMISTRY 12.2
Commercial Thirst Quenchers May Not Be Worth Their Salt

Since the 1970s, coaches have urged athletes to drink only water or highly dilute glucose solutions while exercising. However, many professional athletes and sports enthusiasts can still be seen drinking commercial thirst quenchers. Are these drinks better than ordinary water? Many studies suggest they are not. Because commercial thirst quenchers contain carbohydrates and electrolytes, they remain in the stomach longer than pure water. Medical authorities also question the need for electrolytes in these drinks since a proper diet easily replaces them. Although the carbohydrates may be useful, the carbohydrate concentration is usually too high because drinks with more than 2.5 percent carbohydrates inhibit hydration during exercise. Yet the sugar content of most sports drinks is between 6 and 8 percent. Sports drinks are hypertonic; that is, they have electrolyte concentrations greater than that of normal body fluids. Because of this, sports drinks can cause dehydration by initially drawing body fluids through cell membranes into the stomach by osmosis. This is also what happens to people who drink ocean water in emergencies. Because ocean water is excessively hypertonic, people actually die from dehydration when they drink ocean water. Sports drinks are far less hypertonic than ocean water, but sports drinks can still dehydrate drinkers, and that in turn can decrease athletic performance.

Many athletes still believe that sports drinks are better than water to replace fluids, but even this is usually not the case. The body has developed ways of conserving electrolytes, so you lose water faster than you lose electrolytes when you sweat. Therefore, you are better off drinking water because that is what you are losing, primarily. Since the salt concentration in your body increases when you lose water, you don't want to drink something salty.

More recently, one study has shown that sports drinks did boost performance better than water for marathoners running 25 miles in 95°F weather. Strenuous activity such as this may greatly deplete stores of carbohydrates and electrolytes, and sports drinks can replenish these. However, drinking water is usually the best way to replace the body fluids lost during exercise.

Sources: Chemical & Engineering News, December 24, 1990, page 40; Ronald DeLorenzo, *Journal of Chemical Education*, February 1982, page 153.

12.7 Colloids

Types of Colloids

In Chapter 1 we defined a solution as a homogeneous mixture of two or more components. In contrast, in a heterogeneous mixture we can recognize two or more different phases. Suppose, however, that the particles of a heterogeneous mixture are progressively subdivided into smaller particles until the mixture no longer appears to be heterogeneous, but has not yet become truly homogeneous. This mixture is known as a **colloidal dispersion**, a **colloidal suspension**, or a **colloid**. The components in a colloid cannot be separated by filtration or centrifugation, and the particles are too small to be seen with a microscope, ranging in size between about 1 and 10^3 nm. Colloidal particles do not settle out of the suspension on standing.

If a colloid has two components, the phase that corresponds to the solute in a true solution is called the *dispersed phase*. The other phase, which is similar to the solvent in a true solution, is called the *dispersion medium*. A colloid can be distinguished from a true solution by shining light through it. The relatively large particles in a colloidal dispersion scatter light rays. As a result, we can see a beam of light passing through a colloidal dispersion, a phenomenon known as the **Tyndall effect** (Figure 12.17). In a true solution, the beam of light cannot be seen when viewed at right angles to the axis of the beam because the particles are too small to scatter light.

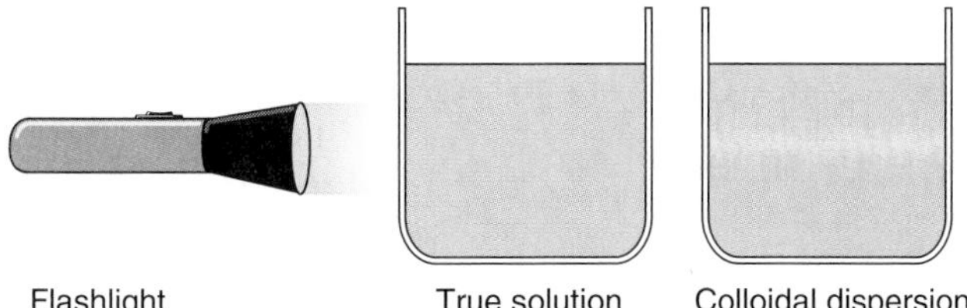

Figure 12.17

Tyndall effect. A device to distinguish between a true solution and a colloidal dispersion. The Tyndall effect is observed in the colloidal solution as a scattering of light rays on the surfaces of relatively large solute particles. This effect is also observed when sunlight passes through a dusty room where the dust forms a colloidal dispersion with air. Automobile headlights in fog produce a similar effect.

One example of a colloid is fog, which consists of finely divided droplets of water suspended in air (Figure 12.18). Fog scatters light from the headlights of an automobile, an example of the Tyndall effect. Another example of a colloidal dispersion is dust suspended in air (Figure 12.19). Sunlight shining through a window is scattered by the dust particles, and we see a beam of light in the room. Cigarette smoke has the same effect. Light shining on cigarette smoke has provided some of the most memorable moments in the black-and-white films of the 1930s. Table 12.4 lists some common types of colloids, their components, and examples of each.

Figure 12.18

Fog is a colloidal dispersion of a liquid in a gas.

Figure 12.19

Dust suspended in air is a colloidal dispersion of a solid in a gas.

Examples of Colloids

Many biological fluids are colloids. For example, blood is an extremely complex mixture that contains several kinds of cells and large molecular complexes that circulate as a colloidal suspension. For example, cholesterol is transported in the blood by several kinds of molecular complexes known collectively as lipoproteins.

The large molecules of a colloidal suspension can be separated from smaller molecules of a true solution by **dialysis**. In dialysis, a colloid is placed in a "bag" made of a membrane of a particular pore size. Such a membrane can be purchased as open-ended tubing. The pore size of the membrane is chosen to permit small solute molecules to pass through, but not the larger colloidal particles, which remain in the bag.

Dialysis is often used in medical practice to treat kidney disorders. Human kidneys remove the waste products of metabolism from blood. One of these waste products is urea, $(NH_2)_2CO$. Blood is passed through a dialysis machine, which contains a long, coiled cellophane tube suspended in a solution that is isotonic with blood. Urea and other low-molecular-weight components in blood pass through the dialysis tubing into the surrounding medium and are removed. The dialyzed blood flows out of the machine into the veins that lead to the heart.

Colloids are also used in the cleaning industry. Many cleaning operations remove grease from fabrics. "Grease" consists of large nonpolar molecules that are attracted to one another more strongly than to water molecules, which are strongly hydrogen bonded to each other. Grease therefore does not mix with water but forms a layer or "slick" on the surface of water. Such a slick is dispersed into a colloidal suspension by a soap or a detergent. The colloid is then washed away with water.

A soap or a detergent consists of large anions having a long nonpolar hydrocarbon "tail" and a polar "head," shown in boldface below for stearate anion, which is the anion of a typical soap, sodium stearate:

```
    H H H H H H H H H H H H H H H H H O
    | | | | | | | | | | | | | | | | | ‖
  H—C—C—C—C—C—C—C—C—C—C—C—C—C—C—C—C—C—C—O⁻
    | | | | | | | | | | | | | | | | |
    H H H H H H H H H H H H H H H H H
                 Nonpolar tail                Polar
                                              head
```

The nonpolar tail of a soap or a detergent anion is similar to the molecules of oil or grease. Such a tail is attracted to grease, while the polar head of the soap anion is attracted to water molecules (Figure 12.20). Thus, grease and soap form a colloidal suspension that can be washed away.

Table 12.4

Common Types of Colloids

Colloid Type (Common Name)	Nature and Composition	Example
Sol	Solid particles dispersed through a liquid	Milk of magnesia (solid magnesium hydroxide dispersed in water), most paints, starch dispersed in water
Gel	Continuous network of solid throughout a liquid medium	Jellies, gelatin, and a slimy precipitate such as aluminum hydroxide
Aerosol	Either a solid or a liquid dispersed in a gas	Smoke (solid dispersed in air), fog (liquid dispersed in air), smog (a combination of smoke and fog)
Emulsion	Particles of a liquid dispersed in another liquid	Milk (particles of butterfat and protein in aqueous dispersion), mayonnaise (a dispersion of salad oil and either lemon juice or vinegar)
Foam	Bubbles of gas suspended in a liquid	Whipped cream, beer foam, soapsuds, canned shaving cream

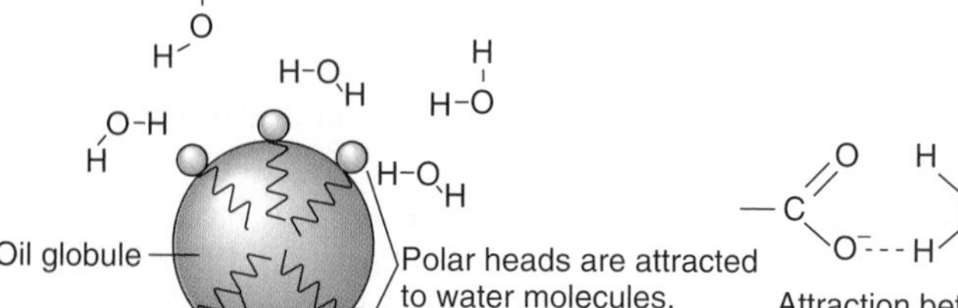

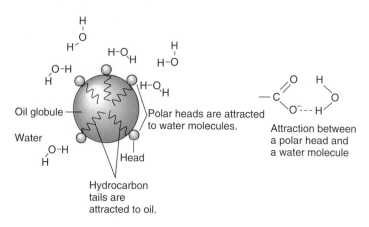

Figure 12.20

A cross section of an oil globule that attracts the hydrocarbon tails of the anions of soap. The polar heads of the anions are attracted to water molecules forming a colloidal particle in an emulsion that can be washed away.

12.8 Water Pollution and Purification

Pollution

Many bodies of water—lakes, streams, bays—contain substantial quantities of organic and inorganic chemicals, detergents, crude and refined oil, household and industrial sewage, radioactive substances, and trash of every description.

Toxic materials discarded into rivers and lakes often include substances such as dichlorodiphenyltrichloroethane (DDT), which is harmful to birds and animals, and polychlorinated biphenyls (PCBs). PCBs are used as coolants in transformers and in the manufacture of paint. Toxic solvents such as carbon tetrachloride, benzene, chloroform, and compounds of lead, mercury, and cadmium are also found in water.

Human waste in rivers and lakes is broken down by bacteria and other microorganisms that require oxygen for their activity. The amount of oxygen required for this purpose is called the **biological oxygen demand** (BOD). When the BOD exceeds the available supply of oxygen, the anaerobic microorganisms (those that function without oxygen) take over. They produce undesirable substances such as hydrogen sulfide, ammonia, and methane. Thus, there are many challenging environmental problems for chemists to solve. Water purification is one such problem. We next discuss current methods of water purification.

Purification

Household and industrial wastewater is subjected to sewage treatment by municipal or industrial sewage treatment plants before the water is returned to lakes and rivers. A complete sewage treatment consists of three stages: primary, secondary, and tertiary treatment. In primary treatment, large, insoluble particles are removed by passing water through debris removal screens. Sand, cinders, and gravel are removed as "grit" in a grit chamber (Figure 12.21). Inorganic and organic solids are allowed to settle out in large sedimentation tanks.

In the secondary stage of sewage treatment, most of the oxygen-demanding waste is removed by treating the water with aerobic microorganisms and

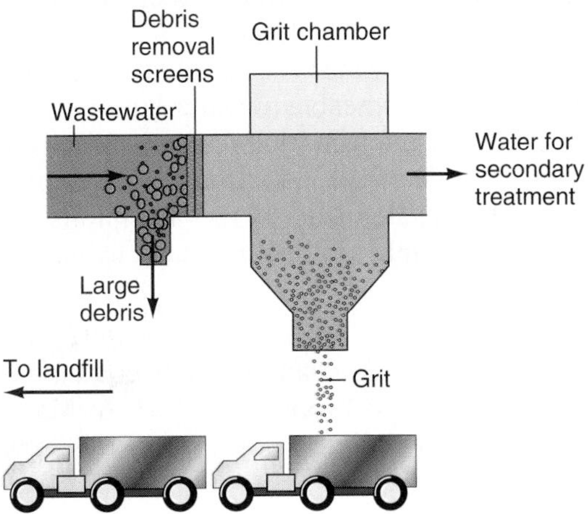

Figure 12.21

Primary sewage treatment.

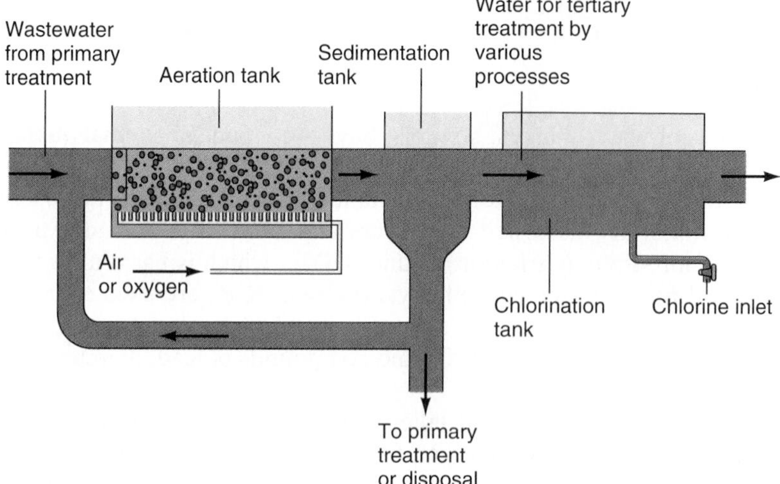

Figure 12.22

Secondary sewage treatment.

with a plentiful supply of oxygen in large aeration tanks, as shown in Figure 12.22. In this stage, disease-causing organisms are killed by chlorinating the water from the aeration tank.

In the tertiary treatment, heavy metal ions, plant nutrients (nitrogen and phosphorus compounds), pesticides and other chemicals that cannot be destroyed by microorganisms, and any radioactive materials remain in the water. These pollutants are removed by specific chemical and physical methods.

To purify water for domestic use, chlorine or ozone is added to kill bacteria and to oxidize any dissolved organic substances. Although chlorine itself is poisonous to humans at high concentrations, a concentration of 1 part per million (1 ppm) is sufficient to kill bacteria without ill effects to humans.

One of the important methods in the tertiary stage of water treatment, as well as in the production of drinking water from seawater, is *desalination*, which removes Na^+, Cl^-, and other ions from water. This effect can be accomplished by distillation, by crystallization (freezing), or by reverse osmosis. Distillation is relatively expensive because it requires considerable energy. It also produces water with a flat taste.

Crystallization involves freezing salt water to a slurry from which the ice particles are separated and remelted. In the freezing process, the salt remains dissolved in the liquid phase of the slurry.

In reverse osmosis, fresh water is separated from seawater by a semipermeable membrane which is permeable to water but not to salts in the seawater. When pressure in excess of the osmotic pressure of seawater (about 26 atm) is applied on the seawater side of the membrane, water molecules pass from the seawater side to the freshwater side of the membrane. Devices for reverse osmosis based on this principle are commercially available.

Water Softening

The water in many wells, lakes, and streams contains Ca^{2+} and Mg^{2+} ions and is known colloquially as **hard water**. The Ca^{2+} or Mg^{2+} ions in hard water form an insoluble, gray precipitate with soap. This precipitate is a salt of stearate anions in soap and either calcium or magnesium ions. The reaction of calcium ions with sodium stearate in soap to produce calcium stearate is

$$\underset{}{Ca^{2+}(aq)} + \underset{\text{soap}}{2NaC_{17}H_{35}CO_2(aq)} \longrightarrow \underset{\text{insoluble salt}}{Ca(C_{17}H_{35}CO_2)_2(s)} + 2Na^+(aq)$$

Because of this reaction, soap does not lather well in hard water, and clothes washed in hard water do not appear to be very clean.

There are two types of water hardness, known as **temporary hardness** and **permanent hardness**, depending on what anions are present. Temporary hard water contains hydrogen carbonate ions (HCO_3^-), which can be removed, along with the cations causing the hardness, by boiling. If the cations in a temporary hard water are calcium ions, the reaction that occurs upon boiling is

$$Ca^{2+}(aq) + 2HCO_3^-(aq) \longrightarrow CaCO_3(s) + CO_2(g) + H_2O(l)$$

The insoluble metal carbonate is the "boiler scale" that forms on the walls of boilers and kettles. Boiler scale is a poor conductor of heat, and its presence lowers the effectiveness of a boiler. Boiler scale removal is expensive and not practical on a large scale.

When water contains sulfates or other anions that do not precipitate upon boiling, it is permanently hard. The ions in permanently hard water can be removed in several different ways. One simple method consists of adding washing soda, $Na_2CO_3(H_2O)_{10}$; the other is called *ion exchange*. Washing soda contains carbonate ions that precipitate the metal ions:

$$Ca^{2+}(aq) + CO_3^{2-}(aq) \longrightarrow CaCO_3(s)$$

Insoluble metal carbonates formed by this reaction are removed by filtration.

In ion exchange, the objectionable cations in hard water are replaced by sodium ions, which do not cause hardness because sodium salts are soluble. In an ion-exchange process, water is passed through a tube packed with natural zeolites or similar synthetic materials. This apparatus is called an ion-exchange column. Zeolites are complex sodium aluminum silicates in which large aluminosilicate anions are arranged in three-dimensional networks of alternating aluminum, oxygen, and silicon atoms. The cavities in this anion network are occupied by sodium ions. When hard water passes through an ion-exchange column containing a zeolite, the Ca^{2+} or Mg^{2+} ions exchange places with the sodium ions and are removed from the water:

$$Ca^{2+}(aq) + 2NaZ(s) \longrightarrow 2Na^+(aq) + CaZ_2(s) \qquad (Z = \text{zeolite anion})$$

Summary

Solution Process

A solution is a homogeneous mixture. An *ideal solution* is a solution in which the solute–solute, solvent–solvent, and solute–solvent interactions are equal. In a real solution, these interactions are never exactly equal, and the properties of real solutions deviate somewhat from ideal behavior. However, the properties of dilute solutions can be predicted with a fair degree of accuracy by assuming ideal behavior.

The *solubility* of one substance in another generally depends on the *intermolecular forces in solute and solvent, temperature, and pressure*. The solubility of most solids in liquids increases with increased temperature, whereas the solubility of gases in liquids, and most liquids in liquids, decreases with increased temperature. The solubility of gases in liquids is governed by **Henry's law**:

$$S = kP$$

where S is the solubility, k a constant, and P the partial pressure.

Properties of Solutions

The properties of solutions that depend on the number of solute particles and not on their nature are called **colligative properties**. These properties include *vapor pressure*, *freezing point depression*, *boiling point elevation*, and *osmotic pressure*.

According to **Raoult's law**, the vapor pressure of an ideal solution is the sum of the vapor pressures of each of the components of the solution:

$$P_{\text{solution}} = P_1 + P_2 + \cdots + P_n$$

The vapor pressure of each component of a solution is the product of the **mole fraction**, X, of that component and the vapor pressure of the pure component at that temperature. The mole fraction of a component is the number of moles of that component divided by the total number of moles. Thus, Raoult's law can be written as

$$P_{\text{solution}} = X_1P_1^\circ + X_2P_2^\circ + \cdots + X_nP_n^\circ$$

The molal concentration of a solution or **molality**, m, is the number of moles of the solute per kilogram of the solvent:

$$m = \frac{\text{moles of solute}}{\text{kg of solvent}}$$

$$= \frac{\text{mass of solute in grams}}{\text{(molar mass of solute)(kg of solvent)}}$$

The boiling point elevation, ΔT_b, of a solvent is the product of the **boiling point constant**, K_b, of the solvent and the molality, m, of the solution:

$$\Delta T_b = K_bm$$

Similarly, the freezing point depression, ΔT_f, of a solvent is the product of the **freezing point constant**, K_f, of the solvent and the molality of the solution:

$$\Delta T_f = K_fm$$

Properties of Electrolyte Solutions

Electrolytes dissociate into ions in water. Therefore, the observed boiling points of electrolyte solutions are higher, and the freezing points are lower, than those observed for nonelectrolyte solutions of equal molality. Data on boiling point elevation or freezing point depression for an electrolyte solution can be used to determine the **percent of dissociation**, or the strength, of the electrolyte.

Osmosis

When two solutions of different concentration are separated by a **semipermeable membrane**, solvent molecules pass through the membrane from the solution of lower concentration to the solution of higher concentration. This process is called **osmosis**. The pressure required to prevent osmosis is the **osmotic pressure**, π. Osmotic pressure can be calculated by the equation

$$\pi = MRT$$

Colloidal Dispersions

Homogeneous mixtures of relatively large particle size (ranging in diameter from approximately 1 to 10^3 nm) which are larger than the particles of "true solutions," but smaller than the particles in a suspension, are called **colloidal dispersions** or **colloids**. In a colloid, one of the components is called the *dispersed phase*, and the other component is the *dispersion medium*. Colloids can be distinguished from true solutions by the *Tyndall effect*.

Water Pollution and Purification

Lakes, rivers, streams, and oceans are solutions that contain many different solutes and sediments, many of which are undesirable. Water is purified by municipal and industrial water treatment plants by first filtering the large, insoluble particles, followed by treatment with aerobic microorganisms, chlorination, and other special processes.

Hard water is caused by Ca^{2+} and Mg^{2+} ions, which precipitate with soaps. *Temporary hard water* contains hydrogen carbonate ions. Temporary hard water can be softened by boiling. *Permanent hard water* contains sulfate or chloride ions. Permanent hard water can be softened by adding sodium carbonate or by passing the water through an ion-exchange column.

New Terms

Aerosol (12.7)
Biological oxygen demand (BOD) (12.8)
Boiling point constant (12.4)
Colligative properties (12.3)
Colloid (colloidal dispersion) (colloid suspension) (12.7)
Degree of dissociation (12.5)
Dialysis (12.7)
Distillation (12.3)
Emulsion (12.7)
Foam (12.7)
Fractional distillation (12.3)
Freezing point constant (12.4)
Gel (12.7)
Hard water (12.8)
Henry's law (12.2)
Hydration (12.1)
Hydrometer (12.4)
Interionic attraction theory (12.5)
Le Châtelier's principle (12.2)
Molal boiling point constant (12.4)
Molal concentration (12.4)
Molal elevation of the boiling point (12.4)
Molal freezing point depression (12.4)
Molality (12.4)
Mole fraction (12.3)
Osmosis (12.6)
Osmotic pressure (12.6)
Percent dissociation (12.5)
Permanent hardness (12.8)
Raoult's law (12.3)
Saturated solution (12.1)
Semipermeable membrane (12.6)
Sol (12.7)
Solubility (12.1)
Supersaturated (12.1)
Temporary hardness (12.8)
Tyndall effect (12.7)

Exercises

General Review

1. Explain the roles of lattice energy and hydration energy in the solubility of an ionic solid in water.
2. Explain the term "ideal solution" in terms of intermolecular forces (solute–solute, solvent–solvent, and solute–solvent interactions).
3. What relative magnitudes of solute–solute, solvent–solvent, and solute–solvent interaction are most favorable for a dissolution process?
4. Which of the following interactions is the least important factor in determining the solubility of a gas in a liquid: solute–solute, solute–solvent, or solvent–solvent interaction? Explain.
5. What is meant by the general statement "like dissolves like"?
6. Why are low-molecular-weight alcohols soluble in water, but high-molecular-weight alcohols insoluble or only sparingly soluble?
7. How does an increase in temperature generally affect the solubility of: **(a)** solids in liquids; **(b)** liquids in liquids; **(c)** gases in liquids?
8. What effect, if any, does an increase in pressure have on the solubility of: **(a)** solids in liquids; **(b)** liquids in liquids; **(c)** gases in liquids?
9. What is Henry's law? Write a mathematical expression for Henry's law and define the symbols used.
10. What is: **(a)** a saturated solution; **(b)** a supersaturated solution?
11. What is Raoult's law? Write a mathematical expression for Raoult's law and define the symbols used.
12. Explain why the vapor pressure of a solution of a solid in a liquid is lower than the vapor pressure of the pure solvent.
13. Show how Raoult's law can be used to calculate the vapor pressure of a solution of two liquid components, assuming ideal behavior.
14. Explain why the boiling point of a solution of a solid in a liquid is higher than that of the pure solvent.
15. Explain why the freezing point of a solution is lower than that of the pure solvent.
16. What is "molality"?
17. Show by examples how the molality of a solution of a nonvolatile nonelectrolyte in a liquid can be used to calculate the boiling point and the freezing point of the solution.
18. What different types of data are needed to determine the molar mass of a nonvolatile nonelectrolyte in a given solvent? Show how you would calculate the molar mass from these data.
19. Explain why the boiling point elevation of a 0.01 *m* solution of sodium chloride is nearly twice that for a 0.01 *m* solution of sucrose, and the boiling point elevation for a 0.01 *m* solution of sodium sulfate is nearly three times that for a 0.01 *m* solution of sucrose.
20. What is interionic attraction theory? How does that theory explain the fact that the boiling point elevation for a dilute sodium chloride solution is *nearly* twice, but not exactly twice, the boiling point elevation for an equally dilute sucrose solution?

21. How does the freezing point depression of a 0.01 *m* solution of acetic acid compare with: **(a)** the freezing point depression of a 0.01 *m* solution of sucrose; **(b)** the freezing point depression of a 0.01 *m* solution of sodium chloride? Explain the difference.
22. What is osmosis, and what is the meaning of osmotic pressure? Give some examples of the importance of osmosis in nature.
23. How can a colloid be distinguished from a true solution?
24. List some major colloids by name (e.g., sol) and describe the composition of each.
25. What are some major water pollutants?
26. Describe some important methods of water purification.
27. What is: **(a)** hard water; **(b)** temporary hard water; **(c)** permanent hard water?
28. How can temporary hard water be softened? Write an equation to illustrate the process.
29. How can permanent hard water be softened? Write equations to illustrate the process.

Solubility and the Factors Affecting Solubility

30. Which of the following compounds would you expect to be water soluble? **(a)** $(NH_4)_3PO_4$; **(b)** H_2S; **(c)** CCl_4; **(d)** C_6H_6 (benzene); **(e)** $HCOOH$; **(f)** CH_3OH; **(g)** NH_3; **(h)** HCl; **(i)** $Ca(NO_3)_2$; **(j)** $CH_2OH(CHOH)_4CHO$; **(k)** $CH_3(CH_2)_6CH_3$; **(l)** P_4; **(m)** CS_2; **(n)** SO_2. Explain your response in each case in terms of intermolecular forces.
31. Which of the following compounds would you expect to be soluble in carbon tetrachloride? **(a)** NH_4NO_3; **(b)** C_6H_6; **(c)** $C_{10}H_8$ (naphthalene); **(d)** SO_2.
32. Explain why Na_2SO_4 is soluble in water but $BaSO_4$ is not.
33. Predict how an increase in temperature might affect the solubility of each of the following substances in water: **(a)** KNO_3; **(b)** $C_{12}H_{22}O_{11}$; **(c)** SO_2; **(d)** O_2.
34. How does the concentration of dissolved oxygen in water in a river at sea level compare with that in a mountain brook? Explain.
35. Explain which of the following substances would you expect to be the most soluble, and which the least soluble, in benzene (C_6H_6) at 25°C: CH_4; C_2H_6; C_5H_{12}.
36. According to the information from the graph in Figure 12.5, is the dissolution of KNO_3 in water endothermic or exothermic? Explain your reasoning.
37. Using the information from the graph in Figure 12.5, approximately how many grams of KNO_3 is needed to prepare a saturated solution of KNO_3 in 40 g of water at 40°C?
38. If 50 g of KNO_3 is dissolved in 100 g of water at 80°C and the solution is cooled to 20°C, how much of the salt crystallizes out of the solution?

Henry's Law

39. When the pressure of nitrogen over water at 20°C is 1.0 atm, the solubility of nitrogen is 6.9×10^{-4} mol L^{-1}. What is the solubility of nitrogen at 20°C when the air pressure is 1.0 atm? The partial pressure of nitrogen in air is 0.79 atm.

Vapor Pressure and Raoult's Law

40. Which has a higher rate of evaporation at 25°C, distilled water or seawater? (Seawater is a solution about 3.5 percent by mass of various salts.) Explain.
41. A 50.0-g sample of ethanol (C_2H_5OH) is mixed with 50.0 g of methanol (CH_3OH) at 20°C. The vapor pressure of pure ethanol at 20°C is 44 torr, and that of methanol is 94 torr. Calculate: **(a)** the mole fraction of ethanol and the mole fraction of methanol in the solution; **(b)** the vapor pressure of ethanol and the vapor pressure of methanol in the solution; **(c)** the vapor pressure of the solution.
42. Explain which of the following 0.100 *m* aqueous solutions has the highest and which has the lowest vapor pressure at 25°C: **(a)** sucrose; **(b)** sodium chloride; **(c)** magnesium chloride.
43. What is the vapor pressure of water over the solution prepared by dissolving 100 g of glucose, $C_6H_{12}O_6$, in 1.00 L of water (the density of water is 1.00 g mL^{-1}) at 25°C? The vapor pressure of water at 25°C is 23.8 torr.
44. The vapor pressure of diethyl ether, $C_2H_5OC_2H_5$, at 10°C is 292 torr. When 4.16 g of solid salicylic acid (a nonvolatile nonelectrolyte) is dissolved in 80.7 g of ether, the vapor pressure is lowered by 8.30 torr. Calculate the molar mass of salicylic acid.

Molality

45. What is the molality of a solution that contains 1.2 mol of a solute per 4.0 L of water? Assume that the density of water is 1.0 g mL^{-1}.
46. What is the molality of a solution that contains 0.320 g of a solute (molar mass 128) in 800 g of carbon tetrachloride?
47. What is the molality of a solution that is prepared by dissolving 25.0 g of sucrose, $C_{12}H_{22}O_{11}$, in 2500 g of water?
48. How many grams of $Al_2(SO_4)_3$ is required to prepare 300 mL of 1.50 *m* solution whose density is 1.40 g mL^{-1}?
49. Explain why the molality of an aqueous solution approaches the molarity when the solution is diluted.

Boiling Point Elevation and Freezing Point Depression

50. Calculate the boiling point and the freezing point of the solution described in problem 47 at 1 atm pressure.

51. Calculate the boiling point and the freezing point of a solution containing 12.5 g of benzoic acid ($C_7H_6O_2$) in 110 g of benzene (C_6H_6). The freezing point of benzene is 5.48°C and its K_f is 4.90°C m^{-1}. The boiling point of benzene is 80.1°C and its K_b is 2.53°C m^{-1}. Benzoic acid is a very weak acid in water but does not ionize in benzene.
52. The radiator of an automobile contains 10 L of water and 1.2 kg of the antifreeze ethylene glycol, $C_2H_4(OH)_2$. Calculate the freezing point and the boiling point of the solution at 1 atm, assuming ideal behavior.
53. A 1.50-g sample of an unknown compound was dissolved in 35.0 g of camphor. The freezing point of the solution was 164.4°C. The freezing point of pure camphor is 178.4°C, and its molal freezing point constant is 40.0°C m^{-1}. Calculate the molar mass of the unknown substance, assuming that it is a nonelectrolyte.
54. A 3.75-g sample of a nonvolatile nonelectrolyte was dissolved in 95.0 g of acetone. The boiling point of the solution was 56.50°C. The boiling point of pure acetone is 55.95°C at 1 atm, and the boiling point constant for acetone is 1.71°C m^{-1}. Calculate the molar mass of the solute.
55. A sample of acetic acid that contains a small amount of water freezes at 15.2°C. Calculate the percent by mass of water in the acetic acid. Pure acetic acid freezes at 16.6°C and its K_f is 3.90°C m^{-1}.

Properties of Electrolyte Solutions

56. Calculate the boiling point and the freezing point of a 0.010 m aqueous solution of each of the following electrolytes, assuming complete dissociation: **(a)** NaCl; **(b)** $MgCl_2$.
57. The experimental freezing point of a 0.100 m solution of NaCl has been found to be −0.348°C. Calculate the apparent percent dissociation of NaCl in this solution.
58. Which of the following 0.10 m aqueous solutions has the lowest freezing point, and which has the highest freezing point? **(a)** sucrose; **(b)** HCl; **(c)** $HC_2H_3O_2$; **(d)** $MgCl_2$. Explain your reasoning.
59. A 0.0100 m acetic acid solution freezes at −0.0194°C. Calculate the percent dissociation of acetic acid in this solution.
60. A 2.563-g sample of a weak monoprotic acid is dissolved in 38.56 g of water. The solution freezes at −1.11°C. A 0.469-g sample of the solution requires 24.36 mL of 0.100 M sodium hydroxide for neutralization. Calculate the molar mass of the acid and its approximate percent dissociation in this solution.

Osmotic Pressure

61. Calculate the osmotic pressure of a 0.100 M glucose solution at 22.0°C.
62. Human blood serum freezes at −0.56°C. Calculate the osmotic pressure of blood at 0.0°C and at 37°C, assuming that 1.0 mL of the blood serum contains 1.0 g of water.
63. The osmotic pressure of horse blood at 10.0°C was found to be 58.75 torr. A 100-mL sample of horse blood contains 5.27 g of hemoglobin. Calculate the molar mass of horse hemoglobin, assuming that it is the only solute in blood.

Colloids

64. What is Tyndall effect?
65. The protein albumin has a molar mass of 6.1×10^4 g mol^{-1}. Calculate the freezing point and osmotic pressure for a 3.0 percent by mass albumin solution in water (density = 1.00 g mL^{-1}) at 25°C, assuming ideal behavior.
66. Explain the action of a soap on an emulsion of oil and water.

Water

67. Write an equation to illustrate what happens: **(a)** when temporary hard water containing calcium hydrogen carbonate is boiled; **(b)** when sodium carbonate is added to permanent hard water containing calcium ions.
68. A sample of municipal water supply was found to contain 0.20 g of Ca^{2+} ions per liter. What mass of sodium carbonate is needed to soften 1000 gallons of this water?

PART 3

Chemical Equilibria and Their Applications

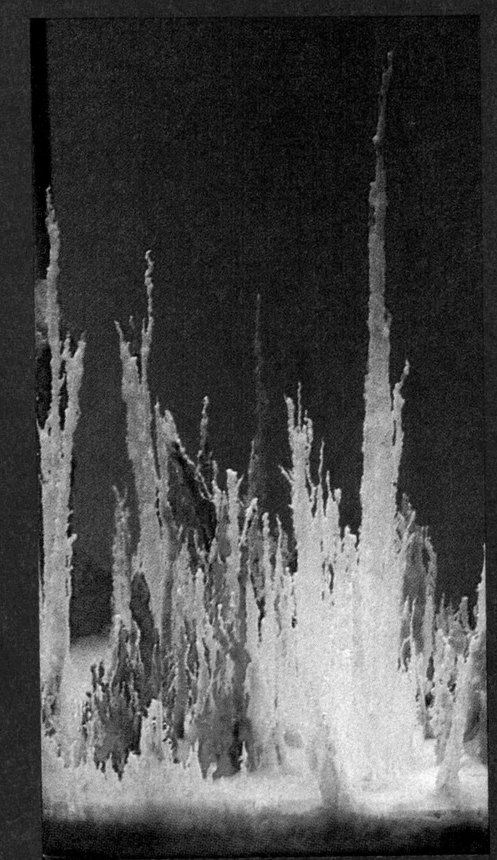

CHAPTER 13

The illustration shows a chemical garden grown by adding salts to a solution of sodium silicate: when the growth of a garden has ceased, the solutions and the solid growths are in dynamic equilibrium.

Chemical Equilibrium

We introduced the idea of dynamic equilibrium in Chapter 11. Let us briefly review the concept of dynamic equilibrium by considering a physical system consisting of a liquid and its vapor in a closed container maintained at constant temperature. A system consisting of liquid water and its vapor is one example.

$$H_2O(l) \rightleftharpoons H_2O(g)$$

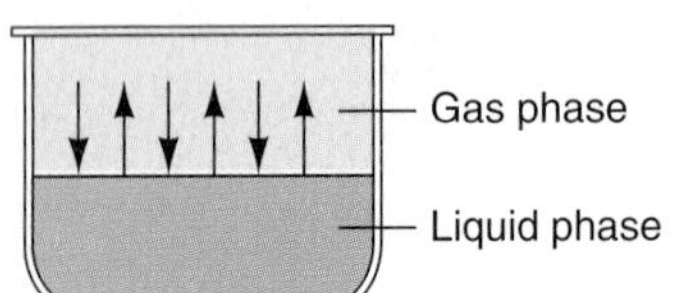

Figure 13.1

Dynamic equilibrium.

When the number of water molecules entering the liquid phase exactly equals the number of water molecules entering the gas phase, there is no net *change in the system, and it is in a state of* dynamic equilibrium *(Figure 13.1). Once equilibrium has been established, the partial pressure of water vapor above the liquid is a constant called the equilibrium vapor pressure. Although there is no net change in the system at dynamic equilibrium, evaporation and condensation occur continuously as water molecules enter and leave the vapor phase. Thus, physical processes or chemical reactions that exist in a state of dynamic equilibrium are* reversible.

A reversible reaction is written with a double arrow. The arrow written from left to right indicates the ***forward reaction****, and the arrow written from right to left indicates the* **reverse reaction**. *The species written to the left of the double arrows are called "reactants," and the species written to the right of the double arrows are "products":*

$$\textit{reactants} \rightleftharpoons \textit{products}$$

13.1 Chemical Equilibrium and Equilibrium Constants

Having considered a physical process in a state of dynamic equilibrium, let us now consider a chemical reaction in a state of dynamic equilibrium. We recall from Chapter 9 that dinitrogen tetroxide (N_2O_4) partially decomposes to form nitrogen dioxide (NO_2); also, nitrogen dioxide partially reacts to form dinitrogen tetroxide in the reverse reaction. This reversible reaction is described by the following equation:

$$N_2O_4(g) \rightleftharpoons 2NO_2(g)$$

A state of dynamic equilibrium can be established starting with $N_2O_4(g)$ or with $NO_2(g)$. If we begin with pure N_2O_4, its concentration decreases to a constant value. During this time, the concentration of NO_2 increases to a constant value. When the concentrations of the two gases are constant, the system is at equilibrium.

Figure 13.2

Brown NO_2 gas in a tube.

NO_2 is a brown gas. The amount of NO_2 in a mixture of NO_2 and N_2O_4 can be determined by measuring the intensity of the brown color (Figure 13.2). N_2O_4 is colorless. When a sample of N_2O_4 is injected into an evacuated container, a brown color soon appears, indicating that NO_2 has been produced. The intensity of the brown color increases with time until it reaches a maximum. At this point NO_2 and N_2O_4 are in equilibrium, and the concentrations of the two gases are constant. (But remember, both the forward and reverse reactions occur continuously, even at equilibrium.) We describe the ratios of the concentrations of N_2O_4 to NO_2 at equilibrium under three different sets of experimental conditions at 383 K to see if any pattern or regularity develops in the values of these ratios.

In the first experiment, 0.100 mol of pure N_2O_4 is injected into a 1.00-L evacuated container, where it partly decomposes until equilibrium is achieved. At this point 0.120 mol NO_2 is present. From the equation for the reaction we know that 2 mol of NO_2 is formed from each mole of N_2O_4 that decomposes. Thus, 0.120 mol of NO_2 is produced by the decomposition of 0.0600 mol of N_2O_4. The number of moles of N_2O_4 remaining at equilibrium equals 0.100 mol − 0.0600 mol = 0.040 mol (Figure 13.3). The mole ratio of NO_2 to N_2O_4 therefore equals 0.120 mol to 0.040 mol, or 3.0:1.0.

In the next experiment we begin with pure NO_2. When 0.100 mol of NO_2 is injected into a 1.00-L evacuated container, it reacts to produce $N_2O_4(g)$. This change is seen as a decrease in the intensity of the brown color of NO_2. At equilibrium, 0.072 mol of NO_2 remains. The number of moles of NO_2 converted to N_2O_4 is 1.00 mol − 0.072 mol; that is, 0.028 mol. Since 2 mol of NO_2 give 1 mol of N_2O_4, the fraction of a mole of N_2O_4 formed from 0.028 mol of NO_2 is

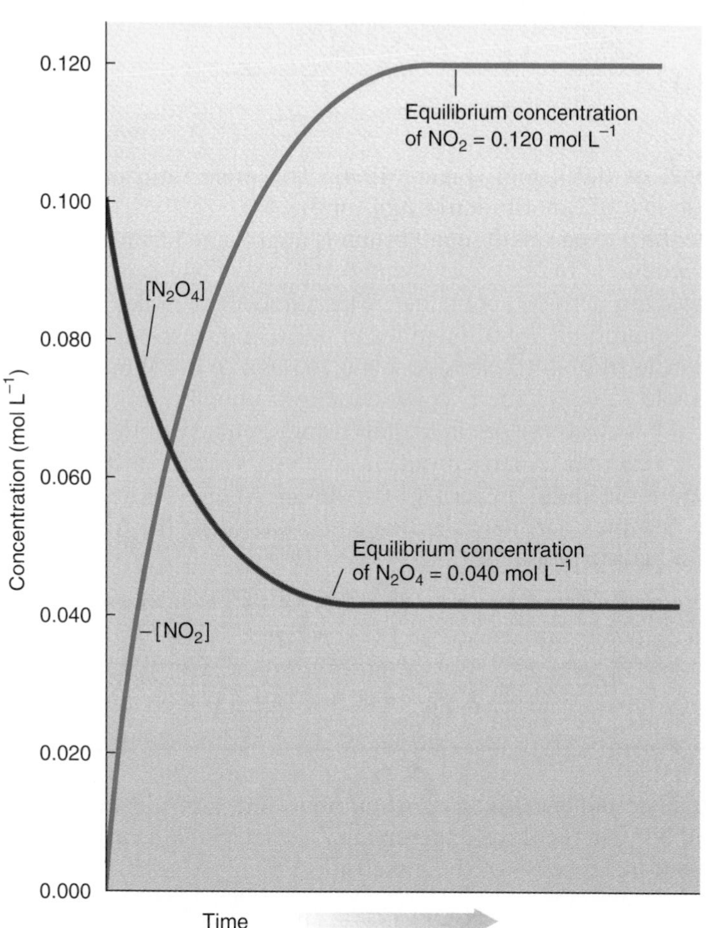

Figure 13.3

Equilibrium established between N_2O_4 and NO_2 at 383 K, as a result of injecting 0.100 mol of N_2O_4 into a 1.00-L evacuated container.

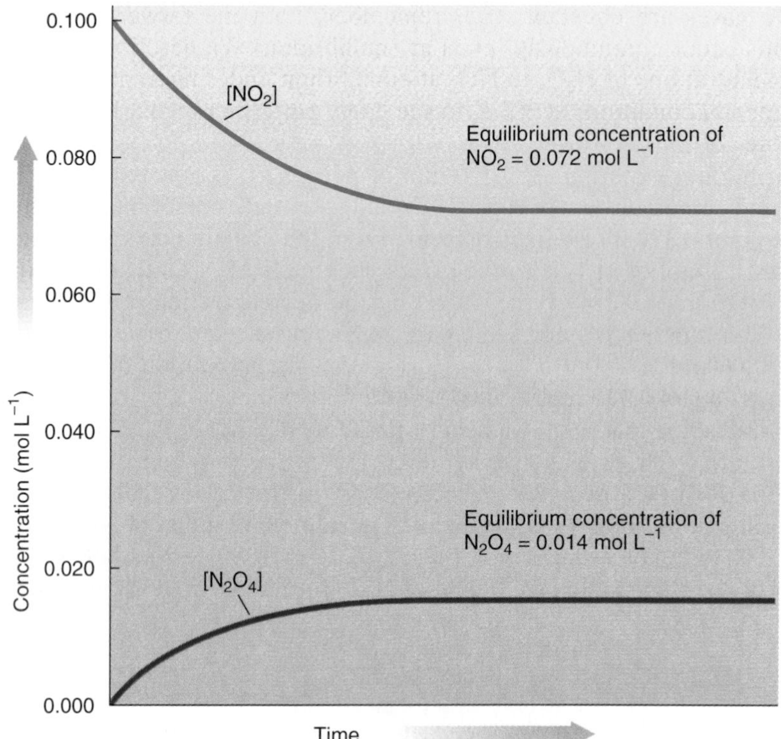

Figure 13.4

Equilibrium concentration of NO_2 and N_2O_4 resulting from injecting 0.100 mol of NO_2 into a 1.00-L evacuated container.

0.028 mol/2, or 0.014 mol (Figure 13.4). The mole ratio of NO_2 to N_2O_4 at equilibrium is 0.072 mol to 0.014 mol, or 5.1:1.0.

In the third experiment, equilibrium is approached from a mixture of reactants and products. In this experiment, 0.100 mol of NO_2 and 0.100 mol of N_2O_4 are injected into a 1.00-L container. The numbers of moles of NO_2 and N_2O_4 present at equilibrium are 0.160 mol and 0.070 mol, respectively (Figure 13.5). The mole ratio of NO_2 to N_2O_4 equals 0.160 mol to 0.070 mol, or 2.3:1.0.

The mole ratios of the two substances at equilibrium do not seem to be related in any way, as we see in the following summary of the results described above. The reaction is carried out in a 1.00-L vessel, so it is convenient to express the mole ratios in terms of moles per liter. The concentration of a substance in moles per liter is indicated by enclosing the formula of the substance with square brackets.

Experiment	**1**	**2**	**3**
$\frac{[NO_2]}{[N_2O_4]}$	$\frac{0.120}{0.040} = 3.0$	$\frac{0.072}{0.014} = 5.1$	$\frac{0.160}{0.070} = 2.3$

There is certainly nothing *constant* about the ratios of the concentrations of NO_2 and N_2O_4 in the three experiments. Let us recall the original equation to see if we can make sense of the mole ratios:

$$N_2O_4(g) \rightleftharpoons 2NO_2(g)$$

Perhaps we can use the coefficients of the balanced equation in a different way. Let us *square* the concentration of NO_2 (2 is the coefficient of NO_2 in the balanced equation) and then divide the squares of the concentration of NO_2 by

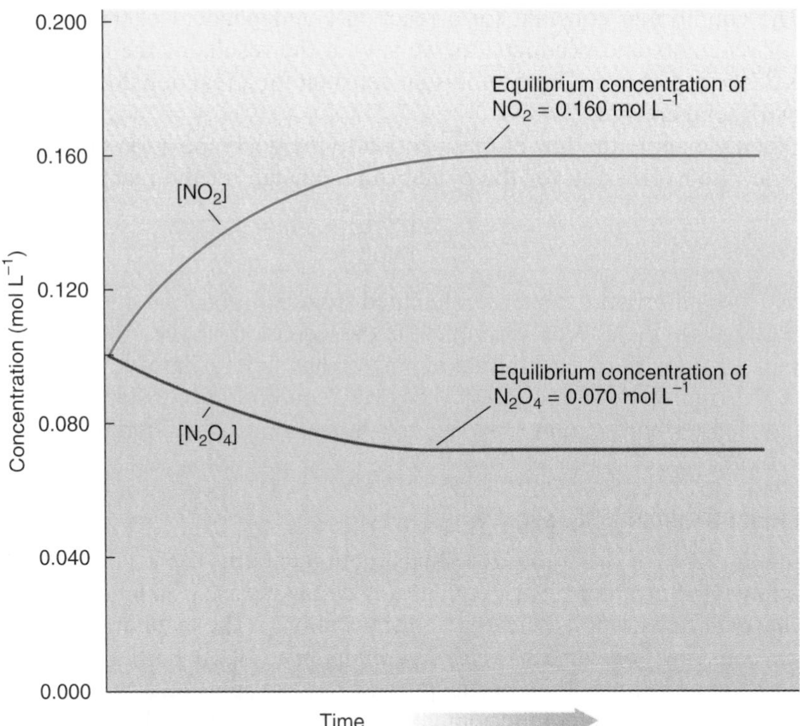

Figure 13.5

Equilibrium concentration of NO_2 and N_2O_4 resulting from injecting 0.100 mol of NO_2 and 0.100 mol of N_2O_4 into a 1.00-L evacuated container.

the concentration of N_2O_4 raised to the power of 1 (1 is the coefficient of N_2O_4 in the balanced equation). When we do this, we find that the ratio $[NO_2]^2/[N_2O_4]$ is nearly identical in all three experiments:

Experiment	1	2	3
$\frac{[NO_2]^2}{[N_2O_4]}$	$\frac{(0.120)^2}{0.040} = 0.36$	$\frac{(0.072)^2}{0.014} = 0.37$	$\frac{(0.160)^2}{0.070} = 0.37$

Many studies of the type described above were carried out by chemists in the nineteenth century. In 1864 the Norwegian chemists Cato Maximilian Guldberg and Peter Waage discovered that at a given temperature *the product of the molar concentrations of the products of a chemical reaction, divided by the product of the molar concentrations of the reactants, is a constant when the concentration of each substance is raised to an exponent equal to its coefficient in the balanced equation*. This generalization is known as the **law of mass action** or the **law of chemical equilibrium**.

Let us apply the law of mass action to the general reaction

$$aA + bB \rightleftharpoons cC + dD$$

At equilibrium, the law of mass action states that the following ratio is a constant:

$$\frac{[C]^c[D]^d}{[A]^a[B]^b} = K \qquad (13.1)$$

where the constant, K, is called the **equilibrium constant**.

The equilibrium constant for a reaction is independent of the concentrations of reactants and products, as we saw in the results of the three experiments described above. The equilibrium constant for a reaction does, however, vary with temperature.

When we apply the law of mass action to the decomposition of dinitrogen tetroxide, the expression for the equilibrium constant of the reaction is

$$K = \frac{[NO_2]^2}{[N_2O_4]}$$

Many equilibrium constants calculated from experimental data have units associated with them. For example, in the equation above, the numerator has units of (mol $L^{-1})^2$ and the denominator has units of mol L^{-1}. The ratio (mol $L^{-1})^2$/(mol L^{-1}) = mol L^{-1}. However, equilibrium constants are often given as dimensionless numbers. We use dimensionless equilibrium constants in this book.

Equilibrium Constants: K_c and K_p

In Equation 13.1 we wrote the law of mass action in terms of the concentrations of reactants and products. Such equilibrium constants are usually expressed as K_c, where the subscript c indicates "concentration." The equilibrium constant for a gaseous reaction can also be expressed in terms of the partial pressures of the components because their partial pressures are proportional to their respective concentrations. An equilibrium constant that is expressed in terms of the partial pressures of the reactants and products is denoted by K_p. For the general reaction

$$a\text{A}(g) + b\text{B}(g) \rightleftharpoons c\text{C}(g) + d\text{D}(g)$$

the equilibrium constant, K_p, is given by

$$K_p = \frac{(P_C)^c(P_D)^d}{(P_A)^a(P_B)^b} \tag{13.2}$$

When the number of moles of gaseous reactants equals the number of moles of gaseous products, K_c equals K_p; otherwise, K_c and K_p are not equal. The relationship between K_c and K_p is developed further in Section 13.5.

Example 13.1

Writing Expressions for Equilibrium Constants from Balanced Equations

The following equations represent various reactions at equilibrium. Write the expression for the equilibrium constant, K_c, for each of these reactions:

(a) $N_2(g) + 3H_2(g) \rightleftharpoons 2NH_3(g)$

(b) $4NH_3(g) + 5O_2(g) \rightleftharpoons 4NO(g) + 6H_2O(g)$

SOLUTION: The coefficients in the balanced equations are used as exponents in the equilibrium constant expressions. The products are placed in the numerator and the reactants in the denominator:

(a) $K_c = \dfrac{[NH_3]^2}{[N_2][H_2]^3}$

(b) $K_c = \dfrac{[NO]^4[H_2O]^6}{[NH_3]^4[O_2]^5}$

Practice Problem 13.1: Write an expression of K_c for each of the following reactions: **(a)** $O_2(g) \rightleftharpoons O_3(g)$; **(b)** $CO(g) + O_2(g) \rightleftharpoons CO_2(g)$. Don't forget to balance the equations.

Example 13.2

Calculating K_c from Equilibrium Concentrations

Gaseous HI decomposes at high temperatures according to the equation

$$2HI(g) \rightleftharpoons H_2(g) + I_2(g)$$

At a temperature of 321°C, HI partially decomposes to produce an equilibrium mixture of HI, H_2, and I_2. Starting with HI, the equilibrium concentrations of HI and I_2 are 1.68 mol L^{-1} and 0.20 mol L^{-1}, respectively. What is the value of K_c?

SOLUTION: The balanced equation for the reaction tells us that 2 mol of HI produces 1 mol of H_2 and 1 mol of I_2. Since only HI is present initially, equimolar amounts of H_2 and I_2 are present at equilibrium. The equilibrium concentration of I_2 is 0.20 mol L^{-1}; therefore, the equilibrium concentration of H_2 is also 0.20 mol L^{-1}. The value of K_c is

$$K_c = \frac{[H_2][I_2]}{[HI]^2} = \frac{(0.20)(0.20)}{(1.68)^2} = 1.4 \times 10^{-2}$$

Practice Problem 13.2: Hydrogen sulfide decomposes to its elements when heated at 1405 K:

$$2H_2S(g) \rightleftharpoons 2H_2(g) + S_2(g)$$

(At higher temperatures, S_8 molecules dissociate into S_2 molecules.)
Some H_2S was placed into a 1.00-L container and heated to 1405 K. At equilibrium, the reaction mixture contained 0.00285 mol of H_2, 0.00715 mol of H_2S, and some S_2. Calculate K_c for this reaction.

Example 13.3

Calculating K_c from Equilibrium Concentrations

At a certain elevated temperature, 4.00 mol of PCl_5 was placed into a 2.00-L flask. At equilibrium, 3.60 mol of PCl_5 remained. Calculate the equilibrium constant, K_c, for the reaction

$$PCl_5(g) \rightleftharpoons PCl_3(g) + Cl_2(g)$$

SOLUTION: To calculate K_c, all concentrations must be expressed in moles per liter. The initial concentration of PCl_5 is 4.00 mol/2.00 L, or 2.00 mol L^{-1}. The equilibrium concentration of PCl_5 is 3.60 mol/2.00 L, or 1.80 mol L^{-1}. From the balanced equation for the reaction we see that the decomposition of 1 mol of PCl_5 produces 1 mol of PCl_3 and 1 mol of Cl_2. The number of moles per liter of PCl_5 that decomposes equals the difference between the initial concentration of PCl_5 and its equilibrium concentration: 2.00 mol L^{-1} − 1.80 mol L^{-1}, or 0.20 mol L^{-1}. Thus, 0.20 mol L^{-1} of PCl_3 and 0.20 mol L^{-1} of Cl_2 are produced. These data are summarized below:

	$PCl_5(g) \rightleftharpoons$	$PCl_3(g)$ +	$Cl_2(g)$
Initial (mol L^{-1})	2.00	0	0
Equilibrium (mol L^{-1})	1.80	0.20	0.20

The equilibrium constant is

$$K_c = \frac{[PCl_3][Cl_2]}{[PCl]^5} = \frac{(0.20)(0.20)}{1.8} = 0.022$$

Sometimes we know the equilibrium constant for a reaction written in one direction, but wish to know the equilibrium constant for the reverse reaction. If K_{c1} is the equilibrium constant for a reaction written in one direction, the equilibrium constant for the reverse reaction equals the reciprocal of K_{c1}, or $1/K_{c1}$.

For example, the expression for the equilibrium constant, K_{c1}, for the decomposition of dinitrogen tetroxide, $N_2O_4(g) \rightleftharpoons 2NO_2(g)$, is

$$K_{c1} = \frac{[NO_2]^2}{[N_2O_4]}$$

The expression for the equilibrium constant for the reverse reaction, K_{c2}, in which nitrogen dioxide converts to dinitrogen tetroxide, $2NO_2(g) \rightleftharpoons N_2O_4(g)$, is

$$K_{c2} = \frac{[N_2O_4]}{[NO_2]^2} = \frac{1}{K_{c1}}$$

Multiple Equilibria

Sometimes we encounter reactions in which the product of one reaction is the reactant for another reaction. When we add the equations for the two reactions, we get the equation for the overall or net reaction. The equilibrium constant for the net reaction, $K_{c(\text{net})}$, is the product of the equilibrium constants K_{c1} and K_{c2} for the two reactions: $K_{c(\text{net})} = K_{c1}K_{c2}$. For example, nitrogen gas reacts with oxygen gas to give nitrogen monoxide, NO, which then reacts with oxygen to give nitrogen dioxide, NO_2. Thus, the product of the first reaction, NO, is the reactant in the second reaction. The following equations and their equilibrium constants summarize the process:

Reaction 1: $N_2(g) + O_2(g) \rightleftharpoons 2NO(g) \qquad K_{c1} = \frac{[NO]^2}{[N_2][O_2]}$

Reaction 2: $2NO(g) + O_2(g) \rightleftharpoons 2NO_2(g) \qquad K_{c2} = \frac{[NO_2]^2}{[NO]^2[O_2]}$

Net: $N_2(g) + 2O_2(g) \rightleftharpoons 2NO_2(g)$

Brown smog over Los Angeles, California. The brown color of this smog is caused by $NO_2(g)$.

The equilibrium constant for the reaction that is the sum of the reactions 1 and 2, $K_{c(\text{net})}$, is the product of K_{c1} and K_{c2}:

$$K_{c1}K_{c2} = \frac{[NO]^2}{[N_2][O_2]} \times \frac{[NO_2]^2}{[NO]^2[O_2]} = \frac{[NO_2]^2}{[N_2][O_2]^2} = K_{c(\text{net})}$$

The above reactions have important consequences for the environment. At the temperature of the combustion chambers of automobile internal combustion engines, nitrogen reacts with oxygen to give some nitrogen monoxide, which is emitted into the atmosphere, where it reacts further with O_2 to give nitrogen dioxide. We recall that NO_2 is a brown gas. The haze of brown "smog" which hovers above the world's major cities is partly the result of this sequence of reactions.

Example 13.4

Calculating K_c for a Reverse Reaction

The equilibrium constant, K_c, for the reaction

$$2SO_2(g) + O_2(g) \rightleftharpoons 2SO_3(g) \quad (1)$$

at 1000 K is 2.8×10^2. Calculate K_c for the reaction

$$2SO_3(g) \rightleftharpoons 2SO_2(g) + O_2(g) \quad (2)$$

SOLUTION: We let K_{c1} be the equilibrium constant for reaction 1. Reaction 2 is the reverse of reaction 1. Therefore, the equilibrium constant, K_{c2}, for reaction 2 is

$$K_{c2} = \frac{1}{K_{c1}} = \frac{1}{2.8 \times 10^2} = 3.6 \times 10^{-3}$$

Example 13.5

Calculating K_c for a Reaction That Is the Sum of Two Reactions

Given the following reactions and their equilibrium constants at 25°C:

$$N_2(g) + O_2(g) \rightleftharpoons 2NO(g) \qquad K_{c1} = 4.1 \times 10^{-31} \quad (1)$$

$$N_2O(g) + \tfrac{1}{2}O_2(g) \rightleftharpoons 2NO(g) \qquad K_{c2} = 1.7 \times 10^{-13} \quad (2)$$

Calculate K_c for the reaction

$$N_2(g) + \tfrac{1}{2}O_2(g) \rightleftharpoons N_2O(g) \quad \text{at } 25°C \quad (3)$$

SOLUTION: To solve this problem we must combine the equations for reactions 1 and 2 to obtain the equation for reaction 3. We see that the reactants in reaction 3 are also the reactants in reaction 1, and that the product of reaction 3 is a *reactant* in reaction 2. Thus we must add reaction 1 and the *reverse* of reaction 2:

$$\begin{array}{ll} N_2(g) + O_2(g) \rightleftharpoons 2NO(g) & \text{(reaction 1)} \\ 2NO(g) \rightleftharpoons N_2O(g) + \tfrac{1}{2}O_2(g) & \text{(reverse of reaction 2)} \\ \hline N_2(g) + \tfrac{1}{2}O_2(g) \rightleftharpoons N_2O(g) & \text{(reaction 3)} \end{array}$$

$K_{c2(rev)}$ for the reverse of reaction 2 is $1/(1.7 \times 10^{-13}) = 5.9 \times 10^{12}$. The equilibrium constant for reaction 3, $K_{c(net)}$, is the product of the equilib rium constants for reaction 1 and the reverse of reaction 2:

$$K_{c(net)} = (4.1 \times 10^{-31})(5.9 \times 10^{12}) = 2.4 \times 10^{-18}$$

13.2 Equilibrium Calculations

Equilibrium Constant and the Position of Equilibrium

The numerical value of an equilibrium constant for a reaction is a measure of how far the forward reaction has progressed when the system has reached equilibrium. If more than 50 percent of the reactants of a reaction have been converted to products when the system has reached equilibrium, the equilibrium is said to lie *to the right*. On the other hand, if less than 50 percent of the reactants in another reaction have been converted to products when the system has reached equilibrium, the equilibrium lies *to the left*. The equilibrium constant for a reaction whose equilibrium lies to the right is greater than the equilibrium constant for another reaction whose equilibrium lies to the left. Thus, the equilibrium constant is a measure of the **position** of an equilibrium or the **point** of an equilibrium.

Let us consider a reaction $A(g) + B(g) \rightleftharpoons C(g) + D(g)$. Suppose that 1.00 mol of A and 1.00 mol B are introduced into a 1.00-L container, and that 0.50 mol of A reacts with 0.50 mol of B to form 0.50 mol of C and 0.50 mol of D. The concentrations of A and B at equilibrium are 1.00 mol L^{-1} − 0.50 mol L^{-1} = 0.50 mol L^{-1}, as follows:

	$A(g)$	+	$B(g)$	$\rightleftharpoons$	$C(g)$	+	$D(g)$
Initial (mol L^{-1})	1.00		1.00		0		0
Equilibrium (mol L^{-1})	1.00 − 0.50		1.00 − 0.50		0.50		0.50

The equilibrium constant for this reaction is

$$K_c = \frac{[C][D]}{[A][B]} = \frac{(0.50)(0.50)}{(0.50)(0.50)} = 1.0$$

In this particular equilibrium, the value of the equilibrium constant indicates that the forward reaction has gone 50 percent to completion. In Example 13.6 we calculate the value of the equilibrium constant for another system in which the forward reaction has gone only 10 percent to completion.

Example 13.6

Calculating the Equilibrium Constant for a System in Which the Forward Reaction Is 10 Percent Complete

Calculate the equilibrium constant, K_c, for the reaction $E(g) + F(g) \rightleftharpoons G(g) + H(g)$ in which the forward reaction is 10 percent complete at equilibrium. Assume that the initial concentrations of E and F are equal.

SOLUTION: Any initial concentration of E and F may be assumed. Let us choose an initial concentration of 1.00 mol L^{-1} for E and 1.00 mol L^{-1} for F. When the forward reaction is 10 percent complete, 0.10 mol (10 percent of 1.00) of E has reacted with 0.10 mol of F to form 0.10 mol of G and 0.10 mol of H. The concentrations of E and F at equilibrium are 1.00 mol L^{-1} − 0.10 mol L^{-1} = 0.90 mol L^{-1}. These data are summarized below.

	E(g)	+ F(g)	⇌ G(g)	+ H(g)
Initial (mol L^{-1})	1.00	1.00	0	0
Equilibrium (mol L^{-1})	1.00 − 0.10	1.00 − 0.10	0.10	0.10

The value of K_c is

$$K_c = \frac{[G][H]}{[E][F]} = \frac{(0.10)(0.10)}{(0.90)(0.90)} = 1.2 \times 10^{-2}$$

In this case the equilibrium position is far to the left, and the equilibrium constant is much less than 1.0.

Example 13.7 also shows how the equilibrium constant is related to the extent of completion of the forward reaction.

Example 13.7

Calculating K_c for a System in Which the Forward Reaction Is 80 Percent Complete

Calculate K_c for the reaction I(*g*) + J(*g*) ⇌ K(*g*) + L(*g*) at equilibrium, assuming that the initial concentration of each reactant is 1.00 mol L^{-1} and that 80 percent of the reactants have converted to products at equilibrium.

SOLUTION: From the balanced equation for the reaction we can see that when 0.80 mol L^{-1} (80 percent of 1.00 mol L^{-1}) of I reacts with 0.80 mol L^{-1} of J, 0.80 mol L^{-1} of K, and 0.80 mol L^{-1} of L are produced at equilibrium. The concentrations of I and J that remain at equilibrium are 1.00 mol L^{-1} − 0.80 mol L^{-1}, that is, 0.20 mol L^{-1}. The initial and equilibrium concentrations of the reactants and products are summarized below.

	I(g)	+ J(g)	⇌ K(g)	+ L(g)
Initial (mol L^{-1})	1.00	1.00	0	0
Equilibrium (mol L^{-1})	1.00 − 0.80	1.00 − 0.80	0.80	0.80

The equilibrium constant, K_c, is

$$K_c = \frac{[K][L]}{[I][J]} = \frac{(0.80)(0.80)}{(0.20)(0.20)} = 16$$

The value of the equilibrium constant for this reaction shows that when more than 50 percent of the reactants have converted to products, the equilibrium constant is much greater than 1.0.

In equilibria with coefficients other than 1, and in which the number of moles of reactants are not equal, the K_c values differ somewhat from the values calculated above at different degrees of completion.

Equilibrium Calculations

Sometimes we wish to know the concentrations of one or all the species present in an equilibrium mixture from the initial concentrations of reactants and the equilibrium constant for the reaction. The following examples illustrate such calculations.

Example 13.8

Calculating the Concentration of a Component of an Equilibrium Mixture

The equilibrium constant for the reaction

$$N_2(g) + O_2(g) \rightleftharpoons 2NO(g)$$

is 4.1×10^{-4} at 2000°C. If the equilibrium concentrations of $O_2(g)$ and $NO(g)$ are 1.2 and 0.011 mol L^{-1}, respectively, what is the equilibrium concentration of $N_2(g)$?

SOLUTION: First, we set up the equilibrium expression for K_c:

$$K_c = \frac{[NO]^2}{[N_2][O_2]}$$

Next, we substitute the value of K_c and the concentrations of $NO(g)$ and $O_2(g)$ into the equilibrium expression.

$$4.1 \times 10^{-4} = \frac{(0.011)^2}{[N_2](1.2)}$$

Finally, we solve for $[N_2]$:

$$[N_2] = \frac{(0.011)^2}{(4.1 \times 10^{-4})(1.2)} = 0.25 \text{ mol L}^{-1}$$

Practice Problem 13.3: At a certain temperature, the equilibrium constant, K_c, for the reaction

$$2HI(g) \rightleftharpoons H_2(g) + I_2(g)$$

is 5.2×10^{-3}. The equilibrium concentrations of $H_2(g)$ and $I_2(g)$ are 1.3×10^{-3} mol L^{-1} and 2.5×10^{-2} mol L^{-1}, respectively. What is the equilibrium concentration of $HI(g)$?

Example 13.9

Calculating the Concentrations of More Than One Component in an Equilibrium Mixture

The equilibrium constant, K_c, for the reaction

$$H_2(g) + I_2(g) \rightleftharpoons 2HI(g)$$

at 793 K is 1.6×10^{-2}. If 1.00 mol of $H_2(g)$ and 1.00 mol of $I_2(g)$ are introduced into a 10.0-L container at 793 K and allowed to come to equilibrium, what concentrations of $H_2(g)$, of $I_2(g)$, and of $HI(g)$ are present?

SOLUTION: In the expression for K_c, all concentrations must be expressed in moles per liter. Since 1.00 mol of each reactant is placed in a 10.0-L container, the initial concentrations of H_2 and I_2 are

$$\frac{1.00 \text{ mol}}{10.0 \text{ L}} = 0.100 \text{ mol L}^{-1}$$

We let x equal the number of moles per liter of H_2 that reacts. We know from the balanced chemical equation that x moles of $H_2(g)$ react with x moles of $I_2(g)$, and that $2x$ moles of HI are produced. The initial and equilibrium concentrations of each reactant and product in this reaction are as follows:

	$H_2(g)$	+	$I_2(g)$	$\rightleftharpoons$	$2HI(g)$
Initial (mol L^{-1})	0.100		0.100		0
Equilibrium (mol L^{-1})	$0.100 - x$		$0.100 - x$		$2x$

The expression for K_c is

$$K_c = \frac{[HI]^2}{[H_2][I_2]}$$

Substituting the data above into this expression yields

$$1.6 \times 10^{-2} = \frac{(2x)^2}{(0.100 - x)^2}$$

This equation, which is a perfect square, is conveniently solved by taking the square root of both sides of the equation to obtain

$$\frac{2x}{0.100 - x} = \sqrt{1.6 \times 10^{-2}} = 0.13$$

Solving for x gives

$$2x = (0.100 - x)(0.13) = 0.013 - 0.13x$$

$$2x + 0.13x = 0.013$$

$$x = \frac{0.013}{2.13} = 0.0061 = [H_2]_{\text{reacted}} = [I_2]_{\text{reacted}}$$

The concentrations of reactants and products present at equilibrium are

$$[H_2] = 0.100 \text{ mol L}^{-1} - 0.0061 \text{ mol L}^{-1} = 0.094 \text{ mol L}^{-1}$$

$$[I_2] = 0.100 \text{ mol L}^{-1} - 0.0061 \text{ mol L}^{-1} = 0.094 \text{ mol L}^{-1}$$

$$[HI] = (2)(0.0061 \text{ mol L}^{-1}) = 0.012 \text{ mol L}^{-1}$$

We can easily check to see if our answers are reasonable. We previously stated that

$$K_c = \frac{[HI]^2}{[H_2][I_2]}$$

If we substitute the values we obtained above for the concentrations of reactants and products into the equilibrium expression, we obtain

$$K_c = \frac{(0.012)^2}{(0.094)(0.094)} = 1.6 \times 10^{-2}$$

Since $K_{c(\text{calculated})} \approx K_{c(\text{measured})}$, our answers are reasonable.

Practice Problem 13.4: The value of K_c for the reaction

$$H_2(g) + Br_2(g) \rightleftharpoons 2HBr(g)$$

at 1768 K is 3.50×10^4. If 1.40 mol of $H_2(g)$ and 1.40 mol of $Br_2(g)$ are placed into a 2.00-L container, what are the molar concentrations of H_2, Br_2, and HBr at equilibrium?

If the initial concentrations of the reactants are not equal, the problem of calculating the equilibrium composition is more complicated, as shown by Example 13.10.

Example 13.10

Calculating the Composition of an Equilibrium Mixture Starting with Unequal Concentrations of the Reactants

Assume that the initial concentrations of H_2 and I_2 in Example 13.9 are 0.200 mol L^{-1} and 0.100 mol L^{-1}, respectively. Calculate the equilibrium concentrations of H_2, I_2, and HI at 793 K.

SOLUTION: Letting x equal the number of moles per liter of H_2 and of I_2 that react, the concentrations of each of the species present initially and at equilibrium are:

	$H_2(g)$	$+ I_2(g)$	$\rightleftharpoons 2HI(g)$
Initial (mol L^{-1})	0.200	0.100	0
Equilibrium (mol L^{-1})	$0.200 - x$	$0.100 - x$	$2x$

Substituting the equilibrium concentrations into the equilibrium expression for the reaction, we obtain

$$1.6 \times 10^{-2} = \frac{(2x)^2}{(0.200 - x)(0.100 - x)}$$

Rearranging this equation, we have

$$1.6 \times 10^{-2} = \frac{4x^2}{0.0200 - 0.200x - 0.100x + x^2}$$

We can rearrange the equation above as follows:

$$4x^2 = 1.6 \times 10^{-2}(0.0200 - 0.300x + x^2)$$

from which

$$4x^2 - 1.6 \times 10^{-2}x^2 + 4.8 \times 10^{-3}x - 3.2 \times 10^{-4} = 0$$

Since 1.6×10^{-2} is negligibly small compared to 4, the equation reduces to

$$4x^2 + 4.8 \times 10^{-3}x - 3.2 \times 10^{-4} = 0$$

This equation is in the form of the general quadratic equation

$$ax^2 + bx + c = 0$$

which can be solved for x by using the quadratic formula

$$x = \frac{-b \pm \sqrt{b^2 - 4ac}}{2a}$$

Substituting the numerical data into this equation yields

$$x = \frac{-4.8 \times 10^{-3} \pm \sqrt{(4.8 \times 10^{-3})^2 - (4)(4)(-3.2 \times 10^{-4})}}{(2)(4)}$$

$$= \frac{4.8 \times 10^{-3} \pm \sqrt{(2.3 \times 10^{-5} + 5.1 \times 10^{-3}}}{8}$$

Note that according to the rules of significant figures, the number 2.3×10^{-5} under the square root sign is very small compared to 3.1×10^{-3} and can therefore be neglected. The equation now simplifies to

$$x = \frac{-4.8 \times 10^{-3} \pm \sqrt{5.1 \times 10^{-3}}}{8}$$

When we solve the quadratic equation, we obtain two values of x: 8.3×10^{-3} and -9.5×10^{-3}. All quadratic equations yield two solutions, but in this case only the positive value has physical meaning—the concentrations of $H_2(g)$ and $I_2(g)$, which reacted to give $HI(g)$.

The number of moles per liter of $H_2(g)$, $I_2(g)$, and $HI(g)$ at equilibrium are

$$[H_2] = 0.200 \text{ mol L}^{-1} - 0.0083 \text{ mol L}^{-1} = 0.192 \text{ mol L}^{-1}$$

$$[I_2] = 0.100 \text{ mol L}^{-1} - 0.0083 \text{ mol L}^{-1} = 0.092 \text{ mol L}^{-1}$$

$$[HI] = (2)(0.0083 \text{ mol L}^{-1}) = 0.017 \text{ mol L}^{-1}$$

Practice Problem 13.5: The value of K_c for the equilibrium

$$N_2O_4(g) \rightleftharpoons 2NO_2(g)$$

at 373 K is 0.36. If 0.100 mol of N_2O_4 is placed into a 1.00-L flask at 373 K, what are the equilibrium concentrations of N_2O_4 and of NO_2?

Predicting the Direction of a Reaction

The equilibrium constant for a reaction can be used to predict its "direction" when known concentrations of reactants and products are mixed. The reaction proceeds to the right if some of the reactants are consumed to form products. The reaction proceeds to the left if some of the products are converted to reactants. To predict the direction of the reaction, we substitute the known concentrations into a ratio that has the form of the equilibrium constant for the reaction. This ratio is known as the **reaction quotient**.

If the reaction quotient is larger than the known value of the equilibrium constant, the reaction proceeds to the left; that is, products convert to reactants until equilibrium is reached. At this point, the reaction quotient equals the equilibrium constant. If the reaction quotient is smaller than the equilibrium constant, the reaction proceeds to the right until equilibrium is established and the reaction quotient once again equals the equilibrium constant.

Example 13.11

Predicting the Direction of a Reaction

The value of K_c for the reaction

$$N_2(g) + O_2(g) \rightleftharpoons 2NO(g)$$

at 2000°C is 0.100. Predict the direction of the reaction when 3.00 mol of N_2 is mixed with 2.00 mol of O_2 and 1.00 mol of NO in a 1.00-L container at 2000°C.

SOLUTION: This problem can be solved by determining the value of the reaction quotient and comparing its value to that of the equilibrium constant.

The expression for the equilibrium constant for the reaction in this problem is

$$K_c = \frac{[NO]^2}{[N_2][O_2]} = 0.100$$

Substituting the given molar concentrations into an analogous expression gives the reaction quotient, Q:

$$Q = \frac{(1.00)^2}{(3.00)(2.00)} = 0.167$$

Since 0.167 is larger than the equilibrium constant, 0.100, the reaction proceeds from right to left. As a result, the concentration of NO decreases and the concentrations of N_2 and O_2 increase until equilibrium is reached.

APPLICATIONS OF CHEMISTRY 13.1
Fish That Change Sex

In some animal species the male/female ratio remains fairly constant. In fact, some animals change their sex in response to environmental stresses. For example, Atlantic silverside fish change sex as the temperature changes; others, like the female African Reed frog and the *Anthias squamipinnis* fish, change sex if there is a decrease in the male population (see photo at right).

These population changes illustrate equilibrium. Let's assume that we have 10 males and 10 females of a given species. Also assume that we must maintain this 1:1 ratio of males to females for maximum reproductive success. If this ratio changes, females can change to males, or males can change to females. Thus there is an equilibrium between the males and females:

$$\overset{10}{\mathrm{F}} \rightleftharpoons \overset{10}{\mathrm{M}}$$

The equilibrium constant, K, for this equilibrium is

$$K = \frac{[\mathrm{M}]}{[\mathrm{F}]} = \frac{10}{10} = 1$$

Now let's remove six males, leaving four behind. We can express this new situation as follows:

$$\overset{10}{\mathrm{F}} \rightleftharpoons \overset{4}{\mathrm{M}}$$

Substituting these new numbers of males and females into the reaction quotient gives us 4/10 = 0.4. Since this reaction quotient is smaller than the equilibrium constant, 1, the reaction proceeds from left to right. As a result, the concentration of females decreases and the concentration of males increases.

Three of the female fish must change into male fish to decrease the female population by three. This will also increase the male population by three.

We wind up with seven males and seven females, and the equilibrium constant K again has a value of 1:

$$K = \frac{[\mathrm{M}]}{[\mathrm{F}]} = \frac{7}{7} = 1$$

Source: *Science News*, March 3, 1990, page 134.

Practice Problem 13.6: The value for K_c for the reaction $2SO_2(g) + O_2(g) \rightleftharpoons 2SO_3(g)$ at 1000 K is 2.4×10^2. Calculate the reaction quotient when $[SO_2] = 4.00 \times 10^{-3}$ mol L^{-1}, $[O_2] = 3.00 \times 10^{-2}$ mol L^{-1}, and $[SO_3] = 1.00 \times 10^{-3}$ mol L^{-1}. Is the system at equilibrium under these conditions? If not, does the reaction proceed to the left or to the right?

13.3 Heterogeneous Equilibria

Thus far we have considered only equilibria in which all the components are in the same phase. These are called **homogeneous equilibria**. Equilibrium systems that consist of two or more phases are **heterogeneous equilibria**. An example of a heterogeneous reaction is the decomposition of calcium carbonate (limestone) to calcium oxide (lime) and carbon dioxide:

$$CaCO_3(s) \longrightarrow CaO(s) + CO_2(g)$$

When this reaction occurs in a closed container at constant temperature, an equilibrium is established between gaseous carbon dioxide, solid calcium carbonate, and calcium oxide:

$$CaCO_3(s) \rightleftharpoons CaO(s) + CO_2(g)$$

For this reaction, the following ratio is a constant which we represent by K':

$$K' = \frac{[CaO(s)][CO_2(g)]}{[CaCO_3(s)]}$$

We can determine the concentration of a gas, or its partial pressure. But what is the meaning of the "concentration of a solid"? The concentration of a solid is related to its density. A given mass of solid—and therefore a given number of moles of solid—occupies a certain volume. When the solid undergoes a chemical reaction, its mass and volume decrease. But the number of moles per unit volume of a solid is constant as long as any solid is present. Therefore, the "concentration" of a solid is a constant. Since $[CaCO_3]$ and $[CaO]$ are constants, the ratio $[CaO]/[CaCO_3]$ is also a constant, K''. Hence, we can rewrite the equilibrium expression for the reaction as

$$K' = \frac{[CaO]}{[CaCO_3]} \times [CO_2] = K''[CO_2]$$

from which

$$K'/K'' = [CO_2] = K_c$$

Thus, $K_c = [CO_2]$ and $K_p = P_{CO_2}$.

The concentration and partial pressure of carbon dioxide at equilibrium with solid calcium carbonate and calcium oxide at a given temperature does not depend on the concentrations of the two solids (as long as some of each is present).

At 900°C, the pressure of CO_2 at equilibrium with $CaCO_3$ and CaO is 1.04 atm. If the volume of the container of the system were decreased at 900°C, some CO_2 would combine with CaO to form $CaCO_3$ until the equilibrium pressure of CO_2 again becomes 1.04 atm. Similarly, if the volume of the container were increased, some $CaCO_3$ would decompose to form more CaO and CO_2 until the pressure of CO_2 again becomes 1.04 atm.

The concentration of a pure liquid in an equilibrium system is also a constant because increasing the number of moles of the liquid increases its volume, and decreasing the number of moles decreases its volume. Thus, the concentrations of *pure solids and pure liquids are omitted from the expression for an equilibrium constant*. This statement applies *only* to pure solids or pure liquids, *not* to gases or solutions since the concentrations of gases and solutes can vary.

Example 13.12

Writing Equilibrium Constant Expressions for Heterogeneous Equilibria

Write the expression of K_c for each of the following reactions at equilibrium:

(a) $2NaHCO_3(s) \rightleftharpoons Na_2CO_3(s) + CO_2(g) + H_2O(g)$
(b) $3Fe(s) + 4H_2O(g) \rightleftharpoons Fe_3O_4(s) + 4H_2(g)$
(c) $CH_4(g) + 4Cl_2(g) \rightleftharpoons CCl_4(l) + 4HCl(g)$
(d) $PbCl_2(s) \rightleftharpoons Pb^{2+}(aq) + 2Cl^-(aq)$

SOLUTION

(a) This equilibrium system includes two solids. Their concentrations are omitted from the equilibrium constant expression. Therefore,

$$K_c = [CO_2][H_2O]$$

(b) Remember to omit the two solids. K_c is

$$K_c = \frac{[H_2]^4}{[H_2O]^4}$$

(c) Omitting the pure liquid, K_c is

$$K_c = \frac{[HCl]^4}{[CH_4][Cl_2]^4}$$

(d) The concentration of pure solid $PbCl_2$ is constant, and the concentrations of Pb^{2+} and Cl^- ions in the solution do not depend on the amount of solid $PbCl_2$ as long as some is present. The expression for K_c is

$$K_c = [Pb^{2+}][Cl^-]^2$$

Practice Problem 13.7: Write an expression for K_c for each of the following equilibria:

(a) $HNO_2(aq) + H_2O(l) \rightleftharpoons H_3O^+(aq) + NO_2^-(aq)$
(b) $Ag_2CrO_4(s) \rightleftharpoons 2Ag^+(aq) + CrO_4^{2-}(aq)$
(c) $P_4(s) + 5O_2(g) \rightleftharpoons P_4O_{10}(s)$

13.4 Factors Affecting Equilibria

We recall that *Le Châtelier's principle* (Chapter 12) states that a stress placed on a system at equilibrium causes the equilibrium to shift to minimize the effect of the stress. Let us consider some factors that affect chemical equilibria.

Change of Concentration

Changing the concentration of one or more of the reactants or products of a reaction at equilibrium disturbs the original equilibrium, and the system changes until equilibrium is restored. If the concentrations of the reactants are increased in a reaction at equilibrium, equilibrium is reestablished by the formation of more products—the equilibrium shifts to the right. If the concentrations of the products are increased in a reaction at equilibrium, the equilibrium shifts to the left by forming more reactants.

For example, the reaction

$$H_2(g) + CO_2(g) \rightleftharpoons H_2O(g) + CO(g)$$

is at equilibrium when 0.50 mol of each of the components is present in a 1.0-L container at 1100 K. The value of K_c at 1100 K is 1.0. When more H_2 is added to this system, it reacts with CO_2 to produce more H_2O and CO until a new equilibrium is established. At this point the concentrations of water vapor and carbon monoxide are greater than 0.50 mol L^{-1} and the concentration of H_2 is greater than the original equilibrium concentration. The concentrations of reactants and products in the new equilibrium system are not the same as the original concentrations, but the value of the equilibrium constant remains 1.0 if the temperature remains constant.

Conversely, when CO is added to the same system at equilibrium, it reacts with water, and more H_2 and CO_2 are produced. The concentration of H_2O decreases, and the concentrations of both H_2 and CO_2 increase until a new equilibrium is established. At this point, the concentration of water is less than 0.50 mol L^{-1}, and the concentrations of H_2 and CO_2 are greater than 0.5 mol L^{-1}. But regardless of the "stress" caused by adding reactants or products, the equilibrium constant is unaltered as long as the temperature stays constant.

When the concentration of one of the components at equilibrium decreases, the equilibrium shifts to replace the component lost. For example, if some H_2O is removed from the equilibrium, $H_2(g) + CO_2(g) \rightleftharpoons H_2O(g) + CO(g)$, H_2 reacts with CO_2 to reestablish equilibrium. At the same time, more CO is also formed.

Change of Pressure

A change in pressure often affects equilibria in which one or more of the reactants and products are gases. The pressure of an equilibrium mixture in a closed container that contains one or more gaseous reactants and products can be changed in three ways.

1. The partial pressure of one or more of the reactants or products can be changed by adding or removing some of it. The effect of changing the partial pressure of a gaseous reactant or product is analogous to the concentration change we discussed above because the partial pressure of a gas is proportional to its concentration.
2. The pressure in a reaction vessel can be altered by changing its volume. Since volume is inversely proportional to pressure, increasing the volume decreases the partial pressure of each gas in the system, and decreasing the volume increases the partial pressure of each gas in the system. Changing the

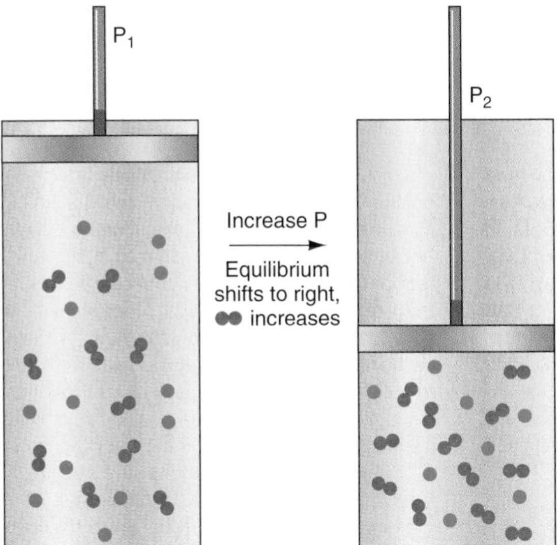

Figure 13.6

Effect of pressure on position of equilibrium.

pressure by changing the volume of the container has no effect on an equilibrium that contains equal numbers of moles of gaseous reactants and products. An example of such an equilibrium system is

$$H_2(g) + CO_2(g) \rightleftharpoons H_2O(g) + CO(g)$$

On the other hand, if a reaction has different numbers of moles of gaseous reactants and products, increasing the pressure by decreasing the volume of the container shifts the equilibrium toward the side that has the smaller volume or lower number of moles (Figure 13.6). For example, the reaction for the Haber process for the manufacture of ammonia is

$$\underset{\text{of reactants}}{\underset{\text{4 mol}}{3H_2(g) + N_2(g)}} \rightleftharpoons \underset{\text{of products}}{\underset{\text{2 mol}}{2NH_3(g)}}$$

Increasing the pressure on this equilibrium system causes the equilibrium to shift to the right. This shift in the equilibrium "relieves the stress" of increased pressure because 2 mol of products exert less pressure at constant temperature and volume than 4 mol of reactants. The Haber process is carried out at a pressure of about 250 atm.

3. The pressure of the system can be increased by adding an inert gas to the system. Increasing the pressure of a system at equilibrium that contains gaseous reactants or products by introducing an inert gas does not change the partial pressures of the gaseous components of the equilibrium system. Therefore, adding an inert gas has no effect on the equilibrium.

Change of Temperature

The effect of a change in temperature on a system at equilibrium depends on whether the forward reaction is exothermic or endothermic. If the forward reaction is *exothermic*, heat can be regarded as a product of the reaction, and increasing the temperature shifts the equilibrium to the *left*:

$$\text{reactants} \rightleftharpoons \text{products} + \text{heat}$$

$$\longleftarrow$$

an increase in temperature shifts the equilibrium to the left

APPLICATIONS OF CHEMISTRY 13.2
The Killer Lake of Cameroon

On August 16, 1984, officials found 37 people lying dead along the roadside by Lake Monoun in the western African Republic of Cameroon (see photo below). Then, in 1986, 1746 human lives were lost and thousands of livestock were killed along the shores of Lake Monoun, the "killer lake." These deaths were caused by CO_2 that had been released by the lake. To understand what had happened, let's first analyze the properties of carbon dioxide dissolved in water, specifically bottled soda.

Bottled soda contains dissolved carbon dioxide. We can represent the reaction of carbon dioxide dissolving in water as follows:

$$CO_2(g) + H_2O(l) \longrightarrow H_2CO_3(aq)$$

As the dissolved carbon dioxide tries to escape in a sealed bottle or can, an equilibrium results:

$$CO_2(g) + H_2O(l) \rightleftharpoons H_2CO_3(aq)$$

While soda remains sealed under 4 atm of pressure, its carbon dioxide stays in solution because high pressure drives the above equilibrium to the right because that side has 0 mol of gas whereas there is 1 mol of gaseous reactant on the left. However, when the bottle is opened, the pressure drops. Le Châtelier's principle tells us that the equilibrium shifts to the side with the greater number of moles of gas. As the gas abruptly leaves the soda can or bottle, it sometimes causes liquid to leave as well.

The killer lake of Cameroon is similar to a can or bottle of soda. Decaying animal and plant life at the bottom of the lake provides a source of carbon dioxide. Since the bottom of a deep body of water is under great pressure, carbon-dioxide is dissolved in water. As long as the carbon dioxide remains under this high pressure, it stays in solution. Because the waters of the killer lake of Cameroon exist in stratified layers with very little mixing, its carbon dioxide stays at the bottom in solution.

Under certain conditions, the stratified layers are disturbed. Volcanic activity, high winds, or an earthquake may cause the layers to mix. Such an event triggered an underwater landslide along the steep slopes near the bottom of Lake Monoun. This disturbance churned the bottom layer of stratified water and carried it closer to the surface, where reduced pressure released the carbon dioxide into the atmosphere. As the carbon dioxide erupted from the lake, it produced a 15-ft-high wave that flattened plants along the shoreline. Since carbon dioxide is heavier than air, it pushed aside all air as it blanketed the ground. Those breathing the carbon dioxide died from suffocation.

Source: *Science News*, December 7, 1985, pages 356–358.

If the forward reaction is *endothermic*, heat can be regarded as a reactant, and increasing the temperature shifts the equilibrium to the *right*:

$$\text{reactants} + \text{heat} \rightleftharpoons \text{products}$$

$$\xrightarrow{\hspace{4cm}}$$

an increase in temperature shifts
the equilibrium to the right

The conversion of oxygen to ozone is an endothermic process:

$$3O_2(g) + \text{heat} \rightleftharpoons 2O_3(g)$$

If the temperature of this equilibrium system increases, the equilibrium shifts to the right, increasing the concentration or pressure of ozone present at equilibrium.

Adding a catalyst to a reaction at equilibrium does not change the position of the equilibrium. A catalyst increases the *rates* of both the forward and reverse reactions equally. Therefore, the position of equilibrium is not changed by a catalyst.

Example 13.13

Influence of Various Factors on an Equilibrium System

Given the following reaction at equilibrium:

$$N_2H_4(g) \rightleftharpoons N_2(g) + 2H_2(g) \qquad \Delta H° = -95.4 \text{ kJ mol}^{-1}$$

How is the position of this system affected by each of the following changes? **(a)** The partial pressure of N_2 is decreased. **(b)** The volume of the container is decreased. **(c)** The temperature is decreased. **(d)** An inert gas is added. **(e)** A catalyst is added.

SOLUTION

(a) Decreasing the partial pressure of N_2 is equivalent to decreasing its concentration. Thus, the equilibrium shifts to the right to replace the N_2 lost. The new equilibrium mixture contains more H_2 and less N_2H_4 than that in the original mixture.

(b) Decreasing the volume of the container increases the partial pressures of all the components of the system. Since 1 mol of reactant, N_2H_4, exerts a smaller partial pressure than 3 mol of products (1 mol of N_2 and 2 mol of H_2), the equilibrium shifts to the left, away from the "stress of increased pressure." The new equilibrium system contains more N_2H_4 and less N_2 and H_2 than that in the original system.

(c) The negative enthalpy change for the forward reaction indicates that it is exothermic. Decreasing the temperature—which is equivalent to removing heat—favors the exothermic reaction, and the equilibrium shifts to the right. The new equilibrium mixture contains more N_2 and H_2 and less N_2H_4 than that in the original system.

(d) Adding an inert gas has no effect on the equilibrium.

(e) Adding a catalyst does not affect the position of any equilibrium.

Practice Problem 13.8: The following reaction occurs at high temperature:

$$PCl_5(g) \rightleftharpoons PCl_3(g) + Cl_2(g) \qquad \Delta H° = 87.9 \text{ kJ}$$

Will the number of moles of PCl_5 in this system increase, decrease, or remain the same when: **(a)** more PCl_3 is added; **(b)** the temperature is increased; **(c)** the volume of the reaction vessel is decreased; **(d)** some Cl_2 is removed; **(e)** a catalyst is added; **(f)** an inert gas is added?

13.5 Relation Between K_p and K_c

Equilibrium constants for reactions in the gas phase are often determined from the partial pressures of the reactants and products rather than from their concentrations. Partial pressures are related to concentrations through the ideal gas equation, $PV = nRT$, assuming ideal behavior. The partial pressure, P, of a gaseous component is

$$P = \frac{nRT}{V}$$

P is in atmospheres when R is 0.0821 L atm mol^{-1} K^{-1}, T is in kelvin, and V is in liters. We know that the equilibrium constant, K_p, for the general reaction

$$a\text{A}(g) + b\text{B}(g) \rightleftharpoons c\text{C}(g) + d\text{D}(g)$$

is given by

$$K_p = \frac{(P_\text{C})^c(P_\text{D})^d}{(P_\text{A})^a(P_\text{B})^b}$$

When we substitute the quantity nRT/V for P, we obtain:

$$\frac{\left(\frac{n_\text{C}RT}{V}\right)^c\left(\frac{n_\text{D}RT}{V}\right)^d}{\left(\frac{n_\text{A}RT}{V}\right)^a\left(\frac{n_\text{B}RT}{V}\right)^b} = \frac{\left(\frac{n_\text{C}}{V}\right)^c\left(\frac{n_\text{D}}{V}\right)^d(RT)^{c+d}}{\left(\frac{n_\text{A}}{V}\right)^a\left(\frac{n_\text{B}}{V}\right)^b(RT)^{a+b}} = \frac{[\text{C}]^c[\text{D}]^d}{[\text{A}]^a[\text{B}]^b} \times (RT)^{c+d-(a+b)} = K_p$$

Since

$$\frac{[\text{C}]^c[\text{D}]^d}{[\text{A}]^a[\text{B}]^b} = K_c$$

the equation simplifies to

$$K_p = K_c(RT)^{\Delta n} \qquad (13.3)$$

where Δn = gaseous moles of products − gaseous moles of reactants. When Δn equals 0, $K_p = K_c$.

Example 13.14

Calculating K_p from K_c

For the reaction $N_2(g) + 3H_2(g) \rightleftharpoons 2NH_3(g)$ at 472°C, K_c is 0.105. Calculate K_p for this reaction at 472°C.

SOLUTION: To solve this problem we use Equation 13.3, which relates K_p to K_c. The data are: $R = 0.0821$ L atm mol^{-1} K^{-1}, $T = 472 + 273 = 745$ K. Since 4 mol of reactants produce 2 mol of products, $\Delta n = 2 - 4 = -2$. Solving Equation 13.3 for K_p gives

$$K_p = K_c(RT)^{\Delta n} = (0.105)(0.0821 \times 745)^{-2} = 2.81 \times 10^{-5}$$

Practice Problem 13.9: Phosphorus pentachloride dissociates upon heating according to the equation

$$PCl_5(g) \rightleftharpoons PCl_3(g) + Cl_2(g)$$

The value of K_c for this reaction at 463 K is 3.25×10^{-2}. Calculate K_p at 463 K.

Summary

Equilibrium Constant and Equilibrium Position

Many reactions form equilibrium mixtures of reactants and products. The product of the molar concentrations of the products of a chemical reaction, divided by the product of the molar concentrations of the reactants, is a constant when the concentration of each substance is raised to an exponent equal to its coefficient in the balanced equation. This constant is called the **equilibrium constant**, symbolized by K_c. When partial pressures are used instead of molar concentrations in the ratio for a gaseous equilibrium, the equilibrium constant is denoted by K_p. K_c can be converted to K_p by the relationship

$$K_p = K_c(TR)^{\Delta n}$$

where R is the gas constant, T is the kelvin temperature, and Δn is the number of moles of gaseous products minus the number of moles of gaseous reactants.

An equilibrium constant varies only with temperature, and does not change when the concentrations or partial pressures of the components of the equilibrium change.

The extent to which the reactants have converted to products in a chemical equilibrium is called the **equilibrium position**. The value of an equilibrium constant thus also indicates the position of the equilibrium. If the components of an equilibrium mixture exist in the same phase, the equilibrium is **homogeneous**. If the components of an equilibrium mixture exist in two or more different phases, the system is **heterogeneous**. In heterogeneous equilibria, the concentrations of pure solids and pure liquids are omitted from the equilibrium constant expression because the concentration of a pure solid or a pure liquid is constant.

Factors Affecting Equilibria

The position of an equilibrium can be shifted to the right or left by changing the *concentrations (or partial pressures)* of some or all of the species present at equilibrium, or by changing the *temperature*. When a reaction is at equilibrium, adding a product shifts the equilibrium to the side of the reactants; adding a reactant shifts the equilibrium to the side of the products. If the forward reaction is exothermic, increasing the temperature shifts the equilibrium to the side of the reactants. If the forward reaction is endothermic, increasing the temperature shifts the equilibrium to the side of the products.

Gaseous equilibria in which the number of moles of reactants is not equal to the number of moles of products can also be shifted by changing the *pressure* by altering the *volume* of the system. When the volume of the container is decreased, the equilibrium shifts to the side having the smaller number of moles of gas. When the volume of the container is increased, the equilibrium shifts to the side that has the larger number of moles of gas.

Changing the total pressure of a mixture of gaseous reactants and products at equilibrium by adding an inert gas has no effect on the equilibrium because it does not alter the partial pressures of the reactants or products. Adding a catalyst increases the rate at which an equilibrium is reached but has no effect on the position of the equilibrium.

New Terms

Equilibrium constant (13.1)
Forward reaction (introduction)
Heterogeneous equilibria (13.3)
Homogeneous equilibria (13.3)
Law of mass action (law of chemical equilibrium) (13.1)
Position (point) of equilibrium (13.2)
Reaction quotient (13.2)
Reverse reaction (introduction)

Exercises

General Review

1. Write an expression for K_c and for K_p for the system

$$2H_2(g) + O_2(g) \rightleftharpoons 2H_2O(g)$$

2. How is an equilibrium constant for a reaction related to the equilibrium constant for the reverse of the reaction?
3. When the equations for two reactions are added to derive an equation that represents the sum of the two reactions, how are the equilibrium constants for the three reactions related?
4. If the numerical value of the equilibrium constant for a reaction $A(g) + B(g) \rightleftharpoons C(g) + D(g)$ equals 1.0, how far has the forward reaction proceeded when equilibrium is established?
5. What does a large numerical value for an equilibrium constant tell us about the relative concentrations of reactants and products present at equilibrium? What does a very small value of an equilibrium constant indicate about the extent of the forward reaction in an equilibrium system?
6. What is meant by the "position" of an equilibrium?
7. What is meant by a "shift" of an equilibrium?
8. What is a "reaction quotient"?
9. If a reaction quotient is larger than the equilibrium constant, in what direction does the reaction proceed before reaching equilibrium?
10. Write an equation relating K_p to K_c. Show how this equation can be rearranged to solve for either of these equilibrium constants from the known value of the other.
11. Write a chemical equation for an equilibrium reaction for which $K_p = K_c$.
12. Explain why the concentrations of pure solids and pure liquids are excluded from the expression for an equilibrium constant.
13. How does an equilibrium shift when the concentration of a component is: **(a)** increased; **(b)** decreased?
14. If the forward reaction for a system at equilibrium is endothermic, how does the equilibrium shift when the temperature is: **(a)** increased; **(b)** decreased?
15. Write an equation for a gaseous equilibrium that shifts to the right when the volume of the container is decreased. Write another equation for a gaseous equilibrium that is not affected by a change in the volume of the container.
16. How is a gaseous equilibrium affected, if at all, by adding an inert gas?
17. What role does a catalyst play in a reaction proceeding toward equilibrium?
18. How does a catalyst affect the position of an equilibrium?

Expression of K_c and K_p for Homogeneous Equilibria

19. Write an expression for K_c for each of the following gaseous equilibria: **(a)** $3O_2(g) \rightleftharpoons 2O_3(g)$; **(b)** $H_2(g) + I_2(g) \rightleftharpoons 2HI(g)$; **(c)** $2HI(g) \rightleftharpoons H_2(g) + I_2(g)$; **(d)** $4NH_3(g) + 3O_2(g) \rightleftharpoons 6H_2O(g) + 2N_2(g)$; **(e)** $2H_2S(g) + 3O_2(g) \rightleftharpoons 2H_2O(g) + 2SO_2(g)$

20. Write an expression for K_p for each of the equilibria in question 19.

Calculation of Equilibrium Constant from Experimental Data

21. In a determination of K_c for the reaction $2HI(g) \rightleftharpoons H_2(g) + I_2(g)$, 0.50 mol of HI is added to an evacuated 1.0-L flask at a certain temperature. When the equilibrium is reached 0.10 mol of HI remains in the flask. **(a)** Calculate the K_c for the system. **(b)** Calculate the value of K_c for the reaction $H_2(g) + I_2(g) \rightleftharpoons 2HI(g)$. **(c)** What is the value of K_c for the reaction $HI(g) \rightleftharpoons \frac{1}{2}H_2(g) + \frac{1}{2}I_2(g)$?

22. In a certain reaction 1.0 mol of reactant A and 1.0 mol of reactant B are placed in a 2.0-L container. After equilibrium is established, 0.10 mol of product C is present in the container. **(a)** Calculate the equilibrium constant for the system $A(g) + B(g) \rightleftharpoons C(g) + D(g)$. **(b)** What would be the value of the equilibrium constant if the reaction were $A(g) + B(g) \rightleftharpoons C(g) + 2D(g)$?

23. In an experiment, 8.10 mol of hydrogen is mixed with 2.94 mol of gaseous iodine in a 1.00-L flask at 448°C. At equilibrium, 5.64 mol of gaseous hydrogen iodide is present in the mixture. Write an equation for the reaction and calculate the value of the equilibrium constant, K_c.

24. A 10-g sample of gaseous PCl_5 is introduced into a 3.0-L flask at 250°C. At equilibrium, 63 percent of the PCl_5 is converted to PCl_3 and Cl_2. Calculate the value of K_c for this reaction at 250°C.

25. What is the value of K_p for the equilibrium in problem 24?

***26.** A mixture of 10.0 mol percent SO_2 and 90.0 mol percent O_2 is added to a reaction flask equipped with a platinum catalyst at 575°C. Ninety percent of the SO_2 is converted to SO_3 at equilibrium. The total pressure of the gaseous mixture at equilibrium is held constant at 1.00 atm. Calculate K_p for the system $2SO_2(g) + O_2(g) \rightleftharpoons 2SO_3(g)$ at 575°C. (*Hint*: The partial pressure in atmospheres of a gaseous component at equilibrium is the same fraction of the total pressure as the number of moles of that component is of the total number of moles.)

* Denotes problems of greater than average difficulty.

Relations of Equilibrium Constants to Equilibrium Equations

27. At 300°C K_p for the reaction

$$2NO(g) + Cl_2(g) \rightleftharpoons 2NOCl(g)$$

is 2.72. Calculate K_p for each of the following reactions at 300°C: **(a)** $2NOCl(g) \rightleftharpoons 2NO(g) + Cl_2(g)$; **(b)** $NO(g) + \frac{1}{2}Cl_2(g) \rightleftharpoons NOCl(g)$.

28. At 425.4°C, K_c for the reaction

$$HI(g) \rightleftharpoons \tfrac{1}{2}H_2(g) + \tfrac{1}{2}I_2(g)$$

is 7.38. What is K_c for each of the following reactions at 425.4°C? **(a)** $2HI(g) \rightleftharpoons H_2(g) + I_2(g)$; **(b)** $H_2(g) + I_2(g) \rightleftharpoons 2HI(g)$.

29. Given the following reactions, which occur at elevated temperatures, and their K_p values:

$$CH_4(g) + H_2O(g) \rightleftharpoons CO(g) + 3H_2(g) \qquad K_p = 25.6$$

$$CO(g) + H_2O(g) \rightleftharpoons CO_2(g) + H_2(g) \qquad K_p = 1.44$$

Calculate K_p for the reaction

$$CH_4(g) + 2H_2O(g) \rightleftharpoons CO_2(g) + 4H_2(g)$$

30. Consider the following reactions at 700°C and their K_c values:

$$2SO_3(g) \rightleftharpoons 2SO_2(g) + O_2(g) \qquad K_c = 2.5 \times 10^{-3}$$

$$NO_2(g) \rightleftharpoons NO(g) + \tfrac{1}{2}O_2(g) \qquad K_c = 0.012$$

Calculate K_c for the reaction

$$SO_2(g) + NO_2(g) \rightleftharpoons SO_3(g) + NO(g)$$

Equilibrium Calculations from Given Values of K_c or K_p

31. For the reaction $N_2(g) + 3H_2(g) \rightleftharpoons 2NH_3(g)$, K_c is 69 at 500°C. Analysis of a 10-L container of the foregoing equilibrium mixture at 500°C showed the presence of 4.0 mol of H_2, and 5.0 mol of NH_3. Calculate the number of moles of N_2 in the container.

32. The equilibrium constant, K_c, for the reaction

$$H_2O(g) + CO(g) \rightleftharpoons H_2(g) + CO_2(g)$$

is 0.63 at 986°C. If 1.00 mol of H_2O and 1.00 mol of CO are added to a 1.00-L container at 986°C, how many moles of each component is present in the equilibrium mixture? What mole percent of CO originally introduced has reacted at equilibrium?

33. At 500 K, K_c for the reaction $PCl_5(g) \rightleftharpoons PCl_3(g) + Cl_2(g)$ is 0.0224. When 1.00 mol of PCl_5 is placed into an empty 1.00-L container at 500 K, what fraction of PCl_5 remains at equilibrium? What is the total pressure in the container at equilibrium?

34. For the reaction $SO_2(g) + NO_2(g) \rightleftharpoons SO_3(g) + NO(g)$ at 973 K the equilibrium constant, K_c, is 9.00. Calculate the equilibrium concentrations of all the species present when: **(a)** 1.00 mol of SO_2 and 1.00 mol of NO_2 are injected into a 1.00-L container at 973 K; **(b)** 1.00 mol of SO_2 and 1.00 mol of NO_2 are injected into a 2.00-L container at 973 K; **(c)** 1.00 mol of SO_3 and 1.00 mol of NO are injected into a 1.00-L container at 973 K.

35. For the reaction $N_2(g) + O_2(g) \rightleftharpoons 2NO(g)$, the value of K_p is 0.050 at 2473 K. If the equilibrium partial pressures of N_2 and NO in this system are 0.20 and 0.045 atm, respectively, what is the partial pressure of O_2?

36. At 1000 K, K_p for the reaction $2NO_2(g) \rightleftharpoons 2NO(g) + O_2(g)$ is 158. When a certain mixture of these reactants and products is at equilibrium at 1000 K, the partial pressure of NO is 0.50 atm and the partial pressure of O_2 is 0.30 atm. Calculate the partial pressure of NO_2 in this equilibrium mixture.

***37.** The value of K_p for the reaction $PCl_5(g) \rightleftharpoons PCl_3(g) + Cl_2(g)$ at 200°C is 0.316. A 3.608-g sample of PCl_5 is injected into a 1.00-L flask containing Cl_2 gas at 20.0°C and 1.00 atm. The flask is then heated to 200°C. Calculate the percent dissociation of PCl_5 at 200°C.

Predicting the Direction of a Reaction

38. The equilibrium constant, K_c, for the reaction

$$N_2(g) + 3H_2(g) \rightleftharpoons 2NH_3(g)$$

at 472°C is 0.105. If 1.00 mol of N_2, 3.00 mol of H_2, and 2.00 mol of NH_3 are injected into a 1.00-L flask at 472°C, in which direction will the reaction proceed? Will the number of moles of N_2 in the flask increase or decrease?

39. For the reaction $SO_2(g) + NO_2(g) \rightleftharpoons SO_3(g) + NO(g)$, K_c is 9.00 at 973 K. If 1.00 mol of SO_2, 1.00 mol of NO_2, 1.00 mol of SO_3, and 1.00 mol of NO are injected into a 1.00-L flask at 973 K, in which direction will the reaction proceed? What are the concentrations of all species at equilibrium?

***40.** The value of K_p for the reaction $CO(g) + H_2O(g) \rightleftharpoons H_2(g) + CO_2(g)$ at 1000°C is 1.6. A mixture of 25 mol percent of CO, 10.0 mol percent of H_2O, 5.0 mol percent of H_2, 3.0 mol percent of CO_2, and 57.0 mol percent of Ne at 1.0 atm total pressure is heated to 1000°C. How is the composition of the mixture changed when equilibrium is reached at 1000°C? Calculate the mole percent of each component present at equilibrium. See the hint given for problem 26.

* Denotes problems of greater than average difficulty.

Heterogeneous Equilibria

41. Write an expression for the equilibrium constant, K_c, for each of the following reactions: **(a)** $Fe(s) + 5CO(g) \rightleftharpoons Fe(CO)_5(g)$; **(b)** $C(s) + H_2O(g) \rightleftharpoons CO(g) + H_2(g)$; **(c)** $2HgO(s) \rightleftharpoons 2Hg(l) + O_2(g)$; **(d)** $HC_2H_3O_2(aq) + H_2O(l) \rightleftharpoons H_3O^+(aq) + C_2H_3O_2^-(aq)$; **(e)** $NH_3(aq) + H_2O(l) \rightleftharpoons NH_4^+(aq) + OH^-(aq)$; **(f)** $2H_2O(l) \rightleftharpoons H_3O^+(aq) + OH^-(aq)$; **(g)** $PbCl_2(s) \rightleftharpoons Pb^{2+}(aq) + 2Cl^-(aq)$; **(h)** $Fe(OH)_3(s) \rightleftharpoons Fe^{3+}(aq) + 3OH^-(aq)$

42. At 125°C the value of K_p for the reaction $NH_4Cl(s) \rightleftharpoons NH_3(g) + HCl(g)$ is 6.0×10^{-9}. Calculate the equilibrium partial pressure of NH_3 and of HCl produced by the partial decomposition of NH_4Cl at 125°C.

43. Ammonium carbamate, $NH_4CO_2NH_2$, decomposes according to the equation $NH_4CO_2NH_2(s) \rightleftharpoons 2NH_3(g) + CO_2(g)$. At equilibrium, the total pressure of NH_3 and CO_2 at 40°C is 0.36 atm. Calculate K_p for this reaction at 40°C.

44. Ethyl acetate, $CH_3COOC_2H_5$ (a good solvent for lacquers) reacts with water to form acetic acid and ethyl alcohol:

$$CH_3COOC_2H_5(aq) + H_2O(l) \rightleftharpoons CH_3COOH(aq) + C_2H_5OH(aq)$$

When 1.0 mol of ethyl acetate is initially mixed with water to make 1.0 L of solution. 0.33 mol of ethyl acetate remains in the solution at equilibrium. Calculate K_c for this reaction.

45. The value of K_c for the system

$$Fe(OH)_2(s) \rightleftharpoons Fe^{2+}(aq) + 2OH^-(aq)$$

is 1.6×10^{-14}. The hydroxide ion concentration in a sample of water from a lake is 5.2×10^{-8} *M*. Calculate the maximum concentration of Fe^{2+} ions that can be present in this water so that no solid $Fe(OH)_2$ precipitates.

Factors Affecting Equilibria

46. In the following system at equilibrium

$$4NH_3(g) + 3O_2(g) \rightleftharpoons 6H_2O(g) + 2N_2(g)$$

how is the amount of nitrogen present affected by each of the following changes? **(a)** The volume of the container is doubled; **(b)** the volume of the container is decreased; **(c)** more water vapor is introduced; **(d)** more ammonia is introduced; **(e)** some oxygen is removed from the system; **(f)** temperature is increased (the forward reaction in this system is exothermic); **(g)** the pressure of the system is increased by introducing helium gas; **(h)** a catalyst is added.

47. For the gaseous equilibrium mixture

$$N_2(g) + O_2(g) \rightleftharpoons 2NO(g) \qquad \Delta H° = +180 \text{ kJ}$$

State the effect on the equilibrium of each of the following changes: **(a)** more oxygen gas is introduced; **(b)** some nitrogen is removed; **(c)** the temperature is increased; **(d)** the volume of the container is decreased without allowing any of the gas to escape; **(e)** the pressure is increased on the system by introducing helium gas; **(f)** a catalyst is added. Explain your reasoning in each case.

48. Consider the Haber process for making ammonia:

$$N_2(g) + 3H_2(g) \rightleftharpoons 2NH_3(g) \qquad \Delta H° = -46 \text{ kJ}$$

Complete the following table to state the effects (increased, decreased, unchanged) the indicated changes will have on the concentrations of the components at equilibrium:

Change	**Concentration of N_2**	**Concentration of H_2**	**Concentration of NH_3**
Increasing pressure by decreasing the volume of the reaction vessel			
Increasing temperature			
Increasing the amount of N_2			
Adding neon gas			
Adding a catalyst			

Relation Between K_p and K_c

49. The value of K_c for the dissociation of phosgene, $COCl_2(g) \rightleftharpoons CO(g) + Cl_2(g)$, at 100°C is 2.19×10^{-10}. What is the value of K_p for this reaction? What are the values of K_c and K_p for the reaction $CO(g) + Cl_2(g) \rightleftharpoons COCl_2(g)$ at 100°C?

50. For the reaction $N_2(g) + O_2(g) \rightleftharpoons 2NO(g)$ the value of K_c at 2400 K is 2.5×10^{-3}. What is the value of K_p at that temperature?

51. The value of K_p for the reaction $N_2(g) + 3H_2(g) \rightleftharpoons 2NH_3(g)$ at 350°C is 7.73×10^{-4}. Calculate K_c at 350°C.

CHAPTER 14

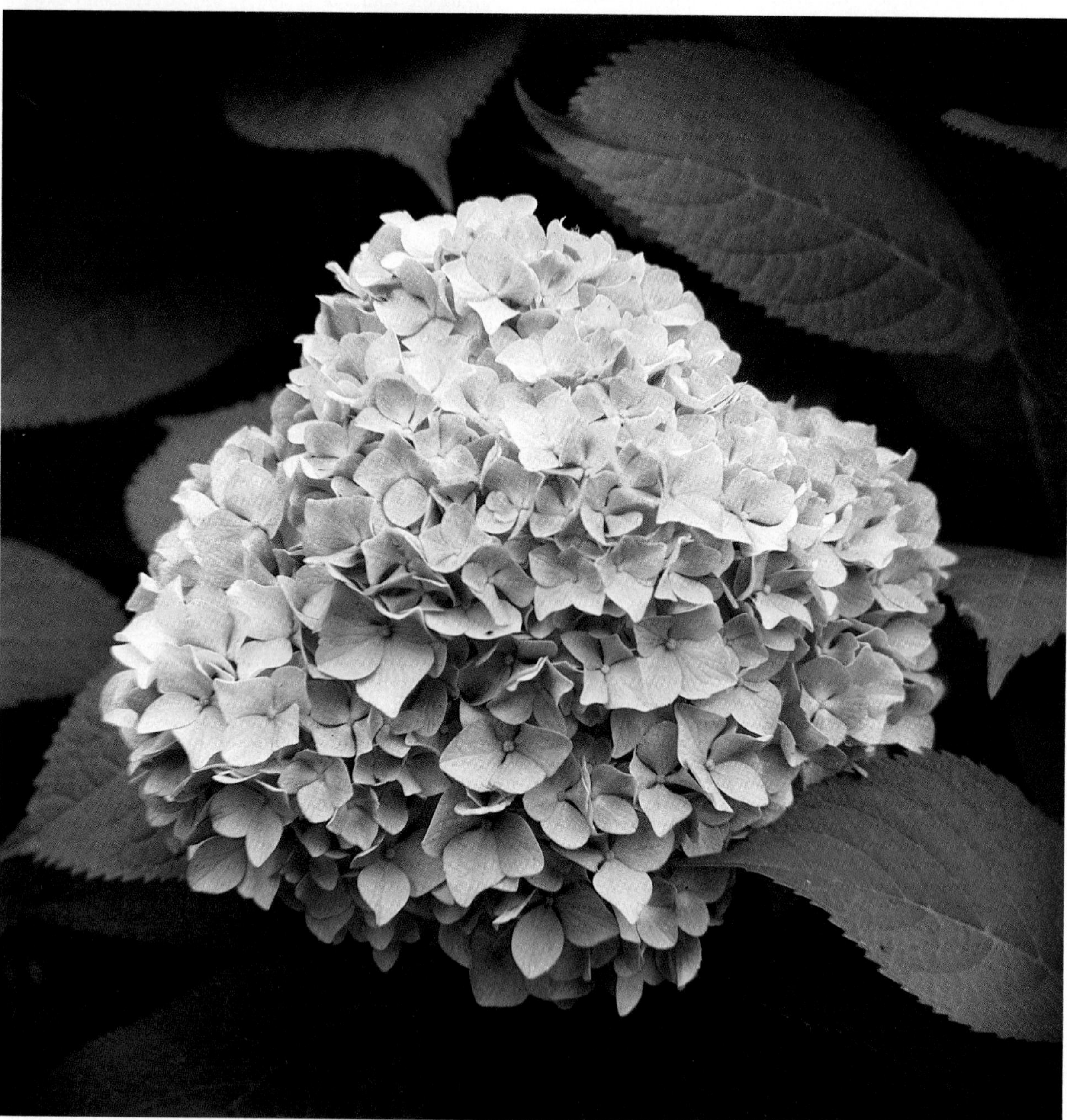

A blue hydrangea is the product of a base soil. A pink hydrangea would be a product of an acid soil.

Acids and Bases

In this chapter we extend our consideration of chemical equilibrium to include reactions of acids with bases. We will extend the discussion of acids and bases that we began in Chapter 4 and merge these ideas with the concepts of chemical equilibrium we introduced in Chapter 13.

14.1 Brønsted Acids and Bases

In Chapter 4 we discussed some properties and reactions of acids and bases. We recall that when an acid dissolves in water, the hydrogen ion concentration increases; when a base dissolves in water, hydroxide ion concentration increases. Substances of this type are called Arrhenius acids and bases. The Arrhenius definition of acids and bases was extended (in 1923) by Johannes Brønsted of Denmark and Martin Lowry of England to include substances that are not necessarily acids and bases in the Arrhenius definition. In the Brønsted–Lowry definition, *an acid is a proton donor* and *a base is a proton acceptor*. We will call the Brønsted–Lowry definition of acids and bases the "Brønsted definition" to be brief.

The reaction of a Brønsted acid, HA, with a base, B, can be represented by the following equilibrium:

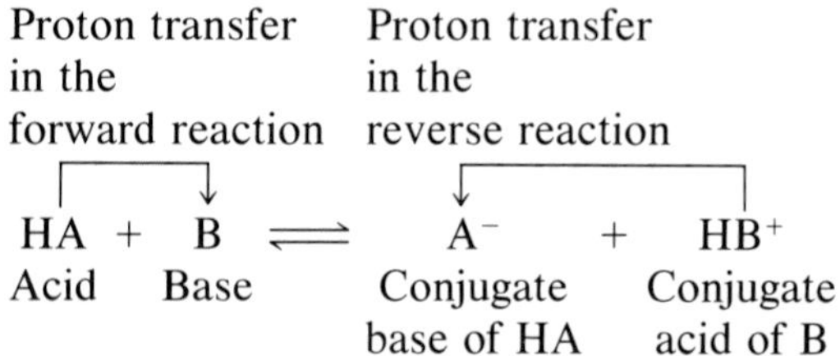

In the forward reaction, acid HA donates a proton to base B to form A^- and HB^+ ions as products. In the reverse reaction, the HB^+ ion is an *acid* that donates a proton to the A^- ion, which thus acts as a base. When a Brønsted acid, HA, loses a proton, the resulting product, A^-, is the **conjugate base** of the original acid, HA. The acid, HA, and its conjugate base, A^-, constitute a **conjugate acid–base pair**. Similarly, when a Brønsted base, B, accepts a proton, the resulting acid, HB^+, is the **conjugate acid** of the original base, B. The base, B, and its conjugate acid, HB^+ comprise another conjugate acid–base pair.

For example, a Brønsted acid–base reaction occurs when gaseous hydrogen chloride is bubbled into aqueous ammonia:

Proton transfer (HCl → NH_3) Proton transfer (NH_4^+ → Cl^-)

$$\underset{\text{Acid}}{HCl(g)} + \underset{\text{Base}}{NH_3(aq)} \rightleftharpoons \underset{\text{Conjugate acid}}{NH_4^+(aq)} + \underset{\text{Conjugate base}}{Cl^-(aq)}$$

In the forward reaction, hydrogen chloride gas is a proton donor. When an NH_4Cl solution is heated, it decomposes to $HCl(g)$ and $NH_3(g)$. In this reverse reaction, ammonium ion is the proton donor. Thus, a hydrogen chloride mole-

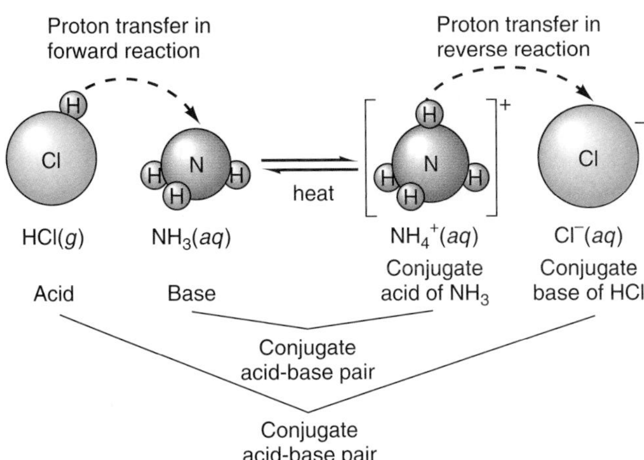

Figure 14.1

A Brønsted-Lowry acid–base reaction. After the transfer of a proton from a hydrogen chloride molecule, the chloride ion that remains is the conjugate base of hydrogen chloride. When an ammonia molecule accepts a proton it forms an ammonium ion, which is the conjugate acid of ammonia.

cule and a chloride ion constitute one conjugate acid–base pair; ammonium ion and an ammonia molecule comprise another conjugate acid–base pair (Figure 14.1).

Now, let us consider the relation of Brønsted acids and bases to Arrhenius acids and bases. According to the Arrhenius definition of acids and bases, water is neither an acid nor a base: it is merely the medium in which reactions occur. However, according to the Brønsted definition, water can react as *either* an acid *or* a base. For example, when acetic acid is dissolved in water, it reacts with water to form an equilibrium mixture of acetic acid molecules ($HC_2H_3O_2$), water molecules, hydronium ions, and acetate ions ($C_2H_3O_2^-$):

$$\underbrace{HC_2H_3O_2(aq)}_{\text{Acid}} + \underbrace{H_2O(l)}_{\text{Base}} \rightleftharpoons \underbrace{H_3O^+(aq)}_{\text{Conjugate acid}} + \underbrace{C_2H_3O_2^-(aq)}_{\text{Conjugate base}}$$

(Proton transfer: from $HC_2H_3O_2$ to H_2O; Proton transfer: from H_3O^+ to $C_2H_3O_2^-$)

In its reaction with acetic acid, water acts as a Brønsted base by accepting a proton from acetic acid.

But let us consider the reaction of aqueous ammonia with water to produce some ammonium ions and hydroxide ions:

$$\underbrace{H_2O(l)}_{\text{Acid}} + \underbrace{NH_3(aq)}_{\text{Base}} \rightleftharpoons \underbrace{NH_4^+(aq)}_{\text{Conjugate acid}} + \underbrace{OH^-(aq)}_{\text{Conjugate base}}$$

(Proton transfer in forward reaction: from H_2O to NH_3; Proton transfer in reverse reaction: from NH_4^+ to OH^-)

In the reaction with aqueous ammonia, water is a proton donor and therefore acts as a Brønsted acid. *A substance that can act either as a proton donor or a proton acceptor is said to be* **amphiprotic**. Thus, water is an amphiprotic substance.

Another amphiprotic substance is aqueous hydrogen carbonate ion, $HCO_3^-(aq)$, which reacts as an acid with aqueous ammonia,

$$HCO_3^-(aq) + NH_3(aq) \rightleftharpoons NH_4^+(aq) + CO_3^{2-}(aq)$$

But, hydrogen carbonate ion reacts as a base with acetic acid,

$$HCO_3^-(aq) + HC_2H_3O_2(aq) \rightleftharpoons H_2O(l) + CO_2(aq) + C_2H_3O_2^-(aq)$$

Water molecules can react with one another to produce a very low concentration of hydronium ions (H_3O^+) and hydroxide ions in water:

$$H_2O(l) + H_2O(l) \rightleftharpoons H_3O^+(aq) + OH^-(aq)$$

A reaction in which molecules of a substance react with one another to form ions is called **autoionization.** The equilibrium expression for the autoionization of water is

$$K = \frac{[H_3O^+][HO^-]}{[H_2O]^2}$$

Since the concentration of pure liquid water is a constant, we can write this equilibrium expression as

$$K[H_2O]^2 = K_w = [H_3O^+][OH^-]$$

where the constant, K_w, is called the **ion product constant for water**. The value of K_w at 25°C is 1.0×10^{-14}. Since hydronium ions and hydroxide ions are present in equal concentrations, $[H_3O^+] = [OH^-] = (1.0 \times 10^{-14})^{1/2} = 1.0 \times 10^{-7}$ mol L^{-1}. Thus, pure water contains very few hydronium ions and hydroxide ions.

Example 14.1

Writing Equations for Brønsted Acid–Base Reactions, and Identifying Acids, Bases, and Conjugate Acid–Base Pairs

Write an equation for the reaction of gaseous ammonia with aqueous acetic acid. Label each substance as a Brønsted acid or base in the forward reaction and as a conjugate acid or conjugate base in the reverse reaction. Also list the conjugate acid–base pairs.

SOLUTION: We know that acetic acid is a stronger acid than water and that ammonia is a stronger base than water. Therefore, when acetic acid and ammonia are mixed, acetic acid molecules donate protons to ammonia molecules producing acetate ions and ammonium ions:

$$\underset{\text{Acid}}{HC_2H_3O_2(aq)} + \underset{\text{Base}}{NH_3(g)} \rightleftharpoons \underset{\text{Conjugate acid}}{NH_4^+(aq)} + \underset{\text{Conjugate base}}{C_2H_3O_2^-(aq)}$$

One conjugate acid–base pair in this reaction consists of acetic acid molecules and acetate ions; the second conjugate acid–base pair consists of ammonia molecules and ammonium ions (Figure 14.2).

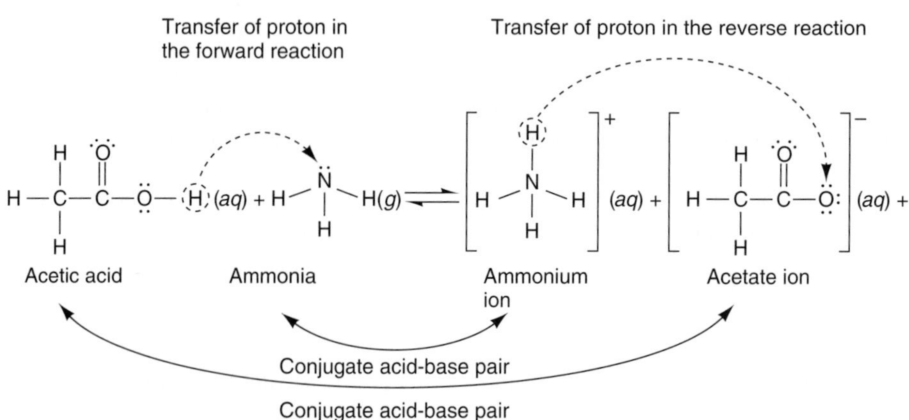

Figure 14.2

The conjugate acid–base pairs in the reaction of acetic acid with aqueous ammonia.

Practice Problem 14.1: Write an equation of the Brønsted acid–base reaction in which hydrogen fluoride gas is bubbled into water. Identify each substance as a Brønsted acid or Brønsted base, and indicate which substances are conjugate acid–base pairs.

14.2 Properties of Acids

Acid Strength

An acid can be characterized as "strong" or "weak" by the extent to which it dissociates in water. We describe an acid as "strong" if it is essentially 100% dissociated in water. In contrast, an acid is "weak" if it is less than 100% dissociated; commonly, an acid that is only slightly dissociated is referred to as a weak acid.

Let us consider the hydrohalic acids, which are binary acids composed of hydrogen and a halogen. They are HF, HCl, HBr, and HI. The relative strengths of these acids are

$$HI > HBr > HCl \gg HF$$

To understand these relative acid strengths, we consider the energy changes associated with the ionization of hydrohalic acids in water.

When a gaseous hydrogen halide, HX, reacts with water, HX molecules donate protons to water producing H_3O^+ and X^- ions:

$$HX(g) + H_2O(l) \longrightarrow H_3O^+(aq) + X^-(aq)$$

We can mentally divide this reaction into several steps, and we can consider the enthalpy change for each step. The net enthalpy change for the reaction equals the sum of the enthalpy changes for the individual steps.

1. We can imagine that in the first step of the reaction gaseous HX dissociates to $H(g)$ and $X(g)$ atoms.
2. In the second step, a gaseous hydrogen atom loses an electron and becomes a hydrogen ion.
3. In the third step, the electron lost by the hydrogen atom is accepted by an X atom, converting it to an X^- ion.
4. In the fourth step, the H^+ ion attaches to a water molecule to form an H_3O^+ ion.
5. In the fifth step, the X^- ions hydrate.

The enthalpy changes accompanying each of these steps at constant pressure are as follows:

(1) $HX(g) \longrightarrow H(g) + X(g)$ $\qquad \Delta H_{dissociation}\ (\Delta H_{dis})$

(2) $H(g) \longrightarrow H^+(g) + e^-$ $\qquad \Delta H_{ionization}\ (\Delta H_i)$

(3) $X(g) + e^- \longrightarrow X^-(g)$ $\qquad \Delta H_{electron\ attachment}\ (\Delta H_{ea})$

(4) $H^+(g) + H_2O(l) \longrightarrow H_3O^+(aq)$ $\qquad \Delta H_{protonation\ of\ water}\ (\Delta H_{pw})$

(5) $X^-(g) \longrightarrow X^-(aq)$ $\qquad \Delta H_{hydration}\ (\Delta H_{hyd})$

Table 14.1

Enthalpy Changes in the Conversion $HX(g) \rightarrow H^+(g) + X^-(g)$

Conversion Reaction	ΔH for $HX(g) \rightarrow H^+(g) + X^-(g)$ (kJ mol^{-1} of HX) $\Delta H_{dis} + \Delta H_i - \Delta H_{ea} = \Delta H_{rxn}$
$HF(g) \longrightarrow H^+(g) + F^-(g)$	569 + 1312 − 333 = 1548
$HCl(g) \longrightarrow H^+(g) + Cl^-(g)$	432 + 1312 − 348 = 1396
$HBr(g) \longrightarrow H^+(g) + Br^-(g)$	366 + 1312 − 324 = 1354
$HI(g) \longrightarrow H^+(g) + I^-(g)$	299 + 1312 − 295 = 1316

The first three steps are the equivalent of converting gaseous HX molecules to gaseous H^+ and X^- ions:

$$HX(g) \longrightarrow H^+(g) + X^-(g)$$

The sum of the enthalpy changes for these three steps for the various hydrogen halides are summarized in Table 14.1. The net enthalpy change is given in the last vertical column in Table 14.1.

How are these enthalpy changes related to acid strength? We can see from Table 14.1 that the bond dissociation energy of hydrogen fluoride is much more endothermic than the bond dissociation energies of the other hydrogen halides. Since hydrogen fluoride has a much higher bond dissociation energy than other hydrogen halides, hydrogen fluoride is a much weaker acid than the other hydrogen halides. The enthalpy of ionization of hydrogen atoms and the protonation enthalpy of water are the same for all four hydrogen halides. The hydration energies of the X^- ions differ only slightly.

Because the enthalpies of conversion of HCl, HBr, and HI to their ions in aqueous solution are less than that for HF, they are stronger acids than HF. According to Brønsted, a strong acid is also a good proton donor. Hydrochloric, hydrobromic, and hydroiodic acids are essentially 100 percent ionized in aqueous solution. Table 14.1 shows that the enthalpy changes for their conversion to gaseous ions differ only slightly.

Since HCl, HBr, and HI are virtually 100 percent dissociated in aqueous solution, the H_3O^+ ion concentrations in 1.0 *M* solutions of these acids are virtually 1.0 *M*. The tendency of a solvent (in this case water) to make strong acids appear to have equal strength is called the *leveling effect*.

Since a strong acid is essentially 100 percent ionized in aqueous solution, its conjugate base has no tendency to regain a proton. The conjugate base of a strong acid is a weak base, but the weaker the acid, the stronger is its conjugate base. Therefore, the order of Brønsted base strength of the halide ions is as follows:

$$F^- \gg Cl^- > Br^- > I^-$$

Acid Dissociation Constant, K_a

In contrast to a dilute solution of a strong acid, which contains virtually no undissociated acid molecules, an aqueous solution of a weak acid, HA, is an *equilibrium mixture* that contains HA molecules, water molecules, A^- anions, and H_3O^+ ions. The strength of a weak acid is related to the value of the equilibrium constant for the dissociation reaction for the acid. For the reaction

$$HA(aq) + H_2O(l) \rightleftharpoons H_3O^+(aq) + A^-(aq)$$

the equilibrium constant expression can be written as

$$K = \frac{[H_3O^+][A^-]}{[H_2O][HA]}$$

Since the concentration of water is constant, we can write

$$K[H_2O] = \frac{[H_3O^+][A^-]}{[HA]}$$

The product of the two constants, K and $[H_2O]$, is itself a constant that we designate K_a. K_a is the **acid dissociation constant** or **acid ionization constant** for a weak acid. Thus, the expression of K_a for the acid HA is

$$K_a = \frac{[H_3O^+][A^-]}{[HA]}$$

The K_a values of some weak acids are listed in Table 14.2.

Table 14.2

Dissociation Constants of Common Weak Acids at 25°C

Name	Formula	K_1	K_2	K_3
Acetic acid	$HC_2H_3O_2$	1.76×10^{-5}		
Ammonium ion	NH_4^+	5.6×10^{-10}		
Arsenic acid	H_3AsO_4	5.6×10^{-3}	1.70×10^{-7}	3.95×10^{-12}
Arsenous acid	H_3AsO_3	6.0×10^{-10}		
Benzoic acid	C_6H_5COOH	6.46×10^{-5}		
Boric acid	H_3BO_3	7.3×10^{-10}	1.8×10^{-13}	1.6×10^{-14}
Carbonic acid	H_2CO_3	4.3×10^{-7}	5.61×10^{-11}	
Chloroacetic acid	$CH_2ClCOOH$	1.4×10^{-3}		
Citric acid	$C_3H_4(OH)(COOH)_3$	7.10×10^{-4}	1.68×10^{-5}	8.4×10^{-6}
Formic acid	$HCOOH$	1.77×10^{-4}		
Hydrazoic acid	HN_3	1.9×10^{-5}		
Hydrocyanic acid	HCN	4.93×10^{-10}		
Hydrofluoric acid	HF	3.53×10^{-4}		
Hydrogen peroxide	H_2O_2	2.4×10^{-14}		
Hydrogen selenate ion	$HSeO_4^-$	1.2×10^{-2}		
Hydrogen sulfate ion	HSO_4^-	1.20×10^{-2}		
Hydrosulfuric acid	H_2S	1.0×10^{-7}	1.3×10^{-13}	
Hypobromous acid	$HBrO$	2.06×10^{-9}		
Hypochlorous acid	$HClO$	3.0×10^{-8}		
Hypoiodous acid	HIO	2.3×10^{-11}		
Iodic acid	HIO_3	1.69×10^{-1}		
Lactic acid	$CH_3CHOHCOOH$	1.4×10^{-4}		
Nitrous acid	HNO_2	4.5×10^{-4}		
Oxalic acid	$H_2C_2O_4$	5.9×10^{-2}	6.4×10^{-5}	
Phosphoric acid	H_3PO_4	7.52×10^{-3}	6.23×10^{-8}	2.2×10^{-13}
Phosphorous acid	H_2PHO_3	1.0×10^{-2}	2.6×10^{-7}	
Selenous acid	H_2SeO_3	3.5×10^{-3}	5×10^{-8}	
Sulfurous acid	H_2SO_3	1.5×10^{-2}	6.3×10^{-8}	

Example 14.2

Calculating the K_a of an Acid from $[H_3O^+]$

The hydronium ion concentration in a 0.100 *M* solution of hydrofluoric acid is 5.7×10^{-3} *M*. Calculate the dissociation constant, K_a, for the acid.

SOLUTION: We first write a chemical equation describing the equilibrium. Beneath each species in the equation we write its initial and equilibrium concentrations, as follows:

	$HF(aq) + H_2O(l)$	$\rightleftharpoons$	$H_3O^+(aq)$	+	$F^-(aq)$
Initial *M*	0.100		~0		0
Equilibrium *M*	$0.100 - 5.7 \times 10^{-3}$		5.7×10^{-3}		5.7×10^{-3}

Note that for the initial hydronium ion concentration we write ~0, implying that the initial hydronium ion concentration is not quite zero, because there are a few hydronium ions present from the autoionization of water. However, the hydronium ion concentration due to the autoionization of water is negligibly small compared with 5.7×10^{-3} mol per liter produced by the dissociation of HF.

According to the balanced chemical equation for the dissociation reaction, $[H_3O^+] = [F^-] = 5.7 \times 10^{-3}$ *M*. To produce 5.7×10^{-3} mol of hydronium ions, an equal number of moles per liter of hydrogen fluoride must dissociate. The concentration of hydrogen fluoride that remains at equilibrium is therefore $0.100\ M - 5.7 \times 10^{-3}\ M = 0.094\ M$. The expression for K_a is

$$K_a = \frac{[H_3O^+][F^-]}{[HF]}$$

Substituting the equilibrium concentration of each species into this expression gives

$$K_a = \frac{(5.7 \times 10^{-3})(5.7 \times 10^{-3})}{0.094} = 3.5 \times 10^{-4}$$

Practice Problem 14.2: The hydronium ion concentration in a 0.200 *M* solution of hydrocyanic acid, HCN, is 9.9×10^{-6} *M*. What is the value of K_a for HCN?

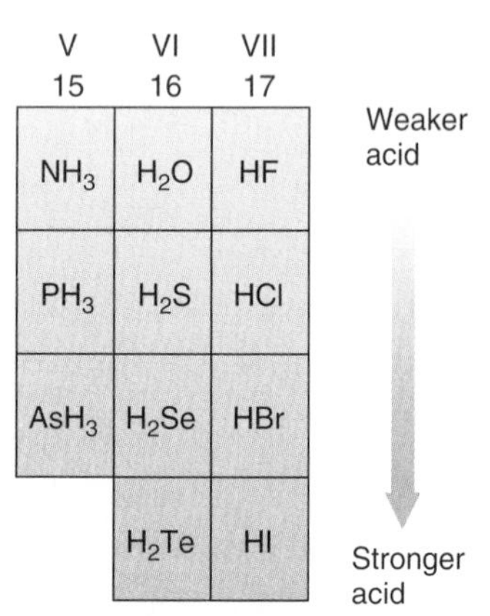

Acid strength increases from top to bottom within a group because the strength of the H-X bond decreases. (X is an element in group V, VI, or VII.)

Figure 14.3

Trend of binary acid strength down the groups.

General Group Trends in Binary Acid Strength

We recall that the acid strength of the binary acids of group VII elements (HF, HCl, HBr, and HI) increases down the group. The acid strength of the binary hydrogen compounds of group VI elements and of group V elements also increases down a group (Figure 14.3). Thus, the acid strength of binary hydrogen compounds of group VI elements—water, hydrogen sulfide, hydrogen selenide, and hydrogen telluride—increases down the groups:

$$H_2O < H_2S < H_2Se < H_2Te$$

We recall that HI is a stronger acid than HF because the H—I bond is weaker than the H—F bond. Similarly, H_2Te is a stronger acid than H_2O because the H—Te bond is weaker than the H—O bond. The H—O bond energy in H_2O is 498 kJ mol^{-1}, and the H—Te bond energy in H_2Te is 238 kJ mol^{-1}.

The acid strengths of the binary hydrogen compounds of group V elements increase from ammonia to phosphine (PH_3) to arsine (AsH_3):

$$NH_3 < PH_3 < AsH_3$$

Phosphine and arsine are weaker acids than water. In fact, NH_3, PH_3, and AsH_3 react with water as proton acceptors, or bases. Arsine is a stronger acid than ammonia because the H—As bond is weaker than the H—N bond. The H—N bond energy in NH_3 is 391 kJ mol^{-1} and the H—As bond energy in AsH_3 is 247 kJ mol^{-1}. We shall see in the next section, however, that bond strength is not the only factor that affects acid strengths.

Variation of Binary Acid Strength Across Periods

We know that hydrogen fluoride is a stronger acid than water and that ammonia is a weaker acid than water. Thus, the relative acid strengths of these three substances are

$$NH_3 < H_2O < HF$$

We can see that the binary acid strength increases from left to right in the second period of the periodic table (Figure 14.4).

Similarly, the strengths of binary hydrogen compounds of third-period elements increase from left to right:

$$PH_3 \ll H_2S \ll HCl$$

These trends in acid strengths within the second and third periods are *not* correlated with bond energies. The N—H bond energy is 391 kJ mol^{-1}, the O—H bond energy is 498 kJ mol^{-1}, and the F—H bond energy 569 kJ mol^{-1}. Similarly, the P—H, S—H, and Cl—H bond energies are 322, 363, and 432 kJ mol^{-1}, respectively.

What, then, is responsible for the trend of increasing acid strengths of binary hydrogen compounds of the second- and third-period elements? The acid strengths of these compounds parallel the increase in electronegativity of the elements from left to right within a period. As the electronegativity of the atom (A) bonded to hydrogen in a binary compound increases, the A—H bond becomes more polar and the H atom is more easily extracted as a proton by a base. For example, oxygen is more electronegative than nitrogen, and H_2O is a stronger acid than NH_3. Similarly, fluorine is more electronegative than oxygen, and HF is a stronger acid than H_2O. In the third period, the electronegativity increases from phosphorus to sulfur to chlorine, and acid strength increases from PH_3 to H_2S to HCl.

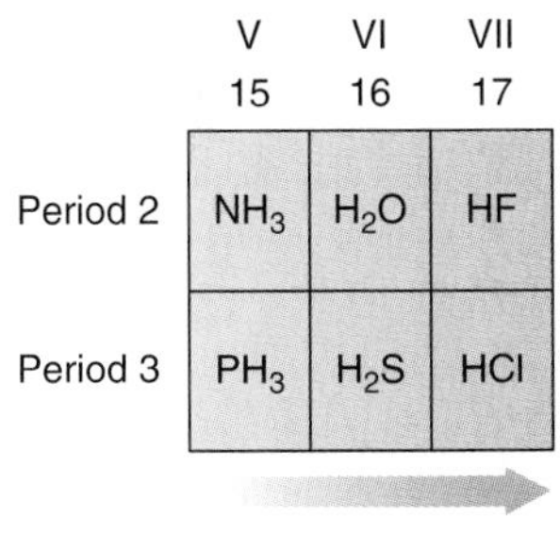

Acid strength increases across a period because the electronegativityof the elements increases across a period, and the H becomes more readily available as a proton.

Figure 14.4

Trend of binary acid strength left-to-right in periods.

Example 14.3

Predicting the Relative Strengths of Binary Acids

Arrange the following acids in the order of increasing acid strength: H_2S, H_2Se, HBr.

SOLUTION: To solve such a problem, remember that the strength of a binary acid (HA) increases across a period (because A becomes more electronegative) and down a group (because the strength of the H—A bond decreases). H_2S and H_2Se are acids of group VI elements. Since selenium is below sulfur in group VI, H_2Se is stronger than H_2S. H_2Se and HBr are acids of the fourth-period elements, selenium and bromine. Bromine lies to the right of selenium, the electronegativity of bromine is higher than the electronegativity of selenium, and therefore HBr is a stronger acid than H_2Se. Thus, the order of increasing acidity is

$$H_2S < H_2Se < HBr$$

Practice Problem 14.3: Arrange the following substances in the order of their increasing acid strength: PH_3, AsH_3, H_2Se.

14.3 Oxoacids and Hydroxo Compounds

An **oxoacid**, as its name implies, contains oxygen. An oxoacid also contains hydrogen and a third element—usually, but not always, a nonmetal. Since an oxoacid contains three elements, it is sometimes called a ternary acid. The relative strengths of oxoacids depend on their molecular structures. In this section we consider the relation of the strengths of oxoacids to their structures.

Structures and Relative Strengths of Oxoacids

The hydrogen atoms in an oxoacid can be bonded to oxygen atoms or to other atoms. Only the hydrogen atoms that are bonded to oxygen atoms can be extracted as protons by a base. These hydrogen atoms are sometimes called ionizable hydrogen atoms or acidic hydrogen atoms.

In the formula for an acid, the ionizable hydrogen atoms are written first, followed by the anion of the acid. The elements in the anion are usually written in the order of their increasing electronegativity. For example, the formula for hypochlorous acid is usually written HClO. However, the hydrogen atom in hypochlorous acid is bonded to the oxygen atom in the arrangement HOCl.

The formula for sulfuric acid is H_2SO_4. The arrangement of the atoms in sulfuric acid is shown by the following structural formula (the lone electron pairs are omitted for clarity):

```
        O
        |
H—O—S—O—H
        |
        O
```

Similarly, the formula for phosphoric acid is usually written as H_3PO_4, but the arrangement of the atoms in phosphoric acid is shown by the structural formula

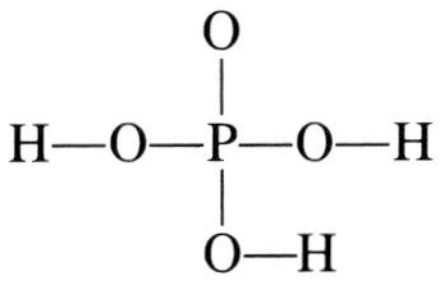

```
        O
        |
H—O—P—O—H
        |
        O—H
```

Each oxoacid we have considered contains one or more OH groups per molecule. Compounds that contain OH groups are therefore collectively called **hydroxo compounds**. The atom in a hydroxo compound to which the OH group is attached greatly influences the strength of the acid. If this atom is highly electronegative, it attracts the bonding electron pair in the O—H bond and makes the hydrogen atom more readily available as a proton. As the electronegativity of the atom to which the OH group is attached increases, acid strength increases. For example, hypochlorous acid, HClO, is stronger than hypobromous acid, HBrO, because chlorine is more electronegative than bromine. Figure 14.5 illustrates how acid strength depends on the electronegativity of the atom to which the OH group is attached.

Let us consider the effect of the electronegativity of the atom X (Figure 14.5) on the acid strengths of two structurally similar acids, phosphoric acid, H_3PO_4, and arsenic acid, H_3AsO_4. In phosphoric acid, three OH groups are attached to a phosphorus atom; in arsenic acid, three OH groups are attached to an arsenic atom. Phosphorus is more electronegative than arsenic. Therefore, phosphoric acid is a stronger acid than arsenic acid.

Example 14.4

Predicting the Relative Strengths of Oxoacids

Which is the stronger acid, sulfurous acid or selenous acid?

SOLUTION: Sulfurous acid, H_2SO_3, and selenous acid, H_2SeO_3, have similar structures. Since sulfur is more electronegative than selenium, sulfurous acid is stronger than selenous acid.

Practice Problem 14.4: Explain whether you expect hypochlorous acid to be a stronger or weaker acid than hypoiodous acid.

Now let us compare the strengths of oxoacids that have the same atom X to which OH groups are attached but which have different structures. The most

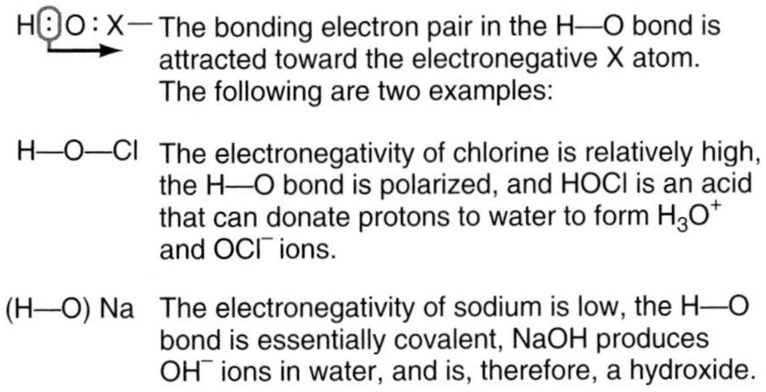

H(:)O : X— The bonding electron pair in the H—O bond is attracted toward the electronegative X atom. The following are two examples:

H—O—Cl The electronegativity of chlorine is relatively high, the H—O bond is polarized, and HOCl is an acid that can donate protons to water to form H_3O^+ and OCl^- ions.

(H—O) Na The electronegativity of sodium is low, the H—O bond is essentially covalent, NaOH produces OH^- ions in water, and is, therefore, a hydroxide.

Figure 14.5

Influence of the electronegativity of the atom X on acid–base character of the compound. The atom X is the atom to which the OH groups are linked in a hydroxo compound.

important factor that determines their relative acid strength is the number of nonprotonated oxygen atoms bonded to the X atom. Nonprotonated oxygen atoms are those which have no hydrogen atoms attached.

The electronegativity of oxygen is the second highest of all the elements. In the chain of atoms, H—O—X—O, the terminal oxygen atom attracts bonding electron pairs throughout the bonding network. Thus, the bonding electron pair in the H—O bond is pulled away somewhat from the H atom by the terminal O atom through the H—O—X—O chain. Increasing the number of nonprotonated oxygen atoms attached to the central X atom increases the electron withdrawing effect from the H atom in the H—O bond, and therefore increases the strength of the acid. For example, nitric acid (HNO_3 or $HONO_2$) is a stronger acid than nitrous acid (HNO_2 or HONO) because nitric acid has two nonprotonated oxygen atoms per molecule, whereas nitrous acid has only one.

Boric acid [H_3BO_3 or $B(OH)_3$], hypochlorous acid (HClO or HOCl), and hypobromous acid (HBrO or HOBr), have *no nonprotonated oxygen atoms* attached to the central atom. The strengths of these acids do not differ greatly from one another. The K_a value of boric acid is about 10^{-10}, K_a for hypobromous acid is approximately 10^{-9}, and K_a for hypochlorous acid is around 10^{-8}.

Acids having *one nonprotonated oxygen atom* per molecule, such as nitrous acid, HNO_2, phosphorous acid, H_2PHO_3, sulfurous acid, H_2SO_3, phosphoric acid, H_3PO_4, and arsenic acid, H_3AsO_4, have K_a values ranging from about 10^{-4} to 10^{-2}. These K_a values are larger than the K_a values for acids with no nonprotonated oxygen atoms by a factor of about 10^5.

Acids with *two nonprotonated oxygen atoms* per molecule are nitric acid, HNO_3 ($K_a \approx 10^1$), chloric acid, $HClO_3$ ($K_a \approx 10^1$), and sulfuric acid, H_2SO_4 ($K_a \approx 10^3$). Perchloric acid, $HClO_4$, which has *three* nonprotonated oxygen atoms per molecule, is a very strong acid whose K_a value is about 10^8.

As the number of oxygen atoms bonded to the X atom increases, the oxidation state of the X atom also increases. Thus, we can relate the oxidation state of the X atom of an oxoacid to the strength of the acid. *Increasing the oxidation state of the X atom increases acid strength.* For example, the oxidation states of S in sulfuric acid (H_2SO_4) and sulfurous acid (H_2SO_3) are +6 and +4, respectively, and sulfuric acid is stronger than sulfurous acid. Similarly, the oxidation state of N in nitric acid (HNO_3) is +5, and the oxidation state of N in nitrous acid (HNO_2) is +3. Nitric acid is a stronger acid than nitrous acid.

We can summarize the factors that affect oxoacid strength as follows.

1. The acid strength of oxoacids that have similar structures increases with increasing electronegativity of the atom X that is bonded to the OH groups of the acid. For example, H_2SO_3 is a stronger acid than H_2SeO_3 because sulfur is more electronegative than selenium.
2. For oxoacids in which the OH groups are bonded to the same atom, X, the acid strength increases as the number of nonprotonated oxygen atoms increases. An increase in the number of nonprotonated oxygen atoms is equivalent to an increase in the oxidation state on the atom X. For oxoacids of chlorine, atom X is chlorine. The relations between acid strength, K_a, the number of nonprotonated oxygen atoms, and the oxidation state on X are illustrated in Table 14.3 for the oxoacids of chlorine, HClO (HOCl), $HClO_2$ (HOClO), $HClO_3$ ($HOClO_2$), and $HClO_4$ ($HOClO_3$):

Table 14.3

Relative Strengths of the Oxoacids of Chlorine

	Increasing acid strength ⟶			
	HOCl	HOClO	$HOClO_2$	$HOClO_3$
Approximate K_a	10^{-8}	10^{-2}	10^{1}	10^{8}
Number of nonprotonated oxygen atoms	0	1	2	3
Oxidation state of Cl	+1	+3	+5	+7

Example 14.5

Predicting the Strengths of Oxoacids

Arrange the following acids in order of increasing strength: sulfurous acid, selenous acid, sulfuric acid, perchloric acid, and arsenous acid.

SOLUTION: Arsenous acid, H_3AsO_3, is the only acid listed that contains no nonprotonated oxygen atoms. Arsenous acid is therefore the weakest acid of this group. Sulfurous acid, H_2SO_3, and selenous acid, H_2SeO_3, each have one nonprotonated oxygen atom attached to the central atom. Sulfuric acid, H_2SO_4, has two nonprotonated oxygen atoms, and perchloric acid, $HClO_4$, has three. Perchloric acid is therefore the strongest of the foregoing acids, followed by sulfuric acid. Sulfurous acid is stronger than selenous acid because sulfur is more electronegative than selenium. The order of increasing acid strength is:

$$H_3AsO_3 < H_2SeO_3 < H_2SO_3 < H_2SO_4 < HClO_4$$

Practice Problem 14.5: Arrange the following acids in order of increasing strength: nitrous acid, nitric acid, hypoiodous acid, and hypochlorous acid.

When the X atom bonded to the OH groups in a hydroxo compound has low electronegativity, as in metal hydroxides such as NaOH, the hydrogen atom of the OH group is not available as a proton (is not acidic). Instead, the compound increases the hydroxide ion concentration in water solution; therefore, it is a base (Figure 14.5). If the electronegativity of the X atom is high, the compound is an acid, if it is low, the compound may be a hydroxide. We next consider some acidic and some basic hydroxo compounds.

Survey of Acidic and Basic Hydroxo Compounds

We recall that metals are less electronegative than nonmetals. If the atom, X, that is connected to one or more hydroxo groups in a hydroxo compound is a metal in a low oxidation state, up to +3, and if only O—H groups are attached to the X atom, the compound is a hydroxide, that is, a base. Some examples of hydroxo compounds that are bases include NaOH and $Mg(OH)_2$.

Table 14.4 lists some hydroxo compounds of representative elements in the second, third, and fourth periods. In each period, most of the hydroxo compounds of the first- and second-group metals are strong bases. Proceeding toward the right in each period, the electronegativity of the atom X increases, and hydroxo compounds become more acidic, particularly when they contain nonprotonated oxygen atoms. Hydroxo compounds of nonmetals at the right of each period are acids.

Table 14.4

Hydroxo Compounds of the Second-, Third-, and Fourth-Period Representative Elements

Group:	I	II	III	IV	V	VI	VII
Second period:	LiOH Strongly basic	$Be(OH)_2$ Amphoteric	H_3BO_3 Weakly acidic $K_1 = 7.3 \times 10^{-10}$	H_2CO_3 (H_2O + CO_2) Acidic $K_1 = 4.3 \times 10^{-7}$	HNO_2 Acidic $K_a = 4.5 \times 10^{-4}$ HNO_3 Strongly acidic K_a = very large		HOF Acidic (unstable)
Third period:	NaOH Strongly basic	$Mg(OH)_2$ Basic	$Al(OH)_3$ Amphoteric	$Si(OH)_4$ or $SiO_2(H_2O)_n$ Weakly acidic $K_1 = 3.2 \times 10^{-10}$	H_2PHO_3 Acidic $K_1 = 1.0 \times 10^{-2}$ H_3PO_4 Acidic $K_1 = 7.5 \times 10^{-3}$	H_2SO_3 Acidic $K_1 = 1.5 \times 10^{-2}$ H_2SO_4 Strongly acidic K_1 = very large	HClO Acidic $K_a = 3.0 \times 10^{-8}$ $HClO_2$ Acidic $K_a = 1.1 \times 10^{-2}$ $HClO_3$ Strongly acidic K_a = very large $HClO_4$ Very strongly acidic K_a = very large
Fourth period:	KOH Strongly basic	$Ca(OH)_2$ Basic	$Ga(OH)_3$ Amphoteric	Ge $(OH)_4$ or $GeO_2(H_2O)_n$ Weakly acidic $K_1 = 2 \times 10^{-9}$	H_3AsO_3 Weakly acidic $K_1 = 6.0 \times 10^{-10}$ H_3AsO_4 Acidic $K_1 = 5.6 \times 10^{-3}$	H_2SeO_3 Acidic $K_1 = 3.5 \times 10^{-3}$ H_2SeO_4 Strongly acidic K_1 = very large	HBrO Weakly acidic $K_a = 2.1 \times 10^{-9}$ $HBrO_3$ Strongly acidic K_a = very large $HBrO_4$ Strongly acidic K_a = very large

Across a period, there is a transition from hydroxides to acids in the middle of a period. For example, in the second period, LiOH and $Be(OH)_2$ are hydroxides, H_3BO_3 is a very weak acid, and the other hydroxo compounds for the elements in this period are weak to strong acids.

There is an interesting variation of the properties in the hydroxo compounds of the group III elements. The hydroxo compound of boron is boric acid, H_3BO_3, which is a weak acid. But the hydroxo compounds of aluminum, $Al(OH)_3$, and of gallium, $Ga(OH)_3$, are bases that have some unique properties. Like all bases, they react with acids. But they also react with strong bases such as sodium hydroxide. *Hydroxides that react with acids and with strong bases* are called **amphoteric hydroxides**. For example, aluminum hydroxide (which is a solid) reacts with aqueous acid in a way that is characteristic of hydroxides:

$$Al(OH)_3(s) + 3H^+(aq) \longrightarrow Al^{3+}(aq) + 3H_2O(l)$$

But since aluminum hydroxide is amphoteric, it also reacts with a concentrated solution of a strong hydroxide:

$$Al(OH)_3(s) + OH^-(aq) \longrightarrow Al(OH)_4^-(aq)$$

The difference in the properties of H_3BO_3 and $Al(OH)_3$ is correlated with the electronegativities of boron and aluminum. The electronegativity of boron (2.0) is higher than the electronegativity of aluminum (1.5). Thus, H_3BO_3 is an acid and $Al(OH)_3$ is an amphoteric hydroxide. The metal atom in an amphoteric hydroxide generally has an electronegativity between 1.5 and 1.9.

As we can see from Table 14.4, beryllium hydroxide is also amphoteric. The electronegativity of beryllium is 1.5, which is not low enough for $Be(OH)_2$ to be strongly basic and not high enough for it to be acidic. Since beryllium hydroxide is amphoteric, it reacts with aqueous sodium hydroxide according to the following equation:

$$Be(OH)_2(s) + 2OH^-(aq) \longrightarrow Be(OH)_4^{2-}(aq)$$

The first element of the fourth group, carbon, forms an unstable, slightly acidic hydroxo compound called carbonic acid, H_2CO_3, which exists in aqueous solution in equilibrium with H_2O and CO_2: $H_2CO_3(aq) \rightleftharpoons H_2O(l) + CO_2(g)$.

Silicon, like carbon, forms a weakly acidic hydroxo compound called silicic acid. The composition of this compound is not known exactly, but is sometimes written as $Si(OH)_4$ or as a hydrated silicon dioxide, $SiO_2(H_2O)_n$, where n is an undetermined number. The electronegativity of silicon (1.8) is between the electronegativities of metals and nonmetals. Silicon is thus a metalloid. Hydroxo compounds of metalloids are either weakly acidic or weakly basic. The weakly acidic character of the hydroxo compound of silicon is probably due to the relatively high, +4 oxidation state of silicon in $Si(OH)_4$.

Tin and lead, the distinctly metallic elements in group IV, form only hydroxides. Tin forms $Sn(OH)_2$, which is amphoteric, and lead forms $Pb(OH)_2$, which is also amphoteric.

Nitrogen, phosphorus, and arsenic, the nonmetallic elements in group V, form distinctly acidic hydroxo compounds. Antimony is a metalloid, which forms a weakly acidic solution of its oxide, Sb_4O_6. The only metal in this group, bismuth, forms a hydroxide, $Bi(OH)_3$, which is basic.

All the non-radioactive elements in groups VI and VII are nonmetals. Their hydroxo compounds are acidic.

Example 14.6

Predicting Relative Acid Strength of Hydroxo Compounds

Which of the following compounds is a stronger acid: $Ga(OH)_3$ or $As(OH)_3$?

SOLUTION: First, we examine the formulas of the compounds to see if they have similar structures. Both of these compounds have similar structures. Neither contains nonprotonated oxygen atoms on the central atom. Because arsenic is more electronegative than gallium, $As(OH)_3$ is a stronger acid than $Ga(OH)_3$. $As(OH)_3$ is arsenous acid, which is usually written as H_3AsO_3, whereas $Ga(OH)_3$ is gallium hydroxide, an amphoteric hydroxide.

Hydroxides and Acids of Transition Metals

Hydroxo compounds of transition metals can be basic hydroxides, amphoteric hydroxides, or acids. Some hydroxo compounds of transition metals which are commonly called "hydroxides" are really hydrated oxides. One such compound is hydrated chromium(III) oxide, $Cr_2O_3(H_2O)_n$ (the value of n can vary). This compound is usually called chromium(III) hydroxide and is represented by the formula $Cr(OH)_3$. It is an amphoteric hydroxide. The transition metals scandium, titanium, and vanadium form hydrated oxides rather than hydroxides.

We pointed out earlier in this section that as the oxidation state of the central atom of a hydroxo compound increases, its acidity increases. This is also true of hydrated oxides. When the oxidation state of a metal in an oxide is high, the metal withdraws electron density from the bonding electron pairs in the O—H bonds of the attached water molecules. We can write this as

$$\begin{array}{c} \quad\quad\quad \mathrm{H} \\ \quad\quad\quad | \\ \mathrm{M}^{x+} \longleftarrow \mathrm{O} \text{—} \mathrm{H} \end{array}$$

where M^{x+} is a metal ion. When the electron density in an O—H bond decreases, the hydrogen atom is readily extracted by a base; that is, the metal oxide is acidic. For example, chromium(VI) oxide, CrO_3, is the anhydride of chromic acid, H_2CrO_4, which is a strong acid. The chromium atom in CrO_3 and in H_2CrO_4 is in +6 oxidation state, which is the highest possible oxidation state for chromium.

Manganese in its +2 oxidation state forms a basic hydroxide, $Mn(OH)_2$. Manganese in its highest oxidation state forms the very strong permanganic acid, $HMnO_4$, the anhydride of which is the unstable and explosive manganese(VII) oxide, Mn_2O_7.

Iron in its lower +2 oxidation state forms iron(II) hydroxide, $Fe(OH)_2$. In the more common +3 state, iron forms a hydrated oxide, $Fe_2O_3(H_2O)_n$, a slightly amphoteric compound that is commonly called iron(III) hydroxide. This hydroxide is represented by the formula $Fe(OH)_3$. The most stable oxidation state for cobalt, nickel, and copper is +2. Each of these metals in its +2 oxidation state forms a well-defined hydroxide: $Co(OH)_2$, $Ni(OH)_2$, and $Cu(OH)_2$. Of these hydroxides, $Co(OH)_2$ and $Cu(OH)_2$ are somewhat amphoteric. Zinc hydroxide and aluminum hydroxide are among the most common amphoteric hydroxides. The reaction of zinc hydroxide with a solution of an alkali metal hydroxide occurs according to the equation

$$Zn(OH)_2(s) + 2OH^-(aq) \longrightarrow Zn(OH)_4^{2-}(aq)$$

We recall that atoms whose electronegativities lie between 1.5 and 1.9 form amphoteric hydroxides. The electronegativity for aluminum is 1.5; for beryllium, 1.5; and for zinc, 1.6. Of the other familiar transition elements in periods 4 and 5, silver and gold form only hydrated oxides.

Mercury in its +1 oxidation state forms neither an oxide nor a hydroxide. Mercury in its +2 state forms mercury(II) oxide, HgO. This compound dissolves very slightly in water to produce a very weakly basic solution that is commonly called mercury(II) hydroxide.

Since an amphoteric hydroxide dissolves in a solution of a strong hydroxide, this property can be used to distinguish an amphoteric hydroxide from a hydroxide which is not amphoteric, as we will see in Example 14.7.

Example 14.7

Distinguishing Between an Amphoteric and a Nonamphoteric Hydroxide

Magnesium hydroxide and zinc hydroxide are white solids. If you were given these two solids in unlabeled containers, what reagent would you use to identify the solids? Write a net ionic equation for the reaction that would occur if the solid were zinc hydroxide.

SOLUTION: A concentrated solution of a strong hydroxide such as an alkali metal hydroxide can be used to distinguish between these two solids. For example, sodium hydroxide solution "dissolves" zinc hydroxide, which is amphoteric. But it does not dissolve magnesium hydroxide, which is not amphoteric. The equation for the reaction of zinc hydroxide in a concentrated solution of an alkali metal hydroxide is

$$Zn(OH)_2(s) + 2OH^-(aq) \longrightarrow Zn(OH)_4^{2-}$$

To carry out this experiment, a small amount of one of the solids is stirred with excess sodium hydroxide solution. If the solid dissolves, it is $Zn(OH)_2$. If it does not dissolve, it is $Mg(OH)_2$.

Practice Problem 14.6: What reagent would you use to identify a solid that can be either iron(II) hydroxide or aluminum hydroxide? Write a net ionic equation for the reaction that would occur if the solid were aluminum hydroxide.

Organic Acids

In organic acids, called **carboxylic acids**, the O—H group is a part of a group of atoms called a **carboxyl group**, $—CO_2H$. The atoms in a carboxyl group are linked as follows:

```
     :O:
      ‖   ..
    —C—O—H
          ..
```

Carboxyl group

The carbon atom of the carboxyl group is attached to a group of atoms called an "R" group. The R group can be a hydrogen atom or a hydrocarbon group such as methyl (CH_3—), ethyl (CH_3CH_2—), propyl ($CH_3CH_2CH_2$—), and others. Thus, the general formula for a carboxylic acid is

$$\text{R—}\overset{\overset{\textstyle :\text{O}:}{\|}}{\text{C}}\text{—}\ddot{\underset{\cdot\cdot}{\text{O}}}\text{—H}$$

Carboxylic acid

In the formulas for carboxylic acids, the R group is written first in the formula of a carboxylic acid, followed by the carboxyl group. The structures of some simple carboxylic acids and their formulas are shown below:

$$\text{H—}\overset{\overset{\textstyle :\text{O}:}{\|}}{\text{C}}\text{—}\ddot{\underset{\cdot\cdot}{\text{O}}}\text{—H} \qquad \text{H—}\underset{\underset{\textstyle \text{H}}{|}}{\overset{\overset{\textstyle \text{H}}{|}}{\text{C}}}\text{—}\overset{\overset{\textstyle :\text{O}:}{\|}}{\text{C}}\text{—}\ddot{\underset{\cdot\cdot}{\text{O}}}\text{—H} \qquad \text{H—}\underset{\underset{\textstyle \text{H}}{|}}{\overset{\overset{\textstyle \text{H}}{|}}{\text{C}}}\text{—}\underset{\underset{\textstyle \text{H}}{|}}{\overset{\overset{\textstyle \text{H}}{|}}{\text{C}}}\text{—}\overset{\overset{\textstyle :\text{O}:}{\|}}{\text{C}}\text{—}\ddot{\underset{\cdot\cdot}{\text{O}}}\text{—H}$$

$HCOOH$	CH_3COOH	CH_3CH_2COOH
Formic acid Causes the "sting" in an ant bite.	Acetic acid (R is CH_3—, methyl group). Ingredient in vinegar.	Propanoic acid (R is CH_3CH_2—, ethyl group). Small amounts in dairy products.

$$\text{H—}\underset{\underset{\textstyle \text{H}}{|}}{\overset{\overset{\textstyle \text{H}}{|}}{\text{C}}}\text{—}\underset{\underset{\textstyle \text{H}}{|}}{\overset{\overset{\textstyle \text{H}}{|}}{\text{C}}}\text{—}\underset{\underset{\textstyle \text{H}}{|}}{\overset{\overset{\textstyle \text{H}}{|}}{\text{C}}}\text{—}\overset{\overset{\textstyle :\text{O}:}{\|}}{\text{C}}\text{—}\ddot{\underset{\cdot\cdot}{\text{O}}}\text{—H}$$

$CH_3CH_2CH_2COOH$

Butanoic acid
(R is $CH_3CH_2CH_2$—,
propyl group).
Responsible
for the odor of
rancid butter.

The only ionizable hydrogen in a carboxylic acid is the one attached to the oxygen atom of the carboxyl group. The ion that remains after the transfer of the proton from a carboxylic acid to a base is called a **carboxylate ion**. A carboxylate ion is the conjugate base of a carboxylic acid. A carboxylic acid reacts with water to establish an equilibrium mixture of carboxylic acid molecules, carboxylate anions, and hydronium ions:

$$\text{R—}\overset{\overset{\textstyle :\text{O}:}{\|}}{\text{C}}\text{—}\ddot{\underset{\cdot\cdot}{\text{O}}}\text{—H}(aq) + H_2O(l) \longrightarrow \left[\text{R—}\overset{\overset{\textstyle :\text{O}:}{\|}}{\text{C}}\text{—}\ddot{\underset{\cdot\cdot}{\text{O}}}:\right]^- (aq) + H_3O^+(aq)$$

Carboxylic acid   Carboxylate anion

The formulas and K_a values of four carboxylic acids—formic, acetic, propanoic, and butanoic acids—are written below.

	HCOOH	CH_3COOH	CH_3CH_2COOH	$CH_3CH_2CH_2COOH$
$K_a =$	1.8×10^{-4}	1.8×10^{-5}	1.3×10^{-5}	1.5×10^{-5}

Formic acid is stronger than acetic acid, but acetic acid, propanoic acid, and butanoic acid do not differ very much in strength, as shown by their K_a values.

We can explain this trend in acid strength in terms of the electronegativity differences of carbon and hydrogen versus carbon and oxygen atoms. Carbon is slightly more electronegative than hydrogen. Therefore, the carbon atom of the methyl group attracts the bonding electron pairs and becomes a relatively electron dense center. The methyl group in acetic acid donates a part of its electron density to the O—H group in the acid. This transfer of electron density increases the covalent character of the O—H bond and makes the hydrogen atom more difficult to remove as a proton. Thus, acetic acid is a poorer proton donor, or a weaker acid, than formic acid, whose R group is a hydrogen atom.

When a hydrogen atom in the hydrocarbon portion of a carboxylic acid is replaced by a more electronegative atom such as a chlorine atom, the highly electronegative Cl atom attracts the bonding electron pair of the O—H bond in the carboxyl group. As a result, the H atom in the carboxyl group becomes more readily available as a proton, and the acid strength increases. For example, the K_a value for monochloroacetic acid, $CH_2ClCOOH$, is 1.4×10^{-3} compared with 1.8×10^{-5} for acetic acid. Substituting additional Cl atoms for H atoms polarizes the O—H bond of the acid even more, and the acid strength increases. Thus, dichloroacetic acid, $CHCl_2COOH$ ($K_a = 3.3 \times 10^{-2}$), is stronger than monochloroacetic acid. Trichloroacetic acid, CCl_3COOH ($K_a = 2 \times 10^{-1}$), is a relatively strong acid.

14.4 Bases

In Chapter 4 we learned that hydroxides of alkali metals and of alkaline earth metals are strong bases. These strong bases are ionic solids that consist of metal ions and hydroxide ions. A weak base such as ammonia, however, produces hydroxide ions in aqueous solution by reacting with water. The reaction is

$$NH_3(aq) + H_2O(l) \rightleftharpoons NH_4^+(aq) + OH^-(aq)$$

A solution in which the hydroxide ion concentration is more than 10^{-7} *M* is said to be basic, or *alkaline*.

Base Ionization Constant, K_b

The general equation for the reaction of a weak base, B, with water is

$$B(aq) + H_2O(l) \rightleftharpoons BH^+(aq) + OH^-(aq)$$

The equilibrium constant for this reaction is given by

$$K = \frac{[BH^+][OH^-]}{[H_2O][B]}$$

The right-hand side of the equation above contains water, whose concentration is virtually constant and is therefore omitted from the expression of the ionization constant:

$$K_b = \frac{[BH^+][OH^-]}{[B]}$$

K_b is called the **base ionization constant**. Table 14.5 lists the K_b values of some weak bases at 25°C.

The base ionization constant of a weak base can be calculated from the known initial molarity of the base solution, and from the measured molar concentration of the hydroxide ions in the solution, as shown in Example 14.8 for aqueous ammonia.

Example 14.8

Calculating the Base Ionization Constant of Ammonia

In a 0.10 *M* solution of ammonia the hydroxide ion concentration is 1.34×10^{-3} *M* at 25°C. Calculate the value of the base ionization constant.

SOLUTION: As before, we first write the chemical equation for the ionization reaction. Then we write the initial and equilibrium concentrations of each species:

	$NH_3(aq) + H_2O(l) \rightleftharpoons$	$NH_4^+(aq)$	+ $OH^-(aq)$
Initial *M*	0.100	0	~0
Equilibrium *M*	$0.100 - 1.34 \times 10^{-3}$	1.34×10^{-3}	1.34×10^{-3}

Autoionization of water produces a few hydroxide ions, whose concentration is negligibly small compared with the concentration of hydroxide ions produced by the ionization of ammonia (1.34×10^{-3} *M*).

The balanced ionization equation shows that ammonium ions and hydroxide ions are present in a 1:1 mole ratio. Therefore, $[NH_4^+] = [OH^-] = 1.34 \times 10^{-3}$ *M*. These ions are produced by partial ionization of the initial 0.10 *M* solution of ammonia, leaving a final ammonia concentration of $(0.10 - 1.34 \times 10^{-3})$ *M*. Since 1.34×10^{-3} is much less than 0.10, the molar concentration of the nonionized ammonia at equilibrium is essentially equal to its initial concentration, 0.10 *M*. That is, according to the rules of significant figures (Chapter 1), subtracting 0.00134 from 0.10 is 0.10:

$$\begin{array}{r} 0.10\ \mathrm{M} \\ -0.00134\ M \\ \hline 0.10\ M \end{array}$$

The value of the base ionization constant is

$$K_b = \frac{[NH_4^+][OH^-]}{[NH_3]} = \frac{(1.34 \times 10^{-3})^2}{0.10} = 1.8 \times 10^{-5}$$

Since K_b is much less than 1.0, the equilibrium lies far to the left, and aqueous ammonia is a relatively weak base (Table 14.5).

Table 14.5

Ionization Constants of Common Weak Bases at 25°C

Name	Ionization Equation	K_b
Ammonia	$NH_3 + H_2O \rightleftharpoons NH_4^+ + OH^-$	1.79×10^{-5}
Aniline	$C_6H_5NH_2 + H_2O \rightleftharpoons C_6H_5NH_3^+ + OH^-$	4.2×10^{-10}
Dimethylamine	$(CH_3)_2NH + H_2O \rightleftharpoons (CH_3)_2NH_2^+ + OH^-$	5.9×10^{-4}
Ethylamine	$C_2H_5NH_2 + H_2O \rightleftharpoons C_2H_5NH_3^+ + OH^-$	6.4×10^{-4}
Hydrazine	$H_2NNH_2 + H_2O \rightleftharpoons H_2NNH_3^+ + OH^-$	1.7×10^{-6}
	$H_2NNH_3^+ + H_2O \rightleftharpoons H_3NNH_3^{2+} + OH^-$	1.3×10^{-15}
Hydroxylamine	$NH_2OH + H_2O \rightleftharpoons NH_3OH^+ + OH^-$	1.1×10^{-8}
Methylamine	$CH_3NH_2 + H_2O \rightleftharpoons CH_3NH_3^+ + OH^-$	4.2×10^{-4}
Pyridine	$C_6H_5N + H_2O \rightleftharpoons C_6H_5NH^+ + OH^-$	1.5×10^{-9}
Trimethylamine	$(CH_3)_3N + H_2O \rightleftharpoons (CH_3)_3NH^+ + OH^-$	6.3×10^{-5}

Practice Problem 14.7: In a 0.20 *M* aqueous ammonia solution the hydroxide ion concentration is 1.9×10^{-3} *M*. Calculate the base ionization constant for 0.20 *M* ammonia. Compare your answer with the value for K_b obtained in Example 14.8.

Many other weak bases are known. These include methylamine, CH_3NH_2 (used in the tanning industry), ethyl amine, $C_2H_5NH_2$ (used in oil refining), and pyridine, C_5H_5N (a solvent and a reagent in analytical chemistry). The reactions of some weak bases with water, and their base ionization constants are listed in Table 14.5.

Basic Anions

Many negative ions are sufficiently basic to extract protons from water. We now consider basic anions and their reactions with water. In Section 14.2 we noted that a weak acid ionizes to produce a relatively strong conjugate base and that a strong acid ionizes to produce a relatively weak conjugate base. For example, hydrochloric acid dissociates virtually completely into hydronium ions and chloride ions in a dilute aqueous solution. Chloride ion is a very weak base that is unable to extract a proton from water. On the other hand, hydrofluoric acid is a weak acid that dissociates only partially in water. Therefore, its conjugate base, the fluoride ion, is a stronger base than chloride ion. Table 14.6 lists some common acids and their conjugate bases.

Table 14.6 also shows that anions such as nitrate and chloride—the conjugate bases of the strong acids nitric acid and hydrochloric acid, respectively—cannot extract protons from water and therefore do not react with water. However, the anions of weak acids can extract protons from water to generate hydroxide ions. For example, the conjugate base of nitrous acid, a weak acid, is nitrite ion. Nitrite ions react with water to form an equilibrium mixture that contains unreacted nitrite ions, nitrous acid molecules, and hydroxide ions:

$$NO_2^-(aq) + H_2O(l) \rightleftharpoons HNO_2(aq) + OH^-(aq)$$

Because nitrite ions react with water to produce hydroxide ions and sodium ions do not react with water, an aqueous solution of sodium nitrite is slightly alkaline.

Table 14.6

Relative Strengths of Some Common Acids and Their Conjugate Bases

	Acid			Base	
Very strong acids, nearly 100 percent dissociated in H_2O	$HClO_4$	↑ Increasing acid strength	Increasing base strength ↓	ClO_4^-	Very weak bases, they do not accept protons from water (not protonated in water)
	H_2SO_4			HSO_4^-	
	HI			I^-	
	HBr			Br^-	
	HCl			Cl^-	
	HNO_3			NO_3^-	
Moderately strong to weak acids, partially dissociated in H_2O	H_3PO_4			$H_2PO_4^-$	Weak bases, partially protonated in water
	HNO_2			NO_2^-	
	HF			F^-	
	HCOOH			$HCOO^-$	
	$HC_2H_3O_2$			$C_2H_3O_2^-$	
	H_2S			HS^-	
	HCN			CN^-	
	HS^-			S^{2-}	
	H_2O			OH^-	
Very weak acids (bases relative to H_2O)	NH_3			NH_2^-	Very strong bases, 100 percent protonated in H_2O
	OH^-			O^{2-}	

Example 14.9

Predicting Relative Basic Strengths of Anions and Writing Equations for Reactions of Anions with Water

Which one of the following solutions is the most basic, 1 *M* sodium nitrate, 1 *M* sodium formate, or 1 *M* sodium cyanide? Write an equation for the reaction responsible for the basicity in each case. Look up the necessary K_a values of weak acids in Table 14.2.

SOLUTION: A 1 *M* solution of sodium nitrate is neutral because neither sodium ions nor nitrate ions react with water. Nitrate ion is the conjugate base of a strong acid, nitric acid, and has virtually no tendency to extract protons from water.

Formate ion, CHO_2^-, is a relatively strong conjugate base of the weak acid, formic acid. Formate ion reacts with water according to the equation

$$CHO_2^-(aq) + H_2O(l) \rightleftharpoons HCHO_2(aq) + OH^-(aq)$$

Cyanide ion, CN^-, is a relatively strong conjugate base of the very weak acid, hydrocyanic acid, HCN. The reaction of cyanide ion with water is

$$CN^-(aq) + H_2O(l) \rightleftharpoons HCN(aq) + OH^-(aq)$$

Table 14.2 shows that hydrocyanic acid is a weaker acid than formic acid. Therefore, cyanide ion is a stronger base than formate ion, and a 1 *M* solution of sodium cyanide is more basic than a 1 *M* solution of sodium formate.

Practice Problem 14.8: Which solution is more basic, 0.50 *M* potassium fluoride or 0.50 *M* potassium nitrite? Write an equation for the reaction of nitrite ions with water.

Hydrolysis

In the preceding section we examined the reactions of some anions, such as nitrite and cyanide ions with water. As we noted, these anions are strong enough bases to extract protons from water to produce an equilibrium mixture containing hydroxide ion.

Now let us consider the reaction of water with a cation such as ammonium ion, that is a conjugate acid of a weak base. Ammonium ions are sufficiently acidic to donate protons to water to form hydronium ions and ammonia molecules:

$$NH_4^+(aq) + H_2O(l) \rightleftharpoons H_3O^+(aq) + NH_3(aq)$$

When a salt of ammonium ions, such as ammonium chloride, is dissolved in water, the resulting solution is slightly acidic because the ammonium ions react with water. *Reactions of ions with water which generate aqueous hydronium ions or aqueous hydroxide ions* are called **hydrolysis reactions**.

The hydrolysis of an anion produces the conjugate acid of the anion. The hydrolysis reaction is therefore related to the ionization of the conjugate acid and to the autoionization of water. This is shown below for the hydrolysis of nitrite ions and the ionization of nitrous acid, the conjugate acid of nitrite ions. The combination of these two reactions gives the reaction for autoionization of water:

Hydrolysis:	$NO_2^-(aq) + H_2O(l) \rightleftharpoons HNO_2(aq) + OH^-(aq)$	K_b
Ionization:	$HNO_2(aq) + H_2O(l) \rightleftharpoons H_3O^+(aq) + NO_2^-(aq)$	K_a
Autoionization:	$2H_2O(l) \rightleftharpoons H_3O^+(aq) + OH^-(aq)$	K_w

We recall from Chapter 13 that when we add two related reactions, the equilibrium constant for the net reaction is the product of the equilibrium constants of the individual reactions. Thus,

$$K_w = (K_a)(K_b)$$

We can write similar equations for the hydrolysis of ammonium ions, and the ionization of its conjugate base, ammonia:

Hydrolysis:	$NH_4^+(aq) + H_2O(l) \rightleftharpoons H_3O^+(aq) + NH_3(aq)$	K_a
Ionization:	$NH_3(aq) + H_2O(l) \rightleftharpoons NH_4^+(aq) + OH^-(aq)$	K_b
Autoionization:	$2H_2O(l) \rightleftharpoons H_3O^+(aq) + OH^-(aq)$	K_w

Once again, we see that $K_w = (K_a)(K_b)$.

We know that the value of K_w at 25°C is 1.0×10^{-14}, and we can look up the values of K_a or K_b for weak acids and weak bases in Tables 14.2 and 14.4, respectively. If we know the value of K_a for a weak acid, we can calculate the value of K_b for its conjugate base. Or if we know K_b for a weak base, we can calculate K_a for its conjugate acid. For example, we find in Table 14.2 that the K_a value for nitrous acid, HNO_2, is 4.5×10^{-4}. The value of K_b for nitrite ion, NO_2^-, is

$$K_b = \frac{K_w}{K_a} = \frac{1.0 \times 10^{-14}}{4.5 \times 10^{-5}} = 2.2 \times 10^{-10}$$

We can perform a similar calculation for any conjugate acid–base pair. For example, K_b for ammonia is 1.8×10^{-5}. The K_a value for the conjugate acid of ammonia (ammonium ion), is

$$K_a = \frac{K_w}{K_b} = \frac{1.0 \times 10^{-14}}{1.8 \times 10^{-5}} = 5.6 \times 10^{-10}$$

Since the values of K_b for nitrite ion and K_a for ammonium ion are very small (much less than 1.0), these ions hydrolyze only to a very slight extent. The equilibrium positions for these hydrolysis reactions lie far to the left. If we know the K_b values of two or more anions, the one with the larger value of K_b hydrolyzes to the greater extent. The anion that hydrolyzes to the greater extent produces the highest concentration of hydroxide ions, assuming that the initial concentrations of the anions are equal.

Practice Problem 14.9: Write equations for the hydrolysis of cyanide and fluoride ions. Calculate the value of K_b for cyanide and for fluoride ions. If one solution has a cyanide ion concentration of 0.5 *M* and a second solution has a fluoride ion concentration of 0.5 *M*, which solution is more basic?

In summary, salts such as sodium chloride, whose ions are derived from strong acids and strong bases, form neutral solutions when dissolved in water because neither their cations nor their anions hydrolyze. On the other hand, salts such as potassium nitrite, which are formed from weak acids and strong bases, form basic solutions because the anions of weak acids hydrolyze. Salts of strong acids and weak bases, such as ammonium chloride, form acid solutions because their cations hydrolyze.

14.5 Lewis Acids and Bases

While the Brønsted definition extends the realm of acids and bases beyond that of the Arrhenius definition, the Lewis definition is even more comprehensive than the Brønsted definition. Gilbert N. Lewis, the American chemist who invented the dot symbols and formulas we discussed in Chapters 7 and 8, gave the following definition of an acid and a base. A **Lewis acid** is an *electron pair acceptor*, and a **Lewis base** is an *electron pair donor*. The Lewis definition thus broadens the realm of acids and bases to include substances that do not contain ionizable hydrogen atoms.

Brønsted bases are also Lewis bases because they donate an electron pair for a proton from an acid. A proton is also a Lewis acid, but not the only one. Many species other than a proton can accept electron pairs for covalent bonding. Many chemical reactions occur by the formation of coordinate covalent bonds (Chapter 8). The Lewis definition of acids and bases is particularly well suited to these reactions.

For example, in the reaction of ammonia with a proton,

$$H^+ + :\overset{\displaystyle H}{\underset{\displaystyle H}{\overset{|}{\underset{|}{N}}}}—H \longrightarrow \left[H—\overset{\displaystyle H}{\underset{\displaystyle H}{\overset{|}{\underset{|}{N}}}}—H \right]^+$$

ammonia is a Lewis base because it acts as an electron pair donor. The proton is a Lewis acid because it accepts the electron pair. The product of a Lewis acid–base reaction is often called an "adduct," because it results from the "addition" of the acid and the base.

Another example of a Lewis acid is boron trichloride, BCl_3. The central boron atom in a BCl_3 molecule does not have a full octet. A boron trichloride

APPLICATIONS OF CHEMISTRY 14.1
Disappearing Books

If you have ever seen an old, brittle, and yellowed newspaper or photograph, you know that paper does not age gracefully (see photo). Paper aging is a very serious concern for the Library of Congress. The Library has about 20 million books in its care, and it receives 7000 additional documents every day. Three-fourths of the Library's books have a life expectancy as short as 25 years. Many nonreplaceable items, such as Alexander Graham Bell's first sketches of a telephone and Freud's letters, face extinction. Why does paper age, and what can be done about it?

Untreated paper made from wood pulp is porous. Ink applied to the surface of untreated paper goes into the pores and spreads to produce a blurred image. To print sharper images, paper processors fill the pores with materials called sizing compounds, such as aluminum sulfate. Aluminum sulfate is the salt of a strong acid (sulfuric acid) and a weak base (aluminum hydroxide). Because paper is between 4 and 7 percent water by weight, the aluminum ions in aluminum sulfate undergo hydrolysis to form an acidic solution. This acidic solution attacks the cellulose fibers in paper, and it decomposes.

The Library of Congress individually treats the pages of its more valuable documents with basic solutions at a cost of hundreds to thousands of dollars per book. Unfortunately, they can only treat 23,000 volumes a year in this way. Also, many of the older works are so brittle that they are beyond treatment. The library preserves these brittle works on microfilm or laser disks.

To combat this problem in the future, the paper-making industry is producing greater quantities of alkaline paper each year. Alkaline paper should last at least 300 years. To help further, alkaline sizing compounds are used to treat paper before it is printed on.

Source: *Chemical & Engineering News,* April 29, 1991, pages 9–11.

molecule can therefore accept a pair of electrons to complete the octet on boron. An ammonia molecule has a lone electron pair on the nitrogen atom; it can therefore act as an electron pair donor. Boron trichloride reacts as a Lewis acid with ammonia as a Lewis base:

$$\begin{array}{ccccccc} & :\ddot{Cl}: & & H & & :\ddot{Cl}: & H \\ & | & & | & & | & | \\ :\ddot{Cl}- & B & + & :N-H & \longrightarrow & :\ddot{Cl}-B & -N-H \\ & | & & | & & | & | \\ & :\underset{..}{Cl}: & & H & & :\underset{..}{Cl}: & H \\ & \text{Lewis acid} & & \text{Lewis base} & & \text{Lewis acid–base adduct} & \end{array}$$

In the course of this reaction, the trigonal planar boron trichloride molecule converts to a tetrahedral structure at the boron atom. Thus, both the boron and the nitrogen atoms in Cl_3BNH_3 are tetrahedral (Figure 14.6).

Figure 14.6

Combination of trigonal planar BCl_3 and pyramidal NH_3 to form Cl_3BNH_3 in which the B and N atoms are tetrahedral.

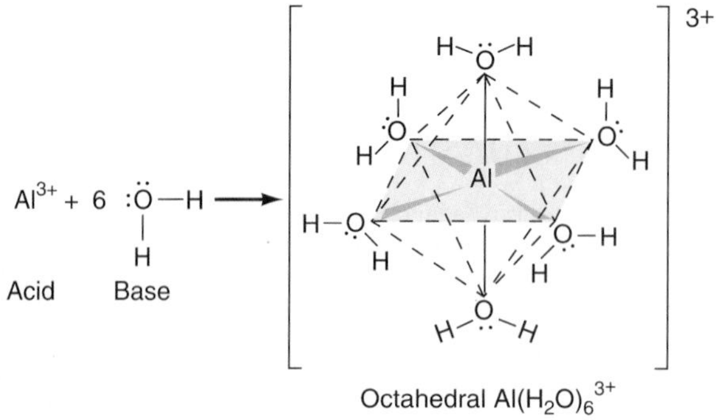

Figure 14.7

Hydration of Al^{3+} ions is an example of a Lewis acid–base reaction. The Al ion is the electron pair acceptor to which water molecules donate their electron pairs.

$$Al^{3+} + 6\ :\ddot{O}H_2 \longrightarrow [Al(H_2O)_6]^{3+}$$

Acid Base

Octahedral $Al(H_2O)_6^{3+}$

Hydration of Metal Ions

Lewis acid–base reactions include the hydration of metal ions. In the hydration of a metal ion, water molecules bond to the ion. The metal ion acts as a Lewis acid, and the water molecules act as Lewis bases. A water molecule has two unshared electron pairs on its oxygen atom. It can donate one of these electron pairs to the cation. The hydration of an aluminum ion is illustrated in Figure 14.7. An aluminum ion can bind six water molecules in an octahedral geometry.

All metal ions in water solution are hydrated to varying extents. Many metal ions, such as Al^{3+}, Fe^{3+}, and Co^{2+} ions, combine with six water molecules. Other ions, such as Zn^{2+} and Cu^{2+} ions, usually bind four water molecules, while Ag^{+} ion binds two water molecules. For many other ions it is impossible to predict the number of water molecules that surround a metal ion.

Hydrolysis of Hydrated Metal Ions

Hydrated metal ions that have a high charge and small radius hydrolyze in water to produce acidic solutions. For example, solutions of beryllium nitrate, aluminum nitrate, chromium(III) nitrate, and iron(III) nitrate are acidic. On the other hand, solutions of sodium nitrate, potassium nitrate, and barium nitrate are neutral. This indicates that hydrated sodium, potassium, and barium ions do not hydrolyze. We recall from our previous discussion that the nitrate ion does not hydrolyze. The acidity of a metal nitrate solution, if any, is due to hydrolysis of the metal ion.

All metal ions in aqueous solution are hydrated. If a highly charged metal ion with a small volume is hydrated, it effectively attracts the bonding electron pairs in the O—H bonds of the water molecules that surround it. As a result, the hydrogen atoms in these water molecules become available as protons that can be extracted by water molecules outside the hydration sphere. Such hydrated metal ions are weak acids that donate protons to water molecules outside the hydration sphere to produce hydronium ions and thus a slightly acidic solution. The higher the charge of the cation and the smaller its volume, the more the O—H bonds are polarized and the more readily the hydrated ion donates protons to water.

Let us consider the hydrolysis of a hydrated aluminum ion. An aluminum ion has a 3+ charge concentrated in a relatively small volume. Hydrated aluminum ions therefore hydrolyze to produce an acidic solution. The hydrolysis of a hydrated aluminum ion proceeds in three steps. These steps occur by the loss of a proton by each of three different water molecules bonded to the aluminum ion. As a result, a hydrated aluminum hydroxide is formed:

Step 1: $Al(H_2O)_6^{3+}(aq) + H_2O(l) \rightleftharpoons Al(H_2O)_5OH^{2+}(aq) + H_3O^+(aq)$

Step 2: $Al(H_2O)_5OH^{2+}(aq) + H_2O(l) \rightleftharpoons Al(H_2O)_4(OH)_2^+(aq) + H_3O^+(aq)$

Step 3: $Al(H_2O)_4(OH)_2^+(aq) + H_2O(l) \rightleftharpoons Al(H_2O)_3(OH)_3(s) + H_3O^+(aq)$

The third step in the hydrolysis of the hydrated aluminum ion produces the insoluble, hydrated aluminum hydroxide. A water solution of an aluminum salt contains equilibrium concentrations of all the species listed in the three equilibria above, except the hydrated aluminum hydroxide, $Al(H_2O)_3(OH)_3$, which is insoluble.

Example 14.10

Hydrolysis of a Hydrated Beryllium Ion

Write an equation for the first step of hydrolysis of $Be(H_2O)_4^{2+}$ ions. Would you expect a solution of beryllium nitrate to be acidic or basic?

SOLUTION: The first step in the hydrolysis of a hydrated beryllium ion is

$$Be(H_2O)_4^{2+}(aq) + H_2O(l) \rightleftharpoons Be(H_2O)_3OH^+(aq) + H_3O^+(aq)$$

The reaction produces hydronium ions. Therefore, an aqueous solution of beryllium nitrate is acidic. (Nitrate ions, we recall, do not hydrolyze.)

Practice Problem 14.10: Write equations for the three steps of hydrolysis of $Fe(H_2O)_6^{3+}$.

Not all hydrated cations hydrolyze. For example, a hydrated barium ion, like a hydrated beryllium ion, has a 2+ charge, but its volume is larger than that of the beryllium ion. Barium ions cannot, therefore, sufficiently polarize the O—H bonds in the water of hydration. Consequently, hydrated barium ions do not hydrolyze. For the same reason, water solutions of alkali metal nitrates do not hydrolyze and their solutions are neither acidic nor basic.

Reactions of Basic Oxides with Acidic Oxides

The reactions of metal oxides with nonmetal oxides can also be classified as Lewis acid–base reactions. Basic metal oxides react with nonmetal oxides to form salts. For example, calcium oxide (lime) reacts with carbon dioxide to form calcium carbonate (limestone):

$$CaO(s) + CO_2(g) \longrightarrow CaCO_3(s)$$

In this reaction, unshared electron pairs of the oxide ions of calcium oxide are donated to the relatively electron-deficient carbon atoms in carbon dioxide molecules. Thus, an oxide ion is a Lewis base, and carbon dioxide is a Lewis acid:

$$Ca^{2+}\ [:\ddot{O}:]^{2-} + O{=}C{=}O \longrightarrow Ca^{2+} + \left[:\ddot{O}{-}\overset{\overset{:O:}{\|}}{C}{-}\ddot{O}:\right]^{2-}$$

Base Acid CaCO$_3$

Another example of a Lewis acid is sulfur trioxide. It can react with oxide ions from a metal oxide to form sulfate ions:

$$[:\ddot{O}:]^{2-} + :\ddot{O}{-}\underset{\underset{:O:}{\|}}{S}{-}\ddot{O}: \longrightarrow \left[:\ddot{O}{-}\overset{\overset{:\ddot{O}:}{|}}{\underset{\underset{:\ddot{O}:}{|}}{S}}{-}\ddot{O}:\right]^{2-}$$

Base Acid

The metal ions associated with the oxide ions combine with the sulfate ions formed to give a metal sulfate as the final product of the reaction. For example, the net reaction of barium oxide with sulfur trioxide is

$$BaO(s) + SO_3(g) \longrightarrow BaSO_4(s)$$

Practice Problem 14.11: Write an equation for the reaction of potassium oxide with sulfur dioxide. Identify the Lewis acid and the Lewis base.

Summary

Definitions of Acids and Bases

In this chapter we discussed the *Arrhenius*, *Brønsted*, and *Lewis* definitions of acids and bases. *An Arrhenius acid produces H^+ ions in water*, and *an Arrhenius base produces OH^- ions in water. A Brønsted acid is a proton donor,* and *a Brønsted base is a proton acceptor. A Lewis acid is an electron pair acceptor*, and *a Lewis base is an electron pair donor.* The substance formed when a Brønsted acid loses a proton is the **conjugate base** of the acid, and the substance formed when a Brønsted base accepts a proton is the **conjugate acid** of the base. *The weaker the acid, the stronger its conjugate base*, and *the weaker the base, the stronger its conjugate acid.*

Acid Strength and Ionization Constants

The strength of binary acids increases down a group of elements, and from left to right across a period. The strength of an oxoacid is determined primarily by the number of nonprotonated oxygen atoms per formula unit of the acid: the greater the number of nonprotonated oxygen atoms, the stronger the acid. For acids with an equal number of nonprotonated oxygen atoms, acid strength generally increases with increasing electronegativity of the atom to which the O—H groups of the acid are attached.

The equilibrium constant for ionization of an acid or a base in aqueous solution is the *ionization constant*, K_a for an acid and K_b for a base. The generalized aqueous equilibria and the expressions for the ionization constants for an acid, HA, and a base, B, are

$$HA(aq) + H_2O(l) \rightleftharpoons H_3O^+(aq) + A^-(aq)$$

$$K_a = \frac{[H_3O^+][A^-]}{[HA]}$$

$$B(aq) + H_2O(l) \rightleftharpoons HB^+(aq) + OH^-(aq)$$

$$K_b = \frac{[HB^+][OH^-]}{[B]}$$

Oxoacids and Hydroxo Compounds

Oxoacids and hydroxides both contain O—H groups attached to a central nonmetal or a metal atom. These compounds are therefore called hydroxo compounds. When the OH groups are attached to a nonmetal atom, the compound is an acid. When the OH groups are attached to a metal atom of low electronegativity and of low oxidation state, the compound is a hydroxide. If the metal atom is of intermediate electronegativity (between approximately 1.5 and 1.9), the compound is likely to be an **amphoteric hydroxide**. When the metal atom is in a high oxidation state, the compound is an acid, for example, H_2CrO_4.

Hydrolysis

Salts formed from strong bases and weak acids, such as sodium fluoride, dissolve in water to produce slightly alkaline solutions because the anions of these salts are able to extract protons from water to produce hydroxide ions. Salts of weak bases and strong acids, such as ammonium chloride, dissolve in water to produce hydronium ions when their cations react with water. *Reactions of ions with water that increase the hydroxide ion or hydronium ion concentration* are called **hydrolysis** reactions. The hydrolysis of an anion A^- can be represented by the equation

$$A^-(aq) + H_2O(l) \rightleftharpoons HA(aq) + OH^-(aq)$$

The base ionization constant, K_b, for A^- ions has the form

$$K_b = \frac{[HA][OH^-]}{[A^-]}$$

K_b for A^- ions is related to K_a for HA and K_w by the relationship

$$(K_a)(K_b) = K_w$$

If either K_a or K_b is known, the other can be calculated from this relationship.

Lewis Acids and Bases

Examples of Lewis acids are molecules in which the central atom has an incomplete octet (such as BCl_3), molecules in which the central atom is relatively electron deficient (for example, CO_2), and metal ions. Molecules such as ammonia and water, which have unshared electron pairs, are Lewis bases. An example of a Lewis acid–base reaction is the hydration of metal ions. The metal ion reacts as a Lewis acid; water acts as a Lewis base. Another example of a Lewis acid–base reaction is the reaction of a nonmetal oxide with an oxide ion. In this process the nonmetal oxide reacts as a Lewis acid and the oxide ion acts as a Lewis base.

New Terms

Acid dissociation (ionization) constant, K_a (14.2)
Amphiprotic substance (14.1)
Amphoteric hydroxide (14.3)
Autoionization (14.1)
Base ionization constant, K_b (14.4)
Brønsted acid (14.1)
Brønsted base (14.1)
Carboxyl group (14.3)
Carboxylate ion (14.3)
Carboxylic acid (14.3)
Conjugate acid (14.1)
Conjugate acid–base pair (14.1)
Conjugate base (14.1)
Hydrolysis (14.4)
Hydrolysis reactions (14.4)
Hydroxo compounds (14.3)
Ion product constant for water (14.1)
Lewis acid (14.5)
Lewis base (14.5)
Oxoacid (14.3)

Exercises

General Review

1. Write an equation to illustrate a Brønsted acid–base reaction. Specify each of the reactants and products in your equation as a Brønsted acid or base. Identify the two conjugate acid–base pairs.
2. Write two equations to illustrate the amphiprotic properties of water.
3. Write an equation for the reaction of ammonia with water, and the expression for K_b.
4. How does the strength of binary acids of elements vary down a group and across a period? Explain.
5. What energy changes can be imagined to occur when a gaseous binary acid, HA, is converted to its gaseous ions, H^+ and A^-?

6. What factors are most important in determining the strength of a binary acid, HA?

7. Write an equation for the dissociation of a weak acid, HA, and the expression for K_a for the acid.

8. If the acid in question 7 were 5.0 percent dissociated in a 1.0 *M* solution, what would be the value for K_a? What would be the values for $[H_3O^+]$, $[A^-]$, and $[HA]$?

9. Why is it easy to find tables of K_a values for weak acids but not for strong acids?

10. Explain and give examples of the relative strengths of some commonly known binary acids of elements within a group of the periodic table and across a period.

11. How does the base strength of conjugate bases vary with the strength of their respective acids? Give some examples that illustrate your answer.

12. Write an equation for the reaction of the conjugate acid of ammonia with the conjugate base of water.

13. If you mixed concentrated solutions of ammonium nitrate and sodium hydroxide, what gas would form as a product of the reaction?

14. Write an equation for the ionization equilibrium of a weak base, B, in water, and the expression for K_b for the base.

15. Write formulas and names for some oxoacids that have zero, one, two, and three nonprotonated oxygen atoms per formula unit.

16. State which of the acids you specified in question 15 is the strongest and which is the weakest. Explain your choice.

17. How does the relative strength of oxoacids that have an equal number of nonprotonated oxygen atoms generally vary with the electronegativity of the atom bonded to the oxygen atoms of the O—H groups? Explain and give examples.

18. Is formic acid a stronger acid or a weaker acid than acetic acid? Explain your reasoning.

19. Why are hydroxides and oxoacids sometimes called hydroxo compounds? Write formulas for some hydroxo compounds that are hydroxides, and for others that are acids.

20. How does the formula for a hydroxo compound reveal whether the compound is a hydroxide or an acid?

21. What is an amphoteric hydroxide? Write the formulas and names of two amphoteric hydroxides.

22. Write a net ionic equation for the reaction of an amphoteric hydroxide with an aqueous acid and with an aqueous solution of a strong base.

23. Explain why aqueous solutions of some salts are neutral, some are acidic, and some are alkaline.

24. Write an equation for a hydrolysis equilibrium for the anion A^- of a weak acid HA, and the expression for K_b of A^- ions.

25. What is the algebraic relationship between the acid dissociation constant for a weak acid and the base ionization constant of its conjugate base?

26. Write formulas of two hydroxo compounds of transition metals which are hydroxides, and the formulas of two others which are acids.

27. Most hydroxo compounds of metals are hydroxides or amphoteric hydroxides. Explain why some are acids.

28. What is: **(a)** a Lewis acid; **(b)** a Lewis base?

29. Write an equation for a typical Lewis acid–base reaction and label each reactant as an acid or base.

30. Write an equation for each of the following types of reaction: **(a)** hydration of a metal ion; **(b)** hydrolysis of a hydrated metal ion; **(c)** combination of a basic metal oxide with a nonmetal oxide.

31. In each of the following reversible reactions, identify the Brønsted acids and bases: **(a)** $H_2SO_4(aq) + H_2O(l) \rightleftharpoons H_3O^+(aq) + HSO_4^-(aq)$; **(b)** $CH_3COOH(l) + HBr(g) \rightleftharpoons CH_3COOH_2^+ + Br^-$; **(c)** $H_2O(l) + H_2O(l) \rightleftharpoons H_3O^+(aq) + OH^-(aq)$; **(d)** $NH_4^+(aq) + OH^-(aq) \rightleftharpoons NH_3(aq) + H_2O(l)$; **(e)** $CO_3^{2-}(aq) + H_2O(l) \rightleftharpoons HCO_3^-(aq) + OH^-(aq)$; **(f)** $Cr(H_2O)_6^{3+}(aq) + H_2O(l) \rightleftharpoons Cr(H_2O)_5OH^{2+}(aq) + H_3O^+(aq)$

32. In each of the reactions in question 31 identify the conjugate acid–base pairs.

33. Write the formula of the conjugate base for each of the following substances: **(a)** HNO_3; **(b)** HSO_4^-; **(c)** NH_3; **(d)** H_2O; **(e)** H_3O^+; **(f)** NH_4^+.

34. Write the formula of the conjugate acid for each of the following substances: **(a)** H_2O; **(b)** Cl^-; **(c)** SO_4^{2-}; **(d)** HPO_4^{2-}; **(e)** $H_2PO_4^-$; **(f)** NH_3; **(g)** OH^-.

Relative Strengths of Acids and Bases

35. Arrange the substances in each of the following parts in the order of decreasing acid strength: **(a)** HF, HCl, HBr, HI; **(b)** H_2Te, H_2S, H_2Se; **(c)** NH_3, PH_3, AsH_3.

36. Arrange the substances in each of the following parts in the order of increasing acid strength: **(a)** SiH_4, PH_3, H_2S, HCl; **(b)** NH_3, H_2O, HF; **(c)** AsH_3, H_2Se, HBr.

37. In each of the following pairs of substances indicate which is the stronger Brønsted base: **(a)** NH_3 or NH_2^-; **(b)** O^{2-} or OH^-; **(c)** F^- or I^-; **(d)** OH^- or HS^-; **(e)** S^{2-} or HS^-; **(f)** HS^- or HSe^-.

Prediction of Brønsted Acid-Base Reactions

38. Predict whether there will be a reaction when each pair of the following substances is mixed, and write an equation for each reaction that occurs: **(a)** ammonia gas and hydrogen bromide gas; **(b)** hydrogen sulfide and water; **(c)** a solution of ammonium bromide and a solution of potassium hydroxide; **(d)** hydrogen chloride and water; **(e)** sodium chloride and water; **(f)** a solution of sodium chloride and a solution of sodium hydroxide; **(g)** sodium oxide and water; **(h)** sodium amide ($NaNH_2$) and water.

Acid and Base Ionization Constants

39. Write an equation for the ionization of each of the following acids or bases in water: **(a)** hydrogen bromide; **(b)** ammonia; **(c)** hydrogen cyanide (HCN); **(d)** methylamine (CH_3NH_2).

40. The chloride ion concentration in a 0.020 *M* solution of hydrochloric acid is found to be 0.020 *M*. What is the hydronium ion concentration?

41. The hydroxide ion concentration of a 0.060 *M* solution of sodium hydroxide is 0.060 *M*. What is the sodium ion concentration?

42. The hypochlorite ion concentration in 0.20 *M* hypochlorous acid is 7.7×10^{-5}. Calculate the K_a for hypochlorous acid.

43. The weak base aniline, $C_6H_5NH_2$ (an important ingredient in the manufacture of dyes), ionizes in water according to the equation

$$C_6H_5NH_2(aq) + H_2O(l) \rightleftharpoons C_6H_5NH_3^+(aq) + OH^-(aq)$$

The hydroxide ion concentration in a 0.15 *M* solution of aniline is 7.9×10^{-6} *M*. What is the value of K_b for aniline?

Oxoacids and Hydroxides

44. Arrange the following substances in the order of their increasing acid strength: H_3AsO_3, H_2SO_3, H_2SO_4, $HClO_4$.

45. Arrange the following substances in the order of their increasing acid strength: $HClO_3$, $HBrO_3$, HIO_3.

46. Which one in each of the following pairs of acids would you predict to be stronger? **(a)** H_3PO_4 or H_3AsO_4; **(b)** H_3PO_3 or H_3AsO_3; **(c)** H_2PHO_3 or H_3BO_3; **(d)** H_3PO_4 or H_3BO_3.

47. Which of the following compounds are acids, which are hydroxides, and which are amphoteric? **(a)** $Be(OH)_2$; **(b)** $B(OH)_3$; **(c)** H_2CO_3; **(d)** $Sn(OH)_2$; **(e)** $Pb(OH)_2$; **(f)** $Ca(OH)_2$; **(g)** HNO_2.

48. Write net ionic equations for the reactions of the amphoteric hydroxide, $Sn(OH)_2$, with: **(a)** hydrochloric acid; **(b)** a solution of sodium hydroxide.

49. Considering the structures of H_3PO_4 and H_2PHO_3 (see Chapter 9 if necessary), explain why the K_a values for these acids are relatively close (Table 14.2).

50. Classify each of the following oxides as acidic, basic, or amphoteric anhydrides: **(a)** CrO_3; **(b)** MnO; **(c)** Mn_2O_7; **(d)** FeO; **(e)** ZnO.

51. Write an equation for the reaction of each of the oxides in question 50 with water, and name the product formed in each case.

52. Describe one test with which you can distinguish solid $Mn(OH)_2$ from solid $Zn(OH)_2$. Write equations for any reactions that occur in your suggested procedure.

53. Which one in the following pairs of acids is the stronger acid? **(a)** Acetic acid or formic acid; **(b)** acetic acid or monochloroacetic acid; **(c)** monochloroacetic acid or monofluoroacetic acid.

Hydrolysis

54. Classify the aqueous solutions of each of the following salts as acidic, neutral, or alkaline: **(a)** $NaCl$; **(b)** NaF; **(c)** Na_2S; **(d)** $NaClO_4$; **(e)** NH_4NO_3; **(f)** $Cr(NO_3)_3$; **(g)** KNO_3.

55. Arrange the following anions in the order of their increasing base strength: SO_4^{2-}, HSO_4^-, HS^-, NH_2^-.

56. Arrange 1 *M* solutions of the following salts in the order of increasing alkalinity: $NaCN$, $NaC_2H_3O_2$, $NaNO_2$.

57. The hydroxide ion concentration of a 0.18 *M* solution of KNO_2 is 2.0×10^{-6} *M*. Calculate the value of K_b for NO_2^- ions.

58. The percent of hydrolysis of fluoride ions in a 0.20 *M* solution of sodium fluoride is $1.2 \times 10^{-3}\%$. Calculate the value of K_b for fluoride ions.

59. The K_a for hypochlorous acid is 3.0×10^{-8}. What is the value of K_b for hypochlorite ions?

60. The K_b for ammonia is 1.8×10^{-5}. What is the value of K_a for ammonium ions?

Lewis Acids and Bases

61. In each of the following reactions, identify the Lewis acid and the Lewis base: **(a)** $AlCl_3(aq) + Cl^-(aq) \rightarrow AlCl_4^-(aq)$; **(b)** $Ag^+(aq) + 2NH_3(aq) \rightarrow Ag(NH_3)_2^+(aq)$; **(c)** $Zn(OH)_2(s) + 2OH^-(aq) \rightarrow Zn(OH)_4^{2-}(aq)$

62. Write the Lewis dot formula for each of the Lewis bases in question 61 and predict the structures of all the products formed in the reactions.

63. Write an equation for the hydration of aluminum ions.

Equation Writing

64. Write a net ionic equation, where appropriate, for each of the following reactions: **(a)** solid aluminum hydroxide is treated with a solution of sodium hydroxide; **(b)** hydration of copper(II) ion to form $Cu(H_2O)_4^{2+}$; **(c)** solid zinc hydroxide is treated with hydrochloric acid; **(d)** solid zinc hydroxide is treated with an excess of concentrated solution of sodium hydroxide.

65. Write an equation for each of the following reactions: **(a)** calcium oxide is heated with sulfur dioxide; **(b)** sodium oxide is treated with sulfur trioxide; **(c)** potassium oxide is treated with carbon dioxide.

CHAPTER 15

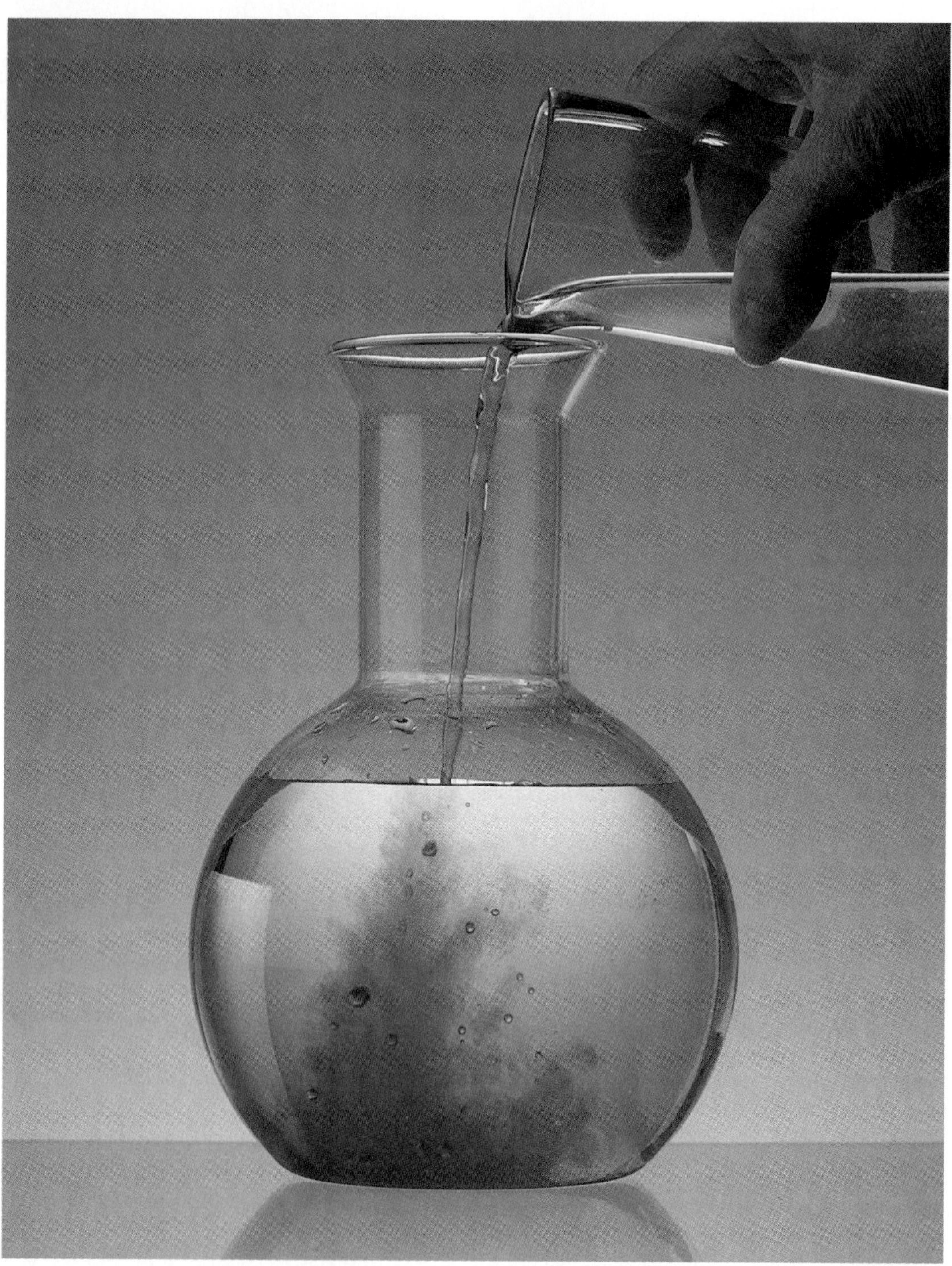

Pouring a base solution into an acid solution containing an indicator. The indicator changes its color in the regions of the solution containing an excess base.

Aqueous Acid–Base Equilibria

In this chapter we consider aqueous equilibria of acids and bases. We describe methods for calculating the hydronium and hydroxide ion concentration in an acid or base solution from the known concentration of an acid or base and its equilibrium constant, K_a *or* K_b*. We will learn how to express the hydronium ion concentrations of aqueous solutions on a scale of acidity called the pH scale. We also discuss buffer solutions, which are solutions whose hydronium ion concentration scarcely alters when a moderate quantity of a strong acid or base is added.*

15.1 Autoionization of Water and K_w

Hydronium and Hydroxide Ion Concentrations in Water

In Chapter 14 we learned that water undergoes autoionization to produce equimolar amounts of hydronium ions and hydroxide ions. The equilibrium equation for the autoionization of water is

$$2H_2O(l) \rightleftharpoons H_3O^+(aq) + OH^-(aq)$$

We recall that the equilibrium constant for the autoionization of water, K_w, equals 1.0×10^{-14} at 25°C, and is expressed as

$$K_w = [H_3O^+][OH^-] = 1.0 \times 10^{-14} \qquad (15.1)$$

Since the concentrations of hydronium ions and hydroxide ions produced by the autoionization of water are equal, we can determine the concentrations of hydronium ions and hydroxide ions in pure water by letting $[H_3O^+] = [OH^-] = x$ in Equation 15.1. Therefore, from the expression of K_w, we have

$$(x)(x) = 1.0 \times 10^{-14} \quad \text{or} \quad x^2 = 1.0 \times 10^{-14}$$

Solving for x by taking the square root of the equation, we have

$$x = [H_3O^+] = [OH^-] = (1.0 \times 10^{-14})^{1/2} = 1.0 \times 10^{-7}\ M$$

Thus, the molar concentration of hydronium ions in pure water at 25°C equals $1.0 \times 10^{-7}\ M$, which is also the molar concentration of hydroxide ions.

Hydronium and Hydroxide Ion Concentrations in Solutions of Strong Acids and Bases

We recall that strong acids, such as HCl, are 100 percent dissociated in their dilute aqueous solutions. The autoionization of water produces a negligibly small concentration of hydronium ions compared to the hydronium ion concentration produced by a strong acid. Similarly, the concentration of hydroxide ions in a solution of a strong base such as NaOH is almost entirely due to the strong base.

When an acid is added to water, the hydronium ion concentration increases and the equilibrium

$$2H_2O(l) \rightleftharpoons H_3O^+(aq) + OH^-(aq)$$

shifts to the left. As a result, the hydroxide ion concentration decreases. Thus, in acid solution, $[H_3O^+] > [OH^-]$. Similarly, when a base is added to water, the hydroxide ion concentration increases and the hydronium ion concentration decreases, so that in a base solution, $[H_3O^+] < [OH^-]$. But in all aqueous solutions, $[H_3O^+][OH^-] = 1.0 \times 10^{-14}$ at 25°C. When either the hydronium ion concentration or the hydroxide ion concentration is known, the other can be calculated from this relationship for acidic or basic solutions (Equation 15.1).

Example 15.1

Calculating the Concentrations of Hydronium and Hydroxide Ions in Strong Acid and Base Solutions

Calculate the hydronium ion concentration and the hydroxide ion concentration in each of the following solutions at 25°C: **(a)** 0.10 *M* nitric acid; **(b)** 0.20 *M* potassium hydroxide; **(c)** 0.0020 *M* calcium hydroxide.

SOLUTION: We first examine the names and formulas of the acids and bases. From our previous study, we recognize all of the above substances as strong acids or bases.

(a) We know that nitric acid is a strong monoprotic acid. Thus, a nitric acid solution labeled 0.10 *M* contains 0.10 mol L^{-1} of hydronium ions and 0.10 mol L^{-1} of nitrate ions. From Equation 15.1, $[H_3O^+][OH^-] = 1.0 \times 10^{-14}$, the concentration of hydroxide ions is

$$[OH^-] = \frac{1.0 \times 10^{-14}}{0.10} = 1.0 \times 10^{-13}\ M$$

(b) Since KOH is a strong base, a KOH solution labeled 0.20 *M* contains 0.20 mol L^{-1} of OH^- ions and 0.20 mol L^{-1} of potassium ions. From Equation 15.1, $[H_3O^+][OH^-] = 1.0 \times 10^{-14}$, the hydronium ion concentration is

$$[H_3O^+] = \frac{1.0 \times 10^{-14}}{[OH^-]} = \frac{1.0 \times 10^{-14}}{0.20} = 5.0 \times 10^{-14}\ M$$

(c) Each mole of $Ca(OH)_2(s)$ that dissolves produces 1 mol of Ca^{2+} ions and 2 mol of OH^- ions:

$$Ca(OH)_2(s) \longrightarrow Ca^{2+}(aq) + 2OH^-(aq)$$

Therefore, in a 0.0020 *M* solution of calcium hydroxide,

$$[OH^-] = 2(0.0020\ M) = 0.0040\ M$$

From the expression for K_w, Equation 15.1, $[H_3O^+][OH^-] = 1.0 \times 10^{-14}$,

$$[H_3O^+] = \frac{1.0 \times 10^{-14}}{[OH^-]} = \frac{1.0 \times 10^{-14}}{0.0040} = 2.5 \times 10^{-12}\ M$$

Practice Problem 15.1: Calculate the hydronium and hydroxide ion concentrations in: **(a)** 0.30 *M* perchloric acid; **(b)** 0.0040 *M* barium hydroxide.

15.2 pH and pOH

pH

The preceding examples show that the concentration of hydronium ions in aqueous solutions is sometimes very small. It is convenient to express the hydronium ion concentrations of solutions on a logarithmic scale. The *negative logarithm of the hydronium ion concentration* is called **pH**. The "p" in the term pH means "power" (that is, exponent) and "H" stands for "hydronium ions." Therefore, pH is defined mathematically as

$$pH = -\log[H^+(aq)] = -\log[H_3O^+] \quad (15.2)$$

The pH of red wine is slightly less than 7.

We recall that a logarithm to the base 10 of a number, n, equals the power or exponent, x, to which 10 must be raised to obtain the number. In terms of pH, the number, n, is the hydronium ion concentration of a solution. If the hydronium ion concentration is 2.00×10^{-3}, it follows from the definition of pH that

$$pH = -\log(2.0 \times 10^{-3})$$

or that

$$10^{-pH} = 2.0 \times 10^{-3}$$

The following steps on a calculator will give the value of the pH: (1) enter 2.0×10^{-3}, (2) press the log key; the value −2.69897 appears on the display; (3) press the (+/−) change sign key, and 2.69897 appears on the display. Thus, the pH of a solution whose hydronium ion concentration is 2.0×10^{-3} *M* equals 2.70. This value, 2.70, has two significant figures, as does 2.0×10^{-3}. We explain this below.

A logarithmic expression has two parts, the characteristic and the mantissa. Digits written to the left of the decimal point make up the characteristic, and digits to the right of the decimal point constitute the mantissa. The characteristic specifies the location of the decimal point in the number to which the log corresponds. This number is called the antilog or inverse log.

The number of significant figures in a logarithmic expression is the number of digits in its mantissa. This number is also the number of significant figures in the antilog. For example, log 2.300 = 0.3617, log 23.00 = 1.3617, and log 230.0 = 2.3617. Each antilog (2.300, 23.00, and 230.0) has four significant figures. The log of each of these numbers has four significant figures (the number of digits in the mantissa). The characteristics 0, 1, and 2 in each log locate the decimal points in their corresponding antilogs. Thus, the hydronium ion concentration of 2.0×10^{-3} has two significant figures and so does its negative log, 2.70, which has two digits in the mantissa.

pOH

Just as the negative logarithm of the hydronium ion concentration is called pH, the *negative logarithm of the hydroxide ion concentration* is called **pOH**:

$$pOH = -\log[OH^-] \quad \text{and} \quad [OH^-] = 10^{-pOH}$$

Now let us consider the relation between K_w, pH, pOH, and the hydronium and hydroxide ion concentrations of a solution. We recall from Equation 15.1 that the product of the hydronium and hydroxide ion concentrations of an aqueous solution equals K_w, or 1.0×10^{-14} at 25°C:

$$[H_3O^+][OH^-] = 1.0 \times 10^{-14}$$

When we take the logarithm of both sides of the equation above we obtain

$$\log[H_3O^+] + \log[OH^-] = \log(1.0 \times 10^{-14})$$

Multiplying both sides of this equation by −1 gives

$$-\log[H_3O^+] + (-\log[OH^-]) = -\log(1.0 \times 10^{-14})$$

Since pH = $-\log[H_3O^+]$, and pOH = $-\log[OH^-]$, the equation above simplifies to

$$\text{pH} + \text{pOH} = 14.00 \qquad (15.3)$$

Thus, for any aqueous solution,

$$\mathbf{[H_3O^+][OH^-] = 1.0 \times 10^{-14}} \quad \text{and} \quad \mathbf{pH + pOH = 14.00}$$

Example 15.2

Calculating the pH and pOH of Neutral, Acidic, and Alkaline Solutions

Calculate the pH and pOH of: **(a)** pure H_2O; **(b)** 0.10 *M* nitric acid; **(c)** 0.20 *M* potassium hydroxide.

SOLUTION

(a) In pure water, $[H_3O^+] = 1.0 \times 10^{-7}$ *M* and $[OH^-] = 1.0 \times 10^{-7}$ *M*. pH = $-\log(1.0 \times 10^{-7}) = 7.00$. The pH value of 7.00 has two significant figures: the two zeros in the mantissa. This is also the number of significant figures in the hydronium ion concentration of 1.0×10^{-7}.

Similarly,

$$\text{pOH} = -\log(1.0 \times 10^{-7}) = 7.00$$

(b) For a 0.10 *M* solution of nitric acid, a strong acid, $[H_3O^+] = 0.10$ *M*, and

$$\text{pH} = -\log(1.0 \times 10^{-1}) = 1.00$$

From the expression for K_w, $[OH^-]$ in a 0.10 *M* solution of nitric acid is

$$[OH^-] = \frac{K_w}{[H_3O^+]} = \frac{1.0 \times 10^{-14}}{0.10} = 1.0 \times 10^{-13}\ M$$

$$\text{pOH} = -\log(1.0 \times 10^{-13}) = 13.00$$

The same result is obtained by solving Equation 15.3 for pOH:

$$\text{pOH} = 14.00 - \text{pH} = 14.00 - 1.00 = 13.00$$

(c) Since potassium hydroxide is a strong base, in a 0.20 *M* solution of potassium hydroxide $[OH^-] = 0.20$ *M*.

$$[H_3O^+] = \frac{1.0 \times 10^{-14}}{0.20} = 5.0 \times 10^{-14}\ M$$

$$\text{pH} = -\log(5.0 \times 10^{-14}) = 13.30$$

From Equation 15.3,

$$\text{pOH} = 14.00 - 13.30 = 0.70$$

We can also get the value of pOH by taking the negative log of $[OH^-]$:

$$\text{pOH} = -\log(2.0 \times 10^{-1}) = 0.70$$

Practice Problem 15.2: Calculate the pH and pOH of: **(a)** 0.25 *M* hydrochloric acid; **(b)** 0.0025 *M* barium hydroxide.

We often need to convert pH to hydronium ion concentration. To do this, we must remember that the hydronium ion concentration equals 10 raised to the exponent of −pH:

$$[H_3O^+] = 10^{-pH}$$

We can easily determine the hydronium ion concentration from a known pH by using a calculator. For a calculator that uses arithmetic logic three steps are required: (1) enter the pH, (2) press the (+/−) key to change the sign, and (3) press the key marked 10^x (or inverse log). The answer appears on the display. The $[OH^-]$ of a solution is found in a similar way from a known value of pOH.

For example, if the pH of a solution is 3.30, $[H_3O^+] = 10^{-3.30}$. To find the value of $[H_3O^+]$: (1) enter 3.30 (the pH), (2) press the (+/−) key to change the sign, (3) press the 10^x key to obtain 5.0×10^{-4} on the display. This is the hydronium ion concentration.

Practice Problem 15.3: **(a)** The pH of a solution is 2.35. What is its hydronium ion concentration? **(b)** The pOH of a solution is 8.96. What is the hydroxide ion concentration?

The preceding discussion can be summarized as follows. For water and any neutral solution, the hydronium ion concentration equals the hydroxide ion concentration:

$$[H_3O^+] = [OH^-] \quad \text{and} \quad pH = pOH$$

For an acid solution,

$$[H_3O^+] > [OH^-] \quad \text{and} \quad pH < 7 < pOH$$

For a basic solution,

$$[H_3O^+] < [OH^-] \quad \text{and} \quad pH > 7 > pOH$$

Table 15.1

Relationships of $[H_3O^+]$, $[OH^-]$, pH, and pOH of Aqueous Solutions

$[H_3O^+]$	$[OH^-]$	pH	pOH	
10	10^{-15}	−1	15	↑
1	10^{-14}	0	14	
10^{-1}	10^{-13}	1	13	Increase in acidity
.	.	.	.	
.	.	.	.	
.	.	.	.	
10^{-7}	10^{-7}	7	7	Neutral solution
.	.	.	.	
.	.	.	.	
.	.	.	.	
10^{-13}	10^{-1}	13	1	
10^{-14}	1	14	0	
10^{-15}	10	15	−1	↓ Increase in alkalinity

APPLICATIONS OF CHEMISTRY 15.1
Can Boiling Water Turn Into an Acid?

We just learned that the pH of pure water at 25°C is 7.00, and we say that water with a pH of 7.00 is neutral. At 25°C, solutions with a pH greater than 7.00 are basic, and solutions with a pH less than 7.00 are acidic. Although this looks to be fairly straightforward, something interesting develops when we boil pure water.

The pH of boiling water is 6.12. Since a water solution with a pH less than 7.00 is acidic, has the water turned acidic upon heating? Calculate the hydronium ion concentration of pure boiling water.

$$[H_3O^+] = 10^{-pH} = 10^{-6.12} = 7.6 \times 10^{-7}$$

Next, let's determine the hydroxide ion concentration, $[OH^-]$, in boiling water. Remember that the autoionization of water produces equimolar amounts of hydronium ions and hydroxide ions. We know this by examinig the equilibrium equation for water's autoionization:

$$2H_2O(l) \rightleftharpoons H_3O^+(aq) + OH^-(aq)$$

This equilibrium equation tells us that if 7.6×10^{-7} mol of hydronium ion forms from the dissociation of boiling water, 7.6×10^{-7} mol of hydroxide ions must also form. By definition, water is neutral because it has equal hydronium and hydroxide ion concentrations. Therefore, boiling water must be neutral because both its hydronium and hydroxide ion concentrations equal 7.6×10^{-7} *M*. Only at 25°C does pure water have a pH of 7.00, and we cannot use the pH = 7.00 definition for neutrality at other temperatures.

In Chapter 13 we learned that the equilibrium constant varies with temperature. At higher temperatures, more water molecules dissociate into hydronium ions and hydroxide ions than at lower temperatures. Therefore, as the temperature of pure water increases, K_w also increases. When we raise the temperature of pure water from 25°C to 100°C, the equilibrium constant, K_w, of water increases from 1.0×10^{-14} to 5.8×10^{-13}.

Table 15.1 shows the variation of hydronium ion concentration, hydroxide ion concentration, pH, and pOH for acid, neutral, and alkaline solutions. The approximate pH values of some common substances are listed in Table 15.2.

Table 15.2

Approximate pH Values of Some Substances

Substance	pH
1 *M* HCl	0.0
0.1 *M* HCl	1.0
Gastric juice	1.4
Lemon juice	2.3
Vinegar	2.9
Orange juice	3.5
Tomato juice	4.5
Coffee	5.0
Urine	5–7
Milk	6.5
Pure water	7.0
Blood	7.4
Seawater	8.1–10
Milk of magnesia	10.5
Household ammonia	11.5
1 *M* NaOH	14.0

15.3 Calculating the $[H_3O^+]$ and pH of Weak Acid and Base Solutions

Weak Acids

We learned in Chapter 14 that the value of K_a for a weak acid and K_b for a weak base is a measure of the extent of its ionization. We also learned how to determine the value of the ionization constant of a weak acid or base from its initial concentration and the concentrations of the ions present in the solution. In this chapter we calculate the ion concentrations from the value of the ionization constant and the initial concentration of the acid or base. Once we know the concentrations of hydronium ions or hydroxide ions, the pH and pOH can be calculated as discussed in Section 15.2. A calculation of this type is illustrated in the following example.

Example 15.3

Calculating the Hydronium Ion Concentration, Hydroxide Ion Concentration, pH, and pOH of a Weak Acid Solution

Calculate the hydronium ion concentration, pH, hydroxide ion concentration, and pOH of a 0.10 *M* solution of acetic acid, $HC_2H_3O_2$. The acid dissociation constant, K_a, for acetic acid is 1.8×10^{-5}.

SOLUTION: Problems of this type can be solved by first writing an equation for the ionization of the weak acid. Then, under each of the species in the equation, we write its initial concentration (before mixing) and its equilibrium concentration. We let x be the number of moles of the weak acid dissociated per liter of solution. The number of moles of acetic acid that dissociates equals the number of moles of hydronium ions and also the number of moles of acetate ions produced. Thus, x is equal to the molar concentration of hydronium ions and the molar concentration of acetate ions:

$$x = [H_3O^+] = [C_2H_3O_2^-]$$

The molar concentration of acetic acid that remains undissociated at equilibrium equals the initial concentration minus x. These data are summarized below:

	$HC_2H_3O_2(aq)$ + $H_2O(l)$ $\rightleftharpoons$	$H_3O^+(aq)$ +	$C_2H_3O_2^-(aq)$
Initial *M*	0.10	~0	0
Equilibrium *M*	$0.10 - x$	x	x

The expression for the dissociation constant of acetic acid, K_a, is

$$K_a = \frac{[H_3O^+][C_2H_3O_2^-]}{[HC_2H_3O_2]}$$

Substituting the equilibrium concentrations of each species into this expression yields

$$1.8 \times 10^{-5} = \frac{(x)(x)}{0.10 - x}$$

This equation can be written in the general form of a quadratic equation, $ax^2 + bx + c = 0$:

$$x^2 + 1.8 \times 10^{-5}x - 1.8 \times 10^{-6} = 0$$

The formula for solving a quadratic equation is

$$x = \frac{-b \pm \sqrt{b^2 - 4ac}}{2a}$$

Substituting the appropriate values into this equation yields

$$x = \frac{-1.8 \times 10^{-5} \pm \sqrt{(1.8 \times 10^{-5})^2 - (4)(1)(-1.8 \times 10^{-6})}}{(2)(1)}$$

$$x_1 = 1.3 \times 10^{-3} \qquad x_2 = -1.4 \times 10^{-3}$$

Both answers are acceptable mathematically, but only the positive answer is meaningful in this case because x represents the number of moles per liter of acetic acid dissociated, and also the concentrations of hydronium ions and acetate ions. Concentrations cannot be negative. Therefore,

$$[H_3O^+] = [C_2H_3O_2^-] = 1.3 \times 10^{-3}\ M$$

$$\text{pH} = -\log(1.3 \times 10^{-3}) = 2.89$$

Since pH is less than 7, the solution is acidic.

$$[OH^-] = \frac{1.0 \times 10^{-14}}{1.3 \times 10^{-3}} = 7.7 \times 10^{-12}\ M$$

$$\text{pOH} = -\log(7.7 \times 10^{-12}) = 11.11$$

Alternatively,

$$\text{pOH} = 14.00 - 2.89 = 11.11$$

When we consider the values of x and the initial concentration, 0.10 M, we can see that x is much less than 0.10 M. The difference between 0.10 M and 0.0013 M is essentially 0.10 M:

$$\begin{array}{r} 0.10\ M \\ -\ 0.0013\ M \\ \hline \approx 0.10\ M \end{array}$$

Because the value of x is extremely small compared to 0.10, x can be neglected in the denominator of the equation,

$$1.8 \times 10^{-5} = \frac{x^2}{0.10 - x}$$

Thus,

$$1.8 \times 10^{-5} = \frac{x^2}{0.10}$$

The simplified equation is then easily solved for x:

$$x^2 = (1.8 \times 10^{-5})(0.10)$$

$$x = \sqrt{(1.8 \times 10^{-5})(0.10)} = 1.3 \times 10^{-3}$$

We can see that the answer obtained here is identical to the answer obtained from the quadratic equation.

In Example 15.3 the acid is so weak that the fraction of a mole which dissociates is negligibly small compared to its initial concentration. Therefore, in that problem we do not need to use the quadratic equation. However, the ionization of a weak acid is not always negligible compared to its initial concentration. Two factors determine whether the value of x is negligibly small: (1) the value of K_a, and (2) the value of the initial concentration of the acid.

Most K_a values are known to an accuracy of about 5 percent. We can therefore first neglect the x and solve the resulting simplified equation for x. Then we can check to see if the expression

$$\left(\frac{x}{\text{initial } M}\right) 100$$

is less than or equal to 5 percent. If it is, neglecting x is normally justified. If the value of x is more than 5 percent of the initial concentration of the acid, the quadratic equation should be used. This 5 percent limit is usually satisfied when K_a is smaller than about 5×10^{-5} and the initial concentration of the acid is 0.01 M or more. As a rule of thumb, the dissociation of an acid is less than 5 percent when the ratio of the initial molar concentration of the acid to the K_a of the acid is more than 100. The equilibrium concentration of the acid is then nearly equal to the initial concentration:

$$\text{When } [\text{acid}]_{\text{initial}}/K_a > 100, \text{ initial } M \approx \text{equilibrium } M.$$

Let us see how these guidelines are applied in Example 15.3, where the initial concentration of the acid is 0.10 M and the value of K_a is 1.8×10^{-5}. The ratio $0.10/(1.8 \times 10^{-5}) = 5.6 \times 10^3$. Since this result is much more than 100, we do not need to use the quadratic equation to solve the problem. As a final check, we see whether the value of x we obtained, 1.3×10^{-3}, is 5 percent or less of the initial concentration of the weak acid:

$$\left(\frac{1.3 \times 10^{-3}}{0.10}\right) 100 = 1.3 \text{ percent}$$

Since the value of x is only 1.3 percent of the initial concentration of the acid, the initial concentration of the acid is essentially equal to its equilibrium concentration.

Example 15.4

Calculating the Hydronium Ion Concentration of a Solution of a Weak Acid and Determining If the Approximation Method Can Be Used

Calculate the hydronium ion concentration and pH in: **(a)** 0.20 M nitrous acid, HNO_2; and **(b)** 0.010 M HNO_2. The acid dissociation constant, K_a, for HNO_2 is 4.5×10^{-4}.

SOLUTION: From the value of K_a, we see that nitrous acid is a weak acid. A solution of nitrous acid contains water, hydronium ions, nitrite ions, and undissociated nitrous acid. We proceed as in Example 15.3 by writing the ionization equation and the initial and equilibrium concentrations of all the species. We let x equal the number of moles per liter of nitrous acid that dissociates; x also equals the equilibrium concentration of hydronium ions and nitrite ions.

	$HNO_2(aq) + H_2O(l) \rightleftharpoons H_3O^+(aq) + NO_2^-(aq)$		
Initial M	0.20	~0	0
Equilibrium M	$0.20 - x$	x	x

The expression for K_a is

$$K_a = \frac{[H_3O^+][NO_2^-]}{[HNO_2]}$$

(a) Substituting the data into this expression gives

$$4.5 \times 10^{-4} = \frac{x^2}{0.20 - x}$$

Next, we check to see if we can neglect the x in the denominator of this expression. The ratio $[\text{acid}]_{\text{initial}}/K_a = 0.20/4.5 \times 10^{-4} = 4.4 \times 10^2$. This is more than 100; therefore, we are justified neglecting x, and the equation simplifies to

$$4.5 \times 10^{-4} = \frac{x^2}{0.20}$$

Solving for x gives

$$x = [H_3O^+] = \sqrt{(4.5 \times 10^{-4})(0.20)} = 9.5 \times 10^{-3}\ M$$

We see that the ratio $(x/[\text{acid}]_{\text{initial}}) = 9.5 \times 10^{-3}/0.20 = 0.048$, or 4.8 percent. Since this is less than 5 percent, we were justified in neglecting x. The pH of the solution is

$$\text{pH} = -\log[H_3O^+] = -\log(9.5 \times 10^{-3}) = 2.02$$

(b) In part (b) we set up the equilibrium expression as we did in part (a).

	$HNO_2(aq) + H_2O(l) \rightleftharpoons H_3O^+(aq) + NO_2^-(aq)$		
Initial M	0.010	~0	0
Equilibrium M	$0.010 - x$	x	x

$$4.5 \times 10^{-4} = \frac{x^2}{0.010 - x}$$

The ratio $0.010/4.5 \times 10^{-4} = 22$. This is less than 100. Therefore, the quadratic equation must be used. The quadratic equation to be solved is

$$x^2 + (4.5 \times 10^{-4}x) - (4.5 \times 10^{-6}) = 0$$

Solving this equation for x gives

$$x = [H_3O^+] = 1.9 \times 10^{-3}\ M$$

The pH of the solution is

$$\text{pH} = -\log[H_3O^+] = -\log(1.9 \times 10^{-3}) = 2.72$$

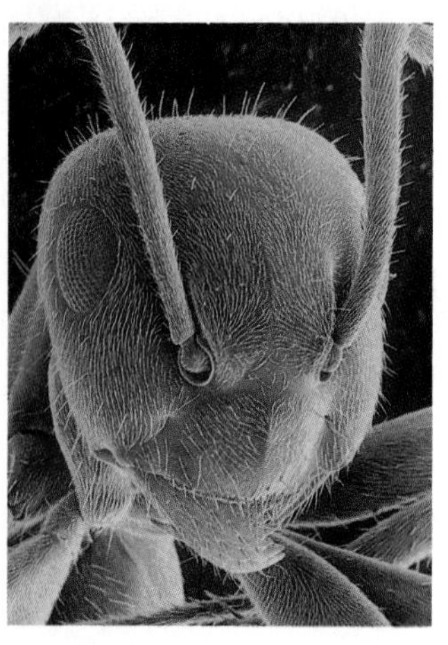

The venom of ants' stingers is formic acid, $HCHO_2$.

Practice Problem 15.4: Calculate the hydronium ion concentration of a 0.020 M formic acid ($HCHO_2$) solution. The K_a of formic acid is 1.77×10^{-4}. Show whether or not you can assume that the initial $M \approx$ equilibrium M.

Weak Bases

We recall from Chapter 14 that a weak base, B, reacts with water to produce hydroxide ions and cations characteristic of the base:

$$B(aq) + H_2O(l) \rightleftharpoons BH^+(aq) + OH^-(aq)$$

Hydroxide ions are formed in this reaction when protons are transferred from water to the base. An example of a weak base is ammonia, which reacts with water to produce ammonium ions and hydroxide ions:

$$NH_3(aq) + H_2O(l) \rightleftharpoons NH_4^+(aq) + OH^-(aq)$$

We also recall that the equilibrium constant for the reaction of a weak base with water is called K_b. We use K_b in a calculation of ion concentrations in an aqueous ammonia solution in Example 15.5.

Example 15.5

Calculating the Ion Concentrations, pH, and pOH of Aqueous Ammonia

Calculate the hydroxide ion concentration, hydronium ion concentration, pH, and pOH of a 0.35 M solution of ammonia. The value of the ionization constant, K_b, of ammonia is 1.8×10^{-5}.

SOLUTION: Problems involving weak bases and weak acids are solved in a similar way. However, in problems involving weak bases, we first solve for the hydroxide ion concentration and then solve for the hydronium ion concentration. In this problem we write the equation for the reaction of ammonia with water. Then we write the initial and equilibrium concentrations of all species. We let x equal the number of moles per liter of ammonia that has reacted when equilibrium is reached (x also equals the concentration of ammonium ions and of hydroxide ions). Then we substitute the equilibrium concentrations for each species into the equilibrium expression and solve for x.

	$NH_3(aq)$ + $H_2O(l)$ ⇌	$NH_4^+(aq)$ +	$OH^-(aq)$
Initial M	0.35	0	~0
Equilibrium M	$0.35 - x$	x	x

The equilibrium expression for K_b is

$$K_b = \frac{[NH_4^+][OH^-]}{[NH_3]}$$

Substituting the equilibrium concentrations of the reactants and products into this equation gives

$$K_b = 1.8 \times 10^{-5} = \frac{x^2}{0.35 - x}$$

Since the ratio of the initial concentration of ammonia to the ionization constant, K_b, is greater than 100, we simplify this equation to

$$1.8 \times 10^{-5} = \frac{x^2}{0.35}$$

Solving for x, we obtain

$$x^2 = (1.8 \times 10^{-5})(0.35)$$

$$x = [OH^-] = \sqrt{(1.8 \times 10^{-5})(0.35)} = 2.5 \times 10^{-3}\ M$$

To be certain that we were justified in assuming that x is negligibly small in the calculation above, we see that the ratio of the concentration of ammonia that ionized to the initial concentration is less than 5 percent. That is,

$$\left(\frac{2.5 \times 10^{-3}}{0.35}\right)(100) = 0.71 \text{ percent}$$

From the expression for K_w,

$$[H_3O^+] = \frac{1.0 \times 10^{-14}}{2.5 \times 10^{-3}} = 4.0 \times 10^{-12}\ M$$

$$pH = -\log(4.0 \times 10^{-12}) = 11.40$$

This pH value is greater than 7, which shows that the solution is alkaline.

$$pOH = 14.00 - 11.40 = 2.60$$

Practice Problem 15.5: Methylamine ionizes according to the equation

$$CH_3NH_2(aq) + H_2O(l) \rightleftharpoons CH_3NH_3^+(aq) + OH^-(aq)$$

The K_b for methylamine is 4.2×10^{-4}. What is the hydroxide ion concentration and pH of a 0.15 M solution of methylamine?

Solutions of Salts of Weak Acids and Bases

We learned in Chapter 14 that salt solutions can be neutral, acidic, or alkaline. A solution of a salt derived from a strong acid and a strong base, such as sodium chloride, is neutral because the ions of the salt do not hydrolyze. A solution of a salt derived from a weak base and a strong acid, such as ammonium chloride, is acidic because the cations of the salt hydrolyze. A salt derived from a strong base and a weak acid, such as sodium acetate, is basic because the anions of the salt hydrolyze. We can calculate the pH of a solution of a salt whose ions hydrolyze by methods similar to those described for weak acids and bases. This is because the cations that hydrolyze react with water as weak acids, and the anions that hydrolyze react with water as weak bases. The following examples illustrate calculations involving hydrolysis reactions.

Example 15.6

Calculating pH of a Solution of an Ammonium Salt

What is the pH of a 0.20 *M* solution of ammonium chloride? The ionization constant, K_b, for ammonia is 1.8×10^{-5}.

SOLUTION: Ammonium chloride is a soluble salt whose cation, NH_4^+, is the conjugate acid of the weak base ammonia. Ammonium ions hydrolyze to produce ammonia and hydronium ions. This reaction and the initial and equilibrium concentrations of the reactants and products are:

	$NH_4^+(aq) + H_2O(l) \rightleftharpoons NH_3(aq) + H_3O^+(aq)$		
Initial *M*	0.20	0	~0
Equilibrium *M*	$0.20 - x$	x	x

The number of moles per liter of the ammonia molecules and hydronium ions produced by hydrolysis is designated as x. The value of x also equals the number of moles per liter of ammonium ions that hydrolyze. Therefore, the equilibrium concentration of ammonium ions is $0.20 - x$. Chloride ions are spectator ions in this process and hence do not appear in the equilibrium. We are given K_b for ammonia. The value of K_a for ammonium ions can be obtained from the expression for K_w: $K_w = K_aK_b$. Thus,

$$K_a = \frac{K_w}{K_b} = \frac{1.0 \times 10^{-14}}{1.8 \times 10^{-5}} = 5.6 \times 10^{-10}$$

The expression for K_a of ammonium ions is

$$K_a = \frac{[NH_3][H_3O^+]}{[NH_4^+]}$$

Substituting the relevant data into this expression gives

$$5.6 \times 10^{-10} = \frac{x^2}{0.20 - x}$$

The value of x is negligibly small compared with 0.20 because the ratio $([\text{acid}]_{\text{initial}})/K_a = [(0.20/5.6 \times 10^{-10})] = 3.6 \times 10^8$, which is much more than 100. Therefore, the equation above can be simplified to

$$5.6 \times 10^{-10} = \frac{x^2}{0.20}$$

Solving for x gives

$$x = [H_3O^+] = \sqrt{(5.6 \times 10^{-10})(0.20)} = 1.1 \times 10^{-5}\ M$$

$$\text{pH} = 4.96$$

Since the pH is less than 7, the solution is acidic. We expected this result because we knew that ammonium ion is a weak acid.

Example 15.7

Calculating the pH of a Solution of a Salt Whose Anion Is Weakly Basic

Calculate the pH of a 0.20 *M* sodium fluoride solution. The acid dissociation constant, K_a, for hydrofluoric acid is 3.5×10^{-4}.

SOLUTION: We will use the same strategy to solve this problem that we used in Example 15.6. Fluoride ion is the conjugate base of hydrofluoric acid, a weak acid. Fluoride ion, acting as a base, reacts with water to form hydrofluoric acid and hydroxide ions. This reaction and the initial and equilibrium concentrations of reactants and products are:

	$F^-(aq) + H_2O(l) \rightleftharpoons$	$HF(aq)$ +	$OH^-(aq)$
Initial *M*	0.20	0	~0
Equilibrium *M*	$0.20 - x$	x	x

In the above expression, x equals the number of moles per liter of hydrofluoric acid formed at equilibrium, the number of moles per liter of hydroxide ions, and also the number of moles per liter of fluoride ions that reacted with water. The expression for the ionization constant, K_b, for fluoride ions is

$$K_b = \frac{[HF][OH^-]}{[F^-]}$$

We are given K_a for hydrofluoric acid. We obtain K_b for fluoride ions from the expression $K_w = K_aK_b$:

$$K_b = \frac{K_w}{K_a} = \frac{1.0 \times 10^{-14}}{3.5 \times 10^{-4}} = 2.9 \times 10^{-11}$$

Substituting the relevant data into this expression gives

$$2.9 \times 10^{-11} = \frac{x^2}{0.20 - x}$$

Since K_b is very small, x is negligibly small compared with 0.20, and the equation simplifies to

$$2.9 \times 10^{-11} = \frac{x^2}{0.20}$$

$$x = [OH^-] = \sqrt{(2.9 \times 10^{-11})(0.20)} = 2.4 \times 10^{-6}\ M$$

$$\text{pOH} = 5.62 \quad \text{and} \quad \text{pH} = 8.38$$

Practice Problem 15.6: Sodium benzoate ($NaC_7H_5O_2$) is a preservative for food products. What is the pH of a 0.15 *M* solution of sodium benzoate? The acid dissociation constant, K_a, for benzoic acid ($HC_7H_5O_2$) is 6.5×10^{-5}.

A salt derived from a weak acid *and* a weak base can give an acidic or basic solution depending on the extent to which the cations and the anions of the salt hydrolyze. For example, ammonium acetate is a salt derived from a weak acid—acetic acid—and a weak base—ammonia. When this salt is dissolved in water, ammonium ions hydrolyze to produce ammonia and hydronium ions, and acetate ions hydrolyze to produce acetic acid and hydroxide ions. The hydronium ions react with the hydroxide ions to form water. The K_a for ammonium ions is equal to K_b for acetate ions; therefore, these two ions hydrolyze to the same extent. Thus, a solution of ammonium acetate is neutral. On the other hand, a solution of ammonium cyanide is basic because cyanide ions hydrolyze to a greater extent than ammonium ions.

15.4 Common Ion Effect and Buffer Solutions

Common Ion Effect

When a solution of a weak acid, HA, is treated with a soluble salt, NaA, whose anion is the same as the anion of the aqueous acid, the acid equilibrium shifts to the left because the anions of the added salt react with the hydronium ions in the equilibrium system. Similarly, when a solution of a weak base is treated with a soluble salt whose cation is the same as the cation of the aqueous base, the base equilibrium shifts to the left because the added anions react with the hydroxide ions in the equilibrium system. *A shift in an equilibrium caused by adding ions common to ions in the equilibrium mixture is called the* common ion effect.

For example, an aqueous solution of acetic acid contains some acetate ions. Acetate ions are also the anions of the soluble salt sodium acetate. Thus, acetate ions are common to acetic acid solution and sodium acetate solution. The effect of adding sodium acetate to an aqueous solution of acetic acid is an example of the common ion effect. In this case, adding sodium acetate disturbs the original acetic acid equilibrium, and the equilibrium shifts to the left to relieve the "stress":

$$NaC_2H_3O_2(s) \xrightarrow{H_2O} Na^+(aq) + C_2H_3O_2^-(aq)$$

concentration of acetate ions increases greatly ↓

$$HC_2H_3O_2(aq) + H_2O(l) \rightleftharpoons H_3O^+(aq) + C_2H_3O_2^-(aq)$$

$\longleftarrow$ so the equilibrium shifts to the left

This shift in the equilibrium position decreases the hydronium ion concentration and the solution becomes less acidic. The pH increases, and the concentration of undissociated acetic acid also increases. If the initial concentration of acetic acid and sodium acetate are known, the pH of the solution can be calculated, as shown in Example 15.8.

Example 15.8

Calculating the pH of an Acetic Acid Solution to Which Sodium Acetate Is Added

What is the pH of a 1.00 *M* solution of acetic acid that is also 0.500 *M* in sodium acetate? The dissociation constant for acetic acid is 1.8×10^{-5}.

SOLUTION: We let x equal the concentration of acetic acid that dissociates. The concentration of hydronium ions is also x, as well as the acetate ions produced by the dissociation of acetic acid. The initial and equilibrium concentration of the various species in this equilibrium system are:

	$HC_2H_3O_2(aq)$ +	$H_2O(l) \rightleftharpoons$	$H_3O^+(aq)$ +	$C_2H_3O_2^-(aq)$
Initial M	1.00		~0	0.50
Equilibrium M	$1.00 - x$		x	$0.50 + x$ (0.50 from sodium acetate; x from acetic acid)

The total acetate ion concentration in the equilibrium mixture equals $0.500 + x$, where x equals the concentration of acetate ions derived from the dissociation of acetic acid and 0.500 *M* is the concentration of acetate ions derived from sodium acetate. Substituting the concentrations of hydronium ions (x), acetate ions ($x + 0.500$), and acetic acid ($1.00 - x$) into the expression for K_a yields

$$K_a = 1.8 \times 10^{-5} = \frac{[H_3O^+][C_2H_3O_2^-]}{[HC_2H_3O_2]} = \frac{(x)(x + 0.500)}{1.00 - x}$$

If we assume that x is very much less than 1.00 or 0.500, the quantities ($x + 0.500$) and ($1.00 - x$) simplify to 0.500 and 1.00, respectively. The equation thus simplifies to

$$1.8 \times 10^{-5} = \frac{(x)(0.500)}{1.00}$$

Therefore,

$$x = [H_3O^+] = \frac{(1.8 \times 10^{-5})(1.00)}{0.500} = 3.6 \times 10^{-5}\ M$$

The value of x is less than 5 percent of either 1.00 or 0.500. Thus, our assumption is correct that x is so small that we can neglect it in terms ($0.500 + x$) and ($1.00 - x$).

$$\text{pH} = -\log(3.6 \times 10^{-5}) = 4.44$$

The $[H_3O^+]$ of a 1.00 *M* solution of acetic acid is 4.2×10^{-3}, and the pH is 2.37. Thus, the addition of sodium acetate decreases the hydronium ion concentration about 100-fold from its value in a 1.00 *M* acetic acid solution. The pH increases by approximately 2 units; the solution is less acidic.

Practice Problem 15.7: Calculate the hydronium ion concentration and pH of a 0.65 *M* solution of benzoic acid, $HC_7H_5O_2$, which is also 1.00 *M* in sodium benzoate, $NaC_7H_5O_2$. The dissociation constant, K_a, for benzoic acid is 6.5×10^{-5}.

Next, let us consider an example of the common ion effect in a solution of a weak base—ammonia—and a salt—ammonium chloride—which provides cations that are the same as the cations of the base solution. When ammonium chloride is added to a solution of ammonia, the aqueous ammonia equilibrium shifts to the left because ammonium ions from the added salt react with hydroxide ions in the equilibrium system of the base:

$$NH_4Cl(s) \xrightarrow{H_2O} NH_4^+(aq) + Cl^-(aq)$$

↓ ammonium ion concentration increases greatly

$$NH_3(aq) + H_2O(l) \rightleftharpoons NH_4^+(aq) + OH^-(aq)$$

⟵ so the equilibrium shifts to the left

Example 15.9

Calculating the pH of a Mixture of Aqueous Ammonia and Ammonium Chloride

What is the pH of a 0.35 *M* solution of ammonia which is also 0.35 *M* in ammonium chloride? The ionization constant, K_b, for ammonia is 1.8×10^{-5}.

SOLUTION: An equation for the equilibrium reaction of ammonia and the molar concentrations of each species are given below. We let x equal the molar concentration of ammonia that ionizes, and also the molar concentration of ammonium ions and of hydroxide ions:

	$NH_3(aq)$ + $H_2O(l)$ ⇌	$NH_4^+(aq)$ +	$OH^-(aq)$
Initial *M*	0.35	0.35	~0
Equilibrium *M*	$0.35 - x$	$0.35 + x$ (from ammonium chloride)	x (from ammonia)

The value of x is expected to be very small compared to 0.35. Therefore, the terms $(0.35 - x)$ and $(0.35 + x)$ simplify to 0.35, and the equilibrium expression becomes

$$1.8 \times 10^{-5} = \frac{(x + 0.35)(x)}{0.35 - x} \approx \frac{0.35x}{0.35}$$

from which

$$x = [OH^-] = 1.8 \times 10^{-5}\ M$$

The value of x is less than 5 percent of 0.35. Thus, our assumption is correct that x is very small and can be neglected in the terms $(0.35 + x)$ and $(0.35 - x)$. Hence,

$$\text{pOH} = 4.74 \quad \text{and} \quad \text{pH} = 9.26$$

In Example 15.5 we found that the hydroxide ion concentration of a 0.35 M ammonia solution is 2.5×10^{-3} M, corresponding to a pH of 11.40. In this example we found that making a 0.35 M ammonia solution 0.35 M in ammonium ions causes the hydroxide ion concentration to decrease from 2.5×10^{-3} M to 1.8×10^{-5} M, which corresponds to a decrease in pH of more than 2 units, from 11.40 to 9.26.

Buffer Solutions

A solution that maintains a nearly constant pH when a small amount of an acid or a base is added to it is called a **buffer solution**. The mixture of solutes that is responsible for the buffering action of a solution is called a **buffer**. Buffer solutions are very important in many areas of chemistry. For example, the pH of blood maintains a constant pH of 7.4 under a very wide range of physiological conditions. Small changes in blood pH lead to serious illness or death. Also, the aqueous phase of all living cells maintains a nearly constant pH of 7.0. Many chemical reactions in aqueous solution change the pH of the solution. These reactions can be studied at constant pH by carrying them out in buffer solutions. In the following paragraphs we consider various aspects of buffer solutions.

A buffer is usually made by mixing a weak acid with its conjugate base, or by mixing a weak base with its conjugate acid. (The acid and base components of the buffer do not react with each other.) An example of a buffer is a mixture of acetic acid and sodium acetate. Acetate ion is the conjugate base of acetic acid, and therefore acetate ions are the base component of the buffer. When an acid is added to a buffer, it reacts with the base component of the buffer; when a base is added to the buffer, it reacts with the acid component of the buffer. Thus, the pH of the buffer solution does not change appreciably, as is shown next.

Buffer Action

To illustrate how a buffer solution resists a change in pH, let us consider a solution of acetic acid and sodium acetate. When a small amount of hydrochloric acid is added to this buffer, acetate ions react with the added hydronium ions to form acetic acid:

$$HCl(aq) + H_2O(l) \longrightarrow H_3O^+(aq) + Cl^-(aq)$$

$\downarrow$ concentration of hydronium ions increases considerably

$$HC_2H_3O_2(aq) + H_2O(l) \rightleftharpoons H_3O^+(aq) + C_2H_3O_2^-(aq)$$

$\longleftarrow$ so the equilibrium shifts to the left

The reaction of hydronium ions with acetate ions is the reverse of the dissociation of acetic acid. As long as the concentration of added hydronium ions is less than the concentration of acetate ions in the solution, acetate ions react with the newly added hydronium ions, and equilibrium will be restored at approximately the same pH. However, if the number of moles of hydrochloric acid added exceeds the number of moles of acetate ions originally in the solution, the pH will decrease considerably.

A buffer solution of acetic acid and sodium acetate also maintains a relatively constant pH when a base such as sodium hydroxide is added. When a small amount of sodium hydroxide solution is added to an acetic acid–sodium acetate buffer solution, the added hydroxide ions react with the hydronium ions in the equilibrium mixture to form water:

$$Na^+(aq) + OH^-(aq)$$

$$\downarrow \text{ hydroxide ion concentration increases, } OH^- \text{ reacts with } H_3O^+\text{, lowering } [H_3O^+]$$

$$HC_2H_3O_2(aq) + H_2O(l) \rightleftharpoons H_3O^+(aq) + C_2H_3O_2^-(aq)$$

$$\xrightarrow{\hspace{6em}}$$

so the equilibrium shifts to the right

As a result of this shift in the equilibrium, the acetic acid concentration decreases and the acetate ion concentration increases. But the decrease in the concentration of acetic acid does not result in a significant change in pH unless the number of moles of sodium hydroxide added exceeds the number of moles of acetic acid present in the buffer solution.

The number of moles of an acid or a base a buffer solution can consume without an appreciable change in its pH is the *capacity* of the buffer. The capacity of a buffer solution depends on the amounts of both the acid and the base components present in the buffer.

Calculating the pH of Buffer Solutions

The pH of a buffer solution can be calculated in a way that is similar to calculations involving the common ion effect that we described earlier. However, calculations of the pH of buffer solutions can also be carried out by the approximation method described below.

Just as we can express the hydronium ion concentration of a solution in terms of pH, we can express the K_a of an acid in terms of its negative logarithm, which is defined as $\mathbf{p}\boldsymbol{K_a}$.

$$-\log K_a = pK_a$$

Let us see how the pK_a of a weak acid is related to the pH of a solution of the weak acid, HA, and its conjugate base, A^-. The equilibrium constant expression for HA is

$$K_a = \frac{[H_3O^+][A^-]}{[HA]}$$

Solving this expression for $[H_3O^+]$ yields

$$[H_3O^+] = K_a \frac{[HA]}{[A^-]}$$

Taking the logarithm of both sides of this equation and multiplying by -1 gives

$$-\log[H_3O^+] = -\log K_a - \log \frac{[HA]}{[A^-]}$$

Since $-\log[H_3O^+]$ equals pH and $-\log K_a$ equals pK_a, we can write

$$pH = pK_a - \log \frac{[HA]}{[A^-]} \quad \text{or} \quad pH = pK_a + \log \frac{[A^-]}{[HA]}$$

Since A^- is a base and HA is an acid, this expression can be written in general terms as

$$\text{pH} = \text{p}K_a + \log \frac{[\text{base}]}{[\text{acid}]} \tag{15.4}$$

This relation is called the *Henderson–Hasselbalch* equation.

An acid buffer is most effective when the concentration of the base component equals the concentration of the acid component. At this point, the pH of the buffer equals the $\text{p}K_a$ value of its acid component. In more general terms, a buffer has the greatest capacity when the [base]/[acid] ratio ranges from 1:10 to 10:1. Under these conditions, $\text{pH} = \text{p}K_a \pm 1$.

In Examples 15.10 and 15.11 we use the Henderson–Hasselbalch equation to calculate the pH of acidic and basic buffers, respectively.

Example 15.10

Calculating the pH of an Acetic Acid–Sodium Acetate Buffer Using the Henderson–Hasselbalch Equation

Calculate the pH of a buffer solution that contains 0.50 *M* acetic acid and 0.80 *M* sodium acetate. The K_a for acetic acid is 1.8×10^{-5}.

SOLUTION: The Henderson–Hasselbalch equation is

$$\text{pH} = \text{p}K_a + \log \frac{[\text{base}]}{[\text{acid}]}$$

We are given the K_a for acetic acid. The $\text{p}K_a = -\log K_a = -\log(1.8 \times 10^{-5}) = 4.74$. We are also given the concentration of the base, 0.80 *M*, and the concentration of the acid, 0.50 *M*. Inserting these data into the Henderson–Hasselbalch equation, we obtain

$$\text{pH} = 4.74 + \log \frac{[0.80]}{[0.50]} = 4.94$$

Example 15.11

Calculating the pH of an Ammonia–Ammonium Chloride Buffer Using the Henderson–Hasselbalch Equation

Calculate the pH of a buffer solution that contains 0.60 *M* ammonia and 0.40 *M* ammonium chloride. The K_b for ammonia is 1.8×10^{-5}.

SOLUTION: The acid component of the buffer consists of ammonium ions, and the base component is ammonia. We are given the K_b for ammonia. We obtain the K_a for ammonium ions from the relation $K_aK_b = K_w$:

$$K_a = \frac{1 \times 10^{-14}}{1.8 \times 10^{-5}} = 5.6 \times 10^{-10}$$

$$\text{p}K_a = -\log(5.6 \times 10^{-10}) = 9.25$$

The concentration of the base is 0.60 *M* and the concentration of the acid is 0.40 *M*. Therefore,

$$\text{pH} = 9.25 + \log \frac{0.60}{0.40} = 9.43$$

In the next example we calculate the change in pH that occurs when an acid is added to a buffer solution.

Example 15.12

Calculating the pH Change When HCl Is Added to a Buffer

Calculate the pH change resulting from adding 0.010 mol of HCl to 1.0 L of a buffer solution that is 0.50 *M* in acetic acid and 0.50 *M* in sodium acetate. The acid dissociation constant, K_a, for acetic acid is 1.8×10^{-5}.

SOLUTION: We calculate the pH before and after adding HCl. We are given the initial concentration of the acid and base components of the buffer. We convert K_a to pK_a and use the Henderson–Hasselbalch equation to obtain the initial pH of the buffer.

$$\mathrm{pH} = \mathrm{p}K_a + \log \frac{[\text{base}]}{[\text{acid}]}$$

$$\mathrm{pH} = -\log (1.8 \times 10^{-5}) + \log \frac{0.50}{0.50} = 4.74$$

Upon adding 0.010 mol of hydrochloric acid, 0.010 mol of hydronium ions react with 0.010 mol of acetate ions, so that the final acetate ion concentration is

$$[C_2H_3O_2^-] = 0.50\ M - 0.010\ M = 0.49\ M$$

The molar concentration of acetic acid increases by 0.010 mol:

$$[HC_2H_3O_2] = 0.50\ M + 0.010\ M = 0.51\ M$$

To determine the pH of the solution after hydrochloric acid has been added, we again use the Henderson–Hasselbalch equation:

$$\mathrm{pH}_{\text{final}} = 4.74 + \log \frac{0.49}{0.51} = 4.72$$

The pH change ($\mathrm{pH}_{\text{final}} - \mathrm{pH}_{\text{initial}}$) = 4.72 − 4.74 = −0.02 pH unit.

If 0.010 mol of HCl were added to a liter of pure water, the H_3O^+ ion concentration would increase from 1.0×10^{-7} *M* to 0.010 *M*. The pH would decrease from 7.00 to 2.00, a decrease of 5 pH units.

We can see from Example 15.12 that when the concentration of the base component of a buffer equals the concentration of the acid component, [base] = [acid], then pH = pK_a. Under these conditions, the buffer is most effective.

Practice Problem 15.8: **(a)** What is the pH of a buffer solution that is 0.85 *M* in benzoic acid, $HC_7H_5O_2$, and 0.85 *M* in sodium benzoate, $NaC_7H_5O_2$? **(b)** What is the pH of 1.0 L of this buffer solution after adding 0.010 mol of nitric acid? **(c)** What is the pH of 1.0 L of this buffer solution after adding 0.020 mol of sodium hydroxide?

Preparation of Buffers

If we wish to prepare a buffer that is effective at a given pH, we choose a weak acid whose pK_a is close to the required pH. (The weak acid can be a molecular acid such as acetic acid or an acidic ion such as ammonium ion.) Then we add the conjugate base of the acid. We make the solution equimolar in the acid and its conjugate base.

When a weak acid with the desired pK_a is not available, an acid whose pK_a is nearest the desired value is chosen. The pH of the buffer can then be set by providing an appropriate [base]/[acid] ratio, as shown in Example 15.13.

Example 15.13

Preparing a Buffer of a Desired pH

What ratio of the concentrations of sodium formate, $NaCHO_2$, and formic acid, $HCHO_2$, is needed to prepare a buffer whose pH is 3.85? The dissociation constant, K_a, for formic acid is 1.8×10^{-4}.

SOLUTION: We use the Henderson–Hasselbalch equation to calculate the ratio of [base]/[acid] required:

$$pH = pK_a + \log \frac{[base]}{[acid]}$$

where the concentration of the base is the formate ion concentration and the concentration of the acid is the formic acid concentration. The pK_a of formic acid is

$$pK_a = -\log K_a = -\log (1.8 \times 10^{-4}) = 3.74$$

Next we substitute the desired pH and the pK_a into the Henderson–Hasselbalch equation and solve for the [base]/[acid] ratio:

$$3.85 = 3.74 + \log \frac{[base]}{[acid]}$$

$$\log \frac{[base]}{[acid]} = 3.85 - 3.74 = 0.11$$

Now we take the antilog of 0.11 to obtain the ratio [salt]/[acid]:

$$\frac{[base]}{[acid]} = \text{antilog } 0.11 = 1.3$$

The concentration of base in this ratio is that of sodium formate and the concentration of acid is that of formic acid. Thus, to make a buffer solution with a pH of 3.85, the sodium formate concentration must be 1.3 times the formic acid concentration. For example, if the formic acid concentration is 0.10 *M*, the concentration of sodium formate must be 0.13 *M*.

Blood Plasma, a Biological Buffer

Blood is a complex liquid that is composed of cells and a fluid called *plasma*, which contains various dissolved solutes. The dissolved solutes in blood contain buffer systems which regulate body pH at a remarkably constant value of 7.4. The principal buffer system in blood consists of carbonic acid

(H_2CO_3), hydrogen carbonate ions (HCO_3^-), carbonate ions (CO_3^{2-}), and carbon dioxide (CO_2). Carbon dioxide is produced by the oxidation of foods in body tissues such as skeletal muscles and the liver. For example, the sugar glucose, $C_6H_{12}O_6$, is oxidized according to the net equation

$$C_6H_{12}O_6(aq) + 6O_2(g) \longrightarrow 6CO_2(g) + 6H_2O(l)$$

Carbon dioxide produced by metabolism diffuses out of various tissues into the blood and is transported to the lungs, where it is exhaled.

The following equilibria are responsible for the buffering action of the carbonic acid–hydrogen carbonate ion buffer. The equilibrium constants for reactions (1), (2), and (3) are K_1, K_2, and K_3, respectively.

$$\begin{array}{llr} (1) & H_2CO_3(aq) + H_2O(l) \rightleftharpoons H_3O^+(aq) + HCO_3^-(aq) & K_1 \\ (2) & CO_2(aq) + H_2O(l) \rightleftharpoons H_2CO_3(aq) & K_2 \\ (3) & CO_2(g) \rightleftharpoons CO_2(aq) & K_3 \\ \hline (4) & CO_2(g) + 2H_2O(l) \rightleftharpoons H_3O^+(aq) + HCO_3^-(aq) & K_1K_2K_3 = K_4 \end{array}$$

The net reaction represented by equation (4) indicates that the $[H_3O^+]$ and the pH of blood depend only on the concentration of hydrogen carbonate ions dissolved in blood and on the partial pressure of gaseous carbon dioxide, $CO_2(g)$, in the air spaces of the lungs.

Strenuous exercise results in an increased production of carbon dioxide by body tissues and ultimately increases the concentration of hydronium ions and hydrogen carbonate ions in the blood.

If the rate of metabolism increases—as it does during strenuous exercise—more carbon dioxide is produced by tissues, where it dissolves. It is transported by blood from the tissues to the lungs, and exhaled as carbon dioxide gas. As a result, the partial pressure of carbon dioxide gas in the air spaces of the lungs increases, and equilibrium (4) shifts to the right. This shift increases the concentrations of hydronium ions and hydrogen carbonate ions in the blood. But hydronium ions react with hydrogen carbonate ions to give carbonic acid in equilibrium (1). As a result, the position of equilibrium (4) scarcely alters.

If, on the other hand, the metabolic rate falls and less carbon dioxide is produced in tissues, less carbon dioxide is exhaled, and equilibrium (4) shifts to the left. Since the reservoir of carbon dioxide gas in the lungs is large and can be changed rapidly by altering the breathing rate, the pH of blood does not change under normal conditions (Figure 15.1).

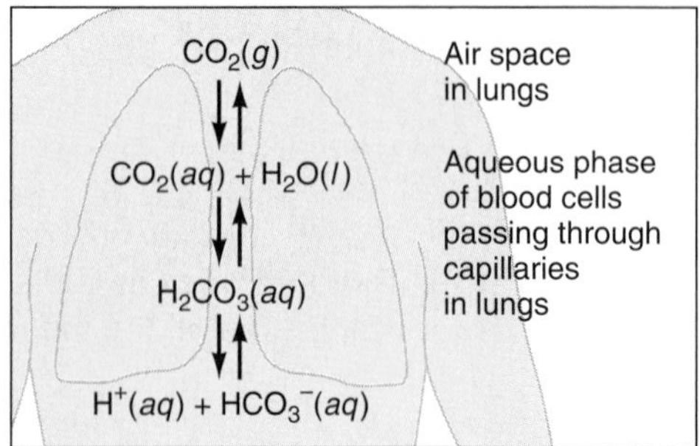

Figure 15.1

pH regulation of blood. The pH of blood is controlled by the ratio of $[HCO_3^-]$ to the partial pressure of CO_2 in the air spaces of the lungs. If excess $H^+(aq)$ ions enter blood, they react with HCO_3^- ions to form more CO_2 gas, and the pH does not change appreciably. If, on the other hand, the hydrogen ion concentration in blood decreases, more CO_2 gas dissolves in the blood, restoring the $H^+(aq)$ ion concentration to its normal level.

When the pH of blood falls below 7.4, the condition is called *acidosis*, and when the pH rises above 7.4, the condition is called *alkalosis*. The pH of blood in severe diabetes can drop as low as 6.8, leading to coma and death.

At first glance it hardly seems possible that a change in pH of 0.6 unit can be fatal. However, many biological processes are exquisitely sensitive to changes in pH, and therefore slight changes in pH radically alter these processes. Furthermore, a change of 0.6 pH unit may not appear to be very large, but it corresponds to a change in the hydronium ion concentration of a factor of 4.

15.5 Polyprotic Acids

Strong Polyprotic Acids

An acid that contains two or more ionizable hydrogens is called a *polyprotic acid*. An acid that contains two ionizable hydrogen atoms in a molecule is called a *diprotic acid*, and an acid that contains three ionizable hydrogen atoms is a *triprotic acid*.

The only common strong polyprotic acid is sulfuric acid. It is a diprotic acid that dissociates in two steps. The first step of dissociation is virtually 100 percent complete:

$$H_2SO_4(aq) + H_2O(l) \longrightarrow H_3O^+(aq) + HSO_4^-(aq) \qquad K_{a1} >> 1$$

A dilute solution of sulfuric acid therefore contains no undissociated sulfuric acid molecules.

Hydrogen sulfate ion, HSO_4^-, is a weak acid, so it dissociates only partially:

$$HSO_4^-(aq) + H_2O(l) \rightleftharpoons H_3O^+(aq) + SO_4^{2-}(aq) \qquad K_{a2} = 1.2 \times 10^{-2}$$

Since the first ionization constant for sulfuric acid, K_{a1}, is much larger than the second ionization constant, K_{a2}, most of the hydronium ions in a solution of sulfuric acid are produced in the first step of dissociation.

Example 15.14

Calculating the Concentrations of Ionic Species in a Sulfuric Acid Solution

Calculate the concentrations of the various ionic species at equilibrium in a 0.10 *M* solution of H_2SO_4. The acid dissociation constant for HSO_4^- is 1.2×10^{-2}.

SOLUTION: The first step of dissociation,

$$H_2SO_4(aq) + H_2O(l) \longrightarrow H_3O^+(aq) + HSO_4^-(aq)$$

is complete and produces 0.10 mol L^{-1} H_3O^+ and 0.10 mol L^{-1} HSO_4^- ions. Hydrogen sulfate ion, however, is a relatively weak acid. The equation for the dissociation equilibrium of hydrogen sulfate ion and the concentrations of the ions in solution after the first and second dissociation steps are given below:

	$HSO_4^-(aq) + H_2O(l) \rightleftharpoons$	$H_3O^+(aq)$ +	$SO_4^{2-}(aq)$
After first step (M)	0.10	0.10	0
Equilibrium M	$0.10 - x$	$0.10 + x$	x

In this equilibrium mixture, x moles per liter of hydrogen sulfate ions dissociate in the second step to produce x moles per liter of hydronium ions and x moles per liter of sulfate ions. The concentration of hydrogen sulfate ions at equilibrium is $0.10 - x$, and the *total* hydronium ion concentration produced by the first and second dissociation steps is $0.10 + x$. The total hydronium ion concentration in the equilibrium mixture must be used in the expression for K_{a2}:

$$K_{a2} = \frac{[H_3O^+][SO_4^{2-}]}{[HSO_4^-]} = 1.2 \times 10^{-2}$$

Substituting the relevant data into this equation gives

$$1.2 \times 10^{-2} = \frac{(0.10 + x)(x)}{0.10 - x}$$

K_{a2} is large compared to 0.10. Therefore, we must solve the quadratic equation to obtain the correct value of x. The solution of the quadratic equation gives 0.010 M as the value of x. The concentrations of various species in a 0.10 M solution of sulfuric acid are

$$[H_2SO_4] = 0\ M$$

$$[H_3O^+] = (0.10 + x)\ M = (0.10 + 0.010)\ M = 0.11\ M$$

$$[HSO_4^-] = (0.10 - x)\ M = (0.10 - 0.010)\ M = 0.09\ M$$

$$[SO_4^{2-}] = x = 0.010\ M$$

Weak Polyprotic Acids

A weak polyprotic acid dissociates only partially, even in the first step. An example of a weak diprotic acid is aqueous hydrogen sulfide, H_2S. Aqueous H_2S dissociates in two steps. The extent of the dissociation is small in both steps, as indicated below by the dissociation constants, K_{a1} and K_{a2}, for steps 1 and 2, respectively:

$$H_2S(aq) + H_2O(l) \rightleftharpoons H_3O^+(aq) + HS^-(aq) \qquad K_{a1} = 1.0 \times 10^{-7}$$

$$HS^-(aq) + H_2O(l) \rightleftharpoons H_3O^+(aq) + S^{2-}(aq) \qquad K_{a2} = 1.3 \times 10^{-13}$$

When hydrogen sulfide gas is bubbled through water to produce a saturated solution, the concentration of undissociated hydrogen sulfide is 0.10 M at 25°C and 1.0 atm. Because both K_{a1} and K_{a2} for H_2S are small, an aqueous solution of hydrogen sulfide has relatively low concentrations of hydronium and hydrogen sulfide ions, and an even lower concentration of sulfide ions, as shown by Example 15.15.

Example 15.15

Calculating the Concentrations of Ions in a Saturated Solution of Hydrogen Sulfide at 25°C

Calculate the concentrations of hydronium ions, hydrogen sulfide ions, and sulfide ions present at equilibrium in a 0.10 *M* solution of H_2S at 25°C. K_{a1} for H_2S is 1.0×10^{-7}; K_{a2} is 1.3×10^{-13}.

SOLUTION: The first step of dissociation and the concentrations of the species present are:

	$H_2S(aq)$ + $H_2O(l)$ ⇌	$H_3O^+(aq)$ +	$HS^-(aq)$
Initial *M*	0.10	~0	0
After step 1 (*M*)	$0.10 - x$	x	x

We let x equal the number of moles per liter of H_2S that dissociates. The molar concentration of hydronium ions is x, and the molar concentration of hydrogen sulfide ions is also x. Next we insert these values into the expression of K_{a1}:

$$K_{a1} = \frac{[H_3O^+][HS^-]}{[H_2S]} = \frac{(x)(x)}{0.10 - x} = 1.0 \times 10^{-7}$$

Because the value of K_{a1} is small, the term $(0.10 - x)$ equals 0.10. The first step of dissociation thus gives the following value of x:

$$x = [H_3O^+] = [HS^-] = \sqrt{(0.10)(1.0 \times 10^{-7})} = 1.0 \times 10^{-4}\ M$$

Some hydrogen sulfide ions dissociate in the second step to produce more hydronium ions plus sulfide ions. The equation below represents the second step of dissociation. We let y equal the number of moles per liter of hydrogen sulfide ions dissociated; this also equals the number of moles per liter of sulfide ions formed:

	$HS^-(aq)$ + $H_2O(l)$ ⇌	$H_3O^+(aq)$ +	$S^{2-}(aq)$
After step 1 (*M*)	1.0×10^{-4}	1.0×10^{-4}	0
Equilibrium *M*	$1.0 \times 10^{-4} - y$	$1.0 \times 10^{-4} + y$	y

Substituting these values into the expression of K_{a2} gives

$$K_{a2} = 1.3 \times 10^{-13} = \frac{(1.0 \times 10^{-4} + y)(y)}{1.0 \times 10^{-4} - y}$$

The value of K_{a2} is extremely small. Therefore, the terms $(1.0 \times 10^{-4} - y)$ and $(1.0 \times 10^{-4} + y)$ each essentially equal 1.0×10^{-4} and the equation simplifies to

$$1.3 \times 10^{-13} = \frac{(1.0 \times 10^{-4})(y)}{1.0 \times 10^{-4}} = y$$

The concentrations of the species present in a 0.10 M solution of H_2S are as follows:

$$[H_2S] = 0.10 - x - y = 0.10 - (1.0 \times 10^{-4}) - (1.3 \times 10^{-13}) = 0.10\ M$$

$$[H_3O^+] = x + y = (1.0 \times 10^{-4}) + (1.3 + 10^{-13}) = 1.0 \times 10^{-4}\ M$$

$$[HS^-] = x - y = (1.0 \times 10^{-4}) - (1.3 \times 10^{-13}) = 1.0 \times 10^{-4}\ M$$

$$[S^{2-}] = y = 1.3 \times 10^{-13}\ M$$

We can see from Example 15.15 that the sulfide ion concentration in a saturated solution of H_2S equals K_{a2} because the concentrations of hydronium ions and hydrogen sulfide ions are essentially equal. The dissociation of hydrogen sulfide ions in the second step is so slight that its equilibrium concentration and that of the hydronium ions are essentially equal.

15.6 Equilibria in Acid–Base Titration

We discussed acid–base titration and stoichiometry in Chapter 4. Now we focus on the equilibria in acid–base titrations and the change of pH during titration.

Titration of a Strong Acid with a Strong Base

Let us suppose that we are titrating 50.0 mL of 0.100 M hydrochloric acid with a solution of 0.100 M sodium hydroxide. The pH of the hydrochloric acid solution before any sodium hydroxide solution is added is $-\log 0.100 = 1.000$. Adding sodium hydroxide increases the pH, but the solution remains acidic until all of the acid has been neutralized, at which point the pH of the solution is 7.00. The point in the titration at which all of the acid has just reacted with the base is called the *equivalence point*. Beyond the equivalence point, the solution becomes alkaline as more base is added and the pH of the solution continues to increase.

When the value of the pH is plotted versus the volume of NaOH added in the titration, the curve shown in Figure 15.2 is obtained. This curve is called a **titration curve**. The titration curve of hydrochloric acid with aqueous sodium hydroxide, or of any strong acid with a strong base, has an "S-shape." This S-shaped curve is often called a *sigmoidal curve*.

Titrations are usually carried out by adding one solution from a buret to another solution in a conical flask called an Erlenmeyer flask. Burets are calibrated in milliliters. It is therefore convenient to express the quantities of acid or base consumed in *millimoles*. One millimole (mmol) equals 0.001 mol, and of course, 1 mL is 0.001 L. Thus a solution whose concentration is 1 M contains 1 mol per liter and also 1 mmol per milliliter. Example 15.16 illustrates the calculation of the pH at different stages of a titration of a strong acid with a strong base.

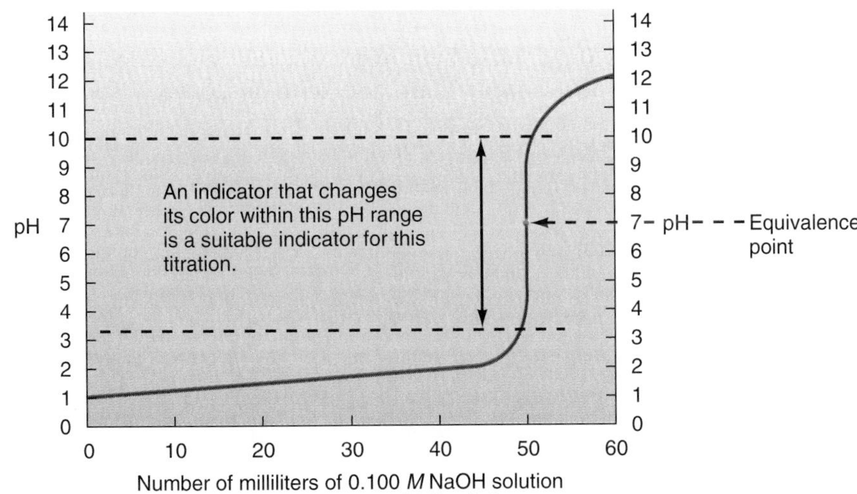

Figure 15.2

The pH change during a titration of HCl with NaOH.

Example 15.16

Calculating pH at Different Stages in the Titration of a Strong Acid with a Strong Base

50.0 mL of 0.100 *M* HCl is titrated with 0.100 *M* NaOH. Calculate the pH after adding: **(a)** 25.0 mL of NaOH; **(b)** 49.9 mL of NaOH; **(c)** 50.0 mL of NaOH; **(d)** 50.1 mL of NaOH.

SOLUTION: In each case we first calculate the number of millimoles of acid and base in the solution after each addition of sodium hydroxide. The concentrations of the acid or base still present and the pH are then calculated.

(a) The number of millimoles (mmol) of hydrochloric acid in 50.0 mL of 0.100 *M* solution is

$$(50.0 \text{ mL})(0.100 \text{ mmol mL}^{-1}) = 5.00 \text{ mmol}$$

The number of millimoles of sodium hydroxide added is

$$(25.0 \text{ mL})(0.100 \text{ mmol mL}^{-1}) = 2.50 \text{ mmol}$$

The number of millimoles of hydrochloric acid remaining is

$$5.00 \text{ mmol} - 2.50 \text{ mmol} = 2.50 \text{ mmol}$$

The total volume of the solution is

$$50.0 \text{ mL} + 25.0 \text{ mL} + 75.0 \text{ mL}$$

$$[H_3O^+] = \frac{2.50 \text{ mmol}}{75.0 \text{ mL}} = 3.33 \times 10^{-2}\ M$$

$$\text{pH} = 1.478$$

(b) Proceeding as in part (a), 0.01 mmol of hydrochloric acid remains in a total volume of 99.9 mL of the solution.

$$[H_3O^+] = \frac{0.01 \text{ mmol}}{99.9 \text{ mL}} = 1 \times 10^{-4} M \quad \text{and} \quad \text{pH} = 4.0$$

(c) After adding 50.0 mL of sodium hydroxide, all the hydrochloric acid has been neutralized. Thus, the solution contains sodium ions and chloride ions. The pH is 7.000.

(d) There is an excess of 0.1 mL of 0.100 *M* sodium hydroxide (0.01 mmol) in the total volume of 100.1 mL of the solution. Therefore,

$$[OH^-] = \frac{0.01 \text{ mmol}}{100.1 \text{ mL}} = 1 \times 10^{-4}\, M$$

$$\text{pOH} = 4.0 \quad \text{and} \quad \text{pH} = 10.0$$

We can see that adding a small amount of sodium hydroxide near the neutralization point causes a large change in pH.

Practice Problem 15.9: For a titration of 50.00 mL of 0.1000 *M* hydrochloric acid with 0.1000 *M* sodium hydroxide, calculate the pH after adding: **(a)** 10.00 mL; **(b)** 45.00 mL; **(c)** 49.50 mL; **(d)** 49.99 mL; **(e)** 50.01 mL; **(f)** 51.00 mL of sodium hydroxide. Plot the pH values versus the number of milliliters of NaOH added.

Titration of a Weak Acid with a Strong Base

Now let us consider the titration of a weak acid, such as acetic acid, with a strong base such as sodium hydroxide. A titration is performed by adding a strong base of known concentration to a solution of the weak acid. At various intervals, the pH of the solution is measured and plotted versus the volume of sodium hydroxide added to obtain the titration curve shown in Figure 15.3.

Acetic acid reacts with sodium hydroxide according to the equation

$$HC_2H_3O_2(aq) + NaOH(aq) \rightleftharpoons NaC_2H_3O_2(aq) + H_2O(l)$$

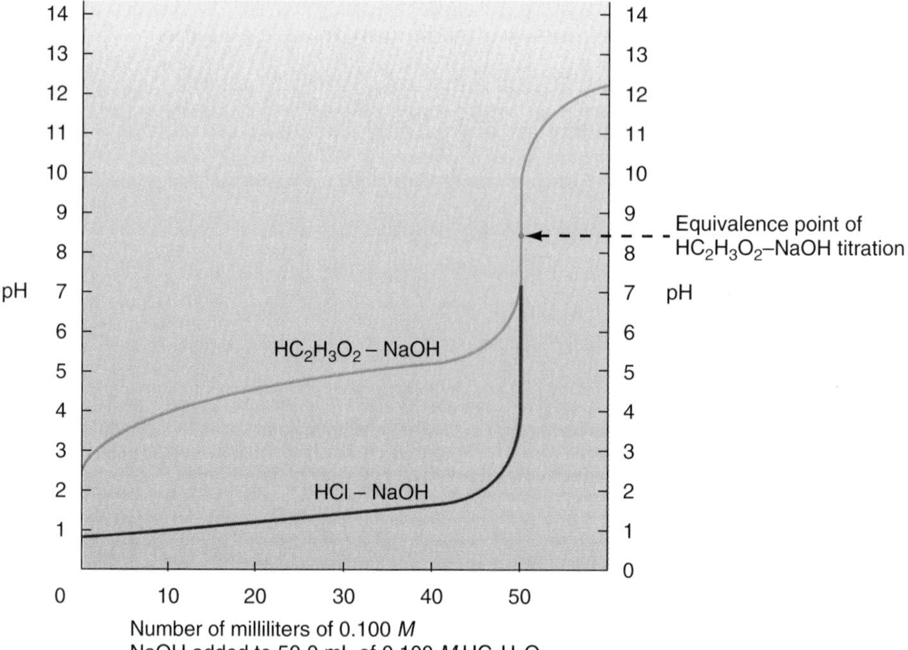

Figure 15.3

Titration curve for the titration of 50.0 mL of 0.100 *M* acetic acid with 0.100 *M* sodium hydroxide. The HCl–NaOH titration curve is also shown for comparison.

Before the equivalence point of the titration is reached, the solution contains unreacted acetic acid and its conjugate base—acetate ions. This solution is a buffer solution and we can calculate its pH by using the Henderson–Hasselbalch equation.

At the equivalence point of the titration, the solution contains sodium ions and acetate ions. The acetate ions hydrolyze, and the pH of the solution is calculated as described in Section 15.3. When sodium hydroxide is added beyond the equivalence point, the pH is determined by the hydroxide ion concentration.

Example 15.17

Calculating the pH at Different Stages of Titration of a Weak Acid with a Strong Base

Calculate the pH in the titration of 50.0 mL of 0.100 *M* acetic acid with 0.100 *M* sodium hydroxide in each of the following instances: **(a)** initial pH of acetic acid; **(b)** after adding 25.0 mL of NaOH; **(c)** after adding 49.9 mL of NaOH; **(d)** after adding 50.0 mL of NaOH; **(e)** after adding 50.1 mL of NaOH. K_a for acetic acid is 1.8×10^{-5}.

SOLUTION: As in Example 15.16, we first calculate the number of millimoles of the acid or base which remains in the solution. Then we determine the molar concentrations of the relevant species from the total volume of the solution. From the molar concentration of the hydronium ions, we calculate the pH.

(a) Using the equilibrium equation

$$HC_2H_3O_2(aq) + H_2O(l) \rightleftharpoons H_3O^+(aq) + C_2H_3O_2^-(aq)$$

as a guide and following our standard method for solving equilibrium problems, we obtain

$$[H_3O^+] = \sqrt{(K_a)[HC_2H_3O_2]} = \sqrt{(1.8 \times 10^{-5})(0.100)}$$

$$= 1.3 \times 10^{-3}\ M$$

$$pH = 2.89$$

(b) The reaction of acetic acid with sodium hydroxide is

$$HC_2H_3O_2(aq) + NaOH(aq) \rightleftharpoons NaC_2H_3O_2(aq) + H_2O(l)$$

After 25.0 mL of sodium hydroxide has been added, acetic acid is half-neutralized.

mmol acetic acid originally present

$$= (50.0\ mL)(0.100\ mmol\ mL^{-1})$$

$$= 5.00\ mmol$$

$$\text{mmol sodium hydroxide added} = (25.0\ mL)(0.100\ mmol\ mL^{-1})$$

$$= 2.50\ mmol$$

$$\text{mmol acetic acid remaining} = 5.00\ mmol - 2.50\ mmol$$

$$= 2.50\ mmol$$

$$\text{mmol acetate ions in the solution} = 2.50\ mmol$$

$$\text{total volume of the solution} = 50.0\ mL + 25.0\ mL = 75.0\ mL$$

$$[HC_2H_3O_2] = \frac{2.50 \text{ mmol}}{75.0 \text{ mL}} = 0.0333\ M$$

$$[C_2H_3O_2^-] = \frac{2.50 \text{ mmol}}{75.0 \text{ mL}} = 0.0333\ M$$

We use the Henderson–Hasselbalch equation to calculate the pH of the solution:

$$\text{pH} = \text{p}K_a + \log \frac{[C_2H_3O_2^-]}{[HC_2H_3O_2]}$$

$$= 4.74 + \log \frac{0.0333}{0.0333} = 4.74$$

(c) Proceeding as in step (b), the number of millimoles of acetic acid remaining is 5.00 mmol − 4.99 mmol = 0.01 mmol, and the number of millimoles of acetate ions produced is 4.99 mmol. The total volume of the solution is 99.9 mL. The [base]/[acid] ratio is larger than it was after adding 25.0 mL of NaOH [part (b)].

$$[C_2H_3O_2^-] = \frac{4.99 \text{ mmol}}{99.9 \text{ mL}} = 0.0499\ M$$

$$[HC_2H_3O_2] = \frac{0.01 \text{ mmol}}{99.9 \text{ mL}} = 1 \times 10^{-4}\ M$$

$$\text{pH} = 4.74 + \log \frac{0.0499}{1 \times 10^{-4}} = 4.74 + 2.7 = 7.4$$

(d) After 50.0 mL of sodium hydroxide has been added, 5.00 mmol of acetate ions are present in 100.0 mL of solution. The solution is slightly alkaline because of hydrolysis of acetate ions. The concentration of acetate ions before hydrolysis is

$$[C_2H_3O_2^-] = \frac{5.00 \text{ mmol}}{100.0 \text{ mL}} = 0.0500\ M$$

Proceeding as usual in solving hydrolysis problems, and letting x equal the number of moles per liter of acetate ions that react, we have:

	$C_2H_3O_2^-(aq) + H_2O(l) \rightleftharpoons$	$HC_2H_3O_2(aq) +$	$OH^-(aq)$
Initial M	0.0500	0	~0
Equilibrium M	$0.0500 - x$	x	x

$$K_b = \frac{[HC_2H_3O_2][OH^-]}{[C_2H_3O_2^-]} = \frac{K_w}{K_a} = \frac{1.0 \times 10^{-14}}{1.8 \times 10^{-5}} = 5.6 \times 10^{-10}$$

Substituting the known concentrations into the expression for K_b gives

$$5.6 \times 10^{-10} = \frac{x^2}{0.0500 - x}$$

Since x is very small compared with 0.0500, the term $0.0500 - x \approx 0.0500$. Solving the equation above for x gives

$$x = [OH^-] = \sqrt{(5.6 \times 10^{-10})(0.0500)} = 5.3 \times 10^{-6}\ M$$

$$pOH = 5.30 \qquad pH = 8.70$$

(e) After adding 50.1 mL of NaOH, the solution contains 0.01 mmol of NaOH in 100.1 mL of solution, and

$$[OH^-] = \frac{0.01\ \text{mol}}{100.1\ \text{mL}} = 1 \times 10^{-4}\ M$$

$$pOH = 4.0 \qquad pH = 10.0$$

Note that here the pH change which results from adding an excess of 0.1 mL of sodium hydroxide is smaller than in the titration of a strong acid with a strong base because the pH at the equivalence point is higher because acetate ions hydrolyze.

Practice Problem 15.10: For a titration of 50.00 mL of 0.1000 M acetic acid with 0.1000 M sodium hydroxide, calculate the pH after adding: **(a)** 10.00 mL; **(b)** 49.00 mL; **(c)** 49.99 mL; **(d)** 50.01 mL; **(e)** 51.00 mL of sodium hydroxide.

Titration of a Weak Base with a Strong Acid

The titration of a weak base with a strong acid follows the same principles as the titration of a weak acid with a strong base. Let us consider the titration of a typical weak base, aqueous ammonia, with a strong acid such as hydrochloric acid. The equation for the titration reaction is

$$NH_3(aq) + HCl(aq) \rightleftharpoons NH_4^+(aq) + H_2O(l)$$

As hydrochloric acid is added at the beginning of the titration, the pH decreases gradually. When one-half of the ammonia has been titrated, the system contains equimolar amounts of ammonia and ammonium ions (the conjugate acid of ammonia). This solution is thus a buffer solution, and the change in pH is slow until the equivalence point is reached. At the equivalence point, the solution contains ammonium chloride in water. Ammonium ion, however, is a weak acid, and it reacts slightly with water to produce hydronium ions. Therefore, the pH at the equivalence point is less than 7.

The titration curve obtained when aqueous ammonia is titrated with aqueous hydrochloric acid is shown in Figure 15.4. It is plotted on the same axes with a sodium hydroxide–hydrochloric acid titration curve for comparison. We can see that the shape of the titration curve for the titration of a base with an acid is similar to the titration curve for the titration of an acid with a base, but the pH change is reversed.

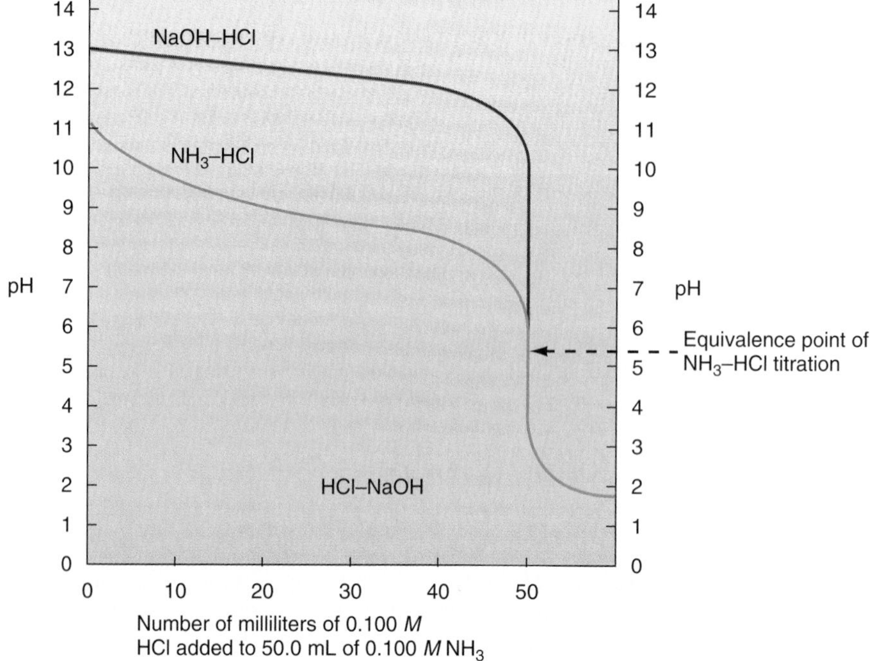

Figure 15.4

Titration curve for the titration of 50.0 mL of 0.100 M NH_3 with 0.100 M HCl. An NaOH–HCl titration curve is also shown for comparison.

Example 15.18

Calculating the pH at Different Stages in the Titration of a Weak Base with a Strong Acid

A 50.0-mL sample of 0.100 M NH_3 is titrated with 0.100 M hydrochloric acid. Calculate the pH after adding: **(a)** 25.0 mL of hydrochloric acid; **(b)** 50.0 mL of hydrochloric acid; **(c)** 52.0 mL of hydrochloric acid.

SOLUTION: In each case we calculate the number of millimoles of the acid or the base remaining, the molar concentrations of the acid or the base, and the pH.

(a) mmol $NH_3(aq)$ initially present

$$= (50.0\ \text{mL})(0.100\ \text{mmol mL}^{-1}) = 5.00\ \text{mmol}$$

$$\text{mmol HCl}(aq) = (25.0\ \text{mL})(0.100\ \text{mmol mL}^{-1}) = 2.50\ \text{mmol}$$

$$\text{mmol NH}_3\ \text{remaining} = 5.00\ \text{mmol} - 2.50\ \text{mmol} = 2.50\ \text{mmol}$$

$$\text{mmol NH}_4^+\ \text{produced} = 2.50\ \text{mmol}$$

The total volume of the solution is 75.0 mL. The molar concentrations of aqueous ammonia and ammonium ions are

$$[\text{NH}_3] = \frac{2.50\ \text{mmol}}{75.0\ \text{mL}} = 0.0333\ M$$

$$[\text{NH}_4^+] = \frac{2.50\ \text{mmol}}{75.0\ \text{mL}} = 0.0333\ M$$

Thus, after 25.0 mL of hydrochloric acid has been added, the solution contains equimolar amounts of ammonia and its conjugate acid, ammonium ion. To determine the pH of the solution we use the Henderson–Hasselbalch equation:

$$pH = pK_a + \log \frac{[NH_3]}{[NH_4^+]}$$

We obtain K_a for ammonium ions from the expression for K_w:

$$K_a = \frac{K_w}{K_b} = \frac{1 \times 10^{-14}}{1.8 \times 10^{-5}} = 5.6 \times 10^{-10}$$

$$pK_a = -\log(5.6 \times 10^{-10}) = 9.25$$

$$pH = 9.25 + \log \frac{0.0333}{0.0333} = 9.25$$

(b) After 50.0 mL of hydrochloric acid has been added, all of the ammonia has been converted to ammonium ions. Ammonium ion is a weak acid that reacts with water to produce hydronium ions and ammonia. Therefore, in this part we must consider the reaction of ammonium ions with water. The initial concentration of ammonium ions is 0.0500 M:

$$[NH_4^+] = \frac{5.00 \text{ mmol}}{100.0 \text{ mL}} = 0.0500\ M$$

We proceed in our usual way of solving equilibrium problems:

	$NH_4^+(aq)$ +	$H_2O(l)$ ⇌	$NH_3(aq)$ +	$H_3O^+(aq)$
Initial M	0.0500		0	~0
Equilibrium M	$0.0500 - x$		x	x

In part (a) we calculated the value of K_a for ammonium ions as 5.6 × 10^{-10}. Because K_a is very small, x is small compared with 0.0500, and

$$K_a = 5.6 \times 10^{-10} = \frac{x^2}{0.0500}$$

$$x = [H_3O^+] = \sqrt{(5.6 \times 10^{-10})(0.0500)} = 5.3 \times 10^{-6}\ M$$

$$pH = 5.28$$

(c) After 52.0 mL of hydrochloric acid has been added, the solution contains excess hydrochloric acid, and the hydronium ion concentration equals the concentration of hydrochloric acid:

$$[H_3O^+] = \frac{0.20 \text{ mmol}}{102.0 \text{ mL}} = 2.0 \times 10^{-3}\ M$$

$$pH = 2.70$$

Practice Problem 15.11: For the titration of 50.0 mL of 0.100 M aqueous ammonia with 0.100 M hydrochloric acid, calculate the pH after adding: **(a)** 10.0 mL of hydrochloric acid; **(b)** 49.9 mL of hydrochloric acid; **(c)** 50.1 mL of hydrochloric acid.

15.7 Acid–Base Indicators

The pH of a solution can be determined by the color of a substance called an **indicator**. An indicator is a weak acid or a weak base whose color changes as a function of pH. An acid indicator can be represented by the formula HInd, and its dissociation in water can be written as follows:

$$\underset{\text{color 1}}{\mathrm{HInd}(aq)} + \mathrm{H_2O}(l) \rightleftharpoons \mathrm{H_3O^+}(aq) + \underset{\text{color 2}}{\mathrm{Ind^-}(aq)}$$

The color of the undissociated molecules, HInd, of an indicator is different from the color of the aqueous indicator ions, Ind^-. These different colors are represented in the above equation as color 1 and color 2, respectively. The colors of the two forms of an indicator (HInd and Ind^-) can be distinguished by the human eye when the ratio of the concentration of one form of the indicator to the concentration of the other is about 10:1. This means that if $[Ind^-]/[HInd] = 10:1$ or greater, we see the color of the Ind^- ions, and if $[Ind^-]/[HInd] = 1:10$ or less, we see the color of the undissociated HInd molecules.

Table 15.3 lists some acid–base indicators, their pH ranges for color transition, and their colors at the pH below and above the interval of transition.

Each substance in Table 15.3 changes color over a relatively narrow pH range. Several different indicators can be impregnated in a paper tape. When this tape, called "pH paper" is tested with an aqueous solution, its color indicates an approximate pH of the solution. pH paper can be used to measure an approximate pH from 0 to 14.

Table 15.3 shows that methyl red indicator changes its color in the pH range 4.4 to 6.2, which is slightly before the equivalence point is reached in the strong acid–strong base titration. Because the pH change is very large near the equivalence point (Figure 15.2), any indicator that changes its color at the pH range represented by the almost vertical portion of the titration curve can be used without serious error. As we can see from Table 15.3, bromphenol blue, methyl orange, bromcresol green, methyl red, litmus, phenol red, or phenolphthalein can be used to indicate the endpoint in a titration of a strong acid with a strong base.

Table 15.3

Some pH Indicators and Their Colors

Indicator	pH Interval for Color Change	Color at Lower pH	Color at Higher pH
Methyl violet	0–2	Yellow	Violet
Thymol blue	1.2–2.8	Red	Yellow
Bromphenol blue	3.0–4.6	Yellow	Purple
Methyl orange	3.1–4.4	Red	Yellow
Bromcresol green	3.8–5.4	Yellow	Blue
Methyl red	4.4–6.2	Red	Yellow
Litmus	4.5–8.2	Red	Blue
Phenol red	6.4–8.2	Yellow	Red
Phenolphthalein	8.3–10.0	Colorless	Pink
Alizarin yellow	10.1–12.0	Yellow	Red

APPLICATIONS OF CHEMISTRY 15.2
Why Do Flowers Change Colors?

For centuries people have noticed that flowers of the same species growing short distances apart frequently have different colors (see photo). Today, it is not uncommon to find that flowers growing close to a house exhibit different colors than their siblings growing just a short distance away. Why is this so?

The acidity or basicity of the soil can influence the color of a flower. A red flower in basic soil may change to blue when moved to acidic soil. In fact, some flowers are natural acid–base indicators. Today, we use many plant extracts in the chemistry laboratory to determine the acidity or basicity of solutions. For example, litmus, present in litmus paper, comes from lichens that grow primarily in the Netherlands. To make litmus paper, a clean strip of paper is dipped into a solution of the litmus extract and dried. Two types of litmus paper, red and blue, can be produced. Red litmus, which detects bases, is made by placing the dried paper into a solution that is slightly acidic. Blue litmus, which detects acids, is made by placing the dried paper into a solution that is slightly basic.

Many other plants are also sources of acid–base indicators. An extract from red rose petals turns bright green in basic solution and pink in acidic solution. Juice extracted from red cherries turns green in basic solutions, red in acidic solution, and blue in neutral solution. Some of the other plants in which this indicator effect exists include red cabbage, blue iris, blue hyacinth, and huckleberries.

All acid–base indicators are weak acids or weak bases. Let's represent the equilibrium between a red hypothetical weak acid indicator, HIn, and its blue ionized form, In^-, as follows:

$$\underset{\text{}}{H_2O(l)} + \underset{\text{Red}}{HIn(aq)} \rightleftharpoons H_3O^+(aq) + \underset{\text{Blue}}{In^-(aq)}$$

All indicators have a special property: Their molecular forms (HIn) are a different color from their ionic forms (In^-).

From Le Châtelier's principle, we know that when we place HIn into an acidic solution, the high hydronium ion concentration drives the equilibrium to the left and the red color of HIn molecules dominates.

In a basic solution, hydroxide ions from the base react with the hydronium ions from the indicator. As hydroxide ions remove hydronium ions to form water molecules, this drives the equilibrium to the right, where the blue color of In^- ions dominates.

Now we can consider our first example: Why are flowers growing alongside a house sometimes a different color than their siblings growing nearby? We know that it is caused by a difference in the pH of the soil, but what makes the pH of the soil by the house different from the pH of the soil in the yard? Calcium carbonate in the brick mortar makes the soil around some houses more basic than the nearby soil.

The titration curve for the titration of a weak acid with a strong base has its vertical section in the pH range from approximately 7 to 10 (Figure 15.3). From Table 15.3 we find that phenolphthalein changes its color in the pH range from 8.3 to 10.0. Therefore, phenolphthalein is a suitable indicator for the titration of acetic acid with sodium hydroxide. Methyl red would be a poor indicator for this titration because its color change occurs in the range pH 4.4 to 6.2. From Figure 15.3 we can see that a large volume of NaOH must be added to change the pH from 4.4 to 6.2, and the indicator changes its color before the equivalence point is reached.

For the titration of a weak base with a strong acid, the vertical portion of the titration curve lies in the pH range from approximately 6 to 4 (Figure 15.4). From Table 15.3 we find that methyl red is a suitable indicator for this titration and that litmus or phenolphthalein would be unsuitable. Figure 15.5 shows the colors of several indicators in acid and basic solution.

Figure 15.5

Colors of methyl orange, bromcresol green, methyl red, litmus, phenol red, and phenolphthalein in basic and acidic solutions.

Summary

Autoionization of Water; Strong Acids and Bases

Pure water, or any neutral aqueous solution, contains very low concentrations of hydronium ions and hydroxide ions, and they exist in equal numbers due to the autoionization of water.

At 25°C the **ion product constant of water**, K_w, equals 1.0×10^{-14}.

$$K_w = [H_3O^+][OH^-] = 1.0 \times 10^{-14}$$

In an acid solution, $[H_3O^+] > [OH^-]$, and in an alkaline solution, $[OH^-] > [H_3O^+]$, but the ion product constant, $[H_3O^+][OH^-]$, equals 1.0×10^{-14} at 25°C. In a dilute solution of a strong monoprotic acid, the hydronium ion concentration equals the concentration of the acid. Similarly, the hydroxide ion concentration of a strong base equals the concentration of the base.

pH and pOH

The acidity of a solution can be expressed as the negative logarithm (base 10) of the $[H_3O^+]$ of the solution. The *negative logarithm of* $[H_3O^+]$ is called **pH**. Similarly, **pOH** is the negative logarithm of the $[OH^-]$ of a solution. For any aqueous solution at 25°C,

$$\text{pH} + \text{pOH} = 14.00$$

Weak Acids and Weak Bases

The hydronium ion concentration of a solution of a weak acid, HA, at equilibrium can be calculated assuming x to be the number of moles per liter of the acid dissociated, which is also the number of moles per liter of H_3O^+ ions and of A^- ions formed:

$$\underset{C - x}{\text{HA}(aq)} + \text{H}_2\text{O}(l) \rightleftharpoons \underset{x}{\text{H}_3\text{O}^+(aq)} + \underset{x}{\text{A}^-(aq)}$$

In the equilibrium above, C is the initial molar concentration of the undissociated acid. Substituting the foregoing values into the expression for the acid dissociation constant, K_a, for the weak acid, we obtain

$$K_a = \frac{x^2}{C - x}$$

In most cases when $C/K_a > 100$, x is negligible compared with the initial concentration, C, and the equilibrium expression simplifies to

$$K_a = \frac{x^2}{C}$$

Similarly, for most weak base solutions, hydroxide ion concentration can be calculated as y from the equation

$$K_b = \frac{y^2}{C}$$

where y is the molar concentration of OH^- and C is the molar concentration of the nonionized base. This simplified expression can usually be used when $C/K_b > 100$.

Calculating the pH of Salt Solutions

If the anion, A^-, of a salt is the conjugate base of a weak acid, HA, it reacts with water, that is, it hydrolyzes:

$$\text{A}^-(aq) + \text{H}_2\text{O}(l) \rightleftharpoons \text{HA}(aq) + \text{OH}^-(aq)$$

If the cation, BH^+, is a conjugate acid of a weak base, B, it hydrolyzes according to the equation

$$\text{BH}^+(aq) + \text{H}_2\text{O}(l) \rightleftharpoons \text{H}_3\text{O}^+(aq) + \text{B}(aq)$$

The base ionization constant, K_b, for A^- ions, or the acid dissociation constant, K_a, for HB^+ ions, can be calculated from the expression for K_w (Chapter 14). The pH of a solution of a salt whose ions hydrolyze can be calculated the way we described for weak acids and bases.

Buffer Solutions

A **buffer solution** *resists a change in pH when a moderate amount of an acid or a base is added* to it. A buffer solution consists of a weak acid and its conjugate base, or of a weak base and its conjugate acid. The anion of an acid is the conjugate base of the acid, and the cation of a base is the conjugate acid of the base. The pH of a buffer is given by the Henderson–Hasselbalch equation,

$$\text{pH} = \text{p}K_a + \log \frac{[\text{base}]}{[\text{acid}]}$$

Polyprotic Acids

A polyprotic acid dissociates in more than one step, and a weak polyprotic acid has more than one dissociation constant. The first step of dissociation of a strong polyprotic acid (such as sulfuric acid) is essentially complete, and hydronium ion concentration produced by this step equals the acid concentration. In calculating the sulfate ion concentration in a solution of sulfuric acid, the second dissociation constant for HSO_4^- ions must be used:

$$K_a = \frac{[H_3O^+][SO_4^{2-}]}{[HSO_4^-]}$$

The hydronium ion concentration used in the expression for K_a of HSO_4^- must be the *total* $[H_3O^+]$ produced in the first and in the second steps of dissociation of the acid. For a weak diprotic acid (such as H_2S), the sulfide ion concentration is nearly equal to K_{a2} of the acid.

Acid–Base Titration and Indicators

An **indicator** used in acid–base titration is usually a weak acid or a weak base whose color changes as a function of pH. An indicator suitable for an acid–base titration is one that exhibits its color change in the region of the vertical portion of the titration curve. A titration curve is a graph in which the pH is plotted (usually on the *y* axis) versus the number of milliliters of the base added on the *x* axis.

At the equivalence point in the titration of a strong acid with a strong base, the solution is neutral and its pH equals 7. In a titration of a weak acid with a strong base, the pH at the equivalence point is more than 7 because the anions of a weak base hydrolyze. In the titration of a weak base with a strong acid, the pH at the equivalence point is less than 7 because the cations of a weak base hydrolyze.

New Terms

Buffer (15.4)
Buffer solution (15.4)
Common ion effect (15.4)
Indicator (acid–base) (15.7)
pH (15.2)
pK_a (15.4)
pK_b (15.4)
pOH (15.2)
Titration curve (15.6)

Exercises

General Review

1. Write a chemical equation for the autoionization of water and a mathematical equation for K_w.
2. Show how K_w can be used to calculate the number of moles of hydronium and hydroxide ions in a liter of pure water.
3. Is the expression for K_w valid in an acidic or alkaline solution?
4. How is the concentration of a strong monoprotic acid in a dilute solution related to the $[H_3O^+]$ of the solution?
5. **(a)** How is the concentration of sodium hydroxide in a dilute solution related to the $[OH^-]$ of the solution? **(b)** How is the concentration of barium hydroxide related to the $[OH^-]$ of the solution?
6. How is the hydronium ion concentration of an acid solution related to the hydroxide ion concentration of the solution?
7. What is: **(a)** pH; **(b)** pOH; **(c)** p*K*?
8. How does the pH of a solution vary with its $[H_3O^+]$?
9. How is the pH of a solution related to the pOH of the solution?
10. How is the degree of dissociation of a monoprotic weak acid related to the initial and equilibrium concentrations of the acid?
11. How is the percent of dissociation of a weak acid related to: **(a)** the magnitude of the K_a of the acid; **(b)** the hydronium ion concentration of the acid solution?
12. If you were given the concentration of a weak base in solution and the hydronium ion concentration of the solution, how would you determine the degree of ionization of the base?
13. Write an equation for hydrolysis of an anion and another for hydrolysis of a cation.

14. Write an expression for K_b of the anion and K_a for the cation you used in question 13.

15. What is meant by "common ion effect" in weak acid–base equilibria? Explain how the hydronium ion concentration of a weak acid solution is affected by adding a salt that contains the conjugate base of the acid. Give an example (other than those in the chapter) of such an acid and a salt.

16. Give an example of a weak base and a salt that contains the conjugate acid of the base. How is the hydroxide ion concentration of this base solution affected by addition of the salt?

17. What is a buffer solution? Give an example (other than those in the chapter) of an acid buffer solution and an alkaline buffer solution.

18. What is meant by "buffer capacity"?

19. Briefly explain how an acid buffer solution resists a change in pH when a small amount of hydrochloric acid is added.

20. Explain how you would choose a buffer system that is effective at a given pH.

21. Use the Henderson–Hasselbalch equation to explain how the [base]/[acid] ratio of a buffer solution affects the pH of the solution.

22. Show that the sulfide ion concentration in a solution of aqueous H_2S, a weak diprotic acid, is equal to K_{a2} of the acid.

23. In a solution of H_2S, the sulfide ion concentration equals K_{a2} for H_2S. Explain why the sulfate ion concentration in a solution of sulfuric acid is not equal to K_{a2} for sulfuric acid.

24. What type of a substance usually serves as an acid–base indicator?

25. What information is needed to choose a suitable indicator for an acid–base titration?

26. State whether the solution at the equivalence point of each of the following titrations is neutral, acidic, or alkaline: **(a)** strong acid–strong base; **(b)** weak base–strong acid; **(c)** weak acid–strong base. Explain your answer in each case.

27. For which of the three titrations in question 26 does the titration curve have the longest vertical portion, almost parallel to the axis representing the pH? Explain how you would use the titration curves for these titrations to select a suitable indicator for each of the titrations.

$[H_3O^+]$ and $[OH^-]$ of Water and of Strong Acids and Bases

28. Calculate the hydronium ion concentration and the hydroxide ion concentration of pure water at 25°C. How many grams of hydronium ions are present in 10 L of water? Could that amount be detected by the balances in your laboratory?

29. What are the hydronium ion concentration and hydroxide ion concentration of a 0.15 M solution of nitric acid?

30. What are the hydronium ion concentration and hydroxide ion concentration of: **(a)** a 0.080 M solution of potassium hydroxide; **(b)** of a 0.0035 M solution of barium hydroxide?

pH and pOH

31. What are the pH and pOH of: **(a)** a 0.010 M solution of nitric acid; **(b)** a 0.25 M solution of hydrochloric acid?

32. What are the pH and pOH of: **(a)** a 0.010 M solution of sodium hydroxide; **(b)** a 0.25 M solution of potassium hydroxide; **(c)** a 0.0056 M solution of barium hydroxide?

33. The pH of a solution is 5.70. What are the pOH, the hydronium ion concentration, and the hydroxide ion concentration of the solution?

34. Calculate the pH of a 3.2×10^{-3} M solution of calcium hydroxide (limewater).

35. If you dissolve 1.58 g of solid potassium hydroxide in water to make 850 mL of solution, what is the pH of the solution?

36. If you dissolve 2.50 g of barium hydroxide octahydrate in water to make 1.20 L of solution, what is the pH of the solution?

37. Calculate the pH of 100 mL of 2.00 M solution of nitric acid. What is the pH of this solution after it is diluted with water to 450 mL?

38. What is the pH of a mixture of equal volumes of 0.10 M hydrochloric acid and 0.100 M hydrobromic acid?

Equilibrium Calculations of Weak Acid and Weak Base Solutions

39. Calculate the hydronium ion concentration, hydroxide ion concentration, pH, and pOH of a 0.400 M solution of HCN.

40. Calculate the hydronium ion concentration of: **(a)** 0.0400 M HCN; **(b)** 0.00400 M HCN.

41. In an experiment, three solutions of acetic acid are prepared. The first solution contains 1.00 mmol of the acid in 13.57 mL of solution, the second contains 1.00 mmol of the acid in 217.1 mL of solution, and the third contains 1.00 mmol of the acid in 1.737×10^3 mL of solution. The percent dissociation in each of these solutions is 1.57, 6.14, and 16.41 percent, respectively. Calculate the dissociation constant, K_a, for each of the three solutions. Does the value of K_a depend on the concentration of the solution?

42. Calculate the pH of 1.2 M hydrofluoric acid and of 0.80 M hypochlorous acid.

43. The pH of a 0.1035 *M* solution of a monoprotic acid is 2.560. Calculate the dissociation constant for the acid.
44. What is the pH of a 0.015 *M* solution of aniline?
45. What is the hydronium ion concentration and hydroxide ion concentration in 0.193 *M* aqueous ammonia?
*46. Calculate the volume of 0.10 *M* aqueous ammonia which is 1.3 percent ionized and contains as many hydroxide ions as 1.0 L of 0.10 *M* sodium hydroxide.
47. The dissociation constants for lactic and benzoic acids are 1.4×10^{-4} and 6.5×10^{-5}, respectively. What is the concentration of a benzoic acid solution that has the same hydronium ion concentration as a 0.10 *M* lactic acid solution?

Hydrolysis

48. Write an equation for hydrolysis of each of the following ions: **(a)** acetate ion; **(b)** ammonium ion; **(c)** nitrite ion; **(d)** formate ion; **(e)** fluoride ion; **(f)** hydrogen carbonate ion; **(g)** sulfide ion (two steps); **(h)** anilinium ion, $C_6H_5NH_3^+$.
49. Indicate whether each of the following 1 *M* solutions is acidic, alkaline, or neutral: **(a)** KBr; **(b)** NH_4Br; **(c)** KCN; **(d)** $RbNO_3$; **(e)** $NaNO_2$; **(f)** KClO; **(g)** $AlCl_3$; **(h)** $NH_4C_2H_3O_2$; **(i)** NH_4CN.
50. Write an expression for K_b for each of the following substances: **(a)** methylamine, CH_3NH_2; **(b)** hypochlorite ion.
51. Using the necessary value of K_a from Table 14.2, calculate the value of K_b for ClO^- ions.
52. Using the necessary value for K_b from Table 14.4, calculate the value of K_a for methylammonium ion, $CH_3NH_3^+$.
53. Calculate the pH of: **(a)** 0.10 *M* NH_4NO_3; **(b)** 0.50 *M* NaBrO; **(c)** 0.30 *M* pyridinium chloride, C_5H_5NHCl; **(d)** 1.6 *M* KNO_2.
54. A 0.0100 *M* solution of potassium cyanide, KCN, has a hydronium ion concentration of 2.22×10^{-11} *M*. Calculate the value of K_b for cyanide ions, the percent of hydrolysis of cyanide ions, and the value of K_a for HCN.
55. A 0.020 *M* solution of anilinium chloride, $C_6H_5NH_3Cl$, has a pH of 3.16. Calculate K_a for anilinium ions, $C_6H_5NH_3^+$, the percent of hydrolysis of anilinium ions, and the K_b for aniline, $C_6H_5NH_2$.

* Denotes problem of greater than average difficulty.

Common Ion Effect and Buffer Solutions

56. Calculate the pH of 0.10 *M* aqueous ammonia, which is also 0.10 *M* in ammonium chloride.
57. A 10.0-mL sample of aqueous ammonia contains 25 percent NH_3 by mass and has a density of 0.91 g cm^{-3}. What is the hydroxide ion concentration of this solution after adding 5.0 g of ammonium chloride and diluting the volume to 100 mL?
58. A 200-mL sample of a 0.10 *M* acetic acid solution is treated with 5.0 g of sodium acetate (assume no volume change). What is the hydronium ion concentration and the pH of the solution before and after adding the sodium acetate?
59. A 2.0-mL sample of 10.0 *M* nitric acid is added to 0.75 L of water. Calculate the change in pH of the acid.
60. Calculate the pH change when 2.0 mL of 10.0 *M* HNO_3 is added to 0.75 L of a buffer solution that is 1.0 *M* in acetic acid and 1.0 *M* in sodium acetate (assume no volume change when HNO_3 is added). Compare the answers in questions 59 and 60.
61. Calculate the pH change when 0.40 g of solid NaOH is added to 0.80 L of a buffer solution that is 0.80 *M* in benzoic acid and 1.2 *M* in sodium benzoate. (Assume no volume change upon adding NaOH.)
62. Calculate the $[HCO_3^-]/[H_2CO_3]$ ratio of blood that has pH 7.40.
63. How many grams of ammonium chloride must be added to 500 mL of 3.0 *M* aqueous ammonia to prepare a buffer solution of pH 10.00? Assume no volume change upon adding ammonium chloride.
*64. What volume of 0.50 *M* sodium benzoate must be added to 500 mL of 2.0 *M* benzoic acid solution to prepare a solution of pH 4.00?

Polyprotic Acids

65. Calculate the hydronium ion concentration, hydrogen sulfate ion concentration, and sulfate ion concentration of a 0.020 *M* solution of sulfuric acid.
66. Citric acid is a triprotic acid that is present in many fruits and vegetables. Its formula is $H_3C_6H_5O_7$. Write the three equilibria for the dissociation of citric acid. For each step, write the expression for the equilibrium constant.
67. Calculate the hydronium ion concentration, hydrogen carbonate ion concentration, carbonate ion concentration, and carbonic acid concentration of a 0.020 *M* solution of carbonic acid, H_2CO_3.

Acid–Base Titration

68. For each of the following titrations, is the pH at the equivalence point equal to 7, more than 7, or less than 7 when the acid has been completely neutralized? **(a)** Nitric acid + potassium hydroxide; **(b)** formic acid + sodium hydroxide; **(c)** hydrofluoric acid + potassium hydroxide; **(d)** hydrobromic acid + rubidium hydroxide; **(e)** sulfuric acid + sodium hydroxide (complete neutralization); **(f)** hydrochloric acid + methylamine, CH_3NH_2. Explain in each case.

69. Write a net ionic equation for the reaction that occurs in each of the titrations in question 68.

70. Calculate the pH at the equivalence point of each of the titrations in question 68, assuming that 50.0 mL of a 1.00 *M* acid is titrated with a solution of 1.00 *M* base to complete the neutralization.

71. Calculate the pH of the solution that results when 50.0 mL of 1.00 *M* HBr is titrated with 1.00 *M* KOH after adding: **(a)** 25.0 mL of KOH; **(b)** 49.9 mL of KOH; **(c)** 50.0 mL of KOH; **(d)** 50.1 mL of KOH; **(e)** 51.0 mL of KOH. Plot the pH versus the volume of KOH added, and list the indicators from Table 15.3 that might be suitable for this titration.

72. Calculate the pH of the solution that results when 25.0 mL of 0.200 *M* hydrofluoric acid is titrated with 0.200 *M* NaOH after adding: **(a)** no NaOH; **(b)** 20.0 mL of NaOH; **(c)** 24.9 mL of NaOH; **(d)** 25.0 mL of NaOH; **(e)** 25.1 mL of NaOH; **(f)** 26.0 mL of NaOH. Plot the pH versus milliliters of NaOH added, and list the indicators from Table 15.3 that might be suitable for this titration.

73. Calculate the pH of the solution that results when 30.0 mL of 0.500 *M* methylamine, CH_3NH_2, is titrated with 0.500 *M* hydrochloric acid after adding: **(a)** no HCl; **(b)** 20.0 mL of HCl; **(c)** 29.0 mL of HCl; **(d)** 29.9 mL of HCl; **(e)** 30.0 mL of HCl; **(f)** 30.1 mL of HCl; **(g)** 32.0 mL of HCl. Plot the pH versus milliliters of HCl added, and list the indicators from Table 15.3 that might be suitable for this titration.

74. The concentration of a weak monoprotic acid (K_a of 1.0×10^{-5}) is determined by titration with sodium hydroxide. At the endpoint of the titration, the molar concentration of sodium ions is 0.040 *M*. What is the pH of the solution at that point? What is the color of litmus at that point?

CHAPTER 16

Rocks and natural deposits of minerals contain metal silicates, phosphates, oxides, and sulfides.

Solubility Equilibria

In Chapter 15 we discussed equilibria in acid–base reactions. In this chapter we continue our discussion of chemical equilibria, but in a rather different context in which we focus on the equilibria between sparingly soluble solids and their ions in aqueous solution. Solubility equilibria exist wherever "insoluble" salts and water are found, that is, everywhere around us. For example, solubility equilibria are related to the effect of fluoride ions on dental health, and the destruction of statues by acid rain. Solubility equilibria also determine the existence or nonexistence of certain compounds in natural deposits. For example, deposits of metal nitrates and most halides are seldom found because these salts are soluble in water. On the other hand, metal silicates, phosphates, oxides, and sulfides are commonly found in natural deposits.

The solubilities of salts in water vary by many orders of magnitude. Some salts are very soluble, and dissolve to the extent of 10 or more moles per liter of water at room temperature. One of the most soluble salts known is lithium chlorate, $LiClO_3$, which has a solubility of about 35 mol per liter of water at 25°C. Other salts dissolve only to the extent of a few tenths of a mole per liter. Such salts are said to be "slightly soluble." For example, mercury(II) chloride, $HgCl_2$, has a solubility of about 0.3 mol per liter. Still other salts dissolve only to the extent of 10^{-20} mol per liter or less. We often use the terms "insoluble" and "sparingly soluble" in the same sense to describe substances that dissolve only to a very small extent. In this chapter we focus on salts whose molar solubilities are less than 0.1 mol per liter of water.

16.1 Molar Solubility and Solubility Product Constant

Solubility Equilibria and Molar Solubility

Solutions of most salts in water contain hydrated ions, not molecules. However, a few salts, including lead(II) acetate, $Pb(C_2H_3O_2)_2$, mercury(II) chloride, $HgCl_2$, and cadmium sulfate, $CdSO_4$, dissolve to form solutions that contain molecules as well as ions. These exceptions are relatively rare and we will not consider them further.

In a saturated solution of a salt above its solid residue, an equilibrium exists between the undissolved solid and its ions in the solution. The equilibrium between aqueous ions and an undissolved solid is called a *solubility equilibrium*. We can write the general equation for the solubility equilibrium of the salt MX as

$$MX(s) \rightleftharpoons M^{n+}(aq) + X^{n-}(aq)$$

The solid is conventionally written on the left (as a reactant).

The *number of moles of a salt that dissolves to form hydrated ions in 1 liter of solution* is called its **molar solubility**. Thus, the number of moles of $MX(s)$ dissolved in a liter of water equals the number of moles of M^{n+} ions, or the number of moles of X^{n-} ions, per liter of a saturated solution of MX, provided that the ions do not hydrolyze:

$$\begin{aligned}\text{moles of MX dissolved in 1 L} &= \text{molar solubility of MX}\\ &= [M^{n+}] = [X^{n-}]\end{aligned}$$

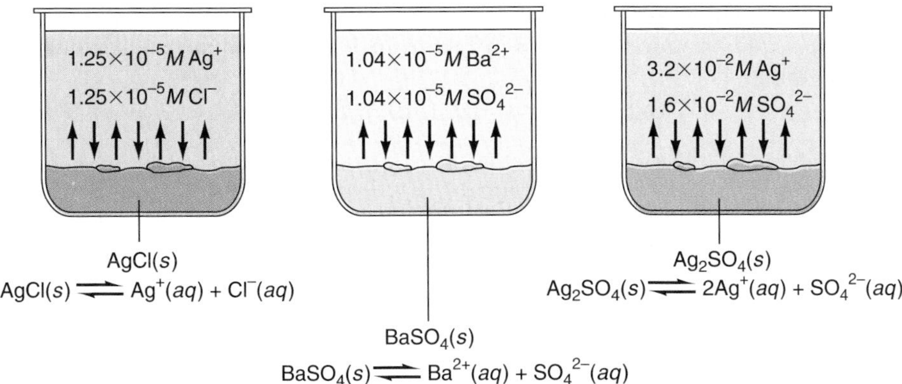

Figure 16.1

Solubility equilibria of AgCl, $BaSO_4$, and Ag_2SO_4.

Three examples of solubility equilibria are summarized in Figure 16.1. When AgCl, $BaSO_4$, and Ag_2SO_4 are separately shaken with water at 25°C to produce a saturated solution of each, an equilibrium is established between the ions in each solution and the solid phase of the salt.

Let us consider the solubility equilibrium of silver chloride. When solid silver chloride is shaken with 1.00 L of water at 25°C, 1.25×10^{-5} mol dissolves. The equilibrium between solid silver chloride and its dissolved ions in a saturated solution can be written as follows:

$$\mathrm{AgCl}(s) \rightleftharpoons \mathrm{Ag}^+(aq) + \mathrm{Cl}^-(aq)$$

The molar solubility of AgCl in 1.00 L of water equals the number of moles of Ag^+ ions, or the number of moles of Cl^- ions, per liter of a saturated solution of AgCl because neither Ag^+ nor Cl^- ions hydrolyze:

$$\begin{aligned}\text{moles of AgCl dissolved in 1.00 L} &= \text{molar solubility of AgCl}\\ &= [\mathrm{Ag}^+] = [\mathrm{Cl}^-]\end{aligned}$$

The Solubility Product Constant

The *equilibrium constant for the solubility equilibrium* of a salt is called its **solubility product constant**, K_{sp}. For example, the solubility equilibrium for silver chloride is

$$\mathrm{AgCl}(s) \rightleftharpoons \mathrm{Ag}^+(aq) + \mathrm{Cl}^-(aq)$$

The equilibrium constant can be written as

$$K = \frac{[\mathrm{Ag}^+][\mathrm{Cl}^-]}{[\mathrm{AgCl}]}$$

But since the concentration of a pure solid is constant, it is not included in the solubility product constant expression, and

$$K_{sp} = [\mathrm{Ag}^+][\mathrm{Cl}^-]$$

The expression for the solubility product constant includes only the molar concentrations of the ions in a saturated solution of the salt, written with their proper exponents if other than one.

The molar concentration of Ag^+ ions and of Cl^- ions in a saturated solution of AgCl at 25°C is 1.25×10^{-5} M. The value of K_{sp} for AgCl at 25°C equals the product of the molar concentrations of the Ag^+ and Cl^- ions in a saturated solution of AgCl:

$$K_{sp} = [\mathrm{Ag}^+][\mathrm{Cl}^-] = (1.25 \times 10^{-5})^2 = 1.56 \times 10^{-10}$$

Barium sulfate and silver sulfate are also sparingly soluble salts whose solubility equilibria are illustrated in Figure 16.1. The solubility equilibrium for barium sulfate is

$$BaSO_4(s) \rightleftharpoons Ba^{2+}(aq) + SO_4^{2-}(aq)$$

The molar solubility of barium sulfate is

$$\text{molar solubility of } BaSO_4 = [BaSO_4]_{dissolved} = [Ba^{2+}] = [SO_4^{2-}]$$

The concentration of Ba^{+2} ions in a saturated solution of barium sulfate at 25°C is 1.04×10^{-5} *M*. The K_{sp} for $BaSO_4$ is

$$K_{sp} = [Ba^{2+}][SO_4^{2-}] = (1.04 \times 10^{-5})^2 = 1.08 \times 10^{-10}$$

When silver sulfate, Ag_2SO_4, dissolves in water, each mole that dissolves produces 1 mol of SO_4^{2-} ions and 2 mol of Ag^+ ions:

$$Ag_2SO_4(s) \rightleftharpoons 2Ag^+(aq) + SO_4^{2-}(aq)$$

Thus, the concentration of Ag^+ ions in a saturated solution of Ag_2SO_4 (prepared by shaking Ag_2SO_4 in water) is twice the concentration of SO_4^{2-} ions (Figure 16.1). The molar solubility of Ag_2SO_4 is

$$\text{molar solubility of } Ag_2SO_4 = [Ag_2SO_4]_{dissolved} = [SO_4^{2-}]$$

The sulfate ion concentration in a saturated solution of silver sulfate is 1.6×10^{-2} *M*. The silver ion concentration in a saturated solution of silver sulfate is twice the solubility of silver sulfate:

$$[Ag^+] = (2)[SO_4^{2-}] = (2)(1.6 \times 10^{-2}) = 3.2 \times 10^{-2}\ M$$

In a solubility product constant expression, as in any equilibrium constant expression, the molar concentration of each ion is raised to the power that equals its coefficient in the balanced equation for the dissolution reaction. Therefore,

$$K_{sp} \text{ for } Ag_2SO_4 = [Ag^+]^2[SO_4^{2-}] = (3.2 \times 10^{-2})^2(1.6 \times 10^{-2})$$
$$= 1.6 \times 10^{-5}$$

Example 16.1

Calculating K_{sp} from Molar Solubility

The solubility of calcium fluoride is 1.6×10^{-2} g L^{-1}. Calculate the solubility product constant for calcium fluoride. Ignore the hydrolysis of fluoride ions.

SOLUTION: First, we convert the solubility of CaF_2 in grams per liter to moles per liter. Then we use the molar solubility in the K_{sp} expression to calculate K_{sp}.

$$\left(\frac{1.6 \times 10^{-2} \text{ g } CaF_2}{1.0 \text{ L}}\right)\left(\frac{1 \text{ mol } CaF_2}{78 \text{ g } CaF_2}\right) = 2.1 \times 10^{-4} \text{ mol L}^{-1}\ CaF_2$$

The solubility equilibrium for CaF_2 is

$$CaF_2(s) \rightleftharpoons Ca^{2+}(aq) + 2F^-(aq)$$

APPLICATIONS OF CHEMISTRY 16.1
Solubility Product and the Osmotic Pill

The concentration of a drug in a tissue or organ changes with time. The drug concentration is higher than necessary shortly after taking a pill, and its concentration falls below effective levels just before we take the next pill. When drug concentrations are too high, the risk of side effects increases. When drug concentrations are too low, the drug is ineffective.

To combat this variation in the time dependence of drug concentrations, "time release" pills have been developed that provide a more steady drug dose. However, when time release pills provide a dose of medication, the same pattern of excessive and inadequate drug concentrations occurs.

An exciting area of drug research today is the development of an osmotic pill. (We recall that osmosis occurs when we separate solutions of different solute concentrations by a semipermeable membrane.) A diagram of an osmotic is shown in the figure at right.

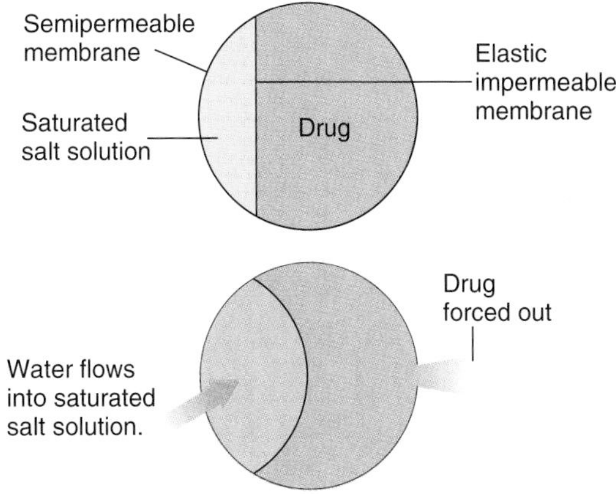

The osmotic pill is divided into two parts. One chamber is filled with a saturated solution of a salt plus some extra undissolved salt. The concentration of ions inside the first chamber is higher than that found in the gastrointestinal tract. Water from body fluids in the gastrointestinal tract passes through the semipermeable membrane. This flow of water from high to low concentration exerts an osmotic pressure on the elastic impermeable membrane, which in turn pushes the drug through a small laser-drilled hole provided at the other end.

The problem of a constant dose delivery appears to remain, but we can use Le Châtelier's principle and knowledge of the solubility constant to show otherwise. As water from the gastrointestinal tract enters the first chamber and dilutes the salt solution, some salt dissolves to reestablish the equilibrium. (We know this from Le Châtelier's principle.) The overall concentration of all ions in the gastrointestinal tract is constant, and the concentration of the solution in the first chamber is also constant. Therefore, the osmotic pressure exerted on the semipermeable membrane also remains constant. This combination assures a constant dose delivery and a constant drug concentration within the body.

Sources: *Chemistry International*, No. 4, pages 3–6; *Chemical & Engineering News*, April 1, 1985, page 30.

From this equilibrium equation we can see that 1 mol of Ca^{2+} ions, and 2 mol of F^- ions are produced in the solution for every mole of CaF_2 that dissolves.

$$\text{molar solubility of } CaF_2 = [Ca^{2+}] = 2.1 \times 10^{-4}\ M$$

$$[F^-] = (2)(2.1 \times 10^{-4}) = 4.2 \times 10^{-4}\ M$$

$$K_{sp} = [Ca^{2+}][F^-]^2 = (2.1 \times 10^{-4})(4.2 \times 10^{-4})^2 = 3.7 \times 10^{-11}$$

Practice Problem 16.1: The solubility of silver chromate is $1.3 \times 10^{-4}\ M$. What is the value of K_{sp} for silver chromate? Ignore hydrolysis.

Table 16.1

Solubility Product Constants for Sparingly Soluble Salts at 25°C

Name	Formula	K_{sp}	Name	Formula	K_{sp}
Aluminum hydroxide	$Al(OH)_3$	2×10^{-32}	Lead(II) sulfide	Pbs	8.2×10^{-28}
Barium carbonate	$BaCO_3$	8.1×10^{-9}	Magnesium carbonate	$MgCO_3$	4.0×10^{-5}
Barium chromate	$BaCrO_4$	2.0×10^{-10}	Magnesium hydroxide	$Mg(OH)_2$	1.5×10^{-11}
Barium fluoride	BaF_2	1.7×10^{-6}	Magnesium oxalate	MgC_2O_4	8.6×10^{-5}
Barium hydroxide	$Ba(OH)_2$	5×10^{-3}	Manganese(II) hydroxide	$Mn(OH)_2$	1.9×10^{-13}
Barium oxalate	BaC_2O_4	1.6×10^{-7}	Manganese(II) sulfide	MnS	1.4×10^{-15}
Barium sulfate	$BaSO_4$	1.08×10^{-10}	Mercury(I) chloride	Hg_2Cl_2	2×10^{-18}
Calcium carbonate	$CaCO_3$	8.7×10^{-9}	Mercury(II) sulfide	HgS	1×10^{-50}
Calcium chromate	$CaCrO_4$	7.1×10^{-4}	Silver bromide	AgBr	7.7×10^{-13}
Calcium fluoride	CaF_2	3.9×10^{-11}	Silver chloride	AgCl	1.56×10^{-10}
Calcium hydroxide	$Ca(OH)_2$	7.9×10^{-6}	Silver chromate	Ag_2CrO_4	9×10^{-12}
Calcium oxalate	CaC_2O_4	2.6×10^{-9}	Silver iodide	AgI	1.5×10^{-16}
Calcium sulfate	$CaSO_4$	2.45×10^{-5}	Silver sulfate	Ag_2SO_4	1.6×10^{-5}
Copper(II) hydroxide	$Cu(OH)_2$	1.3×10^{-20}	Strontium carbonate	$SrCO_3$	1.6×10^{-9}
Copper(II) sulfide	CuS	6.3×10^{-36}	Strontium chromate	$SrCrO_4$	3.6×10^{-5}
Iron(II) hydroxide	$Fe(OH)_2$	1.6×10^{-14}	Strontium hydroxide	$Sr(OH)_2$	3.2×10^{-4}
Iron(II) sulfide	FeS	4.9×10^{-18}	Strontium oxalate	SrC_2O_4	5.6×10^{-8}
Lead(II) carbonate	$PbCO_3$	7.4×10^{-14}	Strontium sulfate	$SrSO_4$	3.8×10^{-7}
Lead(II) chloride	$PbCl_2$	1.6×10^{-5}	Tin(II) hydroxide	$Sn(OH)_2$	6×10^{-27}
Lead(II) chromate	$PbCrO_4$	2.8×10^{-14}	Tin(II) sulfide	SnS	3×10^{-27}
Lead(II) fluoride	PbF_2	3.7×10^{-8}	Zinc hydroxide	$Zn(OH)_2$	1.8×10^{-14}
Lead(II) hydroxide	$Pb(OH)_2$	1.2×10^{-15}	Zinc oxalate	ZnC_2O_4	1.4×10^{-9}
Lead(II) sulfate	$PbSO_4$	1.6×10^{-8}	Zinc sulfide	ZnS	1.1×10^{-21}

Using K_{sp} to Calculate Solubilities of Salts

The K_{sp} values for some sparingly soluble salts at 25°C are listed in Table 16.1. We can use the K_{sp} value for a salt to calculate its molar solubility and the concentrations of its ions in a saturated solution. Examples 16.2 and 16.3 illustrate these calculations for salts of different types of solubility equilibria.

Example 16.2

Calculating the Solubility of a Salt That Produces Equal Numbers of Cations and Anions in Solution

The solubility product constant for lead sulfate is 1.6×10^{-8}. Calculate the molar solubility of lead sulfate in water. How many grams of lead sulfate dissolves in 1.0 L of water?

SOLUTION: Following our previous practice of solving problems in acid–base equilibria, we write an equation for the equilibrium reaction and indicate the equilibrium concentrations of the ions in a saturated solution of the salt. We let x be the unknown concentration:

	$PbSO_4(s) \rightleftharpoons Pb^{2+}(aq) + SO_4^{2-}(aq)$
Equilibrium M	x $\quad$ x

$$\text{molar solubility of } PbSO_4 = [Pb^{2+}] = [SO_4^{2-}] = x$$

$$K_{sp} = 1.6 \times 10^{-8} = [Pb^{2+}][SO_4^{2-}] = (x)(x) = x^2$$

$$x = \text{molar solubility} = \sqrt{1.6 \times 10^{-8}} = 1.3 \times 10^{-4}\ M$$

To convert the molar solubility of a salt to solubility in grams per liter, we multiply the number of moles per liter by the molar mass:

$$\left(\frac{1.3 \times 10^{-4} \text{ mol } PbSO_4}{1.0 \text{ L}}\right)\left(\frac{303 \text{ g } PbSO_4}{1 \text{ mol } PbSO_4}\right) = 3.9 \times 10^{-2} \text{ g L}^{-1}\ PbSO_4$$

Practice Problem 16.2: The K_{sp} of barium chromate is 2.0×10^{-10}. Calculate the solubility of barium chromate in moles per liter and in grams per liter. Ignore hydrolysis.

Example 16.3

Calculating the Solubility of a Compound That Contains Two Anions for Each Cation

The solubility product constant for magnesium hydroxide is 1.5×10^{-11}. Calculate: **(a)** the molar solubility of magnesium hydroxide; **(b)** the molar concentrations of Mg^{2+} and OH^- ions in a saturated solution of magnesium hydroxide; **(c)** the pH of a saturated solution of magnesium hydroxide.

SOLUTION

a. We let x equal the molar solubility of $Mg(OH)_2$. The molar solubility of $Mg(OH)_2$ equals the Mg^{2+} ion concentration. Thus, $[Mg(OH)_2]_{dissolved} = [Mg^{2+}] = x$. But the ratio of Mg^{2+} ion concentration to OH^- ion concentration in solution is 1:2. Therefore, the OH^- ion concentration is $2x$. The solubility equilibrium is

	$Mg(OH)_2(s) \rightleftharpoons$	$Mg^{2+}(aq)$	$+\ 2OH^-(aq)$
Equilibrium M		x	$2x$

$$K_{sp} = 1.5 \times 10^{-11} = [Mg^{2+}][OH^-]^2 = (x)(2x)^2 = 4x^3$$

Solving for x, we obtain

$$x = \text{molar solubility of } Mg(OH)_2$$

$$= \sqrt[3]{\frac{1.5 \times 10^{-11}}{4}} = 1.6 \times 10^{-4}\ M$$

b. $[Mg^{2+}] = \text{molar solubility of } Mg(OH)_2 = 1.6 \times 10^{-4}\ M$

$[OH^-] = (2)(1.6 \times 10^{-4}\ M) = 3.2 \times 10^{-4}\ M$

c. $pOH = -\log(3.2 \times 10^{-4}) = 3.49$ and $pH = 10.51$

Practice Problem 16.3: The K_{sp} for calcium hydroxide is 7.9×10^{-6}. Calculate the solubility of calcium hydroxide in moles per liter and in grams per liter. What is the pH of a saturated solution of calcium hydroxide?

Salts whose formulas have the same ratio of cations to anions have similar solubility equilibria, and their K_{sp} values are a direct measure of their relative molar solubilities. An example of two salts having the same ratio of cations to anions and similar solubility equilibria are AgCl and $PbSO_4$. The K_{sp} values for AgCl and $PbSO_4$ are 1.56×10^{-10} and 1.6×10^{-8}, respectively. Since the K_{sp} for lead(II) sulfate is greater than the K_{sp} for silver chloride, the molar solubility of lead(II) sulfate is greater than that of silver chloride.

The solubility equilibria of $Mg(OH)_2$, $PbCl_2$, and Ag_2SO_4 are similar to one another but different from the solubility equilibria of AgCl and $PbSO_4$. A solution of $Mg(OH)_2$ or $PbCl_2$ contains twice as many anions as cations, and a solution of Ag_2SO_4 contains twice as many cations as anions. The K_{sp} values for these solids are also direct measures of their relative molar solubilities because each forms 3 mol of ions per mole of the dissolved solid.

However, the K_{sp} values for $PbSO_4$ and $Mg(OH)_2$ are not direct measures of their relative solubilities because they have different solubility equilibria; that is, their equilibrium expressions have different mathematical forms. The K_{sp} for $PbSO_4$ (1.6×10^{-8}) is greater than the K_{sp} for $Mg(OH)_2$ (1.5×10^{-11}). But the solubility of $Mg(OH)_2$ is 1.6×10^{-4} *M*, which is greater than the solubility of $PbSO_4$, 1.3×10^{-4} *M*.

16.2 Common Ion Effect in Solubility Equilibria

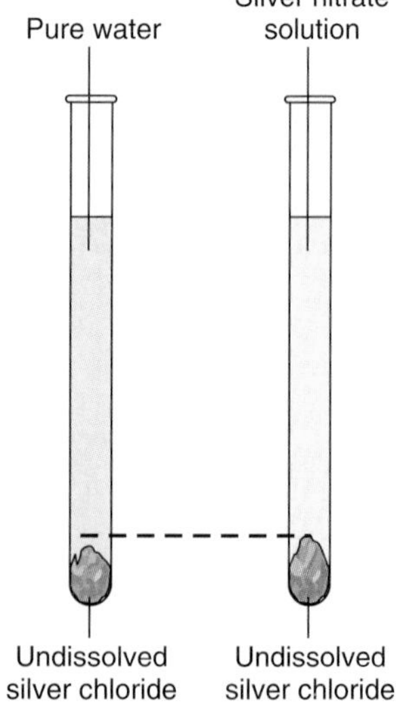

More silver chloride dissolves in pure water than in an equal volume of silver nitrate solution at the same temperature.

Common Ion Effect

We saw in Chapter 15 that adding a common ion to a solution of a weak acid or base shifts the position of the ionization equilibrium of the acid or base. Similarly, adding common ions to a saturated solution of a sparingly soluble salt shifts the position of the solubility equilibrium. Thus, when a saturated solution of a sparingly soluble salt is shaken with a solution of another salt which provides cations or anions common to both salts, some of the sparingly soluble salt originally in solution precipitates. Therefore, the solubility of a sparingly soluble salt is lower in a solution that contains common ions than in pure water. For example, the solubility of silver chloride in a solution of silver nitrate is less than its solubility in pure water.

We can explain the common ion effect for sparingly soluble salts in terms of Le Châtelier's principle. When we add silver nitrate to a saturated solution of silver chloride, silver ions from silver nitrate solution shift the solubility equilibrium of silver chloride to the left. As a result, some silver chloride precipitates, and the chloride ion concentration decreases, as shown schematically below:

$$AgNO_3(aq) \longrightarrow Ag^+(aq) + NO_3^-(aq)$$

additional Ag^+ ions from $AgNO_3$ increase the total Ag^+ ion concentration in the equilibrium system

$$AgCl(s) \rightleftharpoons Ag^+(aq) + Cl^-(aq)$$

so equilibrium shifts to the left

Nitrate ions are also present in the solution, but they are not a component of the solubility equilibrium and therefore do not affect the position of the equilibrium or the solubility of AgCl.

If the concentration of a common ion solution is known, the solubility of a salt can be calculated from its solubility product constant, as shown in the following examples.

Example 16.4

Calculating the Solubility of Silver Chloride in a Solution of Chloride Ions

Calculate the molar solubility of silver chloride in a 0.20 *M* solution of sodium chloride. What is the Ag^+ ion concentration in a saturated solution of AgCl that is also 0.20 *M* in NaCl?

SOLUTION: In all equilibrium problems, the total ion concentrations, regardless of their source, must be used in the expression for the equilibrium constant. The total chloride ion concentration in this equilibrium equals the sum of the concentrations of chloride ions derived from silver chloride, which we designate by x, and the chloride ions derived from sodium chloride, which is 0.20 *M*. The solubility equilibrium for AgCl and the relevant concentration data are:

	$AgCl(s) \rightleftharpoons$	$Ag^+(aq)$ +	$Cl^-(aq)$
Equilibrium *M*		x	$x + 0.20$ (from AgCl, from NaCl)

From Table 16.1, the K_{sp} for AgCl is 1.6×10^{-10}, and we can write

$$K_{sp} = 1.6 \times 10^{-10} = [Ag^+][Cl^-] = (x)(x + 0.20)$$

Since the value of K_{sp} is very small, we assume that x is negligibly small compared with 0.20, and the term $(x + 0.20)$ is essentially 0.20. Therefore,

$$K_{sp} = 1.6 \times 10^{-10} = [Ag^+][Cl^-] = (x)(0.20)$$

and

$$x = [Ag^+] = \text{molar solubility of AgCl} = \frac{1.6 \times 10^{-10}}{0.20} = 8.0 \times 10^{-10}\ M$$

The approximation is valid because 8.0×10^{-10} M is only 4.0×10^{-7} percent of 0.20 M. The molar solubility of AgCl equals the Ag^+ ion concentration in the solution because AgCl is the only source of Ag^+ ions.

The solubility of AgCl in pure water is 1.2×10^{-5} M (Figure 16.1). Thus, the solubility of AgCl in pure water is 15,000 times more than its solubility in 0.20 M NaCl solution.

Practice Problem 16.4: The K_{sp} of gold(I) chloride is 2.0×10^{-13}. What is the molar solubility of gold(I) chloride in a 0.12 M sodium chloride solution?

Example 16.5

Calculating the Molar Solubility of Silver Sulfate in a Silver Ion Solution

What is the molar solubility of silver sulfate in 0.10 M $AgNO_3$ solution?

SOLUTION: We let x equal the molar solubility of Ag_2SO_4. Each mole of Ag_2SO_4 that dissolves produces 1 mol of SO_4^{2-} ions and 2 mol of Ag^+ ions. Thus, $[SO_4^{2-}]$ is x, and $[Ag^+]$ from Ag_2SO_4 is $2x$. The $[Ag^+]$ from $AgNO_3$ is 0.10 M, and the total $[Ag^+]$ is $0.10 + 2x$. These data are summarized below:

	$Ag_2SO_4(s) \rightleftharpoons 2Ag^+(aq) + SO_4^{2-}(aq)$
Equilibrium M	$2x + 0.10$ $\quad x$ (from Ag_2SO ↗, from $AgNO_3$ ↖)

$$K_{sp} = 1.6 \times 10^{-5} = [Ag^+]^2\text{-}[SO_4^{2-}] = (2x + 0.10)^2(x)$$

Since K_{sp} is very small, $x \ll 0.10$, and we can neglect $2x$ in the term $(2x + 0.10)^2$. Therefore,

$$K_{sp} = 1.6 \times 10^{-5} = [Ag^+]^2[SO_4^{2-}] = (0.10)^2(x)$$

and

$$x = [SO_4^{2-}] = \text{molar solubility of } AgSO_4$$

$$= \frac{1.6 \times 10^{-5}}{(0.10)^2} = 1.6 \times 10^{-3}\ M$$

The approximation is valid.

Practice Problem 16.5: The K_{sp} for silver chromate is 9×10^{-12}. What is the molar solubility of silver chromate in a 0.15 M silver nitrate solution? What is the maximum number of grams of silver chromate that can be dissolved in 75 mL of a 0.15 M silver nitrate solution?

Example 16.6

Calculating the Molar Solubility of Lead(II) Chloride in a Lead(II) Nitrate Ion Solution

Calculate the molar solubility of lead(II) chloride in a 0.25 *M* lead(II) nitrate solution. The K_{sp} for $PbCl_2$ is 1.6×10^{-5}.

SOLUTION: Dissolving of 1 mol of lead(II) chloride produces 1 mol of lead ions and 2 mol of chloride ions. If we let x equal the molar solubility of lead(II) chloride, then the $[Pb^{2+}]$ produced from lead(II) chloride that dissolves also equals x. The *total* $[Pb^{2+}]$ in the equilibrium mixture is $0.25 + x$. The $[Cl^-]$ is $2x$. These data are summarized below:

	$PbCl_2(s) \rightleftharpoons Pb^{2+}(aq) + 2Cl^-(aq)$
Equilibrium *M*	$x + 0.25$ $2x$
	from $PbCl_2$ from $Pb(NO_3)_2$

$$K_{sp} = 1.6 \times 10^{-5} = [Pb^{2+}][Cl^-]^2 = (x + 0.25)(2x)^2$$

Because K_{sp} is very small, $x \ll 0.25$, and the equation simplifies to

$$1.7 \times 10^{-6} = (0.25)(4x^2) = x^2$$

$$x = \text{molar solubility of } PbCl_2 = \sqrt{1.6 \times 10^{-5}} = 4.0 \times 10^{-3}\ M$$

Practice Problem 16.6: What are the molar solubility and the solubility in grams per liter of lead(II) hydroxide in a 0.30 *M* solution of lead(II) nitrate? The K_{sp} of lead(II) hydroxide is 1.2×10^{-15}.

The common ion effect in solubility equilibria is applied in quantitative gravimetric analysis (Chapter 4). When a solid precipitates, a solubility equilibrium is established. To minimize the loss of precipitate due to its solubility, an excess of the precipitating reagent is added to shift the equilibrium to the side of the precipitate. For example, the percent by mass of silver in a silver alloy can be determined by gravimetric analysis. A weighed sample of the alloy is dissolved in nitric acid and the silver ions in the solution are precipitated as silver chloride. The precipitate is then filtered, dried, and weighed. From the mass of silver chloride obtained, the mass of silver can be calculated. The percent silver is determined from the mass of silver and the mass of the sample of the original alloy. To decrease the amount of silver chloride that remains in solution, an excess of chloride ions is added when AgCl is precipitated.

The solubility of silver chloride in water is 1.2×10^{-5} mol per liter, or 1.8×10^{-3} g per liter. However, the solubility of silver chloride in 0.20 *M* solution of sodium chloride is 8.0×10^{-10} mol per liter, which is 1.2×10^{-7} g per liter—a decrease to about one fifteen thousandth from the solubility of silver chloride in pure water.

APPLICATIONS OF CHEMISTRY 16.2
The Barium Cocktail

A hand held between a white sheet of paper and a bright light stops the light from reaching the paper and the hand casts a shadow on the sheet of paper. Something similar happens when a hand is held between a photographic film and an x-ray source. The calcium hydroxyapatite, $Ca_5(PO_4)_3OH$ in the bones of the hand prevents x-rays from reaching the photographic film, casting a shadow on the film. When the film is developed, the image of the bones becomes visible. Because skin and other internal structures don't stop x-rays, they leave no shadow on an x-ray photograph.

To diagnose gastrointestinal problems, physicians frequently use x-rays to produce a photographic image of the gastrointestinal tract. Because the stomach and the intestines are flesh, steps must be taken to make the stomach visible. Physicians give their patients barium cocktails or barium enemas made up of a slurry of barium sulfate. Like calcium, barium also stops x-rays. Therefore, a stomach coated with barium sulfate casts a shadow on photographic film (see figure at right).

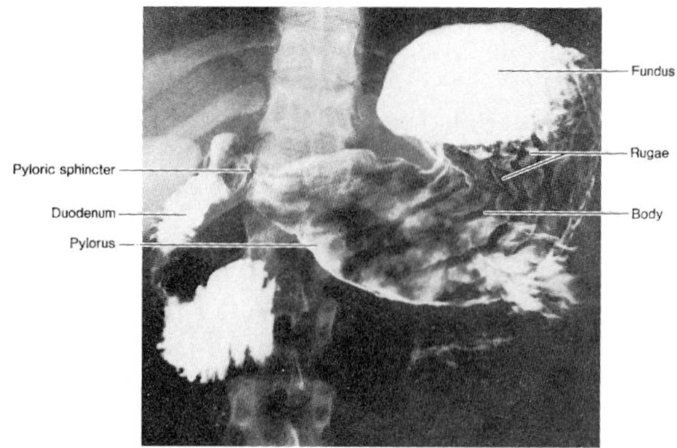

As we learned earlier, barium sulfate is one of the few slightly soluble sulfates. Since barium ions are toxic, even slightly soluble barium sulfate solutions may be unsafe to ingest. Look at the equilibrium that exists between barium sulfate and its ions:

$$BaSO_4(s) \rightleftharpoons Ba^{2+}(aq) + SO_4^{2-}(aq)$$

Physicians take precautions to decrease the barium ion concentration even further by adding magnesium sulfate to the barium sulfate solution. Like most sulfates, magnesium sulfate is soluble in water, and it dissociates completely. This is shown as follows:

$$MgSO_4(s) \longrightarrow Mg^{2+}(aq) + SO_4^{2-}(aq)$$

Since both barium sulfate and magnesium sulfate dissociate to form sulfate ions, we have the common ion effect. Sulfate ions produced by magnesium sulfate shifts the barium sulfate equilibrium to the left. This causes barium ions in the slurry to precipitate as barium sulfate, and this results in a decreased barium ion concentration. Reducing the concentration of toxic barium ions in the cocktail makes the procedure safer.

pH and the Solubility of Hydroxides

The solubility of a sparingly soluble metal hydroxide is influenced by the pH of the solution. Increasing the acidity of the solution increases the solubility of the hydroxide because the added hydrogen ions react with hydroxide ions. As a result, the equilibrium shifts to the right, that is, more of the sparingly soluble hydroxide dissolves to replace the hydroxide ions lost in the reaction. On the other hand, when a base is added to a solution of a sparingly soluble hydroxide, the increased hydroxide ion concentration shifts the equilibrium to the side of the solid causing some of the dissolved metal hydroxide to precipitate.

Example 16.7

Calculating the Solubility of a Metal Hydroxide at a Given pH

What is the solubility of calcium hydroxide in a solution of pH 12.80? The K_{sp} for calcium hydroxide is 7.9×10^{-6}.

SOLUTION: The pOH of the solution is 1.20. This corresponds to an $[OH^-]$ of 0.063 *M*. We let *x* equal the solubility of calcium hydroxide in a 0.063 *M* solution of hydroxide ions. At equilibrium, the calcium ion concentration equals the solubility of calcium hydroxide. The *total* hydroxide ion concentration in the solution is 0.063 *M*. These data are summarized below:

	$Ca(OH)_2(s) \rightleftharpoons$	$Ca^{2+}(aq)$	$+\ 2OH^-(aq)$
Equilibrium *M*		x	0.063

$$K_{sp} = 7.9 \times 10^{-6} = [Ca^{2+}][OH^-]^2 = (x)(0.063)^2$$

$$x = \text{solubility of } Ca(OH)_2 = [Ca^{2+}] = \frac{7.9 \times 10^{-6}}{(0.063)^2} = 2.0 \times 10^{-3}\ M$$

Practice Problem 16.7: Calculate the solubility of calcium hydroxide in pure water and in a solution of pH of 10.90. Compare these results with the results obtained in Example 16.7. How does the solubility of calcium hydroxide vary with pH?

16.3 Forming Precipitates

In the preceding section we saw that adding common ions to a saturated solution of an ionic solid causes some of the salt to precipitate. Adding common ions to a saturated solution of a salt changes the concentrations of the ions in the solution. However, the product of the ion concentrations, each raised to the power as in the K_{sp} for the salt, equals the K_{sp} value for the salt.

When a saturated solution of a sparingly soluble salt is in contact with the solid phase of the salt and the concentration of one of the ions in the solution is decreased, some of the salt will dissolve until a new equilibrium is established with the remaining salt. In the newly formed equilibrium, the ion concentrations are altered, but their product is still equal to the K_{sp} of the salt.

If the concentration of one of the ions in an *unsaturated* solution of a sparingly soluble salt is increased until *the product of the ion concentrations* *exceeds the K_{sp} of the salt*, the salt *starts to precipitate*. After the new equilibrium is reached, the product of the ion concentrations of the supernatant solution then equals the K_{sp} of the salt.

Determining Whether a Precipitate Forms

According to the foregoing discussion, a sparingly soluble salt precipitates when its ion concentrations in a solution are high enough for their product to exceed the K_{sp} of the salt. If the concentration of one of the ions is small, the other must be sufficiently large for their product to exceed the K_{sp} value.

Example 16.8

Determining Whether a Precipitate Forms When Two Solutions Are Mixed

Will a precipitate form when 100 mL of 1.26×10^{-3} *M* barium nitrate is mixed with 200 mL of 0.0200 *M* sodium sulfate?

SOLUTION: From the solubility rules of Chapter 4, we know that barium sulfate is an insoluble salt. The problem here is to determine whether the product of the molar concentrations of barium and sulfate ions in the solution after mixing exceeds the K_{sp} value for barium sulfate. From Table 16.1 we find that K_{sp} for barium sulfate is 1.08×10^{-10}. To calculate the molar concentrations of the ions in the mixture, we first determine the number of moles of each of the ions in the mixture. Then we divide the number of moles by the total volume of the solution to get the molar concentration:

$$\text{moles of } Ba^{2+} \text{ ions in 100 mL of } 1.26 \times 10^{-3}\ M \text{ solution} = (0.100\ \text{L})(1.26 \times 10^{-3}\ \text{mol L}^{-1}) = 1.26 \times 10^{-4}\ \text{mol}$$

$$\text{moles of } SO_4^{2-} \text{ ions in 200 mL of } 0.0200\ M \text{ solution} = (0.200\ \text{L})(0.0200\ \text{mol L}^{-1}) = 4.00 \times 10^{-3}\ \text{mol}$$

Assuming that the volumes are additive, the total volume of the solution = 0.100 L + 0.200 L = 0.300 L. The molar concentrations of barium and sulfate ions are

$$[Ba^{2+}] = \frac{1.26 \times 10^{-4}\ \text{mol}}{0.300\ \text{L}} = 4.20 \times 10^{-4}\ M$$

$$[SO_4^{2-}] = \frac{4.00 \times 10^{-3}\ \text{mol}}{0.300\ \text{L}} = 1.33 \times 10^{-2}\ M$$

The values calculated for $[Ba^{2+}]$ and $[SO_4^{2-}]$ do not assume that a precipitate forms. The product of the calculated molar concentrations of Ba^{2+} and SO_4^{2-} ions in the solution equals $(4.20 \times 10^{-4})(1.33 \times 10^{-2}) = 5.59 \times 10^{-6}$. This value is larger than the value of K_{sp} for $BaSO_4$, 1.08×10^{-10}; therefore, a precipitate of $BaSO_4$ forms.

Practice Problem 16.8: Will a precipitate form when 50 mL of 0.020 *M* calcium nitrate solution is mixed with 50 mL of 0.050 *M* sodium chromate solution?

Example 16.9

Determining Whether a Precipitate of the Type AB_2 Forms

A solution is made 0.0010 *M* in Pb^{2+} and 0.0030 *M* in Cl^- ions. Will a precipitate of $PbCl_2$ form? The K_{sp} for lead chloride is 1.6×10^{-5}.

SOLUTION: Here we are dealing with a salt that has two anions per cation. The K_{sp} expression for lead chloride equals the molar concentration of Pb^{2+} ions in the solution times the square of the molar concentration of Cl^- ions:

$$[Pb^{2+}][Cl^-]^2 = (1.0 \times 10^{-3})(3.0 \times 10^{-3})^2 = 9.0 \times 10^{-9}$$

This value is smaller than the K_{sp} for lead chloride, 1.6×10^{-5}. Therefore, the solution is not saturated, and no precipitate forms. A precipitate forms only if the product of the ion concentrations in the solution exceeds the value of K_{sp}.

Practice Problem 16.9: A solution is made 0.050 *M* in Ag^+ ions and 0.0030 *M* in CrO_4^{2-} ions. Will silver chromate precipitate?

Calculating Concentrations of Ions Needed to Form a Precipitate

In both qualitative and quantitative analysis it is often necessary to calculate the molar concentration of one of the ions that must be provided to precipitate a salt from a solution of an ion whose concentration is known. Example 16.10 illustrates this type of calculation.

Example 16.10

Calculating the Molar Concentration of Chloride Ions Needed to Start the Precipitation of Lead(II) Chloride from a 0.10 *M* Solution of Lead Ions

What minimum concentration of chloride ions is needed to start the precipitation of lead(II) chloride from a 0.10 *M* solution of lead(II) nitrate?

SOLUTION: When precipitation begins, the molar concentration of Pb^{2+} ions times the square of the molar concentration of the Cl^- ions must equal the K_{sp} for $PbCl_2$:

$$K_{sp} = 1.6 \times 10^{-5} = (0.10)[Cl^-]^2$$

$$[Cl^-] = \sqrt{\frac{1.6 \times 10^{-5}}{0.10}} = 1.3 \times 0^{-2}\ M$$

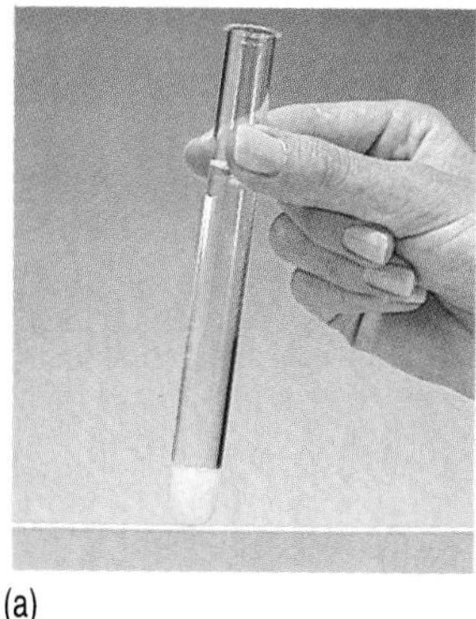

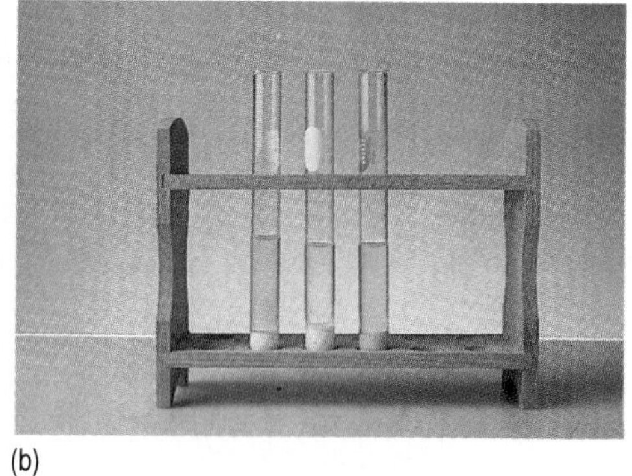

(a) (b)

Figure 16.2

Colors of various precipitates: (a) silver chloride, and (b) barium sulfate, lead(II) chloride, and calcium chromate.

Practice Problem 16.10: What silver ion concentration is required to initiate precipitation of silver sulfate from a 0.085 *M* solution of sodium sulfate?

The colors of various precipitates are illustrated in Figure 16.2.

Calculating Concentrations of Ions Remaining in Solution After Precipitation

When a salt precipitates from a solution, some ions of the salt remain in the supernatant solution. For example, when a solution that contains chloride ions is added to a solution that contains lead ions, $PbCl_2$ precipitates, but some lead ions and chloride ions remain in the solution. As more chloride ions are added, fewer lead ions remain in the solution, but the product of the ion concentrations still equals K_{sp} for $PbCl_2$. Thus, even after excess chloride ions are added to the solution, some lead ions remain in solution.

The ions that remain in solution after precipitation are in a solubility equilibrium with the solid salt and constitute the solubility loss of the salt. This loss of a precipitate due to its solubility causes an error in gravimetric quantitative analysis (Chapter 4). It is important to know the extent of this error so that it can be minimized. Example 16.11 shows how to calculate the percent of ions that do not precipitate.

Example 16.11

Calculating the Percent of Ions Remaining in Solution After Precipitation

What percent of lead ions remain in solution when a 0.10 *M* solution of lead nitrate is made 1.0 *M* in chloride ions?

SOLUTION: The molar concentration of lead ions remaining in the solution can be calculated from the solubility product expression:

$$K_{sp} = 1.6 \times 10^{-5} = [Pb^{2+}][Cl^-]^2 = [Pb^{2+}](1.0)^2$$

Thus, the concentration of Pb^{2+} ions remaining in the solution is

$$[Pb^{2+}] = \frac{1.6 \times 10^{-5}}{(1.0)^2} = 1.6 \times 10^{-5}\ M$$

$$\text{percent } Pb^{2+} \text{ remaining in solution} = \left(\frac{M \text{ remaining in solution}}{M \text{ initially present}}\right) 100$$

$$= \left(\frac{1.6 \times 10^{-5}\ M}{1.0\ M}\right) 100 = 1.6 \times 10^{-3} \text{ percent}$$

To decrease the percent of Pb^{2+} left in solution, we must use a different method. One method uses a solution of chromate ions to precipitate lead(II) ions as lead(II) chromate. Lead(II) chromate has a K_{sp} of 2.8×10^{-14}, which is much smaller than the K_{sp} for lead(II) chloride. Since lead(II) chromate is much less soluble than lead(II) chloride, very few lead(II) ions remain in solution after precipitation with chromate ions.

Practice Problem 16.11: What percent of barium ions remain in solution when a 0.20 M solution of barium nitrate is made 0.40 M in sulfate ions?

16.4 Fractional Precipitation and Simultaneous Equilibria

Fractional Precipitation

Consider a solution that contains two or three different metal ions. If we slowly add a solution that contains anions that form insoluble salts with the cations in the solution, the cations can be selectively precipitated. Cations that form the least soluble salt with the added anions precipitate first. This precipitate can be separated from the solution by filtration or centrifugation. As the anion concentration increases, the salt of second-lowest solubility precipitates. This technique is called *fractional precipitation.* If the two precipitates have similar solubility equilibria, and if their K_{sp} values are sufficiently different, most of the salt of lower solubility precipitates before the second one begins to precipitate. For example, when a solution of sodium sulfate is gradually added to a solution containing barium and calcium ions in equimolar concentrations, barium sulfate precipitates first because its solubility is lower than that of calcium sulfate.

Calculation of the anion concentration needed to precipitate the cations of one metal and not the cations of another metal from an aqueous mixture of the two is illustrated in the following example.

Example 16.12

Calculating the Concentration of Anions to Start the Precipitation of the Cations of One Metal but Not the Cations of Another Metal in an Aqueous Solution

When a solution of sodium chloride is gradually added to a solution that contains 0.10 M Ag^+ and 0.10 M Pb^{2+} ions, at what concentration of chloride ions does the precipitation of AgCl begin? What concentration of chloride ions is required to start the precipitation of $PbCl_2$? What percent of Ag^+ ions remain in the solution when $PbCl_2$ starts to precipitate?

SOLUTION: When AgCl begins to precipitate, the product of the molar concentrations of Ag^+ and Cl^- ions equals the K_{sp} value for AgCl:

$$[Ag^+][Cl^-] = K_{sp} \text{ for AgCl}$$

Solving this equation for $[Cl^-]$ and substituting the known values for $[Ag^+]$ and for K_{sp}, we obtain

$$[Cl^-] = \frac{1.6 \times 10^{-10}}{0.10} = 1.6 \times 10^{-9}\ M$$

When $PbCl_2$ starts to precipitate, the product of the lead ion concentration and the square of the chloride ion concentration equals the K_{sp} of lead chloride:

$$[Pb^{2+}][Cl^-]^2 = K_{sp} \text{ for } PbCl_2$$

Solving for $[Cl^-]$ and substituting the known values into this expression gives

$$[Cl^-] = \sqrt{\frac{K_{sp}}{[Pb^{2+}]}} = \sqrt{\frac{1.6 \times 10^{-5}}{0.10}} = 1.3 \times 10^{-2}\ M$$

Thus, to begin the precipitation of $PbCl_2$ from a 0.10 M Pb^{2+} ion solution, the Cl^- ion concentration must be 1.3×10^{-3} M. At this point, the Ag^+ ion concentration remaining in the solution can be found from the relationship

$$[Ag^+](1.3 \times 10^{-2}) = 1.6 \times 10^{-10} \qquad \text{from which}$$

$$[Ag^+] = \frac{1.6 \times 10^{-10}}{1.3 \times 10^{-2}} = 1.2 \times 10^{-8}\ M$$

The percent of Ag^+ ions *not* precipitated as AgCl when $PbCl_2$ begins to precipitate is

$$\left(\frac{1.2 \times 10^{-8}\ M}{0.10\ M}\right) 100 = 1.2 \times 10^{-5} \text{ percent}$$

This result shows that more than 99.999 percent of the silver ions have precipitated when $PbCl_2$ begins to precipitate.

Practice Problem 16.12: A solution contains 0.12 M barium ions and 0.12 M calcium ions. What concentration of sulfate ions is required to start the precipitation of barium sulfate? What percent of barium ions remain in the solution when calcium ions start to precipitate?

Simultaneous Equilibria in Fractional Precipitation

In the fractional precipitation of cations it is often difficult to control the concentration of added anions. Careful control of anion concentration is particularly important in separating cations from a solution in which their concentrations are low. However, when the anion used as a precipitant is derived from a weak acid, its concentration can be controlled by adjusting the pH of the solution. This method can be used in a selective precipitation of metal sulfides, phosphates, and carbonates. Sulfide, phosphate, and carbonate ions are conjugate bases of weak acids, and their concentrations can be controlled by adjusting the pH. When the pH is decreased, the hydronium ion concentration increases and more of the anions react with the added hydronium ions. As a result, the anion concentration decreases.

The sulfide ion concentration of a solution can easily be controlled by adjusting the pH of a solution of H_2S. A solution of H_2S can be effectively used to selectively precipitate metal ions as metal sulfides.

We can see from Table 16.1 that some metal sulfides have much smaller K_{sp} values than others. Thus, metal ions whose sulfides have low K_{sp} values can be precipitated with a relatively low sulfide ion concentration from a solution of other metal ions whose sulfides have higher K_{sp} values. Metal ions whose sulfides have high K_{sp} values remain in solution until the sulfide ion concentration is increased to the point at which they also precipitate.

The sulfide ion concentration in a solution can be controlled by saturating the solution with H_2S and adjusting the pH of the solution. In a saturated solution of H_2S at 25°C and 1.0 atm, the concentration of undissociated H_2S is 0.10 *M*. This is a constant as long as the solution stays saturated with H_2S at 25°C and 1.0 atm. A useful relationship between the H_3O^+ and S^{2-} ion concentrations in a saturated solution of H_2S can be derived by multiplying the dissociation constant for H_2S, K_{a1} (1.0×10^{-7}), by the dissociation constant for the dissociation of hydrogen sulfide anion, K_{a2} (1.3×10^{-13}):

$$(K_{a1})(K_{a2}) = \left(\frac{[H_3O^+][HS^-]}{[H_2S]}\right)\left(\frac{[H_3O^+][S^{2-}]}{[HS^-]}\right) = \frac{[H_3O^+]^2[S^{2-}]}{[H_2S]}$$

In a saturated solution of H_2S:

$$K_{a1}K_{a2} = \frac{[H_3O^+]^2[S^{2-}]}{0.10} = (1.0 \times 10^{-7})(1.3 \times 10^{-13}) = 1.3 \times 10^{-20}$$

from which

$$[H_3O^+]^2[S^{2-}] = 1.3 \times 10^{-21} \qquad (16.1)$$

Caution: Equation 16.1 cannot be used to calculate the H_3O^+ ion concentration of a pure H_2S solution because equation 16.1 results from adding two equilibria. It is not the equilibrium for acid dissociation of H_2S.

From Equation 16.1 we can see that the S^{2-} ion concentration in a saturated solution of H_2S varies inversely with the square of the H_3O^+ ion concentration. Thus, the S^{2-} ion concentration in a solution saturated with H_2S can be controlled by adjusting the pH to a value at which only the most sparingly soluble sulfide precipitates. The pH of the solution can then be increased to increase the sulfide ion concentration to a point at which the metal sulfide with the next-to-lowest K_{sp} value precipitates. This procedure is repeated until all of the sulfides have precipitated.

Example 16.13

Calculating the Sulfide Ion Concentration in a Saturated Solution of H_2S at Different pH

What is the sulfide ion concentration in a saturated solution of H_2S at 25°C and 1.0 atm in which the pH is adjusted to: **(a)** 10.00; **(b)** 0.30? The K_{a1} and K_{a2} for H_2S are 1.0×10^{-7} and 1.3×10^{-13}, respectively.

SOLUTION

a. The $[H_3O^+]$ corresponding to pH 10.00 is 1.0×10^{-10} *M*. Solving Equation 16.1 for $[S^{2-}]$, we obtain

$$[S^{2-}] = \frac{1.3 \times 10^{-21}}{(1.0 \times 10^{-10})^2} = 0.13\ M$$

This result shows that in an alkaline solution of pH 10.00, the sulfide ion concentration is relatively large.

b. The $[H_3O^+]$ corresponding to pH of 0.30 is 0.50 *M*.

$$[S^{2-}] = \frac{1.3 \times 10^{-21}}{(0.50)^2} = 5.2 \times 10^{-21}\ M$$

This result shows that in a relatively concentrated acid solution saturated with H_2S, the sulfide ion concentration is very small. The results of both parts (a) and (b) of this problem are in accord with Equation 16.1: in a saturated solution of hydrogen sulfide, the sulfide ion concentration varies inversely with the square of the hydronium ion concentration.

Example 16.14

Predicting Which Metal Ion Precipitates as a Metal Sulfide in a Saturated Solution of H_2S at a Given $[H_3O^+]$

A solution that is 0.10 *M* in each of Mn^{2+}, Pb^{2+}, and Zn^{2+} ions is buffered at 0.50 *M* in H_3O^+ and saturated with H_2S. Which cations, if any, precipitate as sulfides under these conditions? The K_{sp} for MnS is 1.4×10^{-15}, for PbS 8.2×10^{-28}, and for ZnS 1.1×10^{-21}.

SOLUTION: Since these sulfides have similar solubility equilibria, and since the K_{sp} of PbS is the smallest of the three, PbS is the least soluble of the three and precipitates before the other two. The sulfide ion concentration needed to start the precipitation of PbS can be calculated from the K_{sp} expression for PbS:

$$[Pb^{2+}][S^{2-}] = K_{sp} \text{ for PbS} = 8.2 \times 10^{-28}$$

$$[S^{2-}] = \frac{K_{sp}}{[Pb^{2+}]} = \frac{8.2 \times 10^{-28}}{0.10} = 8.2 \times 10^{-27}\ M$$

The S^{2-} ion concentration actually present in a saturated solution of H_2S in which $[H_3O^+]$ is 0.50 *M* is 5.2×10^{-21} *M* (Example 16.13b). This sulfide ion concentration is larger than the sulfide ion concentration needed to begin the precipitation of PbS (8.2×10^{-27} *M*) from a 0.10 *M* solution of Pb^{2+} ions. Therefore, PbS precipitates.

To begin the precipitation of ZnS, the S^{2-} ion concentration needed is

$$[S^{2-}] = \frac{K_{sp} \text{ for ZnS}}{[Zn^{2+}]} = \frac{1.1 \times 10^{-21}}{0.10} = 1.1 \times 10^{-20}\ M$$

This value required is larger than the $[S^{2-}]$ actually present in the solution (5.2×10^{-21} *M*), so ZnS does not precipitate under these conditions. Since MnS ($K_{sp} = 1.4 \times 10^{-15}$) is more soluble than ZnS, Mn^{2+} ions will also remain in the solution together with the Zn^{2+} ions. Thus, H_2S in acid solution buffered at 0.50 *M* H_3O^+ will separate Pb^{2+} ions from Mn^{2+} and Zn^{2+} ions.

Practice Problem 16.13: A solution is 0.10 *M* in zinc ions. What sulfide ion concentration is required to start the precipitation of zinc sulfide? What hydronium ion concentration is required to provide the sulfide ion concentration needed to start precipitation of zinc sulfide in a saturated solution of H_2S?

The answer to Practice Problem 16.13 indicates that a larger sulfide ion concentration is needed to precipitate zinc sulfide from a 0.10 *M* solution of zinc ions than is needed to precipitate lead(II) sulfide from a 0.10 *M* solution of lead(II) ions (Example 16.14). According to Equation 16.1, the sulfide ion concentration in a saturated solution of H_2S increases when the acidity of the solution decreases. The acidity can be decreased by adding a base. Thus, when a solution containing Pb^{2+}, Zn^{2+}, and Mn^{2+} ions is saturated with H_2S and a base is slowly added, PbS, ZnS, and MnS precipitate in that order. MnS precipitates last because its K_{sp} is the largest of the three.

Fractional precipitation of PbS, ZnS, and MnS from a solution of these cations can be demonstrated by taking advantage of the different colors of these sulfides: PbS is black, ZnS white, and MnS is light pink (Figure 16.3). When a solution containing Zn^{2+} and Mn^{2+} ions is saturated with H_2S and a base is slowly added, white ZnS precipitates first, followed by the light pink MnS.

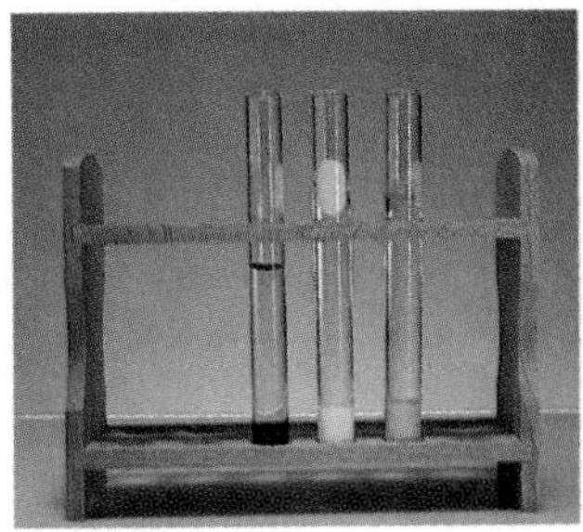

Figure 16.3

Precipitates of PbS, ZnS, and MnS.

In qualitative analysis, a whole group of metal ions can be precipitated as sulfides, leaving another group of metal ions that form more soluble sulfides in the solution. For example, HgS, PbS, Bi_2S_3, CuS, and CdS as a group have small K_{sp} values, ranging from 1.6×10^{-72} for Bi_2S_3 to 8.2×10^{-28} for PbS. Another group of sulfides, MnS, FeS, CoS, NiS, and ZnS, have larger K_{sp} values, ranging from 1.1×10^{-21} for ZnS to 1.4×10^{-15} for MnS.

The first group of metal ions, Hg^{2+}, Pb^{2+}, Bi^{3+}, Cu^{2+}, and Cd^{2+}, can be precipitated as sulfides from an acidic solution saturated with H_2S. In an acidic solution (pH of about 0.5), the sulfide ion concentration is small, but large enough to precipitate the sulfides of that group. The second group of metal ions, Mn^{2+}, Fe^{2+}, Co^{2+}, Ni^{2+}, and Zn^{2+}, if present in the solution, will stay in the solution until the sulfide ion concentration is made large enough for these ions to precipitate as sulfides. This can be accomplished by making the solution alkaline.

After precipitating a group of sulfides or other sparingly soluble salts, the precipitate is separated from the solution by filtration or centrifugation. The precipitate is then dissolved and its ions are separated and identified by various methods. Some methods for dissolving precipitates are discussed in the next section of this chapter. The systematic separation and identification of common cations and anions in solution are considered in Chapter 19, where we discuss qualitative analysis.

16.5 Dissolving Precipitates

Dissolving a precipitate is the opposite of precipitation. In precipitation, the ion product must exceed the K_{sp} of the solid that precipitates. To dissolve a precipitate, the concentration of one or both of the ions in the equilibrium with the solid must be decreased until the product of the ion concentrations is less than the K_{sp} of the precipitate. When the ion concentrations in the solution in contact with a precipitate decreases, some of the precipitate dissolves to replace the ions removed from the solution. As the concentration of the ions in solution continues to diminish, all of the precipitate eventually dissolves. In the following discussion we consider some of the methods used to decrease ion concentrations in solubility equilibria to dissolve precipitates.

Dissolving Hydroxides by Formation of Water

A water-insoluble metal hydroxide can be dissolved in an acid solution. The solubility equilibrium of a solid hydroxide, MOH, in contact with its saturated solution is

$$M(OH)_n(s) \rightleftharpoons M^{n+}(aq) + nOH^-(aq)$$

An important factor in the reaction of a sparingly soluble hydroxide in an acid is the great tendency of H^+ ions to react with OH^- ions to form water. Adding an acid to such a system decreases the OH^- ion concentration by the reaction

$$H^+(aq) + OH^-(aq) \longrightarrow H_2O(l)$$

This reaction has a very large equilibrium constant, K, and goes essentially to completion:

$$K = \frac{1}{K_w} = \frac{1}{1.0 \times 10^{-14}} = 1.0 \times 10^{+14}$$

Adding acid to a metal hydroxide shifts the solubility equilibrium to the right until all of the solid dissolves, provided that a sufficient amount of acid is added. The general net ionic equation for dissolving a solid hydroxide, $M(OH)_n$, in an acid is

$$M(OH)_n(s) + nH^+(aq) \longrightarrow M^{n+}(aq) + nH_2O(l)$$

The net equation for the reaction of a metal hydroxide with an acid can be derived by combining the equation of the solubility equilibrium of the hydroxide with the equation for the reaction of H^+ with OH^- to form water, as shown below for dissolving magnesium hydroxide:

Solubility equilibrium:	$Mg(OH)_2(s) \rightleftharpoons Mg^{2+}(aq) + 2OH^-(aq)$
Formation of water:	$2H^+(aq) + 2OH^-(aq) \longrightarrow 2H_2O(l)$
Net reaction:	$Mg(OH)_2(s) + 2H^+(aq) \longrightarrow Mg^{2+}(aq) + 2H_2O(l)$

Note that the reaction for the formation of water is multiplied by 2 because 1 mol of magnesium hydroxide requires 2 mol of hydrogen ions for neutralization. In an acid solution, the ion product $[Mg^{2+}][OH^-]^2$ is less than the K_{sp} for $Mg(OH)_2$ which means that $Mg(OH)_2$ cannot precipitate under these conditions.

Dissolving Precipitates by Formation of Weak Electrolytes

Just as H^+ ions have a great tendency to react with OH^- ions to form water, H^+ ions also readily react with anions of a weak acid to form molecules of the weak acids. Sparingly soluble salts that contain anions of weak acids may dissolve in strong acids. Hydrogen ions from the acid associate with the anions of the salt in the solubility equilibrium to form molecules of the corresponding weak acid. The salt dissolves to replace the anions removed from the aqueous equilibrium of the salt.

Examples of salts whose anions are conjugate bases of weak acids include most metal carbonates, sulfides, nitrites, fluorides, and phosphates. These salts can often be dissolved in a strong acid by forming HCO_3^-, HS^-, HNO_2, HF, and HPO_4^{2-}, respectively.

Carbonate ion, CO_3^{2-}, is the conjugate base of the weak acid, HCO_3^-, which in turn is the conjugate base of the weak, unstable acid, carbonic acid, H_2CO_3. When HCl is added to solid $CaCO_3$, the solubility equilibrium,

Reaction of calcium carbonate hydrochloric acid.

(1) $CaCO_3(s) \rightleftharpoons Ca^{2+}(aq) + CO_3^{2-}(aq)$

is driven to the right by the reactions

(2) $H^+(aq) + CO_3^{2-}(aq) \rightleftharpoons HCO_3^-(aq)$

(3) $H^+(aq) + HCO_3^-(aq) \rightleftharpoons H_2CO_3(aq)$

(4) $H_2CO_3(aq) \rightleftharpoons H_2O(l) + CO_2(g)$

The sum of equations 1 to 4 is the net equation for dissolving $CaCO_3$ in excess strong acid:

$$CaCO_3(s) + 2H^+(aq) \longrightarrow Ca^{2+}(aq) + H_2O(l) + CO_2(g)$$

When all the $CaCO_3$ has dissolved,

$$[Ca^{2+}][CO_3^{2-}] < K_{sp} \text{ for } CaCO_3$$

Calcium carbonate is the principal ingredient of limestone. Marble is a crystalline, polished limestone. Many statues and the facades of many buildings are made of marble. In industrial areas where acid rain is frequent, the surfaces of marble buildings and statues are damaged because $CaCO_3$ dissolves slowly in acid rain. (See also the discussion of stalagmites and stalactites in Chapter 9.)

Many metal sulfides can also be dissolved in acid solution because sulfide ion is the relatively strong conjugate base of the weak acid, HS^- ion. HS^- ion is the conjugate base of the weak acid H_2S. When a metal sulfide dissolves in acid solution the sulfide ions are converted to HS^- ions and the HS^- ions are converted to H_2S. The solubility equilibrium of the metal sulfide shifts to the right until all of the metal sulfide dissolves. The equilibria for the dissolution of MnS in acid solution are:

$$\begin{aligned} MnS(s) &\rightleftharpoons Mn^{2+}(aq) + S^{2-}(aq) \\ H^+(aq) + S^{2-}(aq) &\rightleftharpoons HS^-(aq) \\ H^+(aq) + HS^-(aq) &\rightleftharpoons H_2S(g) \\ \hline \text{Net equation: } MnS(s) + 2H^+(aq) &\longrightarrow Mn^{2+}(aq) + H_2S(g) \end{aligned}$$

Many metal phosphates are insoluble in water but soluble in aqueous acid. The equilibria for dissolving magnesium phosphate in an acid solution are

$$\begin{aligned} Mg_3(PO_4)_2(s) &\rightleftharpoons 3Mg^{2+}(aq) + 2PO_4^{3-}(aq) \\ 2H^+(aq) + 2PO_4^{3-}(aq) &\rightleftharpoons 2HPO_4^{2-}(aq) \\ 2H^+(aq) + 2HPO_4^{2-}(aq) &\rightleftharpoons 2H_2PO_4^-(aq) \\ 2H^+(aq) + 2H_2PO_4^-(aq) &\rightleftharpoons 2H_3PO_4(aq) \\ \hline \text{Net equation: } Mg_3(PO_4)_2(s) + 6H^+(aq) &\longrightarrow 3Mg^{2+}(aq) + 2H_3PO_4(aq) \end{aligned}$$

Figure 16.4

Tyrannosaurus Rex

Bones (Figure 16.4) and teeth contain hydroxyapatite, $Ca_5(PO_4)_3(OH)$, a solid that has a strong base, the OH^- ion, as an anion. Hydroxyapatite is soluble in acid solution. The dissolution of hydroxyapatite in acidic solution is inhibited by fluoride ions, which replace hydroxide ions in the solid to form some fluorapatite, $Ca_5(PO_4)_3F$. Fluorapatite is less soluble than hydroxyapatite because F^- ion is a weaker base than the OH^- ion. Fluorapatite is also stronger and harder than hydroxyapatite. Thus, adding fluoride ions to toothpaste and drinking water improves the dental health of the population.

Sparingly soluble metal chlorides, bromides, and iodides are generally not soluble in acid because Cl^-, Br^-, and I^- ions are conjugate bases of strong acids that have virtually no tendency to protonate. Salts of these anions can be dissolved by methods discussed in the following sections.

Dissolving Precipitates by Oxidation of Anions

Saturated solutions of sparingly soluble metal sulfides that have extremely small K_{sp} values—such as CuS ($K_{sp} = 6.3 \times 10^{-36}$) and PbS ($K_{sp} = 8.2 \times 10^{-28}$)—have very low sulfide ion concentrations in their solubility equilibria. Therefore, these sulfides cannot be dissolved in hydrochloric acid, but they can be dissolved in oxidizing acids that convert sulfide ions to elemental sulfur. A common oxidizing acid is nitric acid, which contains oxidizing nitrate anions. For example, CuS dissolves in HNO_3 because of the oxidizing action of the NO_3^- ions in an acid medium:

Solubility equilibrium: $$[CuS(s) \rightleftharpoons Cu^{2+}(aq) + S^{2-}(aq)] \times (3)$$

Oxidation of S^{2-}: $$3S^{2-}(aq) + 2NO_3^-(aq) + 8H^+(aq) \longrightarrow 3S(s) + 2NO(g) + 4H_2O(l)$$

Net reaction: $$3CuS(s) + 2NO_3^-(aq) + 8H^+(aq) \longrightarrow 3Cu^{2+}(aq) + 3S(s) + 2NO(g) + 4H_2O(l)$$

Note that the equation for the solubility equilibrium is multiplied by 3 before combining it with the equation for the oxidation of the sulfide ions to obtain the balanced net equation.

Dissolving Precipitates by Complex Ion Formation

If a salt has an extremely low K_{sp} value and if its anions cannot be oxidized, it can often be dissolved by a chemical reaction which forms complex ions. For example, when a copper(II) salt is dissolved in water, the Cu^{2+} ion acts as a Lewis acid in reaction with water, a Lewis base:

$$Cu^{2+}(aq) + 4H_2O(l) \rightleftharpoons Cu(H_2O)_4^{2+}(aq)$$

The $Cu(H_2O)_4^{2+}$ ion consists of a central Cu^{2+} ion linked to four water molecules by coordinate covalent bonds (Section 7.4). A species such as the $Cu(H_2O)_4^{2+}$ ion in which a Lewis acid is attached to one or more Lewis bases, in equilibrium with its components in solution, is called a **complex ion** or a **coordination complex**. The Lewis bases in a complex ion are called **ligands**. The $Cu(H_2O)_4^{2+}$ ions in aqueous solution are slightly dissociated and exist in equilibrium with their components, the Cu^{2+} ions and H_2O molecules. A complex ion whose ligands are water molecules is called an **aqua complex**.

When an aqueous solution of aqua complex ions of copper, $Cu(H_2O)_4^{2+}$, is treated with ammonia, ammonia molecules replace water molecules in the aqua complex to form $Cu(NH_3)_4^{2+}$ ions:

$$\underset{\text{Sky blue}}{Cu(H_2O)_4^{2+}(aq)} + 4NH_3(aq) \longrightarrow \underset{\text{Violet-blue}}{Cu(NH_3)_4^{2+}(aq)} + 4H_2O(l)$$

The aqua complex is sky blue and the copper–ammonia complex is a deeper hue of violet-blue (Figure 16.5). A *complex ion*, such as $Cu(NH_3)_4^{2+}$, *that has ammonia molecules as ligands* is called an **ammine complex**.

Ammonia molecules have a strong tendency to form ammine complex ions with many metal ions, particularly with Ni^{2+}, Cu^{2+}, Zn^{2+}, and Ag^+ ions. Ni^{2+} ion combines with six NH_3 molecules, Cu^{2+} and Zn^{2+} ions each combine with four NH_3 molecules, and Ag^+ ion combines with two NH_3 molecules. Since ammonia has a strong tendency to form complexes with these metal ions, the salts of these ions often dissolve in aqueous ammonia.

Figure 16.5

An aqua complex of Cu^{2+} ions, $Cu(H_2O)_4^{2+}(aq)$, and an aqueous ammine complex of Cu^{2+} ions, $Cu(NH_3)_4^{2+}(aq)$.

For example, silver chloride, which is insoluble in water and acids, readily dissolves in aqueous ammonia to form ammine complex ions of silver, $Ag(NH_3)_2^+$. The dissolution of silver chloride in aqueous ammonia occurs according to the equation that is obtained by combining the equation for the solubility equilibrium of silver chloride with the equation for the reaction of silver ions with ammonia:

AgCl solubility equilibrium:	$AgCl(s) \rightleftharpoons Ag^+(aq) + Cl^-(aq)$
Formation of ammine complex:	$Ag^+(aq) + 2NH_3(aq) \rightleftharpoons Ag(NH_3)_2^+(aq)$
Net equation:	$AgCl(s) + 2NH_3(aq) \longrightarrow Ag(NH_3)_2^+(aq) + Cl^-(aq)$

The Ag^+ ion concentration in a saturated solution of AgCl in contact with the undissolved solid is decreased by the formation of $Ag(NH_3)_2^+$ ions. The solubility equilibrium shifts to the right, increasing the concentration of $Ag(NH_3)_2^+$ ions until all of the solid AgCl dissolves, provided that a sufficient amount of NH_3 is added.

The method of dissolving precipitates by forming complex ions provides a way to separate solids. For example, lead chloride can be separated from silver chloride by treating a mixture of these water-insoluble solids with aqueous ammonia. Silver chloride dissolves by forming ammine complex ions; lead chloride does not react with ammonia and can be recovered from the solution by filtration or centrifugation.

Silver chloride can be reprecipitated from the ammine complex solution by acidifying the solution. Hydrogen ions from the acid react with ammonia molecules in the ammine complex ions to form ammonium ions and silver ions. Silver ions then precipitate with the Cl^- ions in the solution:

$$Ag(NH_3)_2^+(aq) + 2H^+(aq) + Cl^-(aq) \longrightarrow AgCl(s) + 2NH_4^+(aq)$$

Silver chloride can be recovered from the solution by filtration.

Aqueous ions can be separated from one another by exploiting the ability of one of the ions to form complexes. For example, Ni^{2+} ions can be separated from Mg^{2+} ions in solution by adding ammonia to the mixture. The Ni^{2+} ions form stable $Ni(NH_3)_6^{2+}$ ions in the solution, while the Mg^{2+} ions precipitate as $Mg(OH)_2$ from the alkaline solution produced by the ionization of excess ammonia.

Amphoteric hydroxides (Chapter 14) dissolve in excess hydroxide ions to form **hydroxo complex** ions in which hydroxide ions are the ligands. Hydroxo and ammine complex formation can be employed in qualitative analysis, as shown in the following example.

Example 16.15

Identifying Metal Hydroxides by the Formation of Hydroxo Complexes and Ammine Complexes

Three solid hydroxides are placed into three different unlabeled test tubes. One contains $Mg(OH)_2$, the second $Ni(OH)_2$, and the third $Zn(OH)_2$. Describe simple chemical tests that identify each of the hydroxides. Write an equation for the reactions, if any, that occur(s) in the identification of each.

SOLUTION: Zinc hydroxide is amphoteric and forms hydroxo complex ions when treated with a solution of sodium hydroxide that provides a large excess of hydroxide ions. Zinc hydroxide also dissolves in aqueous ammonia to form ammine complex ions of zinc. Nickel hydroxide is not amphoteric; it dissolves in aqueous ammonia but not in sodium hydroxide. Magnesium hydroxide does not dissolve in either sodium hydroxide or in aqueous ammonia.

A sample of the solid from a randomly selected test tube can be treated with a concentrated solution of NaOH. If the sample dissolves, the solid is $Zn(OH)_2$; if it does not dissolve, it is either $Mg(OH)_2$ or $Ni(OH)_2$. Zinc hydroxide dissolves in aqueous NaOH because it is an amphoteric hydroxide.

The reaction is

$$Zn(OH)_2(s) + 2OH^-(aq) \longrightarrow Zn(OH)_4^{2-}(aq)$$

The other two hydroxides do not dissolve in NaOH.

A portion of one of the two remaining solids can be treated with aqueous ammonia. If the solid dissolves, it is $Ni(OH)_2$ because the nickel ion forms a stable ammine complex:

$$Ni(OH)_2(s) + 6NH_3(aq) \longrightarrow Ni(NH_3)_6^{2+}(aq) + 2OH^-(aq)$$

The solid that does not dissolve in NaOH or in NH_3 is $Mg(OH)_2$.

Practice Problem 16.14: How would you identify each of the following hydroxides in different test tubes: aluminum hydroxide, copper(II) hydroxide, calcium hydroxide? Write an equation for the reaction, if any, that occurs in the testing of each.

Dissolving Precipitates by Oxidation and Complexation

Some extremely sparingly soluble sulfides can be dissolved only by oxidizing the anion and forming a complex of the cation. This can sometimes be accomplished in one step by using *aqua regia*, which is a mixture of 3 parts of concentrated hydrochloric acid and 1 part of concentrated nitric acid. For example, mercury(II) sulfide has an extremely small value of K_{sp}, 1×10^{-50}. It does not dissolve in HCl or in HNO_3, but it does dissolve in aqua regia. The nitrate ions from HNO_3 oxidize the S^{2-} ions in the solubility equilibrium of HgS to elemental sulfur, while the chloride ions from HCl form chloro complex ions with Hg^{2+} ions, $HgCl_4^{2-}$. Thus, HgS dissolves in aqua regia according to the following net equation:

$$3HgS(s) + 2NO_3^-(aq) + 12Cl^-(aq) + 8H^+(aq) \longrightarrow 3HgCl_4^{2-}(aq) + 2NO(g) + 3S(s) + 4H_2O(l)$$

To summarize the various ways of dissolving precipitates we have discussed in this section, consider a mixture of the following solids: $Mg(NO_3)_2$, MnS ($K_{sp} = 1.4 \times 10^{-15}$), PbS ($K_{sp} = 8.2 \times 10^{-28}$), and HgS ($K_{sp} = 1 \times 10^{-50}$). We can use the principles discussed in this section to dissolve the components of the mixture selectively to obtain four different metal ion solutions in different containers.

First, we add water and stir. Magnesium nitrate dissolves (metal nitrates are soluble in water, see the solubility rules in Chapter 4). The mixture of the other three solids can be removed by filtration. Next, we add a small amount of HCl to the mixture of the three remaining solids. Manganese sulfide has a relatively large K_{sp} and it therefore dissolves in HCl to form Mn^{2+} ions and the weak acid H_2S. We filter the solution to isolate the two remaining solids. Lead(II) sulfide has an intermediate K_{sp} value and it dissolves in nitric acid, which oxidizes the sulfide ions to elemental sulfur and releases lead(II) ions. The solution is filtered and one solid, HgS, remains. Mercury(II) sulfide has an extremely small K_{sp}. We dissolve the HgS in aqua regia.

Practice Problem 16.15: Write a balanced net ionic equation for the dissolution reaction of each of the following substances: **(a)** magnesium hydroxide in hydrochloric acid; **(b)** manganese(II) sulfide in hydrochloric acid; **(c)** lead(II) sulfide in nitric acid (assume that the nitrate ions are reduced to nitrogen monoxide).

Calculating Solubilities of Solids That Form Complex Ions

We recall that a complex ion in aqueous solution is in equilibrium with the free metal ion and its ligands. For example, $Ag(NH_3)_2^+$ ions are formed from Ag^+ ions and NH_3 molecules in solution where all the species are in equilibrium with one another:

$$Ag^+(aq) + 2NH_3(aq) \rightleftharpoons Ag(NH_3)_2^+(aq)$$

This is an equation for the formation of a complex ion. The equilibrium constant for this reaction is therefore called the **formation constant**, K_f, which has the form

$$K_f = \frac{[Ag(NH_3)_2{}^+]}{[Ag^+][NH_3]^2}$$

The formation constant is sometimes also called the **stability constant** because its magnitude is a measure of the stability of the complex ion; the larger the constant, the more stable the ion. Table 16.2 lists the formation constants for some complex ions.

The overall equation for dissolving AgCl in aqueous ammonia can be obtained by combining the equations for the solubility equilibrium of AgCl and the equilibrium for the formation of $Ag(NH_3)_2{}^+$. The equilibrium constant, K, for the overall equilibrium is the product of the K_{sp} for the solubility equilibrium and K_f for the formation of the complex ion:

Solubility:	$AgCl(s) \rightleftharpoons Ag^+(aq) + Cl^-(aq)$	$K_{sp} = 1.6 \times 10^{-10}$
Complexation:	$Ag^+(aq) + 2NH_3(aq) \rightleftharpoons Ag(NH_3)_2{}^+(aq)$	$K_f = 1.7 \times 10^7$
Overall:	$AgCl(s) + 2NH_3(aq) \rightleftharpoons Ag(NH_3)_2{}^+(aq) + Cl^-(aq)$	

The equilibrium constant for the overall reaction is

$$K = \frac{[Ag(NH_3)_2{}^+][Cl^-]}{[NH_3]^2} = (K_{sp})(K_f) = (1.6 \times 10^{-10})(1.7 \times 10^7)$$

$$= 2.7 \times 10^{-3}$$

We can use the equilibrium constant for the net reaction of complex formation to calculate the molar solubility of a salt in a complexing reagent. In this case the solubility of the salt equals the molar concentration of the complex ion because each mole of salt that dissolves produces 1 mol of complex ion.

Table 16.2

Formation Constants, K_f, of Some Complex Ions in Water at 25°C

Complex Ion	Equilibrium Equation	K_f
$Ag(NH)_2{}^+$	$Ag^+(aq) + 2NH_3(aq) \rightleftharpoons Ag(NH_3)_2{}^+(aq)$	1.7×10^7
$Cd(NH_3)_4{}^{2+}$	$Cd^{2+}(aq) + 4NH_3(aq) \rightleftharpoons Cd(NH_3)_4{}^{2+}(aq)$	1×10^7
$Co(NH_3)_6{}^{2+}$	$Co^{2+}(aq) + 6NH_3(aq) \rightleftharpoons Co(NH_3)_6{}^{2+}(aq)$	1×10^5
$Cu(NH_3)_4{}^{2+}$	$Cu^{2+}(aq) + 4NH_3(aq) \rightleftharpoons Cu(NH_3)_4{}^{2+}(aq)$	5×10^{12}
$Zn(NH_3)_4{}^{2+}$	$Zn^{2+}(aq) + 4NH_3(aq) \rightleftharpoons Zn(NH_3)_4{}^{2+}(aq)$	5×10^8
$Ag(CN)_2{}^-$	$Ag^+(aq) + 2CN^-(aq) \rightleftharpoons Ag(CN)_2{}^-(aq)$	1×10^{21}
$Cu(CN)_4{}^{2-}$	$Cu^{2+}(aq) + 4CN^-(aq) \rightleftharpoons Cu(CN)_4{}^{2-}(aq)$	1×10^{25}
$Fe(CN)_6{}^{4-}$	$Fe^{2+}(aq) + 6CN^-(aq) \rightleftharpoons Fe(CN)_6{}^{4-}(aq)$	1×10^{35}
$Fe(CN)_6{}^{3-}$	$Fe^{3+}(aq) + 6CN^-(aq) \rightleftharpoons Fe(CN)_6{}^{3-}(aq)$	1×10^{42}
$Zn(OH)_4{}^{2-}$	$Zn^{2+}(aq) + 4OH^-(aq) \rightleftharpoons Zn(OH)_4{}^{2-}(aq)$	5×10^{14}

Example 16.16

Calculating the Solubility of a Salt in a Complexing Reagent

What is the molar solubility of AgCl in a solution initially 1.0 *M* in NH_3? How many grams of AgCl can be dissolved in 100 mL of 1.0 *M* NH_3? The formation constant, K_f, for $Ag(NH_3)_2^+$ is 1.7×10^7, and the K_{sp} for AgCl is 1.6×10^{-10}. Ignore the small loss of ammonia due to ionization.

SOLUTION: The initial and equilibrium molar concentrations of the species in the solubility equilibrium are

	$AgCl(s) + 2NH_3(aq)$	$\rightleftharpoons Ag(NH_3)_2^+(aq)$	$+ Cl^-(aq)$
Initial *M*	1.0	0	0
Equilibrium *M*	$1.0 - 2x$	x	x

where x = molar solubility of AgCl = $[Ag(NH_3)_2^+] = [Cl^-]$. Since K_f for $Ag(NH_3)_2^+$ is very large, the complex is very stable, and nearly all of the Ag^+ ions in solution are complexed by NH_3. The equilibrium constant expression is

$$K = \frac{[Ag(NH_3)_2^+][Cl^-]}{[NH_3]^2} = K_f K_{sp} = 2.7 \times 10^{-3}$$

Substituting the concentrations into this expression gives

$$2.7 \times 10^{-3} = \frac{(x)(x)}{(1.0 - 2x)^2}$$

Taking the square root of both sides of this equation yields

$$5.2 \times 10^{-2} = \frac{x}{1.0 - 2x}$$

from which

$$x = 0.047\ M$$

Thus, 1.0 L of 1.0 *M* NH_3 can dissolve 0.047 mol (6.7 g) of AgCl. (The solubility of AgCl in water is only 1.2×10^{-5} mol per liter.) In 100 mL of 1.0 *M* NH_3, the solubility of AgCl is one tenth of its solubility in 1.0 L, or 0.67 g.

Practice Problem 16.16: Calculate the molar solubility of $Cu(OH)_2$ in initially 2.0 *M* NH_3. Ignore the ionization of ammonia.

Summary

Solubility and Solubility Product

The *number of moles of a salt that dissolves to form hydrated ions in 1 L of solution* is called its **molar solubility**. The *equilibrium constant for the solubility equilibrium* of a sparingly soluble salt is called the **solubility product constant** or the **solubility product**, K_{sp}, of the salt. For compounds of similar formulas, the K_{sp} value is a direct measure of their relative solubilities.

The K_{sp} value for a salt can be calculated from the experimentally determined ion concentrations in a saturated solution of the salt. Conversely, the ion concentrations in a saturated solution of a salt, and its solubility, can be calculated from the value of K_{sp}.

Common Ion Effect in Solubility Equilibria

The solubility of a salt decreases when ions common to the equilibrium system of the salt are added. The solubility of a salt in a solution that contains ions common to the salt is calculated from the K_{sp} of the salt using the *total* common ion concentration.

Forming Precipitates

A *salt precipitates if the product of the molar concentrations of its ions in solution (each raised to the power that equals its coefficient in a balanced equation for the dissolution reaction) exceeds the* K_{sp} *value for the salt.*

Fractional Precipitation and Simultaneous Equilibria

Different cations in an aqueous mixture can be successively precipitated by gradually increasing the concentration of anions that form sparingly soluble salts with each of the cations in the solution. For example, many cations can be selectively precipitated as sulfides from a solution in which the sulfide ion concentration is gradually increased. When the sulfide ion concentration is low, the metal sulfides that are the least soluble precipitate first. Increasing the sulfide ion concentration causes metal sulfides of successively higher solubility to precipitate.

Sulfide ion concentration in a solution can be controlled by adjusting the pH of a solution saturated with H_2S. The H_2S concentration in a saturated solution at 25°C and 1.0 atm is 0.10 *M*. As the hydronium ion concentration of such a solution increases, the sulfide ion concentration decreases, as shown by the product of K_{a1} and K_{a2} for H_2S:

$$K_{a1}K_{a2} = \frac{[H_3O^+]^2[S^{2-}]}{[H_2S]}$$

Dissolving Precipitates

A precipitate can be dissolved by decreasing the concentration of the cations, the anions, or both, which are in equilibrium with the precipitate. The precipitate dissolves if the product of the concentrations of the ions in equilibrium with the salt is decreased below the K_{sp} value of the salt. The ion product can be lowered by adding a substance with which the cations or the anions in the solubility equilibrium of the solid react to form water, another weak electrolyte, or complex ions. A salt can also be dissolved by oxidizing its anions in solution, or by both oxidizing its anions and complexing its cations.

New Terms

Ammine complex (16.5)
Aqua complex (16.5)
Complex ion (16.5)
Coordination complex (16.5)
Formation constant (16.5)
Hydroxo complex (16.5)
Ligand (16.5)
Molar solubility (16.1)
Solubility product constant (16.1)
Stability constant (16.5)

Exercises

General Review

1. Define "molar solubility" and the "solubility product constant" for a sparingly soluble salt.
2. Assuming that the molar solubility of a sparingly soluble salt is x, show how x can be calculated from the K_{sp} for each of the following types of salts: **(a)** AB; **(b)** AB_2; **(c)** A_2B; **(d)** AB_3; **(e)** A_2B_3.
3. For each type of the solid listed in question 2, show how the solubility of the salt is related to the cation concentration and to the anion concentration in a saturated solution of the salt.
4. Show how the solubility of each of the sparingly soluble salts listed in question 2 can be calculated in a solution of another substance that provides 1 mol per liter of: **(a)** the cations of the salt; **(b)** the anions of the salt.
5. How does the solubility of a sparingly soluble hydroxide vary with: **(a)** increasing the pH of the solution; **(b)** decreasing the pH of the solution?
6. If you wished to precipitate a sparingly soluble salt by mixing solutions of its ions, how could you decide what the minimum ion concentrations must be to start the precipitation of the salt?
7. If you had a solution containing the cations of two different metals that are known to form insoluble salts with a given anion, X^-, how would you separate these cations from each other by using the anion X^-?
8. How can the sulfide ion concentration in a saturated solution of hydrogen sulfide be varied? Write chemical and mathematical equations relating the sulfide ion concentration in this solution to the hydronium ion concentration of the solution. Show how this equation can be solved for the sulfide ion concentration at any given pH of the solution.
9. After you have precipitated cations of a certain metal as a metal sulfide from a saturated solution of H_2S and other metal cations, would you have to increase or decrease the pH of the solution to start the precipitation of the second metal sulfide that is more soluble than the first one?
10. Explain how the concentrations of the ions in equilibrium with a precipitate should be altered to dissolve the precipitate.
11. Give examples of salts that can be dissolved by using each of the following principles: **(a)** formation of a weak electrolyte; **(b)** oxidation of its anions; **(c)** complex ion formation; **(d)** combined oxidation and complexation.

Solubility and K_{sp}

12. Given the following sulfides and their respective K_{sp} values: CuS, 6.3×10^{-36}; PbS, 8.2×10^{-28}; MnS, 1.4×10^{-15}; HgS, 1×10^{-50}; and ZnS, 1.1×10^{-21}. **(a)** Which of these sulfides is most soluble in water? **(b)** Which has the smallest $[S^{2-}]$ in its saturated solution?
13. Calculate the molar solubility of silver bromide. What are $[Ag^+]$ and $[Br^-]$ in a saturated solution of silver bromide?
14. Calculate the solubility of barium chromate in moles per liter and in grams per liter of water.
15. How many grams of lead(II) sulfate can be dissolved in 3.0 L of water to make a saturated solution of lead(II) sulfate?
16. Calculate the solubility of lead(II) chloride in moles per liter and in grams per liter of water. What are the lead(II) and chloride ion concentrations in a saturated solution of lead(II) chloride?
17. The solubility of silver chloride is 1.25×10^{-5} mol per liter. Calculate K_{sp} for silver chloride.
18. The solubility of magnesium hydroxide is 1.5×10^{-4} mol per liter. Calculate K_{sp} for magnesium hydroxide.
19. The solubility of silver sulfate is 3.3 g per liter of water. Calculate: **(a)** the silver ion concentration and the sulfate ion concentration in a saturated solution of silver sulfate; **(b)** K_{sp} for silver sulfate.
20. The solubility of manganese(II) hydroxide is 3.2×10^{-3} g per liter of water. Calculate: **(a)** the pH of a saturated solution of manganese(II) hydroxide; **(b)** K_{sp} for manganese(II) hydroxide.
21. What is the pH of a saturated solution of strontium hydroxide?
22. How many grams of chromium(III) hydroxide would be required to make 1.4 L of saturated solution of chromium(III) hydroxide? Assume no hydrolysis of Cr^{3+} ions.
23. What volume of solution is needed to dissolve a maximum of 0.20 mg of calcium oxalate? Ignore the hydrolysis of oxalate ions.
24. **(a)** Calculate the molar solubility of $Fe(OH)_3$ in water ($K_{sp} \simeq 4 \times 10^{-38}$. **(b)** The solubility of lanthanum molybdate, $La_2(MoO_4)_3$, is 1.8×10^{-2} g L^{-1}. Calculate the value of K_{sp} for $La_2(MoO_4)_3$. Ignore hydrolysis.

Common Ion Effect on Solubility

25. Calculate the solubility of barium chromate in 0.10 *M* solution of barium nitrate.

26. Calculate the solubility of lead(II) sulfate in 0.20 *M* solution of sodium sulfate.

27. Calculate: **(a)** the solubility of silver chloride in 0.10 *M* hydrochloric acid; **(b)** the solubility of silver bromide in 0.10 *M* hydrobromic acid.

28. **(a)** What is the Pb^{2+} ion concentration in a solution that is 0.080 *M* in NaCl and saturated with $PbCl_2$? **(b)** What is the molar solubility of PbF_2 in a 0.080 *M* solution of NaF?

29. How many grams of $PbCl_2$ can be dissolved in 1.8 L of a 0.40 *M* solution of NaCl?

30. How many grams of silver chromate, Ag_2CrO_4, can be dissolved in 500 mL of 0.50 *M* sodium chromate solution?

31. A portion of a precipitate of barium sulfate is treated with 100 mL of water with stirring, and another portion is treated with 100 mL of a 0.010 *M* sodium sulfate. How many grams of barium sulfate dissolves to form a saturated solution in each case?

32. A 9.5-g sample of magnesium chloride is dissolved in 1.5 L of a saturated solution of lead(II) chloride. What is the molar concentration of Pb^{2+} ions in the solution after precipitation? Assume no change in volume of the solution upon addition of magnesium chloride. How many grams of $PbCl_2$ precipitates?

Formation of Precipitates

33. Write a net ionic equation for the reaction that occurs when solutions of each of the following pairs of substances are mixed: **(a)** sodium chloride and silver nitrate; **(b)** barium nitrate and sodium sulfate; **(c)** lead(II) nitrate and hydrochloric acid; **(d)** potassium chromate and lead(II) nitrate; **(e)** magnesium chloride and sodium hydroxide; **(f)** calcium chloride and sodium phosphate; **(g)** iron(II) nitrate and sodium sulfide; **(h)** silver nitrate and sodium chromate.

34. A solution is 0.10 *M* in barium ions. What is the minimum concentration of sulfate ions that must be supplied to this solution to start the precipitation of barium sulfate?

35. Will a precipitate of mercury(I) chloride form when a 0.20 *M* solution of mercury(I) nitrate, $Hg_2(NO_3)_2$, is made 0.010 *M* in sodium chloride? If a precipitate forms, what are the molar concentrations of Hg_2^{2+} and Cl^- ions remaining in the solution?

36. Will a precipitate form when 0.086 g of solid magnesium chloride is dissolved in 500 mL of 0.050 *M* lead(II) nitrate solution? Assume no volume change when the solid is added to the solution.

37. Will a precipitate form when 200 mL of a 0.10 *M* silver nitrate solution is mixed with 10 mL of 0.010 *M* hydrochloric acid? If a precipitate forms, what is the molar concentration of silver ions remaining in the solution?

38. Calcium fluoride is precipitated from 100 mL of 0.020 *M* solution of calcium nitrate by adding 10 mL of 0.10 *M* sodium fluoride. Calculate the molar concentrations of calcium and of fluoride ions remaining in the solution. Ignore the hydrolysis of fluoride ions.

39. Eighty milliliters of 0.050 *M* magnesium chloride solution is mixed with 20 mL of 0.30 *M* sodium hydroxide. Calculate the molar concentrations of magnesium ions and hydroxide ions remaining in the solution.

40. Eighty milliliters of 0.050 *M* magnesium chloride solution is mixed with 20 mL of 0.10 *M* silver nitrate. Calculate the molar concentration of silver ions and chloride ions remaining in the solution.

Fractional Precipitation and Simultaneous Equilibria

41. Calculate the sulfide ion concentration in an aqueous solution that is 0.30 *M* in HCl and saturated with H_2S at 1 atm and 25°C. The molar concentration of H_2S under these conditions is 0.10.

42. A solution is 0.10 *M* in each of the following cations: Zn^{2+}, Cu^{2+}, Pb^{2+}, and Mn^{2+}. If the sulfide ion concentration were gradually increased in this solution, which sulfide would precipitate first? Which would precipitate last?

43. **(a)** What is the minimum sulfide ion concentration needed to start the precipitation of zinc sulfide from a 0.10 *M* solution of zinc ions? **(b)** What is the maximum concentration of HCl in a saturated solution of H_2S (at 1.0 atm and 25°C) that is required to provide the sulfide ion concentration calculated in part **(a)** of this problem?

44. A solution is 0.10 *M* in Mn^{2+} and 0.10 *M* in Cu^{2+} ions. If this solution were made 1.0×10^{-20} *M* in sulfide ions, would MnS, CuS, or both precipitate? Show by calculation.

45. A solution is 0.10 *M* in each Cu^{2+}, Hg^{2+}, Mn^{2+}, and Zn^{2+}. To separate Cu^{2+} and Hg^{2+} ions from the other ions, they can be precipitated as sulfides in a slightly acid solution saturated with H_2S. Calculate the minimum H^+ ion concentration that permits the precipitation of CuS and HgS, but not MnS or ZnS.

46. A solution containing 0.20 g of lead(II) nitrate in 100 mL is made 1.0 *M* in hydrochloric acid and saturated with hydrogen sulfide to precipitate lead(II) sulfide. What is the percent of lead ions remaining in the solution?

47. One hundred milliliters of a solution was prepared by diluting 10 mL of 25 percent by mass of aqueous ammonia (density 0.910 g cm^{-3}) in which 1.0 g of ammonium chloride was dissolved. Calculate the maximum concentration of magnesium ions that can exist in this solution so that no magnesium hydroxide precipitates.

48. A 50-mL portion of a solution containing 1.0 g of magnesium chloride is to be added to an equal volume of a 2.0 *M* solution of aqueous ammonia. How many grams of solid ammonium chloride must also be added to this solution to prevent the precipitation of magnesium hydroxide?

Dissolving of Precipitates

49. Write the name and the formula of a compound whose aqueous solution can be used to dissolve each of the following solid substances: **(a)** magnesium hydroxide; **(b)** calcium carbonate; **(c)** manganese(II) sulfide; **(d)** strontium phosphate; **(e)** silver chloride; **(f)** aluminum hydroxide (use a compound other than an acid); **(g)** lead(II) sulfide (not soluble in HCl, because K_{sp} is small).

50. Write a net ionic equation for each of the dissolution reactions in question 49.

51. You are given a solid mixture consisting of calcium nitrate, nickel(II) sulfide ($K_{sp} = 3 \times 10^{-19}$), and mercury(II) sulfide ($K_{sp} = 1 \times 10^{-50}$). Devise a scheme of successively dissolving each component of the mixture without essentially affecting the others present. Write an equation for each dissolution reaction you propose to carry out.

Complex Ion Equilibria

52. Calculate the concentrations of $Cu(NH_3)_4^{2+}$, and Cu^{2+} in a solution that is 0.30 *M* in $Cu(NH_3)_4SO_4$ and 0.010 *M* in NH_3 assuming that $Cu(NH_3)_4SO_4$ is completely dissociated into $Cu(NH_3)_4^{2+}$ and SO_4^{2-} ions.

53. Calculate the molar solubility of silver bromide in a 0.80 *M* solution of NH_3.

CHAPTER 17

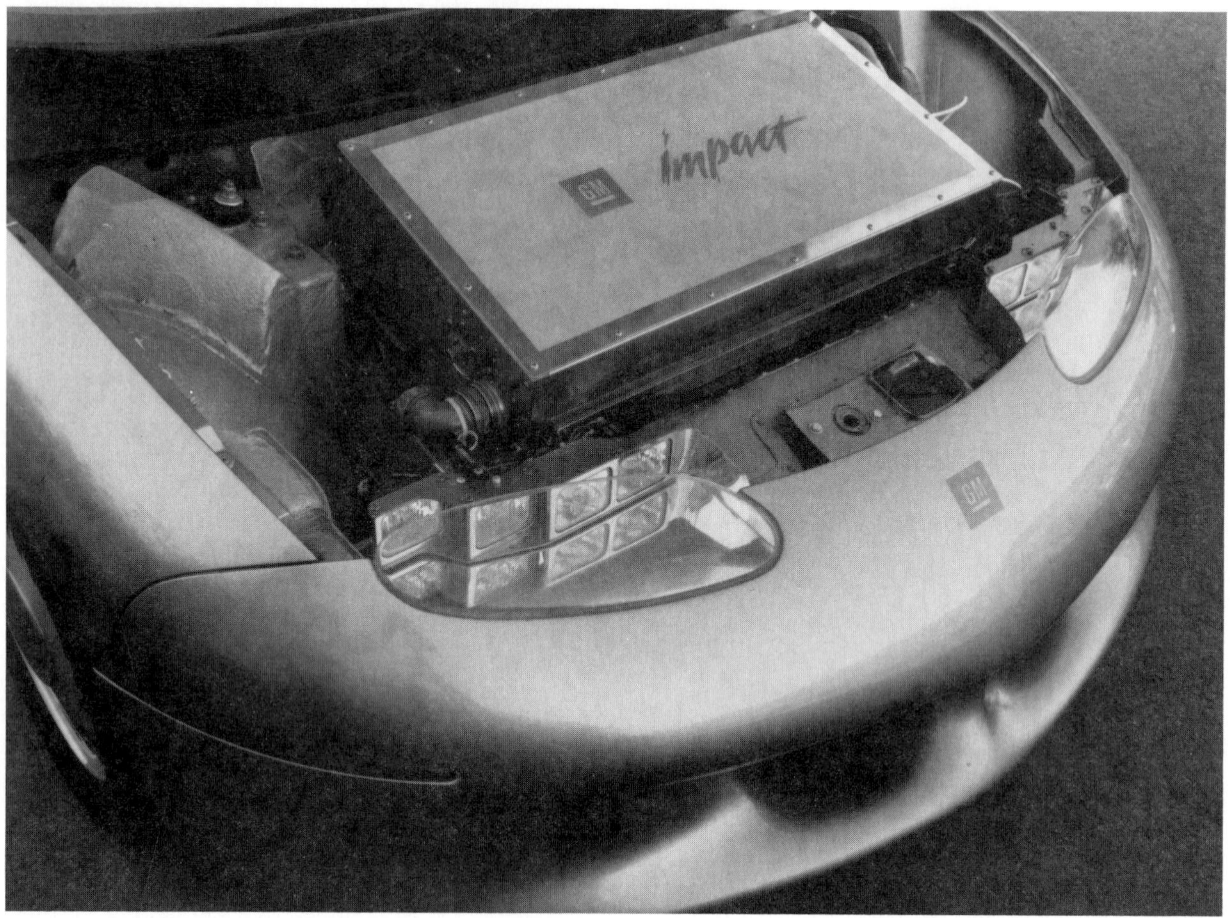

Air pollution caused by automobile exhaust is a major environmental problem in many urban areas. In the near future many automobiles will be powered by electricity rather than by gasoline.

Electrochemistry

In this chapter we discuss spontaneous redox reactions that produce an electric current. We also consider the use of an electric current to drive redox reactions that are not spontaneous. Together these two types of reactions constitute the large field of study called ***electrochemistry****.*

We can use redox reactions to produce electricity. A device in which an electric current is generated by a spontaneous redox reaction *is called a* **cell**. *A* ***battery*** *consists of two or more* cells, *which operate independently to generate electric current by spontaneous redox reactions.* An electrochemical cell that uses a spontaneous redox reaction to generate electric current *is called a* ***voltaic cell*** *or a* ***galvanic cell****. These electrochemical cells are named for the eighteenth-century Italian physicist Alessandro Volta and his contemporary, the physiologist Luigi Galvani. (It was Galvani who first observed that passing an electric current along the nerve of a frog's leg makes the muscle twitch.)*

17.1 Review of Redox Reactions

Figure 17.1

Reaction of magnesium with hydrochloric acid.

We discussed oxidation–reduction (redox) reactions in Chapter 4. We recall that when a species is oxidized, its oxidation state increases. An increase in oxidation state can be caused by loss of electrons. When a species is reduced, its oxidation state decreases. A decrease in oxidation state can be caused by gain of electrons. The reactant that is oxidized acts as a reducing agent, and the reactant that is reduced acts as an oxidizing agent.

A redox reaction in aqueous solution can be separated into two half-reactions. One half-reaction represents oxidation and the other represents reduction. An example of a redox reaction in aqueous solution is the reaction of magnesium with hydrochloric acid (Figure 17.1). This reaction produces a solution of magnesium chloride and hydrogen gas:

$$Mg(s) + 2HCl(aq) \longrightarrow MgCl_2(aq) + H_2(g)$$

In this reaction, magnesium is oxidized; its oxidation state increases from 0 to +2. Hydrogen is reduced from an oxidation state of +1 to an oxidation state of 0. Electrons are transferred from magnesium metal to hydrogen ions. Magnesium metal is the reducing agent, and hydrogen ion is the oxidizing agent. The net ionic equation for this reaction is

$$Mg(s) + 2H^+(aq) \longrightarrow Mg^{2+}(aq) + H_2(g)$$

This redox reaction can be separated into two half-reactions. One half-reaction represents the oxidation of magnesium:

$$Mg(s) \longrightarrow Mg^{2+}(aq) + 2e^-$$

The other half-reaction represents the reduction of aqueous hydrogen ions to hydrogen gas:

$$2H^+(aq) + 2e^- \longrightarrow H_2(g)$$

The electrons lost by magnesium atoms during the reaction are gained by hydrogen ions.

In this reaction the electons are transferred directly from magnesium metal to hydrogen ions in solution. If electron transfer could be made to occur through an external circuit, energy could be harnessed for useful work. In the next section we discuss devices in which spontaneous redox reactions generate an electric current that flows through an external circuit.

17.2 Galvanic Cells

A Simple Galvanic Cell

In Chapter 2 we learned that an active metal can replace the ions of a less active metal. In such a reaction, atoms of the more active metal are oxidized to ions, and ions of the less active metal are reduced to atoms. For example, when a zinc rod is dipped into a copper(II) sulfate solution, zinc atoms are oxidized to zinc ions and copper(II) ions are reduced to copper metal, which deposits on the zinc rod (Figure 17.2):

$$Zn(s) + Cu^{2+}(aq) \longrightarrow Zn^{2+}(aq) + Cu(s)$$

Figure 17.2

A deposit of copper metal on the surface of zinc caused by the reaction of zinc metal with copper(II) ions in solution.

In the reduction of aqueous copper(II) ions by zinc metal, electrons flow from the zinc rod directly to copper(II) ions in the solution. If electron transfer from the zinc rod to the copper ions in solution could be directed through an external circuit, the spontaneous redox reaction could be used to generate an electric current. But when a zinc rod in one container is connected by a copper wire with a copper(II) sulfate solution in a separate container, no current flows through the wire. However, when the two containers are also connected with a tube filled with a solution of an electrolyte such as KCl, current flows through the external circuit (Figure 17.3). The device shown in Figure 17.3 is an example of a *galvanic cell.*

The cell illustrated in Figure 17.3 consists of two **half-cells**. The oxidation half-reaction occurs in one half-cell and the reduction half-reaction occurs in the other half-cell. The sum of the two half-cell reactions is the overall redox reaction that constitutes the *cell reaction*. The *electrode at which oxidation occurs* is called the **anode**, and the *electrode at which reduction occurs* is the

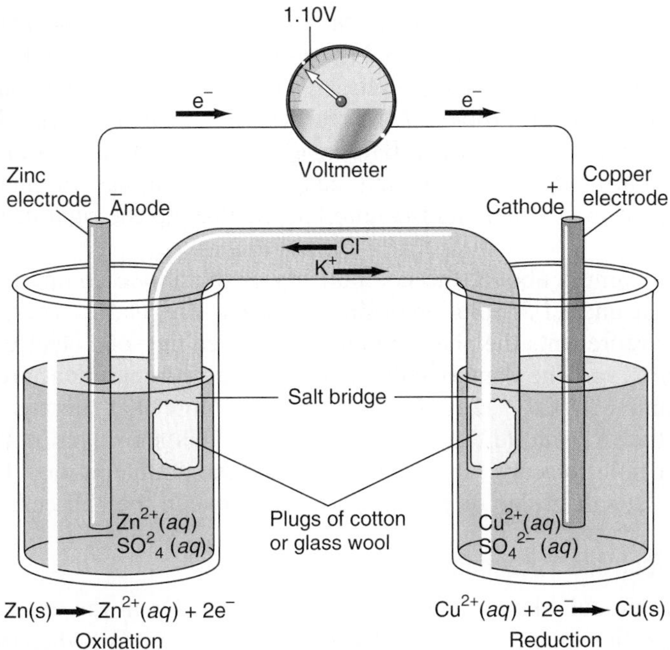

Figure 17.3

A galvanic cell generating electric current by the reaction $Zn(s) + Cu^{2+}(aq) \rightarrow Zn^{2+}(aq) + Cu(s)$.

cathode. In the device shown in Figure 17.3 the zinc rod is the anode. At this electrode the zinc atoms are oxidized to zinc ions:

$$\text{Anode half-reaction:}\quad Zn(s) \longrightarrow Zn^{2+}(aq) + 2e^-$$

Electrons released at the anode travel through the external circuit to the cathode—the copper rod—where they are transferred to copper(II) ions in the solution. Copper(II) ions are reduced to copper atoms that deposit on the copper rod at the cathode:

$$\text{Cathode half-reaction:}\quad Cu^{2+}(aq) + 2e^- \longrightarrow Cu(s)$$

The sum of these half-cell reactions is the cell reaction:

$$Zn(s) + Cu^{2+}(aq) \longrightarrow Zn^{2+}(aq) + Cu(s)$$

In this cell reaction, Zn^{2+} ions form in the zinc half-cell, and Cu^{2+} ions are reduced to Cu° in the copper half-cell. This accumulation of positive charge in the zinc half-cell and depletion of the positive charge in the copper half-cell disturbs the electrical neutrality in both half-cells. Current can pass through the external circuit only if an equivalent amount of charge can pass between the half-cell solutions to maintain the electrical neutrality of the solutions. The electrolyte solution in the U-tube (Figure 17.3), called the *salt bridge*, allows positive and negative ions to pass between half-cells to maintain electrical neutrality. The salt bridge does not allow bulk mixing of the two solutions. A salt bridge can be replaced by a porous plug, which serves the same purpose.

A galvanic cell in which the reactants in both half-cells are in their thermodynamic standard states, that is, 1 *M* solutions, gases at 1 atm external pressure, and a temperature of 25°C, is called a **standard cell**. A galvanic cell can be described as shown below for the cell illustrated in Figure 17.3:

$$Zn \mid Zn^{2+}(1\ M) \parallel Cu^{2+}(1\ M) \mid Cu$$

In this notation for a standard cell, the components of the two half-cells are separated by "double vertical lines" (‖), which represent the salt bridge or the porous partition between the half-cells. A single vertical line represents a *phase boundary*. The concentrations of the reactants (or their pressures, if they are gases) are shown in parentheses. By convention, the components of the anode half-reaction are indicated on the left side of the double vertical lines, and the components of the cathode half-reaction are written to the right of the double vertical lines.

In the example above, the components of the anode compartment are a zinc electrode and a 1 *M* solution of Zn^{2+} ions: $Zn \mid Zn^{2+}(1\ M)$, where the single vertical line represents the phase boundary between the solid electrode and the aqueous solution. The components of the cathode compartment are a copper electrode and 1 *M* Cu^{2+} ions: $Cu^{2+}(1\ M) \mid Cu$. Note that the electrodes are written to the extreme left and extreme right in this shorthand notation, and that the two solutions are "in contact" through the double vertical line, which represents the salt bridge or porous partition between the half-cells.

Standard Electrode Potentials

In the galvanic cell shown in Figure 17.3, a voltmeter in the external circuit indicates that the voltage generated by the cell is 1.10 volt (V) when the concentrations of the Zn^{2+} and Cu^{2+} ions are 1.00 *M* and the temperature is 25°C. The voltage is independent of the sizes of the electrodes or the amount of the

APPLICATIONS OF CHEMISTRY 17.1 Electrochemical Reactions in the Statue of Liberty and in Dental Fillings

Throughout history, we have suffered from our ignorance of basic electrochemical principles. For example, during the Middle Ages, our chemistry ancestors (alchemists) placed an iron rod into a blue solution of copper sulfate. They noticed that bright shiny copper plated out onto an iron rod and they thought that they had changed a base metal, iron, into copper. What actually happened was the redox reaction shown in Equation 1.

$$2Fe(s) + 3Cu^{2+}(aq) \longrightarrow 2Fe^{3+}(aq) + 3Cu(s) \quad (1)$$

This misunderstanding encouraged them to embark on a futile, 1000-year attempt to change base metals into gold.

Some hundred years ago, France presented our country with the Statue of Liberty (see photo). Unfortunately, the French did not anticipate the redox reaction shown in Equation 1 when they mounted the copper skin of the statue on iron support rods. Oxygen in the atmosphere oxidized the copper skin to produce copper ions. Then, since iron is more active than copper, the displacement shown in Equation 1 aided the corrosion of the support bars. As a result of this and other reactions, the statue needed refurbishing before we celebrated its one hundredth anniversary in 1986.

Sometimes dentists also overlook possible redox reactions when placing gold caps over teeth next to teeth with amalgam fillings. The amalgam in tooth fillings is an alloy of mercury, silver, tin, and copper. Atmospheric oxygen oxidizes some of the gold cap to gold ions. Since the metals in the amalgam are more active than gold, contact between the amalgam fillings and gold ions results in redox reactions such as the following.*

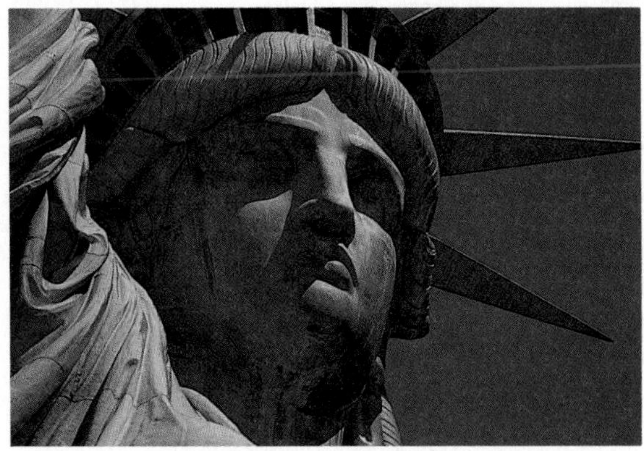

$$3Sn(s) + 2Au^{3+}(aq) \longrightarrow 3Sn^{2+}(aq) + 2Au(s) \quad (2)$$

As a result, the dental fillings dissolve and the patients are left with a constant metallic taste in their mouths.

These examples show that like our ancestors, we continue to experience unfortunate results because of a lack of understanding basic electrochemical principles.

* Equation 2 is oversimplified to illustrate more clearly the basic displacement of gold ions by metallic tin atoms. Actually, only complex ions of gold and tin can exist in aqueous solutions, not the simple cations that are shown.

Source: Ronald DeLorenzo, *Journal of Chemical Education*, May 1985, pages 424–425.

solution in each half-cell. It depends only on the components, the concentrations of the solutions in the half-cells, and the temperature. The voltage of a galvanic cell is the potential difference between the electrodes and is called the **cell potential**.

The magnitude and sign of a cell potential are a direct measure of the spontaneity of the redox reaction that occurs in the cell. A positive cell potential indicates that the cell reaction is spontaneous. The reverse of a spontaneous reaction is not spontaneous, and its cell potential is negative.

The potential of a galvanic *cell* can be measured with a voltmeter. But a single *half-cell* potential cannot be measured directly because one half-cell reaction cannot occur without a simultaneous reaction in another half-cell. However, *relative* half-cell potentials can be determined by arbitrarily designating one half-cell reaction and its potential as the standard to which other half-cell potentials are compared. By international agreement, this reference half-cell is the *standard hydrogen electrode* with the standard potential of 0.00 V. A standard hydrogen electrode consists of a platinum-coated electrode immersed in a 1.00 M H^+ ion solution in contact with hydrogen gas at 1.00 atm

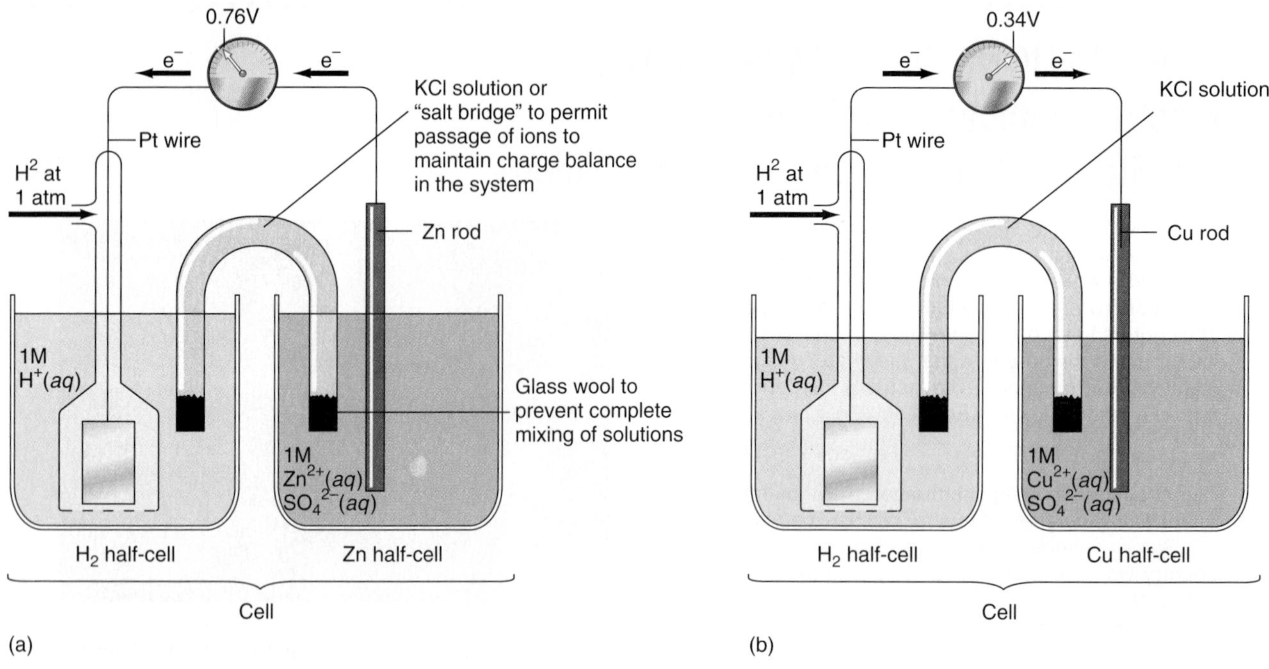

Figure 17.4

Measurement of standard half-cell potentials using a standard hydrogen half-cell.

pressure (Figure 17.4). By convention, the half-cell potential for the reduction of $H^+(aq)$ ions to H_2 gas, or the potential for the oxidation of H_2 to $H^+(aq)$ in the standard hydrogen half-cell, is defined as exactly 0.00 V:

$$\text{Reduction:}\quad 2H^+(aq,\ 1\ M) + 2e^- \longrightarrow H_2(g,\ 1\ \text{atm}) \quad E^\circ_{H^+/H_2} = 0.00\ \text{V}$$

$$\text{Oxidation:}\quad H_2(g,\ 1\ \text{atm}) \longrightarrow 2H^+(aq,\ 1\ M) + 2e^- \quad E^\circ_{H_2/H^+} = 0.00\ \text{V}$$

The symbol E represents electrode potential. The symbol $E°$ designates a **standard potential**; that is, a potential measured under standard conditions (1 M concentrations, 1 atm partial pressure, and 25°C).

When a standard hydrogen half-cell is combined with a standard zinc half-cell as shown in Figure 17.4a, the voltmeter indicates that the cell potential is 0.76 V. When the zinc half-cell is replaced by a standard copper half-cell (Figure 17.4b), the voltmeter indicator needle points in the opposite direction and indicates 0.34 V. The different directions of the voltmeter readings of the cells in Figure 17.4 indicate that the current flows in opposite directions in these cells.

In Figure 17.4a, current flows from the zinc electrode to the platinum electrode. We can predict this because we know that zinc is an active metal that can reduce hydrogen ions. Therefore, the zinc electrode is the anode at which zinc atoms are oxidized to zinc ions. Electrons released by this oxidation half-reaction travel through the external circuit and enter the platinum electrode at which hydrogen ions are reduced to hydrogen gas. The platinum electrode is the cathode because it is the electrode at which reduction occurs. In galvanic cells, electrons flow from the anode to the cathode.

The cell resulting from combining the hydrogen and zinc standard half-cells is described in shorthand form as

$$Zn \mid Zn^{2+}(1\ M) \parallel H^+(1\ M) \mid H_2(1\ \text{atm}) \mid Pt$$

Just as we can combine the half-cell reactions to obtain the cell reaction, we can combine the half-cell potentials to get the cell potential. Because the standard potential for the hydrogen half-cell is zero, the voltmeter reading is the standard potential for the zinc half-cell as shown below:

Anode:	$Zn(s) \longrightarrow Zn^{2+}(aq) + 2e^-$	$E^\circ_{Zn/Zn^{2+}} = ?$
Cathode:	$2H^+(aq) + 2e^- \longrightarrow H_2(g)$	$E^\circ_{H^+/H_2} = 0.00\ V$
Cell:	$Zn(s) + 2H^+(aq) \longrightarrow Zn^{2+}(aq) + H_2(g)$	$E^\circ_{cell} = 0.76\ V$ (from the voltmeter)

The cell potential is the sum of the two half-cell potentials:

$$E^\circ_{cell} = E^\circ_{Zn/Zn^{2+}} + E^\circ_{H^+/H_2}$$

From this expression,

$$E^\circ_{Zn/Zn^{2+}} = E^\circ_{cell} - E^\circ_{H^+/H_2} = 0.76\ V - 0.00\ V = 0.76\ V$$

The cell illustrated in Figure 17.4 can be described as

$$Pt \mid H_2(1\ atm) \mid H^+(1\ M) \parallel Cu^{2+}(1\ M) \mid Cu$$

Copper(II) ions are the ions of a relatively unreactive metal. Copper ions are easier to reduce than hydrogen ions. In this cell, copper(II) ions are reduced to copper metal at the copper electrode, and hydrogen gas is oxidized to hydrogen ions at the hydrogen electrode. The platinum electrode is the anode, and the copper electrode is the cathode. Electrons flow from the anode to the cathode. The half-cell and the cell reactions, and their corresponding potentials in the hydrogen–copper cell, are

Anode:	$H_2(g) \longrightarrow 2H^+(aq) + 2e^-$	$E^\circ_{H_2/H^+} = 0.00\ V$
Cathode:	$Cu^{2+}(aq) + 2e^- \longrightarrow Cu(s)$	$E^\circ_{Cu^{2+}/Cu} = ?$
Cell:	$H_2(g) + Cu^{2+}(aq) \longrightarrow 2H^+(aq) + Cu(s)$	$E^\circ_{cell} = 0.34\ V$ (from the voltmeter)

$$E^\circ_{cell} = E^\circ_{H_2/H^+} + E^\circ_{Cu^{2+}/Cu} = 0.34\ V = 0.00\ V + E^\circ_{Cu^{2+}/Cu}$$

$$E^\circ_{Cu^{2+}/Cu} = 0.34\ V$$

In Figure 17.4a we see that electrons flow from right to left, from the Zn electrode to the hydrogen electrode, whereas in Figure 17.4b, electrons flow from left to right, from the hydrogen electrode to the copper electrode. *In a galvanic cell, electrons flow from the half-cell where oxidation occurs to the half-cell where reduction occurs.*

From the two cells we described above we have obtained the standard half-cell potentials for two half-cells relative to the arbitrarily chosen potential for the standard hydrogen half-cell:

$$Zn(s) \longrightarrow Zn^{2+}(aq) + 2e^- \qquad E^\circ = 0.76\ V$$

$$Cu^{2+}(aq) + 2e^- \longrightarrow Cu(s) \qquad E^\circ = 0.34\ V$$

The *standard half-cell potential for the reverse of a reaction is equal in magnitude but opposite in sign*:

$$Zn(s) \longrightarrow Zn^{2+}(aq) + 2e^- \qquad E^\circ = 0.76\ V$$

$$Zn^{2+}(aq) + 2e^- \longrightarrow Zn(s) \qquad E^\circ = -0.76\ V$$

Standard half-cell potentials are also called standard electrode potentials. By international agreement, standard electrode potentials are tabulated as standard *reduction potentials* for reduction half-reactions. Table 17.1 lists the standard reduction potentials for various half-reactions in order of decreasing potential.

Table 17.1

Standard Reduction Potentials in Water at 25°C

Electrode Reaction	Standard Potential, $E°$ (V)
$F_2(g) + 2e^- \longrightarrow 2F^-(aq)$	2.85
$H_2O_2(aq) + 2H^+(aq) + 2e^- \longrightarrow 2H_2O(l)$	1.776
$PbO_2(s) + SO_4^{2-}(aq) + 4H^+(aq) + 2e^- \longrightarrow PbSO_4(s) + 2H_2O(l)$	1.685
$MnO_4^-(aq) + 8H^+(aq) + 5e^- \longrightarrow Mn^{2+}(aq) + 4H_2O(l)$	1.491
$PbO_2(s) + 4H^+(aq) + 2e^- \longrightarrow Pb^{2+}(aq) + 2H_2O(l)$	1.46
$Cl_2(g) + 2e^- \longrightarrow 2Cl^-(aq)$	1.3583
$Cr_2O_7^{2-}(aq) + 14H^+(aq) + 6e^- \longrightarrow 2Cr^{3+}(aq) + 7H_2O(l)$	1.33
$O_2(g) + 4H^+(aq) + 4e^- \longrightarrow 2H_2O(l)$	1.229
$MnO_2(s) + 4H^+(aq) + 2e^- \longrightarrow Mn^{2+}(aq) + 2H_2O(l)$	1.208
$Br_2(l) + 2e^- \longrightarrow 2Br^-(aq)$	1.065
$NO_3^-(aq) + 4H^+(aq) + 3e^- \longrightarrow NO(g) + 2H_2O(l)$	0.96
$2Hg^{2+}(aq) + 2e^- \longrightarrow Hg_2^{2+}(aq)$	0.905
$Hg^{2+}(aq) + 2e^- \longrightarrow Hg(l)$	0.851
$Ag^+(aq) + e^- \longrightarrow Ag(s)$	0.7996
$Fe^{3+}(aq) + e^- \longrightarrow Fe^{2+}(aq)$	0.770
$MnO_4^-(aq) + 2H_2O(l) + 3e^- \longrightarrow MnO_2(s) + 4OH^-(aq)$[a]	0.588
$I_2(s) + 2e^- \longrightarrow 2I^-(aq)$	0.535
$O_2(g) + 2H_2O(l) + 4e^- \longrightarrow 4OH^-(aq)$[a]	0.401
$Cu^{2+}(aq) + 2e^- \longrightarrow Cu(s)$	0.3402
$SO_4^{2-}(aq) + 4H^+(aq) + 2e^- \longrightarrow SO_2(g) + 2H_2O(l)$	0.20
$Pb^{2+}(aq) + 2e^- \longrightarrow Pb(s)$	−0.1263
$Sn^{2+}(aq) + 2e^- \longrightarrow Sn(s)$	−0.1364
$O_2(g) + 2H_2O(l) + 2e^- \longrightarrow H_2O_2(aq) + 2OH^-(aq)$[a]	−0.146
$Ni^{2+}(aq) + 2e^- \longrightarrow Ni(s)$	−0.23
$PbSO_4(s) + 2e^- \longrightarrow Pb(s) + SO_4^{2-}(aq)$	−0.356
$Fe^{2+}(aq) + 2e^- \longrightarrow Fe(s)$	−0.409
$S(s) + 2e^- \longrightarrow S^{2-}(aq)$	−0.508
$Cr^{3+}(aq) + 3e^- \longrightarrow Cr(s)$	−0.74
$Zn^{2+}(aq) + 2e^- \longrightarrow Zn(s)$	−0.7628
$2H_2O(l) + 2e^- \longrightarrow H_2(g) + 2OH^-(aq)$[a]	−0.8277
$Mn^{2+}(aq) + 2e^- \longrightarrow Mn(s)$	−1.029
$Al^{3+}(aq) + 3e^- \longrightarrow Al(s)$	−1.66
$Be^{2+}(aq) + 2e^- \longrightarrow Be(s)$	−1.70
$Mg^{2+}(aq) + 2e^- \longrightarrow Mg(s)$	−2.375
$Na^+(aq) + e^- \longrightarrow Na(s)$	−2.7109
$Ca^{2+}(aq) + 2e^- \longrightarrow Ca(s)$	−2.76
$Sr^{2+}(aq) + 2e^- \longrightarrow Sr(s)$	−2.89
$Ba^{2+}(aq) + 2e^- \longrightarrow Ba(s)$	−2.90
$Cs^+(aq) + e^- \longrightarrow Cs(s)$	−2.923
$K^+(aq) + e^- \longrightarrow K(s)$	−2.924
$Rb^+(aq) + e^- \longrightarrow Rb(s)$	−2.925
$Li^+(aq) + e^- \longrightarrow Li(s)$	−3.045

Increasing oxidizing ability (↑, upward along the left side of the table)

Increasing reducing ability (↓, downward along the right side of the table)

[a] Half-reaction occurs in basic solution.

Substances at the top of Table 17.1 have the highest reduction potentials, they are easy to reduce, and they are therefore good oxidizing agents. Among them are such strong oxidizing agents as fluorine gas, hydrogen peroxide, permanganate ions, chlorine gas, dichromate ions, oxygen gas, and nitrate ions. The reduced species of good oxidizing agents are difficult to oxidize. They are therefore poor reducing agents.

At the bottom of Table 17.1 we find weak oxidizing agents such as alkali and alkaline earth metal ions. Their reduced counterparts, however, the alkali metal atoms and the alkaline earth metal atoms are easy to oxidize and they are therefore good reducing agents.

Example 17.1

Selecting the Best Oxidizing Agent

Using Table 17.1, determine which of the following species is the best oxidizing agent in aqueous solution under standard conditions: $MnO_2(s)$, $Br_2(l)$, $Cu^{2+}(aq)$.

SOLUTION: The best oxidizing agent has the highest reduction potential, which means that it is easiest to reduce. The standard reduction potentials for the substances in this problem are:

$$MnO_2(s) + 4H^+(aq) + 2e^- \longrightarrow Mn^{2+}(aq) + 2H_2O(l) \qquad E^\circ = 1.208\ \text{V}$$

$$Br_2(l) + 2e^- \longrightarrow 2Br^-(aq) \qquad E^\circ = 1.065\ \text{V}$$

$$Cu^{2+}(aq) + 2e^- \longrightarrow Cu(s) \qquad E^\circ = 0.3402\ \text{V}$$

MnO_2 has the highest reduction potential and is, therefore, the best oxidizing agent, followed by Br_2 and Cu^{2+} in that order.

Practice Problem 17.1: Use Table 17.1 to determine whether $S^{2-}(aq)$, $Fe^{2+}(aq)$, or $Sn(s)$ is the strongest reducing agent in aqueous solution under standard conditions.

17.3 Redox Reactions and Standard Potentials

Cell Reactions and Cell Potentials

We recall that we can divide the net reaction of a galvanic cell into two half-reactions. The oxidation half-reaction occurs at the anode, and the reduction half-reaction occurs at the cathode. When one half-cell reaction is written as oxidation and the other as reduction, the sum of the half-cell reactions is the cell reaction, and the sum of the half-cell potentials is the cell potential:

$$E^\circ_{\text{cell}} = E^\circ_{\text{anode}} + E^\circ_{\text{cathode}} \tag{17.1}$$

However, tables of standard half-cell potentials are written as reduction potentials. *When the two half-reactions in a galvanic cell are written as reduction half-reactions, the one having the lower reduction potential is the anode half-reaction that represents oxidation*. Therefore, *when converting a reduction half-reaction to an oxidation half-reaction, the sign of the reduction potential must be reversed to obtain the potential for the oxidation half-reaction*. The following examples illustrate how these ideas are applied.

Example 17.2

Constructing a Galvanic Cell, Calculating the Standard Potential of the Cell from Standard Reduction Potentials, Identifying the Anode, and Determining the Direction of Electron Flow

The standard reduction potentials for the following half-reactions are:

$$Mn^{2+}(aq) + 2e^- \longrightarrow Mn(s) \qquad E° = -1.029 \text{ V}$$

$$Ni^{2+}(aq) + 2e^- \longrightarrow Ni(s) \qquad E° = -0.23 \text{ V}$$

How can a galvanic cell be constructed of manganese and nickel rods, and of Mn^{2+} and Ni^{2+} ion solutions? Which rod acts as the anode of the cell? Calculate the standard potential of the cell. What is the direction of electron flow in the cell?

SOLUTION: A galvanic cell consists of two half-cells. Each half-cell contains oxidized and reduced forms of a given substance. In this case, one half-cell contains the reduced form of manganese, the manganese rod, and the oxidized form, the manganese(II) ion solution. The other half-cell contains the nickel rod in the nickel(II) ion solution.

The manganese rod acts as the anode because the reduction potential for manganese(II) ions is lower than the reduction potential of nickel(II) ions (Table 17.1). This also means that manganese is more easily oxidized to its ions than nickel. Therefore, the manganese rod is oxidized at the anode, and nickel(II) ions are reduced at the cathode—the nickel rod:

$$\begin{array}{lll} \text{Anode reaction:} & Mn(s) \longrightarrow Mn^{2+}(aq) + 2e^- & E°_{Mn/Mn^{2+}} = 1.029 \text{ V} \\ \text{Cathode reaction:} & Ni^{2+}(aq) + 2e^- \longrightarrow Ni(s) & E°_{Ni^{2+}/Ni} = -0.23 \text{ V} \\ \hline \text{Cell reaction:} & Mn(s) + Ni^{2+}(aq) \longrightarrow Mn^{2+}(aq) + Ni(s) & \end{array}$$

$$E°_{cell} = E°_{Mn/Mn^{2+}} + E°_{Ni^{2+}/Ni}$$

$$= 1.029 \text{ V} + (-0.23 \text{ V}) = 0.80 \text{ V}$$

Note that the manganese half-reaction is written as an oxidation half-reaction. The sign of the potential for the oxidation half-reaction is the reverse of the sign of the reduction potential obtained from Table 17.1 for the reduction half-reaction. Electrons flow from the anode to the cathode, from the manganese rod to the nickel rod.

Practice Problem 17.2: The standard reduction potential for $Ni^{2+}(aq) + 2e^- \longrightarrow Ni(s)$ is -0.23 V and for $Cu^{2+}(aq) + 2e^- \longrightarrow Cu(s)$ is 0.34 V. Describe and draw a diagram to illustrate how you would construct a galvanic cell using a nickel rod, a copper rod, a solution of nickel ions, and a solution of copper(II) ions. Which is the anode of the cell? What is the direction of electron flow? What is the standard potential of the cell?

The magnitude of the cell potential is an index of the spontaneity of the cell reaction. *A positive cell potential indicates that the cell reaction is spontaneous. When the cell potential is zero, the cell reaction is at equilibrium. A negative cell potential indicates that the reaction is not spontaneous in the direction written*; however, the reverse reaction is spontaneous.

When half-reactions are multiplied to balance an equation for a redox reaction, their potentials do not change. The cell potential does not depend on the number of ions, the size of the cell, or the sizes of the electrodes. It depends *only* on the constituents, the concentrations of the ions, the temperature, and the partial pressure of a gas, if present. These principles are illustrated in Example 17.3.

Example 17.3

Describing a Galvanic Cell, Calculating the Standard Potential of a Cell, and Predicting Whether the Cell Reaction Is Spontaneous

Use Table 17.1 to calculate the standard potential for the cell that uses the reaction

$$2MnO_4^-(aq) + 10Cl^-(aq) + 16H^+(aq) \longrightarrow 2Mn^{2+}(aq) + 5Cl_2(g) + 8H_2O(l)$$

Is the reaction spontaneous as written? Describe a cell that generates electricity by this reaction. What is the direction of electron flow? Which electrode is the anode and which is the cathode?

SOLUTION: First, we write the two half-reactions and their standard potentials. Then we add the half-reactions and their standard potentials to obtain the cell reaction and its potential. In this case, Cl_2 has a lower reduction potential than MnO_4^-. Therefore, Cl^- ions are oxidized at the anode, and MnO_4^- ions are reduced at the cathode:

$$\text{Cathode:}\quad 2[MnO_4^-(aq) + 8H^+(aq) + 5e^- \longrightarrow Mn^{2+}(aq) + 4H_2O(l)] \qquad E^\circ_{MnO_4^-/Mn^{2+}} = 1.491\text{ V}$$

$$\text{Anode:}\quad 5[2Cl^-(aq) \longrightarrow Cl_2(g) + 2e^-] \qquad E^\circ_{Cl^-/Cl_2} = -1.3583\text{ V}$$

$$\text{Cell:}\quad 2MnO_4^-(aq) + 10Cl^-(aq) + 16H^+(aq) \longrightarrow 2Mn^{2+}(aq) + 5Cl_2(g) + 8H_2O(l)$$

$$E^\circ_{cell} = E^\circ_{MnO_4^-/Mn^{2+}} + E^\circ_{Cl^-/Cl_2} = 1.491\text{ V} + (-1.3583\text{ V}) = 0.133\text{ V}$$

Because the standard potential is positive, the reaction is spontaneous as written, provided that all reactants are in their standard states.

To generate electricity, the cell must be constructed of two half-cells connected by a salt bridge or a porous partition. One half-cell contains permanganate ions, hydrogen ions, and manganese(II) ions. The other half-cell contains chloride ions and chlorine gas. An inert electrode, such as platinum, is needed in each half-cell. The cell can be described as follows:

$$Pt \mid Cl^- \mid Cl_2 \parallel Mn^{2+}, MnO_4^-, H^+ \mid Pt$$

In this cell, electrons flow from the Cl_2/Cl^- half-cell, where the Pt electrode serves as an anode, to the MnO_4^-/Mn^{2+} half-cell, in which the Pt electrode is the cathode.

Practice Problem 17.3: Use Table 17.1 to calculate the standard cell potential for the reaction $F_2(g) + 2Fe^{2+}(aq) \longrightarrow 2F^-(aq) + 2Fe^{3+}(aq)$. Is the reaction spontaneous? Describe and draw a diagram to illustrate how you would construct this cell. What is the direction of electron flow in this cell? Write the anode and cathode half-reactions of the cell.

Practice Problem 17.4: Using only the substances listed in Table 17.1, describe the galvanic cell that would give the highest cell potential. Would this cell be practical? Explain.

Prediction of Redox Reactions

Often we would like to know whether a redox reaction will occur when various substances are mixed. We can often make such predictions from the relevant standard electrode potential data, assuming that the reactants are in their standard states. We can look up standard reduction potentials to calculate the standard potential for the net redox reaction.

If the standard potential for a redox reaction is positive, the reaction is spontaneous under standard conditions in the direction written. If the standard potential is negative, the *reverse* reaction is spontaneous under standard conditions.

Example 17.4

Predicting Whether a Redox Reaction Occurs

A solution which is 1.0 *M* in iodide ions is treated with chlorine gas at 1.0 atm. Will iodide ions be oxidized to iodine? Using the necessary data from Table 17.1, write the two half-reactions and the overall reaction, if any.

SOLUTION: To check whether this reaction occurs, we write the relevant half-reactions from Table 17.1 and their corresponding standard potentials. If the sum of the standard potentials is positive, the reaction occurs.

$$2I^-(aq) \longrightarrow I_2(s) + 2e^- \qquad E° = -0.535\ V$$

$$Cl_2(g) + 2e^- \longrightarrow 2Cl^-(aq) \qquad E° = 1.3583\ V$$

$$Cl_2(g) + 2I^-(aq) \longrightarrow 2Cl^-(aq) + I_2(s)$$

$$E° = -0.535\ V + 1.3583\ V = 0.823\ V$$

The positive potential for the overall reaction shows that the reaction occurs spontaneously.

Practice Problem 17.5: Liquid bromine is added to a solution that is 1.0 *M* each in Cl^-, Br^-, and I^- ions. Will any reaction occur? Write equations for all reactions that occur.

Corrosion

Corrosion consists of one or more spontaneous redox reactions in which a metal reacts with oxygen and moisture in the atmosphere. For example, several redox reactions occur when iron corrodes or "rusts." A principal reaction of this corrosion and its half-reactions are

$$\begin{array}{r} 2[Fe(s) \longrightarrow Fe^{2+}(aq) + 2e^-] \\ O_2(g) + 2H_2O(l) + 4e^- \longrightarrow 4OH^-(aq) \\ \hline 2Fe(s) + O_2(g) + 2H_2O(l) \longrightarrow 2Fe(OH)_2(s) \end{array}$$

The Fe^{2+} and the OH^- ions produced in the oxidation and reduction half-reactions precipitate as $Fe(OH)_2$. The reduction half-reaction as written occurs in neutral or alkaline medium. In acid medium, the half-reaction is

$$O_2(g) + 4H^+(aq) + 4e^- \longrightarrow 2H_2O(l)$$

Further reaction of $Fe(OH)_2$ with oxygen and water produces iron(III) hydroxide, which decomposes to hydrated iron(III) oxide—the principal ingredient of rust:

$$4Fe(OH)_2(s) + O_2(g) + 2H_2O(l) \longrightarrow 4Fe(OH)_3(s)$$

$$2Fe(OH)_3(s) \longrightarrow \underset{\text{rust}}{Fe_2O_3(H_2O)(s)} + 2H_2O(l)$$

The corrosion of iron can be prevented by painting the metal so that it does not come into contact with oxygen and moisture. The metal can also be covered with a thin layer of another metal more difficult to oxidize than iron. For example, "tin cans" are made by applying a thin layer of tin over steel. The standard potential for oxidation of tin (0.14 V) is lower than that of iron (0.41 V), so tin does not corrode as readily as iron.

The corrosion of iron can also be prevented by covering it with a thin layer of a metal that is easier to oxidize than iron. Iron or steel is covered with a layer of zinc in a process called *galvanization.* The standard potential for oxidation of zinc (0.76 V) is higher than that of iron. Zinc is therefore easier to oxidize than iron. Even when the zinc coating on a galvanized iron is scratched, the iron is still protected from corrosion because the zinc acts as an anode and iron as a cathode at which oxygen is reduced without affecting the surface of iron.

Using the same principle of making iron the cathode, underground iron storage tanks, pipes, and hulls of ships are protected from corrosion by connecting them to bars of magnesium or zinc as the anode material (Figure 17.5).

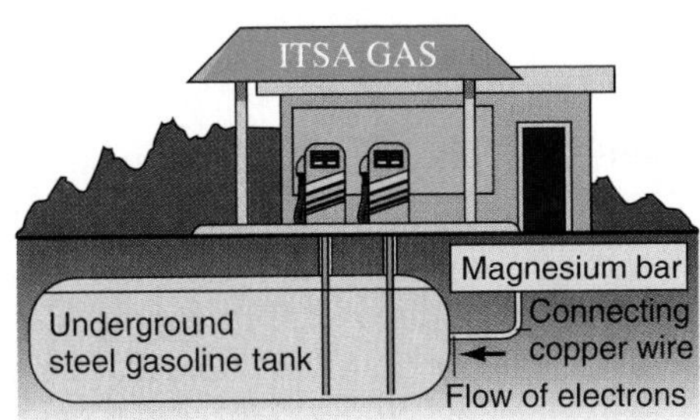

Figure 17.5

Cathodic protection of an underground steel tank using magnesium bars as anode material.

Figure 17.6

Protection of automobile bumpers from corrosion by chromium plating, which also provides visual appeal.

As these bars corrode, they are easier to replace than iron tanks, pipes, and hulls. This technique is known as cathodic protection. The chromium plate on an automobile bumper serves the same purpose and provides visual appeal as well (Figure 17.6).

A Biochemical Application of Electrochemistry

Redox reactions occur in all living cells. For example, photosynthesis in plants is a redox reaction:

$$6CO_2(g) + 6H_2O(l) \xrightarrow{\text{chlorophyll}} C_6H_{12}O_6(s) + 6O_2(g)$$

In animals, redox reactions provide energy during respiration. Respiration is a process in which an organism oxidizes food molecules to provide cells with energy. The net reaction that occurs during respiration is the oxidation of a substance called nicotinamide adenine dinucleotide (NADH) by oxygen. In this reaction, NADH is oxidized to NAD^+ and oxygen is reduced to water:

$$NADH(aq) \longrightarrow NAD^+(aq) + H^+(aq) + 2e^- \qquad E° = 0.32\ V$$
$$\tfrac{1}{2}O_2(g) + 2H^+(aq) + 2e^- \longrightarrow H_2O(l) \qquad E° = 0.82\ V$$
$$NADH(aq) + H^+(aq) + \tfrac{1}{2}O_2(g) \longrightarrow NAD^+(aq) + H_2O(l) \qquad E° = 1.14\ V$$

Since the potential of this reaction is positive, the reaction is spontaneous, and it releases energy that is used for other cell reactions which are not spontaneous.

17.4 Cell Potential, Concentration, and Temperature

The Nernst Equation and Its Applications

Thus far we have considered standard cells. However, galvanic cells are often operated under nonstandard conditions in which the concentrations of reactants are not 1 M, or the gas pressures are not 1 atm. Also, galvanic cells may be operated at temperatures other than the standard temperature (298 K).

The relationship between the potential of a galvanic cell, the concentrations of reactants, and temperature was developed by the German chemist Walther Hermann Nernst (1864–1941):

$$E = E° - \frac{RT}{nF} \ln Q \qquad (17.2)$$

APPLICATIONS OF CHEMISTRY 17.2
Turning the Human Body into a Battery

The heart has its own natural pacemaker that sends nerve impulses (pulses of electric current) throughout the heart approximately 72 times per minute. These electric pulses cause your heart muscles to contract (beat), which pumps blood through the body. The fibers that carry the nerve impulses can be damaged by disease, drugs, heart attacks, and surgery. When these heart fibers are damaged, the heart may run too slowly, stop temporarily, or stop altogether. To correct this condition, artificial heart pacemakers (see figure below) are surgically inserted in the human body. A pacemaker (pacer) is a battery-driven device that sends an electric current (pulse) to the heart about 72 times per minute. Over 300,000 Americans are now wearing artificial pacemakers with an additional 30,000 pacemakers installed each year.

Yearly operations used to be necessary to replace the pacemaker's batteries. Today, pacemakers use improved batteries that last much longer, but even these must be replaced eventually.

It would be very desirable to develop a permanent battery to run pacemakers. Some scientists began working on ways of converting the human body itself into a battery (voltaic cell) to power artificial pacemakers.

Several methods for using the human body as a voltaic cell have been suggested. One of these is to insert platinum and zinc electrodes into the human body as diagrammed in the figure below. The pacemaker and the electrodes would be worn internally. This "body battery" could easily generate the small amount of current (5×10^{-5} ampere) that is required by most pacemakers. This "body battery" has been tested on animals for periods exceeding 4 months without noticeable problems.

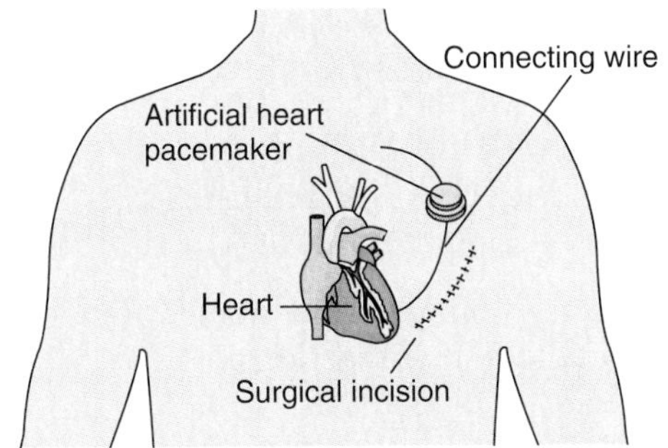

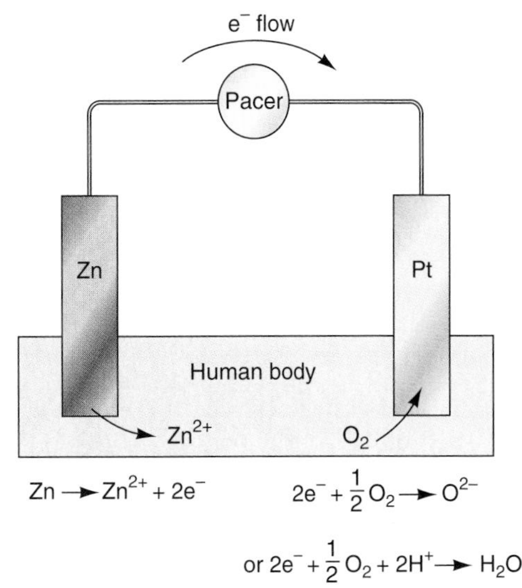

Source: Ronald DeLorenzo, *Problem Solving in General Chemistry*, 2nd ed., Wm. C. Brown, Publishers, Dubuque, Iowa, 1993, pages 406–408.

Equation 17.2 is called the **Nernst equation**. In this equation, E is the cell potential, $E°$ the standard cell potential, R the gas constant 8.31441 J mol^{-1} K^{-1}, T the absolute temperature, n the number of electrons that appears in either half-reaction for a balanced equation for the cell reaction, and F is Faraday's constant, 9.648456×10^4 coulombs (C) per mole of electrons. Q is the reaction quotient we discussed in Chapter 13. The reaction quotient is written in the form of the equilibrium constant for the redox reaction, but the molar concentrations or partial pressures of reactants and products can have any values.

At 25°C, the ratio

$$RT/F = \frac{(8.31441 \text{ J mol}^{-1} \text{ K}^{-1})(298.15 \text{ K})}{9.648456 \times 10^4 \text{ C mol}^{-1}}$$

$$= 0.025693 \text{ J C}^{-1} = 0.025693 \text{ V}$$

We can replace the "ln Q" in Equation 17.2 by "log Q" if we multiply the ratio RT/F by 2.303, because 2.303 log Q equals ln Q. Thus, (2.303)(0.025693 V) = 0.05917 V, which to three significant figures is 0.0592 V. Using this value, a common form of the Nernst equation is

$$E = E° - \frac{0.0592 \text{ V}}{n} \log Q \quad (17.3)$$

Equation 17.3 relates the cell potential to concentrations or partial pressures (included in the factor Q) at 298 K. Some applications of the Nernst equation are illustrated below.

Example 17.5

An Application of the Nernst Equation

Calculate the potential at 298 K of the cell

$$\text{Zn} \mid \text{Zn}^{2+}(1.0 \times 10^{-10}\ M) \parallel \text{Cu}^{2+}(10\ M) \mid \text{Cu}$$

SOLUTION: In this cell the zinc electrode is the anode and the copper electrode is the cathode. The anode and cathode half-reactions, the cell reaction, and their standard potentials are

Anode:	$Zn(s) \longrightarrow Zn^{2+}(aq) + 2e^-$	$E° = 0.76$ V
Cathode:	$Cu^{2+}(aq) + 2e^- \longrightarrow Cu(s)$	$E° = 0.34$ V
Cell:	$Zn(s) + Cu^{2+}(aq) \longrightarrow Zn^{2+}(aq) + Cu(s)$	$E°_{cell} = 1.10$ V

The value of n in this reaction is 2, which is the number of electrons appearing in both anode and cathode half-reactions.

$$Q = \frac{[Zn^{2+}]}{[Cu^{2+}]} = \frac{1.0 \times 10^{-10}\ M}{10\ M} = 1.0 \times 10^{-11}$$

Remember that the concentrations of pure solids and liquids are constant and are not included in the expression of K or Q. Applying the Nernst equation (Equation 17.3), we obtain

$$E = 1.10 \text{ V} - \frac{0.0592 \text{ V}}{2} \log(1.0 \times 10^{-11}) = 1.43 \text{ V}$$

We can see that when Q is less than 1, log Q is negative, the term $[(-0.0592 \text{ V}/n) \log Q]$ is positive, and the value of E is larger than $E°$.

Practice Problem 17.6: Calculate E at 298 K for the reaction

$$Pb(s) + 2Ag^+(aq) \longrightarrow Pb^{2+}(aq) + 2Ag(s)$$

when $[Ag^+] = 1.0 \times 10^{-5}\ M$ and $[Pb^{2+}] = 2.0\ M$.

Example 17.6

Making a Nonspontaneous Reaction Spontaneous by Changing Concentrations

A laboratory method for preparing chlorine gas is the reaction of manganese(IV) oxide with hydrochloric acid:

$$MnO_2(s) + 4H^+(aq) + 2Cl^-(aq) \longrightarrow Cl_2(g) + Mn^{2+}(aq) + 2H_2O(l)$$

Is this reaction spontaneous under standard conditions? What is the potential for the reaction at the instant when $[H^+] = 10\ M$, $[Cl^-] = 10\ M$, $P_{Cl_2} = 1.0$ atm, and $[Mn^{2+}] = 1.0 \times 10^{-10}\ M$? Is the reaction spontaneous under these conditions?

SOLUTION: The standard potential, $E°$, for the reaction is determined in the usual way by adding the two half-reactions and their standard potentials:

$$MnO_2(s) + 4H^+(aq) + 2e^- \longrightarrow Mn^{2+}(aq) + 2H_2O(l) \qquad E° = 1.208\text{ V}$$

$$2Cl^-(aq) \longrightarrow Cl_2(g) + 2e^- \qquad E° = -1.3583\text{ V}$$

$$MnO_2(s) + 2Cl^-(aq) + 4H^+(aq) \longrightarrow Mn^{2+}(aq) + Cl_2(g) + 2H_2O(l) \qquad E° = -0.150\text{ V}$$

The negative value of $E°$ indicates that the reaction is not spontaneous under standard conditions. Therefore, chlorine gas under 1.0 atm Cl_2 pressure cannot be prepared by using 1.0 M HCl for reaction with MnO_2 in 1.0 M Mn^{2+}.

From the data given, the reaction quotient is

$$Q = \frac{[Mn^{2+}]P_{Cl_2}}{[Cl^-]^2[H^+]^4} = \frac{(1.0 \times 10^{-10})(1.0)}{(10)^2(10)^4} = 1.0 \times 10^{-16}$$

$$E = -0.150\text{ V} - \frac{0.0592\text{ V}}{2}\log(1.0 \times 10^{-16}) = 0.32\text{ V}$$

The positive potential under these nonstandard conditions shows that the reaction is spontaneous. Thus, a change in the concentrations of reactants can make a nonspontaneous reaction spontaneous, and vice versa.

Determination of K_{sp}

The Nernst equation can be used to determine K_{sp} values of sparingly soluble salts. For example, the K_{sp} for $PbSO_4$ can be determined by constructing a cell in which one of the half-cells has a Pb electrode in a solution of Pb^{2+} ions in equilibrium with a 1.00 M solution of SO_4^{2-} ions. The other half-cell can have a Cu electrode in a 1.00 M solution of $CuSO_4$. Such a cell can be described as follows:

$$Pb \mid Pb^{2+}(?\ M) \parallel Cu^{2+}(1.00\ M) \mid Cu$$

The cell reaction is

$$Pb(s) + Cu^{2+}(aq, 1.00\ M) \longrightarrow Pb^{2+}(aq, x\ M) + Cu(s)$$

The SO_4^{2-} ions do not appear in the cell description because they do not take part in the electrochemical reaction. However, the Pb^{2+} ion concentration in the cell is determined by the concentration of the SO_4^{2-} ions present (1.00 *M*). The cell potential, *E*, for the cell under the conditions as described, and at 25°C, is 0.697 V. The value for $E°$ based on the data from Table 17.1 is 0.4665 V. Since *n* for the reaction as written is 2, we can write

$$E = E° - \frac{0.0592\text{ V}}{2}\log\frac{[Pb^{2+}]}{[Cu^{2+}]} \quad \text{or} \quad E = E° - 0.0296\text{ V}\log\frac{x}{1.00}$$

Solving this equation for log *x* and substituting the known values for *E* and $E°$, we obtain

$$\log x = -\frac{E - E°}{0.0296\text{ V}} = -\frac{0.697\text{ V} - 0.4665\text{ V}}{0.0296\text{ V}} = -7.787$$

$$x = \text{antilog of } -7.787 = 1.63 \times 10^{-8}$$

The value of *x* equals the molar concentration of Pb^{2+} ions in the cell at equilibrium with 1.00 *M* solution of SO_4^{2-} ions. The value of K_{sp} for $PbSO_4$ is

$$[Pb^{2+}][SO_4^{2-}] = (1.63 \times 10^{-8})(1.00) = 1.63 \times 10^{-8} = K_{sp}$$

This value is in good agreement with the K_{sp} value for $PbSO_4$ in Table 16.1.

Electrochemical methods of the type described above can also be used to determine very low ion concentrations in various solutions, such as a solution of AgI. This is a practical method of measuring concentrations in the range of 10^{-5} *M* or less.

17.5 Commercial Batteries

Many galvanic cells are commercially available for the production of electric energy. Some of these have been on the market for decades and we take them for granted. Others are newly developed and have been used only for special purposes. The older, established category includes the automobile storage battery and the dry cell (or Leclanché cell). The newer types include fuel cells and solar batteries.

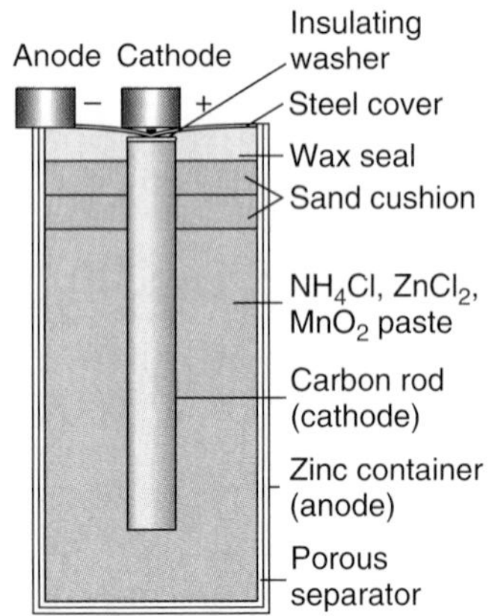

Figure 17.7

A cross section of a typical dry cell.

Primary Cells

Cells that cannot be recharged are called **primary cells**. Once the chemicals in a primary cell are consumed, the cell must be discarded or recycled. Primary cells that do not contain liquid components are called **dry cells**. In the common dry cell used in flashlights, the anode is a zinc container, and the cathode is a carbon rod in the center of the cell (Figure 17.7). The zinc container is lined inside with a porous paper to separate it from the other materials of the cell. The space between the electrodes is filled with a moist paste of NH_4Cl, MnO_2, and $ZnCl_2$.

When a dry cell delivers current, the zinc electrode is the negative electrode (the anode), where zinc is oxidized to zinc ions. The carbon rod is the cathode, at which MnO_2 is reduced to several manganese(III)-containing species. The cell reactions are

$$\begin{array}{rl} \text{Anode:} & Zn(s) \longrightarrow Zn^{2+}(aq) + 2e^- \\ \text{Cathode:} & 2MnO_2(s) + 2NH_4^+(aq) + 2e^- \longrightarrow \\ & \qquad Mn_2O_3(s) + H_2O(l) + 2NH_3(aq) \\ \hline \text{Cell:} & Zn(s) + 2MnO_2(s) + 2NH_4^+(aq) \longrightarrow \\ & \qquad Zn^{2+}(aq) + Mn_2O_3(s) + 2NH_3(aq) + H_2O(l) \end{array}$$

Zinc ions combine to some extent with ammonia to form the ammine complex ions of zinc, $Zn(NH_3)_4^{2+}$. As the cell discharges, the zinc anode is consumed. Since the anode is also the container for the cell, it is protected by a steel case. The potential of this cell is about 1.5 V.

Other primary cells include the mercury cell and the silver oxide cell. These cells can be made very small, which makes them ideal for use in electronic watches, light meters, pacemakers, hearing aids, and pocket calculators. A mercury battery has a zinc anode and mercury(II) oxide cathode in contact with a strongly alkaline solution which is actually a wet paste of KOH and $Zn(OH)_2$ (Figure 17.8).

The mercury cell reactions are

$$\begin{array}{rl} \text{Anode:} & Zn(s) + 2OH^-(aq) \longrightarrow ZnO(s) + H_2O(l) + 2e^- \\ \text{Cathode:} & HgO(s) + H_2O(l) + 2e^- \longrightarrow Hg(l) + 2OH^-(aq) \\ \hline \text{Cell:} & Zn(s) + HgO(s) \longrightarrow ZnO(s) + Hg(l) \end{array}$$

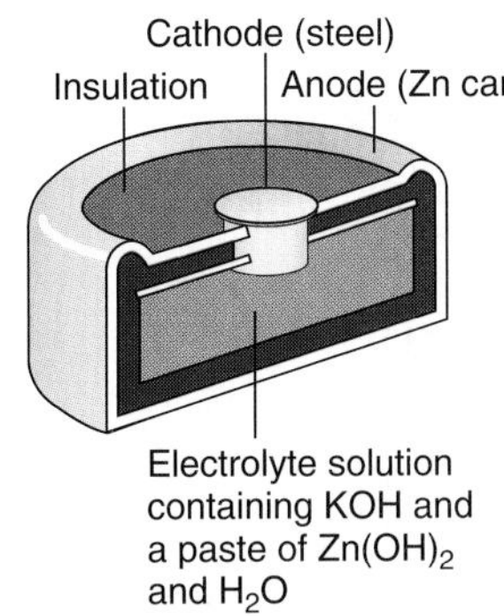

Figure 17.8

A cross section of a mercury battery.

A silver oxide battery has a zinc anode like the mercury battery, but the mercury(II) oxide cathode is replaced by silver oxide. Both mercury and silver oxide batteries deliver 1.3 to 1.5 V.

Secondary Cells or Storage Batteries

A *cell that can be recharged* by passing an electric current from an external source through it to reverse the cell reaction is called a **storage cell**, an accumulator, or a **secondary cell**. Perhaps the most common storage cell is the lead storage cell used in the automobile battery. An automobile battery is recharged by a generator or alternator when the engine is running.

A 12-V automobile storage battery consists of six identical cells (Figure 17.9) connected in series. When the cell discharges, a plate of lead metal acts as the anode and lead(IV) oxide acts as the cathode. Since both the anode and

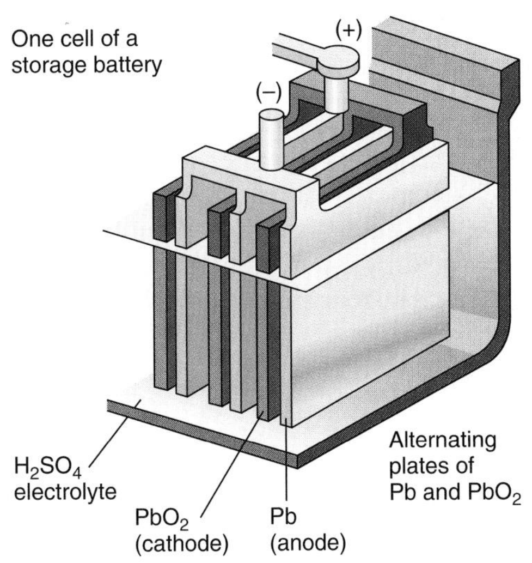

Figure 17.9

One cell of a lead storage battery consisting of a three-plate anode and a three-plate cathode. The more plates per cell, the larger the current. One cell delivers approximately 2 volts. A 12 volt battery consists of six cells connected in series.

cathode contain lead, an automobile battery is often called a lead storage battery. The electrolyte in this battery is sulfuric acid solution of about 30 to 38 percent by mass sulfuric acid.

When a lead storage battery operates, lead metal is oxidized to lead(II) ions, which precipitate with sulfate ions at the electrode surface as lead sulfate:

$$Pb(s) + SO_4^{2-}(aq) \longrightarrow PbSO_4(s) + 2e^- \qquad E° = 0.356 \text{ V}$$

At the cathode, PbO_2 is reduced to lead(II) ions, which also precipitate as $PbSO_4$:

$$PbO_2(s) + 4H^+(aq) + SO_4^{2-}(aq) + 2e^- \longrightarrow PbSO_4(s) + 2H_2O(l) \qquad E° = 1.685 \text{ V}$$

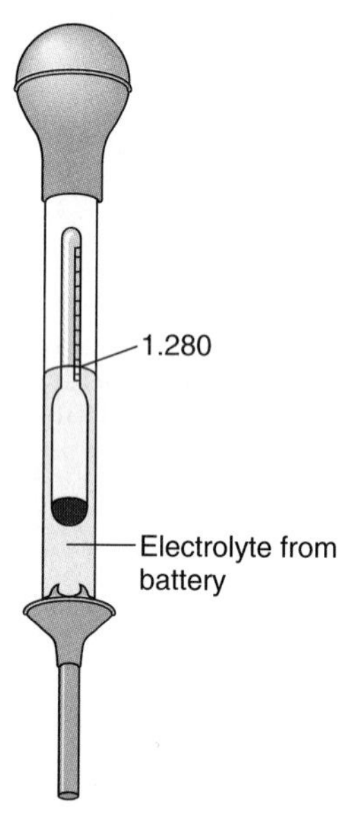

Figure 17.10

A sketch of a hydrometer, an instrument that measures densities of liquids.

When we add the anode and cathode half-reactions and their $E°$ values, we obtain the cell reaction and its $E°$ value:

$$Pb(s) + PbO_2(s) + 4H^+(aq) + 2SO_4^{2-}(aq) \longrightarrow 2PbSO_4 + 2H_2O(l) \qquad E° = 2.041 \text{ V}$$

In the operation of the cell, the Pb electrode is both the anode and the negative terminal by which the current is delivered to the engine.

The $E°$ value obtained above is for one cell. A 12-V battery has six cells, and a 6-V battery has three cells. A 12-V battery can be recharged by a generator which delivers a slightly higher voltage to the battery than the battery itself can produce. This applied voltage drives the cell reaction in reverse.

The actual voltage delivered by a lead storage battery depends on the concentration of sulfuric acid and, to a much lesser extent, on temperature. While the battery discharges, the concentration of sulfuric acid decreases and the voltage drops. As explained in Chapter 4, the concentration of sulfuric acid is related to its density, which can be checked (except for the sealed types) by a hydrometer (Figure 17.10).

Other storage batteries include the *nickel–cadmium battery* called a "NICAD" (Figure 17.11). This battery is used in many cordless appliances, including calculators. A NICAD battery is lighter, lasts longer, and maintains a more nearly constant potential during operation than does a lead storage battery.

Figure 17.11

A nickel–cadmium (Ni–Cd) battery.

The half-reactions during the discharge of a nickel–cadmium battery are:

$$Cd(s) + 2OH^-(aq) \longrightarrow Cd(OH)_2(s) + 2e^-$$

$$NiO_2(s) + 2H_2O(l) + 2e^- \longrightarrow Ni(OH)_2(s) + 2OH^-(aq)$$

A cell recently developed to deliver a large amount of energy to power electric automobiles is the Weber–Kummer cell, developed by the Ford Motor Company. The Weber–Kummer cell is a *sodium–sulfur cell* which has to be operated at 250° to 300°C to maintain both sodium and sulfur in a molten state. The reactants are separated by a porous ceramic partition that allows sodium ions to pass through. The half-reactions that occur during the discharge of this cell are:

$$2Na(l) \longrightarrow 2Na^+(l) + 2e^-$$

$$S(l) + 2e^- \longrightarrow S^{2-}(l)$$

As in any storage cell, these reactions can be reversed by an applied electric current.

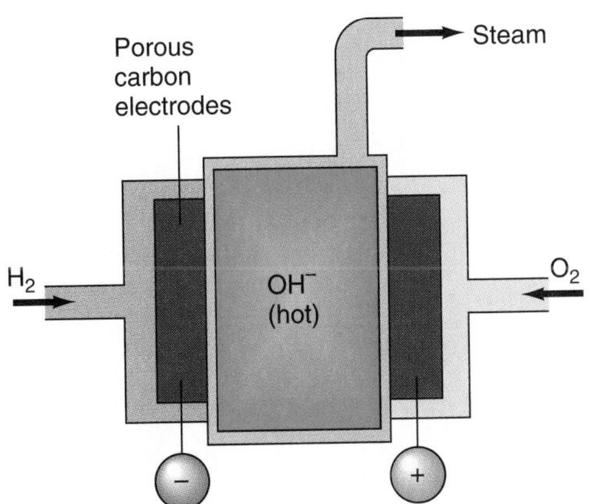

Figure 17.12

A cross section of an H_2–O_2 fuel cell.

Fuel Cells

Cells in which the reactants are continuously supplied to the cell are called **fuel cells**. The reactants in a fuel cell are called the *fuel* and the *oxidizer*. One type of fuel cell, called a hydrogen–oxygen cell, uses hydrogen as the fuel and oxygen as the oxidizer and produces water as a product. In a hydrogen–oxygen fuel cell, energy released from the reaction of hydrogen with oxygen to form water is converted to electrical energy. In the fuel cells currently available, this conversion is less than 50 percent efficient.

A hydrogen–oxygen fuel cell has three compartments separated from one another by porous carbon electrodes (Figure 17.12). Hydrogen gas is passed through the anode compartment and oxygen gas is passed through the cathode compartment. The middle compartment contains a hot solution of KOH.

The cell reactions are

$$\begin{array}{rl} \text{Anode:} & 2[H_2(g) + 2OH^-(aq) \longrightarrow 2H_2O(l) + 2e^-] \\ \text{Cathode:} & O_2(g) + 2H_2O(l) + 4e^- \longrightarrow 4OH^-(aq) \\ \hline \text{Cell:} & 2H_2(g) + O_2(g) \longrightarrow 2H_2O(l) \end{array}$$

Figure 17.13

Apollo spacecraft.

A hydrogen–oxygen cell delivers about 0.9 V. Multiple hydrogen–oxygen cells connected in series were used in *Apollo* and *Gemini* spacecrafts (Figure 17.13), where they provided both electricity and drinking water to the astronauts.

17.6 Electrolysis

We now turn to consider another important aspect of electrochemistry: the use of electric energy to cause nonspontaneous redox reactions to occur. Nonspontaneous redox reactions that are driven by an outside source of electric energy are called **electrolysis** reactions. Some electrolysis reactions are discussed below.

APPLICATIONS OF CHEMISTRY 17.3
Race Car Batteries

As a car battery discharges it eventually produces a decreased voltage. The relationship between battery voltage, its state of charge, and its sulfuric acid concentration within the battery is shown below.

electrodes

$$PbO_2(s) + Pb(s) + 2H^+(aq) + 2HSO_4^-(aq) \longrightarrow 2PbSO_4(s) + 2H_2O(l)$$

$2e^-$ transferred

This is the chemical reaction inside a car battery that produces electricity.

The concentrations of pure solids and liquids are not counted.

$$Q = \frac{[PbSO_4]^2[H_2O]^2}{[PbO_2][Pb][H^+]^2[HSO_2^-]^2} = \frac{(1)^2(1)^2}{(1)(1)[H^+]^2[HSO_4^-]^2} = \frac{1}{[H^+]^2[HSO_4^-]^2}$$

$$E = E° - \frac{0.06}{2} \log \frac{1}{[H^+]^2[HSO_4^-]^2}$$

Voltage delivered by car battery

Concentrations of H^+ and HSO_4^- affect the value of E.

The concentration of sulfuric acid decreases as the chemical reaction in the battery proceeds in the forward direction. This is why auto mechanics frequently determine battery charge with a hydrometer, which is a device that measures the specific gravity (density relative to that of water) of liquids. The specific gravity of the sulfuric acid solution in a fully charged battery is greater than the specific gravity of the sulfuric acid solution in a discharged battery.

Racing cars run more efficiently if their electrical parts always receive an electric current at a constant voltage. This increased efficiency can make the difference between winning and losing a race. Therefore, it is desirable for a racing car to have a battery whose voltage remains constant as the battery's chemical reaction shown above goes in the forward direction.

It's possible to construct a battery whose voltage does not depend on the concentrations of the products and the reactants (upon which the battery's state of charge depends). If all the products and reactants of the battery were pure solids or liquids, their concentrations would remain constant and the battery voltage (E) would always equal the standard voltage ($E°$) at 25°C.

$$E = E° - \frac{0.06}{n} \log 1$$

If all reactants and products are liquids and solids, their concentrations remain constant and their activities equal 1. This makes Q have a value of 1.

log 1 = 0

$$= E° - \frac{0.06}{n}(0) = E° - 0 = E°$$

Some race cars use a silver–zinc battery.

electrodes

$$Ag_2O(s) + Zn(s) + H_2O(l) \longrightarrow 2Ag(s) + Zn(OH)_2(s)$$

$2e^-$ transferred

All products and reactants in the reaction shown above are solids and liquids. The concentration of solids and liquids remains constant as the reaction progresses in the forward direction, thereby maintaining a constant voltage in the battery.

$$Q = \frac{[Ag]^2\,[Zn(OH)_2]}{[Ag_2O]\,[Zn][H_2O]} = \frac{(1)^2(1)}{(1)(1)(1)} = 1$$

$$E = E^\circ - \frac{0.06}{n}\log Q = E^\circ - \frac{0.06}{n}\log 1$$

$$= E^\circ - \frac{0.06}{n}(0) = E^\circ - 0 = E^\circ$$

Source: Ronald DeLorenzo, *Problem Solving in General Chemistry*, 2nd ed., Wm. C. Brown, Publishers, Dubuque, Iowa, 1993.

Electrolysis of Molten Salts

Let us consider the electrolysis of molten sodium chloride. In this process, both sodium metal and chlorine gas can be produced (Figure 17.14). A source of electric current such as a battery acts as an "electron pump" which forces electrons into molten NaCl at one terminal and removes them at the other terminal. Electrons flow into the molten sodium chloride and back to the battery through electrodes constructed of an inert material such as platinum or graphite. The *cell in which electrolysis occurs* is called an **electrolytic cell**. In both a galvanic and an electrolytic cell, oxidation occurs at the anode and reduction at the cathode.

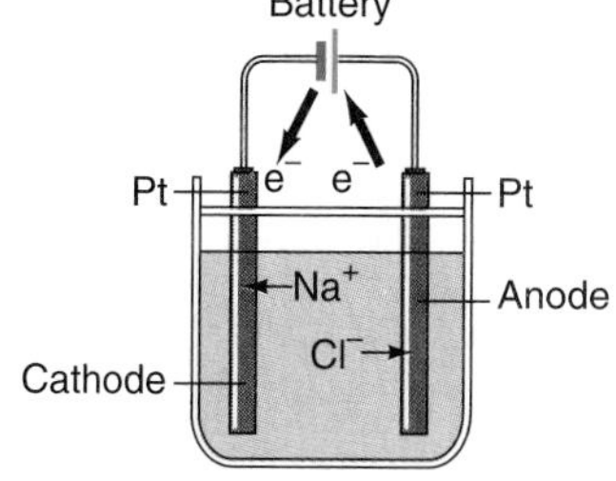

Figure 17.14

Electrolysis of molten sodium chloride. Sodium ions are reduced to elemental sodium at the cathode, and chloride ions are oxidized to chlorine gas at the anode.

As electrons enter the electrolytic cell (Figure 17.14), Na^+ ions are reduced to sodium metal at the cathode, and Cl^- ions are oxidized to chlorine gas at the anode. The cathode and anode half-reactions and the overall reaction that occurs during electrolysis are

$$\text{Reduction, cathode half-reaction:}\quad 2Na^+(l) + 2e^- \longrightarrow 2Na(s)$$
$$\text{Oxidation, anode half-reaction:}\quad 2Cl^-(l) \longrightarrow Cl_2(g) + 2e^-$$

$$\text{Overall cell reaction:}\quad 2Na^+(l) + 2Cl^-(l) \xrightarrow{\text{electrolysis}} 2Na(s) + Cl_2(g)$$

The electrons required for the reduction half-reaction are supplied by the battery; an equal number of electrons return to the battery as a result of the oxidation half-reaction. Note that the cathode is negatively charged because

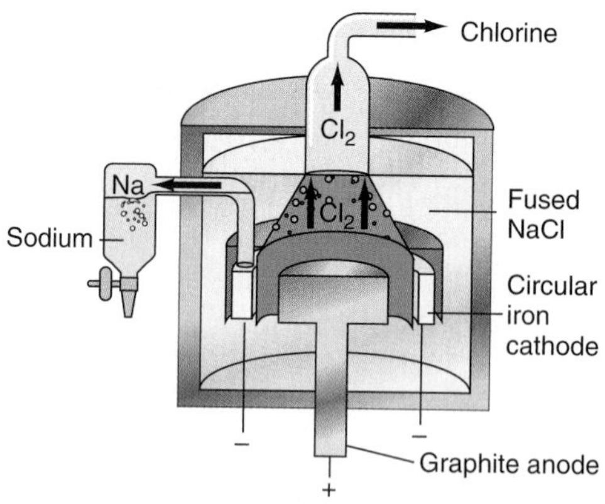

Figure 17.15

Cross section of Downs cell for industrial preparation of sodium.

electrons are forced into it by the battery. The cathode therefore attracts cations. The oxidation half-reaction occurs at the anode. The anode is positively charged, and anions are attracted to it.

Sodium and chlorine are both prepared commercially by the electrolysis of molten sodium chloride in an electrolytic cell called a Downs cell (Figure 17.15). In this cell, which is maintained at high temperatures, sodium forms as a liquid that is drained off, cooled, cast into blocks, and stored in mineral oil to prevent its reaction with oxygen and other components of the atmosphere. Chlorine gas is stored under pressure in corrosion-resistant tanks.

Electrolysis is also used for the industrial preparation of calcium metal and chlorine gas from molten calcium chloride. Calcium ions are reduced at the cathode, and chloride ions are oxidized at the anode:

$$\begin{array}{rl} \text{Cathode:} & Ca^{2+}(l) + 2e^- \longrightarrow Ca(s) \\ \text{Anode:} & 2Cl^-(l) \longrightarrow Cl_2(g) + 2e^- \\ \hline \text{Overall:} & Ca^{2+}(l) + 2Cl^-(l) \longrightarrow Ca(s) + Cl_2(g) \end{array}$$

Electrolysis of Water and Aqueous Solutions

In the electrolysis of water, in the presence of a few ions, some water molecules are reduced to hydrogen gas at the cathode and others are oxidized to oxygen gas at the anode:

$$\begin{array}{rl} \text{Cathode half-reaction:} & 2H_2O(l) + 2e^- \longrightarrow H_2(g) + 2OH^-(aq) \\ \text{Anode half-reaction:} & 2H_2O(l) \longrightarrow O_2(g) + 4H^+(aq) + 4e^- \end{array}$$

When we multiply the cathode half-reaction by 2 and add the half-reactions, the overall electrolysis cell reaction is obtained:

$$2H_2O(l) \longrightarrow 2H_2(g) + O_2(g)$$

In the electrolysis of an aqueous solution of sodium chloride, the sodium ions, which are ions of a very active metal, are not reduced at the cathode. Therefore, the only substance that can be reduced is water. The reduction products of water are hydrogen gas and hydroxide ions:

$$2H_2O(l) + 2e^- \longrightarrow H_2(g) + 2OH^-(aq)$$

The hydroxide ions are only partially neutralized by some hydrogen ions formed by the anode reaction. So the solution becomes increasingly more alkaline as electrolysis proceeds.

The anode reaction is more complicated. There are two possible anode reactions, depending on the concentration of the NaCl solution. Chloride ions can be oxidized to chlorine gas, or water can be oxidized to oxygen gas. Chlorine gas is produced at the anode when a concentrated solution of NaCl is electrolyzed. Electrolysis of a dilute solution of NaCl produces a mixture of Cl_2 gas, O_2 gas, and H^+ ions at the anode. Thus, in the electrolysis of a concentrated solution of sodium chloride, the half-reactions and the overall reaction are

$$\begin{aligned} \text{At cathode:} \quad & 2H_2O(l) + 2e^- \longrightarrow H_2(g) + 2OH^-(aq) \\ \text{At anode:} \quad & 2Cl^-(aq) \longrightarrow Cl_2(g) + 2e^- \\ \hline \text{Overall:} \quad & 2H_2O(l) + 2Cl^-(aq) \longrightarrow H_2(g) + 2OH^-(aq) + Cl_2(g) \end{aligned}$$

As we have noted, during the electrolysis of a sodium chloride solution, sodium ions do not react. Their concentration slowly increases as the amount of water decreases through electrolysis. The hydroxide ion concentration also increases as these ions are formed at the cathode. After the chloride ions have been converted to Cl_2 gas, the solution contains Na^+ and OH^- ions. Evaporation of water from this solution produces relatively pure, solid sodium hydroxide.

Electrolysis of aqueous sodium chloride is used for the industrial production of sodium hydroxide (Figure 17.16). In addition to solid sodium hydroxide, hydrogen and chlorine gases are also obtained in this process as by-products.

Electrolysis of aqueous solutions of some salts is equivalent to electrolysis of water. For example, electrolysis of an aqueous solution of Na_2SO_4 produces hydrogen gas at the cathode and oxygen gas at the anode. The Na^+ and SO_4^{2-} ions are more difficult to reduce than water, and neither of these ions can easily be oxidized. Thus, only water is electrolyzed.

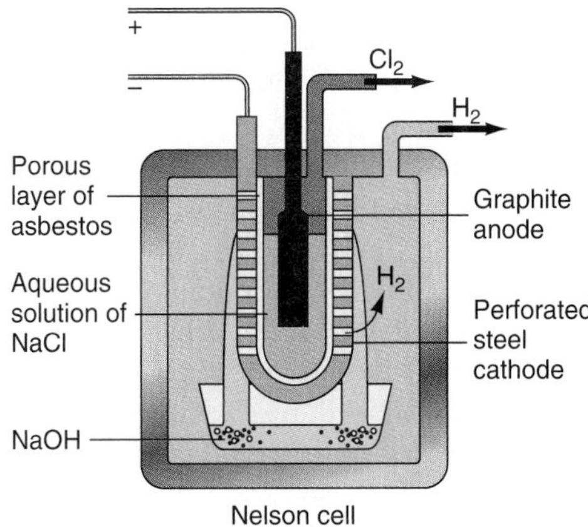

Figure 17.16

Industrial preparation of sodium hydroxide by the electrolysis of an aqueous solution of sodium chloride in a Nelson cell.

Example 17.7

Predicting the Products of Electrolysis Reactions

Write equations for the anode and cathode half-reactions, and for the overall cell reactions, which occur in the electrolysis of each of the following substances using inert electrodes: **(a)** molten copper(II) chloride; **(b)** an aqueous solution of copper(II) sulfate.

SOLUTION:

(a) Molten copper(II) chloride contains only Cu^{2+} and Cl^- ions. The Cu^{2+} ions are reduced to elemental copper at the cathode, and the Cl^- ions are oxidized to chlorine gas at the anode:

$$\text{At cathode: } Cu^{2+}(l) + 2e^- \longrightarrow Cu(s)$$
$$\text{At anode: } 2Cl^-(l) \longrightarrow Cl_2(g) + 2e^-$$
$$\text{Overall cell reaction: } Cu^{2+}(l) + 2Cl^-(l) \longrightarrow Cu(s) + Cl_2(g)$$

(b) We recall that sulfate ions are stable toward oxidation and reduction. Therefore, water is oxidized to oxygen gas at the anode. At the cathode, Cu^{2+} ions could be reduced to Cu metal, or water could be reduced to hydrogen. But we recall that Cu^{2+} ions are easy to reduce. Thus, Cu^{2+} ions are reduced to copper metal at the cathode. The reactions in this electrolysis are

$$\text{At the anode: } 2H_2O(l) \longrightarrow O_2(g) + 4H^+(aq) + 4e^-$$
$$\text{At the cathode: } 2[Cu^{2+}(aq) + 2e^- \longrightarrow Cu(s)]$$
$$\text{Overall: } 2H_2O(l) + 2Cu^{2+}(aq) \longrightarrow O_2(g) + 4H^+(aq) + 2Cu(s)$$

Note that the cathode half-reaction is multiplied by 2 so that the electrons cancel when the half-reactions are added.

Practice Problem 17.7: Write the anode and cathode half-reactions and the overall reactions that occur in the electrolysis of: **(a)** molten barium chloride; **(b)** aqueous solution of lithium fluoride (water is oxidized at the anode).

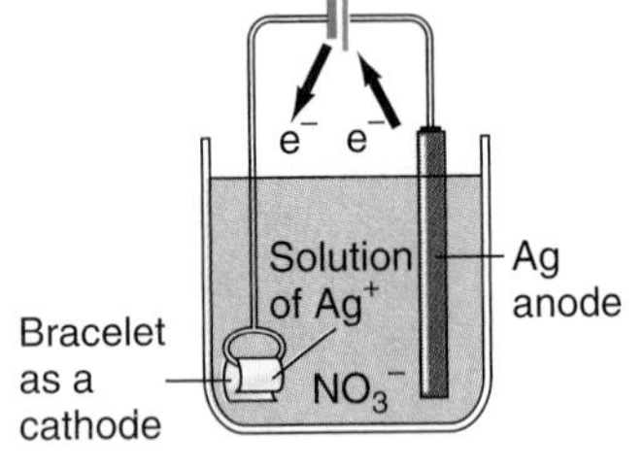

Figure 17.17

A simple cell for silver plating a bracelet. The silver ions in the solution are reduced to silver atoms on the surface of the bracelet. Simultaneously, silver atoms in the silver anode are oxidized to silver ions that go into the solution.

Electroplating and Electrolytic Refining of Metals

Metal objects are often plated with copper, chromium, silver, or gold to protect them against corrosion. For example, a bracelet can be silver plated by making it a cathode in an electrolytic cell containing a solution of silver nitrate and an anode made of silver (Figure 17.17). Silver ions from the $AgNO_3$ solution are reduced to metallic silver on the surface of the bracelet. Silver atoms in the anode are simultaneously oxidized to silver ions, which go into the solution to replace the silver ions reduced onto the bracelet.

In commercial electroplating processes, several electrolytes may be used to make the metal adhere strongly and uniformly to the object being plated. Table 17.2 lists some of the electrolytes used in various electroplating processes.

Table 17.2

Some Electroplating Processes

Metal Used for Plating	Anode	Electrolytes in Solution (percent by mass)	Type of Application
Ag	Ag	4% $KAg(CN)_2$, 4% KCN, 4% K_2CO_3	Tableware, jewelry
Au	Au or Cr	3% $KAu(CN)_2$, 19% KCN, K_2HPO_4 buffer	Jewelry
Cr	Pb	25% CrO_3, 0.25% H_2SO_4	Automobile parts
Zn	Zn	6% $K_2Zn(CN)_4$, 5% NaCN, 8% NaOH, 5% Na_2CO_3	Steel galvanization
Sn	Sn	8% H_2SO_4, 7% $SnSO_4$	Making "tin" cans

We can see from Table 17.2 that metal cyanides are used in many electroplating processes. Cyanide ions form complexes with many metal ions. In a cyanide complex, the concentration of a free metal ion at equilibrium with the complex ion is small. When the metal ion concentration in a solution during an electrolysis process is small, electrolysis is slow, and a smooth deposit of metal forms on the object being plated. When cyanide ions are used in silver plating, the electrode reactions are

$$\text{Anode:}\quad Ag(s) + 2CN^-(aq) \longrightarrow Ag(CN)_2^-(aq) + e^-$$

$$\text{Cathode:}\quad Ag(CN)_2^-(aq) + e^- \longrightarrow Ag(s) + 2CN^-(aq)$$

Cyanide ions are extremely toxic. Therefore, solutions containing cyanide ions must not be dumped into rivers and streams. Metal–cyanide complexes are also used to extract gold and silver from low-grade ores. Solutions containing cyanide ions that were used in mining and electroplating operations and dumped into streams have been responsible for killing fish and other animals.

When electrolysis is used to purify a metal, the impure metal is used as an anode in an electrolytic cell that contains a solution of the metal salt. For example, in the electrolytic refining of copper, an impure copper bar is the anode, a pure copper bar is the cathode, and a $CuSO_4$ solution is the electrolyte (Figure 17.18).

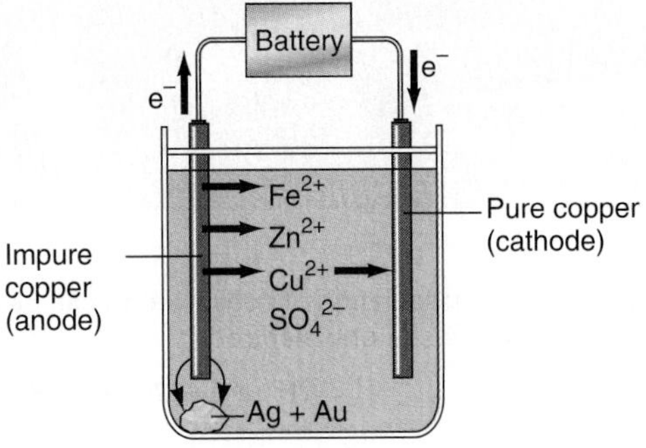

Figure 17.18

Electrolytic refining of copper. With a carefully controlled voltage, zinc, iron, and copper atoms in the impure copper bar are oxidized to ions. Silver and gold fall to the bottom of the cell. The copper ions in solution then redeposit on the surface of the pure copper cathode.

APPLICATIONS OF CHEMISTRY 17.4
Raise the *Titanic*

The *Titanic*, built in 1912 as an unsinkable luxury ocean liner, sank on its maiden voyage after colliding with an iceberg. More than 1500 people died. Today, the *Titanic* lies 2 miles beneath the ocean's surface some 100 miles south of the Great Banks of Newfoundland, Canada. There is considerable interest in raising the *Titanic*. Although there is no major scientific justification for raising the ship, it might have some commercial value because people are interested in seeing things of historical significance. However, the pressure at this depth is about 300 atm, which is too extreme for underwater divers. But some scientists want the challenge of the advanced technological problems that such a feat presents. These scientists want to develop the technology to locate more important objects in the deep ocean. Scientists also want to expand the technology for deep-sea photography.

In the past, sunken ships have been raised by attaching inflatable buoys (big balloons) to them and filling the buoys with air. Compressors located on surface ships supply the air needed to inflate the buoys (see the illustration below). Unfortunately, commercial pumps are not available to pump air at pressure of 300 to 350 atm over distances of 2 to 3 miles.

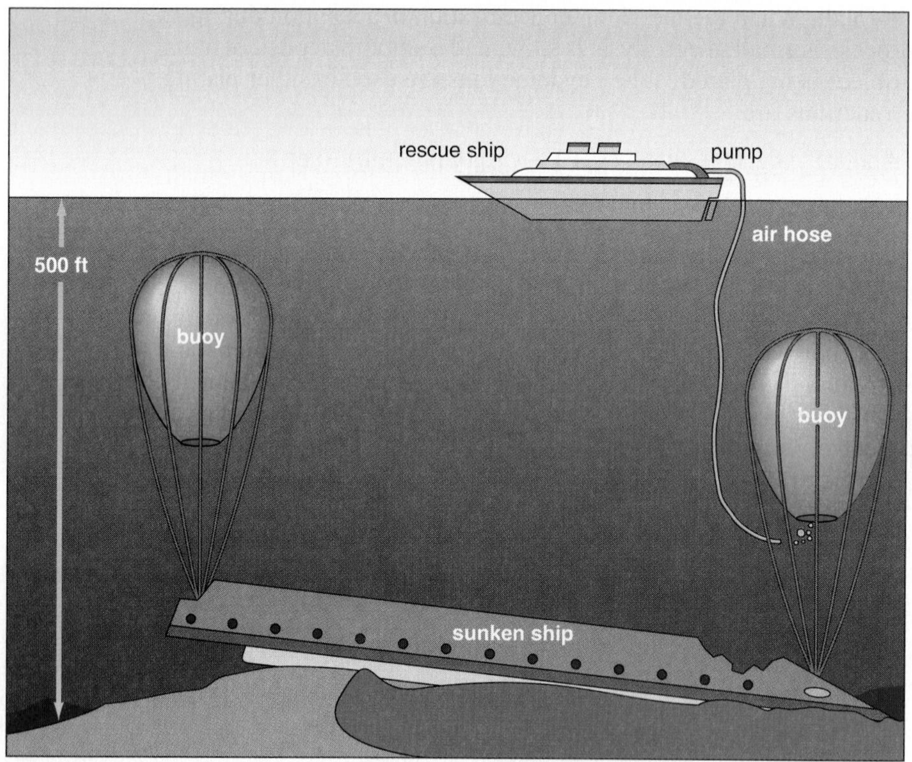

However, there is another way to fill the buoys with gas other than pumping air into them. By substituting a direct current (dc) generator for the pump and a wire for the hose, the hydrogen ions present in the ocean (from the dissociation of ocean water) can be reduced to hydrogen gas at one electrode:

$$2H^+(aq) + 2e^- \longrightarrow H_2(g)$$

This gas can be used to inflate the buoy (see the illustration below).

Scientists have calculated that it would take a 20-MW (20-million-watt) generator operating continuously more than three years to produce enough hydrogen to raise the *Titanic*.

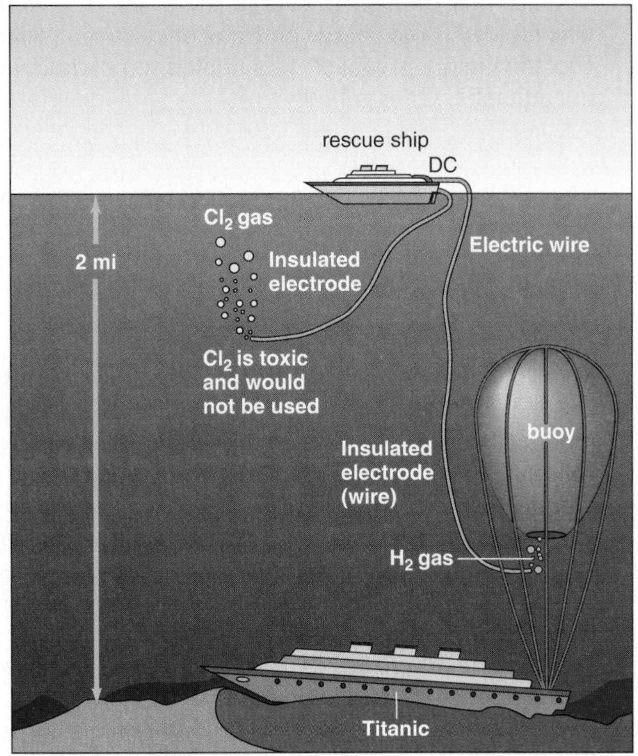

Source: Ronald DeLorenzo, *Problem Solving in General Chemistry*, 2nd ed., Wm. C. Brown Publishers, Dubuque, IA, 1993.

Impure copper frequently contains traces of iron, zinc, silver, and gold. Iron, zinc, and copper atoms in the impure copper anode are oxidized and go into the solution as Fe^{2+}, Zn^{2+}, and Cu^{2+} ions when the voltage is carefully controlled in the electrolysis. Silver and gold are more noble metals than copper, iron, and zinc and therefore are not oxidized at the anode; they fall to the bottom of the cell. In the solution so produced, the Cu^{2+} ions are the easiest to reduce and deposit as pure copper on the copper cathode.

The mixture of silver and gold precipitated in the anode compartment is called "anode slime" or "anode mud" and is a commercial source of silver and gold. The sale of this "mud" often pays for the electrolysis so the refining of copper is highly profitable.

17.7 Quantitative Aspects of Electrolysis

The quantity of electric charge is measured in **coulombs**, C, and **faradays**, F. One faraday is the charge on 1 mol of electrons, which equals 96,487 coulombs. Electric energy in joules, J, is related to potential in volts, V, and the quantity in coulombs, C, as follows:

$$1\ \text{J} = 1\ \text{V C} \tag{17.4}$$

Thus, 1 J of work is done by moving a charge of 1 C through a potential difference; the potential difference is 1 V:

$$1\ \text{V} = 1\ \text{J C}^{-1} \tag{17.5}$$

Faraday's Law and Its Applications

Michael Faraday, an English scientist, showed in 1833 that when a given quantity of electricity is passed through an H_2SO_4 solution, the same amount of H_2 is produced regardless of the size of the electrodes, the concentration of the solution, or the voltage applied. He concluded that *the amount of a substance produced by electrolysis is directly proportional to the quantity of electricity used*. This generalization has become known as **Faraday's law**.

We recall that the *quantity of electric charge* is measured in faradays (F) and coulombs (C). 1 F = charge on 1 mol of electrons = 96,485 C, which to three significant figures is 9.65×10^4 C.

Quantity of electricity is related to rate of charge flow. When *1 coulomb of charge moves through a point in 1 second*, the *rate of charge flow* is 1 **ampere** (A). Thus, the rate of charge flow in amperes equals the number of coulombs that flow through a point in 1 second:

$$\text{A} = \frac{\text{C}}{\text{s}}$$

from which

$$\text{C} = (\text{A})(\text{s}) \tag{17.6}$$

Equation 17.6 relates the amount of electricity in coulombs to the rate of charge flow in amperes and the time in seconds during which current flows.

Let us consider an application of Faraday's law in an electrolysis of a silver salt solution. In this electrolysis, silver ions are reduced at the cathode. The reduction of one silver ion requires one electron. The reduction of 1 mol of silver ions requires 1 mol of electrons, or 1 faraday:

$Ag^+(aq)$	$+\ e^-$	$\longrightarrow$	$Ag(s)$
1 ion	1 electron		1 atom
1 mol ions	1 mol electrons (1 F)		1 mol atoms

By similar reasoning, we conclude that 1 faraday reduces $\frac{1}{2}$ mol of copper(II) ions (Figure 17.19). Note that the reduction of 1 mol of H^+ ions produces $\frac{1}{2}$ mol of H_2 molecules.

We can also apply Faraday's law to oxidation reactions at the anode. For example, the oxidation of 2 mol of chloride ions produces 1 mol of Cl_2 gas and releases 2 mol of electrons (2 faradays):

$2Cl^-$	$\longrightarrow\ Cl_2$	$+$	$2e^-$
2 mol ions	1 mol molecules		2 mol electrons (2 F)

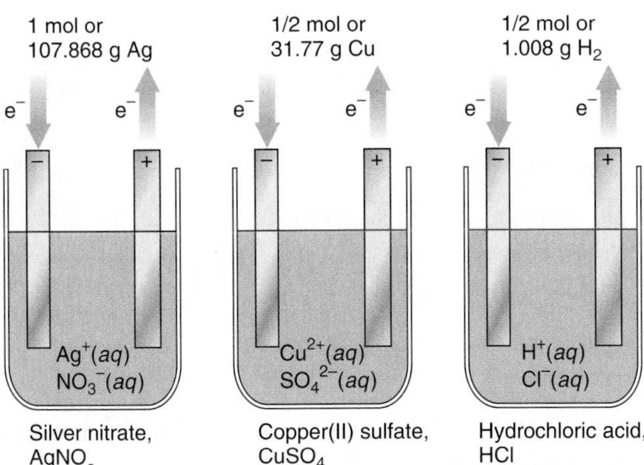

Figure 17.19

The quantities of different substances produced at the cathode in electrolysis by one faraday of electricity (96485 coulombs).

The oxidation of two moles of oxide ions produces one mole of O_2 gas and releases four moles of electrons:

$2O^{2-}$	$\longrightarrow O_2(g)$	+	$4e^-$
2 mol ions	1 mol molecules		4 mol electrons (4 F)

The *quantity of a substance formed by the gain or loss of one faraday during electrolysis* is called its **electrochemical equivalent**. Thus, the electrochemical equivalent of silver is 1 mol of Ag atoms. The electrochemical equivalent of copper is $\frac{1}{2}$ mol of Cu atoms, and the electrochemical equivalent of oxygen is $\frac{1}{4}$ mol of O_2 molecules (because 4 F of charge are lost by oxide ions to produce 1 mol of O_2).

The calculated quantity of a substance obtained in an electrolysis is somewhat larger than the measured quantity because the applied current is never 100 percent efficient. Some of the electric energy can be converted to heat, and some of it may be used to speed up reactions or carry out some other minor unexpected and sometimes unknown reactions that are difficult to predict. Thus, when we apply Faraday's law to calculate the quantity of a substance produced by electrolysis, we shall assume 100 percent current efficiency.

Example 17.8

Mass of a Substance Produced by a Given Number of Coulombs in an Electrolysis

How many grams of oxygen can be produced by 5.00×10^4 C in the electrolysis of water? How many grams of O_2 equals one electrochemical equivalent?

SOLUTION: In the electrolysis of water, oxygen is produced at the anode according to the half-reaction:

$$2H_2O(l) \longrightarrow O_2(g) + 4H^+(aq) + 4e^-$$

Thus, 1 mol of O_2 is formed as a result of the passage of 4 mol of electrons (4 F). The passage of 1 F produces $\frac{1}{4}$ mol of O_2, or 8.00 g of O_2, which is the electrochemical equivalent of O_2. To calculate the mass of O_2 produced by 5.00×10^4 C, we convert coulombs to faradays, faradays to moles of O_2, and moles of O_2 to mass of O_2, according to the following outline:

$$\text{coulombs} \longrightarrow \text{faradays} \longrightarrow \text{moles } O_2 \longrightarrow \text{mass } O_2$$

Substituting numbers into this conversion scheme, we obtain

$$5.00 \times 10^4 \text{ C} \left(\frac{1 \text{ F}}{9.65 \times 10^4 \text{ C}}\right)\left(\frac{1 \text{ mol } O_2}{4 \text{ F}}\right)\left(\frac{32.0 \text{ g } O_2}{1 \text{ mol } O_2}\right) = 4.15 \text{ g } O_2$$

Practice Problem 17.8: How many sodium atoms, and what mass of sodium, can be deposited from molten NaCl by 2.64 F?

Practice Problem 17.9: What mass of Cl_2 gas can be produced by 3.82×10^3 C during an electrolysis of molten sodium chloride?

Example 17.9

Calculating Quantities of Substances Produced in an Electrolysis by Given Amperage and Time

Molten copper(II) chloride is electrolyzed by a current of 2.50 A for 2.00 hours (h). Calculate the mass of copper produced at the cathode and the volume of chlorine produced at the anode when cooled to 20.0°C at 740 torr.

SOLUTION: We multiply the current in amperes by the time *in seconds* to obtain the number of coulombs (Equation 17.6) used in the process. We convert coulombs to faradays and faradays to moles of copper and moles of chlorine produced. Then we convert the moles to mass or volume. The number of coulombs, C, used equals amperes, A, times seconds, s:

$$\text{C} = (2.50 \text{ A})(2.00 \text{ h})\left(\frac{60 \text{ min}}{1\ h}\right)\left(\frac{60 \text{ s}}{1 \text{ min}}\right) = 1.80 \times 10^4 \text{ (A)(s)}$$
$$= 1.80 \times 10^4 \text{ C}$$

The number of faradays is

$$1.80 \times 10^4 \text{ C} \left(\frac{1 \text{ F}}{9.65 \times 10^4 \text{ C}}\right) = 0.187 \text{ F}$$

To convert faradays to moles of copper produced, we need to know the cathode half-reaction. This reaction is

$$\text{Cathode:} \quad Cu^{2+} + 2e^- \longrightarrow Cu(s)$$

According to the cathode half-reaction, one atom of copper is produced by two electrons and 1 mol of copper is produced by 2 F. We use this information to convert faradays to moles of copper, and moles of copper to grams of copper:

$$0.187 \text{ F} \left(\frac{1 \text{ mol Cu}}{2 \text{ F}}\right)\left(\frac{63.5 \text{ g Cu}}{1 \text{ mol Cu}}\right) = 5.94 \text{ g Cu}$$

We can do these calculations in one multistep operation:

$$1.80 \times 10^4 \text{ C} \left(\frac{1 \text{ F}}{9.65 \times 10^4 \text{ C}}\right)\left(\frac{1 \text{ mol Cu}}{2 \text{ F}}\right)\left(\frac{63.5 \text{ g Cu}}{1 \text{ mol Cu}}\right) = 5.92 \text{ g Cu}$$

According to the anode half-reaction,

$$2Cl^- \longrightarrow Cl_2(g) + 2e^-$$

1 mol of Cl_2 is produced by 2 F. To calculate the volume of Cl_2 produced, we first determine the number of moles of Cl_2. Then we convert moles to liters by using the ideal gas law equation, $PV = nRT$. Starting with the previously calculated number of coulombs used in the electrolysis, the number of moles of Cl_2 produced is

$$1.80 \times 10^4 \text{ C} \left(\frac{1 \text{ F}}{9.65 \times 10^4 \text{ C}}\right)\left(\frac{1 \text{ mol } Cl_2}{2 \text{ F}}\right) = 0.0933 \text{ mol } Cl_2$$

$$V = \frac{nRT}{P} = \frac{(0.0933 \text{ mol})(0.0821 \text{ L atm mol}^{-1} \text{ K}^{-1})(293 \text{ K})}{(740/760) \text{ atm}} = 2.31 \text{ L}$$

Practice Problem 17.10: Molten calcium fluoride is electrolyzed with a current of 3.80 A for 3.20 h. Calculate the mass of calcium and the volume of fluorine at STP produced in this electrolysis.

Summary

In this chapter we considered **galvanic** and **electrolytic cells**. In a galvanic cell, electric current is generated by a spontaneous chemical reaction. In an electrolytic cell, a chemical change is produced by electric current forced through the cell.

Galvanic Cells

In an operating galvanic cell, electrons flow through an external circuit from the substance oxidized in one half-cell to the substance reduced in the other half-cell. The **anode** is the electrode at which oxidation occurs, and the **cathode** is the electrode at which reduction occurs. Electrons exit a galvanic cell at the anode, which is the negative electrode. Electrons reenter the cell through the cathode, which is the positive electrode.

The tendency of electrons to flow through the external circuit of a galvanic cell is called the **cell potential**. This potential is measured in volts (V). The quantity of charge is measured in **faradays** (F) or **coulombs** (C). 1 F = 1 mol of electrons = 96,485 C. Electric energy of 1 joule equals the potential of 1 volt times the charge of 1 coulomb: 1 J = 1 V C.

The potential of a galvanic cell depends on the potentials of the half-cell reactions. Adding the half-cell reactions and their potentials gives the cell reaction and its potential if one half-cell reaction is written as reduction and the other as oxidation. Half-cell potentials cannot be directly measured but can be compared with the potential of the *standard hydrogen electrode* whose potential is arbitrarily chosen as exactly zero volts. **Standard reduction potentials** of other half-cells can be measured relative to the potential of the standard hydrogen electrode.

The Nernst Equation and Its Applications

The potential, E, of a cell operating under nonstandard conditions can be calculated by the **Nernst equation**. The Nernst equation can be used to calculate the potential, E, of a cell from its standard potential, $E°$, the number of electrons, n, in the cell reaction as written, and the reaction quotient, Q, at 298 K:

$$E = E° - \frac{0.0592}{n} \log Q \qquad \text{(at 298 K)}$$

Applications of Galvanic Cells

Electrode potentials are important in the construction of commercial batteries and in predicting the spontaneity of redox reactions in chemical synthesis and analysis, as well as in the determination of K_{sp} values for sparingly soluble salts. Commercial batteries are of three main types: **primary cells**, **secondary cells** or **storage cells**, and **fuel cells**.

Electrolysis

According to **Faraday's law**, *the amount of a substance produced by electrolysis is directly proportional to the quantity of electricity used*. The rate of charge flow is measured in **amperes** (A). One ampere is a current of 1 coulomb per second: 1 A = 1 C/s, and 1 C = 1 (A)(s).

One faraday equals the charge of 1 mole of electrons. The quantity of a substance produced by 1 faraday during electrolysis is called its **electrochemical equivalent**. For example, 1 faraday deposits 1 mole of silver from a silver ion solution, or $\frac{1}{2}$ mole of copper from a copper(II) ion solution. If the amperage and the duration of an electrolysis are known, the amount of a substance produced (or decomposed) can be calculated by the following sequence of conversions:

ampere-seconds ⟶ coulombs ⟶ faradays ⟶ moles of substance ⟶ grams of substance.

New Terms

Ampere (17.7)
Anode (17.2)
Battery (introduction)
Cathode (17.2)
Cell (introduction)
Cell potential (17.2)
Coulomb (17.7)
Dry cell (*see* Primary cell)
Electrochemical equivalent (17.7)
Electrochemistry (introduction)
Electrolysis (17.6)
Electrolytic cell (17.6)
Faraday (17.7)
Faraday's law (17.7)
Fuel cell (17.5)
Galvanic cell (introduction and 17.2)
Half-cell (17.2)
Nernst equation (17.4)
Primary cell (17.5)
Secondary cell (17.5)
Standard cell (17.2)
Standard potential (17.2)
Storage cell (17.5)
Volt (17.2 and 17.7)
Voltaic cell (introduction)

Exercises

General Review

1. Explain the difference between galvanic and electrolytic cells.
2. In the operation of a galvanic cell, do the electrons exit the cell by the anode or by the cathode of the cell? Explain.
3. Which electrode, the anode or the cathode, of a galvanic cell is considered the negative electrode? Explain.
4. Which electrode, the anode or the cathode, of an electrolytic cell is the negative electrode? Explain.
5. Explain how you would construct a galvanic cell of the highest possible potential without regard to the length of time the cell could deliver current. What materials would you use in the anode and in the cathode compartments?
6. Describe the construction of, and the materials present in, a standard hydrogen electrode.
7. Does the potential of a galvanic cell depend on: **(a)** the sizes of electrodes; **(b)** the volumes of the electrolytes in the two half-cells; **(c)** the concentrations of the electrolytes; **(d)** temperature?
8. Where (up, down, left, right) in a table of standard reduction potentials do you find species representing: **(a)** the best oxidizing agents; **(b)** the best reducing agents? Give some examples of each from the table.
9. When the half-reactions of a galvanic cell reaction are both given as reduction half-reactions (as in a table of standard electrode potentials), how can you decide which is the anode half-reaction for a standard cell?
10. What is the relationship between cell potential and the spontaneity of the cell reaction?
11. Explain the corrosion of iron and list three different methods to prevent corrosion.

12. Write the Nernst equation and explain its application.
13. Explain the meaning of and give examples of each of the following types of cells: **(a)** storage cells; **(b)** primary cells; **(c)** fuel cells.
14. List two metals and one nonmetal that can best be prepared by electrolysis.
15. When molten sodium chloride is electrolyzed, sodium metal is produced at the cathode compartment, but when an aqueous *solution* of sodium chloride is electrolyzed, hydrogen gas is produced at the cathode. Explain.
16. Discuss the application of electrolysis in: **(a)** electroplating; **(b)** purification of metals.
17. Explain the relationship between the number of moles of a metal deposited at the cathode by 1 F during electrolysis of molten salts of metals in different oxidation states.

Galvanic Cells and Standard Electrode Potentials

18. Describe a cell in which electricity can be generated by the reaction

$$Fe(s) + 2Ag^+(aq) \longrightarrow Fe^{2+}(aq) + 2Ag(s)$$

Answer the following questions concerning this cell: **(a)** What materials are used for electrodes, and what electrolyte solutions can be used in each of the half-cells? **(b)** Which electrode is the anode and which is the cathode in this cell? **(c)** In which direction do the electrons move through the external circuit during the operation of the cell? **(d)** What is the standard potential of the cell? **(e)** Is the reaction as written spontaneous? **(f)** Is the reverse of this reaction spontaneous? **(g)** What is the standard potential for the reverse reaction?

19. A nickel rod in a 1.00 *M* nickel sulfate solution is connected to a copper rod in a 1.00 *M* copper(II) sulfate solution to generate electricity. **(a)** Write equations for the anode and cathode half-reactions and the cell reaction. **(b)** Which electrode is the anode, and which is the cathode? **(c)** Which electrode is the negative terminal, and which is the positive terminal? **(d)** In what direction do electrons move in the external circuit during the operation of this cell? **(e)** What is the standard potential for the cell? **(f)** Is the cell reaction spontaneous? **(g)** What is the standard potential for the reverse of the cell reaction?
20. Describe how you would construct a cell to generate electricity by each of the following reactions: **(a)** $2H^+(aq) + Mg(s) \rightarrow Mg^{2+}(aq) + H_2(g)$; **(b)** $H_2(g) + Cl_2(g) \rightarrow 2H^+(aq) + 2Cl^-(aq)$; **(c)** $Cl_2(g) + 2Br^-(aq) \rightarrow 2Cl^-(aq) + Br_2(l)$. In each case list the materials (electrodes and electrolytes) you would use in each half-cell, and write the half-cell reactions. Also, calculate the standard cell potential, and indicate which electrode is the anode and which is the cathode in each cell.
21. Metals A, B, and C form only divalent aqueous ions. Metal A can reduce B^{2+} ions from their aqueous solution, and metal C can reduce A^{2+} ions. Which one of the three divalent ions, A^{2+}, B^{2+}, or C^{2+}, has the highest standard reduction potential? If you had only these three metals and 1 *M* aqueous solutions of their ions available to you, describe how you would construct a cell of the highest possible potential (indicate the materials you would use in the anode and cathode compartments). For this cell, write the anode and cathode half-reactions, and the cell reaction.
22. Calculate the standard potential, and write the anode and cathode half-reactions, and the cell reaction, for each of the following cells: **(a)** $Fe(s) \mid Fe^{2+}(aq) \parallel Ni^{2+}(aq) \mid Ni(s)$; **(b)** $Pt \mid O_2(g) \mid H_2O_2(aq) \parallel Cu^{2+}(aq) \mid Cu(s)$; **(c)** $Zn(s) \mid Zn^{2+}(aq) \parallel Cl^-(aq) \mid Cl_2(g) \mid Pt$; **(d)** $Pt \mid Fe^{2+}(aq), Fe^{3+}(aq) \parallel H^+(aq), Cr_2O_7^{2-}(aq) \mid Pt$.
23. The standard potential for the cell

$$Sn(s) \mid Sn^{2+}(aq) \parallel H^+(aq), BiO^+(aq) \mid Bi(s)$$

is 0.456 V. Write a balanced equation for the cathode half-reaction and calculate its standard potential.

Prediction of Redox Reactions

24. Predict whether each of the following 2 *M* solutions will be stable toward oxidation or reduction if mixed with an equal volume of 2 *M* nitric acid: **(a)** NaF; **(b)** NaCl; **(c)** NaBr; **(d)** NaI. Where you predict a reaction write a net ionic equation for the reaction, assuming that nitrate ions are reduced to NO gas in the reaction.
25. In some municipalities the water is highly chlorinated during summer, so goldfish will die if you treat your fishpond too long with city water. If the chlorine in the pond were converted to chloride ions, the fish would survive because the chloride ions are not harmful to them. Which of the following substances in a 1 *M* solution can convert chlorine gas (at 1 atm) to chloride ions? **(a)** Iron(II) sulfate; **(b)** iron(III) sulfate; **(c)** sodium sulfate; **(d)** sodium thiosulfate, $Na_2S_2O_3$, [$E°$ for $S_4O_6^{2-}(aq) + 2e^- \rightarrow 2S_2O_3^{2-}(aq)$ is 0.17 V]; **(e)** sodium fluoride.

26. Predict whether a chemical change will occur when equal volumes of the following 2 *M* solutions are mixed: **(a)** mercury(I) nitrate and potassium dichromate; **(b)** iron(II) nitrate and silver nitrate, **(c)** potassium bromide and copper(II) sulfate; **(d)** iron(II) sulfate and sulfuric acid; **(e)** iron(II) sulfate and nitric acid. Where you predict a reaction to occur, write a net ionic equation for the reaction.

27. Predict whether a reaction will occur when the following substances are mixed, and write a net ionic equation for each of the reactions: **(a)** a piece of manganese is placed into 1 *M* hydrochloric acid; **(b)** a 1 *M* solution of iron(II) sulfate is treated with hydrogen peroxide in acid solution; **(c)** a 1 *M* solution of lead(II) nitrate is treated with hydrogen peroxide in basic solution; **(d)** oxygen gas at 1 atm pressure is passed through a 1 *M* acid solution of mercury(I) nitrate; **(e)** a 1 *M* solution of mercury(I) nitrate is treated with powdered aluminum.

28. From Table 17.1 select a suitable substance that can be used to effect each of the following transformations (assuming that all substances are present in 1 *M* concentrations) and write a net ionic equation for each transformation with the substance you select: **(a)** a reducing agent capable of reducing Ag^+ to Ag but not Cu^{2+} to Cu; **(b)** an oxidizing agent capable of oxidizing Fe^{2+} to Fe^{3+} but not Hg_2^{2+} to Hg^{2+}; **(c)** an oxidizing agent capable of oxidizing I^- to I_2 but not Br^- to Br_2; **(d)** a reducing agent capable of reducing I_2 to I^- but not Br_2 to Br^-; **(e)** an oxidizing agent capable of oxidizing Cr^{3+} to $Cr_2O_7^{2-}$ but not Mn^{2+} to MnO_4^- in acid solution.

Applications of the Nernst Equation

29. Predict the effect on the voltage of the reaction

$$Ag^+(aq) + Fe^{2+}(aq) \longrightarrow Ag(s) + Fe^{3+}(aq)$$

caused by each of the following changes: **(a)** the silver ion concentration is increased; **(b)** more elemental silver is added; **(c)** the Fe^{3+} ion concentration is decreased; **(d)** the Fe^{3+} ion concentration is increased.

30. Calculate the cell potential for the reaction in question 29 when $[Ag^+] = 1.0\ M$, $[Fe^{2+}] = 1.0\ M$, and $[Fe^{3+}] = 0.0010\ M$.

31. Calculate the cell potential for the reaction in question 29 when $[Ag^+] = 2.0\ M$, $[Fe^{2+}] = 3.0\ M$, and $[Fe^{3+}] = 0.0010\ M$.

32. If the Ag^+ and Fe^{2+} ion concentrations in question 29 were kept 1.0 *M* each, what should the Fe^{3+} ion concentration be for the cell potential to be zero?

33. The Daniell cell is a cell in which the electrolyte solutions of the anode and cathode compartments are kept in separate layers without a salt bridge or a porous partition. This can be done because one of the solutions has a higher density than the other. This type of a cell is also called a gravity cell. In a Daniell cell, the electrolyte of the cathode compartment is a saturated (1.3 *M*) $CuSO_4$ solution, which is the more dense bottom layer. In this layer, solid copper metal is immersed. The upper anode layer consists of 0.10 *M* $ZnSO_4$ solution with zinc metal as the electrode. Calculate the initial voltage of the cell at 25°C.

34. Calculate the potential of the cell

$$Fe(s) \mid Fe^{2+}(0.010\ M) \parallel Ag^+(0.50\ M) \mid Ag$$

Is the cell reaction spontaneous?

35. A galvanic cell

$$Pt \mid H_2(g, 1\ \text{atm}) \mid H^+(x\ M) \parallel H^+(0.10\ M) \mid H_2(g, 1\ \text{atm}) \mid Pt$$

has a potential of 0.110 V. Calculate the pH of the solution in the anode compartment.

36. The potential of the cell

$$Cu(s) \mid Cu^{2+}(3.80\ M) \parallel Ag^+(x\ M) \mid Ag$$

is 0.380 V. What is the Ag^+ ion concentration in the cathode compartment?

Electrolysis

37. Write the anode and cathode half-reactions and the overall reaction for the electrolysis of each of the following substances: **(a)** molten lithium bromide; **(b)** concentrated aqueous solution of lithium bromide; **(c)** molten strontium chloride; **(d)** water; **(e)** aqueous solution of potassium sulfate; **(f)** aqueous solution of silver nitrate; **(g)** concentrated aqueous solution of copper(II) chloride.

38. In the electrolysis of molten calcium chloride, how many faradays and how many coulombs would produce: **(a)** 1.00 mol of calcium; **(b)** 1.00 mol of chlorine, Cl_2; **(c)** 1.00 g of calcium; **(d)** 1.00 g of chlorine, Cl_2; **(e)** 1.00 L of chlorine at STP; **(f)** 1.00 L of chlorine at 15°C and 700 torr?

39. In the electrolysis of molten gallium(III) bromide, $GaBr_3$, how many faradays and how many coulombs would decompose: **(a)** 1.00 mol of gallium bromide; **(b)** 1.00 g of gallium bromide?

40. Molten gallium bromide is electrolyzed with a current of 3.00 A for 8.00 h. Calculate: **(a)** the mass of gallium produced; **(b)** the mass of bromine produced; **(c)** the mass of gallium bromide decomposed.

41. How long should copper(II) sulfate be electrolyzed with a current of 5.00 A to obtain 5.00 g of copper?

42. If a current of 10.0 A is passed through concentrated hydrochloric acid for 2.00 h, what volume of chlorine gas at 40°C and 765 torr can be obtained?

43. In a nickel plating process involving an Ni(II) salt solution, both nickel metal and some hydrogen gas are produced at the cathode. In this process, 49.2 g of nickel is produced in exactly 2 h using a steady 25.0-A current. What percent of the current is used for the deposition of nickel?

***44.** Chromium(III) sulfate can be converted to chromic acid, H_2CrO_4 (used in chromium plating), by oxidation at the anode in an electrolytic cell. Write an equation for the anode half-reaction and calculate the number of ampere-hours that must be used to oxidize 1.00 L of 0.500 *M* chromium(III) sulfate solution to chromic acid assuming 80 percent current efficiency?

***45.** Aniline, $C_6H_5NH_2$ (used in dyes), can be prepared by electrolytic reduction of nitrobenzene according to the half-reaction

$$C_6H_5NO_2 + 6H^+ + 6e^- \longrightarrow C_6H_5NH_2 + 2H_2O$$

How many ampere-hours is required to produce 0.500 kg of aniline?

* Denotes a problem of greater than average difficulty.

CHAPTER 18

An industrial method of making steel. Molten steel is poured into molds.

Chemistry of the Metals

In Chapter 2 we considered some aspects of the chemistry of metals. We discussed combination reactions of metals with nonmetals. These combination reactions are redox reactions in which the metal is oxidized and the nonmetal is reduced. When metals corrode, they are oxidized by various substances in the atmosphere and hydrosphere. Redox reactions of metals with nonmetals produce the natural deposits of most metals. Pure metals are recovered from these natural deposits, called ores, by reducing them.

In Chapter 4 we discussed the fundamental principles of redox reactions and the methods for writing balanced equations for these reactions. In Chapter 17 we considered redox reactions from the point of view of electrochemistry. In this chapter we concentrate on the redox chemistry of metals.

18.1 Oxidation States of Metals; Oxidizing and Reducing Agents

Oxidation States of Metals

Since the oxidation states of metals change in redox reactions, we now consider the common oxidation states of metals in their compounds. We often need to know whether a substance can be oxidized or reduced. This can be predicted if we know the lowest and highest oxidation states of an element.

Metals cannot usually be reduced to stable negative ions in ordinary chemical reactions. Thus, the zero oxidation state is the lowest oxidation state for metals. Some metals, such as sodium or aluminum, can be oxidized to only one positive, stable oxidation state. Other metals, such as iron and chromium, can be oxidized to two or more positive states.

If an element is in its lowest oxidation state, it cannot be reduced any further, but it can be oxidized to higher states. If the element is in its highest oxidation state, it cannot be oxidized, but it can be reduced. An element in an intermediate oxidation state can be either oxidized or reduced. For example, Fe^{2+} can be oxidized to Fe^{3+} or reduced to elemental iron, Fe. But Al^{3+} can be only reduced to elemental aluminum, Al, because +3 is the maximum oxidation state for aluminum, and there is no intermediate state.

Table 18.1 lists the oxidation states of metals and nonmetals in compounds. The alkali metals (group I) exhibit only a +1 oxidation state in their compounds because an atom of an alkali metal has only one valence electron. The alkaline earth metals (group II) exhibit only a +2 state in their compounds because their atoms have only two valence electrons. Aluminum, in group III, has an oxidation state of +3 in its compounds because an aluminum atom can lose all three of its valence electrons in a redox reaction.

The heavier metals in group III—indium and thallium—can acquire a +3 or a +1 state. The +1 state is more stable than the +3 state for the heaviest member, thallium. Thus, thallium(I) chloride is more common and more stable than thallium(III) chloride. The valence shell electron configuration of an atom of a group III metal is ns^2np^1. When an atom of a group III metal loses its single outer *p* electron, the +1 oxidation state forms; when a group III atom loses its outer *p* electron and its two outer *s* electrons, the +3 oxidation state forms.

The heavier members of group IV, tin and lead, can have oxidation states of +2 or +4. The +2 and +4 states for tin are commonly encountered and have about equal stability. For lead, the +2 state is more stable and more common than the +4 state. An atom of group IV metal has a valence shell electron

Table 18.1

Oxidation States of Elements in Their Compounds[a]

I 1	II 2	3	4	5	6	7	8	9	10	11	12	III 13	IV 14	V 15	VI 16	VII 17	VIII 18
H +1 −1																H +1 −1	He
Li +1	Be +2											B +3	C −4 +2 +4 (−2) (0)	N −3 (−2) (−1) +1 +2 +3 +4 +5	O −2 −1 (−$\frac{1}{2}$) (+2) (+1)	F −1	Ne
Na +1	Mg +2											Al +3	Si (−4) +4	P −3 (+1) +3 +5	S −2 +2 +4 +6	Cl −1 +1 (+2) +3 (+4) +5 (+6) +7	Ar
K +1	Ca +2	Sc 3+	Ti (+2) (+3) +4	V (+2) (+3) +4 +5	Cr (+2) +3 (+4) (+5) +6	Mn +2 (+3) +4 (+6) +7	Fe +2 +3 (+6)	Co +2 (+3)	Ni +2 (+3) (+4)	Cu +1 +2	Zn +2	Ga +3	Ge +4	As −3 +3 +5	Se −2 +2 +4 +6	Br −1 +1 +3 +5 +7	Kr +2 +4
Rb +1	Sr +2	Y +3	Zr +4	Nb (+2) (+3) (+4) +5	Mo (+2) (+3) +4 (+5) +6	Tc +4 (+6) +7	Ru +3 +4 (+5) (+6) (+7) (+8)	Rh (+2) +3 (+4) (+6)	Pd +2 +4	Ag +1 (+2)	Cd +2	In (+1) +3	Sn +2 +4	Sb (−3) +3 +5	Te −2 +2 +4 +6	I −1 +1 +3 +5 +7	Xe +2 +4 +6
Cs +1	Ba +2	La +3	Hf +4	Ta (+4) +5	W (+2) +4 +6	Re (+2) (+3) +4 (+5) (+6) +7	Os (+2) +4 (+5) (+6) (+7) +8	Ir (+2) +3 +4 (+6)	Pt +2 +4 (+6)	Au (+1) +3	Hg +1 +2	Tl +1 +3	Pb +2 +4	Bi +3 (+5)	Po +2 (+4)	At −1	Rn

[a] The oxidation states which are less common are shown in parentheses.

configuration of ns^2np^2. The +2 oxidation state forms when an atom of group IV metal loses its two outer p electrons. In compounds where a group IV metal is in its +4 oxidation state, the bonding is predominantly covalent. For example, lead(II) chloride, $PbCl_2$, is a salt that consists of Pb^{2+} and Cl^- ions and which melts at 501°C. In contrast, lead(IV) chloride, $PbCl_4$, is an oily liquid that freezes at −15°C. (We recall that compounds which are liquid at room temperature are molecular compounds.) Similarly, tin(II) chloride is an ionic compound that is a solid at room temperature, but tin(IV) chloride is a molecular compound, a liquid that freezes at −33°C.

For bismuth, in group V, both +3 and +5 oxidation states exist. Bismuth in the +3 state forms relatively stable ionic compounds such as $BiCl_3$. In the +5 state, bismuth is found only in covalent combination with nonmetals in oxoanions. Of these, one of the most common is bismuthate ion, BiO_3^-. Bismuthate ion is a powerful oxidizing agent that is readily reduced to bismuth(III).

Unlike the oxidation states of most representative elements, the oxidation states of many transition elements cannot be easily predicted from their electron structures. The metals in the 3*d* series, from titanium through zinc, form 2+ ions. All except chromium and copper lose their two 4*s* electrons; chromium and copper lose their single 4*s* electron and one 3*d* electron. A scandium atom loses its 4*s* electrons and its 3*d* electron to form a 3+ ion. The maximum oxidation state of a metal in the scandium through the manganese families equals the group number (Table 18.1).

To the right of manganese, starting with iron in the 3*d* series of transition metals, there is a sharp drop in the commonly observed maximum oxidation states (Table 18.1). Iron exhibits both +2 and +3 states, of which the +3 state is the more stable one. Other relatively unstable oxidation states for iron are +4 and +6. For cobalt, nickel, copper, and zinc, the +2 state is the most stable state. For cobalt and nickel, the +3 state also exists, and the +1 state of copper exists. Nickel also forms relatively unstable +4 oxidation state. The +2 oxidation state for zinc is the only commonly observed oxidation state.

In the 4*d* and 5*d* series of elements, the general trends in oxidation states are similar to that in the 3*d* series with only a few exceptions, which we will not discuss here.

Example 18.1

Predicting Whether a Substance Can Be Oxidized, Reduced, or Both

Predict whether the metal in each of the following substances can be oxidized, reduced, or both oxidized and reduced: **(a)** Sn^{2+}; **(b)** MnO_4^-.

SOLUTION

a. Tin in an Sn^{2+} ion can be oxidized to the +4 state as in $SnCl_4$ or it can be reduced to elemental Sn.

b. The oxidation state of manganese in the MnO_4^- ion is +7, which is the highest state possible for manganese. Therefore, the MnO_4^- ion can be reduced but not oxidized.

Practice Problem 18.1: Predict whether each of the following substances can be reduced, oxidized, or both reduced and oxidized: **(a)** Fe; **(b)** Zn^{2+}; **(c)** Pb^{2+}.

Oxidizing and Reducing Agents

In research and industry it is often necessary to oxidize or reduce a substance. For that purpose we need to find another substance that is an oxidizing or reducing agent. We recall from Chapter 17 that a good oxidizing agent is a substance that has a high reduction potential, and a good reducing agent is a substance that has a high oxidation potential.

If we need an oxidizing agent, we look for a substance that can be reduced. That substance can be an element or a compound. It cannot be an uncombined metal because metals cannot be reduced in ordinary chemical reactions. However, ions of silver and copper are easily reduced and can therefore be used as oxidizing agents, as reflected by their positive standard reduction potentials:

$$Ag^+(aq) + e^- \longrightarrow Ag(s) \qquad E° = 0.80\ V$$

$$Cu^{2+}(aq) + 2e^- \longrightarrow Cu(s) \qquad E° = 0.34\ V$$

An oxidizing agent can also be a nonmetal that can be reduced to its negative ions, or it can be a compound or a polyatomic ion that contains an element in a high oxidation state that can be reduced to lower states.

If we wish to use a nonmetal as an oxidizing agent, we select one that is easy to reduce. These are located in the upper right portion of the periodic table: fluorine, oxygen, and chlorine. Compounds that are good oxidizing agents include those that contain polyatomic ions in which the central element is in a high oxidation state. For example, potassium permanganate, $KMnO_4$, and potassium dichromate, $K_2Cr_2O_7$, are often used as oxidizing agents for reactions in aqueous solution. Permanganate and dichromate ions are good oxidizing agents in aqueous solution because they have relatively high positive reduction potentials:

$$MnO_4^-(aq) + 8H^+(aq) + 5e^- \longrightarrow Mn^{2+}(aq) + 4H_2O(l) \qquad E° = 1.491\ V$$

$$Cr_2O_7^{2-}(aq) + 14H^+(aq) + 6e^- \longrightarrow 2Cr^{3+}(aq) + 7H_2O(l) \qquad E° = 1.33\ V$$

Since the reduction potential for permanganate ion is higher than that for dichromate ion, permanganate ion is a better oxidizing agent than dichromate ion.

A good reducing agent is easily oxidized. Active metals such as the representative metals in groups I and II, aluminum in group III, and transition metals such as zinc and iron are often used as reducing agents. The standard potentials for oxidizing these metals in aqueous solution have relatively large positive values, as shown below for magnesium and aluminum:

$$Mg(s) \longrightarrow Mg^{2+}(aq) + 2e^- \qquad E° = 2.38\ V$$

$$Al(s) \longrightarrow Al^{3+}(aq) + 3e^- \qquad E° = 1.66\ V$$

Of the nonmetals, carbon and hydrogen are often used in industrial processes at high temperatures to reduce metals from their compounds.

Example 18.2

Predicting Whether a Substance Is an Oxidizing Agent or a Reducing Agent

Which of the following species can act only as an oxidizing agent? Which can act only as a reducing agent? Which can act as both an oxidizing agent and a reducing agent? **(a)** Sn; **(b)** BiO_3^-; **(c)** Pb^{2+}.

SOLUTION

a. Elemental tin can be oxidized but it cannot be reduced. Therefore, elemental tin can act as a reducing agent, but not as an oxidizing agent.

b. Bismuth in bismuthate ion, BiO_3^-, is in its highest oxidation state, +5. Bismuthate ion can be reduced but not oxidized. Therefore, bismuthate ion can act as an oxidizing agent, but not as a reducing agent.

c. Lead(II) ion can be oxidized or reduced, and can therefore react either as a reducing agent or as an oxidizing agent.

Practice Problem 18.2: Which of the following substances can act as oxidizing agents? Which can act as reducing agents? Which can act as both oxidizing and reducing agents? **(a)** Nitric acid; **(b)** potassium metal; **(c)** sodium dichromate; **(d)** tin(II) ions.

18.2 Oxidation of Metals

We know that metals can be oxidized, and metal ions can be reduced back to their elemental states. Thus, oxidation of elemental metals and reduction of their ions are two important types of reactions of metals. In this section we discuss the energetics of oxidation of metals to their respective ions. In Section 18.3 we discuss the energetics of reduction of metal ions.

Energetics of Oxidation of Metals

The oxidation of a solid metal to its stable ions in aqueous solution may be imagined to occur in three steps. First, the solid metal converts to gaseous atoms. This process is called *sublimation*. Next, the gaseous atoms convert to gaseous ions in a process called *ionization*. Third, gaseous ions combine with water to form hydrated ions in solution—a process of *hydration*. These steps are outlined in Figure 18.1. The first two processes are endothermic, and the third is highly exothermic.

If the energy changes in these steps are measured at constant pressure, they are the enthalpy changes (Chapter 5). The sum of the enthalpy changes for the three steps equals the enthalpy change for the entire process of oxidation:

$$\Delta H_{\text{sub}} + \Delta H_{\text{i}} + \Delta H_{\text{hyd}} = \Delta H_{\text{ox}}$$

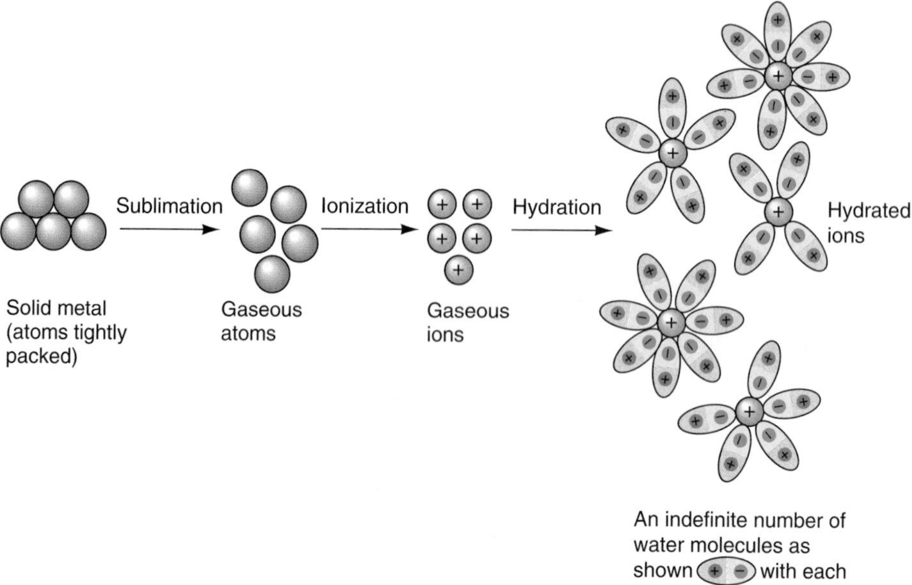

Figure 18.1

Conversion of a solid metal to its aqueous ions. This conversion can be imagined to occur in three steps: (1) The conversion of solid metal to gaseous atoms; (2) the conversion of the gaseous atoms to gaseous ions; and (3) the hydration of the ions in aqueous solution.

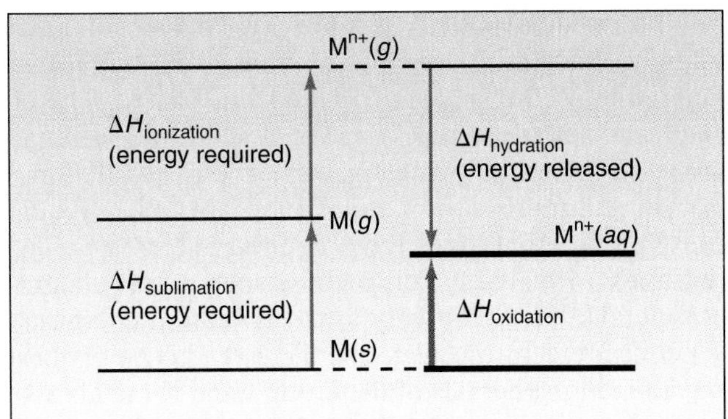

Figure 18.2

Energetics of the oxidation $M(s) \rightarrow M^{n+}(aq)$. Energy is required for sublimation and ionization, and energy is released by hydration. The algebraic sum of these energies is the ΔH of the process (if measured at constant pressure).

In Figure 18.2 the endothermic steps in the oxidation of a metal in aqueous solution are shown by arrows pointed upward and the exothermic step by an arrow pointed downward. The length of each arrow indicates the approximate relative magnitude of the enthalpy change. The net enthalpy change, ΔH_{ox}, is indicated by a different-colored arrow for emphasis.

Figures 18.1 and 18.2 describe the processes and energetics of formation of aqueous ions from a solid metal. If a metal is oxidized to form a solid ionic compound, the energy released in the formation of the solid lattice from the gaseous ions is called the *lattice energy*. In this case the total energy change for the oxidation process is given by

$$\Delta H_{ox} = \Delta H_{sub} + \Delta H_i + \Delta H_{lattice}$$

The sublimation, ionization, and hydration energies of some common metals are listed in Table 18.2. When we examine Table 18.2, we find that the enthalpies of formation of aqueous ions from solid metals are all positive. Thus, the process is endothermic. Table 18.2 also shows that the enthalpy change for the oxidation of a solid metal, $M(s)$, to its aqueous ions, $M^{n+}(aq)$, is lowest for the very active metals such as the alkali metals and the alkaline earth metals.

Table 18.2

Enthalpies of Oxidation of Metals in Their Standard States to Aqueous Ions

Metal	Ion	ΔH_{sub} (kJ mol^{-1})	ΔH_i (kJ mol^{-1})	ΔH_{hyd} (kJ mol^{-1})	ΔH_{ox} for $M(s) \rightarrow M^{n+}(aq)$ (kJ mol^{-1})
K	K^+	89.5	418	−314	194
Na	Na^+	108	495	−397	206
Ca	Ca^{2+}	177	1733	−1598	312
Ba	Ba^{2+}	178	1466	−1322	322
Mg	Mg^{2+}	149	2185	−1941	393
Al	Al^{3+}	324	5132	−4749	707
Zn	Zn^{2+}	126	2637	−2029	734
Cd	Cd^{2+}	111	2496	−1791	816
Fe	Fe^{2+}	416	2321	−1908	829
Ni	Ni^{2+}	431	2485	−2080	836
Pb	Pb^{2+}	197	2164	−1448	913
Cu	Cu^{2+}	339	2700	−2088	951
Hg	Hg^{2+}	61.5[a]	2813	−1807	1068

[a] For Hg this value represents ΔH vaporization since Hg is a liquid.

Metals of intermediate activity, such as iron and nickel, have larger enthalpies of oxidation. And the oxidation enthalpies for "noble metals" such as copper and mercury are the largest of all the metals listed. Therefore, the terms "active metal" and "noble metal" are not "merely descriptive" but have an underlying basis in the enthalpy change required for oxidation.

We can gain still more information from Table 18.2. Examining sublimation enthalpies, we find that the energy required to sublime a metal depends in part on its hardness. Relatively soft metals, such as the alkali metals and the alkaline earth metals, have relatively low sublimation energies. Low sublimation energies contribute to the high chemical reactivity of these metals. The relatively low ionization energies of the group I and II metals also contribute to the high chemical reactivity of these metals.

Of the metals listed in Table 18.2, iron has next to the highest sublimation energy and is one of the hardest metals of the group. However, because of the high hydration energy of Fe^{2+} ions, iron is a metal of intermediate activity. For noble metals, the sum of the sublimation and ionization enthalpies is a relatively large, positive quantity, whereas their hydration enthalpies are smaller negative quantities. The oxidation of a noble metal to its ions in aqueous solution is therefore more endothermic and more difficult than the oxidation of an active or a moderately active metal.

Next, let us examine the trends in hydration energies. The hydration energy of an ion depends on its *charge-to-radius ratio*, also called the **ionic potential**. An ion with a high ionic potential has a high charge concentrated in a small volume. Ions having large ionic potentials strongly attract water molecules and have large negative enthalpies of hydration. Alkali metal ions generally have lower hydration energies than ions of alkaline earth metals, which as a group are smaller and more highly charged than alkali metal ions.

The effect of ionic potential on hydration energy is most prominently displayed by the Al^{3+} ion, which has the highest hydration energy of the ions listed in Table 18.2. Comparing Al^{3+} with Mg^{2+}, we see that not only is the charge of the Al^{3+} ion greater than that of Mg^{2+}, but that an Al^{3+} ion has radius of only 0.057 nm, compared with 0.075 nm for Mg^{2+}.

Example 18.3

Predicting Relative Reducing Powers of Metals from Their Enthalpies of Oxidation

Metal X forms X^+ ions and metal Y forms Y^+ ions. The sublimation and ionization enthalpies of metals X and Y are nearly equal. The radius of an X^+ ion is considerably smaller than the radius of a Y^+ ion. Which of the two metals would you predict to be more readily oxidized to its 1+ ion in aqueous solution? Which of the two metals, X or Y, would you predict to be the better reducing agent?

SOLUTION: Since the charges of X^+ and Y^+ ions are equal, the radius of the ion determines the ionic potential and thus the hydration energy. The smaller the radius, the greater the concentration of positive charge in the ion, and the greater the ionic potential. Because the X^+ ion has a greater ionic potential than the Y^+ ion, the hydration energy of X^+ ions is greater than that of Y^+ ions. Therefore, metal X is more readily oxidized to its aqueous ions than metal Y. Since metal X is more readily oxidized than Y, X is a better reducing agent than Y.

Practice Problem 18.3: The sublimation energy of lithium metal is 155 kJ mol^{-1}, its ionization energy is 519 kJ mol^{-1}, and the hydration energy of lithium ions is −523 kJ mol^{-1}. For cesium metal, these energies are 79, 377, and −285 kJ mol^{-1}, respectively. Which of the two metal ions is likely to have the greater ionic potential? On the basis of the data above, which of the two metals is likely to be the better reducing agent?

Active and Noble Metals

We recall that alkali metals, alkaline earth metals, and aluminum are among the easiest metals to oxidize. They are therefore among the most active metals. Since they are easily oxidized, they do not exist as pure metals in nature. They are found in ionic solids as oxides, halides, sulfates, silicates, and carbonates.

Active metals react readily with most nonmetals, acids, and even water. For example, we recall that sodium reacts violently with water to liberate hydrogen gas:

$$2Na(s) + 2H_2O(l) \longrightarrow 2Na^+(aq) + 2OH^-(aq) + H_2(g)$$

Magnesium or aluminum reacts with cold water only to a small extent, but both react readily with steam or hot water. The reactions of the alkali metals with acids are explosive. Magnesium and aluminum, on the other hand, react at a moderate rate with acids, and these reactions can be used to prepare hydrogen gas in small quantities. Aluminum reacts with acid according to the following net ionic equation:

$$2Al(s) + 6H^+(aq) \longrightarrow 2Al^{3+}(aq) + 3H_2(g)$$

When dilute sulfuric acid reacts with a metal, pure hydrogen evolves as a product. In contrast, nitric acid reacts to produce hydrogen and also oxides of nitrogen as the reduction products of nitrate ions. When a metal reacts with hydrochloric acid, no competing side reactions occur, but the hydrogen produced is contaminated by gaseous hydrogen chloride, whose vapors are present above the solution.

Active metals react with oxygen and with halogens to form oxides and halides, respectively, as we discussed in Chapter 2. In a limited supply of oxygen, all alkali metals form the corresponding oxides. Lithium reacts according to the following equation:

$$4Li(s) + O_2(g) \longrightarrow 2Li_2O(s)$$

In a plentiful supply of oxygen, only lithium forms an oxide. Sodium forms sodium peroxide,

$$2Na(s) + O_2(g) \longrightarrow Na_2O_2(s)$$

Potassium, rubidium, and cesium form *superoxides* with the general formula MO_2 (where M = K, Rb, or Cs). The reaction of potassium with excess oxygen is

$$K(s) + O_2(g) \longrightarrow KO_2(s)$$

All of the alkaline earth metals and aluminum react with oxygen in an open atmosphere to produce their corresponding oxides. In excess oxygen, barium also forms a superoxide, BaO_2. The oxides of both the alkali metals and the alkaline earth metals produce hydroxides when treated with water. These hydroxides are strong bases, although some of them are only sparingly soluble in water. Thus, the oxides of both the alkali metals and the alkaline earth metals are basic anhydrides (Chapter 9).

The reaction of barium oxide with water is a typical example of the reaction of an alkaline earth metal oxide:

$$BaO(s) + H_2O(l) \longrightarrow Ba(OH)_2(s)$$

An alkaline earth metal hydroxide can be converted back to the metal oxide by heating the hydroxide to drive off water:

$$Ba(OH)_2(s) \xrightarrow{\text{heat}} BaO(s) + H_2O(g)$$

Metals of intermediate activity include most of the transition metals and two common representative metals, tin and lead. Like the active metals, the moderately active metals exist in nature mostly as compounds and rarely as pure elements.

Tin and lead react with various nonmetals at elevated temperatures to form binary compounds in which the metals are in their +2 oxidation state. Tin(II) compounds can easily be oxidized to tin(IV) compounds by many oxidizing agents. Thus, tin(II) compounds can be used as reducing agents. Tin(IV) compounds are oxidizing agents because they can be reduced to the tin(II) state or to elemental tin.

Almost all of the commonly encountered stable compounds of lead are Pb(II) compounds such as $PbCl_2$, $PbSO_4$, PbO, PbS, and $PbCO_3$. The few common lead(IV) compounds include lead(IV) oxide, PbO_2 (used in batteries), and tetraethyllead, $Pb(C_2H_5)_4$. Until recently, tetraethyllead was used as an additive in gasoline to prevent the misfiring known as "knocking." But lead released in automobile exhaust is a dangerous environmental pollutant (it causes mental retardation), and gasoline sold in the United States no longer contains lead.

The noble metals (Chapter 2) are the most difficult to oxidize to their ions. The common noble metals are gold, platinum, mercury, silver, and copper. These metals are easy to remember since they are located in the periodic table in the form of an inverted T:

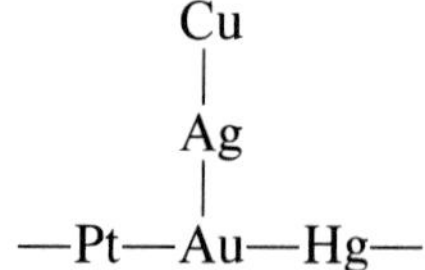

Gold and platinum exist mostly as free elements in nature. These metals are also found to some extent in ores, several of which contain sulfur and tellurium. Copper and silver exist in both free and combined states. Mercury exists mostly in the form of mercury(II) sulfide, called cinnabar, HgS. Some metal ores, as well as native gold and silver, are shown in Figure 18.3.

Activity Series of Metals

In Chapter 17 we learned that the standard reduction potential of the ions of an active metal such as sodium has a highly negative value (Table 17.1). This means that it is difficult to reduce sodium ions to sodium metal, but it is easy to oxidize sodium metal to its ions. On the other hand, the standard reduction potential of the ions of a relatively unreactive metal, copper, has a positive value. This means that it is much easier to oxidize sodium metal than copper metal.

Figure 18.3

Some common metal ores of: (a) gold; (b) silver; (c) iron; (d) lead; (e) zinc.

Metals can be arranged in order of their decreasing ease of oxidation to their respective ions in aqueous solution. This arrangement is called the **activity series** (Table 18.3). We can see from Table 18.3 that the very active metals react with cold water to liberate hydrogen. The less active metals react with steam and with most common acids such as HCl. Metals above hydrogen in the activity series (Table 18.3) react spontaneously with a 1 M nonoxidizing acid such as HCl. That is, the standard potentials of the reactions of these metals with 1 *M* HCl have positive values.

Table 18.3

Activity Series of Common Metals

Li, K, Ba, Sr, Ca, Na	*Very active metals;* react with cold water with the liberation of hydrogen gas; they also react violently with acids (K and Na also react violently with water)
Mg, Al, Mn, Zn, Cr, Fe, Cd	*Metals of intermediate activity;* react with steam or with acids such as HCl with liberation of hydrogen
Co, Ni, Sn, Pb	*Moderately active metals;* react slowly with HCl
H_2	
Cu, Ag, Hg	*Moderately noble metals;* do not react with HCl, but react with oxidizing acids such as HNO_3 and $HClO_4$ that contain oxoanions
Pt, Au	*Very noble metals;* react only with aqua regia

(a)

(b)

Figure 18.4

Reaction of sodium with hydrochloric acid: (a) piece of sodium in hydrochloric acid just before the reaction; (b) explosive reaction causing fire.

Very active metals such as sodium react explosively with acids. The reaction of sodium with hydrochloric acid can liberate so much heat that the hydrogen gas produced reacts explosively with atmospheric oxygen causing fire (Figure 18.4).

Reactions of acids with metals of intermediate activity, such as magnesium and iron, occur relatively smoothly. Magnesium reacts with steam and with hydrochloric acid according to the following equations:

$$\underset{}{Mg(s)} + \underset{\text{Steam}}{2H_2O(g)} \longrightarrow Mg(OH)_2(s) + H_2(g)$$

$$Mg(s) + 2HCl(aq) \longrightarrow MgCl_2(aq) + H_2(g)$$

Note that hydrogen gas is produced in both reactions.

Aluminum reacts with hot water only to a small degree. The exposed surface of aluminum forms a hard, tightly adhering, protective coating of aluminum hydroxide and aluminum oxide which protects the metal from further corrosion. Thus, aluminum is used in storm doors and windows and as siding for houses.

The moderately active metals listed in Table 18.3—cobalt, nickel, tin, and lead—react spontaneously with 1 *M* hydrochloric acid but not with steam.

The metals below hydrogen in the activity series (Table 18.3) do not react with 1 *M* hydrochloric acid. They are not able to reduce hydrogen ions in 1 *M* solution, but their 1 *M* aqueous ions can be reduced by hydrogen gas at 1 atm. These metals are sometimes called noble metals. The moderately noble metals—copper, silver, and mercury—react only with oxidizing acids such as nitric acid or perchloric acid, which contain easily reducible oxoanions. In the reaction of a moderately noble metal with nitric acid, nitrate ions are reduced to brown NO_2 gas, or to colorless NO or N_2O, depending on the concentration of the acid. At high nitric acid concentrations more NO_2 is produced. The reaction of copper with concentrated nitric acid is

$$Cu(s) + 2NO_3^-(aq) + 4H^+(aq) \longrightarrow Cu^{2+}(aq) + 2NO_2(g) + 2H_2O(l)$$

The very noble metals, platinum and gold, react with aqua regia (the Latin term for "royal water," which was so named because of its ability to dissolve gold, the "king" of metals). Aqua regia is a mixture of 1 part of concentrated

(a)

(b)

Figure 18.5

Dissolving of gold in aqua regia. (a) Gold shred being immersed in aqua regia. (b) Most of the shred has dissolved.

HNO_3 and 3 parts of concentrated HCl. Aqua regia dissolves noble metals through the combined oxidizing ability of nitrate ions in strong acid and the formation of stable chloro complex ions with the oxidized metal ion (Figure 18.5). The net ionic equation for dissolving gold in aqua regia is

$$Au(s) + 6H^+(aq) + 4Cl^-(aq) + 3NO_3^-(aq) \longrightarrow AuCl_4^-(aq) + 3NO_2(g) + 3H_2O(l)$$

Example 18.4

Predicting Which Metals React with H_2O, HCl, and HNO_3

Which one of the following metals, Zn, Hg, or Ca, reacts with water at room temperature? Which one reacts with both concentrated hydrochloric acid and with concentrated nitric acid but not with water at room temperature? Which one reacts only with nitric acid but not with water or hydrochloric acid? Write a net ionic equation for each reaction (zinc reduces nitrate ions to ammonium ions).

SOLUTION: From Table 18.3 we see that calcium is an active metal that reacts with water. An equation for the reaction is

$$Ca(s) + 2H_2O(l) \longrightarrow Ca(OH)_2(s) + H_2(g)$$

Zinc is a metal of intermediate activity that does not react with water at room temperature, but reacts with hydrochloric acid:

$$Zn(s) + 2H^+(aq) \longrightarrow Zn^{2+}(aq) + H_2(g)$$

In nitric acid, zinc reduces H^+ ions to H_2 gas and NO_3^- ions to NH_4^+ ions:

$$Zn(s) + 2H^+(aq) \longrightarrow Zn^{2+}(aq) + H_2(g)$$

$$4Zn(s) + 10H^+(aq) + NO_3^-(aq) \longrightarrow 4Zn^{2+}(aq) + NH_4^+(aq) + 3H_2O(l)$$

Mercury does not react with water or with hydrochloric acid, but it reacts with nitric acid, which is an oxidizing acid:

$$Hg(l) + 4H^+(aq) + 2NO_3^-(aq) \longrightarrow Hg^{2+}(aq) + 2NO_2(g) + 2H_2O(l)$$

Practice Problem 18.4: Write a balanced net ionic equation for each of the following reactions: **(a)** aluminum with hot water; **(b)** aluminum with hydrochloric acid; **(c)** lead with nitric acid to form lead(II) ions and nitrogen dioxide gas.

The order of metals in the activity series is related to their $E°$ values, but is only partially related to their positions in the periodic table. The activity of representative metals generally decreases from left to right within a period. For example, the order of decreasing activity of the third period metals is

$$\text{Na} > \text{Mg} > \text{Al}$$

But as Table 18.3 shows, the activities of the transition metals are not related to their positions in the periodic table.

18.3 Reduction of Metal Ions and Metallurgy

Reduction of Metals from Their Compounds

Since active metals are easy to oxidize, it is equally difficult to reduce their ions back to atoms, as shown by the following standard potentials:

$$\text{Na}(s) \longrightarrow \text{Na}^+(aq) + e^- \quad \text{(easy)} \qquad E° = 2.71\text{ V}$$

$$\text{Na}^+(aq) + e^- \longrightarrow \text{Na}(s) \quad \text{(difficult)} \qquad E° = -2.71\text{ V}$$

On the other hand, the noble metals are hard to oxidize, but their ions are easy to reduce:

$$\text{Cu}(s) \longrightarrow \text{Cu}^{2+}(aq) + 2e^- \quad \text{(difficult)} \qquad E° = -0.34\text{ V}$$

$$\text{Cu}^{2+}(aq) + 2e^- \longrightarrow \text{Cu}(s) \quad \text{(easy)} \qquad E° = 0.34\text{ V}$$

The relative ease of reduction of metal ions is an important factor in determining how the metal can be prepared from its compounds. Some metals can be reduced from their binary compounds by heating with a more active metal. In this process, the more active metal is oxidized, and its ions replace the ions of the less active metal in the compound. The ions of the less active metal are reduced to atoms. For example, sodium can reduce aluminum from its compounds, aluminum can reduce iron, and iron can reduce copper:

$$6\text{Na}(s) + \text{Al}_2\text{O}_3(s) \xrightarrow{\text{heat}} 3\text{Na}_2\text{O}(s) + 2\text{Al}(s)$$

$$2\text{Al}(s) + \text{Fe}_2\text{O}_3(s) \xrightarrow{\text{heat}} \text{Al}_2\text{O}_3(l) + 2\text{Fe}(l)$$

$$2\text{Fe}(s) + 3\text{CuO}(s) \xrightarrow{\text{heat}} \text{Fe}_2\text{O}_3(s) + 3\text{Cu}(s)$$

These reactions are all exothermic. A mixture of aluminum and iron(III) oxide, Fe_2O_3, is known as *thermite*. The reaction of aluminum with iron(III) oxide which produces aluminum oxide and iron is called the *thermite reaction*. The thermite reaction is extremely exothermic ($\Delta H = -845$ kJ mol^{-1} Al_2O_3 formed) and generates temperatures in excess of 2000°C. The mixture of Al and Fe_2O_3 must be heated to start the reaction. However, once the reaction starts, so much heat is liberated that it is not only sufficient to sustain the reaction, but

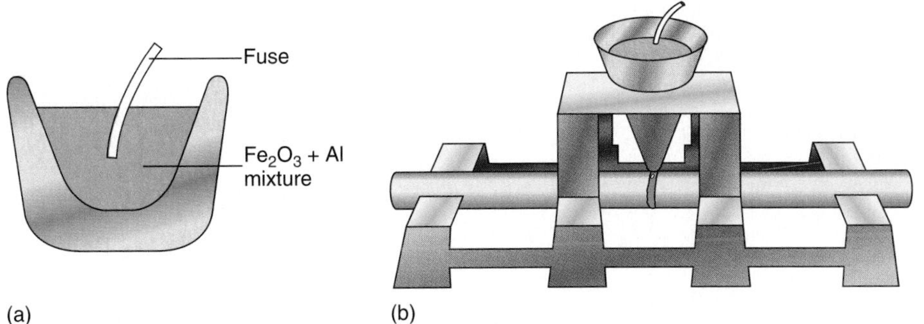

Figure 18.6

Thermite reaction: The reduction of iron(III) oxide with aluminum. (a) The reaction mixture and fuse. (b) Use of the molten iron produced in thermite reaction to weld an iron bar or railing.

the products, aluminum oxide and iron, are formed in their molten states. The molten iron formed by this reaction can be used for welding, as shown in Figure 18.6.

The enthalpy change for reducing a metal from its compound with another more active metal can be divided into two parts: (1) the enthalpy change for oxidizing the more active metal, and (2) the enthalpy change for reducing the less active metal from its compound.

To reduce a metal from its ionic compound, we can imagine that ions in the solid crystal lattice of a compound, MX, must first be converted to gaseous ions:

$$\mathrm{MX}(s) \longrightarrow \mathrm{M}^{n+}(g) + \mathrm{X}^{n-}(g)$$

This is an endothermic process for which the enthalpy change is designated as $\Delta H_{\text{lattice breakage}}$ or $\Delta H_{\text{lattice}}$.

Next, the gaseous metal ions attract electrons to form gaseous metal atoms:

$$\mathrm{M}^{n+}(g) + ne^- \longrightarrow \mathrm{M}(g)$$

This is an exothermic process, and its enthalpy change is called $\Delta H_{\text{electron attachment}}$ or ΔH_{ea}.

The gaseous metal atoms formed then convert to solid metal:

$$\mathrm{M}(g) \longrightarrow \mathrm{M}(s)$$

This process releases energy; its enthalpy change is $\Delta H_{\text{solidification}}$.

An active metal, A, which reduces a less active metal M from its lattice, is oxidized. The oxidation of metal A can be thought of as taking place in two steps. First, the solid metal is converted to gaseous atoms:

$$\mathrm{A}(s) \longrightarrow \mathrm{A}(g)$$

This process is sublimation. Sublimation is an endothermic process and its enthalpy change is represented by ΔH_{sub}. Next, the gaseous A atoms ionize to form gaseous A^{m+} ions:

$$\mathrm{A}(g) \longrightarrow \mathrm{A}^{m+}(g) + me^-$$

Ionization is also an endothermic process for which the enthalpy change is designated as ΔH_{i}.

As the last step in this overall redox process, we can imagine that the gaseous A^{m+} ions combine with the gaseous nonmetal ions, X^{n-}, from the MX lattice to form the new solid $\mathrm{A}_n\mathrm{X}_m$ lattice:

$$n\mathrm{A}^{m+}(g) + m\mathrm{X}^{n-}(g) \longrightarrow \mathrm{A}_n\mathrm{X}_m(s)$$

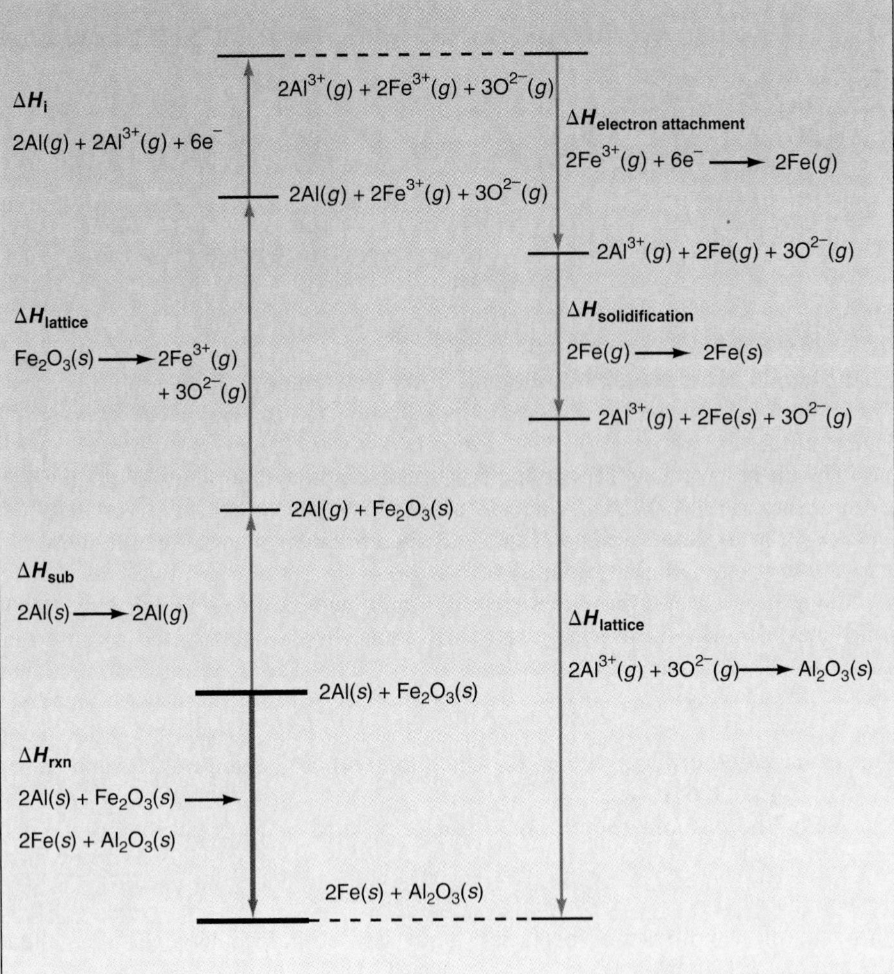

Figure 18.7

Energetics of the reaction $2Al(s) + Fe_2O_3(s) \rightarrow 2Fe(s) + Al_2O_3(s)$. Energy is supplied for the processes of sublimation of solid aluminum, breaking the solid Fe_2O_3 lattice into gaseous ions, and ionization of gaseous aluminum atoms. Energy is released when gaseous Fe^{3+} ions are reduced to Fe atoms. Energy is also released upon solidification of gaseous iron, and when solid lattice of Al_2O_3 forms from gaseous ions. The sum of these energy changes is the enthalpy change for the reaction $2Al(s) + Fe_2O_3(s) \rightarrow 2Fe(s) + Al_2O_3(s)$.

Figure 18.7 illustrates these enthalpy changes for a reaction of elemental iron with aluminum oxide. Endothermic steps are shown by arrows pointed upward, and exothermic steps by arrows pointed downward. The length of each arrow indicates the approximate relative magnitude of the enthalpy change. The sum of all the enthalpy changes in Figure 18.7 is the enthalpy change for the entire reaction, ΔH_{rxn}, which is indicated by a different-colored arrow for emphasis.

Active metals are not the only substances that can reduce less active metals from their oxides. Other reducing agents include carbon (in the form of coke), carbon monoxide, and hydrogen. These reducing agents are relatively inexpensive and readily available. Equations for the reduction of iron from iron(III) oxide with coke, carbon monoxide, and hydrogen, are

$$Fe_2O_3(s) + 3C(s) \longrightarrow 2Fe(s) + 3CO(g)$$

$$Fe_2O_3(s) + 3CO(g) \longrightarrow 2Fe(s) + 3CO_2(g)$$

$$Fe_2O_3(s) + 3H_2(g) \longrightarrow 2Fe(s) + 3H_2O(g)$$

The reducing agent coke is made by heating coal at 350 to 500°C in the absence of air. This treatment drives off the volatile organic compounds that are present in coal. The residue is further heated at 100°C, to produce coke, which is mostly carbon but contains some minerals.

Example 18.5

Predicting Reactions Based on the Activity Series of Metals

Predict whether or not a reaction occurs when each of the following pairs of substances is mixed under the conditions specified. Write a balanced equation for each reaction that occurs. **(a)** Powdered barium is heated with lead(II) oxide; **(b)** tin metal is heated with strontium oxide; **(c)** copper(II) oxide is heated in a stream of hydrogen gas.

SOLUTION

a. Since barium is more active than lead (Table 18.3), barium reacts with lead oxide. Barium is oxidized to form barium oxide and lead is reduced to elemental lead:

$$Ba(s) + PbO(s) \longrightarrow BaO(s) + Pb(s)$$

b. Tin is below strontium in the activity series (Table 18.3), so tin is not able to reduce strontium from its oxide.

c. We can predict that copper oxide reacts with hydrogen because copper is below hydrogen in the activity series (Table 18.3). This means that hydrogen gas can reduce copper(II) ions to elemental copper. The equation for the reaction is

$$CuO(s) + H_2(g) \longrightarrow Cu(s) + H_2O(g)$$

Practice Problem 18.5: Predict whether a reaction occurs when the following pairs of substances are mixed. Write equations for the reactions that do occur. **(a)** Magnesium oxide is heated in a stream of hydrogen gas; **(b)** lead(II) oxide is heated with powdered aluminum.

Reduction of Metal Ions in Aqueous Solution

An active metal can reduce a less active metal from its compounds or from a 1 *M* aqueous solution of its ions. For example, magnesium metal reduces zinc ions, zinc metal reduces iron(II) ions, and metallic iron reduces copper(II) ions:

$$Mg(s) + Zn^{2+}(aq) \longrightarrow Mg^{2+}(aq) + Zn(s)$$

$$Zn(s) + Fe^{2+}(aq) \longrightarrow Zn^{2+}(aq) + Fe(s)$$

$$Fe(s) + Cu^{2+}(aq) \longrightarrow Fe^{2+}(aq) + Cu(s)$$

The reduction of metal ions in solution by a more active metal is easy to demonstrate. For example, when a piece of magnesium ribbon is placed in a solution of zinc nitrate, a deposit of zinc metal forms on the surface of the magnesium ribbon (Figure 18.8a). Similarly, when a strip of zinc is placed in a solution of iron(II) nitrate, it becomes coated by metallic iron (Figure 18.8b), and when a strip of iron is put in a solution of copper(II) nitrate, it soon becomes covered by metallic copper (Figure 18.8c).

The three experiments illustrated in Figure 18.8 enable us to arrange Mg, Fe, Zn, and Cu in order of decreasing activity:

$$Mg > Zn > Fe > Cu$$

This order is the same as that given for these metals in Table 18.3. The same experiments show that aqueous Cu^{2+} ions are easiest to reduce followed by Fe^{2+}, Zn^{2+} and Mg^{2+} ions in that order:

$$Cu^{2+} > Fe^{2+} > Zn^{2+} > Mg^{2+}$$

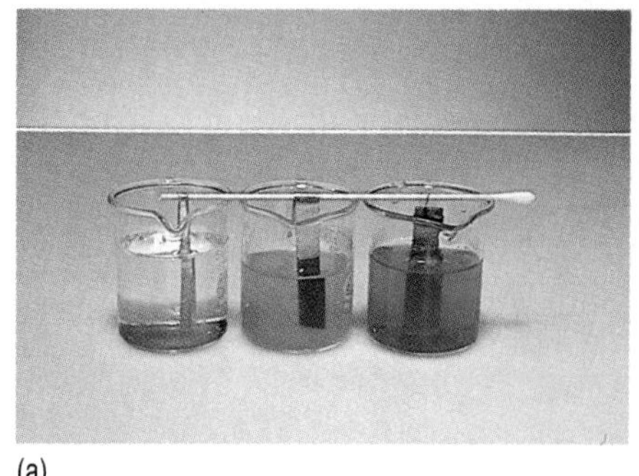

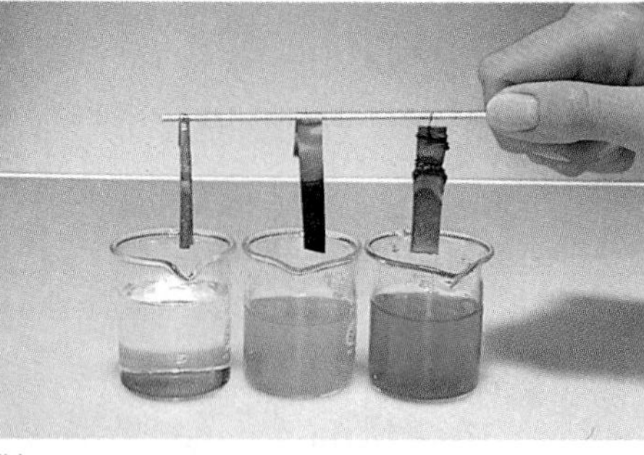

(a) (b)

Figure 18.8

Reactions of aqueous metal ions with more active metals. (a) A strip of magnesium dipped into a zinc sulfate solution, a strip of zinc dipped into an iron(II) sulfate solution, and a strip of iron dipped into a copper(II) sulfate solution. (b) A deposit of zinc on the magnesium strip, a deposit of iron on the zinc strip, and a deposit of copper on the iron strip.

This order of increasing difficulty of reduction of these aqueous metal ions to their elemental states can also be ascertained from Table 17.1, which lists the standard reduction potentials for Cu^{2+}, Fe^{2+}, Zn^{2+}, and Mg^{2+} ions as 0.34, −0.41, −0.76, and −2.38 V, respectively.

We recall that metal ions in aqueous solution are hydrated. So we can imagine that hydrated ions must be dehydrated before they can be reduced to the elemental state. This step in the reduction of aqueous ions is analogous to the step in which the lattice energy is disrupted in the reduction of an ionic

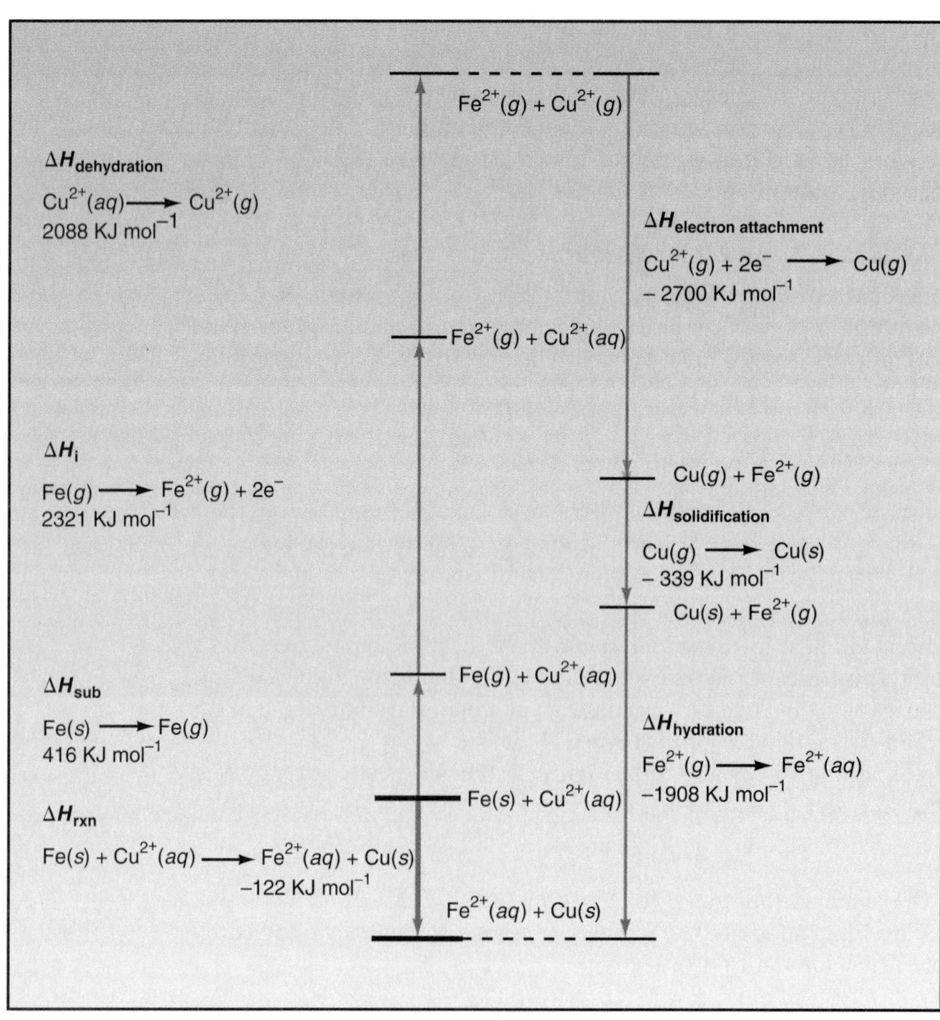

Figure 18.9

Energetics (not drawn exactly to scale) of the reaction $Fe(s) + Cu^{2+}(aq) \rightarrow Cu(s) + Fe^{2+}(aq)$. The enthalpy change for the reaction is $-122\ \text{kJ mol}^{-1}$, which is the sum of all the enthalpy changes for the steps shown in the figure.

solid. The enthalpy of dehydration of an aqueous ion equals in magnitude the enthalpy of hydration of the ion, but has the opposite sign. That is, $\Delta H_{\text{hydration}} = -\Delta H_{\text{dehydration}}$. The enthalpy changes that occur when aqueous copper(II) ions are reduced by metallic iron are illustrated in Figure 18.9.

Example 18.6

Predicting Reduction of Aqueous Metal Ions and Writing Net Ionic Equations

Predict whether a reaction occurs in each of the following cases, and write net ionic equations for the reactions that occur. **(a)** Cadmium metal is placed in a solution of copper(II) sulfate; **(b)** zinc metal is placed in 1 *M* hydrochloric acid; **(c)** barium metal is placed in a solution of iron(II) sulfate.

SOLUTION

a. Cadmium is more active than copper (Table 18.3). Therefore, copper(II) ions are reduced to elemental copper, and cadmium metal is oxidized to cadmium ions:

$$\text{Cd}(s) + \text{Cu}^{2+}(aq) \longrightarrow \text{Cd}^{2+}(aq) + \text{Cu}(s)$$

Sulfate ions do not participate in the reaction and are therefore not included in the net ionic equation.

b. Zinc is above hydrogen in the activity series (Table 18.3). Therefore, zinc metal reduces hydrogen ions in 1 *M* solution to hydrogen gas:

$$\text{Zn}(s) + 2\text{H}^{+}(aq) \longrightarrow \text{Zn}^{2+}(aq) + \text{H}_2(g)$$

c. Barium is one of the most active metals. It reduces iron(II) ions to elemental iron. It also reacts with water with the liberation of hydrogen:

$$\text{Ba}(s) + \text{Fe}^{2+}(aq) \longrightarrow \text{Ba}^{2+}(aq) + \text{Fe}(s)$$

$$\text{Ba}(s) + 2\text{H}_2\text{O}(l) \longrightarrow \text{Ba}^{2+}(aq) + 2\text{OH}^{-}(aq) + \text{H}_2(g)$$

We recall from the solubility rules of Chapter 4 that $BaSO_4$ is insoluble in water. Thus, a third reaction occurs in this case. As the barium ions form by the reactions above, they react with the sulfate ions in the solution to form a precipitate of barium sulfate:

$$\text{Ba}^{2+}(aq) + \text{SO}_4^{2-}(aq) \longrightarrow \text{BaSO}_4(s)$$

The redox reactions in this example could also be predicted from their standard reduction potentials (Table 17.1), as discussed in Chapter 17.

Practice Problem 18.6: Predict whether a reaction occurs in each of the following cases and write net ionic equations for the reactions that occur. **(a)** Magnesium ribbon is placed in a solution of silver nitrate; **(b)** an iron nail is placed in 1 *M* hydrochloric acid.

Metallurgy

The processes of extracting metals from their ores and the subsequent refining of the metals are known collectively as **metallurgy**. Metallurgy may be divided into three major steps: (1) *preliminary treatment of the ore*, (2) *reduction of the metal from its compound*, and (3) *refining the metal.*

In the preliminary treatment of an ore, impurities are removed. This treatment is followed by the conversion of the ore to a suitable form—often an oxide—for reduction. Metal sulfides or carbonates can be converted to oxides by heating in atmospheric oxygen. For example, zinc sulfide converts to zinc oxide,

$$2ZnS(s) + 3O_2(g) \longrightarrow 2ZnO(s) + 2SO_2(g)$$

and magnesium carbonate decomposes to magnesium oxide and carbon dioxide,

$$MgCO_3(s) \longrightarrow MgO(s) + CO_2(g)$$

The metal can then be prepared from its oxide by reducing it by a suitable method.

The sulfide ores of noble metals, such as mercury, copper, and silver, can easily be reduced by heating the metal sulfide in air:

$$HgS(s) + O_2(g) \longrightarrow Hg(l) + SO_2(g)$$

$$Cu_2S(s) + O_2(g) \longrightarrow 2Cu(s) + SO_2(g)$$

Many moderately active transition metals can be reduced from their oxides with carbon or with carbon monoxide. For example, iron is prepared in industry by reducing its oxide with carbon monoxide (obtained from coke) in a blast furnace, as shown in Figure 18.10. This method yields an impure iron that contains carbon, silicon, phosphorus, and sulfur as impurities. The crude product, called *pig iron*, is then purified, and carbon, chromium, manganese, and other ingredients are added to produce *steel.*

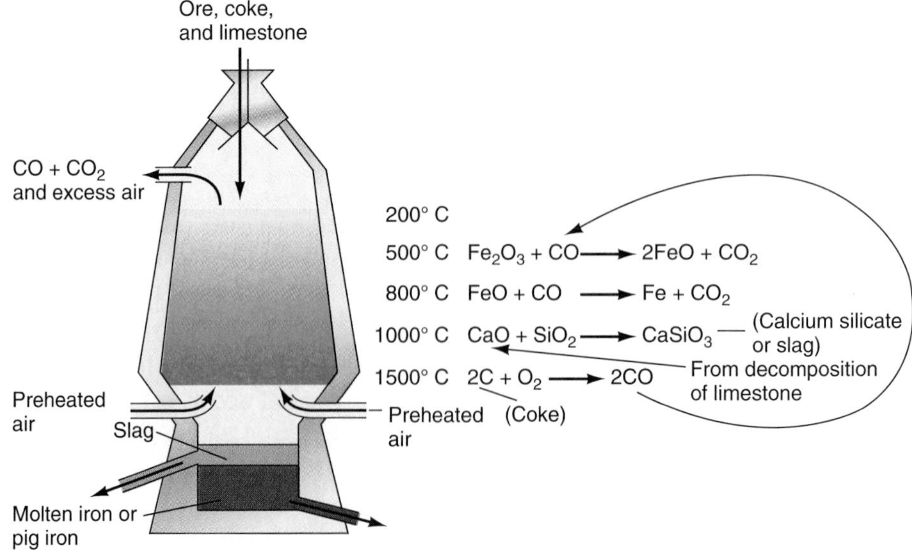

Figure 18.10

Industrial preparation of iron in a blast furnace. The temperature in the furnace increases from top to bottom. The approximate temperatures of the various regions of the furnace are given along with the corresponding reactions which occur in these regions.

Many active or moderately active transition metals, such as manganese, zinc, and chromium, can be reduced by carbon only with difficulty or not at all. Such metals are frequently reduced from their oxides by aluminum:

$$3MnO_2(s) + 4Al(s) \longrightarrow 3Mn(s) + 2Al_2O_3(s)$$

$$3ZnO(s) + 2Al(s) \longrightarrow 3Zn(s) + Al_2O_3(s)$$

$$Cr_2O_3(s) + 2Al(s) \longrightarrow 2Cr(s) + Al_2O_3(s)$$

Reduction with hydrogen produces a relatively pure metal. This method is dangerous, however, if not carried out in an airtight apparatus, because hydrogen reacts explosively with atmospheric oxygen at the high temperatures required for most reduction processes. A laboratory reduction of copper from copper(II) oxide by hydrogen is illustrated in Figure 18.11. The reaction of copper(II) oxide with hydrogen is

$$CuO(s) + H_2(g) \longrightarrow Cu(s) + H_2O(g)$$

Very active metals such as sodium cannot be effectively reduced from their compounds by other active metals. Very active metals are reduced from their molten halides or oxides by electrolysis (Chapter 17).

We recall from Chapter 17 that when molten sodium chloride is electrolyzed, Na^+ ions are reduced at the cathode and Cl^- ions are oxidized at the anode. The combination of these two half-reactions gives the overall redox reaction:

$$\begin{array}{lll} \text{Cathode:} & 2Na^+(l) + 2e^- \longrightarrow 2Na(l) & \text{(reduction)} \\ \text{Anode:} & 2Cl^-(l) \longrightarrow Cl_2(g) + 2e^- & \text{(oxidation)} \\ \hline \multicolumn{2}{l}{2Na^+(l) + 2Cl^-(l) \longrightarrow 2Na(l) + Cl_2(g)} & \text{(redox)} \end{array}$$

Various methods for reducing metals from their oxides are listed in Table 18.4. This table shows that the oxides of silver, gold, mercury and platinum decompose by heating to produce the respective elements. The oxides of these metals and the oxides of several more active metals (Table 18.4) can be reduced by hydrogen or by carbon monoxide. Manganese, zinc, chromium, iron, and cadmium can be reduced from their oxides by carbon or by aluminum. Active metal oxides can be reduced by heating with more active metals or by electrolysis (Table 18.4).

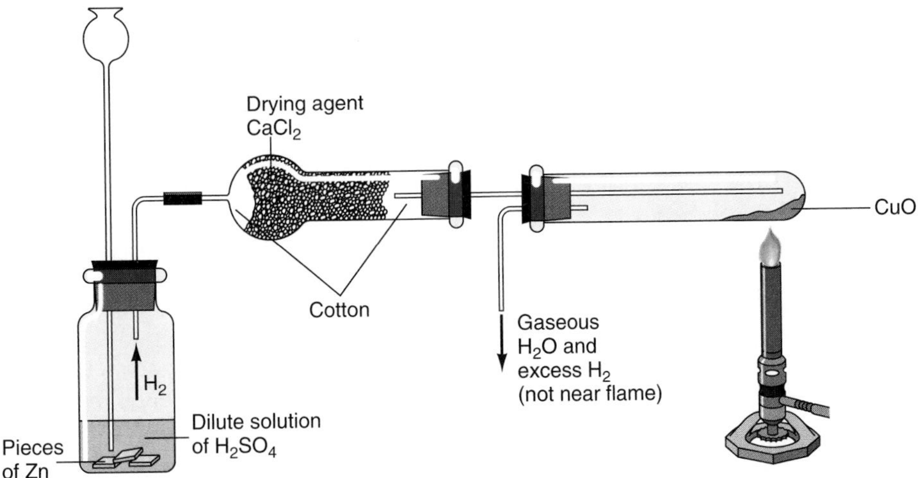

Figure 18.11

Preparation of pure copper from copper(II) oxide by reduction with hydrogen.

Table 18.4

Reduction of Common Metals from Their Oxides

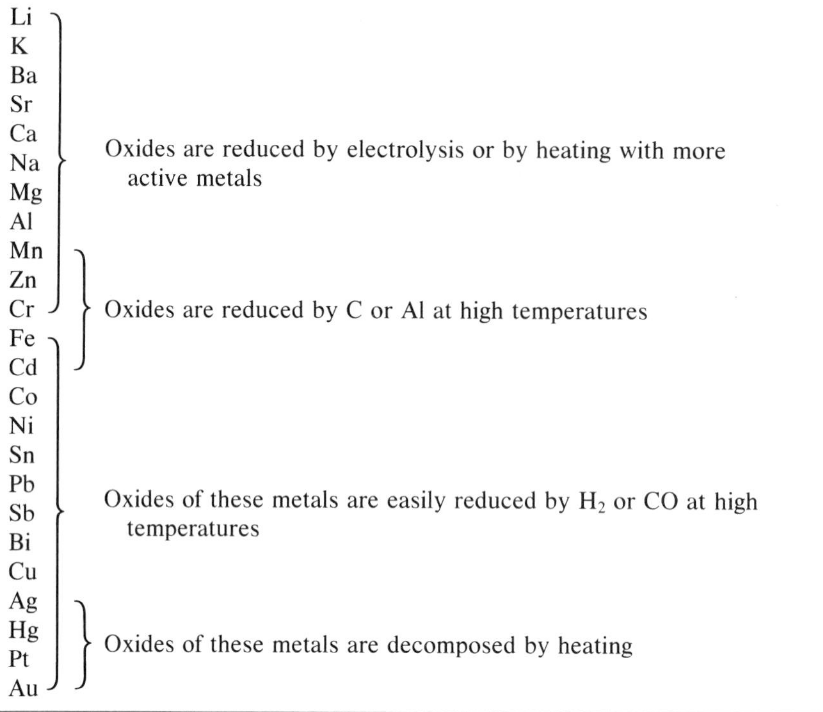

Li, K, Ba, Sr, Ca, Na, Mg, Al	Oxides are reduced by electrolysis or by heating with more active metals
Mn, Zn, Cr, Fe, Cd	Oxides are reduced by C or Al at high temperatures
Fe, Cd, Co, Ni, Sn, Pb, Sb, Bi, Cu	Oxides of these metals are easily reduced by H_2 or CO at high temperatures
Ag, Hg, Pt, Au	Oxides of these metals are decomposed by heating

Example 18.7

Preparing Metals from Their Compounds

Suggest a simple method by which each of the following metals can be prepared from the compound indicated. Write equations for the reactions that occur. **(a)** Chromium from chromium(III) chloride; **(b)** nickel from nickel(II) oxide.

SOLUTION

a. Chromium can be reduced from $CrCl_3$ by a more active metal such as an alkali metal or an alkaline earth metal:

$$3Mg(s) + 2CrCl_3(s) \xrightarrow{\text{heat}} 3MgCl_2(s) + 2Cr(s)$$

b. Nickel can be reduced from its oxide with hydrogen or with carbon monoxide (Table 18.4):

$$NiO(s) + H_2(g) \xrightarrow{\text{heat}} Ni(s) + H_2O(g)$$

$$NiO(s) + CO(g) \xrightarrow{\text{heat}} Ni(s) + CO_2(g)$$

Practice Problem 18.7: How can each of the following metals be prepared from its oxide? **(a)** Zinc; **(b)** tin. Write an equation for each method of preparation.

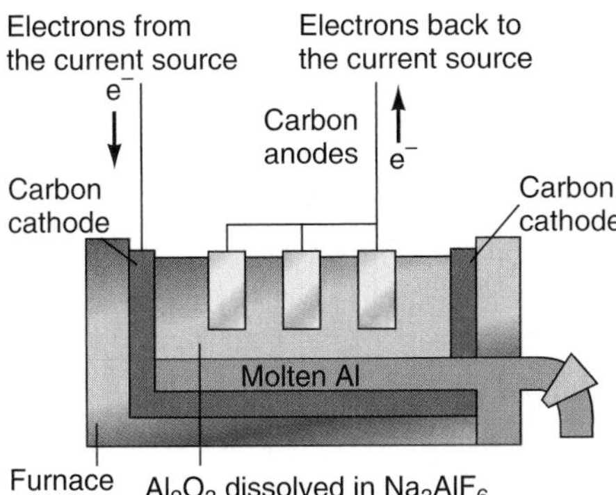

Figure 18.12

Production of aluminum by electrolysis. Aluminum ions are reduced at the cathode, and oxide ions are oxidized at the anode. The carbon anode is consumed in the oxidation of oxide ions to form carbon dioxide. The overall electrolysis reaction is $4Al^{3+}(l) + 6O^{2-}(l) + 3C(s) \rightarrow 4Al(l) + 3CO_2(g)$.

Aluminum is produced commercially by the Hall–Heroult process. Charles Martin Hall was a 22-year-old U.S. inventor who, as an undergraduate at Oberlin College, succeeded in obtaining aluminum from aluminum oxide in a backyard woodshed. The same discovery was made at approximately the same time by Paul Heroult of France who was also 22 years old and worked in a makeshift laboratory.

Today a modified Hall–Heroult process uses naturally occurring impure aluminum oxide, called bauxite, as a starting material. Bauxite is purified, dissolved in molten cryolite, Na_3AlF_6, and electrolyzed at 960°C (Figure 18.12). The cathode and anode in this electrolysis are both made of carbon. The reactions at the electrodes and the overall reaction are:

$$4[Al^{3+}(l) + 3e^- \longrightarrow Al(l)] \quad \text{(cathode reaction)}$$
$$3[C(s) + 2O^{2-}(l) \longrightarrow CO_2(g) + 4e^-] \quad \text{(anode reaction)}$$
$$4Al^{3+} + 6O^{2-} + 3C(s) \longrightarrow 4Al(l) + 3CO_2(g) \quad \text{(overall reaction)}$$

This process produces 99.6 percent pure aluminum, which floats on top of the electrolyte as a solid. The carbon electrodes are consumed during electrolysis, and the heat released in this process maintains the reaction temperature at 960°C.

Tremendous amounts of electricity are required for the industrial production of aluminum. For that reason, aluminum refineries are often located near hydroelectric plants.

18.4 Transition Metals, Their Compounds, and Aqueous Ions

Definition and Properties of Transition Metals

The following discussion provides an overview of the chemistry of common transition metals. We define transition elements as those whose ground-state valence shell configuration is $(n - 1)\, d^x ns^2$ or $(n - 1)\, d^x ns^1$, where x ranges from 1 to 10 and n is the principal quantum number of the highest-energy electrons. For example, for scandium, $n = 4$ and $x = 1$ ($3d^1 4s^2$); for chromium, $n = 4$ and $x = 5$ ($3d^5\, 4s^1$); and for zinc, $n = 4$ and $x = 10$ ($3d^{10} 4s^2$). We will not consider the inner transition metals.

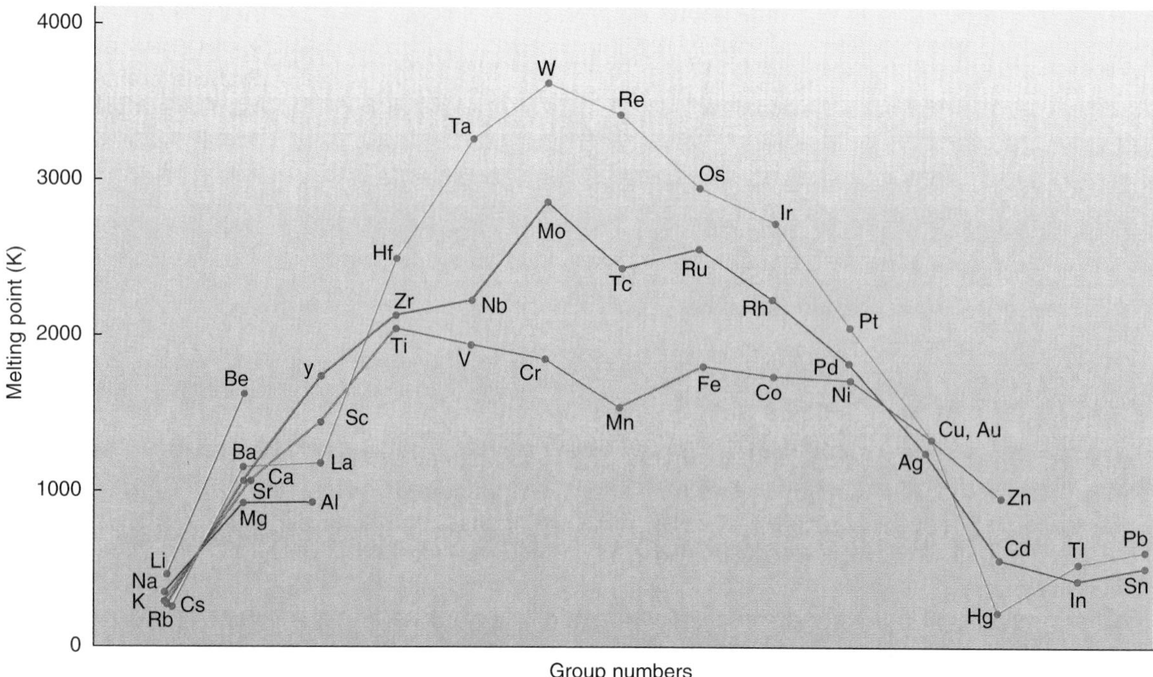

Figure 18.13

Variation of the melting points of metals with their group numbers.

Most transition metals, particularly those in the middle of each of the *d* series, are hard, strong, dense, and high melting (Figure 18.13). Like most other metals, transition metals have a silvery luster, except for copper and gold, which are yellow. Many transition metal compounds and their aqueous ions are colored.

Oxides and Halides of Transition Metals

In Section 18.1 we listed the oxidation states of elements, including the *d* series of transition metals. In this section we give an overview of oxides and aqueous ions of transition metals in their common, stable oxidation states. Table 18.5 lists the oxides of the *d* series of transition metals in their common oxidation states. We can see from Table 18.5 that the number of oxidation states and the number of oxides that exist for a transition metal increase from the scandium family to the middle of each of the *d* series and then decrease to the end of the series. Thus, the elements in the vanadium, chromium, and manganese families exhibit the greatest numbers of oxidation states and the greatest numbers of oxides.

All of the metals in the 3*d* series except scandium form an oxide in which the metal is in its +2 oxidation state. This state can be formed when an atom in the 3*d* series of metals loses two 4*s* electrons (one 4*s* and one 3*d* electron for Cr and Cu). Scandium does not form an ion in the +2 oxidation state because the energies of the 4*s* and 3*d* electrons in a scandium atom are about the same. Thus, scandium forms only one oxide, Sc_2O_3. For similar reasons, yttrium forms only Y_2O_3, and lanthanum forms only La_2O_3.

Proceeding toward the right from scandium in the 3*d* series, we see that titanium, vanadium, chromium, and manganese exhibit oxidation states from +2 up to a maximum that equals the group number (4 for Ti, 5 for V, 6 for Cr,

Table 18.5

Oxides[a] of Transition Metals

Series	Oxidation State	Outer Electron Configuration on the Metal Atom (in Gaseous State)[b]									
		d^1s^2	d^2s^2	d^3s^2	d^5s^1	d^5s^2	d^6s^2	d^7s^2	d^8s^2	$d^{10}s^1$	$d^{10}s^2$
3*d*	1									Cu_2O	
	2		TiO	VO	CrO	MnO	$\mathbf{FeO}$	$\mathbf{CoO}$	$\mathbf{NiO}$	$\mathbf{CuO}$	$\mathbf{ZnO}$
	$2\frac{2}{3}$					$\mathbf{Mn_3O_4}^c$	$\mathbf{Fe_3O_4}^c$	$Co_3O_4{}^c$			
	3	$\mathbf{Sc_2O_3}$	Ti_2O_3	V_2O_3	$\mathbf{Cr_2O_3}$	$\mathbf{Mn_2O_3}$	$\mathbf{Fe_2O_3}$	Co_2O_3			
	4		$\mathbf{TiO_2}$	VO_2		$\mathbf{MnO_2}$					
	5			$\mathbf{V_2O_5}$							
	6				$\mathbf{CrO_3}$						
	7					Mn_2O_7					
	8										
4*d*	1									Ag_2O	
	2			NbO				RhO	PdO		CdO
	3	Y_2O_3				Tc_2O_3		Rh_2O_3			
	4		ZrO_2	NbO_2	MoO_2	TcO_2	RuO_2				
	5			Nb_2O_5	Mo_2O_5						
	6				MoO_3	TcO_3					
	7										
	8						RuO_4				
5*d*	2								PtO		HgO
	3	La_2O_3						Ir_2O_3		Au_2O_3	
	4		HfO_2	TaO_2	WO_2	ReO_2	OsO_2	IrO_2	PtO_2		
	5			Ta_2O_5		Re_2O_5					
	6				WO_3	ReO_3	OsO_3				
	7					Re_2O_7					
	8						OsO_4				

[a] The more common or stable oxides of the 3*d* series of elements appear in bold type.

[b] Exceptions are as follows: Nb, d^4s^1; W, d^4s^2; Tc, d^6s^1; Ru, d^7s^1; Rh, d^8s^1; Pd, $d^{10}s^0$; Pt, d^9s^1.

[c] Oxide contains two-thirds of the metal atoms in the +3 oxidation state and one-third in the +2 state.

and 7 for Mn). The most stable oxide for titanium is TiO_2 and for vanadium V_2O_5. Titanium and vanadium are in their maximum oxidation states in these oxides.

For chromium the most stable oxidation state is +3, which corresponds to the oxide Cr_2O_3. When chromium metal is heated in oxygen, it forms chromium(III) oxide:

$$4Cr(s) + 3O_2(g) \longrightarrow 2Cr_2O_3(s)$$

Chromium(III) oxide is green and is used as a paint pigment. Chromium(VI) oxide is not as stable as chromium(III) oxide. It decomposes to chromium(III) oxide upon heating:

$$4CrO_3(s) \longrightarrow 2Cr_2O_3(s) + 3O_2(g)$$

Unlike chromium, manganese has no single most stable oxide or oxidation state. Three common oxides include Mn_2O_3, Mn_3O_4, and MnO_2. The manganese(VII) oxide, Mn_2O_7, which corresponds to the maximum oxidation state for manganese, exists but is extremely unstable and explodes at the slightest provocation.

The common oxides of iron are FeO, Fe_2O_3, and Fe_3O_4. FeO is easily oxidized to Fe_2O_3, which is therefore the oxide of iron found in nature; it is the principal ingredient of rust. Fe_3O_4 is a "mixed" oxide in which some of the iron is in a +2 oxidation state and some is in a +3 state. Fe_3O_4 is a natural magnet called magnetite. The +3 oxidation state for iron is the most common and stable state, exhibited in compounds such as Fe_2O_3 and $FeCl_3$. For cobalt and nickel, the most common oxides are CoO and NiO, respectively.

For copper, both Cu_2O and CuO exist, but CuO is the more stable one at room temperature because Cu_2O is easily oxidized to CuO. However, at high temperatures, Cu_2O is the more stable one. In the same family, the principal oxides of silver and gold are Ag_2O and Au_2O_3, respectively.

For zinc, cadmium, and mercury, the only oxides are ZnO, CdO, and HgO, respectively. Although the +1 oxidation state exists for mercury, there is no corresponding oxide. Figure 18.14 shows the colors of the principal oxides of each of the 3*d* series of transition metals.

We *cannot* automatically extend our knowledge of the most stable oxidation states of transition metals in their oxides to other types of transition metal compounds. For example, the +5 oxidation state for vanadium is the most stable state in its oxide, but vanadium(V) chloride does not exist because of the steric hindrance that would result from packing five chlorine atoms around the central vanadium atom. For the same reason, chromium exhibits +6 oxidation state in its oxide, CrO_3, but $CrCl_6$ does not exist. For mercury, both mercury(I) chloride, Hg_2Cl_2, and mercury(II) chloride, $HgCl_2$, exist. But the only oxide of mercury is mercury(II) oxide, HgO. Mercury(I) oxide does not exist. Figure 18.15 shows the colors of the principal chlorides of the 3*d* series of transition metals.

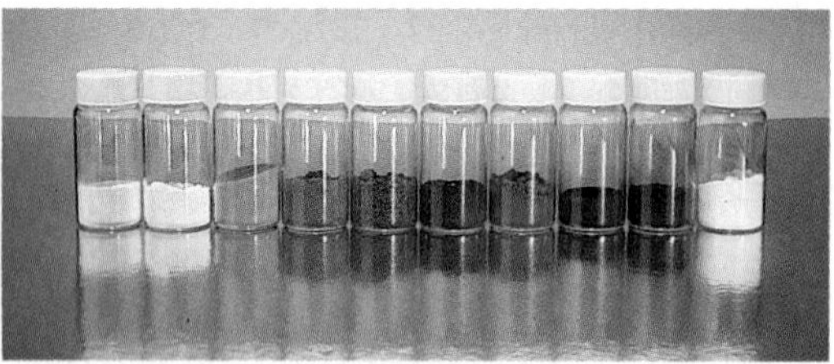

Figure 18.14

Principal oxides of the 3d series of transition metals displayed in their respective positions in the periodic table: Sc_2O_3; TiO_2; V_2O_5; Cr_2O_3, CrO_3; MnO_2; FeO, Fe_2O_3; CoO; NiO; ZnO.

Figure 18.15

Principal chlorides of the 3d series of transition metals displayed in their respective positions in the periodic table: $ScCl_3$, $TiCl_4$, VCl_3, $CrCl_2$, $CrCl_3$, $MnCl_2$, $FeCl_2$, $FeCl_3$, $CoCl_2$, $NiCl_2$, CuCl, $CuCl_2$, $ZnCl_2$.

Mercury forms an Hg^{2+} ion by an atom losing its pair of $6s$ electrons. An Hg_2^{2+} ion consists of two covalently bonded Hg atoms in which each Hg atom has an oxidation state of +1:

$$(Hg—Hg)^{2+} \quad \text{or} \quad Hg_2^{2+}$$

We might expect the +1 oxidation state for copper to be more stable than the +2 state because the outer electron configuration for copper is $3d^{10}4s^1$. This is true at high temperatures, but at room temperature the opposite is true. Copper(I) is easily oxidized to copper(II) at room temperature. Thus, CuO and $CuCl_2$ are more stable than Cu_2O and CuCl. Also, the only simple ion of copper in aqueous solution is copper(II) ion.

To explain the stability of Cu^{2+} ions we must remember that the structure of gaseous atoms and the ionization energy are not the only factors that determine the stability of metal ions in solution or in solid compounds. The stability of an ion in a compound is determined largely by its enthalpy of formation (see Figures 18.7 and 18.9). The enthalpy changes required to convert solid copper to Cu^+ ions in a solid copper(I) compound can be thought of in terms of the following steps:

$$Cu(s) \xrightarrow{\Delta H_{\text{sublimation}}} Cu(g) \xrightarrow{\Delta H_{\text{ionization}}} Cu^+(g) \xrightarrow{\Delta H_{\text{lattice formation}}} Cu^+{}_{\text{in solid lattice}}$$

We recall that sublimation and ionization are endothermic processes and that lattice formation is an exothermic process. Since copper(II) compounds are more stable than copper(I) compounds, the conversion

$$Cu(s) \longrightarrow Cu^{2+}{}_{\text{in solid lattice}}$$

must be more exothermic than the conversion

$$Cu(s) \longrightarrow Cu^+{}_{\text{in solid lattice}}$$

Although the formation of a Cu^{2+} ion from an isolated atom requires more energy than the formation of a Cu^+ ion, the lattice energy released in the formation of a copper(II) compound is greater than the lattice energy for a similar copper(I) compound because the Cu^{2+} ion has a higher ionic potential than the Cu^+ ion. That is, the positive charge of a Cu^{2+} ion is twice the charge of a Cu^+ ion and is concentrated in a smaller volume than the single positive charge of a Cu^+ ion (the ionic radius of Cu^{2+} is 0.069 nm compared with 0.096 nm for Cu^+).

The unusual stability of the chromium(III) ion compared with the chromium(I) ion can be explained in a similar way. The outer electron configuration of a chromium atom is $3d^54s^1$. We might expect the chromium(I) ion to be more stable than the chromium(III) ion because less energy is required to remove the single s electron from an atom than to remove the s electron and two d electrons. But since the chromium(III) ion has a 3+ charge and is relatively small (its radius is 0.076 nm), the lattice energies of chromium(III) compounds are relatively large.

It is not easy to explain why silver does not follow the analogy of copper in the relative stabilities of its ions. Silver has only a +1 oxidation state. For gold, the +1 oxidation state is known, but the +3 state is the more stable one.

Zinc, cadmium, and mercury atoms form stable 2+ ions when they lose their outer s electrons to form a stable outer d^{10} electron configuration for the ions. The outer d orbital energy for zinc, cadmium, and mercury is considerably lower than the outer s orbital energy. Therefore, no d electrons are lost by these atoms in their ordinary chemical reactions, and only 2+ ions exist for zinc and cadmium.

Example 18.8

Predicting Formulas and Naming Stable Transition Metal Chlorides

Write the formula and the name of the most common chloride of each of the following transition metals: **(a)** titanium; **(b)** chromium; **(c)** iron; **(d)** nickel; **(e)** copper.

SOLUTION

a. The most common oxidation state for titanium is +4, and the formula for its corresponding chloride is $TiCl_4$. Its name is titanium(IV) chloride or titanium tetrachloride.

b. The most stable oxidation state of chromium is +3, and the most stable chloride of chromium is $CrCl_3$, called chromium(III) chloride. The older name, still often used, is chromic chloride. (Chromous chloride is $CrCl_2$.)

c. The most stable oxidation state for iron is also +3, and the most stable chloride of iron is $FeCl_3$. The systematic name for this compound is iron(III) chloride. The common name is ferric chloride. Iron(II) chloride, $FeCl_2$, also exists; its common name is ferrous chloride.

d. The common oxidation state for nickel is +2 and its chloride is $NiCl_2$, which is called nickel(II) chloride.

e. For copper, the +2 oxidation state is more common than the +1 state, and the formula for copper(II) chloride is $CuCl_2$. Its common name is cupric chloride. Copper(I) chloride, CuCl, also exists; its common name is cuprous chloride.

Practice Problem 18.8: Write the formula for and name the most stable oxide and fluoride of each of the following metals: **(a)** scandium; **(b)** zinc; **(c)** silver.

Aqueous Monatomic Ions of Transition Metals

Transition metals can exist as monatomic aqueous ions with charges ranging from 1+ to 3+. Table 18.6 lists common monatomic ions of transition metals, their valence shell electron configurations, and their colors in aqueous solution. Ions with d^0 or d^{10} outer electron configuration are colorless. Figure 18.16 displays the colors of stable monatomic aqueous ions of the 3d series of transition metals.

Figure 18.16

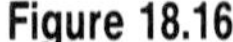

Stable monatomic ions of the 3d series of transition metals: Sc^{3+}; Cr^{3+}; Mn^{2+}; Fe^{3+}; Co^{2+}; Ni^{2+}; Cu^{2+}; Zn^{2+}.

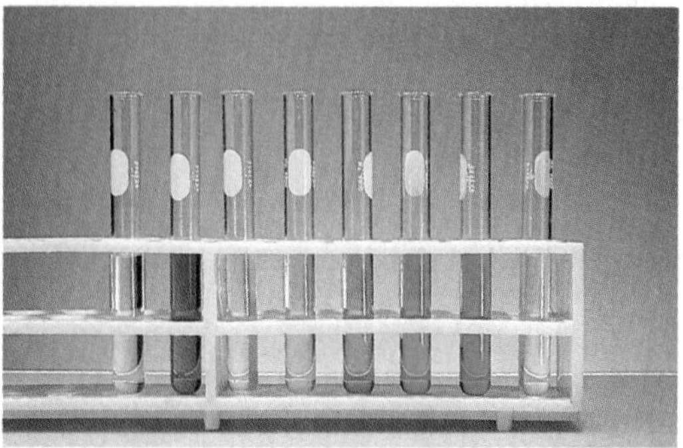

Table 18.6

Common Monatomic Ions of Transition Metals in Aqueous Solution

3*d*	$Sc^{3+}(d^0)$ Colorless	$Ti^{3+}(d^1)$ Violet	$V^{2+}(d^2)$ Violet	$Cr^{2+}(d^4)$ Blue $Cr^{3+}(d^3)$ Green	$Mn^{2+}(d^5)$ Light pink	$Fe^{2+}(d^6)$ Greenish $Fe^{3+}(d^5)$ Yellow	$Co^{2+}(d^7)$ Pink	$Ni^{2+}(d^8)$ Green	$Cu^{2+}(d^9)$ Blue	$Zn^{2+}(d^{10})$ Colorless
4*d*	$Y^{3+}(d^0)$ Colorless								$Ag^{+}(d^{10})$ Colorless	$Cd^{2+}(d^{10})$ Colorless
5*d*	$La^{3+}(d^0)$ Colorless									Hg_2^{2+}[a] Colorless $Hg^{2+}(d^{10})$ Colorless

[a] Consists of two covalently bonded Hg ($d^{10}s^1$) atoms with the total charge of 2+.

The formation of some of the ions listed in Table 18.6 can be predicted from the outer electron configuration of the parent atom; others cannot. Scandium, yttrium, and lanthanum atoms have a d^1s^2 outer electron structure, and they lose all three of these electrons to form 3+ ions. The 2+ ions for vanadium, manganese, iron, cobalt, nickel, and zinc form when the atoms lose their two 4*s* electrons.

Chromium and copper atoms have the outer electron configurations of d^5s^1 and $d^{10}s^1$, respectively. For a chromium atom to form a Cr^{3+} ion, it must lose its single *s* electron and two *d* electrons. The energy required for this ionization is compensated for by the relatively large hydration energy of chromium(III) ions. Similarly, for a copper atom to form Cu^{2+} ion, it must lose its single *s* electron and one *d* electron. The relatively high hydration energy of copper(II) ions is responsible for the stability of these ions in aqueous solution.

Oxoanions of Transition Metals

Although the charge on monatomic aqueous ions of transition metals can range from 1+ to 3+, the oxidation states of transition metals in many of their oxoanions are the maximum for these elements. Among the most commonly encountered oxoanions of transition metals are chromate, CrO_4^{2-}, dichromate, $Cr_2O_7^{2-}$, and permanganate, MnO_4^-, ions. The colors of these ions in aqueous solution are displayed in Figure 18.17.

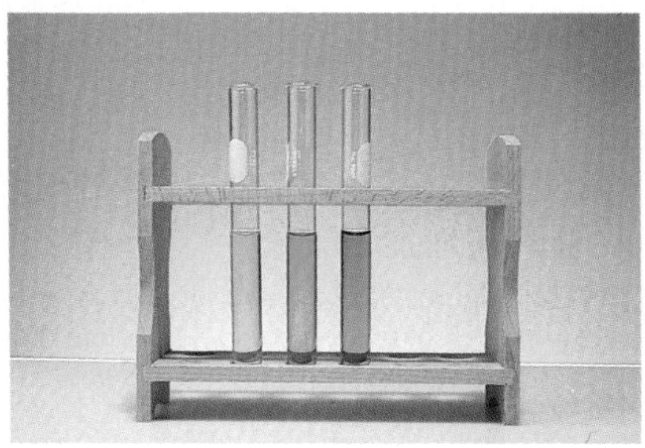

Figure 18.17

Aqueous oxoanions of chromium and manganese: CrO_4^{2-}, $Cr_2O_7^{2-}$, MnO_4.

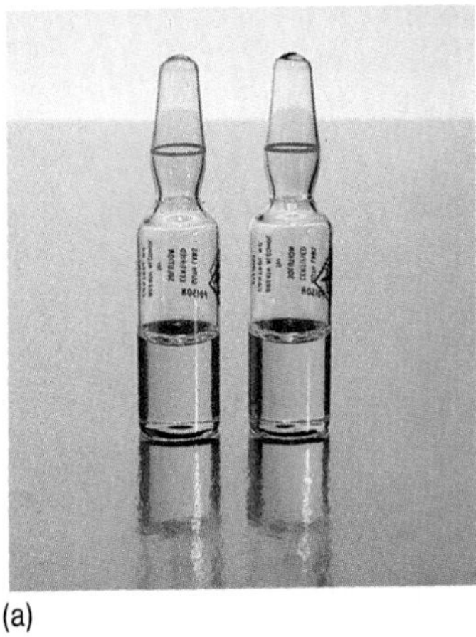
(a)

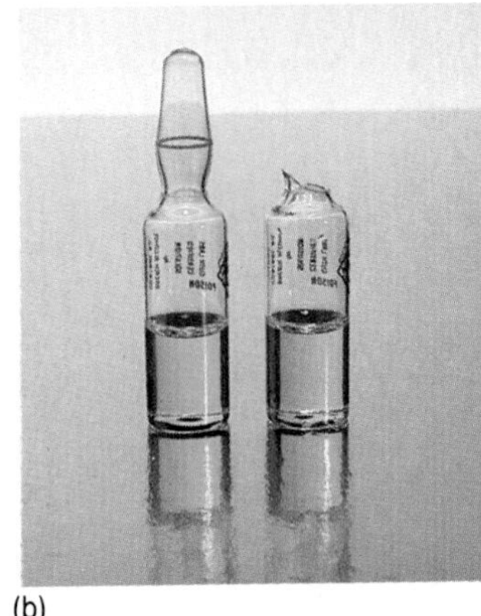
(b)

Figure 18.18

Colors of "breathalizer" solutions before and after exposure to alcohol vapors.

In an aqueous acid solution, chromate and dichromate ions exist in equilibrium with each other, as shown by the following equation:

$$\underset{\text{Yellow}}{2CrO_4^{2-}(aq)} + 2H^+(aq) \rightleftharpoons \underset{\text{Orange}}{Cr_2O_7^{2-}(aq)} + H_2O(l)$$

When a chromate solution is acidified, the equilibrium shifts to the right, and the solution turns orange. When the dichromate solution is made alkaline, the yellow color returns.

We noted previously that the dichromate ion is a common oxidizing agent. When the dichromate ion is used as an oxidizing agent, it is reduced to a chromium(III) ion, the most stable ion of chromium.

The oxidizing action of dichromate ions is used in the breath analyzer or "breathalizer" test for drunk driving suspects (Figure 18.18). Dichromate ions in acid solution oxidize ethyl alcohol to acetic acid:

$$3CH_3CH_2OH(l) + \underset{\text{Orange}}{2Cr_2O_7^{2-}(aq)} + 16H^+(aq) \longrightarrow 3HC_2H_3O_2(aq) + \underset{\text{Green}}{4Cr^{3+}(aq)} + 11H_2O(l)$$

During the test, if vapors containing alcohol are exhaled by the suspect, the color of the solution changes from orange (dichromate) to green, the color of chromium(III) ions. The extent of the color change is a quantitative measure of the amount of alcohol in the breath.

18.5 A Closer Look at Some Common Transition Metals

Scandium, Titanium, and Vanadium Groups

We noted above that scandium, yttrium, and lanthanum form compounds in which these elements are in +3 oxidation states. Although these elements form many compounds, the elements and their compounds have no important common uses.

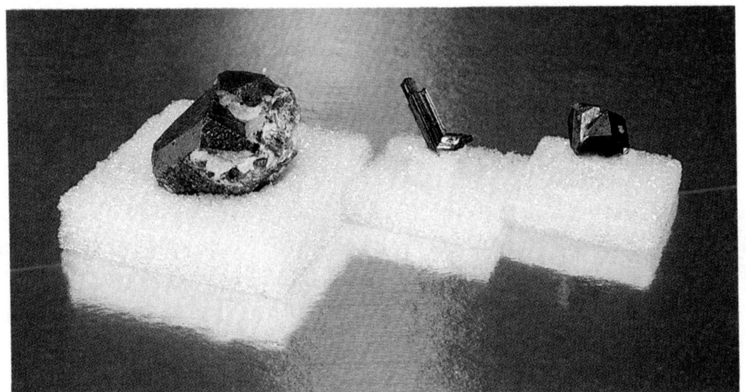

Figure 18.19

Some forms of TiO_2 crystals.

Titanium and its compounds have many uses. We discussed some uses of elemental titanium in Chapter 2. One of the most stable titanium compounds is titanium(IV) oxide or titanium dioxide, TiO_2, which is used as a white paint pigment because of its brilliant white color. When dried and heated at high temperatures, TiO_2 converts to a crystalline form that sparkles like diamonds (Figure 18.19).

Another well-known compound of titanium is titanium(IV) chloride, $TiCl_4$, a liquid (b.p. 136°C) that reacts instantly with moisture in the air to form a dense white cloud of TiO_2:

$$TiCl_4(l) + 2H_2O(l) \longrightarrow TiO_2(s) + 4HCl(g)$$

Because of this property, $TiCl_4$ has been used to produce smoke screens for military purposes (Figure 18.20).

Figure 18.20

Smoke screen caused by the reaction of titanium tetrachloride with water vapor.

The *zirconium* counterpart of titanium oxide is zirconium dioxide, ZrO_2. It, too, is a stable white solid with a very high melting point (2950°C). Zirconium oxide is used as a high-temperature insulator and as a component of refractory (heat-resisting) ceramics. Another zirconium compound, zirconium silicate, $ZrSiO_4$, known as *zircon*, is usually rust-colored. Under certain conditions, however, zircon appears as clear, colorless crystals (Figure 18.21) that sparkle more than diamonds and are therefore prized as gems. However, diamonds are harder and more durable.

The most common oxide of *vanadium* is vanadium(V) oxide, V_2O_5. We recall that vanadium(V) oxide is used as a catalyst in the conversion of SO_2 to SO_3 for the production of sulfuric acid in the contact process (Section 9.7). Vanadium metal and most of its compounds are converted to V_2O_5 when heated in air.

Figure 18.21

A zircon crystal used as a gem.

Chromium and Manganese

We recall that the principal oxides of *chromium* are green Cr_2O_3 and reddish-brown CrO_3. The hydrated form, or water solution of CrO_3, is called *chromic acid*. This acid is a strong acid and is written with the formula H_2CrO_4.

Chromic acid forms salts called *chromates* such as potassium chromate, K_2CrO_4, and lead chromate, $PbCrO_4$. Chromates (and dichromates) contain chromium in its maximum oxidation state, +6. Chromate salts can be decomposed to the very stable green chromium(III) oxide, Cr_2O_3, by heating. Ammonium dichromate burns with a spectacular effect that resembles a miniature volcano producing a green smoke of Cr_2O_3 (Figure 18.22). The reaction is

$$(NH_4)_2Cr_2O_7(s) \longrightarrow N_2(g) + 4H_2O(g) + Cr_2O_3(s)$$

When *manganese* is heated in air, MnO, Mn_2O_3, Mn_3O_4, MnO_2, or a combination of them are produced. We recall that MnO_2 is a catalyst in preparation of oxygen by the decomposition of $KClO_3$. Manganese can be oxidized in pure oxygen to Mn_2O_7, a greenish, semisolid, explosively unstable compound. The acid derived from Mn_2O_7 is *permanganic acid,* $HMnO_4$, which is both a strong acid and a strong oxidizing agent. The salts of this acid, permanganates, are among the best oxidizing agents. In acid solution, permanganate ions can be reduced to manganese(II) ions, which are the most stable ions for manganese.

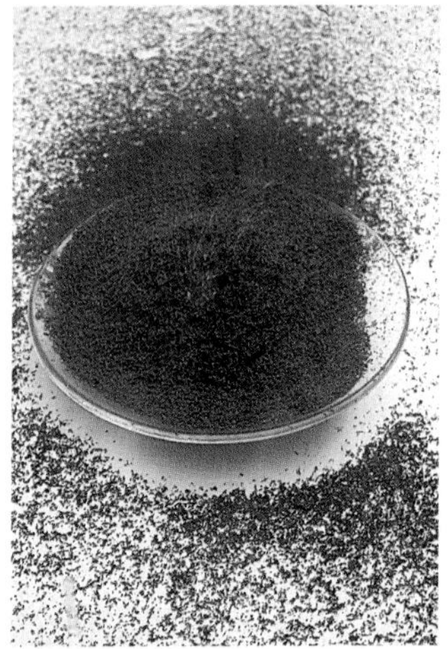

Figure 18.22

A miniature volcano caused by heating ammonium dichromate.

Iron, Cobalt, and Nickel

Iron is a major component of the earth's crust, and it is also believed to be a major component of the earth's core (Figure 18.23). We recall that iron in most of its compounds is in a +2 or a +3 oxidation state. Most of the naturally occurring compounds of iron contain iron in its +3 state because the +2 state is easily oxidized to +3 state. We discussed the uses of elemental iron in Chapter 2, and in Section 18.3 we discussed reduction of iron from its ore.

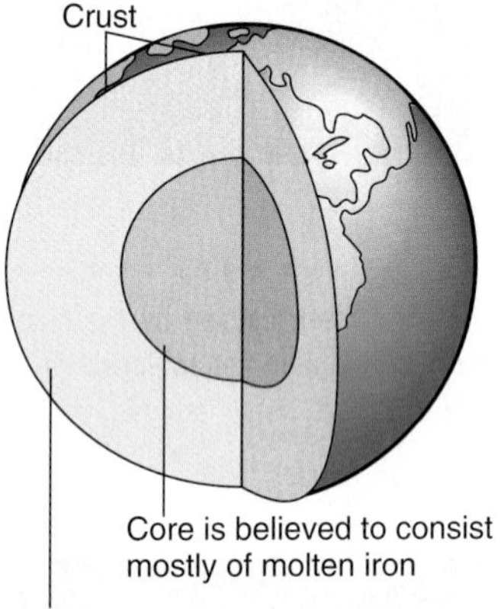
Crust
Core is believed to consist mostly of molten iron
Mantle is believed to contain olivine, a composite of Fe_2SiO_4 and Mg_2SiO_4

Figure 18.23

Major components of earth's crust and core.

Cobalt forms many stable compounds. One compound of cobalt, cobalt(II) chloride hexahydrate, $CoCl_2(H_2O)_6$, is used in simple hygrometers to indicate the approximate relative humidity. When hydrated cobalt(II) chloride is partially dehydrated, its color changes from pink to blue:

$$\underset{\text{Pink}}{CoCl_2(H_2O)_6(s)} \rightleftharpoons \underset{\text{Blue}}{CoCl_2(H_2O)_4(s)} + 2H_2O(g)$$

At high humidity, the cobalt chloride exists mostly as a pink hexahydrate. With decreasing humidity, some hexahydrate is converted to blue tetrahydrate, and the color of the equilibrium mixture of the two hydrates can be used to indicate the percent of relative humidity. At high relative humidity, the hydrate is pink (Figure 18.24), and at low relative humidity the color of the hydrate is blue.

A dilute aqueous solution of $CoCl_2(H_2O)_6$ can also be used as "invisible" ink. A dilute solution of cobalt(II) chloride is almost invisible because the $Co(H_2O)_6^{2+}$ ions are only very pale pink in dilute solution. Writing with such a solution becomes visible when the paper is heated because water evaporates and the blue color of the anhydrous salt becomes easily readable. As heating is discontinued, the writing fades again if the relative humidity is high.

Figure 18.24

Colors of cobalt(II) chloride hexahydrate and cobalt(II) chloride tetrahydrate.

Nickel in its elemental form is more frequently used than any of its compounds. Most nickel compounds, and nickel(II) ions in aqueous solution, are green. Aqueous nickel(II) ions are much lighter green than aqueous chromium(III) ions (Figure 18.25). Finely divided or powdered nickel has the ability to adsorb hydrogen gas. This hydrogen can be easily released upon heating. Nickel metal can therefore be used as a catalyst when liquid vegetable oils are converted to solid fats (Figure 18.26), a process called hydrogenation:

$$\underset{\text{Glycerol trioleate, an edible vegetable oil}}{(C_{17}H_{33}COO)_3C_3H_5(l)} + 3H_2(g) \xrightarrow{\text{Ni}} \underset{\text{Glycerol tristearate, a solid edible fat}}{(C_{17}H_{35}COO)_3C_3H_5(s)}$$

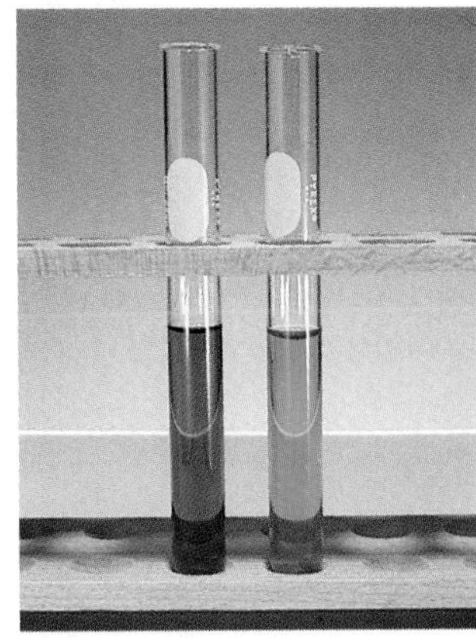

Figure 18.25

Colors of aqueous Cr^{3+} and Ni^{2+} ions.

Copper, Silver, and Gold

One of the most common compounds of *copper*, found in most chemistry laboratories, is copper(II) sulfate, $CuSO_4$. When copper(II) sulfate pentahydrate is heated it is converted to its anhydrous form, $CuSO_4$:

$$\underset{\text{Blue}}{CuSO_4(H_2O)_5(s)} \longrightarrow \underset{\text{White}}{CuSO_4(s)} + 5H_2O(g)$$

Anhydrous copper sulfate is white, but its pentahydrate, $CuSO_4(H_2O)_5$, is blue, as are aqueous copper(II) ions, $Cu(H_2O)_4{}^{2+}$. When the white anhydrous form is exposed to moisture, it converts back to blue pentahydrate. In the pentahydrate, four water molecules are linked to the copper(II) ion by coordinate covalent bonds. The fifth water molecule bridges two of the coordinated water molecules with the sulfate ion (Figure 18.27).

Anhydrous copper(II) sulfate absorbs moisture to form hydrated copper(II) sulfate. Therefore, it is used as a *desiccant* to maintain a dry atmosphere. Anhydrous copper(II) sulfate can also be used to detect moisture in a dry atmosphere. If the white anhydrous salt turns blue, moisture is present. Copper(II) sulfate is also used in electrolytic refining of copper, in copper plating, as a mordant (a substance that fixes a dye to fabric) in the textile industry, and in the manufacture of pigments.

Figure 18.26

Liquid vegetable oil (olive oil) and a solid fat (Crisco).

When a basic solution of a copper(II) salt is heated with a reducing agent, copper(I) oxide precipitates according to the half-reaction

$$2Cu^{2+}(aq) + 2OH^-(aq) + 2e^- \text{ (from reducing agent)} \longrightarrow Cu_2O(s) + H_2O(l)$$

This reaction is the basis for the test for the presence of so-called "reducing sugars" (such as blood sugar, glucose) in the urine in the diagnosis of diabetes. When a sample of urine is mixed with a basic solution of copper(II) salt, a reddish-brown copper(I) oxide precipitates if a reducing sugar is present in the sample; this indicates that the subject may be diabetic.

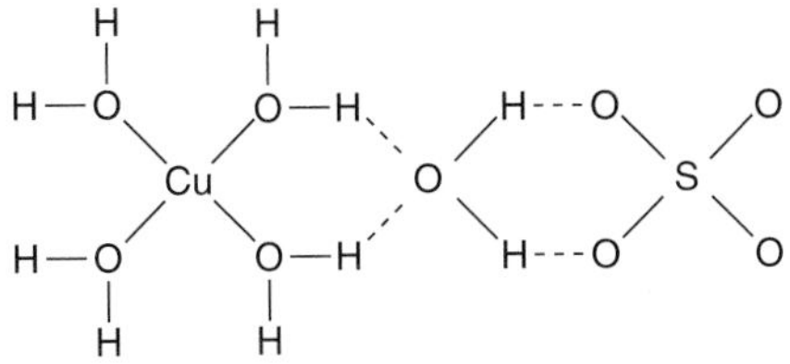

Figure 18.27

Structure of $CuSO_4(H_2O)_5$. Four water molecules are linked to the copper(II) ion by coordinate covalent bonds. The fifth water molecule bridges two of the coordinated water molecules with the sulfate ion through hydrogen bonding (shown by dotted lines).

Silver is often found in nature alloyed with gold, copper, or mercury. It also occurs in nature as silver chloride, $AgCl$, or as silver sulfide, Ag_2S. Silver ions react readily with cyanide ions to form the cyanide complex, $Ag(CN)_2^-$. Silver and its compounds dissolve readily in alkali metal cyanides in the presence of oxygen to form the cyanide complex ions of silver. The silver ion in the cyanide complex can be reduced with zinc to elemental silver. Zinc is much more active than silver and replaces silver in the cyanide complex. The following equations illustrate the production of silver from its sulfide by the formation of cyanide complex and the subsequent reduction of silver with zinc:

$$2Ag_2S(s) + 8CN^-(aq) + O_2(g) + 2H_2O(l) \longrightarrow 4Ag(CN)_2^-(aq) + 2S(s) + 4OH^-(aq)$$

$$2Ag(CN)_2^-(aq) + Zn(s) \longrightarrow 2Ag(s) + Zn(CN)_4^{2-}(aq)$$

Unlike copper, the +1 oxidation state for silver is more stable than the +2 state. Of the compounds of silver, its halides are among the most common. Of these, AgF is soluble in water, while $AgCl$, $AgBr$, and AgI are insoluble. Silver bromide and silver iodide are used in photographic films. Silver nitrate, which is perhaps the most common compound of silver, is used in other photographic materials.

We recall that silver is a noble metal and that noble metal ions are easy to reduce to elemental metal. This easy reducibility of silver ions is exploited in black-and-white photographic film processing.

▢ *Photographic film* for black-and-white pictures is a thin film of cellulose acetate coated with a colloidal suspension of small crystals of $AgBr$ and AgI in gelatin. The size of the silver halide crystals and the relative amounts of $AgBr$ and AgI present determine the sensitivity, or speed, of the film. When light falls on photographic film, silver halide crystals become "activated"; that is, they assume a state that is more easily reduced to elemental silver. The film is developed by placing it in an alkaline solution of an organic reducing agent such as pyrogallol or hydroquinone. The speed of reduction of silver from silver halides, AgX (where $X = Br$ or I), to elemental silver is proportional to the intensity of light during exposure:

$$AgX(s) + e^- \text{ (from reducing agent)} \longrightarrow Ag(s) + X^-(aq)$$

In the fixing process, the developed film is treated with sodium thiosulfate, $Na_2S_2O_3$, solution (called hypo) which dissolves silver halides that were not reduced on those parts of the film that were not exposed to light. The reaction of silver bromide with thiosulfate ions is

$$AgBr(s) + 2S_2O_3^{2-}(aq) \longrightarrow Ag(S_2O_3)_2^{3-}(aq) + Br^-(aq)$$

The film thus produced is called a *negative* because the image is black, finely divided elemental silver, and the unexposed parts are transparent. The negative is printed by a process similar to the one used to make the original image. A piece of photosensitive printing paper is illuminated through the negative and an image appears in which exposed regions of the film are dark and unexposed regions are light.

Gold is sometimes found in its elemental state in natural deposits, although it is usually alloyed with 1 to 50 percent silver. Gold can be found either as veins or dust in quartz rocks, or as nuggets or dust in rocky material deposited by rivers and streams.

None of the binary gold compounds are commonly encountered. Of the existing compounds of gold, the oxides and halides in the +3 oxidation state of the metal are the most prevalent. Gold does not react directly with oxygen, but does react with chlorine at 200°C to form gold(III) chloride, Au_2Cl_6:

$$2Au(s) + 3Cl_2(g) \xrightarrow{200^\circ C} Au_2Cl_6(s)$$

Gold(III) in solution can be detected by a sensitive test in which tin(II) ions reduce gold(III) to colloidal gold (Au°) having a deep purple color known as purple of Cassius.

Zinc, Cadmium, and Mercury

Zinc is often found in nature associated with copper in sulfide ores. Reduction of the mixed metal sulfides gives brass, a metal known to the ancient Romans. Many compounds of zinc, such as ZnO, $ZnSO_4(H_2O)_7$ (white vitriol), and ZnS (zinc blende), are white. They are therefore used in the production of white paint pigments. Zinc oxide is also used as an active ingredient in ointments for skin irritations.

Cadmium resembles zinc in some respects. Like zinc, cadmium exhibits only +2 oxidation state in its compounds, but elemental cadmium is somewhat softer than zinc. Both cadmium and zinc are used in electroplating to provide a corrosion-resistant coating on steel. Yellow cadmium sulfide is used as a paint pigment.

Mercury differs from zinc and cadmium in many ways. Mercury is a liquid noble metal (b.p. 357°C). Unlike zinc and cadmium, mercury does not dissolve in common acids or alkalies. Mercury appears in nature as red cinnabar, HgS, as the free metal, and as the red or yellow oxide, HgO, from which the metal can be recovered by heating the ore in air:

$$2HgO(s) \longrightarrow 2Hg(l) + O_2(g)$$

$$HgS(s) + O_2(g) \longrightarrow Hg(l) + SO_2(g)$$

▢ Like the compounds of lead, mercury vapor and mercury compounds are toxic. Mercury(II) ions react with proteins, destroying their native biological function. Ingestion of food contaminated with mercury(II) ions causes kidney damage. Mercury(II) ions react with the protein tissue of the kidney, which is then unable to remove waste products from the blood. Egg white and milk can be used as antidotes for mercury poisoning; the proteins in eggs and milk precipitate the mercury in the stomach.

Also unlike zinc and cadmium, mercury forms compounds in both +1 and +2 oxidation states. However, only one oxide, mercury(II) oxide, HgO, is known. Another oxide of mercury is sometimes reported with the formula

Hg_2O, but this substance is a mixture of Hg and HgO. Mercury(I) and mercury(II) chlorides both exist. Mercury(I) chloride, Hg_2Cl_2, called *calomel*, has multiple uses. For example, it is used as an electrode in pH meters. Calomel is also used medicinally as a cathartic (purgative) and diuretic (it stimulates secretion of fluids).

The bonding in both mercury(I) and mercury(II) compounds has considerable covalent character. The structures of Hg_2Cl_2 and $HgCl_2$ are

Cl—Hg—Hg—Cl (Hg_2Cl_2)   Cl—Hg—Cl ($HgCl_2$)

As expected of molecular compounds, mercury(II) chloride dissolves in nonpolar solvents such as diethyl ether (CH_3CH_2—O—CH_2CH_3), but its solubility in water is relatively low, 6.9 g per 100 g of H_2O. It dissociates in water into its ions to only a small degree. However, mercury(I) chloride is insoluble in water, ethanol, and diethyl ether. Mercury(II) chloride is also used medicinally. It is called *corrosive sublimate* and is used in dilute solution as an antiseptic.

Summary

Oxidation States of Metals

The maximum oxidation state of most representative metals in their compounds is given by the group numbers of the metals. The heavier metals in groups III, IV, and V exhibit two different oxidation states in their compounds. One of these equals the group number of the metal; the other is the group number minus two. For the heaviest member of each of these groups the lower oxidation state is the more stable one.

Most transition metals exhibit more than one oxidation state. In the 3*d* series of transition metals, scandium in its compounds has only a +3 oxidation state. The maximum oxidation state for titanium is +4, for vanadium +5, for chromium +6, and for manganese +7. Each of these elements also exhibits other oxidation states. Each of the 3*d* series of transition metals except scandium exhibits a +2 oxidation state, which forms when the atoms lose their two 4*s* electrons, except chromium and copper, which lose their single *s* electron and one *d* electron. Iron also exhibits the +3 state and copper the +1 state.

The alkali metals, the alkaline earth metals, and aluminum are good reducing agents because they are easily oxidized. The ions of copper and silver can act as oxidizing agents because they are easily reduced.

Energetics of Oxidation of Metals

The following enthalpy changes occur when a solid metal is oxidized in aqueous solution: $\Delta H_{\text{sublimation}}$, $\Delta H_{\text{ionization}}$, and $\Delta H_{\text{hydration}}$. If the metal is oxidized to form a solid compound, the last enthalpy change in the process is $\Delta H_{\text{lattice formation}}$ instead of $\Delta H_{\text{hydration}}$. The sum of these enthalpy changes is a measure of the activity of the metal. When the total enthalpy change for oxidation of a metal is small, the metal is active and can be easily oxidized to its aqueous ions or to its ions in an ionic solid.

Activity Series of Metals

Metals can be arranged in a series called the **activity series** in order of the relative ease with which they can be oxidized. The metals that are easiest to oxidize are called **active metals**. The active metals can reduce hydrogen ions in a 1 *M* solution. The active metals are good reducing agents. The **noble metals** are the hardest to oxidize, but their ions are easiest to reduce. Noble metals cannot reduce hydrogen ions in a 1 *M* solution. The noble metal ions are relatively good oxidizing agents. The active metal ions are extremely difficult to reduce.

Reduction of Metals and Metallurgy

The methods used in **metallurgy** and in laboratory preparation of metals from their compounds depend on the nature of the metals. The noble metals and metals of intermediate activity may be reduced from their compounds by more active metals, by carbon, or by hydrogen. The most active metals can be reduced from their oxides or halides by electrolysis.

Transition Metals

The common oxidation states exhibited by the 3*d* series of transition metals generally increase from +3 to +7 from scandium through manganese, then decrease to +2 and +3 for iron; +2 for cobalt, nickel, and zinc; and to +1 and +2 for copper. For silver, the +1 state is the only common oxidation state. For mercury, both +1 and +2 states are known, but the +2 state is the more stable one.

Many monatomic aqueous ions of transition metals and their oxoanions, as well as their compounds, are colored. All of the monatomic ions of d^0 or d^{10} structure are colorless. Transition metal ions of d^0, d^5, or d^{10} valence shell structures are usually relatively stable toward oxidation or reduction. However, some monatomic ions having an outer d^3 or d^9 structure, such as Cr^{3+} and Cu^{2+}, are stable because they have relatively high **ionic potentials**.

New Terms

Activity series (18.2) Ionic potential (18.2) Metallurgy (18.3)

Exercises

General Review

1. **(a)** Can elemental metals be oxidized? **(b)** Can they be reduced to stable ions?
2. **(a)** Can elemental nonmetals be oxidized? **(b)** Can they be reduced to stable ions?
3. How can you predict whether an atom, molecule, or ion can act as an oxidizing agent, as a reducing agent, or both?
4. What are some characteristics of: **(a)** active metals; **(b)** noble metals?
5. In which region of the periodic table can we find: **(a)** the most active metals; **(b)** the noble metals; **(c)** the metals of intermediate activity?
6. What is the activity series of metals?
7. What enthalpy changes can be thought to occur in the oxidation of a metal to its ions in an aqueous solution?
8. Explain how each of the enthalpy changes you listed in question 7 affects the activity of a metal.
9. Can free metals act as oxidizing or reducing agents?
10. State whether aqueous monatomic ions of active metals, or aqueous monatomic ions of noble metals, are easier to reduce.
11. What is metallurgy?
12. By what methods are barium, iron, and copper prepared from their compounds? Write equations for these processes.
13. Give an example of: **(a)** a common metal that reacts with water; **(b)** a metal that does not react with water but reacts with 1 *M* hydrochloric acid; **(c)** a metal that does not react with water or hydrochloric acid but reacts with nitric acid; **(d)** a metal that reacts only with aqua regia. Write an equation to illustrate each of these reactions.
14. Can magnesium reduce copper(II) ions from their 1 *M* solution? Can nickel reduce sodium ions in their 1 *M* solution?
15. List three common oxidizing agents.
16. Give examples of some applications of redox reactions.
17. List three common reducing agents.
18. List four main group metals each of which exhibits two different oxidation states in its compounds.
19. List some general properties of transition metals.
20. How do the hardness and melting point of transition metals vary across a period in the periodic table?

21. How does the number of oxidation states exhibited by the transition metals vary across a period of the periodic table?

22. What are the common oxidation states for each of the 3*d* transition metals?

23. Which of the transition metals discussed in this chapter commonly exhibit a +1 oxidation state? For which of these metals is the +1 state the only common oxidation state?

24. List the formulas and names of the most common oxoanions of chromium and manganese.

25. Write the formulas and names of the acids that correspond to chromate ion and to permanganate ion.

26. List three substances that can act only as reducing agents and not as oxidizing agents.

27. Which of the following substances can act only as oxidizing agents? Which can act only as reducing agents? Which can act as either oxidizing or reducing agents? **(a)** Al; **(b)** Cl_2; **(c)** CO; **(d)** Pb; **(e)** PbO; **(f)** PbO_2; **(g)** CrO_4^{2-}; **(h)** $Cr_2O_7^{2-}$; **(i)** MnO_2; **(j)** Ca^{2+}; **(k)** Pb^{2+}; **(l)** Fe^{2+}; **(m)** CrO_3; **(n)** Mg.

Energetics of Oxidation and Reduction

28. The sublimation energy of metal A is 176 kJ mol^{-1} and that of metal B is 339 kJ mol^{-1}. The ionization energy of metal A is lower than that of B. The hydration energies of the ions of A and B are approximately equal. Which metal is more likely to be the better reducing agent in aqueous solution?

29. Referring to question 28, assume that the lattice energies of the oxides of the metals A and B are equal. Is it more likely that metal A can reduce B from its oxide or that B can reduce A from its oxide?

30. The ionization energy of metal A is 495 kJ mol^{-1} and that of metal B is 728 kJ mol^{-1}. The hydration energy of A^+ ions is −197 kJ mol^{-1} and that of B^+ ions is −473 kJ mol^{-1}. Metal A dissolves in 1 *M* HCl, whereas metal B does not. Is metal A likely to be harder than metal B?

Prediction of Reactions

31. Write a balanced equation (a net ionic equation where aqueous ions are present) for each of the following reactions using information derived from the activity series of metals: **(a)** a piece of cesium metal is placed into water (*Danger*: violent reaction); **(b)** aluminum metal is treated with hydrochloric acid; **(c)** copper metal is dissolved in nitric acid (assume that nitrogen dioxide is the reduction product of nitrate ions); **(d)** iron is reduced to Fe from iron(III) oxide with carbon monoxide; **(e)** mercury(II) sulfide is heated in air; **(f)** iron(III) oxide is heated with magnesium; **(g)** lead(II) oxide is heated with powdered zinc; **(h)** nickel wire is placed into a solution of silver nitrate; **(i)** a piece of aluminum is placed in a slightly acid solution of copper(II) sulfate.

Preparation of Metals; Electrolysis

32. Suggest a method for preparing each of the following metals from its oxide, and write an equation for the reaction: **(a)** Mg; **(b)** Ni; **(c)** Ag.

33. Write equations for the anode and cathode reactions and for the overall reaction that occurs in the electrolysis of: **(a)** molten potassium bromide; **(b)** molten calcium chloride.

Equation Writing

34. Write a balanced equation for each of the following chemical changes. Write net ionic equations where appropriate. **(a)** A piece of calcium metal is treated with fluorine gas; **(b)** mercury(II) oxide decomposes upon heating; **(c)** a piece of nickel metal is placed in hydrochloric acid.

35. Iron(II) ions in acid solution are titrated to iron(III) ions with potassium permanganate solution. The permanganate ions are reduced to manganese(II) ions. Write a balanced net ionic equation for this titration reaction.

36. Write a balanced net ionic equation for the oxidation of iron(II) ions in acid solution to iron(III) ions by dichromate ions, which are reduced to chromium(III) ions.

37. Write balanced net ionic equations for the reactions that occur when: **(a)** aluminum wire is dipped into a solution of silver nitrate; **(b)** hydrogen peroxide oxidizes chromium(III) ions to chromate ions in basic solution (hydrogen peroxide is reduced to water in this process); **(c)** magnesium metal is placed in boiling water (magnesium hydroxide is insoluble in water).

38. Write a balanced equation for the reaction that occurs when hydrogen sulfite ions are titrated with permanganate ions in acid solution to produce sulfate ions and manganese(II) ions.

Transition Metals

39. Write the formula and name of the oxide of each of the 3*d* series of transition metals in its highest stable oxidation state.

40. Write the formula of each of the following compounds: **(a)** yttrium chloride; **(b)** iron(III) sulfate; **(c)** vanadium(III) oxide; **(d)** mercury(I) sulfate; **(e)** silver chromate; **(f)** chromium(VI) oxide; **(g)** sodium chromate; **(h)** potassium dichromate; **(i)** cadmium phosphate; **(j)** rubidium permanganate.

41. Write a balanced equation for each of the following reactions (a net ionic equation where appropriate): **(a)** titanium(IV) chloride reacts with water; **(b)** chromium(VI) oxide is decomposed upon heating to oxygen and the green (most stable) oxide of chromium; **(c)** pink cobalt(II) chloride hexahydrate is heated to convert it to blue tetrahydrate; **(d)** blue copper(II) sulfate pentahydrate is heated to produce the white anhydrous sulfate; **(e)** solid silver bromide is dissolved in a solution of potassium thiosulfate to produce $Ag(S_2O_3)_2^{3-}$ ions and bromide ions; **(f)** reaction of manganate ions, MnO_4^{2-}, with one another in acid solution to produce permanganate ions and manganese(II) ions. (A redox reaction in which the same species is oxidized and reduced is called *disproportionation*.)

42. Which of the following substances can act only as oxidizing agents and which can act only as reducing agents? $K_2Cr_2O_7$, Cu^{2+}, Cu, CrO_3, Fe^{2+}.
43. What is the name of the acid derived from each of the following oxides? **(a)** CrO_3; **(b)** Mn_2O_7. **(c)** Write a formula for each of these acids.
44. List the lowest and the highest common oxidation states exhibited by each of the following metals in their compounds: **(a)** scandium; **(b)** chromium; **(c)** manganese; **(d)** iron; **(e)** copper; **(f)** zinc; **(g)** yttrium; **(h)** mercury.
45. Write the formula and the name (both systematic and common where applicable) for the metal oxides corresponding to each of the oxidation states you listed in question 44.
46. Which one in each of the following pairs of chlorides is more common than the other? **(a)** $TiCl_3$ and $TiCl_4$; **(b)** $FeCl_2$ and $FeCl_3$; **(c)** $CoCl_2$ and $CoCl_3$; **(d)** $CuCl_2$ and CuCl; **(e)** $CrCl_2$ and $CrCl_3$.
47. Cite a use for each of the following compounds: **(a)** titanium (IV) oxide; **(b)** titanium(IV) chloride; **(c)** zirconium(IV) silicate; **(d)** potassium dichromate; **(e)** cobalt(II) chloride; **(f)** copper(II) sulfate; **(g)** silver bromide; **(h)** zinc oxide; **(i)** mercury(I) chloride.
48. Write the formula and name of the base that corresponds to each of the following anhydrides: **(a)** MnO; **(b)** Fe_2O_3; **(c)** ZnO.
49. Since mercury(II) oxide can be decomposed to mercury and oxygen by heating, would you predict that CdO, ZnO, and/or Ag_2O can also be decomposed by heating? Use the information in Table 18.4 if necessary.
50. An oxide of chromium is found to contain 55.0 percent chromium. Calculate the formula of the oxide.

Aqueous Chemistry of the *d* Series of Transition Elements

51. List three transition metals that exhibit two different common aqueous monatomic ions. Write the symbols with the correct charges for each of these ions.
52. What is the color of each of the following aqueous ions? **(a)** Sc^{3+}; **(b)** Cr^{3+}; **(c)** Mn^{2+}; **(d)** Fe^{3+}; **(e)** Co^{2+}; **(f)** Cu^{2+}; **(g)** Ag^+; **(h)** Hg_2^{2+}; **(i)** CrO_4^{2-}; **(j)** $Cr_2O_7^{2-}$: **(k)** MnO_4^-.
53. What types of outer electron configurations do the colorless monatomic aqueous ions of the transition metals have in common?
54. Cu^+ ion can easily be oxidized to Cu^{2+} ion, which is the more stable of the two in aqueous solution. Cu^+ ion has a stable d^{10} structure, and its formation from a gaseous Cu atom requires 745 kJ mol^{-1}. The formation of the Cu^{2+} ion (d^9 structure) from a gaseous Cu atom requires 2703 kJ mol^{-1} of energy (the sum of the first and second ionization energies). In the light of these data, how can you explain that an aqueous Cu^{2+} ion is more stable than an aqueous Cu^+ ion? Is your answer consistent with the concept of ionic potential? Do you expect a *gaseous* Cu^{2+} ion to be more stable than a *gaseous* Cu^+ ion?
55. **(a)** Write an equation for the aqueous equilibrium between chromate and dichromate ions. **(b)** Explain why the color of a potassium chromate solution turns from yellow to orange when an acid is added. What change in oxidation state occurs, if any?

CHAPTER 19

The colors of some precipitates and aqueous ions. Many aqueous ions can be identified by the appearance of their precipitates or by the colors of their solutions.

Qualitative Analysis in Aqueous Solution

Qualitative analysis is a branch of chemistry in which we identify unknown substances. Inorganic qualitative analysis is aqueous solution is concerned with the identification of ions present in solution. If the unknown is a solid compound or a mixture of compounds, the solid is usually dissolved by a suitable aqueous reagent. The analysis is then based on characteristic reactions of the ions in solution.

Qualitative analysis in aqueous solution provides us with a way to review the wide variety of chemical reactions we have discussed in this book. In qualitative analysis we can apply our knowledge of redox reactions, acid–base chemistry, solubility equilibria, and complex ion formation. Now that we have considered the chemistry of the elements, including the common transition metals, we can use qualitative analysis to combine chemical theory with practice.

An unknown substance can be identified by systematically studying its chemical reactions with various reagents that are chosen to produce a characteristic color change, a precipitate, or a gas having a specific color or odor. The solubility of an unknown substance in various solvents also provides clues about its identity.

An unknown ionic substance is identified by separate tests of its anion and cation components. Analysis of cations in aqueous solution is usually carried out by separating groups of cations, followed by further separation and identification of individual cations within each group. Anion analysis, on the other hand, does not require the separation of all of the ions because individual tests for the presence or absence of certain anions can be done in the presence of other anions which do not interfere with the tests.

19.1 Precipitation and Separation of Groups of Cations

One of many possible schemes of separating cations into groups in aqueous solution is summarized in Figure 19.1. The general strategy of this (or any) scheme of qualitative analysis is to assume that an unknown substance contains all possible cations and anions.

The qualitative analysis scheme shown in Figure 19.1 is based on the solubilities of metal chlorides, sulfides, hydroxides, and carbonates in aqueous solution. Metal ions in this scheme are divided into five groups.

GROUP I The metal ions in group I are Ag^+, Pb^{2+}, and Hg_2^{2+}. *Group I cations are precipitated as chlorides: $AgCl$, $PbCl_2$, and Hg_2Cl_2, which are all white.* (Figure 19.2).

GROUP II The metal ions in group II are Hg^{2+}, Cu^{2+}, Pb^{2+}, and Sn^{2+}· *Group II cations are precipitated as sulfides* (Figure 19.3) *from an acid solution saturated with H_2S.* These metal sulfides have very small K_{sp} values. In an acid solution saturated with H_2S, the sulfide ion concentration is relatively small. Lead(II) appears in both groups I and II. That is because some Pb^{2+} ions may remain in solution when HCl is added to precipitate the group I cations as chlorides. Hence, Pb^{2+} ions that might remain after adding HCl precipitate as PbS with the group II cations.

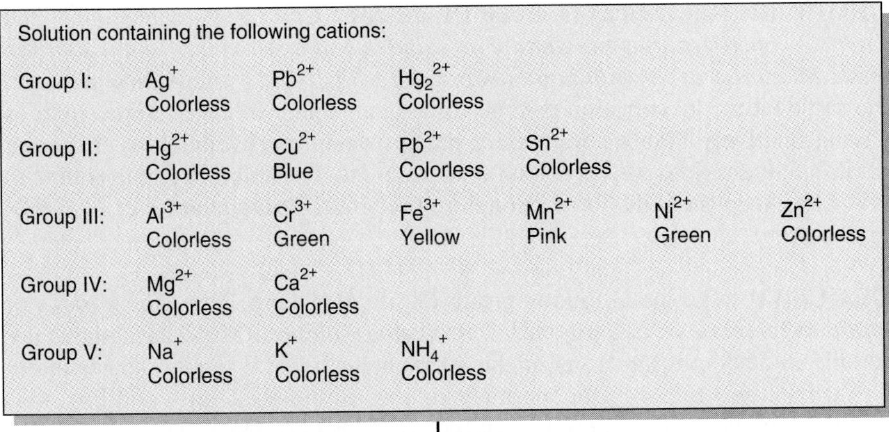

Solution containing the following cations:
Group I: Ag^+ Colorless, Pb^{2+} Colorless, Hg_2^{2+} Colorless
Group II: Hg^{2+} Colorless, Cu^{2+} Blue, Pb^{2+} Colorless, Sn^{2+} Colorless
Group III: Al^{3+} Colorless, Cr^{3+} Green, Fe^{3+} Yellow, Mn^{2+} Pink, Ni^{2+} Green, Zn^{2+} Colorless
Group IV: Mg^{2+} Colorless, Ca^{2+} Colorless
Group V: Na^+ Colorless, K^+ Colorless, NH_4^+ Colorless

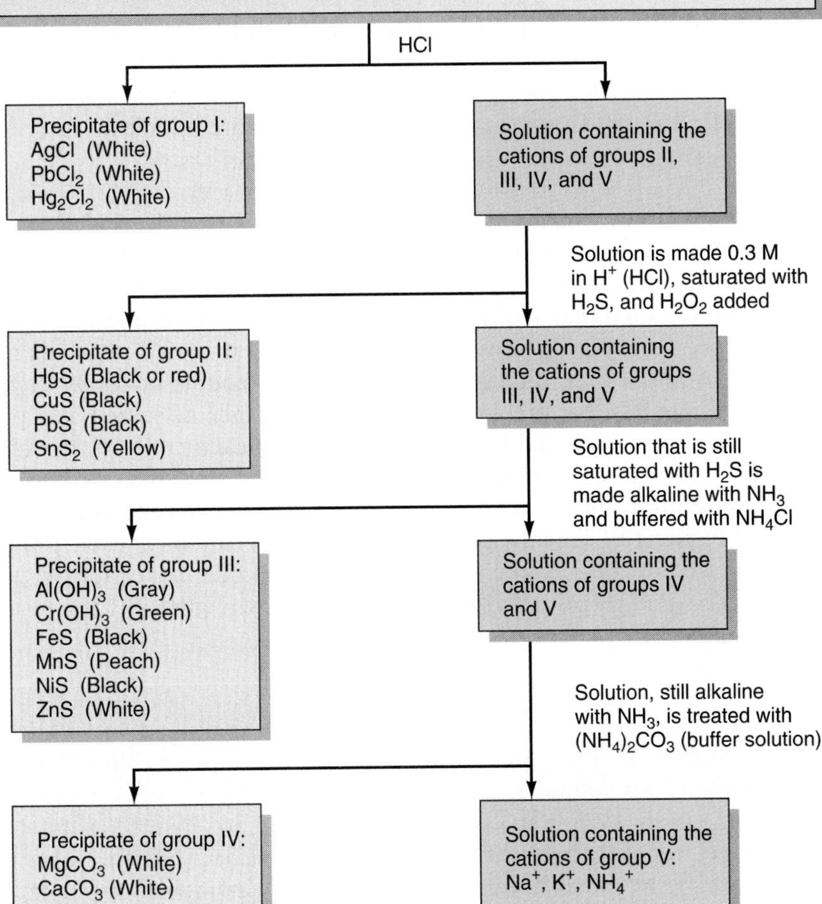

Figure 19.1

A general scheme of qualitative cation analysis for separation of groups of cations.

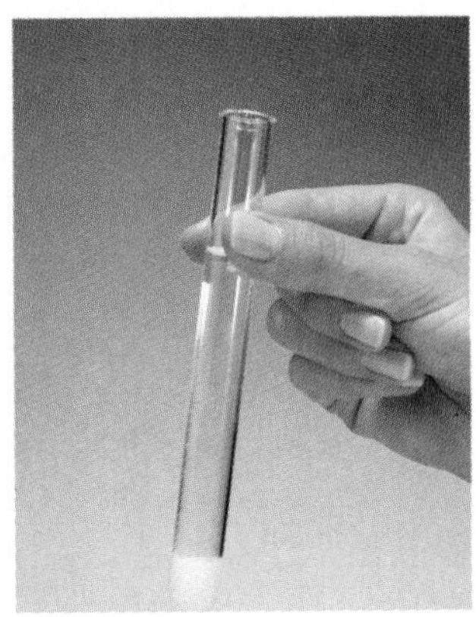

Figure 19.2

Group I precipitate, a mixture of silver chloride, lead(II) chloride, and mercury(I) chloride.

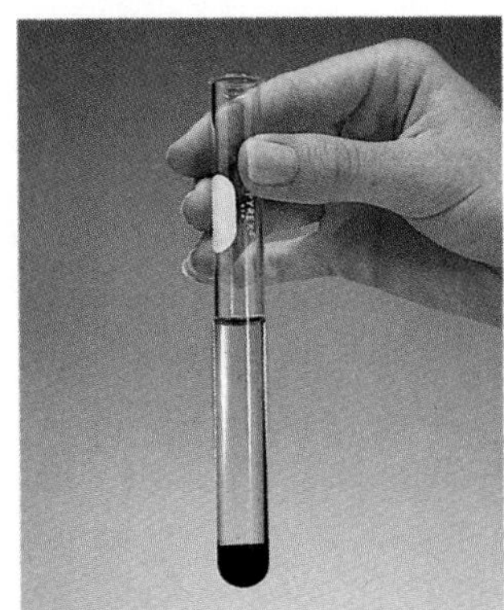

Figure 19.3

Group II precipitate, consisting of black mercury(II) sulfide, black copper(II) sulfide, black lead(II) sulfide, and yellow tin(IV) sulfide. The yellow color of tin(IV) sulfide is masked by the black color of the other sulfides in this group.

GROUP III The cations in group III are Al^{3+}, Cr^{3+}, Fe^{3+}, Mn^{2+}, Ni^{2+} and Zn^{2+}. *Group III cations precipitate as sulfides and hydroxides from a solution made alkaline with ammonia and saturated with H_2S.* In an alkaline solution, the sulfide ion concentration is larger than in an acidic solution. Metal sulfides having relatively "large" K_{sp} values precipitate in this group. Ions that form insoluble hydroxides also precipitate in group III. The solution is buffered with NH_4Cl to keep the $[OH^-]$ low enough to prevent the precipitation of $Mg(OH)_2$ in this group.

GROUP IV The metal ions in group IV are Mg^{2+} and Ca^{2+}· *The cations of group IV precipitate as carbonates in alkaline solution.* Metal carbonates are soluble in acid solution. Ions of the alkaline earth metal family—particularly Ca^{2+}, Sr^{2+}, and Ba^{2+}—from sparingly soluble carbonates. Ba^{2+} and Sr^{2+} ions are often included in this group, but we consider only Mg^{2+} and Ca^{2+}.

GROUP V The metal ions in group V are Na^+, K^+, and NH_4^+ is also included. Most alkali metal salts and ammonium salts are soluble in water. Therefore, *the cations in group V are identified by special tests.*

Figure 19.1 lists most of the metal ions that are usually included in qualitative analysis schemes. After one group of cations has been selectively precipitated, the mixture is centrifuged and the supernatant, which may contain other cations, is decanted into another test tube. The combination of centrifuging and decanting corresponds to filtration. This combination is used rather than filtration because it is often difficult to isolate small amounts of solids from filter paper. Compounds that precipitate as a group are then dissolved, any precipitate remaining is separated from the solution, and each cation is identified.

Some cations can be tentatively identified by the color of the original unknown solution and by the colors of the precipitates in each group. The first block of Figure 19.1 lists the colors of the aqueous cations in each group. The colors of the precipitates in each group are listed in the boxes for the precipitates in each group. Example 19.1 demonstrates how Figure 19.1 can be used to identify unknowns.

Example 19.1

Identifying Cations by Color and by Formation of Precipitates

A solution that may contain Ag^+ Cu^{2+}, Fe^{3+} and Ca^{2+} is blue. When a portion of this solution is treated with hydrochloric acid, a white precipitate forms. When the supernatant from the precipitate is saturated with hydrogen sulfide, a black precipitate appears. Whe the supernatant from this precipitate is made alkaline with ammonia, no change is observed. However, when ammonium carbonate is added to this solution, a white precipitate forms. Which cations are present in the original solution? Use Figure 19.1 as a guide. Write a net ionic equation for the reaction that occurs in each step of the above procedure.

SOLUTION: Of the cations that can be present in the solution, Ag^+ and Ca^{2+} are colorless, Cu^{2+} is blue, and Fe^{3+} is yellow. Because the solution is blue, we conclude that Cu^{2+} is present. However, the blue color of Cu^{2+} can easily mask the yellow color Fe^{3+}, and Fe^{3+} may also be present.

The formation of a white precipitate with HCl demonstrates that Ag^+ is present because it is the only ion in the mixture that forms an insoluble chloride:

$$Ag^+(aq) + Cl^-(aq) \longrightarrow AgCl(s)$$

The formation of a black precipitate from an acidic solution saturated with H_2S is further evidence that Cu^{2+} ions are in the solution. Cu^{2+} ions are the only ions in the mixture that form a black insoluble sulfide with H_2S in an acid solution:

$$Cu^{2+}(aq) + H_2S(aq) \longrightarrow CuS(s) + 2H^+(aq)$$

After Ag^+ and Cu^{2+} ions have been removed, the supernatant solution is made alkaline with ammonia. No precipitate forms. Therefore, Fe^{3+} ions are absent. If Fe^{3+} ions were present, they would precipitate from an alkaline solution as a rust-colored hydroxide, $Fe(OH)_3$, or hydrated oxide, $Fe_2O_3(H_2O)_n$.

When the supernatant is treated with $(NH_4)_2CO_3$, a white precipitate forms, indicating that Ca^{2+} is present:

$$Ca^{2+}(aq) + CO_3^{2-}(aq) \longrightarrow CaCO_3(s)$$

The following flowchart summarizes the results of this analysis (the abbreviation "ppt" stands for "precipitate"):

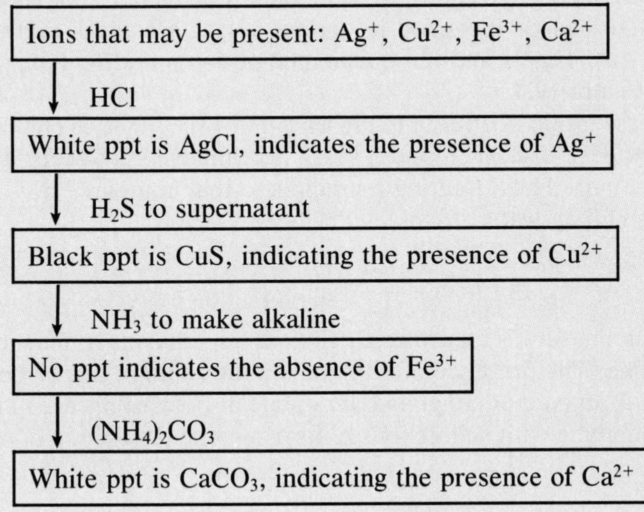

Practice Problem 19.1: A colorless solution may contain one or more of the following ions: Ag^+, Cu^{2+}, Cr^{3+}, and Ca^{2+} When this solution is treated with hydrochloric acid, a white precipitate forms. When the supernatant is made alkaline with ammonia and treated with ammonium carbonate, a white precipitate forms. Which ions are present? Use Figure 19.1 as a guide. Write net ionic equations for the reactions in which precipitate form.

Figure 19.4

Outline of an analysis of group I precipitate.

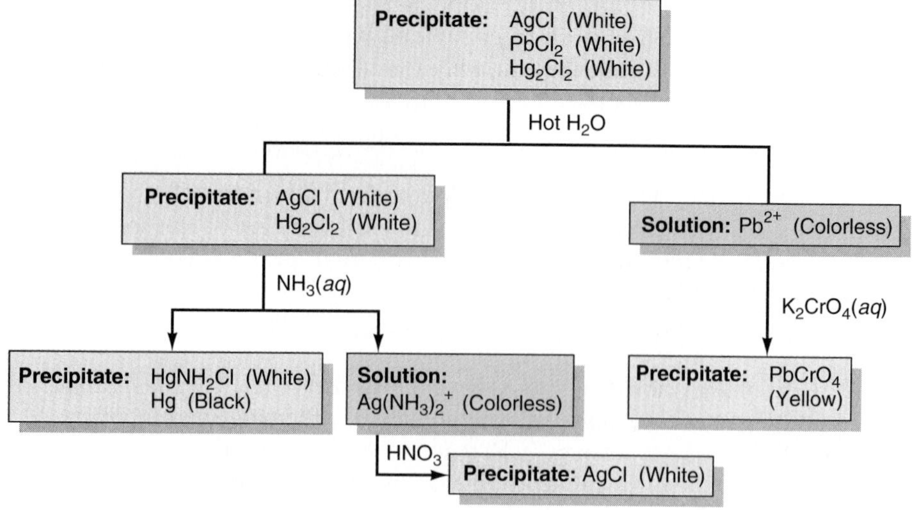

19.2 Group I Analysis

After the cations of a given group have been precipitated, the components of the precipitate can be identified in various ways. We recall that the group I cations are Ag^+, Hg_2^{2+}, and Pb^{2+}. One method for analyzing group I chlorides is given in Figure 19.4.

If all these group I cations are present, they precipitate as chlorides. Of the group I chlorides, lead(II) chloride, $PbCl_2$ is soluble in hot water; the other two chlorides are not. Thus, treating a precipitate that contains $PbCl_2$, AgCl, and Hg_2Cl_2 with boiling water dissolves $PbCl_2$:

$$PbCl_2(s) \xrightarrow{\text{hot } H_2O} Pb^{2+}(aq) + 2Cl^-(aq)$$

The resulting mixture is centrifuged, and the supernatant is decanted into another test tube. The presence of Pb^{2+} ions in the solution can be confirmed by adding a small amount of either sodium sulfate or potassium chromate solution. Lead ions combine with sulfate ions to form a white precipitate of lead sulfate, $PbSO_4$:

$$Pb^{2+}(aq) + SO_4^{2-}(aq) \longrightarrow PbSO_4(s) \quad \text{(white)}$$

Chromate ions react with lead ions to precipitate yellow lead chromate, $PbCrO_4$ (Figure 19.5). The reaction is

$$Pb^{2+}(aq) + CrO_4^{2-}(aq) \longrightarrow PbCrO_4(s) \quad \text{(yellow)}$$

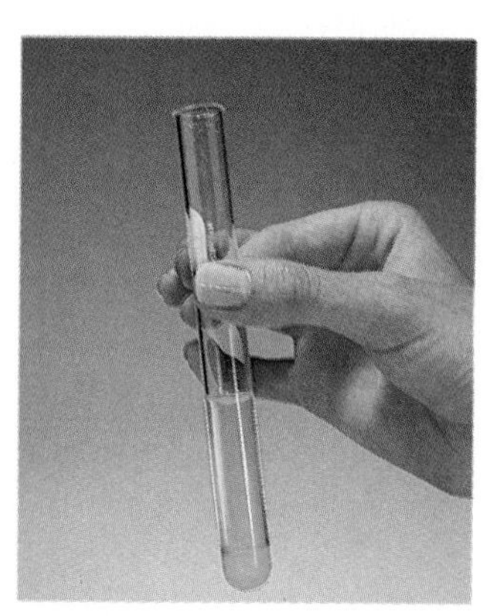

Figure 19.5

A precipitate of lead chromate.

After lead ions have been removed from the group I precipitate, the remaining precipitate can be treated with aqueous ammonia, which dissolves AgCl by forming the ammine complex (Section 16.5):

$$AgCl(s) + 2NH_3(aq) \longrightarrow Ag(NH_3)_2^+(aq) + Cl^-(aq)$$

If the precipitate is a mixture of both silver chloride and mercury(I) chloride, it is virtually impossible to detect whether some fraction of the precipitate dissolves upon treatment with ammonia. Thus, after treating the precipitate with ammonia, the mixture is centrifuged, and the supernatant is tested for $Ag(NH_3)_2^+$ and Cl^- ions by acidifying the solution with nitric acid. Hydrogen ions from nitric acid react with ammonia molecules in $Ag(NH_3)_2^+$ ions to form ammonium ions; the silver ions released react with the chloride ions in solution to form a precipitate of silver chloride:

$$Ag(NH_3)_2^+(aq) + Cl^-(aq) + 2H^+(aq) \longrightarrow AgCl(s) + 2NH_4^+(aq)$$

The formation of a white precipitate of AgCl at this point confirms the presence of Ag^+ ions in the original solution.

Although Hg_2Cl_2 (calomel) does not dissolve in NH_3, it reacts with NH_3 to form black elemental mercury and white mercury(II) amido chloride, $HgNH_2Cl$:

$$Hg_2Cl_2(s) + 2NH_3(aq) \longrightarrow \underset{\text{Black}}{Hg(l)} + \underset{\text{White}}{HgNH_2Cl(s)} + NH_4^+(aq) + Cl^-(aq)$$

This is a redox reaction in which mercury is both oxidized and reduced. (We recall that a redox reaction in which the same substance is oxidized and reduced is called disproportionation.) A gray mixture of a black and a white substance at this point confirms the presence of Hg_2^{2+} ions in the original solution. Mercury normally has a silvery color, but in this case the extremely small size of the precipitated particles makes it appear black.

Example 19.2

Interpreting the Results Obtained in a Group I Cation Analysis

When a solution that contains group I cations is treated with hydrochloric acid, a white precipitate is obtained that may contain one or more of the chlorides, AgCl, $PbCl_2$, and Hg_2Cl_2. The precipitate is treated with hot water. The supernatant obtained gives a yellow precipitate with a solution of sodium chromate. The solid remaining from the hot water treatment dissolves completely in aqueous ammonia. Which group I cations are present in the original solution? Write a net ionic equation for the reaction that occurs at each step of analysis of the precipitate obtained from the solution of group I cations.

SOLUTION: The formation of a yellow precipitate from the hot supernatant confirms the presence of $PbCl_2$ in the precipitate and of Pb^{2+} ions in the original solution. When the precipitate is treated with hot water, $PbCl_2$ dissolves:

$$PbCl_2(s) \xrightarrow{\text{hot } H_2O} Pb^{2+}(aq) + 2Cl^-(aq)$$

The Pb^{2+} ions in the solution react with the chromate ions from sodium chromate to precipitate yellow lead chromate:

$$Pb^{2+}(aq) + CrO_4^{2-}(aq) \longrightarrow PbCrO_4(s)$$

Since the solid remaining from the hot water treatment completely dissolves in aqueous ammonia, AgCl is present. Hg_2Cl_2 is absent because it does not dissolve in ammonia. Silver chloride dissolves in ammonia according to the equation

$$AgCl(s) + 2NH_3(aq) \longrightarrow Ag(NH_3)_2^+(aq) + Cl^-(aq)$$

Thus, Pb^{2+} and Ag^+ ions are present in the original solution, and Hg_2^{2+} ions are absent.

The following flowchart summarizes this analysis:

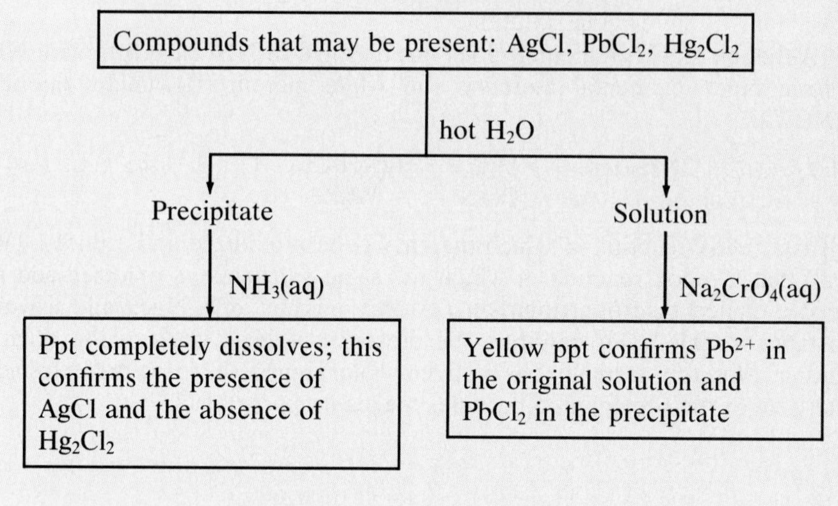

Practice Problem 19.2: A solution of group I cations may contain one or more of the ions Ag^+, Pb^{2+}, and Hg_2^{2+}. When the solution is treated with hydrochloric acid, a white precipitate is obtained that dissolves completely in hot water. Which cations are present, and which are absent?

19.3 Group II Precipitation and Analysis

Group II cations—Hg^{2+}, Cu^{2+}, Pb^{2+}, and Sn^{2+}—precipitate as sulfides from an acidic solution saturated with H_2S. Hydrogen sulfide is a poisonous gas with an unpleasant odor. Hydrogen sulfide is best generated by treating thioacetamide, CH_3CSNH_2, with hot water. The reaction of thioacetamide with water is

$$CH_3CSNH_2(aq) + 2H_2O(l) \xrightarrow{\text{heat}} NH_4^+(aq) + CH_3COO^-(aq) + H_2S(g)$$

This reaction is slow and the gradual formation of H_2S helps to keep most of the gas in the solution. Thus, little noxious H_2S gas enters the atmosphere.

Hydrogen sulfide in aqueous solution is a weak acid, called hydrosulfuric acid, which dissociates in two steps to produce the sulfide ions needed for the precipitation of group II cations:

$$H_2S(aq) + H_2O(l) \rightleftharpoons H_3O^+(aq) + HS^-(aq)$$

$$HS_-(aq) + H_2O(l) \rightleftharpoons H_3O^+(aq) + S^2(aq)$$

The overall dissociation process can be written as the sum of the two reactions above:

$$H_2S(aq) + 2H_2O(l) \rightleftharpoons 2H_3O^+(aq) + S^{2-}(aq)$$

The sulfide ion concentration in acid solution is lower than in neutral solution (see Section 16.4). Thus, only the least soluble metal sulfides precipitate in an acidic solution that contains H_2S.

Inspection of Figure 19.1 shows that a solution of group II cations contains tin in its lower oxidation state as Sn^{2+} ion. In its +2 oxidation state, tin precipitates as tin(II) sulfide, SNS, which is more soluble than tin(IV) sulfide, SnS_2. The K_{sp} for SnS is 3×10^{-27}, whereas the K_{sp} for SnS_2 is 1×10^{-70}! Since SnS_2 is so much more insoluble than SnS, to ensure a virtually complete precipitation of tin as SnS_2 in group II, the tin originally present as Sn^{2+} ions is oxidized to +4 oxidation state using hydrogen peroxide as the oxidizing agent. In the presence of H^+ and Cl^- ions, tin(II) ions are oxidized and form chloro complex ions of tin, $SnCl_6{}^{2-}$, while hydrogen peroxide is reduced to water:

$$Sn^{2+}(aq) + H_2O_2(aq) + 2H^+(aq) + 6Cl^-(aq) \longrightarrow SnCl_6{}^{2-}(aq) + 2H_2O(l)$$

$SnCl_6{}^{2-}$ ions then react with the sulfide ions to form SnS_2:

$$SnCl_6{}^{2-}(aq) + 2S^{2-}(aq) \longrightarrow SnS_2(s) + 6Cl^-(aq)$$

This reaction occurs because SnS_2 is one of the most water-insoluble compounds known. However, this reaction can be shifted to the left by excess chloride ions in acid solution.

Three sulfides of group II cations are black. If one or more of these sulfides is present in the precipitate along with yellow SnS_2, the black color masks the yellow one. However, if the precipitate of the group is yellow, SnS_2 is probably the only sulfide present. An analysis of group II precipitate is outlined in Figure 19.6.

The precipitate of group II sulfides is first treated with HCl, with warming, to dissolve SnS_2. Although the K_{sp} for SnS_2 is very small, SnS_2 is soluble in HCl because tin has a great tendency to form chloro complexes in a solution which contains a high concentration of chloride ions:

$$SnS_2(s) + 4H^+(aq) + 6Cl^-(aq) \longrightarrow SnCl_6{}^{2-}(aq) + 2H_2S(g)$$

The presence of tin can be confirmed after $SnCl_6{}^{2-}(aq)$ has been produced. First, Sn(IV) is reduced to Sn(II) by elemental iron:

$$SnCl_6{}^{2-}(aq) + Fe(s) \longrightarrow Sn^{2+}(aq) + Fe^{2+}(aq) + 6Cl^-(aq)$$

The resulting solution of Sn(II) is treated with $HgCl_2$ solution. The Sn^{2+} ions in solution reduce Hg^{2+} ions to $Hg_2{}^{2+}$ ions. $Hg_2{}^{2+}$ ions then react with chloride

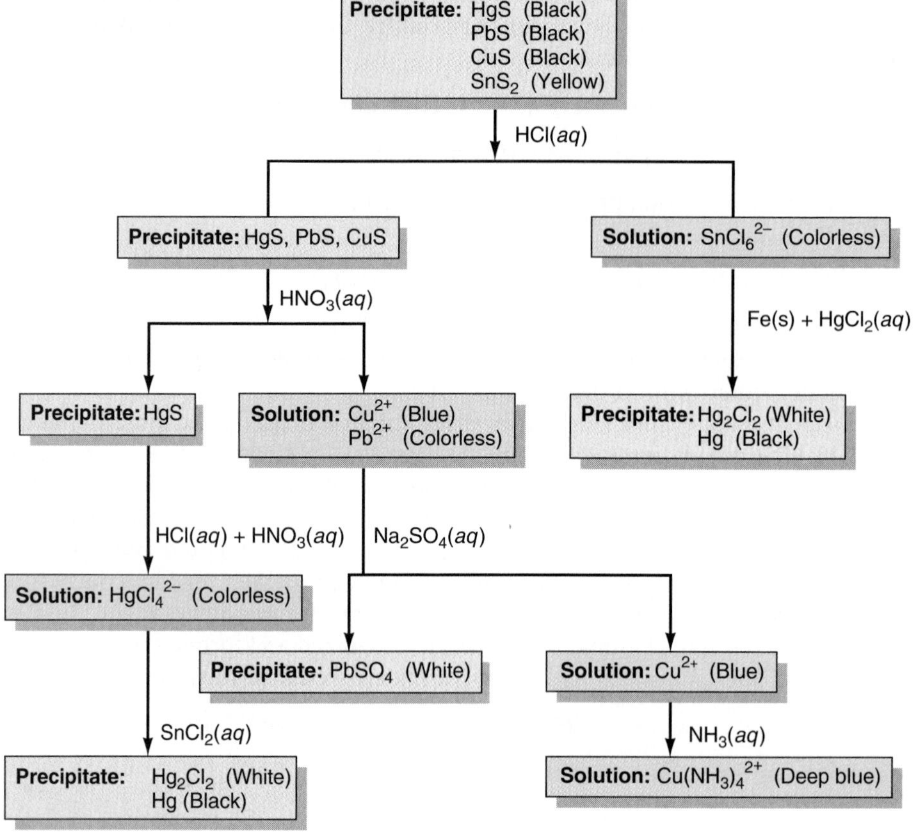

Figure 19.6

Outline of an analysis of group II precipitate.

ions in the solution, precipitating white Hg_2Cl_2. Tin is simultaneously oxidized from +2 oxidation state in Sn^{2+} ions to the +4 oxidation state in $SnCl_6^{2-}$ ions. The overall reaction is

$$Sn^{2+}(aq) + 2Hg^{2+} + 8Cl^-(aq) \longrightarrow \underset{\text{White}}{Hg_2Cl_2(s)} + SnCl_6^{2-}(aq)$$

The precipitate of Hg_2Cl_2 reacts with residual Sn^{2+} ions in the solution to give black elemental mercury:

$$Sn^{2+}(aq) + Hg_2Cl_2(s) + 4Cl^-(aq) \longrightarrow \underset{\text{Black}}{2Hg(l)} + SnCl_6^{2-}(aq)$$

Thus, if Sn^{2+} ions are present in a solution, treatment with $HgCl_2$ solution produces both white and black precipitates. The mixture of these two is likely to appear gray. If the Sn^{2+} concentration is high, a relatively large amount of black Hg forms.

After Sn^{2+} ions have been removed, PbS, CuS, and HgS remain in the precipitate. If the precipitate is treated with nitric acid, an oxidizing acid (see Section 16.5), PbS and CuS dissolve but HgS does not. PbS and CuS can thus be separated from HgS by stirring the mixture with nitric acid. HgS does not dissolve in nitric acid because of its very small K_{sp} value (1×10^{-50}).

The remaining black precipitate of HgS can be dissolved in aqua regia (Section 16.5). Nitrate ions in aqua regia oxidize S^{2-} ions to elemental sulfur, and chloride ions in aqua regia react with Hg^{2+} ions to form chloro complex ions of mercury:

$$3HgS(s) + 8H^+(aq) + 2NO_3^-(aq) + 12Cl^-(aq) \longrightarrow 3HgCl_4^{2-}(aq) + 3S(s) + 4H_2O(l) + 2NO(g)$$

The presence of $HgCl_4^{2-}$ ions in the resulting solution can be confirmed by adding a solution of tin(II) chloride to give a precipitate of white Hg_2Cl_2 and black Hg:

$$2HgCl_4^{2-}(aq) + Sn^{2+}(aq) \longrightarrow Hg_2Cl_2(s) + SnCl_6^{2-}(aq) \quad \text{or}$$

$$HgCl_4^{2-}(aq) + Sn^{2+}(aq) + 2Cl^-(aq) \longrightarrow Hg(l) + SnCl_6^{2-}(aq)$$

Note that this is essentially the same set of reactions as the confirmatory test for Sn^{2+} ions.

Two group II cations may now remain in solution: Pb^{2+} and Cu^{2+}. Pb^{2+} ions are present at his point only if they were incompletely precipitated as lead (II) chloride in group I analysis. The presence of Pb^{2+} ions in the solution can be confirmed by adding a solution of Na_2SO_4, which precipitates white $PbSO_4$:

$$Pb^{2+}(aq) + SO_4^{2-}(aq) \longrightarrow PbSO_4(s)$$

After $PbSO_4$ has precipitated, the supernatant is blue if cooper(II) ions are present. When the solution is made alkaline with ammonia, the blue color of the aqua complex ions of copper deepens when the ammine complex ions of copper are produced:

$$Cu(H_2O)_4^{2+}(aq) + 4NH_3(aq) \longrightarrow Cu(NH_3)_4^{2+}4H_2O(l)$$

This deep blue color confirms the presence of copper(II) ions (Figure 19.7).

Figure 19.7

Aqua complex of copper(II) ions (left) and ammine complex of copper(II) ions (right).

Example 19.3

Interpreting the Results of Analysis of the Precipitate Obtained from Group II Cations

When a solution of group II cations is saturated with H_2S and treated with H_2O_2, a black precipitate is obtained. The precipitate may contain one or more of the following sulfides, HgS, PbS, CuS, and SnS_2. When the precipitate is treated with HCl, part of the precipitate dissolves. When powdered iron and $HgCl^2$ solution are added to this solution, a gray precipitate appears. The precipitate from the HCl treatment dissolves completely in HNO_3, giving a blue solution. When this solution is treated with Na_2SO_4, a white precipitate forms. Which sulfides are present in the original precipitate? Which cations are present in the group II solution?

SOLUTION: The gray precipitate formed in the solution that results from treating the precipitate with HCl indicates that SnS_2 is present in the original precipitate. Since the precipitate that remains after the HCl treatment completely dissolves in HNO_3, HgS is absent because it dissolves only in aqua regia. The blue color of the HNO_3 solution indicates the presence of Cu^{2+} because Cu^{2+} is the only colored group II cation. The white precipitate formed when Na_2SO_4 is added to this solution confirms the presence of Pb^{2+} ions as described above. Thus, the sulfides present in the original precipitate are CuS, PbS, and SnS_2. The group II ions in the original solution are Cu^{2+}, Pb^{2+}, and Sn^{2+}.

The following flowchart is a summary of this analysis:

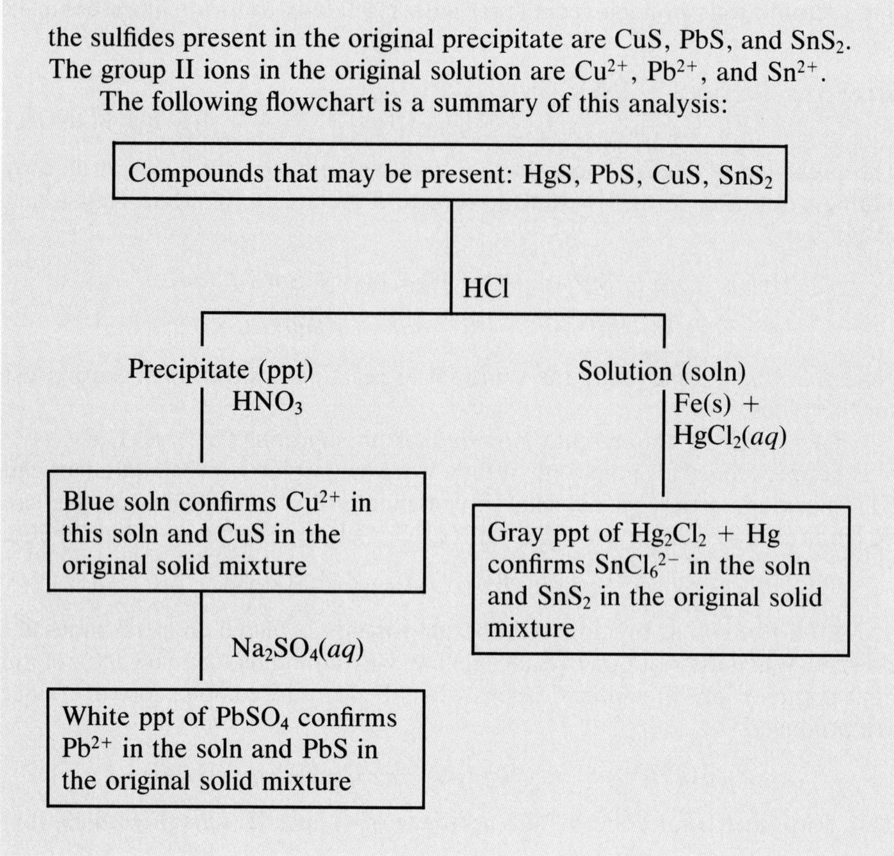

Practice Problem 19.3: A solution that may contain one or more of the group II cations Hg^{2+}, Cu^{2+}, and Sn^{2+} is colorless. When the solution is saturated with H_2S and treated with H_2O_2, a black precipitate forms. Which cation is definitely present? Which cation is definitely absent? Which cation cannot be identified because of insufficient information?

19.4 Analysis of Group III

The sulfides of group III cations—Al^{3+}, Cr^{3+}, Fe^{3+}, Mn^{2+}, and Zn^{2+}—are more soluble than the sulfides of the group II cations. Therefore, to precipitate the group III cations as sulfides, the sulfide ion concentration must be greater than that required to precipitate group II cations. To provide a higher sulfide ion concentration, the equilibrium for the formation of sulfide ions,

$$H_2S(aq) + 2H_2O(l) \rightleftharpoons 2H_3O^+(aq) + S^{2-}(aq)$$

must be shifted to the right. This is done by making the solution alkaline with ammonia, which reacts with hydronium ions:

$$NH_3(aq) + H_3O^+(aq) \longrightarrow NH_4^+(aq) + H_2O(l)$$

When the hydronium ion concentration decreases, the H_2S equilibrium shifts to the right and the sulfide ion concentration increases.

The solution from which the group II sulfides have been precipitated is still saturated with H_2S. When this solution is made alkaline with ammonia, most group III cations precipitate as sulfides. Aluminum ions and chromium(III) ions precipitate as hydroxides because aluminum hydroxide and chromium(III) hydroxide are less soluble than aluminum sulfide or chromium(III) sulfide.

If iron is present in the solution as Fe^{3+}, it is reduced to Fe^{2+} ions by sulfide ions. As a result, FeS precipitates:

$$2Fe^{3+}(aq) + 3S^{2-}(aq) \longrightarrow 2FeS(s) + S(s)$$

Analysis of a group III precipitate is shown in Figure 19.8, and the colors of the components of the precipitate are shown in Figure 19.9.

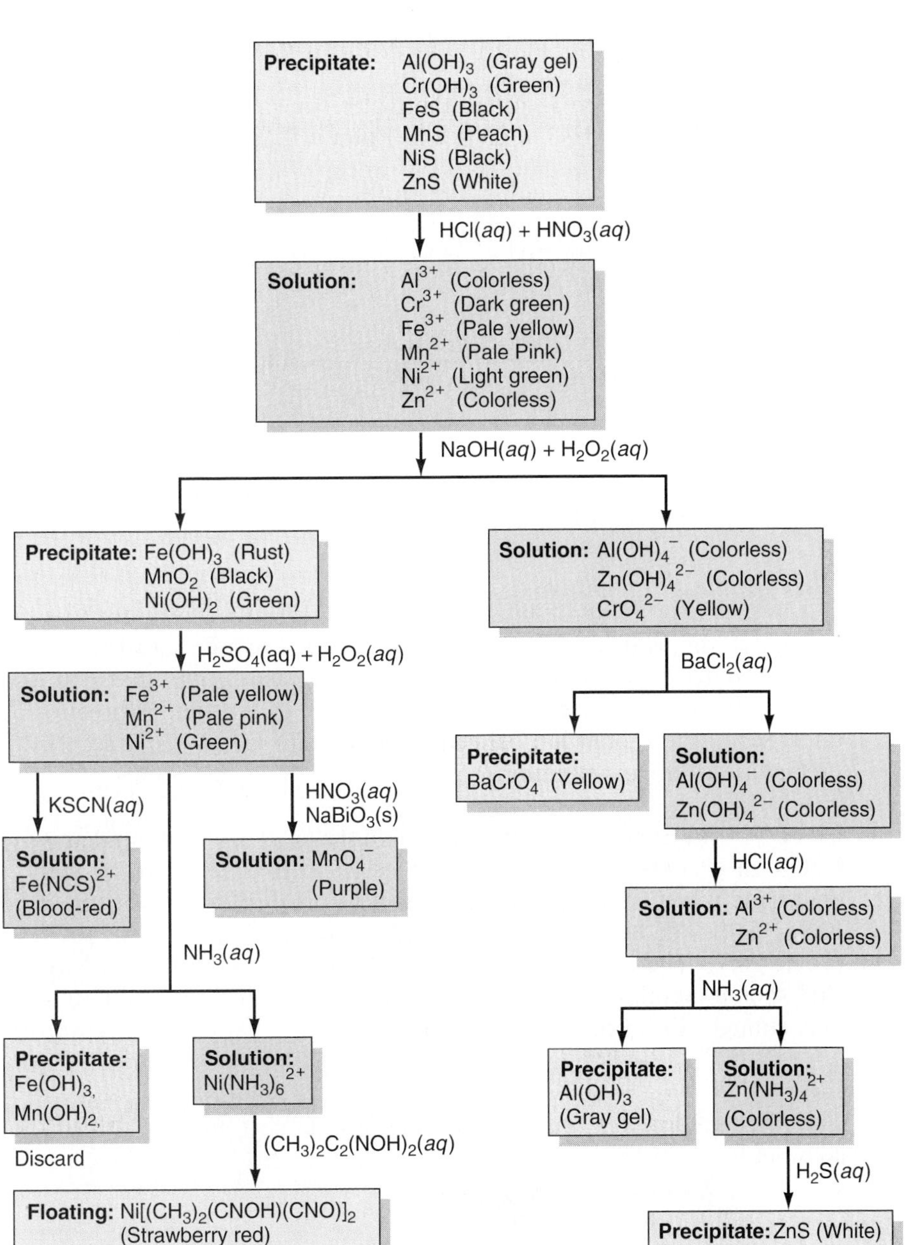

Figure 19.8

An outline of analysis of group III precipitate.

Example 19.4

Calculating the $[S^{2-}]$ in a Saturated Solution of H_2S That is 0.30 *M* in H_3O^+

Calculate the sulfide ion concentration in a 0.30 *M* solution of HCl that is saturated with H_2S at 25°C and 1 atm. K_1 for H_2S is 1.0×10^{-7} and K_2 is 1.3×10^{-13}.

SOLUTION: We will use the overall equation for ionization of H_2S (Section 16.4):

$$H_2S(aq) + 2H_2O(l) \rightleftharpoons 2H_3O^+(aq) + S^{2-}(aq)$$

The equilibrium constant for this system is the product of K_1 and K_2 for H_2S:

$$\frac{[H_3O^+]^2[S^{2-}]}{[H_2S]} = (1.0 \times 10^{-7})(1.3 \times 10^{-13}) = 1.3 \times 10^{-20}$$

We recall from Section 16.4 that the molar concentration of H_2S in a saturated solution of H_2S at 25°C and 1 atm is 0.10 *M*. Thus,

$$[H_3O^+]^2[S^{2-}] = (0.10)(1.3 \times 10^{-20}) = 1.3 \times 10^{-21}$$

We derived this equation as Equation 16.1.

In 0.30 *M* HCl, $[H_3O^+] = 0.30$ *M*. Thus, solving Equation 16.1 for $[S^{2-}]$, we obtain

$$[S^{2-}] = \frac{1.3 \times 10^{-21}}{[H_3O^+]^2} = \frac{1.3 \times 10^{-21}}{(0.30)^2} = 1.4 \times 10^{-20}\ M$$

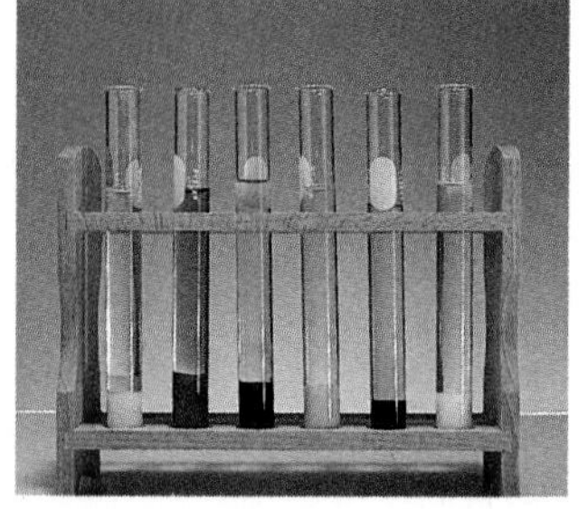

Figure 19.9

Precipitates of aluminum hydroxide (gray gel), chromium(III) hydroxide (dark green), iron(II) sulfide (black), manganese(II) sulfide (peach color), nickel(II) sulfide (black), and zinc sulfide (white).

Practice Problem 19.4: What is the $[S^{2-}]$ in a solution that is 1.0×10^{-4} *M* in $[OH^-]$ and saturated with H_2S at 25°C?

Example 19.5

Deciding Which Cation Precipitates as a Sulfide in 0.30 *M* HCl Solution That is Saturated with H_2S

A solution that is 0.10 *M* in Cu^{2+} ions and 0.10 *M* in Ni^{2+} ions is made 0.30 *M* in HCl and saturated with H_2S. Does Cu^{2+}, Ni^{2+}, or both, precipitate under these conditions? If only one of the cations precipitates, what hydronium ion concentration is required to just start the precipitate of the other cation?

SOLUTION: In Example 19.4 we calculated the $[S^{2-}]$ in a solution that is 0.30 *M* in HCl and saturated with H_2S as 1.0×10^{-20} *M*. Here we will determine whether this $[S^{2-}]$ is sufficient to precipitate either CuS or NiS or both. The K_{sp} for CuS is 6.3×10^{-36} and that for NiS is 3×10^{-19} (Table 16.1). These values indicate that CuS is less soluble than NiS. CuS should therefore precipitate first if the sulfide ion concentration is high enough. From the solubility equilibrium,

$$[Cu^{2+}][S^{2-}] = 6.3 \times 10^{-36}$$

The $[S^{2-}]$ needed to start the precipitation of CuS from a 0.1 *M* Cu^{2+} solution is

$$[S^{2-}] = \frac{6.3 \times 10^{-36}}{0.10} = 6.3 \times 10^{-35}\ M$$

This value is much smaller than the $[S^{2-}]$ actually in the solution (1.4×10^{-20} M); therefore, CuS precipitates.

The $[S^{2-}]$ needed to start the precipitation of NiS is

$$[S^{2-}] = \frac{3 \times 10^{-19}}{0.10} = 3 \times 10^{-18}\ M$$

This value is larger than 1.4×10^{-20} M; therefore, NiS does not precipitate under these conditions. To increase the $[S^{2-}]$ so that NiS can precipitate, the $[H_3O^+]$ must be decreased; that is, the pH must be increased. To calculate the hydronium ion concentration, which provides a sulfide ion concentration of 3×10^{-18} in a saturated solution of H_2S, we solve Equation 16.1 for $[H_3O^+]$:

$$[H_3O^+] = \sqrt{\frac{1.3 \times 10^{-21}}{[S^{-2}]}} = \sqrt{\frac{1.3 \times 10^{-21}}{3 \times 10^{-18}}} = 0.02\ M$$

$$\text{pH} = -\log 0.02 = 1.7$$

Practice Problem 19.5: A solution is 0.10 M in Mn^{2+} ions. If the solution is saturated with H_2S, what hydronium ion concentration is required to initiate the precipitation of MnS?

We see from Example 19.5 that to *start* the precipitation of Ni^{2+} ions in group III, the hydronium ion concentration should be 0.02 M. However, to assure virtually complete precipitation of Ni^{2+} and other group III cations as sulfides and hydroxides, the solution is made alkaline with NH_3 after the separation of group II sulfides (Figure 19.1).

After separation of the group III precipitate from the supernatant solution, the possibly multicolored precipitate is redissolved in aqua regia. In this process, iron(II) is oxidized back to iron(III) by the nitrate ions in aqua regia:

$$FeS(s) + 6H^+(aq) + 3NO_3^-(aq) \longrightarrow Fe^{3+}(aq) + S(s) + 3NO_2(g) + 3H_2O(l)$$

The other sulfides and the hydroxides in this group dissolve readily in acid solution.

Practice Problem 19.6: Write a net ionic equation for dissolving each of the following substances in a nonoxidizing acid solution: **(a)** MnS; **(b)** $Al(OH)_3$.

The solution obtained by dissolving this group III precipitate in acid is made alkaline with sodium hydroxide and is also treated with hydrogen peroxide. As a result, $Fe(OH)_3$, MnO_2, and $Ni(OH)_2$ precipitate. The amphoteric hydroxides, $Al(OH)_3$ and $Zn(OH)_2$, dissolve in excess OH^- ions to form hydroxo complexes, $Al(OH)_4^-$ and $Zn(OH)_4^{2-}$, respectively. Hydrogen peroxide is added to oxidize the Cr^{3+} ions to CrO_4^{2-} ions, and Mn^{2+} ions to MnO_2, which is insoluble in basic solution and precipitates with the hydroxides of iron and nickel. The reactions for the foregoing oxidation processes in basic solution are

$$2Cr^{3+}(aq) + 3H_2O_2(aq) + 10OH^-(aq) \longrightarrow 2CrO_4^{2-}(aq) + 8H_2O(l)$$

$$Mn^{2+}(aq) + H_2O_2(aq) + 2OH^-(aq) \longrightarrow MnO_2(s) + 2H_2O(l)$$

Note that water is the reduction product when hydrogen peroxide acts as an oxidizing agent.

The precipitate consisting of $Fe(OH)_3$, $Ni(OH)_2$, and MnO_2 is dissolved in H_2SO_4 and H_2O_2. Sulfuric acid dissolves the two hydroxides. (Write a net ionic equation for each of these dissolution processes.) Hydrogen peroxide in acid solution reduces MnO_2 to Mn^{2+}:

$$MnO_2(s) + H_2O_2(aq) + 2H^+(aq) \longrightarrow Mn^{2+}(aq) + O_2(g) + 2H_2O(l)$$

Hydrogen peroxide can act either as an oxidizing or as a reducing agent, depending on the pH of the solution and the identity of the second reactant. When H_2O_2 acts as a reducing agent, the oxidation product is oxygen gas.

It is possible to confirm the presence of aqueous Fe^{3+}, Mn^{2+}, and Ni^{2+} ions without separating them. All three ions are colored in aqueous solution, but the green color of the Ni^{2+} ions masks the less intense yellow color of Fe^{3+} and the subtle peach color of Mn^{2+}. The presence of Fe^{3+} ions in the mixture of these three ions can be tested on a portion of the solution by adding thiocyanate ions, SCN^-, in the form of an aqueous alkali metal salt such as KSCN. Fe^{3+} ions react with SCN^- ions to form a blood red complex $FeNCS^{2+}$ (Figure 19.10), confirming the presence of Fe^{3+} ions:

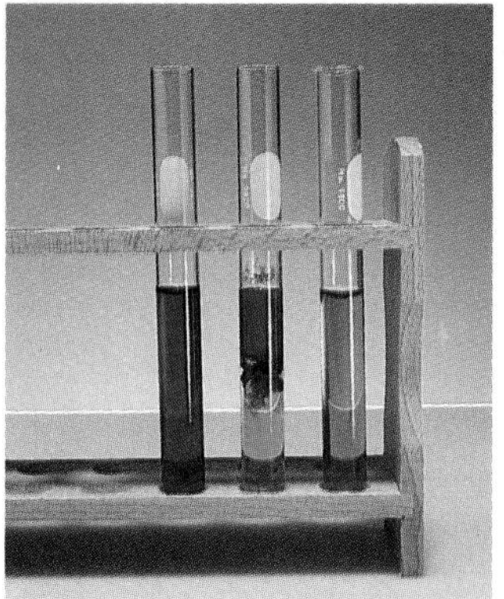

Figure 19.10

A blood red solution of $FeNCS^{2+}$ ions (left), floating strawberry red nickel dimethylglyoxímate (middle), and purple permanganate ion solution.

$$Fe^{3+}(aq) + SCN^-(aq) \longrightarrow FeNCS^{2+}(aq) \quad \text{(blood red)}$$

The presence of Mn^{2+} ions in the solution can be confirmed by oxidizing the Mn^{2+} ions to intensely purple MnO_4^- ions (Figure 19.10). This reaction is carried out with the powerful oxidizing agent, sodium bismuthate, $NaBiO_3$, in an acid solution:

$$2Mn^{2+}(aq) + 5NaBiO_3(s) + 14H^+(aq) \longrightarrow 2MnO_4^-(aq) + 5Na^+(aq) + 5Bi^{3+}(aq) + 7H_2O(l)$$

When another portion of the solution is made alkaline with ammonia, $Fe(OH)_3$ and $Mn(OH)_2$ precipitate, but nickel ions stay in solution as ammine complex ions, $Ni(NH_3)_6^{2+}$. The mixture is then centrifuged and the solution is separated from the precipitate. When the solution is treated with an organic complexing agent called dimethylglyoxime, $(CH_3)_2(CNOH)_2$, a floating strawberry red solid, nickel dimethylglyoximate, $Ni[(CH_3)_2(CNOH)(CNO)]_2$, forms (19.10) and confirms the presence of Ni^{2+} ions in the original solution:

$$Ni(NH_3)_6^{2+}(aq) + 2(CH_3)_2(CNOH)_2(aq) \longrightarrow Ni[(CH_3)_2(CNOH)(CNO)]_2(s) + 2NH_4^+(aq) + 4NH_3(aq)$$

This reaction, written with structural formulas, is

$$\left[\begin{array}{c} H_3N\text{—}Ni(\text{—}NH_3)_4\text{—}NH_3 \end{array}\right]^{2+} + 2\left[\begin{array}{l} H_3C\text{—}C{=}\ddot{N}\text{—}\ddot{O}\text{—}H \\ H_3C\text{—}C{=}\ddot{N}\text{—}\ddot{O}\text{—}H \end{array}\right] \longrightarrow \begin{array}{c} :\ddot{O}\cdots H\cdots :O: \\ H_3C\text{—}C{=}N \quad N{=}C\text{—}CH_3 \\ Ni \\ H_3C\text{—}C{=}N \quad N{=}C\text{—}CH_3 \\ :O\cdots H\cdots \ddot{O}: \end{array} + 2\left[\begin{array}{c} H \\ H\text{—}N\text{—}H \\ H \end{array}\right]^+ + 4\ H\text{—}\ddot{N}(H)\text{—}H$$

If the solution that may contain $Al(OH)_4^-$, $Zn(OH)_4^{2-}$, and CrO_4^{2-} ions is yellow, the CrO_4^{2-} ions are present. These ions can be separated from the solution by adding a solution of barium chloride to precipitate yellow barium chromate:

$$Ba^{2+}(aq) + CrO_4^{2-}(aq) \longrightarrow \underset{\text{yellow}}{BaCrO_4(s)}$$

The formation of this precipitate confirms the presence of $Cr(OH)_3$ in the group III precipitate and of Cr^{3+} ions in the original solution.

The remaining colorless solution containing $Al(OH)_4^-$ and $Zn(OH)_4^{2-}$ is acidified to convert the hydroxo complexes to aqueous Al^{3+} and Zn^{2+} ions:

$$Al(OH)_4^-(aq) + 4H^+(aq) \longrightarrow Al^{3+}(aq) + 4H_2O(l)$$

$$Zn(OH)_4^{2-}(aq) + 4H^+(aq) \longrightarrow Zn^{2+}(aq) + 4H_2O(l)$$

The resulting solution is then made alkaline with ammonia to produce a grayish gelatinous precipitate of $Al(OH)_3$, which confirms the presence of aluminum. $Al(OH)_3$ is amphoteric, but it does not dissolve in the ammonia solution because the hydroxide ion concentration is too low. The ammonia solution contains zinc ions as the ammine complex, $Zn(NH_3)_4^{2+}$. When this solution is saturated with H_2S, a white precipitate of ZnS forms to confirm the presence of zinc:

$$Zn(NH_3)_4^{2+}(aq) + H_2S(aq) \longrightarrow ZnS(s) + 2NH_4^+(aq) + 2NH_3(aq)$$

Example 19.6

Interpreting Results of an Analysis of Group III Precipitate

A solid may contain one or more of the following compounds: $Al(OH)_3$, NiS, and ZnS. The solid dissolves in HNO_3 to produce a colorless solution. The solution is made alkaline with ammonia without any apparent reaction. When this alkaline solution is saturated with H_2S, a white precipitate forms. What is the composition of the solid?

SOLUTION: Because the solution of the solid is colorless, NiS is absent. $Al(OH)_3$ is also absent since no precipitate forms when the solution is made alkaline with ammonia. The formation of a white precipitate with H_2S confirms the presence of ZnS, which is reprecipitated from the solution.

The following flowchart summarizes the analysis:

Compounds that may be present: $Al(OH)_3$, NiS, ZnS

↓ HNO_3

Colorless solution indicates the absence of Ni^{2+} in the solution and of NiS in the original solid.

↓ NH_3 to make alkaline

No precipitate indicates the absence of Al^{3+} in the solution and of $Al(OH)_3$ in the original solid.

↓ H_2S

White precipitate of ZnS confirms the presence of Zn^{2+} in the solution and of ZnS in the original solid.

Practice Problem 19.7: A solution that may contain Cr^{3+}, Fe^{2+}, and Mn^{2+} ions is faintly peach colored. When the solution is made alkaline and saturated with H_2S, a precipitate forms. The precipitate dissolves in HCl. When a portion of this solution is treated with sodium bismuthate, a purple color appears. When another portion of the solution is treated with potassium thiocyanate, no change is observed. Which ions are present in the original solution?

19.5 Analysis of Groups IV and V

Groups IV and V contain cations of alkali metals, of alkaline earth metals, and NH_4^+. These ions are not colored, they do not form colored products with common reagents, and they do not readily form complex ions. Virtually all compounds of the alkali metals are soluble in water. However, carbonates and other compounds of alkaline earth metals are insoluble. Thus, alkaline earth metal ions can be separated from alkali metal ions by precipitating alkaline earth metal ions as carbonates in neutral or basic medium.

The precipitating reagent for group IV cations is a solution of a soluble carbonate such as ammonium carbonate, $(NH_4)_2CO_3$. We recall from Section 16.5 that metal carbonates dissolve in acids. Therefore, the solution of groups IV and V cations is made basic with NH_3 before adding the precipitating agent.

The precipitate of group IV which consists of $MgCO_3$ and $CaCO_3$ is dissolved in HCl. The solution is made alkaline with ammonia, and ammonium oxalate is added to precipitate calcium oxalate, which is white:

$$Ca^{2+}(aq) + C_2O_4^{2-}(aq) \longrightarrow CaC_2O_4(s)$$

Magnesium oxalate is soluble under these conditions; its K_{sp} (8.6×10^{-5}) is larger than the K_{sp} for CaC_2O_4 (2.6×10^{-9}).

Calcium oxalate is precipitated from a buffered solution of NH_3 and $(NH_4)_2C_2O_4$ to keep the pH low enough so that $Mg(OH)_2$ does not precipitate along with CaC_2O_4. Figure 19.11 presents an outline of analysis of group IV precipitate.

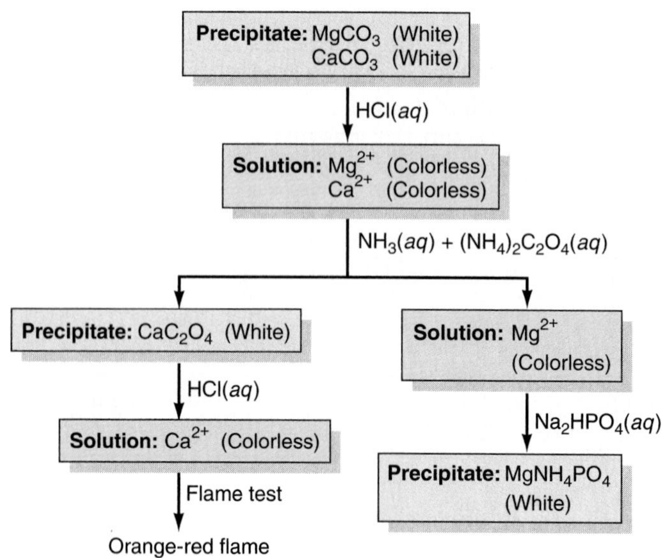

Figure 19.11

An outline of an analysis of group IV precipitate.

Example 19.7

Calculating Whether CaC_2O_4 and $Mg(OH)_2$ Precipitate

A solution that is 0.0050 *M* in Ca^{2+} and 0.0050 *M* in Mg^{2+} ions is made 0.010 *M* in NH_3 and 0.010 *M* in $(NH_4)_2C_2O_4$. Show whether you would expect CaC_2O_4 ($K_{sp} = 2.6 \times 10^{-9}$) and $Mg(OH)_2$ ($K_{sp} = 1.5 \times 10^{-11}$) to precipitate in this solution. K_b for NH_3 is 1.8×10^{-5}.

SOLUTION: A solution that is 0.010 *M* in $(NH_4)_2C_2O_4$ is 0.020 *M* in NH_4^+ ions and 0.010 *M* in $C_2O_4^{2-}$ ions. The product of the calcium ion and oxalate ion concentrations in the solution is

$$[Ca^{2+}][C_2O_4^{2-}] = (0.0050)(0.010) = 5.0 \times 10^{-5}$$

This value is larger than the K_{sp} for CaC_2O_4. Therefore, CaC_2O_4 precipitates.

The OH^- ion concentration in this buffer solution of the weak base NH_3 and its conjugate acid, NH_4^+ ion, can be calculated from the expression for K_b of the system:

	$NH_3(aq) + H_2O(l) \rightleftharpoons$	$NH_4^+(aq)$ +	$OH^-(aq)$
Equilibrium *M*	$0.010 - x$	$(0.020 + x)$	x

where x equals the number of moles per liter of NH_3 ionized.

$$K_b = \frac{[NH_4^+][OH^-]}{[NH_3]}$$

Assuming that x is negligibly small compared with 0.010 or with 0.020, we have

$$1.8 \times 10^{-5} = \frac{(0.020)(x)}{0.010} \quad \text{from which}$$

$$x = [OH^-] = 9.0 \times 10^{-6}$$

The product of the $[Mg^{2+}]$ ion concentration and the square of the $[OH^-]$ ion concentration in the solution is

$$[Mg^{2+}][OH^-]^2 = (0.0050)(9.0 \times 10^{-6})^2 = 4.0 \times 10^{-13}$$

This value is smaller than the K_{sp} for $Mg(OH)_2$. Therefore, magnesium hydroxide does not precipitate.

To confirm the presence of calcium in the precipitate that forms upon adding the oxalate solution, the precipitate is dissolved in HCl, and a platinum or nichrome wire that has been dipped into the solution is placed over flame. An orange-red flame confirms the presence of calcium ion (Figure 19.12).

When a solution or a solid volatile salt of an alkali metal or an alkaline earth metal is vaporized in a flame, the vapor contains atoms whose electrons are in excited states. When the electrons return to the ground state, radiant energy of a characteristic wavelength is emitted in the form of visible light (Chapter 6). The color of the emitted light characterizes the element. As shown

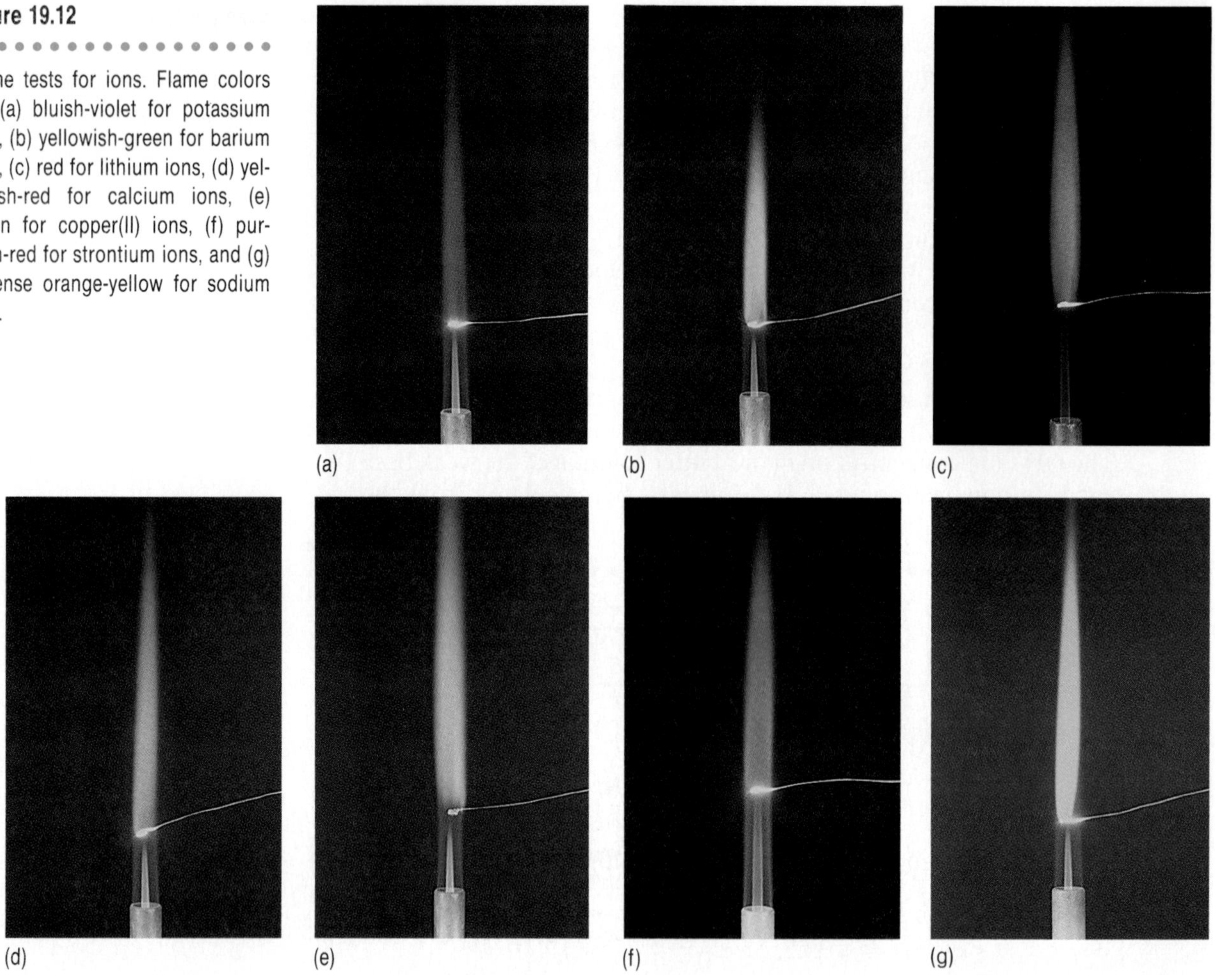

Figure 19.12

Flame tests for ions. Flame colors are (a) bluish-violet for potassium ions, (b) yellowish-green for barium ions, (c) red for lithium ions, (d) yellowish-red for calcium ions, (e) green for copper(II) ions, (f) purplish-red for strontium ions, and (g) a dense orange-yellow for sodium ions.

in Figure 19.12, solutions of calcium salts produce a yellowish-red color; strontium salts impart a crimson or purplish-red color to the flame; barium salts produce a yellowish-green color; lithium salts give a red color; sodium salts have a dense orange-yellow color; potassium, rubidium, and cesium salts have a bluish-violet hue; and copper salts produce a bright green color.

The supernatant from the precipitate of CaC_2O_4 (which contains ammonia) is treated with a solution of sodium monohydrogen phosphate, Na_2HPO_4. If magnesium ions are present, they react with ammonia and monohydrogen phosphate ions to form a white precipitate of magnesium ammonium phosphate:

$$Mg^{2+}(aq) + NH_3(aq) + HPO_4^{2-}(aq) \longrightarrow MgNH_4PO_4(s)$$

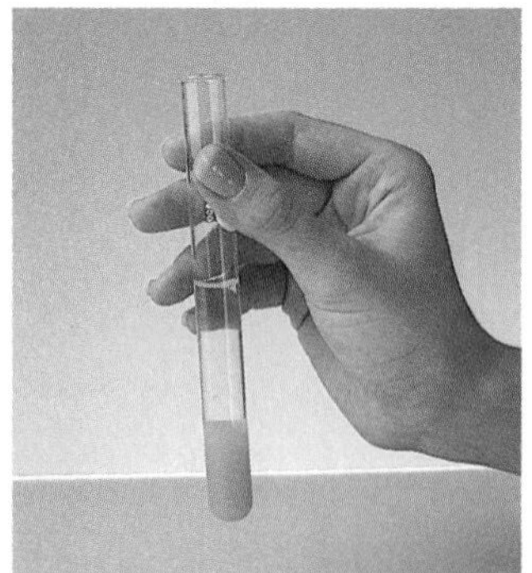

Figure 19.13

Blue suspension ("blue lake") of magnesium hydroxide with the adsorbed dye *para*-nitrobenzeneazoresorcinol.

Another confirmatory test for magnesium is often performed. The precipitate, $MgNH_4PO_4$, is dissolved in hydrochloric acid, and the resulting solution is treated with an organic dye called "magnesium reagent" (*para*-nitrobenzeneazoresorcinol) and with sodium hydroxide until the solution is alkaline. As a result, $Mg(OH)_2$ precipitates and adsorbs the dye on the surface of its particles to form a blue suspension called a blue "lake" (Figure 19.13). The blue lake

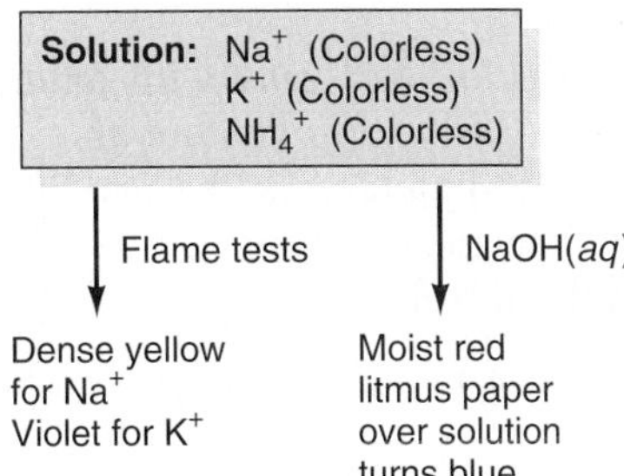

Figure 19.14

Outline of an analysis of a solution containing group V ions.

confirms the presence of Mg^{2+} ions in the original solution. The blue lake suspension can be concentrated to a blue precipitate by centrifuging the mixture.

The supernatant solution from the precipitate of group IV (see Figure 19.1) contains the cations of group V: Na^+, K^+, and NH_4^+. Most salts of these cations are soluble in water. Special tests must therefore be used on this solution to detect the ions of group V.

Testing for NH_4^+ ions is performed on the *original* solution, which contains all of the ions, because ammonium salt solutions were added to precipitate cations in groups III and IV (refer to Figure 19.1). When a solution which contains ammonium ions is treated with sodium hydroxide, the ammonium ions react with the hydroxide ions to form ammonia:

$$NH_4^+(aq) + OH^-(aq) \longrightarrow NH_3(g) + H_2O(l)$$

The evolution of ammonia is tested by holding a moist piece of red litmus paper above the solution. Ammonia reacts with water in the litmus paper to form an alkaline solution, which causes the color of the litmus to change from red to blue.

Sodium and potassium ions can be detected by flame tests performed on the solution from the group IV precipitate (Figure 19.14). If sodium ions and potassium ions are both present, the intense yellow color due to sodium ions masks the short-lived bluish-violet color of potassium ions. However, the blue-violet color of potassium ions in the flame can be seen if the flame is viewed through a cobalt glass plate, which filters out the yellow color caused by sodium.

Example 19.8

Analyzing a Solution Containing Cations from Groups IV and V

A solution may contain one or more of the following cations: Mg^{2+}, Ca^{2+}, Na^+, and NH_4^+. A portion of this solution is treated with an aqueous mixture of ammonia and ammonium oxalate. No precipitate forms. Adding sodium monohydrogen phosphate to another portion of the original solution produces a white precipitate. A third portion of the original solution is treated with sodium hydroxide, and a moist red litmus paper is held above the solution. The litmus paper turns blue. A flame test on a fourth portion of the origial solution does not show any characteristic color. Which ions are present in the solution and which are absent?

SOLUTION: Since no precipitate forms with ammonium oxalate, calcium ions are absent. A white precipitate that forms when sodium monohydrogen phosphate is added in the presence of ammonia indicates the presence of magnesium ions. The Mg^{2+} ions precipitate as magnesium ammonium phosphate:

$$Mg^{2+}(aq) + NH_3(aq) + HPO_4^{2-}(aq) \longrightarrow MgNH_4PO_4(s)$$

The litmus paper test shows the presence of NH_4^+ ions. These ions react with the OH^- ions from NaOH to form NH_3, which reacts with water in the litmus paper, and the color of the paper changes from red to blue. The absence of a color in the flame test indicates the absence of sodium ions and calcium ions.

Practice Problem 19.8: A solution that may contain calcium, magnesium, and potassium ions is treated with solutions of ammonia and ammonium oxalate. A white precipitate forms. The supernatant from the precipitate is treated with sodium monohydrogen phosphate, which produces no precipitate. A flame test performed on another portion of the supernatant solution produces a bluish-violet color. Which ions are present, and which are absent?

19.6 Anion Analysis

Schemes of anion analysis are usually less systematic than schemes of cation analysis. However, in all schemes the anions are divided into groups based on their behavior in major types of reactions such as acid-base reactions, redox reactions, and precipitation reactions.

We consider 10 common anions: SO_4^{2-}, SO_3^{2-}, CO_3^{2-}, NO_3^-, PO_4^{3-}, S^{2-}, Cl^-, Br^-, I^-, and CH_3COO^-. These anions can be divided into five groups. Each anion in a group can be identified by a specific test.

Table 19.1 lists the various groups into which the 10 anions can be divided. The first group contains **anions of volatile acids**. These anions are relatively strong conjugate bases of weak acids, and they are therefore easily protonated to acids whose vapors can be identified by their characteristic odors.

The second group of anions in Table 19.1 is the **sulfate group**. These anions precipitate as barium salts from alkaline solution.

The third group of anions is the **chloride group**. These anions can be precipitated as silver salts from acid solution.

The fourth group contains only nitrate, which is an **oxidizing anion**.

The fifth group contains **reducing anions**. These anions can reduce aqueous permanganate ions, which are purple, to an essentially colorless solution of manganese(II) ions. Note that some anions belong to more than one group.

Table 19.1

Groups of Anions with Common Characteristics

Anions	Group Name	Method of Detection of the Group
SO_3^{2-}, CO_3^{2-}, S^{2-}, $C_2H_3O_2^-$	Anions of volatile acids	With H_2SO_4 these anions form oxides or volatile acids that can be detected by odor or by other means
SO_4^{2-}, SO_3^{2-}, CO_3^{2-}, PO_4^{3-}	Sulfate group anions	Can be precipitated as barium salts from alkaline solution
Cl^-, Br^-, I^-	Chloride group anions	Can be precipitated as silver salts from acid solution
NO_3^-	Oxidizing anion	Can oxidize I^- to I_2, which can be detected in C_6H_{12}
S^{2-}, I^-, Br^-, SO_3^{2-}	Reducing anions	Can reduce and decolorize permanganate solution

Anions of Volatile Acids

The anions of this group, SO_3^{2-}, CO_3^{2-}, S^{2-}, and $C_2H_3O_2^-$, are easily protonated by H_2SO_4 to form weak volatile acids, H_2SO_3, H_2CO_3, H_2S, and $HC_2H_3O_2$, respectively. The vapors of three of these acids can be detected by their odors.

When a solution of sulfite ions is treated with sulfuric acid and warmed, sulfurous acid forms and then decomposes to water and sulfur dioxide, which can be detected by its choking odor:

$$SO_3^{2-}(aq) + 2H^+(aq) \longrightarrow H_2SO_3(aq) \longrightarrow H_2O(l) + SO_2(g)$$

Carbonate ions react with H^+ ions to form carbonic acid, which is in equilibrium with carbon dioxide and water:

$$CO_3^{2-}(aq) + 2H^+(aq) \longrightarrow H_2CO_3(aq) \rightleftharpoons CO_2(g) + H_2O(l)$$

Carbon dioxide is a colorless and odorless gas; its evolution can often be detected by effervescence (bubbling of the reaction mixture as the gas escapes). A more reliable method for detecting CO_2 consists of placing a stirring rod moistened with barium hydroxide solution just above the unknown solution in the reaction flask. Carbon dioxide gas which escapes from the unknown solution reacts with the barium hydroxide solution on the stirring rod to form a white film of insoluble barium carbonate:

$$CO_2(g) + Ba^{2+}(aq) + 2OH^-(aq) \longrightarrow BaCO_3(s) + H_2O(l)$$

Sulfide ions are protonated by sulfuric acid to form hydrogen sulfide:

$$S^{2-}(aq) + 2H^+(aq) \longrightarrow H_2S(g)$$

Hydrogen sulfide gas can easily be detected by its odor of rotten eggs. H_2S can also be detected by a strip of absorbent paper moistened with lead(II) acetate solution. Hydrogen sulfide gas reacts with lead ions in the lead(II) acetate paper, which turns black when black lead sulfide forms:

$$H_2S(g) + Pb^{2+}(aq) \longrightarrow PbS(s) + 2H^+(aq)$$

Acetate ions are protonated by H_2SO_4 to form acetic acid:

$$C_2H_3O_2^-(aq) + H^+(aq) \longrightarrow HC_2H_3O_2(aq)$$

Acetic acid is an ingredient of vinegar, so its vapors smell like vinegar. Another test for acetate ions consists of acidifying the solution and adding ethyl alcohol, with heating. A fruity odor of ethyl acetate, $CH_3COOC_2H_5$, which forms in the reaction, confirms the presence of acetate ions:

$$C_2H_3O_2^-(aq) + H^+(aq) + C_2H_5OH(l) \longrightarrow C_2H_3O_2C_2H_5(aq) + H_2O(l)$$

If a solution contains a mixture of several anions of volatile acids, the odors produced by the H_2SO_4 treatment may mask one another. Only the test for carbonate ions and the lead acetate test for sulfide ions do not depend on odor. If all four of the anions may be present, the sulfite ions, carbonate ions, and sulfide ions can be precipitated from an alkaline solution as silver salts. Silver sulfite and silver carbonate are white and silver sulfide is black. A precipitate that is a mixture of all three of these solids appears gray. If acetate ions are present, they can be detected in the supernatant.

To determine the composition of the precipitate, it is dissolved in an acid. In an acidic solution, carbonate ions are converted to carbon dioxide, which escapes from the solution and can be detected by a stirring rod moistened with barium hydroxide. Sulfide ions in acid solution convert to hydrogen sulfide, which can be detected by lead(II) acetate paper. Sulfite ions convert to sulfurous acid, which is not as volatile as carbonic acid or hydrosulfuric acid. Therefore, the odor of sulfur dioxide can be detected after carbon dioxide and hydrogen sulfide have escaped.

Sulfurous acid can also be identified by oxidizing it to sulfate ions with hydrogen peroxide, followed by precipitation of the sulfate ions as white barium sulfate:

$$H_2SO_3(aq) + H_2O_2(aq) \longrightarrow 2H^+(aq) + SO_4^{2-}(aq) + H_2O(l)$$

$$Ba^{2+}(aq) + SO_4^{2-}(aq) \longrightarrow BaSO_4(s)$$

Example 19.9

Devising a Scheme to Identify the Anions of Volatile Acids

A solution may contain sulfite, carbonate, and sulfide ions. Devise a scheme for identifying each of these anions.

SOLUTION: One portion of the unknown solution can be acidified and treated with hydrogen peroxide and barium nitrate. If a white precipitate forms, sulfite ions are present.

A test for carbonate ions can be done on another portion of the unknown solution by acidifying the solution and testing for carbon dioxide by holding a stirring rod moistened with barium hydroxide above the solution.

Sulfide ions can be tested on a third portion of the unknown by acidifying the solution and testing for H_2S either by its odor or by its reaction with a piece of lead acetate paper held over the solution.

Practice Problem 19.9: Write balanced net ionic equations for the reactions that occur when: **(a)** sulfuric acid is added to a solution of potassium sulfite; **(b)** carbon dioxide is bubbled through a solution of barium hydroxide to form a precipitate of barium carbonate as one of the products; **(c)** hydrogen sulfide gas is bubbled through a lead(II) acetate solution to produce a precipitate of lead sulfide as one of the products; **(d)** hydrogen peroxide is added to sulfurous acid to oxidize it to sulfate ions while hydrogen peroxide is reduced to water.

Sulfate Group Anions

The sulfate group anions are SO_4^{2-}, SO_3^{2-}, CO_3^{2-}, and PO_4^{3-}. These anions are in the same group because like the sulfate ion, they form insoluble barium salts in neutral or alkaline solution. However, in acid solution, only sulfate ions can be precipitated with barium ions. SO_3^{2-}, CO_3^{2-}, and PO_4^{3-} ions are extensively protonated in acid solution, and SO_3^{2-} and CO_3^{2-} ions form volatile acids as discussed in the preceding section. The sulfate group anions also form insoluble (or slightly soluble) silver salts in neutral or alkaline medium.

The presence of SO_4^{2-} ions in a solution of this group of anions can be confirmed by acidifying the solution and by adding barium chloride or barium nitrate solution to precipitate $BaSO_4$. The presence of SO_3^{2-} ions can be confirmed by oxidizing them to SO_4^{2-} ions by H_2O_2 and by subsequent precipitation of the SO_4^{2-} ions as $BaSO_4$ in an acid solution as described earlier. CO_3^{2-} ions are converted to CO_2, which is detected as described in the preceding section.

After SO_4^{2-} and SO_3^{2-} ions have been precipitated from a solution of sulfate group anions, and CO_3^{2-} ions have been expelled as CO_2, the solution is made alkaline to precipitate PO_4^{3-} ions as white $Ba_3(PO_4)_2$ or yellow Ag_3PO_4:

$$3Ba^{2+}(aq) + 2PO_4^{3-}(aq) \longrightarrow Ba_3(PO_4)_2(s)$$

$$3Ag^{+}(aq) + PO_4^{3-}(aq) \longrightarrow Ag_3PO_4(s)$$

A test for PO_4^{3-} can also be done in the presence of other sulfate group anions by acidifying the solution and adding a solution of ammonium molybdate, $(NH_4)_2MoO_4$, which precipitates the PO_4^{3-} ions as yellow ammonium phosphomolybdate, $(NH_4)_3PO_4(MoO_3)_{12}$. In acid solution, PO_4^{3-} ions protonate to form $H_2PO_4^-$ ions and some H_3PO_4,

$$PO_4^{3-}(aq) + 2H^{+}(aq) \longrightarrow H_2PO_4^{-}(aq)$$

A precipitate forms according to the following equation:

$$3NH_4^{+}(aq) + 12MoO_4^{2-}(aq) + H_2PO_4^{-}(aq) + 22H^{+}(aq) \longrightarrow (NH_4)_3PO_4(MoO_3)_{12}(s) + 12H_2O(l)$$

Chloride Group Anions

Like chloride ions, bromide and iodide ions can be precipitated as silver salts from acid solution. Because of this similarity, and because these ions are derived from the same group of elements, Cl^-, Br^-, and I^- are known as the chloride group anions. The colors of silver halides range from white for AgCl, to cream color for AgBr, to pale yellow for AgI (Figure 19.15).

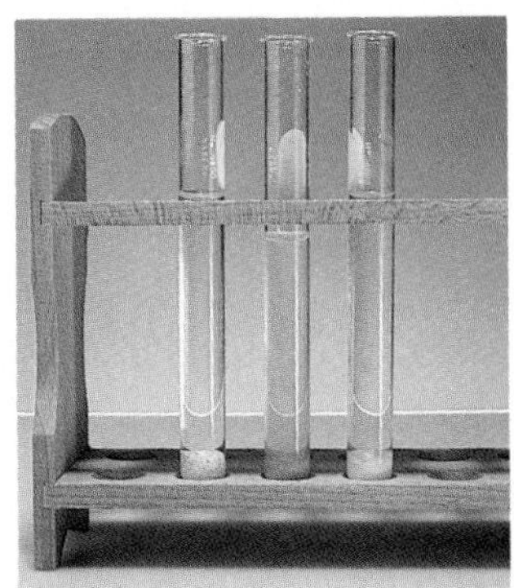

Figure 19.15

Silver halide precipitates: white silver chloride, cream-colored silver bromide, and pale yellow silver iodide.

The color of a silver halide precipitate is not distinctive enough to identify the halide ions in a solution because the color of the precipitate is ambiguous if all three of the halide ions are present. Therefore, a solution suspected of containing halide ions is first treated with a mild oxidizing agent such as potassium nitrite in acid solution to oxidize the most easily oxidizable I^- ions to elemental iodine:

$$2I^{-}(aq) + 2NO_2^{-}(aq) + 4H^{+}(aq) \longrightarrow I_2(aq) + 2NO(g) + 2H_2O(l)$$

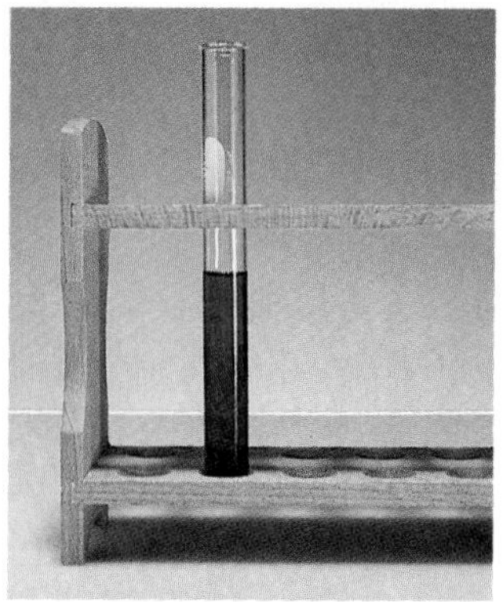

Figure 19.16

The violet solution of iodine in cyclohexane.

The yellowish-brown color of aqueous iodine indicates the presence of I^- ions in the original solution. When the aqueous solution of iodine is shaken with cyclohexane, C_6H_{12} (an organic liquid), the iodine is extracted from the water into the cyclohexane, which forms a distinct upper layer in the mixture. A *violet* color in the cyclohexane layer confirms the presence of I_2 in this layer and of I^- ions in the original solution (Figure 19.16).

After I^- ions have been removed from the solution, Br^- and Cl^- may remain. Br^- ions in aqueous solution can be oxidized to elemental bromine by the permanganate ion, which is a stronger oxidizing agent than the nitrite ion. The reaction of bromide ions with permanganate ions in acid solution is

$$2MnO_4^-(aq) + 10Br^-(aq) + 16H^+(aq) \longrightarrow 2Mn^{2+}(aq) + 5Br_2(aq) + 8H_2O(l)$$

The Br_2 formed by this reaction can also be extracted into cyclohexane to produce an *orange* color in the cyclohexane layer.

After iodide and bromide ions have been removed from the solution, chloride ions may remain in the solution. Chloride ions are more difficult to oxidize than either bromide or iodide ions. Chloride ions cannot be oxidized by a dilute solution of $KMnO_4$. The presence of Cl^- ions in the solution from which the I^- and Br^- ions have been removed is confirmed by adding $AgNO_3$ to precipitate white AgCl.

AgCl has a higher K_{sp} value than AgBr or AgI. AgCl is the only silver halide that is soluble in dilute aqueous ammonia. Thus, AgCl can be separated from a mixture of AgCl, AgBr, and AgI, by treating the mixture with NH_3, then acidifying the supernatant with HNO_3 to reprecipitate AgCl.

Example 19.10

Analyzing a Mixture of Anions of the Sulfate and Chloride Groups

A solution may contain one or more of the following anions: SO_4^{2-}, PO_4^{3-}, and Br^-. When a portion of this solution is made alkaline with ammonia and treated with barium chloride solution, a white precipitate forms. When a second portion of the solution is acidified with nitric acid and treated with barium chloride solution, no precipitate forms. A third acidified portion of the solution is treated with a dilute solution of potassium permanganate and cyclohexane. An orange color is observed in the cyclohexane layer. Which ions are present and which are absent in the original solution?

SOLUTION: A precipitate that forms when barium chloride is added to an alkaline solution indicates the presence of SO_4^{2-}, PO_4^{3-}, or both. The absence of a precipitate when barium chloride is added to an acid solution indicates that SO_4^{2-} ions are absent. Hence, PO_4^{3-} must be present. Barium phosphate, $Ba_3(PO_4)_2$, is insoluble in alkaline solution, but soluble in acid solution. In contrast, $BaSO_4$ is insoluble in both acid and alkaline solutions.

The formation of an orange color in the cyclohexane layer after treatment with $KMnO_4$ confirms the presence of Br_2. Bromine is formed by oxidizing Br^- ions, which therefore must be present in the original solution.

Practice Problem 19.10: A solution may contain Br^- and/or I^- ions. Describe how you would determine the presence or absence of each of these ions.

Oxidizing and Reducing Anions

The only oxidizing anion among the 10 anions we are considering is the nitrate ion. Since the nitrate ion is an oxidizing anion, it can be detected in acid solution by its reaction with iodide ion, which is easily oxidized to elemental iodine:

$$2NO_3^-(aq) + 6I^-(aq) + 8H^+(aq) \longrightarrow 2NO(g) + 3I_2(aq) + 4H_2O(l)$$

The iodine formed by this reaction can be detected by its violet color in cyclohexane, as described earlier.

Another test for nitrate ions is based on their ability to oxidize iron(II) ions in acid solution. The nitrate ions are reduced to nitrogen monoxide, NO, which reacts with residual iron(II) ions to form brown complex ions, $Fe(NO)^{2+}$. The reaction of nitrate ions with iron(II) ions in acid solution is

$$3Fe^{2+}(aq) + NO_3^-(aq) + 4H^+(aq) \longrightarrow 3Fe^{3+}(aq) + NO(aq) + 2H_2O(l)$$

The nitrogen monoxide thus formed reacts with excess iron(II) ions to form the brown complex:

$$Fe^{2+}(aq) + NO(aq) \longrightarrow Fe(NO)^{2+}(aq)$$

In this test, concentrated H_2SO_4 is added to an unknown solution in a test tube. The test tube must be cooled in an ice bath because the $Fe(NO)^{2+}$ complex decomposes at high temperature. Fe^{2+} ions are then slowly added in the form of a solution of $FeSO_4$ that is allowed to slide down along the wall of the test tube so that it will form a second (less dense) layer above the test solution. The color of the $Fe(NO)^{2+}$ complex can then be seen forming slowly as a brown ring between the two layers in the test tube. For that reason, this test is called the *brown ring test* (Figure 19.17).

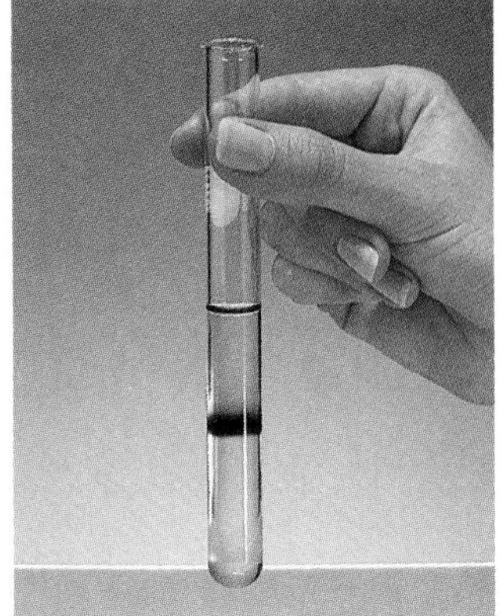

Figure 19.17

The brown ring test for nitrate ions.

The reducing anions we consider are S^{2-}, I^-, Br^-, and SO_3^{2-}. The S^{2-}, I^-, and Br^- ions can be oxidized but not reduced. SO_3^{2-} ion can be both oxidized and reduced. However, since it is more easily oxidized than reduced, it usually acts as a reducing agent. Note that the S^{2-} and SO_3^{2-} ions are also members of the volatile acid group anions.

We have already discussed tests for detecting the reducing anions listed above. Each of these reducing anions can easily reduce purple permanganate ions. Thus, when a small amount of a dilute permanganate solution is added to an acidified solution containing any or all of the reducing anions, the permanganate ions are reduced to nearly colorless Mn^{2+} ions. If too much permanganate is added, the deep purple color of excess permanganate ions makes it impossible to see if some of the ions have been reduced. In acid medium, S^{2-} and SO_3^{2-} ions are converted to H_2S and H_2SO_3, respectively. The reactions of H_2S and H_2SO_3 with permanganate ions in acid solution are:

$$2MnO_4^-(aq) + 5H_2S(aq) + 6H^+(aq) \longrightarrow 2Mn^{2+}(aq) + 5S(s) + 8H_2O(l)$$

$$2MnO_4^-(aq) + 5H_2SO_3(aq) + H^+(aq) \longrightarrow 2Mn^{2+}(aq) + 5HSO_4^-(aq) + 3H_2O(l)$$

The reactions of permanganate ions with I^- and Br^- ions are

$$2MnO_4^-(aq) + 10I^-(aq) + 16H^+(aq) \longrightarrow 2Mn^{2+}(aq) + 5I_2(aq) + 8H_2O(l)$$

$$2MnO_4^-(aq) + 10Br^-(aq) + 16H^+(aq) \longrightarrow 2Mn^{2+}(aq) + 5Br_2(aq) + 8H_2O(l)$$

General Anion Analysis

If an unknown solution is suspected to contain many of the anions discussed in this chapter, a partial separation of the groups of anions, such as that shown in Figure 19.18, is advisable. The presence or absence of each anion in a group can then be confirmed as we have already discussed.

Example 19.11

General Anion Analysis

A solution may contain any or all of the following anions: SO_4^{2-}, SO_3^{2-}, Cl^-, and NO_3^-. A portion of this solution is acidified with HCl and treated with $BaCl_2$ with no apparent reaction. Another portion is made alkaline with NH_3 and treated with $BaCl_2$, forming a white precipitate. Adding HNO_3 and $AgNO_3$ to a fresh portion of the unknown gives a white precipitate. Yet another portion of the unknown, when treated with HCl and dilute KI, and shaken with cyclohexane, gives no evidence of reaction. Which anions are present and which are absent?

SOLUTION: Sulfate ions are absent because no precipitate forms with $BaCl_2$ in acid solution. However, sulfite ions must be present because a precipitate is obtained with $BaCl_2$ in alkaline solution. A white precipitate with HNO_3 and $AgNO_3$ confirms the presence of Cl^- ions. NO_3^- ions are absent since oxidation of I^- to I_2 is not evident when KI and cyclohexane are added to an acidic solution of the unknown.

Practice Problem 19.11: Write a balanced net ionic equation for the reaction that occurs when: **(a)** a solution of barium chloride is added to an alkaline solution containing sulfite ions; **(b)** a solution of silver nitrate is added to an acidic solution containing chloride ions.

Example 19.12

General Anion Analysis

A solution may contain any or all of the anions discussed in this chapter. When a portion of this solution is treated with excess H_2SO_4 and warmed, a gas with an unpleasant odor is evolved from the solution. Warming is continued until the odor can no longer be detected. The solution is then treated with $Ba(NO_3)_2$ solution, whereupon a white precipitate forms. The supernatant from this precipitate is made alkaline with NH_3, producing a precipitate. When the resulting supernatant is acidified with HNO_3 and treated with $AgNO_3$, no precipitate forms.

Another portion of the original solution is acidified with HCl and treated with a solution of potassium iodide and cyclohexane, resulting in a violet color in the cyclohexane layer. When a third portion of the original solution is acidified with HCl and treated with $BaCl_2$, no precipitate forms. Which anions are present, and which are absent? Use Figure 19.18, if necessary.

SOLUTION: The unpleasant odor produced upon adding H_2SO_4 to the solution and warming indicates that some volatile acid group anions are present, probably more than one because no particular characteristic

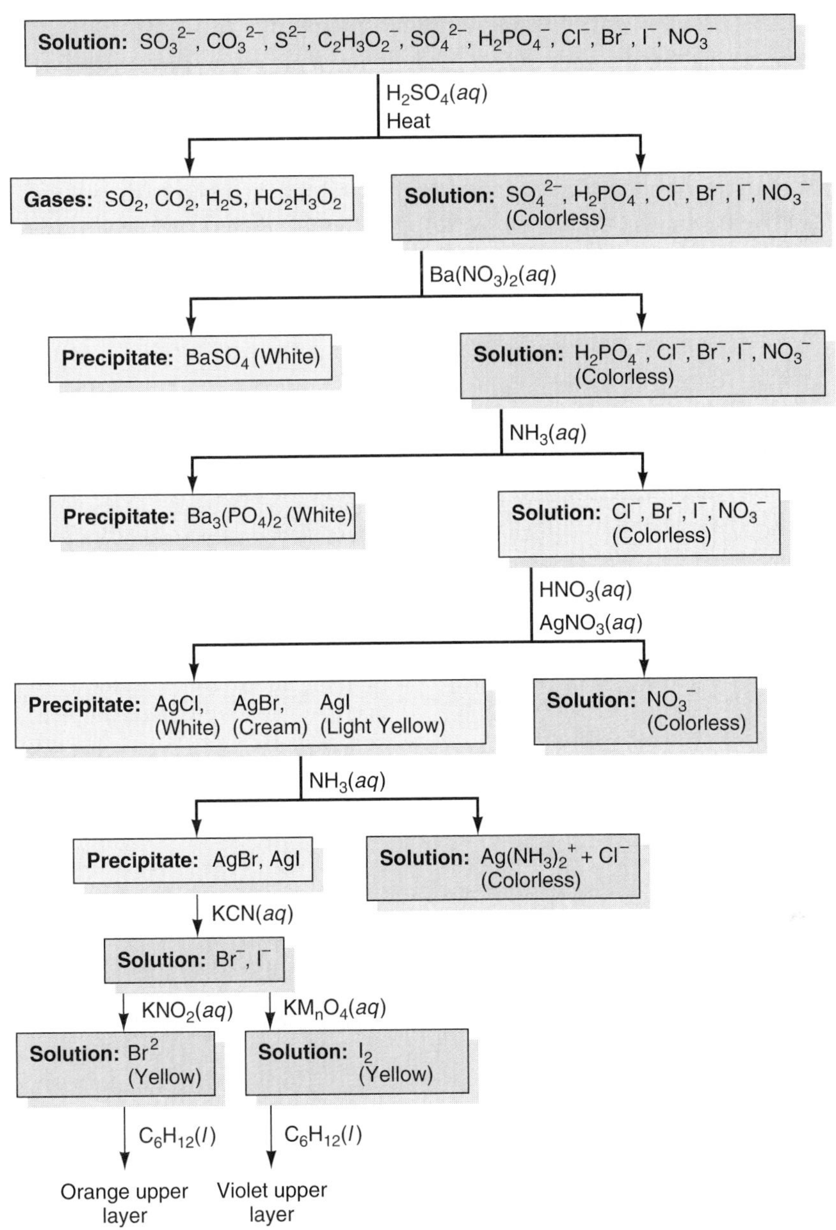

Figure 19.18

A scheme of separating and identifying groups of anions.

odor is detected. More testing is needed to determine whether SO_3^{2-}, S^{2-}, and $C_2H_3O_2^-$ are actually present. The white precipitate formed upon the addition of $Ba(NO_3)_2$ to this acid solution must be $BaSO_4$ because SO_4^{2-} ions were introduced by the addition of H_2SO_4 in the previous step. At this point it is impossible to tell whether SO_4^{2-} ions may have been present in the original solution.

The white precipitate produced when the supernatant from $BaSO_4$ is made alkaline is $Ba_3(PO_4)_2$, indicating that PO_4^{3-} ions are in the original solution. The precipitate here cannot be $BaSO_3$ or $BaCO_3$ because the SO_3^{2-} and CO_3^{2-} ions, if present, were destroyed by the earlier heating with H_2SO_4.

Since the treatment of the supernatant from $Ba_3(PO_4)_2$ with HNO_3 and $AgNO_3$ gives no precipitate, Cl^-, Br^-, and I^- ions are absent.

The violet color produced in the cyclohexane layer when another portion of the original unknown is treated with KI and cyclohexane shows that the unknown contains NO_3^- ions.

Since treating a third portion of the original unknown with HCl and $BaCl_2$ gives no precipitate, SO_4^{2-} ions are absent.

No information is available concerning the presence or absence of CO_3^{2-} ions.

The accompanying flowchart summarizes the foregoing analysis.

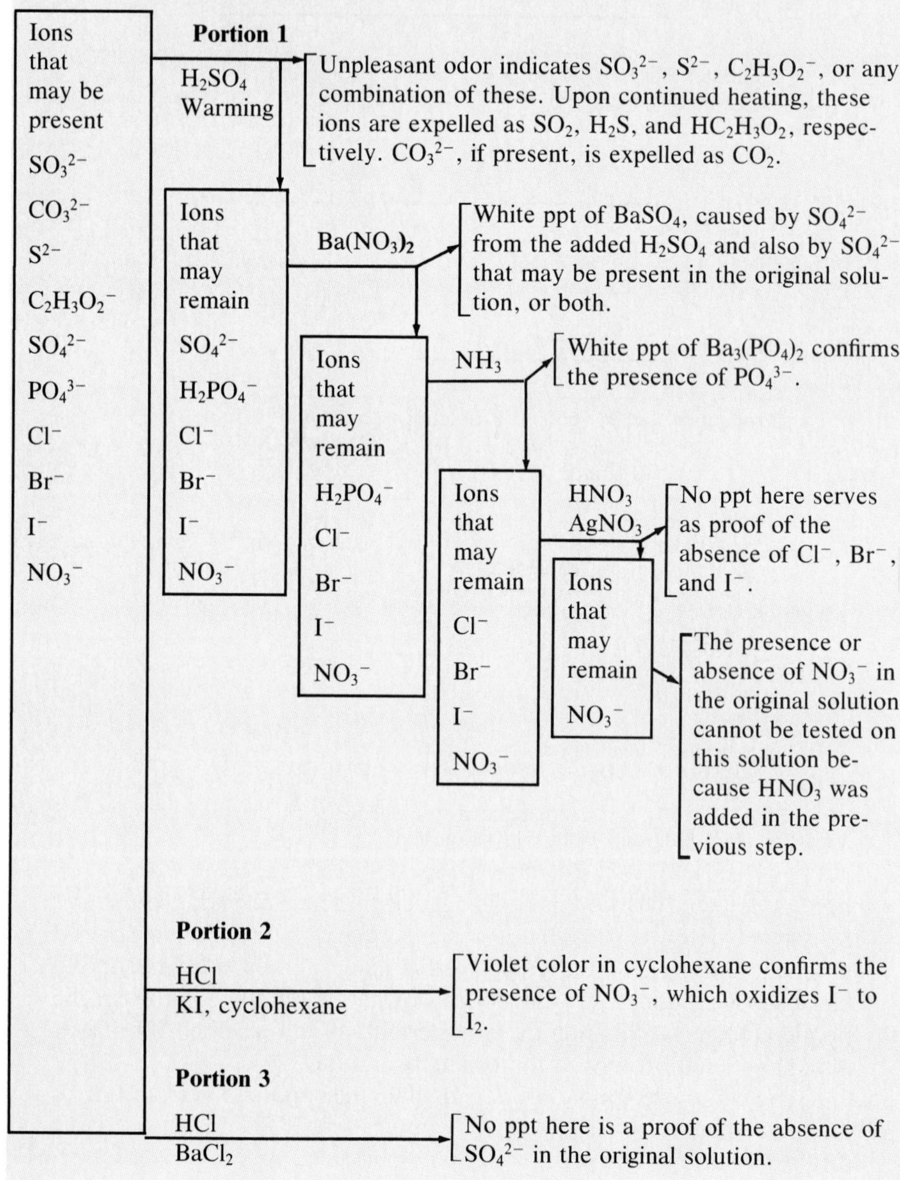

Practice Problem 19.12: A solution may contain sulfate, carbonate, chloride, and/or nitrate ions. Devise a scheme for determining the presence or absence of each of these ions.

19.7 Analysis of Salts

An unknown salt may be a single salt or a mixture of salts. Preliminary examination of an unknown salt along with some preliminary experiments may give valuable clues about the identity of the salt. For example, the color of a salt is an important preliminary indicator of its possible identity. We recall that many transition metal salts are colored and that most salts of main group metals are colorless. However, the colors of the anhydrous and hydrated forms of a salt can differ. For example, anhydrous $CuSO_4$ is white, whereas the hydrated form is blue. Most colored salts also have colored aqueous ions. Some exceptions include HgI_2 that exists in red and yellow crystalline forms, but a solution of aqueous Hg^{2+} and I^- ions is colorless.

A simple flame test performed on a small sample of a solid moistened with HCl can indicate which cation may be present. However, only a few cations (see Section 19.5) yield a characteristic color in the flame of a bunsen burner, and the presence of more than one of these produces a mixture of colors.

The solubility of the salt may give additional clues to its identity. For example, if the salt is soluble in water, it could be an ammonium salt, a salt of an alkali metal, an acetate, or a nitrate of almost any metal (review the solubility rules in Chapter 4 if necessary). If the salt is not soluble in water but dissolves in a nonoxidizing acid such as HCl, the anion of the salt could be a relatively strong conjugate base of a weak acid. Such anions include CO_3^{2-}, PO_4^{3-}, S^{2-}, and SO_3^{2-} (see Section 16.5).

If a salt dissolves in water and produces an acidic or an alkaline solution, the cations or the anions of the salt hydrolyze. We recall that NH_4^+ ions and cations which have high ionic potentials, such as Al^{3+} and Cr^{3+} ions, hydrolyze to produce acidic solutions. Anions such as SO_3^{2-}, CO_3^{2-}, S^{2-}, $C_2H_3O_2^-$, and PO_4^{3-}, which are conjugate bases of weak acids, hydrolyze to produce alkaline solutions (see Sections 14.4 and 15.3).

The presence of any anions of the volatile acid group in an unknown salt or a mixture of salts can be tested by adding H_2SO_4 to a portion of the salt. The vapors produced by such anions in a solid are more concentrated and therefore more easily detectable than the vapors produced by the anions in solution. Even chloride ions present in a solid can be detected as HCl gas, which escapes from the solid when the solid is treated with H_2SO_4 and heated.

Sometimes heating a small portion of a solid on a spatula provides clues to its identity. For example, most metal sulfites decompose upon heating to liberate SO_2, carbonates decompose to give CO_2, sulfides yield SO_2 in air, and heavy metal nitrates evolve brown NO_2. Metal acetates may char (turn black) when heated.

A preliminary analysis like the one outlined above is usually not sufficient to identify the salt. Therefore, the clues gained by preliminary analysis are tested by dissolving the salt in a suitable solvent and analyzing for the suspected cations and anions as described in previous sections.

Example 19.13

Identifying a Salt

A colorless simple salt dissolves in cold water to produce an alkaline solution. Treating the solid with H_2SO_4 and then heating gives no evidence of reaction, and a stirring rod moistened with $Ba(OH)_2$ solution held above the solution of the salt does not turn "milky." When the solution of the solid is treated with HCl, no precipitate forms. When this solution is made alkaline with NH_3 and treated with H_2S, no precipitate forms. Subsequent addition of $(NH_4)_2CO_3$ to this solution gives no evidence of a reaction. When another portion of the original salt solution is made acidic with HCl and treated with $BaCl_2$, no reaction occurs, but when this solution is made alkaline with NH_3, a white precipitate forms. In a flame test, a bluish-violet color is observed in the flame. What is the identity of the salt?

SOLUTION: Since the solution of the salt is alkaline, the anions of the salt hydrolyze. The anion may therefore be SO_3^{2-}, CO_3^{2-}, S^{2-}, $C_2H_3O_2^-$, or PO_4^{3-}. Since no reaction occurs when the solid is heated with sulfuric acid, SO_3^{2-}, CO_3^{2-}, S^{2-}, and $C_2H_3O_2^-$ are absent, and the anion of the salt is PO_4^{3-}. The presence of this anion in the solution of the salt is also indicated by the formation of a white precipitate with NH_3 and $BaCl_2$. Note that CO_3^{2-} would also give a white precipitate with Ba^{2+}, but CO_3^{2-} can be ruled out because there is no effervescence to indicate CO_2 evolution when the salt is heated with H_2SO_4, and the stirring rod moistened with $Ba(OH)_2$ solution does not turn milky.

Since the solution of the salt does not give a precipitate with HCl, with NH_3 and H_2S, or with $(NH_4)_2CO_3$, all of the cations except Na^+, K^+, and NH_4^+ are absent. The bluish-violet color of the flame is characteristic of potassium. The salt is therefore K_3PO_4.

The following flowchart summarizes the foregoing discussion.

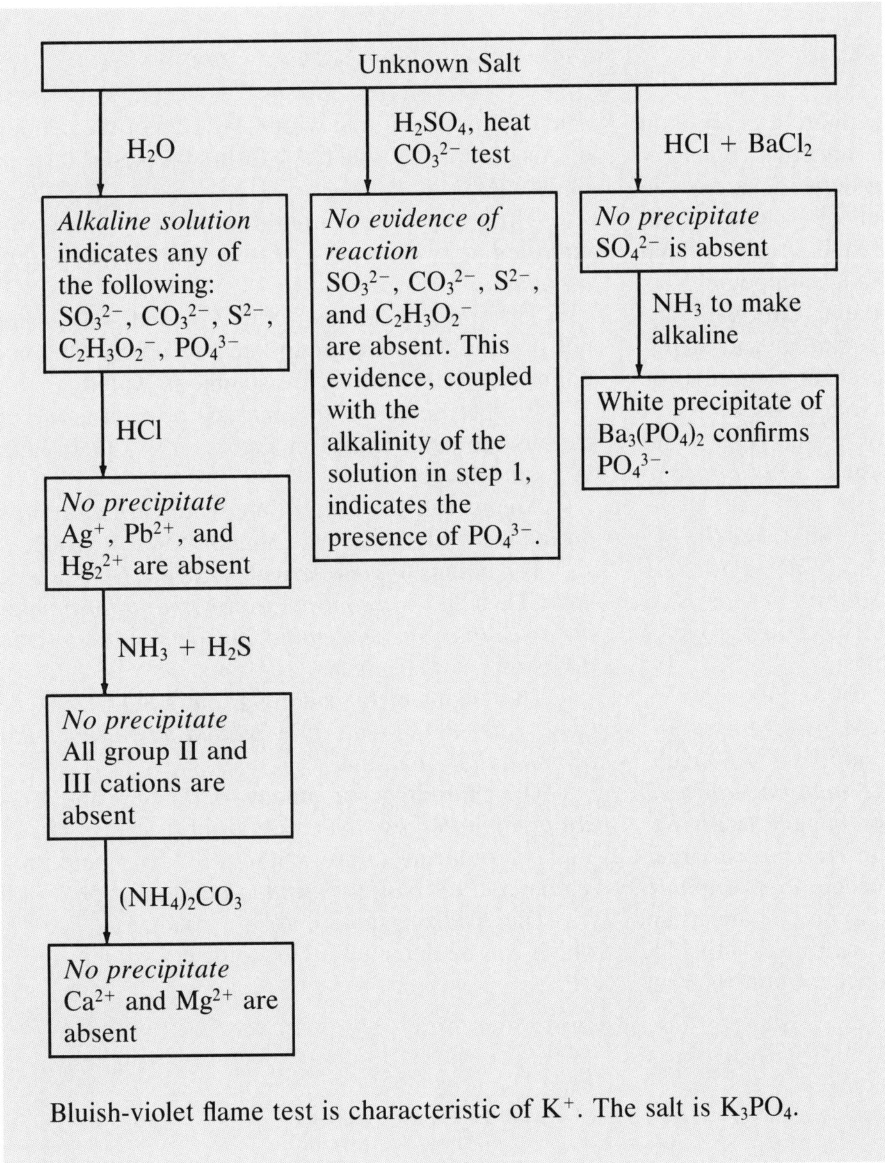
Unknown Salt
H_2O
H_2SO_4, heat
CO_3^{2-} test
HCl + $BaCl_2$
Alkaline solution indicates any of the following: SO_3^{2-}, CO_3^{2-}, S^{2-}, $C_2H_3O_2^-$, PO_4^{3-}
No evidence of reaction SO_3^{2-}, CO_3^{2-}, S^{2-}, and $C_2H_3O_2^-$ are absent. This evidence, coupled with the alkalinity of the solution in step 1, indicates the presence of PO_4^{3-}.
No precipitate SO_4^{2-} is absent
NH_3 to make alkaline
White precipitate of $Ba_3(PO_4)_2$ confirms PO_4^{3-}.
HCl
No precipitate Ag^+, Pb^{2+}, and Hg_2^{2+} are absent
NH_3 + H_2S
No precipitate All group II and III cations are absent
$(NH_4)_2CO_3$
No precipitate Ca^{2+} and Mg^{2+} are absent
Bluish-violet flame test is characteristic of K^+. The salt is K_3PO_4.

Practice Problem 19.13: A simple salt is soluble in water. When HCl is added to the solution, no reaction occurs. When this solution is saturated with H_2S, no precipitate forms. Making the solution alkaline while still saturated with H_2S produces a white precipitate. Treating a fresh sample of the solution with H_2SO_4, followed by heating, gives no reaction. When barium nitrate is added to the solution after heating, a precipitate of barium sulfate forms and is discarded. When the supernatant is made alkaline, no precipitate forms. The alkaline solution is acidified and treated with silver nitrate. A white precipitate forms that dissolves in aqueous ammonia. What is the identity of the salt? Use Figures 19.1, 19.5, and 19.18, if necessary.

Summary

One of the principal goals of qualitative analysis in aqueous solution is to study the aqueous chemistry of cations and anions. Various groups of cations can be separated from one another by selective precipitation. The precipitate of each group—which may contain several different solid compounds—is then dissolved in a suitable solvent. Cations in the solution are separated from one another and identified by various techniques. The scheme of analysis we have chosen to use is summarized below.

Group I cations—Ag^+, Pb^{2+}, and Hg_2^{2+}—are *precipitated as chlorides—$AgCl$, $PbCl_2$, and Hg_2Cl_2—from an acid solution.*

Group II cations—Hg^{2+}, Cu^{2+}, Pb^{2+}, and Sn^{2+}—are *precipitated as sulfides—HgS, CuS, PbS, and SnS_2—from an acid solution saturated with H_2S.* This is possible because the sulfides of this group of cations have very small K_{sp} values.

Group III cations—Al^{3+}, Cr^{3+}, Fe^{3+}, Mn^{2+}, Ni^{2+}, and Zn^{2+}—are *precipitated as hydroxides—$Al(OH)_3$, and $Cr(OH)_3$—and as sulfides—FeS, MnS, NiS, and ZnS—from an alkaline solution saturated with H_2S.* The sulfides of group III have larger K_{sp} values than the sulfides of group II. Thus, a larger sulfide ion concentration is needed to precipitate group III sulfides than is needed to precipitate group II sulfides. An alkaline solution saturated with H_2S provides a larger sulfide ion concentration than an acid solution saturated with H_2S. Two of the cations of this group, Al^{3+} and Cr^{3+}, form hydroxides instead of sulfides.

The **group IV cations**—Mg^{2+} and Ca^{2+}—are *precipitated as carbonates—$MgCO_3$ and $CaCO_3$—from an alkaline solution.*

The **group V cations**—Na^+, K^+, and NH_4^+—are not precipitated as a group because nearly all compounds of these cations are soluble in water.

Further details of the analysis of each group of cations are summarized in Figures 19.1, 19.4, 19.6, 19.8, and 19.11.

Anions can also be classified into various groups on the basis of properties common to each group.

The **anions of volatile acids**—SO_3^{2-}, CO_3^{2-}, S^{2-}, and CH_3COO^-—*are easily protonated to form volatile acids that can be identified by their characteristic odors or by other means.*

The **anions of the sulfate group**—SO_4^{2-}, SO_3^{2-}, CO_3^{2-}, and PO_4^{3-}—*are precipitated as barium salts from alkaline solution.*

The **chloride group anions**—Cl^-, Br^-, and I^-—*are precipitated as silver salts from acid medium.*

The **reducing anions**—SO_3^{2-}, S^{2-}, Br^-, and I^-—*reduce and decolorize permanganate solution.*

The **oxidizing anion**, NO_3^-, oxidizes I^- to I_2, which can be detected by its color in cyclohexane.

Exercises

General Review

1. According to the method of qualitative analysis of cations described in this chapter, which cations belong to: **(a)** group I; **(b)** group II; **(c)** group III; **(d)** group IV; **(e)** group V? Write the symbol and charge for each cation and indicate its color in aqueous solution.
2. Write the formulas of the compounds that precipitate in each group. What is the color of each of these compounds?
3. Name three reagents that can be used to precipitate the group I chlorides.
4. Why are group II sulfides precipitated in acid solution and the group III sulfides in alkaline solution?
5. What reagent can be used to precipitate the carbonates of the group IV cations?
6. Why must carbonates be precipitated in an alkaline solution?
7. What is the color of the flame in the flame test for: **(a)** Na^+ ions; **(b)** K^+ ions?
8. Write a net ionic equation for the precipitation reaction of a cation in each group discussed in this chapter.
9. What chemical properties are common to all the cations of: **(a)** group I; **(b)** group II; **(c)** group III; **(d)** group IV; **(e)** group V?
10. What chemical properties are common to all the anions of the: **(a)** volatile acid group; **(b)** sulfate group; **(c)** chloride group; **(d)** oxidizing anions; **(e)** reducing anions?

11. Select an anion from each group listed in question 10, and write a net ionic equation for the reaction that illustrates the property common to all of the anions in that group.
12. List some preliminary observations and experiments that may be helpful in identifying a salt.
13. List the cations whose salts are generally soluble in water.
14. List the anions whose salts are generally soluble in water.
15. Give some examples of salts which are not soluble in water but which can be expected to dissolve in a non-oxidizing acid such as HCl. Explain why these salts dissolve in HCl and not in H_2O.
16. If a salt does not dissolve in water or in HCl, list some other possible reagents in which it might dissolve. Give an example in each case of the type of salt that might be expected to dissolve in the reagents of your choice (review Section 16.5 if necessary).
17. If a salt dissolves in water and the solution turns red litmus blue, what type of an anion might be present? Give some examples of such anions.
18. If a salt dissolves in water and the solution turns blue litmus red, what type of a cation might be present? Give examples of such cations.

Analysis of Group I Cations

19. A solution that may contain silver ions, and lead(II) ions is treated with hydrochloric acid. A white precipitate forms which does not dissolve in hot water. Which ions are present, and which are absent?
20. A solution that may contain silver ions, mercury(I) ions, and lead(II) ions is treated with hydrochloric acid. A white precipitate forms which is stirred with hot water. The mixture is then centrifuged and the supernatant is separated from the precipitate. The precipitate dissolves completely in aqueous ammonia. The supernatant gives a white precipitate with sulfuric acid. Which ions are present and which are absent in the original solution?
21. Write balanced net ionic equations for the reactions that occur when: **(a)** hydrochloric acid is added to a solution of silver nitrate; **(b)** a solution of sodium sulfate is added to a solution of lead(II) nitrate; **(c)** a precipitate containing both silver chloride and lead chloride is treated with aqueous ammonia; **(d)** a precipitate of mercury(I) chloride is treated with aqueous ammonia.

Analysis of Group II Cations

22. A solution that may contain mercury(II) ions, copper(II) ions, and tin(II) ions is treated with hydrogen peroxide and saturated with hydrogen sulfide at pH 0.50. A precipitate that forms is stirred with hydrochloric acid resulting in a black residue and a colorless supernatant after centrifuging. The residue dissolves completely in aqua regia to produce a colorless solution. The supernatant from the black residue gives a gray precipitate with powdered iron and mercury(II) chloride solution. Which ions are present and which are absent in the original solution? Use Figure 19.6, if necessary.
23. A solid that may contain mercury(II) sulfide, lead(II) sulfide, and tin(IV) sulfide dissolves completely in hydrochloric acid. What is the composition of the solid?
24. A solid that may contain mercury(II) sulfide and copper(II) sulfide partly dissolves in nitric acid. What is the composition of the solid?
25. A blue solution that may contain copper(II) ions and lead(II) ions gives a white precipitate with sulfuric acid. Which cations are present and which are absent in this solution?
26. Write a balanced net ionic equation for the reaction that occurs when: **(a)** a solution of tin(II) nitrate is treated with hydrochloric acid and hydrogen peroxide [one of the products is $Sn(Cl)_6^{2-}$ ion]; **(b)** a solution of $SnCl_6^{2-}$ ions is treated with a solution of sodium sulfide to produce SnS_2 as one of the products; **(c)** a solution of $SnCl_6^{2-}$ ions is treated with iron powder that reduces tin(IV) to tin(II); **(d)** a solution of tin(II) ions is treated with hydrochloric acid and with a solution of mercury(II) chloride to produce a precipitate of mercury(I) chloride and a solution of $SnCl_6^{2-}$ ions; **(e)** a mixture of copper(II) nitrate and lead(II) nitrate solutions is treated with a solution of sodium sulfate; **(f)** a solid mixture of mercury(II) sulfide and coppe-r(II) sulfide is treated with nitric acid; **(g)** a solution of tin(II) chloride is added to a solution of $HgCl_4^{2-}$ ions to produce a precipitate of elemental mercury and a solution of $SnCl_6^{2-}$ ions.

Analysis of Group III Cations

27. A solution is 0.050 M in Pb^{2+} ions and 0.050 M in Zn^{2+} ions. What must be the pH of this solution so that when it is saturated with H_2S, the Pb^{2+} ions precipitate as PbS as completely as possible before any ZnS starts to precipitate? Look up the necessary data.
28. A colorless solution which may contain aluminum ions, nickel(II) ions, and zinc ions is treated with aqueous ammonia, resulting in the formation of a gray, gelatinous precipitate. The supernatant from this precipitate when saturated with hydrogen sulfide forms a white precipitate. Which ions are present and which are absent in the original solution?
29. A solution that may contain Cr^{3+}, Fe^{3+}, and Mn^{2+} ions is treated with NaOH and H_2O_2 forming a precipitate and a yellow supernatant. The precipitate is dissolved in H_2SO_4 and H_2O_2 and the resulting solution is treated with KSCN, which turns the solution blood red. Which ion is present, which is absent, and which may be doubtful because of insufficient information?

30. Name one reagent that would precipitate one of the cations and not the other in each of the following pairs of solutions: **(a)** Al^{3+}, Zn^{2+}; **(b)** Al^{3+}, Fe^{3+}; **(c)** Fe^{3+}, Zn^{2+}.

31. Write a balanced net ionic equation for the reaction that occurs when: **(a)** zinc sulfide is dissolved in hydrochloric acid; **(b)** nickel(II) sulfide is dissolved in nitric acid to produce nickel(II) ions, elemental sulfur, nitrogen dioxide, and water; **(c)** chromium(III) ions are oxidized to chromate ions by hydrogen peroxide in basic solution; **(d)** solid manganese dioxide is reduced to manganese(II) ions by hydrogen peroxide in acidic solution; **(e)** manganese(II) ions in acid solution are oxidized to permanganate ions by solid sodium bismuthate; **(f)** a solution of barium chloride is added to a solution containing both $Al(OH)_4^-$ and CrO_4^{2-} ions; **(g)** a solution of $Zn(NH_3)_4^{2+}$ is saturated with hydrogen sulfide.

Analysis of Cation Groups IV and V

32. Write a net ionic equation for the reaction that occurs when: **(a)** a solution of calcium chloride is mixed with a solution of sodium carbonate; **(b)** solid magnesium carbonate is dissolved in hydrochloric acid; **(c)** a solution of calcium nitrate is mixed with a solution of ammonium oxalate; **(d)** solid calcium oxalate is dissolved in hydrochloric acid; **(e)** a solution of magnesium nitrate is treated with aqueous ammonia and with a solution of sodium monohydrogen phosphate to precipitate magnesium ammonium phosphate; **(f)** a solid pellet of sodium hydroxide is dropped into a solution of ammonium chloride.

General Cation Analysis

33. Name one reagent that would precipitate one of the ions and not the other in each of the following aqueous mixtures of cations: **(a)** Ag^+, K^+; **(b)** Pb^{2+}, Zn^{2+}; **(c)** Cu^{2+}, Al^{3+}; **(d)** Al^{3+}, Fe^{3+}; **(e)** Al^{3+}, Ni^{2+}; **(f)** Fe^{3+}, Mg^{2+}; **(g)** Ca^{2+}, NH_4^+.

34. Each of the following solutions may contain any or all of the cations listed. Devise the simplest possible scheme to separate and identify the ions that may be present, assuming that all of them are present in each case: **(a)** Pb^{2+}, Al^{3+}; **(b)** Al^{3+}, Fe^{3+}; **(c)** Ag^+, Cu^{2+}, Cr^{3+}; **(d)** Mn^{2+}, Ca^{2+}, Na^+; **(e)** Hg_2^{2+}, Hg^{2+}, Zn^{2+}, Ca^{2+}; **(f)** Ag^+, Sn^{2+}, NH_4^+.

35. A solution that may contain silver ions, nickel(II) ions, chromium(III) ions, and calcium ions is colorless and gives a white precipitate with HCl. The supernatant from this precipitate gives a white precipitate upon treatment with NH_3 and $(NH_4)_2CO_3$. Which cations are present and which are absent in the unknown solution?

36. A solution that may contain Na^+, K^+, Fe^{3+}, Zn^{2+}, and Mg^{2+} ions is made alkaline with NaOH and saturated with H_2S, giving a white precipitate. Treatment of the supernatant with $(NH_4)_2CO_3$ gives no evidence of a reaction. A flame test on a separate portion of the unknown solution gives a violet color, but adding KSCN to this solution produces no visible change of color. Which cations are present and which are absent in the original unknown?

37. A colorless solution may contain any or all of the cations in our scheme. Adding HCl to this solution produces a white precipitate that dissolves completely in hot water. When the supernatant is saturated with H_2S, no precipitate forms, but when the solution is made alkaline with NH_3, a grayish gelatinous precipitate appears and is found to be completely soluble in NaOH. The supernatant from this precipitate is treated with $(NH_4)_2CO_3$ with no visible change. A flame test on a portion of the original solution shows no characteristic color, but when NaOH is added to the solution, a moist red litmus paper turns blue when held just above the solution. Which cations are present and which are absent in the original solution?

Anion Analysis

38. Name one reagent that would enable you to distinguish between each of the following pairs of anion solutions in different test tubes: **(a)** SO_4^{2-}, NO_3^-; **(b)** SO_4^{2-}, S^{2-}; **(c)** Cl^-, $C_2H_3O_2^-$; **(d)** NO_3^-, SO_3^{2-}; **(e)** Cl^-, I^- (both in acid solution); **(f)** Br^-, PO_4^{3-}; **(g)** CO_3^{2-}, S^{2-}. What would you observe when you treat each of the solutions with your proposed reagent? Write a net ionic equation for each reaction that occurs in your testing.

39. Devise the simplest possible method of analysis to show the presence or absence of each of the anions listed in each of the following solutions: **(a)** S^{2-}, SO_4^{2-}; **(b)** Cl^-, SO_4^{2-}; **(c)** SO_3^{2-}, SO_4^{2-}; **(d)** $C_2H_3O_2^-$, S^{2-}, CO_3^{2-}; **(e)** Cl^-, Br^-, I^-; **(f)** PO_4^{3-}, NO_3^-, Cl^-; **(g)** Cl^-, Br^-, SO_4^{2-}, S^{2-}.

40. When a solution that may contain NO_3^-, SO_4^{2-}, and CO_3^{2-} ions is treated with barium chloride, a white precipitate forms that completely dissolves in HCl. Another portion of the solution is acidified with HCl and treated with KI, giving a brownish-yellow color. When this solution is shaken with cyclohexane, a violet color appears in the cyclohexane layer. Which anions are present and which are absent in the unknown solution?

41. When a solution that may contain SO_3^{2-}, PO_4^{3-}, Br^-, and Cl^- ions is treated with H_2SO_4 and heated in a water bath, a choking odor is produced. Heating is continued until the odor can no longer be detected. Dilute $KMnO_4$ is then added to the solution, followed by shaking the solution with cyclohexane. An orange color develops in the cyclohexane layer. The aqueous layer is then withdrawn from the cyclohexane layer and treated with $Ba(NO_3)_2$ to remove the sulfate ions previously added in H_2SO_4. Adding silver nitrate solution to the supernatant from the $BaSO_4$ precipitate produces no visible reaction. Another portion of the unknown solution is boiled with HNO_3 and $(NH_4)_2MoO_4$ and then allowed to cool. A yellow precipitate forms. Which anions are present and which are absent in the original unknown?

42. A solution may contain any or all of the anions of our scheme. When a portion of this solution is treated with H_2SO_4 and heated, no odor is detected. A stirring rod moistened with barium hydroxide, as well as a lead acetate paper, both held above the solution during the heating, remain unaffected. Barium nitrate is then added until the precipitation is complete. When the supernatant from this precipitate is made alkaline with NH_3, a white precipitate forms. The supernatant from this precipitate is acidified with HNO_3 and treated with silver nitrate, producing a white precipitate. This precipitate is completely soluble in NH_3.

Another portion of the original solution is made acidic with HCl and treated with a solution of barium chloride. A white precipitate forms. A third portion of the original solution is acidified with HCl, treated with potassium iodide solution, and shaken with cyclohexane. A violet color develops in the cyclohexane layer. Which anions are present and which are absent? Use Figure 19.18, if necessary.

Salt Analysis

43. Consider each of the following pairs of solid compounds and predict which reagent (including water) would dissolve one of the compounds and not the other: **(a)** AgCl, $AgNO_3$; **(b)** $BaCl_2$, $BaSO_4$; **(c)** $PbCl_2$, NH_4NO_3; **(d)** NH_4Cl, $PbSO_4$; **(e)** Na_2S, Ag_2S; **(f)** $BaSO_4$, $BaCO_3$; **(g)** $BaCO_3$, $NiCO_3$; **(h)** $Cu(OH)_2$, $Mg(OH)_2$; **(i)** $Al(OH)_3$, $Fe(OH)_3$; **(j)** $Al(OH)_3$, $Zn(OH)_2$; **(k)** $BaSO_4$, $BaSO_3$; **(l)** MnS ($K_{sp} = 1.4 \times 10^{-15}$), HgS ($K_{sp} = 1 \times 10^{-50}$). [*Hint for parts* **(g)**, **(h)**, *and* **(j)**: Nickel, copper, and zinc have a great tendency to form ammine complexes.]

44. Describe a simple test (preferably one step) that you would perform to distinguish between the two compounds in each of the following pairs of solid compounds (sometimes a simple visual examination of the compounds might identify them): **(a)** $AgNO_3$ and AgCl; **(b)** $Mg(OH)_2$ and $Zn(OH)_2$; **(c)** Na_3PO_4 and $(NH_4)_3PO_4$; **(d)** $CrCl_3$ and $AlCl_3$; **(e)** ZnS and CuS; **(f)** KBr and NH_4I; **(g)** $BaSO_4$ and $CaSO_3$; **(h)** HgI_2 and Hg_2Cl_2; **(i)** Na_2CO_3 and K_3PO_4; **(j)** $Ca(CH_3COO)_2$ and $BaCl_2$; **(k)** ZnO and Al_2O_3. In each case, state the results which you would observe for each of the possibilities.

45. An unknown salt or a mixture of salts that contains carbonate as the only anion is completely soluble in water. What cations may be present?

46. An unknown salt or a mixture of salts is completely soluble in water and contains Ag^+ ions as the only cations. What anions may be present?

47. A solid salt mixture may contain any or all of the following compounds: $FeCl_3$, $NiCl_2$, ZnS, and HgS. When the solid is shaken with hydrochloric acid, a black solid X and a solution Y result. The solid dissolves in aqua regia leaving a small amount of brownish-yellow floating residue. The solution Y is made alkaline with NH_3, producing a blue-violet solution. Indicate which of the compounds above are definitely present, which are absent, and which are undetermined.

48. A solid is known to be a mixture of two or more of the following salts: $CuSO_4$, $AgNO_3$, $(NH_4)_2SO_4$, AgCl, $Ba(NO_3)_2$, $Cr(NO_3)_3$, and $Zn(NO_3)_2$. When this mixture is stirred with water, a white solid, X, and a colorless solution, Y, result. The solid X is unaffected by H_2SO_4 but dissolves in NH_3. The solution Y is slightly acid to litmus, gives no evidence of reaction with H_2SO_4, but when it is made strongly alkaline with NaOH, an odor of ammonia is produced. Which of the compounds above are definitely present, which are absent, and which are undetermined?

PART 4

Thermodynamics and Kinetics

CHAPTER 20

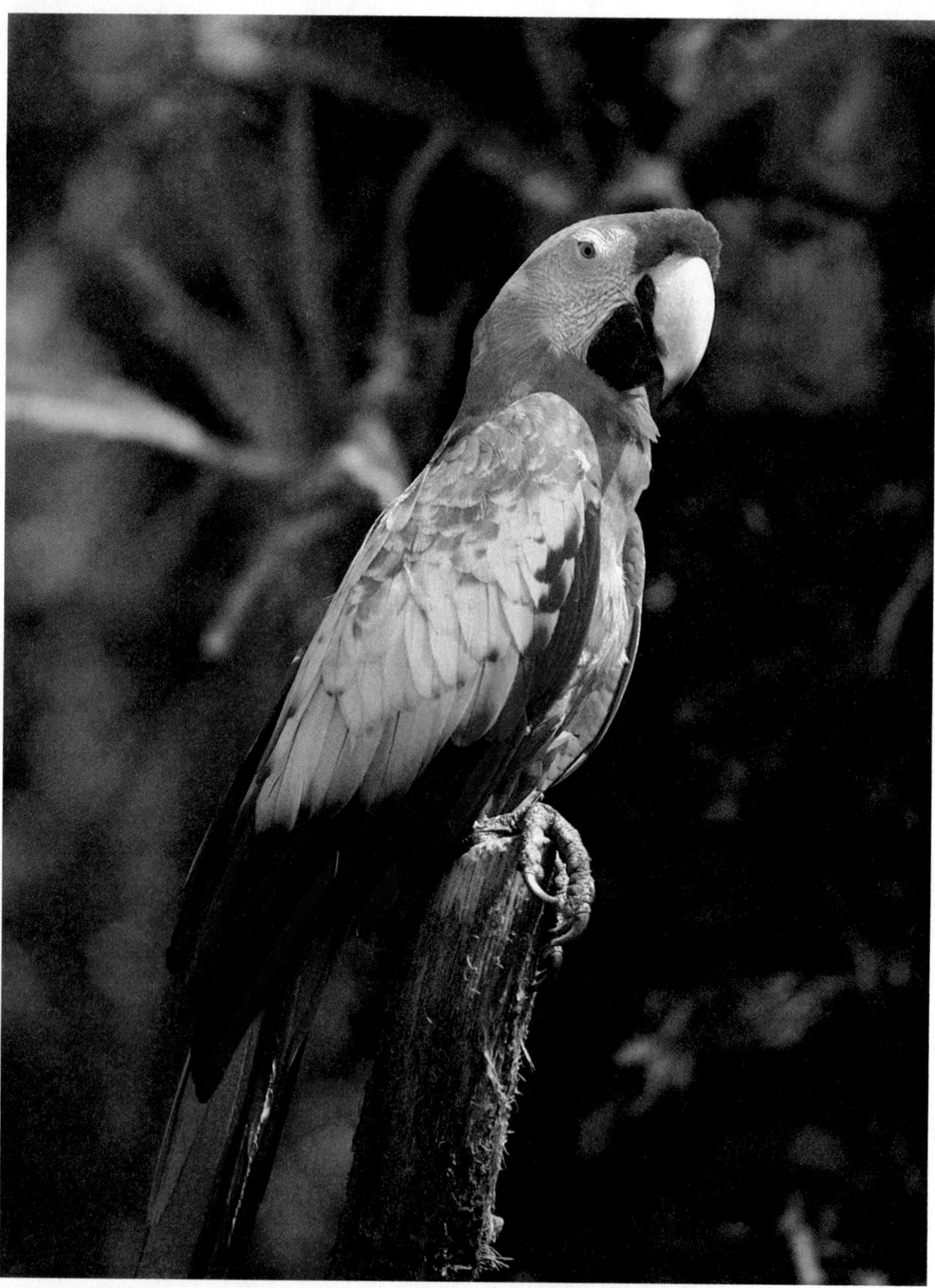

A parrot serves as an example of a possible *system* for study, and its environment as the *surroundings* of the system. The terms system and surroundings are often used in thermodynamics.

Thermodynamics: Spontaneity, Entropy, and Free Energy

In Chapter 5 we discussed heats of reactions. We recall that the heat absorbed or released by a chemical reaction that occurs at constant pressure and does only pressure–volume work is called the enthalpy change for the reaction, enthalpy of reaction, or heat of reaction. The enthalpy of reaction often determines whether or not a reaction occurs spontaneously, but it is not the ultimate criterion of spontaneity. In this chapter we discuss another factor, the change in the degree of randomness, or entropy change, which also plays a role in determining whether or not a reaction or process is spontaneous. We will also learn how enthalpy and entropy changes are related to the position of a chemical equilibrium and how these quantities can be used to determine the values of equilibrium constants.

20.1 Spontaneity and Entropy

Spontaneous Processes

A process that occurs by itself, without outside influence or an external source of work or energy, is said to be **spontaneous**. Examples of spontaneous processes include a ball rolling downhill, iron reacting with oxygen to form iron(III) oxide, ice melting above 0°C, water evaporating in dry air, and glucose dissolving in water. A process that must be driven by an external source of energy is *nonspontaneous*. Thus, a ball can be pushed uphill and iron can be recovered from iron(II) oxide by reduction with carbon. Most spontaneous processes are exothermic, but some spontaneous processes are endothermic. Therefore, *the enthalpy change for a process does not tell us whether or not the process is spontaneous.*

Entropy

Let us consider an example of a spontaneous process that is endothermic: the melting of ice at 1 atm pressure and at 25°C. Ice consists of a crystalline, and therefore highly ordered, array of hydrogen-bonded water molecules. When ice melts, this orderly array is disrupted, and the water molecules in liquid water are less ordered and have a greater freedom of motion, or a greater degree of *randomness or disorder* than the water molecules in ice (Figure 20.1). *The degree of randomness or disorder of a system is called the* ***entropy***, *S, of the system.*

Similar considerations apply to the evaporation of water at 25°C at 1 atm pressure. The water molecules in liquid water, although not as well ordered as those in ice, nevertheless are extensively hydrogen bonded to one another. But when liquid water evaporates, the gaseous water molecules occupy a much larger volume and have greater freedom of motion than they had in the liquid state. Therefore, the evaporation of water is accompanied by a very large increase in the disorder, randomness, or entropy of the system (Figure 20.1). In summary, ice has greater order, more structure, and lower entropy than liquid water, and liquid water has lower entropy than water vapor.

Let us consider another example of a process that occurs with an increase in entropy. When table sugar, sucrose, dissolves in water, the orderly arrangement of sucrose molecules in their crystal lattice converts to a haphazard arrangement of the molecules in a larger volume occupied by the solution

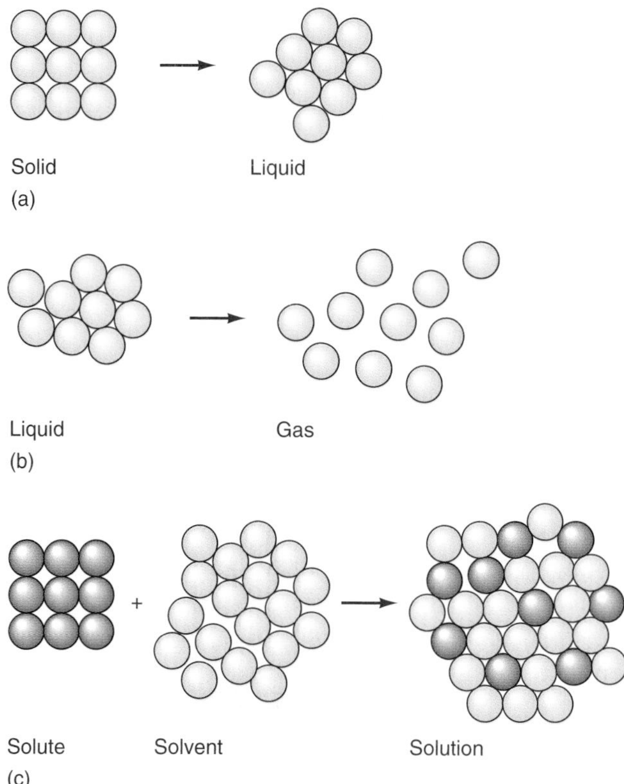

Figure 20.1

Examples of processes occurring with an increase in entropy. (a) Melting: The entropy of the liquid is higher than the entropy of the solid. (b) Vaporization: The entropy is higher in the gas than in the liquid. (c) Dissolving: The entropy is higher in the solution than in the separate solute and the solvent.

(Figure 20.1). The entropy of the sucrose solution is larger than the entropy of the crystalline sucrose and the water used to make the solution. Furthermore, the hydrogen-bonded network of water molecules is somewhat disrupted when sucrose molecules become hydrated, so the entropy of water also increases when the sucrose dissolves.

The Second Law of Thermodynamics

The criterion that determines the spontaneity of all processes, whether exothermic or endothermic, is summarized by the **second law of thermodynamics**: *All spontaneous processes are accompanied by an increase in the entropy of the universe.* This statement may seem surprising, but the "universe" is simply a convenient term that comprises a system and its surroundings (Figure 20.2).

The entropy of the universe includes the entropy of the system and its surroundings. The second law of thermodynamics, like the first law (Chapter 5), is based on observation and experimentation.

The change in entropy for a process, ΔS, equals the difference between the entropies of the final state, S_{final}, and the initial state, S_{initial}:

$$\Delta S = S_{\text{final}} - S_{\text{initial}}$$

When the entropy of the final state is larger than the entropy of the initial state ($S_{\text{final}} > S_{\text{initial}}$), the entropy change is positive, $\Delta S > 0$. When entropy decreases, S_{final} is smaller than S_{initial}, and ΔS is negative.

For a spontaneous process, the entropy change of the *system* can be positive or negative, but the entropy of the *universe* (the system *and* its surroundings) increases. The system and its surroundings constitute the "universe,"

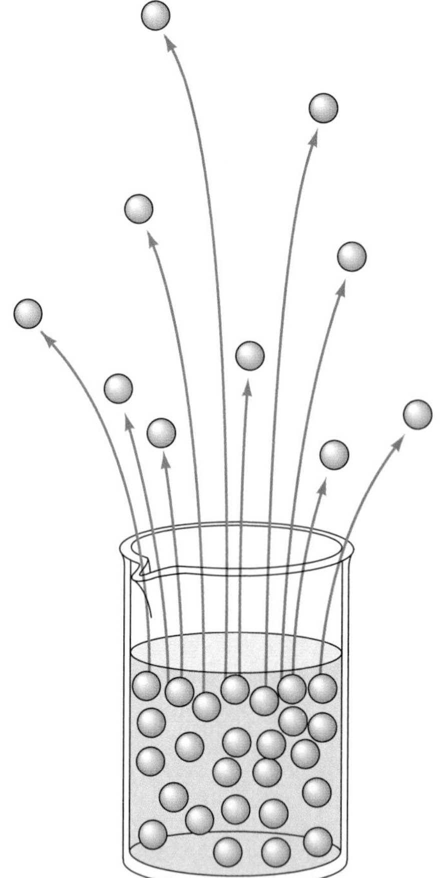

A system: a beaker containing a liquid which evaporates spontaneously. The vapor expands indefinitely in the universe (the system and its surroundings). The entropy of the universe increases.

Figure 20.2

A spontaneous process of evaporation of a liquid is accompanied by an increase in the entropy of the universe.

and the entropy change of the universe, ΔS_{univ}, is the sum of the entropy change of the system, ΔS_{sys}, and the entropy change of its surroundings, ΔS_{sur}:

$$\Delta S_{\text{univ}} = \Delta S_{\text{sys}} + \Delta S_{\text{sur}} \tag{20.1}$$

For any spontaneous process, the entropy change of the universe must be greater than zero. This statement can be expressed mathematically as follows:

For a spontaneous process,

$$\Delta S_{\text{sys}} + \Delta S_{\text{sur}} = \Delta S_{\text{univ}} > 0 \tag{20.2}$$

For a process at equilibrium, ΔS_{univ} equals zero, while ΔS_{sys} and ΔS_{sur} are equal in magnitude, but have opposite signs:

$$\Delta S_{\text{univ}} = \Delta S_{\text{sys}} + \Delta S_{\text{sur}} = 0 \quad \text{or} \quad \Delta S_{\text{sys}} = -\Delta S_{\text{sur}}$$

Changes in entropy that result from heat transfer can be measured. If a system absorbs heat, q, at constant temperature, T, the entropy change of the system near equilibrium equals the heat absorbed per 1 K increase:

$$\Delta S = S_{\text{final}} - S_{\text{initial}} = \frac{q_{\text{rev}}}{T} \tag{20.3}$$

where q_{rev} is the heat absorbed in a **reversible process**. A reversible change occurs in a system that is near equilibrium and occurs in a certain direction infinitely slowly, so that applying an infinitesimal amount of energy stops the process or causes it to go in the opposite direction. Of course, no real processes are reversible in this ideal sense. In the real world, processes can be very *nearly* reversible.

20.2 Calculation of Entropy Changes

Entropy Changes for Phase Changes

The melting of ice at 1 atm pressure and 25°C and the evaporation of water are phase changes that occur with an increase in entropy. Either of these processes can occur in reverse with a decrease in entropy. If we assume that the heat absorbed by ice melting at 0°C and 1 atm is a reversible process, its entropy change, called the *entropy of fusion*, ΔS_{fus}, can be calculated with the equation

$$\Delta S_{fus} = \frac{\Delta H_{fus}}{T_{mp}}$$

where ΔH_{fus} is the enthalpy of fusion of ice and T_{mp} is the Kelvin temperature at which ice melts. The entropy change for the evaporation or vaporization of water can be calculated in a similar way using the values for the enthalpy of vaporization, ΔH_{vap}, and the boiling temperature of water:

$$\Delta S_{vap} = \frac{\Delta H_{vap}}{T_{bp}}$$

Example 20.1

Calculating the Entropy Changes for Fusion of Ice and Evaporation of Water

The enthalpy of fusion of ice at 273 K and 1.00 atm is 6.00 kJ mol^{-1}, and the enthalpy of vaporization of water is 40.7 kJ mol^{-1} at 373 K and 1.00 atm. Calculate the entropy change for each of these processes.

SOLUTION

$$\Delta S_{fus} = \frac{6.00 \text{ kJ mol}^{-1}}{273 \text{ K}} = 0.0220 \text{ kJ mol}^{-1} \text{ K}^{-1} = 22.0 \text{ J mol}^{-1} \text{ K}^{-1}$$

$$\Delta S_{vap} = \frac{40.7 \text{ kJ mol}^{-1}}{373 \text{ K}} = 0.109 \text{ kJ mol}^{-1} \text{ K}^{-1} = 109 \text{ J mol}^{-1} \text{ K}^{-1}$$

We can see from Example 20.1 that the units of entropy change are energy units per mole per kelvin (J mol^{-1} K^{-1}). The entropy change for evaporating a mole of water is larger than that for melting a mole of ice because the increase in the degree of randomness when a liquid evaporates is greater than the increase in the degree of randomness when an equal quantity of a solid melts. This is true not only for water, but for other substances as well, as shown in Table 20.1.

APPLICATIONS OF CHEMISTRY 20.1
Was There a Beginning?

Philosophers and cosmologists have debated for centuries whether the universe had a beginning or whether it simply always existed. The second law of thermodynamics implies that the universe must have had a beginning. The second law of thermodynamics states that in all spontaneous processes, the entropy (disorder) of the universe increases. The entropy of the universe never decreases. As we spontaneously move forward in time, the entropy of the universe increases. One indication of this is that the universe is expanding. As time passes, the stars and galaxies are distributed over greater distances, and their disorder increases.

past	today	future
order	⟶	disorder

It is logical, therefore, to expect that if we could travel backward through time into the past, we would see the universe moving more and more toward an ordered system. It is also logical, however, that there must be a limit to how much something can be ordered (organized). We would eventually reach a point where the universe is as ordered as possible. This would be the point of maximum order for the universe, and we could travel no further into the past. This implies that the universe had a beginning. Today, we call that moment as the Big Bang or the point of maximum order for our universe. According to the Big Bang theory, about 15 to 20 billion years ago an infinitely dense, dimensionless point exploded. All that is today and all that was before the birth of the universe was once concentrated in that dimensionless point.

Today, scientists are continually making measurements to test the accuracy of the predictions made by the Big Bang theory. They have found, for example, that the proportions of hydrogen, helium, and lithium isotopes that we now measure in the universe are exactly what the Big Bang theory predicts they should be.

Optical observations based on the speed of light are also used to test the Big Bang theory. To understand this theory, consider the time it takes for light to travel from the sun to the earth. It takes 8 minutes for light to travel this distance. When we look at the sun we do not see it as it is this instant but rather what it looked like 8 minutes ago. In a similar way, light reaching us from the edge of our universe shows us what that edge looked like 10 to 20 billion years ago, and to look at that edge is to look back through time to the beginning of the universe. In the early 1990s, NASA's Cosmic Background Explorer (CORE) looked back with its optical instrumentation to within a year of the Big Bang. The results of CORE's experiments also support the Big Bang theory.

Sources: Ronald DeLorenzo, *Problem Solving in General Chemistry*, 2nd ed., Wm. C. Brown, Publishers, Dubuque, Iowa, 1993.

Entropy changes of chemical reactions can be positive (entropy increases) or negative (entropy decreases). Chemical reactions in which gaseous or liquid products form from solid reactants occur with an increase in entropy. Reactions in which the reactants and products are of the same phase but the number of moles of the products is greater than the number of moles of the reactants also occur with an increase in entropy.

Table 20.1

Entropies of Fusion and Vaporization of Selected Substances

Substance	Formula	ΔS_{fus} (J mol^{-1} K^{-1})	ΔS_{vap} (J mol^{-1} K^{-1})
Acetic acid	$HC_2H_3O_2$	40.6	61.9
Ammonia	NH_3	28.9	97.5
Benzene	C_6H_6	39.2	87.8
Bromine	Br_2	39.8	85.4
Nitrogen	N_2	11.3	73.6
Water	H_2O	22.0	109.3

Example 20.2

Predicting the Sign of ΔS for Chemical Reactions

Predict whether each of the following reactions occurs with increasing or decreasing entropy: **(a)** $2KClO_3(s) \rightarrow 2KCl(s) + 3O_2(g)$; **(b)** $SO_2(g) + H_2O(l) \rightarrow H_2SO_3(aq)$.

SOLUTION

a. When a solid is converted to another solid, the entropy change is usually small, but when a gas is also formed in the reaction, as in this case, a considerable increase in entropy is expected.

b. Here a decrease in entropy can be predicted because a gaseous reactant converts to an aqueous solution, and the number of moles of products is smaller than the number of moles of reactants. When the number of molecules decreases in a reaction, the degree of randomness in the system also decreases.

Practice Problem 20.1: Predict the sign of ΔS for each of the following reactions: **(a)** $2H_2(g) + O_2(g) \rightarrow 2H_2O(l)$; **(b)** $Ba^{2+}(aq) + SO_4^{2-}(aq) \rightarrow BaSO_4(s)$; **(c)** $3H_2(g) + N_2(g) \rightarrow 2NH_3(g)$; **(d)** $2Na(s) + 2H_2O(l) \rightarrow 2Na^+(aq) + 2OH^-(aq) + H_2(g)$; (e) $4NH_3(g) + 5O_2(g) \rightarrow 4NO(g) + 6H_2O(g)$.

The Third Law of Thermodynamics and Absolute Entropies

In Chapter 5 we noted that it is impossible to determine the absolute value of the enthalpy of a system, although we can determine the change in enthalpy when the system changes from one state to another. However, absolute entropies of substances can be determined using the **third law of thermodynamics**, which can be stated as follows: *The entropy of a perfect crystalline substance* (Figure 20.3) *is zero at zero Kelvin.*

Assuming perfect order in a crystalline substance at 0 K, the amount of heat absorbed by a mole of a substance at different temperatures can be measured from the lowest temperature obtainable to 298 K (the standard temperature in thermodynamics). The sum of the entropies measured this way up to 298 K is the *absolute* standard entropy, $S°$, for a mole of the substance at 1 atm pressure.

Table 20.2 lists values of the absolute entropies ($S°$) of some common substances. This table reveals that crystalline solids have relatively low entropies, liquids have greater entropies, and gases have the highest entropies. These entropy values reflect the relative freedom of molecular motion in different states of matter.

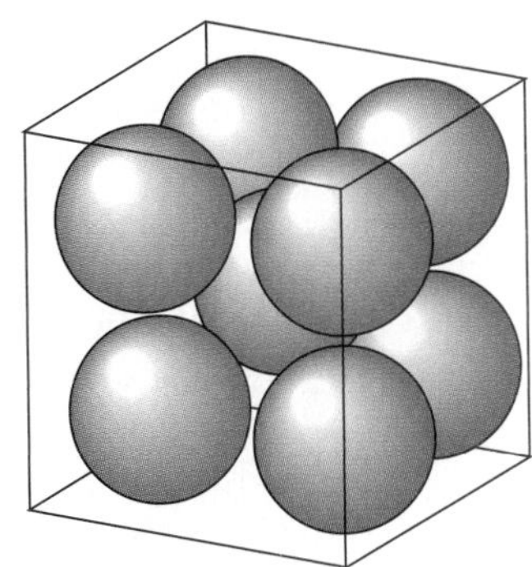

Figure 20.3

A model of a perfect crystal.

Table 20.2

Absolute Entropies of Some Common Substances

Substance	$S°$ (J mol^{-1} K^{-1})
AgCl(s)	96.11
Al(s)	28.32
$Br_2(l)$	152.3
HBr(g)	198.5
$CO_2(g)$	213.6
$H_2O(g)$	188.7
$H_2O(l)$	69.94
KCl(s)	82.7
$KClO_3(s)$	143.0
$O_2(g)$	205.0

Calculating $\Delta S°$ of a Reaction from Absolute Entropies

We can consider the reactants of a chemical reaction as the initial state of a system and the products as the final state of the system. When the reactants in their standard states convert to products in their standard states, the standard entropy change of the reaction equals the standard entropy of the final state (the products) minus the standard entropy of the initial state (the reactants):

$$\Delta S^\circ_{rxn} = \Sigma\, S^\circ_{\text{products}} - \Sigma\, S^\circ_{\text{reactants}} \tag{20.4}$$

For the general reaction

$$aA + bB \longrightarrow cC + dD$$

the standard entropy change, $\Delta S°$, can be given by the equation

$$\Delta S^\circ_{rxn} = (cS^\circ_C + dS^\circ_D) - (aS^\circ_A + bS^\circ_B)$$

We can see that Equation 20.4 for calculating $\Delta S°$ from $S°$ values is similar to Equation 5.6 for calculating $\Delta H°$ of a reaction from ΔH°_f values. However, the value of ΔS°_{rxn} is obtained from *absolute* entropies, $S°$, whereas ΔH°_{rxn} is obtained from enthalpies of *formation*, ΔH°_f.

Example 20.3

Calculating ΔS°_{rxn} from $S°$ Values

Calculate the standard entropy change, $\Delta S°$, for the reaction $2KClO_3(s) \rightarrow 2KCl(s) + 3O_2(g)$ using the absolute entropy values given in Table 20.2.

SOLUTION: We apply Equation 20.4 to obtain

$$\begin{aligned}\Delta S^\circ_{rxn} &= (2)(S^\circ_{KCl}) + (3)(S^\circ_{O_2}) - (2)(S^\circ_{KClO_3})\\ &= (2\text{ mol})(82.7\text{ J mol}^{-1}\text{ K}^{-1})\\ &\quad + (3\text{ mol})(205.0\text{ J mol}^{-1}\text{ K}^{-1}) - (2\text{ mol})(143\text{ J mol}^{-1}\text{ K}^{-1})\\ &= 494\text{ J K}^{-1}\end{aligned}$$

The positive value obtained for ΔS°_{rxn} confirms our earlier qualitative prediction in Example 20.2 that the entropy of this reaction increases.

Practice Problem 20.2: Calculate the standard entropy change for the reaction $2H_2(g) + O_2(g) \rightarrow 2H_2O(l)$. $S°$ for $H_2(g)$ = 130.6 J mol^{-1} K^{-1}.

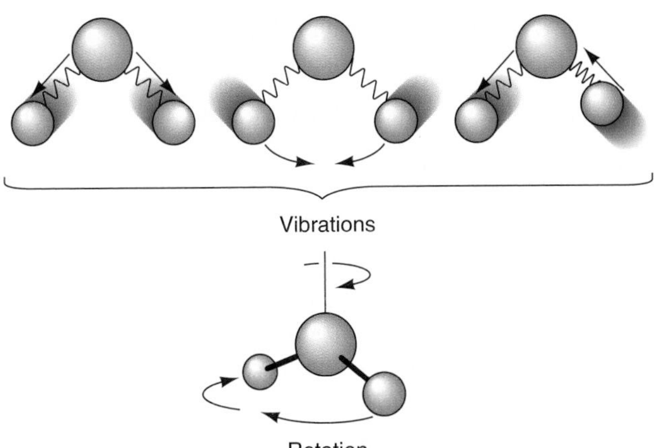

Figure 20.4

Vibration and rotation of a water molecule. Each vibration can be imagined as the motion of the atoms along the arrows and then reversing their directions.

Entropy and Molecular Motion

Different gases and liquids at the same temperature have different entropies because different molecules have different types of motion. Molecules can move from one place to another (this is called *translational motion*), they can *rotate*, and their atoms can *vibrate* in various ways.

Let us briefly consider molecular vibrations. A diatomic molecule continuously vibrates when its two atoms move back and forth along the axis joining the two nuclei. This is the only possible vibration for a diatomic molecule. Hence, diatomic molecules are said to have only one *mode* of vibration. The atoms of polyatomic molecules have more than one mode of vibration. For example, a water molecule has three different modes of vibration: in one mode, the two hydrogen atoms move toward and away from the oxygen in unison; in another mode, the hydrogen atoms undergo a "scissoring" action; and in the third mode, one hydrogen atom moves toward the oxygen atom and the other moves away from the oxygen atom, and then they both reverse (Figure 20.4). A water molecule can also rotate around an axis that passes through the oxygen atom, bisecting the molecule (Figure 20.4).

At a constant temperature, different gases or liquids have different entropies because their molecules have different vibrational and rotational motions. Solids have different entropies due to different degrees of imperfection in their structure. In general, molecules having complex structures have higher entropies than molecules having relatively simple structures. The entropy of a solid, gas, or liquid increases when the substance is heated because molecular motion increases.

20.3 Free Energy and Spontaneity

Free Energy

The heat content, or enthalpy, of a system can be considered in two parts: (1) that which is free to be converted to other forms of energy or to work, and (2) that which is necessary to maintain the system at the specified temperature and is therefore not available for conversion to work. Heat energy that cannot be converted to work is needed for such molecular motions as translation, vibration, and rotation.

This concept of total enthalpy for a system can be expressed in the form of the following equation:

enthalpy = free energy + unavailable energy at a given temperature

This equation expressed in symbols is

$$H = G + TS \tag{20.5}$$

where H is the enthalpy, G the *Gibbs free energy* (after the U.S. physicist Willard Gibbs, 1839–1903, who was one of the founders of thermodynamics), T the Kelvin temperature, and S the entropy of the system. The term representing heat energy that is not available for conversion to work is expressed as the product of absolute temperature and entropy. The total enthalpy change for a process or a chemical reaction at a given temperature, T, can be expressed as

$$\Delta H = \Delta G + T\,\Delta S \tag{20.6}$$

Solving for ΔG yields the conventional form of this equation:

$$\Delta G = \Delta H - T\,\Delta S \tag{20.7}$$

Equation 20.7 relates the free-energy change for a chemical reaction to the enthalpy and entropy changes for the reaction. This equation can be used to calculate the relative contribution of the enthalpy and entropy effects to the free energy. The free-energy change, ΔG, for a process such as a chemical reaction, represents the *maximum work* that can be obtained from the process at constant temperature and pressure.

Spontaneity

For a process to be spontaneous, it must occur with a decrease in free energy (ΔG must be negative) so that the final state has a lower free energy value than the initial state. We can see from the equation $\Delta G = \Delta H - T\,\Delta S$ that if the magnitude of ΔH is greater than the magnitude of $T\,\Delta S$, the sign of ΔG will depend on the sign of ΔH. If the magnitude of $T\,\Delta S$ is greater than the magnitude of ΔH, the sign of ΔG will depend on the sign of ΔS.

We can see from Equation 20.7 that the entropy effect on the free-energy change increases with increasing temperature. This is reasonable because the entropy of a system increases with increasing temperature. Table 20.3 summarizes the effect of the signs of ΔH and ΔS on the sign of ΔG, which indicates whether or not a reaction is spontaneous.

Standard Free-Energy Change

The free-energy change of a process in which reactants in their standard states are converted into products in their standard states at 25°C and 1 atm is called the *standard free-energy change,* $\Delta G°$. (Recall that the standard states of substances are the states in which the substances are most stable at 1 atm pressure and at a specified temperature, commonly 25°C.) Table 20.4 lists thermodynamic data for some selected elements and compounds at 25°C and 1 atm.

The standard free-energy change for a reaction can be calculated from the standard enthalpy and entropy changes for the reaction according to Equation 20.7. The standard enthalpy change is calculated from $\Delta H_f°$ values according to Equation 5.6, and the standard entropy change is calculated from $S°$ values according to Equation 20.4. This type of calculation is illustrated in Example 20.4.

Table 20.3

Criteria for Spontaneity of Reactions

Sign of ΔH	Sign of ΔS	ΔG	Characteristics of Reaction	Example
–	+	Negative	*Reaction is spontaneous at all temperatures*; reverse reaction is not spontaneous	$2O_3(g) \longrightarrow 3O_2(g)$
+	–	Positive	*Reaction is not spontaneous at any temperature*; reverse reaction is spontaneous	$N_2(g) + 3Cl_2(g) \longrightarrow 2NCl_3(l)$
–	–	Negative at low temperature, positive at high temperature	Reaction is spontaneous at low temperatures, not spontaneous at higher temperatures	$2SO_2(g) + O_2(g) \longrightarrow 2SO_3(g)$
+	+	Positive at low temperature, negative at high temperature	Reaction is not spontaneous at low temperatures but becomes spontaneous at high temperatures	$NH_4Cl(s) \longrightarrow NH_3(g) + HCl(g)$

Example 20.4

Calculating ΔG°_{rxn} from ΔH°_f and S° Values

From the appropriate ΔH°_f and S° values obtained from Table 20.4, calculate the standard free-energy change for the reaction

$$2NO(g) + O_2(g) \longrightarrow 2NO_2(g)$$

On the basis of your answer, determine if this reaction is spontaneous under standard conditions.

SOLUTION: The necessary data in Table 20.4 are: ΔH°_f for $NO_2(g)$ is 33.8 kJ mol^{-1}, and for $NO(g)$ 90.4 kJ mol^{-1}; S° for $NO_2(g)$ is 240 J mol^{-1} K^{-1}, for $O_2(g)$ 205 J mol^{-1} K^{-1}, and for $NO(g)$ 211 J mol^{-1} K^{-1}. The standard enthalpy change for the reaction is

$$\begin{aligned}\Delta H^\circ_{rxn} &= (2 \text{ mol})(33.8 \text{ kJ mol}^{-1}) - (2 \text{ mol})(90.4 \text{ kJ mol}^{-1})\\ &= -113 \text{ kJ}\end{aligned}$$

The standard entropy change is

$$\begin{aligned}\Delta S^\circ_{rxn} &= (2 \text{ mol})(240 \text{ J mol}^{-1}\text{ K}^{-1}) - [(2 \text{ mol})(211 \text{ J mol}^{-1}\text{ K}^{-1}) \\ &\quad + (1 \text{ mol})(205 \text{ J mol}^{-1}\text{ K}^{-1})]\\ &= -147 \text{ J K}^{-1}\end{aligned}$$

The free-energy change at 298 K is

$$\begin{aligned}\Delta G^\circ_{rxn} &= \Delta H^\circ_{rxn} - T\,\Delta S_{rxn} = -113 \text{ kJ} - (298 \text{ K})(-0.147 \text{ kJ K}^{-1})\\ &= -69.2 \text{ kJ}\end{aligned}$$

Table 20.4

Thermodynamic Data for Selected Elements and Compounds at 25°C and 1 atm

Substance	ΔH_f° (kJ mol^{-1})	ΔG_f° (kJ mol^{-1})	S° (J mol^{-1} K^{-1})
$Ag(s)$	0	0	42.72
$AgCl(s)$	−127.0	−109.70	96.11
$Al(s)$	0	0	28.32
$Al_2O_3(s)$	−1669.8	−1576.5	51.00
$Br_2(l)$	0	0	152.3
$HBr(g)$	−36.2	−53.2	198.5
C(diamond)	1.88	2.89	2.43
C(graphite)	0	0	5.69
$CO(g)$	−110.5	−137.3	197.9
$CO_2(g)$	−393.5	−394.4	213.6
$CCl_4(l)$	−139.3	−68.6	214.4
$CH_4(g)$	−74.85	−50.8	186.3
$C_2H_2(g)$	226.7	209.2	200.8
$C_2H_4(g)$	52.3	68.11	219.4
$C_2H_6(g)$	−84.67	−32.89	229.5
$C_3H_8(g)$	−103.85	−23.47	269.9
$C_6H_6(l)$ benzene	49.0	124.5	172.8
$C_6H_{12}O_6(s)$ glucose	−1260	−910.6	212.1
$C_{12}H_{22}O_{11}(s)$ sucrose	−2221	−1544	360.2
$CH_3OH(l)$	−238.6	−166.3	126.8
$C_2H_5OH(l)$	−277.6	−174.8	161.0
$HC_2H_3O_2H(l)$ acetic acid	−487.0	−392.4	159.8
$CaCO_3(s)$ calcite	−1207.1	−1128.8	92.88
$Cl_2(g)$	0	0	222.9
$HCl(g)$	−92.30	−95.27	187.0
$F_2(g)$	0	0	203.2
$HF(g)$	−269	−270.7	198.3
$Fe_2O_3(s)$	−822.2	−740.98	89.96
$H_2(g)$	0	0	130.6
$H_2O(l)$	−285.9	−237.2	69.94
$Hg(l)$	0	0	77.4
$I_2(s)$	0	0	116.73
$I_2(g)$	62.3	19.4	260.6
$HI(g)$	25.94	1.30	206.31
$KCl(s)$	−435.9	−408.3	82.7
$KClO_3(s)$	−391.2	−289.2	143.0
$Mg(s)$	0	0	32.51
$MgCl_2(s)$	−641.8	−592.1	89.6
$N_2(g)$	0	0	191.5
$NH_3(g)$	−45.94	−16.66	192.5
$NH_4Cl(s)$	−314.4	−203.0	94.6
$NO(g)$	90.4	86.7	210.62
$NO_2(g)$	33.8	51.84	240.45
$O_2(g)$	0	0	205.0
$O_3(g)$	142.2	163.4	237.6
$Na(s)$	0	0	51.0
$NaCl(s)$	−411.0	−384.0	72.33
$S_8(s)$ rhombic	0	0	31.88
$SO_2(g)$	−296.81	−300.4	248.5
$SO_3(g)$	−395.2	−370.4	256.2
$H_2S(g)$	−20.15	−33.0	205.64
$Zn(s)$	0	0	41.6

This reaction is spontaneous under standard conditions because $\Delta G°$ is negative. Note that the units for both $\Delta H°$ and $\Delta S°$ in the calculation above must both contain either J or kJ. In our example we converted $\Delta S°$ from J K^{-1} to kJ K^{-1} to obtain $\Delta G°$ in kJ. We could just as well convert $\Delta H°$ from kJ K^{-1} to J K^{-1} to obtain $\Delta G°$ in J.

Practice Problem 20.3: Using the appropriate $\Delta H_f°$ and $S°$ values from Table 20.4, calculate $\Delta G°$ for the reaction $4NH_3(g) + 5O_2(g) \rightarrow 4NO(g) + 6H_2O(g)$. Determine if this reaction is spontaneous at 298 K.

Example 20.5

Calculating the Minimum Temperature at Which an Exothermic Reaction with a Negative Entropy Change Is Not Spontaneous Under Standard Conditions

At what temperature would the reaction in Example 20.4 just cease to be spontaneous? Assume that $\Delta H°_{rxn}$ and $\Delta S°_{rxn}$ do not change in the temperature range from 298 K to the temperature in question.

SOLUTION: The data in Example 20.4 show that the negative value for $\Delta G°$ is due to the large negative value of $\Delta H°$, which offsets the negative $\Delta S°$ term. An increase in temperature would increase the entropy effect by increasing the $T\,\Delta S°$ term. For the reaction to just cease to be spontaneous under standard conditions, the value of $\Delta H°$ for the reaction must become equal in magnitude to the value of the $T\,\Delta S°$ term so that $\Delta G°$ is zero in the relationship:

$$\Delta G° = \Delta H° - T\,\Delta S° = 0$$

from which

$$\Delta H° = T\,\Delta S°$$

and

$$T = \frac{\Delta H°}{\Delta S°} = \frac{-113 \text{ kJ}}{-0.147 \text{ kJ K}^{-1}} = 769 \text{ K}$$

Actually, both ΔH and ΔS change somewhat with temperature. However, the answer obtained here is reasonably close to the actual value.

The standard free-energy change for a reaction can also be calculated from the standard free energies of formation of the reactants and products, just as we calculated the standard enthalpy change for a reaction from the standard enthalpies of formation. For an uncombined element in its standard state, the standard enthalpy of formation, $\Delta H_f°$, is zero, and its standard free energy of formation, $\Delta G_f°$, is also zero. For a general reaction

$$aA + bB \longrightarrow cC + dD$$

the standard free-energy change can be calculated by the following equation:

$$\Delta G° = [c\,\Delta G_f°(C) + d\,\Delta G_f°(D)] - [a\,\Delta G_f°(A) + b\,\Delta G_f°(B)] \quad (20.8)$$

Example 20.6

Calculating ΔG°_{rxn} from ΔG°_f Values

From the ΔG°_f values given in Table 20.4, calculate ΔG° for the reaction

$$CH_4(g) + 2O_2(g) \longrightarrow CO_2(g) + 2H_2O(l)$$

Is the reaction spontaneous under standard conditions?

SOLUTION: From Table 20.4, ΔG°_f (in kJ mol^{-1}) for $CH_4(g)$ is -50.8, for $CO_2(g)$ -394, and for $H_2O(l)$ -237. By definition, ΔG°_f for $O_2(g)$ is zero. According to Equation 20.8,

$$\begin{aligned}\Delta G^\circ &= [(1)\ \Delta G^\circ_f(CO_2, g) + (2)\ \Delta G^\circ_f(H_2O, l)] \\ &\quad - [(1)\ \Delta G^\circ_f(CH_4, g) + (2)\ \Delta G^\circ_f(O_2, g)] \\ &= [(1\text{ mol})(-394\text{ kJ mol}^{-1}) + (2\text{ mol})(-237\text{ kJ mol}^{-1})] \\ &\quad - [(1\text{ mol})(-50.8\text{ kJ mol}^{-1}) + 0] \\ &= -817\text{ kJ}\end{aligned}$$

Since ΔG° is less than zero, the products of the reaction are thermodynamically more stable than the reactants at 298 K and 1 atm pressure. The reaction is spontaneous.

Practice Problem 20.4: Using the appropriate ΔG°_f values from Table 20.4, calculate ΔG°_{rxn} for the decomposition of hydrogen peroxide, $2H_2O_2(l) \rightarrow 2H_2O(l) + O_2(g)$. ΔG°_f for $H_2O_2(l) = -120$ kJ mol^{-1}.

20.4 Free Energy and Equilibrium

When a reaction is at equilibrium, no net change occurs, and no free energy is absorbed or released by the system. Thus, *ΔG for a reaction at equilibrium is zero*. (Note that ΔG is not ΔG°.) For ΔG to be zero, the enthalpy change must be equal to the $T\,\Delta S$ term:

$$\Delta H = T\,\Delta S \quad \text{(at equilibrium)}$$

Let us consider the equilibrium between solid and liquid phases of a pure substance. The enthalpy change for a phase change at constant temperature and pressure equals q_p, the heat absorbed or released in the process. Assuming the heat transfer to be a reversible process,

$$\Delta H = q_{rev}$$

Consider, for example, the equilibrium between solid and liquid water when ice melts at 0°C and 1 atm pressure. The heat absorbed by this system increases the entropy of the system without affecting the free-energy change of the system (ΔG remains zero during the melting process).

APPLICATIONS OF CHEMISTRY 20.2
The Cola Space Wars

Equation 20.7 provides interesting insights when viewed as a relationship between two natural forces. Example 20.5 shows that sometimes temperature determines the spontaneity of a chemical reaction. At very high temperatures the $T\,\Delta S$ term is so negative that even endothermic reactions are spontaneous.

All processes in the universe have a natural tendency to strive toward greater disorder and to fall to lower-energy states. Sometimes these tendencies work together and sometimes they work in opposition. When these tendencies work in opposition, temperature determines which of the two tendencies dominates. At very high temperatures, the tendency to increase disorder dominates because the $-T\,\Delta S$ term numerically dominates the ΔH term. At very low temperatures, the tendency to fall to lower states of energy dominates because the $-T\,\Delta S$ term is insignificant compared with the ΔH term.

Entropy changes in chemical reactions can provide the "driving force" for actions we take for granted. This is especially true under conditions of zero gravity such as those encountered by astronauts. Astronauts experience many situations in zero-gravity conditions that are very different from situations here on earth: for example, it is not easy to drink soda pop out of a can. In 1985, NASA made plans to test a special Coca-Cola can on a space shuttle mission. Having a special soda can is essential because drinking carbonated beverages is very difficult in zero gravity. Then Pepsi introduced their own specially designed can, and the Cola Space War began.

Pepsi's can relied on an arrangement in which citric acid dripped onto baking soda to produce carbon dioxide. The carbon dioxide inflated a pouch inside the can, forcing soda out of the can through a special spout. There was only one problem with the arrangement: Citric acid does not "fall" without gravity.

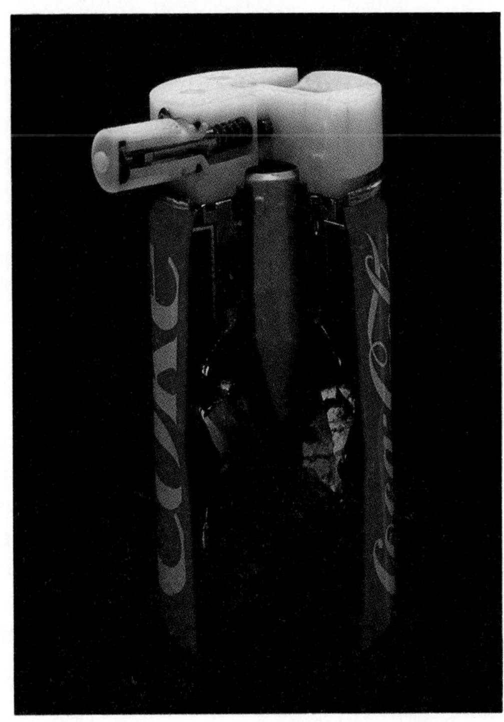

The Coke can stored its soda inside a balloon. Pressure from carbon dioxide outside the balloon forced soda out the balloon and out the can through a specially designed spout (see photo). The final testing of Coke's space can took place successfully in 1991 aboard the Soviet space station Mir.

Source: *Discover*, September 1985, page 9.

Free Energy and Position of Equilibrium

For a spontaneous reaction, the free energy of the reactants, G_R, is greater than the free energy of the products, G_P. The free-energy change for the reaction, ΔG_{rxn}, is therefore less than zero: $\Delta G_{rxn} = G_P - G_R < 0$. When a spontaneous chemical reaction proceeds toward equilibrium, it releases free energy to the surroundings. The free energy of the reactants and the products decreases during the process (Figure 20.5). At equilibrium, the free energy of the reactants equals the free energy of the products at the point at which the free energy of the system is at a minimum, as shown in Figure 20.5.

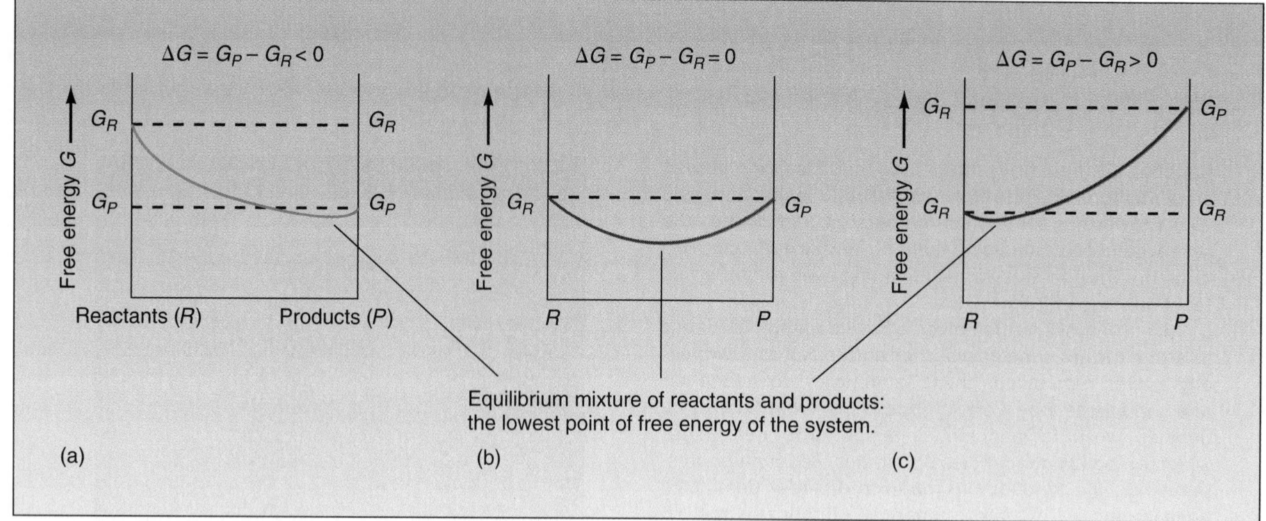

Figure 20.5

Variation of free energy in reversible chemical reactions. An equilibrium can be reached starting with only the reactants, or with only the products. In either case, the free energies of both the reactants and the products decrease until the minimum free energy is reached at equilibrium, where $G_{\text{reactants}} = G_{\text{products}}$.

When a spontaneous reaction reaches equilibrium, the equilibrium mixture contains mostly products and only a small amount of reactants, as shown in Figure 20.5a. Figure 20.5b illustrates a reaction that goes to 50 percent completion. The free energies of the reactants and products of such a reaction are equal before the reaction, $G_R = G_P$. During the reaction, the free energy of both the reactants and the products formed in the reaction decreases until equilibrium is reached. The equilibrium mixture contains equal amounts of the reactants and the products. At equilibrium, $\Delta G_{rxn} = G_P - G_R = 0$.

Figure 20.5c illustrates a nonspontaneous forward reaction ($\Delta G_{rxn} > 0$). The reverse of this reaction is spontaneous. Starting with either the reactants or the products, the free energies of both the reactants and the products decrease during the reaction as the system proceeds toward equilibrium. Since this forward reaction is not spontaneous, it proceeds only to a small extent, and the equilibrium mixture contains mostly reactants.

Figure 20.5 shows that an equilibrium can be reached starting with only the reactants or with only the products. In either case, the free energies of the reactants or the products decrease until the minimum free energy is reached at equilibrium, where $G_R = G_P$.

Relationship Between ΔG and ΔG°

The sign of the free-energy change for a reaction tells us whether or not the reaction is spontaneous. The value of ΔG indicates the extent to which the reaction proceeds. The value of ΔG also indicates the approximate composition of the equilibrium mixture, that is, the position of equilibrium.

Equation 20.9, which we will not derive, relates the free-energy change, ΔG, of a reaction under *any* conditions, with the *standard* free-energy change, ΔG°, for the reaction

$$\Delta G = \Delta G^\circ + RT \ln Q \qquad (20.9)$$

In this reaction, R is the ideal gas constant, 8.314 J mol^{-1} K^{-1}, T the absolute temperature, and Q the reaction quotient (Section 13.2). Recall that the term "ln Q" indicates the natural logarithm of Q. The natural logarithm of a number is the exponent to which the base e (2.718) must be raised to get the number. Thus, $\ln 10.0 = 2.303$ because $(2.718)^{2.303} = 10.0$.

The concentrations or partial pressures expressed by a reaction quotient, Q, are not necessarily equal to the concentrations or partial pressures in an equilibrium mixture. Only for a system at equilibrium is Q equal to K. The free-energy change, ΔG, for a system at equilibrium is zero.

Example 20.7

Calculating ΔG for a Reaction from $\Delta G°$ for the Reaction

Calculate ΔG at 298 K for the reaction

$$3H_2(g) + N_2(g) \longrightarrow 2NH_3(g)$$

at an instant when the partial pressure of H_2 is 2.00 atm, that of N_2 is 1.00 atm, and that of NH_3 is 1.00 atm.

SOLUTION: To apply Equation 20.9, we need to calculate $\Delta G°$ for the reaction. To calculate $\Delta G°$, we use the $\Delta G_f°$ values from Table 20.4. The reaction involves two elements, H_2 and N_2, for which $\Delta G_f°$ is zero. Thus,

$$\Delta G° = 2\ \Delta G_f°(NH_3) - 0 - 0 = (2\ \text{mol})(-16.66\ \text{kJ mol}^{-1})$$
$$= -33.32\ \text{kJ} = -3.332 \times 10^4\ \text{J}$$

Next we need to solve for Q:

$$Q = \frac{(P_{NH_3})^2}{(P_{H_2})^3(P_{N_2})} = \frac{(1.00)^2}{(2.00)^3(1.00)} = 0.125$$

Application of Equation 20.9 yields

$$\Delta G = -3.332 \times 10^4\ \text{J} + (8.314\ \text{J K}^{-1})(298\ \text{K})(\ln 0.125)$$
$$= -3.85 \times 10^4\ \text{J} = -38.5\ \text{kJ}$$

The value obtained for ΔG in Example 20.7 is more negative than the value for $\Delta G°$. This indicates that the forward reaction under the specified conditions proceeds to a greater extent than it would under standard conditions. Under standard conditions, the partial pressure of each component in a gaseous system is 1 atm, and $Q = 1$. When $Q = 1$, $\ln Q = 0$, and $\Delta G = \Delta G°$. For ΔG to be more negative than $\Delta G°$, the term $RT \ln Q$ must be less than zero, that is, a negative number. The $RT \ln Q$ term is negative for any value of Q less than 1.

In Example 20.7 the partial pressure of H_2 is 2 atm, which is more than the standard state pressure. As a result, the equilibrium lies farther to the right than it would under standard conditions.

Practice Problem 20.5: Calculate ΔG for the reaction

$$H_2(g) + Br_2(g) \longrightarrow 2HBr(g)$$

at 350 K when the partial pressure of H_2 is 1.00 atm, that of Br_2 is 2.00 atm, and that of HI is 3.0 atm.

Free-Energy Change and Equilibrium Constant

For a reaction at equilibrium, $\Delta G = 0$ and $Q = K$. We can therefore write

$$\Delta G = \Delta G^\circ + RT \ln K = 0 \quad \text{(at equilibrium)} \tag{20.10}$$

From this equation it follows that

$$\Delta G^\circ = -RT \ln K \tag{20.11}$$

From Equation 20.11 we see that when ΔG° is zero, $\ln K = 0$ and $K = 1$. When the value of ΔG° is negative, $\ln K$ is positive and K is more than 1. When ΔG° is positive, $\ln K$ is negative and K is less than 1. An equilibrium constant that is calculated from Equation 20.11 is called the *thermodynamic equilibrium constant* and is symbolized by K.

Example 20.8

Calculating K from ΔG°

Calculate K for the reaction in Example 20.7 at 298 K.

SOLUTION: To calculate K from ΔG° we apply Equation 20.11. The value for ΔG° for the reaction obtained in Example 20.7 is -33.3 kJ or -3.33×10^4 J. Solving Equation 20.11 for $\ln K$, we obtain

$$\ln K = -\frac{\Delta G^\circ}{RT} = -\frac{-3.33 \times 10^4 \text{ J}}{(8.314 \text{ J K}^{-1})(298 \text{ K})} = 13.441$$

$$K = e^{13.441} = 6.88 \times 10^5$$

Recall that e is the base of natural logarithms. Pocket calculators that do not have a key for e can be used to raise 2.718 (the value for e) to the power of 13.441 to obtain this value for K.

Practice Problem 20.6: The standard free-energy change for the reaction $2SO_2(g) + O_2(g) \rightarrow 2SO_3(g)$ is -140 kJ. Calculate K for this reaction at 298 K.

Example 20.9

Calculating ΔG° from K

For the reaction $N_2O_4(g) \rightleftharpoons 2NO_2(g)$ the equilibrium constant, K_p, is 0.113 at 298 K. Calculate the standard free-energy change, ΔG°, for this reaction.

SOLUTION

$$\Delta G^\circ = -\text{RT} \ln K = -(8.314 \text{ J K}^{-1})(298 \text{ K})(\ln 0.113)$$
$$= 5.40 \times 10^3 \text{ J} = 5.40 \text{ kJ}$$

The result obtained in Example 20.9 shows that a nonspontaneous reaction (with a positive ΔG) can proceed to a small extent reflected by the value of its equilibrium constant. We can show that the decomposition of N_2O_4 to NO_2 does actually take place at room temperature by introducing N_2O_4 at 1 atm and 298 K into an evacuated glass container. After a short time, the initially colorless N_2O_4 develops a brownish tinge, indicating the formation of some NO_2.

Table 20.5

Some Values of $\Delta G°$ and the Corresponding Values of K at 25°C

$\Delta G°$ (kJ)	K
−100	3.4×10^{17}
−50	5.8×10^{8}
−25	2.4×10^{4}
−5	7.5
0	1.0
+5	0.13
+25	4.1×10^{-5}
+50	1.7×10^{-9}
+100	3.0×10^{-18}

Table 20.5 lists some $\Delta G°$ values and their corresponding values of K. We can see that a large negative value for $\Delta G°$ corresponds to a large positive value of K. A reaction with a large positive value of K goes almost to completion. On the other hand, a large positive $\Delta G°$ corresponds to a small value of K. A reaction whose equilibrium constant is less than 1 proceeds only partly to completion.

Practice Problem 20.7: The K_b for the ionization of ammonia, $NH_3(aq) + H_2O(l) \rightleftharpoons NH_4^+(aq) + OH^-(aq)$, is 1.8×10^{-5}. Calculate $\Delta G°$ for this reaction.

20.5 Electrode Potential, Free-Energy Change, and the Equilibrium Constant

We recall from Chapter 17 that a spontaneous redox reaction has a positive cell potential. We also recall from Section 20.3 that a spontaneous reaction has a negative ΔG. The equilibrium constant for a spontaneous reaction is greater than 1.0; for a nonspontaneous reaction it is less than 1.0. Next we discuss the quantitative relationships between the free-energy change for a reaction, its electrode potential, and its equilibrium constant.

$E°$ and $\Delta G°$

When a spontaneous redox reaction occurs at constant temperature and pressure, it can be used to generate electricity or do electrical work. This work can be expressed as the product of the number of moles of electrons transferred, n, the charge per mole of electrons (Faraday's constant, F), and the voltage, E:

$$\text{electrical work} = nFE$$

Recall that F has the units of coulombs per mole of electrons, and E is in volts. One volt equals 1 joule per coulomb. Thus, the units for electrical work are joules:

$$\text{electrical work} = (\text{mol})(\text{C/mol})(\text{J/C}) = \text{J}$$

The change of free energy of a redox reaction equals the maximum electrical work the reaction can do:

$$\Delta G = -nFE \qquad (20.12)$$

Note that for a spontaneous redox reaction, ΔG is negative and E is positive. When a redox reaction is carried out with its reactants and products in standard state conditions, we can write

$$\Delta G° = -nFE° \qquad (20.13)$$

where $\Delta G°$ is the *standard* free-energy change and $E°$ the *standard* potential for the reaction.

Example 20.10

Calculating $\Delta G°$ from $E°$

$E°$ for the reaction

$$2Al^{3+}(aq) + 3Mg(s) \longrightarrow 2Al(s) + 3Mg^{2+}(aq)$$

is 0.715 V. Calculate the standard free-energy change for the reaction.

SOLUTION: We apply Equation 20.13. Since 6 mol of electrons is transferred from 3 mol of Mg to 2 mol of Al^{3+} in the reaction, $n = 6$. Solving Equation 20.13 yields

$$\Delta G° = -(6 \text{ mol})(9.65 \times 10^4 \text{ C mol}^{-1})(0.715 \text{ J C}^{-1}) = -4.14 \times 10^5 \text{ J}$$

Note that $\Delta G°$ depends on the quantity of charge (the number of moles of electrons) transferred. In contrast, $E°$ does not depend on the quantity of charge. The reaction above is spontaneous under standard conditions as shown by its positive $E°$ and negative $\Delta G°$ values. This means that under standard conditions, magnesium metal can reduce aqueous aluminum ions, but aluminum metal cannot reduce magnesium ions from 1 M aqueous solution. This fact is consistent with the positions of aluminum and magnesium in the activity series of metals (Chapter 18).

Practice Problem 20.8: The standard free-energy change for the reaction

$$H_2(g) + Cu^{2+}(aq) \longrightarrow 2H^+(aq) + Cu(s)$$

is −65.6 kJ. Calculate $E°$ for the reaction.

E° and *K*

Since $\Delta G°$ is related to K, and $E°$ is related to $\Delta G°$, K and $E°$ are also related to each other. We recall that Equation 20.12 relates $\Delta G°$ with K:

$$\Delta G° = -RT \ln K$$

Since $\Delta G°$ is also equal to $-nFE°$ according to Equation 20.13, we can write

$$nFE° = RT \ln K \qquad (20.14)$$

from which

$$E° = \frac{RT}{nF} \ln K$$

At 298 K this relationship can be simplified by substituting the numerical values for R, T, and F:

$$E^\circ = \frac{(8.314 \text{ J mol}^{-1} \text{ K}^{-1})(298 \text{ K})}{(n)(96485 \text{ C mol}^{-1})} \ln K$$

$$= \frac{0.0257 \text{ J C}^{-1}}{n} \ln K$$

$$= \frac{0.0257 \text{ V}}{n} \ln K$$

The natural log of a number equals 2.303 times the common log of the number. For example, log 10.0 = 1.000, but ln 10.0 = 2.303. Thus, $\ln K = 2.303 \log K$. Hence, when $\ln K$ is converted to $\log K$, the expression above becomes

$$E^\circ = \frac{0.0592 \text{ V}}{n} \log K \tag{20.15}$$

Example 20.11

Calculating E° from K

The equilibrium constant for the reaction

$$Zn(s) + Cu^{2+}(aq) \rightleftharpoons Zn^{2+}(aq) + Cu(s)$$

is 1.6×10^{37} at 25°C. Calculate E° for the reaction.

SOLUTION: We apply Equation 20.15:

$$E^\circ = \frac{0.0592 \text{ V}}{n} \log K$$

Since $n = 2$,

$$E^\circ = \frac{0.0592 \text{ V}}{2} \log(1.6 \times 10^{37})$$

$$= 1.10 \text{ V}$$

The value obtained for E° in this example is the value measured for the same standard cell illustrated in Figure 17.3.

Practice Problem 20.9: E° for the reaction

$$4Fe^{2+}(aq) + O_2(g) + 4H^+(aq) \rightleftharpoons 4Fe^{3+}(aq) + 2H_2O(l)$$

is 0.459 V. Calculate the value of the equilibrium constant for the reaction.

20.6 Spontaneity and Speed of Reactions

Although ΔG is a measure of the spontaneity of a reaction, it does not specify how *fast* the reaction might go. For example, the standard free-energy change for the reaction

$$H_2(g) + \tfrac{1}{2}O_2(g) \longrightarrow H_2O(l)$$

is -237.2 kJ mol^{-1}, a highly negative value which indicates that the reaction should be spontaneous at room temperature. But a mixture of hydrogen and oxygen can be mixed at room temperature without reaction. However, when a catalyst such as finely powdered palladium metal is added to the mixture, the gases convert readily into water at room temperature. Thus, the reaction of hydrogen with oxygen at 25°C is an example of a spontaneous reaction that is too slow to observe in the absence of a catalyst.

Furthermore, the entropy change of the reaction of hydrogen with oxygen to form water is negative. This makes the $-T\,\Delta S°$ term in the equation $\Delta G° = \Delta H° - T\,\Delta S°$ positive, causing $\Delta G°$ for the reaction to be less negative than $\Delta H°$. Therefore, the conversion of hydrogen and oxygen to water is less favorable at higher temperatures than at lower temperatures. Yet it is well known that hydrogen and oxygen react with explosive violence without a catalyst at the temperature of a flame from a match. This fact is an example of the difference between the speed of a reaction (kinetics) and its tendency to proceed toward the products as predicted by thermodynamics. We will learn in Chapter 21 that by increasing the temperature, we increase the rate (speed) of a reaction. At higher temperatures, the speed of a reaction can compensate for the decreased tendency for the reaction to proceed as predicted by thermodynamics.

Another example of a process that is predicted from thermodynamics to be spontaneous, but is too slow to observe or to be of any significance, is the conversion of diamond to graphite at room temperature and 1 atm. ΔG for the conversion

$$\text{C(diamond)} \longrightarrow \text{C(graphite)}$$

is negative.

From Table 20.4 we find that $\Delta G_f°$(diamond) = 2.89 kJ mol^{-1}. Graphite is the standard (most stable) state for carbon, and its standard free energy of formation is therefore zero. The standard free-energy change for the conversion of diamond to graphite, C(diamond) $\longrightarrow$ C(graphite), is

$$\begin{aligned}\Delta G° = \Delta G_f°\text{C(graphite)} - \Delta G_f°\text{C(diamond)} &= 0 - 2.89 \text{ kJ mol}^{-1}\\ &= -2.89 \text{ kJ mol}^{-1}\end{aligned}$$

We do not observe diamonds changing to graphite at room temperature because the process is much too slow. However, when diamonds are heated to 1500°C in the absence of air, they convert to graphite in a few minutes.

In diamond, the carbon atoms are much closer together than in graphite. For example, the molar volume of diamond is 3.4 cm^3 mol^{-1} and of graphite 5.5 cm^3 mol^{-1}. (See Section 11.7 for C—C bond lengths in diamond and graphite.) Can graphite be converted to diamond? This was first done in 1955 in a laboratory at the General Electric Company. The process requires a pressure of more than 50,000 atm, temperatures above 2800°C, a long time, and a molten metal catalyst. Thus, the cost of synthetic gem-quality diamonds is high.

We discuss rates or speeds of chemical reactions, and the factors affecting them, in Chapter 21.

Summary

Entropy and the Second Law of Thermodynamics

Entropy is a *measure of the degree of randomness or disorder in a system.* According to the **second law of thermodynamics**, *all spontaneous processes are accompanied by an increase in the entropy of the universe* (the system plus its surroundings):

$$\Delta S_{univ} = \Delta S_{sys} + \Delta S_{sur} > 0 \quad \text{for a spontaneous process}$$

The entropy change of a system is given by the equation

$$\Delta S = S_{final} - S_{initial} = \frac{q_{rev}}{T}$$

The Third Law of Thermodynamics; Calculation of ΔS°_{rxn} from S° Values

According to the **third law of thermodynamics**, *the entropy of a perfect crystal at zero kelvin is zero.* We can use this law to determine the absolute entropies of substances. The standard entropy change for a chemical reaction can be calculated from the absolute standard entropies of the reactants and products using the equation

$$\Delta S^\circ = \Sigma\, S^\circ(\text{products}) - \Sigma\, S^\circ(\text{reactants})$$

where the standard entropy of each substance is multiplied by its coefficient in the balanced equation for the reaction because the entropy values per mole are tabulated.

Spontaneity and Free Energy

The contributions of the enthalpy and entropy changes to the spontaneity of a chemical reaction are specified by the equation defining the Gibbs **free energy change, ΔG**:

$$\Delta G = \Delta H - T\,\Delta S$$

For a spontaneous reaction, ΔG is negative.

The standard free-energy change, ΔG°, for a reaction can be calculated from the standard free energies of formation, ΔG°_f, for each reactant and product just as ΔH° for a reaction can be calculated from the values of ΔH°_f:

$$\Delta G^\circ = \Sigma\, \Delta G^\circ_f(\text{products}) - \Sigma\, \Delta G^\circ_f(\text{reactants})$$

where each ΔG°_f value for a substance is multiplied by its coefficient in the balanced equation because ΔG°_f values per mole are tabulated. ΔG° for a reaction can also be determined by calculating ΔH° and ΔS° for the reaction from appropriate data, followed by substitution of these values into the equation

$$\Delta G^\circ = \Delta H^\circ - T\,\Delta S^\circ$$

Free Energy and Equilibrium

The quantities ΔG and ΔG° are related by the equation

$$\Delta G = \Delta G^\circ + RT \ln Q$$

For a reaction at equilibrium, $\Delta G = 0$, $Q = K$, and

$$\Delta G^\circ = -RT \ln K$$

This equation relates the standard free-energy change to the equilibrium constant for a reaction, making it possible to calculate one of these quantities from the value of the other.

ΔG°, E°, and K

The standard free-energy change for a redox reaction is also related to the standard cell potential for the reaction by the equation

$$\Delta G^\circ = -nFE^\circ$$

The standard cell potential is related to the equilibrium constant by the relationship

$$nFE^\circ = RT \ln K$$

Although the value of ΔG for a reaction can tell us whether or not the reaction is spontaneous, this value does not tell us anything about the speed of the reaction.

New Terms

Entropy (20.1)
Free energy (20.3)
Reversible process (in thermodynamics) (20.1)
Second law of thermodynamics (20.1)
Spontaneous (reaction) (20.1)
Third law of thermodynamics (20.2)

Exercises

General Review

1. What is a "spontaneous process"? Give examples of some spontaneous and of some nonspontaneous processes.
2. What is entropy? Give examples of some systems of low entropy and others of higher entropy.
3. Explain how the entropy of a substance changes during a phase change.
4. State the second law of thermodynamics.
5. Many spontaneous reactions occur with decreasing entropy. How does the statement of the second law of thermodynamics apply to such reactions?
6. How does the entropy change for a system at equilibrium compare with the entropy change of the surroundings?
7. How does increasing the temperature of a system usually affect its entropy (S)?
8. In thermodynamics, what is a "reversible" process?
9. Define and write an equation for the entropy change of a system in terms of the temperature of the system and the amount of heat reversibly absorbed or released by the system.
10. State the third law of thermodynamics.
11. How does the entropy of a substance in the solid state usually compare with: **(a)** the entropy of the substance in liquid state; **(b)** the entropy of the substance in the gaseous state?
12. Explain why the entropy of hydrogen gas is lower than the entropy of water vapor under the same conditions.
13. Write an equation that can be used to calculate the entropy change for a reaction from the given entropies of the reactants and products.
14. What is the Gibbs free energy? Write an equation for the Gibbs free energy of a system in terms of the enthalpy change, the entropy change, and the temperature of the system.
15. What is the relationship between the free-energy change of a reaction and the spontaneity of the reaction?
16. List the signs for ΔH and ΔS favorable for a spontaneous reaction. What are the signs for ΔH and ΔS for a reaction that can definitely be predicted to be nonspontaneous?
17. Write an equation for calculating the standard free-energy change for a reaction from the values of the standard free energies of formation for the reactants and products.
18. Show how you would calculate the standard free-energy change for a reaction from given values of standard enthalpies of formation and standard absolute entropies for the reactants and the products.
19. Explain the relationship between free-energy change and equilibrium.
20. What is the value of ΔG for a reaction at equilibrium?
21. How do the values of the reaction quotient, Q, and the equilibrium constant, K, compare for a reaction at equilibrium?
22. What is the relationship between the standard free-energy change for a reaction and the equilibrium constant?
23. What is the relationship between the standard free-energy change for a redox reaction and the standard electrode potential of the reaction?

Predicting the Sign of ΔS

24. Predict the sign of ΔS for each of the following reactions and give a reason for each prediction: **(a)** $AgCl(s) \rightarrow Ag^+(aq) + Cl^-(aq)$; **(b)** $2Na(s) + Cl_2(g) \rightarrow 2NaCl(s)$; **(c)** $H_2(g) + I_2(g) \rightarrow 2HI(g)$; **(d)** $H_2(g) + I_2(s) \rightarrow 2HI(g)$; **(e)** $2H_2O_2(l) \rightarrow 2H_2O(l) + O_2(g)$.

Entropy Changes in Phase Changes

25. The freezing point of ethanol is −117°C and its enthalpy of fusion, ΔH_{fus}, is 7.61 kJ mol^{-1}. Calculate ΔS for the fusion of 1 mol of ethanol.
26. The enthalpy of fusion of ice at 0.00°C and 1.00 atm is 6.00 kJ mol^{-1}. Calculate ΔS for the process of freezing of 1 mol of water at 0.00°C and 1.00 atm.
27. The standard enthalpy of formation of gaseous bromine from its standard liquid state is 30.7 kJ mol^{-1}, and the standard entropy change for the process $Br_2(l) \rightarrow Br_2(g)$ at the normal boiling point of liquid bromine is 93.0 J mol^{-1} K^{-1}. Assuming that $\Delta H°$ of vaporization is independent of temperature, calculate the normal boiling point of bromine from these data.

$S°$ and $\Delta S°$

28. Using $S°$ values given in Table 20.4, calculate $\Delta S°$ for each of the following reactions: **(a)** $N_2(g) + 2O_2(g) \rightarrow 2NO_2(g)$; **(b)** $CO_2(g) + H_2(g) \rightarrow CO(g) + H_2O(g)$; **(c)** $C_6H_{12}O_6(s) + 6O_2(g) \rightarrow 6CO_2(g) + 6H_2O(l)$; **(d)** $4Al(s) + 3O_2(g) \rightarrow 2Al_2O_3(s)$.
29. The standard entropy change for the reaction $2Al(s) + Fe_2O_3(s) \rightarrow Al_2O_3(s) + 2Fe(s)$ is −41.30 J mol^{-1} K^{-1}. Using $S°$ values given in Table 20.4, calculate the standard absolute entropy for solid iron.

ΔG, ΔH, and ΔS

30. A metal M dissolves in an acid and evolves 12.5 kJ of heat per mole at 25.0°C and 1 atm; the entropy change for the reaction is 1.20 J mol^{-1} K^{-1}. Calculate the free-energy change for this reaction.
31. If ΔH for a reaction at 20.0°C is 40.9 kJ mol^{-1} and ΔS is 83.7 J mol^{-1} K^{-1}, what is ΔG for the reaction? Is this reaction spontaneous? If not, calculate the minimum temperature for this reaction to just become spontaneous, assuming that both ΔH and ΔS are independent of temperature.

Calculating ΔG° from ΔH°f and S° Values

32. Calculate $\Delta G°$ for each of the following reactions using the values for $\Delta H_f°$ and $S°$ listed in Table 20.4: **(a)** $N_2(g) + 2O_2(g) \rightarrow 2NO_2(g)$; **(b)** $CH_4(g) + 2O_2(g) \rightarrow CO_2(g) + 2H_2O(l)$; **(c)** $C_6H_{12}O_6(s) + 6O_2(g) \rightarrow 6CO_2(g) + 6H_2O(l)$; **(d)** $2Al(s) + 3ZnO(s) \rightarrow Al_2O_3(s) + 3Zn(s)$.
33. Calculate $\Delta G°$ for the reaction 2C(graphite) + $H_2(g) \rightarrow C_2H_2(g)$ at 25°C. Is this reaction spontaneous at 25°C and 1 atm? If not, what conditions would you change to make it possible to prepare acetylene (C_2H_2) from graphite and hydrogen?

ΔG° and ΔG°f

34. Calculate $\Delta G°$ for each of the following reactions using the values of $\Delta G_f°$ listed in Table 20.4: **(a)** $2H_2O_2(l) \rightarrow 2H_2O(l) + O_2(g)$; **(b)** $NH_4Cl(s) \rightarrow NH_3(g) + HCl(g)$; **(c)** $2CO_2(g) + H_2O(g) \rightarrow C_2H_2(g) + 2\frac{1}{2}O_2(g)$.
35. For the decomposition of calcium carbonate, $CaCO_3(s) \rightarrow CaO(s) + CO_2(g)$, $\Delta G°$ is 130.2 kJ. Calculate the standard free energy of formation for CaO.
36. For the reaction $2HCl(g) + Br_2(l) \rightarrow 2HBr(g) + Cl_2(g)$, $\Delta G°$ is 84.10 kJ and $\Delta G_f°$ for HCl(g) is −95.27 kJ mol^{-1}. Calculate $\Delta G_f°$ for HBr(g). How does the value obtained compare with that given for HCl in Table 20.4? Explain.

Relation of ΔG to ΔG°

37. For the reaction $H_2(g) + I_2(g) \rightarrow 2HI(g)$, $\Delta G°$ is 2.60 kJ. Calculate ΔG for this reaction at 25.0°C at the instant when the partial pressures of the components of the mixture are as follows: $P_{H_2} = 0.250$ atm, $P_{I_2} = 0.500$ atm, and $P_{HI} = 0.100$ atm. Judging from the value you obtained for ΔG, is the forward or the reverse reaction spontaneous at the instant specified?
38. Calculate ΔG for the reaction $KClO_3(s) \rightarrow KCl(s) + 1\frac{1}{2}O_2(g)$ at 25.0° when the pressure of O_2 is 20.0 atm. At what pressure of O_2 would ΔG be equal to zero?
39. For the reaction $N_2O_4(g) \rightarrow 2NO_2(g)$ at 25°C, the value of $\Delta G°$ is 4.858 kJ, indicating that the conversion of N_2O_4 to NO_2 is not spontaneous. Would this conversion be spontaneous if the pressure of N_2O_4 were maintained at 10.0 atm and NO_2 at 1.00 atm? To substantiate your answer, calculate ΔG for the reaction under these conditions.

Free-Energy Change and Equilibrium

40. The equilibrium constant, K, for the reaction $PCl_5(g) \rightleftharpoons PCl_3(g) + Cl_2(g)$ at 25°C is 8.96×10^{-2}. Calculate the standard free-energy change, $\Delta G°$, for this reaction. What do the values of K and $\Delta G°$ for this system imply about the extent of the forward reaction at 25°C?
41. For the conversion of ethylene to ethane, $C_2H_4(g) + H_2(g) \rightleftharpoons C_2H_6(g)$, K_p at 25°C is 5.04×10^{17}. What is the standard free-energy change, $\Delta G°$, for this reaction?
42. The solubility product constant for silver chloride at 25°C is 1.6×10^{-10}. Calculate $\Delta G°$ for the process $AgCl(s) \rightleftharpoons Ag^+(aq) + Cl^-(aq)$.
43. $\Delta G°$ for the reaction $N_2(g) + O_2(g) \rightarrow 2NO(g)$ is 173.1 kJ. Calculate the value of the equilibrium constant for this reaction at 25°C. What do the values of $\Delta G°$ and K imply about the extent of conversion of nitrogen and oxygen to nitrogen(II) oxide at 25°C?
44. Using the appropriate data from Table 20.4, calculate $\Delta G°$ and K for the reaction $H_2(g) + \frac{1}{2}O_2(g) \rightleftharpoons H_2O(g)$ at 25°C.

E°, ΔG°, and K

45. Using the appropriate values from Tables 17.1 and 20.4, calculate $E°$, $\Delta G°$, and the equilibrium constant, K, at 25°C for each of the following reactions: **(a)** $Zn(s) + Cu^{2+}(aq) \rightleftharpoons Cu(s) + Zn^{2+}(aq)$; **(b)** $Pb(s) + Cu^{2+}(aq) \rightleftharpoons Cu(s) + Pb^{2+}(aq)$; **(c)** $Br_2(l) + 2I^-(aq) \rightleftharpoons I_2(s) + 2Br^-(aq)$; **(d)** $5Fe^{2+}(aq) + MnO_4^-(aq) + 8H^+(aq) \rightleftharpoons 5Fe^{3+}(aq) + Mn^{2+}(aq) + 4H_2O(l)$.
46. The equilibrium constant for the reaction

$$Ag^+(aq) + Fe^{2+}(aq) \rightleftharpoons Ag(s) + Fe^{3+}(aq)$$

at 25°C is 3.0. Calculate $E°$ and $\Delta G°$ for this reaction.
47. The standard reduction potential for the half-reaction

$$IO_3^-(aq) + 6H^+(aq) + 5e^- \longrightarrow \tfrac{1}{2}I_2(s) + 3H_2O(l)$$

is 1.20 V. Using additional necessary data from appropriate sources in this book, calculate the standard potential, the equilibrium constant, and the standard free-energy change for the reaction

$$IO_3^-(aq) + 6H^+(aq) + 5I^-(aq) \rightleftharpoons 3I_2(s) + 3H_2O(l)$$

at 25°C. Is this reaction spontaneous?

CHAPTER 21

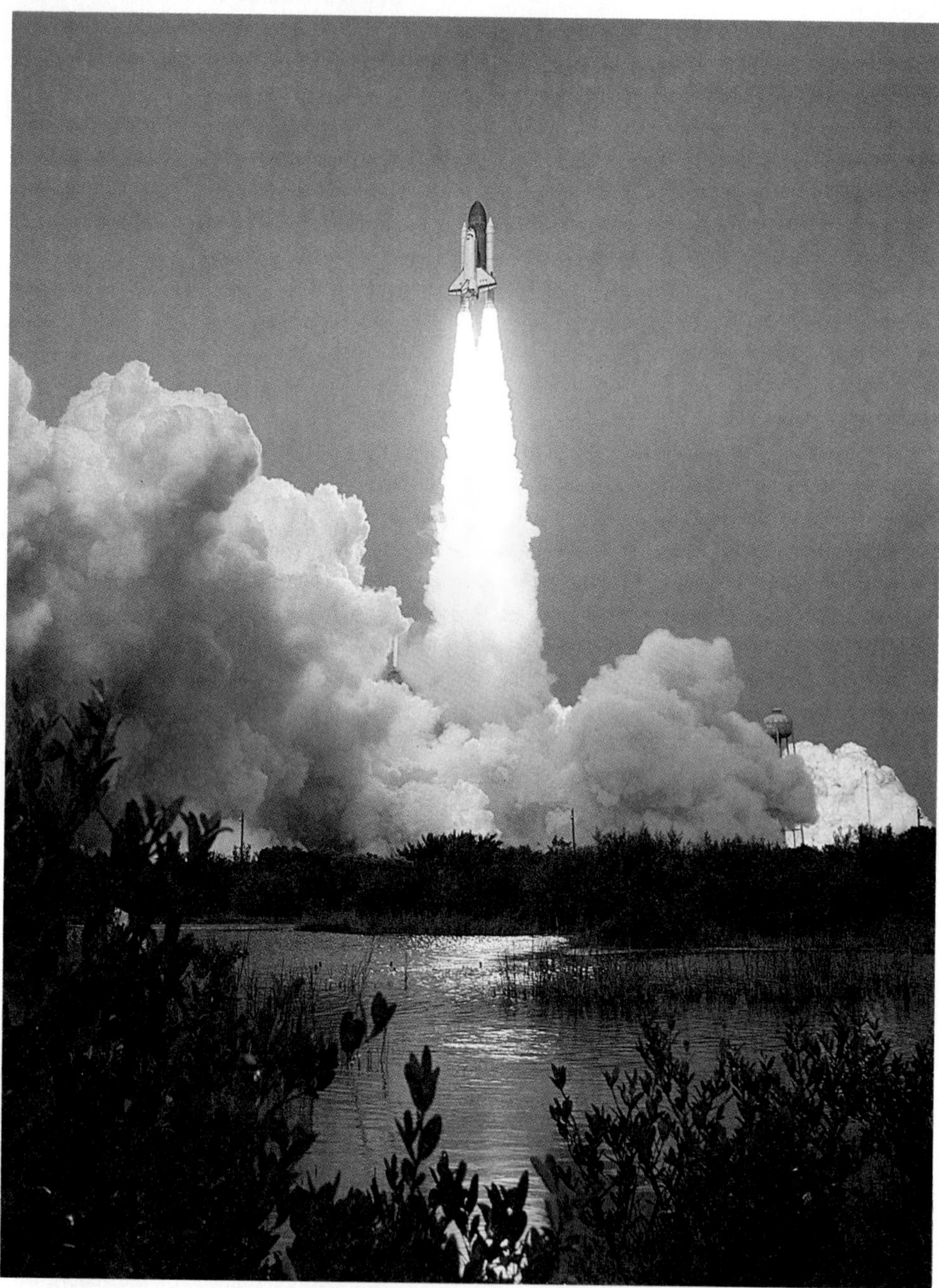

The fuel that propels the Space Shuttle rocket is solid, powdered aluminum mixed with ammonium perchlorate (NH_4ClO_4) and a small amount of iron catalyst in a solid, plastic matrix. When the fuel is ignited, solid Al_2O_3 forms. This Al_2O_3 constitutes the white exhaust material. Various gases also form. The heat of the formation of Al_2O_3 is very high, and this heat causes the gaseous products to expand very rapidly. The expanding gases provide the force that propels the rocket into space.

Chemical Kinetics

In this chapter we discuss the rates *or* velocities *of chemical reactions and the pathways—called* mechanisms*—by which chemical reactions occur. The study of the rates and mechanisms of chemical reactions is known as* **chemical kinetics**. *The study of reaction rates is important in virtually every branch of chemistry, and has many practical applications.*

Some reactions occur in a fraction of a second, others take several minutes, and some take months or years. Reaction rates depend on many variables, of which one of the most important is temperature. Carbon, including diamond, burns at elevated temperatures, but we do not worry about diamond jewelry converting to carbon dioxide at room temperature because the reaction is too slow to observe. Reactions that cause food to spoil occur relatively rapidly at room temperature, but are much slower in a refrigerator at 4°C.

Many common chemical reactions, such as acid–base and precipitation reactions, occur very quickly, as shown by the rapid color change of an indicator or the almost instantaneous formation of a precipitate when some reagents are mixed. However, reactions such as the decomposition of $KClO_3$ *to yield* O_2 *upon heating are slow in the absence of a catalyst.*

21.1 Measurement of Reaction Rates

The **rate of a chemical reaction** equals the change in the number of moles per liter of a product formed, or a reactant consumed, in a given time interval:

$$\text{rate of a chemical reaction} = \frac{\text{change in concentration}}{\text{time}}$$
$$= \frac{\text{moles/liter}}{\text{second}} = \text{mol L}^{-1}\ \text{s}^{-1}$$

Many methods are used to determine reaction rates. For example, when the reactants or products absorb light, the change in the intensity of the light absorbed during the course of the reaction can be measured. The rate of a reaction in the gas phase in which the number of moles of reactants and products are not equal can be measured by the pressure change during the reaction, as shown in Figure 21.1. The change in pressure is related to the change in the number of moles of the products formed or reactants consumed.

Let us consider a simple reaction in which a single reactant, A, decomposes to one product, B:

$$\text{A} \longrightarrow \text{B}$$

The concentration of A decreases during the reaction and the concentration of B increases, as shown graphically in Figure 21.2. As the reaction proceeds, the concentration of reactant decreases and the rate of the reaction decreases.

If we assume that the molar concentration of the product B (Figure 21.2) is M_1 at time t_1, and M_2 at time t_2, the rate of the reaction is

$$\text{rate} = \frac{\text{moles per liter of B formed}}{\text{time interval}} = \frac{M_2 - M_1}{t_2 - t_1} = \frac{\Delta M}{\Delta t}$$

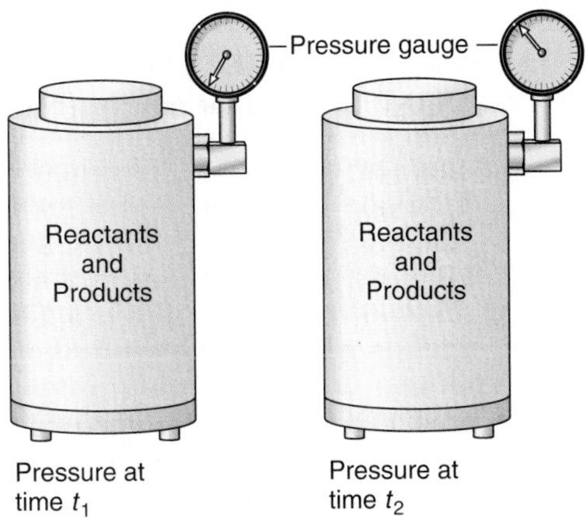

Figure 21.1

Measurement of the rate of the reaction $2N_2O_5(g) \rightarrow 4NO_2(g) + O_2(g)$. The pressure in the reaction vessel is measured at frequent time intervals during the reaction.

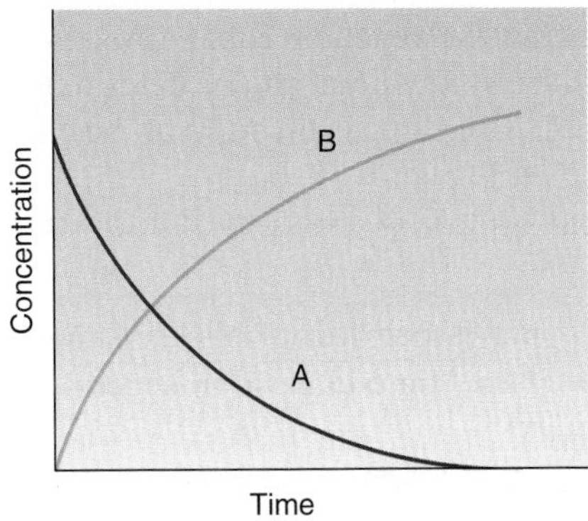

Figure 21.2

Change of concentration of the reactant, A, and the product, B versus time, in the reaction A → B.

The ratio $\Delta M/\Delta t$ equals the change in concentration, ΔM, in the time interval Δt.

An example of a reaction with only one reactant and one product is the conversion of cyclopropane (a fast-acting anesthetic) to propene, a reaction that occurs at elevated temperatures:

$$\underset{\text{Cyclopropane}}{H_2C\overset{CH_2}{\wedge}CH_2(g)} \longrightarrow \underset{\text{Propene}}{H_3C-\overset{H}{\overset{|}{C}}=CH_2(g)}$$

Let us next consider a gas-phase reaction in which one reactant decomposes to two products—the decomposition of nitrogen(V) oxide to nitrogen(IV) oxide and oxygen:

$$2N_2O_5(g) \longrightarrow 4NO_2(g) + O_2(g)$$

In the decomposition of gaseous N_2O_5 to gaseous NO_2 and O_2 at constant temperature, in a container called a "bomb" that maintains a constant volume, the pressure in the bomb increases because the number of moles of the products is greater than the number of moles of the reactant. The increase in pressure is measured at frequent time intervals. If this reaction is carried out in carbon tetrachloride solution, in which both the N_2O_5 and NO_2 are soluble, oxygen bubbles out and the pressure increases as the reaction proceeds. Thus,

Table 21.1

Rate Data for Decomposition of N_2O_5 in CCl_4 at 45°C: $2N_2O_5(g) \longrightarrow 4NO_2(g) + O_2(g)$

Time (s)	$[N_2O_5]$	Time (s)	$[N_2O_5]$
0	1.8×10^{-2}	3000	3.9×10^{-3}
600	1.3×10^{-2}	3600	2.9×10^{-3}
1200	9.3×10^{-3}	4200	2.2×10^{-3}
1800	7.1×10^{-3}	4800	1.7×10^{-3}
2400	5.3×10^{-3}	5400	1.2×10^{-3}

the number of moles of O_2 produced and N_2O_5 decomposed can be calculated in each measured time interval. Table 21.1 lists the molar concentrations of N_2O_5 at different times during its decomposition in carbon tetrachloride solution.

The concentration versus time data listed in Table 21.1 are plotted in Figure 21.3. The rate of disappearance of N_2O_5 (the rate of the reaction) at any given time interval during the reaction equals the slope of the curve at that point. The slope of a curve at any point can be determined by drawing a tangent to the curve at that point. The slope is the ratio of the change along the y axis to the change along the x axis (Figure 21.3).

The rate of disappearance of N_2O_5 during the reaction is given by the expression

$$\text{rate} = \frac{\text{moles per liter of } N_2O_5 \text{ disappearing}}{\text{time interval in seconds}}$$

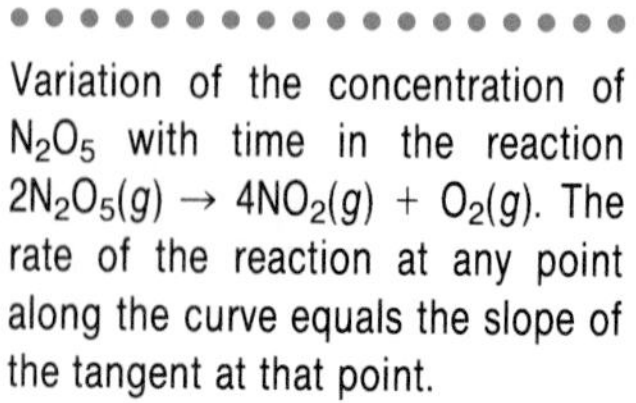

Figure 21.3

Variation of the concentration of N_2O_5 with time in the reaction $2N_2O_5(g) \rightarrow 4NO_2(g) + O_2(g)$. The rate of the reaction at any point along the curve equals the slope of the tangent at that point.

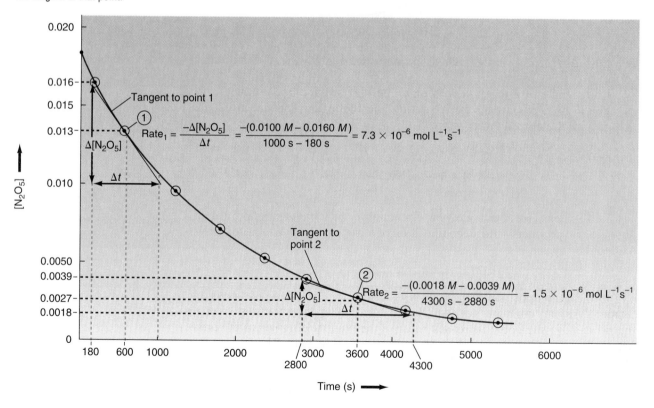

When the rate of a reaction is given as the rate of a reactant whose concentration is decreasing, the rate expression is given a negative sign:

$$\text{rate} = -\frac{\Delta[\text{reactant}]}{\Delta t} = -\frac{[\text{reactant}]_{\text{time 2}} - [\text{reactant}]_{\text{time 1}}}{\Delta t}$$

At point 1 in Figure 21.3, the reaction rate, rate_1, equals the slope of the line at that point:

$$\text{rate}_1 = -\frac{\Delta[N_2O_5]}{\Delta t} = -\frac{(0.0100\ M - 0.0160\ M)}{1000\text{ s} - 180\text{ s}}$$

$$= 7.3 \times 10^{-6}\text{ mol L}^{-1}\text{ s}^{-1}$$

In the expression for the rate of disappearance of a reactant, the difference between the concentration of the reactant at time 2 and that at time 1 is a negative quantity. We therefore write this rate expression with a minus sign so that the rate will be positive.

At point 2, the reaction rate, rate_2, is

$$\text{rate}_2 = -\frac{\Delta[N_2O_5]}{\Delta t} = -\frac{(0.0018\ M - 0.0039\ M)}{4300\text{ s} - 2880\text{ s}}$$
$$= 1.5 \times 10^{-6}\text{ mol L}^{-1}\text{ s}^{-1}$$

As we see from the above calculations and from Figure 21.3, the rate of disappearance of N_2O_5 constantly changes. The rate of a reaction decreases continuously as the reactants are consumed and their concentrations decrease.

Now let us examine the relations of the rate of disappearance of reactants and the rate of formation of products. The rate of formation of a product is written with a positive sign because the concentration of the product at time 2 is greater than the concentration at time 1.

When we consider the reaction, $2N_2O_5(g) \rightarrow 4NO_2(g) + O_2(g)$, we see that when 2 mol of N_2O_5 decompose, 4 mol of NO_2 and 1 mol of O_2 are formed. Thus, at any instant during the reaction, the rate of formation of NO_2 is twice the rate of disappearance of N_2O_5, and the rate of formation of O_2 is one-fourth the rate of formation of NO_2. Also, the rate of formation of O_2 is one-half the rate of disappearance of N_2O_5:

$$\text{rate}_{rxn} = \frac{\Delta[O_2]}{\Delta t} = \left(\frac{1}{4}\right)\frac{\Delta[NO_2]}{\Delta t} = -\left(\frac{1}{2}\right)\frac{\Delta[N_2O_5]}{\Delta t}$$

In general, for a reaction

$$aA + bB \longrightarrow yY + zZ$$

the rate of the reaction is

$$\text{rate} = -\left(\frac{1}{a}\right)\frac{\Delta[A]}{\Delta t} = -\left(\frac{1}{b}\right)\frac{\Delta[B]}{\Delta t} = \left(\frac{1}{y}\right)\frac{\Delta[Y]}{\Delta t} = \left(\frac{1}{z}\right)\frac{\Delta[Z]}{\Delta t} \quad (21.1)$$

Thus, the rate of the reaction $2N_2O_5 \rightarrow 4NO_2 + O_2$ is

$$\text{rate} = -\left(\frac{1}{2}\right)\frac{\Delta[N_2O_5]}{\Delta t} = \left(\frac{1}{4}\right)\frac{\Delta[NO_2]}{\Delta t} = \left(\frac{1}{1}\right)\frac{\Delta[O_2]}{\Delta t}$$

21.2 Reaction Rate and Concentration

Rate Law and Rate Constant

Figures 21.2 and 21.3 show that a change in the rate of a reaction depends on the concentration of the reactants. In general, the rate of the reaction $A \rightarrow B$ is proportional to the molar concentration of the reactant A raised to a power, x. The value of x is determined experimentally:

$$\text{rate} \propto [A]^x$$

This expression of proportionality can be converted to an equality by introducing a proportionality constant, k, called the **rate constant**:

$$\text{rate} = (k)[A]^x \qquad (21.2)$$

An equation that expresses the rate of a reaction in terms of the concentrations of the reactants is called the **rate law** for the reaction. The exponent, x, is called the **order of the reaction** with respect to the reactant in question. When $x = 1$, the reaction is *first order* in the reactant, and when $x = 2$, the reaction is *second order* in that reactant. Third-order reactions are also possible, although they are not frequently encountered. Fractional, zero, and negative orders are also possible. If the rate of a reaction is zero order in a given reactant, the rate of the reaction does not depend on the concentration of that reactant.

For the decomposition reaction of N_2O_5,

$$2N_2O_5(g) \longrightarrow 4NO_2(g) + O_2(g)$$

the rate law is

$$\text{rate} = k[N_2O_5]^x$$

The value of x in this expression must be determined experimentally. This can be done by comparing the concentrations of N_2O_5 and the corresponding reaction rates at different times. *If the ratio of the concentrations at two different times equals the ratio of the corresponding rates, the change in the reaction rate with time is directly proportional to the change in concentration.* Consequently, $x = 1$.

For example, the concentration of N_2O_5 at 600 seconds (point 1 in Figure 21.3) is 4.8 times the concentration at 3600 seconds (point 2):

$$\frac{1.3 \times 10^{-2} \text{ mol L}^{-1}}{2.7 \times 10^{-3} \text{ mol L}^{-1}} = 4.8$$

The ratio of the corresponding rates of the reaction at points 1 and 2 is also nearly the same (within experimental error):

$$\frac{7.3 \times 10^{-6} \text{ mol L}^{-1}\text{ s}^{-1}}{1.5 \times 10^{-6} \text{ mol L}^{-1}\text{ s}^{-1}} = 4.9$$

If we consider the concentrations and rates at other points along the curve in a similar way, we find that the rate varies directly with the concentration of N_2O_5 raised to the first power ($x = 1$). Thus, the decomposition of N_2O_5 is a first-order reaction in N_2O_5,

$$\text{rate} = k[N_2O_5]^1$$

Note that the reaction order of N_2O_5 in this reaction does not equal its coefficient in the balanced equation for the reaction. The coefficient for N_2O_5 in the balanced equation is 2, but the exponent in the rate law is 1. We shall see later that the decomposition of N_2O_5 is not a one-step reaction as the net equation for the reaction might indicate. The net equation for a reaction does not necessarily represent all the steps in the pathway through which the reaction may proceed.

The value of the rate constant, k, can be calculated by solving Equation 21.2 for k if the concentration of the reactant and the order of the reaction with respect to the reactant are known.

Example 21.1

Calculating a Rate Constant for a Reaction

Calculate a rate constant for the reaction $2N_2O_5(g) \rightarrow 4NO_2(g) + O_2(g)$ at 45°C from the information in Figure 21.3.

SOLUTION: A rate constant can be calculated from a known rate law by substituting any known reaction rate and the corresponding reactant concentration into Equation 21.2:

$$\text{rate} = (k)[\text{A}]^x$$

The rate law for the reaction in this problem is

$$\text{rate} = (k)[N_2O_5]^1$$

Substituting the rate at point 1 in Figure 21.3 and the corresponding concentration of N_2O_5 into this expression and solving for the rate constant, k, we obtain

$$k = \frac{\text{rate}}{[N_2O_5]} = \frac{7.3 \times 10^{-6}\ \text{mol L}^{-1}\ \text{s}^{-1}}{1.3 \times 10^{-2}\ \text{mol L}^{-1}} = 5.6 \times 10^{-4}\ \text{s}^{-1}$$

Similarly, at point 2,

$$k = \frac{\text{rate}}{[N_2O_5]} = \frac{1.5 \times 10^{-6}\ \text{mol L}^{-1}\ \text{s}^{-1}}{2.7 \times 10^{-3}\ \text{mol L}^{-1}} = 5.6 \times 10^{-4}\ \text{s}^{-1}$$

The rate constant is independent of the concentrations of the reactants, but varies with temperature. When the rate constant and the rate law for a reaction at a given temperature are known, the rate of the reaction can be calculated at any given concentration of the reactant as shown by Example 21.2.

Example 21.2

Calculating the Rate of a Reaction from the Rate Law

What is the rate of decomposition of N_2O_5 at 45°C at an instant when the concentration of N_2O_5 is 1.5×10^{-3} mol L^{-1}? The rate constant for this first-order reaction at 45°C is $5.7 \times 10^{-4}\ s^{-1}$.

SOLUTION: Substitution of the known values into the first order rate expression yields

$$\begin{aligned}\text{rate} = k[N_2O_5] &= (5.7 \times 10^{-4}\ \text{s}^{-1})(1.5 \times 10^{-3}\ \text{mol L}^{-1}) \\ &= 8.6 \times 10^{-7}\ \text{mol L}^{-1}\ \text{s}^{-1}\end{aligned}$$

Practice Problem 21.1: As in Example 21.2, calculate the rate of decomposition of N_2O_5 at the instant when its concentration is 5.0×10^{-2} mol L^{-1}. Compare your answer with that obtained in Example 21.2. Based on these answers, how does the rate of decomposition of N_2O_5 vary as its concentration decreases?

Determination of Reaction Order

The rate law for a general reaction,

$$aA + bB \longrightarrow cC + dD$$

can be written as follows:

$$\text{rate} = k[A]^x[B]^y \qquad (21.3)$$

In this expression, the reaction is xth order with respect to A and yth order with respect to B. The *overall order of the reaction is the sum of the exponents*, $x + y$. The values of x and y are determined by experiment. In such experiments the concentrations of A and B are varied systematically to learn how the variation affects the reaction rate.

For example, the concentration of one of the reactants, say A, can be doubled while keeping the concentration of B constant. If the rate also doubles, the rate of the reaction varies with the concentration of A to the first power ($x = 1$). Similarly, when the concentration of A is tripled, the rate also triples.

On the other hand, if the concentration of A is kept constant while the concentration of B is doubled and the rate increases four times as a result, the rate varies with the concentration of B raised to the second power ($y = 2$). In this case, the reaction is second order with respect to B. This means that every time the concentration of B is increased by a factor of c, the rate of the reaction increases by the factor of c^2. For example, tripling the concentration of B increases the rate 3^2 or nine times its original value.

Example 21.3

Deriving the Reaction Order and the Rate Law, and Calculating a Rate Constant for a Reaction from Experimental Data

The following data were obtained in a series of experiments for the reaction $2NO(g) + 2H_2(g) \rightarrow N_2(g) + 2H_2O(g)$ at 904°C:

Experiment	Initial [NO]	Initial [H_2]	Rate of Appearance of N_2 (mol L^{-1} s^{-1})
1	0.210	0.122	0.0339
2	0.210	0.244	0.0678
3	0.210	0.366	0.102
4	0.420	0.122	0.136
5	0.630	0.122	0.305

Write the rate law for the reaction.

SOLUTION: To determine the order of a reaction with respect to a given reactant, we examine the relationship between its initial concentration and the rate of the reaction while holding the concentration of the other reactant constant. For this reaction the general expression for the rate law according to Equation 21.3 is

$$\text{rate} = k[\text{NO}]^x[\text{H}_2]^y$$

In experiments 1, 2, and 3, the initial concentration of NO is kept at 0.210 *M*, while the concentration of H_2 in experiment 2 is twice the concentration in experiment 1. As a result, the rate of appearance of N_2 in experiment 2 is twice the rate in experiment 1. In experiment 3, the concentration of H_2 is three times the concentration in experiment 1. The rate of appearance of N_2 in experiment 3 is three times the rate in experiment 1. These results indicate that the rate of the reaction is directly proportional to the concentration of H_2 raised to the first power ($y = 1$) and the reaction is first order in H_2 (Figure 21.4):

$$\text{rate} \propto [\text{H}_2]^1$$

In experiment 4, the initial concentration of H_2 is kept at 0.122 *M* as in experiment 1. The concentration of NO is raised to 0.420 *M* in experiment 4; this is twice the concentration of NO in experiment 1. As a result, the rate of the reaction increases from 0.0339 (the rate in experiment 1) to 0.136. Thus, as the concentration of NO increases by the factor of 2, the rate increases by the factor of 2^2 or 4 (0.136/0.0339 = 4.01).

In experiment 5, the initial concentration of H_2 is kept at 0.122 *M* as in experiment 1. The concentration of NO is raised to 0.630 *M*,

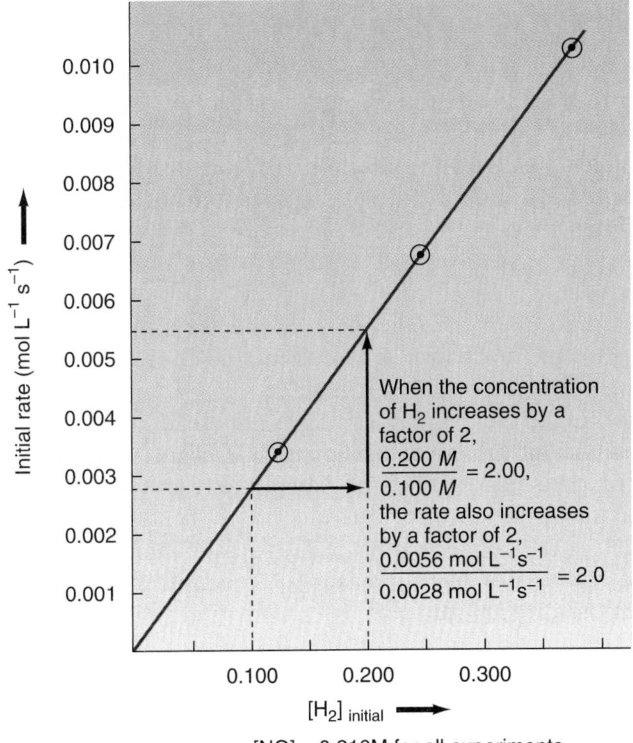

Figure 21.4

A plot of H_2 concentration against the rate of the reaction $2NO(g) + 2H_2(g) \rightarrow N_2(g) + 2H_2O(g)$ when the NO concentration is kept constant.

which is a threefold increase of the concentration of NO in experiment 1. As a result, the rate of the reaction increases by the factor of 3^2, or 9 (0.305/0.0339 = 9.00).

The results in experiments 4 and 5 show that the rate of the reaction is proportional to the *square* of the concentration of NO; $x = 2$ (Figure 21.5). The reaction is therefore second order with respect to NO:

$$\text{rate} \propto [\text{NO}]^2$$

From the information above, the rate law for the reaction is

$$\text{rate} = k[\text{NO}]^2[\text{H}_2]^1$$

Since the sum of the exponents in this rate law is 3, the reaction is third order overall.

The value of the rate constant, k, from the data from experiment 1, is

$$k = \frac{\text{rate}}{[\text{NO}]^2[\text{H}_2]^1} = \frac{0.0339 \text{ mol L}^{-1} \text{ s}^{-1}}{(0.210 \text{ mol L}^{-1})^2(0.122 \text{ mol L}^{-1})}$$
$$= 6.30 \text{ L}^2 \text{ mol}^{-2} \text{ s}^{-1}$$

The rate constant should have an equal value when calculated from other sets of data within the experimental error. For example, the value of k calculated from experiment 2 turns out to be 6.30 $\text{L}^2 \text{ mol}^{-2} \text{ s}^{-1}$, and from experiment 3, 6.32 $\text{L}^2 \text{ mol}^{-2} \text{ s}^{-1}$.

Example 21.4

Using a Rate Law to Predict How the Reaction Rate Changes with Changes in the Concentrations of the Reactants

Consider the rate law in Example 21.3. How is the rate of this reaction affected by each of the following changes? **(a)** The concentration of H_2 is quadrupled; **(b)** the concentration of NO is quadrupled; **(c)** the concentration of H_2 is doubled and simultaneously the concentration of NO is tripled.

SOLUTION

a. Since the rate of the reaction is proportional to the concentration of H_2 to the first power, quadrupling the concentration of H_2 quadruples the rate of the reaction.

b. According to the rate law, the rate of the reaction is proportional to the square of the concentration of NO. Thus, when the NO concentration is quadrupled, the rate of the reaction is increased by 4^2 or 16 times.

c. Doubling the concentration of H_2 and simultaneously tripling the concentration of NO increases the reaction rate by $(2)(3^2)$ or 18 times the original rate.

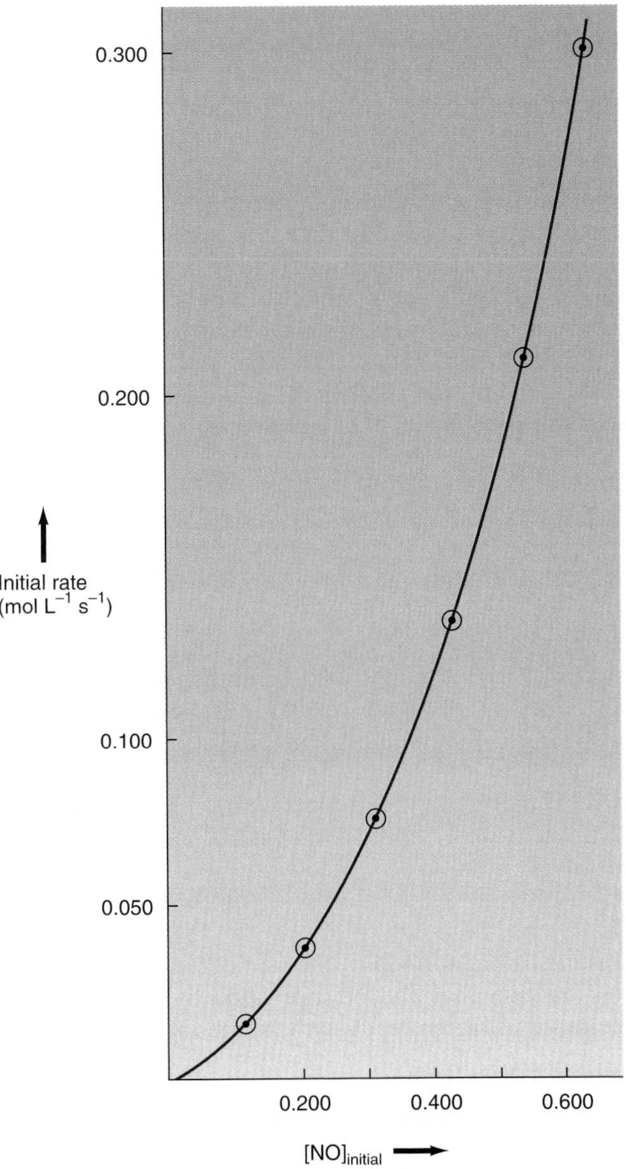

Figure 21.5

A plot of NO concentration against the rate of the reaction $2NO(g) + 2H_2(g) \rightarrow N_2(g) + 2H_2O(g)$ when the H_2 concentration is kept constant.

Practice Problem 21.2: Consider the reaction $2NO(g) + O_2(g) \rightarrow 2NO_2(g)$ and the following experimental data:

Experiment	Initial [NO]	Initial [O_2]	Rate of Appearance of NO_2 (mol L^{-1} s^{-1})
1	1×10^{-3}	1×10^{-3}	7×10^{-6}
2	1×10^{-3}	2×10^{-3}	14×10^{-6}
3	2×10^{-3}	1×10^{-3}	28×10^{-6}

(a) Use these data to write the rate law for the reaction. **(b)** What is the order of the reaction with respect to NO? With respect to O_2? **(c)** What is the overall order of the reaction? **(d)** How does the rate of appearance of NO_2 change when the concentration of O_2 is tripled? When the concentration of NO is tripled?

We can see from the chemical equation and its rate law derived in Example 21.3 that the exponents in a rate law do not necessarily equal the coefficients in a balanced equation. The exponents are determined experimentally. However, the mole ratios of reactants and products in a balanced equation for a reaction can be used to compare the rates of disappearance of reactants or the rates of formation of the products. We recall that the mole ratios of the reactants and products are specified by the coefficients in a balanced equation for the reaction. The use of the coefficients in rate problems is illustrated in Example 21.5b and c.

Example 21.5

Using the Rate Law to Calculate the Rates of Reactants Disappearing and the Rates of Products Forming

In Example 21.3 we determined the rate law for the reaction

$$2NO(g) + 2H_2(g) \longrightarrow N_2(g) + 2H_2O(g)$$

as rate $= k[NO]^2[H_2]$. We also calculated the value of a rate constant, k, as 6.30 $L^2\ mol^{-2}\ s^{-1}$ at 904°C. **(a)** What is the rate of formation of N_2 in the reaction of Example 21.3 when [NO] is 0.155 mol L^{-1} and $[H_2]$ is 0.0986 mol L^{-1}? **(b)** At what rate is NO disappearing when its concentration is 0.155 mol L^{-1} and the concentration of H_2 is 0.0986 mol L^{-1}? **(c)** At what rate is water formed when nitrogen is formed at the rate of 0.136 mol $L^{-1}\ s^{-1}$?

SOLUTION

a. According to Equation 21.1, the rate of formation of N_2 is the rate of the reaction that can be calculated by substituting the given concentrations of the reactants into the rate law expression:

$$\begin{aligned} \text{rate} &= k[NO]^2[H_2] \\ &= (6.30\ L^2\ mol^{-2}\ s^{-1})(0.155\ mol\ L^{-1})^2(0.0986\ mol\ L^{-1}) \\ &= 0.0149\ mol\ L^{-1}\ s^{-1} \end{aligned}$$

b. The balanced equation for the reaction indicates that for each mole of nitrogen formed, 2 mol of NO is consumed. The rate of disappearance of NO is therefore twice the rate of formation of N_2:

$$0.0149\ mol\ N_2\ L^{-1}\ s^{-1}\left(\frac{2\ mol\ NO}{1\ mol\ N_2}\right) = 0.0298\ mol\ NO\ L^{-1}\ s^{-1}$$

c. This part is similar to part (b): For each mole of N_2 formed, 2 mol of H_2O are formed. The rate of formation of H_2O is therefore twice the rate of formation of N_2:

$$0.136\ mol\ N_2\ L^{-1}\ s^{-1}\left(\frac{2\ mol\ H_2O}{1\ mol\ N_2}\right) = 0.272\ mol\ H_2O\ L^{-1}\ s^{-1}$$

Practice Problem 21.3: The rate law for the reaction $2NO(g) + Cl_2(g) \rightarrow 2NOCl(g)$ is rate $= k[NO]^2[Cl_2]$. When [NO] is 0.380 mol L^{-1} and $[Cl_2]$ is 0.380 mol L^{-1}, the rate of the reaction is 5.0×10^{-3} mol L^{-1} s^{-1}. **(a)** Calculate the rate constant for the reaction. **(b)** What is the rate of the reaction (the rate of disappearance of Cl_2) when [NO] is 0.550 mol L^{-1} and $[Cl_2]$ is 0.700 mol L^{-1}? **(c)** At what rate is NOCl appearing when Cl_2 is disappearing at the rate of 4.0×10^{-2} mol L^{-1} s^{-1}?

21.3 Concentration Versus Time Variation in First-Order Reactions

Concentration Versus Time Relationships

In Section 21.1 we showed that the rate of disappearance of a reactant A in a chemical reaction can be expressed by the equation

$$\text{rate} = -\frac{\Delta[A]}{\Delta t}$$

In Section 21.2 we also showed that the rate of disappearance of a reactant A can be related to the concentration of A by the rate law:

$$\text{rate} = k[A]^x$$

We can therefore write

$$\text{rate} = -\frac{\Delta[A]}{\Delta t} = k[A]^x \tag{21.4}$$

If $x = 1$, we can apply calculus to Equation 21.4 (don't worry if you have not learned it yet) to obtain

$$\ln\frac{[A]_0}{[A]_t} = kt \tag{21.5}$$

In Equation 21.5, $[A]_0$ is the initial concentration of A, $[A]_t$ the concentration of A at time t, k the first-order rate constant, and the symbol "ln" represents "natural logarithm." The natural logarithm of a number is the exponent to which the number e (2.71828) has to be raised to obtain the number. Thus, according to Equation 21.5, the natural logarithm of the ratio $[A]_0/[A]_t$ equals kt. This form of the rate law is called an "integrated rate equation" and applies to first-order reactions only.

Using Equation 21.5, we can calculate the concentration of a reactant A in a first-order reaction at any given time, t, if the initial concentration of A, and the rate constant for the reaction, are known. We can also calculate the time for a given concentration of A to decrease to a given lower concentration if we know the rate constant for the reaction.

Example 21.6

Calculating the Concentration of a Reactant That Remains After a Specified Time in a First-Order Reaction

The decomposition of cyclopropane to propene,

$$\underset{H_2C—CH_2(g)}{\overset{CH_2}{\wedge}} \longrightarrow CH_3CH{=}CH_2(g)$$

is a first-order reaction with a rate constant of $1.16 \times 10^{-6}\ s^{-1}$ at 400°C. If the initial concentration of cyclopropane is 5.00×10^{-2} mol L^{-1} and the system is kept at 400°C, what will be its concentration 10.0 h after the beginning of the reaction?

SOLUTION: We apply Equation 21.5. The given data are as follows: $[A]_0 = 5.00 \times 10^{-2}$ mol L^{-1}; $k = 1.16 \times 10^{-6}\ s^{-1}$; $t = 10.0$ h. With these data, the value of ln $[A]_0/[A]_t$ can be calculated:

$$\ln \frac{[A]_0}{[A]_t} = (1.16 \times 10^{-6}\ s^{-1})(10.0\ h)\left(\frac{3600\ s}{1\ h}\right) = 4.176 \times 10^{-2}$$

$$\frac{[A]_0}{[A]_t} = e^{4.176 \times 10^{-2}} = 1.04$$

Solving for $[A]_t$, we obtain

$$[A]_t = \frac{[A]_0}{1.04} = \frac{5.00 \times 10^{-2}\ \text{mol L}^{-1}}{1.04} = 4.81 \times 10^{-2}\ \text{mol L}^{-1}$$

Practice Problem 21.4: The decomposition of N_2O_5 is a first-order reaction. At 40.0°C the rate constant for this reaction in carbon tetrachloride solution is $3.22 \times 10^{-4}\ s^{-1}$. How many seconds will it take for N_2O_5 to decompose from 0.310 M to 1.70×10^{-2} M?

Half-Life

For a first-order reaction, the *time it takes for a reactant to decompose to one-half of its original amount or concentration* is called the **half-life** of the reaction. If the initial concentration of a reactant in a first-order reaction is $[A]_0$, then after the first half-life period, $t_{1/2}$, the concentration is $\frac{1}{2}[A]_0$, after the second half-life period the concentration is $\frac{1}{4}[A]_0$, and so on, as shown in Figure 21.6.

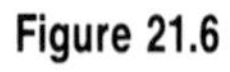

Figure 21.6

The change of the concentration of a reactant A with time in a first-order reaction A → B.

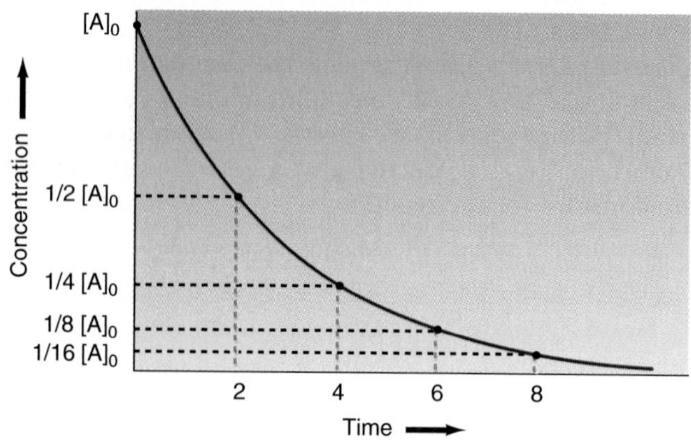

A relationship between the initial concentration of a reactant, $[A]_0$, in a first-order reaction, and the concentration remaining after the first half-life, $\frac{1}{2}[A]_0$, is similar to Equation 21.5:

$$\ln \frac{[A]_0}{\frac{1}{2}[A]_0} = kt_{1/2} \quad (21.6)$$

where $t_{1/2}$ is the half-life. Since the initial concentration of A is twice its concentration after the first half-life, the ratio $[A]_0/\frac{1}{2}[A]_0 = 2$, and ln 2 = 0.693. Equation 21.6 can now be written as

$$0.693 = kt_{1/2} \quad (21.7)$$

Thus, for a first-order reaction the product of the rate constant, k, and the half-life is a constant, 0.693. From this relationship, we can calculate either the rate constant, or the half-life, if the other is known.

Example 21.7

Calculating k, $[A]_t$, and t

The half-life for the first-order reaction $2N_2O_5(g) \rightarrow 4NO_2(g) + O_2(g)$ is 2.40 h at 30.0°C. **(a)** Calculate the rate constant for this reaction at 30.0°C. **(b)** Starting with 2.00 g of N_2O_5, how much of it remains after 9.60 h? **(c)** What length of time will be required for 0.0180 M N_2O_5 to decrease to 0.0100 M?

SOLUTION

a. Solving Equation 21.7 for k, we obtain

$$k = \frac{0.693}{t_{1/2}} = \frac{0.693}{2.40 \text{ h}} = 0.289 \text{ h}^{-1}$$

b. Since one half-life is 2.40 h, 9.60 h corresponds to four half-lives:

$$9.60 \text{ h} \left(\frac{1t_{1/2}}{2.40 \text{ h}}\right) = 4.00t_{1/2}$$

A 2.00-g sample of N_2O_5 will decrease to 1.00 g after the first half-life, to 0.500 g after the second half-life, to 0.250 g after the third half-life, and to 0.125 g after the fourth half-life. However, not all problems of this type lend themselves to this simple step-by-step solution. In most cases, Equation 21.6 must be used. Applying Equation 21.6 to this problem, we obtain

$$\ln \frac{[A]_0}{[A]_t} = kt = (0.289 \text{ h}^{-1})(9.60 \text{ h}) = 2.774$$

$$\frac{[A]_0}{[A]_t} = e^{2.774} = 16.0$$

$$[A]_t = \frac{[A]_0}{16.0} = \frac{2.00 \text{ g}}{16.0} = 0.125 \text{ g}$$

c. In this part of the problem, we are given that $[A]_0 = 0.0180$ M and $[A]_t = 0.0100$ M. Solving Equation 21.5 for t, we obtain

$$t = \frac{\ln([A]_0/[A]_t)}{k} = \frac{\ln(0.0180\ M/0.0100\ M)}{0.289 \text{ h}^{-1}} = 2.03 \text{ h}$$

Practice Problem 21.5: The first-order rate constant for the decomposition of chloroethane to ethene and HCl, $C_2H_5Cl \rightarrow C_2H_4 + HCl$, at 700 K is 2.50×10^{-3} min^{-1}. Calculate: **(a)** the half-life of chloroethane at 700 K; **(b)** the concentration of chloroethane that remains after 1.00 h if the initial concentration is 0.200 mol L^{-1}.

Disintegrations of radioactive substances obey the first-order rate law. We discuss applications of the half-life concept for radioactive substances in Chapter 24.

21.4 The Effect of Temperature on Reaction Rate

We have seen that the rate constant for a reaction is characteristic of the reaction. The rate constant and the rate of a reaction depend on the *nature of the reactants*. The rate of a reaction also depends on the *concentrations* of the reactants, *temperature*, and any *catalyst* present. In this section we consider the temperature dependence of reaction rates, and in the next section we discuss the role of catalysts.

All reaction rates increase with increasing temperature. Foods spoil faster at room temperature than in a refrigerator. Vegetables cook faster in a pressure cooker at 110°C than in an open pot at 100°C. As a rule of thumb, a 10°C increase in temperature approximately doubles the rate of a reaction. The actual variation of reaction rate with temperature must be determined by experiment.

The Arrhenius Equation

An increase in temperature increases the kinetic energy of the reactants (the average speed of the reactant molecules) and the rate constant for a reaction. When the rate constant increases, the rate increases. A quantitative relationship between the rate constant and temperature is provided by the **Arrhenius equation**:

$$k = Ae^{-E_a/RT} \tag{21.8}$$

In this equation, A is a constant, called the frequency factor, which represents the number of collisions of reactant molecules per unit time; e is the base of natural logarithms (2.718); E_a is the activation energy, which we will explain below; R is the gas constant in energy units (8.314 J mol^{-1} K^{-1}); and T is the Kelvin temperature. We can see from Equation 21.8 that an increase in temperature decreases the negative exponent of e. As a result, k increases, and the reaction rate increases.

Collision Theory and Activation Energy

The expression for k in the Arrhenius equation consists of two parts, the frequency factor A, and the quantity $e^{-E_a/RT}$, which represents the fraction of reactant molecules having energy E_a or higher. According to the *collision theory* of reaction rates, the reactant molecules constantly collide with one another, but only a small fraction of the collisions result in the formation of products.

APPLICATIONS OF CHEMISTRY 21.1
Survive Drowning Without Brain Damage

For decades the medical community believed that anyone who stays underwater for more than 3 minutes risks permanent brain damage. The brain, they believed, could not survive more than 3 minutes without oxygen. Also, they said that death was very likely after 6 minutes without oxygen.

In 1975, a Michigan teenager drove his car into an icy pond and remained under water for about 38 minutes. When rescuers found him, medical authorities officially pronounced him dead. His heart and breathing had stopped. However, when rescuers gave him CPR, he regained consciousness and went on to live a normal life.

There are dozens of examples of human beings surviving oxygen deprivation at low temperatures. One recent survivor was a 3-year-old girl in West Virginia. Wearing only a nightgown, she had spent about 5 hours in 27°F weather before friends found her lying in the snow. Because she had no heartbeat and her breathing had stopped, she was clinically dead. But like the Michigan youth, resuscitation efforts paid off: the girl was revived and suffered no brain damage.

Brain damage occurs when a person is deprived of oxygen for more than 3 minutes if normal body temperatures persist during oxygen deprivation. In each of the dozens of documented cases where people emerged unharmed from oxygen deprivation the temperature was less than 0°C. As we just learned, reaction rates (including those involved with life processes) decrease with decreasing temperature. Metabolism slows down, and this reduces the body's need for oxygen.

Most people who survive oxygen deprivation in extreme cold are children. Children's smaller bodies cool down much more quickly than adult bodies. Usually, by the time most larger adult bodies cool down enough to benefit from a decreased oxygen need, they have already been irreversibly damaged from lack of oxygen.

Source: *Science World*, January 29, 1988, page 20.

According to the kinetic-molecular theory (Chapter 10), the colliding molecules of the reactants in a reaction possess different energies at any given temperature. Molecules that have relatively low energies collide and rebound without reacting. Molecules with higher energies may undergo **effective collisions** which result in the formation of products. For collisions to be effective and to lead to products, the colliding reactant molecules must possess a certain *minimum energy in excess of the mean energy of all the reacting molecules.* This minimum *excess energy* is called the **activation energy**, E_a, which is illustrated graphically in Figure 21.7 for a general exothermic reaction A + B → C + D.

Part of the kinetic energy of the colliding molecules A and B (Figure 21.7) is converted to potential energy at the moment of impact. This potential energy, if sufficient for an effective collision, is the activation energy. The curve in Figure 21.7 describes the change in potential energy of the reactants and products during the course of the reaction.

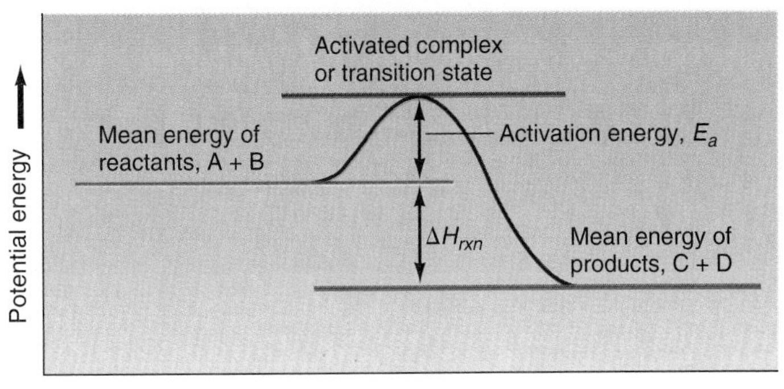

Figure 21.7

Potential energy change during an exothermic reaction A + B → C + D; a graphic illustration of activation energy.

The activation energy can be thought of as the energy needed to break the bonds between the atoms in the reacting molecules so that the atoms can recombine to form new bonds in the product molecules. In this process of bond breaking and bond forming, an unstable combination of atoms is produced. This *unstable combination of atoms* is called an **activated complex**. Its potential energy corresponds to the top of the energy curve in Figure 21.7.

In an activated complex, the bonds in the reactant molecules are about to break while the new bonds in the product molecules are starting to form. An activated complex is a combination of atoms that is in transition between the reactant and the product molecules. This transitory group of atoms is therefore called the **transition state**. The terms "transition state" and "activated complex" are often used interchangeably.

Rate Constants and Reaction Rates at Different Temperatures

The rate of a reaction changes with temperature because the kinetic energy of the reactant molecules changes, and therefore the rate constant changes. We have already mentioned that an increase in temperature, T, decreases the negative exponent of e in Equation 21.8 and, consequently, the rate constant increases. On the other hand, an increase in the activation energy, E_a, increases the magnitude of the negative exponent of e, resulting in a decrease in the rate constant. Since the rate of a reaction is proportional to the rate constant, an increase in temperature or a decrease in activation energy increases the rate of a reaction.

An increase in temperature increases the average kinetic energy and the average speed of the reactant molecules. An increase in kinetic energy of the reactants increases the collision frequency, the number of effective collisions, and the reaction rate. Figure 21.8 illustrates that an increase in temperature increases the fraction of reactant molecules that have the minimum energy needed for a reaction. A decrease in activation energy of a reaction has precisely the same effect: It increases the number of reactant molecules that have enough energy for effective collisions.

The Arrhenius equation can be used to calculate the rate constant for a reaction, and hence the rate of a reaction, at different temperatures. The Arrhenius equation can also be used to determine activation energies of reactions, as shown in the next section.

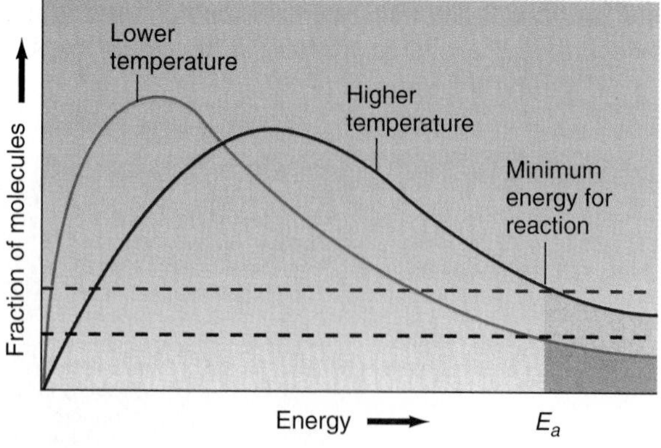

Figure 21.8

The effect of temperature on the number of reactant molecules having the minimum energy needed for reaction.

For most applications, it is convenient to write the Arrhenius equation (Equation 21.8) in a logarithmic form:

$$\ln k = \ln A - \frac{E_a}{RT} \qquad (21.9)$$

Equation 21.9 can be used to calculate the rate constants for a reaction at different temperatures. The factor by which a rate constant increases as a result of temperature increase is also the factor by which the rate of the reaction increases.

Example 21.8

Calculating Increase of Reaction Rate as a Result of Temperature Increase

For a typical first-order reaction, the value of A in the Arrhenius equation is of the order of $1.0 \times 10^{14}\ s^{-1}$ in the temperature range of 0 to 50°C. The activation energy of a certain first-order reaction is 75 kJ mol^{-1}. Calculate the factor by which the rate of this reaction will increase as a result of 10°C increase in temperature from 27°C to 37°C.

SOLUTION: From Equation 21.9 we can calculate the rate constants at the temperatures given. The ratio of the rate constants is also the ratio of the reaction rates because the rate of a reaction is proportional to the rate constant.

$$\text{At 27°C:} \quad \ln k = \ln(1.0 \times 10^{14}) - \frac{7.5 \times 10^4\ \text{J mol}^{-1}}{(8.314\ \text{J mol}^{-1}\ \text{K}^{-1})(300\ \text{K})}$$

$$= 32.24 - 30.07 = 2.17$$

$$k = (2.718)^{2.17} = 8.8\ \text{s}^{-1}$$

Note that k must have the same units as A.

$$\text{At 37°C:} \quad \ln k = \ln(1.0 \times 10^{14}) - \frac{7.5 \times 10^4\ \text{J mol}^{-1}}{(8.314\ \text{J mol}^{-1}\ \text{K}^{-1})(310\ \text{K})}$$

$$= 32.24 - 29.10 = 3.14$$

$$k = 23\ \text{s}^{-1}$$

The value of k at 37°C compared with that at 27°C is

$$\frac{23\ \text{s}^{-1}}{8.8\ \text{s}^{-1}} = 2.6$$

Thus, the rate of the reaction at 37°C is 2.6 times the rate at 27°C.

We can derive an equation that simplifies the type of calculation we performed in Example 21.8. Using the Arrhenius equation (21.8), the ratio of the rate constants, k_1 at temperature T_1 and k_2 at T_2, and thus also the ratio of the reaction rates at these temperatures, can be written as follows:

$$\frac{k_1}{k_2} = \frac{Ae^{-E_a/RT_1}}{Ae^{-E_a/RT_2}} = e^{-E_a/RT_1-(-E_a/RT_2)} = e^{(E_a/R)[(1/T_2-1/T_1)]}$$

Taking the natural logarithm of both sides of the equation, we obtain

$$\ln \frac{k_1}{k_2} = \frac{E_a}{R}\left(\frac{1}{T_2} - \frac{1}{T_1}\right) \quad (21.10)$$

In a similar way we can show that

$$\ln \frac{k_2}{k_1} = \frac{E_a}{R}\left(\frac{1}{T_1} - \frac{1}{T_2}\right) \quad (21.11)$$

Equation 21.10 relates rate constants and their corresponding temperatures to the activation energy. We can see that changing the temperature has the greatest effect on the rates of reactions having large activation energies. Equation 21.10 shows that we do not need to know the frequency factor, A, to calculate the ratio of the rate constants for a reaction at two different temperatures.

Example 21.9

Calculating the Effect of Temperature Change on the Rate of a Reaction

The activation energy of a reaction is 150 kJ mol^{-1}. How does the rate of this reaction change when the temperature increases from 300 K to 310 K?

SOLUTION: Substituting the values given in this problem into Equation 21.11 yields

$$\ln \frac{k_{310}}{k_{300}} = \left(\frac{1.50 \times 10^5 \text{ J mol}^{-1}}{8.314 \text{ J mol}^{-1}\text{ K}^{-1}}\right)\left(\frac{1}{300 \text{ K}} - \frac{1}{310 \text{ K}}\right) = 1.940$$

$$\frac{k_{310}}{k_{300}} = \frac{\text{rate at 310 K}}{\text{rate at 300 K}} = e^{1.940} = 6.96$$

Thus, the rate of a reaction having an activation energy of 150 kJ mol^{-1} increases nearly seven times when the temperature increases by 10 K. The activation energy of the reaction in Example 21.8 was 75 kJ mol^{-1}, which is one-half of the activation energy for the reaction in Example 21.9.

Practice Problem 21.6: The activation energy for the reaction $2N_2O_5(g) \rightarrow 4NO_2(g) + O_2(g)$ is 102 kJ mol^{-1}. The rate constant for the reaction at 318.0 K is 5.0×10^{-4} s^{-1}. Calculate the rate constant at 328.0 K. By what factor does the rate of this reaction change when the temperature increases by 10 K?

Determination of Activation Energy

The Arrhenius equation (21.8) relates the activation energy of a reaction to the rate constant and temperature of the reaction. Thus, the activation energy, E_a, of a reaction can be determined from known values of k and T. This can be done by two different methods. One of these methods is a graphical method (Figure 21.9). The other method consists of calculating the activation energy from Equation 21.10 when k values at two different temperatures are known.

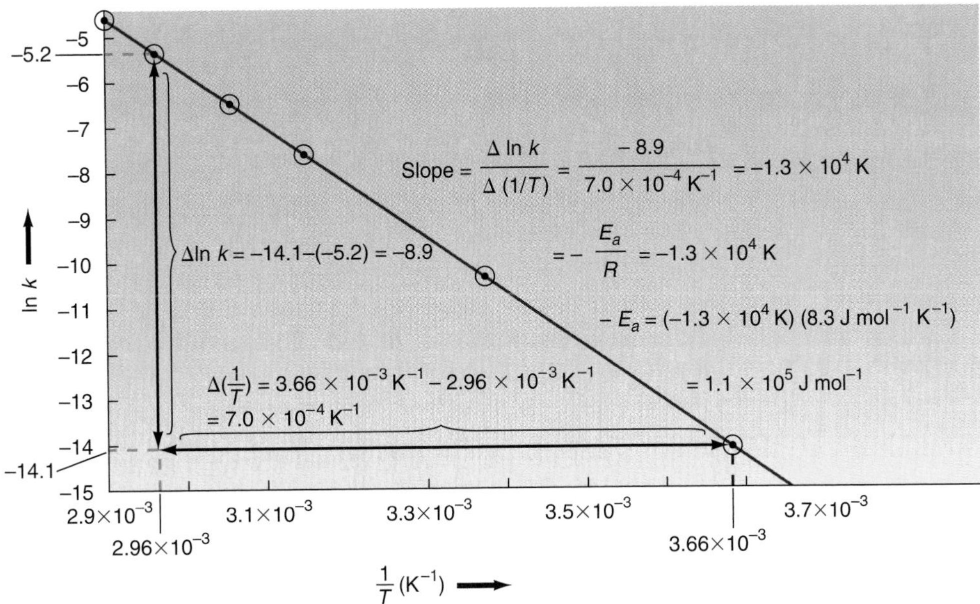

Figure 21.9

A plot of ln k versus 1/T shows how the activation energy can be determined from experimental data. The values shown are for the reaction $2N_2O_5(g) \rightarrow 4NO_2(g) + O_2(g)$.

In a graphic determination of E_a we slightly rearrange the logarithmic form of the Arrhenius equation (21.9):

$$\ln k = -\left(\frac{E_a}{R}\right)\left(\frac{1}{T}\right) + \ln A \qquad (21.12)$$

This rearranged equation has the same form as the equation of a straight line, $y = mx + b$. According to this equation, a straight line is obtained when different values of the variable y are plotted along the y axis and the corresponding values of x are plotted along the x axis. The slope of this line, $\Delta y/\Delta x$, is m, and b is the point at which the line crosses the y axis, the y intercept. In Equation 21.12, $y = \ln k$, $x = 1/T$, m is $-E_a/R$, and b is $\ln A$.

Figure 21.9 illustrates a plot of $\ln k$ along the y axis against $1/T$ along the x axis. The slope of the line obtained is $\Delta(\ln k)/\Delta(1/T)$, which is also $-E_a/R$. The slope is determined from the graph, and E_a is calculated from the value obtained, as shown in Figure 21.9 for the reaction $2N_2O_5(g) \rightarrow 4NO_2(g) + O_2(g)$.

For a nongraphic calculation of E_a, the rate constants for a reaction at two different temperatures must be known. Table 21.2 lists the rate constants for the reaction $2N_2O_5(g) \rightarrow 4NO_2(g) + O_2(g)$ at different temperatures. The activation energy for this reaction can be calculated from two k values and their corresponding temperatures as shown in Example 21.10.

Table 21.2

Rate Constants at Various Temperatures for the Reaction: $2N_2O_5(g) \rightarrow 4NO_2(g) + O_2(g)$

T (K)	$1/T$ (K^{-1})	k (s^{-1})	$\ln k$
273	3.66×10^{-3}	7.78×10^{-7}	−14.067
298	3.36×10^{-3}	3.46×10^{-5}	−10.272
318	3.14×10^{-3}	4.98×10^{-4}	−7.605
328	3.05×10^{-3}	1.50×10^{-3}	−6.502
338	2.96×10^{-3}	4.87×10^{-3}	−5.325

Example 21.10

Calculating E_a for a Reaction from k Values at Two Different Temperatures

The rate constant for the reaction $2N_2O_5(g) \rightarrow 4NO_2(g) + O_2(g)$ at 273 K is $7.78 \times 10^{-7}\ s^{-1}$, and at 338 *K* $4.87 \times 10^{-3}\ s^{-1}$. Calculate the activation energy for the reaction.

SOLUTION: Solving Equation 21.10 for E_a, we obtain:

$$E_a = \frac{\ln(k_1/k_2)(R)}{1/T_2 - 1/T_1}$$

The data given are $k_1 = 7.78 \times 10^{-7}\ s^{-1}$, $k_2 = 4.87 \times 10^{-3}\ s^{-1}$, $T_1 = 273$ K, and $T_2 = 338$ K. Substituting these values into the equation above yields

$$E_a = \frac{\ln\left(\frac{7.78 \times 10^{-7}\ s^{-1}}{4.87 \times 10^{-3}\ s^{-1}}\right)(8.314\ J\ mol^{-1}\ K^{-1})}{\frac{1}{338\ K} - \frac{1}{273\ K}}$$

$$= 1.03 \times 10^5\ J\ mol^{-1} = 103\ kJ\ mol^{-1}$$

This result is in reasonably good agreement with that obtained graphically as shown in Figure 21.9.

Practice Problem 21.7: The rate constant for the reaction $H_2(g) + I_2(g) \rightarrow 2HI(g)$ is $0.0234\ s^{-1}$ at 673 K and $0.750\ s^{-1}$ at 773 K. Calculate the activation energy for this reaction.

21.5 Catalysis

We recall that a catalyst is a substance that increases the rate of a chemical reaction without being consumed by the reaction. The effect of a catalyst is called **catalysis**.

Examples of Catalysis

Potassium chlorate, $KClO_3$, decomposes upon heating to produce solid KCl and O_2 gas. Manganese(IV) oxide, MnO_2, increases the rate of this reaction. After all the $KClO_3$ has decomposed, the original amount of MnO_2 is still present. Thus, MnO_2 is a catalyst. Although none of the catalyst is consumed by the reaction, it must participate in the reaction in some way to increase the reaction rate.

Nearly all chemical reactions in living cells are catalyzed by substances called *enzymes*. Enzyme-catalyzed reactions have rates that are millions of times faster than the corresponding uncatalyzed reactions. For example, an aqueous solution of sucrose—ordinary table sugar—is stable for years (provided that bacterial growth is inhibited). However, when a small amount of an enzyme called sucrase is added to the solution, sucrose is rapidly converted to glucose (blood sugar) and fructose ("fruit sugar," a major component of honey):

$$C_{12}H_{22}O_{11}(aq) + H_2O(l) \xrightarrow{\text{sucrase}} C_6H_{12}O_6(aq) \text{ (glucose)} + C_6H_{12}O_6(aq) \text{ (fructose)}$$

The Role of a Catalyst

A catalyst lowers the activation energy of the reaction that it catalyzes. The details of how some catalysts work are known, those of others are not.

Iodide ions catalyze the decomposition of hydrogen peroxide. This decomposition reaction proceeds in two steps. In the first step, iodide ions react with hydrogen peroxide to form water and hypoiodite ions. The hypoiodite ions formed in the first step are consumed in the second step when they react with excess hydrogen peroxide to form water and oxygen. The sum of these two steps accounts for the decomposition of hydrogen peroxide:

$$H_2O_2(aq) + I^-(aq) \longrightarrow H_2O(l) + IO^-(aq)$$

$$IO^-(aq) + H_2O_2(aq) \longrightarrow H_2O(l) + O_2(g) + I^-(aq)$$

$$2H_2O_2(aq) \xrightarrow{I^-} 2H_2O(l) + O_2(g)$$

The *sequence of individual steps in a net reaction* is called a **reaction mechanism**. In the iodide ion–catalyzed decomposition of hydrogen peroxide, iodide ions are consumed in the first step and regenerated in the second step. Thus, when the reaction is complete, the iodide ions can be recovered. The hypoiodite ion, IO^-, formed in the first step of the reaction cannot be isolated since it is consumed in the second step.

The activation energy for the uncatalyzed decomposition of aqueous hydrogen peroxide is 75.3 kJ mol^{-1}. The activation energy for the same reaction catalyzed by I^- ions is 56.5 kJ mol^{-1}. The activation energies of the iodide ion–catalyzed and uncatalyzed decomposition of hydrogen peroxide are illustrated in Figure 21.10. The ΔH for the reaction is not changed by a catalyst.

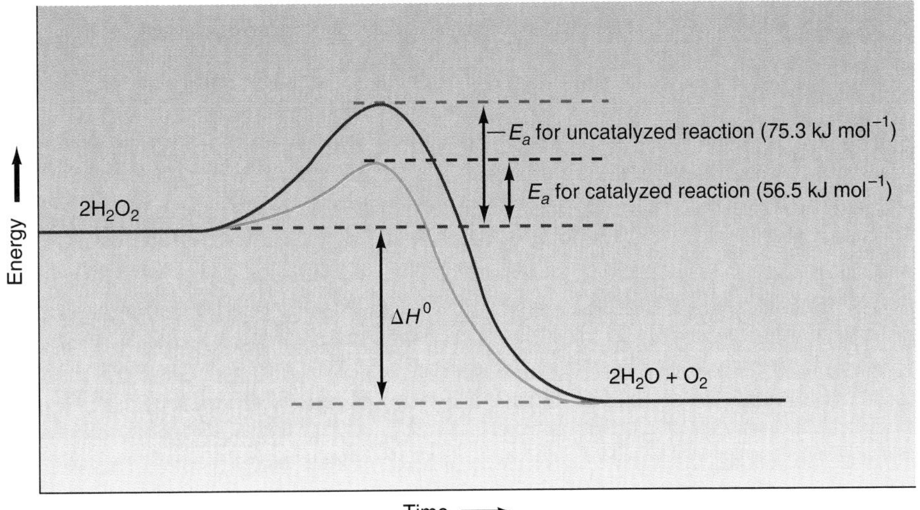

Figure 21.10

Energy change during catalyzed and uncatalyzed decomposition of hydrogen peroxide.

Homogeneous and Heterogeneous Catalysis

When the *reactants and the catalyst for a reaction are in the same phase*, the process is called **homogeneous catalysis**. An example of homogeneous catalysis is the iodide ion–catalyzed decomposition of hydrogen peroxide. The reactant and the catalyst are both present in the same phase—aqueous solution. When the *reactants and the catalyst of a reaction are in different phases*, the catalysis is called **heterogeneous catalysis**.

An example of heterogeneous catalysis is a gaseous reaction in which a finely divided solid metal is used as a catalyst. One example of a metal-catalyzed gaseous reaction is the reaction of hydrogen with oxygen in the presence of finely divided platinum metal. A mixture of hydrogen and oxygen at room temperature can be kept for long periods of time without any noticeable reaction. However, if a stream of hydrogen gas in an open atmosphere is passed over finely divided platinum metal, the platinum starts to glow, indicating that the hydrogen reacts with oxygen from the atmosphere. As the temperature of the glowing platinum increases, the reaction rate increases, and eventually the reaction mixture bursts into flame.

Platinum, nickel, and other metals catalyze gaseous reactions by adsorbing reactant molecules on their surface. As a result, the bonds between the atoms of the reactant molecules weaken, and eventually break, so the atoms can easily regroup to form the product molecules, as illustrated in Figure 21.11 for the nickel-catalyzed reaction of hydrogen with an unsaturated hydrocarbon ethene to convert it to a saturated hydrocarbon ethane (an example of a reaction called hydrogenation). After the reaction, product molecules leave the metal surface, which becomes available for more reactant molecules.

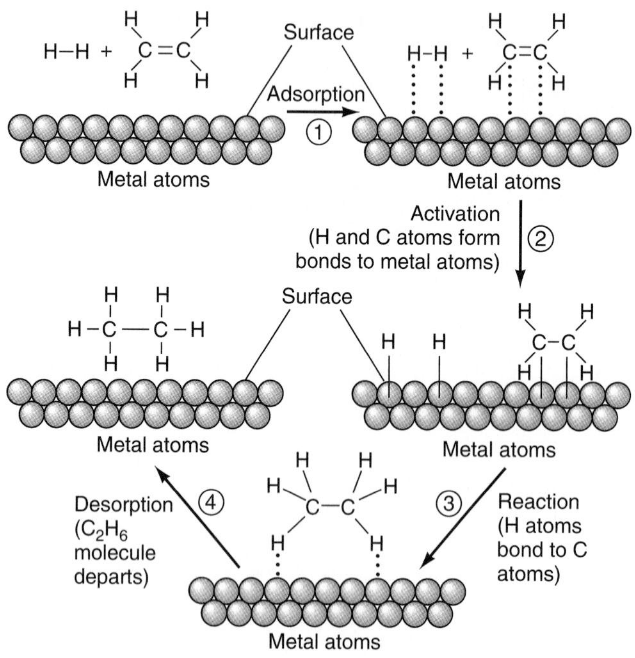

Figure 21.11

Catalysis on a nickel surface for the hydrogenation reaction of ethene to form ethane: $H_2(g) + C_2H_4(g) \rightarrow C_2H_6(g)$. H_2 and C_2H_4 molecules are (1) adsorbed on the nickel surface where the bonds between the H-H atoms are weakened and eventually broken. (2) The carbon atoms of a C_2H_4 molecule become weakly bonded to nickel atoms. (3) The H atoms then add to the carbon atoms of the C_2H_4 molecule to form a C_2H_6 molecule that departs the metal surface, and the surface becomes available for further reactions (4).

Catalytic Converters

Metals that can catalyze gaseous reactions include platinum, nickel, and palladium. Oxides of transition metals such as vanadium can also be used as catalysts. A powdered noble metal such as platinum, or a powdered transition metal oxide is the catalyst in automobile catalytic converters. The powdered catalyst in a catalytic converter is supported on a framework whose shape permits exhaust fumes to pass through and also provides a large surface area for adsorption of gaseous exhaust products. A catalytic converter is illustrated in Figure 21.12.

Catalytic converters react with automobile exhaust gases such as unburned fuel in a gaseous state, including hydrocarbons, carbon monoxide, and nitrogen(II) oxide. Hydrocarbons such as C_8H_{18} (whose common name is isooctane) and carbon monoxide are converted to carbon dioxide in a catalytic converter:

$$2C_8H_{18}(g) + 25O_2(g) \xrightarrow{\text{catalyst}} 16CO_2(g) + 18H_2O(g)$$

$$2CO(g) + O_2(g) \xrightarrow{\text{catalyst}} 2CO_2(g)$$

Nitrogen(II) oxide is converted to nitrogen and oxygen:

$$2NO(g) \xrightarrow{\text{catalyst}} N_2(g) + O_2(g)$$

If a catalyst is to function properly in an automobile, its surfaces must be kept clean. A catalyst that is dirty or "poisoned" is inactive because its surface is coated with adsorbed compounds. Catalytic converters are poisoned by heavy metals. Thus, leaded gasoline cannot be used in modern automobiles.

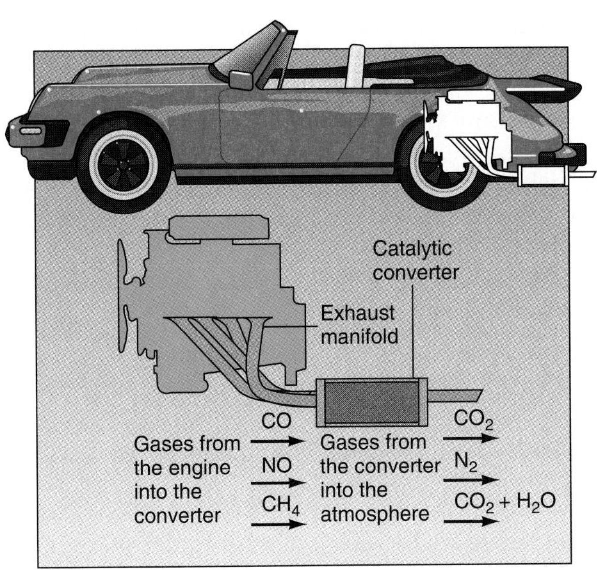

Figure 21.12

A catalytic converter in a Porsche.

21.6 Reaction Mechanisms

We can use the rate law for a reaction to deduce a mechanism for the reaction. The *mechanism* of a reaction consists of one or more steps, called **elementary reactions**. A set of steps, or elementary reactions, constitutes the *pathway* for the overall reaction. Such a pathway describes the order in which bonds are broken and formed, and the positions of atoms and molecules during the reaction.

Elementary Reactions and Molecularity

To propose a mechanism for a reaction, its rate law must first be determined. A mechanism consistent with the rate law can then be proposed, as we explain below.

A reaction that occurs in only one step is an elementary reaction. An example of a one-step reaction is the decomposition of chloroethane to ethene and hydrogen chloride:

$$CH_3CH_2Cl(g) \longrightarrow H_2C{=}CH_2(g) + HCl(g)$$

The rate of this reaction depends only on the concentration of chloroethane:

$$\text{rate} = k[CH_3CH_2Cl]$$

A given chloroethane molecule can decompose without colliding with another molecule of chloroethane. An *elementary reaction in which only one molecule* changes into products is called a **unimolecular reaction**. A unimolecular reaction with the reactant molecule, A, can be written A → products.

An *elementary reaction in which two molecules collide to form products* is a **bimolecular reaction**. A bimolecular reaction can be written A + B → products or A + A → products.

An *elementary reaction in which three molecules simultaneously collide to form products* is a **termolecular reaction**. A termolecular reaction can be written A + B + C → products. Termolecular reactions are not very common because the probability of three molecules colliding at the same time is small. The *number of reactant molecules in an elementary reaction* is called the **molecularity** of the reaction.

Rate Laws and Mechanisms

Let us consider a simple reaction and its mechanism. The reaction

$$O^+(g) + NO(g) \longrightarrow NO^+(g) + O(g)$$

occurs in the upper atmosphere, and its rate law is

$$\text{rate} = k[O^+][NO]$$

According to this rate law, the reaction is first order in each of the reactants, and second order overall. This reaction involves a one-step mechanism that can occur in one of the two ways shown in Figure 21.13.

One of the two possible pathways (1) shown in Figure 21.13 is an atom transfer in which the O^+ ion replaces the O atom in the NO molecule. Another possible pathway (2) is an electron transfer from NO molecule to O^+ ion. Experiments using radioactive $^{18}O^+$ show that the NO^+ ion formed as a product in this reaction does not contain ^{18}O. Instead, the O atom formed as the other

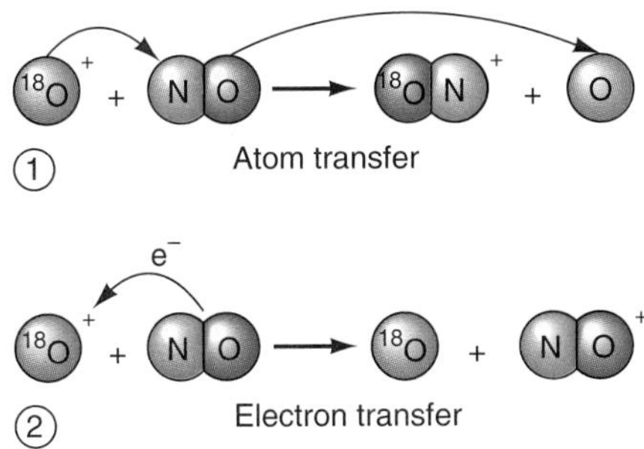

Figure 21.13

Two possible pathways for the reaction $O^+(g) + NO(g) \rightarrow NO^+(g) + O(g)$. The results of experiments using ^{18}O labelling show that the reaction occurs by the atom transfer mechanism, pathway (2).

product of the reaction is the radioactive ^{18}O atom. This atom must be formed from the $^{18}O^+$ ion, and therefore pathway 2 is a possible mechanism for the reaction.

The reaction just described is a bimolecular elementary reaction. It occurs in one step when two particles collide. The rate of the reaction therefore depends on the concentration of each of the two reactants.

We next consider the decomposition reaction of hypochlorite ion, ClO^-. Hypochlorite ions in aqueous solution convert to chlorate ions and chloride ions:

$$3ClO^-(aq) \longrightarrow ClO_3^-(aq) + 2Cl^-(aq)$$

The rate of the reaction is second order in ClO^-:

$$\text{rate} = k[ClO^-]^2$$

According to this rate equation, the reaction cannot be a simple one-step termolecular elementary reaction as the coefficient 3 for ClO^- in the equation for the reaction might suggest.

The following two-step mechanism is consistent with the rate law for the reaction. In the first step, two ClO^- ions collide to form ClO_2^- and Cl^- ions. The ClO_2^- ion formed in the first step is consumed in the second step by collision with a ClO^- ion to form ClO_3^- and Cl^- ions. The sum of these two steps gives the net reaction:

$$\begin{array}{ll} \text{Step 1:} & ClO^-(aq) + ClO^-(aq) \longrightarrow ClO_2^-(aq) + Cl^-(aq) \quad \text{(slow)} \\ \text{Step 2:} & ClO_2^-(aq) + ClO^-(aq) \longrightarrow ClO_3^-(aq) + Cl^-(aq) \quad \text{(fast)} \\ \hline \text{Overall:} & 3ClO^-(aq) \longrightarrow ClO_3^-(aq) + 2Cl^-(aq) \end{array}$$

In a reaction that occurs in two or more steps, the slowest step determines the rate of the overall reaction. The *slowest step of a reaction* is therefore called the **rate-determining step**.

For each elementary reaction (or step) in a reaction mechanism, a rate law can be written. A rate law for an elementary reaction must include the concentration of each of the reactants, raised to the power that equals the coefficient for the reactant in the balanced equation for that step. If the rate law for the rate-determining elementary reaction agrees with the experimentally established rate law for the overall reaction, the proposed mechanism *may* be the correct one.

For the reaction of hypochlorite ions above, the rate law for the first elementary reaction (step 1) is

$$\text{rate} = k[ClO^-][ClO^-] = k[ClO^-]^2$$

This rate law is identical with the experimentally established rate law for the overall reaction. Therefore, if this mechanism is correct, step 1 is the slow, rate-determining step.

When the experimentally established rate law is the same as the rate law for the slowest step in a proposed mechanism, the mechanism may be the correct one, but other possibilities can exist. It is *impossible* to be certain that a proposed mechanism is correct. A mechanism may be correct if it is consistent with the experimentally established rate law and if no experimental evidence demonstrates that it is incorrect.

Example 21.11

Choosing a Possible Mechanism for a Reaction and Determining the Molecularity of the Rate-Determining Step

For the reaction $NO_2(g) + CO(g) \rightarrow NO(g) + CO_2(g)$ below 225°C, the experimental rate law is rate $= k[NO_2]^2$ (note that the reaction is zero order in CO). Which of the following mechanisms is likely to be the correct one for this reaction? What is the molecularity (unimolecular, bimolecular, or termolecular) of the rate-determining step in the likely mechanism?

(1) $NO_2(g) + NO_2(g) \longrightarrow NO_3(g) + NO(g)$ (fast)
$NO_3(g) + CO(g) \longrightarrow NO_2(g) + CO_2(g)$ (slow)

(2) $NO_2(g) + NO_2(g) \longrightarrow N_2O_4(g)$ (slow)
$N_2O_4(g) + CO(g) \longrightarrow N_2O(g) + CO_2(g) + O_2(g)$ (fast)

(3) $NO_2(g) + NO_2(g) \longrightarrow NO_3(g) + NO(g)$ (slow)
$NO_3(g) + CO(g) \longrightarrow NO_2(g) + CO_2(g)$ (fast)

SOLUTION: Mechanism 1 cannot be correct because the slow (rate-determining) step does not agree with the experimental rate law for the reaction.

Mechanism 2 is incorrect because the sum of the elementary reactions:

$$2NO_2(g) + CO(g) \longrightarrow N_2O(g) + CO_2(g) + O_2(g)$$

is not the overall reaction given in this problem.

Mechanism 3 is the only one of the three suggested mechanisms that could be correct because the rate law of the slow, rate-determining step agrees with the experimental rate law, and the sum of the two steps is identical to the overall reaction given. Step 1 is the rate-determining step in this mechanism. This step is a bimolecular elementary reaction because it involves the collision of two reactant molecules.

Practice Problem 21.8: For the reaction $2ICl(g) + H_2(g) \rightarrow 2HCl(g) + I_2(g)$ the rate law is rate $= k[H_2][ICl]$. Which of the following mechanisms is consistent with the proposed rate law? What is the molecularity of the rate-determining step in that mechanism?

(1) $H_2(g) + ICl(g) \longrightarrow HI(g) + HCl(g)$ (fast)
$HI(g) + ICl(g) \longrightarrow HCl(g) + I_2(g)$ (slow)

(2) $H_2(g) + ICl(g) \longrightarrow HI(g) + HCl(g)$ (slow)
$HI(g) + ICl(g) \longrightarrow HCl(g) + I_2(g)$ (fast)

(3) $2ICl(g) + H_2(g) \longrightarrow 2HCl(g) + I_2(g)$ (termolecular elementary reaction)

Example 21.12

Deciding Which Step in a Mechanism Is the Rate-Determining Step and Determining Its Molecularity

For the reaction $2N_2O(g) \longrightarrow 2N_2(g) + O_2(g)$, the rate law is rate = $k[N_2O]$, and a proposed mechanism is

$$\begin{array}{l} N_2O(g) \longrightarrow N_2(g) + O(g) \\ \underline{N_2O(g) + O(g) \longrightarrow N_2(g) + O_2(g)} \\ 2N_2O(g) \longrightarrow 2N_2(g) + O_2(g) \end{array}$$

Which of the two steps in the proposed mechanism is the rate-determining step? What is the molecularity of the rate-determining step?

SOLUTION: The first step is the rate-determining step because it is consistent with the rate law. That is, the rate depends only on the concentration of N_2O. The second step has both dinitrogen oxide and atomic oxygen as reactants. The rate-determining step involves only one molecule and is, therefore, a unimolecular elementary reaction.

Example 21.13

Determining a Rate Law That Is Consistent with a Proposed Mechanism

For the reaction $H_2O_2(aq) + 3I^-(aq) + 2H^+(aq) \rightarrow 2H_2O(l) + I_3^-(aq)$ the following mechanism is proposed:

$$H_2O_2(aq) + H^+(aq) + I^-(aq) \longrightarrow H_2O(l) + HOI(aq) \quad \text{(slow)}$$

$$HOI(aq) + H^+(aq) + I^-(aq) \longrightarrow H_2O(l) + I_2(s) \quad \text{(fast)}$$

$$I_2(s) + I^-(aq) \longrightarrow I_3^-(aq) \quad \text{(fast)}$$

If this mechanism is correct, what is the rate law for the overall reaction? What is the molecularity of the rate-determining step?

SOLUTION: The rate law for the overall reaction must agree with the rate law for the slow, rate-determining step. Thus,

$$\text{rate} = k[H_2O_2][H^+][I^-]$$

If this mechanism is correct, the rate-determining step is a termolecular elementary reaction, that is, first order in each H_2O_2, H^+, and I^-, and third order overall.

Practice Problem 21.9: In Section 21.5 we discussed the iodide ion–catalyzed decomposition of hydrogen peroxide. A possible mechanism for this reaction is

$$H_2O_2(aq) + I^- \longrightarrow H_2O(l) + IO^-(aq) \quad \text{(slow)}$$

$$IO^-(aq) + H_2O_2(aq) \longrightarrow H_2O(l) + O_2(g) + I^-(aq) \quad \text{(fast)}$$

(a) Write an equation for the overall reaction. **(b)** What is the rate law for the overall reaction?

Now that we have learned about reaction mechanisms, we can consider a reaction of great environmental importance. The role of Freon in the conversion of ozone, O_3, to diatomic oxygen, O_2, explains why Freon is no longer used as a propellant in aerosol cans. Freons are a group of stable, noncorrosive compounds used until recently as propellants in aerosol cans. Freons contain

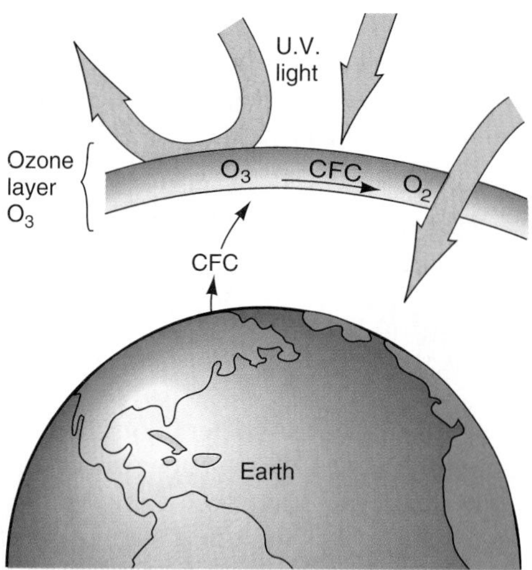

Figure 21.14

Conversion of ozone to diatomic oxygen in the stratosphere. This conversion occurs through several steps and is initiated by the ultraviolet light from the sun and chlorofluorocarbons (CFC) from the earth.

carbon, chlorine, and fluorine, and are therefore called chlorofluorocarbons (CFC).

Freon-11, $CFCl_3$ (trichlorofluoromethane), is used in aerosol cans. Both Freon-11 and Freon-12 (CF_2Cl_2) are used as refrigerants. Freons migrate into the upper atmosphere and cause ozone to be converted to oxygen by a sequence of steps (Figure 21.14). Freon molecules are decomposed by light in the stratosphere (an altitude of about 10 to 50 km) to produce chlorine atoms, shown below for Freon-11:

$$CFCl_3(g) \xrightarrow{\text{light}} CFCl_2(g) + Cl(g)$$

The chlorine atoms have unpaired electrons and are very reactive. They attack ozone molecules, resulting in the following sequence of reactions, which can be repeated over and over again since the chlorine atoms are regenerated in the second reaction:

$$\begin{aligned} &Cl(g) + O_3(g) \longrightarrow ClO(g) + O_2(g) \\ &ClO(g) \xrightarrow{\text{light}} Cl(g) + O(g) \\ &O(g) + O_3(g) \longrightarrow 2O_2(g) \\ \hline &2O_3(g) \longrightarrow 3O_2(g) \end{aligned}$$

A self-sustaining reaction in which a product of one step is a reactant in another step is called a *chain reaction*. The foregoing mechanism for the destruction of ozone is supported by a recent detection of ClO in the stratosphere. The conversion of ozone to oxygen poses an environmental hazard because ozone protects us from ultraviolet light from the sun.

Summary

Measurement of Reaction Rates

Reaction rates can be determined by measuring the concentration of a reactant consumed or a product formed at regular time intervals. The concentration of reactant or product is then plotted against time. If the coefficients for the reactants and products are equal, the reaction rate at any point in time equals the slope of the tangent to the curve at that point:

$$\text{rate} = -\frac{\Delta[\text{reactant}]}{\Delta(\text{time})} = \frac{\Delta[\text{product}]}{\Delta(\text{time})}$$

Factors That Affect Reaction Rates

Factors affecting reaction rates are: (1) the nature of reactants; (2) concentrations of the reactants; (3) temperature; and (4) catalyst.

Rate Law and Order of a Reaction

The rate of a general reaction $aA + bB \rightarrow cC + dD$ can be related to the concentrations of the reactants in an expression called the **rate law** or **rate expression**:

$$\text{rate} = k[A]^x[B]^y$$

where k is a proportionality constant called the **rate constant**, [A] and [B] are the concentrations of the reactants A and B, respectively, and x and y are exponents determined by experiment. The value of x is called the **order of the reaction** with respect to reactant A, and the value of y is the order of the reaction with respect to B. The sum of x and y is the *overall order of the reaction*. The rate of a reaction can be calculated at any given set of concentrations of reactants from the known rate law of the reaction if the value of the rate constant is known.

An expression that relates the concentrations of the reactants with time for a first-order reaction is

$$\ln \frac{[A]_0}{[A]_t} = kt$$

where $[A]_0$ is the initial concentration of the reactant A, $[A]_t$ is the concentration remaining at time t, and k is the rate constant. From this relationship, the expression relating the rate constant, k, with the **half-life**, $t_{1/2}$, of the reactant can be derived:

$$\ln \frac{[A]_0}{\frac{1}{2}[A]_0} = kt_{1/2} \quad \text{or} \quad 0.693 = kt_{1/2}$$

From this relationship, the rate constant for a reaction can be calculated if the half-life of the reactant is known, or the half-life can be calculated from a known value of k.

Arrhenius Equation and Its Applications

The rate constant for a reaction is characteristic of the reaction and depends on the **collision frequency** of the reactant molecules, the **activation energy** of the reaction, and *temperature*.

An expression relating the rate constant to the collision frequency, A, the activation energy, E_a, and the absolute temperature, T, is the **Arrhenius equation**:

$$k = Ae^{-E_a/RT}$$

For most purposes, the Arrhenius equation is used in its logarithmic form:

$$\ln k = \ln A - \frac{E_a}{RT}$$

The Arrhenius equation can be used to determine rate constants at different temperatures, and it can also be used to determine the activation energies of reactions for which the rate constants are known at different temperatures.

Catalysis

A catalyst lowers the activation energy of the reaction it catalyzes. There are two general types of **catalyses**: **homogeneous** and **heterogeneous**.

Reaction Mechanisms

A reaction may occur in one step or in a series of steps called **elementary reactions.** The pathway or the set of steps through which a reaction proceeds is called the **reaction mechanism**. The rate of an overall reaction is determined by the rate of the slowest step in the mechanism, which is, therefore, the **rate-determining step**. An elementary reaction in which only one reactant molecule converts to products is called a **unimolecular reaction**. An elementary reaction in which two reactant molecules collide to form products is called a **bimolecular reaction**, and an elementary reaction in which three reactant molecules simultaneously collide to form products is a **termolecular reaction**.

New Terms

Activated complex (21.4)
Activation energy (21.4)
Arrhenius equation (21.4)
Bimolecular reaction (21.6)
Catalysis (21.5)
Catalyst (21.5)
Effective collision (21.4)
Elementary reaction (21.6)
Half-life (21.3)
Heterogeneous catalysis (21.5)
Homogeneous catalysis (21.5)
Kinetics (introduction)
Molecularity (21.6)

Order of reaction (21.1 and 21.2)
Rate constant (21.2)
Rate-determining step (21.6)
Rate law (or rate expression) (21.2)
Rate of a reaction (21.1)
Reaction mechanism (21.5 and 21.6)
Termolecular reaction (21.6)
Transition state (*see* Activated complex)
Transition state theory (21.4)
Unimolecular reaction (21.6)

Exercises

General Review

1. List three factors that can affect reaction rates.
2. **(a)** What is a rate law? **(b)** Write the general form of the rate law for the reaction $A \rightarrow B$ and for the reaction $A + B \rightarrow C + D$.
3. **(a)** What is meant by the order of a reaction? Considering the general form of the rate expression you wrote in question 2 for the reaction $A + B \rightarrow C + D$, what is the order of this reaction in: **(b)** reactant A; **(c)** reactant B? **(d)** What is the overall order of the reaction?
4. How is the order of a reaction in a given reactant determined?
5. What data would you need to determine the overall order of the reaction $A + B \rightarrow C + D$?
6. What data would you need to determine the rate constant for a reaction?
7. Write an equation for a first-order reaction that relates the initial concentration of a reactant A with its concentration at time t, the rate constant, k, and time t.
8. How can the equation you wrote in question 7 be used? Show how you would rearrange this equation to solve for each of the following: t, k, and the concentration of A remaining at time t.
9. What is the meaning of "half-life"?
10. Write a modified form of the equation in question 7 so that the equation contains $t_{1/2}$ instead of t. What is the numerical value of the product $kt_{1/2}$ in this equation? Show that you can obtain this number on your calculator.
11. Write the general and the logarithmic form of the Arrhenius equation. Identify all the symbols used in the equation, and explain how, according to this equation, the value of the rate constant, and consequently the reaction rate, is affected by varying E_a or by varying T.
12. Explain the term "activation energy."
13. What is: **(a)** collision theory; **(b)** transition state theory?
14. What data would you need to determine the activation energy for a reaction?
15. How does a catalyst increase the rate of a reaction?
16. What two types of catalysis are discussed in this chapter? Give an example of each.
17. What is an elementary reaction?
18. What is the molecularity of a reaction? Illustrate this.
19. Can the rate law for an elementary reaction be written according to the balanced equation for the reaction, or does the rate law have to be established experimentally? Explain.
20. Explain the relationship between the mechanism of a reaction and the rate law for the reaction.

Reaction Rates

21. Acetyl chloride, CH_3COCl, reacts with water to produce acetic acid and hydrogen chloride:

$$CH_3COCl(aq) + H_2O(l) \longrightarrow CH_3COOH(aq) + HCl(aq)$$

The following data were obtained for this reaction:

Time (s)	$[CH_3COCl]$	$[CH_3COOH]$
0	1.20	0
1	1.11	0.08
2	1.05	0.15
3	0.99	0.20
4	0.93	0.27
6	0.81	0.39
8	0.71	0.49
10	0.63	0.57

(a) Make a plot of time versus the concentration of CH_3COCl. **(b)** Determine the rate of disappearance of CH_3COCl at 6 seconds. **(c)** Determine the rate of appearance of CH_3COOH at 6 seconds. **(d)** Determine the average rate of disappearance of CH_3COCl in the time period between 4 and 6 seconds.

22. The rate of disappearance of ozone in the reaction $2O_3(g) \rightarrow 3O_2(g)$ is found to be 6.5 torr s^{-1} during a certain time interval. What is the rate of appearance of oxygen during the same time interval?
23. In the reaction $2NO(g) + O_2(g) \rightarrow 2NO_2(g)$, the rate of appearance of NO_2 is 0.34 torr s^{-1} during a given time interval. What is the rate of disappearance of NO and of O_2 during the same time period?

Rate Law, Rate Constant, and Order of Reaction

24. For the reaction $2HI(g) \rightarrow H_2(g) + I_2(g)$, the following experimental data were obtained at 325°C:

Experiment	[HI]	Rate (mol L^{-1} s^{-1})
1	0.100	2.74×10^{-8}
2	0.200	1.10×10^{-7}
3	0.300	2.47×10^{-7}

(a) From these data, determine the rate law for the reaction, and calculate a value of the rate constant. **(b)** What is the order of the reaction? **(c)** Calculate the rate of this reaction at an instant when the concentration of HI is 4.80×10^{-2} mol L^{-1}. **(d)** If the concentration of HI were increased fivefold, by what factor would the rate increase?

25. In the ozone layer of the upper atmosphere, nitrogen(II) oxide reacts with ozone to form nitrogen(IV) oxide and oxygen: $NO(g) + O_3(g) \rightarrow NO_2(g) + O_2(g)$. The following data were obtained for this reaction at 25°C:

Experiment	[NO]	$[O_3]$	Rate (mol L^{-1} s^{-1})
1	1.00×10^{-6}	9.00×10^{-6}	1.98×10^{-4}
2	2.00×10^{-6}	9.00×10^{-6}	3.96×10^{-4}
3	1.00×10^{-6}	3.00×10^{-6}	6.60×10^{-5}

Use these data to determine: **(a)** the rate law for the reaction; **(b)** the rate constant for the reaction; **(c)** the rate of the reaction when [NO] and $[O_3]$ are each 5.00×10^{-6} *M*.

26. For the reaction $NO_2(g) + CO(g) \rightarrow NO(g) + CO_2(g)$ at 200°C the rate of formation of NO is proportional to the concentration of NO_2 to the second power, and independent of the concentration of CO. **(a)** What is the rate law for the reaction? What is the order of the reaction in NO_2? in CO? What is the overall order? How does tripling the concentration of CO affect the rate of the reaction? tripling the concentration of NO_2? **(b)** Above 225°C the rate of the same reaction is proportional to the concentrations of both NO_2 and CO. Write the rate law for the reaction above 225°C. How would the rate of this reaction be affected by simultaneously doubling the concentrations of both NO_2 and CO?

27. At a certain temperature, the following data are collected for the reaction $2NO(g) + Br_2(g) \rightarrow 2NOBr(g)$:

Experiment	[NO]	$[Br_2]$	Rate (mol L^{-1} s^{-1})
1	0.10	0.10	6.0
2	0.10	0.20	12
3	0.10	0.30	18
4	0.30	0.10	54

Use the data above to determine: **(a)** the rate law for the reaction; **(b)** the order of the reaction in NO, the order in Br_2, and the overall order; **(c)** the rate constant for the reaction; **(d)** the rate of the reaction at an instant when $[NO] = 0.75$ *M* and $[Br_2] = 0.26$ *M*. **(e)** How many times does the rate of this reaction increase when the concentration of NO is tripled? when the concentration of Br_2 is tripled? when the concentrations of both NO and Br_2 are tripled simultaneously?

28. For the reaction $CO(g) + Cl_2(g) \rightarrow COCl_2(g)$, the rate law is rate $= k[CO][Cl_2]^{3/2}$. **(a)** What is the rate constant for this reaction if the concentrations of CO and Cl_2 are each 0.10 *M*, and when the rate of formation of $COCl_2$ is 6.2×10^{-3} mol L^{-1} min^{-1}? **(b)** What is the rate of the reaction when the concentration of each reactant is 0.20 *M*?

Variation of Concentration with Time in First-Order Reactions

29. The decomposition of hydrogen peroxide to water and oxygen is a first-order reaction with a rate constant of 5.0×10^{-6} s^{-1} at a certain temperature. How long will it take at that temperature for a 0.820 *M* solution of H_2O_2 to decrease to 0.500 *M*?

30. The reaction $SO_2Cl_2(l) \rightarrow SO_2(g) + Cl_2(g)$ is a first-order reaction with a rate constant of 2.2×10^{-5} s^{-1} at 320°C. What percent of SO_2Cl_2 is decomposed after heating the substance at 320°C for 1.0 h?

31. A reaction that is first order has 50 percent of the reactant remaining after 2.0 h at a certain temperature. What is the rate constant for the reaction at that temperature?

32. The half-life for the conversion of cyclobutane to ethene

```
   H  H
   |  |                  H  H
H—C—C—H                  |  |
   |  |        ——→      2C═C
H—C—C—H                  |  |
   |  |                  H  H
   H  H
Cyclobutane(g)      Ethene(g)
```

is 8.0×10^{-3} s at 1000°C. **(a)** What is the rate constant of this reaction at 1000°C? **(b)** If you started with 0.90 g of cyclobutane, how much would remain after 0.010 s?

Reaction Rate, Temperature, and Activation Energy

33. In an acid solution, sucrose, $C_{12}H_{22}O_{11}$, converts to the simpler sugars glucose, $C_6H_{12}O_6$, and fructose, $C_6H_{12}O_6$. (Glucose and fructose have different structural formulas.)

$$\underset{\text{Sucrose}}{C_{12}H_{22}O_{11}(aq)} + H_2O(l) \longrightarrow \underset{\text{Glucose}}{C_6H_{12}O_6(aq)} + \underset{\text{Fructose}}{C_6H_{12}O_6(aq)}$$

At 27.0°C the rate constant for this reaction is 2.12×10^{-4} L mol^{-1} s^{-1}, and at 37.0°C it is 8.46×10^{-4} L mol^{-1} s^{-1}. **(a)** Calculate the activation energy for this reaction; **(b)** calculate the frequency factor, A, at 37.0°C. **(c)** By what factor does the rate of this reaction increase when the temperature is raised from 37.0°C to 47.0°C?

34. For the hydrogenation of ethene to produce ethane, $C_2H_4(g) + H_2(g) \rightarrow C_2H_6(g)$, the activation energy is 181 kJ mol^{-1}, and the rate constant at 700 K is 1.3×10^{-3} L mol^{-1} s^{-1}. **(a)** Calculate the rate constant at 720 K. **(b)** By what factor does the rate of the reaction increase when the temperature is increased from 700 K to 720 K?

35. The thermal decomposition of dimethyl ether occurs according to the following equation: $(CH_3)_2O(g) \rightarrow CH_4(g) + H_2(g) + CO(g)$. The frequency factor for this reaction is 8.91×10^{21} min^{-1}, and at 500°C the rate constant is 0.0246 min^{-1}. Calculate the activation energy for this reaction.

36. The rate of a reaction is 1.2 mol L^{-1} min^{-1} at 20°C. **(a)** If the rate of this reaction is exactly doubled by a 10°C increase in temperature, what is the activation energy of the reaction? **(b)** What would be the activation energy if the rate were tripled by a 10°C temperature increase?

Catalysis

37. The activation energy of a reaction is 180 kJ mol^{-1} at 300 K. **(a)** A catalyst decreases the activation energy of the reaction to 150 kJ mol^{-1}. By what factor does the rate of the reaction increase as a result of using this catalyst at 300 K? **(b)** Another catalyst decreases the activation energy of the reaction to 90.0 kJ mol^{-1}. By what factor does the rate increase as a result?

38. The activation energy of a reaction is 50 kJ mol^{-1} at 25°C. Adding a catalyst increases the rate of the reaction by a factor of 10^6. Calculate the activation energy of the reaction in the presence of the catalyst.

Reaction Mechanisms

39. Assuming that each of the following general reactions is an elementary reaction (a reaction occurring in one step exactly as written), what is the molecularity of each of the reactions: **(a)** A + B → C; **(b)** A + B → C + D; **(c)** A → B; **(d)** A → C + D; **(e)** A + A + A → C + D; **(f)** A + B + C → D + E.

40. The reaction of carbon dioxide with hydroxide ions in an aqueous solution in which the OH^- ion concentration is greater than 10^{-4} M occurs in two steps:

Step 1: $CO_2(g) + OH^-(aq) \longrightarrow HCO_3^-(aq)$ (slow)

Step 2: $HCO_3^-(aq) + OH^-(aq) \longrightarrow CO_3^{2-}(aq) + H_2O(l)$ (fast)

Overall: $CO_2(g) + 2OH^-(aq) \longrightarrow CO_3^{2-}(aq) + H_2O(l)$

(a) Write the rate law for the overall reaction; **(b)** What is the rate of disappearance of carbon dioxide in an aqueous solution in which the OH^- ion concentration is 3.0×10^{-4} M, and the CO_2 concentration is 0.40 g L^{-1}? Assume that the reaction occurs at the temperature at which the rate constant is 8.400×10^3 L mol^{-1} s^{-1}. **(c)** What is the molecularity of the rate-determining step in this reaction?

41. Nitrogen(II) oxide from jet engines reacts with ozone in the stratosphere to produce nitrogen(IV) oxide and oxygen:

$$NO(g) + O_3(g) \longrightarrow NO_2(g) + O_2(g)$$

The rate law for this reaction is rate = $k[NO][O_3]$. Which of the following mechanisms is consistent with this rate law?

(a) $NO + NO \longrightarrow N_2O_2$ (slow)
$N_2O_2 + O_3 \longrightarrow N_2O_3 + O_2$ (fast)
$N_2O_3 \longrightarrow NO + NO_2$ (fast)
$NO + O_3 \longrightarrow NO_2 + O_2$

(b) $NO \longrightarrow N + O$ (slow)
$O_3 + O \longrightarrow 2O_2$ (fast)
$O_2 + N \longrightarrow NO_2$ (fast)
$NO + O_3 \longrightarrow NO_2 + O_2$

(c) $NO + O_3 \longrightarrow NO_3 + O$ (slow)
$NO_3 + O \longrightarrow NO_2 + O_2$ (fast)
$NO + O_3 \longrightarrow NO_2 + O_2$

42. A possible mechanism for the reaction $3I^-(aq) + S_2O_8^{2-}(aq) \rightarrow 2SO_4^{2-}(aq) + I_3^-(aq)$ is

(1) $I^- + S_2O_8^{2-} \longrightarrow IS_2O_8^{3-}$ (slow)
(2) $IS_2O_8^{3-} \longrightarrow 2SO_4^{2-} + I^+$ (fast)
(3) $I^+ + I^- \longrightarrow I_2$ (fast)
(4) $I_2 + I^- \longrightarrow I_3^-$ (fast)

What rate law is consistent with this mechanism?

43. The rate law for the reaction $CO(g) + NO_2(g) \rightarrow CO_2(g) + NO(g)$ at 200°C is rate = $k[NO_2]^2$. Is the following mechanism consistent with this rate law?

$2NO_2 \longrightarrow N_2 + 2O_2$ (slow)

$2CO + O_2 \longrightarrow 2CO_2$ (fast)

$N_2 + O_2 \longrightarrow 2NO$ (fast)

PART 5

Some Special Branches of Chemistry

CHAPTER 22

Solutions of some coordination compounds. Coordination compounds are known for the characteristically vivid colors of both their aqueous solutions and their solid states.

Coordination Chemistry

We discussed coordination complexes in Section 16.5. In Chapter 19 we considered applications of coordination complexes in qualitative analysis. We recall that a coordination complex consists of a metal atom or ion which is covalently bonded to one or more atoms or groups of atoms called ligands. In the formation of a coordination complex, the metal ion acts as a Lewis acid and the ligands act as Lewis bases. A coordination complex in solution is in equilibrium with its component metal ions and its ligands. Many stable coordination complexes exist for transition metals. Representative metals form fewer complexes.

22.1 COORDINATION COMPLEXES

Historical Development and Terminology

Coordination complexes of transition metals were studied by the Swiss chemist Alfred Werner around the turn of the century. His work led to modern theories of coordination chemistry. In 1913 Werner received the Nobel Prize in chemistry for this work.

Werner investigated a series of compounds which contained cobalt, chlorine, and ammonia. These colored compounds were thought to have the following empirical formulas:

$CoCl_3 \cdot 6NH_3$ (orange-yellow)
$CoCl_3 \cdot 5NH_3$ (violet)
$CoCl_3 \cdot 4NH_3$ (green)
$CoCl_3 \cdot 3NH_3$ (green)

The dots in these formulas symbolize unspecified bonding between the NH_3 molecules and the rest of the compound.

Coordination complexes in solution can be studied by measuring the electrical conductivity of the solution. The electrical conductivity of a solution depends on the number of moles of ions in the solution. Electrical conductivity measurements of dilute aqueous solutions of $CoCl_3 \cdot 6NH_3$, $CoCl_3 \cdot 5NH_3$, $CoCl_3 \cdot 4NH_3$, and $CoCl_3 \cdot 3NH_3$ suggest that these compounds contain, 4, 3, 2, and 0 mol of ions, respectively, per mole of the compound (Table 22.1). Thus, three of the compounds are probably ionic and one is molecular.

Table 22.1

Data Leading to the Establishment of Structures for Coordination Compounds Containing Cobalt, Chlorine, and Ammonia

Old Formula	Moles of Ions per Mole of Compound as Shown by Conductance	Moles of AgCl Readily Precipitated with $AgNO_3$ per Mole of the Compound	Werner Formula
$CoCl_3 \cdot 6NH_3$	4	3	$[Co(NH_3)_6]Cl_3$
$CoCl_3 \cdot 5NH_3$	3	2	$[Co(NH_3)_5Cl]Cl_2$
$CoCl_3 \cdot 4NH_3$	2	1	$[Co(NH_3)_4Cl_2]Cl$
$CoCl_3 \cdot 3NH_3$	0	0	$[Co(NH_3)_3Cl_3]$

To establish the identities of the components of these compounds and their formulas, the compounds are dissolved in water and the resulting solutions are treated with excess silver nitrate. Any chloride ions in the solution of a given cobalt compound precipitate immediately with silver ions to give solid silver chloride. A mole of $CoCl_3 \cdot 6NH_3$ reacts with 3 mol of silver ions; a mole of $CoCl_3 \cdot 5NH_3$ reacts with 2 mol of silver ions; a mole of $CoCl_3 \cdot 4NH_3$ reacts with 1 mol of silver ions; and $CoCl_3 \cdot 3NH_3$ does not immediately react with silver ions. These data are also summarized in Table 22.1.

Since a mole of $CoCl_3 \cdot 6NH_3$ in solution reacts with 3 mol of silver ions, there must be 3 mol of chloride ions per mole of the compound. Electrical conductivity measurements show that a mole of $CoCl_3 \cdot 6NH_3$ produces 4 mol of ions in solution. Werner proposed that the fourth ion consists of a central cobalt ion surrounded by six ammonia molecules (Figure 22.1). These ammonia molecules are *ligands* that are bonded to the central metal ion by coordinate covalent bonds (Chapter 7). The formula for this complex ion is $Co(NH_3)_6{}^{3+}$. The *central metal ion with its attached ligands* is called the **coordination sphere**.

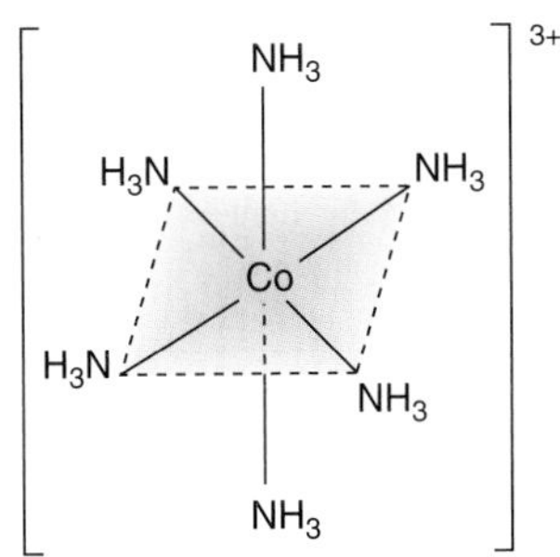

Figure 22.1

Octahedral structure of a $Co(NH_3)_6{}^{3+}$ ion.

In the neutral compound with the empirical formula $CoCl_3 \cdot 6NH_3$, each $Co(NH_3)_6{}^{3+}$ ion is associated with three chloride ions. Thus, the formula for the compound is $[Co(NH_3)_6]Cl_3$. The coordination sphere in a formula of a coordination complex is enclosed in brackets.

Table 22.1 indicates that an aqueous solution of 1 mol of $CoCl_3 \cdot 5NH_3$ readily reacts with excess $AgNO_3$ to produce 2 mol of AgCl. Therefore, two Cl^- ions must be outside of the coordination sphere. A mole of $CoCl_3 \cdot 5NH_3$ therefore consists of 1 mol of $Co(NH_3)_5Cl^{2+}$ ions and 2 mol of Cl^- ions. Since a Cl^- ion and a Co^{3+} ion are a part of the coordination sphere, the charge of the complex ion is 2^+. The formula of the compound is $[Co(NH_3)_5Cl]Cl_2$.

A mole of $CoCl_3 \cdot 4NH_3$ reacts immediately with excess $AgNO_3$ to produce only 1 mol of AgCl (Table 22.1). Therefore, only one Cl^- ion is outside the coordination sphere, and the formula of the complex ion is $[Co(NH_3)_4Cl_2]^+$. Since there are two Cl^- ions in the coordination sphere, the complex ion has a charge of 1+. The formula of the compound is $[Co(NH_3)_4Cl_2]Cl$.

Table 22.1 reveals that an aqueous solution of $CoCl_3 \cdot 3NH_3$ does not conduct electric current, and it does not immediately produce a precipitate with $AgNO_3$. This means that the compound consists of neutral coordination spheres with the formula $[Co(NH_3)_3Cl_3]$.

Example 22.1

Determining the Formula and the Charge of a Complex Cation from Experimental Data

One of the coordination compounds of platinum ion has the empirical formula $PtCl_4(NH_3)_4$. When 1 mol of this compound is dissolved in water and treated with excess $AgNO_3$, 2 mol of AgCl is formed immediately. What are the formula and the charge of the complex cation (the coordination sphere) in this compound?

SOLUTION: Since 1 mol of the coordination compound gives 2 mol of AgCl, two Cl^- ions must be outside the coordination sphere. The other two Cl^- ions are constituents of the coordination sphere. They are covalently bonded to platinum and cannot therefore be easily precipitated with silver nitrate. The two Cl^- ions coordinated to the central platinum(IV) ion decrease the charge of the complex ion from 4+ to 2+. NH_3 is electrically neutral and does not affect the charge of the complex ion. Thus, the formula of this complex cation is $[Pt(NH_3)_4Cl_2]^{2+}$, and the formula of the compound is $[Pt(NH_3)_4Cl_2]Cl_2$. A mole of this compound dissociates in water to 3 mol of ions according to the equation

$$[Pt(NH_3)_4Cl_2]Cl_2(s) \longrightarrow Pt(NH_3)Cl_2^{2+}(aq) + 2Cl^-(aq)$$

Practice Problem 22.1: A coordination compound has a composition that can be expressed by the empirical formula $CuBr_2(H_2O)_4$. When an aqueous solution of this compound is treated with excess silver nitrate, 2 mol of silver bromide immediately precipitate per mole of the compound. Write the formula and the charge of the complex cation in this compound. What is the formula for the compound?

Coordination complexes can be cations such as $Cu(NH_3)_4^{2+}$, anions such as $Fe(CN)_6^{3-}$, or neutral, molecular compounds like $Co(NH_3)_3Cl_3$. Complex cations can combine with simple anions to form salts such as $[Cu(NH_3)_4]Cl_2$. Simple cations can combine with complex anions to give salts like $K_3[Fe(CN)_6]$. Complex cations are also able to combine with complex anions. For example, $Cu(NH_3)_4^{2+}$ ions can combine with $Fe(CN)_6^{3-}$ ions to form $[Cu(NH_3)_4]_3[Fe(CN)_6]_2$. This coordination compound consists of positive and negative coordination spheres. In the cation, $Cu(NH_3)_4^{2+}$, the central Cu(II) ion is surrounded by four ammonia molecules, each attached to the Cu(II) ion by a coordinate covalent bond. Each of these bonds is formed when the nitrogen atom in an ammonia molecule donates its lone electron pair to the Cu(II) ion. The anion in $[Cu(NH_3)_4]_3[Fe(CN)_6]_2$ consists of a central Fe(III) ion covalently bonded to six CN^- ions. The charge on the complex ion is the sum of the oxidation state of the central metal and the charges of all the ligands. Thus, the charge on the complex anion, $Fe(CN)_6^{3-}$, is $(+3) + 6\,(1-) = -3$.

We often refer to the central metal atom in a coordination complex as the central metal "ion" if the complex was formed from the metal ions and ligands. However, when the ion becomes covalently bonded to ligands, it loses its character as an ion. For that reason we do not refer to the central metal "ion" in a complex as having a "charge"; instead, we refer to it as being in a certain oxidation state.

Example 22.2

Determining the Oxidation State of the Central Metal in a Coordination Complex

(a) What is the oxidation state of iron in the complex ion $NiCl_4^{2-}$?
(b) What is the oxidation state of platinum in $[Pt(NH_3)_4]Cl_2$?

SOLUTION: The charge on the complex ion equals the oxidation state of the central metal plus the total charge of all the ligands connected to it.

a. In the $NiCl_4^{2-}$ ion, the central metal ion is surrounded by four Cl^- ions. If the oxidation state of the central metal is x, we can write

$$x + (4)(-1) = -2$$

from which $x = -2 + 4 = +2$.

b. In the complex, $[Pt(NH_3)_4]Cl_2$, the four ligands are neutral ammonia molecules, and the coordination sphere has a charge of 2+ because it is associated with two Cl^- ions. If the oxidation state of Pt is y, we have

$$y + (4)(0) = +2$$

from which $y = +2$.

Practice Problem 22.2: What is the oxidation state of the central metal ion in the coordination spheres of each of the following complexes? **(a)** $Cr(H_2O)_4Cl_2^+$; **(b)** $K_4[Fe(CN)_6]$.

Coordination Number

In a coordination sphere, ligands are bonded to the central metal ion by coordinate covalent bonds. The *number of coordinate covalent bonds that link the central atom or ion to its ligands in a coordination sphere* is called the **coordination number** for the complex. This number equals the number of ligands in the coordination sphere if each ligand is attached to the central atom by only one coordinate covalent bond. Thus, the coordination number for Cu in $Cu(NH_3)_4^{2+}$ is 4 and the coordination number for Pt in $Pt(NH_3)_5Cl^{3+}$ is 6. The most commonly observed coordination numbers are 6, 4, and 2.

In the next section we discuss ligands that can form two or more bonds with the metal ion. Such ligands have more than one unshared electron pair to donate to the metal ion.

22.2 Types of Ligands and Some Important Coordination Complexes

Monodentate Ligands

Ligands can be classified according to the number of coordinate covalent bonds they form with a metal atom or ion. For example, an ammonia molecule has one lone pair on its nitrogen atom. An ammonia molecule can therefore form one coordinate covalent bond with the central atom in a coordination complex. The *atom of a ligand that donates a lone pair for bonding* is called the **donor atom**. Thus, the nitrogen atom in an NH_3 molecule is the donor atom.

A cyanide ion, $:C{\equiv}N:^-$, has two lone pairs of electrons, but it can use only one of these pairs for bonding in a coordination complex. The carbon atom of a cyanide ion is the donor atom in most coordination complexes. *Ligands such as NH_3 and CN^-, which coordinate by only one donor atom*, are called **monodentate ligands** (from Latin *mono*, one, and *dentis*, tooth). Other common monodentate ligands are H_2O, CO, OH^-, Cl^-, Br^-, and thiocyanate ion, SCN^- ($:\ddot{S}{-}C{\equiv}N:^-$).

Polydentate Ligands or Chelating Agents

Ligands that coordinate to a metal ion by two or more donor atoms are called **polydentate ligands**. A *ligand with two donor atoms* is a **bidentate ligand**, a *ligand with three donor atoms* is a **terdentate ligand**, and a *ligand with four donor atoms* is a **tetradentate ligand**. A common example of a bidentate ligand is ethylenediamine, often abbreviated *en* in discussions of coordination compounds:

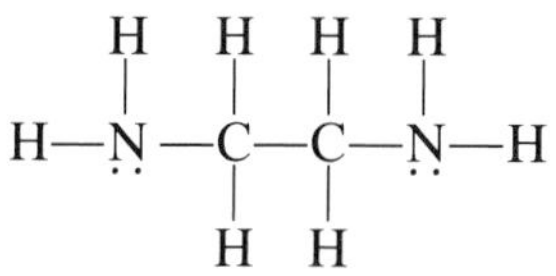

```
    H   H   H   H
    |   |   |   |
H—N—C—C—N—H
    ··  |   |  ··
        H   H
```

Both nitrogen atoms in an ethylenediamine molecule have lone pairs; they are therefore donor atoms.

A polydentate ligand in a coordination complex binds the metal ion through its donor atoms, which act as "claws." A polydentate ligand is therefore called a **chelating agent** (from the Greek word *chele*, claw). Figure 22.2 illustrates a Cr^{3+} ion coordinated with six monodentate CN^- ions and a Cr^{3+} ion bonded to three bidentate ethylenediamine molecules. In both of these structures, the coordinate covalent bonds about the central Cr^{3+} ion are distributed in an octahedral geometry.

A *coordination complex in which the metal ion is coordinated to one or more chelating ligands* is called a **chelate**. An example of a chelate is the $Cr(en)_3{}^{3+}$ ion (Figure 22.2) in which the Cr^{3+} ion is coordinated to three ethylenediamine molecules.

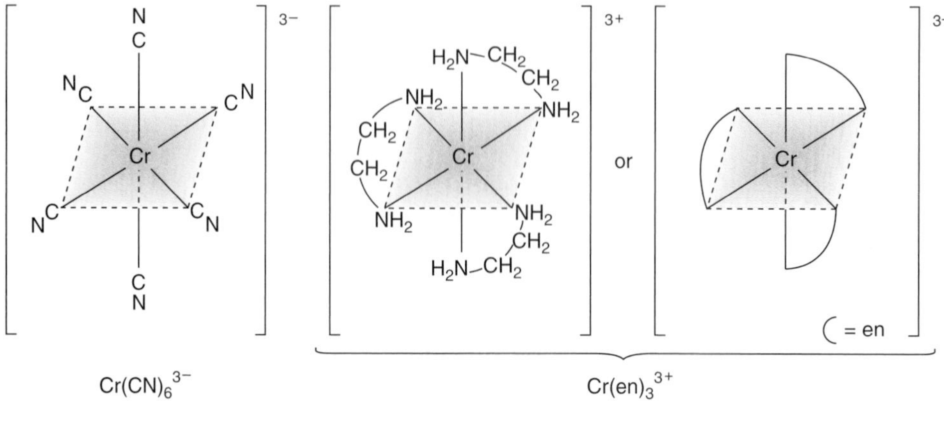

Figure 22.2

Coordination complexes of Cr^{3+} with monodentate CN^- ions as ligands, and with bidentate ethylenediamine (en) molecules.

$Cr(CN)_6{}^{3-}$ — $Cr(en)_3{}^{3+}$ (or; (= en)

Some Important Chelates in Chemistry and Biology

The process of chelate formation, or chelation, has important applications in analytical chemistry. A chelating ligand that is used in analytical chemistry to determine the concentration of metal ions such as Ca^{2+} and Mg^{2+} is ethylenediaminetetraacetate ion, *$EDTA^{4-}$*, which is often written without its charge as EDTA. The structure of this ligand is shown in Table 22.2. A calcium ion coordinated with $EDTA^{4-}$ is shown in Figure 22.3. An $EDTA^{4-}$ ion has six donor atoms—four oxygen atoms and two nitrogen atoms—and is therefore a hexadentate ligand. These donor atoms are attached to a calcium ion in octahedral geometry as illustrated in Figure 22.3.

The $EDTA^{4-}$ ion is derived from ethylenediaminetetraacetic acid, H_4EDTA. A solution of the disodium salt of this acid, Na_2H_2EDTA, can be used to determine quantitatively the concentrations of Ca^{2+} and Mg^{2+} ions in municipal water supplies. A sample to be analyzed is titrated with a standard solution of the sodium salt of H_2EDTA^{2-} ions. The endpoint of this titration is signaled by an indicator color change. Since one H_2EDTA^{2-} ion binds one Ca^{2+} ion (or one Mg^{2+} ion), the total number of moles of these cations in solution can

Table 22.2

Names and Formulas of Common Ligands

Formula	Name	Name as Ligand
H_2O	Water	Aqua
NH_3	Ammonia	Ammine
CO	Carbon monoxide	Carbonyl
$H_2N—CH_2—CH_2—NH_2$	Ethylenediamine	Ethylenediamine (en)
Cl^-	Chloride ion	Chloro
Br^-	Bromide ion	Bromo
CN^-	Cyanide ion	Cyano
SCN^-	Thiocyanate ion	Thiocyanato
$S_2O_3^{2-}$	Thiosulfate ion	Thiosulfato
OH^-	Hydroxide ion	Hydroxo
$C_2O_4^{2-}$	Oxalate ion	Oxalato
$H_3C—C=N—O^-$ | $H_3C—C=N—OH$	Dimethylglyoximate ion	Dimethylglyoximato
O O ‖ ‖ $^-O—C—CH_2$ $CH_2—C—O^-$ \ / $N—CH_2—CH_2—N$ / \ $^-O—C—CH_2$ $CH_2—C—O^-$ ‖ ‖ O O	Ethylenediaminetetraacetate ion	Ethylenediaminetetraacetato (EDTA)
O^- O | ‖ $CH_3—C=C—C—CH_3$ | H	Acetylacetonate ion	Acetylacetonato (acac)

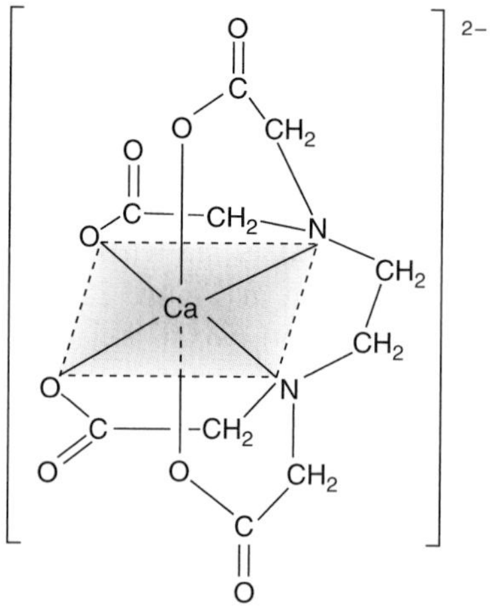

Figure 22.3

$CaEDTA^{2-}$: a chelate of Ca^{2+} ions and $EDTA^{4-}$ ions.

be calculated from the number of moles of H_2EDTA^{2-} required to titrate the solution. (This analysis is valid only if no other ions that bind EDTA are present.)

EDTA is also used in foods to bind metal ions that catalyze the oxidation and spoilage of food. EDTA is present in many shampoos. EDTA removes Ca^{2+} ions present in hard water that inhibit the action of the soap in the shampoo. EDTA also acts as an antidote for lead poisoning by removing Pb^{2+} ions from the body. EDTA is added to whole blood to prevent blood clotting, a complex process that is accelerated by Ca^{2+} ions.

Coordination complexes are important in many living tissues. One biochemically important molecule is *porphin* (Figure 22.4). Porphin has four nitrogen atoms and acts as a tetradentate ligand when two of the nitrogen atoms lose protons.

Derivatives of porphin that contain various side chains are called *porphyrins*. Porphyrins coordinate with metal ions such as Fe^{2+} and Mg^{2+}. Porphyrins are essential components of the oxygen transport protein, *hemoglobin* in blood, and the green plant pigment, *chlorophyll*. Hemoglobin contains structural units called *heme*. The structures of heme and chlorophyll are illustrated in Figure 22.5. In heme, the central metal ion is an Fe^{2+} ion, and in chlorophyll it is an Mg^{2+} ion.

Figure 22.4

The structure of the porphin molecule.

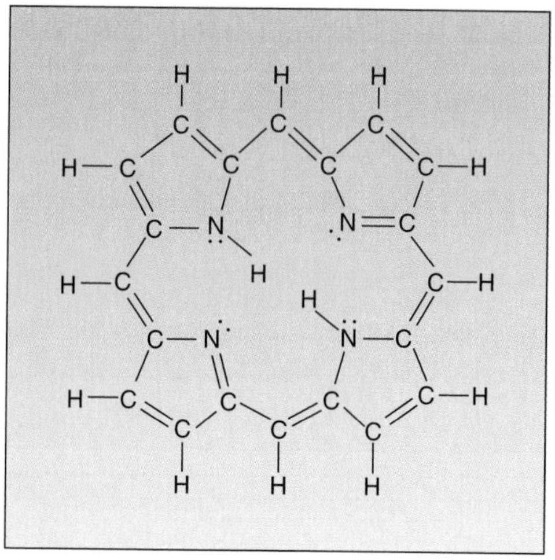

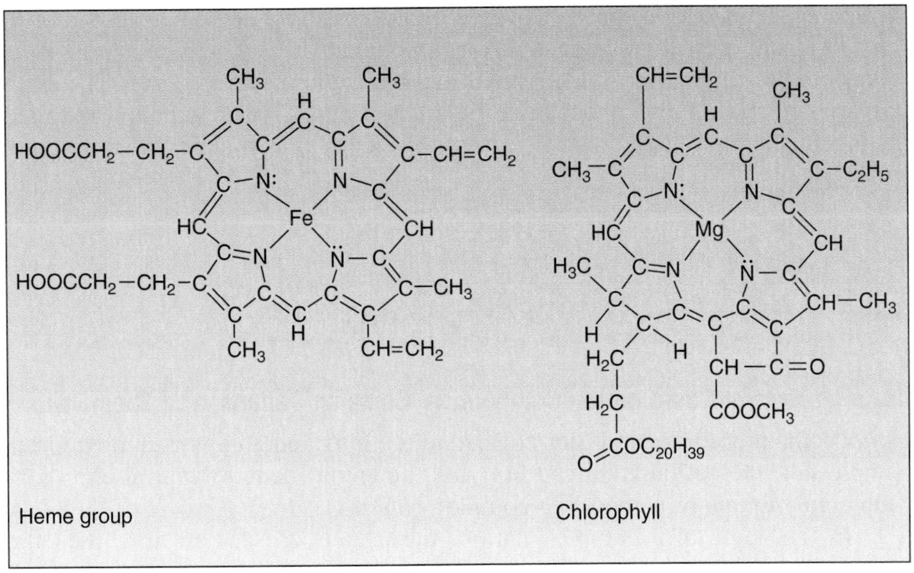

Heme group

Chlorophyll

Figure 22.5

The structures of heme and chlorophyll.

In both heme and chlorophyll, the central metal ion is coordinated to four N atoms in a square planar arrangement. The central metal ion and the four N atoms all lie in the same plane and the N atoms point to the corners of the square. Thus, the basic structures of heme and chlorophyll are similar.

Another structure similar to the porphyrins is *vitamin* B_{12} or *cobalamin* (Figure 22.6). In vitamin B_{12}, a Co^{2+} ion is complexed in a square planar

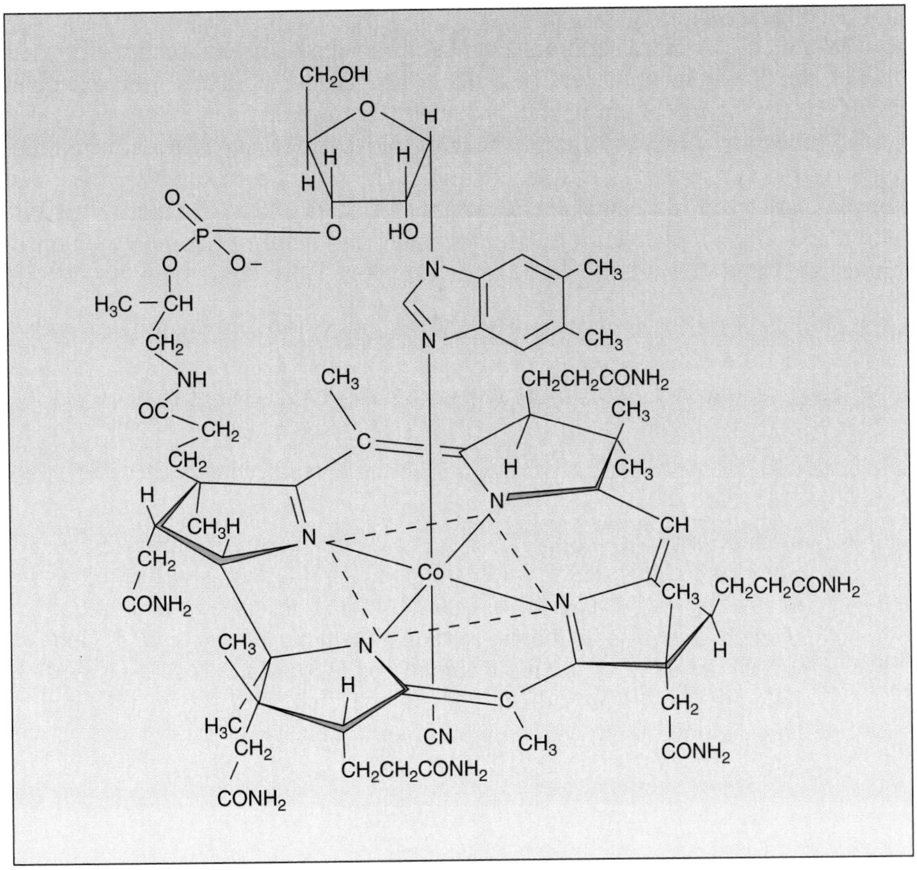

Figure 22.6

The structure of vitamin B_{12} or cyanocobalamin.

geometry by nitrogen atoms in a porphyrinlike ring. The Co^{2+} ion is also coordinated to a nitrogen atom along the axis perpendicular to the square plane that is connected to the porphyrinlike ring through a chain of other atoms. The sixth position along this axis is occupied by a CN group. (This compound is called cyanocobalamin.) Thus the Co^{2+} ion in this complex is in an octahedral environment.

22.3 Naming Coordination Complexes

Naming Neutral Coordination Compounds, Complex Cations, and Their Salts

A coordination compound that consists of cations and anions is named like a simple salt: the cation is named first and the anion second. Neutral coordination compounds are named like complex cations.

In the name of a complex cation, the ligands are named first, then the metal. The oxidation state of the metal is written with a Roman numeral enclosed in parentheses after the name of the metal. The names of the ligands, the metal, and the parentheses are not separated by blank spaces. The names of negative ligands end in −*o*. Most neutral ligands are named as molecules. Two important exceptions are water and ammonia, which are named *aqua* and *ammine*, respectively. If a complex ion contains two or more identical ligands, their number is indicated by a Greek prefix di-, tri-, tetra-, penta-, hexa-, and so on. Thus, the name of the complex cation $Ag(NH_3)_2^+$ is diamminesilver(I) ion. The compound $Ag(NH_3)_2Cl$ is diamminesilver(I) chloride. The prefix mono- is not used for a single ligand.

When two or more different ligands are in the coordination sphere, their names are listed in alphabetical order, disregarding prefixes. For example, $Co(NH_3)_4Cl_2^+$ is tetraamminedichlorocobalt(III) ion.

For ligands with more complex names (particularly names already containing di-, tri-, etc.) such as ethylenediamine, the Greek prefixes bis-, tris-, and tetrakis- are used to designate the number of ligands. Thus, $Cu(en)_2^{2+}$ is bisethylenediaminecopper(II) ion. The formulas and names of some common ligands are listed in Table 22.2.

Example 22.3

Naming Complex Cations, a Salt Consisting of a Complex Cation and a Simple Anion, and a Neutral Coordination Compound

Name each of the following species: **(a)** $Cu(H_2O)_4^{2+}$; **(b)** $[Cr(NH_3)_3Br_3]$; **(c)** $Co(en)_2Br_2^+$; **(d)** $[Fe(H_2O)_5(SCN)]SO_4$.

SOLUTION

a. In this complex, a copper(II) ion is coordinated to four water molecules. We recall that a water molecule as a ligand is called aqua. According to the rules of nomenclature just described, the number of each ligand is listed first, then the name of the ligand, and finally, the name of the metal and its oxidation state. Neutral water molecules as ligands do not affect the charge of the metal ion. Thus, the name of the complex is tetraaquacopper(II) ion.

b. In this complex there are three ammonia molecules and three bromide ions as ligands. These are listed in alphabetical order in the name of the complex. The complex is a neutral coordination sphere. Ammonia molecules are neutral ligands that do not affect the charge of the complex. Since the total charge of the three bromide ions is 3−, the oxidation state of chromium must be +3 for the complex to be neutral. The name of the complex is triamminetribromochromium(III).

c. As before, we list the number of ligands and the names of the ligands first, in alphabetical order, then the name of the metal, followed by its oxidation state. The oxidation state of the cobalt atom in a complex depends on the number and charges of the ligands. Ethylenediamine molecules are neutral and do not affect the charge of the complex ion. The 2− total charge of the two bromide ions neutralizes two plus charges on the cobalt ion. Since the net charge on the complex ion is 1+, the oxidation state of the cobalt ion is 3+. Thus, the complete name of the complex ion is dibromobisethylenediaminecobalt(III) ion. Note that ethylenediamine is a complex ligand and the prefix bis is used to designate the number of these ligands in the complex.

d. $[Fe(H_2O)_5(SCN)]SO_4$ consists of a complex cation and a simple anion. The charge of the complex cation must be 2+ because it is combined with a sulfate ion that has a charge of 2−. Since water molecules are neutral and a thiocyanate ion has a 1− charge, the oxidation state of the central Fe ion is 3+. The name of the compound is pentaaquathiocyanatoiron(III) sulfate.

Practice Problem 22.3: Name the following substances: **(a)** $Co(NH_3)_4Cl_2^+$; **(b)** $[Ni(en)_3](NO_3)_2$.

Naming Complex Anions and Their Salts

A complex anion is named by the same rules as a complex cation, except that the suffix *-ate* is added to the name of the metal. Thus, the names of complex anions are analogous to the names of simple oxoanions such as nitrate, carbonate, and sulfate. For example, the name of $Ni(CN)_4^{2-}$ is tetracyanonickelate(II) ion. When the name of the metal ends with -um, as in chromium and platinum, the suffix *-ate* is added to the stem of the name of the metal to give chromate and platinate, respectively. Thus, $Cr(CN)_6^{3-}$ is hexcyanochromate(III), and $PtCl_4^{2-}$ is tetrachloroplatinate(II). Sometimes the Latin name of the metal is used when the English name is clumsy. For example, cuprate and stannate are used for copper and tin instead of copperate and tinate. Thus, $CuCl_4^{2-}$ is tetrachlorocuprate(II). Metals whose Latin names are generally used are given below.

English Name	Latin Name	Anion Name
Copper	Cuprum	Cuprate
Gold	Aurum	Aurate
Iron	Ferrum	Ferrate
Lead	Plumbum	Plumbate
Silver	Argentum	Argentate
Tin	Stannum	Stannate

Example 22.4

Naming and Writing Formulas for Coordination Compounds Containing Complex Anions

(a) Name $Na[Al(OH)_4]$. **(b)** Name $K_2[SnCl_6]$. **(c)** Write the formula for tetraammineplatinum(II) hexachloroplatinate(IV). **(d)** Write the formula for hexaaquachromium(III) tetracyanonickelate(II).

SOLUTION

a. The oxidation state of aluminum ion is +3, and the name of the compound is sodium tetrahydroxoaluminate(III).

b. The Latin name for tin is stannum. The contribution of 6 chloride ions gives the complex ion a net charge of 2−. Thus, the oxidation state of tin in this compound is +4. The compound has a net charge of zero because the two positive K^+ ions balance the negative charge of the complex anion. Thus, the name of the compound is potassium hexachlorostannate(IV).

c. Tetraammineplatinum(II) ion contains four ammonia molecules coordinated to a Pt^{2+} ion. Therefore, the formula for this ion is $Pt(NH_3)_4{}^{2+}$. A hexachloroplatinate(IV) ion has six Cl^- ions coordinated to a Pt(IV) ion, and the formula for the complex is $PtCl_6{}^{2-}$. Since the complex cation has a 2+ charge and the anion has a 2− charge, the formula for the compound is $[Pt(NH_3)_4][PtCl_6]$.

d. Hexaaquachromium(III) ion has the formula $Cr(H_2O)_6{}^{3+}$, and tetracyanonickelate(II) is $Ni(CN)_4{}^{2-}$. Thus, a neutral combination of these ions has three $Ni(CN)_4{}^{2-}$ ions for every two $Cr(H_2O)_6{}^{3+}$ ions, and the formula of the compound is $[Cr(H_2O)_6]_2[Ni(CN)_4]_3$.

Practice Problem 22.4: **(a)** Name $K[Pt(NH_3)Cl_3]$. **(b)** Write the formula for dicyanobisethylenediaminechromium(III) chloride. **(c)** Write the formula for diamminetetrafluorochromate(III) ion.

22.4 Isomerism in Coordination Complexes

In Section 7.4 we discussed some simple isomers. We recall that isomers are nonidentical compounds which have the same molecular formulas but different structures. *Isomers that differ in the sequence in which the atoms or groups of atoms are bonded together* are called **structural isomers**. For example, cyanic acid, H—O—C≡N, and isocyanic acid, H—N=C=O, are structural isomers because the hydrogen atom is bonded to the oxygen atom in one compound and to the nitrogen atom in the other compound. *Isomers in which the atoms are bonded together in the same sequence but differ in their precise arrangement in space* are called **stereoisomers** (Gr. *stereo*, space).

Structural Isomers

Some important types of structural isomers in coordination complexes include **ionization isomers** and **linkage isomers**. Ionization isomers have different ions inside and outside their coordination spheres:

$$[Co(NH_3)_5Br]SO_4 \quad \text{and} \quad [Co(NH_3)_5SO_4]Br$$
$$\text{Red-violet} \qquad\qquad \text{Red}$$

The sulfate ion in $[Cu(NH_3)_5SO_4]^+$ acts as a monodentate ligand that coordinates to the cobalt ion through an oxygen atom (Figure 22.7).

In linkage isomers a ligand is coordinated to the central metal atom through different donor atoms. For example, the donor atom in a nitrite ion

$$(:\ddot{\underset{..}{O}}-\ddot{N}=\ddot{O}:)^-$$

can be either the nitrogen atom or one of the two oxygen atoms. When the donor atom is nitrogen, the ligand is called *nitro*-. When the donor atom is oxygen, the ligand is called *nitrito*-. The structures and names of two linkage isomers of nitrite ion are shown in Figure 22.8.

Figure 22.7

The sulfate ion as a monodentate ligand in the compound $[Cu(NH_3)_5SO_4]Br$.

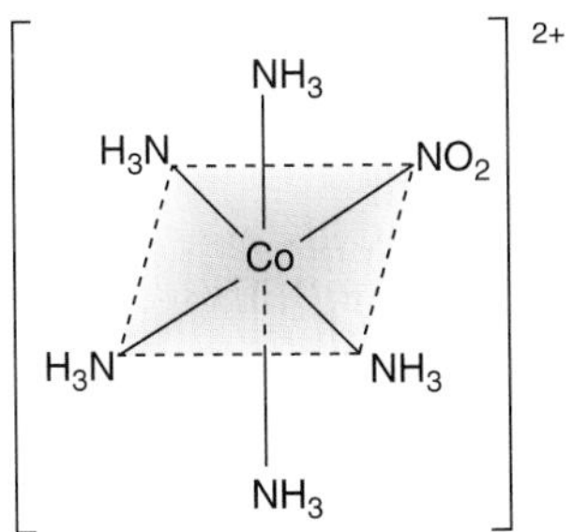

$[Co(NH_3)_5NO_2]^{2+}$
Pentaamminenitrocobalt(III) ion or
Pentaamminenitrito-N-cobalt(III) ion

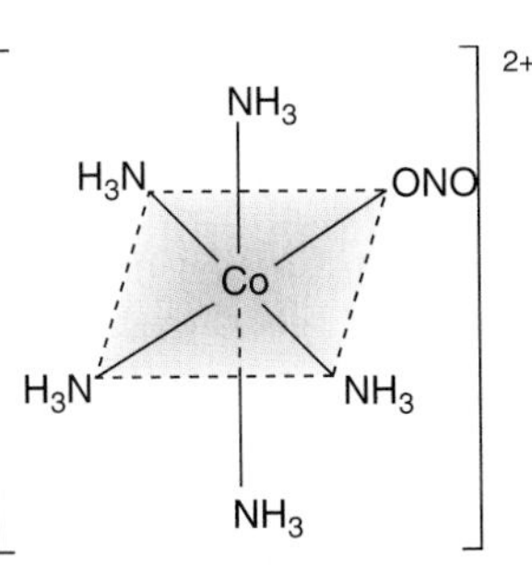

$[Co(NH_3)_5ONO]^{2+}$
Pentaamminenitritocobalt(III) ion or
Pentaamminenitrito-O-cobalt(III) ion

Figure 22.8

Linkage isomers involving different donor atoms of the NO_2^- ion.

In an alternative method of naming linkage isomers, the International Union of Pure and Applied Chemistry (IUPAC) has decreed that the atom through which the ligand is attached to the central atom should be identified. For example, when a nitrite ion is coordinated by the nitrogen atom, the complex is called a nitrito-N-complex; and when it is coordinated by the oxygen atom, it is called a nitrito-O-complex. Thus, $[Co(NH_3)_5NO_2Cl_2$ is pentaamminenitrito-*N*-cobalt(III) chloride, and $[Co(NH_3)_5ONO]Cl_2$ is pentaamminenitrito-*O*-cobalt(III) chloride.

Another example of a ligand that has two donor atoms, and can therefore form linkage isomers, is thiocyanate ion, $:SCN:^-$. The ion is called *thiocyanato-* when it is coordinated by the sulfur atom, and *isothiocyanato-* when it is coordinated by the nitrogen atom. For example, $HgSCN^+$ is thiocyanatomercury(II) ion, and $FeNCS^{2+}$ ion is isothiocyanatoiron(III) ion. According to the IUPAC convention, these complexes are called thiocyanato-*S*-mercury(II) ion and thiocyanato-*N*-iron(III) ion, respectively.

Stereoisomers

We noted above that stereoisomers differ in the arrangement of atoms in space. Stereoisomers are divided into two subcategories, **geometric isomers**, also called **position isomers**, and **optical isomers**. We discuss geometric isomers first.

Geometric isomers can be found in complexes with four ligands in a square planar geometry or with six ligands in an octahedral geometry. In a square planar complex that has two identical ligands of one type and two identical ligands of another type, two identical ligands can lie either on the same side of the central metal atom or on opposite sides of the central atom. When the identical ligands lie on the same side of the central atom, the arrangement is called *cis* (Latin *cis*, on the same side). When two identical ligands lie on opposite sides of a central atom, the isomer is called *trans* (Latin *trans*, across). Examples of geometric isomers in square planar complexes are shown below:

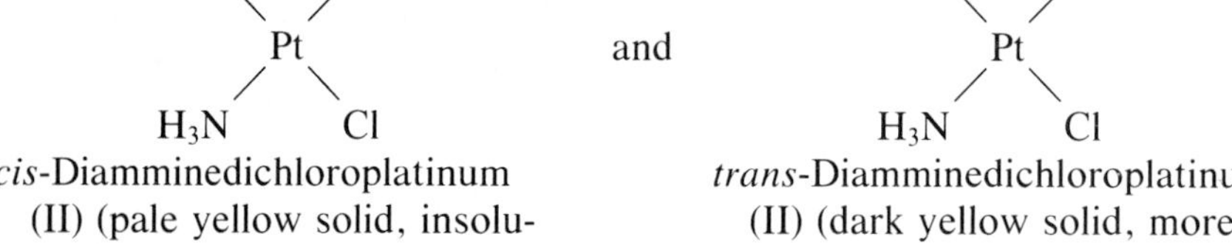

cis-Diamminedichloroplatinum (II) (pale yellow solid, insoluble in water)

trans-Diamminedichloroplatinum (II) (dark yellow solid, more water soluble)

The cis form of these platinum-containing compounds has recently been used as an antitumor agent under the name "cisplatin." This compound is particularly effective in the treatment of testicular cancer and the cancer of the neck area, bladder, and ovary. The trans isomer of this compound is not an antitumor agent.

Cis and trans geometric isomers can also exist in octahedral coordination complexes, as illustrated in Figure 22.9 for $Cr(NH_3)_4Cl_2^+$ ion.

Octahedral complexes that have three identical ligands of one type and three identical ligands of another type can exist as two isomers. The isomer in which all three identical ligands are cis to one another is called the **facial** or **fac- isomer**. The isomer in which two identical ligands are trans to each other is called the **meridional** or **mer- isomer**. Figure 22.10 illustrates the structures of fac- and mer- isomers of triamminetrichlorocobalt(III).

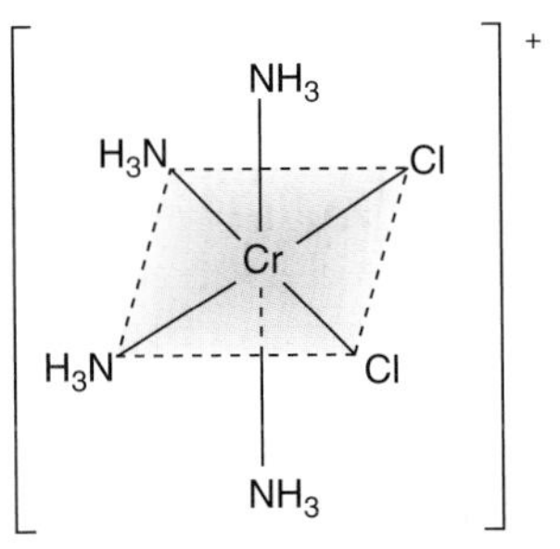

cis-tetraamminedichlorochromium(III) ion (violet)

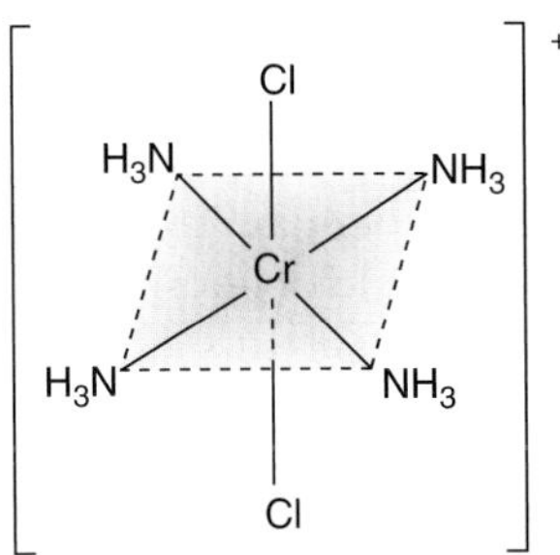

trans-tetraamminedichlorochromium(III) ion (green)

Figure 22.9

Cis-trans isomers of $Cr(NH_3)_4Cl_2^+$ ion.

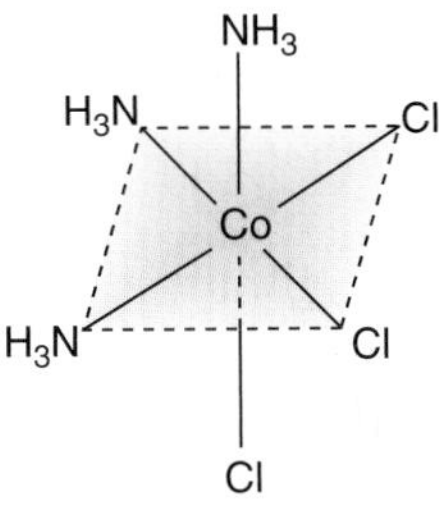

Cis or facial form
fac-triamminetrichlorocobalt(III)

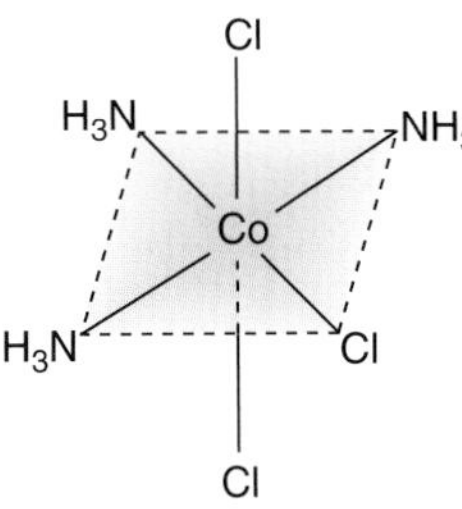

Trans or meridional form
mer-triamminetrichlorocobalt(III)

Figure 22.10

Fac-mer isomers of $Co(NH_3)_6Cl_3$.

Example 22.5

Writing Structural Formulas for Octahedral Complexes of Three Identical Ligands of One Type and Three Identical Ligands of Another Type

Write the structures for all the possible isomers for the neutral complex $[Co(NH_3)_3Cl_3]$.

SOLUTION: We start by positioning two of the ligands, say NH_3, in cis positions at two adjacent corners of an octahedron (a polyhedron with eight equilateral triangular faces) around the central cobalt ion:

H_3N
Co
H_3N

We next place the third NH_3 molecule at each one of the remaining corners (one each time) of the octahedron in each case and examine the structures obtained (we assume that the other three positions are occupied by chloride ions):

(1): H_3N, NH_3, Co, H_3N

(2): NH_3, H_3N, Co, H_3N

(3): H_3N, Co, H_3N, NH_3

(4): H_3N, Co, H_3H, NH_3

In structure (1) two NH_3 molecules are trans to each other, and in structure (2) all the NH_3 molecules are in cis positions. Therefore, structures (1) and (2) are different.

Structure (3) is identical to structure (1) because two NH_3 molecules are trans to each other. To prove that structures (1) and (3) are identical, we can rotate structure (3) clockwise 90° around the axis through the central cobalt ion and perpendicular to the colored square plane. As a result of this rotation, the two structures have become identical.

Structure (4) is identical to structure (2) in which all of the NH_3 molecules are in cis positions. Rotating structure (4) in space shows that it is identical to structure (2). Thus, there are only two isomers, one with identical ligands cis to each other (which is called the *fac* form) and the other with two NH_3 ligands and two Cl ligands trans to each other (the *mer* form). The trans form is called meridional or mer-, and the cis form is called facial or fac-:

```
          Cl                              NH3
  H3N  ..|..  NH3                 H3N  ..|..  Cl
     :\  |  /:                       :\  |  /:
     :   Co  :                       :   Co  :
     :/. | .\:                       :/. | .\:
  H3N    |    Cl                  H3N    |    Cl
          Cl                              Cl
```

trans or meridional form: *mer*-triamminetrichlorocobalt(III)

cis or facial form: *fac*-triamminetrichlorocobalt(III)

We now consider optical isomers. Optical isomers are stereoisomers that are related as *nonsuperimposable mirror images, like right and left hands* (Figure 22.11). Optical isomers may appear at first glance to be identical, but regardless of how we rotate them in space, they cannot be superimposed. An example of a pair of optical isomers is shown for the octahedral complex

Figure 22.11

Left hand and its mirror image.

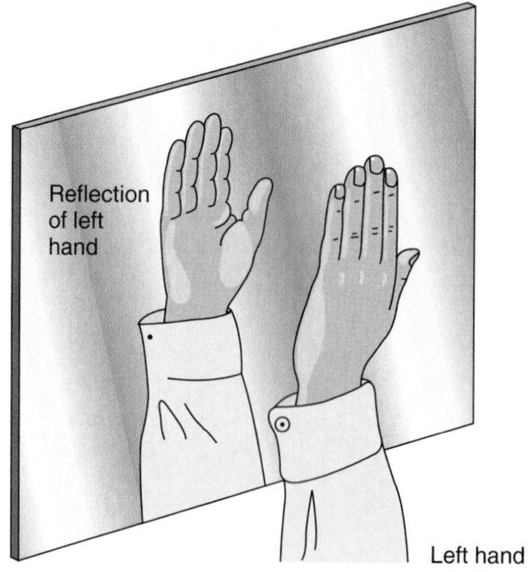

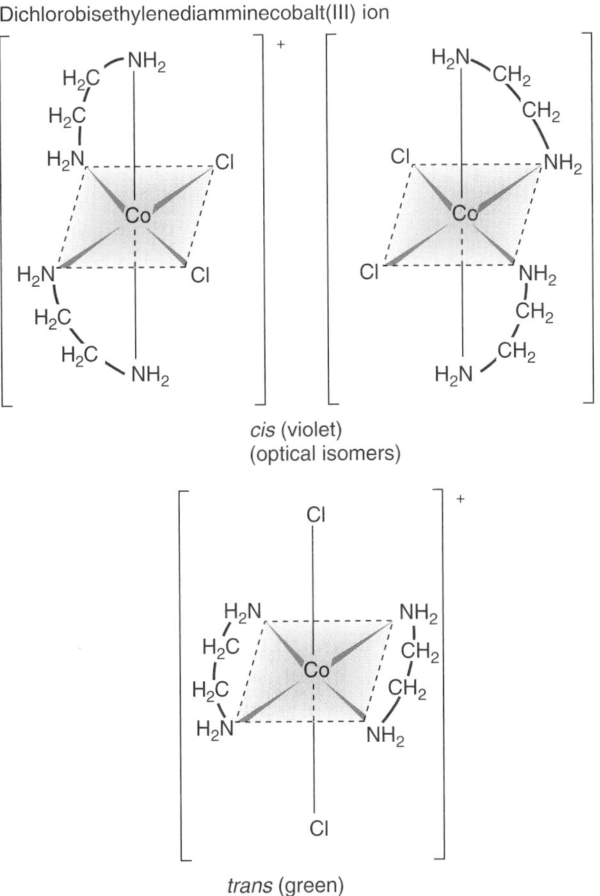

Figure 22.12

Isomers of octahedral complexes with two identical bidentate ligands and two identical monodentate ligands.

$Co(en)_2Cl_2^+$ in Figure 22.12. This complex, which has two identical bidentate ligands and two identical monodentate ligands, can also exist as geometric isomers. The cis and trans isomers of $Co(en)_2Cl_2^+$ are geometrical isomers, but the two cis forms are optical isomers.

Another class of examples of optical isomers is provided by octahedral complexes with three identical bidentate ligands. These isomers are illustrated in Figure 22.13 for $Cr(en)_3^{3+}$ ions, where the ligands are represented by curved lines.

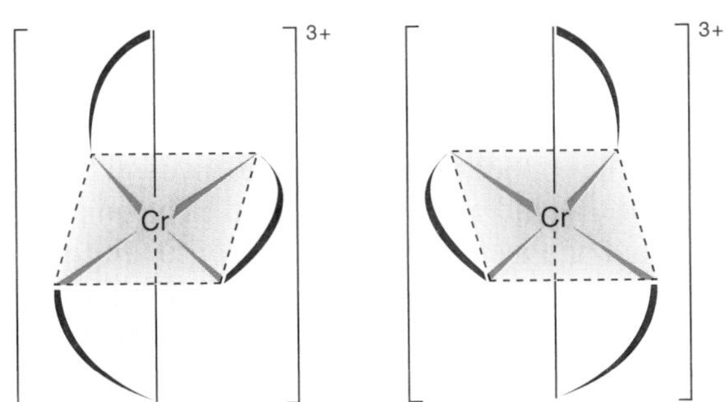

Figure 22.13

Optical isomers of an octahedral complex with three identical bidentate ligands, $Cr(en)^{3+}$. Each ligand is represented by a curved line.

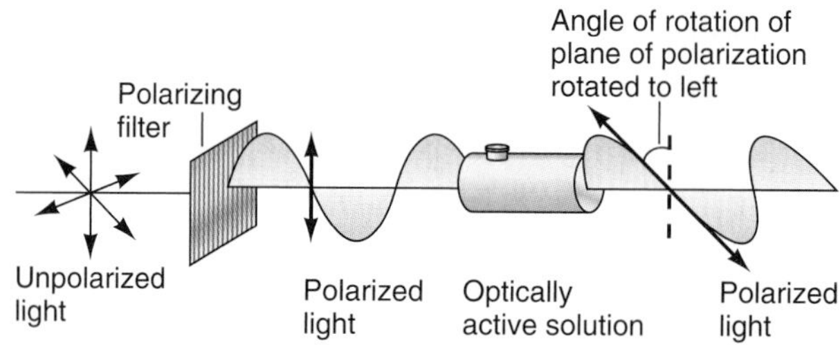

Figure 22.14

Rotation of plane polarized light by an optically active substance (an optical isomer). In this example, the plane of polarized light is rotated to the left (counterclockwise).

Optical isomers have identical physical and chemical properties, but they rotate plane-polarized light in opposite directions. Therefore they are said to be *optically active*. In plane-polarized light, all the electromagnetic waves vibrate in a single plane as shown in Figure 22.14.

Some optically active substances rotate the plane of polarized light to the left (counterclockwise) and others to the right (clockwise). The isomer that rotates the plane of polarized light to the right is called **dextrorotatory** and is called the *dextro,* or *d*, isomer. The mirror image of the dextrorotatory isomer rotates the plane of polarized light to the left and is called **levorotatory**; it is labeled *levo,* or *l*, isomer. An equimolar mixture of two optical isomers is called a **racemic mixture**. A racemic mixture has no net effect on polarized light.

Cis–trans isomerism and optical isomerism are also important in organic and biochemistry. In general, only one of two stereoisomers is physiologically active, as we saw in the example of cis-platin in the treatment of cancer.

22.5 Bonding in Transition Metal Complexes

In Chapter 8 we discussed valence bond theory and molecular orbital theory. These theories can be used to explain the bonding in transition metal complexes. Valence bond theory assumes that the central metal ion (or atom) is linked to its ligands by coordinate covalent bonds. The structure of a transition metal complex is correlated with the hybrid orbitals of the central metal ion. We will not consider the molecular orbital theory of transition metal complexes.

A third theory, called **crystal field theory**, assumes that the bonding between a central metal ion and its ligands in a transition metal complex is essentially electrostatic. The metal ion and the ligands are considered as point charges. This theory explains the magnetic properties, colors, and spectra of coordination complexes. We first consider valence bond theory.

Valence Bond Theory

According to valence bond theory, a coordinate covalent bond forms between a metal atom or ion and a ligand when an unshared electron pair of a ligand enters a vacant hybrid orbital of a metal ion. The hybrid orbitals used by the metal in a coordination complex are determined by the coordination number and the geometry of the complex, or vice versa. Table 22.3 lists some common coordination numbers, geometries of the corresponding complexes, and hybrid orbitals of the metal.

APPLICATIONS OF CHEMISTRY 22.1
The Stereoisomer Clock Within Us

Chemists use many techniques to determine the age of objects. They use carbon dating (see Chapter 24) to find the ages of materials that were once living. Such material includes bones, preserved carcasses, and wooden artifacts. However, there are some limitations with carbon dating. We cannot use it to determine the age of human beings or animals that are still alive. Also, carbon-dating is accurate only for objects less than about 6000 years old.

A new method used for dating objects overcomes some of the limitations of carbon dating. All living organisms contain molecules called amino acids (see Chapter 23). Amino acids exhibit stereoisomerism. Amino acids can exist as levorotatory or dextrorotatory isomers. However, the amino acids in all known life forms are levorotatory isomers. When an organism dies or the body does not replace amino acid stores, some of the levorotatory isomers slowly change into dextrorotatory isomers. Eventually, this change produces a racemic mixture (50 percent levorotatory, 50 percent dextrorotatory).

The change in stereoisomerism occurrs at a constant rate and is therefore a "stereoisomer-clock." Since we know the rate at which the clock "ticks," we can determine the age of objects by measuring the ratio of the levorotatory and dextrorotatory isomers. This technique works for objects up to 500,000 years old.

Now let's see how the stereoisomer clock works in living people and animals. Normally, most tissues produce new amino acids each day. This production is necessary because our bodies use amino acids in proteins which are continually being made to replace damaged proteins in body tissues. However, two areas of our bodies do not produce new amino acids after the moment of birth. The dentin in teeth (that's the layer just under the enamel) and eye lenses contain the same amino acids today that they did at the time of birth. Since our bodies do not replace those amino acids, some of them have been changing into dextrorotatory isomers since birth. Therefore, it is possible to determine the ages of living people and animals by analyzing their dentin or lenses. The method is accurate to within 10 percent.

Using the stereoisomer clock technique, scientists have verified the ages of Georgians who claim to be the oldest living group of people on earth (see photo above). Scientists have also used the method to check the ages of endangered animals to make better management decisions for aiding animal survival.

Source: Science, May 4, 1990, pages 539–540.

Transition metal complexes in which the central atom has a coordination number of 2 include $Ag(NH_3)_2^+$, $Ag(CN)_2^-$, and $Au(CN)_2^-$. In these complexes, the central metal ion has empty *s* and *p* orbitals available for bonding. Thus, in all these complexes, the central metal ion is *sp* hybridized. Two *sp* hybrid orbitals lie along a straight line on opposite sides of the nucleus of the central atom. The complexes are linear (Figure 22.15).

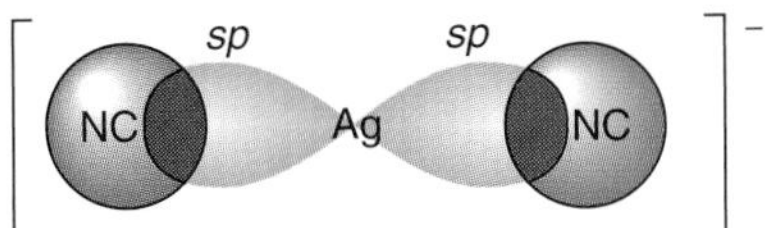

Figure 22.15

The linear structure of an $Ag(CN)_2^-$ ion showing the orbital overlaps of the Ag^+ and the CN^- ions.

Table 22.3

Common Coordination Numbers, Geometries, and Hybridization in Complexes

Coordination Number	Geometry	Hybrid Orbitals on the Central Metal Ion	Examples	
2	Linear	*sp*	$[H_3N—Ag—NH_3]^+$	$[NC—Ag—CN]^-$
4	Tetrahedral	sp^3	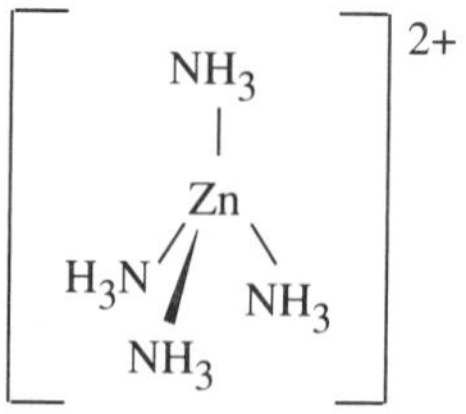 [NH_3, Zn, H_3N, NH_3, NH_3] $^{2+}$	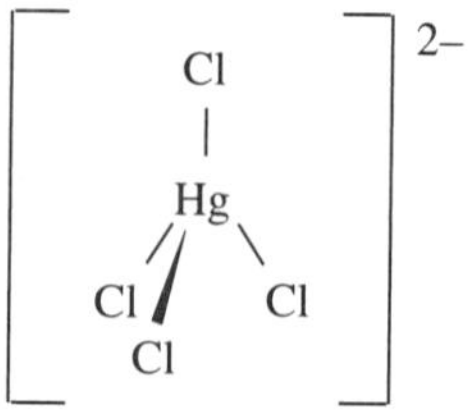 [Cl, Hg, Cl, Cl, Cl] $^{2-}$
	Square planar	dsp^2	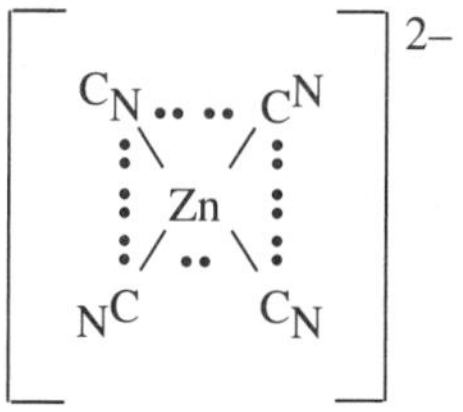 [CN, CN, Zn, NC, CN] $^{2-}$	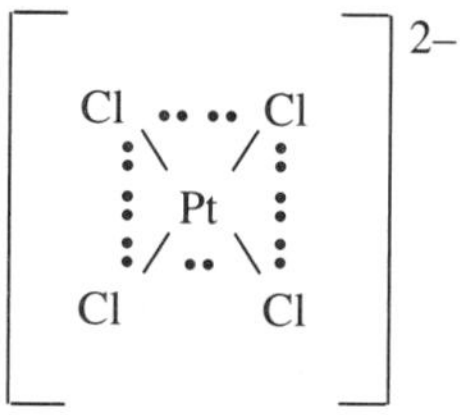 [Cl, Cl, Pt, Cl, Cl] $^{2-}$
6	Octahedral	d^2sp^3	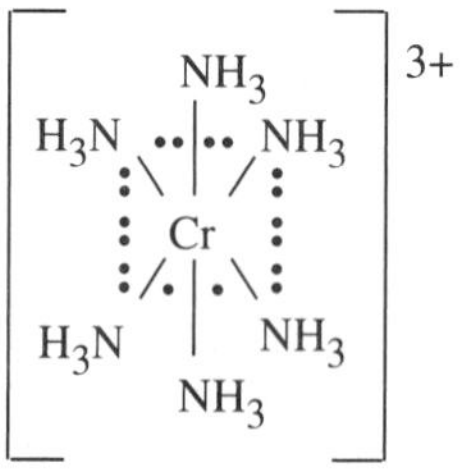 [NH_3, H_3N, NH_3, Cr, H_3N, NH_3, NH_3] $^{3+}$	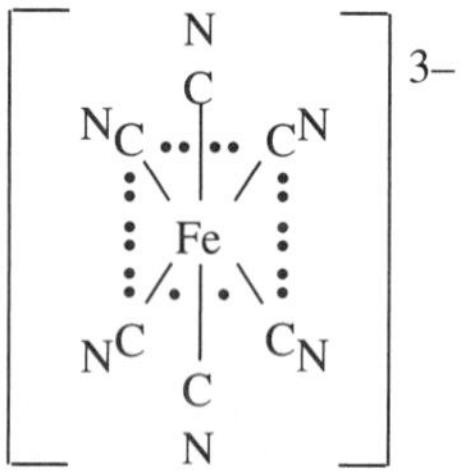 [NC, NC, CN, Fe, NC, CN, CN] $^{3-}$

A coordination complex in which the central metal has a coordination number of 4 is tetraamminezinc ion, $Zn(NH_3)_4^{2+}$. The Zn^{2+} ion has vacant 4*s* and 4*p* orbitals. These orbitals mix to form sp^3 hybrid orbitals. These hybrid orbitals are used to form $Zn(NH_3)_4^{2+}$, as shown diagrammatically below:

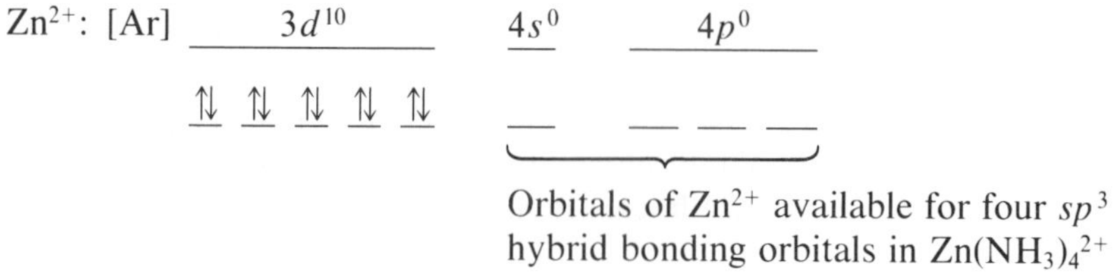

Zn^{2+}: [Ar] $3d^{10}$ $4s^0$ $4p^0$

⥮ ⥮ ⥮ ⥮ ⥮   __   __ __ __

Orbitals of Zn^{2+} available for four sp^3 hybrid bonding orbitals in $Zn(NH_3)_4^{2+}$

The complex is tetrahedral. In a tetrahedral geometry, the four sp^3 hybrid orbitals are as far as possible from one another in space (Figure 22.16).

Coordination complexes in which the central metal ion has a d^8 outer electron configuration can be tetrahedral or square planar. For example, Ni^{2+}, Pd^{2+}, and Pt^{2+} ions have a d^8 outer electron configuration. The eight electrons in the outer *d* orbitals of the central metal ion in a coordination complex can be arranged in two different ways, shown below as configurations (a) and (b):

(a) d^8 ⇅ ⇅ ⇅ ⇅ __ or (b) d^8 ⇅ ⇅ ⇅ ↑ ↑

Configuration (a) has one *d* orbital available for bonding. One *s* orbital and three *p* orbitals in the next-higher principal quantum level are also available. Thus, the central metal ion in configuration (a) can use one *d* orbital, one *s* orbital, and two *p* orbitals to form four dsp^2 hybrid orbitals.

The metal ion in configuration (b) can use one *s* and three *p* orbitals of the next-higher principal quantum level to form four sp^3 hybrid orbitals. Complexes in which the central ion is sp^3 hybridized are tetrahedral. Complexes in which the central ion is dsp^2 hybridized are square planar (Figure 22.17).

In a square planar complex, the *d* orbital used for bonding by the central metal ion is the $d_{x^2-y^2}$ orbital, which has lobes directed toward the corners of a square. In a square planar complex, the p_x and p_y orbitals also participate in bonding. These *p* orbitals lie in the *x*–*y* plane along with the $d_{x^2-y^2}$ orbital. The directional properties of *d* orbitals are shown in Figure 22.18.

The geometry of a square planar or tetrahedral complex with a d^8 electronic configuration can be determined by measuring its interaction with a magnetic field. In a square planar complex in which the central metal ion has a d^8 electronic configuration, all the *d* electrons on the metal ion are paired in four *d* orbitals. These paired electrons are very slightly repelled in a magnetic field, and the complex is *diamagnetic*. As we see in diagram (b) above, a d^8 tetrahedral complex has two unpaired electrons. These unpaired electrons are slightly attracted into a magnetic field. A d^8 tetrahedral complex is therefore *paramagnetic*.

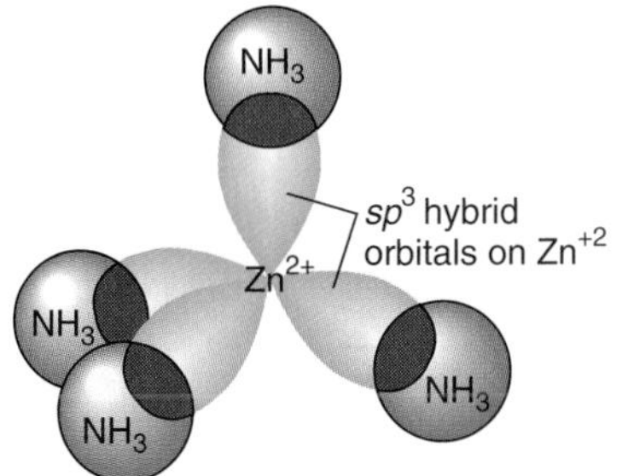

Figure 22.16

The tetrahedral structure of the four sp^3 hybrid orbitals in a $Zn(NH_3)_4^{2+}$ ion.

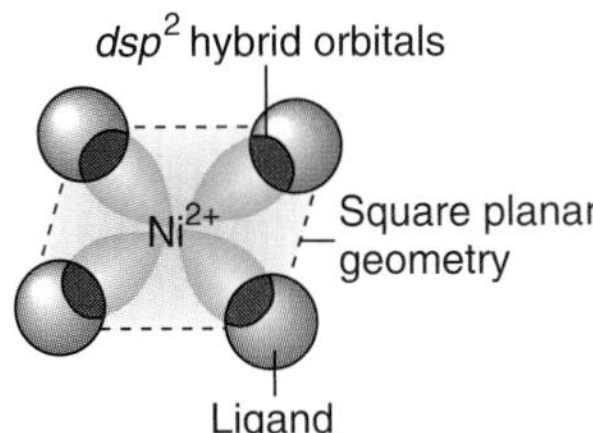

Figure 22.17

The square planar structure of the four dsp^2 hybrid orbitals in a nickel(II) complex.

Figure 22.18

The spatial orientations of *d* orbitals.

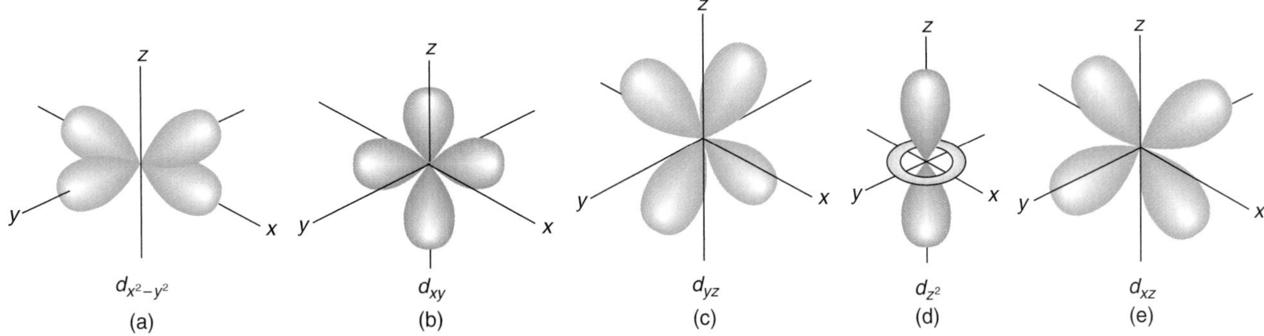

Nickel(II) ion forms a square planar complex with cyanide ions and a tetrahedral complex with chloride ions. The electronic configuration of the Ni^{2+} ion, and the empty orbitals which are hybridized, are diagrammed below for square planar $Ni(CN)_4^{2-}$ and tetrahedral $NiCl_4^{2-}$:

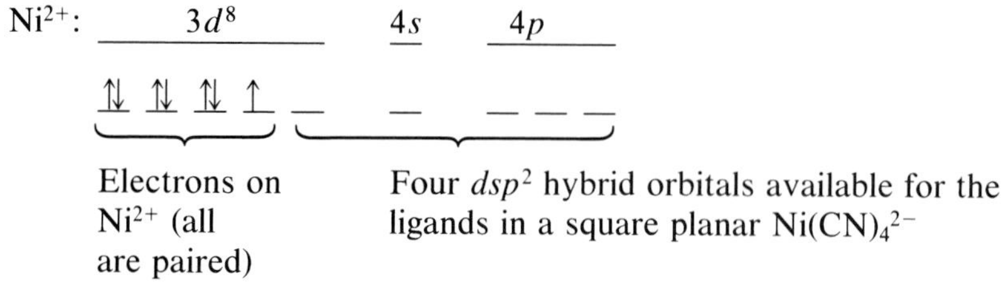

Ni^{2+}: $3d^8$ $4s$ $4p$

⇅ ⇅ ⇅ ↑ _ _ _ _ _

Electrons on Ni^{2+} (all are paired)

Four dsp^2 hybrid orbitals available for the ligands in a square planar $Ni(CN)_4^{2-}$

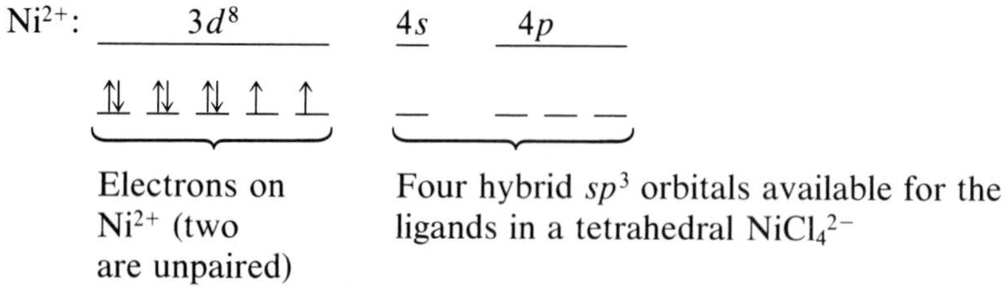

Ni^{2+}: $3d^8$ $4s$ $4p$

⇅ ⇅ ⇅ ↑ ↑ _ _ _ _

Electrons on Ni^{2+} (two are unpaired)

Four hybrid sp^3 orbitals available for the ligands in a tetrahedral $NiCl_4^{2-}$

Complexes of metal ions with a d^9 electronic configuration, such as the Cu^{2+} ion, might be expected to be tetrahedral because all of the *d* orbitals are occupied by one or two electrons. Actually, such coordination complexes are square planar. Valence bond theory explains this observation by assuming that one of the electrons of the metal ion in the coordination complex occupies a 4*p* orbital. As a result, dsp^2 hybrid orbitals can form. The electronic configuration in the square planar $Cu(NH_3)_4^{2+}$ ion is shown below:

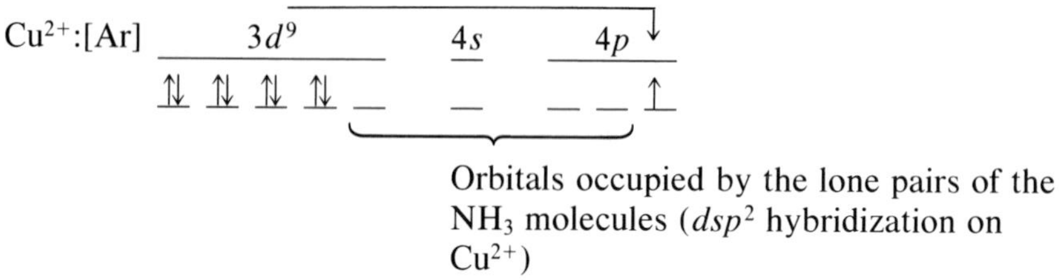

Cu^{2+}:[Ar] $3d^9$ $4s$ $4p$

⇅ ⇅ ⇅ ⇅ _ _ _ _ ↑

Orbitals occupied by the lone pairs of the NH_3 molecules (dsp^2 hybridization on Cu^{2+})

In an octahedral coordination complex such as $Cr(NH_3)_6^{3+}$, the electron arrangement can be described as follows:

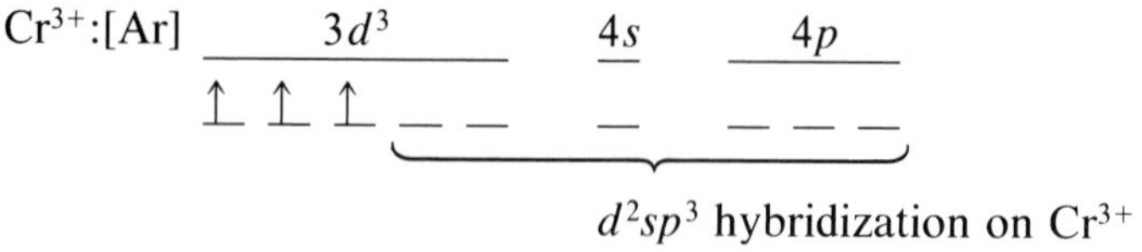

Cr^{3+}:[Ar] $3d^3$ $4s$ $4p$

↑ ↑ ↑ _ _ _ _ _ _

d^2sp^3 hybridization on Cr^{3+}

Experiments have confirmed that $Cr(NH_3)_6^{3+}$ ions are paramagnetic, as valence bond theory predicts.

Example 22.6

Using an Orbital Diagram to Predict the Hybrid Orbitals on the Central Atom and the Magnetic Properties of an Octahedral Manganese(II) Complex

Magnetic measurements show that $Mn(CN)_6^{4-}$ ion has one unpaired electron. Write an orbital diagram for an Mn^{2+} ion and for an $Mn(CN)_6^{4-}$ ion, showing the *d* orbital electronic configuration of the Mn^{2+} ion, and the orbitals available for the lone pairs of the CN^- ions. What is the hybridization of Mn^{2+} in the complex? Is it diamagnetic or paramagnetic?

SOLUTION: The valence shell structure of the Mn^{2+} ion is

Mn^{2+}:	$3d^5$	$4s$	$4p$
	↑ ↑ ↑ ↑ ↑	_	_ _ _

By pairing two of these 3*d* electrons, two empty *d* orbitals become available for bonding:

$3d^5$	$4s$	$4p$
⇅ ⇅ ↑ _ _	_	_ _ _

A total of six orbitals available for bonding

The hybridization on Mn^{2+} is d^2sp^3, and the complex is paramagnetic because one of the *d* electrons on the Mn^{2+} ion is unpaired.

Practice Problem 22.5: Write an orbital diagram for $Co(NH_3)_6^{3+}$ ion showing the *d* orbital population of Co^{3+} ion and the orbitals available for ligand-to-metal bonding. What is the hybridization on cobalt in $Co(NH_3)_6^{3+}$? Is the coordination complex diamagnetic or paramagnetic?

The octahedral coordination complex of Fe^{2+} with cyanide ions as ligands $Fe(CN)_6^{4-}$ is diamagnetic. But when water molecules are the ligands the complex ion $Fe(H_2O)_6^{2+}$ is paramagnetic. The diamagnetism of the hexacyanoferrate(II) ion can be explained by valence bond theory by assuming that the six electrons on the Fe^{2+} ion in the complex occupy three *d* orbitals in pairs, leaving two *d* orbitals to be occupied by the lone pairs of the CN^- ions:

Fe^{2+}:[Ar]	$3d^6$	$4s$	$4p$
	⇅ ⇅ ⇅ _ _	_	_ _ _

d^2sp^3

The paramagnetism of the hexaaquairon(II) ion can be explained by assuming that the six *d* electrons of the Fe^{2+} ion occupy all of the 3*d* orbitals of the ion. One 4*s* orbital, three 4*p* orbitals, and two 4*d* orbitals can then be occupied by the lone electron pairs of the donor atoms of the ligands:

Fe^{2+}:[Ar]	$3d^6$	$4s$	$4p$	$4d$
	⇅ ↑ ↑ ↑ ↑	_	_ _ _	_ _ _ _ _

sp^3d^2

A coordination complex such as $Fe(CN)_6^{4-}$ in which the lower-energy *d* orbitals of the metal are used for ligand bonding is called an **inner orbital complex**. A coordination complex in which the higher-energy *d* orbitals of the metal are used for ligand bonding is called an **outer orbital complex**. An example of an outer orbital complex is $Fe(H_2O)_6^{2+}$. In this complex, the central Fe^{2+} ion is sp^3d^2 hybridized.

Although the explanations of the magnetic properties of different Fe^{2+} complexes above are quite reasonable, valence bond theory cannot adequately explain why some ligands produce inner orbital complexes and other ligands

Figure 22.19

Spatial orientations of the metal and ligand orbitals in the formation of an octahedral complex. (a) Approach of the ligands along the d_{z^2} orbital lobes of the metal. (b) Approach of the ligands along the $d_{x^2-y^2}$ orbital lobes of the metal. (c) Approach of ligands between the d_{yz}, d_{xz}, and d_{xy} orbital lobes.

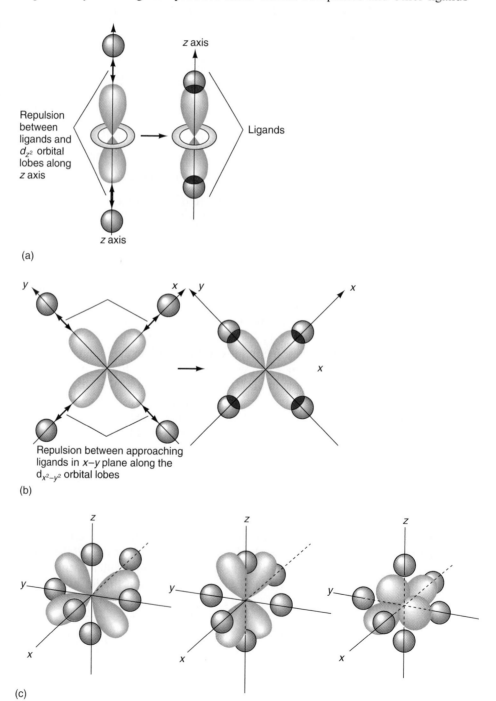

z axis
Repulsion between ligands and d_{z^2} orbital lobes along z axis
Ligands
z axis
(a)

y
x
y
x
x
Repulsion between approaching ligands in x–y plane along the $d_{x^2-y^2}$ orbital lobes
(b)

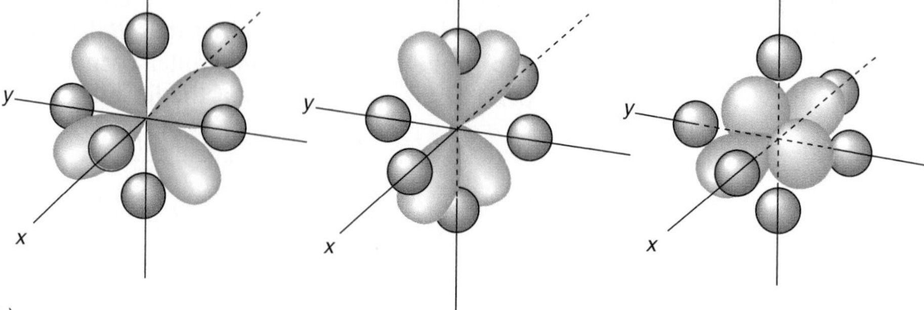

(c)

produce outer orbital complexes. Crystal field theory gives a better account of the magnetic properties of transition metal complexes. It also explains their colors. We discuss crystal field theory next.

Crystal Field Theory

Crystal field theory assumes that the attraction between the central metal ion and the ligands is essentially electrostatic. Ligands that are not negatively charged ions are usually polar molecules. These ligands have an electron pair on the donor atom that is attracted to the positive metal ion.

Crystal field theory also assumes that when ligands approach the central metal ion to form a coordination sphere, the energy of the *d* orbitals of the metal ion splits in a pattern that is characteristic of the geometry of the complex.

Let us consider the formation of an octahedral complex ion such as $Fe(CN)_6^{4-}$. The six ligands in this complex ion are distributed in an octahedral geometry. The d_{z^2} and $d_{x^2-y^2}$ orbital lobes of an isolated Fe^{2+} ion point toward the corners of an octahedron. In building a model for the formation of the complex, we assume by convention that four CN^- ions approach the Fe^{2+} ion along the $d_{x^2-y^2}$ orbital lobes, and two more ligands approach along the d_{z^2} orbital lobes (Figure 22.19). When this happens, the *d* electrons of the Fe^{2+} ion and the lone pairs of the approaching ligands initially repel each other. This repulsion increases the energy of the *d* orbitals of the Fe^{2+} ion as shown in Figure 22.20. Since the ligands approach directly along the lobes of the $d_{x^2-y^2}$ and d_{z^2} orbitals, the repulsion between the electrons in these orbitals and the incoming lone pairs of the ligands is stronger than the repulsion between the

Figure 22.20

The *d* orbital energies and electron populations of an Fe^{2+} ion in the formation of $Fe(CN)_6^{4-}$, an octahedral complex. This diagram illustrates the *d* orbital splitting in the formation of any octahedral complex. Electron arrangement in the split *d* orbitals and the extent of splitting varies in different complexes. The energy difference between the lower and higher energy *d* orbitals in the complex is designated by Δ.

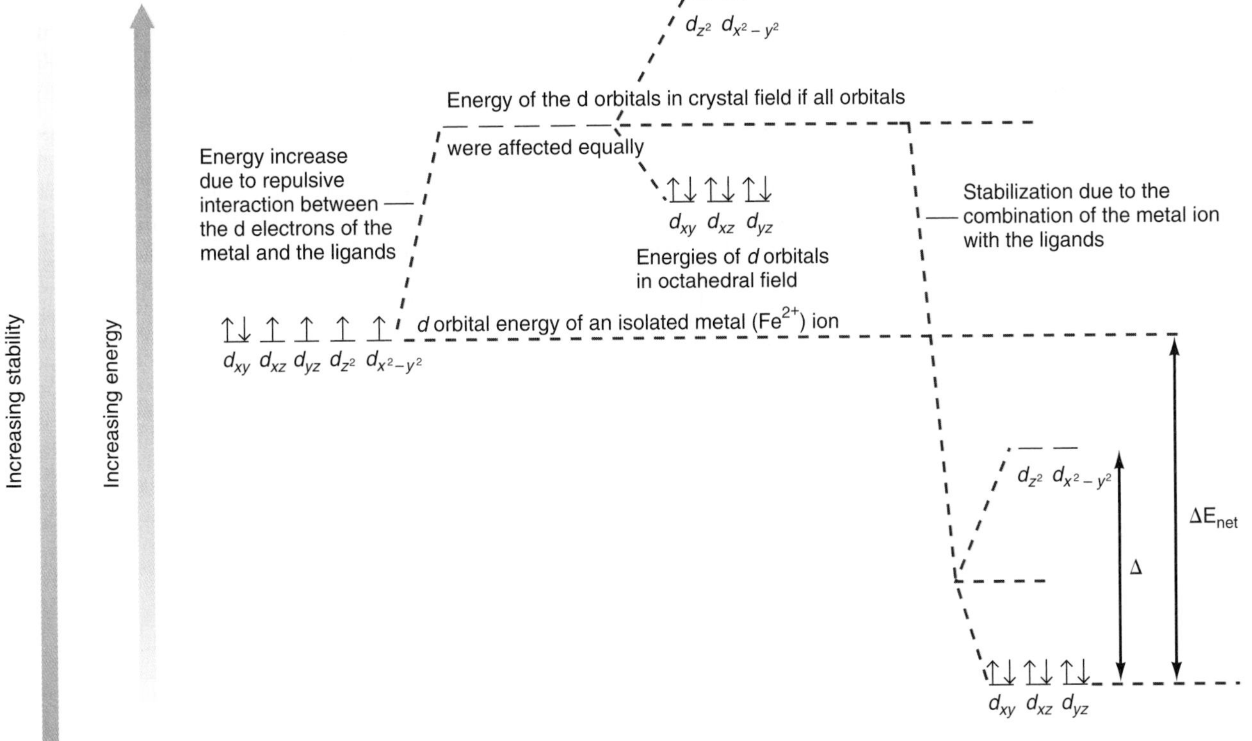

other d orbitals and the incoming ligands. The energies of $d_{x^2-y^2}$ and d_{z^2} orbitals are therefore raised more than the energies of the remaining three d orbitals, whose lobes lie between the lobes of the $d_{x^2-y^2}$ and d_{z^2} orbitals. Thus, the energy of the d orbitals is split (Figure 22.20), and the orbitals no longer have equal energies. However, the energies of $d_{x^2-y^2}$ and d_{z^2} orbitals are equal to each other, as are the energies of d_{xy}, d_{xz}, and d_{yz} orbitals.

The electrostatic field generated by the regular, crystalline array of ligands is known as the **crystal field**. The electrons of the central Fe^{2+} ion in $Fe(CN)_6^{4-}$ occupy the lower-energy d orbitals in pairs. This arrangement is energetically more favorable than the occupancy of both lower- and higher-energy orbitals.

Energy is released when the ligands become attached to the metal ion. The energy of the complex ion is therefore less than the energy of the isolated metal d orbitals, as shown in Figure 22.20. In other words, the complex ion is more stable than the isolated metal ion and its ligands. The complex ion is most stable when the maximum number of electrons are in the lower-energy orbitals (d_{xy}, d_{xz}, and d_{yz}). Electrons in higher-energy d orbitals tend to decrease the stability of the complex.

In a tetrahedral coordination complex, the interaction between the ligands and the d_{xy}, d_{xz}, and d_{yz} orbitals is greater than the interaction between the ligands and the d_{z^2} and $d_{x^2-y^2}$ orbitals because the ligands are positioned between the x, y, and z axes, as shown in Figure 22.21. Therefore, in a tetrahedral complex the d orbital energy splitting pattern (Figure 22.21b) is opposite the pattern for an octahedral complex.

In a square planar coordination complex, the ligands are attached to the central metal ion along the x and y axes (at the corners of a square). Thus, the ligands are repelled more by the $d_{x^2-y^2}$ orbital than by the d_{xy} orbital, whose lobes lie between the x and y axes. The ligands are repelled even less by the d_{z^2} orbital, which has a "doughnut" shape electron density in the x–y plane, but is confined close to the nucleus. The d_{xz} and d_{yz} orbitals are not in the x–y plane and therefore do not repel the ligands as strongly as the other d orbitals do. The d orbital energy splitting pattern for a square planar complex is shown in Figure 22.22.

Ligands differ in their abilities to split the d orbital energies of a metal ion. The following ligands are arranged in the order of their decreasing ability to split the d orbitals energy of the central metal ion in a coordination complex:

$$CN^- > NO_2^- > en > NH_3 > SCN^- > H_2O > OH^- > F^- > Cl^- > Br^- > I^-$$

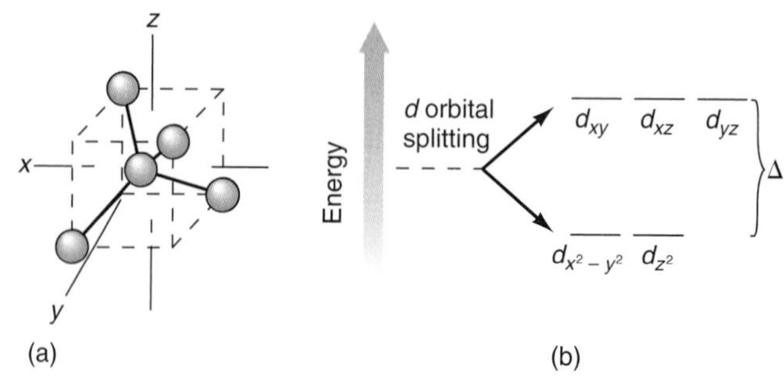

Figure 22.21

d orbital splitting in a tetrahedral complex. (a) Ligand attachment to the central metal ion occurs between the x, y, and z axes, near the lobes of d_{xy}, d_{xz}, and d_{yz} orbitals. (b) d orbital splitting pattern in a tetrahedral complex.

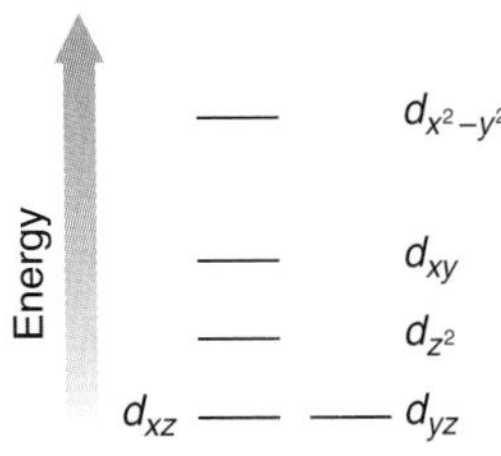

Figure 22.22

d orbital splitting in a square planar complex.

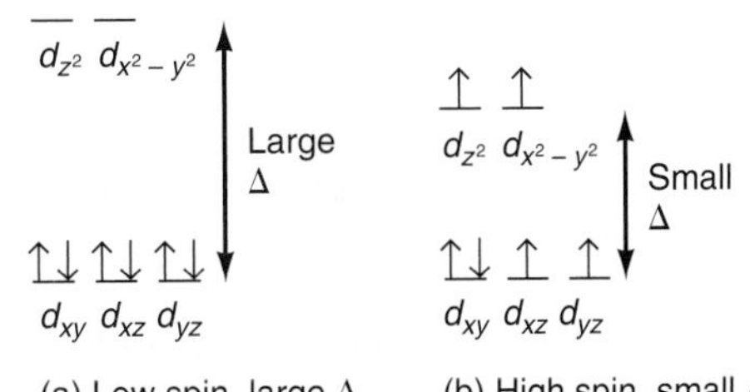

Figure 22.23

d orbital energy splittings and electron population in d^6 complexes: (a) low spin, as in $Fe(CN)_6^{4-}$ and (b) high spin, as in $Fe(H_2O)_6^{2+}$.

A list of ligands arranged in the order of their orbital splitting ability is called the **spectrochemical series.**

The spectrochemical series indicates that the splitting of *d* orbital energy is greater in $Fe(CN)_6^{4-}$ than in $Fe(H_2O)_6^{2+}$, as shown in Figure 22.23. In $Fe(CN)_6^{4-}$, the electrons of the metal ion occupy the lower-energy *d* orbitals in pairs (Figure 22.23a). The $Fe(CN)_6^{4-}$ ion is most stable in this configuration. The CN^- ions cause a large splitting of the *d* orbital energy, so the energy difference between the lower- and higher-energy orbitals is relatively large. The energy required to promote an electron in this complex from a lower- to a higher-energy *d* orbital is greater than the energy associated with the pairing of two electrons in the same orbital. Since the electrons are all paired, the $Fe(CN)_6^{4-}$ ion is diamagnetic.

In the $Fe(H_2O)_6^{2+}$ ion, the H_2O molecules split the *d* orbital energy of the central Fe^{2+} ion less than CN^- ions do. The small energy difference between the lower and the higher-energy *d* orbitals in $Fe(H_2O)_6^{2+}$ permits the electrons of the Fe^{2+} ion to occupy both the lower- and the higher-energy orbital levels according to Hund's rule. This electron arrangement makes the $Fe(H_2O)_6^{2+}$ion paramagnetic because four of the electrons are unpaired (Figure 22.23b).

Complexes such as $Fe(H_2O)_6^{2+}$, in which the *d* orbital energy splitting is relatively small and electrons can occupy both lower- and higher-energy orbitals with maximum number of electrons having parallel spin, are called **high-spin complexes**. Other complexes, such as $Fe(CN)_6^{4-}$, in which the *d* orbital splitting is large and the lower-energy *d* orbitals are filled before any of the electrons enter the higher-energy orbitals, are called **low-spin complexes**. In these complexes the number of unpaired electrons is at a minimum (low spin). Ligands such as CN^- and NO_2^-, which cause large *d* orbital energy splitting, are called **strong field ligands**. Others, such as H_2O and Cl^-, whose *d* orbital splitting ability is relatively small, are called **weak field ligands**.

Example 22.7

Drawing *d* Orbital Splitting Diagrams and Determining Whether Complexes Are High-Spin or Low-Spin

Draw the *d* orbital energy splitting diagrams and the electron arrangements for $Cr(NO_2)_6^{4-}$ (a low-spin complex) and for $CrCl_6^{4-}$ (a high-spin complex).

SOLUTION: We know that both of these complex ions contain chromium(II) because the sum of the charges of the ligands and the metal ion equals the charge of the complex. A Cr^{2+} ion has four *d* electrons. Since NO_2^- is a strong field ligand and Cl^- is a weak field ligand, the energy difference between the lower- and higher-energy *d* orbitals is greater in $Cr(NO_2)_6^{4-}$ than in $CrCl_6^{4-}$. Thus, all four of the electrons in $Cr(NO_2)_6^{4-}$ are in the lower-energy *d* orbitals, while in $CrCl_6^{4-}$ three electrons are in the lower-energy orbitals and one is in a higher-energy orbital, as shown by the orbital splitting diagrams:

For $Cr(NO_2)_6^{4-}$	For $CrCl_6^{4-}$
— — d_{z^2} $d_{x^2-y^2}$	↑ — d_{z^2} $d_{x^2-y^2}$
⇅ ↑ ↑ d_{xy} d_{xz} d_{yz}	↑ ↑ ↑ d_{xy} d_{xz} d_{yz}

Both complexes are paramagnetic, but the paramagnetism of $CrCl_6^{4-}$ is greater than that of $Cr(NO_2)_6^{4-}$ because $CrCl_6^{4-}$ has 4 unpaired electrons, whereas $Cr(NO_2)_6^{4-}$ has only two unpaired electrons.

Practice Problem 22.6: Draw *d* orbital energy splitting diagrams for $Co(CN)_6^{3-}$ and $Cr(H_2O)_6^{3+}$. Which of these two complexes is the high-spin complex? Are both complexes paramagnetic? Explain.

22.6 Color in Transition Metal Complexes

Most coordination complexes of the 3*d* series of transition metals are colored. The major exceptions to this generalization are the complexes of Sc^{3+} and Zn^{2+}, which are colorless. Crystal field theory can explain the colors of coordination complexes. The energy difference between the higher- and lower-energy *d* orbitals, Δ, on the central metal ion in a coordination complex is in the range of energies of visible light. Absorption of energy from white light can therefore excite an electron from a lower- to a higher-energy *d* orbital in a coordination complex.

The extent of *d* orbital energy splitting (the value of Δ) in a complex determines the wavelength of the light absorbed for the transition of an electron from a lower to a higher orbital level. The wavelengths of the radiant energy not absorbed are reflected or transmitted by the complex. These wavelengths determine the color we see. The color of the light transmitted is the complement of the color that is absorbed. For example, $Ti(H_2O)_6^{3+}$ ion is violet because it absorbs yellow-green light with a wavelength of about 500 nm. A solution of $Cu(H_2O)_4^{2+}$ ions absorbs yellow light at about 580 nm, and the transmitted light is blue. Some colors and their complements are: red and bluish green, blue and yellow, green and purple.

Substitution of different ligands in a complex changes the color by changing the *d* orbital energy splitting. $Co(NH_3)_6^{3+}$ is yellow because it absorbs violet light with a wavelength of approximately 473 nm. When a weaker field ligand such as a Cl^- ion is substituted for an NH_3 molecule, the resulting complex, $Co(NH_3)_5Cl^{2+}$, is purple because it absorbs yellow-green light with a wavelength of about 530 nm. The Cl^- ion is a weaker field ligand than the NH_3 molecule, and it decreases the value of Δ so that the electron transition requires the absorption of lower-energy, longer-wavelength radiation from white light.

Coordination complexes of Zn^{2+}, Cd^{2+}, Hg^{2+}, Cu^+, and Ag^+ ions are colorless because all the *d* orbitals of these ions are completely filled with electrons, making an electron transition impossible. Complexes of Sc^{3+}, Y^{3+}, and La^{3+} are also colorless because these ions have no *d* electrons.

Summary

Terminology

A **coordination complex** contains at least one **coordination sphere** in which a central metal ion or atom is attached to one or more **ligands** by coordinate covalent bonding. A coordination complex can be a cation, an anion, or a neutral molecule. The number of coordinate covalent bonds between the central metal ion and its ligands is called the **coordination number**. The charge on a coordination sphere equals the sum of the oxidation state of the central metal and the charges of attached ligands.

A ligand that is attached to the central metal ion in a coordination sphere by only one **donor atom** is a **monodentate ligand**. Ligands that are attached to the central metal by two or more donor atoms are called **polydentate ligands** or **chelating agents**. Ligands that are attached by two, three, or four donor atoms are called **bidentate**, **terdentate**, and **tetradentate** ligands, respectively.

Naming Coordination Complexes

In naming a complex cation or a neutral complex, the ligands are named first, in alphabetical order, then the central metal, followed by its oxidation state in parentheses. The number of ligands of each kind is indicated by prefixes di-, tri-, tetra-, and so on. For ligands with complex names, the prefixes bis-, tris-, and tetrakis- are used.

A complex anion is named similarly, except that the suffix -ate is added to the name of the central metal. When the English name of the metal with suffix -ate is clumsy, the stem of its Latin name is used, followed by the suffix -ate.

Isomerism

Isomeric coordination complexes can be divided into **structural isomers** and **stereoisomers**. Structural isomers include **ionization** and **linkage** isomers. In stereoisomers, the ligands bonded to the central metal ion occupy different positions in space. Stereoisomers include **cis–trans** isomers and **optical isomers**. Optical isomers that rotate plane-polarized light to the right (clockwise) are called **dextrorotatory**, dextro, or ***d*** isomers. Optical isomers that rotate plane-polarized light to the left (counterclockwise) are called **levorotatory**, **levo**, or ***l*** isomers.

Valence Bond Theory

Valence bond theory assumes that in a coordination complex, the central metal ion is linked to its ligands by coordinate covalent bonds. The donor atoms of the ligands contribute electron pairs for these bonds. The electron pairs of the ligands occupy vacant orbitals of the metal. The vacant orbitals used by the metal depend on the coordination number of the metal and the geometry of the complex.

Crystal Field Theory

According to **crystal field theory**, bonding between the central metal ion and its ligands is essentially electrostatic. Crystal field theory also assumes that when ligands approach the central metal ion to form a coordination sphere, the *d* orbital energy of the metal ion splits in a pattern that is characteristic of the geometry of the complex. Some ligands can split the orbital energy more than other ligands. Ligands that cause large *d* orbital splitting are called **strong field ligands**, and those that cause small splitting are **weak field ligands**. The greater the *d* orbital energy difference, the greater is the likelihood that electrons will occupy the lower-energy orbitals with spins paired.

The color of a transition metal complex is due to its absorbing some component of white light. The color we see results from the components of white light *transmitted* by the complex. That color is the complement of the color absorbed. The energy of the component of white light absorbed by a complex must be just sufficient to excite the electrons from lower- to higher-energy *d* orbitals in the complex. Thus, the color of a complex depends on the energy difference between the *d* orbitals that are split.

New Terms

Bidentate ligand (22.2)
Chelate (22.2)
Chelating agent (22.2)
Coordination number (22.1)
Coordination sphere (22.1)
Crystal field (22.5)
Crystal field theory (22.5)
Dextrorotatory isomer (22.4)
Donor atom (22.2)
Facial isomer (22.4)
Geometric isomers (22.4)
High-spin complex (22.5)
Inner orbital complex (22.5)
Ionization isomers (22.4)
Levorotatory isomer (22.4)
Linkage isomers (22.4)
Low-spin complex (22.5)
Meridional isomer (22.4)
Monodentate ligand (22.2)
Optical isomers (22.4)
Outer orbital complex (22.5)
Polydentate ligand (22.2)
Position isomers (22.4)
Racemic mixture (22.4)
Spectrochemical series (22.5)
Stereoisomers (22.4)
Strong field ligand (22.5)
Structural isomers (22.4)
Terdentate ligand (22.2)
Tetradentate ligand (22.2)
Weak field ligand (22.5)

Exercises

General Review

1. What is: **(a)** a ligand; **(b)** a coordination sphere?
2. How can you determine whether negative ions in a coordination complex are inside or outside the coordination sphere of the complex?
3. Can a coordination complex contain more than one coordination sphere? If your answer is yes, give the formula of one such complex.
4. Write a formula to illustrate a complex compound that is composed of: **(a)** monatomic cations and negatively charged coordination spheres; **(b)** positively charged coordination spheres and monatomic anions; **(c)** neutral molecules that are coordination spheres; **(d)** positive coordination spheres and negative coordination spheres.
5. What is meant by "coordination number" in a coordination complex?
6. Give the formula of: **(a)** a monodentate ligand; **(b)** a bidentate ligand.
7. What is the difference between a chelating agent and a chelate? Illustrate your answer by writing a formula for each.
8. Explain how you can determine the oxidation state of the central metal ion in a coordination sphere from the formula of the complex.
9. What are: **(a)** structural isomers; **(b)** stereoisomers?
10. Write formulas to illustrate each of the following types of structural isomers: **(a)** ionization isomers; **(b)** linkage isomers.
11. Write structural formulas to illustrate the following types of stereoisomers: **(a)** cis–trans isomers; **(b)** optical isomers.
12. Referring to optical isomers, what is meant by "dextro" and "levo"?
13. What is a "racemic mixture"?
14. Explain and illustrate using specific examples how the valence bond theory correlates the structure of a coordination complex with the hybridization on the central metal ion in: **(a)** a linear complex; **(b)** a tetrahedral complex; **(c)** a square planar complex; **(d)** an octahedral complex.

15. What is the hybridization on a metal ion with a d^6 valence shell electron population in an: **(a)** inner orbital octahedral complex; **(b)** outer orbital octahedral complex?
16. What are some of the major advantages of the crystal field theory over the valence bond theory in explaining the properties of transition metal complexes?
17. Why are orbital energies of the central metal ion "split" in a coordination complex?
18. Arrange several common ligands in the order of their decreasing ability to split *d* orbitals of a metal ion in a complex. What is the name of such a list of ligands?
19. Is a low-spin complex generally formed by strong field or by weak field ligands? Explain and illustrate.
20. Explain why complexes of metal ions with d^0 or d^{10} electron population are colorless.
21. If a complex is red, does it absorb components of white light closer to 400 nm or closer to 700 nm?
22. Why is $Ag(NH_3)_2^+$ called a complex ion?

Formulas and Charges of Complexes

23. A coordination compound of chromium contains 1 mol of chromium, 3 mol of chlorine, and 5 mol of ammonia per mole of the compound. Treatment of 0.5 mol of this compound with silver nitrate gives an immediate precipitate of 1 mol of silver chloride. Write the formula of the compound and indicate the coordination sphere in brackets. What is the oxidation state of chromium in this compound?
24. What is the oxidation state of the central metal in each of the following coordination complexes? **(a)** $Ag(NH_3)_2^+$; **(b)** $FeCl_4^-$; **(c)** $Fe(CN)_6^{3-}$; **(d)** $K_4[Fe(CN)_6]$; **(e)** $[Cr(H_2O)_4Cl_2]^+$; **(f)** $[Co(NH_3)_3Cl_3]$; **(g)** $[Co(NH_3)_6]Br_3$.
25. What are the charges on the complex ions in question 24d and g?
26. What is the coordination number of the metal ion in each of the complexes in question 24 and in question 31k?
27. Write the formula of the compound that results from the combination of $Cu(NH_3)_4^{2+}$ and $Fe(CN)_6^{3-}$ ions.

Types of Ligands

28. Which of the following species can act as ligands and which cannot? **(a)** H_2; **(b)** CH_4; **(c)** Be^{2+}; **(d)** F^-; **(e)** AlH_3; **(f)** PH_3; **(g)** NH_4^+.
29. Classify each of the following ligands as monodentate, bidentate, terdentate, and so on: **(a)** H_2O; **(b)** CN^-; **(c)** $H_2NCH_2CH_2NHCH_2CH_2NH_2$; **(d)** acetylacetonate ion (Table 22.2); **(e)** Cl^-.
30. Which of the following ligands can act as chelating agents? **(a)** OH^-; **(b)** $EDTA^{4-}$; **(c)** Br^-; **(d)** dimethylglyoximate ion; **(e)** porphyrin.

Nomenclature of Coordination Complexes

31. Name each of the following: **(a)** $Ag(NH_3)_2^+$; **(b)** $Cu(H_2O)_4^{2+}$; **(c)** $Fe(CN)_6^{3-}$; **(d)** $Fe(CN)_6^{4-}$; **(e)** $Co(NH_3)_6Br_3$; **(f)** $[Cr(H_2O)_4Cl_2]NO_3$; **(g)** $K_2[PtCl_6]$; **(h)** $Na_2[Zn(OH)_4]$; **(i)** $Al(OH)_4^-$; **(j)** $[Cu(NH_3)_4]_3[Fe(CN)_6]_2$; **(k)** $Mn(NH_2CH_2CH_2NH_2)_3^{2+}$.
32. Write a formula for each of the following: **(a)** tetraamminecopper(II) ion; **(b)** tetraaquadichlorochromium(III) ion; **(c)** tetrahydroxozincate ion; **(d)** sodium tetrahydroxozincate; **(e)** tetraamminedichlorocobalt(III) chloride; **(f)** trisethylenediaminenickel(II) bromide; **(g)** diamminesilver(I) hexacyanoferrate(II); **(h)** aquapentaammineruthenium(III) ion.

Isomerism in Coordination Complexes

33. Indicate which type of isomers (ionization, or linkage) each of the following pairs of compounds represents: **(a)** $[Cr(H_2O)_6]Cl_3$ and $[Cr(H_2O)_4Cl_2]Cl(H_2O)_2$; **(b)** $[Co(NH_3)_4Cl_2]NO_2$ and $[Co(NH_3)_4Cl(NO_2)]Cl$; **(c)** $[Fe—SCN]^{2+}$ and $[Fe—NCS]^{2+}$.
34. Are the isomers in question 33 structural isomers or stereoisomers?
35. Sketch the structures of cis and trans isomers for each of the following complexes: **(a)** square planar $[Pt(NH_3)_2(NO_2)_2]$; **(b)** square planar $[NiCl_2(PR_3)_2]$, where R = $—CH_3$ group (these groups are not in the plane with Ni, Cl, and P atoms); **(c)** $[Co(NH_3)_4Br_2]^+$; **(d)** $[Cr(NH_3)_2(C_2O_4)_2]^-$ ($C_2O_4^{2-}$ ion is a bidentate ligand).
36. For the cis isomer of $[Cr(NH_3)_2(C_2O_4)_2]^-$ (see question 35d) two different structures are possible. Sketch these structures. What type of isomerism do they represent?
37. Sketch the isomeric structures for $[Co(en)_3]^{3+}$ (en = ethylenediamine, $NH_2CH_2CH_2NH_2$). What type of isomerism is represented by these structures?
38. What is the general name of the type of isomerism illustrated by all the structures in questions 35, 36, and 37?
*39. In an experiment, the compound $Co(en)_2(NO_2)_2Cl$ is prepared in three different isomeric forms. One of the forms does not react with either $AgNO_3$ or with additional ethylenediamine (en) and is optically inactive. Another form reacts with $AgNO_3$ to give a white precipitate but does not react with ethylenediamine; this form is optically inactive. The third form is optically active and reacts with both $AgNO_3$ and with ethylenediamine. Sketch the structures of each of these three isomeric forms, including the optical isomers of the third form. (*Hint*: A bidentate ligand can occupy cis positions but not trans positions.)

* Denotes a problem of greater than average difficulty.

Valence Bond Theory

40. What is the hybridization on the central metal ion in each of the following complexes? **(a)** MnF_6^{3-}; **(b)** $Co(CN)_6^{3-}$; **(c)** $Cr(NH_3)_6^{3+}$; **(d)** $Ag(CN)_2^-$; **(e)** $Cu(H_2O)_4^+$; **(f)** $Zn(NH_3)_4^{2+}$; **(g)** $Zn(CN)_4^{2-}$.

41. What is the structure of each of the complexes listed in question 40?

42. Using valence bond theory, predict whether each of the following complexes is likely to be diamagnetic or paramagnetic: **(a)** $Cr(NH_3)_6^{3+}$; **(b)** $[CrCl_4(H_2O)_2]^-$; **(c)** $Co(H_2O)_6^{2+}$; **(d)** $Ag(CN)_2^-$; **(e)** $Cu(H_2O)_4^{2+}$.

43. Using valence bond theory, explain why $Co(CN)_6^{3-}$ is diamagnetic but $CoCl_6^{3-}$ is paramagnetic.

44. Draw diagrams showing the outer electron configuration and the hybridization on the central metal ion of each of the following four-coordinate complexes: **(a)** $Zn(CN)_4^{2-}$, tetrahedral; **(b)** $Cu(NH_3)_4^{2+}$, square planar; **(c)** $Pt(NH_3)_4^{2+}$, square planar; **(d)** $Ni(CN)_4^{2-}$, square planar; **(e)** $NiCl_4^{2-}$, tetrahedral.

45. Would you expect each of the following complexes to be diamagnetic or paramagnetic? **(a)** $Zn(CN)_4^{2-}$; **(b)** $Cu(NH_3)_4^{2+}$; **(c)** $NiCl_4^{2-}$.

46. Explain how magnetic measurements can establish whether a four-coordinate nickel(II) complex is planar or tetrahedral.

47. Show that magnetic measurements cannot be used to determine whether a copper(II) complex is planar or tetrahedral.

***48.** There is some evidence for the existence of a compound of the type $Li[Fe(CO)Cl_3]$. Would you expect this compound to be diamagnetic or paramagnetic assuming that it has a planar arrangement of CO and the three Cl^- groups around the iron?

* Denotes a problem of greater than average difficulty.

Crystal Field Theory

49. Draw a *d* orbital energy splitting pattern in an octahedral crystal field and describe the electron distribution in the lower- and higher-energy *d* orbitals of each of the following ions: **(a)** Mn^{2+} in a weak field; **(b)** Mn^{2+} in a strong field; **(c)** Fe^{2+} in a strong field; **(d)** Fe^{2+} in a weak field; **(e)** Fe^{3+} in a strong field.

50. Give an example of each complex in question 49 and write the formula and the name for it. Which of these complexes are high-spin and which are low-spin complexes?

51. Using crystal field theory, compare the magnetic properties of $CrCl_6^{4-}$ and $Cr(CN)_6^{4-}$ ions. Which one of these ions would be attracted more strongly to a magnet?

52. The complex $Ni(CN)_4^{2-}$ is diamagnetic but $NiCl_4^{2-}$ is paramagnetic with two unpaired electrons. Explain these observations using crystal field theory. On the basis of your explanation, what geometry would you predict for each of these complexes?

Color in Complexes

53. Which of the following complexes would you expect to be colorless? **(a)** $Fe(CN)_6^{3-}$; **(b)** $Zn(CN)_4^{2-}$; **(c)** $Cu(NH_3)_4^{2+}$; **(d)** $Ag(NH_3)_2^+$; **(e)** $NiCl_4^{2-}$; **(f)** $V(H_2O)_6^{3+}$; **(g)** $La(H_2O)_6^{3+}$.

54. All the following complexes are blue or violet: $Cu(H_2O)_4^{2+}$, $Cu(NH_3)_4^{2+}$, and $CuCl_4^{2-}$. Which of these three complexes would you expect to absorb the component of visible light of shortest wavelength? Explain.

55. The complex $Co(NH_3)_6^{3+}$ absorbs most strongly at 470 nm. Would you predict the color of this complex to be blue or some other color?

CHAPTER 23

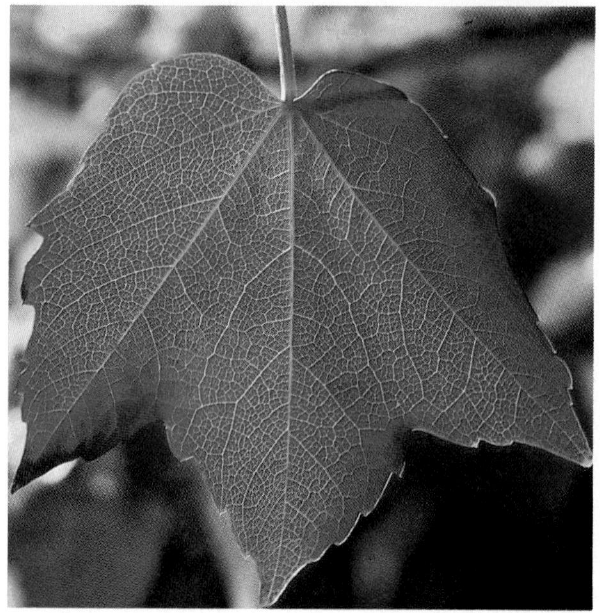

In mid-autumn, red pigments such as *cyanidin* are formed from sugars trapped in the leaf. Mixed yellow and red pigments may make the leaf orange.

In winter all pigments decompose revealing the brown of wood compounds *cellulose* (top) and *lignin* (bottom).

Organic Chemistry

Until the early nineteenth century it was believed that the substances in living cells were fundamentally different from the substances of nonliving matter. Compounds found in living organisms were called organic compounds, *and compounds found in nonliving matter were* inorganic compounds. *In 1828, Friedrich Wöhler, a German chemist, heated an aqueous solution of the inorganic compound ammonium cyanate, NH_4NCO, and obtained the organic compound urea, a known constituent of the urine of mammals:*

$$\underset{\text{Ammonium cyanate}}{NH_4N{=}C{=}O(aq)} \longrightarrow \underset{\text{Urea}}{H_2N{-}\overset{\overset{\displaystyle O}{\|}}{C}{-}NH_2(aq)}$$

Following the work of Wöhler, many other organic compounds were prepared from inorganic substances. As a result, the belief that matter in living organisms is inherently different from inanimate matter was gradually abandoned. Today, organic chemistry may be defined as the branch of chemistry that deals with compounds containing carbon–hydrogen bonds, carbon–carbon bonds, or both.

Millions of organic compounds have been discovered. Many organic compounds are derived from coal, natural gas, and petroleum; others are produced by the chemical industry. Synthetic organic compounds are used to make plastics, drugs, clothing, paints, and many other products. In this chapter we discuss some important types of organic compounds and their reactions. Since organic chemistry is closely related to biochemistry, some important biological molecules are also considered.

23.1 Hydrocarbons

Hydrocarbons are among the most commonly encountered organic compounds. We discussed the structures and bonding of methane, ethane, ethene, and ethyne—the smallest and simplest organic molecules—in Chapter 8. In this chapter we consider hydrocarbons in more detail.

Classification of Hydrocarbons

Hydrocarbons can be divided into two main classes: **aliphatic hydrocarbons** and **aromatic hydrocarbons**. Aliphatic hydrocarbon molecules can consist of chains of carbon atoms or rings of carbon atoms (Figure 23.1). The hydrocarbons we discussed in Chapter 8—alkanes, alkenes, and alkynes—are examples of aliphatic hydrocarbons. Aromatic hydrocarbons consist of cyclic molecules or "rings" of carbon atoms that are linked by sigma bonds and by delocalized pi bonding, as shown in Figure 23.1. Various aromatic hydrocarbons contain one or more carbon rings, which are often connected to various side chains. Examples of aromatic hydrocarbons include benzene (a good solvent—though a toxic one—for fats and paints), naphthalene (present in some mothballs), and anthracene (used in making color designs on cotton fabric).

The word "aliphatic hydrocarbon" originally referred to molecules related to fat, and the term "aromatic hydrocarbon" originally alluded to the aromas of

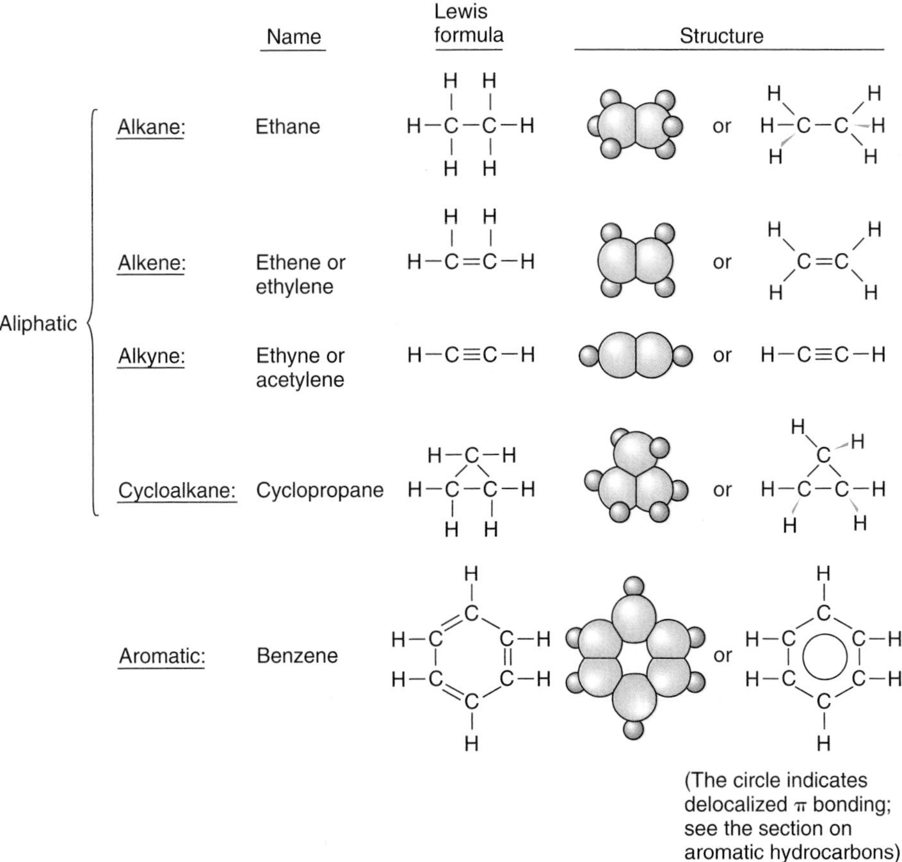

Figure 23.1

Types of hydrocarbons. Alkanes are hydrocarbons with the general formula C_nH_{2n+2}. Alkenes contain double bonds and are represented by the general formula C_nH_{2n}. Alkynes contain triple bonds; they are represented by the general formula C_nH_{2n-2}. Cycloalkanes consist of cyclic molecules. Aromatic hydrocarbons contain cyclic molecules in which the carbon atoms are linked by sigma bonds and partial pi bonding.

these compounds. But these terms have lost their original meanings. The diagram below summarizes the classification of hydrocarbons.

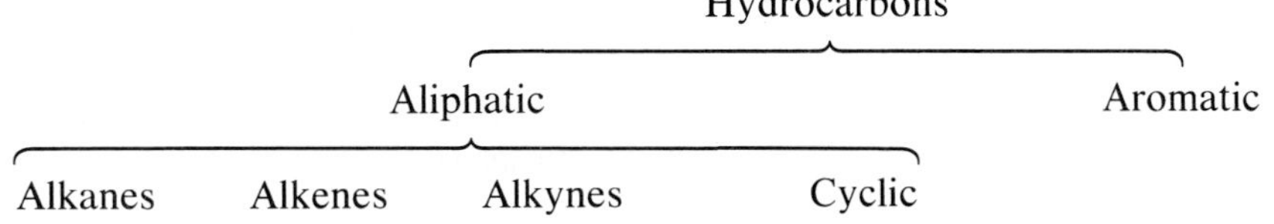

A review of the structures and bonding in alkanes, alkenes, and alkynes in Chapter 8 will help you to understand this chapter.

Alkanes

Alkanes are components of crude oil, a complex mixture of aliphatic and aromatic hydrocarbons whose constituents vary from methane to molecules containing 50 or more carbon atoms. The components of crude oil can be separated by fractional distillation (see Section 12.3). The boiling points of alkanes increase with increasing molar mass as shown in Table 23.1. When crude oil is distilled, the vapors of hydrocarbons that boil at lower temperatures are condensed and separated from the higher-boiling residue. The fractions obtained at different temperatures are listed in Table 23.2.

Table 23.1

Straight-Chain Alkanes and Their Boiling Points

Name	Molecular Formula	Condensed Structural Formula	Boiling Point (°C)
Methane	CH_4	CH_4	−164.0
Ethane	C_2H_6	CH_3CH_3	−88.6
Propane	C_3H_8	$CH_3CH_2CH_3$	−42.1
Butane	C_4H_{10}	$CH_3CH_2CH_2CH_3$ or $CH_3(CH_2)_2CH_3$	−0.5
Pentane	C_5H_{12}	$CH_3CH_2CH_2CH_2CH_3$ or $CH_3(CH_2)_3CH_3$	36.1
Hexane	C_6H_{14}	$CH_3CH_2CH_2CH_2CH_2CH_3$ or $CH_3(CH_2)_4CH_3$	68.9
Heptane	C_7H_{16}	$CH_3CH_2CH_2CH_2CH_2CH_2CH_3$ or $CH_3(CH_2)_5CH_3$	98.4
Octane	C_8H_{18}	$CH_3CH_2CH_2CH_2CH_2CH_2CH_2CH_3$ or $CH_3(CH_2)_6CH_3$	125.7
Nonane	C_9H_{20}	$CH_3CH_2CH_2CH_2CH_2CH_2CH_2CH_2CH_3$ or $CH_3(CH_2)_7CH_3$	150.8
Decane	$C_{10}H_{22}$	$CH_3CH_2CH_2CH_2CH_2CH_2CH_2CH_2CH_2CH_3$ or $CH_3(CH_2)_8CH_3$	174.1

Table 23.2

Petroleum Fractions

Approximate Boiling Range (°C)	Principal Product	Approximate Number of Carbon Atoms per Molecule	Physical State of Fraction at 25°C	Use
Below 20	Petroleum gas or natural gas	C_1–C_5	Gas	Heating
20–60	Petroleum ether or light petroleum	C_5–C_6	Liquid	Industrial solvent
60–120	Light naphtha or ligroin	C_6–C_7	Liquid	Industrial solvent
40–200	Gasoline	C_5–C_{10}	Liquid	Fuel for motor vehicles
175–300	Kerosene	C_{10}–C_{18}	Liquid	Aircraft fuel and home heating oil
250–400	Diesel fuel	C_{12}–C_{20}	Liquid	Diesel fuel
Above 400	Lubricating oils	C_{20}–C_{30}	Viscous liquid	Lubricant
	Paraffin wax	C_{16}–C_{40}	Semisolid to solid	Manufacture of candles
	Asphalt	C_{20}–C_{40}	Semisolid to solid	Roofing and road construction

We can see from Table 23.1 that the straight-chain alkanes form a series of compounds in which *each member has one more CH_2 unit than its preceding member*. The molecules in this series have the general formula C_nH_{2n+2}, where n is the number of carbon atoms in the molecule.

Methane, the simplest alkane, is the principal ingredient of natural gas, which is widely used for cooking and home heating. Minor components of natural gas include ethane, propane, and butane. Propane is the major component of bottled gas, also used for home heating and cooking where municipal natural gas is not available. Butane is used in disposable pocket lighters and in the fuel canisters of camping lanterns and stoves. Alkanes with about 5 to 12 carbon atoms are constituents of gasoline. Different ways of representing the

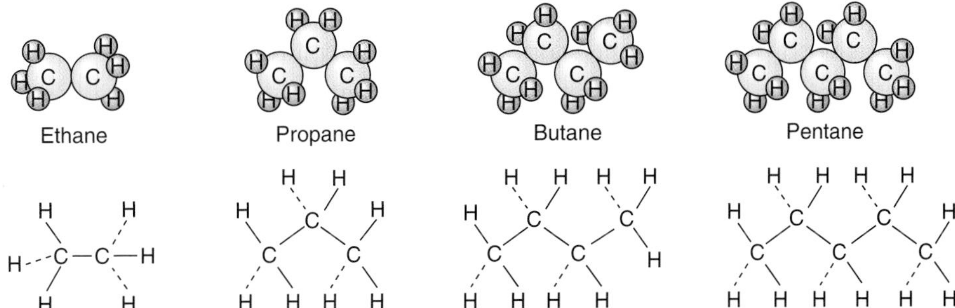

Figure 23.2

Formulas of alkanes. Note how each carbon atom is linked to other atoms in tetrahedral structure.

structures of alkanes are shown in Figure 23.2. (Recall that in an alkane, each carbon atom has a tetrahedral structure.)

The components of gasoline include both straight-chain and branched-chain alkanes. The term "straight chain" means that the carbon atoms are arranged "one after another" with no branches. We can see from Figure 23.2 that a "straight" chain actually has a "zigzag" arrangement because the carbon atoms in the chain are tetrahedral. The alkanes listed in Table 23.1 are straight-chain alkanes. An alkane can be represented by a Lewis formula or by a condensed formula, as shown below for butane:

$$\begin{array}{c} \begin{array}{ccccccccc} & & H & & H & & H & & H \\ & & | & & | & & | & & | \\ H & — & C & — & C & — & C & — & C & — H \\ & & | & & | & & | & & | \\ & & H & & H & & H & & H \end{array} \\ \text{(Lewis formula)} \end{array} \quad \text{or} \quad \begin{array}{c} CH_3(CH_2)_2CH_3 \\ \text{(Condensed formula)} \end{array}$$

The formula for a larger molecule such as octane can be written $CH_3—CH_2—CH_2—CH_2—CH_2—CH_2—CH_2—CH_3$ or $CH_3(CH_2)_6CH_3$.

It is possible to link four or more carbon atoms together in different ways to produce branched-chain hydrocarbons with the same molecular formula but different structural formulas; the molecules are structural isomers (see Section 22.4). For example, two substances that have the same molecular formula, C_4H_{10}, but different structural formulas are butane and methylpropane:

$$\begin{array}{ccccccccc} & H & & H & & H & & H & \\ & | & & | & & | & & | & \\ H— & C & — & C & — & C & — & C & —H \\ & | & & | & & | & & | & \\ & H & & H & & H & & H & \end{array} \qquad \begin{array}{ccccc} & & H & & \\ & & | & & \\ H & & H—C—H & & H \\ | & & | & & | \\ H—C & — & C & — & C—H \\ | & & | & & | \\ H & & H & & H \end{array}$$

or	or
$CH_3(CH_2)_2CH_3$	$CH(CH_3)_3$
Butane	Methylpropane
m.p.: −135°C	m.p.: −145°C
b.p.: −0.5°C	b.p.: −10°C

Although the molecular formulas for these two isomers are the same, C_4H_{10}, their structural formulas and their physical properties are different.

One important alkane is the component of gasoline that best prevents engine "knocking" (caused by premature combustion). It has the molecular formula C_8H_{18} and is commonly called "octane." Its correct name is 2,2,4-trimethylpentane and it has the structural formula

$$\begin{array}{ccccccccc} & & CH_3 & & & & CH_3 & & \\ & & | & & & & | & & \\ CH_3 & — & C & — & CH_2 & — & C & — & CH_3 \\ & & | & & & & | & & \\ & & CH_3 & & & & H & & \end{array}$$

2,2,4-Trimethylpentane
(commonly called "octane")

This compound is assigned an arbitrary octane number (rating) of 100, whereas heptane, $CH_3CH_2CH_2CH_2CH_2CH_2CH_3$, the component of gasoline that has the lowest ability to prevent knocking, is assigned an octane number of 0. Other alkanes, or mixtures thereof, are assigned octane numbers within this range. Molecules with many branches have high octane numbers. The "antiknock" properties of gasoline improve when gasoline contains a high fraction of branched alkanes.

Naming Alkanes

The International Union of Pure and Applied Chemistry (IUPAC) has developed a systematic method for naming organic compounds. This nomenclature system is called *IUPAC nomenclature*. The IUPAC system for naming alkanes is outlined below.

1. Find the *longest continuous carbon chain* in the molecule. The number of carbon atoms in this chain dictates the *parent name* of the compound (see Table 23.1 for the systematic name based on the length of the carbon chain). For example, in the structural formula

$$\begin{array}{ccccccccc} & & CH_3 & & & & & & \\ & & | & & & & & & \\ CH_3 & — & CH & — & CH_2 & — & CH_2 & — & CH_3 \end{array}$$

the longest chain contains five carbon atoms. Consulting Table 23.1, we find that the parent name is pentane.

2. A substituent on the longest continuous chain echoes the name of the alkane from which the substituent is derived. Thus, a $—CH_3$ group is called a *methyl* group because it is derived from the name methane. To name the substituent we drop the ending *-ane* and add *-yl*. Thus, a $CH_3CH_2—$ group is called an *ethyl* group because it is derived from the parent alkane ethane. Such substituents are collectively called **alkyl groups**. Thus, in the compound above we have a methyl substituent:

$$\begin{array}{ccccccccc} & & CH_3 & \longleftarrow & & & & & \text{methyl} \\ & & | & & & & & & \\ CH_3 & — & CH & — & CH_2 & — & CH_2 & — & CH_3 \end{array}$$

3. The longest continuous chain is numbered from one end to the other so that the carbon atoms to which the substituents are attached to the chain have the *smallest possible numbers*:

$$\begin{array}{ccccccccc} & & CH_3 & & & & & & \\ & & | & & & & & & \\ CH_3 & — & CH & — & CH_2 & — & CH_2 & — & CH_3 \\ 1 & & 2 & & 3 & & 4 & & 5 \end{array}$$

The location of the substituent, a methyl group, is indicated by the number of the carbon atom to which it is attached, followed by the name of the substituent (the methyl group), which is followed by the name of the parent alkane, pentane in this case. Thus, the IUPAC name is developed as follows:

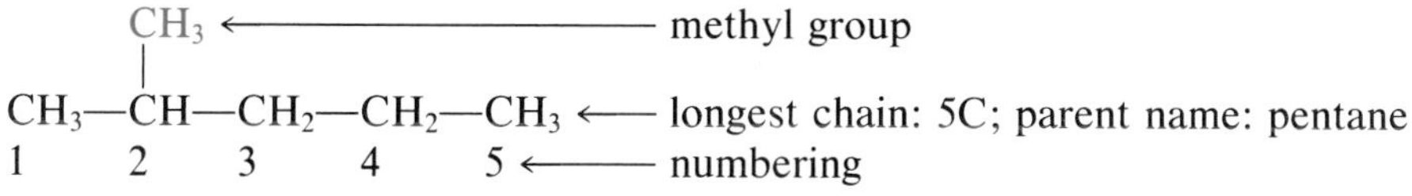

```
CH3 ←——————————— methyl group
 |
CH3—CH—CH2—CH2—CH3 ←— longest chain: 5C; parent name: pentane
1    2    3    4    5 ←—— numbering
```

IUPAC name: 2-Methylpentane (*not* 4-methylpentane)

Another similar example is

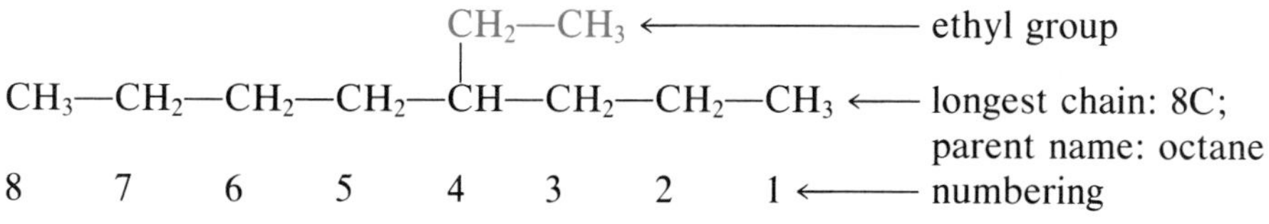

```
                         CH2—CH3 ←——————— ethyl group
                          |
CH3—CH2—CH2—CH2—CH—CH2—CH2—CH3 ←— longest chain: 8C;
                                    parent name: octane
8    7    6    5    4    3    2    1 ←—— numbering
```

IUPAC name: 4-Ethyloctane

Numbering from the other end of the chain would give the name 5-ethyloctane. This name is unacceptable according to IUPAC rules because the substituent has a higher number (5) than in the correct name, 4-ethyloctane.

The next example shows that the carbon atoms of the longest chain do not have to be drawn in a straight, horizontal line (the molecule as written can be redrawn if desired to show more clearly the longest chain):

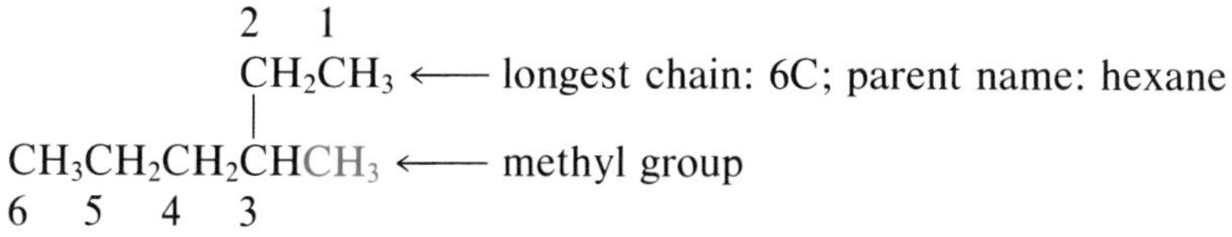

```
              2  1
              CH2CH3 ←— longest chain: 6C; parent name: hexane
              |
CH3CH2CH2CHCH3 ←— methyl group
6   5   4   3
```

IUPAC name: 3-Methylhexane

4. Two or more identical alkyl groups attached to the longest chain are indicated by prefixes di-, tri-, tetra-, etc. The set of numbers indicating the positions of the alkyl groups is chosen to be the lowest possible and the numbers in the set are separated from one another by commas. The last number and the name are separated by a dash. For example, the compound

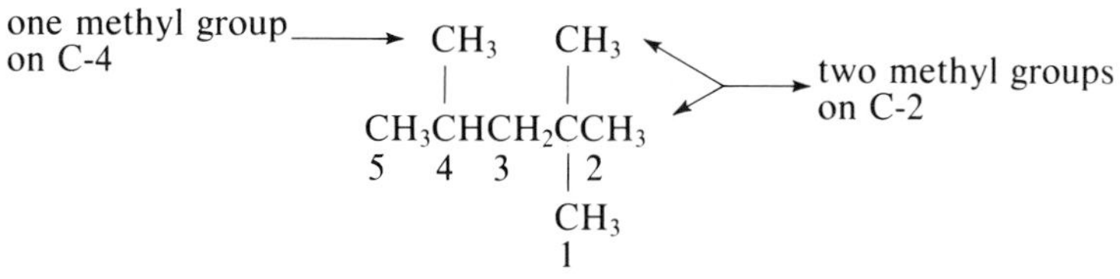

```
one methyl group ——→ CH3   CH3 ↖
on C-4                |     |     >——→ two methyl groups
                   CH3CHCH2CCH3   ↙      on C-2
                   5   4  3  | 2
                             CH3
                             1
```

is named 2,2,4-trimethylpentane (not 2,4,4-trimethylpentane, which has a higher set of position numbers).

Example 23.1

Naming a Branched Alkane

Name the compound

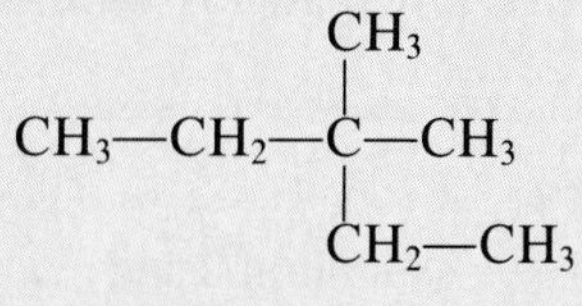

$$CH_3-CH_2-\underset{\displaystyle CH_2-CH_3}{\overset{\displaystyle CH_3}{\underset{|}{\overset{|}{C}}}}-CH_3$$

SOLUTION: The longest chain in this compound contains five carbon atoms:

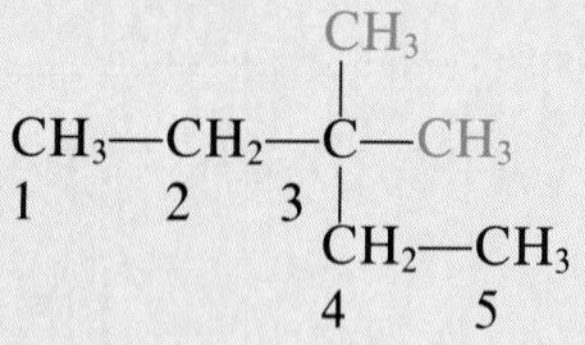

$$\underset{1}{CH_3}-\underset{2}{CH_2}-\underset{3}{\overset{\displaystyle CH_3}{\underset{|}{\overset{|}{C}}}}-CH_3 \quad \underset{4}{CH_2}-\underset{5}{CH_3}$$

In this formula the two methyl groups are attached to the carbon number 3. Thus, the IUPAC name of the compound is 3,3-dimethylpentane.

Practice Problem 23.1: Write structural formulas for: **(a)** 2,3-dimethylhexane; **(b)** 2,2-dimethyl-4-ethylheptane.

Example 23.2

Writing Structural Formulas for Isomers of an Alkane

Write the structural formulas for all the isomers of pentane, C_5H_{12}.

SOLUTION: Pentane has five carbon atoms. Each of the two terminal atoms is connected to three hydrogen atoms and one other carbon atom. The nonterminal carbon atoms are each linked to two hydrogen atoms and two other carbon atoms:

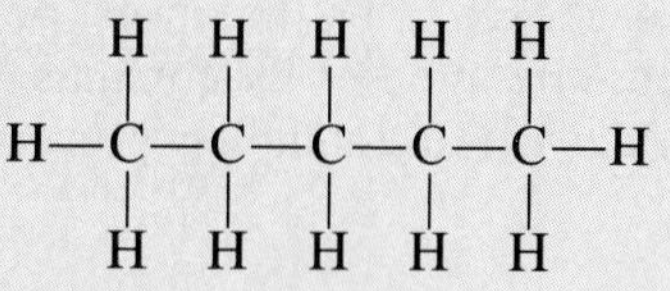

$$H-\underset{\displaystyle H}{\overset{\displaystyle H}{\underset{|}{\overset{|}{C}}}}-\underset{\displaystyle H}{\overset{\displaystyle H}{\underset{|}{\overset{|}{C}}}}-\underset{\displaystyle H}{\overset{\displaystyle H}{\underset{|}{\overset{|}{C}}}}-\underset{\displaystyle H}{\overset{\displaystyle H}{\underset{|}{\overset{|}{C}}}}-\underset{\displaystyle H}{\overset{\displaystyle H}{\underset{|}{\overset{|}{C}}}}-H$$

In writing structures for possible isomers, one of the terminal $-CH_3$ groups (for brevity we call it a C atom) can be moved to various nonterminal carbon atoms. The procedure is repeated until all possible nonequivalent positions are exhausted, as shown below.

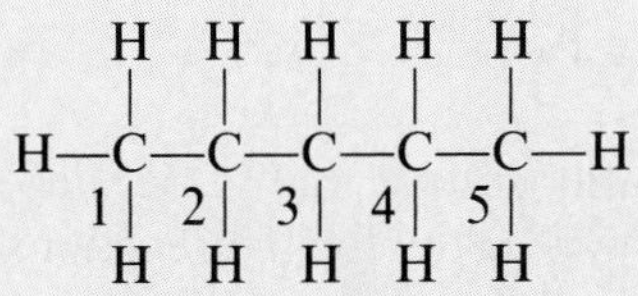

```
   H  H  H  H  H
   |  |  |  |  |
H—C—C—C—C—C—H
  1| 2| 3| 4| 5|
   H  H  H  H  H
```

Structure 1:
Pentane

Original structure of pentane

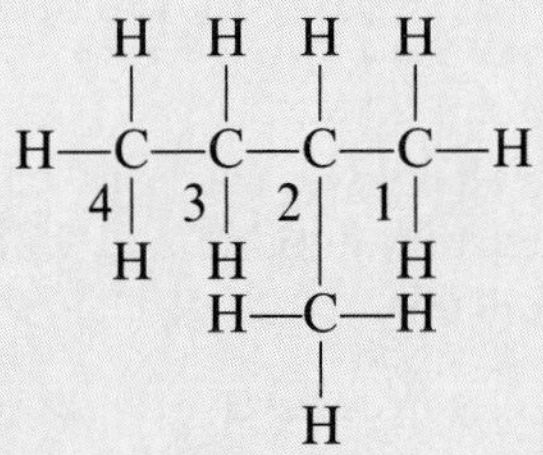

```
   H  H  H  H
   |  |  |  |
H—C—C—C—C—H
  4| 3| 2| 1|
   H  H  |  H
       H—C—H
          |
          H
```

Structure 2:
2-Methylbutane

One terminal C atom is moved to position 2 in the remaining carbon chain. This structure is clearly different from structure 1.

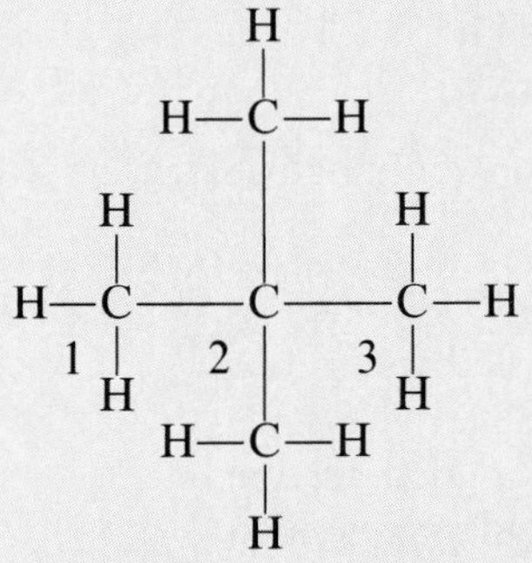

```
         H
         |
      H—C—H
         |
  H      |      H
  |      |      |
H—C——C——C—H
 1|  2   |  3   |
  H      |      H
      H—C—H
         |
         H
```

Structure 3:
2,2-Dimethylpropane

Two terminal C atoms are moved to the middle position of the remaining chain of three C atoms. This structure differs from both of the structures 1 and 2.

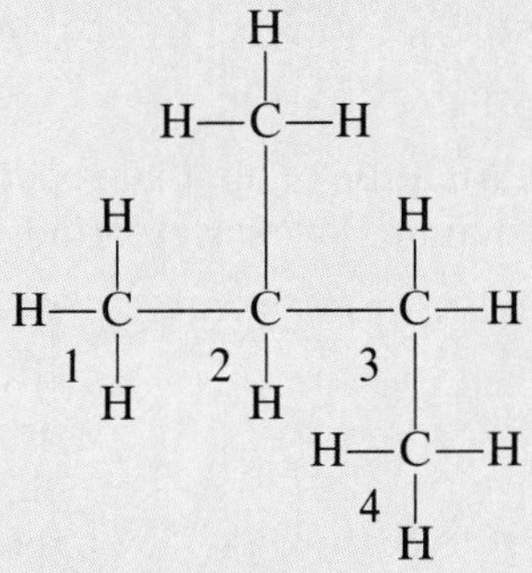

```
         H
         |
      H—C—H
         |
  H      |      H
  |      |      |
H—C——C——C—H
 1|   2|    3   |
  H     H       |
             H—C—H
               4|
                H
```

Structure 4:
2-Methyl butane

In this structure the longest carbon chain can be considered to consist of either atoms 1,2,3,4 or 4,3,2, and the unnumbered C atom. In either case, the structure is identical to structure 2.

The structures and discussion above show that there are only three isomers of C_5H_{12}, represented by the structures 1, 2, and 3. Structure 4 is identical to structure 2. All other attempts to reposition the five carbon atoms lead only to structures identical to 1, 2, or 3.

Practice Problem 23.2: Draw and name all five isomers of hexane.

23.2 Reactions of Alkanes

We recall that hydrocarbons burn in an atmosphere that is rich in oxygen to produce carbon dioxide and water, as shown below for the reactions of methane and ethane.

$$CH_4(g) + 2O_2(g) \longrightarrow CO_2(g) + 2H_2O(g)$$

$$2CH_3CH_3(g) + 7O_2(g) \longrightarrow 4CO_2(g) + 6H_2O(g)$$

Alkanes also react with halogens at temperatures above 200°C, or when the reaction mixture is irradiated with ultraviolet or strong white light, to produce halogenated alkanes, called *alkyl halides*, as shown below for chlorination of methane:

$$CH_4(g) + Cl_2(g) \longrightarrow CH_3Cl(g) + HCl(g)$$

CH_3Cl: Methyl chloride or chloromethane (used as a refrigerant)

Further chlorination of methane occurs upon continued irradiation of a reaction mixture containing excess chlorine:

$$CH_3Cl(g) + Cl_2(g) \longrightarrow CH_2Cl_2(l) + HCl(g)$$

CH_2Cl_2: Dichloromethane or methylene chloride (a good solvent for oils and waxes, also used as a local anesthetic in dentistry)

$$CH_2Cl_2(l) + Cl_2 \longrightarrow CHCl_3(l) + HCl(g)$$

$CHCl_3$: Trichloromethane or chloroform (used as an anesthetic and as a solvent for fats)

$$CHCl_3(l) + Cl_2(g) \longrightarrow CCl_4(l) + HCl(g)$$

CCl_4: Tetrachloromethane or carbon tetrachloride (also a good solvent for fats)

In practice, when a halogenation reaction is carried out, a mixture of alkyl halides is generally obtained. The components of the mixture can be separated by fractional distillation since the various chloromethanes have different boiling points. The boiling points of the chlorination products of methane are

CH_3Cl	CH_2Cl_2	$CHCl_3$	CCl_4
b.p. −24°C	b.p. 40°C	b.p. 62°C	b.p. 77°C

Except for reactions with oxygen and halogens, alkanes are relatively unreactive compared with alkenes and alkynes.

23.3 Unsaturated Hydrocarbons

Hydrocarbons with double or triple bonds between carbon atoms are said to be *unsaturated* because they contain less hydrogen than the corresponding alkanes, or *saturated* hydrocarbons. Saturated hydrocarbons contain the maximum possible number of hydrogen atoms per carbon atom. Alkanes are saturated hydrocarbons; alkenes and alkynes are unsaturated. Alkenes have the general formula C_nH_{2n}, and alkynes, C_nH_{2n-2}.

Alkenes

Alkenes can be made by heating alkanes in the presence of a suitable catalyst in the absence of air. The carbon–carbon bonds in an alkane tend to rupture when heated, producing an alkene and either hydrogen or another alkane. The products formed when butane is heated in the presence of powdered iron and in the absence of air are:

$$\underset{\text{Butane}}{C_4H_{10}} \xrightarrow{Fe(s)} \underset{\text{1-Butene}}{\begin{array}{ccccccccccc} & & H & & H & & H & & H & & \\ & & \mid & & \mid & & \mid & & \mid & & \\ H & — & C & — & C & — & C & = & C & — & H \\ & & \mid & & \mid & & & & & & \\ & & H & & H & & & & & & \end{array}} + \underset{\text{2-Butene}}{\begin{array}{ccccccccccc} & & H & & H & & H & & H & & \\ & & \mid & & \mid & & \mid & & \mid & & \\ H & — & C & — & C & = & C & — & C & — & H \\ & & \mid & & & & & & \mid & & \\ & & H & & & & & & H & & \end{array}}$$

$$+ \underset{\text{Propene}}{\begin{array}{ccccccccc} & & H & & H & & H & & \\ & & \mid & & \mid & & \mid & & \\ H & — & C & — & C & = & C & — & H \\ & & \mid & & & & & & \\ & & H & & & & & & \end{array}} + \underset{\text{Ethene}}{\begin{array}{ccccccc} & & H & & H & & \\ & & \mid & & \mid & & \\ H & — & C & = & C & — & H \end{array}} + CH_4 + CH_3CH_3 + H_2$$

Naming Alkenes

Like alkanes, alkenes can be named by the systematic, IUPAC nomenclature. When naming an alkene, the suffix *-ane* of the parent alkane is changed to *-ene* for the alkene. For alkenes that contain four or more carbon atoms, the position of the double bond is indicated by a numerical prefix, counting from the end of the carbon chain closest to the double bond:

$$\underset{\text{Ethene}}{CH_2{=}CH_2} \qquad \underset{\text{Propene}}{CH_3{—}CH{=}CH_2}$$

$$\underset{\text{1-Butene}}{\begin{array}{ccccccc} CH_3 & — & CH_2 & — & CH & = & CH_2 \\ 4 & & 3 & & 2 & & 1 \end{array}} \qquad \underset{\text{2-Butene}}{\begin{array}{ccccccc} CH_3 & — & CH & = & CH & — & CH_3 \\ 4 & & 3 & & 2 & & 1 \end{array}}$$

In naming branched-chain alkenes, the numbering is based on the longest continuous chain of carbon atoms *containing the double bond*:

$$\underset{\text{2,3-Dimethyl-1-butene}}{\begin{array}{ccccccc} CH_3 & — & CH & — & C & = & CH_2 \\ 4 & & 3\ \mid & & 2\ \mid & & 1 \\ & & CH_3 & & CH_3 & & \end{array}} \qquad \underset{\text{4-Methyl-2-heptene}}{\begin{array}{ccccccccc} CH_3 & — & CH_2 & — & CH_2 & — & CH & — & CH_3 \\ 7 & & 6 & & 5 & & 4\ \mid & & \\ & & & & & & 3\ CH & & \\ & & & & & & \Vert & & \\ & & & & & & 2\ CH & & \\ & & & & & & \mid & & \\ & & & & & & 1\ CH_3 & & \end{array}} \qquad \underset{\substack{\text{2-Ethyl-1-butene}\\ \text{(not a pentane)}}}{\begin{array}{ccccccc} CH_3 & — & CH_2 & — & C & = & CH_2 \\ 4 & & 3 & & 2\ \mid & & 1 \\ & & & & CH_2 & & \\ & & & & \mid & & \\ & & & & CH_3 & & \end{array}}$$

Example 23.3

Naming a Branched Alkene

Name the following compound:

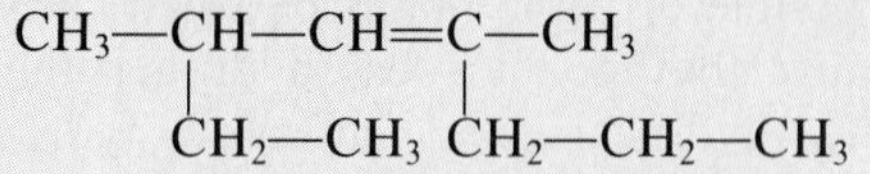

$$\begin{array}{l} CH_3{-}\underset{|}{CH}{-}CH{=}\underset{|}{C}{-}CH_3 \\ \quad\;\; CH_2{-}CH_3 \;\; CH_2{-}CH_2{-}CH_3 \end{array}$$

SOLUTION: The longest carbon chain containing the double bond in this compound has eight carbon atoms. These are numbered in the formula

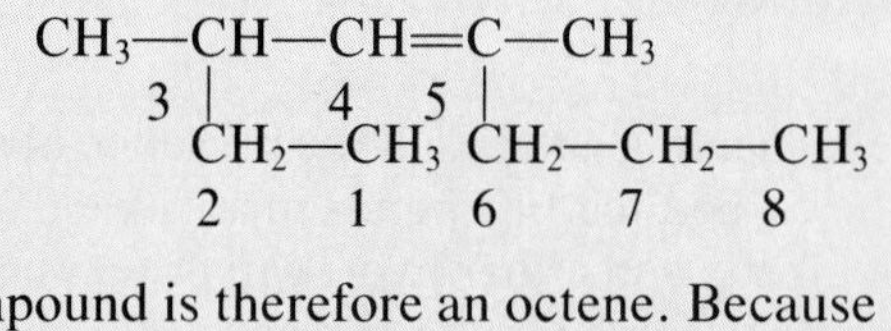

$$\begin{array}{l} CH_3{-}\underset{3\,|}{CH}{-}\overset{4}{CH}{=}\underset{5\,|}{C}{-}CH_3 \\ \quad\;\; \underset{2}{CH_2}{-}\underset{1}{CH_3} \;\; \underset{6}{CH_2}{-}\underset{7}{CH_2}{-}\underset{8}{CH_3} \end{array}$$

The parent compound is therefore an octene. Because the double bond is on the fourth carbon atom from either end of the eight-carbon chain, the parent compound is a 4-octene. Methyl groups are attached to the third and fifth carbon atoms. Therefore, the name for the compound is 3,5-dimethyl-4-octene.

Practice Problem 23.3: Write structural formulas for: **(a)** 2,3-dimethyl-3-heptene; **(b)** 5,6-dimethyl-3-heptene; **(c)** 2,5-dimethyl-2-hexene.

Some alkenes contain two or more double bonds. The names of alkenes with two double bonds end in *-diene*, and those with three double bonds end in *-triene*:

$$\underset{1}{CH_2}{=}\underset{2}{CH}{-}\underset{3}{CH}{=}\underset{4}{CH_2} \qquad \underset{1}{CH_2}{=}\underset{2}{CH}{-}\underset{3}{CH}{=}\underset{4}{CH}{-}\underset{5}{CH}{=}\underset{6}{CH}{-}\underset{7}{CH_3}$$

1,3-Butadiene　　　　1,3,5-Heptatriene

Alkenes containing alternating single and double bonds as shown are valuable as monomers in the polymerization reactions described in the next section.

Cycloalkanes and cycloalkenes are named as their open-chain counterparts, except the prefix cyclo- is used with the name (Figure 23.3). The structure of a cyclic molecule can be simplified by letting each corner of a polygon represent a carbon atom and enough hydrogen atoms to provide an octet for the carbon atom (Figure 23.3). Each side of the polygon drawn as a single line represents a carbon–carbon σ bond. A side of a polygon together with a line drawn parallel to the side inside the polygon represents one double bond. For simplicity, the hydrogen atoms attached to each carbon atom are omitted.

The number of hydrogen atoms connected to a carbon atom in a hydrocarbon molecule is determined by the rule that each carbon atom must have four bonding electron pairs (a full octet). Thus, a carbon atom connected to two other carbon atoms by single bonds must have two hydrogen atoms. A carbon atom connected to one other carbon atom by a single bond and to another carbon atom by a double bond can only have one hydrogen atom. A carbon atom connected to two other carbon atoms by double bonds cannot have any hydrogen atoms attached to it.

Naming Cis–Trans Isomers

No rotation occurs around a carbon–carbon double bond. Therefore, cis–trans isomers (see Section 22.4) are possible, as shown below for 2-butene:

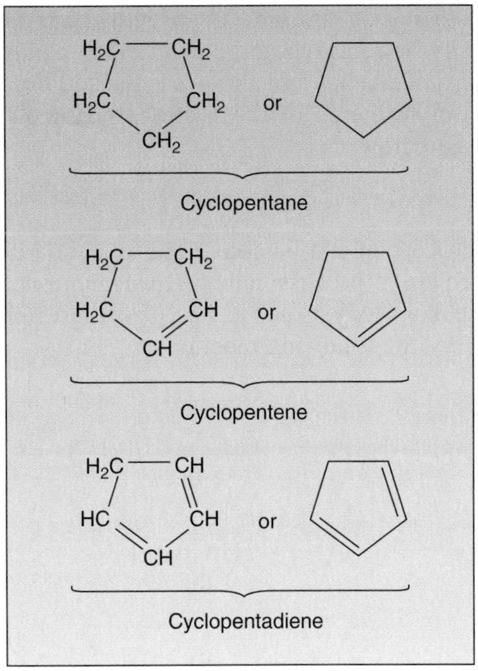

Figure 23.3

Structures of cyclopentane, cyclopentene, and cyclopentadiene.

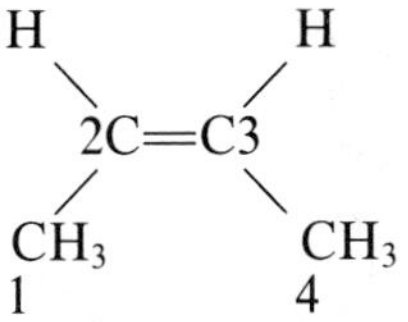

$$\underset{\substack{1}}{CH_3}(H)\,{}^{2}C{=}C^{3}\,(H)\underset{\substack{4}}{CH_3}$$

cis-2-Butene
(b.p. 3.7°C)

$$\underset{\substack{1}}{CH_3}(H)\,{}^{2}C{=}C^{3}\,(\overset{4}{CH_3})H$$

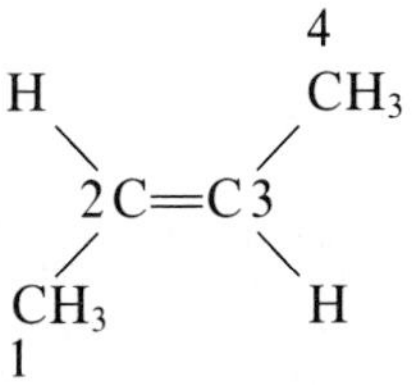

trans-2-Butene
(b.p. 0.9°C)

Cis and *trans*-2-Butenes have different boiling points, which indicates that these isomers are truly different compounds.

Alkynes

Alkynes are named according to rules analogous to those for naming alkanes. The suffix *-ane* of the corresponding alkane is changed to *-yne*:

$H{-}C{\equiv}C{-}H$ — Ethyne (acetylene)

$H{-}C{\equiv}C{-}CH_3$ — Propyne (methyl acetylene)

$H{-}\underset{1}{C}{\equiv}\underset{2}{C}{-}\underset{3}{CH_2}{-}\underset{4}{CH_3}$ — 1-Butyne

$\underset{1}{CH_3}{-}\underset{2}{C}{\equiv}\underset{3}{C}{-}\underset{4}{CH_3}$ — 2-Butyne

$\underset{7}{CH_3}{-}\underset{6}{CH_2}{-}\underset{5}{CH_2}{-}\underset{4}{C}{\equiv}\underset{3}{C}{-}\underset{2}{CH_2}{-}\underset{1}{CH_3}$ — 3-Heptyne

23.4 Reactions of Alkenes and Alkynes

Alkenes and alkynes are more reactive than alkanes. In the presence of a catalyst, alkenes and alkynes react readily with hydrogen, hydrogen halides, halogens, acids, oxidizing agents, and even with water because double and triple bonds can easily be ruptured and converted to bonds of lower order. A characteristic reaction of alkenes and alkynes is the addition of small molecules or atoms to the multiple bonds of the hydrocarbon molecules. Addition of hydrogen is discussed below.

Hydrogenation

An unsaturated hydrocarbon can be converted to a saturated hydrocarbon by the addition of hydrogen, a process called *hydrogenation*. The reaction takes place in the presence of a catalyst such as finely powdered platinum, palladium, or nickel, as shown by the following reactions:

$$\underset{\text{Ethene}}{CH_2{=}CH_2(g)} + H_2(g) \xrightarrow{Pt} \underset{\text{Ethane}}{\overset{H\quad\; H}{\overset{|\quad\;\; |}{CH_2{-}CH_2(g)}}} \quad \text{or} \quad CH_3CH_3(g)$$

$$\underset{\text{Ethyne}}{HC{\equiv}CH(g)} + 2H_2(g) \xrightarrow{Pt} \underset{\text{Ethane}}{CH_3CH_3(g)}$$

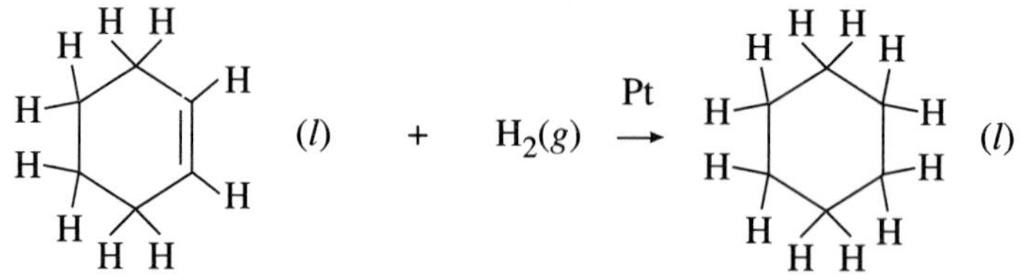

$$\underset{\text{Cyclohexene}}{C_6H_{10}(l)} + H_2(g) \xrightarrow{Pt} \underset{\text{Cyclohexane}}{C_6H_{12}(l)}$$

Hydrogenation is commonly used to convert liquid oils, such as corn oil, peanut oil, and safflower oil, which contain carbon–carbon double bonds, to saturated fats known as margarines:

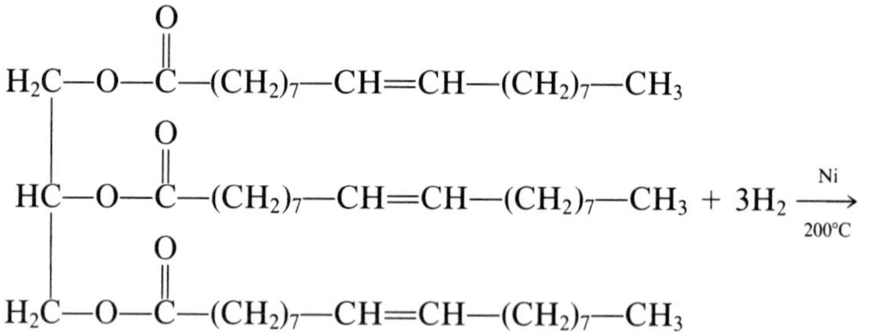

$$\begin{array}{l} H_2C{-}O{-}\overset{O}{\overset{\|}{C}}{-}(CH_2)_7{-}CH{=}CH{-}(CH_2)_7{-}CH_3 \\ \;|\\ HC{-}O{-}\overset{O}{\overset{\|}{C}}{-}(CH_2)_7{-}CH{=}CH{-}(CH_2)_7{-}CH_3 + 3H_2 \xrightarrow[200°C]{Ni} \\ \;| \\ H_2C{-}O{-}\overset{O}{\overset{\|}{C}}{-}(CH_2)_7{-}CH{=}CH{-}(CH_2)_7{-}CH_3 \end{array}$$

Triolein (a liquid oil)

$$\begin{array}{l} H_2C{-}O{-}\overset{O}{\overset{\|}{C}}{-}(CH_2)_7{-}CH_2{-}CH_2{-}(CH_2)_7{-}CH_3 \\ \;|\\ HC{-}O{-}\overset{O}{\overset{\|}{C}}{-}(CH_2)_7{-}CH_2{-}CH_2{-}(CH_2)_7{-}CH_3 \\ \;| \\ H_2C{-}O{-}\overset{O}{\overset{\|}{C}}{-}(CH_2)_7{-}CH_2{-}CH_2{-}(CH_2)_7{-}CH_3 \end{array}$$

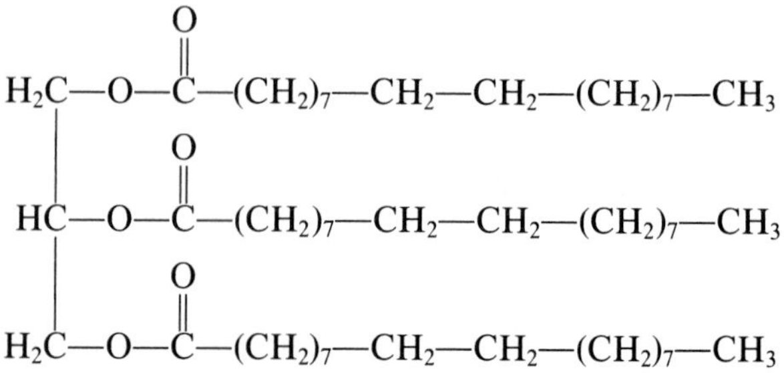

Tristearin (a solid fat)

Addition of Hydrogen Halide, HX

Hydrogen halides (HCl, HBr, and HI) add to double and triple bonds to produce alkyl halides. For example, HCl adds to ethene to produce ethyl chloride, which is also called chloroethane:

$$CH_2{=}CH_2(g) + HCl(g) \xrightarrow{\text{catalyst}} CH_3CH_2Cl(g) \qquad \text{(a local anesthetic)}$$

Alkynes such as ethyne react with hydrogen halides in a similar way. While only one molecule of hydrogen or hydrogen halide adds to a double bond, two molecules can add to a triple bond, as shown below for the reaction of ethyne with hydrogen chloride:

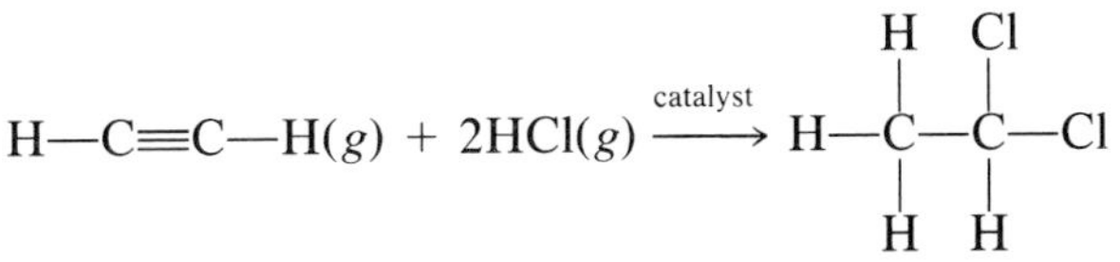

$$H{-}C{\equiv}C{-}H(g) + 2HCl(g) \xrightarrow{\text{catalyst}} H{-}\overset{\displaystyle H}{\underset{\displaystyle H}{\overset{|}{\underset{|}{C}}}}{-}\overset{\displaystyle Cl}{\underset{\displaystyle H}{\overset{|}{\underset{|}{C}}}}{-}Cl$$

1,1-Dichloroethane (in high concentrations, a narcotic)

Halogenation and Halogenated Hydrocarbons

Elemental halogens also add to a double or triple bond in a reaction called *halogenation*. Alkenes and alkynes can be halogenated even at room temperature and in the dark. The reaction of ethene with chlorine is

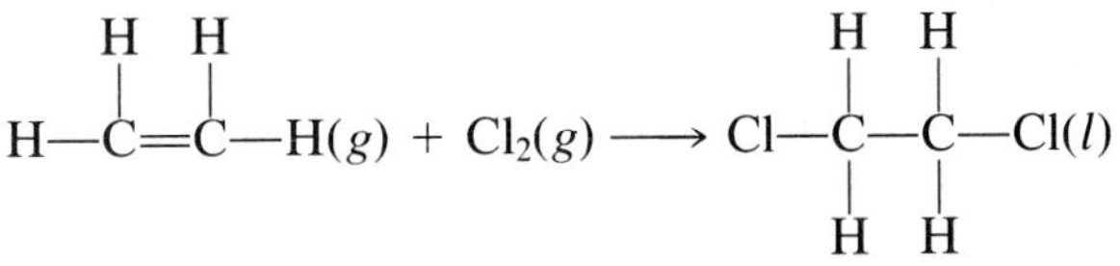

$$H{-}\overset{\displaystyle H}{\overset{|}{C}}{=}\overset{\displaystyle H}{\overset{|}{C}}{-}H(g) + Cl_2(g) \longrightarrow Cl{-}\overset{\displaystyle H}{\underset{\displaystyle H}{\overset{|}{\underset{|}{C}}}}{-}\overset{\displaystyle H}{\underset{\displaystyle H}{\overset{|}{\underset{|}{C}}}}{-}Cl(l)$$

1,2-Dichloroethane (a solvent for fats, oils, and waxes)

Chlorine reacts with ethyne to give 1,1,2,2,-tetrachloroethane:

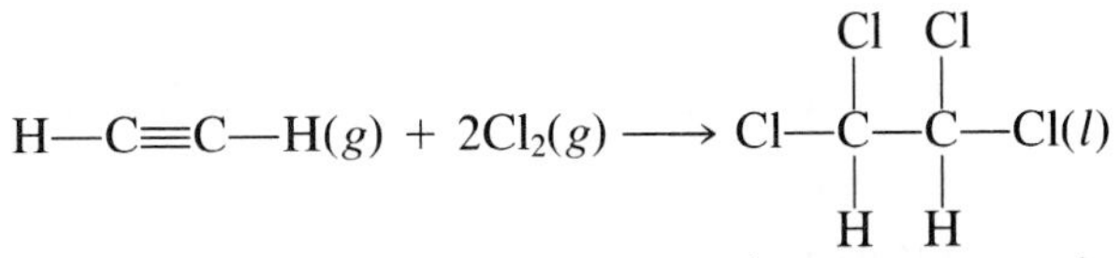

$$H{-}C{\equiv}C{-}H(g) + 2Cl_2(g) \longrightarrow Cl{-}\overset{\displaystyle Cl}{\underset{\displaystyle H}{\overset{|}{\underset{|}{C}}}}{-}\overset{\displaystyle Cl}{\underset{\displaystyle H}{\overset{|}{\underset{|}{C}}}}{-}Cl(l)$$

(a paint remover and ethyl alcohol denaturant)

Many halogenated hydrocarbons have important commercial uses. Methyl chloride, CH_3Cl, is a gas at room temperature, whereas dichloroethane, CH_2Cl_2, trichloroethane or chloroform, $CHCl_3$, and tetrachloroethane or carbon tetrachloride, CCl_4, are liquids. These three liquids can be used as solvents for grease, oils, and other organic substances. Chloroform is used as an anesthetic, and ethyl chloride, CH_3CH_2Cl, as a local anesthetic. Ethyl chloride boils at 12.3°C, and when it is sprayed on an injured area of the body, its rapid evaporation cools the sprayed area and deadens the pain. Carbon tetrachloride has been used as a dry-cleaning liquid, but because of its toxic and suspected carcinogenic effects at high concentration, it has been replaced in the dry-cleaning industry by trichloroethene ($CHCl{=}CCl_2$).

Freons are hydrocarbons that are fully halogenated with both chlorine and fluorine. Dichlorodifluoromethane (CCl_2F_2, Freon-12) and trichlorofluoromethane (CCl_3F, Freon-11) are used as refrigerants in air conditioners, freezers, and refrigerators, and also as propellants in aerosol spray cans. Freons are very unreactive and may exist in the atmosphere for years. They diffuse into the stratosphere, where they are broken down by ultraviolet light from the sun to produce free Cl atoms. The Cl atoms have unpaired electrons and are therefore highly reactive. They react with ozone, thereby decreasing the ozone layer in the upper atmosphere. (We discussed the mechanism of this reaction in Chapter 21.) Ozone absorbs ultraviolet light from the sun, thus protecting life on earth from radiation damage. Substitutes for Freons as refrigerants and as propellants in aerosol cans are therefore being used whenever possible.

Polymerization

In Chapter 9 we discussed some inorganic polymers. We recall that polymers consist of characteristic repeating covalently bonded units called monomers. In this section we consider some polymers derived from hydrocarbons. When ethene is heated at 100 to 400°C at high pressure and in the presence of a catalyst, many molecules unite end to end to form long-chain polymeric molecules containing approximately 1000 carbon atoms each:

$$\underset{\text{Ethene}}{n\mathrm{CH_2{=}CH_2}(g)} \longrightarrow \underset{\text{Ethylene units}}{\mathrm{+\!CH_2{-}CH_2\!+}_n(s)}$$

where n is the number of ethylene units. The product of this reaction is a translucent solid called *polyethylene*, with a melting point of 110 to 137°C, depending on the method of manufacture. It is a tough, flexible material, a "plastic," that resists most chemical reagents. Polyethylene is used to make toys, water pipes, wrapping material, and many other commercial products.

A *reaction in which many identical units (monomers) join to form a polymer* is called **polymerization**. Many different polymers have been developed in recent years, and many manufactured products consist wholly or partly of polymeric materials. Polymers are also found in many naturally occurring substances, such as cellulose, a major structural component of plants.

The polymer *polypropylene* was first prepared from propene in 1954. The production of polypropylene can be represented by the following equation:

$$\underset{\text{Propene}}{n\mathrm{CH_3CH{=}CH_2}(g)} \longrightarrow \underset{\text{Propylene units}}{\mathrm{+\!\underset{\underset{\displaystyle CH_3}{|}}{CH}{-}CH_2\!+}_n(s)}$$

Polypropylene has a higher melting point (170°C) than polyethylene and is in many ways superior to the latter. Polypropylene is used to make clothing such as long underwear, ropes, and the synthetic grass Astroturf.

Polymerization of tetrafluoroethene gives the product Teflon:

$$\underset{\text{Tetrafluoroethene}}{n\mathrm{CF_2{=}CF_2}(g)} \longrightarrow \underset{\text{Teflon}}{\mathrm{+\!CF_2{-}CF_2\!+}_n(s)}$$

Teflon, which is exceptionally resistant to heat and chemicals, is used as tubing and sheets in chemical laboratories, for lining reaction vessels and cooking utensils, to make artificial heart valves, and so on.

Natural rubber is a polymer of isoprene monomers (see below). In the polymerization of isoprene, there is a shift of double bonds (indicated by curved arrows) in each monomer, as shown by the following equation for the reaction of three isoprene units to form a trimeric portion of the polymer:

$$3\ CH_2{=}C(CH_3){-}CH{=}CH_2(l) \rightarrow {-}CH_2{-}C(CH_3){=}CH{-}CH_2{-}CH_2{-}C(CH_3){=}CH{-}CH_2{-}CH_2{-}C(CH_3){=}CH{-}CH_2{-}\ (s)$$

Isoprene (2-methyl= 1,3-butadiene) — A trimeric portion of natural rubber

A polypropylene molecule in rubber may contain up to 2000 monomeric units.

Natural rubber is obtained from rubber trees, but isoprene can be polymerized in a laboratory to form *synthetic rubber* that has characteristics nearly identical to those of natural rubber. Rubber molecules are long-chain polymers of carbon and hydrogen atoms. These long chains can easily be bent, twisted, compressed, or stretched to a reasonable extent without breaking the bonds.

Figure 23.4 shows the structures of some common polymers, and Table 23.3 lists some common synthetic polymers and their uses. (See Figures 23.5 and 23.6 for the meaning of the hexagons.)

Table 23.3

Some Synthetic Polymers and Their Uses

Name and Structure of Monomer	**Name of the Polymer**	**Uses of the Polymer**
Ethylene, $CH_2{=}CH_2$	Polyethylene	Tubing, wrapping material, films, molded objects
Propylene, $CH_2{=}CHCH_3$	Polypropylene	Fibers for clothes and carpeting, knitted surgical mesh, material for hip-joint repair
Vinyl chloride, $CH_2{=}CHCl$	Polyvinyl chloride (PVC)	Floor coverings, records, imitation leather
Vinylidine chloride, $CH_2{=}CCl_2$	Polyvinylidine chloride	Saran Wrap
Acrylonitrile, $CH_2{=}CH{-}C{\equiv}N$	Polyacrylonitrile	Fibers for Orlon and Acrilan
Methyl methacrylate, $CH_2{=}C(CH_3){-}C({=}O){-}O{-}CH_3$	Polymethyl methacrylate	Plexiglas, Lucite (transparent sheets and rods)
Tetrafluoroethylene, $CF_2{=}CF_2$	Polytetrafluoroethylene (Teflon)	Electrical insulation, artificial heart valves, housing for heart pacemakers, laboratory ware, nonstick coatings for kitchen ware, Gore-Tex garments
Styrene, $CH_2{=}CH{-}C_6H_5$	Polystyrene	Styrofoam insulation, molded objects
Vinyl acetate, $CH_2{=}CH{-}O{-}C({=}O){-}CH_3$	Polyvinyl acetate	Adhesives, coatings, chewing gum

Figure 23.4

Structures of some polymeric materials. The corners of the hexagons represent carbon atoms with sufficient hydrogen atoms to satisfy the octet at each carbon atom (see Figure 23.5 and the relevant discussion).

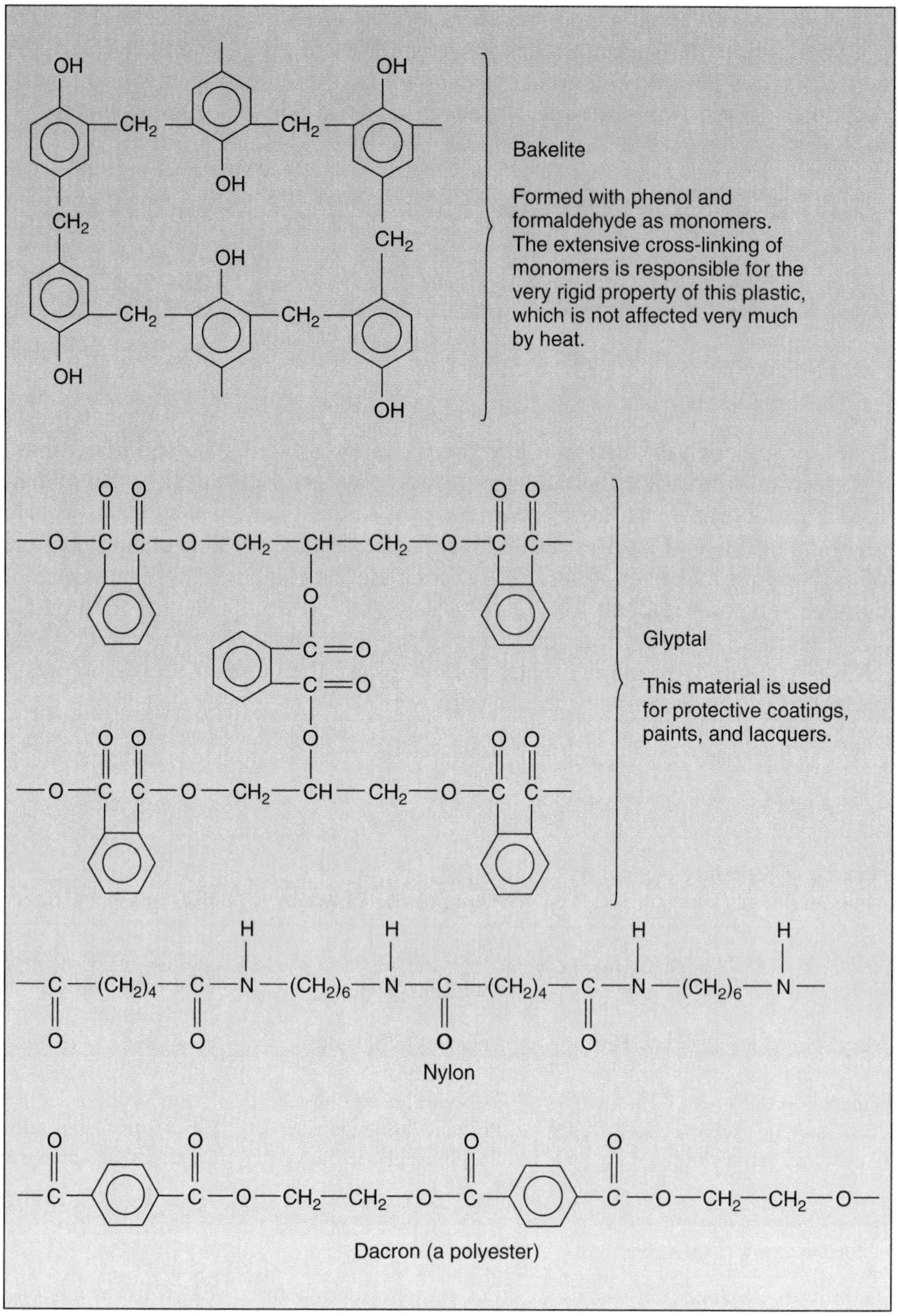

OH OH CH$_2$ CH$_2$ OH CH$_2$ CH$_2$ OH CH$_2$ CH$_2$ OH OH

Bakelite

Formed with phenol and formaldehyde as monomers. The extensive cross-linking of monomers is responsible for the very rigid property of this plastic, which is not affected very much by heat.

Glyptal

This material is used for protective coatings, paints, and lacquers.

Nylon

Dacron (a polyester)

Figure 23.5

The structure of benzene.

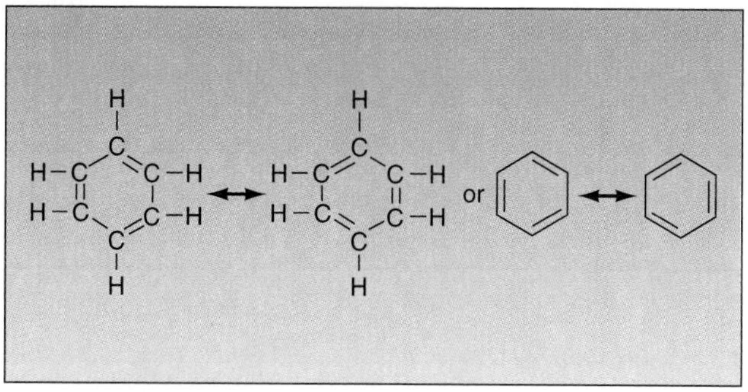

or

APPLICATIONS OF CHEMISTRY 23.1
What's the Difference Between an Artificial Ice Skating Rink and Plastic Food Wrap?

There is a relationship between the number of carbon atoms in a hydrocarbon molecule and the physical properties of the hydrocarbon such as viscosity and volatility. For example, hydrocarbons with between one and four carbon atoms per molecule (methane, ethane, propane, and butane) are gases. Gasoline, a volatile liquid, is a mixture of hydrocarbon molecules containing 6 to 12 carbon atoms. Kerosene, less volatile than gasoline, is another mixture of hydrocarbons containing 9 to 14 carbon atoms per molecule. Because of its decreased volatility, kerosene is a safer fuel than gasoline in room heaters.

As the number of carbon atoms per hydrocarbon molecule increases, the properties change from kerosene to fuel oil to lubricating oils. Each mixture in this progression becomes more viscous and less volatile. Eventually, as we continue to increase the number of carbon atoms, we obtain grease, asphalt, and tar.

A similar trend persists when chemists polymerize hydrocarbons such as ethene. As the size of the polyethylene molecule increases, so does its viscosity. These physical properties also depend on other factors, such as polymerization conditions, and the presence of other chemicals, such as catalysts.

When chemists polymerize ethene to produce molecules with about 40,000 carbon atoms, the result is material we use in plastic food wrap. If we increase the number of carbon atoms per polyethylene molecule even further, we produce even tougher products. One of these is suitable for use in flexible milk bottles (60,000 carbon atoms/molecule), and another in more rigid bleach bottles (80,000 carbon atoms/molecule).

When the number of carbon atoms per polyethylene molecule reaches 800,000, the material becomes very viscous and abrasion resistant. Such material can be used as a substitute for metal bearings and as artificial ice in skating rinks.

The polyethylene surface used as artificial ice in skating rinks (see photo) is 60 percent less expensive to maintain than real ice. After 4 years of daily vacuuming and weekly silicone treatments, maintenance workers flip the rink over to expose the unused bottom surface. This fresh surface is good for another 4 years.

Source: Earl F. Pearson, Curtis C. Wilkins, and Norman W. Hunter, *Journal of Chemical Education*, August 1988.

23.5 Aromatic Hydrocarbons and Their Reactions

Structures and Examples

The simplest representative of aromatic hydrocarbons is benzene. Lewis resonance structures of benzene are illustrated in Figure 23.5. Many other aromatic hydrocarbons are derivatives of benzene. In a benzene molecule, the C—C bond lengths and bond energies are equal. The C—C bond energy in benzene is greater than the C—C single-bond energy but less than the C=C double-bond energy. Similarly, the C—C bond length (0.140 nm) in benzene is intermediate between the C—C single-bond length (0.154 nm) and C=C double-bond length (0.134 nm).

For each C—C σ bond in a benzene molecule, there is a π bonding contribution. This contribution arises from the sideways overlap of the p orbitals of the carbon atoms lying perpendicular to the plane of the benzene ring (Figure 23.6). The single electron in each of these p orbitals can move freely throughout all six of the p orbitals. Such electrons are said to be *delocalized* throughout the

Figure 23.6

The orbital representation of σ and π bonding in a benzene molecule.

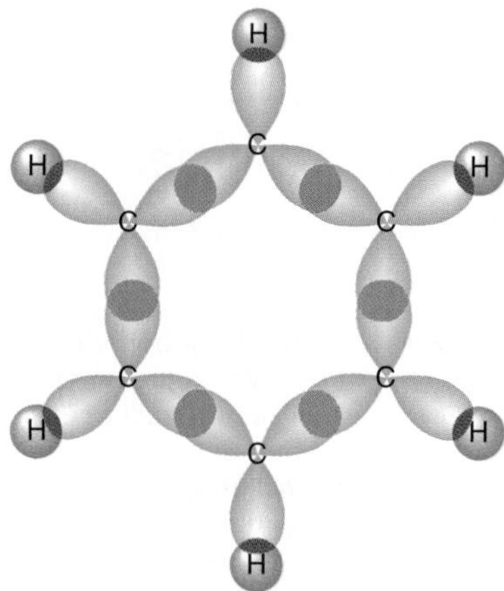

(a) The σ bond structure of benzene. Each of the six C atoms has a *p* orbital perpendicular to the plane of paper. The parallel overlapping of these *p* orbitals forms a delocalized π electron cloud in front of and behind the plane of paper. This π bonding is shown in part (b) of this figure.

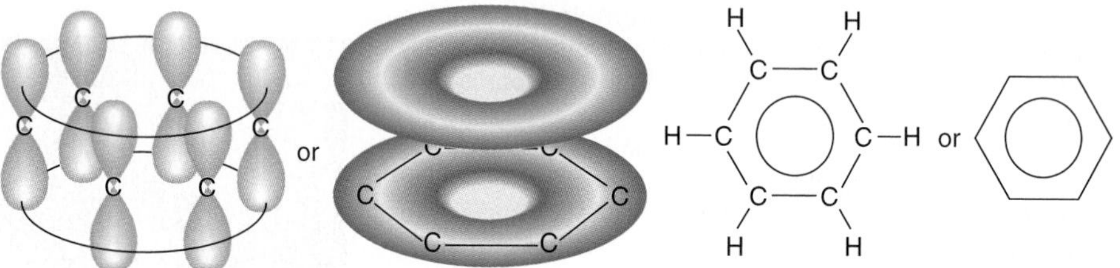

(b) The π bond structure of sideways overlapping of *p* orbitals of the carbon atoms in benzene molecule.

(c) Conventional representation for benzene; the circle represents delocalized π electrons.

molecule. The concept of delocalization, illustrated in Figure 23.6, is similar to that discussed in Chapter 9 for the resonance hybrids of the carbonate ion, nitrate ion, and the sulfur dioxide molecule. We recall that delocalized π electrons in a chemical bond are sometimes indicated by dotted lines. The delocalized π electrons in a benzene ring are usually represented by a solid circle inside the hexagon, as illustrated in Figure 23.6.

Benzene is a toxic, colorless, flammable liquid (b.p. 79.7°C). Benzene is used in vast quantities as a starting material for the synthesis of a variety of products, such as dyes, plastics, drugs, perfumes, explosives, and detergents.

Reactions of Benzene

Benzene is unusually stable because of its delocalized π electrons. Benzene is therefore chemically less reactive than unsaturated aliphatic hydrocarbons, which readily undergo various addition reactions to their π bonds.

Thus, when benzene is treated with a halogen such as bromine, no reaction occurs unless a Lewis acid such as $FeBr_3$ is added as a catalyst to the reaction mixture. Under these conditions, a bromine atom replaces a hydrogen atom in the benzene ring to produce bromobenzene:

Br

$$(l) + Br_2(l) \xrightarrow{FeBr_3} (l) \quad + \quad HBr(g)$$

Bromobenzene (used as a solvent and for organic synthesis)

Using similar techniques, one or more hydrogen atoms in a benzene ring can be substituted with other atoms or groups of atoms. For example, benzene reacts with nitric acid in a sulfuric acid medium to produce nitrobenzene:

NO_2

$$(l) + HNO_3 \xrightarrow{H_2SO_4} (l) \quad + \quad H_2O(l)$$

Nitrobenzene (an oily liquid used for manufacture of other compounds)

If the reaction mixture is heated for several hours, a second nitro ($-NO_2$) group is added to the ring to produce dinitrobenzene:

NO_2 NO_2

$$(l) + HNO_3(l) \xrightarrow{H_2SO_4} (s) \quad + \quad H_2O(g)$$

NO_2

Dinitrobenzene (used for synthesis and in dyes)

Dinitrobenzene has three isomers, called *ortho-*, *meta-*, and *para*-dinitrobenzene (Figure 23.7). In the ortho isomer, the nitro groups are on adjacent carbon atoms of the benzene ring. The meta isomer has an unsubstituted carbon atom between the two substituent groups, and the para isomer has the two substituents opposite each other.

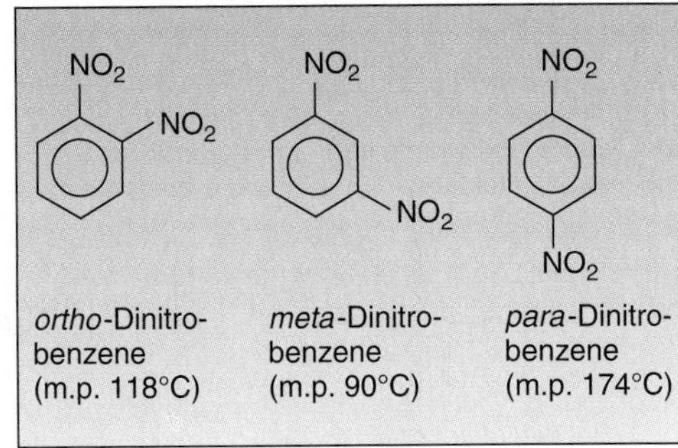

Figure 23.7

Ortho-, meta-, and para- isomers of dinitrobenzene.

An alkyl group can be substituted onto an aromatic ring by treatment of benzene or its derivative with an alkyl halide, using aluminum chloride as a catalyst:

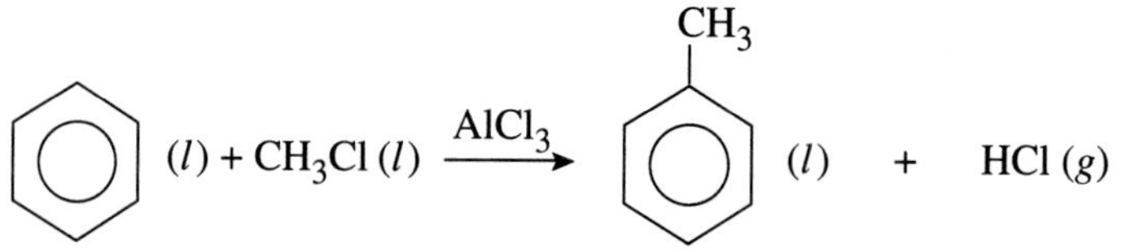

$$C_6H_6\,(l) + CH_3Cl\,(l) \xrightarrow{AlCl_3} C_6H_5CH_3\,(l) + HCl\,(g)$$

Toluene (a solvent for oils; used for the preparation of dyes, perfumes, saccharin, and explosives—TNT)

By similar techniques, a great variety of substituted aromatic hydrocarbons (Figure 23.8) can be made.

Figure 23.8

Some examples of benzene derivatives.

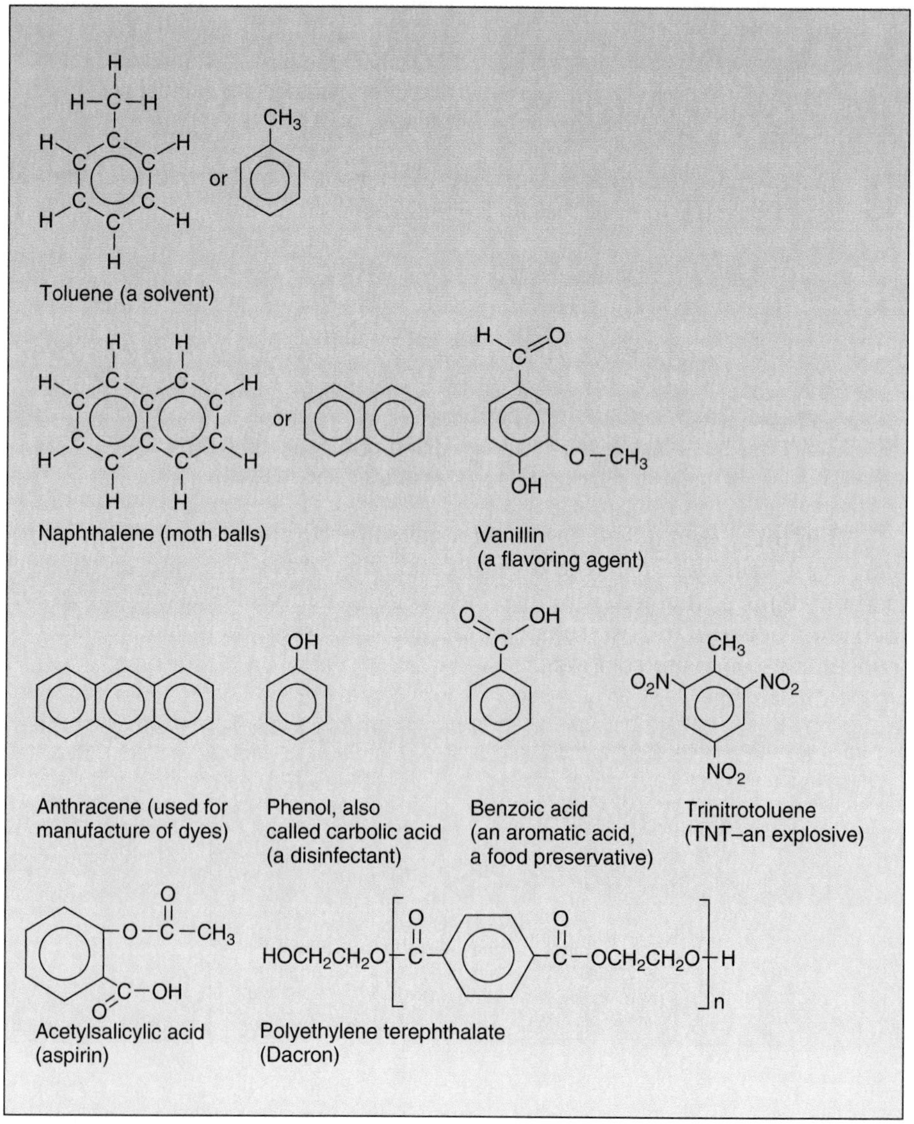

Toluene (a solvent)

Naphthalene (moth balls)

Vanillin (a flavoring agent)

Anthracene (used for manufacture of dyes)

Phenol, also called carbolic acid (a disinfectant)

Benzoic acid (an aromatic acid, a food preservative)

Trinitrotoluene (TNT–an explosive)

Acetylsalicylic acid (aspirin)

Polyethylene terephthalate (Dacron)

23.6 Alcohols, Ethers, and Amines

Alcohols

Many organic compounds contain oxygen, and many contain nitrogen. We next discuss some organic compounds that contain oxygen—alcohols and ethers. An alcohol has a hydroxyl group (—OH) covalently bonded to a carbon atom. For example, in methyl alcohol, CH_3OH, the —OH group is bonded to the carbon atom of a methyl group, CH_3OH. In ethyl alcohol, CH_3CH_2OH, the —OH group is bonded to the carbon atom of an ethyl group. A *characteristic group of atoms common to a class of organic compounds* is called a **functional group**. The functional group of alcohols is the —OH group.

Alcohols can be classified as *primary*, *secondary*, or *tertiary alcohols*, depending on the number of carbon atoms connected to the carbon bearing the functional group. In a primary alcohol, the carbon atom that is linked to the —OH group is connected to one other carbon atom and two hydrogen atoms, as in CH_3CH_2OH, or to three hydrogen atoms, as in CH_3OH. In a secondary alcohol, the carbon atom bearing the —OH group is bonded to two other carbon atoms and one hydrogen atom. In a tertiary alcohol, the carbon atom bearing the —OH group is bonded to three other carbon atoms. General formulas for primary, secondary, and tertiary alcohols are shown below (R, R′, and R″ are any alkyl groups):

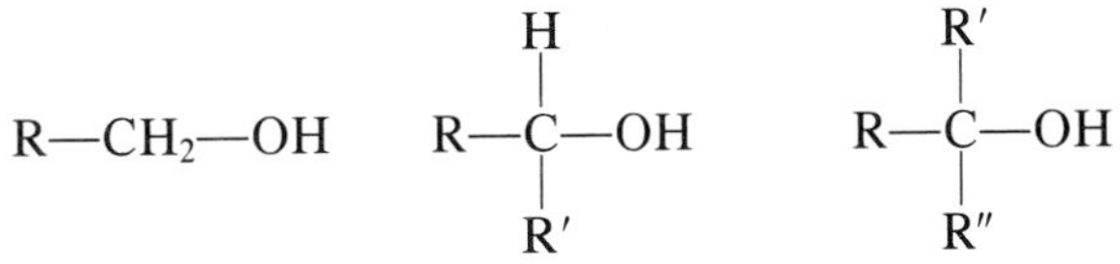

```
                        H                    R′
                        |                    |
R—CH2—OH            R—C—OH               R—C—OH
                        |                    |
                        R′                   R″

A primary          A secondary           A tertiary
alcohol            alcohol               alcohol
```

Simple examples of primary, secondary, and tertiary alcohols (assuming that $R = R' = R'' = —CH_3$) are

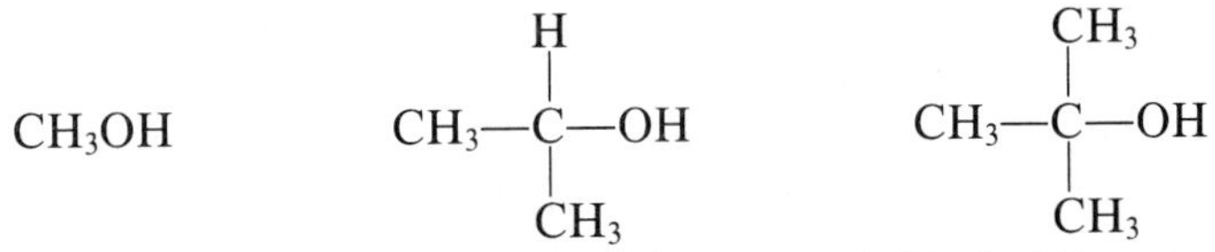

```
                          H                        CH3
                          |                        |
CH3OH                CH3—C—OH                 CH3—C—OH
                          |                        |
                          CH3                      CH3

Methanol              2-Propanol              2-Methyl-2-propanol
(methyl alcohol)      (isopropyl alcohol)     (tertiary butyl alcohol)
```

Alcohols are named in the IUPAC system by dropping the *−e* ending from the corresponding alkane and substituting the ending *−ol*. Thus, CH_3OH is named methan*ol*, and CH_3CH_2OH is called ethan*ol*. The name of a branched alcohol is based on the longest unbranched chain of carbon atoms that contains the —OH group, and the numbering begins at the end of the chain closer to the —OH group:

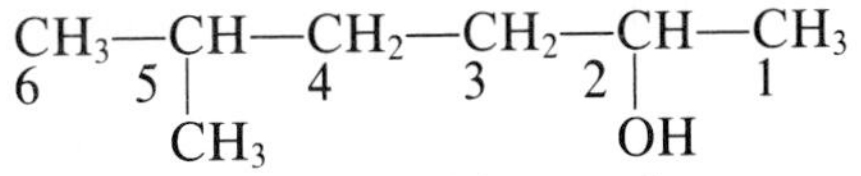

```
CH3—CH—CH2—CH2—CH—CH3
 6   5 |  4    3   2 |   1
      CH3            OH
     5-Methyl-2-hexanol
```

For the three alcohols given above as simple examples of primary, secondary, and tertiary alcohols, the IUPAC name is given first, followed by the common name in parentheses.

Methanol (whose common name is methyl alcohol) was at one time prepared in large quantities by heating wood in absence of air. Therefore, another name for methanol is wood alcohol. Methanol is synthesized industrially by the reaction of carbon monoxide and hydrogen in the presence of a catalyst such as Fe_2O_3 at high temperature and pressure:

$$\underset{}{CO(g)} + 2H_2(g) \xrightarrow[\text{catalyst}]{Fe_2O_3} \underset{\text{Methanol}}{CH_3OH(l)}$$

Methanol is widely used in organic synthesis, in the manufacture of formaldehyde (see the next section), to "denature" ethyl alcohol (make it nondrinkable), and as a fuel. Methanol is highly toxic; ingestion of only a few milliliters can cause nausea and blindness. Imbibing large quantities of methanol may be fatal. Ethanol that is sold as a disinfectant or fuel is denatured to prevent the loss of alcohol tax revenue when people use the "fuel" for drinking purposes.

Ethanol can be prepared by the bacterial action of yeast on sugar in the absence of oxygen, as shown by the following reaction of glucose:

$$\underset{\text{Glucose}}{C_6H_{12}O_6(aq)} \xrightarrow{\text{yeast}} \underset{\text{Ethanol}}{2CH_3CH_2OH(l)} + 2CO_2(g)$$

The conversion of sugar to alcohol and carbon dioxide is called *fermentation*. Most of the ethanol produced industrially in the United States is prepared by a catalyzed reaction of ethene with water:

$$H-\underset{}{\overset{H}{\overset{|}{C}}}=\overset{H}{\overset{|}{C}}-H(g) + H-O-H(l) \xrightarrow[\substack{200\text{ atm}\\ 300°C}]{H_3PO_4} H-\underset{H}{\underset{|}{\overset{H}{\overset{|}{C}}}}-\underset{H}{\underset{|}{\overset{H}{\overset{|}{C}}}}-O-H(l)$$

Adding water to a double bond is similar to the addition of HCl and Cl_2 discussed previously. The H atom from water adds to one of the carbon atoms, and the —OH group adds to another carbon atom. As a result, the C=C double bond converts to a single bond.

When ethanol is consumed, it is oxidized in the human body to acetaldehyde (a cause of "hangover") and ultimately to acetic acid. Acetic acid is a normal constituent of cells and is not toxic at low concentrations.

The various commercially available alcoholic beverages have the following approximate percentages of ethanol by volume:

Whiskey	40–50 (80–100 proof)	Wine	7–20
Brandy	45–55 (90–110 proof)	Beer	3–4

In the alcohol industry, the concentration of ethanol in a beverage is indicated by "proof," which is twice the percent ethanol by volume.

High concentrations of alcohol in the blood lead to impaired physical ability and mental judgment. Consuming too much alcohol can cause temporary blindness (hence the expression "blind drunk"). Ethanol acts on the central nervous system as a depressant, and extreme overdoses are sometimes fatal.

Other familiar aliphatic alcohols include 2-propanol (also called isopropyl alcohol and "rubbing alcohol"),

```
   H  OH H
   |  |  |
H—C—C—C—H
   |  |  |
   H  H  H
```

and ethylene glycol,

```
    H  H
    |  |
 H—C—C—H
    |  |
   OH OH
```

Ethylene glycol (also called dialcohol or diol) is used as an antifreeze.

Ethers

Ethers are compounds with the general formula R—O—R′, where R and R′ can be identical or different aliphatic or aromatic groups of atoms. The functional group of ethers is the —O— group. Ethers can be prepared by elimination of a molecule of water from two molecules of alcohol using an acid catalyst:

$$CH_3CH_2O{-}H(l) + HO{-}CH_2CH_3(l) \xrightarrow[\text{heat}]{H^+} \underset{\text{Diethyl ether}}{CH_3CH_2{-}O{-}CH_2CH_3(l)} + H_2O(l)$$

Dimethyl ether, $CH_3{-}O{-}CH_3$, is a gas that is used as a refrigerant. The higher-molecular-weight aliphatic ethers are liquids. Diethyl ether, $CH_3CH_2{-}O{-}CH_2CH_3$ (often simply called "ether"), is a volatile liquid (b.p. 35°C) which is an excellent solvent for many organic substances. It was used as an anesthetic for many years because it depresses the activity of the central nervous system. But ether irritates the respiratory system and causes nausea and vomiting. Diethyl ether has been replaced as an anesthetic by methyl propyl ether, $CH_3{-}O{-}CH_2CH_2CH_3$, which is relatively free of side effects.

Amines

Amines are nitrogen-containing organic compounds. Alcohols and ethers can be thought of as water molecules in which one or both of the hydrogen atoms are replaced by alkyl groups. Similarly, **amines** can be regarded as ammonia molecules in which one, two, or all three of the hydrogen atoms are replaced by alkyl groups. Amines are classified as *primary*, *secondary*, and *tertiary*, depending on the number of alkyl groups attached to the nitrogen atom. Primary amines have the general formula RNH_2, secondary amines, RR′NH, and tertiary amines, RR′R″N. The R groups can be identical or different. The functional group of primary amines is the $—NH_2$ group.

Examples of primary amines include methylamine (a flammable gas at 25°C and 1 atm that is used in the tanning industry) and ethylamine (a flammable liquid used in oil refining and in organic synthesis). The formulas for these amines are

$$\mathrm{H_3C{-}\ddot{N}(H){-}H} \quad \text{or} \quad \mathrm{CH_3\dot{N}H_2}$$
Methylamine

$$\mathrm{CH_3CH_2{-}\ddot{N}(H){-}H} \quad \text{or} \quad \mathrm{CH_3CH_2NH_2}$$
Ethylamine

The simplest secondary amine is dimethylamine, a gas at room temperature, which, like many low-molecular-weight amines, smells like dead fish. Dimethylamine is used in the manufacture of soaps; it is also an insecticide that attracts boll weevils, exterminating them on contact. Diethylamine is a flammable, strongly alkaline liquid used in the rubber and petroleum industries, as well as in making dyes and drugs. The formulas for dimethylamine and diethylamine are

$$\mathrm{CH_3{-}\ddot{N}(H){-}CH_3} \quad \text{or} \quad \mathrm{CH_3NHCH_3}$$
Dimethylamine

$$\mathrm{CH_3CH_2{-}\ddot{N}(H){-}CH_2CH_3} \quad \text{or} \quad \mathrm{CH_3CH_2\dot{N}HCH_2CH_3} \quad \text{or} \quad \mathrm{(CH_3CH_2)_2\dot{N}H}$$
Diethylamine

Common *tertiary amines* include trimethylamine,

$$\mathrm{H_3C{-}N(CH_3)_2{:}}$$

a gas with a pungent, fishy, ammoniacal odor (used as an insect attractant), and triethylamine,

$$\mathrm{CH_3CH_2{-}N(CH_2CH_3)_2{:}}$$

a liquid with a strong odor of ammonia used in the synthesis of other compounds.

Amines can contain more than one $-NH_2$ group per molecule. For example, we recall from Chapter 22 that ethylenediamine, $H_2NCH_2CH_2NH_2$, is a chelating ligand. Other examples of polyamines include putrescine, $H_2NCH_2CH_2CH_2CH_2NH_2$, and cadaverine, $H_2NCH_2CH_2CH_2CH_2CH_2NH_2$. Both of these compounds are present in decaying meat and fish and are responsible for their foul odors.

There are also many aromatic amines. The simplest of these is aniline,

$$\mathrm{C_6H_5NH_2}$$

Aniline is a poisonous, oily liquid with a characteristic odor; it is used for the manufacture of dyes, drugs, varnishes, perfumes, and shoe polish.

23.7 Carbonyl Compounds

Carbonyl compounds constitute a group of oxygen-containing organic compounds that contain the **carbonyl group**,

$$\overset{\displaystyle O}{\overset{\|}{-C-}}$$

Important classes of carbonyl compounds include aldehydes, ketones, carboxylic acids, and esters. Below are general formulas of these compounds, where R and R′ can be identical or different alkyl groups:

$$\underset{\text{Aldehyde}}{R-\overset{\displaystyle O}{\overset{\|}{C}}-H} \qquad \underset{\text{Ketone}}{R-\overset{\displaystyle O}{\overset{\|}{C}}-R'} \qquad \underset{\text{Carboxylic acid}}{R-\overset{\displaystyle O}{\overset{\|}{C}}-OH} \qquad \underset{\text{Ester}}{R-\overset{\displaystyle O}{\overset{\|}{C}}-O-R'}$$

These compounds are discussed below in more detail.

Aldehydes

Aldehydes contain the characteristic functional group

$$-\overset{\displaystyle O}{\overset{\|}{C}}-H$$

In an aldehyde molecule, either an alkyl group and a hydrogen atom or two hydrogen atoms are bonded to a carbonyl group:

$$R-\overset{\displaystyle O}{\overset{\|}{C}}-H$$

Aldehydes can be prepared by oxidation of primary alcohols using either $KMnO_4$ or $K_2Cr_2O_7$ as an oxidizing agent. Continued oxidation converts the aldehyde to a carboxylic acid:

$$\underset{\text{Alcohol}}{R-\overset{\overset{\displaystyle H}{|}}{\overset{\displaystyle O}{\overset{|}{C}}}-H \atop \quad\ \ |\ \ \atop \quad\ \ H} \xrightarrow{KMnO_4} \underset{\text{Aldehyde}}{R-\overset{\displaystyle O}{\overset{\|}{C}}-H} \xrightarrow{KMnO_4} \underset{\text{Acid}}{R-\overset{\displaystyle O}{\overset{\|}{C}}-OH}$$

Note that in the conversion of an alcohol to an aldehyde, one hydrogen atom is removed from the —OH group of the alcohol, and another hydrogen atom is removed from the carbon atom bearing the —OH group. In a conversion of an aldehyde to an acid, an oxygen atom is added to the functional group characteristic of the aldehyde to obtain the functional group characteristic of the carboxylic acids, called a *carboxyl group*, $-\overset{\displaystyle O}{\overset{\|}{C}}-OH$.

Alcohols can also be oxidized to aldehydes by atmospheric oxygen in the presence of a catalyst at high temperature. For example, methanol can be converted to formaldehyde, and ethanol to acetaldehyde:

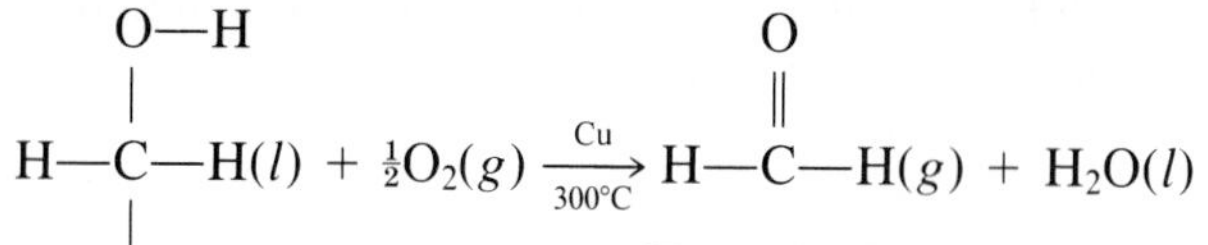

$$\begin{array}{c} \text{O—H} \\ | \\ \text{H—C—H}(l) \\ | \\ \text{H} \end{array} + \tfrac{1}{2}O_2(g) \xrightarrow[300°\text{C}]{\text{Cu}} \begin{array}{c} \text{O} \\ \| \\ \text{H—C—H}(g) \end{array} + H_2O(l)$$

Formaldehyde

$$\begin{array}{c} \text{H}\ \ \text{O—H} \\ |\ \ \ \ | \\ \text{H—C—C—H}(l) \\ |\ \ \ \ | \\ \text{H}\ \ \text{H} \end{array} + \tfrac{1}{2}O_2(g) \xrightarrow[300°\text{C}]{\text{Cu}} \begin{array}{c} \text{H}\ \ \text{O} \\ |\ \ \ \ \| \\ \text{H—C—C—H}(l) \\ | \\ \text{H} \end{array} + H_2O(l)$$

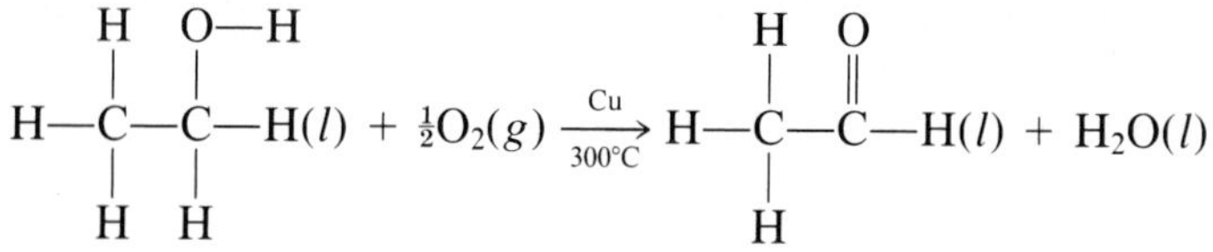

Acetaldehyde

Ethanol in wine can be oxidized to either acetaldehyde or acetic acid if the wine is in contact with air for long time intervals. A wine bottle is stored on its side to prevent air from drying the cork and eventually oxidizing the alcohol in the bottle.

Formaldehyde is a colorless gas at room temperature with a suffocating, pungent odor. Its water solution (about 37 percent by mass) is called formalin, which is a poisonous liquid that is used as a disinfectant, as a preservative of biological specimens, and as an embalming fluid. Acetaldehyde is a colorless gas or a volatile liquid (b.p. 20.8°C) with a pungent, fruity odor. It is used mainly in organic synthesis and as a mild reducing agent.

Alcohols burn in atmospheric oxygen to produce carbon dioxide and water vapor as shown below for the burning of ethanol:

$$C_2H_5OH(l) + 3O_2(g) \longrightarrow 2CO_2(g) + 3H_2O(g)$$

Methanol is often used as an additive in gasoline to enhance the octane rating of the gasoline and to conserve petroleum.

Ketones

The oxidation of secondary alcohols produces **ketones.** Tertiary alcohols are not oxidized under similar conditions. The characteristic functional group of ketones is the carbonyl group. In a ketone, the carbonyl group is bonded to two alkyl or aryl groups—R and R′:

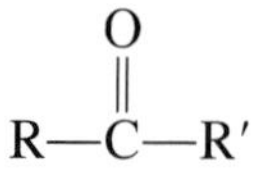

$$\begin{array}{c} \text{O} \\ \| \\ \text{R—C—R}' \end{array}$$

where R and R′ can be different or identical alkyl or aryl groups. An **aryl group** is a benzene ring or its derivative, less one hydrogen atom.

The oxidation of the simplest secondary alcohol, 2-propanol, yields the simplest ketone, propanone, commonly known as acetone:

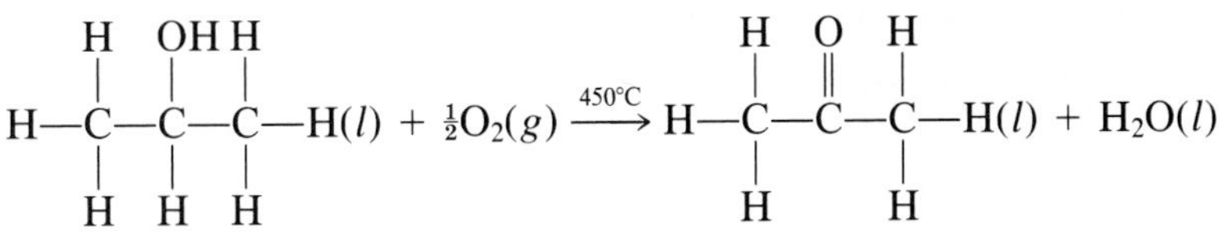

$$\mathrm{H{-}\overset{\overset{\displaystyle H}{|}}{\underset{\underset{\displaystyle H}{|}}{C}}{-}\overset{\overset{\displaystyle OH}{|}}{\underset{\underset{\displaystyle H}{|}}{C}}{-}\overset{\overset{\displaystyle H}{|}}{\underset{\underset{\displaystyle H}{|}}{C}}{-}H}(l) + \tfrac{1}{2}O_2(g) \xrightarrow{450^\circ C} \mathrm{H{-}\overset{\overset{\displaystyle H}{|}}{\underset{\underset{\displaystyle H}{|}}{C}}{-}\overset{\overset{\displaystyle O}{\|}}{C}{-}\overset{\overset{\displaystyle H}{|}}{\underset{\underset{\displaystyle H}{|}}{C}}{-}H}(l) + H_2O(l)$$

The formula of acetone is usually written $CH_3{-}\overset{\overset{\displaystyle O}{\|}}{C}{-}CH_3$. Acetone is a colorless, volatile, and flammable liquid (b.p. 56.2°C). It dissolves fats, adhesives, and paints. Because of its grease-dissolving ability and fast drying action (volatility), acetone is often used in science laboratories for rinsing glassware. It is also used to remove fingernail polish. Acetone has a "sweet" odor. The presence of acetone in the breath is characteristic of diabetes.

Carboxylic Acids

We discussed some simple carboxylic acids in Chapter 14. We recall that the characteristic functional group of carboxylic acids is the carboxyl group. This group may be regarded as a combination of a carbonyl group and a hydroxyl group:

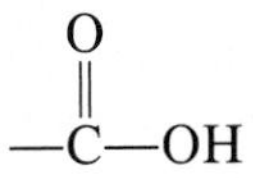

$$-\overset{\overset{\displaystyle O}{\|}}{C}{-}OH$$

Carboxyl group

We also recall that an aldehyde is the first product of the mild oxidation of a primary alcohol. If the aldehyde is not removed from the reaction mixture (by distillation or by other means), the oxidation of the alcohol continues to the corresponding carboxylic acid as shown below for the oxidation of ethanol:

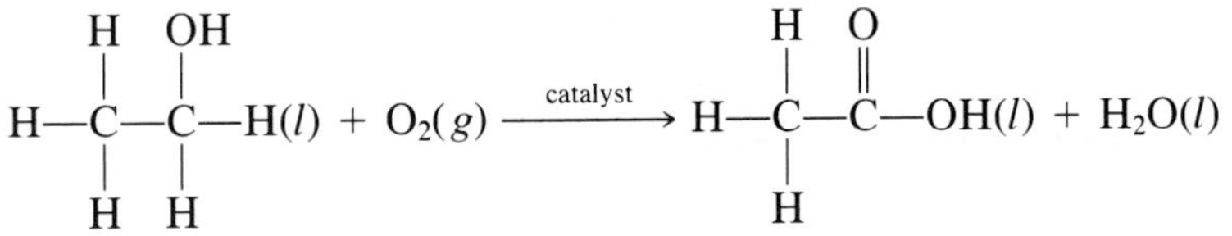

$$\underset{\text{Ethanol}}{\mathrm{H{-}\overset{\overset{\displaystyle H}{|}}{\underset{\underset{\displaystyle H}{|}}{C}}{-}\overset{\overset{\displaystyle OH}{|}}{\underset{\underset{\displaystyle H}{|}}{C}}{-}H}(l)} + O_2(g) \xrightarrow{\text{catalyst}} \underset{\text{Acetic acid or ethanoic acid}}{\mathrm{H{-}\overset{\overset{\displaystyle H}{|}}{\underset{\underset{\displaystyle H}{|}}{C}}{-}\overset{\overset{\displaystyle O}{\|}}{C}{-}OH}(l)} + H_2O(l)$$

If ethanol produced by fermentation is exposed to air, it is oxidized to acetic acid according to the above equation. Some homemade wines and whiskey ("moonshine"), if not made in airtight systems, contain a considerable amount of acetaldehyde, acetic acid, or both. These two compounds cause "hangovers."

The oxidation of methanol yields methanoic acid, HCOOH (also called formic acid). The oxidation of propanol gives propanoic acid, CH_3CH_2COOH. The structures and relative strengths of the four simplest organic acids are discussed in Chapter 14.

APPLICATIONS OF CHEMISTRY 23.2
Plastic Bags Versus Paper Bags

Every year Americans pollute the environment with millions of tons of plastic (polymer) waste. A complaint commonly heard is that, unlike paper bags, plastic bags "sit around indefinitely" because they are not biodegradable. Let's take a look at some of the interesting fallacies in this argument.

First, plastic bags do not sit around indefinitely. Chemists can alter plastics (polymers) to make them biodegradable. For example, chemists mix starch compounds with polyethylene to produce a decomposable product. Also, even unmodified plastic bags eventually degrade over a period of several decades.

Chemists can also make polymers such as polyethylene biodegradable by periodically inserting ketone groups throughout the hydrocarbon chain. These ketone functional groups are photosensitive, which means that ultraviolet light from the sun causes the modified polymers to break into small segments. Then microorganisms can easily attack these smaller segments to finish the degradation process.

This leads us to a second fallacy: Paper is biodegradable. Actually, neither paper nor "biodegradable plastics" are biodegradable under common disposal conditions. Most of our waste, both paper and plastic, ends up in a landfill, where we cover it with dirt. Protected by the dirt, neither sunlight nor air can reach the waste to aid in its decomposition. Paper-decomposing microorganisms need oxygen to survive.

Recently, scientists uncovered a completely legible 40-year-old newspaper from a landfill in Florida (see photo). Also, scientists found 40-year-old food samples and lawn clippings that were as fresh as the day they were discarded.

Which bag is better for the environment, paper or plastic? There isn't a clear-cut answer to this question. The best that we may be able to do for our environment is to recycle. Many people throughout the world carry their own reusable bags to grocery stores. By reusing these bags dozens of times, they may have made the wisest choice of all.

Sources: *Science World*, April 5, 1991, pages 6–7 and April 20, 1990, pages 11–16; Michael Chejlava, *Chemical & Engineering News*, July 2, 1990, page 2; *Chemecology*, October 1990, pages 2–5.

Esters

An **ester** is a compound in which the hydrogen atom of the carboxyl group of a carboxylic acid is replaced by an alkyl or an aryl group:

$$\text{R}\text{—}\overset{\overset{\displaystyle\text{O}}{\|}}{\text{C}}\text{—O—R}'$$

The R and R′ groups can be identical or different. In an ester derived from formic acid, R = H. Thus, the functional group for esters is

$$\text{—}\overset{\overset{\displaystyle\text{O}}{\|}}{\text{C}}\text{—O—R}$$

An ester can be produced by the reaction of a carboxylic acid with an alcohol. For example, the reaction of acetic acid with ethanol produces the ester ethyl acetate:

$$CH_3-\overset{\overset{\large O}{\|}}{C}-OH(l) + H-O-CH_2CH_3(l) \longrightarrow$$

$$CH_3-\overset{\overset{\large O}{\|}}{C}-O-CH_2CH_3(l) + H_2O(l)$$

Ethyl acetate

In this reaction of a carboxylic acid with an alcohol, the —OH group of the acid is replaced by the —OR group of the alcohol. The —OH group of the acid combines with the —H of the alcohol to form water.

In the equation above we can see that the name of the ester is derived from the names of the alkyl group of the alcohol and the carboxyl group of the acid. Thus, the ester that forms from ethyl alcohol and acetic acid is ethyl acetate, and the ester that forms from methyl alcohol and formic acid is methyl formate.

Many esters are volatile liquids with pleasing odors. They are widely distributed in nature and are responsible for the pleasant odors of many fruits and flowers. For example, ethyl acetate and methyl butyrate are responsible for the odor of pineapples, butyl acetate is present in bananas, and octyl acetate is found in oranges.

Fats and oils such as lard, butter, cottonseed oil, and olive oil are esters of complex acids and an alcohol called glycerol. Glycerol has three —OH groups:

$$\begin{array}{ccccccc} & & H & & H & & H & \\ & & | & & | & & | & \\ H & - & C & - & C & - & C & - & H \\ & & | & & | & & | & \\ & & OH & & OH & & OH & \end{array}$$

For the structure of a typical oil and a fat, see the discussion on hydrogenation in Section 23.4. Table 23.4 summarizes the various types of oxygen-containing aliphatic organic compounds discussed in this section.

Table 23.4

Types of Aliphatic Oxygen-Containing Compounds

General Name	Functional Group	Example	Name
Alcohol	—OH	C_2H_5-OH	Ethyl alcohol
Aldehyde	$-\overset{\overset{O}{\|}}{C}-H$	$CH_3-\overset{\overset{O}{\|}}{C}-H$	Acetaldehyde
Acid	$-\overset{\overset{O}{\|}}{C}-OH$	$CH_3-\overset{\overset{O}{\|}}{C}-OH$	Acetic acid
Ketone	$-\overset{\overset{O}{\|}}{C}-$	$CH_3-\overset{\overset{O}{\|}}{C}-CH_3$	Acetone
Ester	$-\overset{\overset{O}{\|}}{C}-O-$	$CH_3-\overset{\overset{O}{\|}}{C}-O-CH_3$	Methyl acetate
Ether	—O—	$C_2H_5-O-C_2H_5$	Diethyl ether

Example 23.4

Boiling Points and Structures of Organic Compounds

Acetic acid and methyl formate are isomers with the molecular formula $C_2H_4O_2$. However, acetic acid boils at 118°C, while methyl formate boils at 32°C. Suggest an explanation for this large difference in boiling points.

SOLUTION: As we discussed in Chapter 11, a liquid has a high boiling point when the intermolecular forces in it are strong. Since acetic acid and methyl formate have the same molar masses, they have approximately equal London forces. However, acetic acid (CH_3COOH) can form a hydrogen-bonded dimer. Hydrogen bonding between two molecules of acetic acid is shown below by dotted lines:

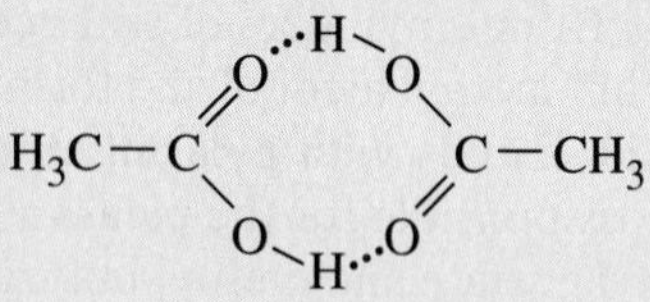

$$H_3C-C\begin{smallmatrix}\nearrow O\cdots H-O\searrow \\ \searrow O-H\cdots O\nearrow\end{smallmatrix}C-CH_3$$

In methyl formate,

$$H-\overset{\overset{\Large O}{\|}}{C}-O-CH_3$$

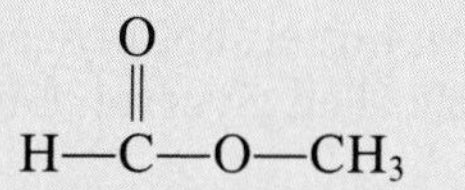

all of the hydrogen atoms are attached to carbon atoms, and hydrogen bonding is impossible. Therefore, acetic acid has a much higher boiling point than methyl formate.

23.8 Introduction to Biochemistry

Biochemistry is the study of the chemical reactions and molecular structures of living cells. In this section we provide only a bird's eye view of this vast subject.

The basic unit of all living organisms is the *cell.* A cell is a smallest unit of living matter that is capable of functioning independently. Cells combine to form *tissues*; tissues make up *organs*; and a system of organs constitutes an *organism*. Cells differ greatly in size (10^{-12} to 10^{-6} cm^3 of volume), shape, and function. A *bacterium* is a microscopic example. As examples of the approximate composition of living matter, the compositions of the bacterium *Escherichia coli* (*E. coli*) and of rat's liver are listed in Table 23.5.

All cells contain a high percentage of water (about 70 percent by mass), which acts as a transport system for food and waste and as the medium for chemical reactions. In animals, water also helps to control body temperature with its high heat capacity and its high heat of vaporization.

Cells contain four major types of compounds: **proteins**, **nucleic acids**, **carbohydrates**, and **lipids**. The following discussion will focus on these substances.

Table 23.5

Approximate Composition of the Cells of the Bacterium *E. coli* and the Cells of Rat Liver (Percent by Mass)

Component	*E. coli*	Rat Liver
Water	70	69
Protein	15	21
Nucleic acids	7	1.2
Carbohydrates	3	3.8
Lipids	2	6
Inorganic ions	1	0.4
Amino acids	0.4	—
Nucleotides	0.4	—

Carbohydrates

Carbohydrates are the most abundant organic molecules in the biosphere (the part of the world in which life can exist). The term **carbohydrate** embraces a wide range of molecular structures, from relatively simple substances such as glucose to polymers of glucose—among which are starch and cellulose—and other molecules with extremely complex structures. Carbohydrates are also called **saccharides**, which exist as monomers called **monosaccharides** (*simple sugars*) and as polymers—**polysaccharides**. The empirical formula for many common carbohydrates is $(CH_2O)_n$, hence the name "hydrate of carbon." The empirical formula, however, does not hint at the structural complexity of carbohydrates nor at the wide range of their chemical and biochemical reactivity.

Carbohydrates have very diverse biochemical functions. Carbohydrates are an important food for most organisms, and carbohydrate metabolism provides a significant fraction of the energy available to most organisms. The interactions of cells with one another often occurs through their membranes, which are covered with characteristic carbohydrates.

Carbohydrates also have structural roles. They are a major component of bacterial cell walls. The exterior shells of animals such as crabs and snails are made of complex carbohydrates. The polysaccharide cellulose is the most abundant organic molecule in plants; indeed, it is the second most abundant molecule (after water) in the biosphere. Many complex carbohydrates are found in animal tissues as well.

The first reference to "wine sugar," glucose, appeared in Moorish writings of the twelfth century, a time when Europe languished in the Dark Ages. But we can date modern carbohydrate chemistry to the late nineteenth century, when the German chemist Emil Fischer carried out his research, including a proof of the structure of glucose, one of the great achievements in chemistry.

Glucose is the principal sugar in blood (about 0.1 percent by mass), and it is therefore also called blood sugar. Glucose is also abundant in fruits such as grapes and is therefore sometimes called grape sugar (or "wine sugar"). In hospitals, glucose in saline solution (about 0.6 to 0.9 percent NaCl solution) is administered intravenously before and after an operation, or whenever oral nutrition is not possible. Glucose and most other simple sugars exist predominantly in a cyclic structure. In the cyclic structure of glucose, an oxygen atom

Figure 23.9

Cyclic structures of α-D- and β-D-forms of glucose. These forms differ in the orientation of the OH group on the number 1 carbon atom. In α-D-glucose this OH group is perpendicular to a plane through all the carbon atoms in the ring, assuming these atoms to be planar. Actually, they are puckered as shown. In β-D-glucose this OH group is roughly in the plane through the ring.

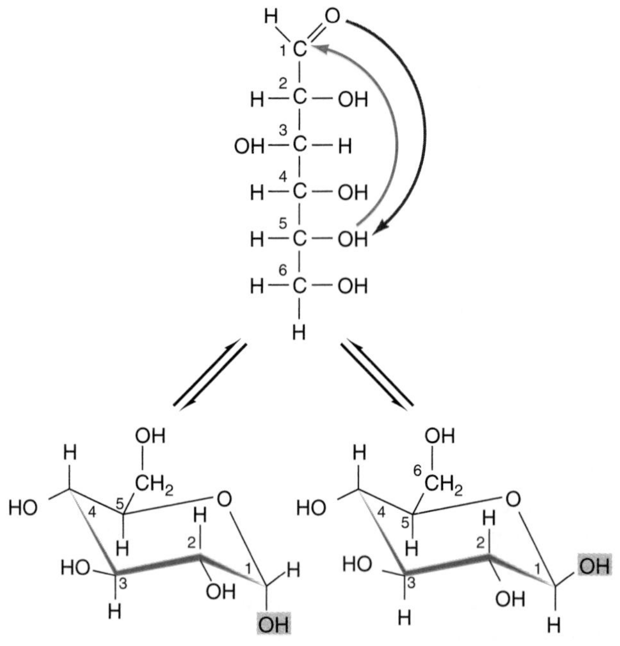

is a part of a puckered six-membered ring. Figure 23.9 illustrates the straight-chain structure and two cyclic structures of glucose. In the two isomeric cyclic structures of glucose, the —OH group on the carbon atom marked number 1 (Figure 23.9) can point perpendicular to a plane of the carbon atoms in the ring, or it can be positioned approximately in that plane. The isomer in which the —OH group is perpendicular to the plane is called α-D-glucose, and the other isomer is β-D-glucose, as shown in Figure 23.9.

Another important monosaccharide is called *fructose*. Fructose is found in many fruits (hence its name; the Latin name for fruit is *fructus*). When fructose is dissolved in aqueous solution, it rotates the plane of polarized light to the left. As a result, fructose is sometimes called *levulose*. The straight-chain and cyclic structure for fructose are shown in Figure 23.10. This cyclic structure is a five-membered ring that contains an oxygen atom.

Figure 23.10

Cyclic structure of fructose.

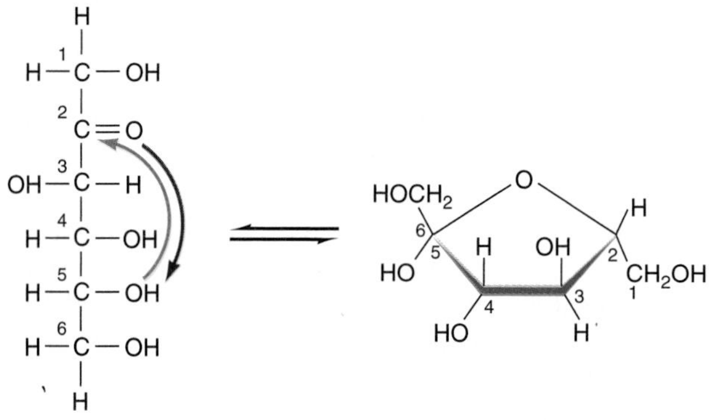

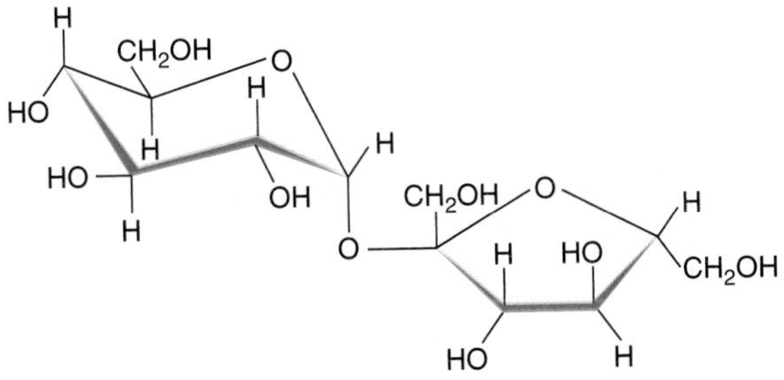

Figure 23.11

Structure of sucrose, a disaccharide. A molecule of sucrose consists of two monosaccharide units, glucose and fructose. A sucrose molecule is formed from an glucose molecule and a fructose molecule with the elimination of a water molecule.

Sucrose (table sugar) is a *disaccharide* consisting of two monosaccharide units, α-D-glucose and fructose (Figure 23.11). Note that a molecule of sucrose is formed by the elimination of a molecule of H_2O from the —OH groups of the glucose and the fructose molecules.

Other important disaccharides are *lactose* and *maltose*. Lactose is a major carbohydrate component of milk (about 5 percent in cow's milk and 7 percent in human milk) and is also called "milk sugar." Maltose occurs in malt, the grain used to brew beer and other alcoholic beverages, and is therefore known as "malt sugar."

Starch is a polysaccharide that is made from many α-D-glucose units (Figure 23.12). Starch is found in plants such as corn (50 percent), rice (75 percent), and potatoes (20 percent). When starch is heated, it is converted to *dextrin*, a sweet substance with good adhesive properties that is used as an adhesive for stamps and as wallpaper glue.

Figure 23.12

Structures of starch and cellulose. Starch is formed by linking 1000 to 4000 glucose units through oxygen atoms. Cellulose is formed by linking about 3000 β-D-glucose units through oxygen atoms.

Repeating unit

Starch

Repeating unit

Cellulose

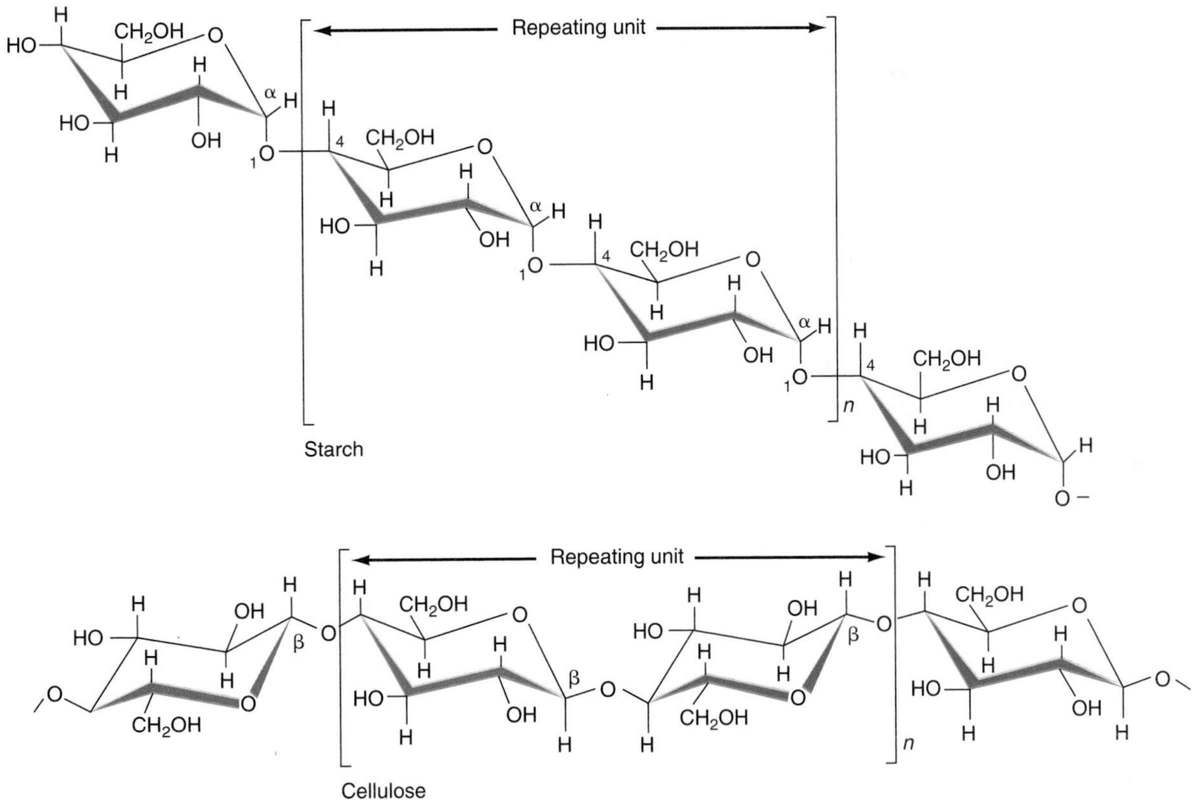

Glycogen is an animal polysaccharide whose structure is very similar to that of starch. Glycogen is a carbohydrate found in liver and muscle tissue of animals.

Cellulose is another important polysaccharide consisting of β-D-glucose units (Figure 23.12). This small structural difference between starch and cellulose has important consequences on the properties of these substances. Starch is slightly water-soluble and is used as food. Cellulose is insoluble and cannot be digested by human beings, although termites find it quite nutritious. Cows and other hoofed animals are able to use cellulose as a food because their stomachs contain bacteria that convert cellulose to glucose. Cotton is almost pure cellulose. Wood is about 50 percent cellulose, but wood is a very complex material that contains many other polysaccharides and nearly 10 percent inorganic salts.

Lipids

Most **lipids** are naturally occurring esters that contain at least one long-chain carboxylic acid called a fatty acid. *Fats*, oils, and *waxes* are examples of lipids. We recall that fats are esters of glycerol (see Section 23.7) with saturated fatty acids, whereas oils contain unsaturated fatty acids. Some naturally occurring fatty acids are listed in Table 23.6.

Waxes are esters of fatty acids and long-chain monohydroxy alcohols (alcohols that contain only one —OH group per molecule). For example, the wax of the sperm whale is derived from cetyl alcohol, $CH_3(CH_2)_{14}CH_2OH$. Waxes form protective coatings on several parts of both animals and plants. Wax coats the feathers of ducks and other birds, and also protects the surfaces of plant tissues. Fats are major energy reserves in humans and animals. The generic term "lipid" includes several other types of molecules such as *phospholipids* and *steroids*. Phospholipids are esters derived from glycerol, fatty acids, and a phosphoric acid derivative:

Table 23.6

Some Naturally Occurring Fatty Acids

Name and Example of Origin		Formula	Melting Point (°C)
Saturated	Lauric acid (in laurel trees or shrubs, coconut or palm)	$CH_3(CH_2)_{10}COOH$	44
	Myristic acid (in nutmeg fat)	$CH_3(CH_2)_{12}COOH$	54
	Palmitic acid (in many vegetable fats)	$CH_3(CH_2)_{14}COOH$	63
	Stearic acid (in many animal and vegetable fats)	$CH_3(CH_2)_{16}COOH$	70
Unsaturated	Linoleic acid (in linseed oil)	$CH_3(CH_2)_4CH{=}CHCH_2CH{=}CH(CH_2)_7COOH$	−5
	Oleic acid (in olive oil)	$CH_3(CH_2)_7CH{=}CH(CH_2)_7COOH$	13.4
	Palmitoleic acid (in butter fat)	$CH_3(CH_2)_5CH{=}CH(CH_2)_7COOH$	−1

$$\begin{array}{l} \text{H—C—O—}\overset{\text{O}}{\overset{\|}{\text{C}}}\text{—R} \quad \longleftarrow \text{from fatty acids} \\ \text{from glycerol} \longrightarrow \text{H—C—O—}\overset{\text{O}}{\overset{\|}{\text{C}}}\text{—R} \\ \text{H—C—O—}\overset{\text{O}}{\overset{|}{\underset{\text{O}}{\underset{|}{\text{P}}}}}\text{—O—R}' \quad \longleftarrow \text{from phosphoric acid derivative} \end{array}$$

A phospholipid

Phospholipids are important constituents of the membranes that surround every living cell.

Steroids contain a basic four-ring hydrocarbon unit shown in Figure 23.13. Some important steroids and their functions are shown in Figure 23.14.

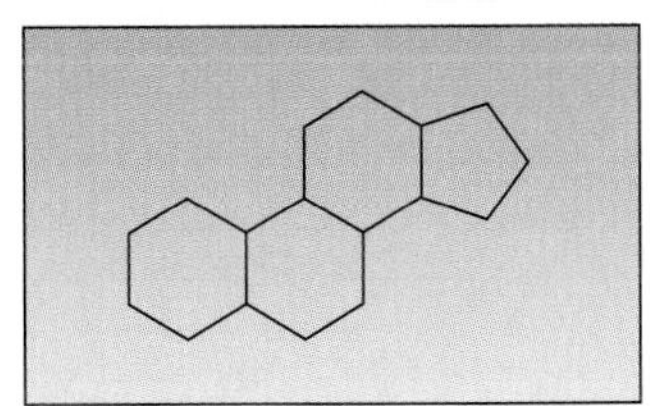

Figure 23.13

Basic unit of steroids. Note the three hexagons and a pentagon.

Proteins

Proteins are polymers that consist of monomers called **α-amino acids**. An amino acid consists of a carboxylic acid that contains an amine group, $—NH_2$. The structures of two simple amino acids are shown below:

$$\underset{\text{Glycine}}{\underset{\underset{\text{NH}_2}{|}}{\text{CH}_2}\text{—}\overset{\overset{\text{O}}{\|}}{\text{C}}\text{—OH}} \qquad \underset{\text{Alanine}}{\text{CH}_3\text{—}\underset{\underset{\text{NH}_2}{|}}{\text{CH}}\text{—}\overset{\overset{\text{O}}{\|}}{\text{C}}\text{—OH}}$$

Glycine and alanine are both called α-amino acids because the amine group is connected to the carbon atom adjacent to the carboxyl group (the α-carbon atom). An α-amino acid can be written with a general formula

$$\text{R—}\underset{\underset{\text{NH}_2}{|}}{\overset{\alpha}{\text{CH}}}\text{—}\overset{\overset{\text{O}}{\|}}{\text{C}}\text{—OH}$$

One amino acid can react with another by linking the carbon atom of the carboxyl group of one of the acids with the nitrogen atom of the amino group of the other acid, as shown by the equation below. Such a linkage is called a **peptide linkage**. In the formation of a peptide linkage, a molecule of water is eliminated.

$$\text{CH}_3\text{—}\underset{\underset{\text{NH}_2}{|}}{\text{CH}}\text{—}\overset{\overset{\text{O}}{\|}}{\text{C}}\text{—OH} + \text{H—}\underset{\underset{\text{H}}{|}}{\text{N}}\text{—CH}_2\text{—}\overset{\overset{\text{O}}{\|}}{\text{C}}\text{—OH} \longrightarrow$$

peptide linkage

$$\text{CH}_3\text{—}\underset{\underset{\text{NH}_2}{|}}{\text{CH}}\text{—}\overset{\overset{\text{O}}{\|}}{\text{C}}\text{—}\underset{\underset{\text{H}}{|}}{\text{N}}\text{—CH}_2\text{—}\overset{\overset{\text{O}}{\|}}{\text{C}}\text{—OH} + \text{H}_2\text{O}$$

Alanylglycine, a *dipeptide*

Figure 23.14

Some important steroids and their functions.

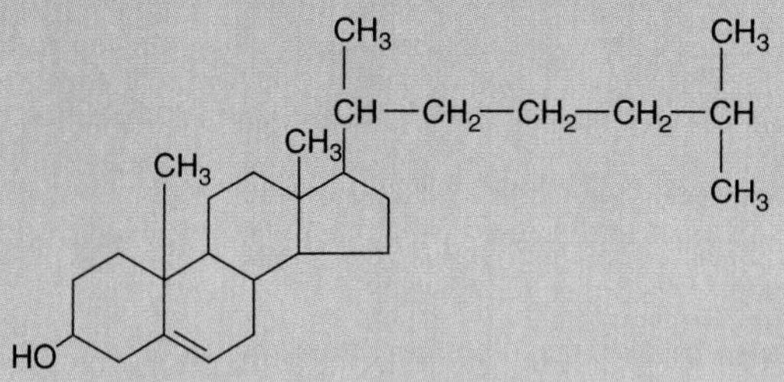

Cholesterol

A constituent of blood, brain tissue, bile, and gallstones. Cholesterol regulates fat metabolism and makes it possible for cells to hold large quantities of water. However, too much cholesterol might lead to atherosclerosis.

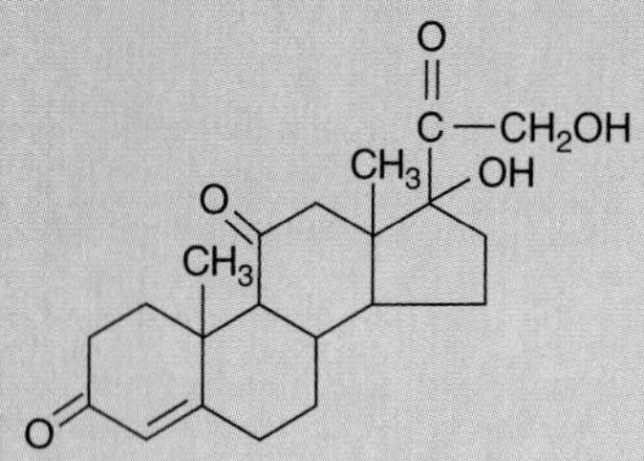

Cortisone

Affects protein metabolism.

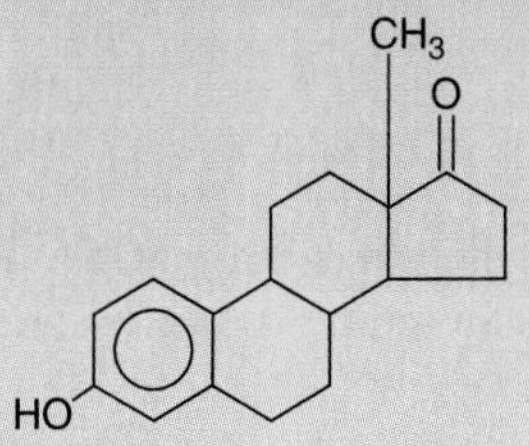

Estrone

A female sex hormone

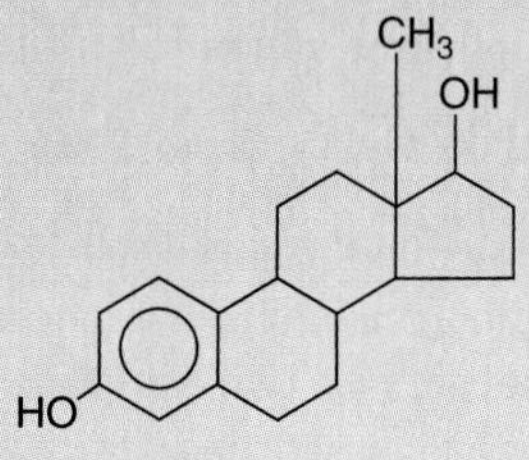

Estradiol

A female sex hormone

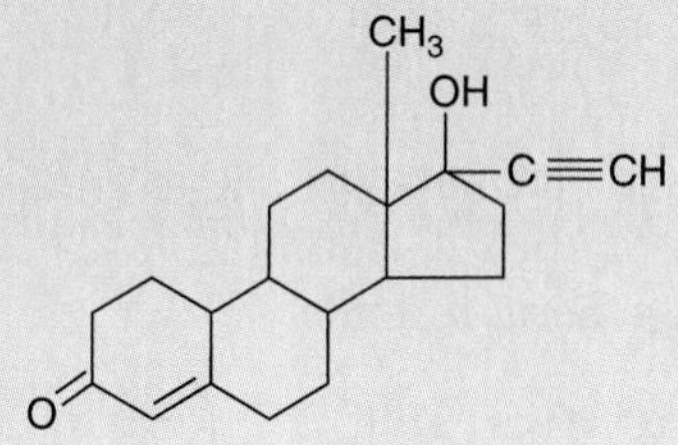

Norethindrone

An oral contraceptive

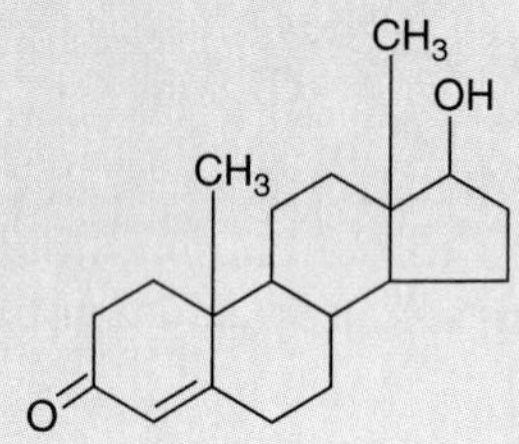

Testosterone

A male sex hormone

Long chains of amino acids can be formed in this way. A peptide that contains two amino acid units is called a *dipeptide*, one that contains three amino acid units is a *tripeptide*, and so on.

Proteins are polypeptide chains that contain a minimum of 50 up to more than 1000 amino acid units. Proteins have diverse functions in all cells. Naturally occurring proteins of plants and animals contain about 20 different amino acids (that is, 20 different R groups). Table 23.7 lists some important classes of proteins and their functions.

Table 23.7

Classes of Proteins and Their Functions in the Body

Protein	Function
Enzymes	Catalyze chemical reactions in the body
Hormones	Regulate metabolism and growth
Storage proteins	Store metal ions and release amino acids as needed
Structural proteins	Form bones, cartilage, and connective tissue
Contractile proteins	Form muscle tissue
Transport proteins	Help to bind and transport various other molecules in bloodstream
Protective proteins	Act as antibodies and blood-clotting agents

Nucleic Acids

Nucleic acids are giant polymers present in nearly all cells. Nucleic acids are of two types: **deoxyribonucleic acids** (DNA) and **ribonucleic acids** (RNA). DNA is primarily responsible for transmitting genetic information from one generation to the next during cell division. Strings of DNA constitute genes, the individual units of heredity. Most RNA is present in the fluid called cytoplasm outside the nucleus of a cell.

The molar masses of DNA molecules range from a few million for small bacteria to the billions in higher animals. RNA molecules are much smaller, their molar masses range from 20,000 to 40,000.

The repeating units in both DNA and RNA polymers are called **nucleotides**. Each nucleotide consists of three parts:

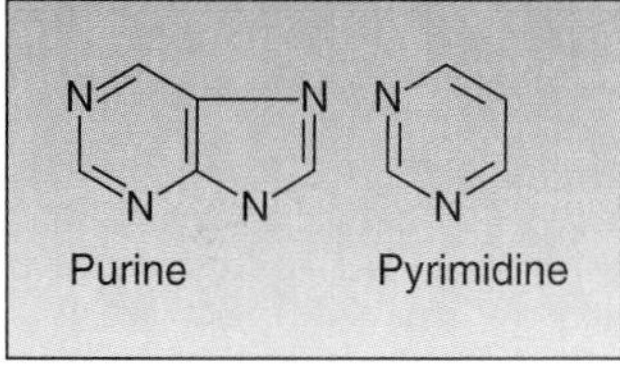

Figure 23.15

Structures of purine and pyrimidine.

1. A cyclic ring system (called a heterocyclic ring system because the rings contain nitrogen in addition to carbon). The heterocyclic ring systems of nucleic acids are derived from molecules of a substance called purine or another substance called pyrimidine (Figure 23.15). The ring systems of purine and pyrimidine are Lewis bases and are often simply called "bases."
2. A sugar that is either D-ribose or 2-deoxy-D-ribose (Figure 23.16).
3. One or more phosphate groups.

The sugar of DNA nucleotides is 2-deoxy-D-ribose. In a molecule of 2-deoxy-D-ribose, one of the carbon atoms is attached to two H atoms, while the corresponding carbon atom in the ribose molecule in RNA is attached to an H atom and an OH group (Figure 23.16).

In DNA, the bases derived from purine are adenine (A) and guanine (G), and those derived from pyrimidine are cytosine (C) and thymine (T). These bases are incorporated in nucleotides (Figure 23.17). A nucleotide unit and the linkage of two of these units in DNA is illustrated in Figure 23.18.

Cells also contain uncombined nucleotides, which play a major role in metabolism. For example, adenosine triphosphate, called ATP, is the primary

Ribose (in RNA)

Deoxyribose (in DNA)

Figure 23.16

Structures of ribose and 2-deoxy-D-ribose.

Figure 23.17

The four bases present in the nucleotide units of DNA molecules and their usual symbols.

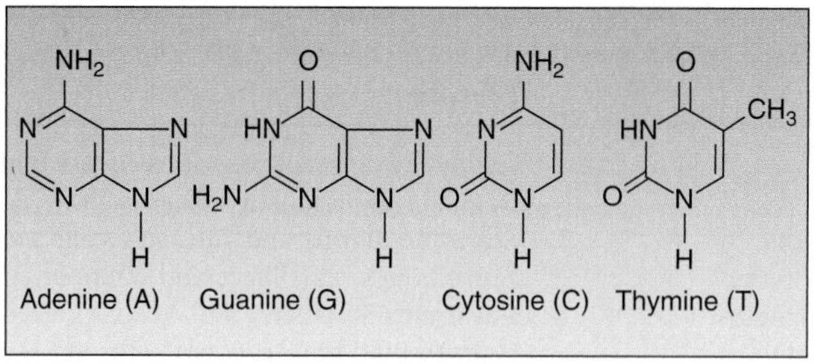

Adenine (A) Guanine (G) Cytosine (C) Thymine (T)

Figure 23.18

A nucleotide, and the linkage of two nucleotides in a DNA molecule.

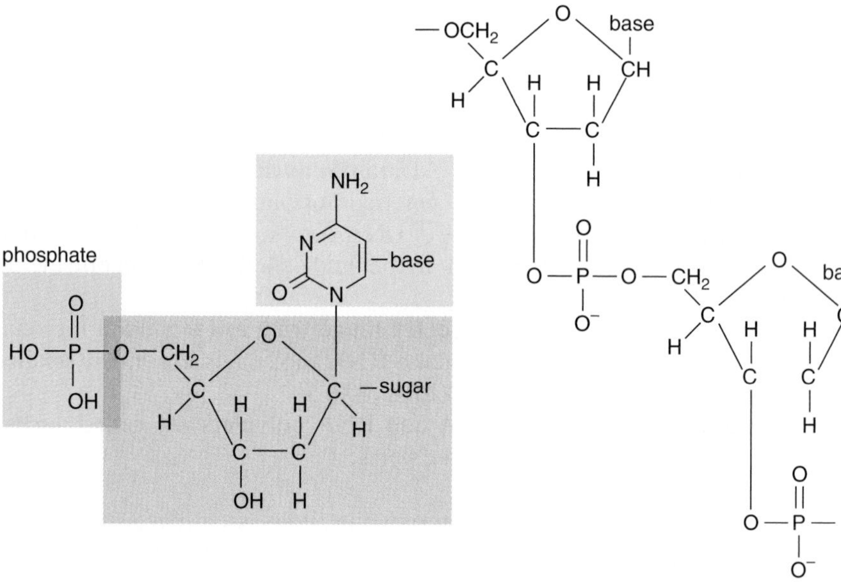

phosphate — base — sugar

A nucleotide (deoxycytidylic acid)

Two nucleotide units in a DNA molecule

Figure 23.19

Conversion of ATP to ADP with the release of energy.

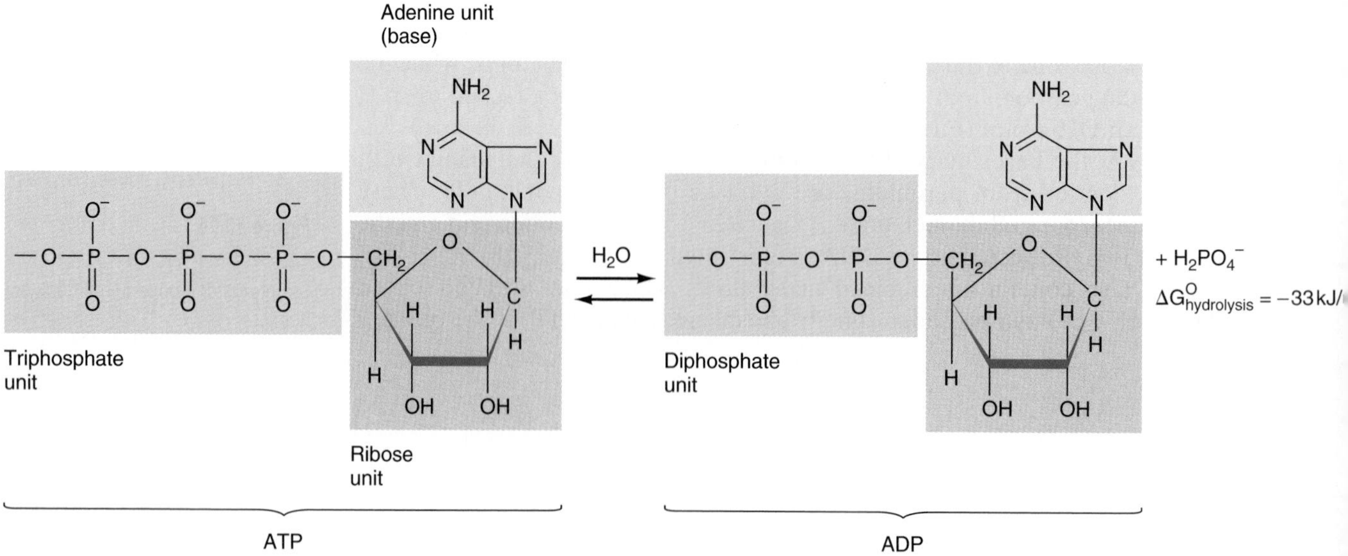

Adenine unit (base)

Triphosphate unit

Ribose unit

H_2O

Diphosphate unit

$+ H_2PO_4^-$

$\Delta G^{O}_{hydrolysis} = -33\,kJ/$

ATP

ADP

source of energy—the "energy currency"—in cells. The hydrolysis of adenosine triphosphate (ATP) to adenosine diphosphate (ADP) releases about 33 kJ per mole ($\Delta G^\circ = -33$ kJ mol^{-1}) (see Figure 23.19). This energy is used to drive processes such as muscle contraction and the synthesis of various polymers in cells.

DNA consists of two intertwined helical polymeric strands composed of nucleotide units. The intertwined strands interact by hydrogen bonds between bases on each strand. The bases form the "base pairs" adenine–thymine and guanine–cytosine, as shown by the dotted lines representing hydrogen bonding in Figure 23.20. In a double-helical DNA molecule A is always paired with T, and G is always paired with C. Figure 23.21a is a schematic illustration of the double helix of DNA, and part (b) illustrates a space-filling model of DNA.

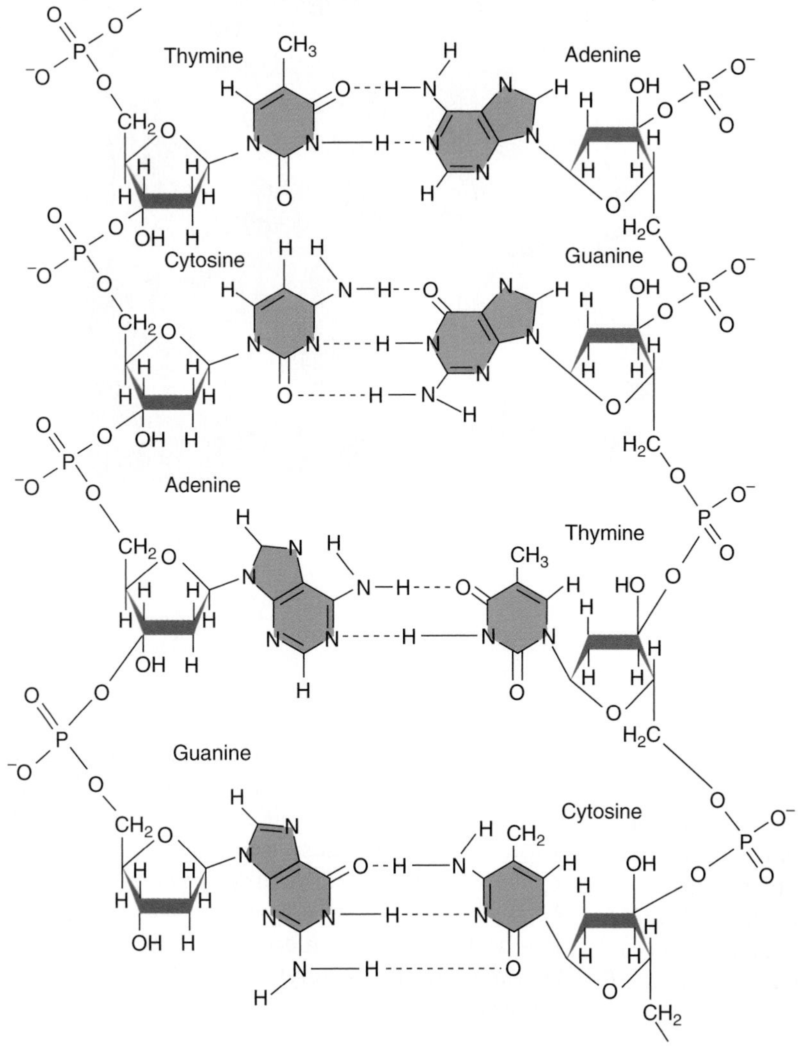

Thymine Adenine Cytosine Guanine Adenine Thymine Guanine Cytosine

Figure 23.20

Hydrogen bonding between the two polymeric strands in DNA.

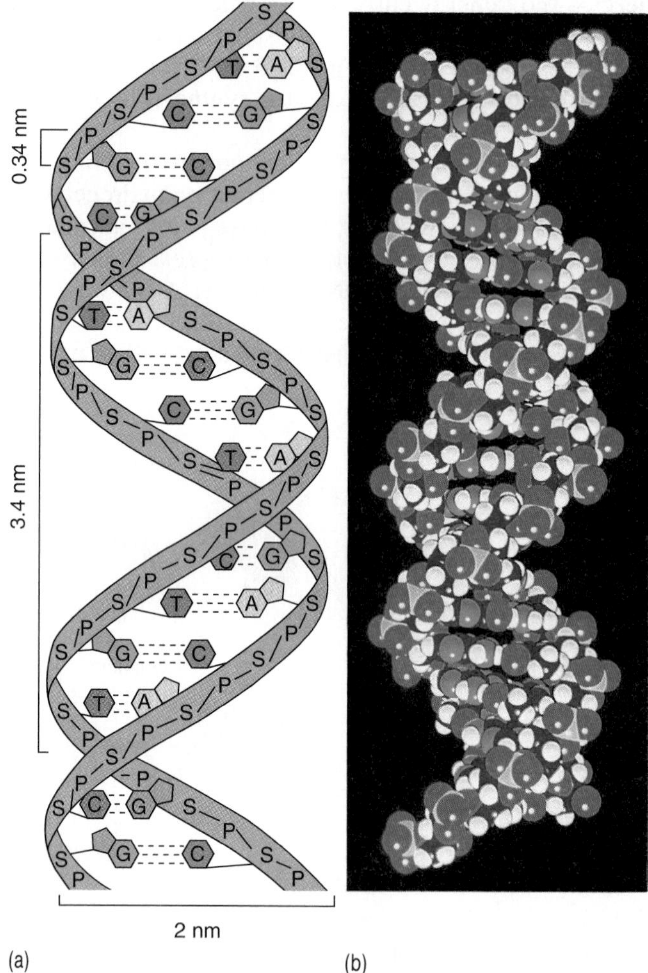

Figure 23.21

A double-stranded helical structure of DNA. (a) A schematic representation. S = sugar, P = phosphate, C = cytosine, G = guanine, A = adenine, and T = thymine. Note that A is always aligned with T, and G with C. The dotted lines represent hydrogen bonds (see Figure 23.20). (b) A space-filling model.

Summary

Hydrocarbons

Organic chemistry is the branch of chemistry that deals with compounds that contain carbon–hydrogen bonds, carbon–carbon bonds, or both. Compounds that contain only carbon and hydrogen are called hydrocarbons. **Aliphatic hydrocarbons** include *alkanes*, *alkenes*, *alkynes*, and *cyclic analogs* of these compounds. **Aromatic hydrocarbons** include *benzene and its derivatives*.

Alkanes are *saturated* hydrocarbons. The general formula of an alkane is C_nH_{2n+2}. Alkenes (C_nH_{2n}) and alkynes (C_nH_{2n-2}) are *unsaturated* hydrocarbons. Alkanes react with halogens to form *alkyl halides*.

Hydrogenation of alkenes or alkynes produces corresponding alkanes. Reactions of alkenes or alkynes with hydrogen halides or halogens (*halogenation*) yield halogenated hydrocarbons. Many alkenes can be made to polymerize to form such important polymers as polypropylene, Teflon, and synthetic rubber.

Oxygen- and Nitrogen-Containing Organic Compounds

Some important oxygen-containing organic compounds are listed below by their names and general formulas:

R—OH	R—C(=O)—H	R—C(=O)—OH	
Alcohols	**Aldehydes**	**Carboxylic acids**	
R—C(=O)—R′	R—C(=O)—O—R′	R—O—R′	R—N(R″)—R′
Ketones	**Esters**	**Ethers**	**Amines**

Aldehydes can be made by oxidation of primary alcohols, and ketones can be prepared by the oxidation of secondary alcohols. Aldehydes can be oxidized to carboxylic acids. Esters can be prepared by reacting alcohols with carboxylic acids. Ethers can be prepared by heating alcohols with an acid catalyst.

Biochemistry

Biochemistry is the study of the chemical reactions and molecular structures found in living organisms. Cells contain four major types of organic compounds: **carbohydrates**, **lipids**, **proteins**, and **nucleic acids**. Carbohydrates are also referred to as saccharides, which may be simple sugars (**monosaccharides**) or polymers of monosaccharides called **polysaccharides**. Most lipids are naturally occurring esters in which at least one of the acid component is a long-chain carboxylic acid called a **fatty acid**. Fats, oils, and waxes are examples of lipids. Proteins are composed of **α-amino acids** linked by **peptide linkages**. **Nucleic acids** are giant polymers present in nearly all cells. These polymers are classified into two groups: **deoxyribonucleic acids** (DNA) and **ribonucleic acids** (RNA). The repeating units of both DNA and RNA are called **nucleotides**. DNA is a double helical molecule in which A is always paired with T and G is always paired with C (see Figure 23.20). DNA is primarily responsible for transmitting genetic information from one generation to the next during cell division. RNA plays a key role in protein synthesis.

New Terms

α-amino acids (23.8)
Alcohol (23.6)
Aldehyde (23.7)
Aliphatic hydrocarbon (23.1)
Alkyl group (23.1)
Amine (23.6)
Amino acid (23.8)
Aromatic hydrocarbon (23.1 and 23.5)
Aryl group (23.7)
Carbohydrate (23.8)
Carbonyl group (23.7)
Deoxyribonucleic acid, DNA (23.8)
Ester (23.7)
Ether (23.6)
Fatty acid (23.7)
Functional group (23.6)
Halogenation (23.4)
Hydrogenation (23.4)
Ketone (23.7)
Lipid (23.8)
Monosaccharide (23.8)
Nucleic acid (23.8)
Nucleotide (23.8)
Peptide linkage (23.8)
Polymerization (23.4)
Polypeptide (23.8)
Polysaccharide (23.8)
Protein (23.8)
Ribonucleic acid, RNA (23.8)
Saccharide (*see* Carbohydrate).

Exercises

General Review

1. What are some differences between aliphatic and aromatic hydrocarbons? Write the formula for and the name as an example of each of these two types of compounds.
2. What is: **(a)** a saturated hydrocarbon; **(b)** an unsaturated hydrocarbon?
3. Write the formula and the name for: **(a)** an alkane; **(b)** an alkene; **(c)** an alkyne; **(d)** a cyclic alkane; **(e)** a cyclic alkene.
4. What are the general formulas for alkanes, alkenes, and alkynes?
5. Draw structures for the isomers of butane.
6. What are the systematic names for ethylene and for acetylene?
7. What are the products of: **(a)** combustion of a hydrocarbon in excess oxygen; **(b)** halogenation of an alkane in a limited supply of the halogen?
8. How can alkenes be prepared from alkanes?
9. What is the hydrogenation product of an alkene?
10. Are alkenes less or more reactive with halogens than benzene is?
11. Give two examples of commercially important halogenated hydrocarbons and their uses.
12. Give an example of and the use of a polymer that consists of alkene monomers.
13. What is a functional group?
14. What is the characteristic functional group of each of the following classes of compounds? **(a)** Alcohols; **(b)** aldehydes; **(c)** carboxylic acids; **(d)** ketones; **(e)** esters; **(f)** ethers; **(g)** amines.
15. Write a structural formula for a primary, secondary, and tertiary alcohol.
16. What is the product of a mild oxidation of a primary alcohol? If this product is further oxidized, what will be produced?
17. What is the oxidation product of a secondary alcohol?
18. How can an ester be prepared?
19. How can an ether be prepared?
20. What class of hydrocarbons has delocalized π electrons?
21. How can benzene be converted to chlorobenzene?
22. Explain the difference between cell, tissue, organ, and organism in a living creature.
23. What are the four major compounds present in a living cell? Which of these is the most abundant?
24. Write a formula for a common monosaccharide and for a common disaccharide.
25. Name two common polysaccharides.
26. Name three classes of compounds that are lipids.
27. What are the monomeric units of proteins? Describe how these units are linked to one another. What is this type of linkage called?
28. Which has a higher molar mass, DNA or RNA?
29. What are the major functions of DNA and of RNA?
30. What are the repeating units that make up both DNA and RNA molecules?

Naming Alkanes

31. Name each of the following alkanes:

```
(a) CH3—CH2—CH2—CH2—CH2—CH3        (b) CH3—CH2—CH2
                                                  |
                                                 CH2
                                                  |
                                                 CH3

(c) CH3—CH2—CH2—CH—CH3        (d)                         CH3
                 |                                         |
                CH2                                       CH2
                 |                                         |
                CH3              CH3—CH2—CH2—CH—CH—CH3
                                              |
                                             CH3

(e)      CH3      (f)      CH3         (g)    CH3
          |                 |                  |
    CH3—CH            CH3—C—CH3               CH—CH2—CH3
          |                 |                  |
         CH3               CH3            CH3—C—CH3
                                               |
                                              CH3
```

32. Write structural formulas for: **(a)** methylpropane; **(b)** 4-ethyloctane; **(c)** 2,3-dimethylhexane; **(d)** 3-ethyl-3-methylheptane; **(e)** 2,2,4-trimethyl-4-ethylheptane; **(f)** cyclopentane; **(g)** cyclohexane.

Naming Alkenes, Alkynes, and Halosubstituted Hydrocarbons

33. Name each of the following compounds:

(a) $CH_3—CH{=}CH_2$ **(b)** $CH_3—C{\equiv}C—CH_3$ **(c)** 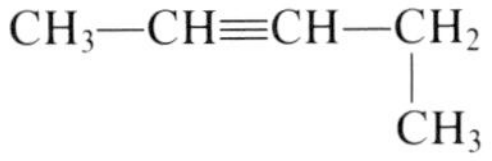 $CH_3—CH{\equiv}CH—CH_2$ (with CH_3 on the last CH_2)

(d) 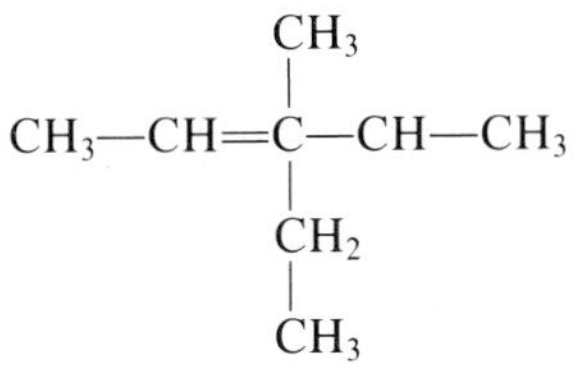
```
          CH3
          |
CH3—CH=C—CH—CH3
          |
          CH2
          |
          CH3
```

(e) 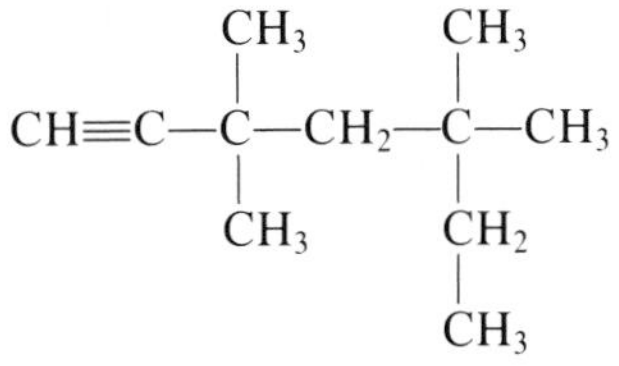
```
         CH3      CH3
         |        |
CH≡C—C—CH2—C—CH3
         |        |
         CH3      CH2
                  |
                  CH3
```

(f) $CH_2Cl—CH_2Cl$

(g) 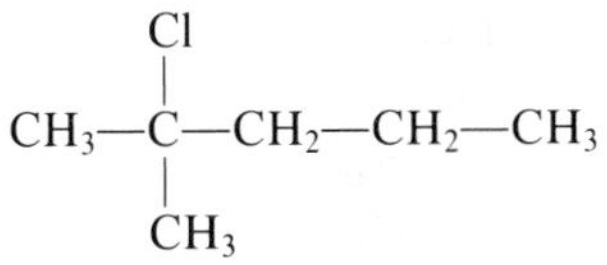
```
      Cl
      |
CH3—C—CH2—CH2—CH3
      |
      CH3
```

(h) $CHCl_3$ **(i)** $CHCl{=}CHF$

34. Write a structural formula for each of the following compounds: **(a)** 1,1-dibromoethane; **(b)** *cis*-2-butene; **(c)** *trans*-1,2-difluoro-1-butene; **(d)** 2,4-dimethyl-3-heptene; **(e)** cyclohexene; **(f)** 1,3-butadiene; **(g)** 1,3-cyclopentadiene.

Naming Common Oxygen-Containing Organic Compounds and Amines

35. Name each of the following compounds:

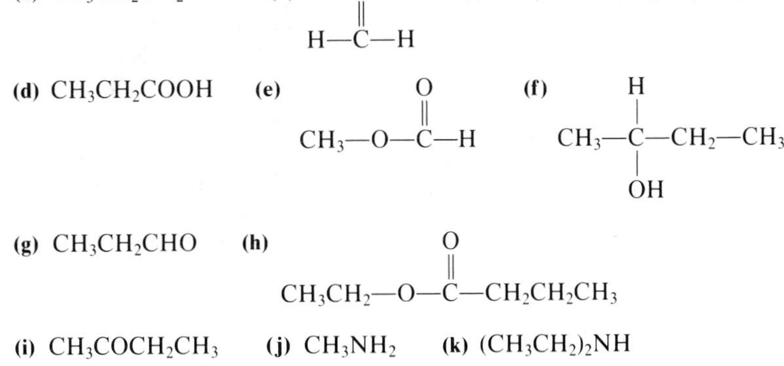

36. Write a structural formula for each of the following compounds: **(a)** acetic acid; **(b)** ethanol; **(c)** 2-propanol; **(d)** ethylene glycol; **(e)** acetaldehyde; **(f)** acetone; **(g)** methyl ethyl ether; **(h)** ethyl acetate; **(i)** butyl propionate; **(j)** ethylamine; **(k)** trimethylamine.

Structure and Bonding

37. What is the geometry (linear, trigonal planar, or tetrahedral) of the starred atoms in each of the following compounds? Review the discussion on bonding in Chapter 8 if necessary.

(a) $CH_3—\overset{*}{C}H_2—\overset{*}{C}H_3$ **(b)** $CH_3—\overset{*}{C}H=\overset{*}{C}H_2$ **(c)** $CH_3—\overset{*}{C}\equiv\overset{*}{C}H$ **(d)** $CH_3—\overset{*}{O}H$

(e) $CH_3—\overset{*}{C}(=O)—CH_3$ **(f)** $CH_3—\overset{*}{C}(=O)—\overset{*}{O}H$ **(g)** $CH_3—\overset{*}{O}—CH_2—CH_3$

(h) $CH_3—\overset{*}{C}H_2—\overset{*}{O}—\overset{*}{C}(=O)—H$ **(i)** $\overset{*}{C}H_2(CH_2)_5$ (ring) **(j)** $\overset{*}{C}H=CH(CH_2)_4$ (ring)

38. What is the hybridization (sp, sp^2, etc.) on each of the starred atoms in question 37?

39. How many σ and how many π bonds are there in each of the molecules listed in question 37? (Consider one resonance structure for any of the species where delocalized π bonding is involved.)

Isomerism

40. Draw all the structural isomers for each of the following compounds, excluding any cyclic structures: **(a)** C_4H_{10}; **(b)** C_5H_{12}; **(c)** C_6H_{14}; **(d)** C_4H_8; **(e)** C_5H_8; **(f)** $C_2H_4Br_2$; **(g)** $C_3H_6Cl_2$; **(h)** C_2H_6O.

41. Draw the structures for all isomers for each of the following compounds: **(a)** C_4H_8; **(b)** $C_2H_2F_2$.

Reactions of Organic Compounds; Equation Writing

42. Write an equation for each of the following reactions, and name the organic product, if any, formed in each case: **(a)** complete combustion of propane; **(b)** complete chlorination of methane; **(c)** hydrogenation of 1-propene; **(d)** complete hydrogenation of 2-butyne; **(e)** addition of HCl to ethene; **(f)** complete bromination of ethyne; **(g)** fermentation of glucose; **(h)** industrial preparation of methanol from carbon monoxide and hydrogen at high temperature and pressure; **(i)** industrial preparation of ethanol from ethane and water at high temperature and pressure; **(j)** mild oxidation of ethanol; **(k)** oxidation of formaldehyde; **(l)** oxidation of $CH_3CHOHCH_2CH_3$; **(m)** reaction of methanol with formic acid; **(n)** heating ethanol with an acid catalyst; **(o)** reaction of benzene with chlorine in the presence of $FeBr_3$ as a catalyst; **(p)** reaction of benzene with methyl chloride using aluminum chloride as a catalyst.

Biochemistry

43. Write a structural formula for each of the following compounds: **(a)** glucose; **(b)** sucrose; **(c)** glycine; **(d)** alanine.

44. Using structural formulas, write an equation for the reaction of alanine with glycine to form alanylglycine. Point out the peptide linkage in the product.

45. What type of organic compound (alcohols, ethers, etc.) does each of the following substances belong to? **(a)** Glucose; **(b)** acetone; **(c)** alanine; **(d)** a fat.

46. List some major sources and uses of each of the following polysaccharides: **(a)** starch; **(b)** dextrin; **(c)** glycogen; **(d)** cellulose.

47. **(a)** What is the major function of fats in the bodies of humans and animals? **(b)** What is the major biological function of waxes?

48. **(a)** To what class of substances (carbohydrates, proteins, lipids) do enzymes belong? **(b)** What is the major function of enzymes and of hormones?

49. What is the full name of: **(a)** DNA; **(b)** RNA?

50. How does the structure of DNA differ from the structure of RNA?

51. What type of forces hold the two strands of double helical DNA together?

52. An uncombined nucleotide in living cells is ATP. What is the full name of this nucleotide, and what is its major function?

CHAPTER 24

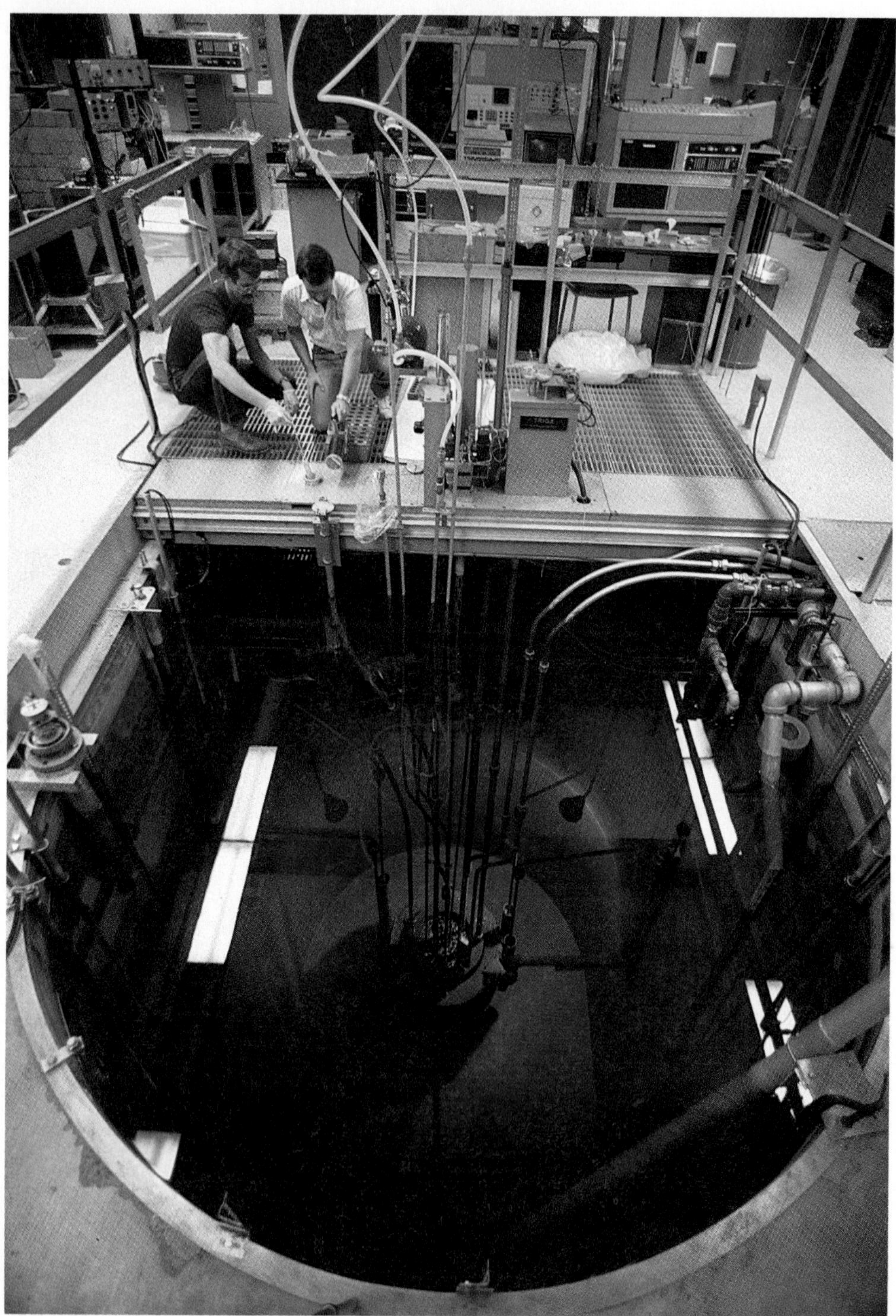

The core of a nuclear reactor.

Nuclear Chemistry

In the preceding chapters we have discussed the electronic structures of atoms and molecules and the correlation of the electronic structures to physical properties and chemical reactivity. We have said very little about atomic nuclei. In Chapter 1 we briefly discussed the principal components of nuclei: protons and neutrons. We also discussed radioactivity and the nature of alpha, beta, and gamma rays emitted by the nuclei of radioactive substances. In this chapter we expand on these topics.

24.1 Radioactivity

Some naturally occurring elements, such as radium, uranium, and an isotope of carbon (carbon-14) are **radioactive**; that is, they spontaneously emit radiant energy, material particles, or both. This spontaneous radioactivity is called **natural radioactivity**. Any nuclear species, or **nuclide**, that is naturally radioactive is known as a **radionuclide**. The emission of radiant energy and material particles by a radionuclide is also called **radioactive decay**.

There are four major types of radioactive decay: (1) *alpha emission*, (2) *beta emission*, (3) *positron emission*, and (4) *electron capture*. Gamma rays may also be emitted in a radioactive disintegration. These four processes are discussed below.

Alpha Emission and Nuclear Equations

In radioactive decay, the *nucleus that disintegrates* is called the **parent nuclide**, and the *nucleus that forms* is called the **daughter nuclide**. The formation of the daughter nuclide is often accompanied by the production of smaller material particles, which we discuss below. Radioactive decay processes also release energy.

Radioactive decay processes can be described by *nuclear equations*. In a nuclear equation, the symbols for the reacting nuclei are the symbols for the nuclides of the corresponding atoms. The *mass number* (the total number of protons and neutrons) of the nucleus is indicated by a superscript at the left of the symbol. The atomic number is written as a subscript, also at the left of the symbol. For example, a radon nucleus, with 86 protons and 136 neutrons, has a mass number of 222 and an atomic number of 86; thus, its nuclear symbol is written

$$^{222}_{86}\text{Rn}$$

An example of a radioactive decay process is the spontaneous decay of uranium-238 into thorium-234 and an alpha particle. This process is written in a nuclear equation as follows:

$$^{238}_{92}\text{U} \longrightarrow {}^{234}_{90}\text{Th} + {}^{4}_{2}\text{He}$$

A *nuclear disintegration that occurs by emission of alpha particles* is called **alpha emission**. We can see from the above equation that alpha emission by the parent ^{238}U nuclide creates a smaller daughter nuclide having two fewer protons and two fewer neutrons than the parent nuclide (Figure 24.1).

In a nuclear equation we must balance both mass and nuclear charge. The total mass number at the left of the arrow must equal the total mass number at the right of the arrow. The sum of atomic numbers must also be equal on both sides of the arrow. Thus, in the equation for alpha emission by uranium, the

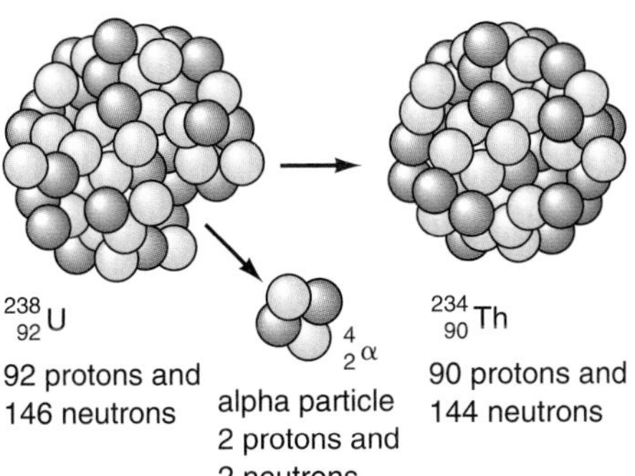

Figure 24.1

Alpha emission by uranium-238. An alpha emission produces an alpha particle and a daughter nuclide having two fewer protons and two fewer neutrons than the parent nuclide.

total mass number to the left of the arrow is 238, and the total to the right is also 238. Similarly, the atomic number to the left is 92, which equals the total number of protons in the products to the right of the arrow. Thus, in a nuclear equation, the mass numbers and nuclear charges must be balanced.

Example 24.1

Predicting the Product of a Nuclear Reaction and Writing an Equation for the Reaction

An isotope of radon, ^{222}Rn, decays by alpha emission. What daughter nuclide forms in this process? Write an equation for this process.

SOLUTION: First, we write the symbol for the parent nuclide of radon including its mass number. Then we indicate its atomic number, which we find in the periodic table. Next we write an arrow and the symbol for helium, the element whose nucleus is an alpha particle. We also write the mass number of the alpha particle as a superscript and the charge as a subscript. The daughter isotope whose identity is unknown at this point we represent by X:

$$^{222}_{86}Rn \longrightarrow {}^{4}_{2}He + X$$

The mass number of the daughter isotope X must be $222 - 4 = 218$, and the atomic number is $86 - 2 = 84$. X can now be found from the periodic table: it is the element whose atomic number is 84—polonium, Po. The complete nuclear equation for the decay of radon-222 by alpha emission is

$$^{222}_{86}Rn \longrightarrow {}^{4}_{2}He + {}^{218}_{84}Po$$

Practice Problem 24.1: Write a nuclear equation for the alpha emission of thorium-230 (^{230}Th). What is the name of the daughter nuclide?

Nuclear disintegrations can also emit beta particles, neutrons, protons, and positrons. Table 24.1 lists the symbols, masses, and charges of particles emitted by nuclear disintegrations. Gamma rays that are also emitted are a form of radiant energy.

Beta particles are electrons. They are represented by either β or e. The proton is also the nucleus of a hydrogen atom, so the symbols p and ^{1}H are used interchangeably. The positron has a mass equal to that of an electron but has a *positive* charge whose magnitude equals that of the negative charge of an electron. The symbol for a positron is therefore $^{0}_{1}e$, whereas an electron or a beta particle is $^{0}_{-1}e$.

Table 24.1

Common Types of Radiation Emitted by Radioactive Substances

Radiation	Nature	Symbol	Mass Number of Particle	Charge
Alpha	Particles	α, ${}^{4}_{2}\alpha$, ${}^{4}_{2}He$	4	2
Beta	Particles	β, ${}^{0}_{-1}\beta$, ${}^{0}_{-1}e$	0	−1
Gamma	Radiant energy	γ		
Neutron	Particles	n, ${}^{1}_{0}n$	1	0
Proton	Particles	p, ${}^{1}_{1}p$, ${}^{1}_{1}H$	1	1
Positron	Particles	${}^{0}_{1}e$, ${}_{+}\beta$, ${}_{+1}\beta$	0	1

Of all the particles listed in Table 24.1, only neutrons and protons are known to exist as discrete entities in the nuclei of atoms. The other particles, including the gamma rays, are products of nuclear reactions.

Beta Emission, Positron Emission, and Electron Capture

Radioactive decay that occurs by the emission of beta particles is called **beta emission** or **beta decay**. For example, carbon-14 decays to nitrogen-14 by beta emission (Figure 24.2):

$$ {}^{14}_{6}C \longrightarrow {}^{14}_{7}N + {}^{0}_{-1}e $$

In this reaction, the atomic number (nuclear charge) of the daughter nuclide is one unit greater than that of the present nuclide, while the mass number remains the same. This is somewhat surprising, but it can be explained if we assume that a neutron consists of a proton and an electron, which can decay as follows:

$$ {}^{1}_{0}n \longrightarrow {}^{1}_{1}p + {}^{0}_{-1}e $$

Thus, beta emission is equivalent to converting a neutron to a proton and a beta particle.

Radioactive decay in which positrons are emitted is called **positron emission**. In positron emission, the atomic number of the parent nuclide decreases by one unit, while the mass number remains unchanged. An example of positron emission is the decay of carbon-11 to boron-11 (Figure 24.3):

$$ {}^{11}_{6}C \longrightarrow {}^{11}_{5}B + {}^{0}_{+1}e $$

In positron emission, a proton converts into a neutron and a positron:

$$ {}^{1}_{1}p \longrightarrow {}^{1}_{0}n + {}^{0}_{1}e $$

Some nuclear reactions occur by *capture of an electron in the K shell of an atom by the nucleus of the atom*, a process called **electron capture** or **K-**

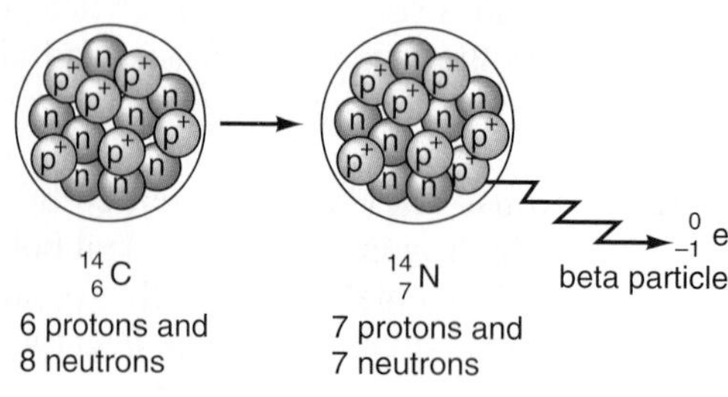

Figure 24.2

The decay of carbon-14 to nitrogen-14 by beta emission.

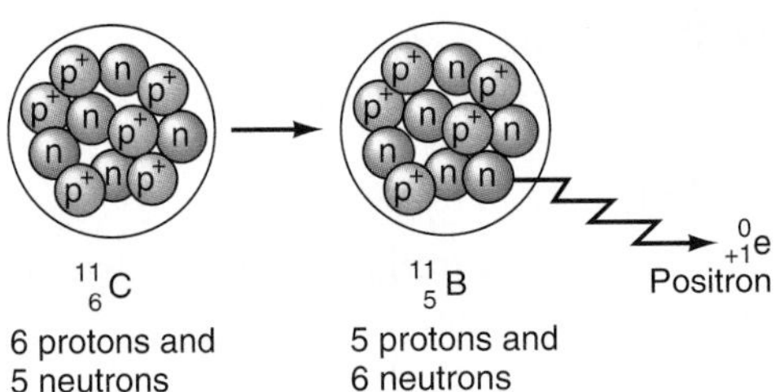

Figure 24.3

Positron emission by carbon-11 to produce boron-11.

capture. K- capture converts a proton into a neutron. For example, the conversion of potassium-40 into argon-40 occurs by electron capture:

$$^{40}_{19}K + {}^{0}_{-1}e \longrightarrow {}^{40}_{18}Ar$$

As with positron emission, electron capture produces a nucleus with a charge one unit lower than that of the parent, while the mass number remains unchanged.

Neutron emission can also occur in nuclear reactions, but it is not a common mode of natural radioactive decay.

Example 24.2

Writing Nuclear Equations

Write balanced nuclear equations for the following reactions: **(a)** positron emission by phosphorus-30; **(b)** decay of neptunium-238 to plutonium-238.

SOLUTION:

a. The information given in this part of the problem can be written by an incomplete equation

$$^{30}_{15}P \longrightarrow {}^{0}_{1}e + X$$

where X is the daughter nuclide. To balance the mass, the mass number for the daughter nuclide X must be 30. To balance the charge, the nucleus of X must have a charge of 14. From the periodic table we find that the element having an atomic number of 14 is silicon. The complete nuclear equation is

$$^{30}_{15}P \longrightarrow {}^{0}_{1}e + {}^{30}_{14}Si$$

b. From the periodic table we find the atomic numbers for neptunium (Np) and plutonium (Pu). An incomplete nuclear equation for this reaction is

$$^{238}_{93}Np \longrightarrow {}^{238}_{94}Pu + X$$

where X is a particle or radiation emitted. To balance both mass and charge in this equation, the mass number of X must be zero and its charge -1. Thus, X is a beta particle, and the complete nuclear equation is

$$^{238}_{93}Np \longrightarrow {}^{238}_{94}Pu + {}^{0}_{-1}e$$

This is an equation for the beta decay of ^{238}Np.

Practice Problem 24.2: Write an equation for each of the following processes: **(a)** electron capture by rubidium-81; **(b)** decay of polonium-212 to lead-208.

24.2 Nuclear Stability

Empirical Guidelines of Stability

In this section we examine some criteria that enable us to predict which nuclei are stable and which are radioactive.

1. Nuclei with 83 protons or fewer are stable, and nuclei with more than 83 protons are unstable. This rule has two surprising exceptions. The nuclei of technetium, Te, atomic number, 43, and promethium, Pm, atomic number 61 (elements with atomic numbers much lower than 83) are also unstable. These two elements are found in nature only in trace quantities.
2. Nuclei having 2, 8, 20, 28, 50, and 82 protons or neutrons are generally more stable than other nuclei. These numbers are called **magic numbers**. We recall that the chemical stabilities of atoms are related to the atoms having a total of 2, 10, 18, 36, 54, or 86 electrons, which are the numbers of electrons in noble gas atoms from helium through radon. However, these numbers are not the same as magic numbers that are associated with nuclear stability. Nuclei that contain a magic number of protons and a magic number of neutrons are said to be "doubly magic"; they are extra stable. Examples of such nuclei are $^{4}_{2}He$, $^{16}_{8}O$, $^{40}_{20}Ca$, and $^{208}_{82}Pb$.
3. Nuclei with even numbers of protons and even numbers of neutrons are generally more stable than those having an odd number of protons, an odd number of neutrons, or both an odd number of protons and an odd number of neutrons. Only five stable nuclei are in the odd–odd category. The following table relates the numbers of protons and neutrons in nuclei with numbers of stable nuclei.

Number of protons Number of neutrons	Even Even	Even Odd	Odd Even	Odd Odd
Number of stable nuclei	157	52	50	5

Example 24.3

Predicting Stabilities of Nuclei

Decide which of the following nuclei might be unstable and therefore radioactive: **(a)** $^{210}_{85}At$; **(b)** $^{120}_{50}Sn$; **(c)** $^{20}_{11}Na$.

SOLUTION

(a) The nucleus of astatine-210 is predicted to be radioactive because it contains more than 83 protons. According to rule 1, nuclei of more than 83 protons are unstable. The nucleus of astatine-210 also contains an odd number of protons (85) and an odd number of neutrons (125).

(b) An $^{120}_{50}Sn$ nucleus contains an even number of protons (50) and an even number of neutrons (70). It is therefore likely to be stable.

(c) $^{20}_{11}Na$ is expected to be radioactive because it contains an odd number of protons (11) and an odd number of neutrons (9).

Practice Problem 24.3: Would you predict the following nuclei to be radioactive? **(a)** Helium-4; **(b)** francium-220.

Belt of Stability

Small, stable nuclei have about equal numbers of neutrons and protons. For larger stable nuclei, the neutron-to-proton ratio increases with increasing atomic number, as shown in Figure 24.4. A *plot of the number of neutrons versus the number of protons in stable nuclei forms a band* called the **belt of stability** (Figure 24.4). Nuclei outside the belt of stability are radioactive and decay to more stable nuclei having neutron-to-proton ratios characteristic of nuclei inside the belt.

A small nucleus whose neutron-to-proton ratio is too high lies above the belt of stability (Figure 24.4). Such a nucleus can decay by beta emission, which decreases the number of neutrons by one and increases the number of protons by one. Neutron emission also occurs in some cases, but it is much less common than beta emission.

Nuclei whose neutron-to-proton ratio is too low lie below the belt of stability. Such nuclei can decay either by positron emission or by electron capture. Both processes decrease the number of protons and increase the number of neutrons. Small nuclei whose neutron-to-proton ratio is less than 1 usually decay by positron emission.

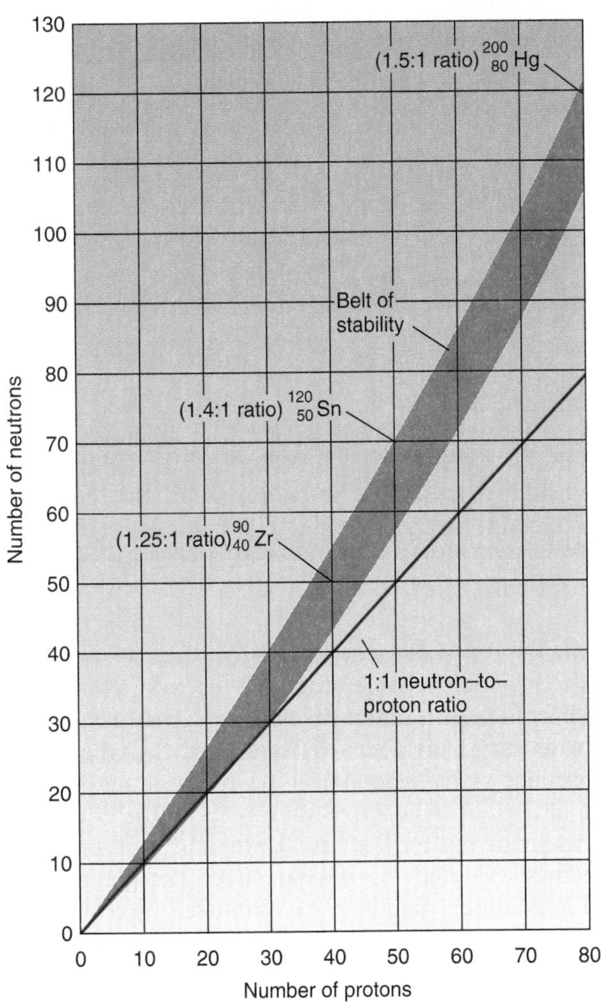

Figure 24.4

Belt of stability: A plot of the number of neutrons versus the number of protons in stable nuclei. Most of the radioactive nuclei occur outside of this belt.

Large nuclei, such as ^{200}Hg, have a neutron-to-proton ratio of 1.5 or higher. These nuclei are indicated at the upper regions of the belt of stability (Figure 24.4). Large, unstable nuclei generally decay by alpha emission, which decreases both the number of neutrons and the number of protons by two and produces smaller nuclei which often lie within the belt of stability.

Example 24.4

Predicting Stability of Nuclei and the Mode of Decay of Unstable Nuclei

Predict whether each of the following nuclei is radioactive according to Figure 24.4, and if it is, write an equation for a possible mode of decay: (a) $^{47}_{20}Ca$; (b) $^{11}_{6}C$; (c) $^{235}_{92}U$; (d) $^{16}_{8}O$.

SOLUTION

a. The neutron-to-proton ratio (27/20 = 1.35) in a $^{47}_{20}Ca$ nucleus is more than indicated by the belt of stability for a nucleus with 20 protons. It decays by beta emission:

$$^{47}_{20}Ca \longrightarrow ^{47}_{21}Se + ^{0}_{-1}e$$

b. Nuclei of ^{11}C lie below the belt of stability because their neutron-to-proton ratio is less than 1. Since these nuclei are small, they decay by positron emission:

$$^{11}_{6}C \longrightarrow ^{11}_{5}B + ^{0}_{1}e$$

c. Nuclei of ^{235}U lie outside of the belt of stability toward the upper region of the belt (Figure 24.1), where the principal mode of decay is alpha emission:

$$^{235}_{92}U \longrightarrow ^{231}_{90}Th + ^{4}_{2}He$$

d. ^{16}O nuclei are not radioactive. They are "doubly magic" because they contain magic numbers of 8 protons and 8 neutrons.

Practice Problem 24.4: Write an equation for most likely decay of ^{208}Po.

The empirical rules discussed in Section 24.1 are only general guidelines, so information obtained from the belt of stability should be used with caution. For example, $^{231}_{90}Th$ might be predicted to decay by alpha emission. Instead, ^{231}Th decays by beta emission.

Some large nuclei with high neutron-to-proton ratios undergo a series of nuclear reactions to produce a stable product lying within the belt of stability. Such a series is referred to as a **radioactive decay series**, or *nuclear disintegration series*. Four such series have been discovered. One of these is the *decay of uranium-238* to produce stable lead-206 as the end product (Figure 24.5).
◘ The existence of certain radioactive substances in the environment can be explained by the radioactive decay series. Uranium-238 is a minor component of many rocks and occurs in the ores of gold, silver, and platinum, and in other minerals of certain localities. Radon-222 is produced by a series of reactions starting with uranium-238. Radon-222 causes lung cancer among some miners

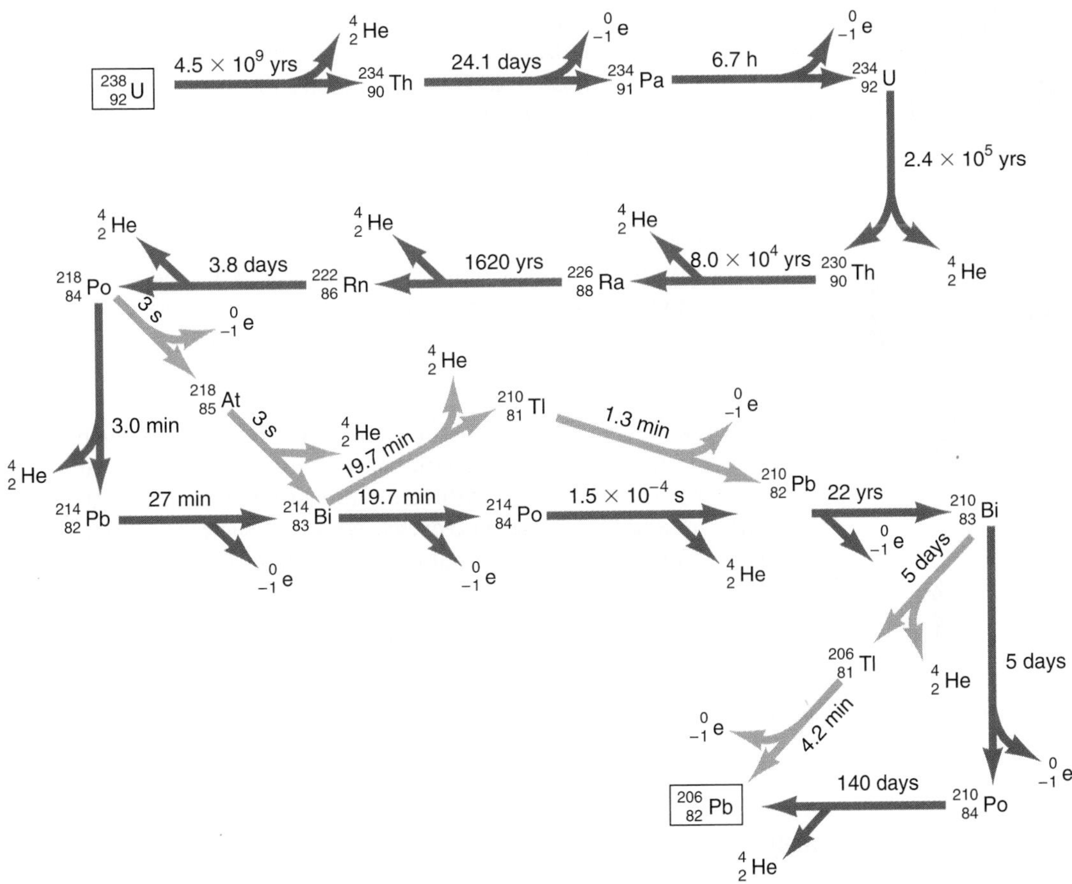

Figure 24.5

Radioactive decay series. Series of steps of radioactive decay of uranium-238 to lead-206 as the stable end product. The dotted arrows indicate alternate paths of decay. The over-all equation for the disintegration of uranium-238 is:

$$^{238}_{92}U \xrightarrow{4.51 \times 10^9 \text{ yrs}} {}^{206}_{82}Pb + 8\,^{4}_{2}He + 6\,^{0}_{-1}e.$$

The times listed are half-lives (see Section 24.4).

who work with uranium-containing ore. Traces of radioactive ^{238}U are found in phosphate fertilizers used in tobacco fields. Recent evidence suggests that ^{238}U and other nuclides produced in its decay series might be one of the causes of cancer among cigarette smokers. There has been a recent concern about the level of radioactive radon gas found in homes of various localities, particularly where uranium ores are plentiful.

24.3 Preparation of New Nuclei; Induced Radioactivity

Nuclear Transmutation

In 1919, Lord Rutherford bombarded nitrogen-14 with fast-moving alpha particles emitted by radium. As a result, nitrogen-14 was converted into oxygen-17:

$$^{14}_{7}N + ^{4}_{2}He \longrightarrow ^{17}_{8}O + ^{1}_{1}H$$

This was the first *artificial conversion of one nucleus into another*, a process called **nuclear transmutation**.

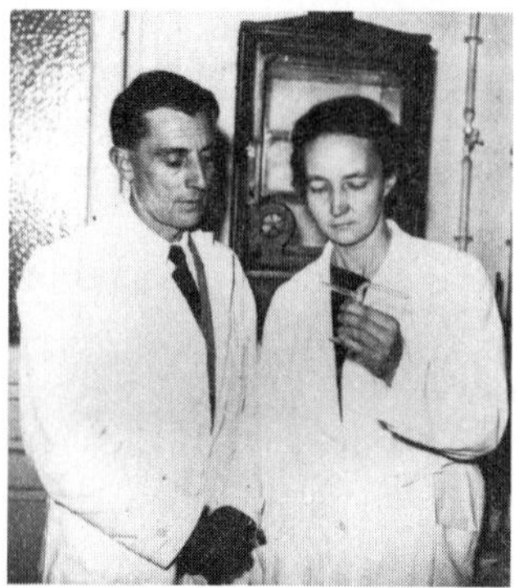

Figure 24.6

Marie Curie's daughter, Irene, and Irene's husband, Frederic Joliot.

Equations for nuclear reactions, like the nuclear transmutation reaction described above, are often written in a shorthand form. In this form, the major target nucleus is written first, followed in parentheses by the smaller bombarding particle and the ejected particle, separated with a comma, and ending with the product nucleus to the right of the closing parentheses:

$$^{14}_{7}N(^{4}_{2}He,^{1}_{1}p)^{17}_{8}O$$

In simplified nuclear equations, alpha particles, beta particles, protons, and neutrons are usually written with symbols α, β, p, and n, respectively. For simplicity, the masses and charges of these particles are omitted. Thus, the reaction above can be written as

$$^{14}_{7}N(\alpha,p)^{17}_{8}O$$

In 1934, Marie Curie's daughter, Irene Curie, and her husband, Frederic Joliot of France (Figure 24.6), bombarded aluminum, boron, and magnesium with alpha particles from a naturally radioactive substance. All the elements were transmuted to product nuclei that were radioactive, decaying by positron emission. For example, aluminum is first converted to ^{30}P by bombardment with α particles:

$$^{27}_{13}Al(\alpha,n)^{30}_{15}P$$

The ^{30}P formed as a product is radioactive; it decays to ^{30}Si by positron emission:

$$^{30}_{15}P \longrightarrow ^{30}_{14}Si + ^{0}_{1}e$$

Radioactivity produced in a laboratory is called **induced** or **artificial radioactivity**.

Example 24.5

Writing a Complete Nuclear Equation from a Shorthand Equation

Write the complete and balanced nuclear equation for the following shorthand form $^{27}_{13}Al(\alpha,n)^{30}_{15}P$.

SOLUTION: In the shorthand form of the equation, the α represents the bombarding particle or $^{4}_{2}He$, and n or $^{1}_{0}n$ is the ejected particle. The complete equation is

$$^{27}_{13}Al + ^{4}_{2}He \longrightarrow ^{30}_{15}P + ^{1}_{0}n$$

Practice Problem 24.5: Write complete and shorthand nuclear equations for the reaction that occurs when ^{45}Sc nuclei are bombarded by neutrons to produce ^{45}Ca nuclei and protons.

Particle Accelerators and Bombardment with Positive Particles

Alpha particles emitted from naturally radioactive sources have relatively low energies—about 10 million electron volts (MeV). (Electron volts are energy units often used in nuclear chemistry; 1 eV = 1.6×10^{-19} J.) Therefore only light nuclei can be converted to other nuclei by bombarding them with alpha particles from natural sources. Particle accelerators produce high-energy particles for the study of nuclear reactions and for the production of artificial radionuclides.

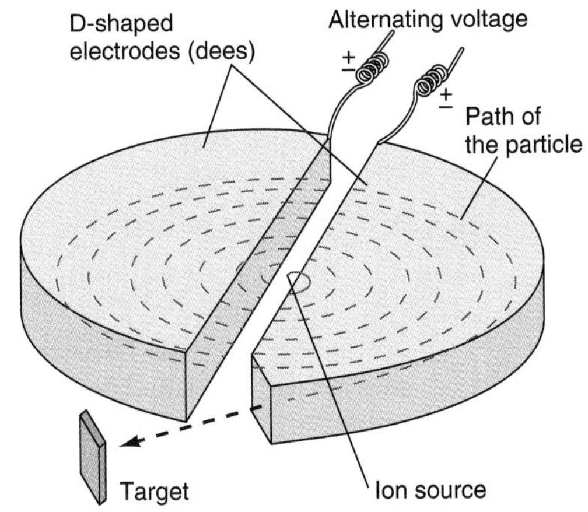

Figure 24.7

Diagram of a cyclotron. Charged particles are accelerated along the spiral path by the application of alternating voltage to the D-shaped electrodes.

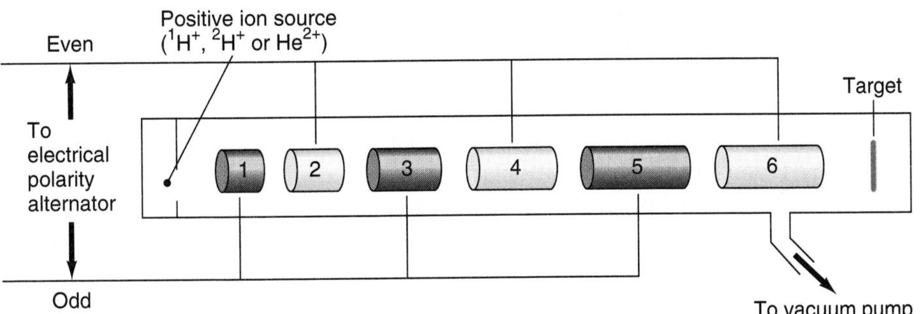

Figure 24.8

Diagram of a simple linear particle accelerator. A positive particle is attracted to the negatively charged cylinder 1. As the particle passes through the cylinder, the cylinder's charge converts to positive, repelling the positive particle toward the second cylinder, which at that instant is negatively charged. This process continues through several cylinders.

Particle accelerators for charged particles such as protons, electrons, and deuterons (the nuclei of deuterium atoms) were first built in the 1930s by E. O. Lawrence in California. These early instruments were called **cyclotrons**. In a cyclotron, positively charged particles are accelerated in a spiral path between D-shaped magnets, illustrated in Figure 24.7. The acceleration is accomplished by rapidly alternating the polarity of the electromagnets from negative to positive so that a positive particle is repelled by the positive pole and is attracted by the negative pole. Ions introduced at the center of the cyclotron are accelerated in an expanding spiral path until they acquire sufficient energy to cause a nuclear reaction when they strike a target material.

Another type of particle accelerator is a **linear accelerator** in which the particles are made to flow through the center of a series of cylinders in which the negative and positive charges constantly alternate as illustrated in Figure 24.8. As a positive particle leaves its source at the left, the first cylinder is negatively charged and the particle flies toward it. As the particle passes through the first cylinder, the charge of the cylinder converts to positive, repelling the particle, which is now accelerated toward the second cylinder which at that instant is negatively charged. This process continues through several cylinders of increasing length until the particle has acquired its maximum energy as it hits the target. A linear accelerator (which cost $100 million to build) at Stanford University in California is 2 miles long and can accelerate electrons to 20 billion electron volts (Figure 24.9).

Figure 24.9

Stanford linear accelerator.

During the period of 1937 to 1940, three new elements—not yet discovered but predicted at that time—were prepared by bombarding certain nuclei with positive particles in particle accelerators. These elements were technetium, francium, and astatine, discovered in 1937, 1939, and 1940, respectively. The equations for the reactions by which these elements were prepared are

$$^{96}_{42}\text{Mo} + ^{2}_{1}\text{H} \longrightarrow ^{97}_{43}\text{Tc} + ^{1}_{0}\text{n}$$

$$^{230}_{90}\text{Th} + ^{1}_{1}\text{H} \longrightarrow ^{223}_{87}\text{Fr} + 2^{4}_{2}\text{He}$$

$$^{209}_{83}\text{Bi} + ^{4}_{2}\text{He} \longrightarrow ^{210}_{85}\text{At} + 3^{1}_{0}\text{n}$$

Bombardment with Neutrons

Neutrons are uncharged and therefore cannot be accelerated. However, because neutrons are electrically uncharged, they are not repelled by nuclei as positive particles are. Neutrons can thus be captured by nuclei without having to be accelerated.

Neutrons are used as the bombarding particles to produce isotopes for research and medicine. The necessary neutrons can be generated in many different ways; a common method is the bombardment of beryllium-9 with alpha particles from radium-222:

$$^{9}_{4}\text{Be} + ^{4}_{2}\text{He} \longrightarrow ^{12}_{6}\text{C} + ^{1}_{0}\text{n}$$

Neutrons produced by this reaction can be used, for example, to make cobalt-60 that is used in radiation treatment of cancer patients. To produce cobalt-60, iron-58 is bombarded with neutrons in a nuclear reactor to give iron-59. Iron-59 decays to cobalt-59 and beta particles. The cobalt-59 then captures a neutron to produce cobalt-60. This sequence of reactions is:

$$^{58}_{26}\text{Fe} + ^{1}_{0}\text{n} \longrightarrow ^{59}_{26}\text{Fe}$$

$$^{59}_{26}\text{Fe} \longrightarrow ^{59}_{27}\text{Co} + ^{0}_{-1}\text{e}$$

$$^{59}_{27}\text{Co} + ^{1}_{0}\text{n} \longrightarrow ^{60}_{27}\text{Co}$$

Cancerous cells are more sensitive than normal cells to radiation and are therefore destroyed by a selective dose of radiation.

In 1947, neutron bombardment was used to discover element 61, promethium. This element was produced by bombarding neodymium-142 with neutrons:

$$^{142}_{60}\text{Nd} + ^{1}_{0}\text{n} \longrightarrow ^{143}_{61}\text{Pm} + ^{0}_{-1}\text{e}$$

Elements whose atomic numbers are higher than that of uranium, 92, are called **transuranium elements**. These elements have been made by high-energy bombardment techniques in nuclear reactors and accelerators.

24.4 Half-Life of Radioactive Elements

We mentioned in Chapter 21 that radioactive decay processes obey a first-order rate law. We also recall that the half-life of a reactant is the time it takes for its concentration or mass to decrease to one half of its original value.

Each radionuclide has a characteristic half-life that ranges from a fraction of a second to billions of years. For example, the half-life of sulfur-35 is 88 days. Thus, if 10.0 g of sulfur-35 were present at a given time, only 5.00 g would

APPLICATIONS OF CHEMISTRY 24.1
Nuclear Chemistry in Political Science

Neutrons provide us with a tool to determine what chemical elements are present in materials and how much of each element is present. When a sample is bombarded with neutrons, elements in the sample absorb some of the neutrons and turn into radioactive isotopes. Each radioactive isotope emits its own characteristic radiation, and by making careful measurements to determine the type of radiation being emitted, the identity of the isotopes can be determined. Since the intensity of radiation depends only on the amount of radioactive material present, it is also possible to determine how much of each radioactive element is present in the unknown sample. If the unknown sample is human hair, much information about the person can be deduced. Hair is a "living diary" because hair contains deposits of all chemicals that a person has ever consumed. If you drank soda from an aluminum can last week, your body deposited some of the aluminum ions that were dissolved in the soda in your hair. Each day your hair grows about $\frac{1}{2}$ mm, and each day's growth contains trace amounts of whatever you have eaten (see illustration).

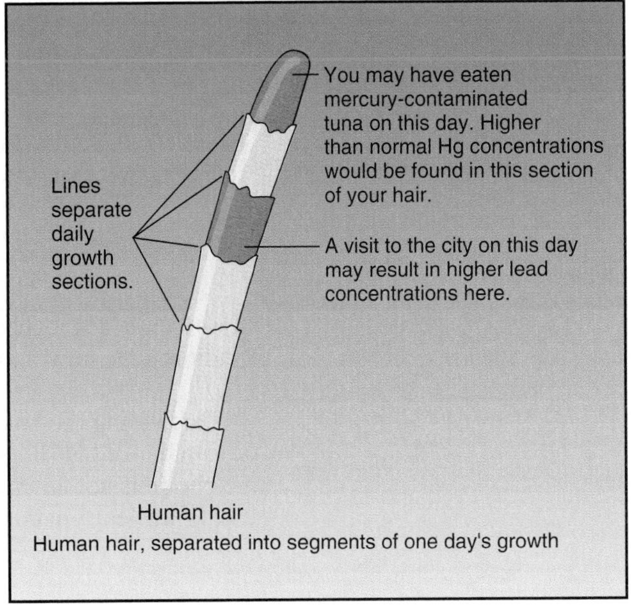
You may have eaten mercury-contaminated tuna on this day. Higher than normal Hg concentrations would be found in this section of your hair.
Lines separate daily growth sections.
A visit to the city on this day may result in higher lead concentrations here.
Human hair
Human hair, separated into segments of one day's growth

In 1991, scholars suggested that Zachary Taylor, the twelfth president of the United States, had been assassinated. It was believed that because Taylor opposed slavery in new states seeking admission to the Union, his enemies may have added arsenic to Taylor's food. To test this hypothesis, samples of Taylor's hair were bombarded with neutrons. Any arsenic in his hair would have undergone the following change:

$$^{75}As + 5n \longrightarrow {}^{80}As$$

Since arsenic-80 is very radioactive, it is easy to detect even if present in amounts as little as 1×10^{-17} g. However no arsenic was detected in Taylor's hair samples, and the arsenic assassination theory crumpled.

Source: Chemical & Engineering News, July 1, 1991, page 40.

remain after the first half-life period of 88 days. After another 88-day period, one-half of 5.00 g, or 2.50 g, of ^{35}S would remain, and so on. Sulfur-35 decays by beta emission:

$$^{35}_{16}S \longrightarrow {}^{35}_{17}Cl + {}^{0}_{-1}e$$

Table 24.2 lists the half-lives of some common radioisotopes.

Example 24.6

Calculating the Amount of a Substance That Remains After Several Half-Lives

The half-life of $^{131}_{53}I$ is 8.00 days. How many grams of a 4.60-g sample of iodine-131 would remain after 24.0 days?

SOLUTION: After the first half-life of 8.00 days, 2.30 g would remain, after another 8.00 days (16.0 days total), one-half of 2.30, or 1.15 g, would remain; and after 24.0 days (three half-lives), 0.575 g would remain.

Table 24.2

Half-Lives of Some Common Radioisotopes

Isotope	Mode of Decay	Half-Life (yr)
$^{239}_{94}Pu$	Alpha emission	24,000
$^{238}_{92}U$	Alpha emission	4.51×10^9
$^{235}_{92}U$	Alpha emission	7.1×10^8
$^{232}_{90}Th$	Alpha emission	1.4×10^{10}
$^{228}_{88}Ra$	Beta emission	6.7
$^{226}_{88}Ra$	Alpha emission	1590
$^{90}_{38}Sr$	Beta emission	28.8
$^{60}_{27}Co$	Beta emission	5.26
$^{40}_{19}K$	Alpha emission	1.3×10^9
$^{14}_{6}C$	Beta emission	5730

Practice Problem 24.6: The half-life of radium-228 is 6.70 years. If you had 0.580 g of radium-228, how many grams would remain after 20.1 years?

Calculating the amount of a radioactive substance remaining after an integral number of half-lives is a fairly easy task, as shown in Example 24.6. If the time period is not an integral number of half-lives, we apply Equations 21.5 and 21.7. In nuclear chemistry, Equation 21.5 is usually written as

$$\ln \frac{N_0}{N_t} = kt \qquad (24.1)$$

where N_0 is the number of radioactive nuclei at time 0, and N_t is the number of nuclei remaining at time t. After one half-life, the ratio $N_0/N_t = 2$, and $\ln(N_0/N_t) = \ln 2 = 0.693 = kt_{1/2}$ (Equation 21.7). The rate constant, k, can thus be calculated from a given half-life:

$$k = \frac{0.693}{t_{1/2}} \qquad (24.2)$$

From the known value of the rate constant, k, we can calculate N_0, N_t, or t given from the values of the other two variables. A calculation of this type is illustrated in Example 24.7.

Example 24.7

Calculating the Fraction of a Radioactive Substance Remaining After a Given Time

The half-life of strontium-90 is 28.8 years. What fraction of a sample of strontium-90 remains after 10.0 years?

SOLUTION: This problem can be solved in three steps: **(1)** We calculate the rate constant k from Equation 24.2; **(2)** we calculate N_0/N_t from Equation 24.1; **(3)** we determine the reciprocal of N_0/N_t, which is the fraction that remains (N_t/N_0).

(1) $k = \dfrac{0.693}{28.8\ \text{yr}} = 0.0241\ \text{yr}^{-1}$

(2) $\ln \frac{N_0}{N_t} = kt = (0.0241\ \text{yr}^{-1})(10.0\ \text{yr}) = 0.241$

(3) $\frac{N_0}{N_t} = e^{0.241} = 1.27$

$$\frac{N_t}{N_0} = \frac{1.00}{1.27} = 0.787$$

Practice Problem 24.7: The half-life of carbon-14 is 5.73×10^3 years. How many grams of a 0.286-g sample of carbon-14 remains after 1000 (1.00×10^3) years? (*Hint*: The mass of a substance is proportional to its number of nuclei.)

24.5 Uses of Radioactivity

Age of Rocks, the Earth, and the Moon

The uranium-238 decay series (Figure 24.5) provides a method of determining the ages of uranium-containing minerals, and ultimately also the age of the earth and the moon. Uranium-containing minerals always contain some lead-206, which is the stable end product of the radioactive decay of uranium-238. Assuming that no lead-206 was present at the time when a rock was formed and that no lead-206 was lost during its formation from uranium-238, the age of the rock can be calculated from the half-life of uranium-238 and from the measured quantities of uranium and lead in the rock, as shown in Example 24.8.

Example 24.8

Calculating the Age of a Rock from Given Masses of Uranium and Lead in the Rock

A sample of a rock contains 0.312 g of ^{206}Pb and 1.502 g of ^{238}U. The half-life of ^{238}U is 4.51×10^9 years. Calculate the age of the rock.

SOLUTION: The problem can be solved in four steps: **(1)** We calculate the mass of uranium that decayed to lead; **(2)** we determine the total mass of uranium that was present when the rock was formed; **(3)** we calculate the decay rate constant for uranium from Equation 24.2; and **(4)** we apply Equation 24.1 to obtain *t*, the age of the rock.

1. The mass of ^{238}U that produced 0.312 g of ^{206}Pb is

$$0.312\ \text{g}\ ^{206}\text{Pb}\left(\frac{1\ \text{mol}\ ^{206}\text{Pb}}{206\ \text{g}\ ^{206}\text{Pb}}\right)\left(\frac{1\ \text{mol}\ ^{238}\text{U}}{1\ \text{mol}\ ^{206}\text{Pb}}\right)\left(\frac{238\ \text{g}\ ^{238}\text{U}}{1\ \text{mol}\ ^{238}\text{U}}\right) = 0.360\ \text{g}\ ^{238}\text{U}$$

2. The total mass of ^{238}U present when the rock was formed at time 0 equals the amount of ^{238}U present at time *t*, plus the amount that produced 0.312 g of ^{206}Pb, or 1.502 g + 0.360 g = 1.862 g.

3. $$k = \frac{0.693}{t_{1/2}} = \frac{0.693}{4.51 \times 10^9 \text{ yr}} = 1.54 \times 10^{-10} \text{ yr}^{-1}$$

4. Since the mass of uranium is proportional to the number of its nuclei, we solve Equation 24.1 for t using the known mass of uranium when the rock was formed at time 0 and the mass of uranium remaining at time t:

$$t = \frac{\ln(N_0/N_t)}{k} = \frac{\ln(1.862 \text{ g}/1.502 \text{ g})}{1.54 \times 10^{-10} \text{ yr}} = 1.40 \times 10^9 \text{ yr}$$

Rocks that do not contain uranium usually have potassium-40, which decays to argon-40 by electron capture:

$$^{40}_{19}\text{K} + {}^{0}_{-1}\text{e} \longrightarrow {}^{40}_{18}\text{Ar} \qquad t_{1/2} = 1.3 \times 10^9 \text{ yr}$$

The age of such rocks can be determined from the ratio of potassium-40 to argon in a way similar to that described for uranium/lead ratio. However, gaseous argon can escape from minerals, and this method may not always be reliable.

The oldest rocks found, granite from the west coast of Greenland are 3.7×10^9 (3.7 billion) years old. This indicates that the crust of the earth has been solid for over 3.5 billion years. It has been estimated that it took 1 billion to 1.5 billion years for the earth's surface to solidify from the molten mass in which it was probably formed.This means that the age of the earth is about 4.0 billion to 4.5 billion years.

Analysis of rock samples taken from the moon indicates that the age of the moon is about the same as that of the earth. The age of a large meteorite that fell in Mexico in 1969 was found to be 4.6×10^9 years. This is the oldest object ever found, and its age indicates that the solar system must be at least 4.6 billion years old.

Age of Carbon-Containing Matter

Substances that contain carbon also contain a radioactive isotope of carbon, carbon-14. The carbon-14 content of a substance can be used to determine its age. Carbon-14 is formed in the upper atmosphere when nitrogen-14 is bombarded with neutrons from cosmic rays:

$$^{14}_{7}\text{N} + {}^{1}_{0}\text{n} \longrightarrow {}^{14}_{6}\text{C} + {}^{1}_{1}\text{H}$$

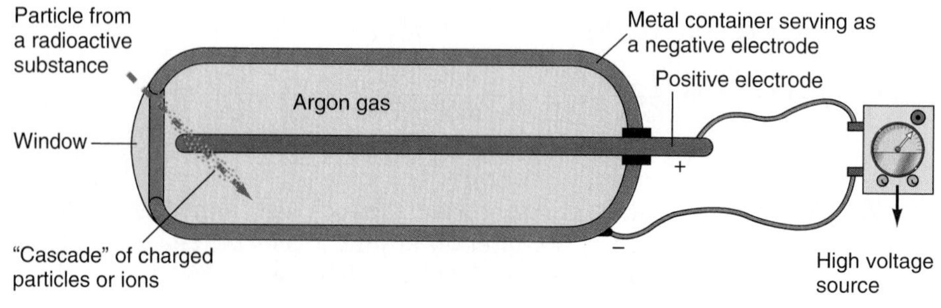

Figure 24.10

Diagram of a Geiger counter. A particle from a radioactive decay process (or a "cosmic ray") passes through a window and ionizes argon gas in the tube. This produces a "cascade" of electrically charged particles of argon gas in the tube. As argon ions hit the electrodes, a "pulse" is produced in the current that is automatically recorded. It is heard as a "click."

The carbon-14 produced reacts with oxygen to form $^{14}CO_2$, which is incorporated in plants and other living organisms. Thus, a certain fraction of all the carbon atoms in living organisms is ^{14}C, which decays to ^{14}N by beta emission:

$$^{14}_{6}C \longrightarrow {}^{14}_{7}N + {}^{0}_{-1}e \qquad t_{1/2} = 5730 \text{ yr}$$

The decay of ^{14}C to ^{14}N occurs at approximately the same rate as the conversion of ^{14}N to ^{14}C by cosmic rays, although some variation in the intensity of cosmic rays has been noted from time to time. The calculated ages of objects are corrected for such variations.

Living organisms have a constant ^{14}C content. When a plant dies, it no longer takes in carbon dioxide by photosynthesis, and its ^{14}C activity decreases in time. Plant-eating animals also have a constant ^{14}C content, which starts to decrease when they die.

The amount of radioactivity in a given quantity of matter can be determined by a device called a **Geiger counter** (Figures 24.10 and 24.11). A Geiger counter has a window that allows the passage of alpha, beta, and gamma rays from a radioactive source. These rays convert argon atoms in the tube to ions. When ions are produced by entering radiation, current flows between the positive electrode and the metal container. The number of particles entering the tube can be electronically recorded and is a measure of the amount of a radioactive substance present in the source. Various other sophisticated instruments are now available to measure radioactivity.

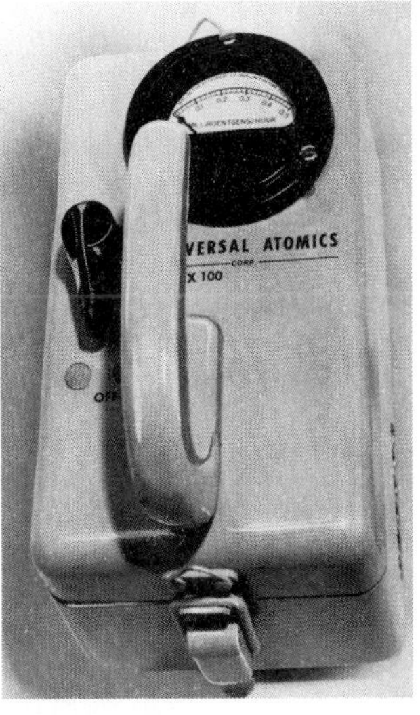

Figure 24.11

A Geiger counter.

To determine the age of a carbon-containing object, the radioactivity of a given quantity of carbon in the object is measured with a Geiger counter and compared with the radioactivity emitted by the same quantity of carbon from living matter. The age of the object can then be calculated as shown in Example 24.9.

Example 24.9

Calculating the Age of a Tree Trunk from Its Carbon-14 Content

The count of radioactivity of a gram of carbon obtained from an old tree trunk is found to be 0.887 times the count measured for the same quantity of carbon from a living tree. Calculate the age of the tree trunk ($t_{1/2}$ of ^{14}C is 5730 years).

SOLUTION: First, we calculate carbon-14 decay rate constant, k, from Equation 24.2. Then we apply Equation 24.1 to obtain the value of t, the age of the tree trunk.

$$k = \frac{0.693}{5730 \text{ yr}} = 1.21 \times 10^{-4} \text{ yr}^{-1}$$

The value of N_0 is not given in the problem, but we know that $N_t =$ $0.887N_0$. Thus, according to Equation 24.1,

$$\ln \frac{N_0}{0.887N_0} = kt$$

$$t = \frac{\ln(1/0.887)}{1.21 \times 10^{-4} \text{ yr}^{-1}} = 991 \text{ yr}$$

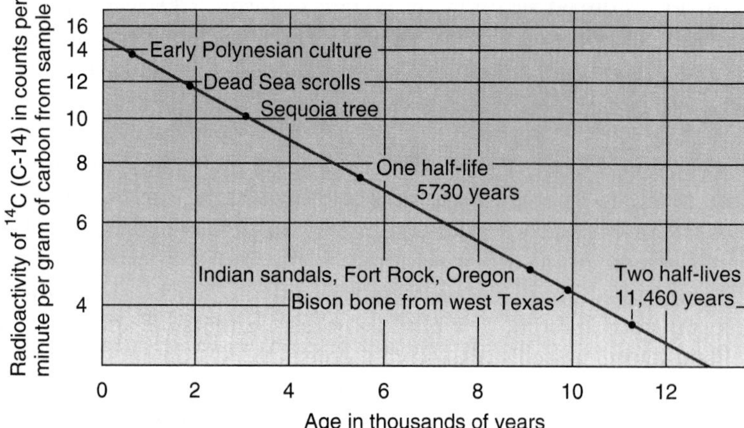

Figure 24.12

Age of substances that contain carbon. The amount of radioactivity (number of counts per minute) is determined by a Geiger counter and the corresponding age is read from the curve.

^{14}C-dating is frequently used in archeology. By carefully measuring the count of radioactivity obtained from an archeological object, it is possible to determine the age of the object from a plot such as that shown in Figure 24.12, or from a calculation such as that in Example 24.9.

Practice Problem 24.8: A wooden Indian sandal found in an archeological site shows a carbon-14 decay rate of 12 disintegrations per minute per gram of carbon. The disintegration rate of living trees is 15 disintegrations per minute per gram of carbon. The half-life of carbon-14 is 5730 years. What is the age of the sandal?

Radioisotopes in Research and Medicine

Several radionuclides are used as *tracers* to "label" an element in a substance. The pathway of such a substance in plant photosynthesis or in animal metabolism can be traced with radiation detectors. Such a study is possible because all isotopes of an element have the same chemical properties. For example, Melvin Calvin and his associates at the University of California used $^{14}CO_2$ to trace the pathway of the photosynthesis reaction through various intermediates to establish the mechanism of this reaction, which has the net equation

$$6CO_2(g) + 6H_2O(l) \longrightarrow C_6H_{12}O_6(s) + 6O_2(g)$$

Radioisotopes can help us determine reaction mechanisms. For example, consider the reaction of an alcohol with a carboxylic acid:

$$R{-}OH + HO{-}\overset{\overset{\displaystyle O}{\|}}{C}{-}R' \longrightarrow R{-}O{-}\overset{\overset{\displaystyle O}{\|}}{C}{-}R' + H_2O$$

A question here is: Does the oxygen in water come from the —OH of the alcohol or that of the acid? To answer that question, "labeled" oxygen-18 can be introduced into the —OH group of either the alcohol or the acid. When the labeled oxygen is in the alcohol, it is found to be present in the ester that forms, and not in water, as shown by the following equation, where the labeled oxygen is marked with an asterisk.

$$R{-}\overset{*}{O}{-}H + H{-}O{-}\overset{\overset{\displaystyle O}{\|}}{C}{-}R' \longrightarrow R{-}\overset{*}{O}{-}\overset{\overset{\displaystyle O}{\|}}{C}{-}R' + H_2O$$

APPLICATIONS OF CHEMISTRY 24.2
Predicting Earthquakes in California

Two major earthquakes have occurred in California in the last 300 years, and both of those occurred when the state had a small population. California is the most populated state in the union, and a major earthquake would be devastating economically and might cause a large loss of human life (see photo at right). Were the two largest California earthquakes part of a regular repeating series of earthquakes? If so, can we predict the next major California earthquake?

Scientists excavated a 13-ft trench about 60 miles northeast of Los Angeles. They found that the sedimentary layers had shifted and separated at several locations by several feet. Each shift of several feet represents the occurrence of a major earthquake. By carbon-dating the remains of once-living organisms at each of these breaks, the approximate time when each earthquake took place was determined. In 1990, the activities of the organic matter in these separated sedimentary layers ranged from 12.90 to 15.06 counts per minute per gram of carbon sample tested.

Since a living organism has an activity of 15.3 counts per minute per gram of carbon sample, the ages of the organisms range from 130 to 1410 years. (Try to duplicate these results by calculating the ages as shown in Example 24.9.) We now know that over the last 1400 years, major California earthquakes have taken place about every 150 years. Since the last large earthquake took place in California in 1857 (the 1989 earthquake was relatively tame), some think that the next major one is due around the year 2010. However, the degree of strata separation suggests to others that the 1857 quake was much smaller than previous major earthquakes. This would make the 1720 earthquake the last major quake. If this is true, the next big quake has been overdue for the last 130 years. Since the largest previous interval between quakes was 275 years, a major earthquake in California appears imminent.

Sources: *Science*, July 1, 1988, page 413; *Science News*, July 6, 1985, page 5; Ronald DeLorenzo, *Journal of Chemical Education*, August 1980, page 601.

This result shows that the oxygen in the water molecule comes from the —OH of the acid. On the other hand, if the reaction is carried out using labeled, ^{18}O, in the acid, the radioactive oxygen ends up in the water, showing again that the reaction involves the cleavage of the O—H bond in the alcohol and the C—O bond of the acid:

Cleavage of these bonds

$$R{-}O{-}H + H{-}\overset{*}{O}{-}\overset{\overset{\displaystyle O}{\|}}{C}{-}R' \longrightarrow R{-}O{-}\overset{\overset{\displaystyle O}{\|}}{C}{-}R' + H_2\overset{*}{O}$$

A small amount of iodine-131, which is radioactive, can be used to check the condition of the thyroid gland (in the throat). This gland is necessary for the complete development and function of every cell in the body. Most of the iodine ingested by the body is used up by this gland. When one drinks a solution of NaI containing a small amount of iodine-131, a counter placed close to the thyroid region of the neck can be used to trace the rate of absorption of NaI by the gland. A normal thyroid absorbs about 12 percent of the iodine within a few hours. Deviation from this rate indicates abnormality.

Brain tumors have a tendency to absorb indium and copper from the bloodstream. Tracing the paths of indium-111 and copper-64 isotopes can enable a surgeon to locate such tumors quite accurately. Gallium-67 has been shown to be 95 percent accurate in locating lesions due to Hodgkin disease, which are deep inside the body.

24.6 Nuclear Fission and Fusion

Nuclear Stability and Binding Energy

Figure 24.13

Albert Einstein.

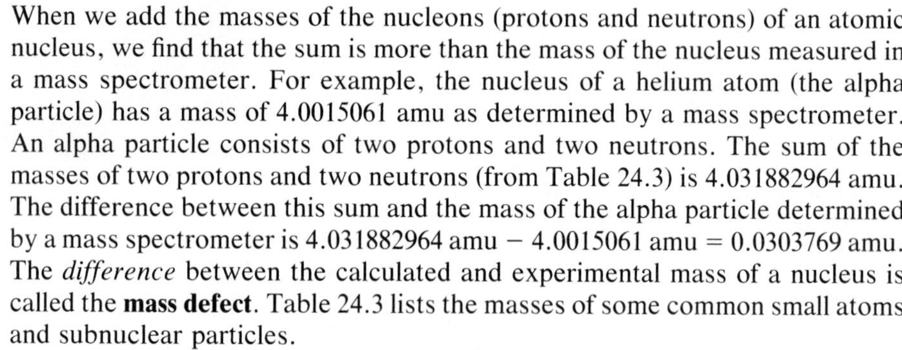

When we add the masses of the nucleons (protons and neutrons) of an atomic nucleus, we find that the sum is more than the mass of the nucleus measured in a mass spectrometer. For example, the nucleus of a helium atom (the alpha particle) has a mass of 4.0015061 amu as determined by a mass spectrometer. An alpha particle consists of two protons and two neutrons. The sum of the masses of two protons and two neutrons (from Table 24.3) is 4.031882964 amu. The difference between this sum and the mass of the alpha particle determined by a mass spectrometer is 4.031882964 amu − 4.0015061 amu = 0.0303769 amu. The *difference* between the calculated and experimental mass of a nucleus is called the **mass defect**. Table 24.3 lists the masses of some common small atoms and subnuclear particles.

The energy equivalent of the mass that is lost when a nucleus forms from its constituent protons and neutrons can be thought of as the **nuclear binding energy**—the *energy that holds the nucleus together*. Thus, the energy equivalent of the mass defect is the nuclear binding energy. The binding energy of a nucleus can be calculated from the well-known equation of Albert Einstein (Figure 24.13) that relates mass and energy:

$$E = mc^2 \tag{24.3}$$

where E is the energy, m the mass, and c the speed of light. Using this equation, the binding energy of a nucleus per nucleon can be calculated from the known mass defect.Example 24.10 illustrates how the mass defect and the binding energy can be calculated for the nucleus of a helium-4 atom (the alpha particle).

Table 24.3

Masses of Common Small Atoms and Particles of Atoms

Particle	Mass (amu)
$^{1}_{1}H$	1.007825[a]
$^{2}_{1}H$	2.0140[a]
$^{3}_{1}H$	3.01605[a]
$^{4}_{2}He$	4.0026031[a]
$^{1}_{0}n$	1.008665012
$^{1}_{1}p$	1.007276470
$^{4}_{2}\alpha$	4.0015061
$^{0}_{-1}e$ or $^{0}_{1}e$	0.00054858026

[a] Represents atomic mass (nucleus + electrons).

Example 24.10

Calculating the Mass Defect and the Binding Energy for the Nucleus of a Helium-4 Atom

The experimental mass of the nucleus of a helium-4 atom is 4.0015061 amu. Using the data from Table 24.3, calculate the mass defect for a helium-4 nucleus and the binding energy per nucleon. 1 amu = 1.66056×10^{-24} g, and the speed of light is 2.997925×10^{8} m s^{-1}.

SOLUTION: We have already calculated the mass defect of a helium-4 nucleus, but we show again how this is done in more detail below using the data from Table 24.3:

Mass of two protons	= (2)(1.007276470 amu)	= 2.014552940 amu
Mass of two neutrons	= (2)(1.008665012 amu)	= 2.017330024 amu
	Calculated mass of He-4 nucleus	= 4.031882964 amu

The difference between the calculated and experimental mass is the mass defect:

Calculated mass of ^{4}He:	4.031882964 amu
Experimental mass of ^{4}He:	4.0015061 amu
Mass defect:	0.0303769 amu

We use Einstein's equation to calculate the binding energy. If we express the mass in kilograms and the speed of light in meters per second, the energy will emerge in joules. Thus we must convert the mass defect in atomic mass units to kilograms:

$$0.0303769 \text{ amu} \left(\frac{1.66056 \times 10^{-24} \text{ g}}{1 \text{ amu}}\right)\left(\frac{1 \text{ kg}}{1000 \text{ g}}\right) = 5.04426 \times 10^{-29} \text{ kg}$$

$$E = mc^2 = (5.04426 \times 10^{-29} \text{ kg})(2.997925 \times 10^{8} \text{ m s}^{-1})^2$$
$$= 4.53355 \times 10^{-12} \text{ kg m}^2 \text{ s}^{-2} \text{ or } 4.53355 \times 10^{-12} \text{ J}$$

Since there are four nucleons in a nucleus of a helium-4 atom, the binding energy per nucleon is

$$\frac{4.53352 \times 10^{-12} \text{ J}}{4 \text{ nucleons}} = 1.13338 \times 10^{-12} \text{ J nucleon}^{-1}$$

In nuclear chemistry, binding energies are usually expressed in millions of electron volts (MeV). 1 MeV = 1.6021×10^{-13} J. Thus, the binding energy per nucleon of a helium nucleus in million electron volts is

$$1.13338 \times 10^{-12} \text{ J} \left(\frac{1 \text{ MeV}}{1.602189 \times 10^{-13} \text{ J}}\right) = 7.07395 \text{ MeV}$$

Practice Problem 24.9 The mass of a deuteron (the nucleus of a deuterium atom, ^{2}H) is 2.0135 amu. Using the data from Table 24.3, calculate the mass defect and the binding energy for a deuteron. What is the binding energy per nucleon in MeV?

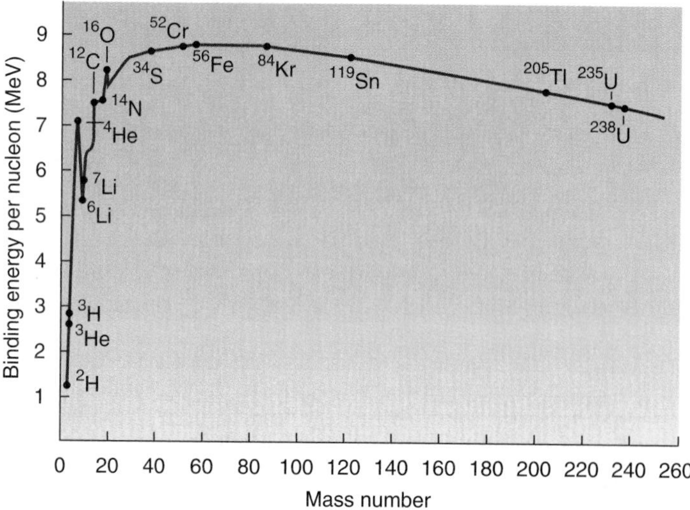

Figure 24.14

Variation of binding energy per nucleon with increasing mass numuer of nuclei.

The binding energy per nucleon is relatively low for small nuclei such as ${}^{2}_{1}H$ and ${}^{6}_{3}Li$, but increases sharply with increasing mass number, reaches a maximum at mass numbers of about 40 to 60, and then decreases slowly toward the heavier nuclei (Figure 24.14).

The plot in Figure 24.14 indicates that the most stable nuclei (the nuclei with highest binding energies) are in the vicinity of ${}^{52}Cr$ and ${}^{56}Fe$. It is therefore reasonable to predict from Figure 24.14 that if small nuclei could be fused into more stable, larger nuclei, the difference in the binding energy would be released to the surroundings. Similarly, splitting very large nuclei, such as ${}^{238}U$ and ${}^{235}U$, could yield more stable nuclei of intermediate size. This process of splitting large nuclei would also release energy. The fusion of small nuclei to larger ones, and the splitting of large nuclei to smaller ones, have been carried out as predicted. Both of these processes release large amounts of energy. We discuss these processes in more detail below.

Nuclear Fission

The splitting of large nuclei is called *fission*. The first nuclear fission reaction was carried out in 1938 by Otto Hahn and Fritz Strassman at the Kaiser Wilhelm Institute in Berlin. They found that when uranium-235 was exposed to neutrons, one of the products was barium, an element of intermediate mass number. The significance of this discovery was communicated to Niels Bohr by Lise Meitner (a former colleague of Hahn) and Otto Frisch, then refugees from Germany who were working in Copenhagen.

Figure 24.15

Enrico Fermi.

At the time Bohr was just preparing to visit the United States. Arriving in January 1939, Bohr discussed the Hahn and Strassman discovery with Einstein and others. A group of refugee physicists in the United States, Leo Szilard, Edward Teller, and Eugene Wigner, all three from Hungary, Victor Weisskopf of Austria, and Enrico Fermi (Figure 24.15) of Italy drafted a letter, signed by Einstein, to President Roosevelt. The letter explained the potential of nuclear fission and warned that Germany might already be working to build a nuclear bomb.

In February 1940, a modest fund of $6000 was made available to start research on nuclear fission in the United States, and later the Manhattan Pro-

Figure 24.16

Detonation of the first atomic bomb in Alamogordo, N.M., in 1945.

ject was established. This initial effort grew to enormous proportions involving many scientists. The first controlled nuclear fission reaction was carried out at the University of Chicago on December 2, 1942. On July 16, 1945, the first nuclear bomb was detonated on the Alamogordo air base, 120 miles south of Albuquerque, New Mexico (Figure 24.16). By this time, the original $6000 authorized had grown to $2,000,000,000. On August 6 (Japanese time), 1945, a nuclear bomb was dropped on Hiroshima, Japan, and on August 9 another bomb was dropped on Nagasaki. As a result, World War II ended.

In a nuclear fission reaction, uranium-235 nuclei are bombarded by neutrons. As a result, the uranium nucleus splits into two smaller nuclei, and more neutrons are also produced. The smaller fractions of ^{235}U may be $^{139}_{56}Ba$ and $^{94}_{36}Kr$ or other similar nuclides. The reaction of splitting U-235 to produce ^{139}Ba and ^{94}Kr is

$$^{235}_{92}U + ^{1}_{0}n \longrightarrow ^{139}_{56}Ba + ^{94}_{36}Kr + 3^{1}_{0}n$$

The extra neutrons produced in this reaction split more ^{235}U nuclei producing more neutrons, and so on (Figure 24.17). A *fission reaction that is self-sustaining and occurs by the production of neutrons that split other nuclei in continued succession* is called a **chain reaction**, illustrated in Figure 24.17.

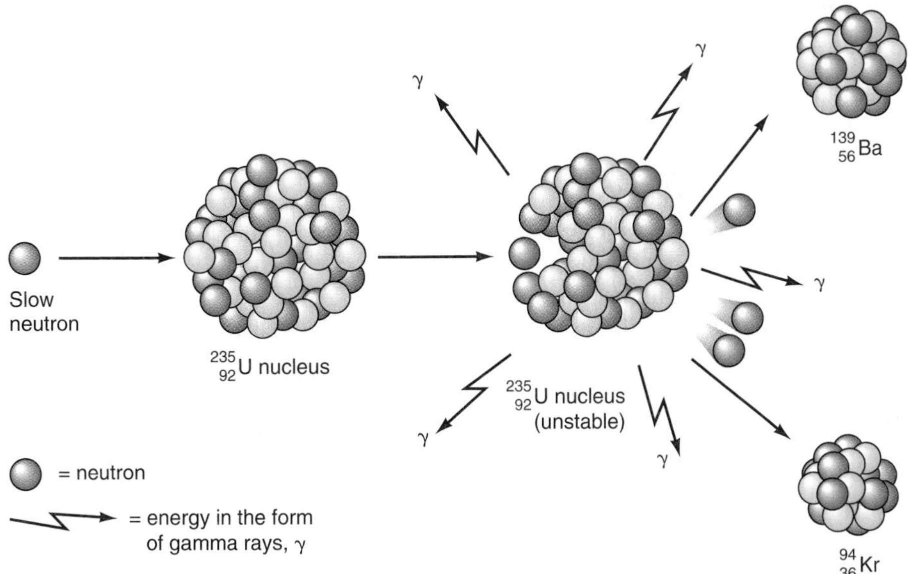

Figure 24.17

A chain reaction. A neutron splits an (unstable) U-235 nucleus to smaller nuclei and neutrons. The neutrons produced in the first step split additional U-235 nuclei, and so on.

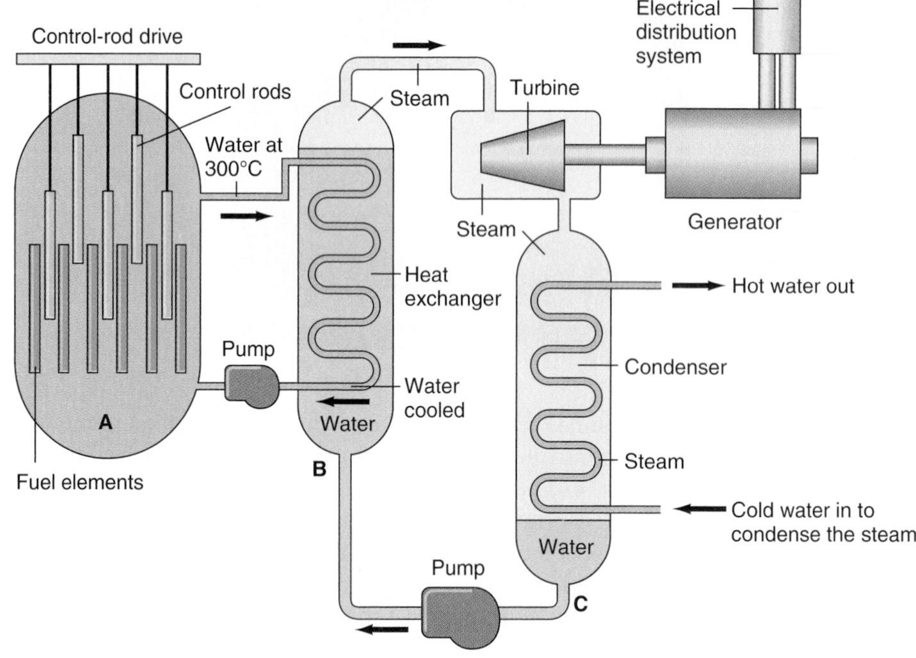

Figure 24.18

Diagram of a nuclear fission reactor used to generate electricity. The speed of the reaction is controlled by cadmium or boron rods. When the rods are fully inserted, fission may stop or it may occur at a low level. Fewer neutrons are absorbed by the rods when they are partially withdrawn, and a more vigorous fission occurs.

Figure 24.19

A nuclear power plant.

Experiments have shown that both uranium-235 and plutonium-239 can be used successfully in fission reactions. Natural uranium consists mostly of ^{238}U and only about 0.7 percent of ^{235}U. Plutonium can be made only by nuclear transmutation reactions. Problems of separating ^{235}U from ^{238}U were staggering at first, and in 1942, with the world at war, only about 0.000001 lb (about half a milligram) of fissionable material was available. As a result of intensive efforts, these problems were solved, and a large-scale plant for separation of ^{235}U from ^{238}U was built at a 70-square-mile tract near Knoxville, Tennessee, now known as Oak Ridge. Large-scale plutonium production reactors were built on an isolated 1000-square-mile tract on the Columbia river north of Pasco, Washington—the Hanford engineering works:

For a nuclear chain reaction to occur, there must be a minimum amount, or **critical mass**, of fissionable material (^{235}U or ^{239}Pu) present. If the amount is too small, the loss of neutrons to the surroundings will prevent a chain reaction.

In a **nuclear reactor** (Figures 24.18 and 24.19), the reaction must occur rapidly enough to produce usable heat energy, but it should not overheat. Control of such a reactor is achieved by means of cadmium or boron rods that absorb neutrons. When the rods are fully inserted, fission may stop or it may occur at a low level. As the rods are withdrawn, fewer neutrons are absorbed by the control rods, and a more vigorous reaction occurs.

As with all other fuels, the supply of uranium-235 is limited, and according to some estimates, the supply of low-cost ^{235}U will be exhausted in 40 years or less. For that reason, *reactors that produce energy and more fissionable material than they consume*—called **breeder reactors**—have been developed. In these reactors, some of the control rods are replaced with rods containing ^{238}U.

In a breeder reactor, some ^{235}U is consumed for its operation, but at the same time, the nonfissionable ^{238}U is converted to fissionable ^{239}Pu as shown by the following equations:

$$^{238}_{92}U + ^{1}_{0}n \longrightarrow ^{239}_{92}U$$

$$^{239}_{92}U \longrightarrow ^{239}_{93}Np + ^{0}_{-1}e$$

$$^{239}_{93}Np \longrightarrow ^{239}_{94}Pu + ^{0}_{-1}e$$

Thus, breeder reactors provide both power and fissionable plutonium.

Nuclear Fusion

Nuclear fusion involves the merger of small nuclei to form larger, more stable species, whose binding energy is represented by a point higher in the nuclear binding energy curve (Figure 24.14). The conversion of hydrogen to helium by this pathway is the primary source of solar energy. Spectroscopic evidence indicates that the sun is composed of 73 percent by mass H, 26 percent He, and only 1 percent of other elements. Some fusion reactions that occur in the sun are the following:

(1) $^{1}_{1}H + ^{1}_{1}H \longrightarrow ^{2}_{1}H + ^{0}_{1}e$

(2) $^{1}_{1}H + ^{2}_{1}H \longrightarrow ^{3}_{2}He$

The $^{3}_{2}He$ nuclei formed in the second step fuse to form $^{4}_{2}He$ and two protons in the third step:

(3) $^{3}_{2}He + ^{3}_{2}He \longrightarrow ^{4}_{2}He + 2\,^{1}_{1}H$

As a result of these processes, γ radiation is emitted (Figure 24.20). This sequence of steps is summarized by the equation

$$4\,^{1}_{1}H \longrightarrow ^{4}_{2}He + 2\,^{0}_{1}e$$

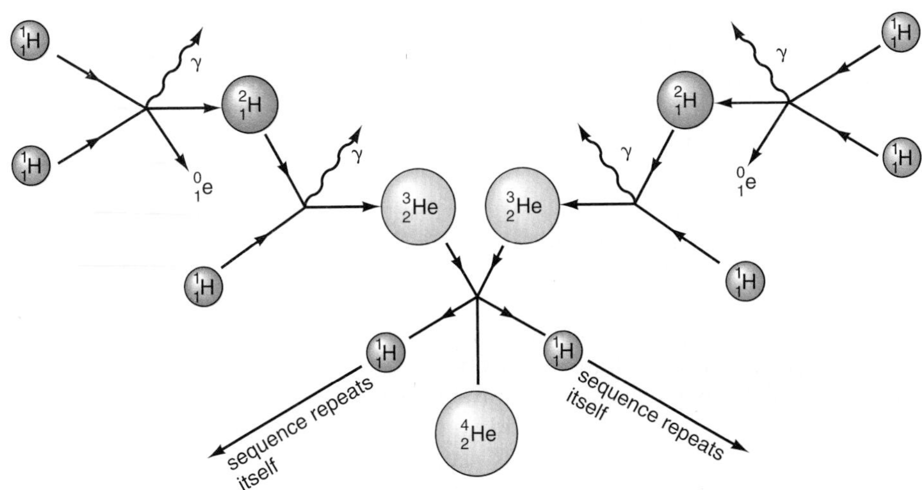

Figure 24.20

A possible sequence of steps for the fusion of hydrogen to helium, which can be summarized by the equation $4\,^{1}_{1}H \rightarrow ^{4}_{2}He + 2\,^{0}_{1}e$ + energy.

The temperature in the center of the sun has been estimated to be about 1.5 $\times$ 10^7°C. At such a high temperature, atoms dissociate into positive ions and electrons. A *neutral mixture of positive ions and electrons at high temperature* is called **plasma**. In a plasma, small nuclei have sufficiently high energy to overcome their mutual repulsion so that collisions between them can lead to fusion. *Nuclear reactions at very high temperatures* (more than 1 million K) are called **thermonuclear reactions**.

To start a fusion reaction, an extremely high temperature is needed. The fusion of deuterons with triterons provides the energy released by a hydrogen bomb:

$$^{2}_{1}H + ^{3}_{1}H \longrightarrow ^{4}_{2}He + ^{1}_{0}n$$

This reaction requires the lowest possible temperature for a fusion reaction, about 40,000,000 K. To explode a hydrogen bomb, the high temperature needed can be achieved by using a conventional fission bomb to trigger the fusion reaction.

Fusion reactions provide more energy per gram of fuel than fission reactions, and they are also "cleaner" than fission reactions; that is, the products of a fusion reaction are stable. The end products of fusion reactions are stable isotopes of helium, compared with dangerous radioactive isotopes formed by ssion reactions. A further advantage of fusion is the much greater abundance of light isotopes suitable for such reactions than the heavier isotopes of uranium and plutonium needed for fission. For example, deuterium is present in natural waters, and various methods have been developed for its production from water.

Although fusion can be carried out in thermonuclear weapons by initiating the explosion with the use of an atomic bomb, a controlled fusion reaction for energy production is difficult to achieve because of the extremely high temperatures required for activation of such reactions. On December 24, 1982, researchers at Princeton University's $314 million Tokamak Fusion Test Reactor carried out the first U.S. test of a controlled fusion reaction. In this experiment, superheated nuclei and electrons derived from deuterium and tritium formed a plasma that was confined by two magnetic fields as illustrated by Figure 24.21. The small-scale reaction lasted only 50 ms, but some scientists hope that fusion reactions can be developed to generate usable energy some time in the twenty-first century.

24.7 Effects of Radiation on Matter

When matter, living or inanimate, is exposed to alpha, beta, and gamma rays, the molecules of the matter can become ionized, they can be excited to higher energy, or they may dissociate into fragments. Our environment is constantly exposed to radiation from cosmic rays, ultraviolet radiation from the sun, and radiation from radioactive elements such as uranium in rocks. This radiation varies with location and depends on altitude, on the amount of sunshine, and on the percentage of uranium in rocks.

Alpha particles have the greatest ionizing power, followed by beta particles, and then gamma rays. However, gamma rays are the most penetrating to tissues, followed by beta and alpha rays, in that order. Exposure of human skin

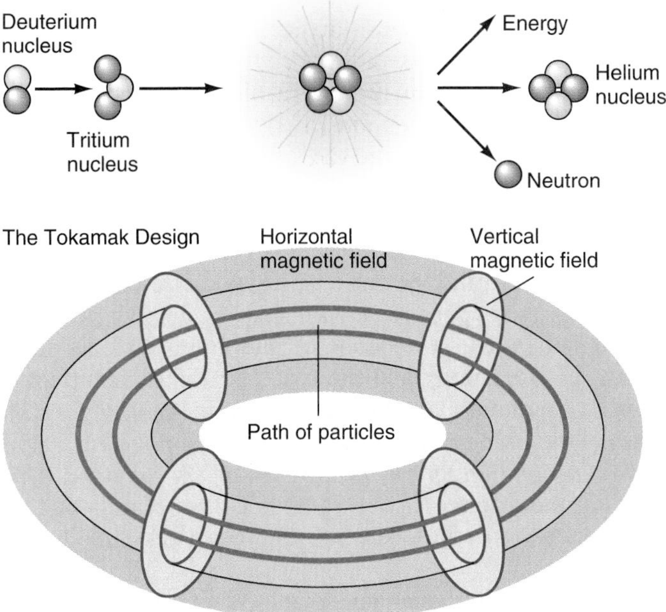
A nucleus of hot deuterium (known as heavy hydrogen) fuses with a tritium nucleus to produce an unstable nucleus that immediately decays into helium and an energy-bearing neutron.
Deuterium nucleus
Tritium nucleus
Energy
Helium nucleus
Neutron
The Tokamak Design
Horizontal magnetic field
Vertical magnetic field
Path of particles
In Princeton's device, super-heated nuclei, confined by two magnetic fields, speed around and collide to produce fusion.

Figure 24.21

The design of the nuclear fusion reaction carried out in the Tokamak Fusion Test Reactor of Princeton University on December 24, 1982.

to alpha rays is fairly harmless unless the particles get into the body, in which case they may be fatal. To effectively shield human tissue from the harmful effects of radiation, a few sheets of paper are sufficient to stop alpha rays, but several feet of concrete is needed to prevent gamma radiation from reaching the tissue (Figure 24.22).

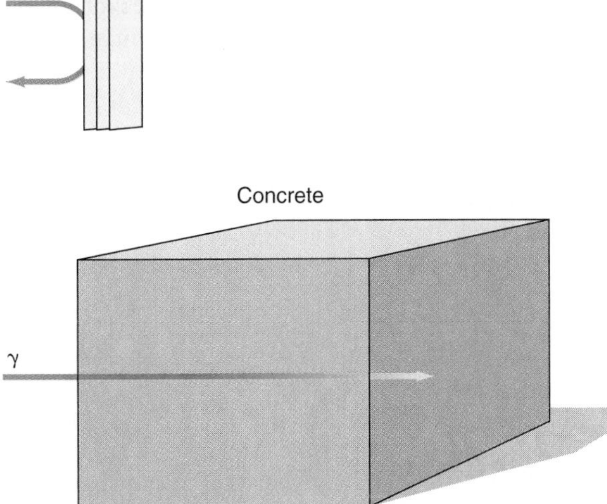
α
Three sheets of paper
Concrete
γ
3 ft

Figure 24.22

Relative penetrating abilities of alpha and gamma rays.

Radiation in high doses kills living organisms, whereas lower doses of ionizing radiation can cause leukemia, bone cancer, or birth defects. Damage to living tissues by radiation can be divided into two general types: (1) somatic effects—immediate or delayed changes in cell structure that do not affect future generations; (2) genetic effects—changes in the genes that produce physical changes in future generations.

The degree of radiation damage to organisms depends on many factors, including the extent of penetration and energy of the radiation, the total dose received, the length of exposure, and the type of tissue exposed. It is not easy to determine the level of radiation that definitely causes damage to tissues, because the damage may be delayed for several days or years.

Radiation dosage can be measured by several different units. The fundamental unit of radioactivity is the **curie** (Ci). One curie corresponds to 3.70×10^{10} nuclear disintegrations per second, which is the decay rate of 1 g of radium.

The intensity of radiation depends on the number of disintegrations per second, and also on the energy and the type of radiation emitted. Other commonly used units of radiation include the *rad* and the *roentgen*. A **rad** (radiation absorbed dose) is the *amount of ionizing radiation equivalent to* 1×10^{-2} *J of energy absorbed by one kilogram of body tissue*. A **roentgen** (after Wilhelm Conrad Roentgen, the discoverer of x-rays) is a dose of radiation that delivers 8.4×10^{-3} J of energy to 1 kg of air. A roentgen is essentially equivalent to a rad.

There is a difference in the damage that a rad of radiation can do to tissues (a rad of alpha rays can cause more damage than a rad of beta rays). Therefore, a unit of radiation that relates the intensity of radiation and its effect on body tissues is the roentgen equivalent for humans—**rem.** A rem is the product of the energy of ionizing radiation in rads and its *relative biological effectiveness* (RBE):

$$1 \text{ rem} = (1 \text{ rad})(1 \text{ RBE})$$

The value for RBE varies with the type of tissue involved, the dose rate, and its total amount. For many tissues, the RBE is approximately 1 for beta rays or gamma rays, and 10 for alpha rays.

The probable effects of short-time radiation on human beings are listed in Table 24.4. The effects of long-term lower-level radiation cannot be exactly specified, but they are under careful study by scientists, lawmakers, and others. It is believed that a dose of about 1 rem might cause about 1 case of cancer out of each 10,000 population within 20 to 30 years. The U.S. National Council on Radiation Protection and Measurements has recommended that the radiation dosage for general population be limited to 0.17 rem (170 millirems) per year from all sources above background level.

Table 24.4

Effects of Short-Time Radiation on Humans

Radiation Dose (rem)	Effects
0–25	No detectable clinical effects
25–50	Slight decrease in the count of white blood cells
100–200	Nausea and a considerable decrease in white blood cells
500	Death of about half the population within 1 month of the exposure
1000	All the population exposed would be killed

It has been estimated that the background radiation (from cosmic rays, sun, rocks, etc.) contributes about half of the estimated safe dosage received by an average person in the United States. This amounts to approximately 100 to 130 millirems (mrem) per year. The exposure in a chest x-ray examination is about 20 mrem. People associated with nuclear power production are exposed to about 5 mrem per year, and those in uranium mining are exposed to 4 mrem per year. Annual dosage received from color TV, watch dials, and so on, is approximately 0.04 mrem.

Summary

Natural Radioactivity

Substances that spontaneously emit material particles, radiant energy, or both, are called **radioactive substances**. Spontaneous radioactivity of naturally occurring substances is called **natural radioactivity**. Nuclides of radioactive elements are known as **radionuclides**.

The disintegration of radioactive substances, known as **radioactive decay**, can occur by **alpha emission**, **beta emission**, **positron emission**, and **electron capture**. A decaying isotope is known as the **parent isotope**, and the new isotope that is produced by the decay process is the **daughter isotope**.

Nuclear Equations

In nuclear equations, the mass number is written as a superscript at the upper left of the symbol for a nuclide, and the atomic number (the nuclear charge) may be written as a subscript at the lower left. In balancing a nuclear equation, the mass number and nuclear charge must be balanced.

Prediction of Stability of Nuclei

The following empirical guidelines help us predict whether or not a substance is radioactive: (1) All nuclei of more than 83 protons are radioactive; (2) nuclei that have 2, 8, 20, 28, 50, or 82 (the **magic numbers**) protons or neutrons are less likely to be radioactive than nuclei having other numbers of protons or neutrons; (3) nuclei that have an even number of protons and an even number of neutrons are generally less likely to be radioactive than nuclei that have an odd number of protons, an odd number of neutrons, or both; and (4) nuclei that lie inside the **belt of stability** (illustrated in Figure 24.4) are generally stable, with only a few exceptions.

Nuclear Transmutation

Radionuclides can be prepared by **nuclear transmutation**, in which certain nuclei are bombarded by small positive particles in *particle accelerators* such as **cyclotrons** and **linear accelerators**. Radionuclides can also be prepared by bombarding nuclei with neutrons. Equations for nuclear transmutations can be written in a shorthand form. For example, the equation

$$^{14}_{7}\text{N} + ^{4}_{2}\text{He} \longrightarrow ^{17}_{8}\text{O} + ^{1}_{1}\text{p}$$

in a shorthand form is

$$^{14}_{7}\text{N}(^{4}_{2}\alpha,\ ^{1}_{1}\text{p})^{17}_{8}\text{O}$$

or even more simply

$$^{14}_{7}\text{N}(\alpha,\text{p})^{17}_{8}\text{O}$$

Half-Life of Radioactive Elements and Related Applications

Radioactive decay obeys first-order reaction kinetics that can be expressed by the equation

$$\ln \frac{N_0}{N_t} = kt$$

where N_0 is the number of radioactive nuclides at time 0 and N_t is the number of nuclides at time t. The rate constant for such a reaction can be calculated from the known half-life, $t_{1/2}$:

$$k = \frac{0.693}{t_{1/2}}$$

Using the knowledge of the first-order kinetics and the value of the half-life of ^{238}U, the ages of uranium-containing minerals, the earth, and the moon can be calculated from the experimentally determined uranium and lead content. The ages of organic materials can similarly be determined by the method of *^{14}C dating*. Radionuclides are also used in research, medicine, and in many other fields.

Nuclear Fission and Fusion

Small and large nuclei are less stable than those of intermediate mass. Small nuclei can therefore be fused at extremely high temperature to produce larger, more stable nuclei, and energy. This is the principle of **nuclear fusion** or **thermonuclear reactions**. Nuclear fusion in the sun produces energy. Humans use the fusion process to produce thermonuclear weapons. The splitting of large nuclei to smaller ones, and the simultaneous production of energy, is known as **nuclear fission**, used in nuclear reactors and in atomic bombs.

Mass Defect and Nuclear Binding Energy

The difference between the calculated and experimental mass of a nuclide is called the **mass defect**. The energy equivalent to the mass defect is the **nuclear binding energy**. The energy produced by nuclear fission, or fusion, can be calculated from the mass defect using Einstein's equation, $E = mc^2$.

Units and Effects of Radiation

Radiation effects on human beings can be divided into somatic effects (immediate or delayed changes in the body) and genetic effects (effects that produce changes in future generations). Some common units used to measure radiation dosage include the **curie** (3.70×10^{10} nuclear disintegrations per second), the **rad** (0.01 J of energy absorbed by 1 kg of body tissue), and the **rem**, which is the product of radiation in rads and a factor called *relative biological effectiveness, RBE*. It has been suggested that the radiation dosage for general population be limited to 0.17 rem per year.

New Terms

Alpha emission (24.1)
Belt of stability (24.2)
Beta decay (24.1)
Beta emission (24.1)
Breeder reactor (24.6)
Chain reaction (24.6)
Critical mass (24.6)
Curie (24.7)
Cyclotron (24.3)
Daughter nuclide (24.1)
Electron capture (24.1)
Geiger counter (24.5)
Induced (artificial) radioactivity (24.3)
K-capture (24.1)
Linear accelerator (24.3)
Magic numbers (24.2)
Mass defect (24.6)
Natural radioactivity (24.1)
Nuclear binding energy (24.6)
Nuclear fission (24.6)
Nuclear fusion (24.6)
Nuclear reactor (24.6)
Nuclear transmutation (24.3)
Nuclide (24.1)
Parent nuclide (24.1)
Plasma (24.6)
Positron emission (24.1)
Rad (24.7)
Radioactive (24.1)
Radioactive decay (24.1)
Radioactive decay series (24.2)
Radionuclide (24.1)
Rem (24.7)
Roentgen (24.7)
Thermonuclear reaction (24.6)
Transuranium element (24.3).

Exercises

General Review

1. Describe the nature of alpha, beta, and gamma rays. If the rays consist of particles, give their names, symbols, masses, and charges.
2. What is the difference between natural radioactivity and induced radioactivity?
3. What is meant by "radioactive decay"?
4. What are parent and daughter nuclides?
5. Explain quantitatively what happens to the mass and charge of a nuclide undergoing each of the following modes of decay: **(a)** alpha emission; **(b)** beta emission; **(c)** positron emission; **(d)** electron capture.
6. Where in the periodic table do you find elements that are all radioactive?
7. What are "magic numbers"? Give some numerical values of magic numbers.
8. Explain the significance of odd and even numbers of protons and neutrons in nuclei.
9. Explain the significance of the neutron-to-proton ratio in nuclei.
10. What is the "belt of stability" described in this chapter?
11. What is a radioactive decay series? What is the first element of one of these series described in this chapter, and what is the last (stable) element of this series?
12. What is nuclear transmutation? Give examples of particles used to bombard a target nucleus to effect nuclear transmutation.

13. What is the purpose of cyclotrons, and of linear accelerators? How is a cyclotron different from a linear accelerator?
14. What are transuranium elements? Do they exist in nature? If not, how were they produced?
15. There are two radioactive elements with atomic numbers less than 83. What are their names and atomic numbers?
16. Write a logarithmic equation that relates the rate constant of a radioactive decay with the half-life of the decaying substance. Show how the rate constant can be calculated from the equation.
17. Explain how the age of a uranium-containing rock can be determined.
18. Explain how the age of an old wooden sandal can be determined by radiocarbon dating.
19. Give examples of some uses of radioisotopes.
20. What is meant by "tracers" and radioisotope "labeling"?
21. What data are needed to calculate the binding energy of a nucleus?
22. Draw a curve to show how the binding energy per nucleon changes with increasing mass number of nuclei.
23. What is: **(a)** nuclear fission; **(b)** nuclear fusion?
24. Explain how a fission reaction can be controlled in a nuclear reactor (without an explosion).
25. What two elements are commonly used as fuels in nuclear reactors?
26. What is a breeder reactor?
27. Why is a hydrogen bomb called a thermonuclear weapon?
28. What is the meaning of "plasma" in nuclear chemistry?
29. What data are needed to calculate the energy obtainable from a nuclear fission or fusion?
30. Name and define three common units used to measure radiation dosage.

Nuclear Equations for Natural Decay Processes

31. Write a balanced equation for each of the following radioactive decay processes: **(a)** alpha emission by ^{225}Ac; **(b)** beta emission by ^{40}K; **(c)** positron emission by ^{13}N; **(d)** electron capture by ^{51}Cr.
32. The uranium-235 radioactive series starts with the emission of an alpha particle in the first step, followed by a beta emission in the second step, and an alpha emission in the third step. What daughter isotope is formed at this point? Write an equation for each of the foregoing steps of decay.

Prediction of Nuclear Stability

33. Predict whether each of the following nuclides is likely to be radioactive or not; explain the basis of your prediction in each case: **(a)** ^{16}O; **(b)** ^{17}O; **(c)** ^{208}Pb; **(d)** ^{18}F; **(e)** ^{222}Rn.
34. Decide whether each of the following nuclides lies within the belt of stability illustrated in Figure 24.4: **(a)** ^{23}Na; **(b)** ^{24}Na; **(c)** ^{102}Ag; **(d)** ^{238}Np.
35. What would you predict to be the most likely decay mode of each of the nuclides in question 34 that does not lie within the belt of stability? Write an equation for each of these decay processes.

Nuclear Transmutation by Particle Bombardment

36. The first artificially produced element was technetium. Write an equation for the production of ^{97}Tc by the bombardment of $^{96}_{42}Mo$ with deuterons, $^{2}_{1}H$.
37. In each of the following nuclear reactions, replace X by its correct symbol, mass, and charge, and balance the equation, if not already balanced: **(a)** $^{8}_{4}Be + ^{4}_{2}He \rightarrow X + \gamma$; **(b)** $^{239}_{94}Pu + ^{1}_{0}n \rightarrow X + ^{140}_{54}Xe + 3^{1}_{0}n$; **(c)** $^{27}_{13}Al + ^{1}_{1}H \rightarrow ^{27}_{14}Si + X$; **(d)** $^{32}_{16}S + ^{2}_{1}H \rightarrow ^{30}_{15}P + X$; **(e)** $X + ^{4}_{2}He \rightarrow ^{17}_{8}O + ^{1}_{1}p$; **(f)** $2X + 2\,^{1}_{0}n \rightarrow ^{4}_{2}He$; **(g)** $^{12}_{6}C + ^{1}_{1}H \rightarrow X + \gamma$.
38. Complete each of the following induced nuclear reactions: **(a)** $^{10}_{5}B(n,\alpha)X$; **(b)** $^{43}_{20}Ca(\alpha,p)X$; **(c)** $^{82}_{35}Br(n,\gamma)X$; **(d)** $^{238}_{92}U(^{22}Ne,4n)X$.
39. Write a complete nuclear equation for each of the abbreviated versions in question 38.

Half-Life

40. Radium-226 has a half-life of 1590 years. Starting with 3.00 g of radium-226, how much will remain after: **(a)** 795 years; **(b)** 1590 years; **(c)** 3180 years?
41. The half-life of phosphorus-30 is about 3.0 minutes. How much of a 0.010-mg sample of phosphorus-30 remains after 1.5 minutes? after 10 minutes?
42. The half-life of cobalt-60 is 5.26 years. **(a)** Calculate the decay rate constant for cobalt-60. **(b)** How much of a 10.0-g sample of cobalt-60 remains after 20.0 years? **(c)** How long will it take for a 10.0-g sample of cobalt-60 to decrease to 9.00 g?
43. The activity of a sample of a radioisotope decreases to 94.4 percent of its original activity in 5.0 minutes. Calculate the half-life of this radioisotope.
44. The half-life of plutonium-239 is 24,000 years. Of all the plutonium waste produced today, what fraction will remain in the year 4000?

Determination of Age by U/Pb and K/Ar Ratios

45. What is the age of a rock in which the mole ratio of uranium-238 to lead-206 is 1.10? The half-life of uranium-238 is 4.51×10^9 years.

46. A sample of uraninite (a uranium-containing ore) from the Black Hills of South Dakota was found to contain 0.228 g of lead-206 for each gram of uranium-238. On the basis of this information, estimate the minimum age of the uraninite.

47. A sample of rock was found to contain 2.10×10^{-5} mol of potassium-40 and 1.10×10^{-5} mol of argon-40. Assuming that all the argon-40 came from the decay of potassium-40 and none escaped, what is the age of the rock ($t_{1/2}$ for potassium-40 is 1.3×10^9 years)?

Carbon-14 Dating

48. An ancient wooden object found in a cave has only one-fourth of the carbon-14 activity per gram of carbon of that of a growing tree. What is the age of the object ($t_{1/2}$ for carbon-14 is 5730 years)?

49. Trees and other living organisms contain a mixture of carbon-12 and carbon-14 and produce 15.3 disintegrations (counts on a Geiger counter) of carbon-14 atoms per minute per gram of carbon. A sample of wood from an Egyptian mummy case gives 9.4 counts per minute as the carbon-14 activity per gram of carbon in the mummy case. Calculate the age of the mummy case.

Matter–Energy Equivalence

50. Calculate the amount of energy that can be obtained from the conversion of 0.0200 g of matter to energy.

51. Assuming gasoline to be pure octane, C_8H_{18} ($\Delta H_f^\circ = -208.4$ kJ mol^{-1}, density = 2.65 kg gallon^{-1}), how many gallons of gasoline must be burned to $CO_2(g)$ and $H_2O(g)$ to release enough energy equivalent to that produced by the conversion of 0.0200 g of matter to energy? See question 50.

Mass Defect and Nuclear Binding Energy

52. Calculate the mass defect and the binding energy per nucleon for each of the following nuclides: **(a)** $^{9}_{4}Be$; **(b)** $^{94}_{40}Zr$; **(c)** $^{235}_{92}U$. The experimentally determined masses of these nuclides are 9.01218 amu for $^{9}_{4}Be$, 93.9061 amu for $^{94}_{40}Zr$, and 235.0439 amu for $^{235}_{92}U$. The masses of nucleons are listed in Table 24.3.

APPENDIX A

Answers to Odd-Numbered Problems

CHAPTER 1

Practice Problems

1.1 1.22×10^{-10} m, 1.22×10^{-8} cm, 122 pm

1.3 328°C, 601 K

1.5 5×10^2 g

1.7 89.0% oxygen

1.9

Atom	Nucleons	Protons	Neutrons	Electrons
(a) $^{11}_{5}B$	11	5	6	5
(b) $^{52}_{24}Cr$	52	24	28	24
(c) $^{197}_{79}Au$	197	79	118	79

Exercises

1 Scientists seek knowledge of the world around us. They do this by using a systematic approach to solving a scientific problem called the scientific method.

3 Matter has mass and occupies space.

5

Quantity	SI unit	Symbol
length	meter	m
mass	kilogram	kg
time	second	s
temperature	kelvin	K
energy	joule	J

7 2.58×10^3 g, 2.58×10^7 mg, 5.68 lb

9 27°C, 300 K

11 25°C

13 15.8 mL

15 **(a)** 3 **(b)** 2 **(c)** 4 **(d)** 1 **(e)** 2 **(f)** 5 **(g)** 4

17 **(a)** 0.40 m^2 **(b)** 2.87 g/mL **(c)** 25.2 g **(d)** 32.85 cm

19 Dalton viewed atoms as hard, tiny, indivisible particles which were alike for the same element, but different for atoms of other elements. Atoms did not change to become different atoms when combining to make compounds, and they combined in definite proportions. In modern theory, we know that the atom is divisible into smaller particles, such as protons, neutrons, and electrons.

21 9 protons, 10 neutrons, 9 electrons

23 The atomic weight of an element is the weighted average of all the isotopes of that element according to their abundance in nature. The atomic mass of an atom is the sum of the number of protons and neutrons for the specific isotope involved.

25 **(a)** 751 **(b)** 0.0751 **(c)** 0.000018

27 **(a)** 2.5×10^2 cm **(b)** 2.5×10^3 mm **(c)** 2.5×10^{-3} km **(d)** 98 in. **(e)** 8.2 ft

29 **(a)** 5.0×10^{-2} L **(b)** 0.053 qt **(c)** 50 cm^3 **(d)** 5.0×10^{-5} m^3

31 40°C, 313 K

33 1.1×10^2 gal

35 0.13% error, 1.3 ppt

37 4.000 g $\pm$ 0.0001 g

39 **(a)** $V = 39.209$ cm^3 **(b)** $d = 0.86592$ g/cm^3

41 0.05 L

43 A hypothesis is an educated guess based on observed phenomena. A theory is a network of interrelated hypotheses and models consistent with experimentally established laws. A law is a statement that has been shown to have no exceptions under a given set of conditions.

45 Inorganic chemistry–investigates non-carbon-containing substances.
Organic chemistry–investigates carbon-containing substances.
Physical chemistry–investigates how the laws of physics are applied to the structure and changes of matter.
Analytical chemistry–investigates methods for determining the composition of matter.
Biochemistry–investigates the chemical and physical changes occurring in living organisms.

47 **(a)** compound **(b)** element **(c)** compound **(d)** compound **(e)** mixture **(f)** mixture **(g)** element **(h)** mixture **(i)** compound **(j)** compound **(k)** element **(l)** mixture **(m)** mixture

49 **(a)** physical change **(b)** chemical change **(c)** physical change **(d)** physical change **(e)** physical change **(f)** chemical change **(g)** chemical change

51 **(a)** CO_2 contains 1 carbon atom and 2 oxygen atoms.
(b) H_2SO_4 contains 2 hydrogen atoms, 1 sulfur atom and 4 oxygen atoms
(c) C_3H_8 contains 3 carbon atoms and 8 hydrogen atoms.
(d) $C_6H_{12}O_6$ contains 6 carbon atoms, 12 hydrogen atoms, and 6 oxygen atoms.

53 **(a)** 120.0 g bromine **(b)** AuBr, $AuBr_3$

55 **(a)** first oxide of tin: 1.35 g oxygen for every 5.00 g tin; second oxide of tin: 0.674 g oxygen for every 5.00 g tin **(b)** 1.35 g O/ 0.674 g O = 2/1

57 1.0×10^{-13}%

59

Particle	Mass	Charge
(a) proton	1 amu	+1
(b) neutron	1 amu	0
(c) electron	~0 amu	−1

61

Particle	Protons	Neutrons	Electrons
(a) ^{12}C	6	6	6
(b) ^{4}He	2	2	2
(c) ^{35}Cl	17	18	17
(d) ^{23}Na	11	12	11

63 **(a)** 2.01584 amu **(b)** 16.12672 amu **(c)** 20.15840 amu

65 ^{107}Ag = 51.85%; ^{109}Ag = 48.15%

67 7.23 km/L

69 0.01000 cm

71 2.65 g/mL

73 1.20×10^3 lb gold

CHAPTER 2

Practice Problems

2.1 **(a)** Li^+ **(b)** Be^{2+}

2.3 (a) AlF_3 (b) Li_3P (c) $CaBr_2$

2.5 (a) sulfur trioxide (b) sulfur difluoride (c) disulfur dichloride (d) iodine heptafluoride (e) dinitrogen tetroxide

2.7 (a) potassium carbonate (b) barium nitrate
(c) aluminum sulfite (d) lithium phosphate
(e) strontium hydrogen sulfate (f) magnesium chlorate
(g) ammonium arsenate (h) iron(III) nitrate

2.9 (a) $4Li(s) + O_2(g) \longrightarrow 2Li_2O(s)$
(b) $6Mg(s) + P_4(s) \longrightarrow 2Mg_3P_2(s)$
(c) $16Al(s) + 3S_8(s) \longrightarrow 8Al_2S_3(s)$

2.11 (a) $Na_2O(s) + H_2O(l) \longrightarrow 2NaOH(aq)$
(b) $N_2O_5(g) + H_2O(l) \longrightarrow 2HNO_3(aq)$

2.13 $2Al(s) + 6HCl(aq) \longrightarrow AlCl_3(aq) + 3H_2(g)$

2.15 (a) $4Li(s) + O_2(g) \longrightarrow 2Li_2O(s)$
(b) $2Rb(s) + 2H_2O(l) \longrightarrow 2RbOH(aq) + H_2(g)$

2.17 (a) $4Al(s) + 3O_2(g) \longrightarrow 2Al_2O_3(s)$
(b) $8Sn(s) + S_8(s) \longrightarrow 8SnS(s)$
(c) $PbCl_2(s) + Cl_2(g) \longrightarrow PbCl_4(l)$

Exercises

1 Mendeleev classified the elements according to their chemical and physical properties. Elements with similar chemical properties were placed in a "family." For example, the alkali metals, Li, Na, K, Rb, Cs, all react violently with water, combine with halogens in a 1:1 ratio, and combine with oxygen in a 2:1 ratio. Physical properties, such as melting point, boiling point, and density, showed consistent trends on Mendeleev's periodic table. In addition, the elements were arranged in order of increasing atomic weight. Although this left gaps in the table, Mendeleev realized these gaps represented elements yet to be discovered, and he was able to predict their properties with amazing accuracy.

3 The word "periodic" refers to the repetition of chemical properties observed.

5 An example of periodicity of properties of elements in the periodic table would be the combining ratios of metals with halides. Group I metals combine in a 1:1 ratio, such as NaCl; Group II metals combine in a 1:2 ratio, such as $MgCl_2$; Group III metals combine in a 1:3 ratio, such as $AlCl_3$.

7

Group	Group Name	Elements
I	alkali metals	Li, Na, K, Rb, Cs, Fr
II	alkaline earth metals	Be, Mg, Ca, Sr, Ba, Ra
VII	halogens	F, Cl, Br, I, At
VIII	noble gases	He, Ne, Ar, Kr, Xe, Rn

9 There are 42 main group elements in the first six periods.

11 There are 14 elements in each of the lanthanide and actinide series.

13 Metals tend to be lustrous, malleable, and conduct electricity; nonmetals tend to be dull, brittle, and nonconducting.

15 For most representative metal ions, the magnitude of the positive charge is equal to the group number. For monatomic nonmetal ions, the charge is equal to the group number minus eight.

17 The *-ide* ending is used for monatomic negative ions.

19 A "binary compound" is a substance composed of only two elements.

21 Ionic bonding is the electrostatic attraction between ions of opposite charge. This type of bonding predominates between metals and nonmetals. Covalent bonding consists of a sharing of electrons between two nonmetals.

23

Formula	Name	Formula	Name
HCl	hydrochloric acid	HNO_2	nitrous acid
HBr	hydrobromic acid	H_2SO_4	sulfuric acid
HI	hydroiodic acid	H_3PO_4	phosphoric acid
HNO_3	nitric acid	$HClO_4$	perchloric acid

25 The elements that exist as gases at room temperature are hydrogen, H_2, nitrogen, N_2, oxygen, O_2, fluorine, F_2, chlorine, Cl_2, and the noble gases, helium, He, neon, Ne, argon, Ar, krypton, Kr, xenon, Xe, and radon, Rn. The elements that exist as liquids at room temperature are cesium, Cs, francium, Fr, mercury, Hg, gallium, Ga, and bromine, Br_2.

27 Allotropy results when an element can exist in more than one form. Examples of allotropes are carbon, which can exist as graphite and diamond, or phosphorus, which can exist as red or white phosphorus.

29 The compound formed by magnesium and phosphorus will be ionic since a metal is combining with a nonmetal. Magnesium is a representative metal found in group II, and can form only one ion, Mg^{2+}. Phosphorus is a representative nonmetal found in group V and will form a phosphide ion, P^{3-}. These ions must combine in a ratio in which the charges cancel. This ratio must be 3Mg:2P. With this ratio, three Mg^{2+} ions result in a total positive charge of +6, and two phosphide ions result in a total negative charge of −6. The overall net charge of the compound is zero. Magnesium phosphide, Mg_3P_2, is the result.

31 The term "noble" applied to gases and metals means they do not react under normal conditions.

33 (a) 54 neutrons (b) 42 electrons (c) 40 electrons (d) 42 protons

35 (a) not isoelectronic (b) isoelectronic (c) isoelectronic (d) isoelectronic (e) isoelectronic

37 Hydrogen is able to form both a +1 and a −1 ion; thus it appears in both group I and group VII.

39 metal A, +2; metal B, +3 and +5

41 (a) NH_4^+ (b) AsO_4^{3-} (c) ClO_3^- (d) ClO_4^-
(e) MnO_4^- (f) $Cr_2O_7^{2-}$

43 (a) KBr (b) $CaBr_2$ (c) Li_3N (d) Mg_3P_2 (e) Al_2O_3
(f) HBr (g) $KMnO_4$ (h) $Mg_3(PO_4)_2$ (i) $(NH_4)_2SO_4$
(j) CsH_2PO_4 (k) SnF_2 (l) $SnSe_2$ (m) $Ca(OH)_2$
(n) Cr_2S_3 (o) CrO_3 (p) $Mn(NO_3)_2$ (q) $Fe_2(CrO_4)_3$
(r) ScI_3 (s) Zn_3As_2 (t) $TiCl_4$ (u) CCl_4 (v) As_2S_3
(w) P_4O_6

45 Ionic bonding involves an electrostatic attraction of ions of opposite charge resulting from a transfer of electrons. Covalent bonding involves a sharing of electrons between atoms. Ionic bonding predominates between a metal and a nonmetal. Covalent bonding predominates between two nonmetals.

47 Sodium chloride is an ionic compound commonly known as table salt. Look at some table salt and see that many of the crystals are cubic in shape. The shape of the sodium chloride lattice is due to the way sodium ions and chloride ions pack into the crystal. A sodium chloride lattice has alternating sodium ions and chloride ions. Each sodium ion is surrounded by six chloride ions. Each chloride ion is surrounded by six sodium ions. Thus, it is impossible to say which sodium ion goes with each chloride ion, and one cannot describe NaCl as a molecule, but rather as a formula unit. In addition, when sodium chloride crystals are dissolved in water, individual sodium ions and chloride ions are present in solution, and are not found as units of NaCl.

49 (a) $2K(s) + F_2(g) \longrightarrow 2KF(s)$
(b) $Ba(s) + I_2(s) \longrightarrow BaI_2(s)$
(c) $6Ca(s) + P_4(s) \longrightarrow 2Ca_3P_2(s)$
(d) $16Al(s) + 3S_8(s) \longrightarrow 8Al_2S_3(s)$
(e) $6Li(s) + N_2(g) \longrightarrow 2Li_3N(s)$
(f) $2Al(s) + 3Br_2(l) \longrightarrow 2AlBr_3(s)$
(g) $Mg(s) + H_2(g) \longrightarrow MgH_2(s)$

51 (a) decomposition (b) combination (c) combination
(d) combination (e) single displacement (f) combination
(g) double displacement (h) single displacement
(i) decomposition

53 (a) $Mg(s) + CuO(s) \longrightarrow MgO(s) + Cu(s)$; single displacement

(b) $BaBr_2(aq) + K_2SO_4(aq) \longrightarrow BaSO_4(s) + 2KBr(aq)$; double displacement
(c) $2Cu_2O(s) + O_2(g) \longrightarrow 4CuO(s)$; combination
(d) $Li_2O(s) + CO_2(g) \longrightarrow Li_2CO_3(s)$; combination
(e) $PtO_2(s) + \text{heat} \longrightarrow Pt(s) + O_2(g)$; decomposition
(f) $3Ca(NO_3)_2(aq) + 2Na_3PO_4(aq) \longrightarrow Ca_3(PO_4)_2(s) + 6NaNO_3(aq)$; double displacement
(g) $Zn(s) + 2HCl(aq) \longrightarrow ZnCl_2(aq) + H_2(g)$; single displacement
(h) $2Al(s) + 3H_2SO_4(aq) \longrightarrow 2Al_2(SO_4)_3(aq) + 3H_2(g)$; single displacement

55 Alkali metals are softer than alkaline earth metals.

57 Group VII; halogens

59 Sodium reacts vigorously with water: $2Na(s) + 2H_2O(l) \longrightarrow 2NaOH(aq) + H_2(g)$; chromium does not react with water: $Cr(s) + H_2O(l) \longrightarrow$ no reaction.

61 Fe and C, as iron carbide

63

Element	Allotrope	Properties
Carbon	graphite	shiny, slippery, conducting, brittle
	diamond	shiny, very hard, high melting
	soot	dull, black solid, soft
Phosphorus	white	white, poisonous, ignites spontaneously in air, corrosive, reacts with most metals and nonmetals
	red	red, less reactive, air stable
Oxygen	O_2	colorless, odorless gas, supports combustion
	O_3	pale blue gas, characteristic odor of operating electrical equipment, absorbs cosmic rays from the sun

65 An amalgam is an alloy of two or more metals that cannot be separated by physical means.

67

Element	Bromide	Sulfide
Na	NaBr	Na_2S
Mg	$MgBr_2$	MgS
Al	$AlBr_3$	Al_2S_3

69 (a) $FeCl_2$, $FeCl_3$ (b) CuCl, $CuCl_2$ (c) Hg_2Cl_2, $HgCl_2$ (d) $SnCl_2$, $SnCl_4$ (e) $PbCl_2$, $PbCl_4$ (f) $CrCl_3$, $CrCl_6$

CHAPTER 3

Practice Problems

3.1 1.68×10^{-24} g

3.3 0.01439 mol Au, 8.666×10^{21} Au atoms

3.5 20.1% C, 2.82% H, 23.5% N, 53.6% O

3.7 Empirical formula: CH_2; molecular formula: C_4H_8

3.9 3.12 mol O_2

3.11 81.84% $CaCO_3$

3.13 Limiting reactant: Au; 0.337 g Cl_2 reacted, 0.535 g Cl_2 unreacted; 0.961 g $AuCl_3$ obtained

3.15 (a) 0.0379 mol L^{-1} (b) 0.144 mol L^{-1}

3.17 94.4 mL

3.19 0.750 M

Exercises

1 The atomic weight of an element is the weighted average of the mass number of the isotopes of the element according to their relative abundance. The mass number gives the mass of an individual isotope.

3 The atomic mass unit is based on the mass of a single carbon-12 atom as a standard. Carbon-12 has a mass of exactly 12 amu.

5 Atomic weights are obtained from a weighted average of all the isotopes according to their natural abundance.

7 One mole of any particle, atoms or molecules, contains Avogadro's number, 6.02×10^{23} of particles.

9 Chemical reactions are based on the combining ratios of particles. If one wishes to investigate reactions in the laboratory, one cannot weigh out a single atom or molecule of a substance. By scaling up what is happening on the molecular level with the mole quantity, laboratory measurements take on meaning.

11 (a) A molecule of acetic acid, $HC_2H_3O_2$, contains 4 H atoms. (b) A mole of acetic acid contains 4 moles of H atoms.

13 4.04 g H

15 $\dfrac{\text{at. wt}}{\text{Avogadro's number}}$ = mass of 1 average atom in grams

17 (a) 1.01 g H/mol (b) 18.0 g H_2O/mol (c) 132.2 g $(NH_4)_2SO_4$/mol (d) 48.0 g O_3/mol (e) 216.0 g $Cr_2O_7^{2-}$/mol

19 To find the simplest formula of a compound from the given percent composition:

1. Assume 100 grams of the compound. The percentage of each element of the compound is thus the grams of that element.
2. Convert the mass of each element to moles using the atomic weight of the element. Divide the mass of the element by its atomic weight. It is important to use atomic weights to at least the same number of significant figures as the percentage data.
3. Divide the moles of each element by the mole value which is smallest. This gives the simplest ratio of how the elements have combined. With reliable data, the ratio often works out to be whole numbers. If the ratio does not work out to whole numbers, multiply through by a factor which will yield whole number ratios.
4. Write the simplest formula using the ratio values of each element as the subscripts in the formula.

21 A hydrate is a salt in which water molecules are chemically bound within the crystal lattice. These water molecules may be driven off by heating the solid hydrate.

23 The limiting reactant is the reactant which is completely consumed during a reaction. It is called the limiting reactant because it limits the amount of product which can form.

25 (a) A dilute solution contains a small amount of solute dissolved in a given amount of solvent. (b) A concentrated solution contains a large amount of solute dissolved in a given amount of solvent.

27 1. First find the mass of sodium chloride needed: 52.6 g NaCl

$$500\text{ mL} \times \frac{1\text{ L}}{1000\text{ mL}} \times \frac{1.80\text{ mol NaCl}}{1\text{ L}} \times \frac{58.4\text{ g NaCl}}{1\text{ mol NaCl}}$$

2. Weigh out 52.6 g NaCl.
3. Dissolve the NaCl in a 500 mL volumetric flask using less than 500 mL of water.
4. Add water up to the 500 mL mark on the volumetric flask.

29 (a) An electrolyte is a substance that dissolves in water and is able to conduct an electric current. (b) A strong electrolyte is a substance that almost completely dissociates into ions in water. Examples of strong electrolytes are sodium chloride, hydrochloric acid and sodium hydroxide. (c) A weak electrolyte is a substance that only partly dissociates into ions in water. Examples of weak electrolytes are acetic acid, ammonia, and water itself. (d) A nonelectrolyte is a substance that does not dissociate in water, and hence does not conduct electricity. Examples of nonelectrolytes are sugar, alcohol, and gasoline.

31 (a) 4.00 g/mol (b) 2.02 g/mol (c) 107.0 g/mol (d) 143.4 g/mol (e) 150.3 g/mol (f) 106.0 g/mol (g) 262.9 g $Mg_3(PO_4)_2$ g/mol

33 **(a)** 0.50 mol **(b)** 0.061 mol **(c)** 0.56 mol **(d)** 0.29 mol **(e)** 0.17 mol

35 **(a)** 3.3×10^{-12} mol O_2 **(b)** 6.6×10^{-12} mol O atoms

37 **(a)** 2.74×10^{23} molecules **(b)** 6.68×10^{22} molecules

39 3 mol C atoms, 6 mol O atoms, 9 mol atoms (total)

41 **(a)** 3.50×10^2 g N **(b)** 212.0 g N

43 $K_2Cr_2O_7$

45 $Na_2S_2O_8$

47 $NaClO_3$

49 $Na_2CO_3(H_2O)_{10}$

51 %C = 91.27%; %H = 8.749%; empirical formula = C_7H_8

53 Molecular formula = N_2O; $NH_4NO_3(s) \longrightarrow N_2O(g) + 2H_2O(l)$

55

	NH_3	O_2	NO	H_2O
(a)	2.02 g	0.149 mol	3.75 g	0.178 mol
(b)	0.0364 mol	2.74×10^{22} molecules	0.0364 mol	0.938 g
(c)	7.82×10^{23} molecules	41.6 g	7.82×10^{23} molecules	1.95 mol

57 $2\,KClO_3(s) \longrightarrow 2KCl(s) + 3\,O_2(g)$; 2.400 g O_2; 0.1500 mol O_2; 9.030×10^{22} O_2 molecules

59 $2C_2H_6(g) + 7O_2(g) \longrightarrow 4CO_2(g) + 6H_2O(g)$
(a) 6 mol CO_2 **(b)** 9 mol H_2O **(c)** 50 g CO_2 **(d)** 0.997 mol CO_2

61 1.6×10^4 g HNO_3 solution; 3.1×10^3 g H_3PO_4

63 **(a)** 10.0 mol H_2O **(b)** 1.11 mol C_8H_{18} **(c)** 127 g C_8H_{18} **(d)** 181 mL C_8H_{18}

65 Iodine is the limiting reactant; 10.9 g Al remains unreacted; 0.0667 moles and 27.2 grams of AlI_3 is produced.

67 314 g CCl_2F_2

69 19.5 g BF_3, 52.3% yield

71 **(a)** Cr_3O_8 **(b)** $48\,CrO_3(s) + S_8(l) \longrightarrow 16Cr_3O_8(s) + 8SO_2(g)$ **(c)** 1.80 g Cr_3O_8, 94.5% yield

73 20.4 g $KHC_8H_4O_4$

75 0.434 L

77 70.8 mL

79 3.26 M HNO_3

81 0.556 g $PbCl_2$

CHAPTER 4

Practice Problems

4.1 **(a)** 18.3 *M* **(b)** 27.7 mL soln

4.3 **(a)** 0.020 *M* OH^- **(b)** 0.040 *M* OH^-

4.5 Total equation: $H_2SO_4(aq) + 2KOH(aq) \longrightarrow Na_2SO_4(aq) + 2H_2O(l)$
Ionic equation: $H^+(aq) + HSO_4(aq) + 2K^+(aq) + 2OH^-(aq) \longrightarrow 2K^+(aq) + SO_4^{2-}(aq) + 2H_2O(l)$
Net ionic equation: $H^+(aq) + HSO_4^-(aq) + 2OH^-(aq) \longrightarrow SO_4^{2-}(aq) + 2H_2O(l)$

4.7 **(a)** Total equation: $HI(aq) + KOH(aq) \longrightarrow KI(aq) + H_2O(l)$.
Net ionic equation: $H^+(aq) + OH^-(aq) \longrightarrow H_2O(l)$
(b) Total equation: $2HNO_3(aq) + Mg(OH)_2(s) \longrightarrow Mg(NO_3)_2(aq) + 2H_2O(l)$
Net ionic equation: $2H^+(aq) + Mg(OH)_2(s) \longrightarrow Mg^{2+}(aq) + 2H_2O(l)$

4.9 Total equation: $2NH_3(aq) + H_2SO_4(aq) \longrightarrow (NH_4)_2SO_4(aq)$
Net ionic equation: $2NH_3(aq) + H(aq) + HSO_4^-(aq) \longrightarrow 2NH_4^+(aq) + SO_4^{2-}(aq)$

4.11 36.48 mL NaOH solution

4.13 0.9023 g KHPh

4.15 **(a)** There are two possible combinations of ions which might produce a precipitate: Mercury(I) ions with chloride ions and hydrogen ions with nitrate ions. Consulting the solubility rules, one finds most chlorides are soluble except lead chloride, silver chloride, and mercury(I) chloride. Thus, mercury(I) chloride will precipitate. The other product, nitric acid, is a strong acid and remains dissociated. The net ionic equation shows the formation of mercury(I) chloride from its constituent ions:

$$Hg_2^{2+}(aq) + 2Cl^-(aq) \longrightarrow Hg_2Cl_2(s)$$

(b) There are two possible combinations of ions which might produce a precipitate: magnesium ions with hydroxide ions and sodium ions with chloride ions. Consulting the solubility rules, one finds most hydroxides are insoluble except those of alkali metals; calcium hydroxide and strontium hydroxide are slightly soluble. Since magnesium hydroxide is not among this list of soluble hydroxides, it will form a precipitate. Sodium chloride is soluble since most salts of alkali metals and ammonium ions are soluble. The net ionic equation shows the formation of magnesium hydroxide from its constituent ions:

$$Mg^{2+}(aq) + 2OH^-(aq) \longrightarrow Mg(OH)_2(s)$$

4.17 **(a)** $Pb(NO_3)_2(aq) + K_2SO_4(aq) \longrightarrow PbSO_4(s) + 2KNO_3(aq)$
$Pb^{2+}(aq) + SO_4^{2-}(aq) \longrightarrow PbSO_4(s)$
(b) $3Ca(NO_3)_2(aq) + 2Na_3PO_4(aq) \longrightarrow Ca_3(PO_4)_2(s) + 6NaNO_3(aq)$
$3Ca^{2+}(aq) + 2PO_4^{3-}(aq) \longrightarrow Ca_3(PO_4)_2(s)$
(c) $2AgNO_3(aq) + Na_2S(aq) \longrightarrow Ag_2S(s) + 2NaNO_3(aq)$
$2Ag^+(aq) + S^{2-}(aq) \longrightarrow Ag_2S(s)$

4.19 33.33% $BaCO_3$; 41.44 mL soln

4.21 Step 1: Identify the oxidation states.
Step 2: Determine which species has been oxidized and which species has been reduced.
Step 3: The number of electrons lost must equal the number of electrons gained. Find the common multiple and assign the proper coefficients to the oxidized pair and reduced pair of compounds.
Step 4: Balance the remaining atoms by inspection.

4.23 25.6 mg I_2

Exercises

1 An Arrhenius acid is a substance which produces H^+ ions (as H_3O^+ ions) in water solution. An Arrhenius base is a substance which produces OH^- ions in water solution.

3 A Brønsted-Lowry acid is a proton donor; a Brønsted-Lowry base is a proton acceptor.

5 A weak acid will dissociate only to a small extent in water; few H^+ ions will be produced. A weak base will dissociate only to a small extent in water; few OH^- ions will be produced.

7 Two common weak acids are acetic acid, $HC_2H_3O_2$, and nitrous acid, NHO_2. A common weak base is ammonia, NH_3.

9 **(a)** $H^+(aq) + NO_3^-(aq) + K^+(aq) + OH^-(aq) \longrightarrow K^+(aq) + NO_3^-(aq) + H_2O(l)$
(b) $H^+(aq) + Cl^-(aq) + NH_3(aq) \longrightarrow NH_4^+(aq) + Cl^-(aq)$
(c) $HNO_2(aq) + Na^+(aq) + OH^-(aq) \longrightarrow Na^+(aq) + NO_2^-(aq) + H_2O(l)$
(d) $HCN(aq) + NH_3(aq) \longrightarrow NH_4^+(aq) + CN^-(aq)$

11 The steps for determining the concentration of an acid solution using a standard solution of base are discussed in detail in Section 4.3. These steps are briefly outlined below:

1. A known volume of acid solution is pipetted into an Erlenmeyer flask, and a few drops of indicator solution are added.

2. A solution of standardized base is placed into a buret, and an initial volume reading is taken.
3. The standardized base solution is delivered from a buret into the flask containing the acid solution until a color change of the indicator signals the endpoint has been reached.
4. A final volume reading is taken off the buret, and the acid concentration may be calculated. The volume of base added is the difference between the final and initial volumes read off the buret.

$$M_{\text{base}} \times V_{\text{base}} \text{ in L} \times \frac{1 \text{ mol acid}}{1 \text{ mol base}} \times \frac{1 \text{ L}}{V_{\text{acid}} \text{ in L}} = M_{\text{acid}}$$

The data needed for this determination include: concentration of base, initial and final volumes of base (read off the buret), and volume of acid titrated.

13 The stages of neutralization for the polyprotic acid, H_3X, using the base, MOH, are:

$$H_3X + MOH \longrightarrow MH_2X + H_2O$$

$$MH_2X + MOH \longrightarrow M_2HX + H_2O$$

$$M_2HX + MOH \longrightarrow M_3X + H_2O$$

15 **(a)** $2H^+(aq) + 2Cl^-(aq) + Pb^{2+}(aq) + 2NO_3^-(aq) \longrightarrow PbCl_2(s) + 2H^+(aq) + 2NO_3^-(aq)$
(b) $2K^+(aq) + SO_4^{2-}(aq) + Ba^{2+}(aq) + 2NO_3^-(aq) \longrightarrow BaSO_4(s) + 2K^+(aq) + 2NO_3^-(aq)$

17 Oxidation–reduction reactions involve a change in oxidation states of two reactant species. One reactant will be reduced by gaining electrons, another reactant will be oxidized by losing electrons. A single displacement reaction will always be an oxidation–reduction reaction.

19 **(a)** Oxidation involves the gain of oxygen, loss of hydrogen, or loss of electrons. **(b)** Reduction involves the loss of oxygen, gain of hydrogen, or gain of electrons.

21 A simple redox reaction shown in net ionic form is:

$$Cu(s) + 2Ag^+(aq) \longrightarrow Cu^{2+}(aq) + 2Ag(s)$$

ox. state 0 +1 +2 0

Copper is oxidized; it lost 2 electrons.
Silver(I) ion is reduced; each ion gained 1 electron.

23 **(a)** HNO_3, nitric acid (a monoprotic acid)
(b) H_2SO_4, sulfuric acid (a diprotic acid)
(c) H_3PO_4, phosphoric acid (a triprotic acid)

25 $H_3PO_4(aq) + H_2O(l) \longrightarrow H_2PO_4^-(aq) + H_3O^+(aq)$
dihydrogen phosphate ion; hydronium ion

$H_2PO_4^-(aq) + H_2O(l) \longrightarrow HPO_4^{2-}(aq) + H_3O^+(aq)$
hydrogen phosphate ion; hydronium ion

$HPO_4^{2-}(aq) + H_2O(l) \longrightarrow PO_4^{3-}(aq) + H_3O^+(aq)$
phosphate ion; hydronium ion

27 Strength of acids in water is determined by the degree of dissociation in water. If dissociation is complete, a large "amount" of hydronium ions is produced, and the acid is considered to be strong. Partial dissociation in water indicates a weak acid. **(a)** HNO_3 is a stronger acid than $HC_2H_3O_2$ in water because HNO_3 dissociates completely. $HC_2H_3O_2$ only partially dissociates in water. **(b)** HCl is a stronger acid in water than HNO_2 because HCl dissociates completely. HNO_2 only partially dissociates.

29 **(a)** $HCl(aq) + NaOH(aq) \longrightarrow NaCl(aq) + H_2O(l)$
(b) $2HCl(aq) + Ba(OH)_2(aq) \longrightarrow BaCl_2(aq) + 2H_2O(l)$
(c) $Mg(OH)_2(s) + 2HCl(aq) \longrightarrow MgCl_2(aq) + 2H_2O(l)$
(d) $2KOH(aq) + H_2SO_4(aq) \longrightarrow K_2SO_4(aq) + 2H_2O(l)$
(e) $2Al(OH)_3(s) + 3H_2SO_4(aq) \longrightarrow Al_2(SO_4)_3(aq) + 6H_2O(l)$
(f) $Sr(OH)_2(aq) + 2HC_2H_3O_2(aq) \longrightarrow Sr(C_2H_3O_2)_2(aq) + 2H_2O(l)$
(g) $Ca(OH)_2(s) + 2HC_2H_3O_2(aq) \longrightarrow Ca(C_2H_3O_2)_2(aq) + 2H_2O(l)$
(h) $NH_3(aq) + HNO_2(aq) \longrightarrow NH_4NO_2(aq)$
(i) $2NH_3(aq) + H_2SO_4(aq) \longrightarrow (NH_4)_2SO_4(aq)$

31 Neutralization is the complete reaction of an acid and a base to yield a salt and water. The equivalence point is the pH at which an acid and a base have been neutralized. The endpoint refers to the pH and volume at which an indicator in a titration changes color, signaling the end of a titration. If the indicator has been chosen properly, the volume of the equivalence point and the volume of the endpoint differ only slightly.

33 20.0 mL NaOH solution; 30.0 mL NaOH solution

35 0.01405 *M* NaOH

37 0.31 g $Ba(OH)_2$

39 0.09544 *M* NaOH

41 0.139 *M* H_3PO_4

43 0.121 *M* NaOH

45 90.7% N

47 **(a)** AgCl **(b)** $PbCl_2$ **(c)** Hg_2Cl_2 **(d)** $CaCO_3$ **(e)** No reaction **(f)** $BaSO_4$

49 **(a)** $BaCl_2$ and Na_2SO_4 **(b)** $AgNO_3$ and HCl **(c)** $CaCl_2$ and K_2CO_3 **(d)** $Ba(NO_3)_2$ and Na_3PO_4

51 First, test solubility in water. Of these three compounds, only silver chloride is water insoluble. To distinguish between barium chloride and sodium nitrate, add a source of sulfate ions such as potassium sulfate. Barium chloride will react with potassium sulfate to form insoluble barium sulfate and soluble potassium chloride:

$$BaCl_2(aq) + K_2SO_4(aq) \longrightarrow BaSO_4(s) + 2KCl(aq)$$

On the other hand, sodium nitrate will not react with potassium sulfate.

53 12.69% S

55 10.4% P

57 0.324 g Ag

59 **(a)** $Ca(s) + 2HCl(aq) \longrightarrow CaCl_2(aq) + H_2(g)$
(b) $2K(s) + H_2SO_4(aq) \longrightarrow K_2SO_4(aq) + H_2(g)$
(c) $3Zn(s) + 2FeCl_3(aq) \longrightarrow 3ZnCl_2(aq) + 2Fe(s)$
(d) $Mg(s) + 2AgNO_3(aq) \longrightarrow Mg(NO_3)_2(aq) + 2Ag(s)$

61 **(a)** $Ca(s) + 2H^+(aq) \longrightarrow Ca^{2+}(aq) + H_2(g)$
(b) $2K(s) + 2H^+(aq) \longrightarrow 2K^+(aq) + H_2(g)$
(c) $3Zn(s) + 2Fe^{3+}(aq) \longrightarrow 3Zn^{2+}(aq) + 2Fe(s)$
(d) $Mg(s) + 2Ag^+(aq) \longrightarrow Mg^{2+}(aq) + 2Ag(s)$

63 **(a)** Cr = +6 **(b)** P = +5 **(c)** Fe = +3 **(d)** N = +5 **(e)** N = −3 **(f)** S = +2

65 **(a)** $14H^+(aq) + 2MnO_4^-(aq) + 10Cl^-(aq) \longrightarrow 2Mn^{2+}(aq) + 5Cl_2(g) + 8H_2O(l)$
(b) $5H^+(aq) + Cr_2O_7^{2-}(aq) + 3HNO_2(aq) \longrightarrow 2Cr^{3+}(aq) + 3NO_3^-(aq) + 4H_2O(l)$
(c) $2Fe^{2+}(aq) + 2H^+(aq) + H_2O_2(aq) \longrightarrow 2Fe^{3+}(aq) + 2H_2O(l)$
(d) $18H^+(aq) + 6MnO_4^-(aq) + SI^-(aq) \longrightarrow 6Mn^{2+}(aq) + SIO_3^-(aq) + 9H_2O(l)$
(e) $8CdS(s) + 16NO_3^-(aq) + 32H^+(aq) \longrightarrow 8Cd^{2+}(aq) + 16NO_2(g) + S_8(s) + 6H_2O(l)$
(f) $3H_4IO_6^-(aq) + 2Cr^{3+}(aq) \longrightarrow 3IO_3^-(aq) + Cr_2O_7^{2-}(aq) + 8H^+(aq) + 2H_2O(l)$

(g) $2HgCl_2(s) + Sn^{2+}(aq) + 4Cl^-(aq) \longrightarrow Hg_2Cl_2(s) + SnCl_6^{2-}(aq)$
(h) $I_2(s) + 2S_2O_3^{2-}(aq) \longrightarrow 2I^-(aq) + S_4O_6^{2-}(aq)$

67 $6H^+(aq) + 2MnO_4^-(aq) + 5H_2C_2O_4(aq) \longrightarrow 2Mn^{2+}(aq) + 10CO_2(g) + 8H_2O(l)$
Molarity $= 0.01904\ M\ MnO_4^-$

69 $I_2(s) + 2S_2O_3^{2-}(aq) \longrightarrow 2I^-(aq) + S_4O_6^{2-}(aq)$
0.531 g I_2

CHAPTER 5

Practice Problems

5.1 927 kJ
5.3 −2991 kJ
5.5 -2.17×10^3 J
5.7 -2.90×10^3 J
5.9 1.95×10^5 J/mol

Exercises

1 **(a)** An exothermic reaction releases heat to the surroundings. For example, the combustion of wood produces heat. The materials used in heat packs, sodium acetate trihydrate and water, produce heat when they are mixed.
(b) An endothermic reaction absorbs heat from the surroundings. The melting of ice is an endothermic process. Ice feels cold because it is absorbing the heat from its surroundings (your hand) to undergo the phase change to a liquid. The materials used in cold packs, ammonium cyanate and water, result in an endothermic process when they are mixed.

3 **(a)** 4.184 J **(b)** 4.184 kJ in 1 kcal (each unit increased by a factor of 1000)

5 One calorie is the heat/energy required to raise 1 g of water by 1°C.

7 The equation, $\Delta E = q + w$, states that the difference in internal energy, ΔE, of a system is given by the sum of the heat, q, and work, w, which occurred during the change. Work done on the system by the surroundings is negative in sign. Heat flow into the system from the surroundings is positive in sign.

9 The "state" of a system is defined by variables, called state functions, which make up the system. Examples of state functions include pressure, temperature, and energy. A change in state of state function does not depend on the path of the change. For example, consider the energy change for the burning of a gram of sugar. One can burn sugar in a calorimeter and find the energy released during the reaction. This same amount of energy is released when the body metabolizes ("burns") a gram of sugar using a biochemical pathway.

11 Pressure–volume work will be negligibly small for solids and liquids since no appreciable volume change occurs.

13 The measured heat of a reaction equals the enthalpy change for the reaction under constant pressure conditions.

15 Enthalpy of formation is by definition the enthalpy change resulting from the formation of a substance from its elements as found in nature. Enthalpy of reaction is more general and gives the enthalpy change of any chemical reaction.

17 The term ΔH_f° refers to the standard enthalpy change. Standard conditions are 1 atmosphere pressure and 298°K. ΔH_f refers to the enthalpy change under nonstandard conditions.

19 Energy is required to break bonds and the same amount of energy is released when that type of bond forms. If the bond strength is large, the molecule is more stable.

21 An ionic bond is an electrostatic attraction between ions of opposite charge. The terms "bond strength" or "bond energy" refer to covalent bonds. A more appropriate term for ionic compounds is "lattice energy."

23 **(a)** Heat capacity is the amount of heat required to raise the temperature of 1 mole of substance by one degree Celsius.
(b) Specific heat is the heat capacity of 1 gram of a substance.

25 Under constant pressure conditions, the enthalpy change equals the internal energy change because no work is being done (volume change is negligible).

27 A constant pressure calorimeter is open to the atmosphere; a bomb calormeter is sealed.

29 Major fuels currently available are fossil fuels (coal, oil, and natural gas) and nuclear energy.

31 3.59 cal; 3.59×10^{-3} kcal

33 2.1×10^5 J

35 +144 J absorbed and 6 J work done on the system; $-150\text{ J} = \Delta E_{surr}$. Thus $\Delta E_{sys} + \Delta E_{surr} = 0$ and the first law of thermodynamics is obeyed.

37 A negative ΔH indicates an exothermic reaction. The products are thermodynamically more stable than the reactants.

39	355 Cal	**51**	−137 kJ	**61**	11.1°C
41	−405 kJ	**53**	−315 kJ/mol	**63**	0.483 J/g°C
43	−24.2 kJ	**55**	−185 kJ	**65**	5.70×10^3 J/°C
45	−169.0 kJ	**57**	498 kJ/mol	**67**	2851 kJ
47	−3267 kJ	**59**	−391 kJ	**69**	7.13×10^4 J/mol
49	99.1 kJ/mol				

71 **(a)** 2.43×10^3 kJ **(b)** 60.9 ft^3 propane **(c)** 143 ft^3 methane

CHAPTER 6

Practice Problems

6.1 96.5 MHz

6.3 **(a)** $\Delta E = 1.64 \times 10^{-18}$ J; $\nu = 2.47 \times 10^{15}$ Hz; $\lambda = 1.22 \times 10^{-7}$ m, UV region
(b) $\Delta E = 3.03 \times 10^{-19}$ J; $\nu = 4.57 \times 10^{14}$ Hz; $\lambda = 6.57 \times 10^{-7}$ m, visible region
(c) $\Delta E = 1.06 \times 10^{-19}$ J; $\nu = 1.60 \times 10^{14}$ Hz; $\lambda = 1.88 \times 10^{-6}$ m, IR region
(d) $\Delta E = 4.90 \times 10^{-20}$ J; $\nu = 7.40 \times 10^{13}$ Hz; $\lambda = 4.06 \times 10^{-6}$ m, IR region

6.5 For a 4d orbital, $n = 4$ and $l = 2$

6.7

	$1s^2$	$2s^2$	$2p$			$3s^2$	
Mg atom:	↑↓	↑↓	↑↓	↑↓	↑↓	↑↓	0 unpaired e^-
	$1s^2$	$2s^2$		$2p^6$		$3s^2$	
Mg^{2+} ion:	↑↓	↑↓	↑↓	↑↓	↑↓		0 unpaired e^-
	$1s^2$	$2s^2$		$2p^6$			
N atom:	↑↓	↑↓	↑	↑	↑		3 unpaired e^-
	$1s^2$	$2s^2$		$2p^3$			
N^{3-} ion:	↑↓	↑↓	↑↓	↑↓	↑↓		0 unpaired e^-
	$1s^2$	$2s^2$		$2p^6$			
Cl atom:	↑↓	↑↓	↑↓	↑↓	↑		1 unpaired e^-
	$1s^2$	$2s^2$		$2p^5$			

6.9 **(a)** Sr; [Kr]$5s^2$ **(b)** Kr; [Ar]$3d^{10}4s^24p^6$
(c) Br; [Ar]$3d^{10}4s^24p^5$ **(d)** Rb; [Kr]$5s^1$
(e) Mn; [Ar]$3d^54s^2$ **(f)** Cu; [Ar]$3d^{10}4s^1$
(g) Zn; [Ar]$3d^{10}4s^2$ **(h)** Ba; [Xe]$6s^2$

Exercises

1 Evidence for the presence of electrons in atoms includes:
1. A stream of negative particles was produced in a cathode ray tube by William Crookes in 1879.
2. The mass to charge ratio of an electron was measured by J. J. Thomson in 1879.
3. The charge of an electron was measured by R. A. Millikan in 1909. Using the mass to charge ratio and the charge, Millikan calculated the mass of an electron to be 0.00055 amu.

3

Particle	Mass, amu	Charge
proton	1	+1
neutron	1	0
electron	~0	−1
alpha	4	2

5 *Curie*–Marie Curie and her husband, Pierre, studied Becquerel rays. They demonstrated that radioactivity is a property of certain atoms.
Rutherford–established that the structure of an atom consists of a dense nucleus surrounded by a diffuse cloud of electrons.
Bohr–established that electrons are found in "orbits" around the nucleus. Each orbit occurs at a discrete energy and distance from the nucleus.
DeBroglie–established that photons have both particle and wave properties.
Heisenberg–established that both the momentum and position of a subatomic particle cannot be known simultaneously.
Schrödinger–developed an equation to describe the standing waves associated with electron motions.
Pauli–established that no two electrons in an atom can have the same set of four quantum numbers.

7 Alpha rays are positive particles having a mass of 4 amu and a charge of +2. They are the least energetic particles emitted by a radioactive substance. Alpha particles are easily absorbed by a few centimeters of air.

Beta rays are negative particles with a mass of approximately 0 and a charge of −1. They are high-velocity electrons and can penetrate a sheet of aluminum several millimeters thick.

Gamma rays are neutral and are the most energetic of the three rays emitted by radioactive substances. Only lead metal will stop the penetration of gamma rays.

9 $E = h\nu$ where h is Planck's constant and ν is the frequency.

11 Quantum theory is the basis of modern atomic theory. Radiant energy is emitted or absorbed by matter in discrete units called quanta. The magnitude of a quantum is directly proportional to the frequency of light.

13 Quantum theory states that energy is emitted or absorbed by matter in small, discrete units called quanta. If electrical current is passed through, for instance, hydrogen gas, the hydrogen electrons are "excited" from lower energy levels to higher energy levels according to how many quanta of energy are absorbed. When these electrons drop back down to lower energy levels, they release the quanta of energy in the form of light at the wavelength equivalent to the energy difference, producing a line spectrum. Energy and wavelength are inversely proportional: $\Delta E = hc/\lambda$.

15 $\nu = 6.00 \times 10^{14}$ Hz; $\Delta E = 3.98 \times 10^{-19}$ J
$\lambda = c/\nu$; $\nu = c/\lambda$; $\Delta E = h\nu$

17 **(a)** Transitions from an orbit $n > 2$ to an orbit $n = 2$ produce lines in the visible region of the electromagnetic spectrum.
(b) Transitions from $n > 1$ to $n = 1$ produce lines in the ultraviolet region of the electromagnetic spectrum.

19 $\Delta E_{n_h \rightarrow n_l} = R_h\left[\dfrac{1}{n_h^2} - \dfrac{1}{n_l^2}\right]$

21 The electron configuration of boron is: $1s^2 2s^2 2p^1$. The valence shell contains 2 electrons in the $2s$ orbital and 1 electron in the $2p$ orbital. Possible numeric values for the quantum numbers are:

	n	l	m_l	m_s
$2s\ e^-$	2	0	0	$+\frac{1}{2}$
	2	0	0	$-\frac{1}{2}$
$2p\ e^-$	2	1	−1, 0 or +1	$+\frac{1}{2}$ or $-\frac{1}{2}$

23 $n = 1$: $1s$ only; $n = 2$: $2s$ and $2p$; $n = 3$: $3s$, $3p$ and $3d$; $n = 4$: $4s$, $4p$, $4d$, and $4f$

25 s subshell can hold $2e^-$; p subshell can hold $6e^-$; d subshell can hold $10e^-$; f subshell can hold $14e^-$

27 $s < p < d < f$

29

	Elements	Location of Highest-Energy Electrons
(a)	B to Ne	$4p$
(b)	Na to Mg	$3s$
(c)	K to Ca	$4s$
(d)	Ti to Zn	$3d$
(e)	Ga to Kr	$4p$

31 $4d$ electrons

33

Element	Electron Configuration		# unpaired e^-
Li	↑↓ ↑	$1s^2\ 2s^1$	1
Be	↑↓ ↑↓	$1s^2\ 2s^2$	0
B	↑↓ ↑↓ ↑ _ _	$1s^2\ 2s^2\ 2p^1$	1
C	↑↓ ↑↓ ↑ ↑ _	$1s^2\ 2s^2\ 2p^2$	2
N	↑↓ ↑↓ ↑ ↑ ↑	$1s^2\ 2s^2\ 2p^3$	3
O	↑↓ ↑↓ ↑↓ ↑ ↑	$1s^2\ 2s^2\ 2p^4$	2
F	↑↓ ↑↓ ↑↓ ↑↓ ↑	$1s^2\ 2s^2\ 2p^5$	1
Ne	↑↓ ↑↓ ↑↓ ↑↓ ↑↓	$1s^2\ 2s^1\ 2p^6$	0
K	↑	[Ar] $4s^1$	1
Ca	↑↓	[Ar] $4s^2$	0
Sc	↑ _ _ _ _ ↑↓	[Ar] $3d^1\ 4s^2$	1
Ti	↑ ↑ _ _ _ ↑↓	[Ar] $3d^2\ 4s^2$	2
V	↑ ↑ ↑ _ _ ↑↓	[Ar] $3d^3\ 4s^2$	3
Cr	↑ ↑ ↑ ↑ ↑ ↑	[Ar] $3d^5\ 4s^1$	6
Mn	↑ ↑ ↑ ↑ ↑ ↑↓	[Ar] $3d^5\ 4s^2$	5
Fe	↑↓ ↑ ↑ ↑ ↑ ↑↓	[Ar] $3d^6\ 4s^2$	4
Co	↑↓ ↑↓ ↑ ↑ ↑ ↑↓	[Ar] $3d^7\ 4s^2$	3
Ni	↑↓ ↑↓ ↑↓ ↑ ↑ ↑↓	[Ar] $3d^8\ 4s^2$	2
Cu	↑↓ ↑↓ ↑↓ ↑↓ ↑↓ ↑	[Ar] $3d^{10}\ 4s^1$	1
Zn	↑↓ ↑↓ ↑↓ ↑↓ ↑↓ ↑↓	[Ar] $3d^{10}\ 4s^2$	0
Ga	↑↓ ↑ _ _	[Ar]$3d^{10}\ 4s^2\ 4p^1$	1
Ge	↑↓ ↑ ↑ _	[Ar]$3d^{10}\ 4s^2\ 4p^2$	2
As	↑↓ ↑ ↑ ↑	[Ar]$3d^{10}\ 4s^2\ 4p^3$	3
Se	↑↓ ↑↓ ↑ ↑	[Ar]$3d^{10}\ 4s^2\ 4p^4$	2
Br	↑↓ ↑↓ ↑↓ ↑	[Ar]$3d^{10}\ 4s^2\ 4p^5$	1
Kr	↑↓ ↑↓ ↑↓ ↑↓	[Ar]$3d^{10}\ 4s^2\ 4p^6$	0

35 **(a)** Ionization energy decreases down a group.
(b) Ionization energy increases from left to right within a period.

37 **(a)** orbital–a region in space where the probability of finding an electron is the greatest.
(b) *s* orbital–a spherical region in space with $l = 0$.

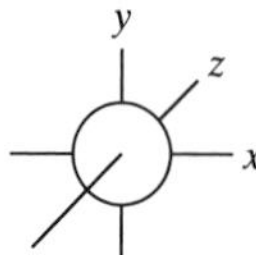

(c) p_y orbital–a spatial region along the *y*-axis with $l = 1$.

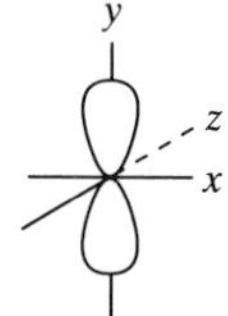

(d) d_{xy} orbital–a spatial region with $l = 2$.

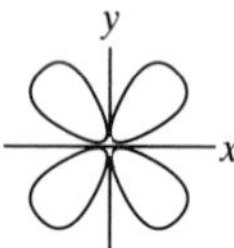

(e) d_{z^2} orbital–a spatial region with $l = 2$.

(f) $d_{x^2-y^2}$ orbital–a spatial region with $l = 2$.

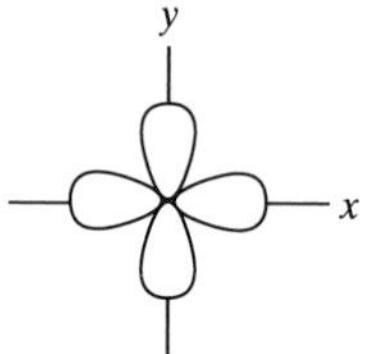

(g) principal energy level in an atom–the energy level given by the quantum number, *n*.
(h) For visible spectra, electron transitions from $n \longrightarrow 2$; for ultraviolet spctra, electron transitions from $n \longrightarrow 1$; for infrared spectra, electron transitions from $n \longrightarrow 3$.
(i) Pauli exclusion principle–any two electrons may not have the same set of quantum numbers.
(j) Hund's rule–half-filled orbitals with parallel spin and/or completely filled orbitals are especially stable.
(k) atomic shell–a region located at the discrete distance from the nucleus having a discrete energy.
(l) atomic subshell–a region within a shell occupied by orbitals.
(m) atomic radius–one-half the distance between nuclei of two identical atoms bonded by a single covalent bond.
(n) unpaired electron–a single electron occupying an orbital.
(o) ionization energy–the energy required to remove an electron.
(p) electron affinity–the energy released upon the gain of an electron.

39 The wavelength of x-ray radiation is much shorter than the wavelength of blue light.

41 Infrared radiation has longer wavelengths than x-ray, blue, or red radiation and thus is on the lower energy side of the electromagnetic spectrum. Energy and wavelength are inversely proportional.

43 The line spectrum of an element is characteristic of that element. Line spectra have been used to identify elements found in our sun and other stars.

45 If electrons were to be found at arbitrary locations around the nucleus, a gas discharge tube would produce a continuous spectrum. The fact that a line spectrum is produced indicates the presence of discrete energy levels.

47 Light is a form of energy and is characterized by its frequency and wavelength.

49 The difference in energy between quantum levels decreases as the number of the quantum level increases.

51 5.44×10^{-19} J

53 $\nu = 7.32 \times 10^{14}$ Hz; $\Delta E = 4.85 \times 10^{19}$ J

55

Element	n	l	m_l	m_s	subshell
Ca	4	0	0	$+\frac{1}{2}$ or $-\frac{1}{2}$	*s*
Zn	3	2	−2, −1, 0, +1 or +2	$+\frac{1}{2}$ or $-\frac{1}{2}$	*d*
Ga	4	1	−1, 0 or +1	$+\frac{1}{2}$ or $-\frac{1}{2}$	*p*

57 For the second energy level, $n = 2$; l may have values of 0 or 1 ($l \leq n - 1$). Since a *d* orbital has $l = 2$, *n* must be 3 or greater.

59

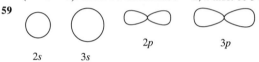

2*s* 3*s* 2*p* 3*p*

61

	Element	n	l	m_l
(a)	Li	2	0	0
	Na	3	0	0
	K	4	0	0
	Rb	5	0	0
	Cs	6	0	0
(b)	Be	2	0	0
	Mg	3	0	0
	Ca	4	0	0
	Sr	5	0	0
	Ba	6	0	0
(c)	F	2	1	−1, 0 or +1
	Cl	3	1	−1, 0 or +1
	Br	4	1	−1, 0 or +1
	I	5	1	−1, 0 or +1
	As	6	1	−1, 0 or +1

63 $8e^-$; $2e^-$ in 2*s* and $6e^-$ in 2*p*.

65

Element	Electronic Structure
H	↑ $1s^1$
He	⇅ $1s^2$
Li	⇅ ↑ $1s^2\ 2s^1$
Be	⇅ ⇅ $1s^2\ 2s^2$
B	⇅ ⇅ ↑ _ _ $1s^2\ 2s^2\ 2p^1$
C	⇅ ⇅ ↑ ↑ _ $1s^2\ 2s^2\ 2p^2$
N	⇅ ⇅ ↑ ↑ ↑ $1s^2\ 2s^2\ 2p^3$
O	⇅ ⇅ ⇅ ↑ ↑ $1s^2\ 2s^2\ 2p^4$

Element	Electronic Structure
F	⇅ ⇅ ⇅ ⇅ ↑
	$1s^2$ $2s^2$ $2p^5$
Ne	⇅ ⇅ ⇅ ⇅ ⇅
	$1s^2$ $2s^2$ $2p^6$
Na	⇅ ⇅ ⇅ ⇅ ⇅ ↑
	$1s^2$ $2s^2$ $2p^6$ $3s^1$
Mg	⇅ ⇅ ⇅ ⇅ ⇅ ⇅
	$1s^2$ $2s^2$ $2p^6$ $3s^2$
Al	⇅ ⇅ ⇅ ⇅ ⇅ ⇅ ↑ __ __
	$1s^2$ $2s^2$ $2p^6$ $3s^2$ $3p^1$
Si	⇅ ⇅ ⇅ ⇅ ⇅ ⇅ ↑ ↑ __
	$1s^2$ $2s^2$ $2p^6$ $3s^2$ $3p^2$
P	⇅ ⇅ ⇅ ⇅ ⇅ ⇅ ↑ ↑ ↑
	$1s^2$ $2s^2$ $2p^6$ $3s^2$ $3p^3$
S	⇅ ⇅ ⇅ ⇅ ⇅ ⇅ ⇅ ↑ ↑
	$1s^2$ $2s^2$ $2p^6$ $3s^2$ $3p^4$
Cl	⇅ ⇅ ⇅ ⇅ ⇅ ⇅ ⇅ ⇅ ↑
	$1s^2$ $2s^2$ $2p^6$ $3s^2$ $3p^5$
Ar	⇅ ⇅ ⇅ ⇅ ⇅ ⇅ ⇅ ⇅ ⇅
	$1s^2$ $2s^2$ $2p^6$ $3s^2$ $3p^6$

67

Element	# valence e^-	# se^-	# pe^-	# de^-	# fe^-
N	5	2	3	0	0
Mg	2	2	0	0	0
S	6	2	4	0	0
Sc	3	2	0	1	0
I	7	2	5	0	0

69

	Element	# se^-	# pe^-	# de^-	# fe^-
(a)	Br	2	5	0	0
(b)	Al	2	1	0	0
(c)	Ag	2	0	10	0

71 Sulfide ion and scandium(III) ion

73

Species	# unpaired e^-	Species	# unpaired e^-
(a) Al	1	**(h)** Sc	1
(b) Al^{3+}	0	**(i)** Cu	1
(c) Na	1	**(j)** Mn	5
(d) Na^+	0	**(k)** As	3
(e) O	2	**(l)** Sn	2
(f) O^{2-}	0	**(m)** Zn^{2+}	0
(g) F^-	2		

75 Hund's rule accounts for each of these deviations. According to Hund's rule, half-filled orbitals with parallel spin and/or completely filled orbitals are more stable. For boron and aluminum, the first ionization involves the removal of a single *p* electron. Removal of this electron leaves a filled *s* orbital and empty *p* orbitals. Since the loss of the *p* electron results in a more stable electronic state, it is favored and should take less energy. Similarly, the first ionization energy for sulfur and selenium involves the removal of the fourth electron in *p* orbitals. Removal of this electron leaves a filled *s* orbital and half-filled *p* orbitals. Once again, a more stable electronic state is achieved and thus should take less energy.

77 $P^{3-} > Cl^- > Ca^{2+} > Sc^{3+}$

79 Fe–largest; Fe^{3+}–smallest
For Fe^{3+}, the total negative charge has decreased while the effective nuclear charge has remained constant. The remaining electrons are pulled in closer to the nucleus.

81 K < Li < I < F

CHAPTER 7

Practice Problems

7.1 When an element forms an ion, it does so by losing or gaining the number of electrons needed to have a noble gas electron configuration or a filled subshell. For thallium, the elemental electron configuration is: [Xe]: $5d^{10}6s^26p'$. By losing one electron to form Tl^+, the 6*p* orbitals are empty and the 6*s* orbitals are filled, resulting in a stable ion. By losing three electrons, the 6*s* and the 6*p* orbitals will be empty and a stable Tl^{3+} ion results.

7.3 Br—Cl is the weakest bond.

7.5 N_2 :N≡N: N—N bond order = 3
C_2H_4 H—C(H)=C(H)—H C—C bond order = 2
C_2H_2 H—C≡C—H C—C bond order = 3

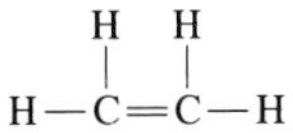

7.7 :O=S(—O:)—O: ⟷ O=S(—O:)—O: ⟷ :O—S(—O:)=O
Bond order for S—O is 4/3.

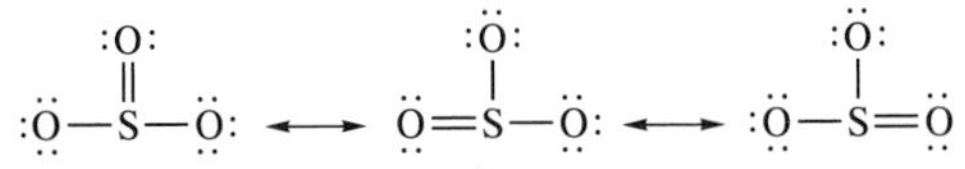

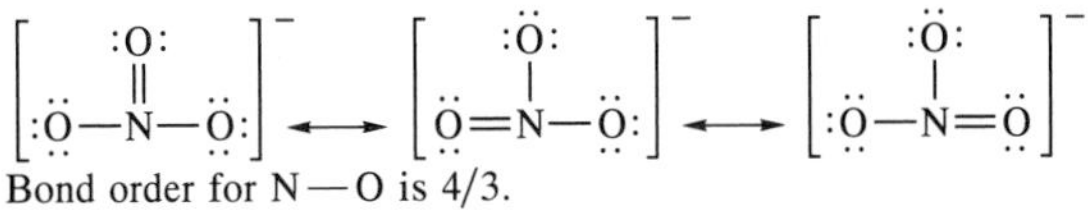

$[$:O—N(=O:)—O:$]^-$ ⟷ $[$O=N(—O:)—O:$]^-$ ⟷ $[$:O—N(—O:)=O$]^-$
Bond order for N—O is 4/3.

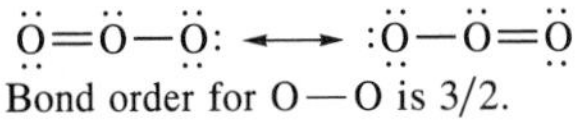

O=O—O: ⟷ :O—O=O
Bond order for O—O is 3/2.

7.9 **(a)** linear, nonpolar **(b)** tetrahedral, polar **(c)** trigonal bipyramidal, nonpolar **(d)** for each C, tetrahedral, nonpolar

7.11

	Electron Pair Structure	Molecular Structure	Polarity
(a)	tetrahedral	trigonal pyramidal	polar
(b)	linear	linear	nonpolar
(c)	tetrahedral	bent	polar

Exercises

1 Ionic compounds form from the reaction of metals with nonmetals. They are crystalline and have high melting points. An ionic solid is held together by the electrostatic attraction between ions of opposite charge (ionic bonds). Molecular compounds tend to be gases, liquids, and low melting solids. Covalent bonds are present in molecular solids.

3 "Completion of an octet" generally refers to the tendency of a covalently bonded atom to acquire eight electrons by the presence of nonbonding electrons and by the sharing of electrons with neighboring atoms.

5 Two elements forming a covalent bond complete their octets by sharing electrons.

7 A polar bond results from unequal sharing of electrons due to electronegativity differences in the atoms forming bonds. Examples of polar diatomic molecules are CO, HCl, NO, BrCl.

9 Dipole moment measures the magnitude and direction of polarity.

11 **(a)** 4.0 **(b)** 3.5 **(c)** 0.7–1.0 **(d)** 2.1

13 An electronegativity difference relates the ionic character of a bond. The greater the electronegativity difference, the greater the ionic character of that bond. The greater the ionic character, the greater the bond energy.

15

Lewis Dot Formula	Lewis Structure
:Cl: above, :Cl : C : Cl:, :Cl: below	:Cl: above, :Cl—C—Cl:, :Cl: below
:O: above, H : C : H (with :: to O)	:O: above, H—C(=O)—H
H : C ::: N:	H—C≡N:

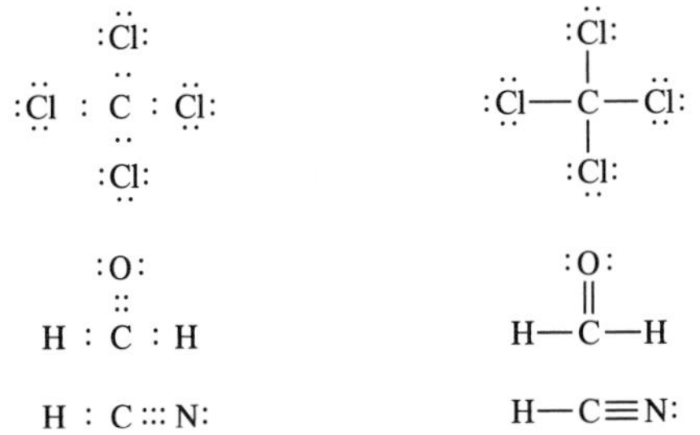

17 Valence shell electron pair repulsion (VSEPR) theory is used to predict the geometry of a molecule based on the shape that minimizes the electron pair repulsions around a central atom.

19 **(a)** linear **(b)** trigonal planar **(c)** tetrahedral **(d)** trigonal bipyramidal **(e)** octahedral

21 Molecular structures are experimentally determined.

23 VSEPR theory predicts the electronic pair structure by assigning the geometry that minimizes electronic repulsions of all regions of electron density around the central atom.

25 Polar triatomic molecule: H—C≡N:
Nonpolar triatomic molecule: O=C=O

27 CCl_4, CO_2, BF_3

29 Greatest covalent character: N—O
This bond is covalent because the electronegativity difference is less than 1.5.

31 Greatest ionic character: Rb—O
This bond is ionic because the electronegativity difference is greater than 1.5.

33 NO, BrCl, and HI

35 HF

37 Calculated Br—F bond energy = 176 kJ mol^{-1}; 58 kJ "extra" energy

39 Fr—F

41
(a) CCl_4: C bonded to four :Cl:
(b) :Cl—N—Cl: with N bonded to :Cl:
(c) :Br—Cl:
(d) :Cl—S—Cl:
(e) H—N—N—H, each N bonded to H
(f) $[BCl_4]^-$: B bonded to four :Cl:
(g) $[H—C(=O)—O:]^-$
(h) H—C(=O)—H
(i) H(H)C=C(H)H
(j) H—C≡C—H
(k) H—C(H)(H)—O—H
(l) H—C(H)(H)—N(H)—H
(m) O=N—O:
(n) $[O=N—O:]^-$
(o) H—O—O—H
(p) SF_6: S bonded to six :F:
(q) $[NH_4]^+$: N bonded to four H
(r) :Cl—S(—O:)—Cl:
(s) O=N—O—N=O
(t) $[PCl_4]^+$: P bonded to four :Cl:
(u) $[PCl_6]^-$: P bonded to six :Cl:
(v) $[O=N=O]^+$
(w) H—N(H)—O—H

43 $[H—C(=O)—O:]^- \longleftrightarrow [H—C(—O:)=O]^-$

$[:O—S=O] \longleftrightarrow [O=S—O:]$

45

	Lewis Structure	Electron Pair Structure	Molecular Structure
(a)	BF_3	trigonal planar	trigonal planar
(b)	$SiCl_4$	tetrahedral	tetrahedral
(c)	$[NH_4]^+$	tetrahedral	tetrahedral
(d)	$[PH_4]^+$	tetrahedral	tetrahedral
(e)	CH_3—CH_2—CH_3	tetrahedral around each C	tetrahedral around each C
(f)	H—C≡C—H	linear around each C	linear around each C
(g)	O=C=O	linear	linear
(h)	$H_2C=O$	trigonal planar	trigonal planar
(i)	$[PF_6]^-$	octahedral	octahedral
(j)	SF_6	octahedral	octahedral
(k)	IF_5	octahedral	square pyramidal
(l)	$[CO_3]^{2-}$	trigonal planar	trigonal planar
(m)	NO_3	trigonal planar	trigonal planar
(n)	$[O—N=O]^-$	trigonal planar	bent

(o)	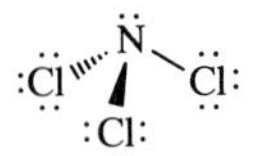 :Cl—N(—Cl:)—Cl:	tetrahedral	trigonal pyramidal
(p)	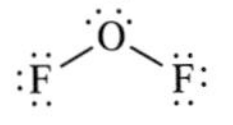 :F—O—F:	tetrahedral	bent

47

	Lewis Structure	Electron Pair Structure	Molecular Structure
(a)	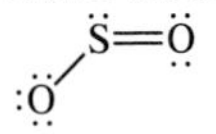 :O—S=O	trigonal planar	bent
(b)	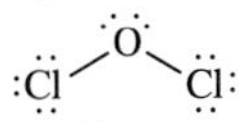 :Cl—O—Cl:	tetrahedral	bent
(c)	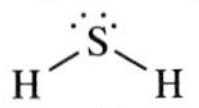 H—S—H	tetrahedral	bent
(d)	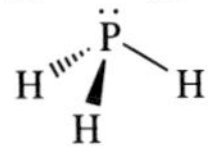 PH_3 (H—P(—H)—H)	tetrahedral	trigonal pyramidal
(e)	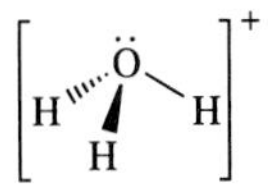 $[H_3O]^+$	tetrahedral	trigonal pyramidal
(f)	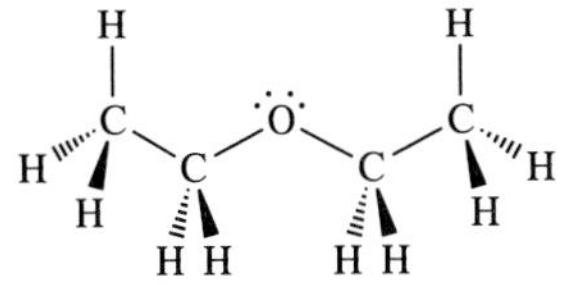 H_3C—CH_2—O—CH_2—CH_3	tetrahedral around each C; tetrahedral around O	tetrahedral around each C; bent around O
(g)	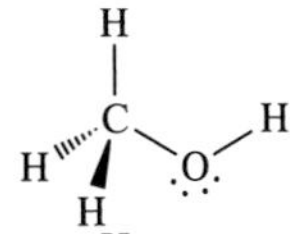 H_3C—O—H	tetrahedral around C; tetrahedral around O	tetrahedral around C; bent around O
(h)	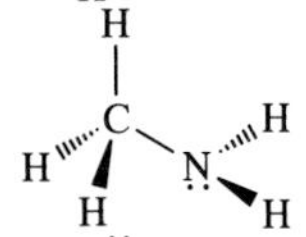 H_3C—NH_2	tetrahedral around C; tetrahedral around N	tetrahedral around C; trigonal pyramidal around N
(i)	 :O—S(—O:)—O:	tetrahedral	trigonal pyramidal
(j)	 SF_4	trigonal bipyramidal	seesaw
(k)	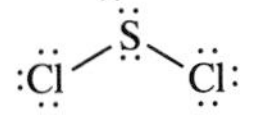 :Cl—S—Cl:	tetrahedral	bent
(l)	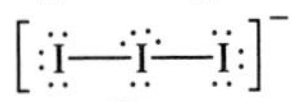 [:I—I—I:]$^-$	trigonal bipyramidal	linear
(m)	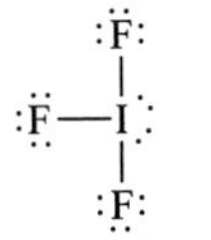 IF_3	trigonal bipyramidal	T-shaped
(n)	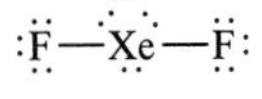 :F—Xe—F:	trigonal bipyramidal	linear
(o)	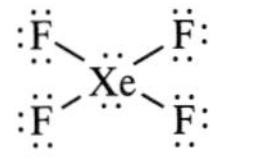 XeF_4	octahedral	square planar

CHAPTER 8

Practice Problems

8.1 A sp^2 hybrid orbital of Al overlaps with a $4p$ orbital of Br.

8.3 Each Si atom in disilane is tetrahedral and sp^3 hybridized. A sp^3 hybrid orbital of Si overlaps with a $1s$ orbital of H.

8.5 H—C(H)(H)—C≡C—C(H)(H)—H (carbons labeled a, b, b, a)

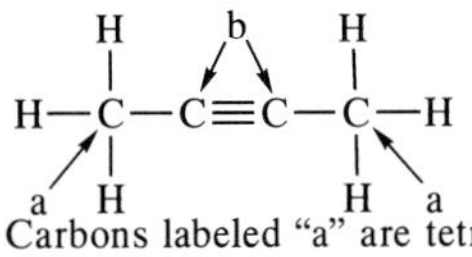

Carbons labeled "a" are tetrahedral and sp^3 hybridized. Carbons labeled "b" are linear and sp hybridized.

8.7 **(a)** electron pair structure: trigonal bipyramidal; molecular structure: T-shaped; hybridization: sp^3d **(b)** electron pair structure: octahedral; molecular structure: square pyramidal; hybridization: sp^3d^2 **(c)** electron pair structure: trigonal bipyramidal; molecular structure: linear; hybridization: sp^3d **(d)** electron pair structure: octahedral; molecular structure: square planar; hybridization: sp^3d^2

8.9

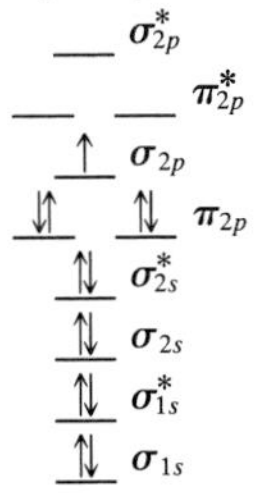

σ^*_{2p} ( )
π^*_{2p} ( ) ( )
σ_{2p} (↑) Bond order = 2½
π_{2p} (⇅) (⇅) N_2^+ is paramagnetic.
σ^*_{2s} (⇅)
σ_{2s} (⇅)
σ^*_{1s} (⇅)
σ_{1s} (⇅)

Exercises

1 Valence bond theory assumes atomic orbitals mix to form hybrid orbitals. Molecular orbital theory assumes atomic orbitals overlap to form molecular orbitals.

3 The electron pair geometry gives the shape of the molecule based on the number of regions of electron density around a central atom.

Electronic Geometry	# of regions of e^- density	Hybridization
linear	2	sp
trigonal planar	3	sp^2
tetrahedral	4	sp^3
trigonal bipyramidal	5	sp^3d
octahedral	6	sp^3d^2

5 A saturated hydrocarbon contains only C—C and C—H bonds.

7 A sigma bond is a head-on overlap of orbitals.

9 A double bond consists of 1σ and 1π bond. A triple bond consists of 1σ and 2π bonds.

11 N: $2s$ (⇅) $2p$ (↓)(↓)(↓) 3 bonding sites

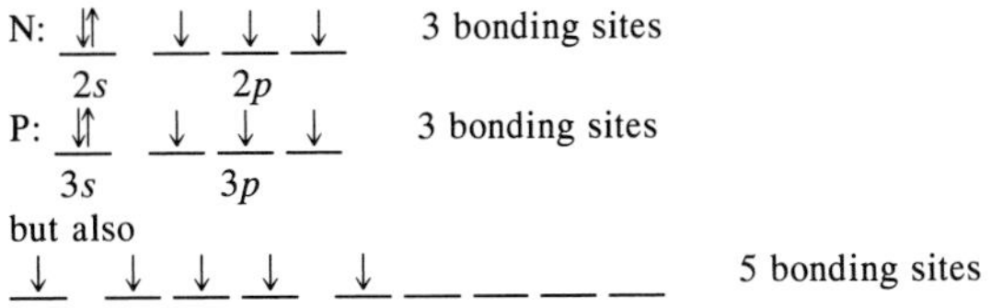

P: $3s$ (⇅) $3p$ (↓)(↓)(↓) 3 bonding sites

but also

$3s$ (↓) $3p$ (↓)(↓)(↓) $3d$ (↓)( )( )( )( ) 5 bonding sites

13 S: $3s$ (⇅) $3p$ (⇅)(↓)(↓) 2 bonding sites

$3s$ (⇅) $3p$ (↓)(↓)(↓) $3d$ (↓)( )( )( )( ) 4 bonding sites

$3s$ (↓) $3p$ (↓)(↓)(↓) $3d$ (↓)(↓)( )( )( ) 6 bonding sites

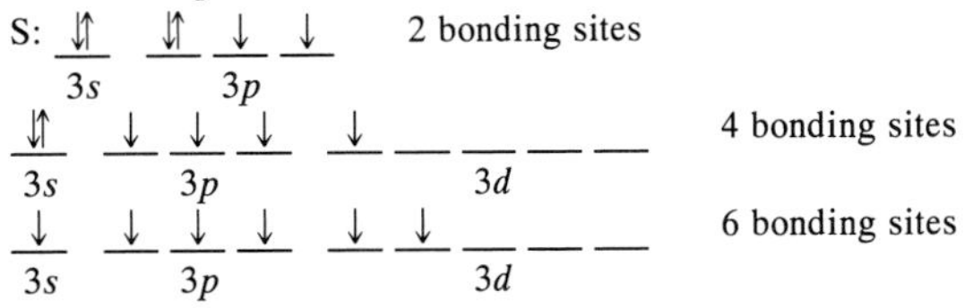

15 Bond order = (# e^- in bonding MOs − #e^- in antibonding MOs)/2

17 Three atomic orbitals; one s orbital and two p orbitals

19 **(a)** tetrahedral, sp^3 **(b)** tetrahedral, sp^3 **(c)** trigonal planar, sp^2 **(d)** linear, sp

21 **(a)** sp^3 **(b)** sp^3 **(c)** sp^3 **(d)** C, sp^3; O, sp^3 **(e)** C, sp^3; N, sp^3 **(f)** sp^3 **(g)** sp^2 **(h)** left to right: sp^3, sp^2, sp^2 **(i)** left to right: sp^3, sp, sp **(j)** both N, sp^3 **(k)** sp^2 **(l)** sp^3

23 A sigma bond is stronger than a pi bond.

25

	N_2	O_2	F_2
σ^*_{2p}	—	—	—
π^*_{2p}	— —	↓ ↓	⇅ ⇅
π_{2p} / σ_{2p}	⇅ σ_{2p}; ⇅ ⇅ π_{2p}	⇅ ⇅ π_{2p}; ⇅ σ_{2p}	⇅ ⇅ π_{2p}; ⇅ σ_{2p}
σ^*_{2s}	⇅	⇅	⇅
σ_{2s}	⇅	⇅	⇅
σ^*_{1s}	⇅	⇅	⇅
σ_{1s}	⇅	⇅	⇅

E ↑

The σ_{2p} orbital energy is lower in O_2 and F_2 than in N_2.

27 **(a)** $KK(\sigma_{2s})^2(\sigma^*_{2s})^2(\pi_{2p})^4$
(b) $KK(\sigma_{2s})^2(\sigma^*_{2s})^2(\sigma_{2p})^2(\pi_{2p})^4(\pi^*_{2p})^2$
(c) $KK(\sigma_{2s})^2(\sigma^*_{2s})^2(\sigma_{2p})^2(\pi_{2p})^4(\pi^*_{2p})^3$
(d) $KK(\sigma_{2s})^2(\sigma^*_{2s})^2(\sigma_{2p})^2(\pi_{2p})^4(\pi^*_{2p})^4$
(e) $KK(\sigma_{2s})^2(\sigma^*_{2s})^2(\sigma_{2p})^2(\pi_{2p})^4(\pi^*_{2p})^1$
(f) $KK(\sigma_{2s})^2(\sigma^*_{2s})^2(\sigma_{2p})^2(\pi_{2p})^4$
(g) $KK(\sigma_{2s})^2(\sigma^*_{2s})^2(\sigma_{2p})^2(\pi_{2p})^3$
(h) $KK(\sigma_{2s})^2(\sigma^*_{2s})^2(\sigma_{2p})^2(\pi_{2p})^4(\pi^*_{2p})^1$
(i) $KK(\sigma_{2s})^2(\sigma^*_{2s})^2(\sigma_{2p})^2(\pi_{2p})^4$
(j) $KK(\sigma_{2s})^2(\sigma^*_{2s})^2(\sigma_{2p})^2(\pi_{2p})^4(\pi^*_{2p})^3$

29 **(a)** diamagnetic **(b)** paramagnetic **(c)** paramagnetic
(d) diamagnetic **(e)** paramagnetic **(f)** diamagnetic
(g) paramagnetic **(h)** paramagnetic **(i)** diamagnetic
(j) paramagnetic

CHAPTER 9

Practice Problems

9.1 The electronic configuration of S is $3s^23p^4$; that of P is $3s^23p^3$. By using their unpaired electrons in their ground states, S can form two bonds, and P can form 3. Through promotion of one electron into a *d* orbital, S can form 4 bonds and P, 5. Through promotion of a second electron, S can form 6 bonds. The number of half-filled orbitals must be even for S but odd for P.

9.3 O, O=N—N=O, O (two O atoms on each N)

Structure is trigonal planar and hybridization is sp^2 at each N atom.

9.5 The Lewis structure is a resonance hybrid:

O=O—O ⟷ O—O=O

Each bond corresponds to 1 σ bond and ½ π bond, for a total bond order of 1½.

9.7 $S_8(s) + 8O_2(g) \longrightarrow 8SO_2(g)$
$SO_2(g) + H_2O(l) \longrightarrow H_2SO_3(aq)$

Exercises

1

Nonmetal (N) Group →	V	VI	VII
Metal (M) Group I	M_3N	M_2N	MN
II	M_3N_2	MN	MN_2
III	MN	M_2N_3	MN_3
IV	M_3N_4 or M_3N_2	MN_2 or MN	MN_4 (M^{4+}) or MN_2 (M^{2+})
V	M_3N_5 or MN	M_2N_5 or M_2N_3	MN_5 (M^{5+}) or MN_3 (M^{3+})

3 **(a)** Not exactly. In addition to MX_n (n = group #), the formulae MX_{n-2} are found for some combinations in groups III and IV.
(b) Not exactly. In addition to MX_n ($n = 8$ − group #), you can have MX_{n+2}, MX_{n+4}, or MX_{n+6} for heavier elements in groups V, VI, and VII with certain halides.

5

	EPG	Hybridization	Structure
PF_3	tetrahedral	sp_3	trigonal pyramid
PF_5	trigonal bipyramid	dsp^3	trigonal bipyramid
SF_2	tetrahedral	sp^3	bent
SF_4	trigonal bipyramid	dsp^3	seesaw
SF_6, S_2F_{10}	octahedral	d^2sp^3	octahedral

7 XeF_2: xenon difluoride or xenon(II) fluoride; XeF_4: xenon tetrafluoride or xenon(IV) fluoride; XeF_6: xenon hexafluoride or xenon(VI) fluoride

9 A molecule consisting of two or more identical subunits

11 F is a much smaller atom than I. There isn't enough space around a S atom for 6 I atoms.

13 B: B_2O_3 boron(III) oxide or diboron trioxide; C: CO carbon monoxide, CO_2 carbon dioxide; Si: SiO_2 silicon dioxide; N: N_2O dinitrogen monoxide or nitrous oxide, NO nitrogen monoxide or nitric oxide, N_2O_3 dinitrogen trioxide, NO_2 nitrogen dioxide, N_2O_4 dinitrogen tetroxide, N_2O_5 dinitrogen pentoxide; P: P_4O_6 tetraphosphorus hexoxide; P_4O_{10} tetraphosphorus decoxide; S: SO_2 sulfur dioxide, SO_3 sulfur trioxide.

15 B: no simple "sodium borate"; C: $NaHCO_3$ sodium hydrogen carbonate, Na_2CO_3 sodium carbonate; Si: $NaSiO_3$ sodium silicate; N: $NaNO_2$ sodium nitrite, $NaNO_3$ sodium nitrate; P: $NaH(HPO_3)$ sodium dihydrogen phosphite, $Na_2(HPO_3)$ sodium hydrogen phosphite, NaH_2PO_4 sodium dihydrogen phosphate, Na_2HPO_4 sodium hydrogen phosphate, Na_3PO_4 sodium phosphate; S: $NaHSO_3$ sodium hydrogen sulfite, Na_2SO_3 sodium sulfite, $NaHSO_4$ sodium hydrogen sulfate, Na_2SO_4 sodium sulfate; Cl: NaOCl sodium hypochlorite, $NaClO_2$ sodium chlorite, $NaClO_3$ sodium chlorate, $NaClO_4$ sodium perchlorate.

17 The FXeF bond angle in XeF_6 will be less than 90°, since it has *seven* electron pairs and will be a "crowded" octahedral geometry.

19 I is a larger atom than Br, so 7 F atoms can fit around the I but not the Br. Since Cl is even smaller than Br, ClF_7 would not be expected to be stable.

21

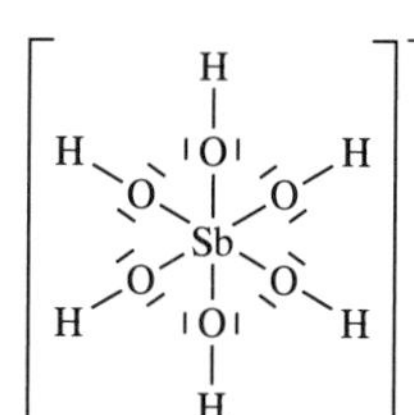

Sb is a larger atom than As, so it can accommodate 6 surrounding atoms as opposed to 4 for As.

23 **(a)** $PCl_3(l) + Cl_2(g) \xrightarrow{\Delta} PCl_5(s)$
(b) $S_8(s) + 4Cl_2(g) \xrightarrow{\Delta} 4S_2Cl_2(l)$
(c) $S_8(s) + 8O_2(g) \xrightarrow{\Delta} 8SO_2(g)$
(d) $Br_2(l) + 3F_2(g) \xrightarrow{\Delta} 2BrF_3(l)$
(e) $CO_2(g) + H_2O(l) \xrightarrow{\Delta} H_2CO_3(aq)$
(f) $CaO(s) + H_2O(l) \longrightarrow Ca(OH)_2(s)$
(g) $2B(OH)_3 \xrightarrow{\Delta} B_2O_3(s) + 3H_2O(l)$
(h) $CaCO_3(s) + CO_2 + H_2O(l) \longrightarrow Ca(HCO_3)_2(aq)$
(i) $Ca(HCO_3)_2(aq) \longrightarrow CaCO_3(s) + CO_2(g) + H_2O(l)$
(j) $SiO_2(s) + 4HF(aq) \xrightarrow{\Delta} SiF_4(g) + 2H_2O(l)$
(k) $SiO_2(s) + 2NaOH(aq) \longrightarrow Na_2SiO_3(aq) + H_2O(l)$
(l) $NH_4NO_3(s) \xrightarrow{\Delta} N_2O(g) + 2H_2O(l)$
(m) $2NO(g) + O_2(g) \xrightarrow{\Delta} 2NO_2(g)$
(n) $N_2O_3(s) + H_2O(l) \longrightarrow 2HNO_2(aq)$
(o) $2HNO_2(aq) + Ca(OH)_2(aq) \longrightarrow Ca(NO_2)_2(aq) + 2H_2O(l)$
(p) $P_4O_6(s) + 6H_2O(l) \longrightarrow 4H_3PO_3(aq)$
(q) $H_3PO_3(aq) + 2NaOH(aq) \longrightarrow Na_2(HPO_3)(aq) + 2H_2O(l)$
(r) $H_3PO_4(aq) + 3NaOH(aq) \longrightarrow Na_3PO_4(aq) + 3H_2O(l)$
(s) $H_2SO_4(l) \longrightarrow H_2O(g) + SO_3(g)$
(t) $(CaSO_4)_2H_2O(s) + 3H_2O(l) \longrightarrow 2CaSO_4(H_2O)_2(s)$
(u) $PCl_3(l) + 3H_2O(l) \longrightarrow H_3PO_3(aq) + 3HCl(g)$

25 −1: HCl, NaCl, $AlCl_3$, etc.; +1: HOCl; +3: $NaClO_2$; +4: ClO_2; +5: $KClO_3$; +6: Cl_2O_6; +7: $HClO_4$

27 S has 6 valence electrons: $3s^23p^4$ (2 unpaired electrons). By promotion of one or two electrons and use of the 3*d* orbitals, it can have 4 or 6 unpaired electrons. It can therefore form 2, 4, or 6 bonds. It cannot form 1, 3, or 5 bonds without having an unpaired electron "left over."

29 N has no available *d* orbitals ($n = 2$ level), so can't form more than 4 bonds. P has *d* orbitals available ($n = 3$ level), so can expand its octet.

31 Na_3CPO_5

CHAPTER 10

Practice Problems

10.1 1.25 L
10.3 17.8 L
10.5 66.7% $CaCO_3$
10.7 **(a)** 0.714 g L^{-1} **(b)** 0.643 g L^{-1}
10.9 94.8 g mol^{-1}
10.11 16 g mol^{-1}; CH_4

Exercises

1 gas pressure: Force exerted divided by area over which force is exerted.
torr: Pressure needed to hold up a column of mercury 1 mm high.
1 atm pressure: Average pressure exerted by the atmosphere at sea level; pressure needed to support a column of mercury 760 mm high.
atmospheric pressure: Actual pressure exerted by the atmosphere at a given point.
barometer: Pressure-measuring device in which the height of liquid in a single arm is measured.
manometer: Pressure-measuring device in which the heights of liquids in two arms are compared.
Boyle's law: Pressure and volume of a gas are inversely proportional if the amount of gas and the temperature are kept constant.
Charles' law: Volume and temperature of a gas are directly proportional if the amount of gas and the pressure are kept constant.
ideal gas: A substance that perfectly obeys all the gas laws under all circumstances.
Gay-Lussac's law: The volumes of gaseous reactants and products at constant temperature and pressure can be expressed in whole number ratios.
Avogadro's law: Equal volumes of all gases at the same temperature and pressure contain equal numbers of molecules.
Dalton's law of partial pressures: The total pressure of a mixture of gases equals the sum of the partial pressures exerted by the individual gases in the mixture.
Graham's law of effusion: The rate of effusion of a gas is inversely proportional to the square root of its density (or molar mass).
effusion: Passage of gas spontaneously from a region of higher pressure to a region of lower pressure.
diffusion: Passage of one gas through another to cause mixing.
equation of state: A mathematical equation relating pressure, temperature, and volume for a sample of matter.

3 The water column would be higher by a factor of 13.6 (the ratio of their densities).

5 Absolute temperature is the temperature measured on a scale on which the lowest temperature (i.e., zero degrees) corresponds to absolute zero, the lowest possible temperature, where all molecular motions cease.

7 $P_t = P_1 + P_2 + P_3 + \ldots P_n$
If P_t and all but one (say P_1) of the partial pressures are known, then $P_1 = P + P_2 + P_3 + \ldots + P_n$.

9 E.g., $C_3H_8(g) + 5\ O_2(g) \longrightarrow 3\ CO_2(g) + 4\ H_2O(g)$.
Suppose C_3H_8 were the limiting reactant. Then

$$x\text{ g }C_3H_8 \times \frac{1\text{ mol }C_3H_8}{44.0\text{ g }C_3H_8} \times \frac{3\text{ mol }CO_2}{1\text{ mol }C_3H_8} = \text{mol }CO_2$$

Then put moles CO_2 into the ideal gas equation with the specified temperature and pressure to calculate the volume.

$$or\ x\text{ L }O_2 \times \frac{3\text{ L }CO_2}{5\text{ L }O_2} = \text{L }CO_2$$

(assuming that the reactants and products were at the same temperature and pressure).

11 He added terms to account for decreased effective volume due to the non-zero volume of the molecules and to account for decreased effective pressure due to intermolecular attractions.

13 If equal volumes of gas contain equal numbers of molecules, then gases will react in whole number ratios of volumes because they react in whole number ratios of molecules.

15 78% N_2, 21% O_2, 1% Ar, traces of CO_2, SO_2, . . . , with variable amounts of water vapor.

17 The temperature at which liquid water begins to condense from a sample of "wet" gas; the temperature at which the relative humidity becomes 100%.

19 **(a)** $n_{CO_2} = n_{O_2} = n_{N_2}$ (Avogadro's law)
(b) g $CO_2 >$ g $O_2 >$ g N_2 ($MM_{CO_2} > MM_{CO_2} > MM_{N_2}$)
(c) $KE_{O_2} = KE_{O_2} = KE_{O_2}$ (from kinetic molecular theory)
(d) $V_{N_2} > V_{O_2} > V_{CO_2}$ $\left(V \frac{1}{MM}\right)$
(e) All equal since number of moles all equal.

21 1.03×10^4 g; 2.06×10^4 g

23 13.8 mL

25 0.123 L

27 1.98 atm

29 67.0 K

31 6.26 L

33 389 K

35 **(a)** 503 mL **(b)** 0.719 g

37 1.21 L

39 28.0 g mol^{-1}

41 44.3 g mol^{-1}

43 0.541 g L^{-1}

45 78.0 g mol^{-1}

47 **(a)** 30.0 g mol^{-1} **(b)** CH_3 **(c)** C_2H_6

49 113 g mol^{-1}

51 1.59 atm

53 21 atm

55 **(a)** 0.921 atm **(b)** 0.0374 mol

57 0.276 g

59 0.0240 L

61 1.15 g

63 $v_{He} > v_{Ne} > v_{Ar} > v_{Kr} > v_{Xe}$

65 1.99×10^5 cm s^{-1}

67 3 mL s^{-1}

CHAPTER 11

Practice Problems

11.1 Water can form 4 H-bonds, CH_3CH_2OH only 2, and CH_3CH_3 none. Therefore, we would expect that the vapor pressures could increase $H_2O < CH_3CH_2OH < CH_3CH_3$, and the boiling points would decrease $H_2O > CH_3CH_2OH > CH_3CH_3$.

11.3 55.6 kJ

11.5 **(a)** solid **(b)** gas **(c)** gas **(d)** liquid **(e)** liquid

Exercises

1 viscosity: Resistance of a liquid to a flow.
surface tension: Energy required to increase the surface area of a liquid.
equilibrium vapor pressure: Pressure exerted by a vapor in contact with the corresponding liquid.
boiling point: Temperature at which the vapor pressure of a liquid equals the surrounding atmospheric pressure.
freezing point: Temperature at which a solid and the corresponding liquid are in equilibrium.
hydrogen bonding: Intermolecular force resulting from the attraction of a hydrogen atom bonded to a small, electronegative atom (O, N, or F) in one molecule to an electron-rich atom (usually O, N, or F) in an adjacent molecule.
dipole–dipole interaction: Intermolecular force resulting from the electrostatic attraction between two polar (unsymmetrical) molecules.
London forces: Intermolecular force resulting from the attraction between two instantaneous (temporary) dipoles.
ionic solid: Solid held together by forces between oppositely charged particles.
molecular solid: Solid which consists of molecules (neutral) held together by relatively weak intermolecular forces.
covalent solid: Solid consisting of atoms held together by covalent bonds.
unit cell: Smallest unit of a solid. Structure of the solid can be built up by putting together many unit cells.
phase diagram: A graph of pressure *vs.* temperature showing the regions in which the three phases (solid, liquid, gas) are most stable.
critical temperature: Temperature above which a substance can exist only as a gas.
triple point: Combination of pressure and temperature at which the three phases coexist.

3 As intermolecular interactions increase, both the viscosity and the surface tension increase. Stronger intermolecular forces mean that the molecules will flow less readily (higher viscosity) and that the outer "skin" of the liquid will be more tightly held (higher surface tension).

5 The stronger the intermolecular forces, the lower the vapor pressure, since the molecules have a more difficult time escaping from the liquid. As the temperature is increased, the molecules have more energy and can more easily overcome the forces of attraction, and the vapor pressure increases.

7 A dipole–dipole interaction results from the attraction of two molecules which are polar—that is, unsymmetrical with respect to the distribution of electrons and charges. Liquids such as PCl_3 or $(CH_3)_2CO$ are held together predominantly by this force.
Hydrogen bonding results from the attraction of a hydrogen atom bonded to a small, electronegative atom (F, N, O) to the electronegative atom in an adjacent molecule. Liquids such as HF or $HOCH_3$ are held together by this force.
London forces result from the interaction of two instantaneous (temporary) dipoles. This happens as a result of temporary imbalances in electron distribution in the molecules. Nonpolar liquids such as Br_2 or CH_4 are held together by this force.

9 In general, a larger heat of vaporization results from stronger intermolecular forces.

11 In general, a high boiling point is caused by large intermolecular forces.

13 The boiling point of a liquid is the temperature at which the vapor pressure equals the surrounding atmospheric pressure. Since vapor pressure increases with temperature, the boiling point will increase with pressure also.

15 *ionic*—ions held together by ionic bonds. Examples: NaCl, CaF_2, $BaSO_4$
metallic—atomic "kernels" held together in a "sea" of electrons. Examples: Mg, Fe
covalent—atoms held together by covalent bonds. Examples: SiO_2, C, SiC
molecular—molecules held together by dipole–dipole, London, or hydrogen bonding forces. Examples: SCl_2, H_2O, I_2

17 NaCl—fcc cell of Cl^- with Na^+ in the octahedral holes.
CsCl—simple cubic cell of Cl^- with Cs^+ in the cubic holes.

19 A simple cubic cell has eight particles at the corners of a cube. A body-centered cubic cell has the eight corner particles, plus one particle in the center of the cell. A face-centered cubic cell has the eight corner particles, plus one particle in the center of each face.

21 **(a)** Kinetic energy of the particles and the temperature both increase. **(b)** Kinetic energy of the particles and the temperature both remain constant. **(c)** Both increase. **(d)** Both remain constant. **(e)** Both increase.

23 As the intermolecular forces get stronger, the critical temperature increases.

25 Gases: CO, CO_2, N_2O, NO, NO_2, SO_2, OF_2, O_2F_2, Cl_2O, ClO_2
Liquids: SO_3, H_2O, H_2O_2, Br_2O, BrO_2

27 In general, the smaller the atom or molecule, the more likely the substance will be a gas. Materials with intermediate-size particles are liquids, and large-size particles generally give solids.

29 London forces

31 Water has the higher boiling point, since its intermolecular forces are stronger (hydrogen bonds *vs.* London forces).

33 In a pressure cooker, the atmospheric pressure is higher, making the boiling point higher and allowing the vegetables to cook faster.

35 (a) nonpolar (b) London forces (c) water—hydrogen bonds *vs.* London forces (d) carbon disulfide—heavier molecule, larger London forces (e) CS_2—higher vapor pressure due to weaker intermolecular forces (f) H_2O—higher boiling point due to stronger forces (g) No—ionic solids are not very soluble in nonpolar liquids (h) H_2O—heat of vaporization increases with strength of forces

37 (a) increases (b) increases (c) decreases (d) increases (e) increases (f) increases

39 $HC_2H_3O_2 > SO_2 > PH_3 > CO_2 > H_2$

41 (a) and (d) in water; (b) and (c) in carbon tetrachloride.

43 A net total of 2 particles: 8 corner particles × 1/8 particle per cell, plus the one in the center.

45 (a) 2 (b) 0.533 nm (c) 0.152 nm^3 (d) 45.7 cm^3

47 (See fig. 11.26(b) for CO_2 phase diagram.) CO_2 is a solid at 1 atm below −78°C; above that temperature it is a gas. Above 5.2 atm, it can be converted to a liquid. Above 31°C it can only exist as a gas.

49 Ice is less dense than water because the forces holding it together (i.e., hydrogen bonds) are very directional, resulting in "holes" in the structure. This open structure means a lower density. When pressure is applied, the structure collapses to a denser, more fluid one (i.e., liquid). The liquid film allows easy gliding of a skate blade over the surface. Glass has a more dense solid than liquid, so doesn't melt under pressure and could not be a good skating surface.

51

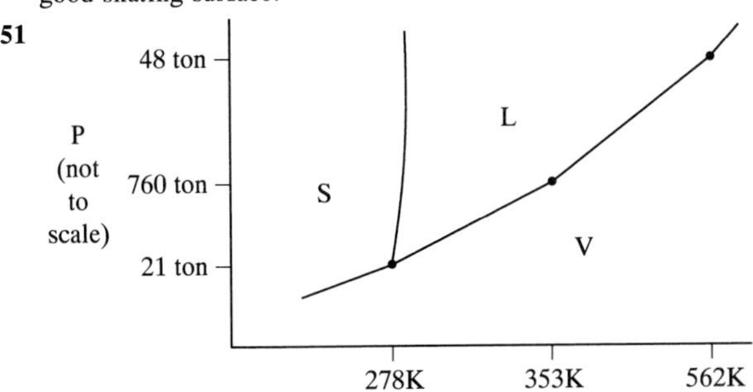

53 (a) 6.00 kJ (b) 40.6 kJ

55 0.430 g

57 The molecules of oxygen (O_2) are smaller than those of sulfur (S_8), so the intermolecular forces are smaller.

59 As a group, (OF_2, Cl_2O, ClO_2) are smaller than (Br_2, BrO_2), and these are smaller than IO_2. Intermolecular forces rise as the size of the molecule does.

CHAPTER 12

Practice Problems

12.1 $Ba(NO_3)_2$, being an ionic compound, would be expected to be water soluble. Methanol, CH_3OH, being a low-molecular-weight alcohol, would be expected to be soluble due to its H-bonding interactions with water.

12.3 0.304 atm

12.5 100.266°C

12.7 284 g mol^{-1}

12.9 30.1 atm

Exercises

1 Energy is required to break apart the crystalline lattice of a solid. Energy is given off when the ions interact with the water molecules. The relative sizes of the lattice energy and the hydration energy determine whether the solution process is exo- or endothermic. Materials with an exothermic heat of solution are generally more soluble than those with endothermic heats of solution.

3 It is most favorable if the solvent–solute instructions are greater in magnitude (i.e., more exothermic) than the solvent–solute or solute–solute interactions.

5 In general, the more alike the solvent–solute and solute–solute interactions (e.g., London forces and London forces), the greater the solubility of the solute in the solvent.

7 Increasing the temperature generally increases the solubility of a solid in a liquid, but decreases the solubility of a liquid or gas in a liquid.

9 The solubility of a gas in a liquid is directly proportional to the partial pressure of the gas above the solution:

$$s = kP$$

where s = solubility of the gas
k = constant characteristic of the gas, liquid, and temperature
P = partial pressure of the gas

11 The vapor pressure due to a given material above a solution containing that material is directly proportional to the mole function of that material in the solution.

$$P_i = X_i P_i^\circ$$

where P_i = vapor pressure of component i above the solution
X_i = mole fraction of component i in the solution
P_i° = vapor pressure of pure component i

13 Ideal solution behavior means all components obey Raoult's law. Then

$$P_T = P_i + P_2 + \cdots P_n$$

$$P_T = X_1 P_1^\circ + X_2 P_2^\circ + \cdots X_n P_n^\circ$$

15 The molecules of the solute "block" the solvent molecules and make it more difficult for them to form their lattice. Energy must be removed (i.e., the temperature must be lowered) to "lock" the solvent molecules into place.

17 The changes in boiling and freezing points are directly proportional to the molality of the solution:

$$\Delta T_f = K_f m \quad \text{and} \quad \Delta T_b = K_b m$$

If the molality of the solution is known, using the K_f or K_b for the solvent allows calculation of the change in boiling or freezing point, ΔT_b or ΔT_f. Adding ΔT_b to the boiling point for the pure solvent, or subtracting ΔT_f from the freezing point of the pure solvent, gives the actual boiling or freezing point for the solution.

19 Sodium chloride and sodium sulfate are strong electrolytes, meaning that they are virtually 100% dissociated into ions in dilute solutions. NaCl gives 2 ions per mole, and Na_2SO_4 gives 3 ions per mole. Each solute particle (molecule *or* ion) has the same effect on the boiling point of the solution.

21 The freezing point depressions of NaCl would be greatest, of acetic acid next, and of sucrose the least, in approximately a 2:1.01:1 ratio. This is because they are, respectively, a strong, weak, and non-electrolyte. Stronger electrolyte means more particles in the solution and larger freezing point depression.

23 By shining a light through the mixture, the Tyndall effect can be observed in a colloid but not a solution. In a colloid, the light is scattered by the suspended particles and will be visible at right angles to the light beam.

25 Dissolved materials such as heavy metal ions, Cl^-, SO_4^{2-}, etc.; organic matter; synthetic materials such as plastic, solvents, pesticides, etc.; human wastes; disease-producing organisms; radioactive materials; etc.

27 **(a)** Hard water contains dissolved ions such as Ca^{2+} and Mg^{2+}.
(b) Hard water which contains HCO_3^- is temporary hard water, since the hardness can be removed by boiling.
(c) Permanent hard water contains no HCO_3^- and cannot be softened by boiling.

29 Permanent hard water can be softened by ion exchange, in which NA^+ attached to a large anion (in a natural or synthetic zeolite or resin) is exchanged for Ca^{2+}:

$$2NaZ(s) + Ca^{2+}(aq) \longrightarrow CaZ_2(s) + 2Na^+(aq)$$

31 **(a)** no **(b)** yes **(c)** yes **(d)** no

33 **(a)** increase **(b)** increase **(c)** decrease **(d)** decrease

35 The solubility would increase $CH_4 < C_2H_6 < C_5H_{12}$. The first two are gases; the third, a liquid (at 25°C). The intermolecular forces in C_5H_{12} would be most like those in C_6H_6 due to the similarity in molar mass and molecular size.

37 24 g

39 5.5×10^{-4} mol L^{-1}

41 **(a)** $X_{ethanol} = 0.411$; $X_{methanol} = 0.589$
(b) $P_{ethanol} = 18$ torr; $P_{methanol} = 55$ torr
(c) 73 torr

43 23.5 torr

45 0.30 *m*

47 0.0292 *m*

49 The density of an aqueous solution of a solid is greater than 1.00 g mL^{-1} due to the added solute. As it is diluted, the density becomes closer to 1.00 g mL^{-1} or (1.00 kg L^{-1}) and $\frac{\text{mol solute}}{\text{kg solvent}}$ approaches $\frac{\text{mol solute}}{\text{L solution}}$

51 $T_b = 82.5°C$; $T_f = 0.92°C$

53 122 g mol^{-1}

55 0.65% H_2O

57 87% dissociated

59 4% dissociated

61 2.42 atm

63 1.58×10^4 g mol^{-1}

65 $T_f = -0.00094°C$; $\pi = 0.012$ atm

67 **(a)** $Ca^{2+}(aq) + 2HCO_3^-(aq) \longrightarrow CaCO_3(s) + CO_2(g) + H_2O(l)$
(b) $Ca^{2+}(aq) + CO_3^{2-}(aq) \longrightarrow CaCO_3(s)$

CHAPTER 13

Practice Problems

13.1 **(a)** $3O_2(g) \rightleftharpoons 2O_3(g)$

$$K_c = [O_3]^2/[O_2]^3$$

(b) $2CO(g) + O_2(g) \rightleftharpoons 2CO_2(g)$

$$K_c = [CO_2]^2/[CO]^2[O_2]$$

13.3 7.9×10^{-2} *M*

13.5 $[N_2O_4] = 0.040$ *M*; $[NO_2] = 0.12$ *M*

13.7 **(a)** $K_c = [H_3O^+][NO_2^-]/[HNO_2]$ **(b)** $K_c = [Ag^+]^2[CrO_4^{2-}]$
(c) $K_c = 1/[O_2]^5$

13.9 1.24

Exercises

1 $K_c = \dfrac{[H_2O]^2}{[H_2]^2[O_2]}$ $\quad K_p = \dfrac{P_{H_2O}{}^2}{P_{H_2}{}^2 P_{O_2}}$

3 If equations (1) and (2) are added to produce equation (3), then $K_1 \times K_2 = K_3$.

5 A large value of *K* means the reaction lies to the right—that is, that the equilibrium concentrations of the products are larger than those of the reactants. A small value of *K* means the reaction lies to the left.

7 An equilibrium shifts when the concentrations of the reactants and products change to establish a new equilibrium position.

9 The reaction will shift to the left.

11 Any reaction in which the number of moles of gas is the same on the left and right sides, e.g., $N_2(g) + O_2(g) \rightleftharpoons 2NO(g)$.

13 The reaction will shift **(a)** to remove an excess of a material or **(b)** to produce more of a material in deficit.

15 The reaction $2H_2(g) + O_2(g) \rightleftharpoons 2H_2O(g)$ would be shifted to the right by decreasing the volume (to smaller number of moles of gas). The reaction $CO_2(g) + H_2(g) \rightleftharpoons CO(g) + H_2O(g)$ would not be affected by volume changes (equal numbers of moles of gas on the right and left).

17 It allows the reaction to reach the same equilibrium position faster.

19 **(a)** $K_c = [O_3]^2/[O_2]^3$ **(b)** $K_c = [HI]^2/[H_2][I_2]$
(c) $K_c = [H_2][I_2]/[HI]^2$ **(d)** $K_c = [H_2O]^6[N_2]^2/[NH_3]^4[O_2]^3$
(e) $K_c = [H_2O]^2[SO_2]^2/[H_2S]^2[O_2]^3$

21 **(a)** 4.0 **(b)** 0.25 **(c)** 0.50

23 $K_c = [HI]^2/[H_2][I_2] = 50$

25 0.73

27 **(a)** 0.368 **(b)** 1.65

29 36.9

31 0.57 mol

33 46.7 atm

35 0.20 atm

37 22.9%

39 shift to the right; $[SO_3] = [NO] = 1.50$ *M*; $[SO_2] = [NO_2] = 0.50$ *M*

41 **(a)** $K_c = [Fe(CO)_5]/[CO]^5$ **(b)** $K_c = [CO][H_2]/[H_2O]$
(c) $K_c = [O_2]$ **(d)** $K_c = [H_3O^+][C_2H_3O_2^-]/[HC_2H_3O_2]$
(e) $K_c = [NH_4^+][OH^-]/[NH_3]$ **(f)** $K_c = [H_3O^+][OH^-]$
(g) $K_c = [Pb^{2+}][Cl^-]^2$ **(h)** $K_c = [Fe^{3+}][OH^-]^3$

43 6.9×10^{-3}

45 5.9 *M*

47 **(a)** shift right **(b)** shift left **(c)** shift right **(d)** no effect
(e) no effect **(f)** no effect
See Section 13.4 for explanations.

49 K_p for $COCl_2 \rightleftharpoons CO + Cl_2 = 6.71 \times 10^{-9}$
K_p for $CO + Cl_2 \rightleftharpoons COCl_2 = 1.49 \times 10^8$
K_c for $CO + Cl_2 \rightleftharpoons COCl_2 = 4.57 \times 10^9$

51 2.02

CHAPTER 14

Practice Problems

14.1 $HF(aq) + H_2O(l) \rightleftharpoons H_3O^+(aq) + F^-(aq)$
acid base acid base
Conjugate pairs: HF/F^-; H_3O^+/H_2O

14.3 AsH_3 is a stronger acid than PH_3 since acid strength increases down a group. H_2Se is a stronger acid than AsH_3 since acid strength increases to the right along a period. Therefore the order of acidity is $PH_3 < AsH_3 < H_2Se$.

14.5 Nitrous (HNO_2), nitric (HNO_3), hypoiodous (HIO), and hypochlorous (HClO) acids have one, two, zero, and zero nonprotonated oxygen atoms, respectively. Nitric will therefore be the strongest and nitrous the second-strongest. Between HClO and HIO, chlorine is more electronegative than iodine, so you would expect HClO to be stronger than HIO. The order would be HIO < HClO < HNO_2 < HNO_3.

14.7 1.8×10^{-5} (same as Example 14.8)

14.9 $CN^-(aq) + H_2O(l) \rightleftharpoons HCN(aq) + OH^-(aq)$

$$K_b = 2.0 \times 10^{-5}$$

$F^-(aq) + H_2O(l) \rightleftharpoons HF(aq) + OH^-(aq)$

$$K_b = 2.8 \times 10^{-11}$$

0.50 *M* CN^- more basic than 0.50 *M* F^-.

14.11 $K_2O(s) + SO_3(g) \longrightarrow K_2SO_4(s)$
base acid

Exercises

1 $HF(aq) + N_2H_4(aq) \rightleftharpoons N_2H_5^+(aq) + F^-(aq)$
acid base acid base
Conjugate pairs: HF/F^- and $N_2H_5^+/N_2H_4$

3 $NH_3(aq) + H_2O(l) \rightleftharpoons NH_4^+(aq) + OH^-(aq)$

$$K_b = \frac{[NH_4^+][OH^-]}{[NH_3]}$$

5 Dissociation of HA: $HA(g) \longrightarrow H(g) + A(g)$
Ionization of H: $H(g) \longrightarrow H^+(g) + e^-$
Formation of A^-: $A(g) + e^- \longrightarrow A^-(g)$

7 $HA(aq) + H_2O(l) \rightleftharpoons H_3O^+(aq) + A^-(aq)$

$$K_a = \frac{[H_3O^+][A^-]}{[HA]}$$

9 Strong acids are essentially 100% dissociated to $H^+ + A^-$ in solution, so their K_a values are so large as to be considered infinite.

11 The stronger the acid, the weaker the conjugate base. HCl is a stronger acid than HF, and F^- is a stronger base than Cl^-.

13 $NH_4^+(aq) + OH^-(aq) \rightleftharpoons NH_3(aq) + H_2O(l)$
NH_3 has high, but not unlimited, solubility in water, so the smell of NH_3 would be the indication of reaction.

15

zero:	H_3BO_3 boric	H_3AsO_3 arsenous	HClO hypochlorous
one:	HNO_2 nitrous	$HClO_2$ chlorous	H_2SO_3 sulfurous
two:	H_2SO_4 sulfuric	HNO_3 nitric	$HClO_3$ chloric
three:	$HClO_4$ perchloric	$HBrO_4$ perbromic	

17 As the electronegativity of the central element increases, the strength of the acid increases. The central atom draws electron density to itself, polarizing (and weakening) the O—H bond(s).

HClO > HBrO $H_2SO_3 > H_2SeO_3$

19 Hydroxides and oxoacids both contain O—H bonds. NaOH, $Ca(OH)_2$, $Fe(OH)_2$ are hydroxides (bases). $(HO)_2SO_2$, $(HO)_3PO$, $HOClO_3$ are oxoacids.

21 An amphoteric hydroxide is one which dissolves in (i.e., reacts with) either an acid or a hydroxide base. $Al(OH)_3$, aluminum hydroxide, and $Zn(OH)_2$, zinc(II) hydroxide, are examples.

23 It depends on the strengths of the acid and base of which the ions of the salt are conjugates. If the cation is the conjugate acid of a weak base, the solution will be acidic due to hydrolysis of this ion. If the anion is the conjugate base of a weak acid, the solution will be basic due to hydrolysis of this ion. If the cation is a small metal ion which is highly charged (≥3+), the solution will be acidic due to hydrolysis of this ion. If the cation and anion are both conjugates of strong electrolytes, neither ion will hydrolyze and the solution will be neutral.

25 $K_w = K_a K_b$

27 If the oxidation number of the metal is high (e.g., +6 in H_2CrO_4 or +7 in $HMnO_4$), the effective electronegativity of the central atom is very high. This polarizes and weakens the O—H bond and makes the compound a stronger acid.

29 $NH_3(g) + BF_3(g) \longrightarrow H_3N—BF_3$
base acid

31 **(a)** $H_2SO_4(aq) + H_2O(l) \rightleftharpoons H_3O^+(aq) + HSO_4^-(aq)$
acid base acid base
(b) $CH_3COOH(l) + HBr(g) \rightleftharpoons CH_3CO_2H_2^+ + Br^-$
base acid acid base
(c) $H_2O(l) + H_2O(l) \rightleftharpoons H_3O^+(aq) + OH^-(aq)$
base acid acid base
(d) $H_3O^+(aq) + I^-(aq) \rightleftharpoons H_2O(l) + HI(aq)$
acid base base acid
(e) $NH_4^+(aq) + OH^-(aq) \rightleftharpoons NH_3(aq) + H_2O(l)$
acid base base acid
(f) $CO_3^{2-}(aq) + H_2O(l) + HCO_3^-(aq) + OH^-(aq)$
base acid acid base

33 **(a)** NO_3^- **(b)** SO_4^{2-} **(c)** NH_2^- **(d)** OH^- **(e)** H_2O **(f)** NH_3

35 **(a)** HI > HBr > HCl > HF **(b)** $H_2Te > H_2Se > H_2S$ **(c)** $AsH_3 > PH_3 > NH_3$

37 **(a)** NH_2^- **(b)** O^{2-} **(c)** F^- **(d)** OH^- **(e)** S^{2-} **(f)** HS^-

39 **(a)** $HBr(g) + H_2O(l) \longrightarrow H_3O^+(aq) + Br^-(aq)$
(b) $NH_3(aq) + H_2O(l) \rightleftharpoons NH_4^+(aq) + OH^-(aq)$
(c) $HCN(aq) + H_2O(l) \rightleftharpoons H_3O^+(aq) + CN^-(aq)$
(d) $CH_3NH_2(aq) + H_2O(l) = CH_3NH_3^+(aq) + OH^-(aq)$

41 0.060 *M*

43 4.2×10^{-10}

45 $HIO_3 < HBrO_3 < HClO_3$

47 **(a)** amphoteric **(b)** acid **(c)** acid **(d)** amphoteric **(e)** amphoteric **(f)** hydroxide **(g)** acid

49 H_3PO_4 and H_3PO_3 have the structures $(HO)_3PO$ and $(HO)_2PO$ (with H bonded to P), respectively. Since they both have one nonprotonated oxygen atom, it isn't surprising that their K_a values are similar.

51 **(a)** $CrO_3(s) + H_2O(l) \longrightarrow H_2CrO_4(aq)$ chromic acid
(b) $MnO(s) + H_2O(l) \longrightarrow Mn(OH)_2(s)$ manganese(II) hydroxide
(c) $Mn_2O_7(s) + H_2O(l) \longrightarrow 2HMnO_4(aq)$ permanganic acid
(d) $FeO(s) + H_2O(l) \longrightarrow Fe(OH)_2(s)$ iron(II) hydroxide
(e) $ZnO(s) + H_2O(l) \longrightarrow Zn(OH)_2(s)$ zinc(II) hydroxide

53 **(a)** formic **(b)** monochloroacetic **(c)** monofluoroacetic

55 $SO_4^{-2} < HSO_4^- < HS^- < NH_2^-$

57 2.2×10^{-11}

59 3.3×10^{-7}

61 **(a)** $AlCl_3(aq) + Cl^-(aq) \longrightarrow AlCl_4^-(aq)$
acid base
(b) $Ag^+(aq) + 2NH_3(aq) \longrightarrow Ag(NH_3)_2^+(aq)$
acid base
(c) $Zn(OH)_2(s) + 2OH^-(aq) \longrightarrow Zn(OH)_4^{2-}(aq)$
acid base

63 $Al^{3+}(g) + 6H_2O(l) \longrightarrow Al(H_2O)_6^{3+}(aq)$

65 **(a)** $CaO(s) + SO_2(g) \longrightarrow CaSO_3(s)$
(b) $Na_2O(s) + SO_3(g) \longrightarrow Na_2SO_4(s)$
(c) $K_2O(s) + CO_2(g) \longrightarrow K_2CO_3(s)$

CHAPTER 15

Practice Problems

15.1 **(a)** $[H_3O^+] = 0.30\ M$, $[OH^-] = 3.3 \times 10^{-14}\ M$
(b) $[H_3O^+] = 1.2 \times 10^{-2}\ M$, $[OH^-] = 0.0080\ M$

15.3 **(a)** $[H_3O^+] = 4.5 \times 10^{-3}\ M$ **(b)** $[OH^-] = 1.1 \times 10^{-9}\ M$

15.5 $[OH^-] = 7.9 \times 10^{-3}\ M$, pH = 11.90

15.7 $[H_3O^+] = 4.2 \times 10^{-5}\ M$, pH = 4.37

15.9 **(a)** 1.18 **(b)** 2.28 **(c)** 3.30 **(d)** 5.00 **(e)** 9.00
(f) 11.00

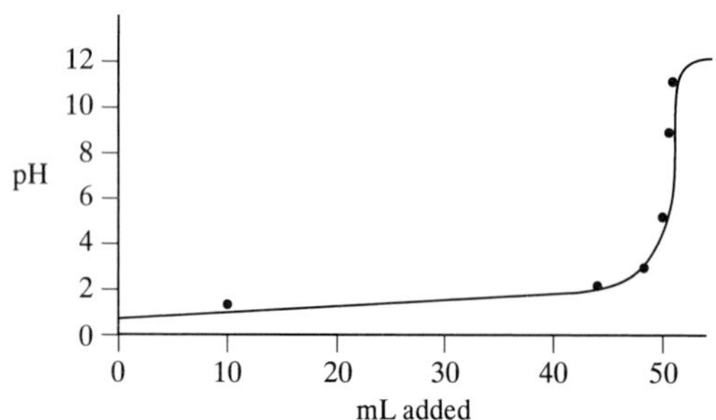

15.11 **(a)** 9.876 **(b)** 6.57 **(c)** 4.00

Exercises

1 $2H_2O(l) \rightleftharpoons H_3O^+(aq) + OH^-(aq)$
$K_w = [H_3O^+][OH^-] = 1.0 \times 10^{-14}$ (at 25°C)

3 It is valid in *all* aqueous solutions, regardless of any solute.

5 **(a)** For NaOH, $[OH^-]$ = the molar concentration of NaOH.
(b) For $Ba(OH)_2$, $[OH^-]$ = twice the molar concentration of $Ba(OH)_2$ (the formula contains 2 OH^-).

7 **(a)** pH = log $[H_3O^+]$ **(b)** pOH = log $[OH^-]$
(c) pK = −log K

9 pH + pOH = 14.00

11 **(a)** The larger the K_a, the larger the percent dissociation.
(b) The larger the $[H_3O^+]$, the larger the percent dissociation.

13 $OCl^-(aq) + H_2O(l) \rightleftharpoons HOCl(aq) + HOCl(aq)$
$CH_3NH_3^+(aq) + H_2O(l) \rightleftharpoons CH_3NH_2(aq) + H_3O^+(aq)$

15 The position of an acid–base equilibrium is shifted by adding the conjugate species of the one already in the solution. If we consider the system

$$HCO_2H(aq) + H_2O(l) \rightleftharpoons H_3O^+(aq) + CO_2H^-(aq)$$

adding $NaCO_2H$ (sodium formate) to the solution would shift the equilibrium to the left, decreasing the $[H_3O^+]$.

17 A buffer solution is one that resists changes in pH when strong acid or base is added. An acid buffer might be HCO_2H + $NaCO_2H$; a basic buffer might be N_2H_4 + $[N_2H_5]NO_3$.

19 The buffer contains the weak acid HA and its salt, furnishing A^-. If strong acid is added, it reacts with the A^- and produces HA. As long as the total quantity of HA + A^- is large and that of strong acid is small, the ratio $[HA]/[A^-]$ will remain fairly constant, and so will the pH.

$$\left(\text{pH} = \text{p}K_a + \log\frac{[\text{anion}]}{[\text{acid}]}\right)$$

21 pH = pK_a + log([base]/[acid])
If [base]/[acid] remains constant, so does pH. If more base is added, the pH increases; if more acid is added, the pH decreases.

23 For a *weak* diprotic acid, $[A^{2-}] = K_{a2}$ for H_2A. This is because $[H_3O^+] \approx [HA^-]$ from the first step of the dissociation. The second step in the reaction (to form a second H_3O^+ and A^{2-}) proceeds only very slightly.

$$K_{a2} = \frac{[H_3O^+][A^{2-}]}{[HA^-]}$$

If $[H_3O^+] = [HA^-]$, then $K_{a2} = [A^{2-}]$. For H_2SO_4, the first dissociation is complete and the second step is pretty large (over 10% dissociated), so the above argument is not valid and $K_{a2} \neq [SO_4^{2-}]$.

25 You need to know the approximate pH at the equivalence point, which means you need to know what is being titrated *vs.* what (strong acid + strong base, strong acid + weak base, etc.).

27 The strong acid–strong base titration curve would have the largest vertical section. To select an appropriate indicator for a titration, you need to select an indicator whose transition range is as close to the center of the vertical section of the titration curve as possible. This would typically be about 6–8 for strong acid–strong base, about 4–6 for weak base–strong acid, and about 8–10 for strong base–weak acid.

29 HNO_3 is a strong acid, so $[H_3O^+] = 0.15\ M$
$[OH^-] = K_w/[H_3O^+] = 6.7 \times 10^{-14}\ M$.

31 **(a)** pH = 2.00, pOH = 12.00 **(b)** pH = 0.60, pOH = 13.40

33 pOH = 8.30, $[H_3O^+] = 2.0 \times 10^{-6}$, $[OH^-] = 5.0 \times 10^{-9}$

35 12.52

37 0.35

39 $[H_3O^+] = 1.4 \times 10^{-5}\ M$, pH = 4.85
$[OH^-] = 7.1 \times 10^{-10}\ M$, pOH = 9.15

41 $K_a = 1.84 \times 10^{-5}$, 1.85×10^{-5}, and 1.86×10^{-5} in the three solutions. For practical purposes, K_a is independent of initial conditions, as all equilibrium constants should be.

43 7.5×10^{-5}

45 $[H_3O^+] = 5.38 \times 10^{-12}\ M$, $[OH^-] = 1.86 \times 10^{-3}\ M$.

47 0.22 *M*

49 **(a)** neutral **(b)** acidic **(c)** basic **(d)** neutral **(e)** basic
(f) basic **(g)** acidic **(h)** neutral **(i)** basic
(a) and **(d)** salt of strong acid + strong base **(b)** salt of weak base and strong acid **(c) (e),** and **(f)** salt of strong base and weak acid **(g)** due to hydrolysis of $Al(H_2O)_6^{3+}$ **(h)** salt of weak acid and weak base, with $K_a \approx K_b$ **(i)** salt of weak acid and weak base with $K_a > K_b$

51 3.3×10^{-7}

53 **(a)** 5.13 **(b)** 11.19 **(c)** 2.85 **(d)** 8.78

55 $K_a = 2.4 \times 10^{-5}$, $K_b = 4.2 \times 10^{-10}$, % hydrolysis = 3.4%

57 $2.5 \times 10^{-5}\ M$

59 5.42 pH units

61 zero

63 14 g

65 $[H_3O^+] = 0.026M$, $[SO_4^{2-}] = 0.0065\ M$, $[HSO_4^-] = 0.013\ M$

67 $[H_3O^+] = [HCO_3^-] = 9.3 \times 10^{-5}\ M$, $[CO_3^{2-}] = 5.6 \times 10^{-11}\ M$, $[H_2CO_3] = 0.020\ M$

69 **(a)** $H_3O^+(aq) + OH^-(aq) \longrightarrow 2H_2O(l)$
(b) $HCO_2H(aq) + OH^-(aq) \longrightarrow H_2O(l) + CO_2H^-(aq)$
(c) $HF(aq) + OH^-(aq) \longrightarrow H_2O(l) + F^-(aq)$
(d) $H_3O^+(aq) + OH^-(aq) \longrightarrow 2H_2O(l)$
(e) $H_3O^+(aq) + HSO_4^-(aq) + 2OH^-(aq) \longrightarrow 3H_2O(l) + SO_4^{2-}$
(f) $CH_3NH_2(aq) + H_3O^+(aq) \longrightarrow CH_3NH_3^+(aq) + H_2O(l)$

71 **(a)** 0.48 **(b)** 3.0 **(c)** 7.00 **(d)** 11.0 **(e)** 12.00

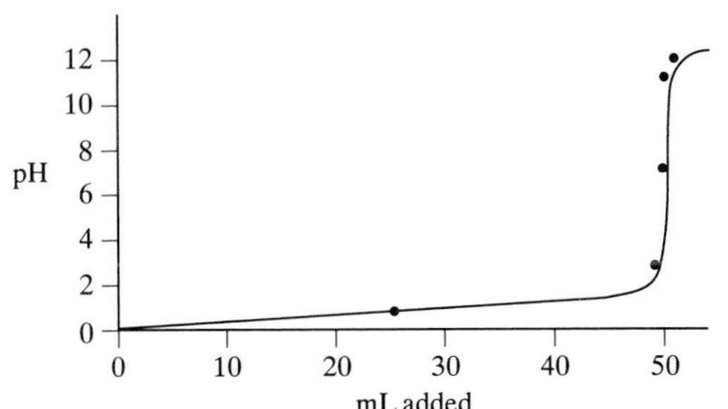

Indicators: methyl orange, bromcresol green, methyl red, litmus, phenol red (best), phenolphthalein

73 **(a)** 12.62 **(b)** 10.32 **(c)** 9.15 **(d)** 8.14 **(e)** 5.61 **(f)** 3.08 **(g)** 1.79

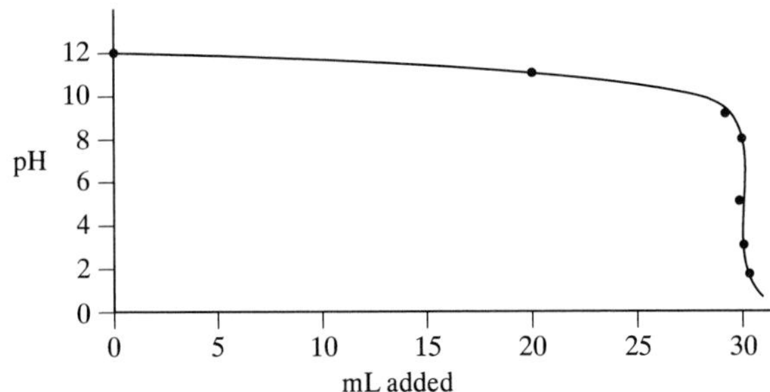

CHAPTER 16

Practice Problems

16.1 1.1×10^{-15}

16.3 $1.3 \times 10^{-2}\ M$, 9.3 g L^{-1}; 12.41

16.5 $4 \times 10^{-10}\ M$; 1×10^{-8} g

16.7 $1.3 \times 10^{-2}\ M$ in water; $2.0 \times 10^{-4}\ M$ at pH 13.30. Solubility of $Ca(OH)_2$ decreases with increasing pH.

16.9 Yes

16.11 1.4×10^{-7}%

16.13 $[S^{2-}] = 1.1 \times 10^{-20}\ M$; $[H_3O^+] = 0.34\ M$

16.15 **(a)** $Mg(OH)_2(s) + 2H_3O^+(aq) \longrightarrow 4H_2O(l) + 2Mg^{2+}(aq)$
(b) $MnS(s) + 2H_3O^+(aq) \longrightarrow Mn^{2+}(aq) + 2H_2O(l) + H_2S(g)$
(c) $3PbS(s) + 8H_3O^+(aq) + 2NO_3^-(aq) \longrightarrow 3Pb^{2+}(aq) + 3S(s) + 2NO(g) + 12H_2O(l)$

Exercises

1 Molar solubility is the number of moles of the salt which can be dissolved in 1 L of solution. The solubility product constant is the product of the concentrations of the ions in equilibrium with the solid material, all raised to powers equal to their stoichiometric ratios in the formula for the salt.

3 **(a)** molar solubility $= x = [A^{n+}][B^{n-}]$ **(b)** $x = [A^{n+}] = \frac{1}{2}[B^{n-}]$ **(c)** $x = \frac{1}{2}[A^{n+}] = [B^{n-}]$ **(d)** $x = [A^{n+}] = \frac{1}{3}[B^{n-}]$ **(e)** $x = \frac{1}{2}[A^{n+}] = \frac{1}{3}[B^{n-}]$

5 **(a)** As the pH increases, the solubility decreases. **(b)** As the pH decreases, the solubility increases. This is due to the shift of the solubility equilibrium by adding or removing OH^-.

7 By adding sufficient X^- to cause the precipitation of the cation with the smaller K_{sp}, but not so much X to cause the precipitation of the cation with the larger K_{sp}.

9 You would need to increase the pH, to increase the $[S^{2-}]$ in the solution.

11 **(a)** $CaCO_3$, as in $CaCO_3(s) + 2H_3O^+(aq) \longrightarrow Ca^{2+}(aq) + 3H_2O(l) + CO_2(g)$
(b) CuS, as in $3CuS(s) + 8H_3O^+(aq) + 2NO_3^-(aq) \longrightarrow 3Cu^{2+}(aq) + 2S(s) + 12H_2O(l) + 2NO(g)$
(c) AgCl, as in $AgCl(s) + 2NH_3(aq) \longrightarrow Ag(NH_3)_2^+(aq) + Cl^-(aq)$
(d) HgS, as in $3HgS + 8H_3O^+(aq) + 2NO_3^-(aq) + 12Cl^-(aq) \longrightarrow 3PbCl_4^{2-}(aq) + 3S(s) + 12H_2O(l) + 2NO(g)$

13 $8.8 \times 10^{-7}\ M$

15 0.12 g

17 1.56×10^{-10}

19 **(a)** $[Ag^+] = 2.1 \times 10^{-2}\ M$, $[SO_4^{2-}] = 1.1 \times 10^{-2}\ M$; **(b)** 4.7×10^{-6}

21 12.93

23 31 mL

25 $2.0 \times 10^{-9}\ M$

27 **(a)** $1.6 \times 10^{-9}\ M$ **(b)** $7.7 \times 10^{-12}\ M$

29 5.0×10^{-2} g

31 2.43×10^{-4} g in the H_2O; 2.51×10^{-7} g in the Na_2SO_4

33 **(a)** $Ag^+(aq) + Cl^-(aq) \longrightarrow AgCl(s)$
(b) $Ba^{2+}(aq) + SO_4^{2-}(aq) \longrightarrow BaSO_4(s)$
(c) $Pb^{2+}(aq) + 2Cl^-(aq) \longrightarrow PbCl_2(s)$
(d) $Pb^{2+}(aq) + CrO_4^{2-}(aq) \longrightarrow PbCrO_4(s)$
(e) $Mg^{2+}(aq) + 2OH^-(aq) \longrightarrow MgOH_2(s)$
(f) $3Ca^{2+}(aq) + 2PO_4^{3-}(aq) \longrightarrow Ca_3(PO_4)_2(s)$
(g) $Fe^{2+}(aq) + S^{2-}(aq) \longrightarrow FeS(s)$
(h) $2Ag^+(aq) + CrO_4^{2-}(aq) \longrightarrow Ag_2CrO_4(s)$

35 Yes. $[Hg_2^{2+}] = 0.18\ M$, $[Cl^-] = 3 \times 10^{-9}\ M$

37 Yes. 0.095 M

39 $[Mg^{2+}] = 0.020\ M$; $[OH^-] = 2.7 \times 10^{-5}\ M$

41 $1.4 \times 10^{-20}\ M$

43 **(a)** $1.1 \times 10^{-20}\ M$ **(b)** 0.34 M

45 0.34 M

47 $9.9 \times 10^{-4}\ M$

49 **(a)–(d)** HCl, hydrochloric acid (or any other strong acid) **(e)** NH_3, ammonia **(f)** NaOH, sodium hydroxide (KOH also would work) **(g)** HNO_3, nitric acid.

51 Add water first. The $Ca(NO_3)_2$ will dissolve, leaving the NiS and HgS unreacted. To the residue, add HCl, which will dissolve NiS but not HgS. To the residue, add a mixture of HCl and HNO_3, which dissolves HgS.

$$Ca(NO_3)_2(s) \longrightarrow Ca^{2+}(aq) + 2NO_3^-(aq)$$

$$NiS(s) + 2H_3O^+(aq) \longrightarrow H_2S(g) + Ni^{2+}(aq)$$

$$3HgS(s) + 12Cl^-(aq) + 2NO_3^-(aq) + 8H_3O^+(aq) \longrightarrow 3HgCl_4^{2-}(aq) + 2NO(g) + 3S(s) + 12H_2O(l)$$

53 $2.9 \times 10^{-3}\ M$

CHAPTER 17

Practice Problems

17.1 The best reducing agent has the most negative standard reduction potential. Thus, sulfide ion ($E° = -0.508$ V) is the best reducing agent among the species listed in this problem. Iron(II) ions ($E° = -0.409$ V) would follow sulfide, and tin ($E° = -0.1364$ V) would follow iron(II) ions.

17.3 Anode: $2Fe^{2+}(aq) \longrightarrow 2Fe^{3+}(aq) + 2e^-$ $\quad E^\circ_{Fe^{2+}/Fe^{3+}} = -0.770$ V
Cathode: $F_2(g) + 2e^- \longrightarrow 2F^-(aq)$ $\quad E^\circ_{F^-/F_2} = 2.85$ V
Cell: $2Fe^{2+}(aq) + F_2(g) \longrightarrow 2Fe^{3+}(aq) + 2F^-(aq)$ $\quad E^\circ_{cell} = 2.08$ V
A positive cell potential indicates the reaction is spontaneous.

Cell Construction:

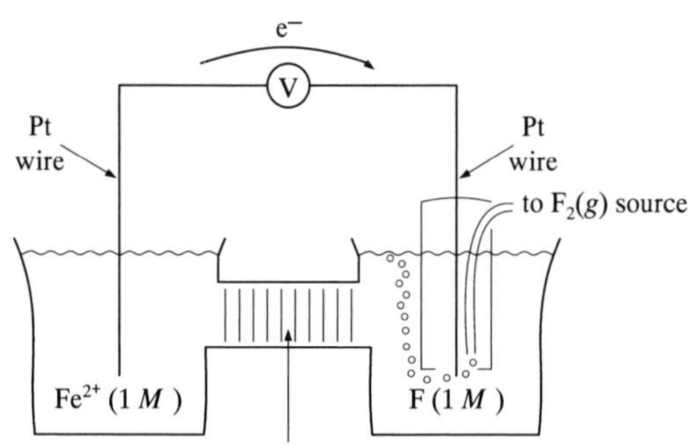

17.5 Bromine is able to oxidize iodide ions, but not chloride ions:

$Br_2(l) + 2I^-(aq) \longrightarrow 2Br^-(aq) + I_2(s)$

17.7 **(a)** Anode: $2Cl^-(l) \longrightarrow Cl_2(g) + 2e^-$
Cathode: $Ba^{2+}(l) + 2e^- \longrightarrow Ba(s)$
Cell: $Ba^{2+}(l) + 2Cl^-(l) \longrightarrow Ba(s) + Cl_2(g)$

(b) Anode: $2H_2O(l) \longrightarrow O_2(g) + 4H^+(g) + 4e^-$
Cathode: $4H_2O(l) + 4e^- \longrightarrow 2H_2(g) + 4OH^-(aq)$
Cell: $2H_2O(l) \longrightarrow 2H_2(g) + O_2(g)$

17.9 1.40 g Cl_2

Exercises

1 A galvanic cell has a positive cell potential and is spontaneous. An electrolytic cell has a negative cell potential and is nonspontaneous.

3 For a galvanic cell, the anode is negative and the cathode is positive. Oxidation occurs at the anode. The electrons lost leave the anode and move toward the cathode in a spontaneous fashion.

5 This cell would produce 5.90 V but is not "practical" and would last only a few seconds:

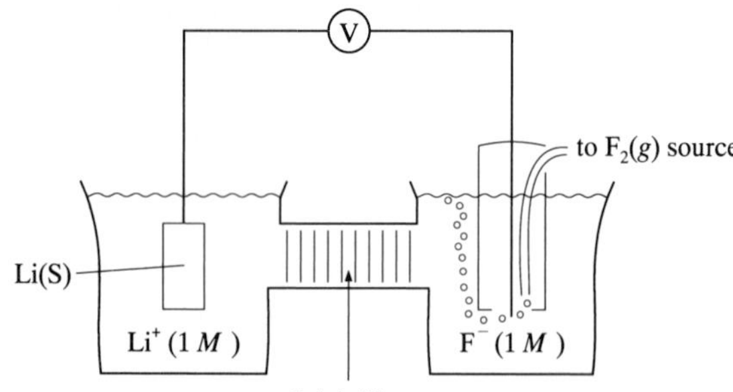

7 **(a)** no **(b)** no **(c)** yes **(d)** yes

9 For the anode, choose compounds easily oxidized. They are found on the bottom right portion of Table 17.1. For example, Li(*s*), Mg(*s*), Ba(*s*).

11 Corrosion of iron occurs in a series of three reactions:

$2Fe(s) + O_2(g) + 2H_2O(l) \longrightarrow 2Fe(OH)_2(s)$

$34Fe(OH)_2(s) + O_2(g) + 2H_2O(l) \longrightarrow 4Fe(OH)_3(s)$

$2Fe(OH)_3(s) \longrightarrow Fe_2O_3(H_2O)(s) + 2H_2O(l)$
[rust]

Methods of preventing corrosion are: painting the metal surface, plating the metal with a thin layer of another metal more resistant to oxidation, galvanization, and cathodic protection.

13 **(a)** storage cell—a cell which can be recharged; for example, an automobile battery.
(b) primary cell—a cell which cannot be recharged; for example, a flashlight battery.
(c) fuel cell—a cell in which the reactants are continuously supplied; for example, a hydrogen-oxygen fuel cell.

15 In molten sodium chloride, water is absent, and the sodium formed may be collected. In aqueous solution, water is more easily reduced to hydrogen gas than sodium ions to sodium metal.

17 One faraday is required for each mole of electrons needed to form each mole of metal.

19 **(a)** Anode: $Ni(s) \longrightarrow Ni^{2+}(aq) + 2e^-$; cathode: $Cu^{2+}(aq) + 2e^- \longrightarrow Cu(s)$ **(b)** anode: Ni; cathode: Cu **(c)** negative terminal: Cu; positive terminal: Ni **(d)** Electrons move toward the cathode. **(e)** 0.696 V **(f)** spontaneous **(g)** −0.696 V

21 $E^\circ_{C/C^{2+}} > E_{A/A^{2+}} {}^\circ E^\circ_{B/B^{2+}}$

Anode: $B(s) \longrightarrow B^{2+}(aq) + 2e^-$
Cathode: $C^{2+} + 2e^- \longrightarrow C(s)$
Cell: $B(s) + C^{2+}(aq) \longrightarrow B^{2+}(aq) + C(s)$

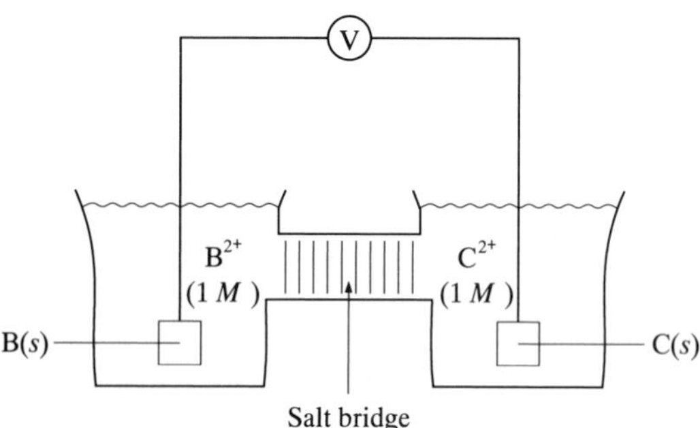

23 $BiO^+(aq) + 2H^+(aq) + 3e^- \longrightarrow Bi(s) + H_2O(l)$; $E^\circ_{BiO^+/Bi} = 0.320$ V

25 **(a)–(d)** Cl_2 converted to Cl^- **(e)** Cl_2 not converted to Cl^-

27 **(a)** $Mn(s) + 2H^+(aq) \longrightarrow Mn^{2+}(aq) + H_2(g)$

(b) $2Fe^{2+}(aq) + H_2O_2(aq) + 2H^+(aq) \longrightarrow 2Fe^{3+}(aq) + 2H_2O(l)$

(c) $Pb^{2+}(aq) + 2OH^-(aq) \longrightarrow Pb(OH)_2(s)$

(d) $2Hg_2^{2+}(aq) + O_2(g) + 4H^+(aq) \longrightarrow 4Hg^{2+}(aq) + 2H_2O(l)$

(e) no reaction

29 **(a)** voltage decreases **(b)** voltage unchanged **(c)** voltage increases **(d)** voltage decreases

31 0.25 V

33 1.136 V

35 pH = 0.86

37 **(a)** Anode: $2Br^-(l) \longrightarrow Br_2(l) + 2e^-$
Cathode: $2Li^+(l) + 2e^- \longrightarrow 2Li(s)$
Cell: $2Li^+(l) + 2Br^-(l) \longrightarrow 2Li(s) + B_2(l)$

(b) Anode: $2Br^-(aq) \longrightarrow Br_2(l) + 2e^-$
Cathode: $2H_2O(l) + 2e^- \longrightarrow H_2(g) + 2OH^-(aq)$
Cell: $2H_2O(l) + 2Br^-(aq) \longrightarrow H_2(g) + Br_2(l) + 2OH^-(aq)$

(c) Anode: $2Cl^-(l) \longrightarrow Cl_2(g) + 2e^-$
Cathode: $Sr^{2+}(l) + 2e^- \longrightarrow Sr(s)$
Cell: $Sr^{2+}(l) + 2Cl^-(l) \longrightarrow Sr(s) + Cl_2(g)$

(d) Anode: $2H_2O(l) \longrightarrow O_2(g) + 4H^+(aq) + 4e^-$
Cathode: $4H_2O(l) + 4e^- \longrightarrow 2H_2(g) + 4OH^-(aq)$
Cell: $2H_2O(l) \longrightarrow O_2(g) + 2H_2(g)$

(e) Anode: $2H_2O(l) \longrightarrow O_2(g) + 4H^+(aq) + 4e^-$
Cathode: $4H_2O(l) + 4e^- \longrightarrow 2H_2(g) + 4OH^-(aq)$
Cell: $2H_2O(l) \longrightarrow O_2(g) + 2H_2(g)$

(f) Anode: $2H_2O(l) \longrightarrow O_2(g) + 4H^+(aq) + 4e^-$
Cathode: $4Ag^+(aq) + 4e^- \longrightarrow 4Ag(s)$
Cell: $2H_2O(l) + 4Ag^+(aq) \longrightarrow O_2(g) + 4Ag(s) + 4H^+(aq)$

(g) Anode: $2H_2O(l) \longrightarrow O_2(g) + 4H^+(aq) + 4e^-$
Cathode: $2Cu^{2+}(aq) + 4e^- \longrightarrow 2Cu(s)$
Cell: $2H_2O(l) + 2Cu^{2+}(aq) \longrightarrow O_2(g) + 4H^+(aq) + 2Cu(s)$

39 (a) 3.00 F, 2.90×10^5 C (b) 9.70×10^{-3} F, 9.36×10^2 C

41 3.04×10^3 s

43 90.0%

45 $Cr^{3+}(aq) + 4H_2O(l) \longrightarrow H_2CrO_4(aq) + 6H^+(aq) + 3e^-$; 80.4 amp-hr

CHAPTER 18

Practice Problems

18.1 (a) oxidized only (b) reduced only (c) both

18.3 Li^+ has the greater ionic potential and is the better reducing agent.

18.5 (a) no reaction
(b) $2Al(s) + 3PbO(s) \xrightarrow{\Delta} 3Pb(s) + Al_2O_3(s)$

18.7 (a) $2ZnO(s) + C(s) \xrightarrow{\Delta} 2Zn(s) + CO_2(g)$
(b) $SnO(s) + CO(g) \xrightarrow{\Delta} Sn(s) + CO_2(g)$

Exercises

1 They can be oxidized, but not reduced under ordinary conditions.

3 A species in its highest oxidation state can act only as an oxidizing agent. A species in its lowest oxidation state can act only as a reducing agent. A species in an intermediate oxidation state can be either.

5 (a) groups I and II (b) the later, heavier transition elements (e.g., Os, Au, Ag) (c) the earlier transition elements and the metals in groups III–V

7 The three conceptual steps are the sublimation of the metal, ionization of the gaseous metal atoms, and hydration of the ions.

9 Free (elemental) metals can (and often do) act as reducing agents. It is extremely rare for them to act as oxidizing agents.

11 Metallurgy is the study and practice of the processes involved in producing metals and alloys from their ores.

13 (a) $2Na(s) + 2H_2O(l) \longrightarrow 2NaOH(aq) + H_2(g)$
(b) $Zn(s) + 2H_3O^+(aq) \longrightarrow Zn^{2+}(aq) + 2H_2O(l) + H_2(g)$
(c) $Cu(s) + 4H_3O^+(aq) + 2NO_3^-(aq) \longrightarrow Cu^{2+}(aq) + 2NO_2(g) + 6H_2O(l)$
(d) $Au(s) + 3NO_3^-(aq) + 6H_3O^+(aq) + 4Cl^-(aq) \longrightarrow AuCl_4^-(aq) + 3NO_2(g) + 9H_2O(l)$

15 HNO_3, $K_2Cr_2O_7$, $KMnO_4$, F_2, etc.

17 Virtually any metal—I^-, C, H_2, CO, Fe^{2+}, etc.

19 They generally are hard, high melting, dense, strong, of intermediate reactivity, exhibit multiple oxidation states, and have colored ions.

21 The early members (Sc, Y, La) show essentially only one state (3+). The number of states increases as you move to the right, reaches a maximum about the (Mn, Tc, Re) group, then drops off until at or near the end only one state may be exhibited (Zn, Ag, Cd, etc.).

23 Cu^+, Ag^+, Au^+, and Hg_2^{2+} are the only common ions in the 1+ oxidation state. Only for Ag^+ is it the only stable state.

25 H_2CrO_4, chromic acid; $HMnO_4$, permanganic acid.

27 (a) reducing (b) either (c) either (d) reducing (e) either (f) oxidizing (g) oxidizing (h) oxidizing (i) either (j) oxidizing (k) either (l) either (m) oxidizing (n) reducing

29 Since A is the better reducing agent (ΔH_{subl} and ΔH_{ion} are less endothermic for A), you would expect that A could reduce B from its oxide.

31 (a) $2Cs(s) + 2H_2O(l) \longrightarrow 2Cs^+(aq) + 2OH^-(aq) + H_2(g)$
(b) $2Al(s) + 6H_3O^+(aq) \longrightarrow 2Al^{3+}(aq) + 3H_2(g) + 6H_2O(l)$
(c) $Cu(s) + 2NO_3^-(aq) + 4H_3O^+(aq) \longrightarrow Cu^{2+}(aq) + 2NO_2(g) + 6H_2O(l)$
(d) $Fe_2O_3(s) + 3CO(g) \xrightarrow{\Delta} 2Fe(s) + 3CO_2(g)$
(e) $HgS(s) + O_2(g) \xrightarrow{\Delta} Hg(l) + SO_2(g)$
(f) $2Fe_2O_3(s) + 6Mg(s) \xrightarrow{\Delta} 4Fe(s) + 6MgO(s)$
(g) $PbO(s) + Zn(s) \xrightarrow{\Delta} ZnO(s) + Pb(s)$
(h) $Ni(s) + 2Ag^+(aq) \longrightarrow 2Ag(s) + Ni^{2+}(aq)$
(i) $2Al(s) + 3Cu^{2+}(aq) \longrightarrow 3Cu(s) + 2Al^{3+}(aq)$

33 (a) Anode: $2Br^- \longrightarrow Br_2 + 2e^-$
Cathode: $K^+ + e^- \longrightarrow K(l)$
Overall: $2K^+ + 2Br^- \longrightarrow 2K(l) + Br_2(g)$
(b) Anode: $2Cl^- \longrightarrow Cl_2(g) + 2e^-$
Cathode: $Ca^{2+} + 2e^- \longrightarrow Ca(l)$
Overall: $Ca^{2+} + 2Cl^- \longrightarrow Ca(l) + Cl_2(g)$

35 $5Fe^{2+}(aq) + MnO_4^-(aq) + 8H_3O^+(aq) \longrightarrow 5Fe^{3+}(aq) + Mn^{2+}(aq) + 12H_2O(l)$

37 (a) $Al(s) + 3Ag^+(aq) \longrightarrow 3Ag(s) + Al^{3+}(aq)$
(b) $2Cr^{3+}(aq) + 3H_2O_2(aq) + 10OH^-(aq) \longrightarrow 2CrO_4^{2-}(aq) + 8H_2O(l)$
(c) $Mg(s) + 2H_2O(l) \xrightarrow{\Delta} Mg(OH)_2(s) + H_2(g)$

39 Sc_2O_3, TiO_2, V_2O_5, CrO_3, Mn_2O_7, Fe_2O_3, CoO, NiO, CuO, ZnO

41 (a) $TiCl_4(l) + 2H_2O(l) \longrightarrow TiO_2(s) + 4HCl(g)$
(b) $4CrO_3(s) \xrightarrow{\Delta} Cr_2O_3(s) + 3O_2(g)$
(c) $CoCl_2(H_2O)_6(s) \xrightarrow{\Delta} CoCl_2(H_2O)_4(s) + 2H_2O(g)$
(d) $CuSO_4(H_2O)_5(s) \xrightarrow{\Delta} CuSO_4(s) + 5H_2O(g)$
(e) $AgBr(s) + 2S_2O_3^{2-}(aq) \longrightarrow Ag(S_2O_3)_2^{3-}(aq) + Br^-(aq)$
(f) $5MnO_4^{2-}(aq) + 8H_3O^+(aq) \longrightarrow 4MnO_4^-(aq) + Mn^{2+}(aq) + 12H_2O(l)$

43 (a) chromic acid, H_2CrO_4 (b) permanganic acid, $HMnO_4$

45 (a) Sc_2O_3, scandium(III) oxide
(b) CrO, chromium(II) oxide (chromous oxide) and CrO_3, chromium(VI) oxide
(c) MnO, manganese(II) oxide (manganous oxide) and Mn_2O_7, manganese(VII) oxide
(d) FeO, iron(II) oxide (ferrous oxide) and Fe_2O_3, iron(III) oxide (ferric oxide)
(e) Cu_2O, copper(I) oxide (cuprous oxide) and CuO, copper(II) oxide (cupric oxide)
(f) ZnO, zinc(II) oxide
(g) Y_2O_3, yttrium(III) oxide
(h) HgO, mercury(II) oxide (no mercury(I) oxide exists) (mercuric oxide)

47 (a) white pigment (b) generation of smoke screens (c) gemstone (d) oxidizing agent (e) humidity indicator (f) desiccant (g) photographic film (h) ointment (i) purgative

49 Ag and Hg are very close together in the activity series, so it would be likely that Ag_2O, like HgO, would decompose on heating. However, Cd and Zn are very far above Hg (much "less noble"), so it is unlikely that their oxides would decompose on heating.

51 chromium forms Cr^{2+} and Cr^{3+}; mercury forms Hg_2^{2+} and Hg^{2+}; iron forms Fe^{2+} and Fe^{3+}

53 They are either d^0 (e.g., Sc^{3+}) or d^{10} (e.g., Cu^+, Zn^{2+}).

55 (a) $2CrO_4^{2-}(aq) + 2H_3O^+(aq) \rightleftharpoons Cr_2O_7^{2-}(aq) + 3H_2O$
(b) When acid is added, the increased $[H_3O^+]$ causes the above equilibrium to shift to the right, favoring the orange $Cr_2O_7^{2-}$ over the yellow CrO_4^{2-}. No change in oxidation state takes place.

CHAPTER 19

Practice Problems

19.1 Ag^+ present; $Ag^+(aq) + Cl^-(aq) \longrightarrow AgCl(s)$
Ca^{2+} present; $Ca^{2+}(aq) + C_2O_4^{2-}(aq) \longrightarrow CaC_2O_4(s)$

19.3 Cu^{2+} absent, Hg^{2+} present, Sn^{2+} uncertain

19.5 9.6×10^{-4} *M*

19.7 Mn^{2+} present

19.9 (a) $SO_3^{2-}(aq) + 2H_3O^+(aq) \longrightarrow H_2SO_3(aq) + 3H_2O(l)$
$H_2SO_3(aq) \longrightarrow SO_2(g) + H_2O(l)$
(b) $CO_2(g) + Ba^{2+}(aq) + 2OH^-(aq) \longrightarrow BaCO_3(s) + H_2O(l)$
(c) $Pb^{2+}(aq) + H_2S(g) + 2H_2O(l) \longrightarrow 2H_3O^+(aq) + PbS(s)$
(d) $H_2SO_3(aq) + H_2O_2(aq) + H_2O(l) \longrightarrow SO_4^{2-}(aq) + 2H_3O^+(aq)$

19.11 (a) $Ba^{2+}(aq) + SO_3^{2-}(aq) \longrightarrow BaSO_3(s)$
(b) $Ag^+(aq) + Cl^-(aq) \longrightarrow AgCl(s)$

19.13 $AlCl_3$

Exercises

1 (a) Ag^+, Pb^{2+}, Hg_2^{2+}—all colorless (b) Cu^{2+} (blue), Pb^{2+} (colorless), Hg^{2+} (colorless), Sn^{2+} (colorless) (c) Al^{3+} (colorless), $Fe^{2+/3+}$ (pale green/yellow), Mn^{2+} (pale pink), Zn^{2+} (colorless), Cr^{3+} (blue), Ni^{2+} (green) (d) Ca^{2+}, Mg^{2+}—both colorless (e) Na^+, K^+, NH_4^+—all colorless.

3 NaCl, KCl, HCl

5 $(NH_4)_2CO_3$

7 (a) yellow-orange (b) violet

9 (a) They form insoluble Cl^- salts. (b) They form S^{2-} salts insoluble in acid solution. (c) They form S^{2-} or OH^- compounds insoluble in alkaline solution. (d) They form insoluble CO_3^{2-} salts. (e) They form no common insoluble salts.

11 (a) $SO_3^{2-}(aq) + 2H_3O^+(aq) \longrightarrow SO_2(g) + 3H_2O(l)$
(b) $2PO_4^{3-}(aq) + 3Ba^{2+}(aq) \longrightarrow Ba_3(PO_4)_2(s)$
(c) $Ag^+(aq) + I^-(aq) \longrightarrow AgI(s)$
(d) $2NO_3^-(aq) + 6I^-(aq) + 8H_3O^+(aq) \longrightarrow 3I_2(aq) + 2NO(g) + 12H_2O(l)$
(e) $10Br^-(aq) + 2MnO_4^-(aq) + 16H_3O^+(aq) \longrightarrow 2Mn^{2+} + 5Br_2(aq) + 24H_2O(l)$

13 Na^+, K^+, NH_4^+ (all); Ca^{2+}, Mg^{2+} (many)

15 $CaCO_3$, $Ba_3(PO_4)_2$, CaC_2O_4. These dissolve in HCl because they contain the anion of a weak acid. This anion is protonated by HCl, shifting the solubility equilibrium $MX \rightleftharpoons M^{2+}(aq) + X^{2-}(aq)$ to the right.

17 An anion which is the anion (conjugate base) of a weak acid, which can undergo hydrolysis to produce OH^-. Examples might be SO_3^{2-}, $C_2H_3O_2^-$, CO_3^{2-}.

19 Only Ag^+ is present, since its chloride does not dissolve in hot water.

21 (a) $Cl^-(aq) + Ag^+(aq) \longrightarrow AgCl(s)$
(b) $Pb^{2+}(aq) + SO_4^{2-}(aq) \longrightarrow PbSO_4(s)$
(c) $AgCl(s) + 2NH_3(aq) \longrightarrow Ag(NH_3)_2^+(aq) + Cl^-(aq)$ ($PbCl_2$ gives no reaction with NH_3)
(d) $Hg_2Cl_2(s) + 2NH_3(aq) \longrightarrow HgNH_2Cl(s) + Hg(l) + NH_4^+(aq) + Cl^-(aq)$

23 The solid must be just SnS_2, since PbS and HgS are insoluble in HCl.

25 Both Pb^{2+} and Cu^{2+} are present. The precipitate confirms the Pb^{2+} and the color confirms the Cu^{2+}.

27 At the point where ZnS begins to precipitate

$$K_{sp} = [Zn^{2+}][S^{2-}] = (.050)[S^{2-}] = 1.1 \times 10^{-21}$$

or $[S^{2-}] = 2.2 \times 10^{-20}$

$$[H_3O^+]^2[S^{2-}] = 1.3 \times 10^{-21}$$

or $[H_3O^+]^2 = (1.3 \times 10^{-21})/(2.2 \times 10^{-20})$

so $[H_3O^+] = 0.24$ *M* and pH = 0.57.

29 Cr^{3+} and Fe^{3+} are present; Mn^{2+} is doubtful. The yellow color of the first supernatant implies Cr^{3+} in the original solution; the blood-red color later implies Fe^{3+}. Mn^{2+}, if present, would precipitate as MnO_2 with the $Fe(OH)_3$ and dissolve in the H_2SO_4/H_2O_2. No confirming test was done for Mn^{2+}, however, so no conclusion can be reached.

31 (a) $ZnS(s) + 2H_3O^+(aq) \longrightarrow Zn^{2+}(aq) + 3H_2O(l)$
(b) $NiS(s) + 2NO_3^-(aq) + 4H_3O^+ \longrightarrow Ni^{2+}(aq) + S(s) + 2NO_2(g) + H_2O(l)$
(c) $2Cr^{3+}(aq) + 100H^-(aq) + 3H_2O_2(aq) \longrightarrow 2CrO_4^{2-}(aq) + 8H_2O(l)$
(d) $MnO_2(s) + 2H_3O^+(aq) + H_2O_2(aq) \longrightarrow Mn^{2+}(aq) + O_2(g) + 3H_2O(l)$
(e) $2Mn^{2+}(aq) + 5NaBiO_3(s) + 14H_3O^+(aq) \longrightarrow 2MnO_4^-(aq) + 5Bi^{3+}(aq) + 21H_2O + 5Na^+(aq)$
(f) $Ba^{2+}(aq) + CrO_4^{2-}(aq) \longrightarrow BaCrO_4(s)$ ($Al(OH)_4^-$ does not react)
(g) $Zn(NH_3)_4^{2+}(aq) + H_2S(aq) \longrightarrow ZnS(s) + 2NH_3(aq) + 2NH_4^+(aq)$

33 (a) HCl will precipitate AgCl but not K^+. (b) HCl will precipitate $PbCl_2$ but not Zn^{2+}. (c) NH_3 will precipitate $Al(OH)_3$ but not Cu^{2+}. (d) NaOH will precipitate $Fe(OH)_3$ but not Al^{3+}. (e) NH_3 will precipitate $Al(OH)_3$ but not Ni^{2+}. (f) NH_3 will precipitate $Fe(OH)_3$ but not Mg^{2+}. (g) $(NH_4)_2CO_3$ will precipitate $CaCO_3$ but not NH_4^+.

35 Ag^+, Ca^{2+} present; Cr^{3+}, Ni^{2+} absent

37 Pb^{2+}, Al^{3+}, NH_4^+ present; all others absent

39 (a) Acidify with HCl. Bad smell indicates H_2S. Add Ba^{2+} to the solution. A white precipitate indicates SO_4^{2-}.
(b) Acidify with HNO_3, then add Ba^{2+}. A white precipitate indicates SO_4^{2-}. To the supernatant, add Ag^+. A white precipitate indicates Cl^-.
(c) Acidify with HCl and add Ba^{2+}. A white precipitate indicates SO_4^{2-}. To the supernatant, add H_2O_2. A white precipitate indicates SO_3^{2-}.
(d) Add Ba^{2+}. A white precipitate indicates CO_3^{2-}. To the supernatant, add Ag^+. A black precipitate indicates S^{2-}. To the supernatant, add H_2SO_4. The odor of vinegar indicates $C_2H_3O_2^-$.
(e) Acidify and add NO_2^- and C_6H_{12}. A violet color in the C_6H_{12} layer indicates I^-. To the water layer, add MnO_4^- and more C_6H_{12}. An orange color in the C_6H_{12} layer indicates Br^-. To the water layer, add Ag^+. A white precipitate indicates Cl^-.

(f) Add Ba^{2+}. A white precipitate indicates the presence of PO_4^{3-}. To the supernatant, add Ag^+ (*not* as $AgNO_3$!) A white precipitate indicates Cl^-. To the supernatant, add acid and I^-. A brown color indicates the presence of NO_3^-.
(g) Acidify the solution with HNO_3. A rotten-egg odor indicates S^{2-}. Then add Ba^{2+}. A white precipitate indicates SO_4^{2-}. To the supernatant, add MnO_4^-. A brown coloration indicates Br^-. Finally, add Ag^+. A white precipitate indicates Cl^-.

41 SO_3^{2-}, Br^-, PO_4^{3-} present; Cl^- absent

43 **(a)** Water would dissolve $AgNO_3$, not AgCl. **(b)** Water would dissolve $BaCl_2$, not $BaSO_4$. **(c)** Water (cool) would dissolve NH_4NO_3, not $PbCl_2$. **(d)** Water would dissolve NH_4Cl, not $PbSO_4$. **(e)** Water would dissolve Na_2S, not Ag_2S. **(f)** HCl would dissolve $BaCO_3$, not $BaSO_4$. **(g)** NH_3 would dissolve $NiCO_3$, not $BaCO_3$; *or:* H_2SO_4 would dissolve $NiCO_3$, not $BaCO_3$. **(h)** NH_3 would dissolve $Cu(OH)_2$, not $Mg(OH)_2$. **(i)** NaOH would dissolve $Al(OH)_3$, not $Fe(OH)_3$. **(j)** NH_3 would dissolve $Zn(OH)_2$, not $Al(OH)_3$. **(k)** HCl would dissolve $BaSO_3$, not $BaSO_4$. **(l)** HCl would dissolve MnS, not HgS.

45 Na^+, K^+, NH_4^+

47 HgS, $FeCl_3$, $NiCl_2$ present; ZnS uncertain

CHAPTER 20

Practice Problems

20.1 **(a)** $\Delta S < 0$ **(b)** $\Delta S < 0$ **(c)** $\Delta S < 0$ **(d)** $\Delta S > 0$ **(e)** $\Delta S > 0$

20.3 $\Delta G° = -958.2$ kJ

20.5 $\Delta G = 1.02 \times 10^5$ J

20.7 $\Delta G° = 2.71 \times 10^4$ J

20.9 $K = 2.94 \times 10^{13}$

Exercises

1 A spontaneous process is a process which occurs by itself, without outside influence or an external source of work or energy. Examples of spontaneous processes: precipitation reactions, corrosion, gas expanding into a vacuum; nonspontaneous processes: metal refining, boiling of water, silverplating.

3 A solid has a well-ordered lattice. Upon melting, the lattice is disrupted into a less ordered state. Upon vaporization, the liquid becomes gaseous and has the highest degree of disorder.

5 The second law of thermodynamics refers to an increase in entropy of the universe. The universe consists of a system and its surroundings. If the increase in entropy of the surroundings exceeds the decrease in entropy of the system, the process is spontaneous.

7 Increasing the temperature usually increases the entropy.

9 $\Delta S = \dfrac{q_{rev}}{T}$

11 $S_{solid} < S_{liquid} < S_{gas}$

13 $\Delta S = \Sigma\, nS_{products} - \Sigma\, nS_{reactants}$

15 If $\Delta G < 0$, the process is spontaneous; if $\Delta G > 0$, the process is nonspontaneous; if $\Delta G = 0$, the process is at equilibrium.

17 $\Delta G° = \Sigma\, n\Delta G_f°$ of products $- \Sigma\, n\Delta G_f°$ of reactants

19 At equilibrium $\Delta G = 0$

21 $Q = K$ at equilibrium

23 $\Delta G° = -n\text{FE}°$

25 $\Delta S_{fus} = 0.0488$ kJ mol^{-1} K^{-1}

27 $\Delta S_{vap} = \dfrac{\Delta H_{vap}}{T_b}$

29 $S°$ of Fe(s) $= 25.17$ J mol^{-1} K^{-1}

31 $\Delta G = 16.4$ kJ mol^{-1}; nonspontaneous; $T = 489$ K

33 $\Delta G° = 209.2$ kJ mol^{-1}

35 $\Delta G_f°$ of CaO(s) $= -604.2$ kJ mol^{-1}

37 $\Delta G = -3.66 \times 10^3$ J; spontaneous

39 spontaneous: $\Delta G = -8.47 \times 10^2$ J

41 $\Delta G° = -1.01 \times 10^6$ J

43 $K = 1.06 \times 10^{-30}$; a large, positive $\Delta G°$ value and small K value indicate the forward reaction is not favored.

45

	$E°$, V	$\Delta G°$, J	K
(a)	1.103	-2.13×10^5	2.17×10^{37}
(b)	0.4665	-9.00×10^4	5.97×10^{15}
(c)	-0.725	1.40×10^5	2.97×10^{-25}
(d)	0.721	-3.48×10^5	9.56×10^{60}

47 $E° = 0.66$ V; $\Delta G° = -3.2 \times 10^5$ J; $K = 6.6 \times 10^{55}$; spontaneous

CHAPTER 21

Practice Problems

21.1 2.8×10^{-5} Ms^{-1}. Rate increases with increasing concentration.

21.3 **(a)** 9.1×10^{-2} M^{-2} s^{-1} **(b)** 1.9×10^{-2} Ms^{-1} **(c)** 8.0×10^{-2} Ms^{-1}

21.5 **(a)** 27.7 min **(b)** .172 M

21.7 1.4×10^2 kJ mol^{-1}

21.9 **(a)** $2H_2O_2(aq) \longrightarrow 2H_2O(l) + O_2(g)$
(b) Rate $= k[H_2O_2][I^-]$

Exercises

1 Temperature, concentration, catalyst

3 **(a)** The order is the power to which the concentration is raised in the rate law. In Rate $= k[A]^n[B]^m$, **(b)** the order in A is n, **(c)** the order in B is m, and **(d)** the overall order is $n + m$.

5 You would need to determine the rate at a number of combinations of concentrations of A and B, and see how the rate varied as a function of the concentrations.

7 $\ln([A]_0/[A]_t) = kt$

9 The half-life of a reaction is the time required for the concentration of the reactant to drop to one-half of its initial value.

11 $k = Ae^{-E_a/RT}$ or $\ln k = \ln A - E_a/RT$ where k = rate constant, A = frequency factor, e = base of natural logarithms (2.71828...), E_a = activation energy, R = gas constant, and T = absolute temperature. If E_a is larger, k will be smaller, since the negative exponent will be smaller in magnitude. If T is larger, k will be larger, since the negative exponent will be larger in magnitude.

13 **(a)** Collision theory of reaction rate works from the assumption that molecules in the reaction mixture are constantly colliding with one another, but that only a fraction (given by $e^{-E_a/RT}$) of the collisions have sufficient energy to lead to reaction.
(b) In this theory, the transition state (or activated complex) is an unstable combination of atoms which forms temporarily on the reaction pathway between reactants and products. It represents the point of maximum energy on the reaction coordinate.

15 A catalyst gives the reactants an alternate pathway (mechanism) of lower activation energy.

17 An elementary reaction is a reaction that occurs in one step, exactly as it is written.

19 If a reaction is an elementary one, it is assumed that it occurs exactly as written, so its order and molecularity are the same. This means the orders equal the stoichiometric coefficients. No experiments are necessary.

21 **(a)**

[CH_3COCl]: 1.2, 1.1, 1.0, .9, .8, .7, .6

t, s: 0, 2, 4, 6, 8, 10

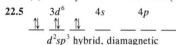

(b) -5.4×10^{-2} Ms^{-1} **(c)** $+5.4 \times 10^{-2}$ Ms^{-1}
(d) -6.0×10^{-2} Ms^{-1}

23 $\Delta[NO]/\Delta t = -0.34$ torr s^{-1}; $\Delta[O_2]/\Delta t = 0.17$ torr s^{-1}

25 **(a)** Rate $= k[NO][O_3]$ **(b)** $k = 2.20 \times 10^7\ M^{-1}\ s^{-1}$
(c) 5.50×10^{-4} Ms^{-1}

27 **(a)** Rate $= k[NO]^2[Br_2]$ **(b)** 1st order in Br_2, 2nd order in NO, 3rd order overall **(c)** $k = 6.0 \times 10^3\ M^{-2}\ s^{-1}$
(d) 1.2×10^3 Ms^{-1} **(e)** 9 times, 3 times, 27 times

29 9.9×10^4 s

31 0.35 hr^{-1}

33 **(a)** 106 kJ mol^{-1} **(b)** 6.0×10^{14} L mol^{-1} s^{-1}
(c) 3.6 times

35 349 kJ mol^{-1}

37 **(a)** 1.7×10^5 **(b)** 4.7×10^{15}

39 **(a), (b)** bimolecular; **(c), (d)** unimolecular;
(e), (f) termolecular

41 Only **(c)**

43 Yes

CHAPTER 22

Practice Problems

22.1 **(a)** $Cu(H_2O)_4^{2+}$; $[Cu(H_2O)_4]Br_2$

22.3 **(a)** tetraammine dichlorocobalt(III) ion
(b) trisethylenediammine nickel(II) nitrate

22.5 $3d^6$ ⇅ ⇅ ⇅ $4s$ __ $4p$ __ __ __
d^2sp^3 hybrid, diamagnetic

Exercises

1 **(a)** A ligand is an atom or group of atoms acting as a Lewis base (i.e., donating a pair of electrons) towards a metal ion acting as a Lewis acid. **(b)** A coordination sphere is a metal ion with the ligands bonded directly to it.

3 A coordination complex *can* contain more than one coordination sphere as in $[Ag(NH_3)_2][Co(NH_3)_2(CN)_4]$

5 The coordination number is the number of donor atoms attached to the metal.

7 A chelating agent is the ligand which can bond to the metal to form a ring; a chelate is the complex formed by that interaction. $H_2NCH_2CH_2NH_2$ is a chelating agent; $[Co(H_2NCH_2CHNH_2)_3]^{3+}$ is a chelate.

9 **(a)** Structural isomers have the same formula but a different pattern of bonds and bonded atoms. Stereoisomers have the same formula, the same pattern of bonds and bonded atoms, but a different arrangement of the bonds in space.

11 **(a)** H_3N, H_3N / Pt / Cl, Cl and H_3N, Cl / Pt / Cl, NH_3

(b)

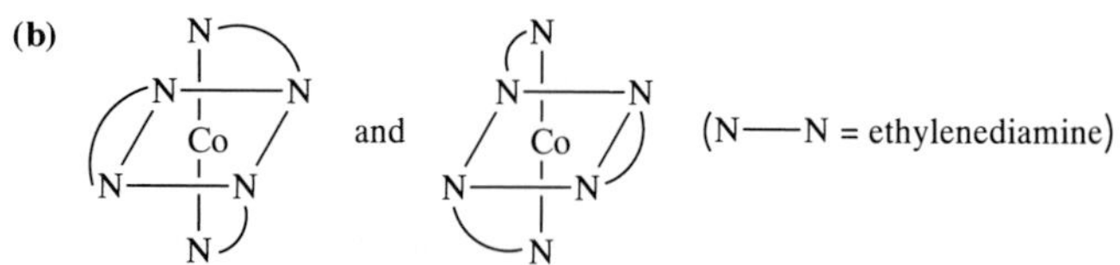

13 A racemic mixture is a mixture of equal amounts of the dextro and levo isomers of an optically active compound.

15 **(a)** d^2sp^3 **(b)** sp^3d^2

17 The electrons donated by the ligands repel the electrons in the *d* orbitals of the metal ion. The stronger the repulsion, the more destabilized (i.e., the higher the energy of) the metal electrons. Electrons in metal orbitals which point most directly at the incoming ligands are most affected.

19 The stronger the ligand, the greater the splitting of the *d* orbitals, increasing the chances that it will be a low-spin complex. For example, consider $Fe(CN)_6^{3-}$ and $Fe(H_2O)_6^{3+}$. Both contain Fe(III), d^5, but $Fe(CN)_6^{3-}$ is low-spin and $Fe(H_2O)_6^{3+}$ is high-spin:

$Fe(CN)_6^{3-}$	$Fe(H_2O)_6^{3+}$
$d_{x^2-y^2}$ __ d_{z^2} __	$d_{x^2-y^2}$ ↑ d_{z^2} ↑
d_{xy} ⇅ d_{xz} ⇅ d_{yz} ↑	d_{xy} ↑ d_{xz} ↑ d_{yz} ↑

21 Near 700 nm

23 $[Cr(NH_3)_5Cl]Cl_2$; Cr(III)

25 **(d)** -2 **(g)** $+3$

27 $[Cu(NH_3)_4]_3[Fe(CN)_6]_2$

29 **(a), (b), (e)** monodentate **(c)** terdentate **(d)** bidentate

31 **(a)** diamminesilver(I) ion **(b)** tetraaquacopper(II) ion
(c) hexacyanoterrate(III) ion **(d)** hexacyanoferrate(II) ion
(e) hexaammine cobalt(III) bromide
(f) tetraaquadichlorochromium(III) nitrite **(g)** potassium hexachloroplatinate(IV) **(h)** sodium tetrahydroxogincate(II)
(i) dichlorodifluorocobaltate(II) ion
(j) tetraamminecopper(II) hexacyanoferrate(III)
(k) trisethylenediamine manganese(II) ion

33 **(a)** and **(b)** are ionization isomers; **(c)** is a linkage isomer.

35

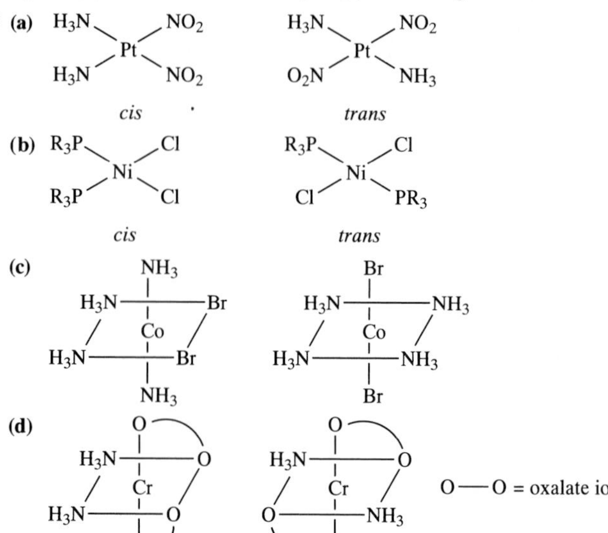

37 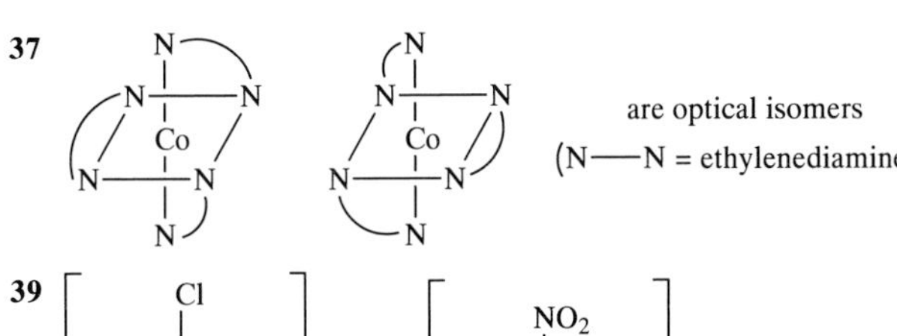

39 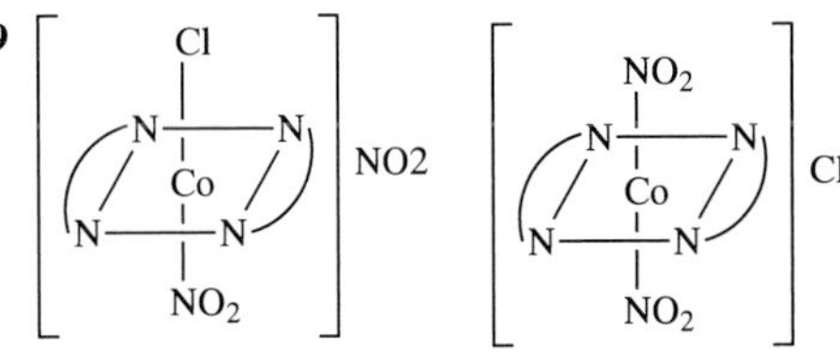

no reaction with Ag⁺ or en — no reaction with en; reacts with Ag⁺

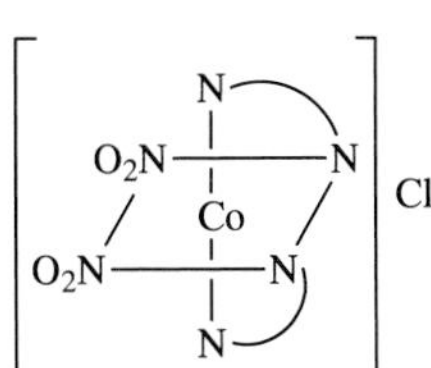 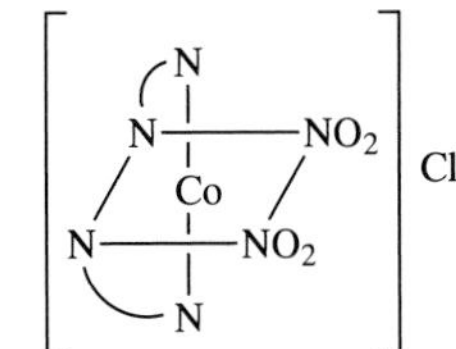

reacts with Ag⁺ and en

41 **(a), (b), (c)** octahedral **(d)** linear **(e)** square planar **(f), (g)** tetrahedral

43 $Co(CH)_6^{3-}$ is an inner-orbital complex, $COCl_6^{3-}$ is an outer-orbital complex.

45 **(a)** diamagnetic **(b)** and **(c)** paramagnetic

47 CU(II) has an odd total number of electrons and 9 in the *d* orbitals. All CU(II) complexes have one unpaired electron, so magnetic measurements would not be of help here.

49
```
(a)  __  __              (b)  ↑   ↑              (c)  __  __
    ⇅   ⇅   ↑                 ↑   ↑   ↑              ⇅   ⇅   ⇅
(d)  ↑   ↑               (e)  __  __
    ⇅   ↑   ↑                 ⇅   ⇅   ↑
```

51 Both contain Cr(II), d^4
```
 ↑  __            __  __
 ↑   ↑   ↑        ⇅   ↑   ↑
  CrCl6 4−         Cr(CN)6 4−
```
$CrCl_6^{4-}$ would be more strongly attracted to a magnet, since it has four unpaired electrons as compared with two in $Cr(CN)_6^{4-}$

53 **(b), (d),** and **(g)** should be colorless, since they have either a d^{10}(**b** and **d**) or d^0(**g**) configuration.

55 Since the complex absorbs in the blue region of the spectrum, it would reflect or transmit some other color or blend of colors.

CHAPTER 23

Practice Problems

23.1 **(a)**
```
       CH3
       |
CH3—CHCHCH2CH3
         |
         CH3
```

(b)
```
      CH3     CH2CH3
      |       |
CH3—C—CH2CHCH2CH2CH3
      |
      CH3
```

23.3 **(a)**
```
       CH3
       |
CH3—CH—C=CHCH2CH2CH3
          |
          CH3
```

(b)
```
                    CH3
                    |
CH3CH2CH=CHCHCH—CH3
                 |
                 CH3
```

(c)
```
      CH3      CH3
      |        |
CH3—C=CHCH2CHCH3
```

Exercises

1 Aliphatic hydrocarbons consist of carbon compounds such as alkanes, alkenes, and alkynes. They may be "open chain" or "cyclic." Aromatic hydrocarbons are cyclic and are linked by sigma bonds and delocalized pi bonds. The chemical properties of aromatic hydrocarbons are very different from those of aliphatic hydrocarbons. An example of an aliphatic hydrocarbon is butane, $CH_3CH_2CH_2CH_3$, the fuel used in camp stoves. An example of an aromatic hydrocarbon is benzene, C_6H_6, a common organic solvent.

3 **(a)** C_6H_{14}, hexane **(b)** C_6H_{12}, hexene **(c)** C_6H_{10}, hexyne **(d)** C_6H_{12}, cyclohexane **(e)** C_6H_{10}, cyclohexene

5 The two isomers of butane are:
```
                        CH3
                        |
CH3CH2CH2CH3 and CH3CHCH3
```

7 **(a)** $2C_4H_{10}(g) + 13O_2(g) \longrightarrow 8CO_2(g) + 10H_2O(g)$
(b) $CH_4(g) + Cl_2(g) \longrightarrow CH_3Cl(g) + HCl(g)$

9 alkane

11 CH_3Cl, chloromethane—used as a refrigerant; CCl_4, carbon tetrachloride—used in dry cleaning

13 A functional group results from the replacement of one or more hydrogen(s) on an alkane with a multiple bond or a heteroatom.

15 primary alcohol: $CH_3CH_2—OH$
```
secondary alcohol:    OH
                      |
                   CH3CHCH3
tertiary alcohol:     OH
                      |
                 CH3—C—CH3
                      |
                      CH3
```

17 a ketone

19 An ether can be prepared by the elimination of a water molecule from two molecules of alcohol using a suitable catalyst.

21 $C_6H_6 + Cl_2 \xrightarrow{FeCl_3} C_6H_5Cl + HCl$

23 nucleic acids, carbohydrates, proteins, and lipids

25 starch and cellulose

27 The monomeric units of proteins are amino acids. They are linked by amide bonds. This amide bond is called a peptide linkage.

29 DNA is found in the nucleus of a cell, and is primarily responsible for transmitting genetic information from one generation to the next during cell division. RNA is found in the cytoplasm outside the nucleus and is responsible for protein synthesis in the cell.

31 **(a)** h-hexane **(b)** h-pentane **(c)** 3-methylhexane **(d)** 3,4-dimethylheptane **(e)** methylpropane **(f)** dimethylpropane **(g)** 2,2,3-trimethylpentane

33 **(a)** propene **(b)** 2-butyne **(c)** 2-pentyne **(d)** 3-ethyl-4-methyl-2-pentene **(e)** 3,3,5,5-tetramethyl-1-hexyne **(f)** 1,2-dichloroethane **(g)** 2-chloro-2-methylpentane **(h)** trichloromethane **(i)** 1-chloro-2-fluoroethene

35 **(a)** 1-propanol **(b)** methanol **(c)** 1-methoxypropane **(d)** propanoic acid **(e)** methyl methanoate **(f)** 2-butanol **(g)** propanol **(h)** ethylbutanoate **(i)** butanone **(j)** methylamine **(k)** diethylamine

37 **(a)** both carbons are tetrahedral **(b)** both carbons are trigonal planar **(c)** both carbons are linear **(d)** oxygen is bent **(e)** carbon is trigonal planar **(f)** carbon is trigonal planar; oxygen is bent **(g)** oxygen is bent **(h)** carbon on left is tetrahedral; oxygen is bent; carbon on right is trigonal planar **(i)** carbon is tetrahedral **(j)** carbon is trigonal planar

39 **(a)** 10σ **(b)** $8\sigma, 1\pi$ **(c)** $6\sigma, 2\pi$ **(d)** 5σ **(e)** $9\sigma, 1\pi$ **(f)** $7\sigma, 1\pi$ **(g)** 11σ **(h)** $10\sigma, 1\pi$ **(i)** 18σ **(j)** $16\sigma, 1\pi$

41 **(a)**

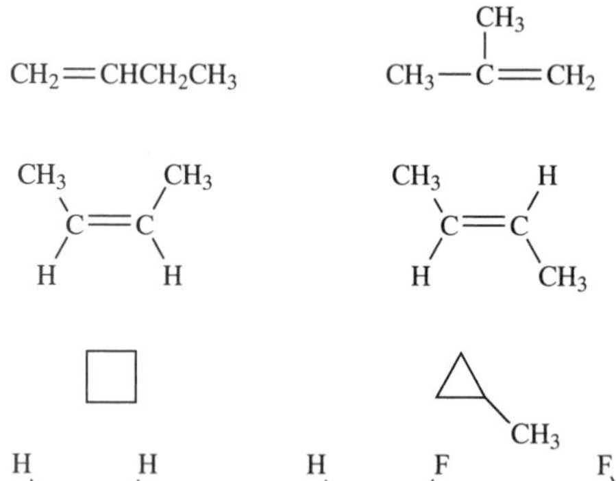

$CH_2{=}CHCH_2CH_3$ $CH_3{-}C(CH_3){=}CH_2$

$(CH_3)(H)C{=}C(CH_3)(H)$ (cis) $(CH_3)(H)C{=}C(H)(CH_3)$ (trans)

□ (cyclobutane) △–CH_3 (methylcyclopropane)

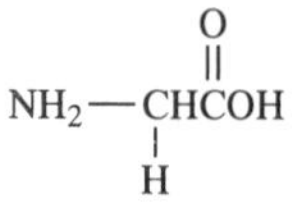

(b) $(H)(F)C{=}C(H)(F)$ $(H)(F)C{=}C(F)(H)$ $(F)(F)C{=}C(H)(H)$

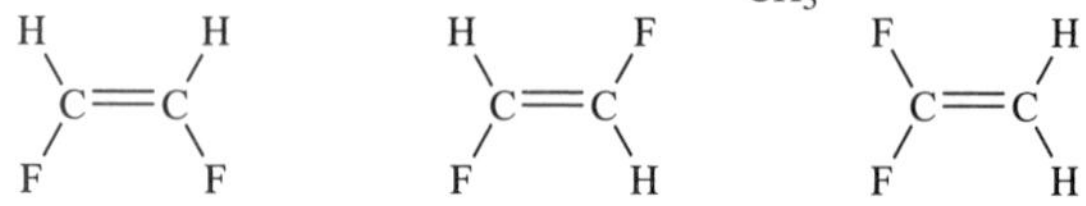

43 **(a)** CH_2OH; H, O, OH; H; OH, H; HO, H; H, OH (pyranose ring)

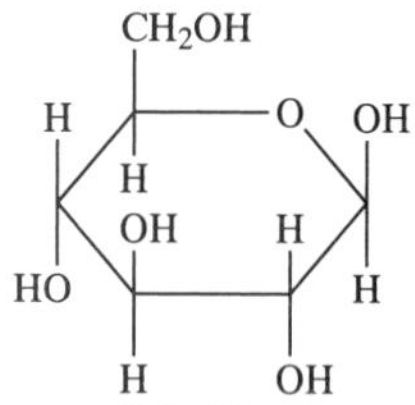

(b) CH_2OH; H, O, H; H; OH, H; HO; H, OH —O— $HOCH_2$, O, H; H, HO; CH_2OH; OH, H (pyranose ring linked through O to furanose ring)

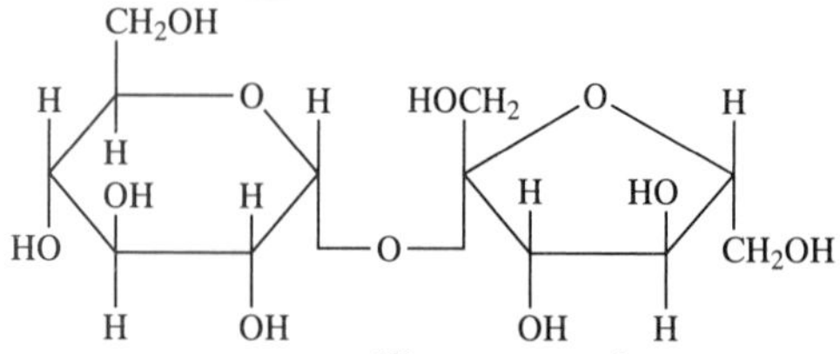

(c) $NH_2{-}\underset{\displaystyle H}{\underset{|}{C}}H{-}\overset{\displaystyle O}{\overset{\|}{C}}OH$ **(d)** $NH_2{-}\underset{\displaystyle CH_3}{\underset{|}{C}}H{-}\overset{\displaystyle O}{\overset{\|}{C}}OH$

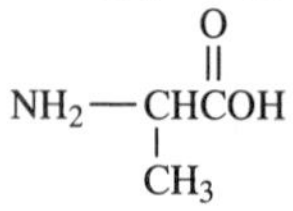

45 **(a)** aldehyde, alcohol **(b)** ketone **(c)** amine, carboxylic acid **(d)** ester

47 **(a)** Fats are the major energy reserve of humans and animals.
(b) Waxes provide protective coats on the surfaces of plants and many animals.

49 **(a)** DNA—deoxyribonucleic acid
(b) RNA—ribonucleic acid

51 H-bonding forces hold two strands of double helical DNA together.

CHAPTER 24

Practice Problems

24.1 ${}^{230}_{90}Th \longrightarrow {}^{4}_{2}He + {}^{226}_{88}Ra$; the name of the daughter nuclide is radium-226

24.3 **(a)** stable; **(b)** radioactive

24.5 ${}^{27}_{13}Al + {}^{4}_{2}He \longrightarrow {}^{30}_{15}P + {}^{1}_{0}n$
Shorthand notation: ${}^{27}_{13}Al(\alpha, n){}^{30}_{15}P$

24.7 0.253 g

24.9 mass defect = 0.0005 amu; binding energy = 7×10^{-14} J; 0.5 MeV

Exercises

1

Ray	Mass	Charge
α	4	+2
β	~0	−1
γ	0	0

3 Radioactive decay is the emission of radiant energy and material particles by a radionuclide.

5 **(a)** mass decreases, charge decreases **(b)** mass unchanged, charge increases **(c)** mass unchanged, charge decreases **(d)** mass unchanged, charge decreases

7 A "magic number" is the number of protons or neutrons found in a nucleus having exceptional stability. Nuclei having 2, 8, 20, 28, 50, or 82 protons or neutrons are more stable than other nuclei.

9 Nuclei with a 1:1 neutron-to-proton ratio lie in a "belt of stability." Nuclei outside this belt of stability are unstable.

11 radioactive decay series is a series of nuclear reactions undergone by a large, unstable nucleus to produce a stable product lying within the belt of stability. The first element described in this chapter to undergo a radioactive decay series is uranium-238; this series ends with stable lead-206.

13 Cyclotrons and linear accelerators produce high-energy particles for the study of nuclear reactions and to produce artificial radionuclides. Cyclotrons are circular in design, and linear accelerators are linear in design.

15 technetium, atomic number = 43; promethium, atomic number = 61

17 **(1)** Calculate the mass of uranium that decayed to lead.
(2) Determine the mass of uranium that was present when the rock was formed.
(3) Calculate the decay rate constant.
(4) Using the rate equation for first-order decay, calculate the age of the rock.

19

Isotope	Use
${}^{14}C$	dating archeological artifacts which are carbon based
${}^{60}Co$	treatment of cancer
${}^{131}I$	diagnosis of the condition of the thyroid gland
${}^{111}In$	location of brain tumors

21 The experimental mass and calculated mass of a nucleon are needed to determine the mass defect. The mass defect and speed of light are needed to calculate the binding energy.

23 **(a)** Nuclear fission is the splitting of large nuclei; **(b)** Nuclear fusion is the joining of small nuclei.

25 Uranium-235 and plutonium-239

27 A hydrogen bomb is called a thermonuclear weapon because detonation requires a very high temperature. The high temperature (40,000,000 K) is achieved by using a conventional fission bomb to trigger the fusion reaction.

29 Data needed: calculated mass of nucleon; experimental mass of nucleon; mass defect

31 **(a)** $^{225}_{89}Ac \longrightarrow ^{4}_{2}He + ^{221}_{87}Fr$
(b) $^{40}_{19}K \longrightarrow ^{0}_{-1}e + ^{40}_{20}Ca$
(c) $^{13}_{7}N \longrightarrow ^{0}_{-1}e + ^{13}_{6}C$
(d) $^{51}_{24}Cr + ^{0}_{-1}e \longrightarrow ^{51}_{23}V$

33 **(a)** stable; has a "magic number" of protons and neutrons, hence "doubly magic." **(b)** unstable and radioactive; has an odd number of neutrons and an even number of protons. **(c)** stable; has an even number of neutrons and an even number of protons. **(d)** unstable and radioactive; has an even number of neutrons and an odd number of protons **(e)** unstable and radioactive; has more than 83 protons

35 **(a)** $^{23}_{11}Na \longrightarrow ^{4}_{2}He + ^{19}_{9}F$ **(b)** $^{24}_{11}Na \longrightarrow ^{0}_{-1}e + ^{24}_{12}Mg$
(c) $^{102}_{47}Ag \longrightarrow ^{4}_{2}He + ^{98}_{45}Rh$ **(d)** $^{238}_{93}Np \longrightarrow ^{0}_{-1}e + ^{238}_{92}Cl$

37 **(a)** $^{12}_{6}C$ **(b)** $^{96}_{40}Zr$ **(c)** $^{1}_{0}n$ **(d)** $^{4}_{2}He$ **(e)** $^{14}_{7}N$ **(f)** $^{1}_{1}H$
(g) $^{13}_{7}N$

39 **(a)** $^{10}_{5}B + ^{0}_{1}n \longrightarrow ^{4}_{2}He + ^{6}_{4}Be$
(b) $^{43}_{20}Ca + ^{4}_{2}He \longrightarrow ^{1}_{1}H + ^{46}_{21}Sc$
(c) $^{82}_{35}Br + ^{1}_{0}n \longrightarrow \gamma + ^{83}_{35}Br$
(d) $^{238}_{92}U + ^{22}_{10}Ne \longrightarrow 4^{1}_{0}n + ^{256}_{102}No$

41 7.1×10^{-3} mg; 9.9×10^{-4} mg

43 $t_{1/2} = 60.1$ min

45 4.20×10^{9} yr

47 8.58×10^{8} yr

49 1.35×10^{4} yr

51 1.52×10^{4} gallons

APPENDIX B

Ionization Energies, Electron Affinities, and Electronegativities

First and Second Ionization Energies of Elements (kJ mol^{-1})

Key: Li 520 ← First ionization energy; 7297 ← Second ionization energy

H 1311																	He 2372 5250
Li 520 7297	Be 899 1757											B 801 2427	C 1087 2353	N 1402 2855	O 1313 3388	F 1681 3376	Ne 2081 3963
Na 495 4563	Mg 737 1450											Al 577 1816	Si 786 1577	P 1011 1903	S 1000 2258	Cl 1255 2297	Ar 1521 2665
K 419 3070	Ca 590 1145	Sc 631 1235	Ti 658 1310	V 650 1414	Cr 652 1591	Mn 717 1509	Fe 762 1561	Co 758 1645	Ni 736 1751	Cu 745 1958	Zn 906 1733	Ga 579 1979	Ge 760 1537	As 947 1798	Se 941 2075	Br 1143 2084	Kr 1351 2370
Rb 403 2653	Sr 549 1064	Y 616 1180	Zr 660 1267	Nb 664 1382	Mo 685 1558	Tc 703 1473	Ru 710 1617	Rh 720 1744	Pd 804 1874	Ag 731 2073	Cd 868 1631	In 588 1820	Sn 708 1412	Sb 834 1592	Te 869 1795	I 1008 1842	Xe 1171 2046
Cs 375 2422	Ba 503 965	La 541 1103	Hf 676 1438	Ta 760 1563	W 770 1708	Re 759 1602	Os 840 1640	Ir 868	Pt 868 1795	Au 888 1978	Hg 1006 1809	Tl 590 1971	Pb 716 1450	Bi 703 1610	Po 813 1834	At 917 1940	Rn 1037
Fr 370 2123	Ra 510 976	Ac 666 1168															

Ce 540 1187	Pr 529	Nd 531	Pm	Sm 540	Eu 547	Gd 594	Tb 577	Dy 656	Ho	Er 587	Tm 561	Yb 600	Lu 593
Th 671	Pa	U 587	Np	Pu 560	Am 579	Cm	Bk	Cf	Es	Fm	Md	No	Lr

Electron Affinities of Main Group Elements (kJ mol^{-1}) (Values from H. Hotop and W. C. Lineberger. *J. Phys. Chem. Ref. Data*, 1975, 4, 539.)

H -73							He 0
Li -60	Be 0	B -27	C -122	N +7	O -141	F -328	Ne 0
Na -53	Mg 0	Al -44	Si -134	P -71.7	S -200	Cl -349	Ar 0
K -48	Ca 0	Ga -29	Ge -120	As -77	Se -195	Br -325	Kr 0
Rb -47	Sr 0	In -29	Sn -121	Sb -101	Te -190	I -295	Xe 0
Cs -45	Ba 0	Tl -30	Pb -110	Bi -110	Po -180	At -270	Rn 0

Pauling's Electronegativities of Elements

H 2.1																
Li 1.0	Be 1.5											B 2.0	C 2.5	N 3.0	O 3.5	F 4.0
Na 0.9	Mg 1.2											Al 1.5	Si 1.8	P 2.1	S 2.5	Cl 3.0
K 0.8	Ca 1.0	Sc 1.3	Ti 1.5	V 1.6	Cr 1.6	Mn 1.5	Fe 1.8	Co 1.9	Ni 1.9	Cu 1.9	Zn 1.6	Ga 1.6	Ge 1.8	As 2.0	Se 2.4	Br 2.8
Rb 0.8	Sr 1.0	Y 1.2	Zr 1.3	Nb 1.6	Mo 1.8	Tc 1.9	Ru 2.2	Rh 2.2	Pd 2.2	Ag 1.9	Cd 1.7	In 1.7	Sn 1.8	Sb 1.9	Te 2.1	I 2.5
Cs 0.8	Ba 1.0	La 1.1	Hf 1.3	Ta 1.5	W 1.7	Re 1.9	Os 2.2	Ir 2.2	Pt 2.2	Au 2.4	Hg 1.9	Tl 1.8	Pb 1.9	Bi 1.9	Po 2.0	At 2.2
Fr 0.8	Ra 1.0	Ac 1.1	Th 1.3	Pa 1.4												

APPENDIX C

Atomic and Ionic Radii

Atomic and ionic radii (in Å) of the main group elements. 1 Å = 0.1 nm = 10^{-10} m.

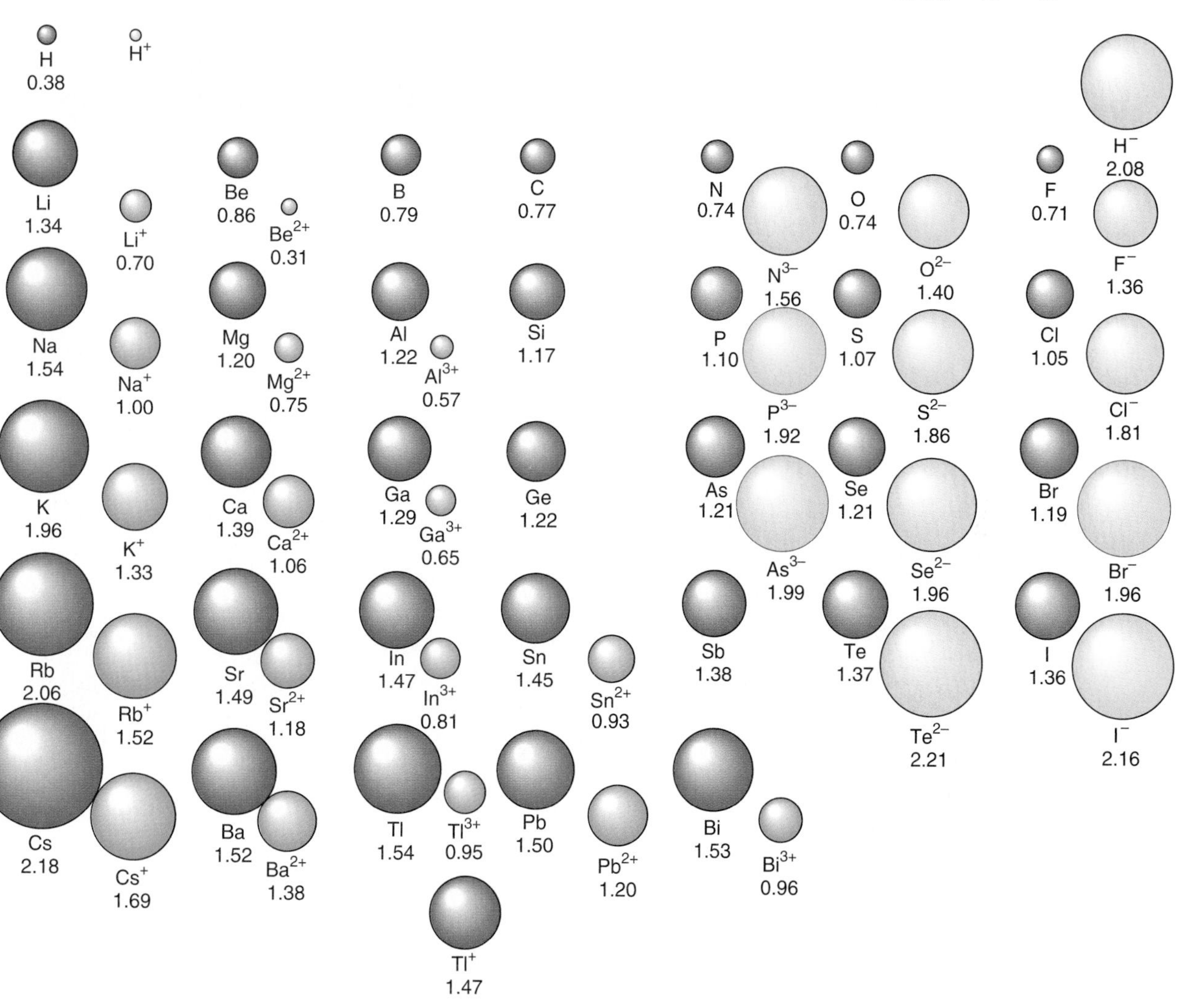

Atomic and ionic radii (in Å) of transition elements.

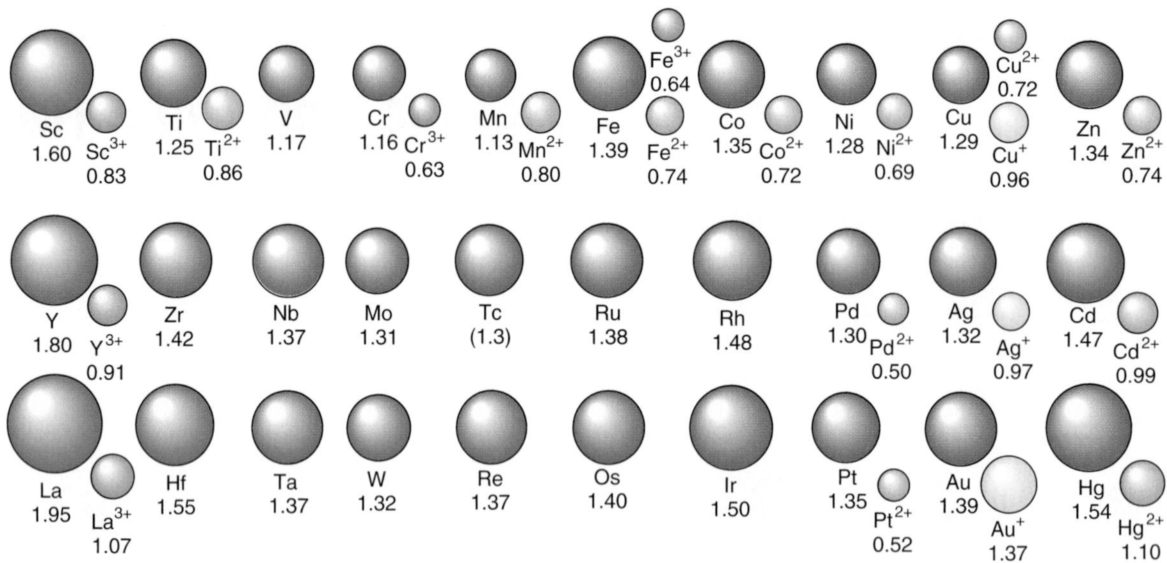

APPENDIX D

Equilibrium Constants

Dissociation Constants of Common Weak Acids at 25°C

Name	Formula	K_1	K_2	K_3
Acetic acid	$HC_2H_3O_2$	1.76×10^{-5}		
Ammonium ion	NH_4^+	5.6×10^{-10}		
Arsenic acid	H_3AsO_4	5.6×10^{-3}	1.70×10^{-7}	3.95×10^{-12}
Arsenous acid	H_3AsO_3	6.0×10^{-10}		
Benzoic acid	C_6H_5COOH	6.46×10^{-5}		
Boric acid	H_3BO_3	7.3×10^{-10}	1.8×10^{-13}	1.6×10^{-14}
Carbonic acid	H_2CO_3	4.3×10^{-7}	5.61×10^{-11}	
Chloroacetic acid	$CH_2ClCOOH$	1.4×10^{-3}		
Citric acid	$C_3H_4(OH)(COOH)_3$	7.10×10^{-4}	1.68×10^{-5}	8.4×10^{-6}
Formic acid	$HCOOH$	1.77×10^{-4}		
Hydrazoic acid	HN_3	1.9×10^{-5}		
Hydrocyanic acid	HCN	4.93×10^{-10}		
Hydrofluoric acid	HF	3.53×10^{-4}		
Hydrogen peroxide	H_2O_2	2.4×10^{-14}		
Hydrogen selenate ion	$HSeO_4^-$	1.2×10^{-2}		
Hydrogen sulfate ion	HSO_4^-	1.20×10^{-2}		
Hydrosulfuric acid	H_2S	1.0×10^{-7}	1.3×10^{-13}	
Hypobromous acid	$HBrO$	2.06×10^{-9}		
Hypochlorous acid	$HClO$	3.0×10^{-8}		
Hypoiodous acid	HIO	2.3×10^{-11}		
Iodic acid	HIO_3	1.69×10^{-1}		
Lactic acid	$CH_3CHOHCOOH$	1.4×10^{-4}		
Nitrous acid	HNO_2	4.5×10^{-4}		
Oxalic acid	$H_2C_2O_4$	5.9×10^{-2}	6.4×10^{-5}	
Phosphoric acid	H_3PO_4	7.52×10^{-3}	6.23×10^{-8}	2.2×10^{-13}
Phosphorous acid	H_2PHO_3	1.0×10^{-2}	2.6×10^{-7}	
Selenous acid	H_2SeO_3	3.5×10^{-3}	5×10^{-8}	
Sulfurous acid	H_2SO_3	1.5×10^{-2}	6.3×10^{-8}	

Ionization Constants of Common Weak Bases at 25°C

Name	Ionization Equation	K_b
Ammonia	$NH_3 + H_2O \rightleftharpoons NH_4^+ + OH^-$	1.79×10^{-5}
Aniline	$C_6H_5NH_2 + H_2O \rightleftharpoons C_6H_5NH_3^+ + OH^-$	4.2×10^{-10}
Dimethylamine	$(CH_3)_2NH + H_2O \rightleftharpoons (CH_3)_2NH_2^+ + OH^-$	5.9×10^{-4}
Ethylamine	$C_2H_5NH_2 + H_2O \rightleftharpoons C_2H_5NH_3^+ + OH^-$	6.4×10^{-4}
Hydrazine	$H_2NNH_2 + H_2O \rightleftharpoons H_2NNH_3^+ + OH^-$	1.7×10^{-6}
	$H_2NNH_3^+ + H_2O \rightleftharpoons H_3NNH_3^{2+} + OH^-$	1.3×10^{-15}
Hydroxylamine	$NH_2OH + H_2O \rightleftharpoons NH_3OH^+ + OH^-$	1.1×10^{-8}
Methylamine	$CH_3NH_2 + H_2O \rightleftharpoons CH_3NH_3^+ + OH^-$	4.2×10^{-4}
Pyridine	$C_5H_5N + H_2O \rightleftharpoons C_5H_5NH^+ + OH^-$	1.5×10^{-9}
Trimethylamine	$(CH_3)_3N + H_2O \rightleftharpoons (CH_3)_3NH^+ + OH^-$	6.3×10^{-5}

Solubility Product Constants for Sparingly Soluble Salts at 25°C

Name	Formula	K_{sp}	Name	Formula	K_{sp}
Aluminum hydroxide	$Al(OH)_3$	2×10^{-32}	Lead(II) sulfide	Pbs	8.2×10^{-28}
Barium carbonate	$BaCO_3$	8.1×10^{-9}	Magnesium carbonate	$MgCO_3$	4.0×10^{-5}
Barium chromate	$BaCrO_4$	2.0×10^{-10}	Magnesium hydroxide	$Mg(OH)_2$	1.5×10^{-11}
Barium fluoride	BaF_2	1.7×10^{-6}	Magnesium oxalate	MgC_2O_4	8.6×10^{-5}
Barium hydroxide	$Ba(OH)_2$	5×10^{-3}	Manganese(II) hydroxide	$Mn(OH)_2$	1.9×10^{-13}
Barium oxalate	BaC_2O_4	1.6×10^{-7}	Manganese(II) sulfide	MnS	1.4×10^{-15}
Barium sulfate	$BaSO_4$	1.08×10^{-10}	Mercury(I) chloride	Hg_2Cl_2	2×10^{-18}
Calcium carbonate	$CaCO_3$	8.7×10^{-9}	Mercury(II) sulfide	HgS	1×10^{-50}
Calcium chromate	$CaCrO_4$	7.1×10^{-4}	Silver bromide	AgBr	7.7×10^{-13}
Calcium fluoride	CaF_2	3.9×10^{-11}	Silver chloride	AgCl	1.56×10^{-10}
Calcium hydroxide	$Ca(OH)_2$	7.9×10^{-6}	Silver chromate	Ag_2CrO_4	9×10^{-12}
Calcium oxalate	CaC_2O_4	2.6×10^{-9}	Silver iodide	AgI	1.5×10^{-16}
Calcium sulfate	$CaSO_4$	2.45×10^{-5}	Silver sulfate	Ag_2SO_4	1.6×10^{-5}
Copper(II) hydroxide	$Cu(OH)_2$	1.3×10^{-20}	Strontium carbonate	$SrCO_3$	1.6×10^{-9}
Copper(II) sulfide	CuS	6.3×10^{-36}	Strontium chromate	$SrCrO_4$	3.6×10^{-5}
Iron(II) hydroxide	$Fe(OH)_2$	1.6×10^{-14}	Strontium hydroxide	$Sr(OH)_2$	3.2×10^{-4}
Iron(II) sulfide	FeS	4.9×10^{-18}	Strontium oxalate	SrC_2O_4	5.6×10^{-8}
Lead(II) carbonate	$PbCO_3$	7.4×10^{-14}	Strontium sulfate	$SrSO_4$	3.8×10^{-7}
Lead(II) chloride	$PbCl_2$	1.6×10^{-5}	Tin(II) hydroxide	$Sn(OH)_2$	6×10^{-27}
Lead(II) chromate	$PbCrO_4$	2.8×10^{-14}	Tin(II) sulfide	SnS	3×10^{-27}
Lead(II) fluoride	PbF_2	3.7×10^{-8}	Zinc hydroxide	$Zn(OH)_2$	1.8×10^{-14}
Lead(II) hydroxide	$Pb(OH)_2$	1.2×10^{-15}	Zinc oxalate	ZnC_2O_4	1.4×10^{-9}
Lead(II) sulfate	$PbSO_4$	1.6×10^{-8}	Zinc sulfide	ZnS	1.1×10^{-21}

APPENDIX E

Standard Reduction Potentials

Standard Reduction Potentials in Water at 25°C

Electrode Reaction	Standard Potential, $E°$ (V)
$F_2(g) + 2e^- \longrightarrow 2F^-(aq)$	2.85
$H_2O_2(aq) + 2H^+(aq) + 2e^- \longrightarrow 2H_2O(l)$	1.776
$PbO_2(s) + SO_4^{2-}(aq) + 4H^+(aq) + 2e^- \longrightarrow PbSO_4(s) + 2H_2O(l)$	1.685
$MnO_4^-(aq) + 8H^+(aq) + 5e^- \longrightarrow Mn^{2+}(aq) + 4H_2O(l)$	1.491
$PbO_2(s) + 4H^+(aq) + 2e^- \longrightarrow Pb^{2+}(aq) + 2H_2O(l)$	1.46
$Cl_2(g) + 2e^- \longrightarrow 2Cl^-(aq)$	1.3583
$Cr_2O_7^{2-}(aq) + 14H^+(aq) + 6e^- \longrightarrow 2Cr^{3+}(aq) + 7H_2O(l)$	1.33
$O_2(g) + 4H^+(aq) + 4e^- \longrightarrow 2H_2O(l)$	1.229
$MnO_2(s) + 4H^+(aq) + 2e^- \longrightarrow Mn^{2+}(aq) + 2H_2O(l)$	1.208
$Br_2(l) + 2e^- \longrightarrow 2Br^-(aq)$	1.065
$NO_3^-(aq) + 4H^+(aq) + 3e^- \longrightarrow NO(g) + 2H_2O(l)$	0.96
$2Hg^{2+}(aq) + 2e^- \longrightarrow Hg_2^{2+}(aq)$	0.905
$Hg^{2+}(aq) + 2e^- \longrightarrow Hg(l)$	0.851
$Ag^+(aq) + e^- \longrightarrow Ag(s)$	0.7996
$Fe^{3+}(aq) + e^- \longrightarrow Fe^{2+}(aq)$	0.770
$MnO_4^-(aq) + 2H_2O(l) + 3e^- \longrightarrow MnO_2(s) + 4OH^-(aq)$[a]	0.588
$I_2(s) + 2e^- \longrightarrow 2I^-(aq)$	0.535
$O_2(g) + 2H_2O(l) + 4e^- \longrightarrow 4OH^-(aq)$[a]	0.401
$Cu^{2+}(aq) + 2e^- \longrightarrow Cu(s)$	0.3402
$SO_4^{2-}(aq) + 4H^+(aq) + 2e^- \longrightarrow SO_2(g) + 2H_2O(l)$	0.20
$Pb^{2+}(aq) + 2e^- \longrightarrow Pb(s)$	−0.1263
$Sn^{2+}(aq) + 2e^- \longrightarrow Sn(s)$	−0.1364
$O_2(g) + 2H_2O(l) + 2e^- \longrightarrow H_2O_2(aq) + 2OH^-(aq)$[a]	−0.146
$Ni^{2+}(aq) + 2e^- \longrightarrow Ni(s)$	−0.23
$PbSO_4(s) + 2e^- \longrightarrow Pb(s) + SO_4^{2-}(aq)$	−0.356
$Fe^{2+}(aq) + 2e^- \longrightarrow Fe(s)$	−0.409
$S(s) + 2e^- \longrightarrow S^{2-}(aq)$	−0.508
$Cr^{3+}(aq) + 3e^- \longrightarrow Cr(s)$	−0.74
$Zn^{2+}(aq) + 2e^- \longrightarrow Zn(s)$	−0.7628
$2H_2O(l) + 2e^- \longrightarrow H_2(g) + 2OH^-(aq)$[a]	−0.8277
$Mn^{2+}(aq) + 2e^- \longrightarrow Mn(s)$	−1.029
$Al^{3+}(aq) + 3e^- \longrightarrow Al(s)$	−1.66
$Be^{2+}(aq) + 2e^- \longrightarrow Be(s)$	−1.70
$Mg^{2+}(aq) + 2e^- \longrightarrow Mg(s)$	−2.375
$Na^+(aq) + e^- \longrightarrow Na(s)$	−2.7109
$Ca^{2+}(aq) + 2e^- \longrightarrow Ca(s)$	−2.76
$Sr^{2+}(aq) + 2e^- \longrightarrow Sr(s)$	−2.89
$Ba^{2+}(aq) + 2e^- \longrightarrow Ba(s)$	−2.90
$Cs^+(aq) + e^- \longrightarrow Cs(s)$	−2.923
$K^+(aq) + e^- \longrightarrow K(s)$	−2.924
$Rb^+(aq) + e^- \longrightarrow Rb(s)$	−2.925
$Li^+(aq) + e^- \longrightarrow Li(s)$	−3.045

↑ Increasing oxidizing ability (left side, upward)

↓ Increasing reducing ability (right side, downward)

[a] Half-reaction occurs in basic solution.

APPENDIX F

Thermodynamic Data

Thermodynamic Data for Selected Elements and Compounds at 25°C and 1 atm

Substance	ΔH_f° (kJ mol^{-1})	ΔG_f° (kJ mol^{-1})	S° (J mol^{-1} K^{-1})
$Ag(s)$	0	0	42.72
$AgCl(s)$	−127.0	−109.70	96.11
$Al(s)$	0	0	28.32
$Al_2O_3(s)$	−1669.8	−1576.5	51.00
$Br_2(l)$	0	0	152.3
$HBr(g)$	−36.2	−53.2	198.5
C(diamond)	1.88	2.89	2.43
C(graphite)	0	0	5.69
$CO(g)$	−110.5	−137.3	197.9
$CO_2(g)$	−393.5	−394.4	213.6
$CCl_4(l)$	−139.3	−68.6	214.4
$CH_4(g)$	−74.85	−50.8	186.3
$C_2H_2(g)$	226.7	209.2	200.8
$C_2H_4(g)$	52.3	68.11	219.4
$C_2H_6(g)$	−84.67	−32.89	229.5
$C_3H_8(g)$	−103.85	−23.47	269.9
$C_6H_6(l)$ benzene	49.0	124.5	172.8
$C_6H_{12}O_6(s)$ glucose	−1260	−910.6	212.1
$C_{12}H_{22}O_{11}(s)$ sucrose	−2221	−1544	360.2
$CH_3OH(l)$	−238.6	−166.3	126.8
$C_2H_5OH(l)$	−277.6	−174.8	161.0
$HC_2H_3O_2H(l)$ acetic acid	−487.0	−392.4	159.8
$CaCO_3(s)$ calcite	−1207.1	−1128.8	92.88
$Cl_2(g)$	0	0	222.9
$HCl(g)$	−92.30	−95.27	187.0
$F_2(g)$	0	0	203.2
$HF(g)$	−269	−270.7	198.3
$Fe_2O_3(s)$	−822.2	−740.98	89.96
$H_2(g)$	0	0	130.6
$H_2O(l)$	−285.9	−237.2	69.94
$Hg(l)$	0	0	77.4
$I_2(s)$	0	0	116.73
$I_2(g)$	62.3	19.4	260.6
$HI(g)$	25.94	1.30	206.31
$KCl(s)$	−435.9	−408.3	82.7
$KClO_3(s)$	−391.2	−289.2	143.0
$Mg(s)$	0	0	32.51
$MgCl_2(s)$	−641.8	−592.1	89.6
$N_2(g)$	0	0	191.5
$NH_3(g)$	−45.94	−16.66	192.5
$NH_4Cl(s)$	−314.4	−203.0	94.6

$NO(g)$	90.4	86.7	210.62
$NO_2(g)$	33.8	51.84	240.45
$O_2(g)$	0	0	205.0
$O_3(g)$	142.2	163.4	237.6
$Na(s)$	0	0	51.0
$NaCl(s)$	−411.0	−384.0	72.33
$S_8(s)$ rhombic	0	0	31.88
$SO_2(g)$	−296.81	−300.4	248.5
$SO_3(g)$	−395.2	−370.4	256.2
$H_2S(g)$	−20.15	−33.0	205.64
$Zn(s)$	0	0	41.6

APPENDIX G

Vapor Pressure of Water

Vapor Pressure of Water at Different Temperatures

Temperature (°C)	Vapor Pressure (mm Hg)	Temperature (°C)	Vapor Pressure (mm Hg)
0	4.58	29	30.04
5	6.54	30	31.82
10	9.21	31	33.70
11	9.84	32	35.66
12	10.52	33	37.73
13	11.23	34	39.90
14	11.99	35	42.18
15	12.79	40	55.32
16	13.63	45	71.88
17	14.53	50	92.51
18	15.48	55	118.04
19	16.48	60	149.38
20	17.54	65	187.54
21	18.65	70	233.7
22	19.83	75	289.1
23	21.07	80	355.1
24	22.38	85	433.6
25	23.76	90	525.8
26	25.21	95	633.9
27	26.74	100	760.0
28	28.35		

GLOSSARY

Numbers in parentheses indicate sections in which the terms were first introduced.

A

Absolute zero (10.2) the lowest temperature that can be attained theoretically (−273.15°C).

Accuracy (1.3) the agreement between a result of a measurement and the accepted value.

Acid, Arrhenius definition (2.5) a substance which increases the hydrogen ion concentration in water (*see also* Brønsted acid and Lewis acid).

Acid dissociation (ionization) constant, K_a (14.2) the equilibrium constant for the dissociation of a weak acid.

Acid–base reaction (4.2) a reaction of an acid with a base.

Acidic anhydride (9.4) *see* acidic oxide.

Acidic oxide (2.7 and 9.4) an oxide that reacts with water to form an acidic solution.

Actinide (2.1) an inner transition metal preceded by the element actinium in the periodic table.

Activated complex (21.4) an unstable combination of the atoms of the reactants at the top of the potential energy barrier as a reaction proceeds from reactants to products.

Activation energy (21.4) the minimum energy needed for a reaction to occur.

Active metal (2.7) a metal that readily reacts with nonmetals, or with 1 M hydrochloric acid to liberate hydrogen gas.

Activity series (18.2) a list of metals in the order of decreasing ease of oxidation.

Actual yield (3.4) the quantity of a product actually obtained in a reaction.

Aerosol (12.7) a colloid in which a solid is dispersed in a gas.

Alcohol (23.6) an organic compound containing the hydroxyl group, —OH.

Aldehyde (23.7) an organic compound with the functional group —CHO, and with the general formula RCHO where R is an H atom, an alkyl group, or an aryl group.

Aliphatic hydrocarbon (23.1) a hydrocarbon that does not contain benzene rings.

Alkali metal (2.1) a metal in Group I (1) in the periodic table.

Alkaline earth metal (2.1) a metal in Group II (2) in the periodic table.

Alkane (8.2) a hydrocarbon consisting of molecules in which all the carbon atoms are linked by single bonds.

Alkene (8.2) a hydrocarbon whose molecules contain one or more double bonds.

Alkyl group (23.1) an alkane molecule minus one hydrogen atom.

Alkyne (8.2) a hydrocarbon whose molecules contain one or more triple bonds.

Allotropic forms or allotropes (2.8) two or more forms of an element in the same physical state which differ in their physical and chemical properties.

Alloy (2.7) a homogeneous mixture of two or more metals.

Alpha emission (24.1) a nuclear disintegration emitting alpha particles.

Alpha particle (6.1) the nucleus of a helium atom.

Alpha rays (6.1) rays that consist of alpha particles.

Amine (23.6) an organic compound that can be regarded as derived from ammonia by replacing one, two, or three H atoms by alkyl or aryl groups. [A primary amine has the general formula RNH_2, secondary amine RR′NH, the tertiary amine RR′R″N, where R, R', and R'' are alkyl or aryl groups.]

Amino acid (23.8) a carboxylic acid containing the amine group, $-NH_2$.

Ammine complex (16.5) a coordination complex in which the ligands are ammonia molecules.

Amorphous substance (1.1) a substance in which there is no definite geometrical arrangement of atoms or molecules.

Ampere (17.7) the rate of charge flow when one coulomb of charge moves through a point in one second.

Amphiprotic substance (14.1) a substance that can act either as a proton donor or a proton acceptor.

Amphoteric hydroxide (14.3) a hydroxide which can react either with an acid or with a base.

Amplitude of a wave (6.2) one half of the vertical distance between a crest and a trough of a wave.

Analytical chemistry (1.1) the branch of chemistry concerned with the methods and instruments for determining the composition of matter.

Angular molecule (7.6) *see* bent molecule.

Anhydrous compound (3.3) a compound without any water absorbed into it.

Anion (2.5) a negatively charged ion.

Anode (6.1) the electrode at which oxidation occurs.

Antibonding molecular orbital (8.4) a molecular orbital in which the electron density is concentrated outside of the region between the nuclei of the two bonded atoms.

Aqua complex (16.5) a coordination complex in which the ligands are water molecules.

Aromatic hydrocarbon (23.1 and 23.5) a hydrocarbon which contains one or more benzene rings.

Arrhenius equation (21.4) an expression which relates the rate constant of a reaction with absolute temperature:

$$k = Ae^{-E_a/RT}$$

where k is the rate constant, A is a constant, E_a is the activation energy, R is the gas constant, and T is the absolute temperature.

Aryl group (23.7) benzene ring or its derivative minus one hydrogen atom.

Atom (1.5) the basic unit of an element that retains its identity during chemical reactions.

Atomic mass unit (1.6) the mass of exactly one twelfth of a C-12 atom.

Atomic number (1.6) the number of protons in the nucleus of an atom.

Atomic theory (1.5) a theory that explains the nature of matter in terms of atoms.

Atomic weight (1.6) the weighted average mass of all the isotopes of an element.

Aufbau process (6.6) a hypothetical process to arrive at the ground-state elec-

Aufbau process (6.6) (*continued*) tron configurations of atoms by adding protons and electrons to a hydrogen atom to build successively larger atoms.

Autoionization (14.1) a reaction in which the molecules of a substance react with one another to form ions.

Avogadro's law (10.2) a law stating that equal volumes of all gases at the same temperature and pressure contain equal numbers of molecules.

Avogadro's number (3.1) 6.02×10^{23}; the number of particles in a mole.

Axial atoms (7.5) the atoms that lie in the opposite sides of the central trigonal plane in a trigonal bipyramidal structure.

Azimuthal quantum number (6.5) *see* subshell quantum number.

B

Barometer (10.1) an instrument used to measure atmospheric pressure.

Base, Arrhenius definition (2.7) a substance that increases the hydroxide ion concentration in water solution (*see also* Brønsted base and Lewis base).

Base ionization constant, K_b (14.4) the equilibrium constant for the ionization of a weak base.

Basic anhydride (9.4) basic oxide.

Basic oxide (2.7) an oxide that reacts with water to form a basic solution.

Battery (chapter 17, introduction) a device used to generate electric current by spontaneous redox reaction.

Belt of stability (24.2) a plot of the number of neutrons versus the number of protons in stable nuclei.

Bent molecule (angular molecule) (7.6) a triatomic molecule in which the atoms do not lie along a straight line.

Beta decay (24.1) a radioactive decay of nuclei emitting beta particles.

Beta emission (24.1) an emission of a beta particle (a high speed electron) from an unstable nucleus.

Beta rays (6.1) high speed electrons emitted by radioactive nuclei.

Bidentate ligand (22.2) a ligand which bonds to the metal ion by two atoms.

Bimolecular reaction (21.6) an elementary reaction in which two molecules collide to form products.

Binary compound (2.2) a compound that consists of two elements.

Biological chemistry (biochemistry) (1.1) a branch of chemistry primarily concerned with the study of the physical and chemical changes that occur in living organisms.

Biological oxygen demand (BOD) (12.8) the amount of oxygen required for microorganisms to be active in breaking down human waste in rivers and lakes.

Body-centered cubic unit cell (11.5) a cubic unit cell in which there is a lattice point in each corner of the cell, and one in the center of the cube.

Boiling point (11.3) the temperature at which the vapor pressure of a liquid equals the atmospheric pressure.

Boiling point constant (12.4) the elevation of the boiling point caused by dissolving one mole of a nonionizing solute in 1 kg of solvent.

Bond angle (7.5) the angle between two lines drawn from the central atom of a molecule to two adjacent atoms bonded to the central atom.

Bond dissociation energy (5.4) the energy required to break a covalent bond.

Bond order in a Lewis model (7.4) the number of bonding electron pairs per bond.

Bond order in the molecular orbital theory (8.4) one half of the difference between the number of bonding electrons and the number of antibonding electrons.

Bonding molecular orbital (8.4) a molecular orbital in which the electron density is concentrated in the region between the nuclei of the two bonded atoms.

Borane (9.1) a binary compound of boron and hydrogen.

Boyle's law (10.2) the volume of a sample of gas varies inversely with pressure at constant temperature.

Breeder reactor (24.6) a nuclear reactor that produces more fissionable material than it consumes.

Brønsted acid (4.1) a proton donor.

Brønsted base (4.1) a proton acceptor.

Buffer capacity (15.4) the amount of acid or base that can be added to a buffer solution before the pH changes significantly.

Buffer solution (15.4) a solution that resists the change of pH when a small amount of an acid or a base is added to it.

C

Calorie (1.2) a unit of energy equivalent to the energy needed to raise the temperature of one gram of water by one degree celsius.

Calorimeter (5.5) an instrument used to measure heat changes.

Carat (2.8) a measure used to express the purity of gold. A 24 carat gold is pure gold. Also, a unit of weight of precious stones; 1 carat = 200 mg.

Carbohydrate (23.8) a polyhydroxyaldehyde or ketone. The general formula for carbohydrates is $C_x(H_2O)_y$. Examples are glucose, sucrose, and starch.

Carbonyl group (23.7) the $>C{=}O$ group.

Carboxyl group (14.3) a group of atoms —COOH in a carboxylic acid.

Carboxylic acid (14.3) an acid that contains the carboxyl group (—COOH).

Catalysis (21.5) a process in which a catalyst increases the rate of a reaction.

Catalyst (2.8) a substance that increases the rate of a reaction without being consumed by the reaction.

Cathode (6.1) the electrode at which reduction occurs.

Cathode rays (6.1) electrons emitted by the cathode (negative electrode) when high voltage is applied to electrodes in an evacuated tube.

Cation (2.5) a positively charged ion.

Cell in electrochemistry (chapter 17, introduction) a compartment in which electricity is produced by a spontaneous redox reaction or in which a nonspontaneous redox reaction is driven by an electric current.

Cell potential (17.2) the potential difference between the electrodes of a cell.

Celsius temperature scale (1.2) a temperature scale that defines the freezing point of water as zero degrees and the boiling point of water as 100 degrees at one atmosphere pressure.

Centigrade temperature scale (1.2) is identical to the Celsius scale.

Chain reaction, nuclear (24.6) a self-sustaining nuclear fission reaction in which the neutrons produced by splitting one nucleus split other nuclei.

Charles' law (10.2) states that at constant pressure the volume of a sample of a gas is directly proportional to the absolute temperature.

Chelate (22.2) a coordination complex in which the metal ion is coordinated to one or more chelating ligands.

Chelating agent (22.2) a polydentate ligand.

Chemical change (1.1 and 1.4) process in which the composition of a substance is altered.

Chemical equilibrium (chapter 13 introduction and 13.1) the state of a chemical reaction in which the rate of the forward reaction equals the rate of the reverse reaction.

Chemical property (1.1 and 1.4) a property of matter that determines how a substance changes into one or more other substances.

Chemical reaction (1.4) a change of matter in which one or more substances convert into different substances.

Chemistry (1.1) a science that examines the properties, composition, structure, and changes of matter.

Cis isomer (22.4) a stereoisomer in which two atoms or groups of atoms are on the same side of some reference line or plane in the molecule.

Colligative properties (12.3) properties of solutions that depend on the number of solute particles but not on the nature of these particles.

Collision frequency (21.4) the frequency with which molecules collide.

Colloid (colloidal dispersion) (12.7) a homogeneous mixture in which the constituent particles are larger than those of a true solution, but not large enough to be separated by filtration or centrifugation.

Combination reaction (2.7) a reaction in which elements or compounds combine to form a new compound.

Combustion (2.8) the reaction of a substance with oxygen, usually with the production of heat and flame.

Common ion effect (15.4) the shift in equilibrium caused by adding a compound containing an ion that is identical (common) to one of the components of the equilibrium system.

Complex ion (16.5) a negatively or positively charged coordination complex.

Compound (1.4) a pure substance that consists of two or more elements held together by chemical bonds.

Concentrated solution (3.5) a solution which contains a relatively large quantity of a solute in a given quantity of solvent.

Conjugate acid (14.1) a substance that forms when a Brønsted base gains a proton.

Conjugate acid-base pair (14.1) a Brønsted acid-base pair in which the only difference between the members is that one of them has one more proton than the other.

Conjugate base (14.1) a substance that forms when a Brønsted acid loses a proton.

Continuous spectrum (6.2) a distribution of radiant energies of gradually changing wavelengths.

Conversion factor (1.2) a ratio that can be used to convert one quantity and its units to another equivalent quantity and its units.

Coordinate covalent bond (7.4) a covalent bond in which the bonding electron pair is contributed by only one of the bonded atoms.

Coordination complex (16.5) a molecule or ion in which a Lewis acid such as a metal ion is attached to one or more Lewis bases called ligands, in equilibrium with its components in solution.

Coordination number (11.5) in a crystal, the number of nearest neighbor atoms or ions of an atom or ion. In a coordination complex, the number of bonds the metal ion or atom forms with its ligands.

Coordination sphere (22.1) in a coordination complex, the central metal ion with its attached ligands.

Coulomb (17.2) the SI unit of electric charge: the quantity of electricity equal to the current of one ampere flowing for one second; also the charge on 6.24×10^{18} electrons.

Covalent bond (2.4) a bond between two atoms which is formed by sharing of electrons between the atoms.

Covalent solid (network solid) (11.5) a solid which consists of a three dimensional network of covalently bonded atoms.

Critical mass (24.6) the minimum mass of fissionable material in which a chain reaction can occur.

Critical pressure (11.6) the minimum pressure that must be applied to liquify a gas at its critical temperature.

Critical temperature (11.6) the highest temperature at which a gas can be liquified.

Crystal field theory (22.5) a theory which assumes that the bonding between the central metal ion and its ligands in a coordination complex is essentially electrostatic.

Crystalline solid (1.1) a solid that consists of atoms, molecules, or ions arranged in a regular geometric pattern.

Cubic close-packing structure (ccp) (11.5) a structure in which the layers of atoms are in an ABCABCABCA . . . arrangement.

Curie, Ci (24.7) a unit of radioactivity equal to 3.70×10^{10} nuclear disintegrations per second (this is the decay rate of one gram of radium).

Cyclotron (24.3) a device for accelerating nuclear particles, used in nuclear transmutation.

D

d-Block elements (6.6) elements in which the d orbitals are being filled (elements in groups 3 through 12).

Dalton's law of partial pressures (10.4) states that the total pressure of a gaseous mixture equals the sum of the partial pressures of the components of the mixture.

Daughter nuclide (24.1) a nuclide formed by the decay of another nuclide called the parent nuclide.

Decomposition reaction (2.7) a reaction in which a substance breaks down into two or more simpler substances.

Deduction (1.1) a reasoning process in which general ideas or theories lead to specific conclusions.

Degenerate orbitals (6.5) orbitals of the same energy.

Degree of dissociation of a weak electrolyte (12.5) the extent to which an electrolyte dissociates into ions in aqueous solution.

Density (1.2) the mass of a unit volume of an object.

Deoxyribonucleic acid, DNA (23.8) a polynucleotide in which the sugar component is deoxyribose.

Deuterium (1.6) an isotope of hydrogen that has one neutron in its nucleus.

Dew point (10.7) the temperature at which water just begins to condense out of air.

Dextrorotatory isomer (22.4) an optical isomer whose solution rotates plane polarized light to the right.

Dialysis (12.7) the separation of small solute particles from colloidal particles by a semipermeable membrane.

Diamagnetic substance (8.3) a substance that contains no unpaired electrons and is slightly repelled by a magnetic field.

Diffusion (10.5) mixing one substance with another as a result of random molecular motion.

Dilute solution (3.5) a solution which contains a relatively small quantity of a solute in a given quantity of solvent.

Dimer (9.2) a molecule formed by combination of two identical molecules.

Dipole moment (7.2) bond dipole moment: the product of the partial charges on two atoms and the distance between the nuclei. Molecular dipole moment: the vector sum of all the bond dipole moments in the molecule.

Dipole-dipole interaction (11.4) attractive forces between polar molecules.

Diprotic acid (4.1) an acid that can produce two moles of protons per mole of acid.

Displacement reaction (2.7) a reaction in which one atom or group of atoms displaces an atom or group of atoms of another substance.

Disproportionation reaction (9.5) a reaction in which a substance is simultaneously oxidized and reduced.

Distillation (12.3) a process in which a liquid is vaporized and the more volatile component is condensed to separate it from the less volatile component.

Donor atom (22.2) an atom of a ligand which donates its electron pair to the metal ion in a coordination complex.

Double bond (7.4) a covalent bond formed by two shared electron pairs.

Double displacement reaction (2.7) a reaction in which there is an exchange of components $AB + CD \rightarrow AD + CB$.

Dry cell *see* primary cell.

Dynamic equilibrium (11.2) a state in which two opposite processes occur at equal rates.

E

Effective collision (21.4) a collision of reactant molecules which leads to the formation of products.

Effusion (10.5) the flow of gas through a small opening or a porous material into a region of lower pressure.

Electrochemical equivalent (17.7) the quantity of a substance formed or decomposed by the loss or gain of one faraday during electrolysis.

Electrochemistry (chapter 17, introduction) the branch of chemistry that deals with the study of the relations between electricity and redox reactions.

Electrolysis (17.6) a process in which a nonspontaneous redox reaction is forced to occur by an electric current.

Electrolyte (3.5) a substance whose water solution conducts electric current.

Electrolytic cell (17.6) a cell in which a nonspontaneous redox reaction is forced to occur by an electric current.

Electron (1.6) a subatomic particle which has a very small mass and carries a negative charge.

Electron affinity (6.7) the energy change that occurs when an electron adds to a gaseous atom.

Electron capture (24.1) nuclear decay by capture of an electron from an inner orbital of an atom.

Electron density (6.5) the probability of finding an electron in a particular region in an atom.

Electron pair structure (7.5) the geometry of all the orbitals surrounding the central atom of a molecule or a polyatomic ion, or the concentration of electrons in a particular region of an atom or a molecule.

Electronegativity (7.3) a measure of the ability of an atom in a molecule to attract electrons surrounding it.

Element (1.4 and 1.6) a substance whose atoms all have the same atomic number.

Elementary reaction (21.6) one step in the mechanism of a reaction.

Empirical formula, also called the simplest formula (3.3) the formula of a compound written with the smallest integer subscripts.

Emulsion (12.7) a colloid in which a liquid is dispersed throughout another liquid.

Endpoint (4.3) the point in a titration reaction at which an indicator signals the completion of the reaction.

Endothermic reaction (5.1) a reaction that absorbs heat.

Energy (1.2) the ability to do work.

Enthalpy (5.2) a thermodynamic quantity used to describe heat changes that occur at constant pressure.

Enthalpy change, ΔH, of reaction (5.2) the heat absorbed or evolved by the reaction at constant pressure.

Enthalpy of formation (5.3) the enthalpy change for the formation of one mole of a compound from its elements.

Enthalpy of reaction (5.3) enthalpy change of a reaction.

Entropy (20.1) the degree of randomness or disorder of a system.

Equation of state (10.3) ideal gas law equation.

Equatorial atoms (7.5) the atoms that lie in the corners of the central trigonal plane in a trigonal bipyramidal structure.

Equilibrium (11.2) a state in which no change occurs as time elapses.

Equilibrium constant (13.1) the ratio of the product of the molar concentrations of the products of a reaction to the product of the molar concentrations of the reactants, each raised to the power of its coefficient in the balanced equation for the reaction.

Equivalence point (4.3) the point at which a given quantity of a reactant is totally consumed in a reaction with an equivalent quantity of another reactant.

Ester (23.7) an organic compound formed by the reaction between a carboxylic acid and an alcohol; it has the general formula RCOOR′.

Ether (23.6) an organic compound with the general formula R—O—R′ where R and R′ can be identical or different alkyl or aryl groups.

Excited state (6.3) an energy state that is higher than the ground state.

Exothermic reaction (5.1) a reaction that releases heat.

F

***f*-Block elements (6.6)** elements in which the *f* orbitals are being filled (the lanthanides and actinides).

Face-centered cubic unit cell (11.5) a cubic unit cell which has a lattice point at the center of each face of the cell, in addition to the points at the corners.

Fahrenheit temperature scale (1.2) a temperature scale that defines the freezing point of water as 32 degrees and the boiling point of water as 212 degrees at one atmosphere pressure.

Faraday, F (17.2) the charge of one mole of electrons. 1 faraday = 96,487 coulombs.

Faraday's law (17.7) a law stating that the quantity of a substance produced or decomposed by electrolysis is directly proportional to the quantity of electricity used.

Fatty acid (23.7) a long-chain carboxylic acid.

First law of thermodynamics (5.2) the change of the internal energy, ΔE, of a system is the sum of the heat, q, and work, w, transferred during the change: $\Delta E = q + w$.

First-order reaction (21.2) a reaction whose rate is proportional to the concentration of a single reactant, raised to the first power.

Foam (12.7) a colloid in which a gas is dispersed in a liquid.

Formula (1.4) a combination of symbols of the elements that make up a compound.

Formula weight (3.1) the sum of the

atomic weights of the atoms in a formula unit of a compound.

Forward reaction (chapter 13, introduction) the reaction which is written from left to right.

Free energy (20.3) the energy that is available to do useful work.

Freezing point (11.3) the temperature at which the solid and liquid phases of a substance are in equilibrium.

Freezing point constant (12.4) the freezing point depression caused by dissolving one mole of a nonionizing solute in one kilogram of solvent.

Frequency of a wave (6.2) the number of times per second a complete wave passes through a point.

Fuel cell (17.5) a cell to which the reactants are continuously supplied.

Functional group (23.6) a group of atoms in the molecule of an organic compound which is responsible for the characteristic chemical behavior of the molecule.

G

Galvanic cell (chapter introduction and 17.2) a voltaic cell.

Gamma rays (6.1) high energy radiation emitted by nuclei of radioactive atoms.

Gas (1.4) a sample of matter that spreads throughout the whole volume of its container and can expand indefinitely.

Gay-Lussac's law of combining volumes (10.2) states that the volumes of gaseous reactants and products at constant temperature and pressure can be expressed in whole number ratios.

Geiger counter (24.5) a device that measures the rate of radioactive decay processes.

Gel (12.7) a colloid in which a liquid is dispersed in a solid.

Geometric isomers (22.4) isomers in which atoms are linked together in the same sequence, but which differ in the arrangement of some atoms, or groups of atoms, in space.

Graham's law of effusion (10.5) states that the rate of effusion of a gas is inversely proportional to the square root of its molar mass or density.

Gravimetric analysis (4.4) a branch of quantitative analysis that involves weighing the precipitate produced in a reaction.

Ground state (6.3) the lowest energy state in a system.

Group in a periodic table (2.1) a column in the periodic table.

H

Half-cell (17.2) the part of an electrochemical cell in which either the oxidation or reduction occurs.

Half-life of a reaction (21.3) the time it takes for a reactant to decrease to one half of its original amount or concentration.

Half-reaction (4.5) One of two parts of a redox reaction, oxidation (loss of electrons) or reduction (gain of electrons).

Halide ion (2.2) a negatively charged halogen ion.

Halogen (2.1) an element in Group VII (17) in the periodic table.

Halogenation (23.4) a reaction in which halogen is added to a compound.

Hard water (12.8) contains calcium and magnesium ions.

Heat of fusion (11.3) the amount of heat required to melt a specified quantity of a solid. The heat of fusion of ice at 0°C is 333 J g^{-1}.

Heat of reaction *see* enthalpy of reaction.

Heat (1.2) a form of energy.

Heat capacity (5.5) the amount of heat required to raise the temperature of an object by one degree Celsius (or 1 K).

Heat of combustion (5.6) the enthalpy change for a combustion reaction.

Heat of formation *see* enthalpy of formation.

Heat of vaporization (11.2) the amount of heat required to vaporize a specified quantity of a liquid. The heat of vaporization of water is 2.26 kJ g^{-1}.

Heisenberg uncertainty principle (6.4) it is impossible to simultaneously determine the precise position and momentum of an electron in an atom.

Henderson-Hasselbalch equation (15.4) the relationship between pH, pK_a, and the concentrations of the acid and base components in a buffer solution.

Henry's law (12.2) states that the solubility of a gas is directly proportional to the partial pressure of the gas above the solution.

Hertz (6.2) the SI unit of frequency.

Hess's law of heat summation (5.3) a law stating that if a reaction goes through two or more steps, the enthalpy of reaction is the sum of the enthalpies of all the steps.

Heterogeneous catalysis (21.5) a catalysis in which the catalyst is in a different phase than the reactants.

Heterogeneous equilibrium (13.3) a system at equilibrium in which all the components are not of the same phase.

Heterogeneous matter (1.4) matter that contains two or more visibly different components.

Heteronuclear diatomic molecule (2.4) a diatomic molecule containing atoms of different elements.

Hexagonal close-packing structure (hcp) (11.5) a structure with hexagonal unit cells; the layers of atoms in this structure are arranged in an ABABAB . . . sequence.

High-spin complex (22.5) a coordination complex in which the maximum number of electrons have parallel spin.

Homogeneous catalysis (21.5) the catalyst is in the same phase as the reactants.

Homogeneous equilibrium (13.3) a system at equilibrium in which all the components are in the same phase.

Homogeneous matter (1.4) matter that appears the same throughout, without any visibly different components.

Homologous series (23.1) a series of compounds in which one compound differs from the preceding one by having one more $—CH_2—$ group.

Homonuclear diatomic molecule (2.4) a diatomic molecule containing atoms of the same element.

Hund's rule (6.6) states that the most stable arrangement of electrons in orbitals of equal energy is the one in which there is a maximum number of electrons with parallel spins.

Hybrid orbital (8.1) an orbital formed by mixing two or more atomic orbitals.

Hydrate (3.3) a compound that contains water molecules chemically bonded to it.

Hydration (3.3) a process in which a compound absorbs water molecules to form a hydrate.

Hydroacid (2.7) an acid that does not contain oxygen.

Hydrocarbon (2.7) a binary compound of hydrogen and carbon.

Hydrogen bonding (11.4) the attraction of two molecules so that an H atom

Hydrogen bonding (11.4) *(continued)* bonded to an F, O, or N atom in one molecule is attracted to an F, O, or N atom in another molecule.

Hydrogenation (23.4) a reaction in which hydrogen is added to an unsaturated hydrocarbon to convert it to a saturated hydrocarbon.

Hydrolysis of ions (14.4) a reaction of ions with water resulting in either an increase or a decrease of pH.

Hydrometer (12.4) a device that measures the density of a liquid.

Hydronium ion (4.1) H_3O^+, the predominant form of the hydrogen ion, H^+, in aqueous solution.

Hydroxo complex (16.5) a coordination complex in which the ligands are hydroxide ions.

Hypothesis (1.1) a tentative explanation of a phenomenon.

I

Ideal gas (10.3 and 10.5) a hypothetical gas which obeys the ideal gas laws perfectly.

Ideal gas equation (10.3) an equation describing the relations between pressure (P), volume (V), temperature (T), and quantity (n) of a gas: $PV = nRT$, where R is the gas constant.

Indicator (4.3) a substance whose color changes at the endpoint of a titration.

Induced (artificial) radioactivity (24.3) radioactivity produced in a laboratory as opposed to natural (spontaneous) radioactivity.

Induction (1.1) a reasoning process that leads from specific observations to general statement(s).

Infrared spectrum (6.2) the spectral region of 700–1500 nm.

Inner orbital complex (22.5) according to the valence bond theory, a coordination complex in which the electron pairs of the ligands occupy the lowest energy orbitals ("inner orbitals") of the metal ion.

Inner transition metals (6.6) the *f*-block metals (lanthanides and actinides).

Inorganic chemistry (1.1) a branch of chemistry primarily concerned with non-carbon-containing substances.

Interhalogen compound (8.3) a binary compound consisting of two different halogens.

Interionic attraction (12.5) attraction between oppositely charged ions in solution.

Intermolecular forces (11.4) attractive forces between molecules.

Internal energy (5.2) the total of all types of energy in a system.

Ion (2.2) an electrically charged atom or group of atoms.

Ion product constant for water (14.1) the product of the molar concentrations of hydronium and hydroxide ions in pure water.

Ionic bonding (2.4) attraction between positively and negatively charged ions.

Ionic character of a covalent bond (7.3) the extent to which the bonding electrons are attracted toward one of the bonded atoms and away from the other.

Ionic compound (2.2) a compound that consists of ions.

Ionic equation (4.2) an equation that specifies the ionic species as reactants or products in an aqueous reaction.

Ionic potential (18.2) the charge-to-radius ratio of an ion.

Ionic solid (11.5) a solid that consists of ions.

Ionization (3.5) the formation of ions from neutral atoms or molecules.

Ionization energy (6.7) the energy required to remove the least tightly bound electron from a gaseous atom or ion.

Ionization isomers (22.4) isomers of a coordination complex that differ by the anion that is coordinated to the metal atom or ion.

Isoelectronic species (2.2) atoms or ions that have the same number and arrangement of electrons.

Isomers (7.4) compounds of the same molecular formula, but with different arrangements of atoms.

Isotopes (1.5 and 1.6) atoms of an element that have different masses.

J

Joule (1.2) a unit of energy equivalent to the energy of a two-kilogram mass moving with the velocity of one meter per second.

K

K-capture (24.1) *see* electron capture.

Kelvin temperature (1.2) is the Celsius temperature plus 273.15 (*see* absolute zero).

Ketone (23.7) an organic compound with the general formula RR′C=O.

Kinetic energy (5.1) energy of motion.

Kinetic molecular theory (10.5) the theory that a gas consists of molecules in constant random motion.

Kinetics of chemical reactions (chapter 21, introduction) the study of reaction rates and mechanisms.

L

Lanthanide (2.1) an inner transition metal preceded by the element lanthanum in the periodic table.

Lattice (1.1) a repeated pattern of particles.

Lattice energy (5.4) the energy required to break the lattice of a mole of an ionic compound.

Law in science (1.1) a verbal or mathematical statement of a relationship that is always true under certain circumstances.

Law of conservation of matter (1.5) matter can neither be created nor destroyed.

Law of constant composition (1.5) the elements in a compound are present in definite proportions by mass.

Law of definite proportions (1.5) law of constant composition.

Law of mass action (law of chemical equilibrium) (13.1) states that the value of the equilibrium constant expression for a reaction is always a constant at a given temperature.

Law of multiple proportions (1.5) in different binary compounds of elements A and B, the masses of the element B in combination with a fixed mass of A in each of these compounds can be expressed in the ratios of small whole numbers.

Le Châtelier's principle (12.2) if a stress is applied to a system at equilibrium, the system shifts to minimize the stress.

Levorotatory isomer (22.4) an optical isomer whose solution rotates plane polarized light to the left.

Lewis acid (14.5) an electron pair acceptor.

Lewis base (14.5) an electron pair donor.

Lewis structure (7.4) a combination of Lewis symbols to represent a molecule or a polyatomic ion. Shared electron pairs are drawn as dashes connecting the atoms, and unshared electron pairs are shown as dot pairs.

Lewis symbol (7.4) a symbol of an atom surrounded by dots which represent the valence electrons of the atom.

Ligand (16.5) a Lewis base attached to a metal atom or ion in a coordination complex.

Limiting reactant (3.4) the reactant in a chemical reaction that limits the quantity of a product obtained.

Line spectrum (6.2) a spectrum which contains radiation of certain specific wavelengths and is observed as a pattern of lines.

Linear accelerator (24.3) a linear device for accelerating nuclear particles, used in nuclear transmutation.

Linear structure (7.5) a structure in which all the points lie along a straight line.

Linkage isomers (22.4) isomers in a coordination complex that differ in the atom by which a ligand is coordinated to the metal atom or ion.

Lipid (23.8) a naturally occurring ester which contains at least one long-chain carboxylic acid called a fatty acid.

Liquid (1.4) a sample of matter that has a tendency to flow and to assume the shape of its container.

London dispersion forces (11.4) attractive forces between temporarily polar molecules.

Lone electron pair (7.4) a pair of valence electrons on an atom not used for bonding in a molecule.

Low-spin complex (22.5) a coordination complex which contains the lowest possible number of unpaired electrons.

M

Magic numbers (24.2) the numbers of protons or neutrons in stable nuclei.

Magnetic quantum number, m_l (6.5) the quantum number that relates to the orientation of an orbital in space relative to the other orbitals with the same l and n values (orbitals of the same energy and shape). It can have integral values from l to $-l$, including 0.

Manometer (10.1) a device used to measure gas pressure.

Mass (1.2) the quantity of matter in an object.

Mass defect (24.6) the difference between the calculated and experimental mass of a nucleus.

Mass number (1.6) the total number of protons and neutrons in an atom.

Mass spectrometer (3.1) an instrument that separates ions by their mass-to-charge ratio and indicates their masses and abundance.

Matter (1.1) anything that occupies volume and has mass.

Metal (2.1 and 2.2) an element which has a characteristic luster or shine and is a good conductor of heat and electricity.

Metallic bonding (11.5) the attraction between positive metal ions and the surrounding mobile electrons in a metallic solid.

Metalloid (2.1) an element with properties intermediate between those of metals and nonmetals.

Metallurgy (18.3) the process of producing metals from their ores.

Metathesis reaction (2.7) a double displacement reaction.

Mixture (1.4) a sample of matter that contains two or more different components that retain their identity.

Model (1.1) a hypothetical entity used to explain the properties and changes of matter.

Molal boiling point elevation (12.4) boiling point constant.

Molal freezing point lowering (12.4) freezing point constant.

Molality (12.4) the number of moles of solute per kilogram of solvent.

Molar mass (3.1) the mass of one mole of a substance.

Molar solubility (16.1) the number of moles of a solute in one liter of a saturated solution.

Molar volume (10.2) the volume occupied by one mole of a substance. The molar volume of a gas at 0°C and 1 atm is 22.4 L.

Molarity (3.5) the number of moles of a solute in one liter of solution.

Mole (3.1) the quantity of matter that contains as many atoms, molecules, or ions as the number of atoms in exactly 12 grams of carbon-12 isotope.

Mole fraction (12.3) the ratio of the number of moles of a component of a solution to the total number of moles of all the components.

Molecular formula (3.3) a formula specifying the exact number of atoms of each element in a molecule of a compound.

Molecular orbital (8.1 and 8.4) a spatial region in a molecule where one or two electrons can most probably be found.

Molecular orbital (MO) theory (8.1 and 8.4) a theory that considers a molecule as a "multinuclear atom" in which all electrons belong to molecular orbitals that extend over the entire molecule.

Molecular solid (11.5) a solid that consists of molecules.

Molecular structure (7.5) geometric arrangement of atomic nuclei in a molecule or a polyatomic ion.

Molecular weight (3.1) the sum of the atomic weights of the atoms in a molecule of a compound.

Molecularity (21.6) the number of molecules reacting in an elementary reaction.

Molecule (1.5) a neutral group of two or more atoms of the same or different elements held together by natural forces called chemical bonds.

Monatomic ion (2.5) an atom that carries an electric charge.

Monodentate ligand (22.2) a ligand that bonds to the metal ion by only one atom.

Monomer (9.2) the smallest repeating unit in a polymer.

Monoprotic acid (4.1) an acid that can produce one mole of protons per mole of the acid.

Monosaccharide (23.8) a simple sugar; common simple sugars contain six carbon atoms per molecule; they include glucose and fructose.

N

Natural radioactivity (24.1) radioactivity that is caused by spontaneous nuclear disintegration in nature.

Nernst equation (17.4) an equation relating cell potential with standard cell potential, temperature, the number of electrons in a balanced equation for the reaction, and the reaction quotient.

Net ionic equation (4.2) an ionic equation in which the spectator ions are omitted.

Network solid (11.5) a covalent solid.

Neutralization point (4.3) the equivalence point in an acid-base reaction.

Neutralization reaction (4.2) an acid-base reaction to form water.

Neutron (1.6) a neutral particle in a nucleus of an atom with a mass of slightly more than that of a proton.

Noble gas (2.1) an element in Group VIII (18) in the periodic table.

Noble metal (2.7) a metal which does not readily react with nonmetals and does not react with 1 M hydrochloric acid.

Nonbonding electron pair (7.4) *see* lone electron pair.

Nonelectrolyte (3.5) a substance whose water solution does not conduct electric current.

Nonmetal (2.1 and 2.2) an element that is usually a poor conductor of heat and electricity.

Nonpolar bond (7.2) a covalent bond in which the bonding electron pair is shared equally between the bonded atoms.

Normal boiling point (11.3) the temperature at which the vapor pressure of a liquid equals the atmospheric pressure of 1 atm.

Normal freezing point (11.3) the temperature at which the solid and liquid phases of a substance are in equilibrium at 1 atm pressure.

Nuclear binding energy (24.6) the energy that holds the nucleus together.

Nuclear fission (24.6) a nuclear reaction in which large nuclei are split into smaller nuclei with the simultaneous emission of energy.

Nuclear fusion (24.6) a nuclear reaction in which small nuclei are fused together into larger nuclei with the simultaneous emission of energy.

Nuclear reactor (24.6) a device in which nuclear reactions are carried out.

Nuclear transmutation (24.3) a conversion of one nucleus into another as a result of bombardment by neutrons or other particles.

Nucleic acid (23.8) a very high molecular weight polymer that carries genetic information.

Nucleon (1.6) a proton or a neutron.

Nucleotide (23.8) a repeating unit in nucleic acid polymer.

Nuclide (24.1) the nucleus of a specified isotope of an element.

O

Octahedral structure (7.5) the structure of a molecule in which the central atom lies in the center of an octahedron and six atoms connected to it lie at the corners of the octahedron.

Octet rule (7.4) a rule specifying that with a few exceptions, the central atom in a molecule or a polyatomic ion tends to have eight electrons on its valence shell.

Optical isomers (22.4) stereoisomers which are related as nonsuperimposable mirror images, like right and left hands.

Orbit (6.3) the path of the electron in a hydrogen atom about the nucleus of the atom according to Bohr theory.

Orbital (6.5) a three-dimensional region around the nucleus of an atom in which the probability of finding an electron is highest.

Order of reaction (21.1 and 21.2) the sum of the exponents in the rate law of a reaction.

Organic chemistry (1.1) a branch of chemistry concerned with carbon-containing substances.

Osmosis (12.6) the diffusion of a solvent through a semipermeable membrane.

Osmotic pressure (12.6) the minimum pressure required to prevent osmosis.

Outer orbital complex (22.5) according to the valence bond theory, a coordination complex in which the electron pairs of the ligands do not fully occupy all the lowest energy orbitals of the metal atom or ion.

Oxidation (4.5) an increase in oxidation state.

Oxidation number (4.5) oxidation state.

Oxidation-reduction reaction (4.5) a reaction in which oxidation and reduction occur simultaneously.

Oxidation state (4.5) the real or apparent charge on an atom assigned according to a set of rules.

Oxidizing agent (4.5) a reactant that oxidizes another reactant.

Oxoacid (2.5) an acid that contains oxygen.

Oxoanion (2.5) an anion derived from oxoacid.

P

***p*-Block elements (6.6)** elements in which the atoms have *p* electrons as their highest energy electrons.

Paramagnetic substance (8.3) a substance that contains one or more unpaired electrons and is attracted by a magnet.

Parent nuclide (24.1) a nuclide before decay to another nuclide called the daughter nuclide.

Partial pressure (10.4) the pressure exerted by a component of a gaseous mixture.

Pascal, Pa (10.1) the SI unit of pressure, a pressure of one newton, N, per square meter, N m^{-2} (one newton is the force required to accelerate one kilogram mass one meter per second each second, kg m s^{-2}). 1 atm = 101,325 Pa.

Pauli exclusion principle (6.6) no two electrons in an atom can have the same set of four quantum numbers.

Peptide (23.8) a substance made from two or more amino acids.

Peptide linkage (23.8) a bond linking two amino acid molecules; it is formed between the carbon atom of the carboxyl group of one amino acid and the nitrogen atom of the amino group of another amino acid.

Percent by mass (3.5) the number of mass units of a component in a collection in one hundred mass units of the collection.

Percent by volume (3.5) the number of volume units of a component of a liquid in one hundred volume units of the liquid.

Percent dissociation (12.5) degree of dissociation × 100.

Percent yield (3.4) the ratio of the mass of a product obtained in a reaction to the calculated mass of the product, multiplied by 100.

Period in a periodic table (2.1) a row in the periodic table.

Periodic law (2.1) when elements are arranged in the order of their increasing atomic numbers, their physical and chemical properties vary periodically.

Permanent hard water (12.8) hard water that cannot be softened by boiling.

pH (15.2) the negative logarithm of the hydrogen ion concentration.

Phase (1.4) a homogeneous portion of a heterogeneous mixture, often used synonymously with the term "state of matter".

Phase diagram (11.6) a graph of pressure-versus-temperature showing the conditions at which a substance exists as a solid, liquid, or gas.

Photon (6.2) a quantum of radiant energy.

Physical change (1.1) a change in which the basic nature or the composition of a substance is not altered.

Physical chemistry (1.1) a branch of chemistry primarily concerned with the application of physics in the study of the structure and changes of matter.

Physical property (1.4) a property of matter that can be directly observed by our senses, such as color and odor, or that can be measured with a scientific instrument.

Pi (π) bond (8.2) a covalent bond formed by sideways overlap of two *p* orbitals; its electron density is concentrated on opposite sides of the internuclear axis.

Piezoelectric substance (9.4) a substance that generates an electric potential when it is mechanically deformed in a certain direction.

pK_a (15.4) the negative logarithm of the acid dissociation constant.

pK_b (15.4) the negative logarithm of the base ionization constant.

Plasma (24.6) a neutral mixture of positive ions and electrons at high temperature.

pOH (15.2) the negative logarithm of the hydroxide ion concentration.

Polar bond (7.2) a covalent bond between atoms with different electronegativities.

Polar molecule (7.2) a molecule which possesses a net dipole moment.

Polyatomic ion (2.5) a group of two or more covalently bonded atoms carrying an electric charge.

Polydentate ligand (22.2) a ligand that bonds to a metal ion by two or more atoms.

Polymer (9.2) a substance which has a high molar mass and consists of many repeating units of low molar mass.

Polymerization (23.4) a reaction which forms a polymer.

Polypeptide (23.8) a polymer formed from amino acid molecules joined together by peptide linkages.

Polyprotic acid (4.1) an acid that can produce two or more moles of protons per mole of acid.

Polysaccharide (23.8) a carbohydrate which contains two or more monosaccharide units; an example is starch.

Position isomers (22.4) *see* geometric isomers.

Position (point) of equilibrium (13.3) a property of a system at equilibrium measured by the value of the equilibrium constant.

Positron (24.1) a particle with a mass equal to that of an electron, but with a positive charge.

Positron emission (24.1) an emission of a positron from an unstable nucleus.

Potential energy (5.1) the energy possessed by an object by virtue of its position or composition.

Precipitate (2.7) a product of an aqueous reaction that settles to the bottom of the reaction container.

Precipitation reaction (4.4) a reaction in which a precipitate forms.

Precision (1.3) the agreement between two or more results of a measurement.

Pressure (10.1) the force exerted per unit area.

Primary alcohol (23.6) an alcohol with the general formula RCH_2OH where R is an H atom or an alkyl group.

Primary cell (17.5) a cell that cannot be recharged.

Principal quantum number, *n* (6.5) the quantum number that defines the energy and the size of an atomic orbital.

Product of a reaction (2.6) a substance formed as a result of a reaction.

Protein (23.8) a polymer consisting of amino acid monomers.

Proton (1.6) a positively charged particle of an atomic nucleus, its mass is slightly less than that of a neutron.

Pure substance (1.4) an element or a compound.

Q

Qualitative analysis (4.4) determination of the identities of the elements of a compound or the components of a mixture.

Quantitative analysis (4.3 and 4.4) chemical analysis of various quantities of substances; it includes volumetric and gravimetric analysis.

Quantum (6.3) the smallest quantity of radiant energy that can be absorbed or emitted by an atom.

Quantum number (6.5) an integer in a set of integers that describe the energy and other properties of an atomic orbital.

R

Racemic mixture (22.4) a mixture of equal amounts of optical isomers.

Rad (24.7) *r*adiation *a*bsorbed *d*ose—the amount of ionizing radiation equivalent to 1×10^{-2} joule of energy absorbed by one kilogram of body tissue.

Radiant energy (6.1) energy of electromagnetic waves propagated by simultaneous variation of electric and magnetic field intensities.

Radiation (6.1) emission of radiant energy or nuclear particles or both.

Radioactive decay (24.1) disintegration of unstable nuclei producing other nuclear particles, energy, or both.

Radioactive decay series (24.2) a sequence of nuclear disintegrations starting with a long-lived, unstable nuclide and ending with a stable isotope of lower atomic number.

Radioactive substance (24.1) a substance whose nuclei disintegrate to other nuclei by emission of nuclear particles, energy, or both.

Radioactivity (6.1) spontaneous emission of subatomic particles, or electromagnetic radiation, or both, by unstable nuclei.

Radionuclide (24.1) a nuclide of a radioactive element.

Raoult's law (12.3) states that the partial pressure of a component of a solution is directly proportional to the mole fraction of that component.

Rate constant (21.2) the proportionality constant in the rate law for a chemical reaction.

Rate-determining step (21.6) the slowest step in a multistep reaction.

Rate law or rate expression (21.2) an expression relating the rate of a reaction with the rate constant and the concentrations of the reactants.

Rate of a reaction (21.1) the change in the concentration of a reactant or product of a reaction per unit time.

Reactant (2.6) a substance which undergoes chemical change.

Reaction mechanism (21.5 and 21.6) the sequence of steps in a net reaction.

Reaction quotient (13.2) a ratio that has the same form as the equilibrium constant, but whose concentration values are not necessarily the same as those in the equilibrium expression.

Redox reaction (4.5) a reaction in which oxidation and reduction occur.

Reducing agent (4.5) a reactant that reduces another reactant.

Reduction (4.5) a decrease in oxidation state.

Relative biological effectiveness (RBE) (24.7) a measure of the biological effects of various types of radiation; a factor used to convert rads to rems.

Relative humidity (10.7) the ratio of the partial pressure of water vapor in air to the vapor pressure of water in saturated air at the same temperature.

Rem (24.7) *r*oentgen *e*quivalent for *m*an—the product of the energy of ionizing radiation in rads and its relative biological effectiveness (RBR): 1 rem = (1 rad)(1 RBE).

Representative element (2.1) an element whose properties are representative of its position in the periodic table—an element in Groups I through VIII (groups 1 and 2 and 13 through 18) according to the periodic table used in this text.

Resonance hybrid (7.4) the average of all the resonance structures that can be written for a molecule.

Resonance structure (7.4) one of the two or more Lewis structures that can be written for a molecule.

Reverse reaction (chapter 13 introduction) in an equilibrium system, the reaction written from right to left.

Reversible process in thermodynamics (20.1) a change that occurs in a system that is near equilibrium, and occurs in a certain direction infinitely slowly, so that applying an infinitesimal amount of energy stops the process or causes it to go in the opposite direction.

Ribonucleic acid, RNA (23.8) a polymer of ribonucleotide units.

Roentgen (24.7) a dose of radiation that delivers 8.4×10^{-3} joule of energy to one kilogram of air.

S

***s*-Block elements (6.6)** the elements in which the atoms have *s* electrons as their highest energy electrons.

Saccharide *see* carbohydrate.

Salt (2.2) an ionic compound other than a hydroxide or an oxide.

Saturated hydrocarbon (8.2) a hydrocarbon whose molecules contain the maximum number of hydrogen atoms.

Saturated solution (3.5 and 12.1) a solution which contains the maximum amount of solute at a given temperature and the dissolved and undissolved solutes are in equilibrium.

Schrödinger wave equation (6.4) an equation that relates the wave properties of a particle to its mass, energy, and spatial coordinates.

Scientific method (1.1) a method of problem-solving based on experimentation and reasoning that leads from observations to the construction of unifying principles for explanation and prediction.

Second law of thermodynamics (20.1) states that spontaneous processes are accompanied by an increase in the entropy of the universe.

Secondary alcohol (23.6) an alcohol in which the hydroxyl group is bonded to a carbon atom that is also bonded to two other carbon atoms and a hydrogen atom.

Secondary cell (17.5) a cell that can be recharged.

Semipermeable membrane (12.6) a membrane that allows passage of solvent molecules such as water molecules, but not larger solute molecules.

Shared electron pair (7.4) a bonding electron pair in a covalent bond.

Shell (6.5) the set of orbitals in an atom having the same principal quantum number, *n*; a shell is also referred to as the "principal quantum level."

SI units (1.2) basic units adopted by the General Conference of Weights and Measures.

Sigma (σ) bond (8.1) a covalent bond formed by orbital overlap along the axis of the bonded nuclei.

Significant figures (1.3) the digits in a recorded measurement that are known with certainty, plus the estimated final digit.

Silane (9.1) a binary compound of silicon and hydrogen.

Silicate (9.4) a compound consisting of metal ions and polyatomic anions which contain silicon and oxygen atoms.

Silicone (9.4) a polymer containing —O—Si—O— chain with various hydrocarbon groups attached to the silicon atoms.

Simplest formula (3.3) empirical formula.

Single displacement reaction (2.7) a reaction in which a substance displaces a component of another substance: $A + BC \rightarrow AC + B$.

Sol (12.7) a colloid in which a liquid is dispersed in another liquid.

Solid (1.4) a rigid sample of matter whose volume is not greatly influenced by changes in temperature and pressure.

Solubility (12.1) the amount of solute needed to make a saturated solution in a given amount of solvent at a given temperature.

Solubility product constant, K_{sp} (16.1) the equilibrium constant for the solubility equilibrium of a sparingly soluble ionic compound.

Solute (3.5) a substance dissolved in a solvent; usually the component of a solution present in the smaller amount.

Solution (1.4) a homogeneous mixture.

Solvent (3.5) a substance which dissolves another substance; usually the component of a solution present in the larger amount.

***sp* hybrid orbital (8.1)** a hybrid orbital formed by mixing an *s* and a *p* orbital.

***sp*^{2} hybrid orbital (8.1)** a hybrid orbital formed by mixing an *s* orbital with two *p* orbitals.

***sp*^{3} hybrid orbital (8.2)** a hybrid orbital formed by mixing an *s* orbital with three *p* orbitals.

***sp*^{3}d hybrid orbital (8.3)** a hybrid orbital formed by mixing an *s* orbital with three *p* orbitals and one *d* orbital.

***sp*^{3}d^2 hybrid orbital (8.3)** a hybrid orbital formed by mixing an *s* orbital with three *p* orbitals and two *d* orbitals.

Specific heat (5.5) the amount of heat required to raise the temperature of one gram of a substance by one degree Celsius.

Spectator ions (4.2) the ions that are present but are not changed in a reaction.

Spectrochemical series (22.5) a list of ligands arranged in the order of their orbital splitting ability.

Spectrum (6.2) an array of radiation arranged in the order of some varying characteristic such as the wavelength.

Spin quantum number, m_s (6.5) the quantum number that designates the direction of the spin of an electron relative to another electron. Its possible values are $+\frac{1}{2}$ and $-\frac{1}{2}$.

Spontaneous reaction (20.1) a reaction which occurs without outside influence or supply of energy.

Square planar structure (7.6) the structure of a molecule in which the central atom lies in the center of a square and four other atoms connected to the central atom lie in the corners of the square so that all of the atoms lie in a plane.

Square pyramidal structure (7.6) the structure of a molecule in which one atom lies at the apex of a pyramid and is connected to four other atoms that lie in the corners of the square base of the pyramid.

Standard cell (17.2) a galvanic cell in which the concentrations are 1 M, the temperature is 25°C, and the pressure is 1 atm.

Standard cell potential (17.3) the potential of a cell operating under standard conditions (1 M concentrations, 1 atm pressure, and a specified temperature, usually 25°C).

Standard conditions (standard temperature and pressure, STP) of gases (10.1) 0°C and 1 atmosphere.

Standard enthalpy (heat) of formation (5.3) the enthalpy change when one mole of a compound in its standard state forms from its elements in their standard states at 1 atm and 298 K.

Standard electrode potential (17.2) standard reduction potential.

Standard hydrogen electrode (17.2) an electrode which contains a platinum wire immersed in a 1 M H^+ solution and H_2 gas at 1 atm. The potential of a standard hydrogen electrode is defined as 0 V.

Standard reduction potential (17.2) the potential of a standard reduction half-cell measured relative to the standard hydrogen electrode.

Standard state of a system (5.2) 1 atm pressure and specified temperature, usually 298 K.

Standard solution (4.3) a solution whose concentration is accurately known.

Standardization (4.3) a determination of an accurate concentration of a solution.

State function (5.2) a property that is determined by the state of the system; these include variables like temperature and pressure.

States of matter (1.4) solid, liquid, and gas.

Stereoisomers (22.4) isomers in which the atoms are bonded in the same sequence but differ in their precise arrangement in space.

Steric hindrance (9.2) the space limitation that makes it difficult or impossible to fit large atoms or groups of atoms into a given space.

Stoichiometry (chapter 3, introduction) quantitative aspects of chemistry such as the calculation of the quantities of reactants and products in a chemical reaction.

Storage cell (17.5) a secondary cell (a cell that can be recharged).

Strong acid (4.1) an acid that is completely dissociated into ions in a dilute aqueous solution.

Strong base (4.1) a base that is completely dissociated into ions in a dilute aqueous solution.

Strong electrolyte (3.5) an electrolyte that completely dissociates into ions in a dilute solution.

Strong field ligand (22.5) a ligand that causes a large splitting of *d* orbital energy of the metal ion in a coordination complex.

Structural isomers (22.4) isomers which differ in the sequence in which the atoms or groups of atoms are bonded together.

Subatomic particle (6.1) a particle smaller than an atom (a component of an atom).

Subshell (6.5) a subdivision within a shell of an atom designated by the quantum number *l*.

Subshell quantum number, *l* (6.5)—also called azimuthal quantum number and angular momentum quantum number—the quantum number that designates the shape of an atomic orbital; the values of *l* can be integers from 0 to n − 1.

Supernatant solution (2.7) the solution above a precipitate.

Supersaturated solution (12.1) a solution that contains more solute than the saturated solution at the same temperature; any added solute to a supersaturated solution will not be in equilibrium with the dissolved solute but will cause some dissolved solute to precipitate to form a saturated solution.

Surface tension (11.1) the energy required to increase or stretch a unit area of a liquid's surface.

Surroundings in thermodynamics (5.2) the part of the universe outside of a system.

Symbol (1.4) a letter or a combination of letters used to represent an element.

System in thermodynamics (5.2) any part of the universe of interest or under observation in an experiment.

T

Temperature (1.2) the property of an object that determines whether heat flows to it or away from it spontaneously.

Temporary hard water (12.8) hard water that can be softened by boiling, it contains HCO_3^- ions and Ca^{2+} and Mg^{2+} ions.

Termolecular reaction (21.6) an elementary reaction in which three molecules simultaneoulsy collide to form products.

Tertiary alcohol (23.6) an alcohol in which the hydroxyl group is bonded to a carbon atom that is also bonded to three other carbon atoms.

Tetrahedral structure (7.5) the structure of a molecule in which the central atom is in the center of a tetrahedron and four atoms bonded to the central atom lie at the corners of the tetrahedron.

Theoretical yield (3.4) the calculated quantity of a product of a reaction.

Theory (1.1) a network of interrelated hypotheses and models consistent with experimentally established laws.

Thermochemistry (chapter 5, introduction) the study of heat changes in chemical reactions.

Thermodynamics (chapter 20, introduction) the study of all types of energy changes accompanying chemical reactions.

Thermonuclear reaction (24.6) a nuclear fusion reaction that occurs at very high temperature.

Third law of thermodynamics (20.2) the entropy of a perfect crystalline substance is zero at zero Kelvin.

Titration (4.3) a procedure in which one solution is added from a buret to another solution in a separate container for a reaction.

Titration curve (15.6) a plot of pH versus the number of milliliters of a base added in a titration of an acid with a base.

Torr (10.1) a pressure unit equal to 1 mm Hg.

Total equation (4.2) an equation in which the formulas of all the reactants and products are written whether or not they are present as ions or molecules in solution.

Trans isomer (22.4) a stereoisomer in which two atoms or groups of atoms are on the opposite sides of some reference line or plane in the molecule.

Transition metal (2.1) an element in groups 3 through 12 in the periodic table.

Transition state *see* activated complex.

Transition state theory (21.4) a theory that a chemical reaction proceeds through a transition state before products are formed.

Transuranium element (24.3) an element with an atomic number greater than that of uranium.

Trigonal bipyramidal structure (7.5) the structure of a molecule in which the central atom lies in the center of an equilateral triangle (a trigonal plane), three atoms connected to the central atom occupy the corners of the triangle, and two atoms lie on opposite sides of the central atom along a straight line perpendicular to the triangle.

Trigonal planar structure (7.5) a structure of a tetratomic molecule in which the central atom and the three atoms bonded to it lie in a plane and all the bond angles are 120°.

Trigonal pyramidal structure (7.6) a structure in which the central atom is in the apex of a pyramid and three atoms connected to it lie in the corners of the triangular base of the pyramid.

Triple bond (7.4) a covalent bond formed by three shared electron pairs.

Triple point (11.6) the point in the phase diagram of a substance showing the temperature and pressure at which the solid, liquid, and gaseous states of the substance exist at equilibrium.

Triprotic acid (4.1) an acid that can produce three moles of protons per mole of the acid.

Tritium (1.6) hydrogen atom that contains two neutrons.

Tyndall effect (12.7) the scattering of light by colloidal particles.

U

Ultraviolet spectrum (6.2) the spectral region of 200–400 nm.

Unimolecular reaction (21.6) an elementary reaction (a step in a reaction mechanism) in which only one molecule is the reactant.

Unit cell (11.5) the smallest characteristic structural unit of a crystalline substance.

Unpaired electron (6.6) an electron occupying an orbital by itself.

Unsaturated hydrocarbon (8.2) a hydrocarbon whose molecules contain double or triple bonds and thus do not contain the maximum number of hydrogen atoms.

Unshared electron pair (7.4) *see* lone electron pair.

V

Valence bond (VB) theory (8.1–8.3) a theory that explains the covalent bond as an electron pair bond formed between two atoms by overlapping of orbitals of the two atoms. The valence bond theory in coordination complexes (22.5) assumes that a coordinate covalent bond is formed between a metal ion and a ligand when an unshared electron pair of the ligand enters a vacant hybrid orbital of the metal ion.

Valence electrons (6.6) the outermost electrons in an atom that participate in chemical bonding.

Valence shell electron pair repulsion (VSEPR) theory (7.5) the theory that the valence shell electron pairs of the central atom of a molecule keep as far as possible from one another.

Van der Waals forces (11.4) intermolecular forces that include both dipole-dipole interaction and London forces.

Vapor pressure, equilibrium vapor pressure (11.2) the pressure exerted by vapor in equilibrium with its liquid or solid phases.

Viscosity (11.1) a measure of a fluid's resistance to flow.

Volt (17.2) the SI unit of potential difference. The potential difference when one joule of energy is required to move one coulomb of charge from a lower potential to a higher potential.

Voltaic cell (chapter 17 introduction) an electrochemical cell which uses a spontaneous redox reaction to generate electric current.

Volumetric analysis (4.3) a branch of quantitative analysis in which volumes of reactant solutions are measured.

W

Water of hydration (3.3) the water contained in a hydrate.

Wavelength (6.2) the distance from a point in one wave to an equivalent point in the next wave.

Weak acid (4.1) an acid that is only slightly ionized in aqueous solution.

Weak base (4.1) a base that is only slightly ionized in aqueous solution.

Weak electrolyte (3.5) an electrolyte that dissociates into ions in water solution only to a small degree.

Weak field ligand (22.5) a ligand that causes a small splitting of *d* orbital energy of the metal ion in a coordination complex.

Weight (1.2) the force that gravity exerts on an object.

Work (5.1) the movement of an object against some force, measured as the product of the force and distance.

CREDITS

PHOTOGRAPHS

Longcore Maciel Studio © Wm. C. Brown Publishers:
Figures: 1.2, 1.4(a–c), 1.5, 1.6, 1.9, p. 14 Margin, 1.13(a, b), 1.14, 1.15(a–c), 1.20, 2.3, 2.4, 2.7(a–b), p. 56 Margin, 2.13, 2.15, 2.16, 2.17, 2.18(a–b), 2.19, p. 75 Margin, p. 80 Margin Courtesy of Courtney Hughes, p. 83 Margin Courtesy of Liina Ladon, 3.3, p. 102 Margin, 3.5, p. 105 Margin, 3.6, 3.7(a–f), 3.8(a–c), 4.1(a–b), 4.2, 4.3, 4.5(a–b), 4.6(a–e), 4.7(a–b), 4.8(a–b), 4.9, 4.10(a–b), 9.21, 9.30(a–b), 9.32, 10.3, 10.14, 10.16(a–b), 11.2, 11.12, 12.16(a–b), 13.2, 15.5, 16.2(a–b), 16.3, p. 619 Margin, 16.5, 17.1, 17.2, 17.11, 18.4(a–b), 18.5(a–b), 18.8(a–b), 18.14, 18.15, 18.16, 18.17, 18.18(a–b), 18.19 Courtesy of Alan Pribula, 18.21 Courtesy of Courtney Hughes, 18.22, 18.24, 18.25, 18.26, 19.2, 19.3, 19.5, 19.7, 19.9, 19.10, 19.12(a–g), 19.13, 19.15, 19.16, 19.17.

EXPANDED CONTENTS

p. vii: Philip de Loutherbourg/Science Museum of London; **p. viii (top):** Longcore Maciel Studio © Wm. C. Brown Publishers; **p. viii (bottom):** Jasper Johns/Albright-Knox Art Gallery/VAGA; **p. ix (top):** Longcore Maciel Studio © Wm. C. Brown Publishers; **p. ix (bottom):** © Emil Muench/Photo Researchers, Inc.; **p. x (top):** Uniphoto, Inc.; **p. x (bottom):** From *Biochemistry,* J. David Rawn. © 1989 J. David Rawn and courtesy of Richard J. Feldmann; **p. xi:** © Larry West/Photo Researchers, Inc.; **p. xii (top):** © Dave Watts/ Tom Stack & Associates; **p. xii (bottom):** © 1987 Matt Bradley/ © Southern Stock Photos; **p. xiii:** © 1991 A.H. Rider/Photo Researchers, Inc.; **p. xiv (top):** © Douglas Faulkner/Photo Researchers, Inc.; **p. xiv (bottom):** From *General Chemistry,* Second Edition, by P. W. Atkins and J. A. Beran. Copyright © 1992 by P. W. Atkins and J. A. Beran. Reprinted by permission of W. H. Freeman and Company; **p. xv (top):** James Mayhew/National Audubon Society/Photo Researchers, Inc.; **p. xv (bottom):** Longcore Maciel Studio © Wm. C. Brown Publishers; **p. xvi (top):** © Comstock, Inc./Bob Grant; **p. xvi (bottom):** Courtesy of General Motors; **p. xvii (top):** Courtesy of McEngelvan; **p. xvii (bottom):** Longcore Maciel Studio © Wm. C. Brown Publishers; **p. xviii (top):** © Pat Canova/© Southern Stock Photo Agency; **p. xviii (bottom):** © Harvey Olsen/© 1991 Southern Stock Photos; **p. xix (top):** Longcore Maciel Studio © Wm. C. Brown Publishers; **p. xix (bottom):** Uniphoto, Inc.; **p. xx (top):** Photo Researchers, Inc.; **p. xx (bottom):** © Spencer Grant/Photo Researchers, Inc.

PART OPENERS

Part One: "This Must Be the Place," © 1965 Roy Lichtenstein.
Part Two: From *Biochemistry,* J. David Rawn. © 1989 J. David Rawn and courtesy of Richard J. Feldmann.
Part Three: From *General Chemistry,* Second Edition, by P. W. Atkins and J. A. Beran. Copyright © 1992 by P. W. Atkins and J. A. Beran. Reprinted by permission of W. H. Freeman and Company.
Part Four: Comstock, Inc.
Part Five: From *Biochemistry,* J. David Rawn. © 1989 J. David Rawn and courtesy of Richard J. Feldmann.

Chapter 1

Opener: Philip de Loutherbourg/Trustees of the Science Museum of London; **Figure 1.1**: © Phillip Hayson/Photo Researchers, Inc.; **p. 18, Application**: © Dr. Charles Belinky/Photo Researchers, Inc.; **1.13(c)**: Fletcher & Baylis/Photo Researchers, Inc.; **1.13(d)**: © 1986 Ray Ellis/Photo Researchers, Inc.; **1.16**: The Bettmann Archive; **1.18**: The Granger Collection, New York.

Chapter 2

Opener: Longcore Maciel Studio © Wm. C. Brown Publishers; **Figure 2.1**: The Bettmann Archive; **p. 65, Application**: © H. Fichner/ Southern Stock Photo Agency; **p. 73, Application**: © Gary Milburn/ Tom Stack & Associates; **2.20(a)**: © Lawrence Migdale/Photo Researchers, Inc.; **2.20(b–d)**: © Russ Lappa/Photo Researchers, Inc.; **2.22(a–d)**: © Russ Lappa/Photo Researchers, Inc.; **p. 79 top, Application**: © Will McIntyre/Photo Researchers, Inc.; **p. 79 bottom, Application**: © Kenneth Eward/Photo Researchers, Inc.; **2.23(a–c)**: © Russ Lappa/Photo Researchers, Inc.; **2.23(d)**: © 1987 Rich Treptow/ Photo Researchers, Inc.

Chapter 3

Opener: Jasper Johns, *Numbers in Color,* 1958–59, encaustic and newspaper on canvas, 66 ½ × 49 ½", Albright-Knox Art Gallery, Buffalo, New York, Gift of Seymour H. Knox, 1959; © Jasper Johns/ VAGA, New York 1992; **p. 92., Margin**: Geoff Tompkinson/Science Photo Library/Photo Researchers, Inc.; **p. 93, Margin**: The Bettmann Archive; **p. 111, Application**: © Les Riess/Southern Stock Photo Agency.

Chapter 4

Opener: Longcore Maciel Studio © Wm. C. Brown Publishers; **p. 152, Margin top**: John Shaw/Tom Stack & Associates; **p. 152, Margin bottom**: © B. Drake/Southern Stock Photo Agency.

Chapter 5

Opener: © Emil Muench/Photo Researchers, Inc.; **Figure 5.9**: © Henry Fichner/Southern Stock Photo Agency; **p. 190, Margin**: Courtesy of Parr Instrument Company.

Chapter 6

Opener: Uniphoto, Inc.; **p. 208, Application**: Courtesy of International Atomic Energy Agency, Vienna, Austria; **Figure 6.5**: The Bettmann Archive; **p. 218, Margin**: The Bettmann Archive.

Chapter 7

Opener: From *Biochemistry,* J. David Rawn. © 1989 J. David Rawn and courtesy of Richard J. Feldmann; **Figure 7.1**: The Bettmann Archive; **7.6**: The Bettmann Archive.

Chapter 8

Opener: © Larry West/Photo Researchers, Inc.; **p. 302, Application**: Southern Stock Photo Agency.

Chapter 9

Opener: © Dave Watts/Tom Stack & Associates; **Figure 9.15**: © Blair Howard/Southern Stock Photo Agency; **p. 342, Margin**: Roberto De Gugliemo/Science Photo Library/Photo Researchers, Inc.; **9.19**: © Gary Retherford/Photo Researchers, Inc.; **9.20**: © Michael Abbey/Science Source/Photo Researchers, Inc.; **9.23(a)**: Roberto De Gugliemo/Science Photo Library/Photo Researchers, Inc.; **9.23(b)**: M. Claye Jacana/Photo Researchers, Inc.; **9.23(c)**: Field Museum, Chicago/Tom McHugh/Photo Researchers, Inc.; **p. 346 top, Application**: © Allen Green/Photo Researchers, Inc.; **p. 346 bottom, Application**: NASA/Science Source/Photo Researchers, Inc.; **9.31**: Springer/Bettmann Film Archive.

Chapter 10

Opener: © 1987 Matt Bradley/© Southern Stock Photos; **Figure 10.7**: The Bettmann Archive; **p. 376, Application**: AP/Wide World Photos; **p. 387, Application**: © 1984 Ralph Krubner/Southern Stock Photo Agency; **10.20**: © Jim Tuten/Southern Stock Photo Agency; **10.23**: © 1981 Dave Baird/Tom Stack & Associates; **10.24**: AP/Wide World Photos; **10.25(a–b)**: Courtesy of The Metropolitan Museum of Art. All rights reserved.

Chapter 11

Opener: © 1991 A.H. Rider/Photo Researchers, Inc.; **Figure 11.4**: © Everett Johnson/Southern Stock Photo Agency; **11.5**: © Jon Feingersh/Tom Stack & Associates; **p. 439, Application**: © Timothy O'Keefe/Southern Stock Photo Agency.

Chapter 12

Opener: © Douglas Faulkner/Photo Researchers, Inc.; **Figure 12.6**: © 1987 Timothy O'Keefe/Southern Stock Photo Agency; **12.7**: © 1990 Jon Feingersh/Tom Stack & Associates; **12.8**: © 1987 Steven Frink/© 1989 Southern Stock Photos; **p. 464, Application**: © Chuck St. John/Southern Stock Photo Agency; **12.18**: © Greg Vaughn/Tom Stack & Associates; **12.19**: © Richard Weymouth Brooks/Photo Researchers, Inc.

Chapter 13

Opener: From *General Chemistry*, Second Edition, by P. W. Atkins and J. A. Beran. Copyright © 1992 by P. W. Atkins and J. A. Beran. Reprinted by permission of W. H. Freeman and Company. **p. 498, Margin**: © Tom McHugh/Photo Researchers, Inc.; **p. 507, Application**: © Mike Mesgleski/Southern Stock Photo Agency; **p. 512, Application**: Thierry Orban/Sygma Photo News Agency.

Chapter 14

Opener: James Mayhew/National Audubon Society/Photo Researchers, Inc.; **p. 545, Application**: © Joseph Nettis 1986/Photo Researchers, Inc.

Chapter 15

Opener: Longcore Maciel Studio © Wm. C. Brown Publishers; **p. 556, Margin**: © Roberto Santos/© 1991 Southern Stock Photos; **p. 564, Margin**: Dr. Jeremy Burgess/Science Photo Library/Photo Researchers, Inc.; **p. 576, Margin**: Reuters/Bettmann; **p. 589, Application**: Photo Researchers, Inc.

Chapter 16

Opener: © Comstock, Inc./Bob Grant; **Figure 16.4**: Courtesy of the Department of Library Services, American Museum of Natural History, V/C 2827; **p. 608, Application**: Courtesy of Kent Van De Graaff.

Chapter 17

Opener: Courtesy of General Motors; **p. 635, Application**: © Pete Salouttos/Southern Stock Photos; **Figure 17.6**: Courtesy of Cadillac Motor Car Division, GM; **17.13**: © 1973 Hank Morgan/Photo Researchers, Inc.

Chapter 18

Opener: Courtesy of McEngelvan; **Figure 18.3(a)**: Brian Parker/Tom Stack & Associates; **18.3(b)**: © 1976 Thomas R. Taylor/Photo Researchers, Inc.; **18.3(c)**: © 1985 Phillip Hayson/Photo Researchers, Inc.; **18.3(d)**: © 1975 Paolo Koch/Photo Researchers, Inc.; **18.3(e)**: © A. W. Ambler/Photo Researchers, Inc.; **18.20**: AP/Wide World Photos.

Chapter 19

Opener: Longcore Maciel Studio © Wm. C. Brown Publishers.

Chapter 20

Opener: © Pat Canova/© Southern Stock Photo Agency; **p. 763, Application**: © Scott Berner/NASA/© Southern Stock Photo Agency; **p. 829, Application**: © The Coca-Cola Co. "Coca-Cola" and "Coke" are registered trademarks of The Coca-Cola Co.

Chapter 21

Opener: © Harvey Olsen/© 1991 Southern Stock Photos.

Chapter 22

Opener: Longcore Maciel Studio © Wm. C. Brown Publishers; **p. 711, Application**: Courtesy of Dannon Yogurt.

Chapter 23

Opener: Uniphoto, Inc./Photo Researchers, Inc. (Adapted from *ChemMatters*, Oct. 1986, pp. 8–9, American Chemical Society); **Figure 23.21(b)**: Figure 15.5 on page 223 of *Biology*, 3d ed., by Sylvia S. Mader. © 1989 by Wm. C. Brown Publishers; **p. 863, Application**: © Jerry Wachter/Photo Researchers, Inc.; **p. 874, Application**: Adapted from *ChemEcology*, Oct. 1990, Chemical Manufacturers' Association/© 1985 Hasking/Uniphoto, Inc.

Chapter 24

Opener: © Spencer Grant/Photo Researchers, Inc.; **Figure 24.6**: The Bettmann Archive; **24.9**: Tom McHugh/Photo Researchers, Inc.; **24.11**: The Bettmann Archive; **p. 911, Application**: Paul Richards/UPI; **24.13**: The Bettmann Archive; **24.15**: UPI/Bettmann Newsphotos; **24.16**: UPI/Bettmann Newsphotos; **24.19**: © David M. Doody/Tom Stack & Associates.

ILLUSTRATIONS/LINE ART

Chapter 12

Figure 12.21: Adapted from page 346 of *Chemistry*, 3d ed., by John C. Bailar, Jr., Therald Moeller, et al. © 1989 by Harcourt Brace Jovanovich; **12.22**: Adapted from page 346 of *Chemistry*, 3d ed., by John C. Bailar, Jr., Therald Moeller, et al. © 1989 by Harcourt Brace Jovanovich.

Chapter 17

Figure 17.3: Adapted from figure 19-9 on page 628 of *General Chemistry*, 2d ed., by Kenneth W. Whitten and Kenneth D. Gailey. © 1984 by Saunders College Publishing; **17.7**: Adapted from page 624 of *Chemistry*, 3d ed., by John C. Bailar, Jr., Therald Moeller, et al. © 1989 by Harcourt Brace Jovanovich; **17.8**: Adapted from figure 19.6 on page 796 of *Chemistry*, 4th ed., by Raymond Chang. © 1991 by McGraw Hill; **17.9**: Adapted from figure 18.23 on page 610 of *General Chemistry: Principles and Structure*, by James E. Brady. © 1990 by John Wiley & Sons; **17.10**: Adapted from figure 19.6 on page 571 of *Chemistry: The Central Science*, 5th ed., by Theodore L. Brown, H. Eugene LeMay, Jr., et al. © 1990 by Prentice-Hall; **17.12**: Adapted from figure 17.21 on page 554 of *General Chemistry: Principles and Structure*, by James E. Brady and Gerard Humiston. © 1982 by John Wiley & Sons; **17.19**: Adapted from figure 19-6 on page 522 of *General Chemistry*, 1st ed., by Kenneth W. Whitten and Kenneth D. Gailey. © 1981 by Saunders College Publishing.

Chapter 20

Figure 20.4: Adapted from figure 17.4 on page 548 of *Chemistry: The Central Science*, 5th ed., by Theodore L. Brown, H. Eugene LeMay, Jr., et al. © 1990 by Prentice-Hall.

Chapter 21

Figure 21.11: Adapted from figure 14.17 on page 441 of *General Chemistry*, 9th ed., by Henry F. Holtzclaw, Jr., William R. Robinson, et al. © 1991 by D. C. Heath and Company; **21.12**: Adapted from figure 14.20 on page 508 of *Chemistry: The Central Science*, 5th ed., by Theodore L. Brown, H. Eugene LeMay, Jr., et al. © 1990 by Prentice-Hall.

Chapter 22

Figure 22.12: Adapted from figure 28.4 on page 809 of *Chemistry*, 3d ed., by John C. Bailar, Jr., Therald Moeller, et al. © 1989 by Harcourt Brace Jovanovich; **22.21**: Adapted from figure 22-7 on page 557 of *General Chemistry: Principles and Modern Applications*, 4th ed., by Ralph H. Petrucci. © 1985 by Macmillan Publishing Company.

Chapter 23

Figure 23.21(a): Adapted from figure 15.5 on page 223 of *Biology*, 3d ed., by Sylvia S. Mader. © 1989 by Wm. C. Brown Publishers.

Chapter 24

Figure 24.7: Adapted from figure 21.4 on page 697 of *Chemistry: The Central Science*, 5th ed., by Theodore L. Brown, H. Eugene LeMay, Jr., et al. © 1990 by Prentice-Hall; **24.8**: Adapted from page 878 of *Chemistry: A Systematic Approach*, by Harry H. Sisler, Richard D. Dresdner, et al. © 1980 by Oxford University Press; **24.14**: Adapted from figure 21.9 on page 958 of *Chemistry*, 2d ed., by Steven S. Zumdahl. © 1989 by D. C. Heath and Company; **24.18**: Adapted from figure 31.7 on page 889 of *Chemistry: A Systematic Approach*, by Harry H. Sisler, Richard D. Dresden, et al. © 1980 by Oxford University Press; **24.21**: Adapted from *Baltimore Sun*, January 2, 1983.

INDEX

A

B

C

F

G

H

I

J

K

L

M

N

O

P

S

T

U

V

W

X

Y

Z

SI Units and Conversion Factors

Length: 1 meter = 100 centimeters
= 1.0936 yards
= 3.2808 feet
= 39.370 inches
1 inch = 2.54 centimeters (exactly)
1 kilometer = 1000 meters
= 0.62137 mile
1 mile = 5280 feet
= 1.6093 kilometers

Mass: 1 kilogram = 1000 grams
= 2.2046 pounds
1 pound = 16 ounces
= 453.59 grams
= 0.45359 kilograms
1 ounce = 28.349 grams
1 ton = 2000 pounds
= 907.18 kilograms
1 metric ton = 1000 kilograms
= 2204.6 pounds
1 atomic mass unit = 1.66056×10^{-27} kilogram

Volume: 1 liter = 1000 milliliters
= 1 cubic decimeter
= 10^{-3} cubic meter
= 1.0567 quarts
1 gallon = 4 quarts
= 8 pints
= 3.7854 liters
1 quart = 32 fluid ounces
= 0.94633 liter
1 fluid ounce = 29.573 milliliters

Energy: 1 joule = 0.2390 calorie
1 calorie = 4.184 joules
1 electron volt = 1.6021×10^{-19} joule

Temperature:
$$\text{K} = {}^\circ\text{C} + 273.15$$
$$ {}^\circ\text{C} = ({}^\circ\text{F} - 32)/1.8 = ({}^\circ\text{F} - 32)(5/9) = [({}^\circ\text{F} + 40)/(1.8)] - 40$$
$$ {}^\circ\text{F} = ({}^\circ\text{C})(1.8) + 32 = ({}^\circ\text{C})(9/5) + 32 = [({}^\circ\text{C} + 40)(1.8)] - 40$$

Pressure: 1 atmosphere = 101.325 kilopascals
= 760 torr (mm Hg)
= 14.70 pounds per square inch